W0079118

Advances in Cryogenic Engineering

Materials

VOLUME 32

A Continuation Order Plan is available for this series. A continuation order will bring delivery of each new volume immediately upon publication. Volumes are billed only upon actual shipment. For further information please contact the publisher.

Advances in Cryogenic Engineering *Materials*

VOLUME 32

Edited by

R. P. Reed and A. F. Clark

National Bureau of Standards
U.S. Department of Commerce
Boulder, Colorado

SPRINGER SCIENCE+BUSINESS MEDIA, LLC

The Library of Congress cataloged the first volume of this title as follows:

Advances in cryogenic engineering. v. 1–
New York, Cryogenic Engineering Conference; distributed by Plenum Press, 1960–
v. illus., diagrs. 26 cm.
Vols. 1– are reprints of the Proceedings of the Cryogenic Engineering Conference, 1954–
Editor: 1960- K. D. Timmerhaus.

1. Low temperature engineering—Congresses. I. Timmerhaus, K. D., ed. II. Cryogenic Engineering Conference.

TP490.A3 660.29368 57-35598

Proceedings of the Sixth International Cryogenic Materials Conference (ICMC), held August 12–16, 1985, in Cambridge, Massachusetts

ISBN 978-1-4613-9873-8 ISBN 978-1-4613-9871-4 (eBook)
DOI 10.1007/978-1-4613-9871-4

© 1986 Springer Science+Business Media New York
Originally published by Plenum Press New York in 1986
Softcover reprint of the hardcover 1st edition 1986

All rights reserved

No part of this book may be reproduced, stored in a retrieval system, or transmitted in any form or by any means, electronic, mechanical, photocopying, microfilming, recording, or otherwise, without written permission from the Publisher

CONTENTS

AUSTENITIC STEEL WELDING

COMPOSITES AND POLYMERS

REFRIGERATION MATERIALS

MECHANICAL PROPERTIES

PHYSICAL PROPERTIES

NONDESTRUCTIVE EVALUATION

MATERIALS AND DEVICES FOR SUPERCONDUCTING ELECTRONICS

SUPERCONDUCTING NITRIDE FILMS

FINE FILAMENTARY SUPERCONDUCTORS

SUPERCONDUCTORS — AC LOSSES;

MECHANICAL, ELECTROMAGNETIC, AND PHYSICAL PROPERTIES

RADIATION DAMAGE AND FLUX PINNING IN SUPERCONDUCTORS

MULTIFILAMENTARY Nb_3Sn SUPERCONDUCTORS

NEW SUPERCONDUCTING MATERIALS AND METHODS

INDEXES

FOREWORD

The Sixth International Cryogenic Materials Conference (ICMC) was held on the campus of Massachusetts Institute of Technology in Cambridge in collaboration with the Cryogenic Engineering Conference (CEC) on August 12–16, 1985. The complementary program and the interdependence of these two disciplines foster the conference. Its manifest purpose is sharing the latest advances in low temperature materials science and technology. Equally important, areas of needed research are identified, priorities for new research are set, and an increased appreciation of interdisciplinary, interlaboratory, and international cooperation ensues.

The success of the conference is the result of the able leadership and hard work of many people: S. Foner of M.I.T. coordinated ICMC efforts as its Conference Chairman. A. I. Braginski of Westinghouse R & D Center planned the program with the assistance of Cochairmen E. N. C. Dalder of Lawrence Livermore National Laboratory, T. P. Orlando of M.I.T., D. O. Welch of Brookhaven National Laboratory, and numerous other committee members. A. M. Dawson of M.I.T., Chairman of Local Arrangements, and G. M. Fitzgerald, Chairman of Special Events, skillfully managed the joint conference. The contributions of the CEC Board, and particularly its conference chairman, J. L. Smith, Jr. of M.I.T., to the organization of the joint conference are also gratefully acknowledged.

The combined conferences are held biennially; the next CEC/ICMC will be hosted by Fermi National Accelerator Laboratory, Batavia, Illinois, on June 14–18, 1987. The ICMC Board sponsors special-topic conferences in other years. Proceedings of these conferences are published as the Cryogenic Materials Series by Plenum. Nonmetallic Materials and Composites at Low Temperatures IV will be held in Heidelberg, West Germany on June 14–18, 1986. The Institute of Metal Research in Shenyang, China will be the location for another special ICMC emphasizing cryogenic metals research; its tentative dates are June 21–23, 1988.

Participation in the conference once again surpassed that of previous conferences: 634 registrants, 121 contributed papers (70 oral and 51 poster), and 41 invited papers. The program committee initiated sessions of invited papers to highlight research in three areas: New Methods to Characterize Cryogenic Materials, Nb_3Al and Superconductivity Worldwide, and Nb_3Sn. The Second ICMC Symposium on Materials and Processing for Superconducting Electronics was sponsored jointly with the Magnetics Society of the Institute of Electronics and Electrical Engineers. This timely and growing subject area was summarized by W. J. Gallagher of IBM in the plenary session for that day. In the other ICMC plenary speech, J. W. Morris, Jr. outlined how materials can be designed for their low temperature properties, in particular, for high field superconducting magnets.

There has been excellent progress in the development of stronger, tougher, austenitic steels for low temperature applications, especially in Japan. Alloy systems, including Fe-Mn, Fe-Mn-Cr, and Fe-Cr-Ni, have been studied. Advanced thermomechanical processing techniques, such as cross-rolling and rapid cooling following hot-rolling, have been applied to produced uniform, fine-grain structures. Refining practices have been modified to control inclusion density. As a result, a number of stronger, tougher, austenitic steels have been produced and are reported in this volume.

The possibilities of using Carnot- or Ericsson-type refrigerators at very low temperatures has led to research on possible magnetic refrigeration materials. These materials require either high heat transfer with a magnetocaloric effect (Carnot type) or ferromagnetism with a large magnetic moment (Ericsson type). Studies of magnetic and thermal properties of candidate materials were presented and compose an entire section.

Thermal and electrical insulators will have to withstand irradiation in fusions reactors. Studies are currently under way to assess the effects of neutron irradiation at 4 K on polymer-based composites, the primary candidate insulation materials. Review papers and research results from preliminary studies are included.

In the development of superconductors, significant progress is reported: enhanced critical current densities for NbTi, very fine filaments from processing improved starting materials, and improved properties from modified processing techniques such as hot isostatic pressing. Also, successful new processing techniques for some of the more advanced superconductors, such as B1 and C15, are reported from around the world. The superconducting supercollider (SSC) accelerator is providing a large impetus for new conductor designs, many of which are discussed in these papers. The push for higher critical temperatures and more stable films for junctions and devices continues, with notable achievements in the nitrides and in some integrated circuits.

This volume is the largest ever published by ICMC, reflecting the continual increasing interest and research related to low temperature materials and the increasing quality of the conference research papers.

R. P. Reed
A. F. Clark

BEST PAPER AWARDS

Awards for the best papers of the 1983 ICMC proceedings, Advances in Cryogenic Engineering — Materials, volume 30, were presented at the 1985 conference. These papers were selected by the awards committee from nominations of the editors in the categories of superconducting and structural materials. We thank the authors for their exemplary contributions.

MAGNETIC IMPURITY SCATTERING IN *in situ* SUPERCONDUCTORS

D.K. Finnemore, H.C. Yang, V.G. Kogan, and S.L. Miller

Ames Laboratory-USDoE, Ames, Iowa

CORRELATION OF DIELECTRIC AND MECHANICAL DAMPING AT LOW TEMPERATURES

G. Hartwig and G. Schwarz

Nuclear Research Center Karlsruhe
Karlsruhe, Federal Republic of Germany

1985 INTERNATIONAL CRYOGENIC MATERIALS CONFERENCE BOARD

A. F. Clark, Chairman . . .	National Bureau of Standards Boulder, Colorado, USA
A. I. Braginski	Westinghouse Research and Development Center Pittsburgh, Pennsylvania, USA
E. W. Collings	Battelle Memorial Institute Columbus, Ohio, USA
D. Evans	Rutherford Laboratory Chilton, Didcot, England
S. Foner	Francis Bitter National Engineering Laboratory Cambridge, Massachusetts, USA
H. C. Freyhardt	University of Göttingen Göttingen, Federal Republic of Germany
G. Hartwig	Institut für Technische Physik Karlsruhe, Federal Republic of Germany
T. Horiuchi	Kobe Steel Kobe, Japan
D. C. Larbalestier	University of Wisconsin Madison, Wisconsin, USA
J. W. Morris, Jr.	University of California Berkeley, California, USA
R. P. Reed	National Bureau of Standards Boulder, Colorado, USA
K. Tachikawa	National Research Institute for Metals Ibaraki, Japan
G. K. White	CSIRO Division of Applied Physics Sydney, Australia
K. A. Yushchenko	E. O. Paton Institute of Electrowelding Kiev, USSR

STRUCTURAL ALLOYS FOR HIGH FIELD SUPERCONDUCTING MAGNETS

J. W. Morris, Jr.

Materials and Molecular Research Division
Department of Materials Science and Mineral Engineering and
Lawrence Berkeley Laboratory,
University of California, Berkeley, California

ABSTRACT

Research toward structural alloys for use in high field superconducting magnets is international in scope, and has three principal objectives: the selection or development of suitable structural alloys for the magnet support structure, the identification of mechanical phenomena and failure modes that may influence service behavior, and the design of suitable testing procedures to provide engineering design data. This paper reviews recent progress toward the first two of these objectives. The structural alloy needs depend on the magnet design and superconductor type and differ between magnets that use monolithic and those that employ force-cooled or ICCS conductors. In the former case the central requirement is for high strength, high toughness, weldable alloys that are used in thick sections for the magnet case. In the latter case the need is for high strength, high toughness alloys that are used in thin welded sections for the conductor conduit. There is productive current research on both alloy types. The service behavior of these alloys is influenced by mechanical phenomena that are peculiar to the magnet environment, including cryogenic fatigue, magnetic effects, and cryogenic creep. The design of appropriate mechanical tests is complicated by the need for testing at 4 K and by rate effects associated with adiabatic heating during the tests.

INTRODUCTION

Many of the most challenging needs for cryogenic materials relate to the design and construction of large, high field superconducting magnets that operate at temperatures of 4 K or below. Three distinct types of materials are used in these magnets: superconducting wire to carry the current that creates the field, high strength structural alloys to position and support the superconductor, and electrical insulators to separate the conductors from one another. There are important problems with each of these material types. The present review summarizes problems and progress that relate to the cryogenic structural alloys that will be used in the magnet case and support structure.

Research on cryogenic structural alloys primarily supports the development of magnetic confinement systems for magnetic fusion. These systems will employ magnetic fields of 8 to 20 T generated by large-diameter magnets. Since the Lorentz force on the conductor increases with both the field and the dimensions of the magnet, the structure of such a magnet must support very high forces. High field superconducting magnets are also central components in a number of other technologies, such as particle accelerators, but the magnetic fields and magnet dimensions that are currently projected for such systems are smaller, and the known structural materials needs are correspondingly less severe.

The materials research that supports Magnetic Fusion Energy has three major objectives: (1) to provide structural alloys that satisfy the design requirements of large, high field superconducting magnets; (2) to identify unusual mechanical phenomena that may influence the behavior of the alloys in the magnet environment and anticipate failure modes; and (3) to provide probative testing procedures and test data to guide magnet design. Recent progress toward each of these objectives is covered in specific technical sessions within this conference. The present paper provides a brief overview of selected aspects of research toward the first two.

In assessing the magnet materials needs of fusion systems it must be remembered that these devices are ultimately intended to produce electric power in competition with fossil and fission devices. The materials contribution to system cost and reliability is a relevant issue. The materials used in fusion reactors must be inexpensive as well as functional and reliable.

STRUCTURAL ALLOYS

The increasingly stringent specifications for magnet structural alloys result from the increasing size and field of the magnets, both of which raise the mechanical load on the magnet structure. While current systems, such as the Mirror Fusion Test Facility (MFTF-B) at LLNL and the Large Coil Test Facility at Oak Ridge, were built with available structural alloys, future designs will almost certainly require alloys that offer superior combinations of strength and toughness at 4 K.[1]

The precise property requirements for magnet structural alloys depend on the structural support scheme as well as on the detailed magnet design. Two alternative methods of support are now in use. The most common is an exoskeletal configuration in which the superconductor receives its primary support from the external magnet case. This design method was used for the MFTF-B and for several of the coils in the Large Coil Project. The alternate method is to place the superconductor within a high-strength conduit that provides internal support. In this method the superconductor is a loose winding that is cooled by flowing helium. The exoskeleton can then be a relatively low-strength construction of aluminum or stainless steel. The "internally cooled cabled superconductor" (ICCS) concept was used in the Westinghouse Nb_3Sn coil for the Large Coil Project, and has been chosen for the "outsert coil" for the MIT-LLNL Multipurpose Coil now in design at MIT.[2]

The material properties required for the case of a magnet differ from those needed for an ICCS conduit. It is useful to divide the research toward new structural alloys on this basis. The alloys that are under development for the magnet case are high strength structural steels, including nitrided austenitic Fe-Ni-Cr and Fe-Mn-Cr alloys and ferritic Fe-Ni cryogenic steels. Aluminum alloys may also be useful in specific applications, despite their lower strength. The alloys that are being considered for the ICCS superconductor conduit are Fe-based and Ni-based superalloys whose heat treatments are compatible with those of the superconducting wire. Both sets of alloys will probably find other applications as well. For example, the Fe-based superalloy A286 is used for the magnet support structure forgings for MFTF-B, and conventional stainless steels have been used for the ICCS conductor conduits of NbTi magnets that do not require heat treatment after fabrication.

MAGNET CASE ALLOYS

An example of the case configuration in a high field superconducting magnet is shown in Figure 1, which is a photograph of the Ying-Yang coil set from the MFTF-B at LLNL. The physical size of the coil set is apparent in the photograph, as is the fact that the case is a welded construction.

Conventional austenitic stainless steels were used in the construction of the MFTF coil case, as in other magnets that depend on exoskeletal support. As shown in Figure 2, these alloys offer a combination of strength and toughness at 4 K that easily meets the design requirements for the MFTF.[3,4] They can be welded with available techniques and filler metals so that they retain the needed strength and toughness.[4] It is likely, however, that future fusion magnets will require superior materials. Figure 2 also shows the design specifications developed by the Japan Atomic Energy Research Institute (JAERI) for a future tokomak machine.[1] These specifications call for a strength-toughness combination that is significantly beyond the capability of the existing stainless steels. Preliminary specifications for the case alloys for future U.S. machines are somewhat less challenging, as shown in Figure 3, but will certainly require the development of structural steels with improved properties in the welded condition.

These future needs have been addressed through metallurgical research that uses the relations between alloy microstructure and mechanical properties to select or design suitable materials.

The Microstructure-Property Relations

The essential properties of structural alloys in the magnet case are strength and toughness; the alloys must not fail in service. Other characteristics are also important to optimize the performance of the magnet system. The relevant properties of the magnet case alloy include its fatigue resistance, its sensitivity to radiation, its coefficient of thermal expansion, and its cost.

The structure-property relations governing the interplay between strength and toughness are shown in Figure 4. At moderate temperature all of the alloys of interest fracture in a ductile mode and have reasonably high fracture toughness, as indicated by the toughness-temperature plot at upper left. However, the fracture toughness decreases as the alloy is made stronger, yielding the strength-toughness characteristic shown at upper right. As the temperature is lowered the yield strength increases, as illustrated in the plot at lower left. The rate of increase depends on the crystal structure and on the density of microstructural features such as interstitial solute atoms that represent thermally activated barriers to dislocation glide. If the yield strength becomes too high relative to the critical stress for brittle fracture then the alloy becomes brittle, often over a fairly narrow temperature range that defines the ductile-brittle transition temperature. The dominant mode of brittle fracture below the transition is the easier of transgranular cleavage, in which the metal grains fracture, and intergranular fracture, in which the grains separate from one another across their boundaries.

To create an alloy that is both strong and tough at 4 K one must first ensure that the alloy is ductile at that temperature. If the alloy is not ductile, the ductile-brittle transition can be suppressed by lowering the strength of the alloy, by moderating the increase in strength at low temperature, or by increasing the alloy's resistance to brittle fracture, as must be done in ferritic steels and in some austenitic steels. The proper metallurgical approach to improve resistance to brittle fracture depends on the

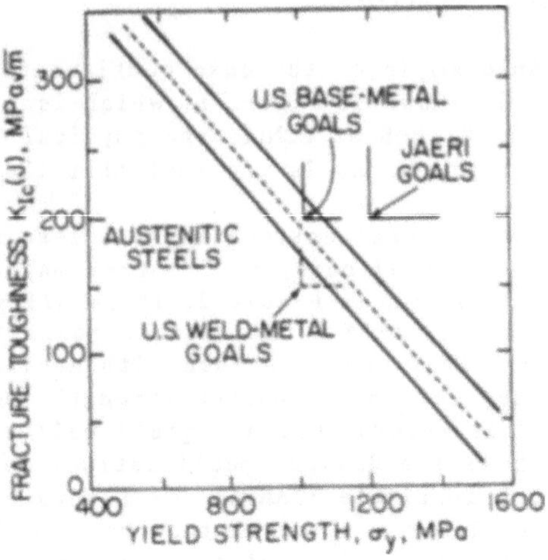

Fig. 3. U.S. specifications for cryogenic structural steels.

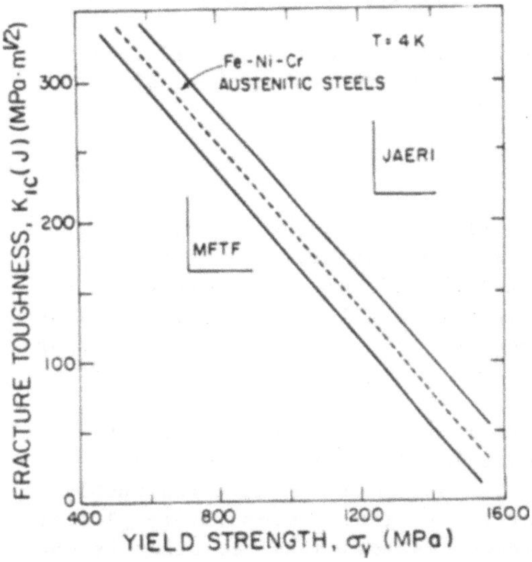

Fig. 2. Strength-toughness properties of stainless steels with MFTF and JAERI specifications.

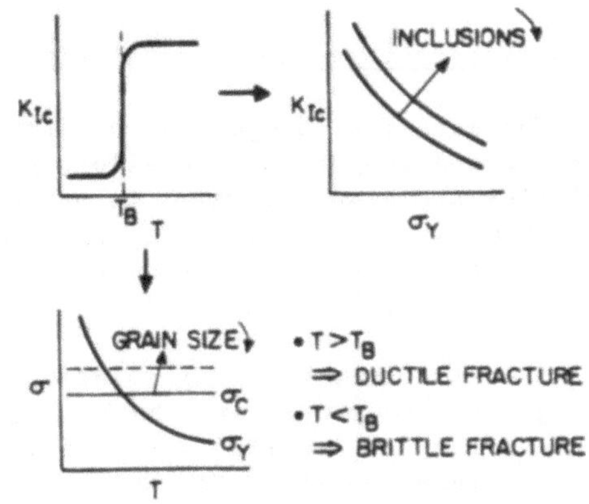

Fig. 4. Strength-toughness behavior.

fracture mode. If the brittle mode is intergranular, the alloy should be treated to eliminate intergranular surfactants that cause embrittlement, as sulfur and phosphorus do in many steels, or changed to incorporate intergranular surfactants that promote cohesion, as boron does in ferritic Fe-Mn steels and Ni_3Al intermetallics. If the brittle mode is transgranular the alloy can usually be made more resistant to brittle fracture by refining its grain size.

Once the alloy is made tough at 4 K one is left with the problem of maximizing its toughness in the ductile mode. The toughness is a monotonically decreasing function of the yield strength, as illustrated in the graph at the upper right in Figure 4; the toughness at given strength can only be increased by raising the strength-toughness characteristic as a whole. The strength-toughness characteristic is affected by the composition of the steel, in ways that are not well understood, and is also strongly influenced by the purity of the alloy. The effect of alloy purity has its source in the mechanism of ductile fracture, which occurs through the nucleation and coalescence of voids that usually form at inclusion particles within the matrix. Decreasing the density of inclusions, by improving the purity of the alloy, inhibits ductile fracture at given yield strength and improves the strength-toughness characteristic.

Three classes of alloys have been suggested as candidates for the magnet case. The most important of these includes the austenitic steels, of which two types are under active development: Fe-Ni-Cr alloys (modified 300 grade stainless steels) and Fe-Mn-Cr alloys. Promising ferritic cryogenic steels have also been developed, along with new aluminum alloys that may be suitable for selected applications.

Austenitic Stainless Steels

The strength-toughness characteristic of the conventional Fe-Ni-Cr stainless steels is presented in Figure 2.[3] These alloys have adequate strength and toughness to satisfy the specifications for current systems, but do not have the properties that may be needed for future machines, for example, the JAERI specifications given in the figure.

To satisfy this need, the Japanese Atomic Energy Research Institute (JAERI) has instituted a substantial alloy development program involving several steel companies.[1,6] The promising alloys that have emerged from that program include both Fe-Ni-Cr and Fe-Mn-Cr grades.

Fe-Ni-Cr Austenitic Steels

The results of basic research[7] suggest that there are three critical elements in designing an austenitic Fe-Ni-Cr alloy to achieve properties within the "JAERI Box" (Figure 2). The first is solution hardening for alloy strength. The most efficient solutes are the interstitial species carbon and nitrogen. Of the two, nitrogen is preferred since carbon precipitates with chromium during heat treatment and decreases alloy toughness. However, relatively high nitrogen contents, near 0.35 weight percent, are required to achieve strength above 1200 MPa (Figure 5). This nitrogen level exceeds the solubility of Fe-Ni-Cr unless the chromium and nickel contents are raised to high levels or unless a significant quantity of manganese is added.

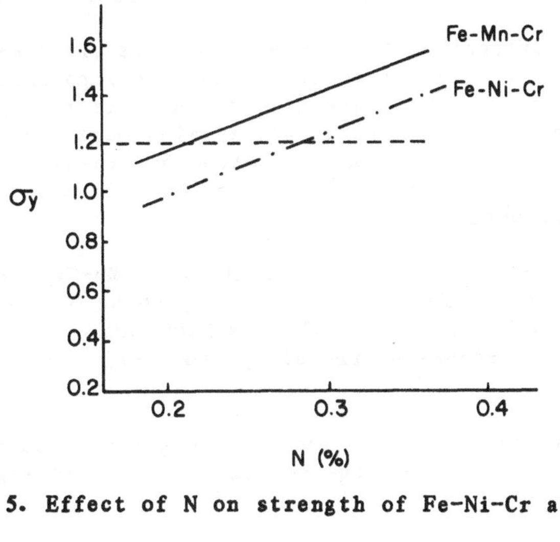

Fig. 5. Effect of N on strength of Fe-Ni-Cr alloys (4 K).

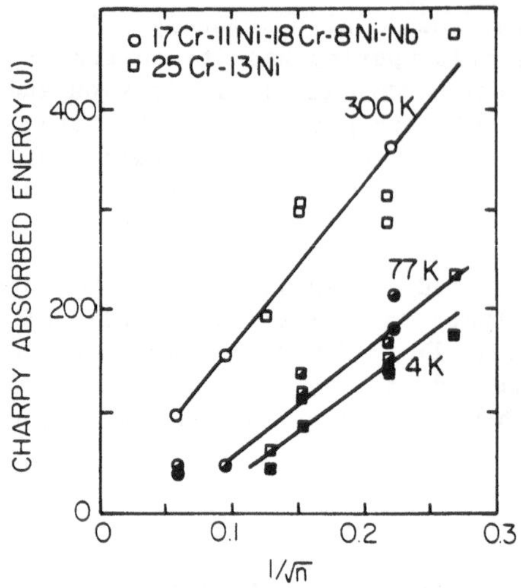

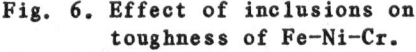

Fig. 6. Effect of inclusions on toughness of Fe-Ni-Cr.

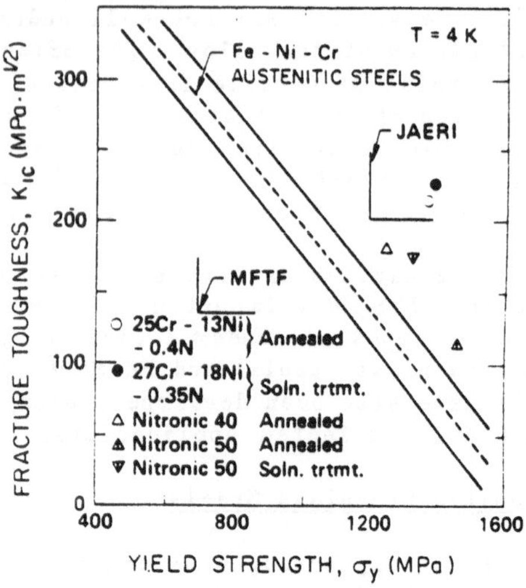

Fig. 7. Strength-toughness relation for Fe-Ni-Cr alloys.

The next two requirements relate to alloy toughness. For exceptional toughness the alloy must be thermally stable in the fcc austenitic structure at 4 K; the presence of thermally induced martensite or retained δ-ferrite introduces islands of brittle phase that promote fracture. Empirical results suggest that the austenite phase should be mechanically stable as well. This requires a high alloy content of chromium and nickel. In addition, the alloy should be melted to high purity to reduce the concentration of inclusions that act as nucleation sites for ductile fracture (Figure 6[8]).

These criteria have been used by the Nippon Steel Company[8] to melt alloys that satisfy the JAERI criteria, as shown in Figure 7.[9] The most promising of those appearing in the figure is a 27Cr-18Ni-2Mn-0.35N-0.02C alloy that is melted to high purity. More recent results were presented at the 1985 International Cryogenic Materials Conference (ICMC).[10]

Fe-Mn-Cr Austenitic Steels

The second important class of austenitic cryogenic steels is based on the Fe-Mn-Cr ternary. Such steels have been under intensive development for a variety of applications in recent years, particularly in Japan.

The principal advantages of the Fe-Mn alloys are their excellent strengthening response to nitrogen additions and their potentially lower cost. The main disadvantages of these alloys are their susceptibility to mechanical transformation to the hexagonal, epsilon-martensite phase, which decreases the fracture toughness, and their tendency toward intergranular fracture at low temperature. Both of these problems have been overcome in the development of production alloys.[11-15]

The influence of nitrogen on the strength of Fe-Mn-Cr alloys is shown schematically in Figure 8. The addition of only about 0.2 wt.% nitrogen suffices to achieve a 4 K yield strength near 1200 MPa. Additional strength is imparted by processing the alloys to a relatively fine grain size.

The Fe-Mn-Cr-N alloys have good fracture toughness when they fracture in the ductile mode, provided that the content of the alloying species is

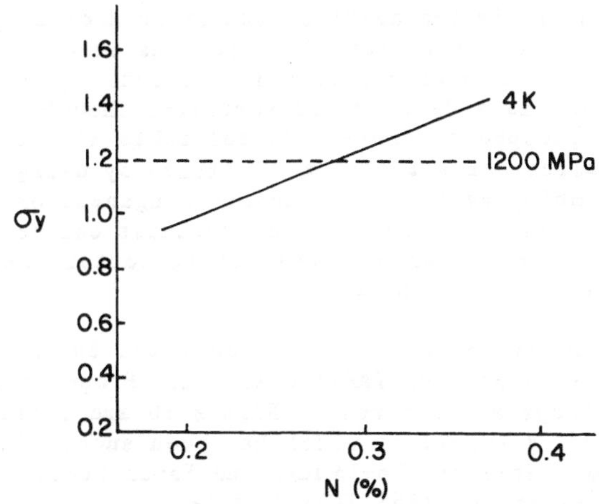

Fig. 8. Effect of N on strength of Fe-Mn-Cr alloys (4 K).

high enough to stabilize the fcc matrix against transformation to the hexagonal, ε-martensite phase. However, these alloys are sometimes susceptible to intergranular cracking, which causes them to exhibit a ductile-brittle transition in which the brittle, low-temperature mode is intergranular. Some examples are given in Figure 9.[16] Interestingly, the intergranular cracking is most pronounced in alloys melted from very pure starting materials. It is not yet clear whether the intergranular weakness is inherent to the alloy or is associated with the segregation of manganese, and possibly nickel. Recent research has shown that the intergranular fracture can be eliminated by slowing the cooling rate of the alloy, which allows the intergranular chemistry to adjust during cooling, or by introducing small additions of carbon and silicon, which interact with the embrittling species in ways that are not yet understood.[26] Carbon and silicon are invariably present in some concentration in commercial alloys.

Fe-Mn-Cr alloys may also be liable to temper embrittlement if they are held for too long a time at intermediate temperatures. Carbide precipitation has been identified as a source of temper embrittlement and can be minimized, if necessary, by adding a small amount of molybdenum.[15]

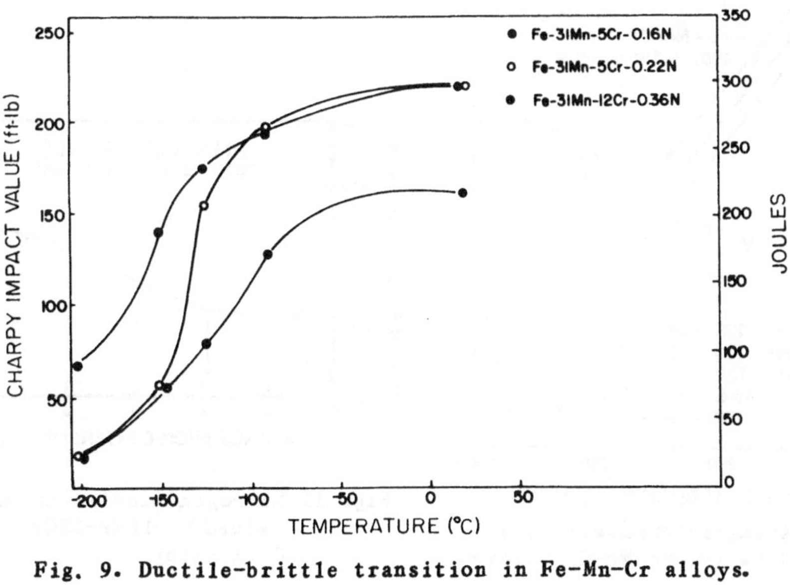

Fig. 9. Ductile-brittle transition in Fe-Mn-Cr alloys.

To maximize toughness in the ductile mode it is necessary to minimize the inclusion content. A major source of inclusions in Fe-Mn alloys are the MnS precipitates that form when sulfur is present. It is also widely believed that phosphorus is a deleterious impurity, though recent research suggests that a small phosphorus content is tolerable if the sulfur content is low.[15] The sulfur content can be minimized by using a low-sulfur melting practice and employing either elemental manganese or low-sulfur ferromanganese as a starting material. The toughness can be further improved by adding trace elements such as calcium to getter the remaining sulfur into hard, spherical inclusions.[15]

These alloy design techniques have been combined to produce structural alloys that approach or exceed the JAERI criteria. Properties of some of the more successful alloys are plotted in Figure 10 and include contributions from Kobe Steel[11,14] (Fe-22Mn, which has been successfully produced in thick plates), Nippon Steel[12] (Fe-25Mn), and Japan Steel Works.[5] Additional data were presented at the 1985 ICMC.[6,14,15].

Welding

The work summarized above virtually ensures that austenitic steel plates with very high strength-toughness combinations at 4 K will be available for the construction of large magnets. However, the magnet case must also be welded, and will probably have to be designed to the properties of the weldment. Research toward suitable welding procedures for these alloys has only recently begun.

The welding requirements for magnet structural alloys concern both the welding procedure and the properties of the welded plate. The welding procedure should be practical for field fabrication of the magnet case, and should be a high-rate welding process to minimize construction costs. On the other hand, the welded construction should have a combination of strength and toughness that is competitive with that of the base plate.

The metallurgical problems that are most likely to intrude during the welding of the new austenitic steels include nitrogen loss during welding,

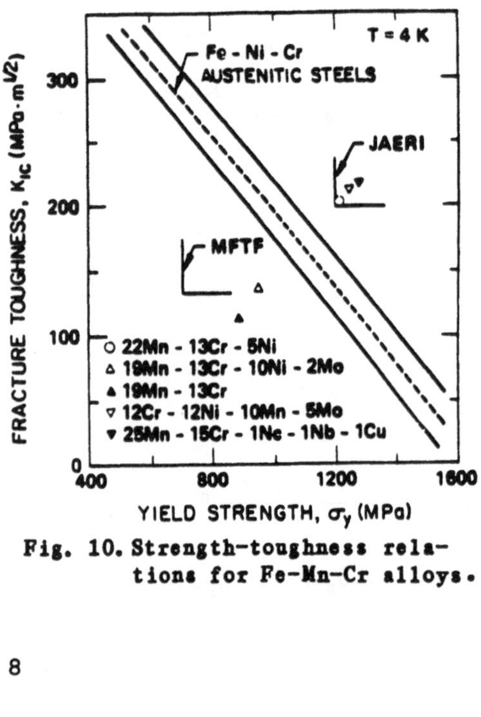

Fig. 10. Strength-toughness relations for Fe-Mn-Cr alloys.

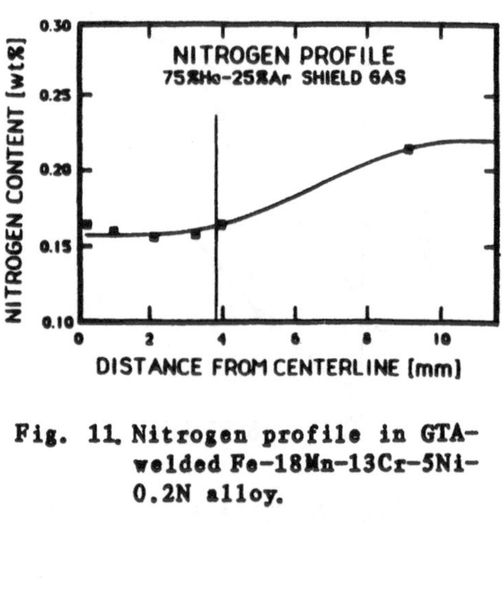

Fig. 11. Nitrogen profile in GTA-welded Fe-18Mn-13Cr-5Ni-0.2N alloy.

which may decrease the strength of the weldment, and hot cracking or the development of a coarse, heterogeneous structure in the weld metal or heat affected zone, which may cause a loss of toughness.

The former problem is illustrated in Figure 11, taken from work on the gas-tungsten arc (GTA) welding of a Kobe 18Mn-13Cr-5Ni-0.2N steel at LBL[17] Welding causes a loss of nitrogen in both the weld metal and heat affected zone. The nitrogen content in the weld metal can be restored by adding nitrogen to the shielding gas, which raises the strength to base metal values, but results in a deterioration in the toughness. On the positive side, however, the strength-toughness combination of the plate welded without nitrogen gas addition approaches the tentative U.S. requirements of 1000 MPa yield, 150 MPa$\sqrt{m}$ fracture toughness.

There are also encouraging recent results from Kobe Steel[18], showing that the nitrogen content and strength of weldments in Fe-22Mn-13Cr-5Ni-0.2N are retained when electron beam welding is used. Since the welded plates also appear to have good toughness, the EB welding process is promising.

Unfortunately, GTA welding is a low-rate welding process that is tedious and expensive when it is applied to thick plates, and EB welding usually requires a high vacuum that prevents its use in the field. Much developmental work remains to be done on the welding of the new high-strength steels to achieve reliable strength-toughness combinations with welding procedures that can be used for the field welding of large magnet cases.

Ferritic Cryogenic Steels

The preference for austenitic steels for the superconducting magnet case has two origins: their favorable low-temperature ductility in the as-cooled condition and their low magnetic susceptibility. The latter requirement has largely disappeared as studies have shown that the presence of saturated ferromagnetic materials will not adversely affect performance. In fact, the modified A286 superalloy used for the conductor conduit in the Westinghouse large coil is ferromagnetic at 4 K, and ferritic alloys are serious candidates for the reactor first wall that sits within the magnetic field. The low-temperature brittleness of the ferritic alloys remains a problem. Even the best of the ferritic alloys that are commercially available have ductile-brittle transition temperatures above 4 K.

On the other hand, the ferritic alloys offer specific advantages. They are relatively inexpensive, familiar alloys that naturally reach strength levels above 1300 MPa at 4 K. Alloy design efforts have been undertaken in both the United States and Japan to create ferritic steels that are suitable for at least special applications in the magnet structure.

The problem with the ferritic steels is the ductile-brittle transition, and the metallurgical challenge is to devise alloy treatments that suppress the transition temperature (T_B) to below 4 K. The starting alloys are the Fe-Ni ferritic steels that have been successfully used at temperatures of 77 K or below, e.g., Fe-(9-12)Ni.

The source of the ductile-brittle transition in 9-12Ni steels is relatively straightforward[19] and can be understood by reference to the 'Yoffee diagram' in the lower left of Figure 4. These alloys naturally quench into a lath martensitic structure whose basic element is a 'packet' of parallel laths (Figure 12). While the substructure of a packet appears refined in

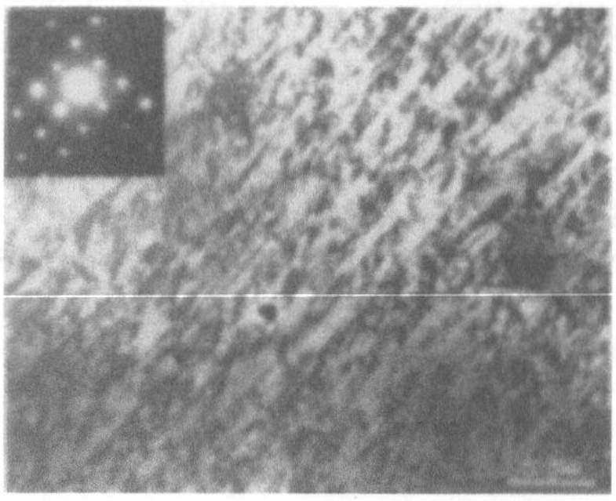

Fig 12. Parallel Martensite laths within packet in Fe-Ni Steel.

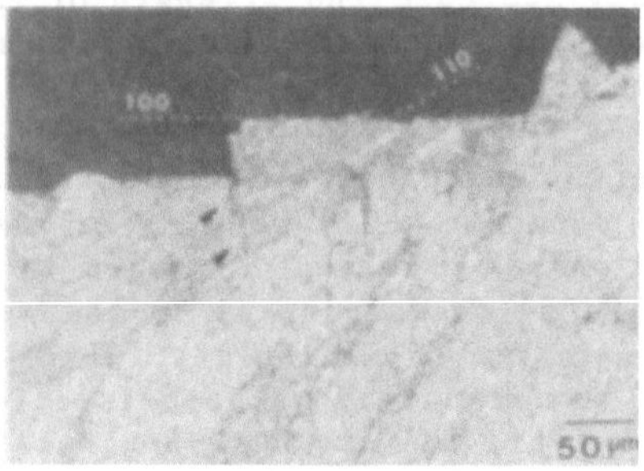

Fig. 13. Profile SEM fractograph showing cleavage fracture in Fe-Ni Steel.

an optical or bright-field transmission electron micrograph, crystallographic studies show that the laths share a common crystallography. Hence, they have a common {100} cleavage plane and cleave as a unit under high stress. When the alloy is cooled its strength increases dramatically until the cleavage stress for the martensite packet is exceeded in the highly stressed region ahead of the crack tip. The alloy then fractures in a brittle manner, predominantly through the successive cleavage of martensite packets along {100} planes. The crystallography of cleavage fracture is documented in the profile fractograph shown in Figure 13.[20]

The most direct way to decrease T_B in a ferritic steel is to refine the effective grain size, which is the martensite packet size in the best Fe-Ni cryogenic steels. In conventional 6-9Ni steels intended for use at 77 K or higher, this is accomplished by giving the alloy an intercritical temper (in the two-phase $\alpha+\gamma$ region) to introduce a controlled distribution of precipitated austenite phase along the martensite lath boundaries.[21,23] This austenite breaks up the crystallographic alignment of the martensite packets (in a somewhat subtle way), increases the resistance to cooperative cleavage fracture, and decreases the ductile-brittle transition temperature. However, this method has not been successful in suppressing T_B to below 4 K.

An alternative approach that has produced promising 4 K properties uses cyclic heat treatment to refine the packet size. There are two methods for doing this, diagrammed in Figure 14[24]. Both involve heating

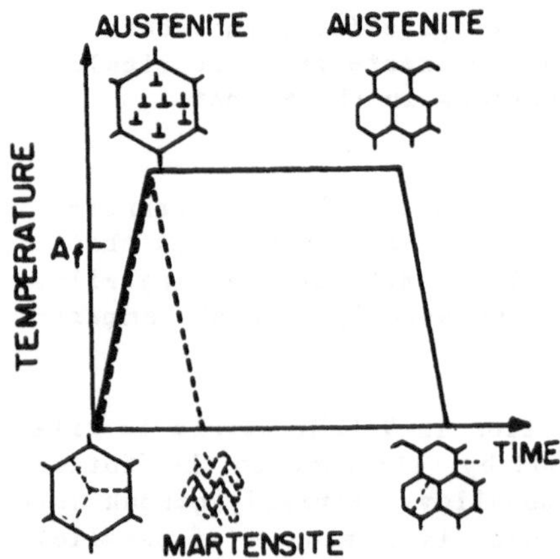

Fig. 14. Microstructural refining processes showing grain vs packet refinement for Fe-Ni steel.

the alloy to above the austenite reversion temperature to revert it into the γ (fcc) structure. Unless it is done very slowly, the $\alpha \rightarrow \gamma$ transformation involves some shear or mechanical deformation that causes the fresh γ phase to have a high density of dislocations. If the alloy is held at a temperature in the γ region, it will recrystallize to eliminate these defects. The result is to refine the austenite grain size, so that the martensite packet size is smaller after the alloy is returned to room temperature. This technique is termed "grain refinement" and can be used in the processing of thick plates. If, on the other hand, the alloy is quenched immediately after reversion to γ, recrystallization does not occur. The highly defective γ phase then transforms to a lath martensite in which the martensite packets are not well aligned. The result is a very small effective grain size. This technique is termed "packet refinement" and is useful in welding.

The optimum thermal treatments for the base plate alternate austenite reversion with intercritical annealing treatments to condition the microstructure. It was found some years ago that a four-step cycle of this type, designated the '2B' treatment, could produce an exceptional combination of strength and toughness near 4 K in an Fe-12Ni-0.25Ti alloy.[25] Two more recent versions of the thermal cycling treatment are diagrammed in Figure 15[26].

The 4 K strength-toughness combinations obtained in Fe-12Ni-0.25Ti that was processed through the thermal cycle schedules described above are plotted in Figure 16[26] and compared with the strength-toughness characteristic for the conventional stainless steels and to the JAERI goal. Ferritic alloys can be reproducibly made that have 4 K strength-toughness combinations above those of the conventional stainless steels. While these alloys do not yet have the toughness required by the JAERI goal, detailed fractography shows that the fracture mode in the best alloys is still not completely ductile. It follows that further improvements in processing should yield a substantial addition to the toughness. Research on these alloys is continuing.

A surprising and important property of the ferritic alloys is their weldability for 4 K service.[27] Fundamental research has shown that the best heat treatment for the Fe-Ni alloys is a rapid thermal cycle into the γ field. As was illustrated in Figure 14, a rapid reversion cycle directly

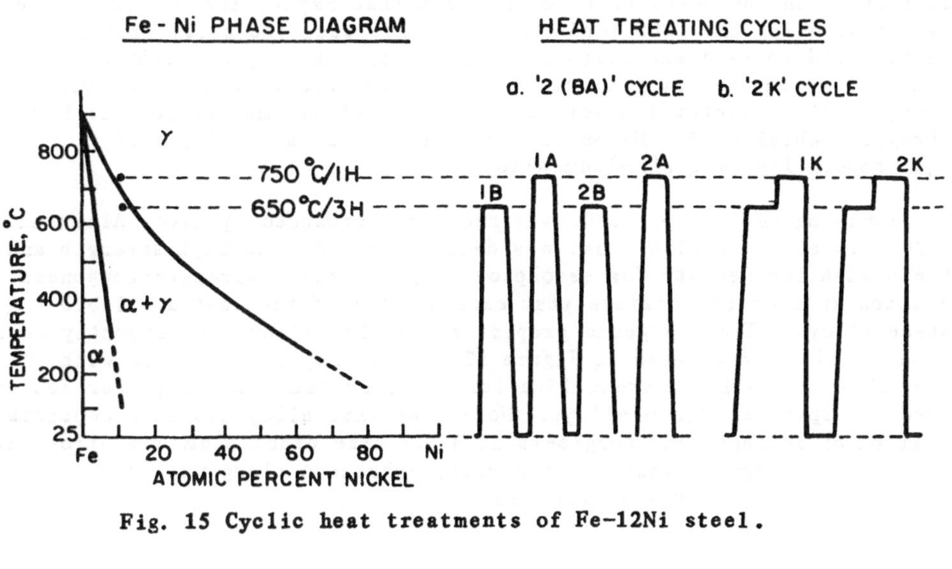

Fig. 15 Cyclic heat treatments of Fe-12Ni steel.

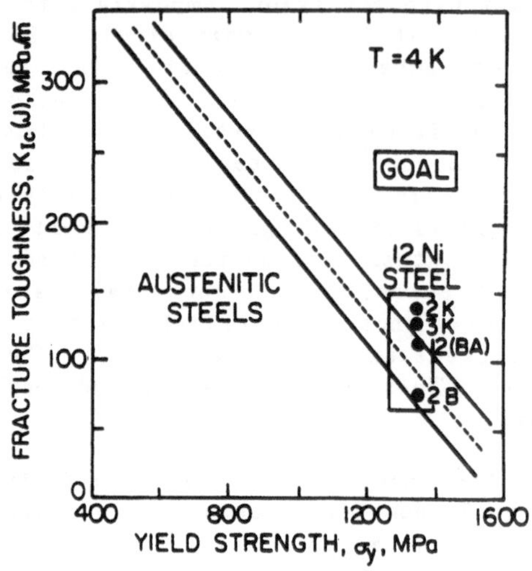

Fig. 16. Strength-toughness combinations for Fe-Ni steels.

decomposes the lath alignment within a martensite packet to create an extremely fine effective grain size. The rapid heating and cooling rates employed are impractical for the processing of bulk alloys, but are naturally imparted if the alloy is welded in a fine wire, multipass process in which each deposited bead is repeatedly heated by subsequent passes. An appropriate GTA welding technique was developed some years ago in a joint project between Kobe Steel and Nippon Kokan to weld 9Ni steel for use at 77 K and above. A suitable modification of this process has been shown to yield weldments that have exceptional toughness at 4 K.

It follows from these results that ferritic cryogenic steels can be made to have an attractive combination of strength and toughness at 4 K. These alloys should also be useful for high field magnet structures.

Aluminum Alloys

Compared with the best cryogenic steels, aluminum alloys have low strength and toughness at 4 K. Nonetheless, aluminum alloys may find applications in the cryogenic structure of high field superconducting magnets because of their light weight and their physical properties. A particular application is in the outer cases of magnets that employ internal structure, such as those using the ICCS conductor. In this case the outer case is not always required to have exceptional strength, and the light weight of the aluminum alloys may be advantageous in magnet construction and assembly. For example, the structural alloy for the Westinghouse magnet for the Large Coil Project, which used a Nb_3Sn conductor shrouded in JBK-75, a high strength superalloy, was aluminum alloy 2219.

A promising new alloy, 2090, was recently announced by Alcoa Aluminum. Alloy 2090 is an Al-Li alloy that was designed to combine high strength and toughness with low density for aerospace use, and has a strength-toughness combination at room temperature that exceeds that of the best available aerospace alloys. The cryogenic properties of this alloy have recently been measured[27] and are presented in Figure 17, where they are compared with those of 2219 and with the trend line for the room temperature properties of advanced aerospace aluminum alloys. Note that this alloy has an exceptional combination of strength and toughness at 4 K. Like other aluminum alloys it has the unusual property that its toughness increases along with its strength as the temperature is lowered.

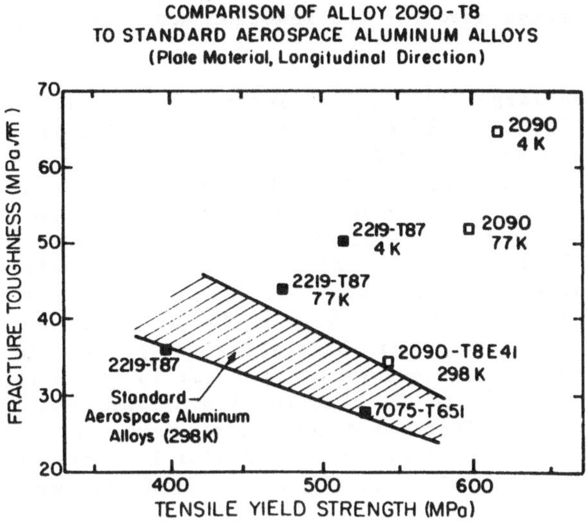

Fig. 17. Strength-toughness combinations of aluminum alloys.

STRUCTURAL ALLOYS FOR ICCS CONDUCTORS

Figure 18 is a cross section through an ICCS conductor of the type used in the Westinghouse Large Coil. It consists of a relatively loose bundle of superconducting wire inside a metal conduit. Helium flows through the interstices of the wire bundle to cool the superconductor. The ICCS geometry offers advantages in magnet construction that have made it the configuration of choice for several large systems, including the Westinghouse Large Coil, the MIT 12 T coil, and the MIT outsert coil for the DOE Multipurpose Coil.

The conduit for the ICCS conductor is typically made by wrapping a sheet of metal around the wire bundle, seam-welding it to form a tube, and squaring the tube so that the conductor can be stacked in a compact winding.[28] If the superconductor is NbTi, the conductor sheath can be made of any suitable cryogenic steel. Conventional stainless steels have been used. But if the conductor is Nb_3Sn that must be reacted after assembly, the conductor sheath must be made of an alloy that can tolerate the long-time, high-temperature reaction treatment and still preserve a good strength-toughness combination at 4 K.

The conventional stainless steels are generally unsuitable for Nb_3Sn conductor conduits because they form brittle precipitate phases after prolonged heating at high temperature. The materials of choice have been superalloys[29], since these are normally processed by aging for long times at

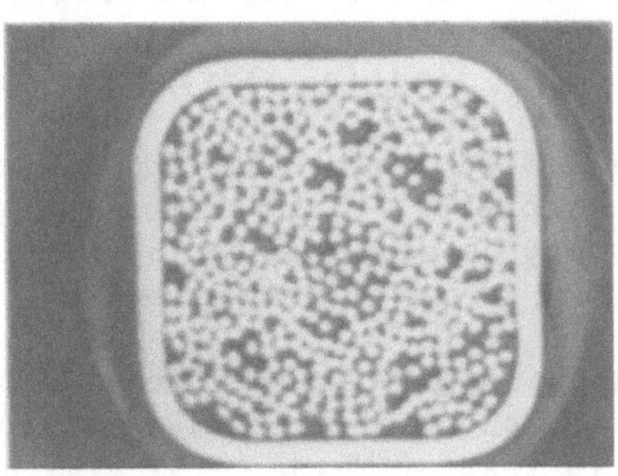

Fig. 18. Cross section of an internally cooled, cabled superconductor (ICCS).

temperatures in the range used for the reaction heat treatment of Nb_3Sn, 700-750°C.

The Westinghouse large coil used a bronze-processed Nb_3Sn conductor that was reacted at 700°C for 30 h. The alloy chosen for the conduit case was JBK-75, a modified version of A286[30]. Its nominal composition is Fe-29Ni-15Cr-2Ti-1.25Mo-0.25Al-0.25V-0.016C. It hardens during heat treatment through the precipitation of the ordered intermetallic phase γ' Ni_3(Ti,Al,-Mo), an ordered precipitate that coarsens slowly. Like most precipitation-hardened alloys JBK-75 has a yield strength that depends only slightly on test temperature, but exceeds 1000 MPa (150 ksi) at 4 K after a slight cold work followed by heat treatment for 30 h. at 700°C. Since the alloy ductility and thin sheet toughness remain high after autogenous GTA welding, it meets the obvious mechanical requirements for use in the conductor sheath. Moreover, work done parallel to the construction of the Westinghouse coil led to the development of alternative heat treatments for the alloy that are potentially compatible with the superconductor heat treatment and lead to superior property combinations in the alloy and in its weldments.[31,32]

However, it is not clear that JBK-75 is suitable for advanced magnets such as the MIT multipurpose coil. The advanced magnets require tailored material properties whose precise nature is under active investigation by magnet designers. The issues of greatest interest at this time concern the strain introduced into the superconductor by the conduit and the most critical magnet failure modes.

Most superalloys have thermal expansion coefficients that are much larger than that of Nb_3Sn. As a consequence, the superconductor is compressed on cooling from the reaction temperature (near 1000 K) to 4 K. Since the critical current within the superconductor (the maximum current that it can carry) decreases with strain, the thermal expansion mismatch may cause a loss in magnet performance. To overcome this problem there should be an appropriate match between the thermal expansion coefficients of the conduit alloy and the superconductor.

The problem of choosing the optimum thermal expansion coefficient is, however, complicated by the fact that the system is also strained by the winding operation and by the Lorentz force developed in the coil when the magnet is operated. The Lorentz strain depends at least on the operating field, the precise design of the composite conductor, and the elastic modulus of the conduit alloy. Both theoretical and experimental studies are currently underway to specify these effects more precisely and identify suitable combinations of thermal expansion coefficient, modulus, strength, and conduit configuration.

The most threatening failure modes for magnets that employ ICCS conductors are also uncertain as of this writing. It is clear that the conduit alloy must have some minimum strength and toughness to resist fracture under the internal pressures developed in service, but it is possible that fatigue strength may be a more limiting service requirement.[34] Strength-toughness specifications have recently been selected to guide alloy development and testing in the United States: 750 MPa yield strength and 100 $MPa\sqrt{m}$ fracture toughness. However, these criteria may well change as the magnet designs evolve and may be supplemented by specific requirements on fatigue resistance and other physical properties.

At this writing it appears that the preferred alloy for the MIT Outsert Coil would have a coefficient of thermal expansion substantially below that of JBK-75 and close to that of Nb_3Sn.[2] Several candidate low-expansion alloys (Incoloy 903, 905 and 909) have been identified and tested after heat treatments similar to those required for multifilamentary Nb_3Sn.[35]

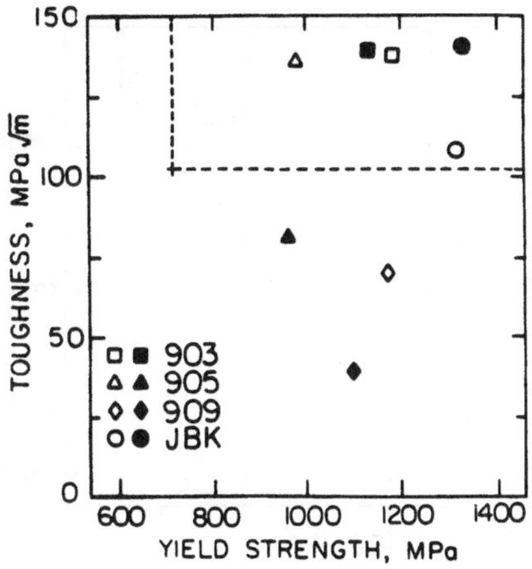

Fig. 19. Strength-toughness combinations for ICCS sheath alloys.

Some of the recent results are plotted in Figure 19. They show that Incoloy 903 and JBK-75 exceed strength-toughness specifications in the welded condition. Incoloy 905 fails in the welded condition, but may meet requirements Fig. 19. Strength-toughness combinations for ICCS sheath alloys if alternate heat treatments are used. However, the Incoloy alloys 903 and 905 are susceptible to embrittlement when exposed to oxygen at high temperature and must be treated with care.

The results of recent research on ICCS sheath alloys suggest that there are suitable, available alloys for current magnets. However, future ICCS magnets will almost certainly require new alloys with tailored properties. The optimal properties of future alloys are not yet clear.

CRYOGENIC MECHANICAL PHENOMENA AND FAILURE MODES

A second objective of cryogenic structural materials research is to identify and understand those mechanical phenomena that are important to the behavior of structural alloys at cryogenic temperatures. By way of illustration we shall consider three phenomena: fatigue, magnetic field effects, and cryogenic creep.

Fatigue

Metal fatigue is a potential failure mode in all high field magnet structures and is the most threatening failure mode in at least some components and designs. The modern procedure for fail-safe design against fatigue is based on the fatigue crack growth rate: one assumes the existence of small flaws or cracks in the initial structure and adjusts the cyclic load so that these cannot grow to critical size during the service life of the structure.

The data base that underlies fatigue design is the relation between the fatigue crack growth rate, da/dn (the increase in crack growth rate per stress cycle) and the cyclic stress intensity, ΔK, which is a measure of the effective cyclic stress at the crack tip. The general relation between the crack growth rate and the cyclic stress intensity is shown in Figure 20[36] and can be divided into three regions: The first, region I, is the "threshold" region, which is characterized by a very low crack growth rate that increases rapidly with ΔK. There is usually a practical "threshold value" of the stress intensity below which the crack growth rate can be ignored. The threshold value is sensitive to the environment and the microstructure and can be controlled, within bounds, by metallurgical treatment of the

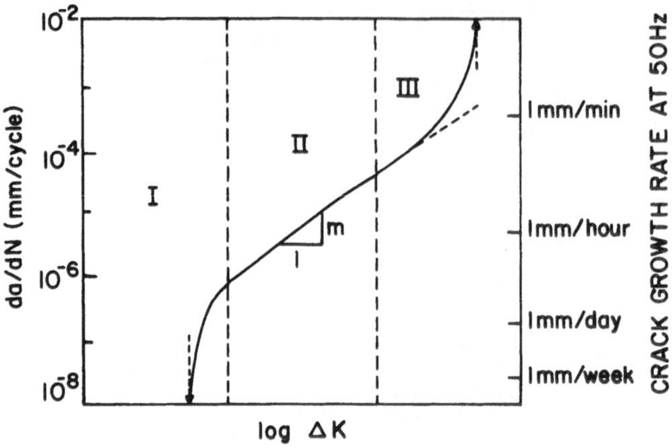

Fig. 20 Fatigue crack growth relation (courtesy R. Ritchie).

alloy. At higher values of ΔK, in region II (the "Paris" region), the crack growth rate obeys the simple power law

$$da/dn = A(\Delta K)^m \tag{1}$$

where the constants A and m depend on the material, but are only slghtly dependent on the microstructure. At still higher ΔK the maximimum stress intensity, K_{max}, approaches the plane strain fracture toughness, K_{Ic}. This leads to region III behavior, in which the crack growth rate increases dramatically, culminating in failure when

$$K_{max} = K_{Ic} \tag{2}$$

The portion of the fatigue crack growth curve that is most important to design depends on the load spectrum that the device will experience, and is particularly sensitive to the number of stress cycles it must withstand.

If a device must survive a very large number of stress cycles, as may be the case for the support structure for the poloidal coils in a tokamak fusion device, then it must be designed so that the cyclic stress intensity, ΔK, remains in the threshold region during operation. Since the stress intensity increases with the square root of the crack length, and since the crack growth rate in the threshold region is far below that which would be predicted by extrapolating the crack growth rate in the Paris region, the design is severely penalized unless the crack growth rate in the threshold region is known with reasonable precision. Because it is difficult and expensive to obtain such data at 4 K, the existing data base is sparse, and relevant data is only now beginning to appear[37]. The development of a satisfactory design data base for threshold fatigue at 4 K will almost certainly require the use of closed-loop, helium recondensation systems along the lines of the system recently constructed in Japan.[37,39]

If the device is subject to fewer load cycles, as may be the case for the toroidal coils in a fusion device, the structure may be designed to loads that lie in the Paris region of the crack growth curve. In this case it is desirable to find metallurgical techniques to minimize the crack growth rate. While the crack growth rate in the Paris region is relatively insensitive to the microstructure, it is influenced by phase transformations that may occur during fatigue, with the consequence that the fatigue crack growth rate is relatively low for both the metastable austenitic steels (e.g., 304 and 304L), which undergo a transformation to α-martensite near

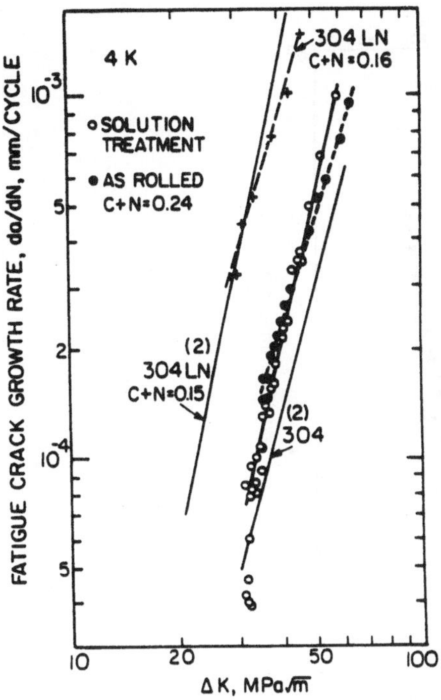

Fig. 21. Fatigue crack growth rates for austenitic steels. Data points for Kobe Fe-18Mn[13]. Solid lines from Ref. 46.

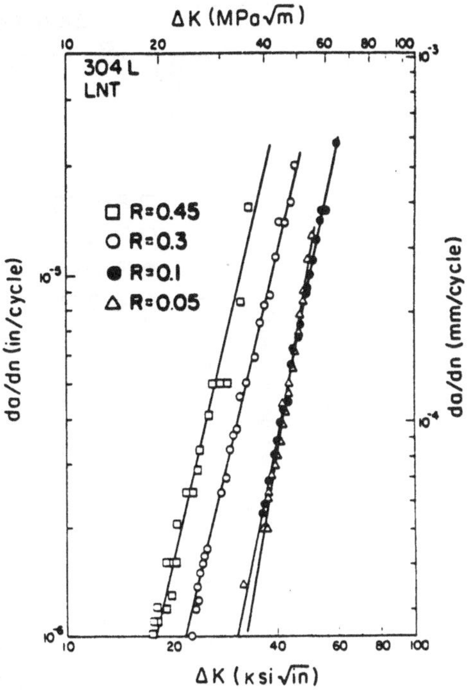

Fig. 22. Dependence of fatigue crack growth rate on load ratio.

the growing crack, and some of the Fe-Mn steels, which undergo a transformation to ε-martensite near the growing crack. The suppression of the fatigue crack growth rate for these metastable austenitic steels is illustrated in Figure 21, in which the 4 K crack growth rates of 304 stainless steel and Kobe Fe-18Mn-13Cr-5Ni-0.2N are compared with that of 304LN, a more stable austenitic steel.

The favorable fatigue crack growth rates of the metastable austenitic steels are due to the stress redistribution near the tip of the growing crack that accompanies the phase transformation. Though details of the process remain poorly understood, the effect of the transformation is to make the crack behave as if it closed on itself during relaxation, decreasing the effective value of ΔK[36]. This behavior not only decreases the crack growth rate for given ΔK, but has the consequence that the crack growth rate depends on the "load ratio"

$$R = K_{min}/K_{max} \tag{3}$$

The R-dependence of the fatigue crack growth rate of the metastable austenitic steel 304L at 77 K is illustrated in Figure 22.[40]

It is desirable to find alternate representations of the fatigue crack growth behavior that eliminate the R-dependence in order to minimize the data that must be measured to characterize fatigue properties. One simple method follows from the observation that the extent of transformation near the crack tip, and hence the transformation effect, depends on the maximum stress intensity, K_{max}[40]. If the crack growth rate is plotted against K_{max} rather then against ΔK the fatigue curves coalesce onto a single curve, as shown in Figure 23. This alternative representation is only valid for R

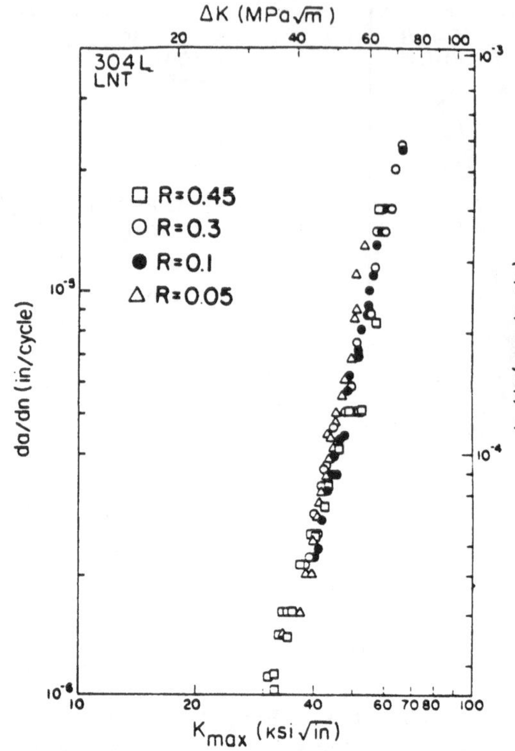

Fig. 23. Fatigue crack growth rate plotted against K_{max}.

values less than about 0.5. For higher values of R the transformation has a lesser influence, and the fatigue crack growth rates approach those of table austenitic steels.

Magnetic Field Effects

The structure of a superconducting magnet is subjected to high magnetic fields, which raises the question of whether the magnetic field may influence the mechanical properties of the structural alloy. Such an effect is most likely in metastable austenitic stainless steels and in ferritic steels that are toughened by the introduction of precipitated austenite, since the transformation from the paramagnetic austenite to the ferromagnetic α-martensite phase is thermodynamically promoted by the presence of a magnetic field.

To test this possibility cryogenic tensile tests have been conducted in magnetic fields up to 19 T on the metastable austenitic steels 304L and 304LN[41,42]. The results show that the magnetic field has a measurable influence on the tensile behavior, but that the effect is not large enough to affect mechanical design. A typical result is shown in Figure 24, which compares the stress-strain curves of 304L at 77 K in fields of 0T and 19 T. The magnetic field causes a slight decrease in the yield stress and the early stage flow stress, increases the ultimate tensile strength, and decreases the tensile elongation. However, each of these changes is relatively small. The fatigue crack growth rates of 304L and 304LN were also measured in an 8 T magnetic field at 77 K but did not change measurably from the zero-field values. The influence of magnetic fields on the mechanical properties of tempered 9Ni steel was also small.

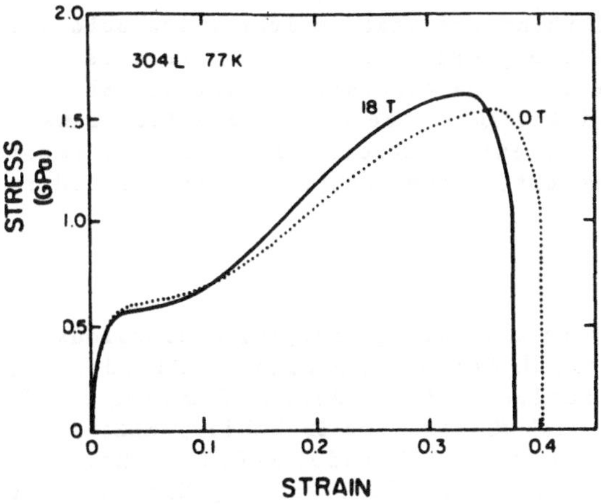

Fig. 24. Effect of magnetic field on stress-strain curve of 304L.

It should be noted, however, that theoretical work at Tohoku University[43] has predicted a significant magnetic field effect on the fracture toughness of certain alloys that has not yet been tested.

Cryogenic Creep

Creep, or the continuous plastic deformation of a material under load, is generally regarded as a high temperature phenomenon. Until quite recently creep was considered irrelevant to the design of cryogenic structures, such as high field magnets, despite fairly extensive work by Startzev and coworkers in the Soviet Union[44] that documents the possibility of significant creep at low temperature. Recent work at Westinghouse[45] suggests that creep not only occurs at 4 K in the austenitic steels that are candidates for magnet structures, but that the magnitude of the 4 K creep rate may be large enough to cause failure unless it is specifically considered in design.

As of this writing the data on 4 K creep in cryogenic structural alloys must be considered preliminary. The existing data are, however, sufficient to identify cryogenic creep as an important subject for further study.

CONCLUSIONS

This paper has briefly treated the status of research and development toward materials for the structures of high field superconducting magnets. The status may be summarized as follows:

1. A series of new austenitic steels with exceptional 4 K strength-toughness combinations has been developed and should be available for use in future magnets. However, the practical weldability of these alloys remains to be demonstrated.

2. There has also been substantial progress in the development of other classes of structural alloys that may have applications in high field magnets, including ferritic steels and high strength aluminum alloys.

3. Promising alloys have been identified for the conduits of ICCS conductors. However, the optimum property needs for the conduit alloys of future ICCS conductors have not been specified with the precision needed to guide new alloy design.

4. The mechanical phenomena that influence the behavior of structural alloys at 4 K are reasonably well understood. There is a significant need for additional fatigue data, particularly in the threshold region, and for better understanding of the influence of martensitic transformations on the fatigue of metastable austenitic steels. The possible importance of cryogenic creep has only recently been recognized, and should receive serious attention.

ACKNOWLEDGMENTS

The author wishes to acknowledge helpful discussions with coworkers at LBL, including B. Fultz, M. Strum, J. Glazer, D. Dietderich, G.-M. Chang and J. Chan, and at LLNL, including E. Dalder, L. Summers and J. Miller. He has also used discussions with R. Reed and R. Tobler, NBS; M. Hoenig and H. Becker, M.I.T.; S. Shimamoto and E. Tada, JAERI; T. Horiuchi, S. Tone, M. Shimada, and R. Ogawa, Kobe Steel; T. Sakamoto, Nippon Steel; J. Wells and L. Roth, Westinghouse; and V. Der, DOE. This work was supported by the Director, Office of Energy Research, Office of Development and Technology, Magnetic Systems Division of the U.S. Department of Energy under Contract No. DE-AC03-76SF00098.

REFERENCES

1. S. Shimamoto, K. Yoshida, H. Nakajima, Y. Ogane, K. Koizumi, and M. Oshikiri, Requirements for Cryogenic Structural Alloys for Superconducting Magnet Cases, in "Advances in Cryogenic Engineering Materials," vol. 32, Plenum Press, New York (1986).
2. M. O. Hoenig, Multipurpose Coil Design Requirements, Presented at DOE Workshop on Conductor-Sheath Issues for ICCS, Germantown, Md., (July, 1985),
3. D. T. Read and R. P. Reed, in "Materials Studies for Magnetic Fusion Energy Applications - II," NBSIR79-1609, National Bureau of Standards Boulder, Colorado, (1979), p. 81.
4. C. D. Henning, in "Proceedings, 9th International Cryogenic Engineering Conference", K. Yasukochi and H. Nagano, Eds., Butterworths, London, (1982), p. 7.
5. C. D. Henning and E. N. C. Dalder, in "Proceedings of the 5th International Conference on Structural Mechanics for Reactor Technology," vol. 10, Berlin (1979).
6. H. Nakajima, K. Yoshida, M. Oshikiri, K. Okuno, K. Kawano and S. Shimamoto, Fracture Toughness of Newly Developed Cryogenic Steels for Fusion Experimental Reactor, in "Advances in Cryogenic Engineering," vol. 32, Plenum Press, New York (1986).
7. T. Sakamoto, Y. Nakagawa, I. Yamauchi, T. Zaizen, H. Nakajima and S. Shimamoto, in "Advances in Cryogenic Engineering - Materials," vol. 30, Plenum Press, New York (1984), p. 137.
8. T. Sakamoto, Nippon Steel, Work presented at the Japan-U.S. Meeting on Structural Alloys for Fusion Magnets, JAERI, Tokyo (December, 1985).
9. J. W. Morris, Jr. and E. N. C. Dalder, "Cryogenic Structural Materials for High Field Superconducting Magnets," Lawrence Livermore National Laboratory Rept. 92681, Livermore, California (1985).
10. T. Sakamoto and Y. Nakagawa, Effects of Mn on the Properties of High Nitrogen Austenitic Stainless Steels, in "Advances in Cryogenic Engineering," vol. 32, Plenum Press, New York (1986).
11. S. Tone, M. Shimada, T. Horiuchi, Y. Kasamatsu, H. Nakajima and S. Shimamoto, in "Advances in Cryogenic Engineering - Materials", vol. 30, Plenum Press, New York (1984), p. 145
12. H. Masumoto, K. Suemune, H. Nakajima and S. Shimamoto, in "Advances in Cryogenic Engineering Materials," vol. 30, Plenum Press, New York (1984), p. 169.

13. R. Ogawa and J. W. Morris, Jr., in "Advances in Cryogenic Engineering Materials," vol. 30, Plenum Press, New York (1984), p. 177.
14. T. Horiuchi, Cryogenic Fe-Mn Austenitic Steels, in "Advances in Cryogenic Engineering Materials," vol. 32, Plenum Press, New York (1986).
15. K. Suemune and K. Sugino, A High-Strength, High-Manganese Stainless Steel for Cryogenic Use, in "Advances in Cryogenic Engineering Materials," vol. 32, Plenum Press, New York (1986).
16. M. J. Strum, Lawrence Berkeley Laboratory, Unpublished Research, (1985).
17. J. W. Chan and J. W. Morris, Jr., The Influence of Welding on the Mechanical Properties of Nitrogen-Strengthened High Manganese Cryogenic Steel, in "Advances in Cryogenic Engineering Materials," vol. 32, Plenum Press, New York (1986).
18. S. Tone, M. Hiromatsu, J. Numata and T. Horiuchi, Cryogenic Properties of Electron Beam Welded Joints in a 22Mn-13Cr-5Ni-0.22N Austenitic Stainless Steel, in "Advances in Cryogenic Engineering Materials," vol. 32, Plenum Press, New York (1986).
19. J. I. Kim, C. K. Syn and J. W. Morris, Jr., Metall. Trans. 14A:93 (1983).
20. Y. H. Kim, Ph.D. Thesis, Dept. Materials Science and Mineral Engineering, University of California, Berkeley (1985).
21. C. W. Marshall, R. F. Heheman and A. R. Triano, Trans. ASM 55:135 (1962).
22. S. Nagashima, T. Ooka, S. Sakino, H. Mimura, T. Fuzishima, S. Yano and H. Sakurai, Trans. ISIJ 11:402 (1971).
23. J. W. Morris, Jr., C. K. Syn, J. I. Kim and B. Fultz, in "Proceedings, International Conference on Martensitic Transformations," Cambridge, Mass. (1979).
24. H. J. Kim, Y. H. Kim and J. W. Morris, Jr., Metall. Trans. (in press).
25. S. Jin, J. W. Morris, Jr. and V. F. Zackay, Metall. Trans. 6A:129 (1975).
26. H. J. Kim and J. W. Morris, Jr., Lawrence Berkeley Laboratory, Unpublished Research (1985).
27. J. Glazer, S. L. Verzasconi, E. N. C. Dalder, W. Yu, R. A. Emigh, R. O. Ritchie and J. W. Morris, Jr., Cryogenic Mechanical Properties of Al-Cu-Li-Zr Alloy 2090, in "Advances in Cryogenic Engineering Materials," vol. 32, Plenum Press, New York (1986).
28. R. E. Gold, W. A. Logsdon, G. E. Grotke and B. Lustman, in "Advances in Cryogenic Engineering Materials," vol. 28, Plenum Press, New York (1982), p. 759.
29. M. M. Steeves, M. O. Hoenig and C. J. Cyders, in "Advances in Cryogenic Engineering Materials," vol. 30, Plenum Press, New York (1984) p. 883.
30. J. A. Brooks and R. W. Krenzer, Welding J. 53(6):242 (1974).
31. M. J. Strum, M.S. Thesis, Dept. Materials Science and Mineral Engineering, University of California, Berkeley (1982).
32. L. T. Summers, Ph.D. Thesis, Department of Materials Science and Mineral Engineering, University of California, Berkeley (1985).
33. L. T. Summers and J. W. Morris, Jr., Improvements in the Weldability of a Superconductor Sheath Material, in "Advances in Cryogenic Engineering Materials," vol. 32, Plenum Press, New York (1986).
34. P. Rezza, M. O. Hoenig, M. M. Steeves and Y. Iwasa, Mechanical Behavior of an Internally Cooled Cabled Superconductor (ICCS) under Cyclic Loading, in "Advances in Cryogenic Engineering Materials," vol. 31, Plenum Press, New York (1986).
35. L. T. Summers and E. N. C. Dalder, An Investigation of the Cryogenic Mechanical Properties of Low Thermal Expansion Superalloys, in "Advances in Cryogenic Engineering Materials," vol. 32, Plenum Press, New York (1986).
36. R. O. Ritchie, in "Fatigue Thresholds," vol. 1, J. Backlund, A. Blum and C. J. Beevers, Eds., EMAS, Ltd., Warley, U.K. (1982), p. 503.
37. R. L. Tobler, Near-Threshold Fatigue Crack Growth Behavior of Austenitic Stainless Steels at Cryogenic Temperatures, in "Advances in Cryogenica Engineering Materials," vol. 32, Plenum Press (1986).

38. K. Nagai, T. Ogata, T. Yuri and K. Ishikawa, Fatigue Testing at 4 K with Helium Recondensation System, in "Advances in Cryogenic Engineering Materials," vol. 32, Plenum Press (1986).
39. M. Akamatsu, M. Taneda, Y. Ohtsu, T. Tsukuda, S. Kataoka and T. Horiuchi, A Coolant Distributing Refrigerator for Machines to Test Materials at Liquid Helium Temperature, in "Advances in Cryogenic Engineering Materials," vol. 32, Plenum Press, New York (1986).
40. G.-M. Chang, M.S. Thesis, Department of Materials Science and Mineral Engineering, University of California, Berkeley (1984).
41. J. W. Morris, Jr. and B. Fultz, in "Proceedings, International Cryogenic Materials Conference, Kobe, Japan, 1982," K. Tachikawa and A. Clark, Eds., Butterworths, London (1982) p. 343.
42. B. Fultz, G. Fior, R. Kopa and J. W. Morris, Jr., Magneto-Mechanical Effects in Steels with Metastable Austenites, in "Advances in Cryogenic Materials," vol. 32, Plenum Press, New York (1986).
43. T. Shoji, Y. Shindo and M. Saka, Magneto-Fracture Mechanics and Structural Integrity of Superconducting Magnets, presented at the Japan-U.S. Workshop on Structural Materials for Fusion Magnets, JAERI, Tokyo (December, 1984).
44. V. I. Startzev, V. P. Soldatov, V. D. Natsik and V. V. Abraimov, Phys. Stat. Solidi 59(A):377 (1980).
45. L. D. Roth and E. N. C. Dalder, Creep of 304LN and 316L Stainless Steels at Cryogenic Temperatures, in "Advances in Cryogenic Engineering Materials," vol. 32, Plenum Press, New York (1986).
46. R. L. Tobler, D. T. Read and R. P. Reed, in "Materials Studies for Magnetic Fusion Energy Applications at Low Temperatures - II," NBSIR79-1609, NBS, Boulder, Colorado (1979) p. 81.

REQUIREMENTS FOR STRUCTURAL ALLOYS FOR SUPERCONDUCTING MAGNET CASES

S. Shimamoto, H. Nakajima, K. Yoshida, and E. Tada

Division of Thermonuclear Fusion Research
Japan Atomic Energy Research Institute
Naka-Machi, Naka-Gun, Ibaraki-Ken, Japan

ABSTRACT

This paper describes the requirements of cryogenic structural alloys for superconducting magnets in fusion reactors and their status of development. The mechanical property targets are a yield strength over 1,200 MPa and a fracture toughness over 200 MPa$\sqrt{m}$ at 4 K. These targets are deduced from strain characteristics of Nb_3Sn materials, stress analysis, and crack propagation. Several alloys that satisfy these criteria have already been developed in a collaboration of JAERI and Japanese industries. This year, we are studying the 4 K mechanical properties of the welded parts on these materials.

INTRODUCTION

The superconducting magnet is indispensable for plasma engineering and for fusion reactors in order to create a high magnetic field in a large space with realistic and economical power consumption. The engineering development has been accomplished, and now the magnet systems to confine the high-temperature plasma are being constructed. However, at the present state of the art, a fusion reactor magnet system is not possible. Both component development and the scaling-up project are necessary, and recently developed material should be used in a real magnet system for practical demonstration. It is generally thought that superconducting magnet technology development depends only on superconducting material research, but that is not sufficient. Just after the discovery of high-field material, this understanding was true. However, once the high-field materials reach the target level, developments in other technical areas of the system, such as structural material, instrumentation, and refrigeration, are necessary in order to establish a high level of operation for the entire magnet system. We have succeeded in generating 12 T in a large bore with a high-current Nb_3Sn conductor. Therefore, in material development, cryogenic structural alloy development has become the primary task.

REQUIREMENTS AND STATUS OF SUPERCONDUCTING MAGNET DEVELOPMENT

After the achievement of critical plasma in the present large Tokamak (JT-60), from the viewpoint of plasma technology, the next step in fusion development is a realization of burning. It will be done in the Tokamak Fusion Experimental Reactor (FER), whose magnetic parameters are shown in Table 1. It is impossible to operate the FER with a conventional magnet system because

Table 1. Key Parameters of the Fusion Experimental Reactor Magnet System

	Toroidal	Poloidal
Coil Bore	~10 m	2~20 m
Field	12 T	8 T
Current	~20 kA	50~100 kA
Operation	DC	Pulsed
Energy	~30 GJ	8 GJ

it demands a great amount of electrical power and high efficiency. The entire magnet system must be superconducting. In the design of the FER, 14 toroidal coils and 20 poloidal coils are required. The total weight of the magnet system will be over 10,000 tons. The structural material will be more than half of the total weight; the rest will be mainly conductor. Progress on plasma technology and cost analysis may change the geometry and coil numbers of the FER. However, the order of magnitude does not change so much. Figure 1 shows the requirements and the status of the superconducting toroidal coil. Since 12 T has been realized with JAERI's (Japan Atomic Energy Research Institute's) Test Module Coil using Nb_3Sn, the next key advance in technology is total system engineering for the magnet of several GJ. The superconducting technology is advancing smoothly in Japan and is becoming reliable. Therefore, such kind of system development will be successful and fruitful.

Figure 2 shows the requirements and status of the superconducting pulsed coil for a tokamak poloidal coil. Owing to the difficulty of putting barriers in wire and cabling, this developmental work started much later than that of the toroidal coil. Naturally, the scale-up of this development is also much further behind. However, the development of pulsed coils in Japan is

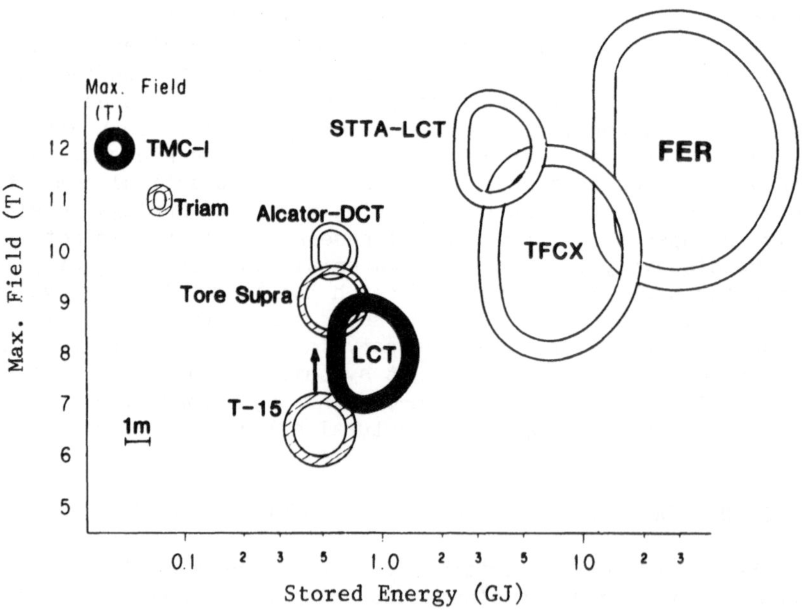

Fig. 1. Superconducting toroidal coil: requirements and status (black: realized; hatched: under fabrication; blank: future project).

also advancing, and JAERI is fabricating a high-current pulsed coil of the 30-MJ class. The pulse power for the testing will be supplied from a generator for the JT-60 operation.

The component development level of the magnet system has not advanced uniformly owing to the intrinsic characteristics of the components and interest in development. Figure 3 shows the developmental status of the components in the superconducting magnet. The high-field material and the superconducting stability technologies are the most advanced. In this situation, much more effort should be given to other items, such as the large refrigerators and structural materials. If not, it is impossible to realize or to operate a large magnet system forever. In a magnet system each component is equally important, since malfunction of one component stops the operation of the total system. With uniform development, good quality control, and excellent system checks, a superconducting magnet system can be very reliable and practical. What is missing up to now is this fundamental understanding and philosophy.

REASONS FOR NEW ALLOY DEVELOPMENT

As already described, excellent superconducting materials are not sufficient for realization of a large superconducting magnet system. To generate a high field in a large space, an excellent mechanical structure is required to support the superconducting material. The technical comparison of superconducting material and structural material follows:

Measurement Technique

Since superconductivity is an electromagnetic behavior, its measurement is very accurate, and an accuracy as high as one percent can be pursued. Superconducting behavior can be measured anywhere in a magnet; there is no geometrical restriction. The stress measurement still relies on strain gauge techniques, whose accuracy is not so excellent as electromagnetic measurements. In addition, geometrical restrictions prevent stress measurements at all points of the magnet. Therefore, at present, we cannot fully evaluate stress.

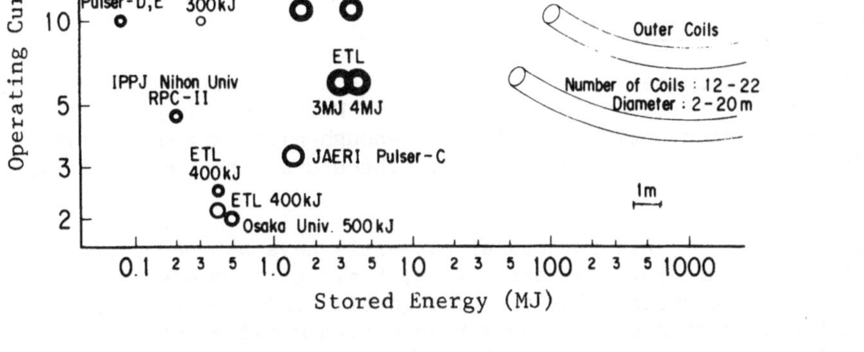

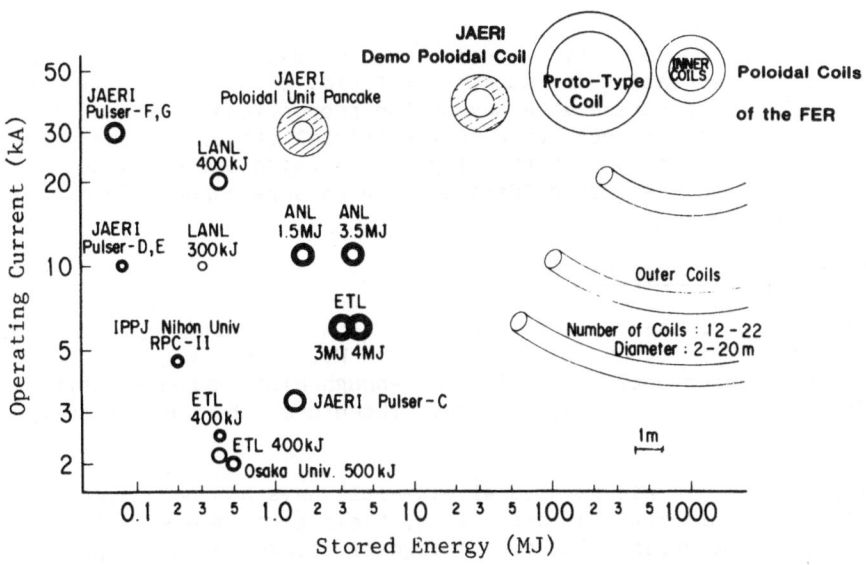

Fig. 2. Superconducting pulsed poloidal coil: requirements and status (black: realized; hatched: under fabrication; blank: future project).

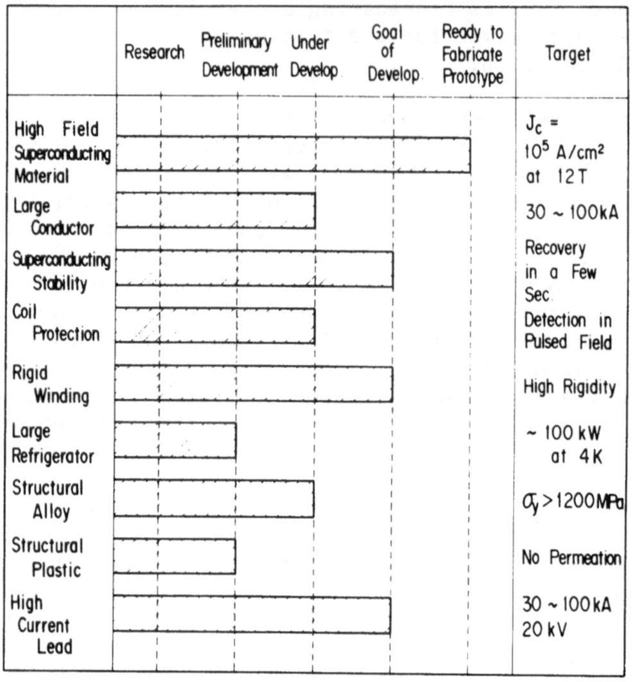

Research
Preliminary Development
Under Develop
Goal of Develop
Ready to Fabricate Prototype
Target
High Field Superconducting Material
J_c = 10^5 A/cm² at 12T
Large Conductor
30 ~ 100kA
Superconducting Stability
Recovery in a Few Sec
Coil Protection
Detection in Pulsed Field
Rigid Winding
High Rigidity
Large Refrigerator
~ 100 kW at 4K
Structural Alloy
σ_y > 1200MPa
Structural Plastic
No Permeation
High Current Lead
30 ~ 100kA 20 kV

Fig. 3. Degree of progress in key technique for superconducting magnet system.

Limit of Operating Zone

Superconductivity shows a drastic change at the critical point, and the critical characteristic can be well defined. The stress-strain curve does not show drastic change, and the operational zone cannot be definitely defined by a physical property.

Recovery to Operational State

Even if superconductivity reverts to the normal state for some reason, by cooling or by decreasing the current or field, it can be easily recovered. On the other hand, if stress goes beyond the proportional limit of the structure, the material cannot come back to the initial state. In the worst case, the material is stressed to the breaking point, which causes catastrophic failure. This is an absolute difference between superconducting material and structural material.

Design Margins in a Large Coil

In a large coil the current density is usually limited by the stability of the superconductivity or coil protection. With the present technique, however, the superconducting material has enough high current density that it is unaffected by these limiting factors. Therefore, the current margin is automatically high.

The present cryogenic structural materials have not been specially developed for cryogenic applications. Thus, their performance is limited by their physical properties. The electromagnetic force increases proportionally with the size of a coil. The stress design margins for a large coil are automatically limited by present material properties.

Necessity of Structural Alloy Development

It can be now understood that for realization of a large, high-field superconducting coils, especially coils for fusion, structural material development is as important as superconducting material development.

REQUIREMENT FOR NEW CRYOGENIC ALLOYS

The high-current Nb_3Sn conductor technique is well established for the high-field generation in a large bore, since 12-T generation was demonstrated with JAERI's Test Module Coil in June 1985. Therefore, the design of a large, high-field coil must be based on the performance of Nb_3Sn.

However, it is well-known that there is a degradation of critical current in Nb_3Sn at a certain level of strain. This degradation characteristic is more pronounced in higher fields. The design limit of strain at the level of 12 T must be 0.50%, and the design value should be around 0.25%. The coil case, as well as the conductor, should respect these values.

The peak stress in a large coil depends on coil support, type of winding, and operational mode, including fault mode. From our experience of stress calculation, the peak value is 500 to 800 MPa. The 304LN (N = 0.15%) stainless steel, currently one of the most popular alloys for cryogenic use, has a yield strength of ~800 MPa and an ultimate strength of ~1,700 MPa at 4 K. The 316LN (N = 0.15%) has a yield strength of ~900 MPa and an ultimate strength of ~1,650 MPa at 4 K. Since these materials do not have a sufficient safety margin, new materials with enough margin are required. Figure 4 shows required stress-strain performance of a new alloy in comparison with that of 304LN stainless steel and a superconductor. The required yield strength should be 1200 MPa.

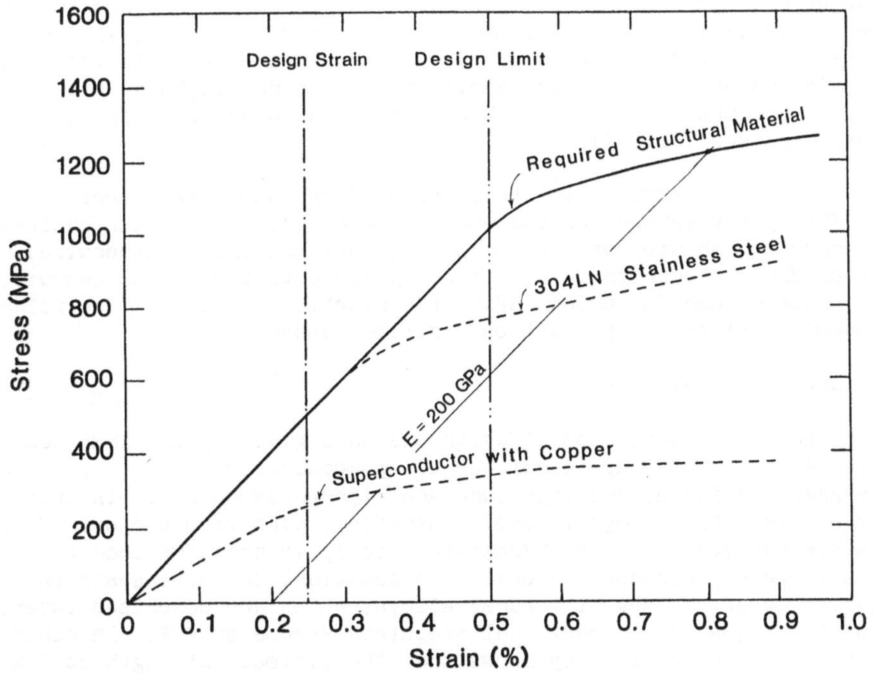

Fig. 4. Stress-strain curve of structural alloys and conductor at 4 K.

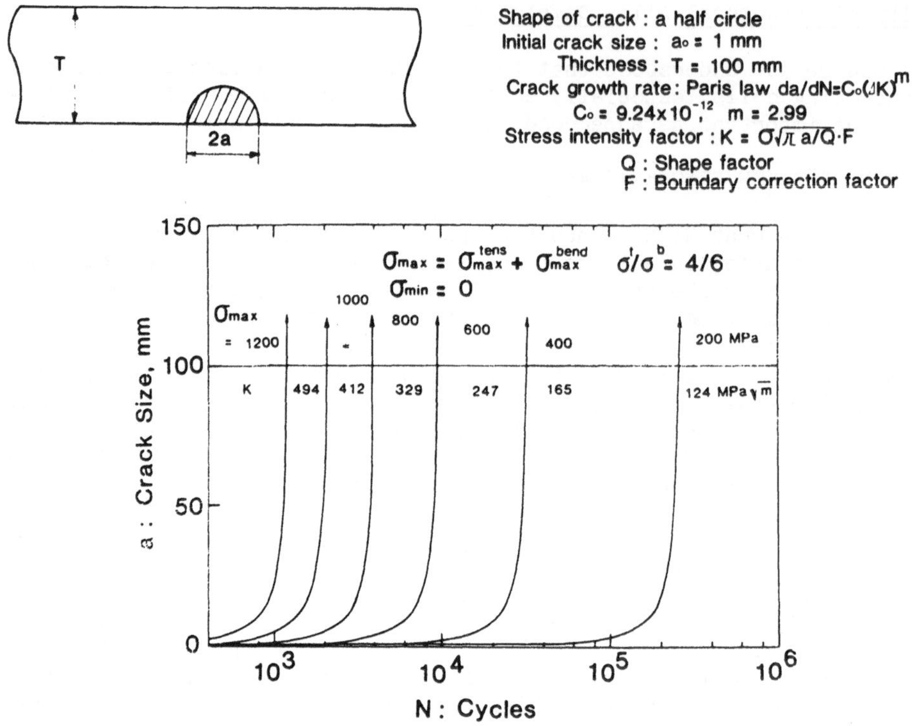

Fig. 5. Relation of crack propagation due to repeated cycles.

The crack propagation analysis[1] includes two assumptions: a plate thickness of 100 mm, which is a reasonable value in a large coil, and an initial crack radius of 1 mm. Figure 5 shows the result. At 500 MPa and 10^4 cycles, a fracture toughness, K_{Ic}, of 200 MPa$\sqrt{m}$ is required. For most alloys, if the yield strength increases, the fracture toughness decreases. In the conventional stainless steels described above, the fracture toughness goes sufficiently over 200 MPa$\sqrt{m}$. However, these materials do not satisfy the yield characteristic (see Table 2).

The irradiation effect will be studied after candidate materials are selected. The heat treatment of the sheath material for the Nb_3Sn cable-in-conduit conductor should not affect the property required. Otherwise, we should consider a conductor fabrication technique that puts the conduit at the last stage of manufacturing, after the reaction of Nb_3Sn. The welded joints should satisfy the properties described above.

PROPOSED NEW DESIGN STANDARD

Up to now, the conventional design standard given by the ASME code has been referred to for the design of large superconducting coils. The ASME code was prepared for high-temperature and high-pressure use. In this code, there is no basis for cryogenic use. Therefore, even with the use of conventional stainless steel, it is inconsistent to apply the ASME code to the design of a large superconducting coil. In addition, the stress-strain characteristics of the new structural alloys, which are described later, are different from those of conventional stainless steels at 4 K. In other words, the new structural alloy shows that the ultimate strength at 4 K is much nearer to the yield strength than that of conventional stainless steel (see Figure 6). Thus, if we apply the ASME code to the new structural alloy

Table 2. Developmental Targets for New Cryogenic Structural Alloys

Yield strength	– More than 1,200 MPa
Fracture toughness	– More than 200 MPa$\sqrt{m}$
Fatigue characteristics	– Similar to those of 316 austenitic stainless steel
Magnetic permeability	– Not specified (nonmagnetic material is preferred)
Corrosion resistance	– Good rust resistance
Others	– Good workability and weldability

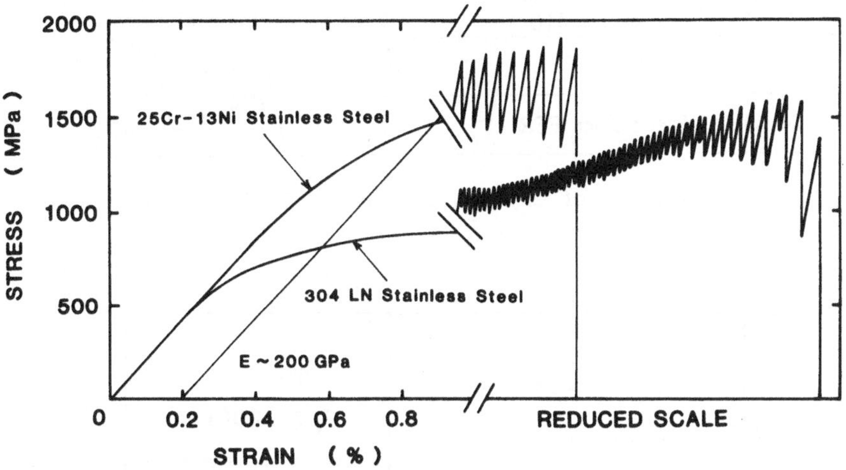

Fig. 6. Measured stress-strain curve of high Cr-Ni stainless steel and 304LN stainless steel.

Table 3. ASME Code and Proposed New Design Standard

		ASME Code	Proposed New design Standard
Base Temperature		300 K	4 K
Strain Limit		no indication	0.5 % *
Sm		(2/3)Sy, (1/3)Su	(2/3)Sy, (1/2)Su **
Fatigue	Design Stress Amplitude (Sa)	(1/2) Failure Stress	(1/2) Failure Stress
	Design Cycle (Nf)	(1/20) Cycle to Failure	(1/10) Cycle to Failure

* due to Nb_3Sn limit
** New structural material shows that Su is much nearer to Sy than the classic materials.

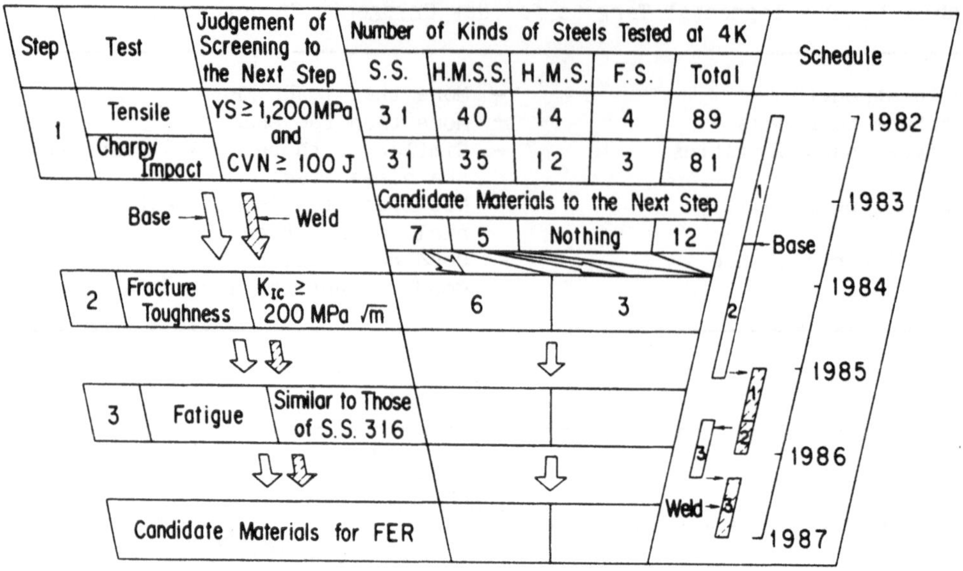

Fig. 7. JAERI's steps of cryogenic structural alloy development.

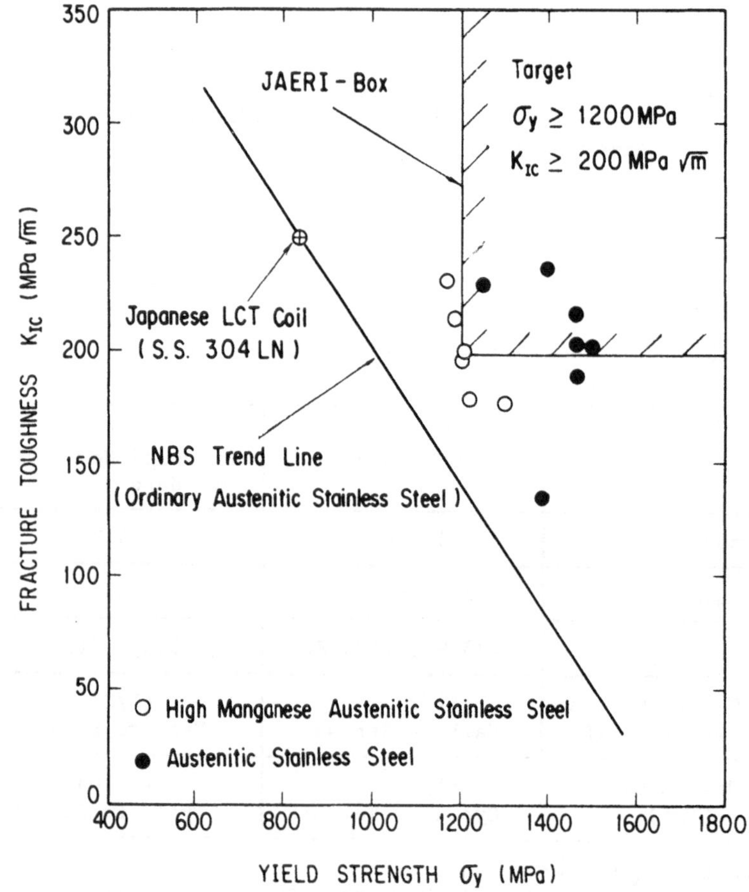

Fig. 8. Fracture toughness and yield strength of recently developed alloys at 4 K (the status in August 1985).

the design limitation is only that of ultimate strength. In this case, the range of the linear portion of the stress-strain curve is sufficient. The design standard should be modified from this viewpoint. Our new, proposed stress value, Sm, is either 2/3 of the yield strength or 1/2 of the ultimate strength.

In the ASME code, there is no restriction on the strain limit. Since we use Nb_3Sn conductor for high-field generation, a strain limitation should be written into the new code. The design limit of the strain should be 0.50% owing to beginning of critical current degradation.

There is no data base on fatigue characteristics at 4 K. If fatigue characteristics at 4 K are similar to those at room temperature in the low-cycle range, the ASME design code limit of 1/20 of the cycles to failure restricts performance, and in the high-cycle range, the design limit of 1/2 of the amplitude of the failure stress. In the Tokamak reactor, repeated stress amplitude due to normal pulse operation is not so high, but the stress value due to an upset or emergency condition will be high. However, the number of cycles in the latter condition is not very high. Therefore, we propose that in the low-cycle range, the limit should be relaxed to 1/10 of the cycles to failure. The proposed design standard is listed in Table 3 and compared with the ASME code. This proposal is not a definite one and should be discussed with international experts.

In addition to the proposals described above, we should take into account the serration effect at 4 K, which occurs in the zone over the yield strength value. In this zone, the structural alloy can be never practically used due to its distinguished repeated performance. The technical issue is how we include this effect in a design standard.

We have some other issues on weldments: safety margins and partial penetration of welding. In ASME, the safety margins in a welded part are the same as for the alloy, and full welding is always required. Solutions to these problems should be proposed later, after more experience in design and a larger data base.

STATUS OF NEW ALLOY DEVELOPMENT

After the completion of the Japanese LCT coil fabrication, JAERI established collaboration contracts with four Japanese industries that prepare newly developed alloy samples. With the planning shown in Fig. 7, JAERI carried out cryogenic testing (tensile, Charpy and fracture toughness) on these materials. At first we had the impression that our targets were too high. However, after several studies of testing and fabrication, we obtained materials that satisfy the JAERI Box. Figure 8 shows the fracture toughness-yield strength characteristics of the new materials and 304LN stainless steel. The promising candidates for future magnets are high-Mn stainless steel and high-Cr-Ni stainless steel.[2] These two materials have already been melted on an industrial scale of several tens of tons. The results obtained are very promising, and the new materials will be used in JAERI's demonstration pulse coil program. Thus, we want to experience actual practical application of the new materials.

Besides this projected work, this year we are developing welding techniques and evaluating welds. Some of this work was presented at the 1985 ICMC.[3]

CONCLUSIONS

On the basis of our past experience of hardware and the conceptual design of the fusion reactor, we proposed requirements for cryogenic structural alloys to be used in superconducting magnet systems in the reactor. Several candidate alloys have been obtained after a few years work. Of these alloys, high-Mn stainless steel and high-Cr-Ni stainless steel, which were melted on an industrial scale, are very promising. This year we will evaluate welds. Then, fatigue testing will be started in one year. Weld evaluation and fatigue testing are the keys to the last stage. In addition, we should establish design standards for a large superconducting magnet system, which will require a lot of discussion.

ACKNOWLEDGMENT

The authors thank Drs. S. Mori, A. Tomabechi, and M. Tanaka for their continuous encouragement of this work. The contributions of Nippon Steel Co., Kobe Steel, Japan Steel Works, and Kawasaki Steel Co. are hereby acknowledged.

REFERENCES

1. K. Yoshida, H. Nakajima, K. Koizumi, M. Shimada, Y. Sanada, Y. Takahashi, E. Tada, H. Tsuji, and S. Shimamoto, Development of cryogenic structural materials for tokamak reactor, in "Austenitic Steels at Low Temperatures," R. P. Reed and T. Horiuchi, eds., Plenum Press, New York (1983), pp. 29-39.
2. H. Nakajima, K. Yoshida, K. Okuno, M. Oshikiri, E. Tada, S. Shimamoto, R. Miura, M. Shimada, S. Tobe, K. Suemune, T. Sakamoto, and K. Nohara, Results of fracture toughness tests on newly developed structural materials for superconducting coils of fusion experimental reactor, in: "Advances in Cryogenic Engineering - Materials," vol. 32, Plenum Press, New York (1986).
3. S. Tone, M. Hiromatsu, J. Numata, T. Horiuchi, H. Nakajima, and S. Shimamoto, Cryogenic properties of electron beam welding joint in a 22Mn-13Cr-5Ni-0.22N austenitic stainless steel, in: "Advances in Cryogenic Engineering - Materials," vol. 32, Plenum Press, New York (1986).

CRYOGENIC Fe-Mn AUSTENITIC STEELS

T. Horiuchi, R. Ogawa and M. Shimada

Kobe Steel, Ltd.
Chuo-ku, Kobe, Japan

ABSTRACT

It is taken for granted that structural materials used for superconducting magnets of a fusion reactor must combine strength, toughness and fatigue resistance especially in case of welded structures at cryogenic temperatures. Austenitic Fe-Mn alloys were expected to be promising candidates to meet the above requirements. However, they proved to have some problems to be overcome in terms of cryogenic brittle fractures. To optimize the mechanical properties, chromium, nickel, nitrogen and other elements were added to Fe-Mn alloy. Consequently, 22Mn-13Cr-5Ni steel has been developed as the candidate steel. Optimum manufacturing conditions have been established and the steel has been tested at cryogenic temperatures. The results show that the steel maintains satisfactory strength of more than 1.2 GPa and fracture toughness of more than 200 MPa$\sqrt{m}$ at 4 K.

INTRODUCTION

Many researchers are interested in the potential of Fe-Mn alloys for cryogenic use in the high field superconducting magnets. The conventional Fe-Mn alloys have been usually used in casted or forged forms because of their poor formability. The Hadfield steel (13Mn-1C steel) and 18Mn-5Cr steel are the typical examples among the Fe-Mn alloys. A new Fe-Mn alloy with excellent formability has recently been developed by Kobe Steel and used in the fabrication of a fusion experimental reactor, the JT-60, because of their higher non-magnetic stability and lower cost than conventional stainless steel.[1] This successful usage stimulates the industries to develop new cryogenic Fe-Mn base alloys.

On the other hand, there is a need to develop techniques for reducing the nickel content of conventional cryogenic steels which have a high nickel content because the nickel significantly increases the cost.[2,3,4,5] Nickel could be replaced by manganese because they have similar metallurgical effects on microstructures of steels. Consequently, Fe-Mn-Cr base alloys, which are modifications of Fe-Mn systems, have been investigated.

In the following, recent studies on the basic properties of the Fe-Mn alloy will be reviewed briefly. An on-going alloy development project will be described with examples primarily taken from the research at Kobe Steel.

The essential property of the structural alloys is considered to be a high strength accompanied by a high toughness. It is a matter of course that other properties should be optimized in order to fulfill the performance of the cryogenic system.

Phase Diagram

Figure 1 shows a structural diagram of Fe-Mn-Cr ternary alloy at 4 K.[6] The influence of Cr on phases is minimal for Cr content of less than 15%, while δ-ferrite is retained for higher Cr content. At high Mn and Cr levels, the σ-phase is also present. Though the diagram is very similar to that at 300 K reported by Kato et al.,[7] the $(\gamma+\alpha'+\varepsilon)/(\gamma+\varepsilon)$ and $(\gamma+\varepsilon)/\gamma$ phase boundaries shift to a higher Mn content at 4 K.

Toughness

Fe-Mn alloys containing more than 30% Mn content are austenitic even at 4 K and are expected to have high cryogenic toughness. Nevertheless, their toughness decreases with the addition of Mn, providing its content exceeds approximately 30%, as shown in Fig. 2.[8] The phases present in Fe-Mn alloys are plotted in Fig. 3.[9,10] For Mn content of more than 36%, austenite of the Fe-Mn alloys is very stable, but its cryogenic toughness decreases with Mn content. Therefore, the transformations are not directly connected with this anomaly, though it might appear to be due to a strain-induced transformation of the residual austenite in the alloy.

The alloy in this composition range fractures in an intergranular mode at cryogenic temperatures.[9,10] Mn addition to Fe-Mn alloys at high Mn content causes an increase in stacking fault energy.[11,12,13] High stacking fault energy reduces the width of an expanded dislocation and enhances the

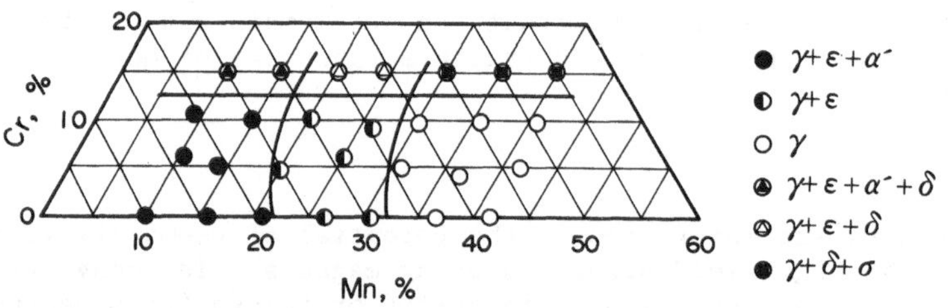

Fig. 1. A structural diagram of the Fe-Mn-Cr System at 4 K.[6]

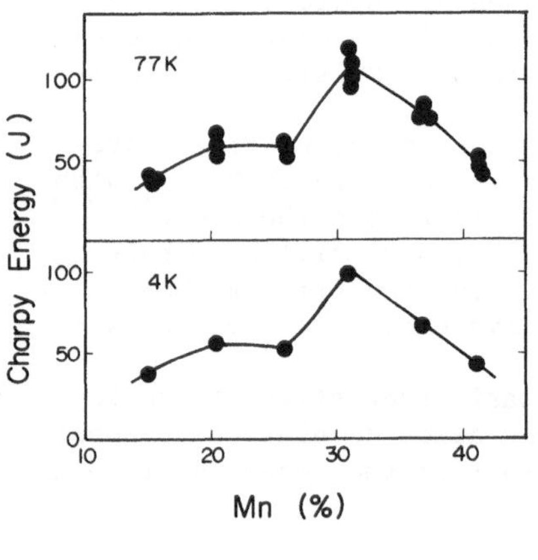

Fig. 2. Toughness versus Mn content for Fe-Mn alloys at 77 K and 4 K.[8]

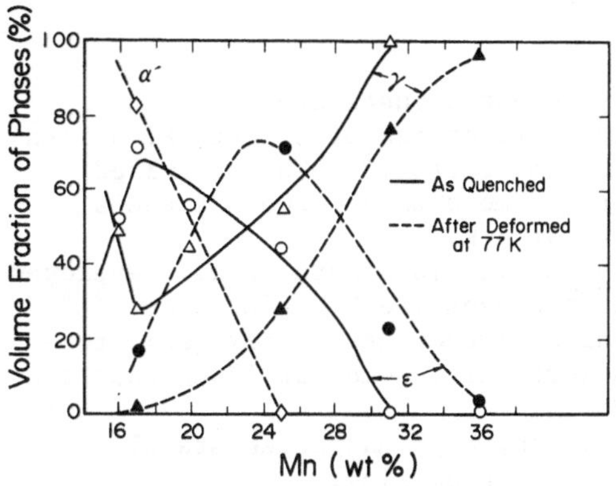

Fig. 3. The phases present in Fe-Mn alloys as a function of Mn content.[10]

concentration of the stress around dislocations. Therefore, it may become difficult to relieve a stress concentration at grain boundaries and an intergraunular fracture may result.[11] However, the metallurgical reason for this behavior has not been fully clarified as yet.

Improvement of Toughness

Two techniques proposed in order to suppress the ductile-brittle transition of the alloys are grain size control[14] and a combination of chromium additive and cooling control, respectively.[15,16]

Figure 4 shows the grain size dependence on the transition temperature of Fe-40Mn alloy.[14] The embrittlement is suppressed at 4 K in the alloys with a grain size of less than 30 μm. However, Tone et al. reported the reverse grain size dependence of both Charpy absorbed energy and fracture toughness at 4 K in austenitic Fe-Mn-Cr-Ni alloys.[26] The effects of grain boundaries in fcc metals on toughness are not well understood.

Shibata et al. found that the cryogenic toughness of solutioned Fe-32Mn-7Cr-0.3N alloy was improved by reheating it at 773 K though no significant change in its microstructure was observed.[15] Morris also found the same phenomenon and Cr and N segregation at grain boundaries. Then he suggested that heating at 873 K for 30 min. might cause remedial segregation of the Cr-N pairs near grain boundaries.[16]

At any rate, an application of the simple Fe-Mn austenitic alloy to cryogenic structures will produce some problems due to embrittlement, the reason for which is still under investigation. Therefore, Fe-Mn alloys must be improved by practical and reliable techniques.

A proportionate addition of chromium, nickel, carbon, and nitrogen is one of the practical and promising methods to be discussed.

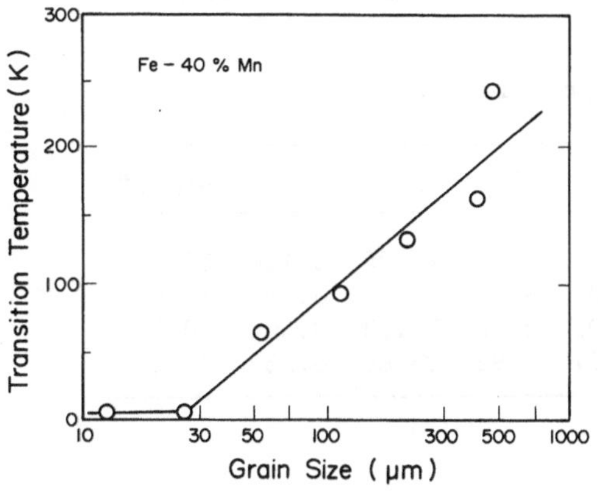

Fig. 4. The grain size dependence of ductile-brittle transition temperature in Fe-40Mn alloy.[14]

Recently JAERI (Japan Atomic Energy Research Institute) has made public the requirements for the mechanical properties of structural alloys used for superconducting magnets in a fusion reactor.[17] They required rustproof alloys to have a yield strength of more than 1,200 MPa and a fracture toughness of more than 200 MPa√m at 4 K.

Before these requirements were made public, Yoshimura et al. developed the 25Mn-5Cr-1Ni steel[2] and Miura et al. proposed the 32Mn-7Cr steel.[3,4] While these alloys exhibited good mechanical properties at cryogenic temperatures, their strength and toughness were still somewhat lower than the requirements, and they were not rustproof.

In order to fulfill JAERI's requirements, high manganese stainless steels containing nitrogen were thought to be promising.

Alloying Effects

The authors have studied alloying effects on the cryogenic mechanical properties of high manganese austenitic stainless steels.[18,19,20] The studies were conducted on a series of steels containing 15 to 28% manganese, 1 to 7% nickel, 12 to 18% chromium, and 0.14 to 0.26% nitrogen.

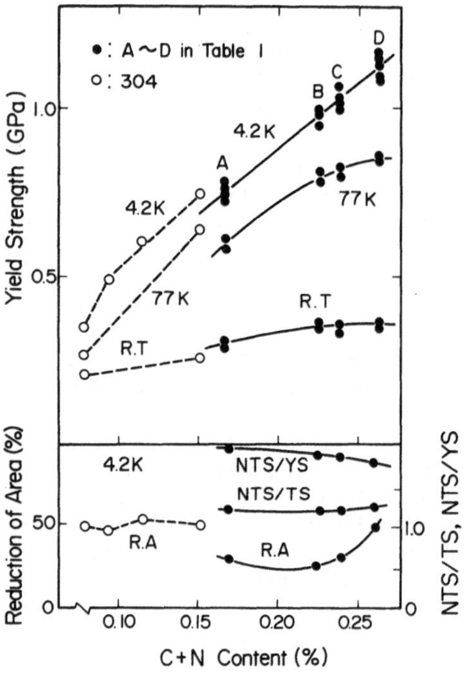

Fig. 5. The effects of C+N content on yield strength, reduction of area, notch yield strength and notch tensile strength of Fe-Mn-Cr base alloys.[20]

Table 1. Chemical Compositions (%)

Alloy	C	Si	Mn	P	S	Ni	Cr	N	$Ni_{eq.}$*
A	0.030	0.11	15.5	0.004	0.01	4.00	12.98	0.140	16.9
B	0.035	0.10	16.2	0.001	0.01	3.94	14.67	0.193	18.9
C	0.036	0.10	18.3	0.001	0.01	3.98	11.93	0.204	20.3
D	0.032	0.01	18.4	0.001	0.01	4.92	13.80	0.233	22.1

* $Ni_{eq.}$ = Ni + 30C + 30N + 0.5Mn

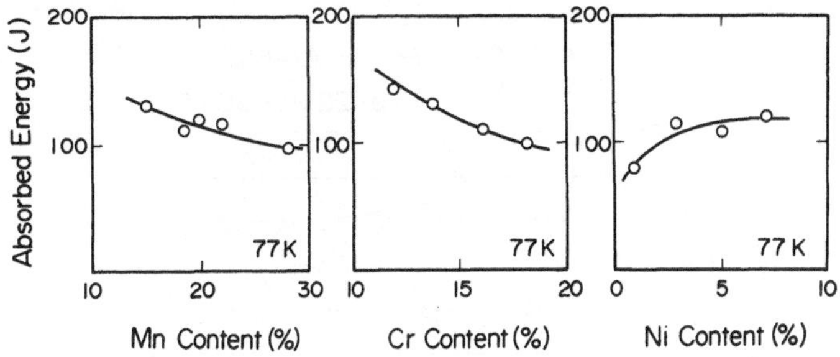

Fig. 6. The effects of Mn, Cr, and Ni content on the Charpy absorbed energy at 77 K.[20]

Figure 5 shows the experimental results of tensile tests. For comparison, data of 304 stainless steel are plotted in the figure. The chemical compositions of the alloys corresponding to the alloy in Fig. 5 are tabulated in Table 1. Yield strength at cryogenic temperatures of the solutioned alloys is strongly affected by the addition of carbon and nitrogen just like the 304 stainless steel.[18,21] The other main elements (Cr, Mn, Ni) in the alloys do not contribute much to the strength because their atomic sizes are very similar to that of iron. However, addition of Mn and Cr has the advantage of increasing the nitrogen solubility limit.[22]

Figure 6 shows the effects of Mn, Cr and Ni on Charpy absorbed energy at 77 K. As shown in the graphs, the absorbed energy decreased slightly with manganese and chromium content and increased gradually with nickel content. Their effects were not significant as far as their austenicity is concerned. Figure 7 shows the effects of $Ni_{eq.}$ on Charpy absorption energy and magnetic stability. As the increase in permeability is due to induced α' martensite, $Ni_{eq.}$ must be more than 20% if a stable nonmagnetic property is desired. Figure 8 illustrates in more detail the phase stability against deformation at 4 K. For comparison, the results of 316 stainless steel are also plotted in the figure. The Fe-Mn system differs

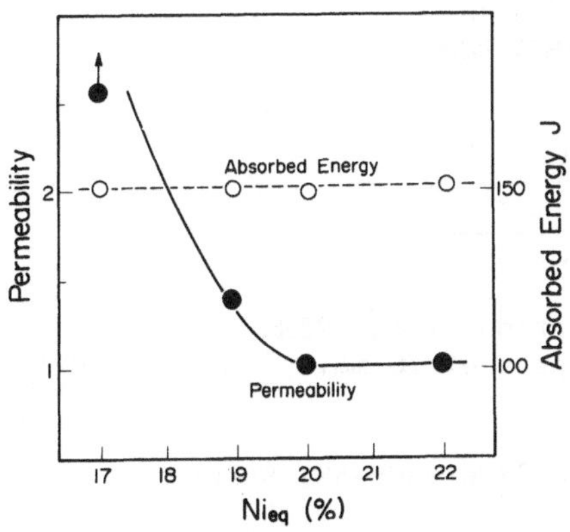

Fig. 7. The effects of $Ni_{eq.}$ on Charpy absorbed energy and permeability at fracture surface at 77 K.[20]

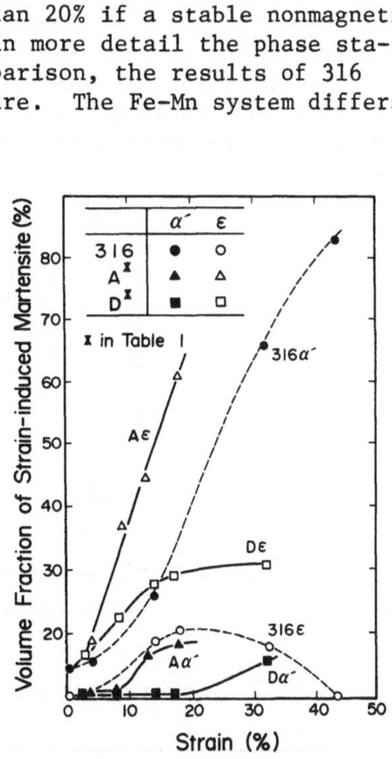

Fig. 8. The relation between strain and volume fraction of strain-induced martensite.[20]

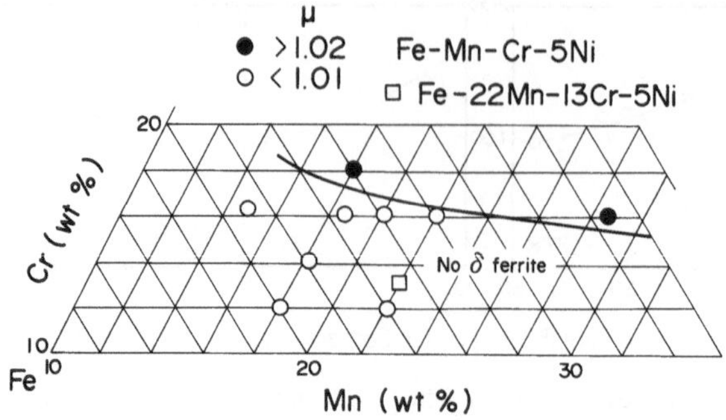

Fig. 9. The measurement of δ-ferrite.

from the Fe-Cr-Ni system mainly by its propensity toward transformation to the ε phase rather than the α' phase. This means that the magnetic stability of the Fe-Mn system is superior to that of the Fe-Cr-Ni system. Moreover, the chromium effect on phase stability was also investigated.[23] It was found that an excess addition of chromium produces δ-ferrite, as shown in Fig. 9.

In order to determine the chemical composition of the new alloy, properties and conditions, such as strength, toughness, phase stability, weldability, corrosion resistance, manufacturing process, and cost, were taken into consideration.[24] Table 2 shows the chemical composition of the selected alloy.

Manufacturing Conditions

As one can see in Fig. 5, the required yield strength at 4 K can be obtained by an addition of carbon and nitrogen of 0.26% or more. On the other hand, carbon content must be minimal and nitrogen content much less than the solubility limit for satisfactory weldability. These contradictory requirements could be reconciled to some extent by employing the "Thermo-Mechanical Control Process" (TMCP), which also reduces production costs.

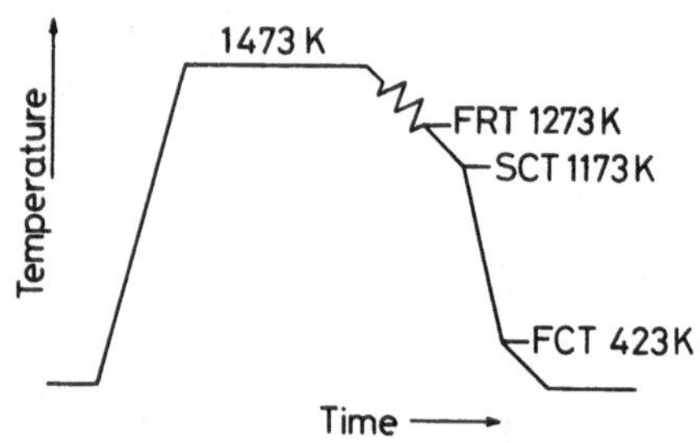

Fig. 10. The manufacturing condition of the 22Mn steel plate with thickness of 70 mm.

Table 2. Chemical Composition of the 22Mn-13Cr-5Ni Steel

C	Si	Mn	P	S	Ni	Cr	N
0.05	0.10	22.0	0.01	0.01	5.0	13.0	0.220

The 22Mn steel with the nominal chemical composition shown in Table 2 was prepared on a trial commercial basis using a 15-ton electric furnace. Using a large ingot, the effects of manufacturing conditions on cryogenic properties of the 70 mm thick plate were investigated in order to optimize the conditions.[25,26,27] The results suggest, for optimal processing, to finish rolling at 1,273 K (1000°C) and to start water cooling at 1,173 K (900°C), followed by air cooling at 423 K (150°C), as shown in Fig. 10.

The alloy manufactured through this process has been studied and evaluated as follows:

Evaluation

Figure 11 shows the tensile properties against temperature. The yield strength of more than 1,200 MPa and high ductility at 4 K have been confirmed. At the same time, the Charpy impact toughness was determined to be satisfactory at 4 K, as shown in Fig. 2. The 22Mn steel has a strength-toughness combination superior to the 304 stainless steels.

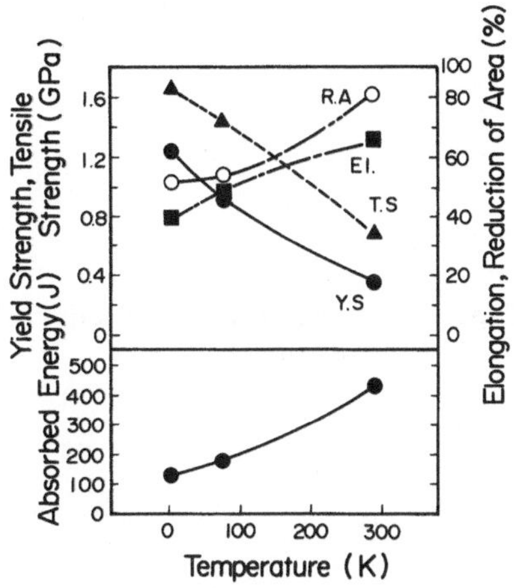

Fig. 11. The tensile properties and Charpy absorbed energy as a function of temperature.[25]

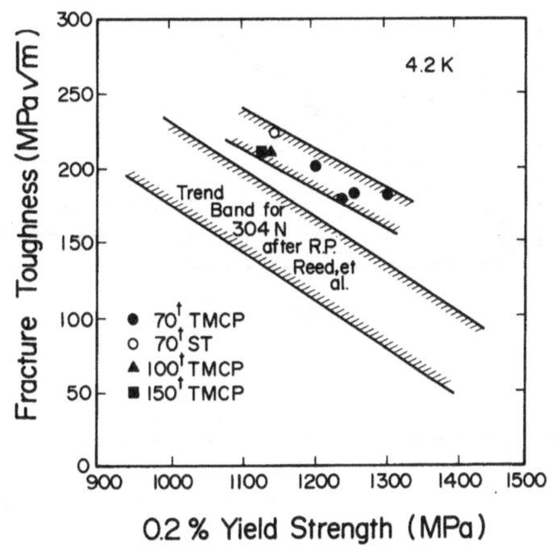

Fig. 12. The relation between yield strength and fracture toughness at 4 K.

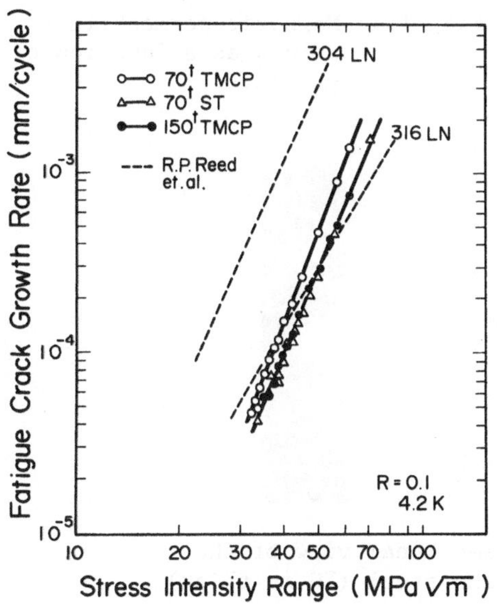

Fig. 13. The fatigue crack growth rate of the 22Mn steel at 4 K.

Moreover, the fatigue crack growth rate at 4 K is also shown in Fig. 13.[28] While a transformation-induced crack closure was not expected in this alloy because α' phase, whose positive volume change caused the crack closure, did not form at the crack tip,[29] satisfactory fatigue crack growth rate properties were observed. An electron microscope revealed that the crack in high manganese steel propagated zigzag on the [111] plane.[30] This behavior may be related to the satisfactory fatigue crack growth rate of the 22Mn alloy because contact of rugged surfaces is expected to cause a crack closure.

Figure 14 shows the magnetization curves at 4 K. For comparison, conventional austenitic stainless steels are plotted in the figure. As expected, 22Mn steel exhibits excellent nonmagnetic stability. And Fig. 15

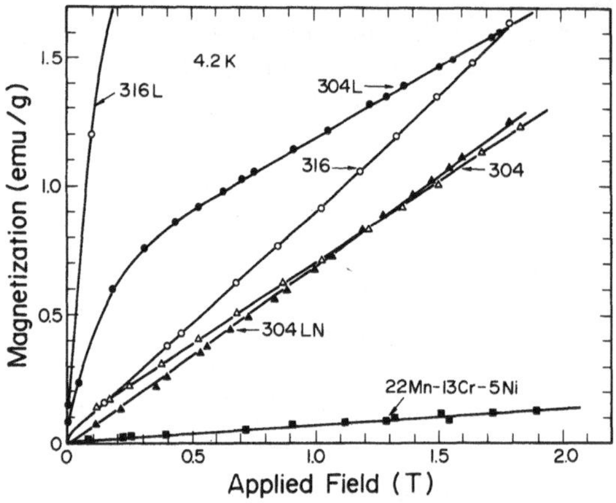

Fig. 14. Magnetization curves at 4 K.

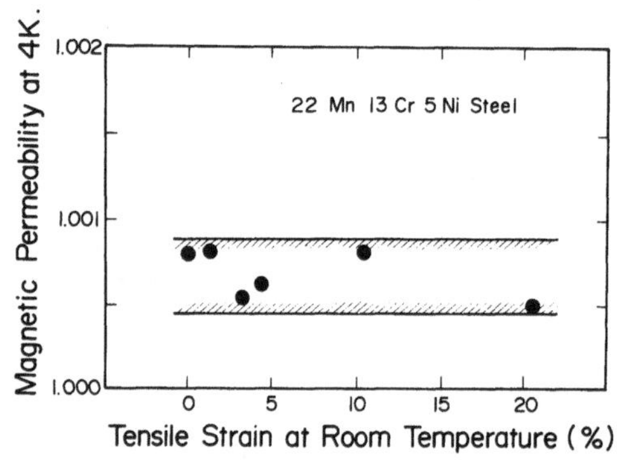

Fig. 15. Magnetic permeability at 4 K as a function of strain.

Photo 1. The macroscopic cross-sectional view of the electron-beam-welded 22Mn steel (70 mm thick).

shows the permeability at 4 K of the alloy deformed at ambient temperatures. These data proved that the alloy was highly stable in a magnetic sense against deformation. This makes the alloy most suitable for application in a high magnetic field.

Photo.1 shows the cross-sectional view of the welded plate. The 22Mn steel was confirmed to be weldable by electron beam welding (EBW) though the steel contained an amount of nitrogen which might cause voids in welded metal.

The mechanical evaluations of the weldment have been carried out and have yielded promising results. New welding consumables are also being developed specifically for the alloy. Studies on the welding will be reported in another section of this volume[31].

SUMMARY

The recent progresses of researches on cryogenic Fe-Mn alloys are reviewed briefly. As a typical example of developing cryogenic iron-manganese base alloy, so called 22Mn steel is introduced with various kinds of data, because it is the only alloy that is manufactured on an industrial level. It proves to possess superior properties as a cryogenic steel, such as high strength, high toughness, excellent magnetic stability, and good weldability.

REFERENCES

1. "Kobe Steel's Non-Magnetic Steel Plate NONMAGNE 30" Report of Kobe Steel, Ltd. (1979)
2. H. Yoshimura, H. Masumoto and T. Inoue, "Adv. Cryo. Eng." vol.28, Plenum Press, N.Y. (1981), p.115
3. K. Ohnishi and R. Miura, Tetsu-to-Hagane 60: S1989 (1980)
4. R. Miura, H. Nakajima, Y. Takahashi and K. Yoshida, "Adv. Cryo. Eng." vol.30, Plenum Press, N.Y. (1983), p.245
5. J. W. Morris, Jr., S. K. Hwang, K. A. Yushchenko, V. I. Belotzerkovetz and O. G. Kvasnevskii, "Adv. Cryo. Eng." vol.24, Plenum Press, N.Y. (1978), p.91
6. J. Namekata and K. Higashi, Tetsu-to-Hagane 66: S355 (1980)
7. T. Kato, S. Fukui, M. Fujikura and K. Ishida, Trans. Iron Steel Inst. Jpn 16: 16 (1976)
8. J. Namekata and Y. Kondo, Tetsu-to-Hagane 68: S1464 (1982)
9. Y. Tomota and J. W. Morris, Jr., Tetsu-to-Hagane 70: S503 (1984)
10. J. W. Morris, Jr. and E. N. C. Dalder, R & D - Kobe Steel Engineering Reports 34: 1 (1984)
11. J. Namekata, Y. Kondo and T. Matsumoto, Tetsu-to-Hagane 69: S1482 (1983)
12. H. Schumann, Krist. Tech 10: 1141 (1974)
13. P. Y. Volosevich, V. N. Gridnev and Y. N. Petrov, Phys. Met. Metalligr. 42: 372 (1976)
14. T. Matsumura, Y. Kondo and J. Namekata, Tetsu-to-Hagane 70: S1292 (1984)
15. K. Shibata and T. Fujita, Tetsu-to-Hagane 71: S601 (1985)
16. J. W. Morris, Jr., Private Communication
17. K. Yoshida, K. Koizumi, H. Nakajima, M. Shimada, Y. Sanada, Y. Takahashi, E. Tada, H. Tsuji and S. Shimamoto, "Proc. ICMC Kobe Japan" Butterworths, Guildford, (1982) p.417
18. T. Horiuchi, R. Ogawa, M. Shimada, S. Tone, M. Yamaga and Y. Kasamatsu, "Adv. Cryo. Eng." vol.28, Plenum Press, N.Y., (1981), p.93
19. R. Ogawa, M. Shimada, Y. Kanetsuki and T. Horiuchi, "Proc. 33th meeting Cryogenic Association Japan" (1981) C2-4
20. T. Horiuchi, R. Ogawa and M. Shimada, R & D - Kobe Steel Engineering Reports" 34: 53 (1984)

21. Y. Takahashi, K. Yoshida, M. Shimada, E. Tada, R. Miura and S. Shimamoto, "Adv. Cryo. Eng." vol.28, Plenum Press, N.Y. (1981), p.73
22. W. M. Small and R. D. Pehlke, Trans Met. Soc. AIME 242: 2501 (1968)
23. R. Ogawa and J. W. Morris, Jr., Tetsu-to-Hagane 70: S498 (1984)
24. H. Nakajima, K. Yoshida, Y. Takahashi, M. Shimada, S. Tone, R. Ogawa, M. Yamaga, H. Kaji, T. Horiuchi and Y. Kasamatsu, "Proc. ICMC Kobe Japan" Butterworths, Guildford, (1982), p.405
25. H. Kaji, M. Hiromatsu, S. Tone, M. Shimada and S. Shimamoto, R & D - Kobe Steel Engineering Reports 34: 57 (1984)
26. S. Tone, M. Shimada, T. Horiuchi, Y. Kasamatsu, H. Nakajima and S. Shimamoto, "Adv. Cryo. Eng." vol.30, Plenum Press, N.Y. (1983) p.145
27. S. Tone, M. Hiromatsu, H. Kaji, M. Shimada, H. Nakajima and S. Shimamoto, Tetsu-to-Hagane 70: S502 (1984)
28. S. Tone, M. Hiromatsu, H. Kaji, R. Ogawa and S. Shimamoto, Tetsu-to-Hagane 71: S506 (1985)
29. K. Katagiri, R. Ogawa, G. M. Chang and J. W. Morris, Jr., Private Communication
30. R. Ogawa and J. W. Morris, Jr., in "Fatigue at Low Temperatures" ASTM STP857 R. I. Stephens ed., American Society for Testing and Materials, Philadelphia (1985), p.47
31. S. Tone, M. Hiromatsu, J. Numata, T. Horiuchi, H. Nakajima, and S. Shimamoto, to be published in "Adv. Cryo. Eng." vol.32

NICKEL AND NITROGEN ALLOYING EFFECTS ON THE STRENGTH AND TOUGHNESS OF AUSTENITIC STAINLESS STEELS AT 4 K

R. P. Reed and P. T. Purtscher
Fracture and Deformation Division
National Bureau of Standards
Boulder, Colorado

K. A. Yushchenko
E. O. Paton Institute of Electrowelding
Kiev, USSR

ABSTRACT

The tensile strength and fracture toughness at 4 K were studied as a function of Ni (6-15 wt.%) and N (0.90-0.28 wt.%) contents for eight austenitic stainless steels. Results indicate that Ni increases the tensile yield strength and decreases the fracture toughness, $K_{Ic}(J)$, and Ni has little effect on tensile yield strength but increases the fracture toughness. The temperature dependence of the yield strength is given by $\sigma_y = \sigma_o e^{-AT}$, where σ_o is the yield strength at 0 K, and A is the slope of ln σ_y vs. T. The parameter A is proportional to the stacking fault energy. Lower Ni alloys exhibited brittle facets on fracture surfaces. The quality index, a new parameter = $\sigma_y \cdot K_{Ic}(J)$, relates to the capacity of the alloy to achieve greater strength or toughness, but not at the expense of the other parameter. Nickel alloying increases the quality factor; nitrogen has little effect.

INTRODUCTION

To constrain magnetic forces developed in high-field superconducting magnets, structural alloys that are strong and tough at 4 K are required. It has been demonstrated that N is a potent solid-solution strengthening agent in austenitic stainless steels at 4 K.[1] However, the toughness of the high-N alloys decreases as the strength increases.[2] In this study, Ni content was varied from 6 to 15 wt.% in alloys containing 19% Cr, 3 to 5% Mn, low C, and 0.09 to 0.28% N. Tensile strength and fracture toughness of the alloy series were measured in liquid helium and correlated with Ni and N contents.

EXPERIMENTAL PROCEDURES AND RESULTS

The eight laboratory heats of steel included in this study were supplied by E. O. Paton Institute of Electrowelding, Kiev, USSR (see Table 1). All alloys were completely austenitic and in the form of 25-mm-thick plates. They were tested after solution treatment at 1060°C for 1 h and water quenching. The procedures and accuracies for tensile testing of round specimens and fracture toughness tests (J-integral) using compact-tension specimens have been discussed elsewhere.[2-4]

Table 1. Chemical Composition and Metallurgical Characterization

Alloy	Composition (wt.%)								Grain Size	Hardness
	Cr	Ni	Mn	C	N	S	P	Si	(μm)	(R_B)
413	18.8	5.6	4.1	0.020	0.256	0.005	0.014	0.25	122	84
414	18.4	8.9	4.2	0.018	0.281	0.005	0.014	0.28	165	82
415	20.5	12.8	5.5	0.024	0.265	0.004	0.015	0.34	200	81
416	20.8	14.9	5.2	0.028	0.277	0.004	0.015	0.41	160	82
520	19.8	5.8	3.2	0.014	0.260	0.004	0.012	0.41	180	85
739	18.7	8.7	3.7	0.014	0.093	0.005	0.013	0.23	220	72
740	19.7	11.5	3.9	0.012	0.141	0.005	0.013	0.27	250	72
741	20.8	14.7	4.4	0.016	0.197	0.004	0.014	0.33	207	76

Table 2. Mechanical Property Results at 4 K and Physical Parameters

Alloy	Tensile Strength, σ_y (MPa)	σ_u (MPa)	Elongation (%)	Fracture Toughness, $K_{Ic}(J)$ (MPa$\sqrt{m}$)	Estimated Stacking Fault Energy, γ (mJ/m^2)	Estimated T_{md} (K)
413	1075	1418	11	75	8	309
414	950	1650	42	143	28	224
415	1000	1580	32	204	58	97
416	1160	1630	22	239	71	47
520	960	1216	7	87	7	316
739	460	1477	40	273	26	275
740	720	1370	42	244	45	186
741	870	1425	29	254	67	80
AISI 304					24	
AISI 316					55	
AISI 310					98	

The averages of the tensile strength and fracture toughness (2 or 3 measurements for each alloy) at 4 K are listed in Table 2. The high-N alloy stress-strain curves are characterized by high yield strengths and a lack of work hardening at large deformations. The lower N, less-stable alloys exhibit the characteristic three stages of stress-strain behavior identified earlier.[5] Nitrogen increases σ_I (the elastic limit); N and Ni reduce and finally eliminate stage II [associated with hexagonal close-packed (ε) martensite and body-centered cubic (α') martensite at slip-band intersections] and stage III (associated with α'-induced work hardening). Discontinuous yielding occurred in all measurements, and the load drops near fracture ranged from about 25% for the stronger alloys to about 10% for the more ductile alloys.

The temperature dependence of the tensile yield strength, σ_y, was measured for three alloys from 4 to 295 K. The results are plotted in Fig. 1. There is a linear relationship between $\ell n\ \sigma_y$ and temperature, similar to that of other austenitic steels.[3] The least stable alloy, 739, exhibits little change of yield strength in the region between about 100 to 200 K. This region has been identified previously for metastable alloys[6] and is associated with the α' and ε martensitic transformations.[7]

The tensile yield strength is a strong function of the C and N content, as shown in Fig. 2. Here the function C + 2N is used. Reed and Simon[1] found this to be a representative function when they used regression analysis to describe all σ_y data on AISI 304 type alloys at 4 K. The data of this paper closely correspond to those from earlier research on Fe-19Cr-10Ni base alloys.[8]

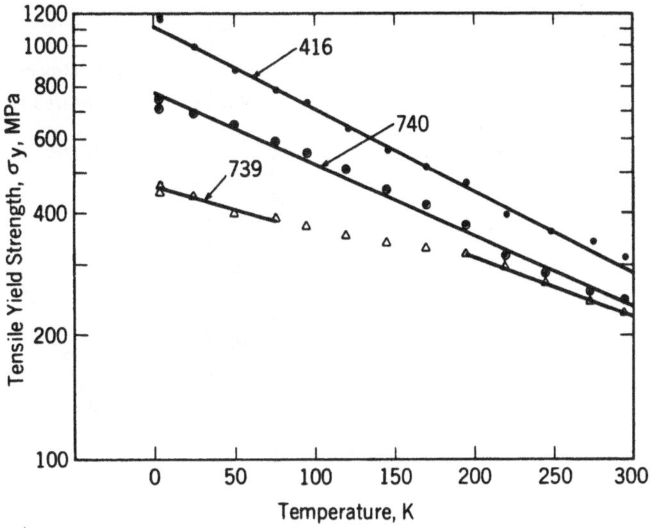

Fig. 1. Temperature dependence of the yield strength for 416, 739, and 740 alloys.

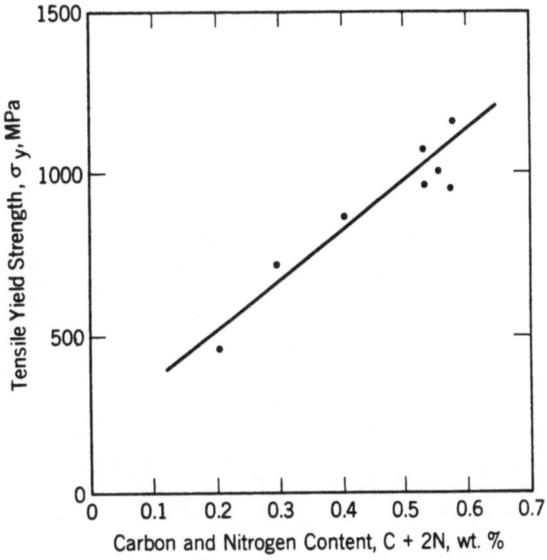

Fig. 2. Tensile yield strength at 4 K as a function of C and N content (C + 2N). Line indicates trend of a series of nine 19Cr-10Ni steels with varying C and N contents.[8]

The effect of Ni content on fracture toughness is shown in Fig. 3. Notice the trend of increasing toughness with increasing Ni in the19Cr-4Mn-0.02C-(0.26-0.28)N alloys (413-416). From Table 2 it is clear that the alloys with lower N contents form no systematic series with respect to dependence of toughness on Ni content; presumably the N varies too much to permit comparison solely on the basis of Ni. Analyses of the data (Table 2) reveal that, at constant Ni content, toughness tends to be less at higher N content. In Fig. 3 the values of $K_{Ic}(J)$ reported for the lower N content alloys (739, 740, 741) were adjusted to include consideration of the effect of N variability.[9]

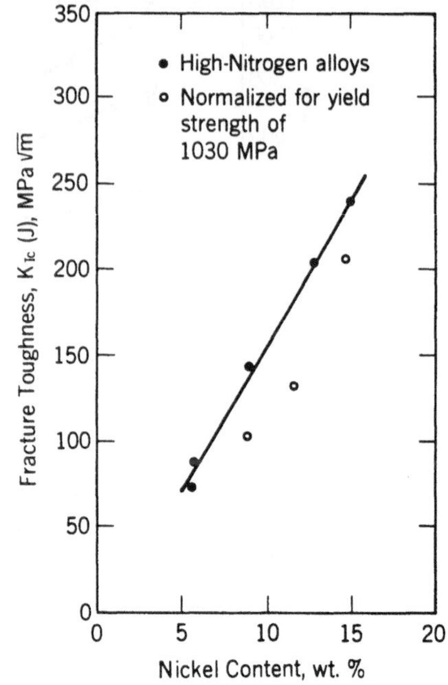

Fig. 3. Fracture toughness at 4 K, estimated from J-integral measurements, versus Ni content.

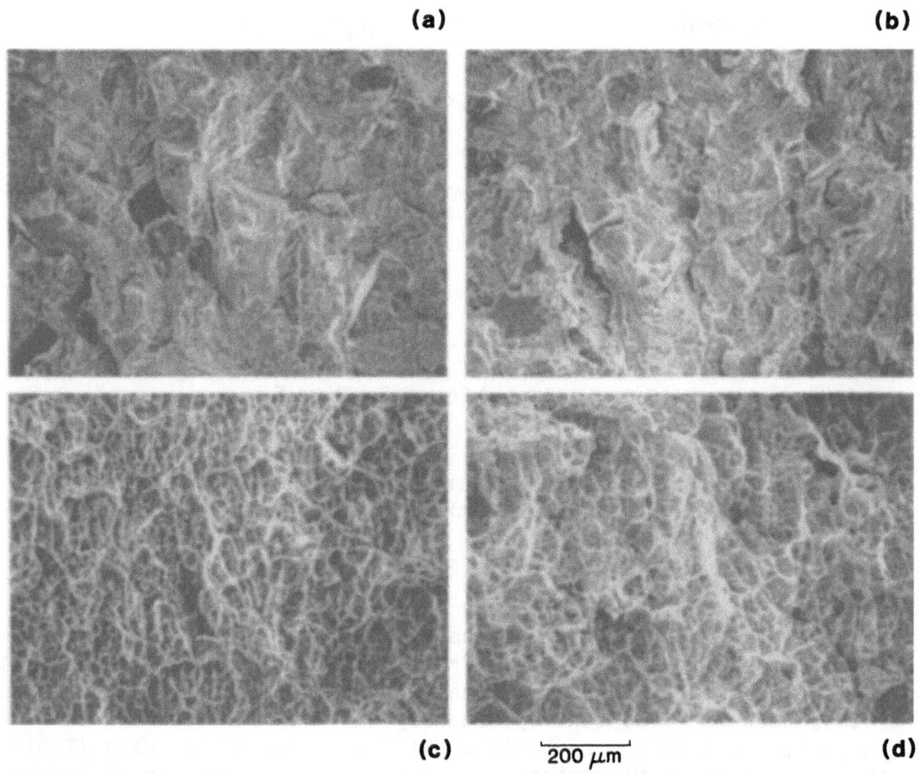

Fig. 4. Fracture surfaces at crack tip of alloys (a) 413, (b) 414, (c) 415, and (d) 416 (SEM photomicrograph).

The fracture appearances in the surface area associated with the critical J-integral measurements were characterized by scanning electron microscopy (SEM). The fracture surfaces of the high-N alloys (413-416) are shown in Fig. 4. As the Ni content increases, the fracture morphology becomes more dimpled (microvoid coalescence). Nonmetallic inclusions are associated with most dimples. These large (3-4 μm) inclusions are Mn silicates, according to energy-dispersive x-ray analysis.

At low-Ni content [low $K_{Ic}(J)$], the fracture surfaces are faceted. There are small, step-like facets and large (size of grain) flat facets. The large facets, shown in Fig. 5a, are apparently {111} planes of austenite. Their appearance is very similar to those recently identified as {111} by Tobler and Meyn[10] who used back-scattered Laue x-ray analysis and described them as regions of "slip-band decohesion." Linear traces on the faceted surfaces, such as those shown in Fig. 5b, correspond to other {111} traces.

DISCUSSION

All alloys, except 416, partially transformed to body-centered cubic (α') martensite during plastic deformation. The average volume percent of α' that formed throughout the gage length, normalized by dividing by the percent elongation, correlates well with T_{md} of Williams et al.[11] (see Table 2 and Fig. 6.)

From earlier work,[3] the temperature dependence of the tensile yield stress of austenitic stainless steels is best portrayed as

$$\sigma_y = \sigma_o e^{-AT} \quad (1)$$

where σ_o is the yield strength at 0 K, and A is a constant.

Reed and Arvidson[3] pointed out that there may be a correlation between the temperature dependence of the tensile yield strength (constant A) and the stacking fault energy (γ); this is evident in Fig. 7. For estimation of γ we have chosen to use the Schramm-Reed analysis,[12] compiled in Table 2, which produces admittedly high γ values, but we think accurate relative values of γ, considering only the major alloying elements Cr, Ni, and Mn.

The primary deformation mechanism in stage I of polycrystalline austenitic steels at low temperatures may be associated with cross-slip, cutting of forest dislocations, or solid-solution strengthening mechanisms. Some theoretical approaches for these possible rate-controlling deformation processes imply a linear dependence of ln σ on T [see Schrock and Seeger[13] (cross-slip) and Haasen[14] (solid-solution strengthening)]. Clearly, we must continue studies to identify and characterize the rate-controlling deformation mechanism in austenitic steels.

There is strong dependence of fracture toughness on Ni content (Fig.3) wnen the N content is held constant. SEM photomicrographs of the fracture surfaces at the initial fatigue-sharpened crack tip are shown in Fig. 4. As Ni content increased, the corresponding fracture surfaces exhibited a transition from faceted to dimpled surfaces. These faceted surfaces, shown at higher magnification in Fig. 5, are cleavage-like in appearance but do not show the characteristic river patterns found on the cleavage facets of body-centered cubic alloys. The surface is probably a {111} austenite plane, considering the orientations of {111} slip-band traces and the similarity to regions in another high-N austenitic stainless steel, which were identified as {111} austenite by Tobler and Meyh.[10] Similar fracture features have been observed in austenitic stainless steels charged with H and

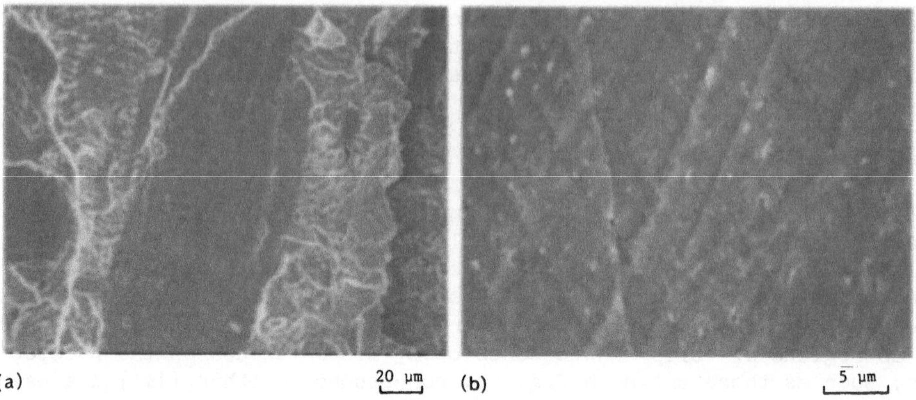

Fig. 5. Cleavage-like areas that become apparent in the more brittle alloys: (a) alloy 413, low magnification SEM photomicrograph; (b) alloy 413, high magnification SEM photomicrograph.

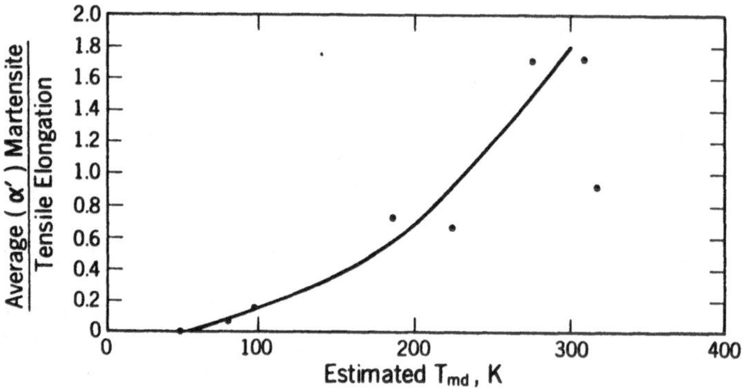

Fig. 6. The normalized average amount of α' martensite within the gage length of tensile specimens (after deforming at 4 K) plotted against the calculated md temperatures (based on chemical composition).

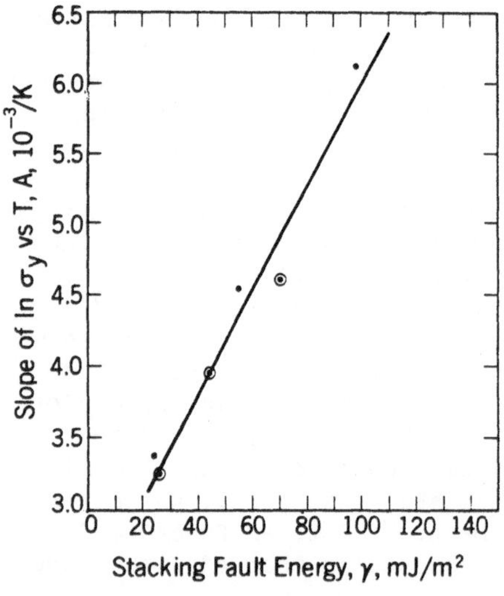

Fig. 7. Temperature dependence of yield strength (A of Eq. 1) versus estimated stacking fault energy, based on chemical composition. Three circled data points represent AISI 304, 316, and 310 alloys from a previous study.[3] The temperature range used for alloy 739 was 200-300 K.

fractured at room temperatures.[15,16] The study of Hannula et al.[15] shows a similar trend: brittleness increased with decreasing austenite stability.

Further study of the possible origin of apparently brittle {111} austenite was pursued with optical microscopy.[9] The microstructure of alloy 740, containing 11.5% Ni, is characterized by lath-like α' martensite formation, contained within {111} austenite sheets and by lath-like α' martensite that formed at {111} slip-band intersections. The microstructure of alloy 413, containing only 5.6% Ni, is composed of many {111} slip-band traces. These slip-band traces are broader than those observed in alloy 740 and more distinctly etched. Less α' formation was noticed at slip-band intersections. One may conclude that there is more extensive faulting in the lower Ni alloy. Possible conclusions are: more extensive stacking-fault clusters exist, more hexagonal close-packed martensite is formed, or deformation twins are produced within the {111} slip-band packets.

Nickel and nitrogen contribute quite differently to the low temperature mechanical behavior of austenitic stainless steels. Nickel improves toughness and does not affect strength; nitrogen improves strength and reduces toughness. For most structural applications, both strength and toughness are required. It is useful, we believe, to consider both of these parameters together, as the product $\sigma_y \cdot K_{Ic}(J)$, and to examine the effects of alloying on this product, called the <u>quality index</u>. In Fig. 8, Ni and C plus 2N contents are plotted versus the quality index. Increasing Ni content increases this index. Stated another way, increasing Ni content leads to improved low temperature mechanical behavior with increased toughness at nearly constant strength. Additions of C and N have little effect on the quality index. Therefore, one adjusts N content to affect strength and suffers from an attendant change of toughness.

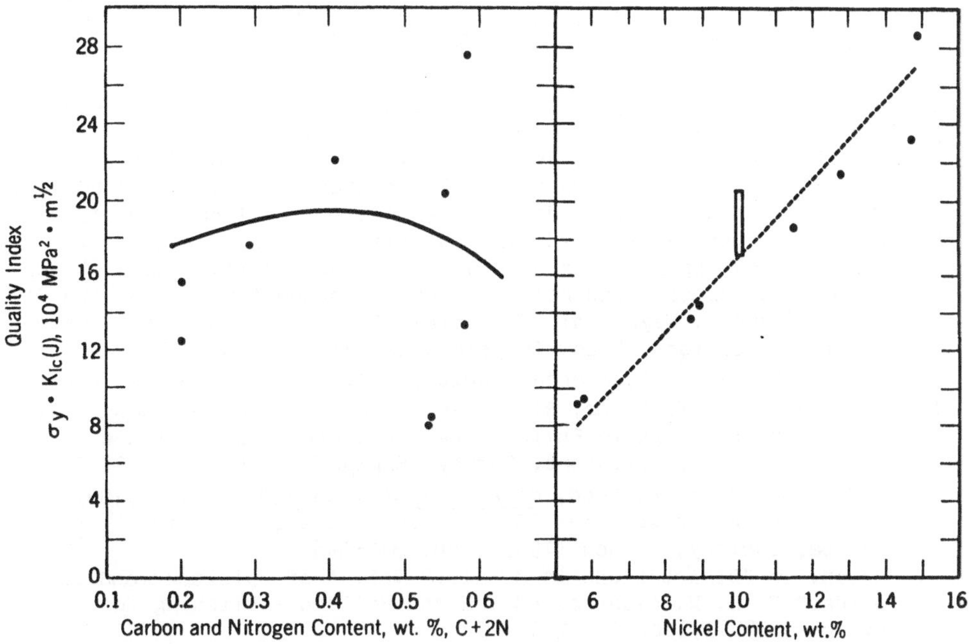

Fig. 8. The quality index $\sigma_y \cdot K_{Ic}(J)$ at 4 K plotted against C + 2N and Ni content. Data represented by solid line (C + 2N) and band (Ni) are a series of nine 19Cr-10Ni steels with varying C and N contents.[8]

SUMMARY

A series of eight austenitic stainless steels with varying Ni and N contents were measured at 4 K to determine tensile strength and fracture toughness. The results of the study are:

1. Nitrogen addition raises the tensile yield strength and reduces fracture toughness. Nickel addition raises the fracture toughness and has little effect on strength.

2. The temperature dependence follows the relation $\sigma_y = \sigma_o e^{-AT}$. The parameter A is directly proportional to the stacking fault energy, and tensile yield strength is strongly affected by N content.

3. A quality index, defined as the product $\sigma_y \cdot K_{Ic}(J)$, has been proposed to portray mechanical behavior. Nickel is shown to increase the quality index; nitrogen has little effect.

ACKNOWLEDGMENTS

The Office of Fusion Energy, Dr. Victor Der, Project Monitor, sponsored this research. We owe special thanks to Robert Walsh for performing tensile measurements and to Lonnie Scull for conducting the chemical analyses.

REFERENCES

1. R. P. Reed and N. J. Simon, Adv. Cry.Eng.--Mater. 30:127-136 (1984).
2. R. L. Tobler, D. T. Read, and R. P. Reed, in: "Fracture Mechanics: Thirteenth Conference," ASTM STP 743, American Society for Testing and Materials, Philadelphia (1981), pp. 250-268.
3. R. P. Reed and J. M. Arvidson, Adv. Cryo. Eng.--Mater. 30:263-270 (1984).
4. D. T. Read and R. L. Tobler, Adv. Cryo. Eng.--Mater. 28:17-28 (1982).
5. R. P. Reed and R. L. Tobler, Adv. Cryo. Eng.--Mater. 28:49-56 (1982).
6. R. L. Tobler, R. P. Reed, and D. S. Burkholter, Adv. Cryo. Eng.--Mater. 26:107-119 (1980).
7. R. P. Reed, in: "Materials at Low Temperatures," American Society for Metals, Metals Park, Ohio (1983), pp. 295-341.
8. R. L. Tobler and R. P. Reed, in: "Materials Studies for Magnetic Fusion Energy Applications at Low Temperatures--III," NBSIR 80-1627, National Bureau of Standards, Boulder, Colorado (1980), pp. 17-48.
9. P. T. Purtscher and R. P. Reed, in "Materials Studies for Magnetic Fusion Energy Applications at Low Temperatures-VIII," NBSIR 85-3025, National Bureau of Standards, Boulder, Colorado (1985), pp. 123-144.
10. R. L. Tobler and D. Meyn, in: "Materials Studies for Magnetic Fusion Energy Applications at Low Temperatures--VIII," NBSIR 85-3025, National Bureau of Standards, Boulder, Colorado (1985), pp. 167-179.
11. I. Williams, R. G. Williams, R. C. Capellaro, in: "Proceedings of Sixth International Cryogenic Engineering Conference," IPC Science and Technology Press, Guildford, Surrey, England (1976), pp. 337-341.
12. R. E. Schramm and R. P. Reed, Metall. Trans. 6A:1345-1351 (1975).
13. G. Schoeck and A. Seeger, in: "Defects in Crystalline Solids," The Physical Society, London (1955), pp. 340-346.
14. P. Haasen, in "Dislocations in Metals," vol. 4 of "Dislocations in Solids," F. R. N. Nabarro, ed., North-Holland Publishing Co., New York (1979), pp. 155-190.
15. S.-P. Hannula, H. Hanninen, and S. Tahtinen, Metall. Trans. 15A:2205-2211 (1984).
16. G. R. Caskey, in: "Fractography and Materials Science," ASTM STP 733, American Society for Testing and Materials, Philadelphia (1981), pp. 86-97.

IMPROVEMENT OF TOUGHNESS OF A HIGH-STRENGTH, HIGH-MANGANESE STAINLESS STEEL FOR CRYOGENIC USE

K. Suemune, K. Sugino, and H. Masumoto
Central R & D Bureau
Nippon Steel Corporation
Tokyo, Japan

H. Nakajima and S. Shimamoto
Japan Atomic Energy Research Institute
Ibaraki, Japan

ABSTRACT

Steel that possesses proof stress not less than 1200 MPa, fracture toughness not less than 200 MPa$\sqrt{m}$ at 4 K and good rust resistance is required as the structural material for superconducting magnets (SCM). Some of the authors previously reported that 0.05C-0.3Si-25Mn-15Cr-1Cu-1Ni-0.2N steel satisfied nearly all the requirements and was a promising structural material for SCM. The present report concerns the results of investigation into the effects of several elements on the low temperature toughness of this steel. It became clear that a decrease of S content to less than 0.005% and an addition of a small amount of Ca remarkably improved the toughness of this steel. Moreover, an addition of about 1% Mo was observed to prevent a decrease in toughness by sensitization treatment. It is concluded that 0.05C-0.3Si-22/25Mn-13/15Cr-1Cu-1/3Ni-1Mo-0.2N-Ca steel can satisfy the requirements.

INTRODUCTION

The structural material of SCM is required to possess high-strength, high-toughness and good fatigue behavior at liquid helium temperature. This material is also required to be corrosion resistant so as to maintain a high vacuum because its surface is exposed to a vacuum for thermal insulation. The authors have attempted to develop a high-manganese stainless steel to satisfy the following requirements proposed by JAERI:

1. Proof stress at 4 K : 1200 MPa minimum
2. Toughness at 4 K : (1) K_{IC}:200 MPa$\sqrt{m}$ minimum
   (2) vE :100 J minimum
3. Corrosion resistance : No rust from dew

Some of the authors previously reported that 0.05C-0.3Si-25Mn-15Cr-1Cu-1Ni-0.2N steel was a promising candidate material through an investigation of the effects of various alloying elements on the mechanical properties and corrosion-resistance of high-manganese steel[1]. In the present report, the influence of several elements on the low temperature toughness of this steel and its improvement are discussed.

Table 1. Mechanical Properties of 0.05C-0.3Si-25Mn-15Cr-1Ni-1Cu-0.2N Steel

1) Manufactured Process
150 kg VIM → 150 kg ingot → Heating at 1150°C → Rolling into 100 mm thick slab → Heating at 1150°C → Rolling into 16 mm thick plate → Solution-treating (1100°C×30 min. WQ)

2) Chemical Composition

(wt.%)

C	Si	Mn	P	S	Cu	Ni	Cr	Nb	N	Aℓ
0.064	0.32	25.0	0.019	0.009	1.03	0.99	15.4	0.063	0.212	0.005

3) Mechanical Properties (Transverse direction)

Testing Temp.	PS (MPa)	TS (MPa)	Eℓ (%)	RA (%)	vE (J)
77K	854	1243	29	58	89
4K	1275	1652	33	46	70

EFFECTS OF S AND Ca ON TOUGHNESS

The chemical composition and the mechanical properties of 0.05C-0.3 Si-25Mn-15Cr-1Cu-1Ni-0.2N steel plate are shown in Table 1.[1] As the absorbed energy (vE) of the steel was a little too low, it remained a problem to improve it. It is well known that a decrease of S content and an addition of Ca reduce the amount of the elongated manganese sulfide inclusions and hence improve the transverse direction toughness of rolled carbon steels.[2] Therefore, the authors attempted to improve the toughness of the high-manganese stainless steel by applying this principle.

Seven high-manganese stainless steels were melted in a vacuum melting furnace. Every steel ingot of 150 or 300 kg was hot-rolled into 100 mm thick slab; then the slab was reheated at 1150°C and rolled into 30 mm thick plate. The plate was solution-treated by heating at 1100°C for 30 min followed by water quenching. A part of every steel plate was subsequently sensitized at 700°C for 2 h in consideration of the welding HAZ and stress relief annealing. The chemical compositions of these high-manganese steels are shown in Table 2. Steels M1, M2 and M3, to which Ca was not added, contain 0.020, 0.010 and 0.005%S, respectively. A small amount of Ca is added to steels M4, M5 and M6. Steel M6 contains 1.0%Mo. Tensile tests and Charpy impact tests were carried out on all steels at 77 K. Then, on the promising steels selected by the test results at 77 K, tensile tests, Charpy impact tests and fracture toughness tests were performed at 4 K. All test specimens were cut in the transverse direction. Moreover, all plates were examined for nonmetallic inclusions by an optical microscope and computer-aided microanalyzer (CMA) on the longitudinal section. Precipitates also were observed by optical microscope and electron microscope on the above steels.

Table 2. Chemical Composition of High-Mn Stainless Steels

(wt.%)

Steel	C	Si	Mn	P	S	Cu	Ni	Cr	Mo	N	Aℓ	Ca
M1	0.031	0.38	24.3	0.012	0.020	0.84	1.02	13.2	-	0.220	0.030	-
M2	0.031	0.38	24.4	0.012	0.010	0.84	1.02	13.1	-	0.231	0.014	-
M3	0.040	0.42	25.4	0.013	0.005	0.76	1.03	12.9	-	0.230	0.022	-
M4	0.040	0.50	25.4	0.013	0.006	0.76	1.03	13.0	-	0.225	0.025	0.0026
M5	0.022	0.52	24.9	0.014	0.003	0.79	3.10	13.0	-	0.219	0.017	0.0034
M6	0.030	0.40	22.4	0.014	0.003	0.77	3.06	13.0	1.00	0.222	0.021	0.0025

Tensile test specimens were 7 mm in diameter and 45 mm in length of round reduced section. A strain rate of 7×10^{-5}/s was applied until yielding occurred, and then the rate was changed to 1×10^{-3}/s until failure occurred. The 0.2% proof stress (PS) was measured by strain gauges directly attached to the specimens. Tensile strength (TS), elongation (El), reduction of are (RA), and Young's modulus (E) were also measured. The Charpy impact test specimens were full size. The Charpy impact tests at 4 K were performed in the following way[3] A specimen was inserted in a small glass Dewar which was filled with liquid helium, and the specimen was struck while in the Dewar. The absorbed energy (vE) was calculated by subtracting that for the Dewar. Fracture toughness tests were performed by the J-integral method in accordance with ASTM E813-81 using 25.4 mm thick compact tension specimens. The fracture surfaces of the Charpy impact and fracture toughness test specimens were observed by SEM, and the ferromagnetic phase was measured by ferrite scope. Precipitates extracted by the replica method were observed by an electron microscope.

Tensile and Charpy impact test results at 77 K are listed in Table 3. The relation between PS and vE at 77 K is shown in Fig. 1. The data shown

Table 3. Mechanical Properties of High-Mn Stainless Steels at 77 K

Steel	Heat Treatment*	PS (MPa)	TS (MPa)	Eℓ (%)	RA (%)	vE (J)
M1	S	838	1358	46	38	62
M2	S	873	1297	36	49	89
M3	S	860	1349	52	52	117
	T	915	1360	47	34	68
M4	S	834	1352	57	44	164
	T	898	1361	30	43	76
M5	S	852	1348	53	53	166
	T	893	1354	54	52	112
M6	S	851	1350	55	52	183
	T	905	1294	40	59	171

* S ; 1100°C × 30 min, WQ
T ; 1100°C × 30 min, WQ → 700°C × 2 h, AC

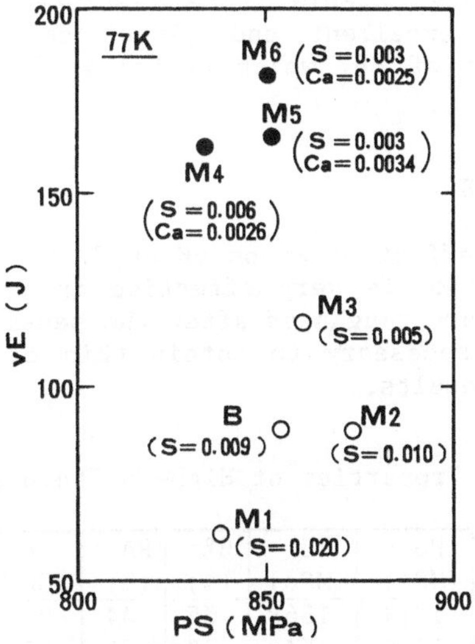

Fig. 1. Relation between PS and vE of high-manganese stainless steels at 77 K. Steel B; the steel shown in Table 1.

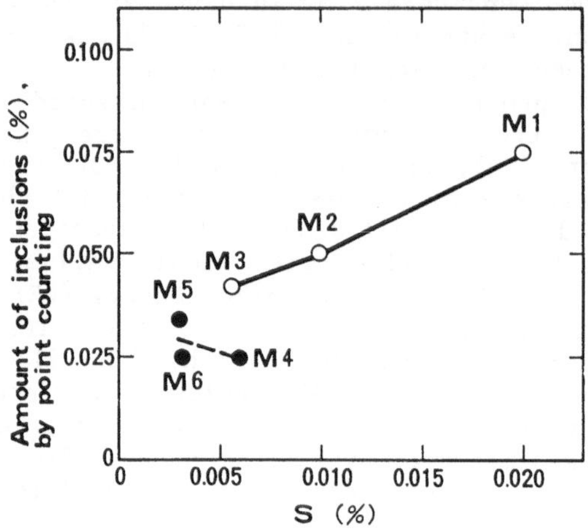

Fig. 2. Cleanliness of high-manganese stainless steel.

in Table 1 are also given in the figure for comparison. The vE increases with a decrease in S content, and remarkably with the addition of Ca to lower S steels. A change in the amount of nonmetallic inclusions by S content is shown in Fig. 2. Inclusions, which consist mainly of elongated MnS or A-type inclusions by JIS, decrease with a decrease in S content. Therefore, this increase in vE is thought to be due to the decrease of A-type inclusions. In Ca-added steels, A-type inclusions were reduced in quantity and spherical inclusions or C-type inclusions by JIS appeared, and then vE was highly improved. Most of the C-type inclusions of these Ca-added steels were found to consist of the mixture of Al_2O_3 and Ca (O, S) by CMA.

Table 4 shows the results of tensile, Charpy impact and fracture toughness tests at 4 K on the three promising steels selected by the results shown in Table 3. The PS is nearly equal and vE is higher than the target values in these steels. The fracture toughnesses (K_{IC}) of steels M4 and M6 are excellent and almost reach the target value. Especially the toughness of M6, which contains 1%Mo, is high enough to satisfy the objective.

EFFECT OF Mo ON TOUGHNESS

Figure 3 shows the effect of Mo on vE at 77 K after sensitization at 700°C. The addition of Mo is very effective in protecting against the decrease of low temperature toughness after the sensitization treatment at 700°C. The Mo content necessary to obtain this effect was found to be about 1% by other test results.

Table 4. Mechanical Properties of High-Mn Stainless Steels at 4 K

Steel	Heat Treatment*	PS (MPa)	TS (MPa)	Eℓ (%)	RA (%)	E (GPa)	vE (J)	K_{IC} (MPa$\sqrt{m}$)
M4	S	1214	1568	35	44	208	133	180
M5	S	1176	1548	38	51	229	137	
M6	S	1185	1565	38	47	207	163	216

* S = 1100°C × 30 min, WQ

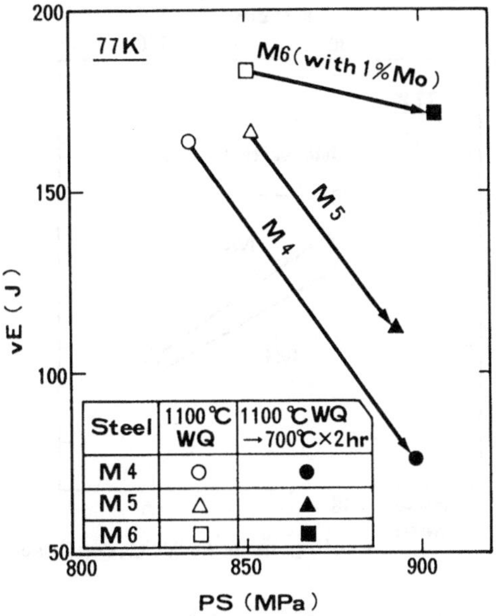

Fig. 3. Effect of Mo on the toughness of high-Mn stainless steel at 77 K.

Many precipitates were observed along grain boundaries in steel without Mo after the sensitization treatment, but not in Mo-added steel. This effect of Mo is thought to be due to its prevention of C diffusing. Examples of microstructure are shown in Fig. 4.

Figure 5 shows the toughness of steels M4, M5 and M6 after the simulated welding thermal cycles. Steel M6 with 1% Mo has higher toughness than the others irrespective of weld heat input.

Conclusively, Mo is very effective in preventing embrittlement of the HAZ by stress relief annealing.

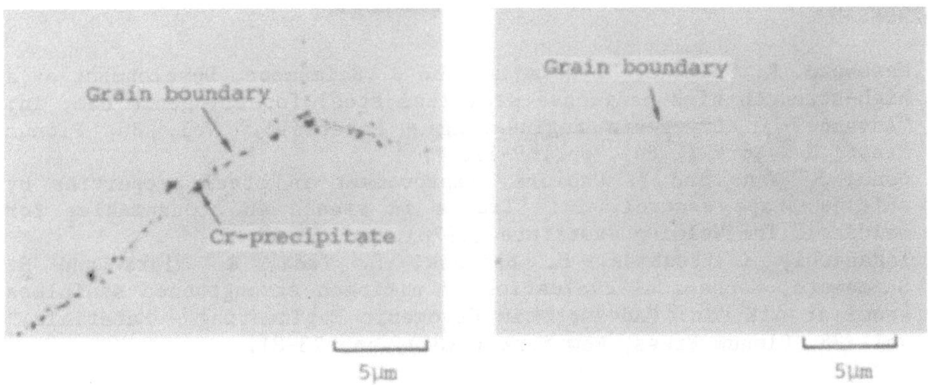

(a) Steel M4 (Mo free) (b) Steel M6 (1% Mo-added)

Fig. 4. Electron micrographs of high-Mn stainless steels after sensitization treatment at 700°C.

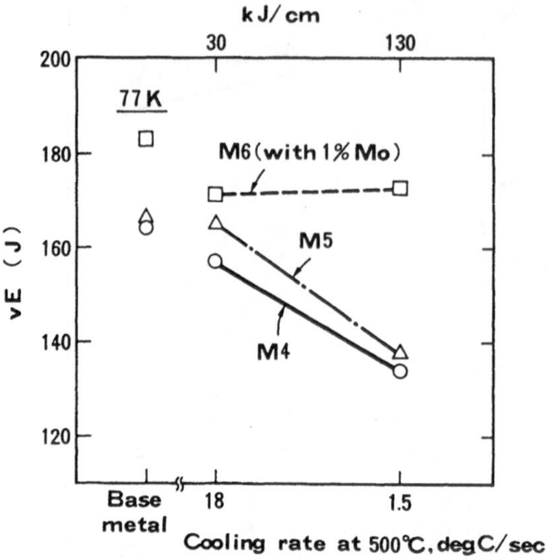

Fig. 5. Change of vE by simulated weld thermal cycle. (maximum temperature 1250°C)

CONCLUSIONS

1. The low temperature toughness of 0.05C-0.3Si-25Mn-15Cr-1Cu-1Ni-0.2N steel reported previously was remarkably improved by decreasing the S content to less than 0.005% and adding a small amount of Ca.
2. The addition of 1%Mo to the high-manganese steel is very effective in preventing the decrease in the toughness due to sensitization treatment; it also improves the toughness of HAZ.
3. Consequently, 0.05C-0.3Si-22/25Mn-13/15Cr-1/3Ni-1Cu-1Mo-0.2N-Ca steel proved to be a suitable material for SCM. This steel is going to be manufactured on a trial commercial scale.

REFERENCES

1. H. Masumoto, K. Suemune, H. Nakajima and S. Shimamoto, Development of a high-strength high manganese stainless steel for cryogenic use, in: "Advances in Cryogenic Engineering - Materials," Vol. 30, Plenum Press, New York (1984), pp.169-176.
2. Y. Oono, S. Yano and Y. Okamura, Improvement in steel properties by sulfide shape control, in: "Trends in steels and consumables for welding," The Welding Institute (1978), pp. 55-68.
3. Y. Takahashi, K. Yoshida, M. Shimada, E. Tada, R. Miura and S. Shimamoto, Mechanical evaluation of nitrogen strengthened stainless steel at 4 K, in: "Advances in Cryogenic Engineering - Materials," vol. 28, Plenum Press, New York (1982), pp. 73-81.

EFFECT OF METALLURGICAL VARIABLES ON STRENGTH AND TOUGHNESS OF Mn-Cr AND Ni-Cr STAINLESS STEELS AT 4.2 K

S. Yamamoto, N. Yamagami, and C. Ouchi

Technical Research Center
Nippon Kokan K. K.
Kawasaki, Japan

ABSTRACT

For the structural materials used for fusion reactors, high strength and superior toughness are required at liquid helium temperature in heavy plate (over 100 mm thick). Both Mn-Cr and Ni-Cr austenitic steels are candidate materials for the next Fusion Reactor Project at JAERI (Japan Atomic Energy Research Institute), which has specifications of yield strength and K_{IC} fracture toughness of over 1200 MPa and 200 MPa$\sqrt{m}$, respectively. This paper systematically examines the effect of metallurgical variables and thermomechanical processing parameters on strength and toughness obtained at 4.2 K for both types of stainless steels. Strength obtained at 4.2 K is influenced principally by chemical composition and austenitic grain size. Carbon and nitrogen exert a much greater influence on strength than substitutional elements in both Mn-Cr and Ni-Cr stainless steels. The value of Ky, expressing the grain size dependence of yield strength in the Hall-Petch relation, is 50 MPa$\sqrt{mm}$ for both steels. Toughness evaluated by Charpy impact energy and the K_{IC} value at 4.2 K is controlled by strength, Ni content, cleanliness of the steel, and austenitic grain size. The increase of Ni content and a reduction of S content improve both values markedly while maintaining high strength. For an optimization of the relationship between strength and toughness, austenitic grain refinement through thermomechanical processing is the most important process in heavy plate production. Candidate materials satisfying JAERI specifications have been determined for both types of stainless steels; their chemical compositions are 0.02C-22Mn-13Cr-7.5Ni-0.22N and 0.01C-1.5Mn-18Cr-12Ni-3.5Mo-0.21N.

INTRODUCTION

For the structural materials used for fusion reactors with large superconducting magnets, very high strength and superior toughness are required at liquid helium temperature in heavy plate (over 100 mm thick). Both Mn-Cr and Ni-Cr austenitic steels are candidate materials[1] for the next Fusion Reactor Project at JAERI, which has yield strength, Charpy impact energy, and K_{IC} fracture toughness specifications of over 1200 MPa (122.4 kg/mm^2), 100 J (10.2 kg-m) and 200 MPa$\sqrt{m}$, respectively.

Metallurgical variables influencing the strength at 4.2 K are interstitial elements, such as C and N, substitutional elements, such as Mo and V, and austenitic grain size.[2] In particular, it has been reported that increased N content was very effective for strengthening austenitic steel at 4.2 K. It has been well-known that solubility of N in the austenitic phase increased with Cr and Mn contents[3] in steel, and thus, to utilize strengthening by N addition fully, stainless steels of both high Cr-Ni[4] and high Mn-Cr[5] systems have been investigated. As for the toughness at cryogenic temperatures, it has been reported that Charpy impact energy was correlated with the mean free path of nonmetallic inclusions.[4] Impurity content, such as S or P, Ni content, and austenitic grain size are the main controlling factors of the toughness at liquid helium temperature.

It is also important to establish the processing method that satisfies such a severe strength and toughness requirements for heavy plates of stainless steel. For this goal, in-line solution treatment combined with thermomechanical processing in plate mills is the most promising process; it has been successfully industrialized for HSLA steel or line pipe steel. Major advantages of this process are grain refinement through the control of recrystallization during hot rolling as well as saving energy compared with conventional off-line solution treatment.

In the first part of this paper, effects of the metallurgical factors described above on the strength and toughness at 4.2 K are systematically investigated for both the Mn-Cr and Ni-Cr stainless steels. In the remaining half, the relationships between thermomechanical processing variables and mechanical properties in heavy plates are examined. Finally, the optimum alloy compositions, as well as their properties in both types of stainless steels that satisfy the JAERI specification, are demonstrated.

EXPERIMENTAL PROCEDURE

The ranges of alloying compositions of the Mn-Cr and Ni-Cr stainless steels investigated are listed in Table 1. The basic compositions are 0.02C-25Mn-13Cr-5.5Ni-0.25N and 0.005C-18Cr-12Ni-2.5Mo-0.17N, and the content of interstitial elements, such as C or N, and substitutional elements, such as Mn, Cr, or Mo, were varied independently. All heats were melted in a 50 kg or 150 kg vacuum induction furnace. Ingots were hot rolled down to 15 mm-thick plates. The reheating and finish-rolling temperatures were 1200 and 1100°C, respectively. For thermomechanical processing, reheating temperature and finish-rolling temperature were varied from 1100 to 1250°C and from 1000 to 1150°C, respectively. The plate thickness was also varied up to 100 mm. Hot rolled plates were directly water quenched immediately after hot rolling. Some of the hot rolled plates were solution treated at temperature between 1050 and 1250°C followed by water quenching. By these changes in the

Table 1. The Ranges of Alloying Compositions (wt%) in the Mn-Cr and Ni-Cr Stainless Steels Investigated.

	C	Si	Mn	P	S	Cr	Ni	Alloying Elements	sol.Al	T.N
	0.02	0.30	25.0	0.003	0.001	13.5	5.5		0.01	0.25
Mn-Cr	0.02 ~ 0.25	0.30	20.0 ~ 30.0	0.003 ~ 0.026	0.0010 ~ 0.0150	10.0 ~ 20.0	0 ~ 12.0	Mo : 0 ~ 4.0 Cu, V : 0 ~ 0.5	0.01	0.15 ~ 0.45
	0.005	0.30	1.5	0.003	0.001	18.0	12.0	2.5	0.01	0.17
Ni-Cr	0.005 ~ 0.05	0.30	1.5	<0.005	<0.001	18.0	12.0 ~ 14.0	Mo : 2.5 ~ 4.5	0.01	0.02 ~ 0.21

hot rolling and solution treatment conditions, austenitic grain size in as-rolled and solution-treated plates was varied from 10 to 120 μm.

For investigations of both the thermomechanical processing parameters and the effects of plate thickness, two production heats of stainless steels were used. One is 22Mn-13Cr-5.5Ni steel melted in 5 Mg vacuum induction furnace. The other is a 18Cr-13Ni-3.7Mo 26 Mg ingot prepared from the Electro Furnace - Vacuum Oxygen Decarburizing process. The 150 mm-thick slabs prepared from these large ingots were rolled to 15 to 100 mm-thick plate in a laboratory plate mill. Tensile specimens (6.0 mm in diameter and 21 mm in gauge length), 2 mm Charpy V-notch impact specimens and 13 mm compact tension specimens were prepared from the plates parallel to the rolling direction. Tensile tests were carried out using a 10 Mg Instron servohydraulic machine and multi-type cryostat with single pull rod. In the Charpy impact test, Morris's method[6] is adopted. Fracture toughness values of K_{IC} were calculated from J_{IC} values obtained by the compliance method using a 20 Mg Instron servohydraulic machine and cryostat.

RESULTS

Strength-Controlling Factors

Figure 1 shows strength variations at ambient temperature and 4.2 K with C and N contents in Mn-Cr steels with a fixed austenitic grain size of 15 μm. C and N are extremely effective for strengthening, especially at 4.2 K. The increments in yield strength per 1 wt% C and 1 wt% N addition are 1960 MPa/wt% and 2940 MPa/wt%, respectively, which are 3 or 5 times larger than those at the ambient temperature. Figure 2 summarizes the strengthening effect due to various alloying elements, where the increments of yield strength per 1 wt% alloying element at 4.2 K, 77 K, and 293 K are represented. The increment of yield strength increases with the decrease of test temperature in any alloying element. The marked strengthening due to C and N at 4.2 K was also observed in Ni-Cr stainless steel. Cu, V, and Mo exhibited the largest strengthening among the substitutional alloying elements, and Mn, Cr, and Ni exerted no influence on the strength.

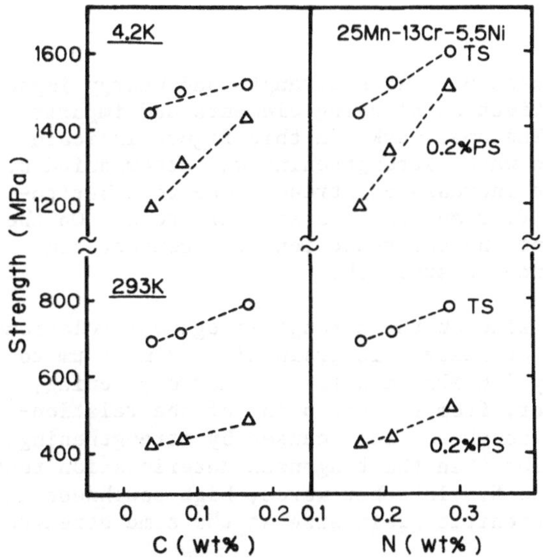

Fig. 1. Strength variations at the ambient temperature and 4.2 K with C and N content in Mn-Cr stainless steels.

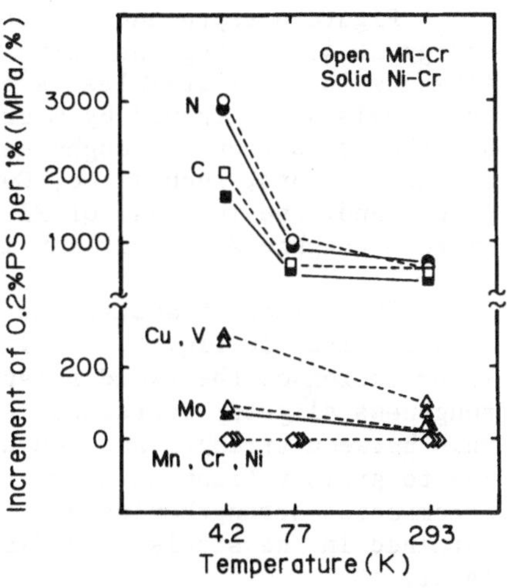

Fig. 2. The increment of yield strength per 1 wt% alloying element at 4.2 K, 77 K and 293 K.

It is important to note that an increase of carbon content over 0.10 wt% in both stainless steels caused the precipitation of Cr carbide along the austenitic grain boundaries in as-hot-rolled plate. These grain boundary carbides not only deteriorate low temperature toughness but also enhance surface rusting in high-Cr steels. Thus, strengthening due to carbon cannot be utilized in both types of stainless steel. On the other hand, an increase of N content over 0.20 wt% in steel did not form any nitride during thermomechanical processing. Therefore, the increase of N content is the most effective strengthening route at 4.2 K, and Cu, V, or Mo can be used as a supplemental strengthening alloying element.

Figure 3 shows the change of yield strength with austenitic grain size in 22Mn-13Cr-5.5Ni steels with the different contents of C and N, where the yield strength is plotted as a function of reciprocal root of austenitic grain size. Austenitic grain size was varied by changing both conditions of hot rolling and solution treatment. The values of Ky, expressing the grain size dependence of yield strength in the Hall-Petch relation, are 25 and 50 at the ambient temperature and liquid helium temperature, respectively. That is, strengthening due to grain refinement is very remarkable at 4.2 K. A similar trend was observed in Ni-Cr stainless steel, in which the values of Ky at the ambient temperature and 4.2 K were 17 and 50, respectively. The quantitative strengthening effects due to various alloying elements and grain size in both stainless steels can be summarized by the following equations:

in Mn-Cr stainless steel

$$0.2\%PS(MPa) \text{ at } 293K = 98 + 590(\%C) + 570(\%N) + 20(\%Mo) + 80(\%Cu) + 40(\%V) + 25d\gamma^{-1/2} \quad (1)$$

$$0.2\%PS(MPa) \text{ at } 4.2K = 290 + 1960(\%C) + 2940(\%N) + 100(\%Mo) + 290(\%Cu) + 260(\%V) + 50d\gamma^{-1/2} \quad (2)$$

in Ni-Cr stainless steel

$$0.2\%PS(MPa) \text{ at } 293K = 20 + 470(\%C) + 690(\%N) + 50(\%Mo) + 17d\gamma^{-1/2} \quad (3)$$

$$0.2\%PS(MPa) \text{ at } 4.2K = 150 + 1700(\%C) + 2870(\%N) + 80(\%Mo) + 50d\gamma^{-1/2} \quad (4)$$

Controlling Factors of Toughness

Figure 4 represents the relationship between strength and Charpy impact energy at 4.2 K, together with the effect of alloying elements and impurity elements in Mn-Cr stainless steel. The open marks in this figure indicate the steels strengthened by C or N, in which strengthening was accompanied by a marked reduction of toughness. The increase of strength due to substitutional elements, such as Mo, Cu, and V, resulted in a similar trend. On the other hand, the increase of Ni content and the reduction of S content can improve toughness without any influence on strength.

The effect of austenitic grain size on the strength-toughness relationship is given in Fig. 5. Refinement of austenitic grain size from 45 µm to 15 µm increases the yield strength by 170 MPa, and for this strengthening, toughness slightly decreases. However, from the viewpoint of the relationship between strength and toughness, toughness loss caused by strengthening due to grain refinement is much smaller than the toughness deterioration that accompanies strengthening due to C and N. In other words, high toughness is obtained in the steels with finer austenitic grain size at the same strength level.

In the relationship of strength and fracture toughness, K_{IC}, the same trend is observed with the strength-Charpy impact energy relationship, as shown in Fig. 6. The value of K_{IC} decreases with strengthening due to N in

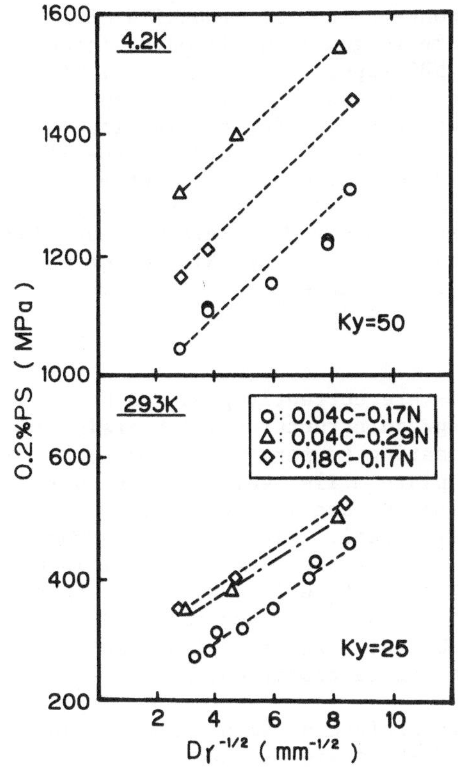

Fig. 3. The change of yield strength with austenitic grain size in Mn-Cr stainless steels.

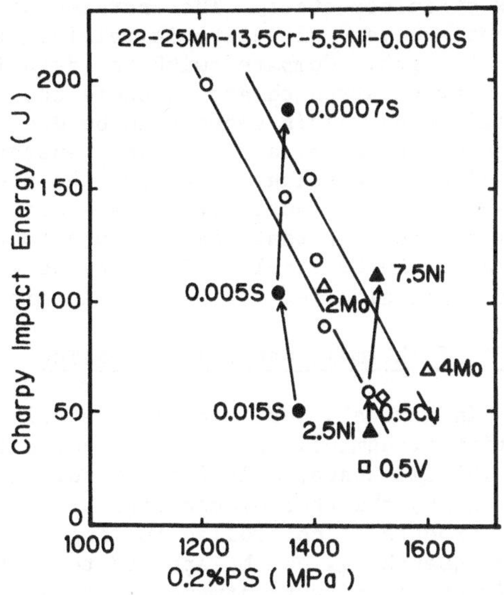

Fig. 4. The relationship between strength and Charpy impact energy at 4.2 K, together with the effect of alloying elements and impurity elements in Mn-Cr stainless steel.

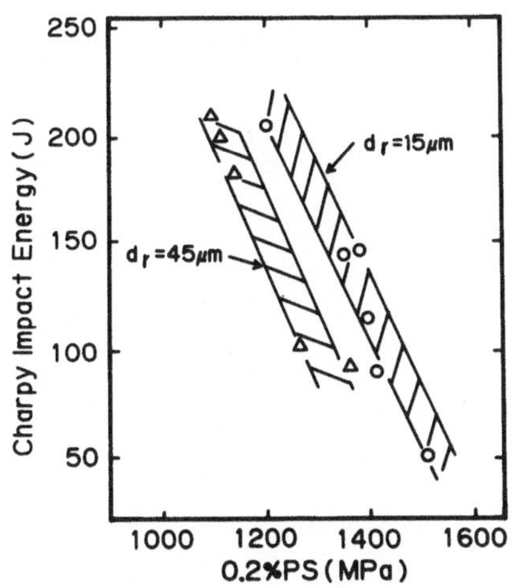

Fig. 5. The effect of austenitic grain size on strength-toughness relationship in Mn-Cr stainless steel.

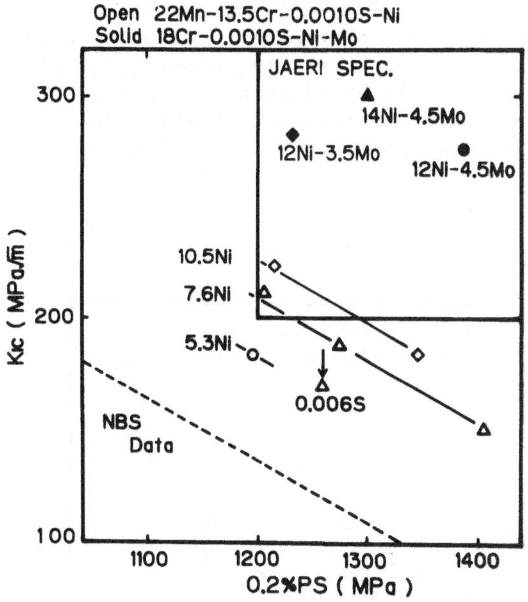

Fig. 6. The relation of strength and fracture toughness, together with the effect of alloying elements and impurity elements.

Mn-Cr stainless steel. The increase of Ni content and the reduction of S content are very effective in raising fracture toughness while maintaining high strength. Compared with the data in a NBS report,[7] a better relationship between strength and fracture toughness is obtained in the present investigation. This appears to be due mainly to the difference in cleanliness of the steel and the finer austenitic grain size achieved by thermomechanical processing. That is, the steels used here contain S content less than 0.0010%, and austenitic grain size was reduced to 50 μm. It is also important to note that the fracture toughness of Ni-Cr-Mo steels was superior to that of Mn-Cr steels. This may be attributable to higher Ni content and the beneficial effect of Mo addition in the former steel.

Effect of Thermomechanical Processing

In general, most stainless steel plates are currently produced by solution treatment, where coarse austenitic grain structures (with grain size over 100 μm) unavoidably form at solution treating temperatures. On the other hand, the thermomechanical process in plate mills, which is the combined process of controlled rolling and subsequent accelerated cooling or direct quenching, can be applied to austenitic steel. This may be referred to as direct solution treatment. That is, high temperature controlled rolling that is carried out above recrystallization temperature in austenitic stainless steel can produce very refined austenitic grains through repeated static recrystallization during hot rolling, and subsequent rapid cooling can suppress the carbide precipitation in as-hot-rolled condition.

Figure 7 gives the change of austenitic grain size with plate thickness and reheating temperature in both Mn-Cr and Ni-Cr stainless steels. In both types of steels, austenitic grain size decreases with the reduction of plate thickness. Fine grain size (smaller than 30 μm) is obtained in plates thinner than 20 mm, and even in the heavy plate (thicker than 75 mm), austenitic grain size ranged from 40 to 70 μm, which is about a half that in a conventional solution treating condition. Austenitic grain size decreases further by lowering the reheating temperature, but one must be very careful not to reduce rolling temperature below recrystallized temperature of stainless steels by adopting an extremely low reheating temperature. Recrystallization temperatures in Mn-Cr steel, type 304, and type 316 obtained by hot rolling experiments were 950°C, 970°C, and 1050°C, respectively. Austenitic

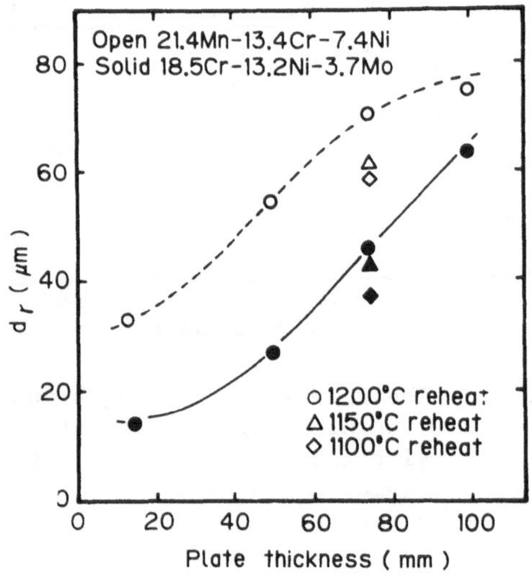

Fig. 7. The change of austenitic grain size with plate thickness and reheating temperature.

grain size in 18Cr-13Ni-3.7Mo stainless steel is smaller by about 20 μm than that in 22Mn-13Cr stainless steel under the same hot-rolling conditions as shown in Fig. 7. This appears to be caused by the grain refinement effect due to Mo addition.

ALLOY DESIGNING AND DISCUSSION

Strength obtained at 4.2 K can be controlled principally by N content and austenitic grain size, as shown in Fig. 2 and Fig. 3. In Mn-Cr stainless steels, the solubility of N is more than 0.35 wt%, and strengthening by N addition is fully utilized, although excessive strengthening tended to decrease toughness, as shown in Fig. 4 and Fig. 6. Therefore, an optimum addition of nitrogen is important. Cleanliness of the steel, which depends on S and O contents, plays an important role in the improvement of toughness, and it was confirmed that reduction of S content below 10 ppm markedly improved both Charpy impact energy and K_{IC} values at 4.2 K. Austenitic grain refinement, which improves the strength-toughness relationship, can be achieved through a thermomechanical process, and austenitic grain sizes smaller than 50 μm are obtained in heavy plates. Taking these factors mentioned above into consideration, optimum alloying compositions for the heavy plate that can satisfy the specifications proposed by JAERI were determined in Mn-Cr and Ni-Cr stainless steels.

By assuming austenitic grain size is 50 μm in heavy plates (over 100 mm thick) and using equation (2), 0.22 wt% N is necessary to attain the proposed yield strength greater than 1200 MPa in Mn-Cr stainless steel. The amount of Ni content should be determined by toughness. It is implied from comparison of Fig. 4 and Fig. 6 that the K_{IC} fracture toughness requested by JAERI may be more difficult to achieve than the specified value of Charpy impact energy. Ni content over 7.0 % is necessary to attain a fracture toughness greater than 200 MPa$\sqrt{m}$ in Mn-Cr stainless steel. In Ni-Cr stainless steel the solubility of N is 0.2 wt% at most for the common Cr level about 18 wt%,[3] and under this limitation, an additional increase of strength by Mo is necessary to obtain a yield strength over 1200 MPa at 4.2 K. On the basis of equation (4), Mo content greater than 3.0 wt% is needed in Ni-Cr stainless steel containing 18 wt% Cr. It is important to note that fracture toughness in these Ni-Cr stainless steels is high enough to fulfill the specification by JAERI, as demonstrated in Fig. 6.

Alloy compositions of the candidate Mn-Cr and Ni-Cr steels as well as their mechanical properties at 4.2 K are listed in Table 2. These two types of stainless steels satisfy the values of yield strength, Charpy impact

Table 2. Alloying Composition and Mechanical Properties of the Candidate Materials at 4.2 K.

	C	Si	Mn	P	S	Cr	Ni	Mo	sol.Al	T.N
A	0.02	0.30	22.0	0.003	0.001	13.5	7.5	–	0.01	0.22
B	0.01	0.30	1.5	0.003	0.001	18.5	12.0	3.5	0.01	0.21

	0.2%PS (MPa)	TS (MPa)	$vE_{4.2}$ (J)	K_{IC} (MPa$\sqrt{m}$)
A	1203	1458	192	212
B	1232	1529	208	282
spec.	≧1200		≧100	≧200

energy, and fracture toughness proposed by JAERI. Compared with steel A in Mn-Cr type stainless steel, steel B in Ni-Cr type stainless steel exhibits better toughness, although the alloying cost in steel B might be higher than that in steel A.

CONCLUSIONS

1. Strength obtained at 4.2 K is influenced principally by chemical composition and austenitic grain size. Carbon and nitrogen exert a greater influence on strength than substitutional elements, and the strengthening coefficient of these elements at 4.2 K is 3 to 5 times that at ambient temperature in both types of steels. The quantitative strengthening effects due to various alloying elements and grain size at 4.2 K can be summarized by the following equations:

$$0.2\%PS(MPa) \text{ in Mn-Cr stainless steel} = 290 + 1960(\%C) + 2940(\%N) + 100(\%Mo) + 290(\%Cu) + 260(\%V) + 50d\gamma^{-1/2}$$

$$0.2\%PS(MPa) \text{ in Cr-Ni stainless steel} = 150 + 1700(\%C) + 2870(\%N) + 80(\%Mo) + 50d\gamma^{-1/2}$$

2. Toughness evaluated by Charpy impact energy and K_{IC} values at 4.2 K is controlled by strength, alloying elements such as Ni, cleanliness, and austenitic grain size. The increase of Ni content and a reduction of S content in both Mn-Cr and Ni-Cr stainless steels improve markedly both values while maintaining high strength. A better combination of strength and toughness is obtained by grain refinement through thermomechanical processing than by high strengthening due to C and N.

3. Candidate materials that fulfill the proposal of JAERI were determined in both Mn-Cr and Cr-Ni stainless steels. The optimum compositions are 0.02C-22Mn-13Cr-7.5Ni-0.22N and 0.01C-1.5Mn-18Cr-12Ni-3.5Mo-0.21N, respectively.

REFERENCES

1. H. Nakajima, K. Yoshida, Y. Takahashi, E. Tada, M. Oshikiri, K. Koizumi, S. Shimamoto, R. Miura, M. Shimada, S. Tone, H. Masumoto, T. Sakamoto, in: "Advances in Cryogenic Engineering - Materials," vol. 30, Plenum Press, New York (1984), p. 219.
2. R.L. Tobler, D.H. Beekman, R.P. Reed, in: "Austenitic Steels at Low Temperatures," Plenum Press, New York (1983), p. 135.
3. P.D. Hodgson, R. Jackson, Metals Forum, 4:192 (1981),
4. T. Sakamoto, Y. Nakagawa, I. Yamauchi, T. Zaizen, in: "Advances in Cryogenic Engineering - Materials," vol. 30, Plenum Press, New York (1984), p. 137.
5. R. Ogawa, J.W. Morris, Jr., in: "Advances in Cryogenic Engineering-Materials," vol. 30, Plenum Press, New York (1984), p. 177.
6. S. Jin, W.A. Horwood, J.W. Morris, Jr., "Advances in Cryogenic Engineering," vol. 19, Plenum Press, New York (1974), p. 373.
7. D.T. Read, R.P. Reed, Cryogenics, 31:415 (1981), p. 415.

EFFECT OF Mn ON THE CRYOGENIC PROPERTIES OF

HIGH NITROGEN AUSTENITIC STAINLESS STEELS

T. Sakamoto, Y. Nakagawa, and I. Yamauchi

Stainless Steel Laboratory
Nippon Steel Corporation
Kanagawa, JAPAN

ABSTRACT

High nitrogen containing austenitic stainless steel is a promising structural material for use in superconducting coils. However, high Cr and Ni contents are required to obtain sufficient strength and toughness at the same time (25% Cr and 13% Ni are necessary to attain the target values proposed by JAERI). As there is a possibility of decreasing Ni and Cr contents by the addition of Mn without any change of nitrogen content in solid solution, the effect of Mn on the cryogenic properties was investigated and the following results were found:
1) Mn is nearly as effective as Cr in increasing N solubility.
2) Mn is more effective than Ni in increasing the strength at cryogenic temperatures when the amount of N is equal.
3) Mn has an austenite stabilizing ability when Cr content is low, and it works as a ferrite forming element when Cr content is high.
4) Toughness deteriorates to some extent by the addition of Mn.
The optimum Mn content varies with the contents of Cr and Ni, and some suitable Cr-Ni-Mn combinations are possible, depending on the circumstances of application.

INTRODUCTION

High proof stress, fracture toughness and nonmagnetism at liquid He temperature are the principal properties required for the structural material used in construction of superconducting magnets.

High-nitrogen Ni-Cr stainless steels with reduced second phase particles are promising material for this purpose, as previously reported.[1,2] For example, 25Cr-14Ni-4Mn-0.35N steel has a 0.2% proof stress of 1390 MPa, a Charpy V-notch energy of 151 J and a K_{IC} of 238 MPa$\sqrt{m}$ at 4K.

This kind of steel tends to be expensive, since it contains large amounts of Ni and Cr. High Mn austenitic steels with the addition of Cr to enhance the corrosion resistance are, on the other hand, also candidate materials for this purpose; their cryogenic properties are similar to those of high-N Ni-Cr stainless steels.[3,4] Therefore, there exists a possibility of a material of new composition, in which Ni and Cr are replaced by Mn.

Mn can be a substitute for Ni, as it is an austenite forming element: it can also be a substitute for Cr, as it increases the solubility of N, which is indispensable for the attainment of sufficient proof stress.

Table 1. Composition Range of Specimens in Weight Percent

Mn	0 ~ 20 %	
Cr	13 ~ 23 %	when Mn ≥ 7 %
	17 ~ 31 %	when Mn ≤ 7 %
Ni	2 ~ 19 %	
N	0 ~ 0.6 %	mainly 0.25 ~ 0.35 %
C	0.03 %	

The purpose of this paper is to study fundamentally the effect of Mn on the solubility of N, the strength, the toughness, and the phase stability at cryogenic temperature, in order to make use of this element effectively for structural use.

EXPERIMENTAL METHOD

About 80 kinds of steels having the composition ranges denoted in Table 1 were used. Mainly, 50 kg ingots were prepared by vacuum-induction melting, but some high Mn containing ingots were prepared by melting in air.

An average Ca content of 0.003% was added, and Al content was limited to less than 0.03% to avoid harmful inclusions. The inclusion content of the specimen ranged from 0.05% to 0.17%.

All specimens were taken from the 30-mm-thick plate after solution heat-treatment at 1100°C. Tensile properties were determined using 7 mm diameter-35 mm gauge length specimens and impact toughness was measured as the absorbed energy value by Charpy V-notch testing.

These tests were all conducted at 77 K, since the proof stress and impact toughness at 4 K has a close relationship with those at 77 K. For example, 0.2% proof stress of 1200 MPa at 4 K corresponds to 960 MPa at 77 K, and vE of 100 J at 4 K corresponds to 150 J at 77 K.

Austenite phase stability was measured by the magnetic permeability (μ) and by inspection of microstructure.

RESULTS

Effect of Mn on the Solubility of N

The solubility of N with varied Cr and Mn content was determined by using the specimens with constant Ni content around 15%.

Here, we define soluble where no bubbling occurs within the ingot after vacuum melting, and the solubility limit as the boundary N content beyond which the bubbling occurs. Solubility of N in the Ni-Cr matrix does not vary much with the temperature, so that N remains in solid solution after cooling at a normal cooling rate.

Fig. 1 shows the solubility limits which were experimentally determined by this method. The solubility of N increases with Cr and Mn remarkably. The following experimental equation was derived from the result.

$$N(\%) = 0.021(Cr + 0.9Mn) - 0.204 \qquad (1)$$

Mn is nearly as effective as Cr in increasing the solubility of N.

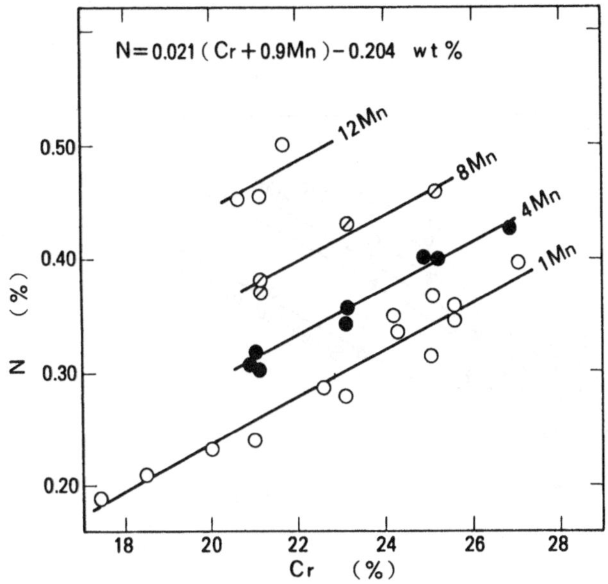

Fig. 1. Effect of Cr and Mn on solubility.

Effect of Mn on the Solution Hardening of N

Mn takes part in increasing the proof stress by its substitutional solid solution hardening. Consequently, Mn-Cr steel shows higher proof stress than Ni-Cr steel when N content is equal.

However, the contribution of Mn in increasing the proof stress is slight when compared with that of N, and the fairly large amount of N is also necessary for Mn-Cr steel to obtain high proof stress at low temperatures.

The relation between N content and the proof stress at 77 K is shown in Fig. 2. The amount of N is the major factor in increasing the proof stress, but the rate of increase seems smaller when Mn content is high

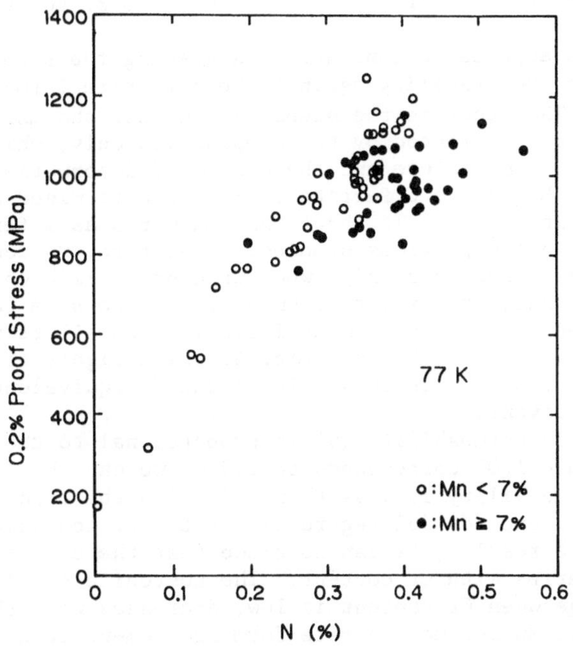

Fig. 2. Relation between N content and 0.2% proof stress at 77 K.

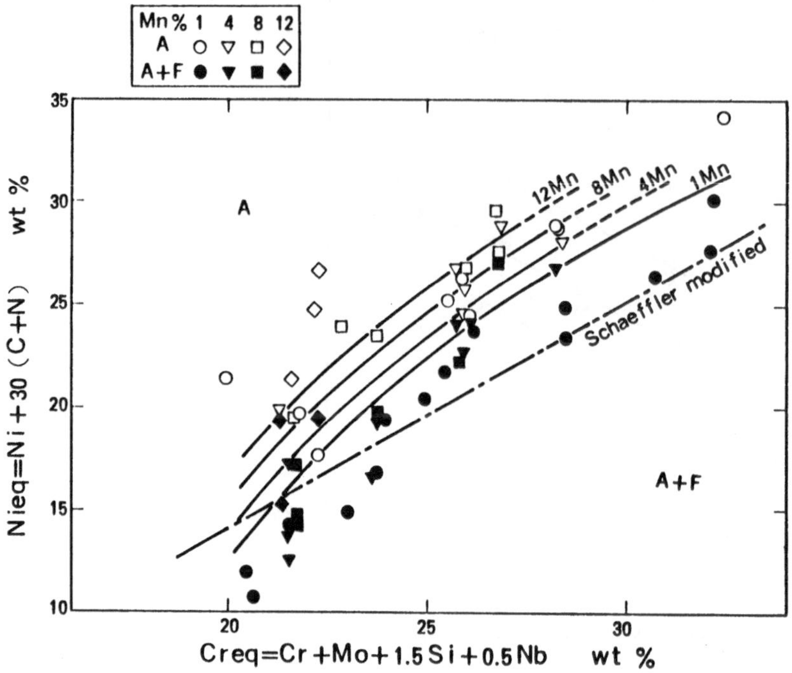

Fig. 3. Effect of Mn on the austenite-ferrite phase boundary.

(>7% in Fig. 2). And finally the proof stress does not increase beyond the critical N value of about 0.3%. Therefore, further addition of N in high Mn steel is useless from the viewpoint of the proof stress.

Effect of Mn on the Stability of Austenite

Mn has been considered to be an austenite stabilizing element. Delong[5] alloted Mn 0.5; Hammer,[6] 0.31; and Espy,[7] 0.87 for their Ni equivalent.

There are two aspects to consider in assessing the stability of austenite. One is the stability against the δ ferrite formation when Cr content is high. The other is the stability against the martensite formation. The former is affected by the composition only, while the latter also depends on the temperature and the degree of deformation.

The effect of Mn on the δ ferrite formation determined by our experiment is shown in Fig. 3. It shows that Mn works as a ferrite-forming element, contrary to the previous studies. Therefore, Mn content must be low to obtain single phase austenite when Cr content is high.

On the other hand, the effect of Mn on the martensite formation is complicated. Adjusting our experimental results and allotting a Ni equivalent of 0.6 to Mn, we obtained Fig. 4, which signifies that the martensite forming tendency depends only on the Ni equivalent, and the Cr equivalent is irrelevant.

As the magnetic permeability (μ) is proportional to the quantity of α martensite and μ = 1.02 corresponds to 0.3%α, we can obtain material whose magnetic permeability is less than 1.02 with the condition that its Ni equivalent value exceeds 28% regardless of the Cr equivalent value.

From the above results, we can conclude that the effect of Mn on the phase stability varies with Cr content. The austenite stabilizing effect of Mn is very large when Cr content is low, decreases with the increase of Cr, and finally, Mn becomes ferrite-forming element at high Cr content.

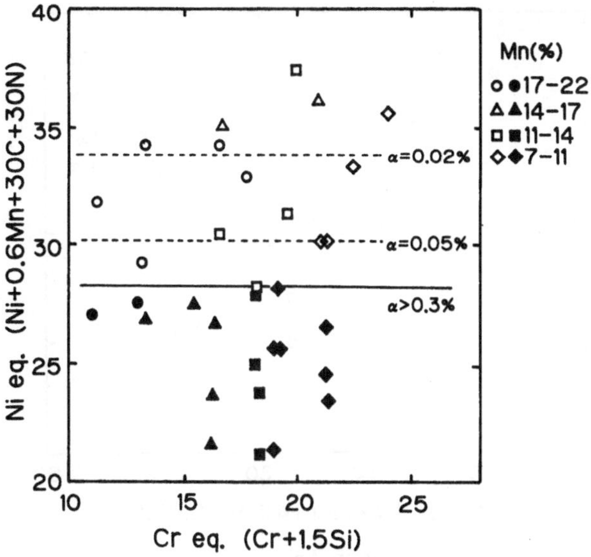

Fig. 4. Austenite-Martensite phase boundary of high Mn-Cr-Ni steel.

Effect of Mn on the Impact Toughness

Before investigating the effect of Mn on the impact toughness, there is one important fact worth noticing, that is, the effect of S: When Mn is added in the metallic form of Mn, S content inevitably increases because a considerable amount of S is usually contained in metallic Mn. S decreases the impact toughness remarkably, as shown in Fig. 5. Accordingly, some consideration must be given to content when selecting materials to avoid this important defect.

Comparing the impact toughness of Ni-Cr-N steel with that of Mn-Cr-N steel, we find out the former is a little superior to the latter when the proof stress is equal.[2] Mn seems to decrease the impact toughness to some extent.

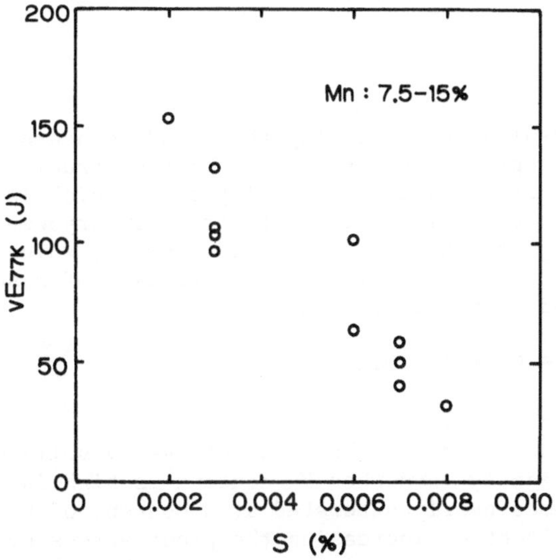

Fig. 5. Relation between S content and vE at 77 K.

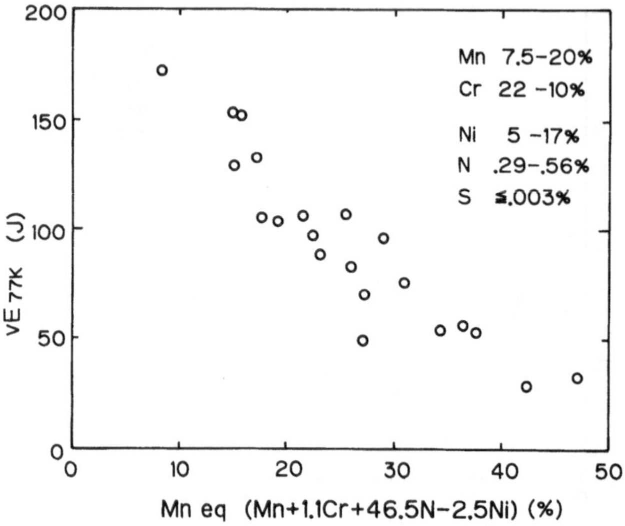

Fig. 6. Relation between Mn eq. and vE at 77 K.

A multiple linear regression analysis was made with 21 specimens which contain less than 0.003% S, and the effect of elements on the impact toughness relative to that of Mn was calculated.

The result is represented by the following equation:

$$\text{Mn eq.} = \text{Mn} + 1.1\text{Cr} + 46.5\text{N} - 2.5\text{Ni} \ (\%) \qquad (2)$$

The effects of Mn and Cr are nearly equal and N decreases the impact toughness 46.5 times more than Mn. Ni is the only element that improves the impact toughness.

Fig. 6 shows the relation between this Mn equivalent and vE at 77 K. The impact toughness of Mn-containing materials can be predicted using this equivalent.

Inclusions and small second phase particles decrease the impact toughness of Ni-Cr-N steels markedly, as we reported previously.[1] However, the effect of these small particles was not clear on the materials of this experiment, whose inclusion content ranges from 0.05% to 0.08%.

CONCLUSIONS

The following results were found by this study concerning the effect of Mn on the cryogenic properties necessary for the structural material:

1. Mn is useful in increasing the solubility of N, which is the most effective element in enhancing the proof stress at cryogenic temperatures. In this regard, Mn can be replaced with Cr, which also effectively increases N solubility. A rough standard of N solubility in a small ingot formed by vacuum melting is:

$$\text{N}(\%) = 0.021(\text{Cr} + 0.9\text{Mn}) - 0.204$$

when Ni content is around 15%.

2. The effect of N in increasing the proof stress by solid solution strengthening is slightly less when Mn content is high in the Cr-Ni-Mn matrix, and it becomes nearly saturated at N content of 0.3%. However, as Mn has its own effect of increasing the proof stress by substitutional solid solution strengthing, the necessary N content can be lower than that in low Mn Cr-Ni steels.

3. The behaviour of Mn with regard to the stability of austenite is very complicated. Mn is a ferrite-forming element at the austenite-ferrite boundary of the high Cr content region.
   On the contrary, Mn works as an austenite stabilizing element at the austenite-martensite boundary of the low Cr region. And by allotting a Ni equivalent of 0.6 to Mn, this boundary line becomes dependent only on the Ni equivalent, regardless of Cr equivalent.
4. The impact toughness at 77 K deteriorates with the addition of Mn. This effect is similar to that of Cr. Ni is the sole element that improves the impact toughness. As S deteriorates the impact toughness markedly, some consideration must be given to decreasing the S content.

The optimum Mn content varies with the amount of Cr and Ni for the reasons mentioned above. Therefore, several suitable Cr-Ni-Mn combinations are possible according to the circumstances of application.

REFERENCES

1. T. Sakamoto et al., Nitrogen-containing 25Cr-13Ni stainless steel as a cryogenic structural material, in "Advances in Cryogenic Engineering-Materials" vol. 30, Plenum Press, New York (1984), pp. 137-144.
2. H. Nakajima et al., Test Results of Cryogenic Structural Materials of FER in Viewgraphs No. M-6, DOE-JAERI joint planning Work Shop Dec. (1984).
3. S. Tone et al., The development of a nitrogen-strengthened high-Manganese austenitic stainless steel for a large superconducting magnet, in "Advances in Cryogenic Engineering-Materials" vol. 30 Plenum Press, New York (1984), pp. 145-152.
4. H. Masumoto et al., Development of a high-strength high-Manganese stainless steel for cryogenic use, ibid. pp. 169-176.
5. W. T. Delong et al., Weld. J. 35(11): 84s-90s (1956).
6. F. C. Hull, Weld. J. 52(5): 193s-203s (1973).
7. R. H. Espy, Weld. J. 61(5): 149s-156s (1982).

AN INVESTIGATION OF THE CRYOGENIC MECHANICAL PROPERTIES OF LOW THERMAL-EXPANSION SUPERALLOYS

L. T. Summers and E. N. C. Dalder

Lawrence Livermore National Laboratory
Livermore, California

ABSTRACT

Four Fe-based superalloys, JBK-75, Incoloy 903, Incoloy 905, and Incoloy 909 were evaluated as tube materials for ICCS Nb_3Sn superconductors. Evaluation consisted of 4-K tensile and elastic-plastic fracture-toughness testing, and a microstructural characterization of unwelded and autogenously gas-tungsten-arc welded sheet given a simulated postweld processing treatment of 15% cold reduction by rolling followed by a Nb_3Sn-reaction heat treatment of 96 hours at 700 ^oC plus 48 hours at 730 ^oC. Results indicate that JBK-75 and Incoloy 903 showed satisfactory combinations of strength and toughness for ICCS tube use requiring long Nb_3Sn-reaction heat treatments. Incoloy 905 welds and 909 showed unacceptable fracture toughness. Results are discussed in terms of microstructural changes caused by the extended Nb_3Sn-reaction heat treatment.

INTRODUCTION

Many fusion coil designs[1-3] use unalloyed or alloyed Nb_3Sn superconductor, surrounded by a thin-walled tube, and cooled by the forced flow of LHe inside this tube; the so-called "internally cooled coiled superconductor" or ICCS. This report describes the investigation of the suitability of four alloys proposed for use as the tube.

The high-strain sensitivity of A15 conductors[4] requires that the conductor be fully fabricated prior to reaction heat treatment. Fabrication includes wrapping with the tube alloy, seam welding, and swaging to final size. The tube alloy sees the high-temperature heat treatment that forms Nb_3Sn in the conductor. Minimizing thermal-expansion mismatch between the Nb_3Sn conductor and tube is beneficial, and use of alloys with controlled thermal expansion can substantially increase the current-carrying capacity of short lengths of ICCS conductors.[5]

Tube-alloy performance targets in this program are those set forth in the Vail Workshop[6] in October 1984: a 4-K yield strength (σ_y) of at least 700 MPa, a 4-K fracture toughness (K_{IC}) of at least 100 MPa $\sqrt{m}$, and a tensile elongation of at least 10%. In this investigation, the 4-K K_{IC} and tensile properties of four candidate alloys were determined on

Table 1. Compositions, weight percent

Alloy	Fe	Cr	Ni	Co	Nb	Ti	Al	Mo	Mn	C	Si
JBK-75	Bal	15.0	29.0	----	----	2.15	1.25	1.25	0.05	0.016	--
I-903	Bal	----	38.0	15.0	3.0	1.4	0.9	0.5	0.2	0.03	--
I-909	Bal	----	38.2	13.0	4.7	1.5	0.4	0.5	0.2	0.03	0.25
I-905	Bal	----	48.9	0.33	4.7	1.59	0.04	----	0.05	0.01	--

unwelded sheet and autogenous (without filler material being added) gas-tungsten-arc (GTA) welds after a simulated Nb_3Sn-reaction heat treatment of 96 hours at 700°C plus 48 hours at 730°C. Work performed by Wu et al.[7] showed that this treatment yielded the maximum critical current density for internal bronze-process Nb_3Sn superconductor.

Four alloys were selected for evaluation. The first, JBK-75, is the same precipitation-hardening stainless steel as that used for the tube for the Westinghouse Large Coil.[8] The three remaining alloys (Incoloy 903, 905, and 909) are controlled thermal-expansion alloys, with compositions given in Table 1. All four alloys harden by the precipitation of an ordered intermetallic phase of the "Ni_3X" type, where "X" may be Ti, Al, or Nb, either alone or in combination.

PROCEDURE

The starting material was 1.54- to 1.77-mm-thick sheets. Sheets were sealed in stainless steel bags and solution treated in air for one hour at 1253 K. Following descaling, weld specimens were prepared using the GTA process. All welds were full penetration "bead-on-plate" welds.

Metallographic specimens were sealed in stainless steel bags and heat treated in air at temperatures from 923 to 1073 K for up to 500 hours. Temperatures were controlled to ± 2 K, as determined by a Chromel-Alumel thermocouple. Specimens for mechanical testing were aged using the previously described heat treatment.[7]

Aging curves were generated using microhardness measurements obtained from metallographic specimens. Mechanical test specimens were described in Ref. 9. The K_{IC} was determined using the unloading-compliance J-integral method of ASTM E-813[10] modified for testing at 4 K.[11] The specimen thickness is lower than prescribed by ASTM E-813, but is similar to the thickness of an actual conductor sheath. Therefore, the estimates of K_{IC} based on these tests may be conservative. Tensile tests were run at a crosshead speed of 1.27 mm/min in a "hard" machine and testing conformed to procedures set forth in ASTM E-8[12], modified for testing at 4 K.[13] Specimens were oriented such that the applied loads were taken both along and transverse to the rolling/welding direction .

RESULTS

Mechanical Testing

Results appear in Table 2. The average ultimate tensile strengths (σ_u), 0.2% offset yield strengths, and elongations to failure (e_f) of the four alloys are given. The 4-K σ_y of all unwelded and GTA-welded

Autogenous GTA Welds at 4.2 K

Alloy	Condition	Orientation	Elastic modulus, GPa	Ultimate strength, MPa	Yield strength, MPa	Elongation in 12.7 mm, %	Plane-strain fracture toughness MPa $\sqrt{m}$	NTR[c]	NYR[d]
Incoloy 903	Unwelded	Longitudinal[a]	138.6	1530	1016	28.9	165.4	0.97	1.46
		Transverse[b]	156.2	1611	1184	27.9	110.1	0.93	1.28
	Welded	Longitudinal	151.5	1499	1139	9.1	127.4	0.83	1.09
		Transverse	-----	1439	1070	9.3	150.9	0.88	1.19
Incoloy 905	Unwelded	Longitudinal	171.4	1453	906	11.1	133.0	0.80	1.20
		Transverse	215.7	1527	1024	15.8	140.5	0.85	1.26
	Welded	Longitudinal	195.6	1472	942	7.4	81.5	----	----
		Transverse	-----	1502	984	14.2	136.8	0.71	1.02
Incoloy 909	Unwelded	Longitudinal	174.2	1432	1115	4.6	-----	0.73	0.94
		Transverse	161.9	1606	1236	7.0	69.7	0.65	0.85
	Welded	Longitudinal	140.9	1506	1119	4.0	39.5	0.55	0.74
		Transverse	-----	1454	1084	3.5	38.9	0.56	0.74
JBK-75	Unwelded	Longitudinal	232.7	1884	1398	34.2	108.3	0.93	1.26
		Transverse	274.3	1710	1238	30.0	----	0.96	1.33
	Welded	Longitudinal	243.9	1855	1295	17.5	148.8	0.85	1.26
		Transverse	-----	1831	1415	9.7	133.0	0.81	1.06

[a] Longitudinal - parallel to rolling direction of sheet in direction of welding.
[b] Transverse - perpendicular to rolling direction of sheet in direction of welding.
[c] Ratio of notched to unnotched tensile strength at a stress concentration of 4.0.
[d] Ratio of notched to unnotched yield strength at a stress concentration of 4.0.

Table 3. Summary of Microstructural Observations on Fully Reacted Alloys

	Incoloy 903		Incoloy 905		Incoloy 909		JBK 75	
Sampling location	Base metal	Weld metal	Base metal	Weld metal	Base metal	Weld metal	Base metal	Weld metal
Fracture surfaces	Ductile rupture	Ductile rupture, separation along dendrites	Cleavage & intergranular fracture	Separation along dendrites	Intergranular fracture	Separation along dendrites	Ductile rupture plus intergranular fracture	Ductile rupture separation along dendrites
General Microstructure	Austenite, discontinuous grain-boundary phase, coarse intragranular precipitate		Coarse, intergranular precipitate, serrated grain boundaries		Semicontinuous grain-boundary phase, fine intragranular precipitate		Austenite plus cellular phases growing from grain boundaries	

sheet surpass the 700-MPa goal. Elongation values of Incoloy 903, 905, and JBK-75 sheet exceed the 10.0% requirement. Elongation values of Incoloy 909 sheet were unacceptably low, with values of 4.6% (L) and 7.0% (T). Elongation values of Incoloy 905 welds tested in the T direction and JBK-75 welds tested in the L direction exceeded the 10.0% requirement. However, e_f values for Incoloy 903 welds tested in both directions and JBK-75 welds tested in the T direction were between 9.0 and 10.0%. Incoloy 905 welds tested in the L direction and Incoloy 909 welds tested in both directions fell well below the 10.0% requirement, being 7.4%, 4.0%, and 3.5%, respectively. Incoloy 909 had a weld e_f of 3.75% while Incoloy 905 and 903 welds showed e_f values of 7.4 to 14.2% and 9.1 to 9.3%, respectively.

Notched tensile data at a stress concentration of 4 are summarized in Table 2. Incoloy 903 and JBK-75 had ratios of notched tensile strength ($\sigma_{notched}$) to unnotched tensile strength, or "NTR", no lower than 0.81. Welded Incoloy 905 had a NTR ratio of 0.71. Incoloy 909 had significantly lower NTR values, 0.65 to 0.73 for sheet and 0.55 to 0.56 for welds.

Fracture toughnesses of all alloys are given in Table 2. Incoloy 909 sheet and welds have unacceptably low K_{IC} of 69.7 MPa $\sqrt{m}$, and 38.9 to 39.5 MPa $\sqrt{m}$, respectively. Incoloy 905 welds tested in the L direction had a K_{IC} of 81.5 MPa $\sqrt{m}$, but this value rose to 136.8 MPa $\sqrt{m}$ when tested in the T direction.

The 4-K K_{IC} values showed a qualitative correlation to NTR and ratio of σ_{un}/σ_y, or "notched yield ratio," NYR. Those materials with high K_{IC} displayed high NTR and NYR values, while those with low K_{IC} showed low NTR and NYR ratios. For example, Incoloy 909, which has a weld K_{IC} of about 39 MPa $\sqrt{m}$, has an NTR of about 0.55.

Modulus of elasticity (E) values for the four alloys are in Table 2. These values were determined from the elastic-strain region of the tensile stress-strain curves. Data are reported for E values obtained for sheet L and T orientations and weld L orientations. Note that the two Co-free alloys (JBK-75 and Incoloy 905) have significantly higher (in excess of 170 GPa) E values than Incoloy 903 and 909, which have E values of 138 GPa to 156 GPa and 141 GPa to 174 GPa, respectively.

MICROSTRUCTURAL CHANGES

For tensile and K_{IC} rupture surfaces of sheet, microstructural observations are summarized in Table 3. Incoloy 903 showed ductile rupture. JBK-75 showed ductile rupture in grain interiors with embrittlement of grain boundaries. Incoloy 905 showed a mixed-mode failure with regions of transgranular cleavage and of intergranular fracture. Incoloy 909 is embrittled, with a rough, intergranular, fracture surface.

Microstructural observations on fracture surfaces of heat-treated welds are summarized in Table 3. Incoloy 903 and JBK-75 showed failures that are mostly ductile. A dendritic weld microstructure was evident, indicating that some embrittlement occurred at dendrite interfaces. Incoloy 903 showed evidence of the fracture following the columnar weld-metal grain-structure. Incoloy 905 and 909 showed pronounced evidence of separation both along weld-metal-grain and dendrite boundaries.

JBK-75 formed grain-boundary phases by a cellular precipitation mechanism.[14] The cellular product was seen in the weld at aging times of less than 8 hours. Continued aging resulted in rapid coarsening of the cellular product in the weld and decoration of sheet grain boundaries by the cellular product. The common overaging product in Fe-Cr-Ni superalloys is Ni_3Ti eta phase[9] which forms by a cellular mechanism and is probably responsible for the weld and sheet grain-boundary decoration in JBK-75.

Fully reacted Incoloy 903 weld and sheet showed discontinuous precipitation along grain boundaries and a coarse intragranular precipitate. Semiquantitative SEM-EDS analysis showed that the intergranular precipitates are slightly enriched in Nb and Ti with respect to the matrix, indicating that these precipitates are of the $Ni_3(Nb,Ti)$ type.

Fully reacted Incoloy 905 sheet showed serrated grain boundaries resulting from grain-boundary migration caused by growth of the coarse, intergranular, precipitate plates. Incoloy 905 welds showed intragranular precipitates. Incoloy 909 sheet and welds showed heavy, small, intergranular precipitation early in aging. SEM-EDS analysis showed enrichment with Nb and Ti. These precipitates are probably an intermetallic overaging product, such as $Ni_3(Nb, Ti)$.

Weld and sheet specimens of these alloys showed a difference in aging response. The welds age more slowly and to lower peak-hardness values than does sheet.[9]

DISCUSSION

Susceptibility of Fe-base superalloys to formation of embrittling precipitates is complex, and depends on both processing history and composition. Holding other factors constant, the alloy with the higher concentration of hardening elements such as Ti, Al, Nb, Ta, and C, will be most likely to form unwanted phases. These assumptions provide guidelines for material selection and understanding of microstructural changes.

Low K_{IC} and rapid grain-boundary precipitation in Incoloy 909 is expected, since the combined Al, Ti and Nb content of this alloy is nominally 6.6 wt%. The high Nb/Ti ratio favors the substitution of Nb for Ti in γ', or the possible formation of γ''.[15] Substitution of Nb for Ti increases lattice mismatch and the strain energy that drives the formation of equilibrium phases.

Incoloy 905 contains large amounts of elements that participate in hardening-phase formation (6.33 wt%). Replacement of Co with Ni in this alloy affects the precipitate morphology. Overaging products, rather than forming as grain-boundary precipitates, form as intragranular Widmanstatten precipitates that contribute to transgranular fracture and reduction of K_{IC}.

Both JBK-75 and Incoloy 903 contain lesser reduction of the hardening elements: 3.4 and 5.3 wt%, respectively. Also significant is the lower Nb/Ti ratio in Incoloy 903 as compared to other Incoloys, resulting in alloys that, while showing grain-boundary precipitation, do not do so to an extent that K_{IC} is compromised. Welds in JBK-75 are sensitive to formation of a cellular phase mixture at grain and dendrite interfaces, due to segregation of Ti to dendrite interfaces during weld solidification.[9] Hardness mismatches during isothermal heat treatments of these alloys are caused by this Ti segregation, which forms "precipitate-free zones" (pfz) in welds. Multistep aging treatments reduce the extent of pfz. Similar σ_y for welds and sheets in this investigation results from beneficial effects of the long aging at 973 K.[9]

Based on K_{IC} and σ_y requirements, Incoloy 903 and JBK-75 meet the specified targets. The low e_f of Incoloy 903 weld is below the needed 10%. Owing to the weld's high K_{IC}, this may not be a major concern. Incoloy 905 and 909 have good σ_u and σ_y, but insufficient K_{IC}, particularly in their welds.

Incoloy 903 and 909 have low E values, about 156 GPa. Incoloy 905 and JBK-75 have values of 204 GPa and 238 GPa, respectively. If the tube is a load-bearing member, designers must account for differences in moduli of tube alloys. Use of a higher-E alloy, such as JBK-75, may allow reductions in tube-wall thicknesses and a higher fraction of superconductor in the coil pack. However this advantage must be weighed against effects of thermally induced strains caused by the higher thermal-contraction coefficient of JBK-75.

Alloys with large amounts of Co (Incoloy 903 and Incoloy 909) had lowest E values. It is assumed that replacement of Ni and Fe with Co is responsible for E changes. If this is true, development of E-composition correlations based on published information may provide a means for planning a focused alloy-design program to produce an alloy with both high E and low thermal-expansion coefficient.

CONCLUSIONS

1. All alloys formed extensive coarse precipitates during a long (973 K:96 h plus 1003 K:48 h) Nb_3Sn-reaction heat treatment. Incoloy 909 and 905 sheet and GTA welds showed heavy intergranular and intragranular precipitation. JBK-75 welds showed extensive formation of an Ni_3Ti-based phase that did not adversely affect K_{IC}. Both JBK-75 sheet and Incoloy 903 sheet and welds showed much less sensitivity to overaging than did Incoloy 905 and 909.

2. Yield strengths of all alloys are above the 700-MPa requirement. However, Incoloy 905 welds and Incoloy 909 welds and sheet had K_{IC} values that are well below the 100 MPa $\sqrt{m}$ target. Intergranular failure and reduced K_{IC} resulted from heavy precipitation of brittle equilibrium phases that occurred intragranularly in Incoloy 905 and at the dendrite and grain boundaries of Incoloy 909.

3. Tensile elongation is acceptable in JBK-75 weld and sheet. Incoloy 903 and 905 display satisfactory e_f in sheet but marginal e_f in welds. Incoloy 909 had poor e_f in both weld and sheet.

4. The E values of Incoloy 903 and 909 are considerably lower than Incoloy 905 and JBK-75, possibly due to the replacement of Ni and Fe with Co in these alloys.

RECOMMENDATIONS

Based upon 4-K mechanical properties, Incoloy 903 and JBK-75 are acceptable for ICCS conductor-tube alloy usage in applications requiring extended Nb_3Sn heat treatments. The E value of Incoloy 903 is low and its effect on any design must be examined. Incoloy 905 and 909 should be excluded from consideration on the basis of low K_{IC} values.

ACKNOWLEDGEMENTS

The assistance of Dennis Freeman with mechanical testing, Robert Kershaw and Richard Gross with metallography, and John Holthius with thermomechanical processing, are all gratefully acknowledged. This work was performed under the auspices of the U.S. Department of Energy by the Lawrence Livermore National Laboratory under contract number W-7405-ENG-48.

REFERENCES

1. J. R. Miller, "Design of Aggressive Superconducting TFCX Magnet Systems," Lawrence Livermore National Laboratory, Livermore, California, UCRL-92371 (1985).
2. B. Blackwell, M. Greenwald, D. Gwinn, B. Lipschultz, E. Marmer, B. Montgomery, R. Parker, P. Politzer, M. Porkolab, J. Schultz, S. Wolfe, "Alcator DCT Proposal," Plasma Fusion Center, Massachusetts Institute of Technology, (Sept. 1983).
3. K. Yoshida, H. Nakajima, K. Koizumi, M. Shimada, Y. Sanada, Y. Takahashi, E. Tada, H. Tsuji, S. Shimamoto, "Development of Cryogenic Structural Materials for Tokamak Reactors," in "Austenitic Stainless Steels at Low Temperatures," R. Reed and T. Horiuchi, eds., Plenum Press, N.Y., (1983), pp. 29-39.
4. J. W. Ekin, in "Filamentary A-15 Superconductors," M. Suenaga and A. F. Clark, eds. (Plenum, NY, 1980).
5. M. M. Steeves and M. O. Hoenig, IEEE Trans MAGN-19, pg. 374 (1983).
6. Task Group on Conductor Sheath Alloys, NBS-DOE Workshop on Materials at Low Temperatures, Vail, Colorado, October 1984.
7. I. W. Wu, D. R. Dietderich, W. V. Hassenzahl, and J. W. Morris, Jr., "The Influence of Microstructure on the Properties of Bronze-Processed Multifilamentary Wire," United States-Japan Workshop on High Field Superconducting Materials for Fusion, San Diego CA, May 23-25, 1983.
8. P. Sanger et al., Adv Cryo Eng. 28, 751 (1981).
9. L. T. Summers and J. W. Morris, Jr., Adv Cryo Eng. 30 23 (1983).
10. ASTM Designation E-813-81, "Standard Test Method for J_{IC}, A Measure of Fracture Toughness."
11. E.N.C. Dalder and R. Brady, "Standard Practice for Tension Testing of Stainless Steels at Cryogenic Temperatures," Appendix A to MEL 83-001675, Lawrence Livermore National Laboratory, April 1983.
12. ASTM Designation E-8-82, "Standard Method of Tension Testing of Metallic Materials."
13. E.N.C. Dalder, R. Street, and D. Freeman, "Standard Practice for Fracture Testing of Stainless Steels at Cryogenic Temperatures," Appendix B to MEL 83-001675, Lawrence Livermore National Laboratory, April 1985.
14. H. J. Beattie, Jr., and W. C. Hagel, Trans AIME 215, 911 (1957).
15. J. H. N. van Vucht, J. Less-Common Met. 11, 308 (1966).

WELDING ADVANCES IN CRYOGENICS

K. A. Yushchenko

E. O. Paton Institute of Electrowelding
Ukr.SSR Academy of Sciences
Kiev, USSR

ABSTRACT

Recent developments in welding technology are reviewed. Mechanical properties of recently developed alloys and welds produced by various processes are tabulated.

INTRODUCTION

In modern cryogenics, the developmental level of welding materials and techniques determines, to a large extent, the success of space exploration programs, the reliability of containers for liquefied gases, and the capabilities ities of cryogenerators, cryomotors, superconducting cables, and magnetic systems for thermonuclear fusion plants. In the 1960s, the dominant research effort was in welds for pipeline shells and pipes, which were manufactured from relatively thin, small pieces of metal. Problems that arose in the 1980s required the development of new (in principle) materials, processes, and equipment.

Welded structures are used in all types of applications at cryogenic temperatures. Along with the development of systems that must operate in this temperature range, requirements of cryogenic welding materials science are continuously increasing. The experience of the last twenty years showed that development of structural materials must consider the service conditions at cryogenic temperatures.[1,2]

What are these conditions? The principal ones are:

- brittle fracture resistance in the loading condition down to 4 K and low sensitivity to stress raisers
- good weldability and flexibility in metallurgical processing, stamping, and bending
- high strength
- corrosion resistance
- low thermal conductivity and magnetization

Depending on the service conditions, other requirements may be a low coefficient of thermal expansion, fracture resistance during thermocycling or irradiation, and satisfactory performance in strong magnetic fields, to name a few. Both structural materials and welding materials must meet all these requirements. Welding materials are primarily welding wires, electrodes, and fluxes.

RECENT DEVELOPMENTS

In the last two decades, the development of welding science and engineering for cryogenic applications has been characterized primarily by the development and industrial production of good, weldable structural steels and alloys. Examples of alloys that were developed for service over a wide range of cryogenic temperatures are given in Table 1; they are the results of research in the USSR, United States, Japan, Great Britain, and the Federal Republic of Germany. Note especially the austenitic nitrogen-strengthened steels. Alloying 18 to 22%Cr and 10 to 16%Ni steels with nitrogen sharply increases their strength properties in the temperature range 4.2 to 77 K.

Table 1. Properties of Steels and Alloys for Cryogenic Applications

Type of Alloy	Temperature, K	Yield Strength, MPa	Ultimate Strength, MPa	Impact Toughness (CVN), kgm/cm^2	Fracture Toughness (K_{Ic}), $MPa\sqrt{m}$
Fe-18Cr-10Ni-0.5Ti	293	220	540	30	*
Fe-18Cr-13Ni 2.5Mo-0.5Ti (and similar alloys)	4.2	580	1300	26	350
Fe-18Cr-10Ni-0.15N	293	320	600	30	*
Fe-18Cr-13Ni-2.5Mo-0.15N (and similar alloys)	4.2	900	1500	22	280
Fe-0.3C-13Cr-19Mn-0.18N	293	370	650	23	240
Fe-0.03C-13Cr-20Mn-5Ni-0.2N (and similar alloys)	4.2	960	1400	17	130
Fe-0.03C-20Cr-16Ni-0.25N	293	370	650	30	300
(and similar alloys)	4.2	1150	1500	25	160
Fe-0.03C-12Cr-10Ni-0.6Mo-	293	830	1050	12.0	200
0.25Ti (and similar alloys)	20	1400	1500	5.0	140

*not determined

Fig. 1. Cryogenic vessel from 18Cr-10Ni stainless steel (1400 m^3).

This discovery is very important in the recent development of materials for structures operating in strong magnetic fields. The combination of high strength (more than 1500 MPa), yield strength (more than 1200 MPa), and a high ductility enabled the development of such unique structures as the magnetic power systems of thermonuclear fusion plants--tokamak, Mirror Fusion Test Facility (MFTF), and Large Coil Project (LCP); load-carrying elements of rotors of cryoelectric machines; high pressure vessels for the storage of liquefied gases (Fig. 1); and elements of a cryogenic aerodynamic tunnel.

Table 2. Properties of High Nickel Alloys

Type of Alloy	Temperature, K	Yield Strength, MPa	Ultimate Strength, MPa	Impact Toughness (CVN), kgm/cm^2	Fracture Toughness (K_{Ic}), $MPa\sqrt{m}$
60Ni-20Cr-9Mo-6W-3Al	293	1080	1400	6.0	130
	4.2	1300	1600	4.0	120
Inconel 706	293	950	1150	5.0	120
	4.2	1100	1470	4.0	176
Inconel 718	293	1090		-	126
	4.2	1440		-	198

Table 3. Mechanical Properties of Fe-0.03C-20Cr-16Ni-6Mn-0.25N Steel Weldments Made by Different Methods with Fe-0.01C-19Cr-16Ni-6Mn-2Mo-N-2W Wire

Welding Method	Temperature, K	Yield Strength, MPa	Ultimate Strength, MPa	Elongation, %	Impact Toughness (CVN), kgm/cm^2	Fracture Toughness (K_{Ic}),* $MPa\sqrt{m}$
Electrodes AHB 40,# manual	293 20	372 1150	657 1420	35 28	16.0 7.5	220 155
Submerged arc welding with AHK-45 flux⁺	293 20	360 1200	620 1405	38 26.0	14.5 8.0	- 168
Flux-cored wire of 20Cr-16Ni-6Mn-2Mo-2W type	293 20	340 1080	610 1320	30 22	12.0 4.6	- 140
Electroslag welding, flux AHK-45⁺	293 20	325 1016	585 1380	48 36.0	22.0 10.6	- 180
Electron beam welding with filler wire	293 20	396 1150	670 1400	36.0 24.0	18.9 10.8	- 172
Automatic argon-arc welding with tungsten electrode	293 20	380 1120	650 1430	44.0 27.0	18.0 9.6	- 162

* at 77 K
\# 0.04C, 7.0Mn, 0.60Si, 19.6Cr, 16.2Ni, 0.15N, 3.0Mo, 0.02S, 0.022P
$^+$ CaF_2, Al_2O_3, MgO, SiO, ZrO_2

Great improvements were made in the weldability of high-strength high-nickel steels.[3,4,5] Table 2 gives typical alloys and their properties. Their improved properties are essential when devices are subjected to heavy loads over a broad temperature range, from normal to cryogenic temperatures. Improvements in the weldability of serial alloys, such as Inconel 716, 708, alloy 286, and XH60MB,* qualified them for use in components and models of cryogenerator rotors. It is feasible to predict their widespread use in industrial power plants of the future, having 1.2 to 2.0 MW or even higher capacity.

Aluminum alloys, having good weldability, are commonly used in cryogenic structures. Many new structures use particularly thick plate; in some cases

*Ni base, 0-0.05C, 0.3Si, 0.01S, 0.01P, 0.5Mn, 19-21Cr, 8-10Mo, 4-6W, 3.2-3.6Al, 5.0Fe

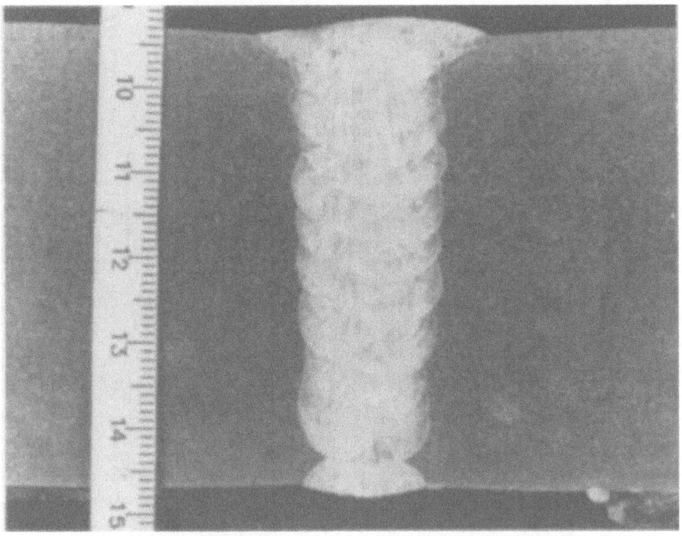

Fig. 2. Macrostructure of weld made by narrow-gap welding.

the wall thickness is 150 to 200 mm. The joining of such thick aluminum alloys requires highly efficient welding procedures as well as the very best welding materials. A high degree of automation is necessary for the welding procedures, not only to increase productivity, but also to obtain the highest quality welds by minimizing the technological and metallurgical defects. Welding sites must now be large enough to handle the large pieces of metal needed for storage tanks of cryogenic liquids.

For these reasons, considerable effort is being directed to the development and improvement of the basic principles of weld alloying and welding materials. Among the new developments, studies of the welding of 9%Ni ferritic steels[6] and stable, nitrogen-alloyed austenitic steels[7] are worthy of mention. Use of ferritic wires on 12%Ni base for the automatic welding of storage tanks for liquefied natural gas results in a yield strength of more than 550 MPa at 293 K; austenitic systems Fe-20Cr-(14-16)Ni-(2-4)Mo-2W-(0.20-0.30)N have yield strengths of more than 400 MPa. The development of the theory of alloying stable austenitic welds[2,7] led to the development and use of the following processes:

- manual welding with coated electrodes in all positions
- submerged arc welding
- electroslag welding
- gas electric welding in shielding gases with a flux-core wire
- electron beam welding with a filler wire
- automatic argon-arc welding with a filler wire

Table 3 gives the mechanical properties of Fe-0.03C-20Cr-16Ni-6Mn-0.25N steel weldments made by these different processes.

For thick plate, the narrow-gap welding method with AHK-45 flux* was used in a previous study.[8] The technique that was developed provided a quality weld (Fig. 2) of 150-mm-thick metal, although the welding of 400- to 600-mm-thick plate is feasible. The alloying system developed, materials, and the welding processes are suitable for a wide range of austenitic and ferritic steels, and even maraging steels when some reduction of weld strength

*CaF_2, Al_2O_3, MgO, SiO, ZrO_2

(a)

(b)

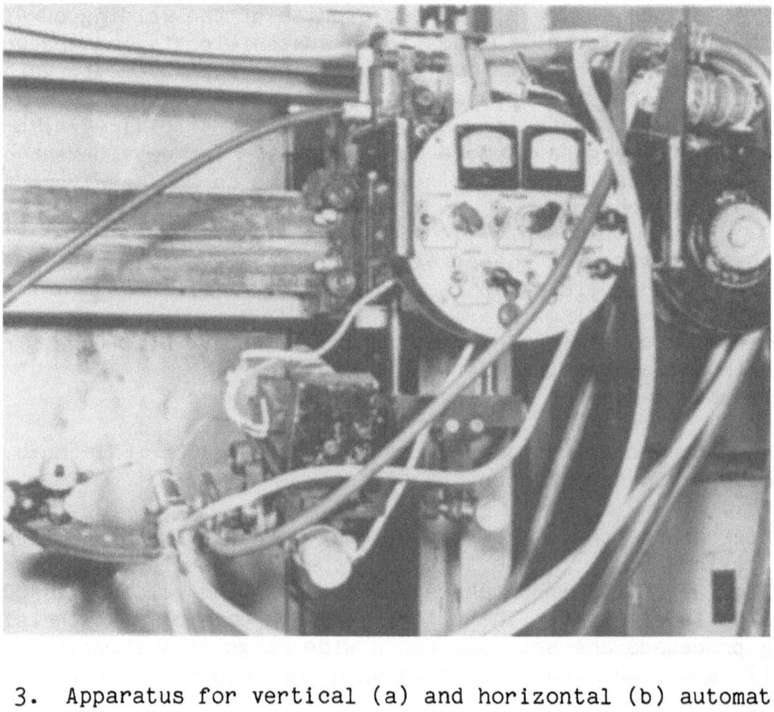

Fig. 3. Apparatus for vertical (a) and horizontal (b) automatic TIG welding of LNG storage vessels from 6 to 9%Ni steel.

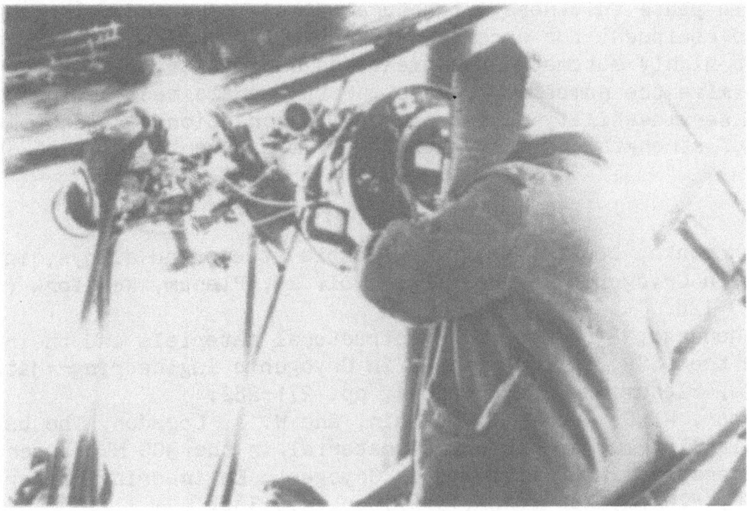

Fig. 4. Welding 6 to 9%Ni steel in the overhead position using flux-covered wire.

is tolerable at room temperature. Note that in welding ferritic Fe-9Ni and maraging Fe-12Cr-10Ni-0.6Mo-0.2Ti steels with a stable austenitic wire of the type mentioned above, a matching weld at cryogenic temperatures is attained. Achievements in the field of cryogenic machine building depend largely upon the welding equipment. As a rule, not only are specialized automatic machines required, but also special power sources and control systems. Remarkable progress has already been made in the USSR, United States, and Japan. Highly efficient equipment has been developed for shop and site welding of unique structures. Figures 3 through 5 show examples of this equipment.

What are the prospects of welding development for cryogenic engineering? First of all, materials will be improved with regard to their susceptibility to pore and crack formation, process stability, and overall quality. With

Fig. 5. Machine for submerged arc welding of 6 to 9%Ni steel.

the increase in plate thickness, structural dimensions, and weld length, the technology and equipment for on-site welding will grow drastically, especially that of highly automated equipment and associated welding techniques that will minimize the number of passes. Future goals must include improvements in weld serviceability in actual service conditions, reliability, and optimization of strength, which together will minimize the cost of structures.

REFERENCES

1. K. A. Yushchenko, Low temperature weldable steels and alloys, in: "Advances in Cryogenic Engineering," vol. 24, Plenum, New York (1978), pp. 120-128.
2. K. A. Yushchenko, Progress in cryostructural materials and their welding in the USSR, in: "Advances in Cryogenic Engineering--Materials," vol. 30, Plenum, New York (1984), pp. 271-282.
3. G. D. Hooper, W. G. Moore, T. T. Shin, and W. A. Logsdon, The use of Inconel Alloy 706 as structural material in the 300 MVA superconducting generator, in: "Advances in Cryogenic Engineering--Materials," vol. 30, Plenum, New York (1984), pp. 359-366.
4. S. Tone, M. Ogawa, M. Yamaga, H. Kaji, T. Horiuchi, Y. Kasamatus, H. Nakajima, K. Yoshida, Y. Takahashi, and M. Shimada, Preliminary study on structural material selection for large superconducting magnets, in: "Austenitic Steels at Low Temperature," Plenum, New York (1983), pp. 263-275.
5. A. Ya. Ishchenko, E. O. Paton Institute of Electrowelding, Kiev, USSR, Collection of papers.
6. K. Agusa, M. Kosho, and N. Nishiyama, Matching ferritic filler MIG welding of 9% Ni Steel, in "Proceedings of the International Cryogenic Materials Conference, Kobe, Japan," Butterworths, London (1983), pp. 368-371.
7. N. I. Kakhovskii, K. A. Yushchenko, and G. G. Monko, Fundamentals of steel and weld metal alloying in structures working for a long time at low temperatures, Strength Mater. 8 (1970).
8. K. A. Yuschenko, G. G. Monko, A. M. Solokha, and I. V. Dubovetskii, Welding of 03X20H16A 6 thick steel, in: "Cryogenic Materials and Their Welding," Naukova Dumka (1985), pp. 471-477.

CRYOGENIC PROPERTIES OF ELECTRON-BEAM WELDED JOINTS IN A 22Mn-13Cr-5Ni-0.22N AUSTENITIC STAINLESS STEEL

S. Tone, M. Hiromatsu, J. Numata, and T. Horiuchi
Kobe Steel, Ltd.
Hyogo, Japan

H. Nakajima and S. Shimamoto
Japan Atomic Energy Research Institute
Ibaraki, Japan

ABSTRACT

The electron-beam (EB) welding process is being considered for welding structural materials for the large superconducting magnets, because it is a highly effective process for welding thick, heavy, steel plates. A 70-mm-thick 22Mn-13Cr-5Ni-0.22N austenitic stainless steel plate, recently developed for cryogenic use, was welded by the EB welding process, and the characteristics of the welded joint were investigated. The results are: (1) The welded joint had sound cross sections without any weld defects, such as cracks and blowholes, although the weld metal had a fully austenitic microstructure and a high nitrogen content. (2) The tensile strength of the welded joint was about 1570 MPa at 4 K, and its joint efficiency was about 95 percent. (3) The fracture toughness of the weld metal was about 240 MPa$\sqrt{m}$ at 4 K, a value higher than that of the base metal. (4) The results, on a laboratory scale, suggest that the EB welding process is suitable for welding the 22Mn-13Cr-5Ni-0.22N austenitic stainless steel.

INTRODUCTION

Several kinds of welding processes, such as the electron-beam (EB), gas tungsten-arc, and gas metal-arc welding processes, will be used to construct a large superconducting magnet. Among these welding processes, the EB welding process is strongly considered, because it is very effective in welding thick, heavy, steel plates.

Previous studies[1,2] indicate that 22Mn-13Cr-5Ni-0.22N austenitic stainless steel (22Mn steel) plates have desirable cryogenic properties, i.e., high yield strength, good fracture toughness and fatigue crack growth rate properties, and a stable austenitic structure at 4 K.

In this study, the 22Mn steel plate with 70-mm thickness was welded by the EB process in the laboratory, and characteristics of the welded joint were investigated. Applicability of the EB welding process to the 22Mn steel plate was evaluated.

Table 1. Chemical composition* of steel plate tested.

C	Si	Mn	P	S	Ni	Cr	N
0.04	0.34	21.82	0.013	0.004	4.94	12.84	0.212

*Wt. % in check analysis

EXPERIMENTAL PROCEDURE

Material and Welding

The test plate of the 22Mn steel, whose chemical composition is shown in Table 1, was made on a trial commercial scale from a 15-ton electric furnace heat. A bottom-pored 15-ton ingot was reheated to 1473 K and hot-rolled into 200-mm-thick slabs. The slab was reheated to 1473 K and hot-rolled into 70-mm-thick plate. In this case, the plate was accelerated-cooled immediately after hot-rolling. Mechanical properties at 4 K of the 22Mn steel plate with 70-mm thickness are shown in Table 2.

A pair of small pieces, 70-mm-thick, 150-mm-wide, and 400-mm-long, taken from the hot-rolled and accelerated-cooled steel plate, were EB welded with the conditions shown in Table 3. The welded joints were inspected with X-rays.

Test Methods

Tensile properties of the welded joint were determined using 7-mm-diameter, 35-mm-gage-length specimens. Tensile specimens were aligned transverse to the centerline of the weld and were taken from the quarter-thickness of the welded joint, as shown in Fig. 1. The tensile test was carried out in liquid helium at 4 K. Strain rates before and after yielding were $7 \times 10^{-5}\ s^{-1}$ and $1 \times 10^{-3}\ s^{-1}$, respectively.

Fracture toughness of the welded joint was measured by the J-integral method in accordance with ASTM E813-81, using 25-mm-thick and 51-mm-wide compact tension specimens. Specimens were oriented as shown in Fig. 1 and were taken from the quarter-thickness of the welded joint. Specimens with side-grooves were fatigue-precracked at the center of the weld metal, at the fusion line, and 5 mm from the fusion line. The test was conducted at 4 K using the single-specimen unload-compliance technique.

The macrostructures and microstructures of the welded joint were examined and its Vickers hardness number was measured.

TEST RESULTS AND DISCUSSION

Macrostructure

No defects were observed in the X-ray inspection of the welded joints. The macrostructure of the cross section of the welded joint is shown in Fig. 2. No weld defects, such as cracks and blowholes, were observed, although the weld metal has a fully austenitic microstructure and a high nitrogen content (0.195 percent.) Furthermore, the width of the weld metal is very narrow and uniform.

Table 2. Mechanical properties of 22Mn-13Cr-5Ni-0.22N steel plate.

Plate Thickness (mm)	Test Temp. (K)	Tensile Properties a)			Fracture Toughness b)
		0.2% Yield Strength (MPa)	Tensile Strength (MPa)	Elongation (%)	$K_{Ic}(J)$ ($MPa\sqrt{m}$)
70	4	1241	1660	37	189

a) 7-mmϕ round specimen (GL:35 mm), 1/4t, T direction.
b) 25-mmt compact tension specimen, 1/4t, TL direction

Table 3. Electron-beam welding conditions.

Plate Thickness	70 mm
Welding Position	Horizontal
Welding Current	150 mA
Welding Voltage	150 kV
Welding Speed	15 cm/min.
Heat Input	90 kJ/cm
Preheating	None
Number of Passes	1
Groove Preparation (Unit:mm)	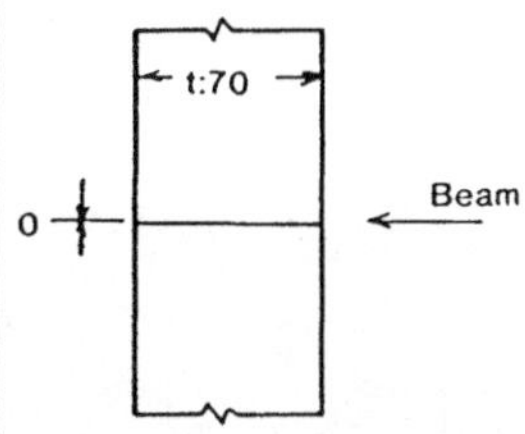

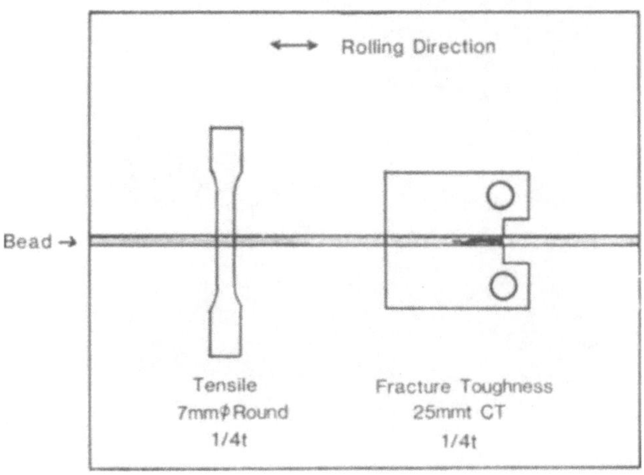

Fig. 1. Test specimen locations and orientations.

Fig. 2. Macrostructure of electron-beam welded joint.

Table 4. Tensile properties of electron-beam welded joint.*

Plate Thickness (mm)	Test Temp. (K)	0.2% Yield Strength (MPa)	Tensile Strength (MPa)	Elongation (%)	Location of Fracture
70	4	1187	1559	17	Weld Metal ~ Fusion Line
		1172	1584	16	Weld Metal ~ Fusion Line

*As-welded condition

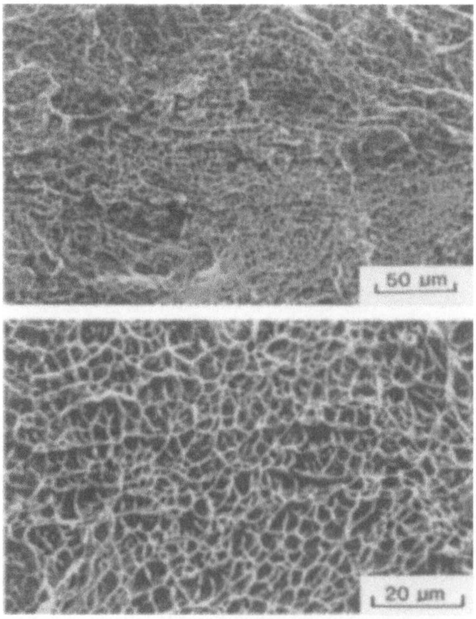

Fig. 3. SEM fractographs of welded joint tensile specimen tested at 4 K.

Tensile Properties

The tensile test results of the welded joint are summarized in Table 4. Its yield strength and tensile strength at 4 K were about 1180 MPa and 1570 MPa respectively, values lower than those of the base metal. However, the joint efficiency has a high value, about 95 percent. The main reason for the lower yield strength and tensile strength is probably the decrease in nitrogen content from 0.212 percent in the base metal to 0.195 percent in the weld metal.

The fracture surface of a representative tensile specimen was examined by scanning electron microscopy (SEM). Figure 3 shows the SEM fractograph. The fracture surface revealed that the weld metal fractured in a ductile manner.

Fracture Toughness

The results of the fracture toughness test at 4 K are shown in Fig. 4. The $K_{Ic}(J)$ at 4 K for the weld metal was about 240 MPa$\sqrt{m}$, higher than that of the base metal. The $K_{Ic}(J)$ values at 4 K at the fusion line and 5 mm from fusion line were approximately 180 MPa$\sqrt{m}$, the same as that of the base metal.

The fracture surfaces of representative fracture toughness specimens examined by SEM are shown in Fig. 5. All positions of the welded joint fractured in a ductile manner. The fracture surfaces of the base metal, at the fusion line, and 5 mm from fusion line had a relatively coarse texture. In contrast, the fracture surface of the weld metal had fine dimples, which suggests the higher fracture toughness in the weld metal.

Predominant factors affecting fracture toughness of weld metal at cryogenic temperatures may be sensitization, ferrite content, nitrogen pickup, and oxide inclusions.[3] In the case of the EB welding process used

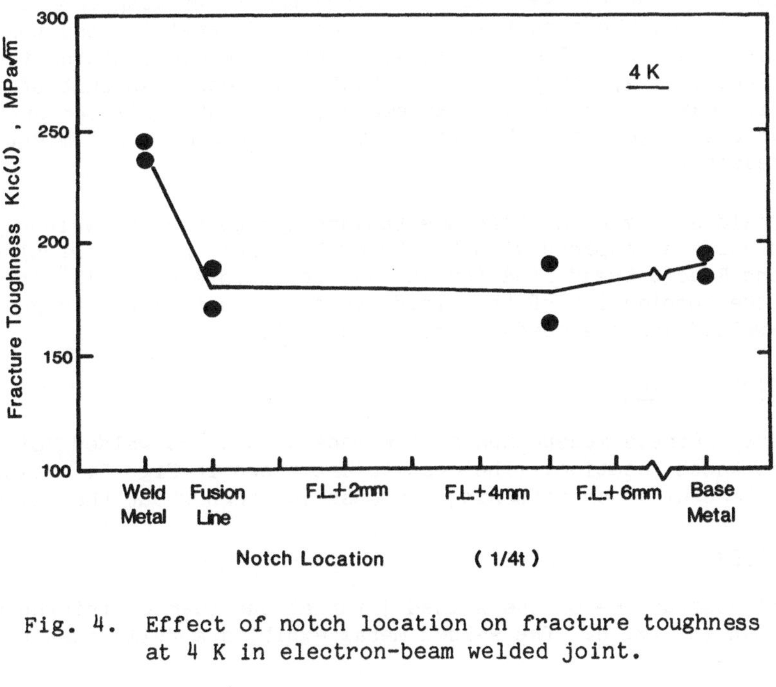

Fig. 4. Effect of notch location on fracture toughness at 4 K in electron-beam welded joint.

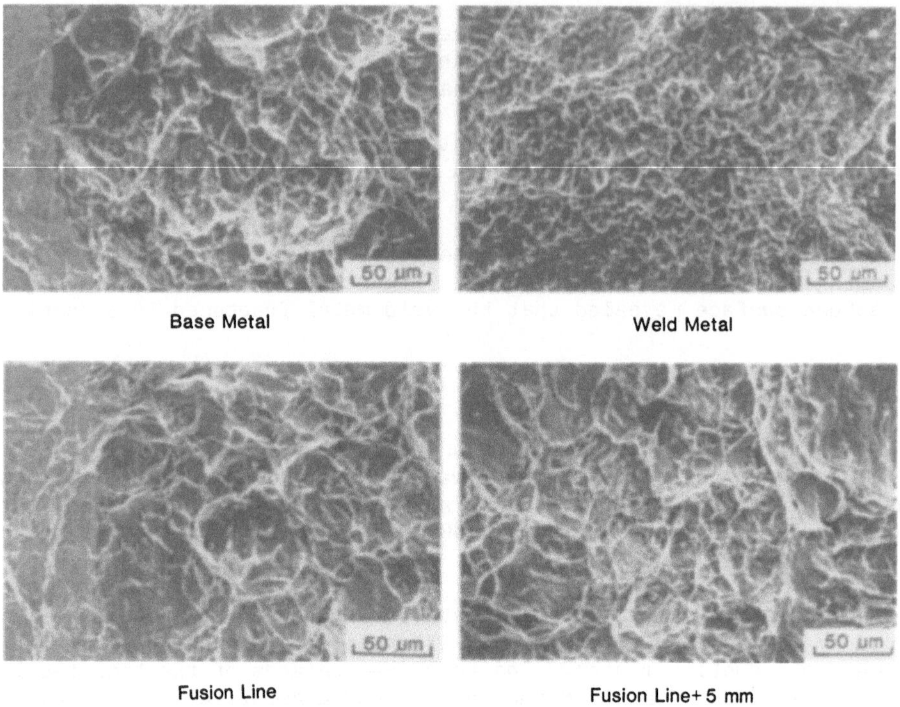

Fig. 5. SEM fractographs of base metal and welded joint fracture toughness specimens tested at 4 K.

on the 22Mn steel, these factors apparently had no harmful influence on the fracture toughness of the weld metal. Since the EB welding process is a single-pass method, the sensitization of the weld metal is less than that of multipass weldments. The weld metal of the 22Mn steel had no ferrite content. Furthermore, its nitrogen content was lower than that of the base metal and its oxygen content was extremely low, about 0.001 percent. The higher fracture toughness of the weld metal in the 22Mn steel might be due to these reasons.

The yield strength and fracture toughness data for the welded joint of the 22Mn steel are compared with the trend line for the 22Mn steel plates[2] and Reed and Read's trend line for austenitic stainless steels[4] in Fig. 6. Note that the combination of the yield strength and fracture toughness at 4 K in the EB-welded joint exceeds these trend lines.

Hardness Distribution

Vickers hardness measurements were made across the welded joint at the quarter-thickness location. The results are shown in Fig. 7. Neither remarkable hardening nor softening was observed throughout the welded joint.

Microstructure

The microstructure of the welded joint at the quarter-thickness location is shown in Fig. 8. The welded metal exhibits a fully austenitic microstructure.

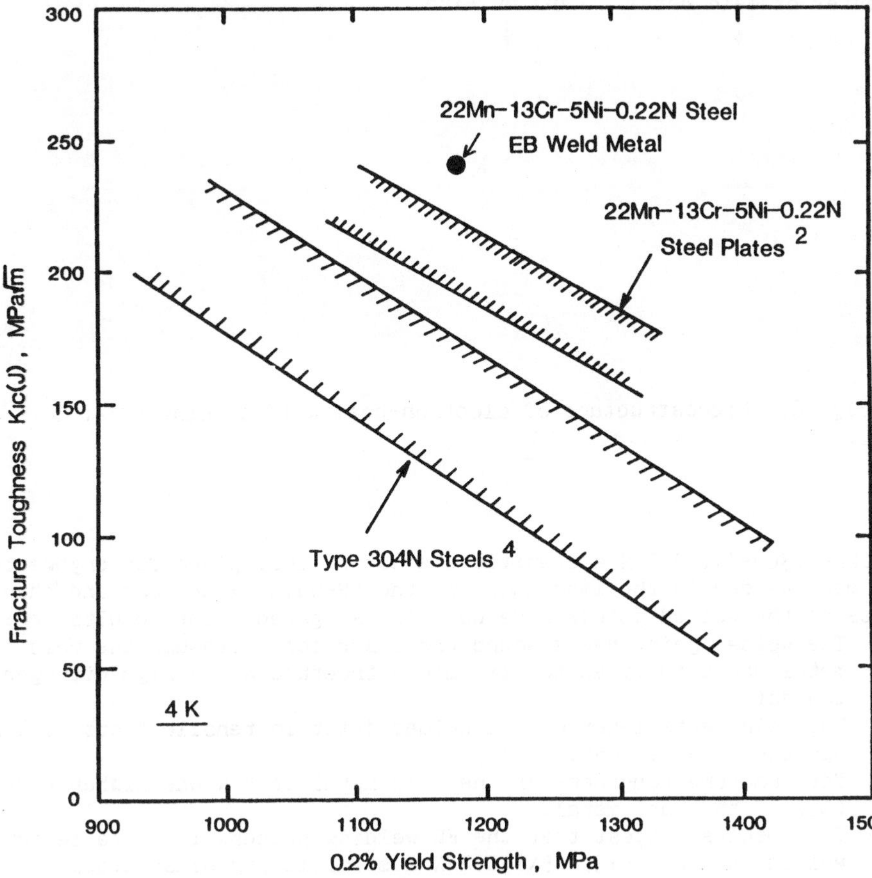

Fig. 6. Relationship between fracture toughness and yield strength at 4 K in electron-beam welded joint and base metals of 22Mn-13Cr-5Ni-0.22N steel and Type 304N steels.

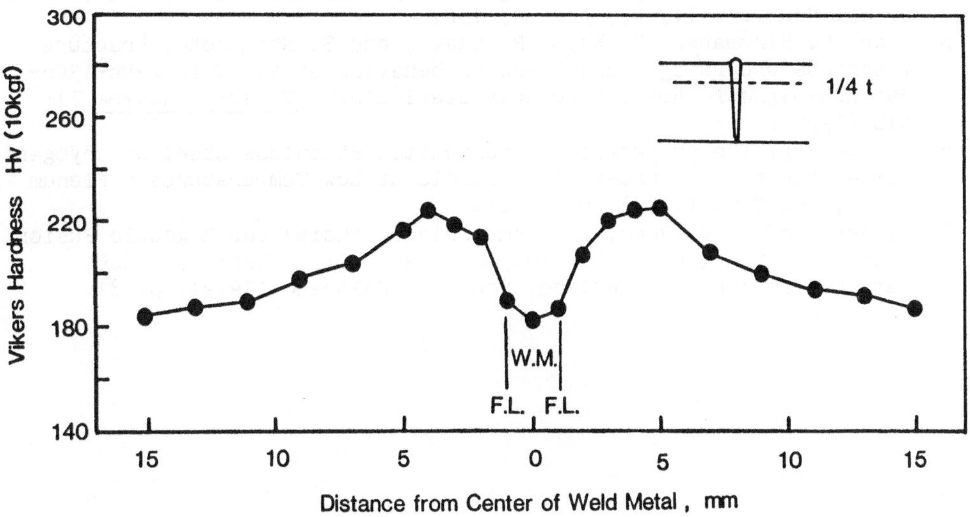

Fig. 7. Hardness distribution of electron-beam welded joint.

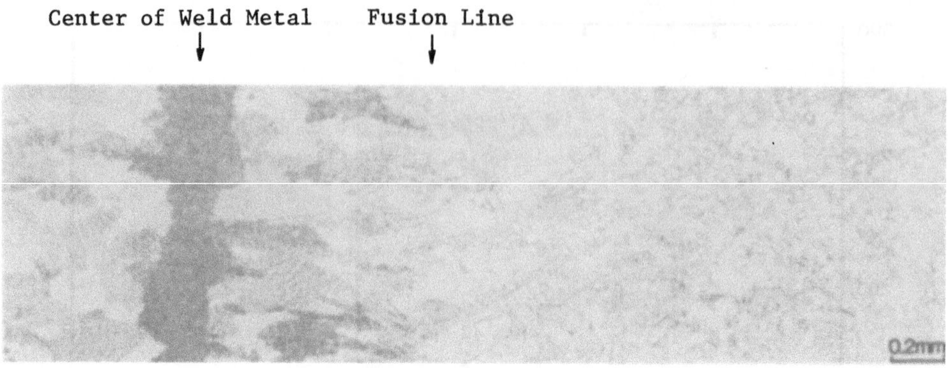

Fig. 8. Microstructure of electron-beam welded joint (1/4t).

SUMMARY

A 22Mn-13Cr-5Ni-0.22N austenitic stainless steel plate for cryogenic use has been welded in the laboratory by the EB-welding process and characteristics of the welded joints have been investigated. The results are:

1. The welded joint has a sound cross section, although the weld metal has a fully austenitic microstructure and a high nitrogen content.
2. The joint efficiency of the welded joint in tensile tests at 4 K was about 95 percent.
3. The fracture toughness of the weld metal at 4 K was higher than that of the base metal.
4. The results suggest that the EB welding process is suitable for welding the 22Mn-13Cr-5Ni-0.22N austenitic stainless steel.

REFERENCES

1. S. Tone, M. Shimada, T. Horiuchi, Y. Kasamatsu, H. Nakajima, and S. Shimamoto, The development of a nitrogen-strengthened high-manganese austenitic stainless steel for a large superconducting magnet, in: "Advances in Cryogenic Engineering - Materials," vol. 30, Plenum Press, New York (1984), pp. 145-152.
2. S. Tone, M. Hiromatsu, H. Kaji, R. Ogawa, and S. Shimamoto, Fracture toughness and fatigue crack growth behavior at 4.2 K in 22Mn-13Cr-5Ni non-magnetic heavy thickness steel plate, Tetsu-to-Hagane 71: 310 (1985).
3. H. I. McHenry, The properties of austenitic stainless steel at cryogenic temperatures, in: "Austenitic Steels at Low Temperatures," Plenum Press, New York (1983), pp. 1-27.
4. R. P. Reed and D. T. Read, in: "Materials Studies for Magnetic Fusion Energy Applications at Low Temperatures - III," NBSIR 79-1609, National Bureau of Standards, Boulder, Colorado (1979), p. 81.

CRYOGENIC MECHANICAL PROPERTIES OF HIGH-MANGANESE STEEL WELDMENTS

J.W. Chan and J.W. Morris, Jr.

Materials and Molecular Research Division,
Lawrence Berkeley Laboratory, and
Department of Materials Science and Mineral Engineering,
University of California, Berkeley, California

ABSTRACT

The structural alloys used in fusion reactor magnets of the next generation are required to have a 4.2 K yield strength and fracture toughness combination superior to that of alloys used currently. An alloy of nominal composition 18Mn-5Ni-16Cr-0.22N after proper thermomechanical treatment approaches the JAERI projected requirements and exceeds the proposed U.S. requirements for 4.2 K yield strength and fracture toughness in the base metal. However, the properties in the welded condition are still uncertain. This work investigates the effect of autogenous GTAW on the cryogenic properties of this alloy. It was found that the 4.2 K yield strength of the weld can be increased through the addition of nitrogen to a 75% He/25% Ar shield gas; specifically, a yield strength comparable to that of the base metal was achieved for a 6 vol% nitrogen addition. Fracture toughness data are still preliminary; however, autogenous welds made on this alloy approach the U.S. requirements for both yield strength and K_{Ic}.

INTRODUCTION

The next generation of fusion reactors will require high field superconducting magnets for plasma confinement. The low allowable strain limit for the winding material, service at liquid helium temperature and other design considerations impose stringent requirements on the cryogenic mechanical properties of the alloys used as structural materials for these magnets.

The projected requirements as proposed by JAERI[1] are 1200 MPa yield strength and a fracture toughness of 200 MPa$\sqrt{m}$ for the base metal at 4.2 K and similar properties in the welded condition. The tentative U.S. requirements[2] are less stringent, calling for a yield strength of 1000 MPa and a fracture toughness of 200 MPa$\sqrt{m}$ for the base metal with 1000 MPa yield strength and a K_{Ic} of 150 MPa$\sqrt{m}$ at 4.2 K in the welded condition.

Previous research[3] indicates that with proper thermomechanical treatment a nitrogen-strengthened high-manganese austenitic alloy of nominal composition 18Mn-5Ni-16Cr-0.22N approaches the JAERI specifications and more than satisfies the U.S. requirements for the base plate. However, these properties were achieved in the as-rolled state; it is not certain that these properties can be retained in the welded condition.

Table 1. Alloy Composition

Mn	Ni	Cr	N	C	Si	S	P
17.98	4.96	16.26	0.216	0.024	0.53	0.010	0.004

The purpose of this work is to establish the effects of welding on the cryogenic mechanical properties of this alloy, and to examine how suitable process modifications can be used to obtain properties matching those of the base metal.

EXPERIMENTAL PROCEDURE

The composition and thermomechanical history of the alloy used in this work are given in Tables 1 and 2, respectively. The yield strength of the base alloy is 1140 MPa.[3] Autogenous GTA welds were made on the 8 mm thick plates using the parameters listed in Table 3. As shown in Figure 1, the weld centerline is in the transverse direction of the base plate. Shield gases of 75% He/25% Ar and a 75% He/25% Ar mixture plus nitrogen additions of 1, 3, and 6 vol% were used. Tensile and compact tension specimens were machined from the welded plates. Figure 1 illustrates the orientation of these specimens with respect to the weld. The tensile specimens were 25.4 mm gauge length flat composite specimens with the weld at the center of the gauge section, and the compact tension specimen dimensions were W=50.8 mm and B=6 mm. The mechanical testing was conducted at 4.2 K using an Instron model 1332 load frame. J_{Ic} values were obtained by the compliance method and the K_{Ic} values calculated from the J_{Ic} data according to ASTM E813-81. The fracture surfaces of the compact tension specimens were examined by scanning electron microscopy. Interstitial nitrogen contents were determined by a commercial laboratory from 1 mm wide through-thickness slices removed from the welds.

RESULTS AND DISCUSSION

The autogenous welds made using the 75% He/25% Ar shield gas have a 4.2 K yield strength of 1040 MPa. This value is 100 MPa lower than the 4.2 K yield strength of the base metal. A profile of the interstitial nitrogen content across the weld bead is shown in Figure 2. The interstitial nitrogen content drops from 0.22 wt% in the base metal to 0.16 wt% across the fusion zone. Since the 4.2 K yield strength of this alloy is strongly dependent on its interstitial content,[4] part of the loss in strength can be attributed to the depletion of interstitial nitrogen from the weld fusion zone.

Table 2. Processing

VACUUM INDUCTION MELTING
HOT-FORGED TO 80 mm
SOAKED AT 1250°C FOR 1 HOUR
HOT-ROLLED TO 30 mm
MACHINED TO 8 mm THICKNESS
AUTOGENOUS GTAW

Table 3. Welding Parameters

SHIELDING GASES	75% He / 25% Ar 75/25 MIXTURE + 1% N_2 3% N_2 6% N_2
BACKGROUND CURRENT	80-100A
PULSE CURRENT	150-185A
DUTY CYCLE	75%
ELECTRODE	2.38 mm diameter, 60° truncated tip
STICKOUT	4.76 mm
TRAVEL SPEED	50.8 mm/min
BACKING GAS	100% Ar

The addition of nitrogen to the shield gas changes the interstitial nitrogen profile by significantly raising the nitrogen content in the weld fusion zone. A profile of a weld made with 3 vol% nitrogen addition to the shield gas is shown in Figure 3. In this case, the concentration rises to above 0.25 wt% at the weld centerline as compared to 0.16 wt% across the fusion zone of the nitrogen free shield gas weld.

Figure 4 shows the change in the average interstitial nitrogen content of the fusion zone as a result of nitrogen additions to the shield gas. The average interstitial nitrogen level can be raised to above 0.25 wt%, slightly higher than the 0.22 wt% of the base metal, with the addition of 6 vol% nitrogen to the shield gas.

The increase in 4.2 K yield strength as a function of the average interstitial nitrogen content is shown in Figure 5. The yield strength increases with increasing average interstitial nitrogen content from a low of 1040 MPa for welds made with the nitrogen free shield gas to a high of 1130 MPa for welds made with a 6 vol% nitrogen addition to the shield gas. The latter value approaches the yield strength of the base metal.

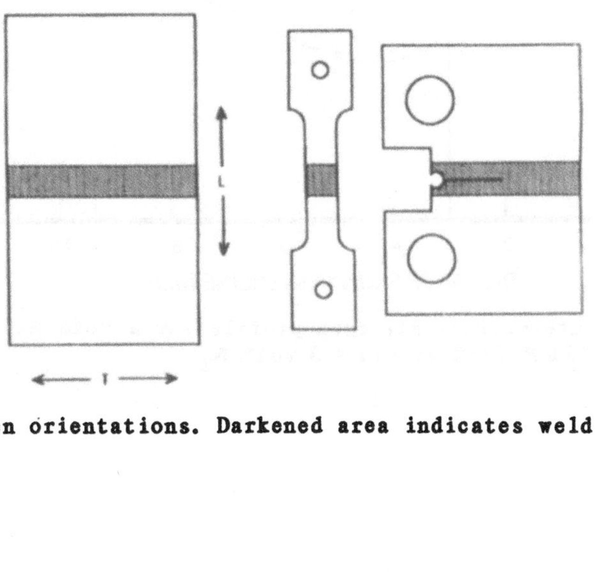

Fig. 1. Specimen orientations. Darkened area indicates weld metal region.

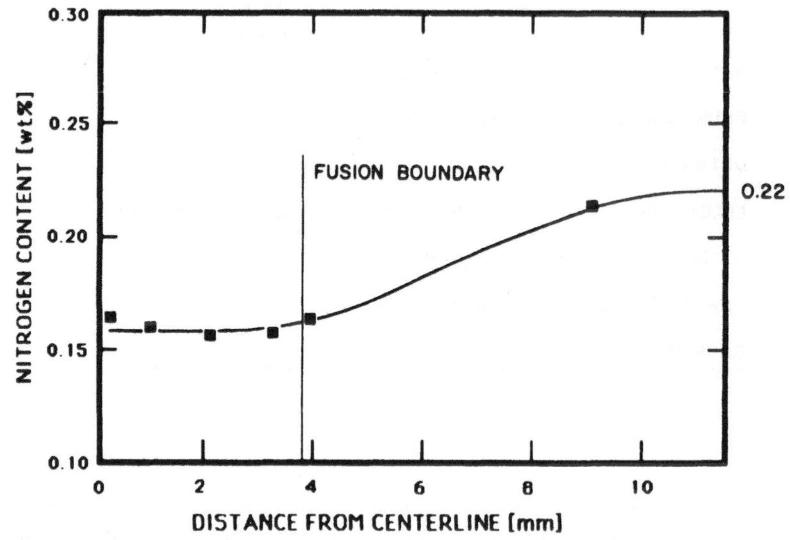

Fig. 2. Interstitial nitrogen profile for a weld made using 75% He/25% Ar shield gas.

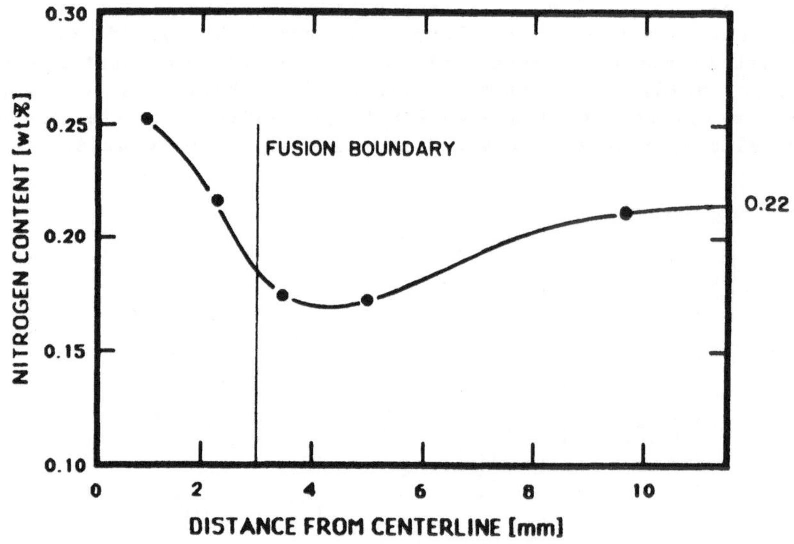

Fig. 3. Interstitial nitrogen profile for a weld made using 75% He/25% Ar mix + 3 vol% N_2.

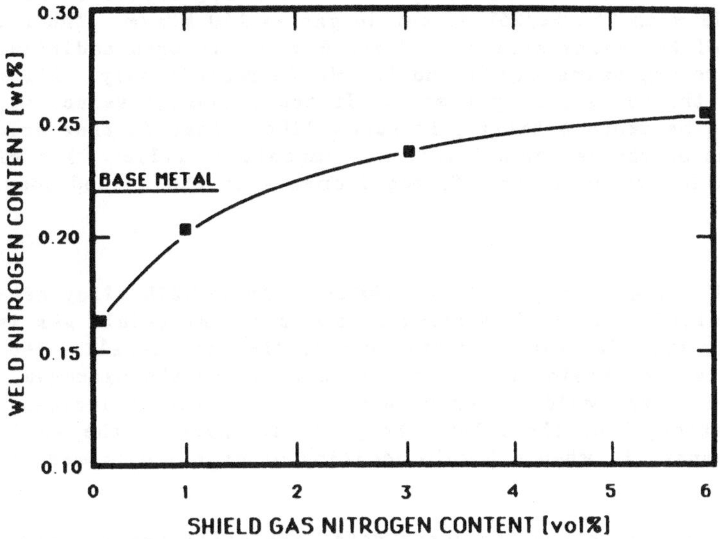

Fig. 4. Average interstitial nitrogen content in the weld as a function of nitrogen content of the shield gas.

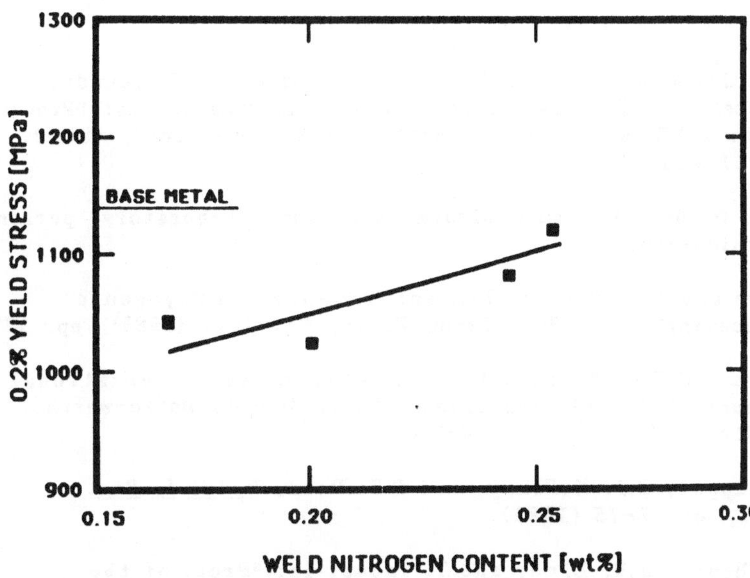

Fig. 5. Dependence of 4.2 K yield strength on average interstitial nitrogen content.

K_{Ic} values were obtained for a single set of welds. However, there was significant scatter in the raw data, and, since fracture toughness measurements of weldments are generally subject to large scatter,[5,6] these data must be considered preliminary. The 4.2 K K_{Ic} fracture toughness of the weld made with 75% He/25% Ar shield gas is 140 MPa√m. The fracture toughnesses of the welds made with 3 and 6 vol% nitrogen addition to the shield gas are approximately 70 and 110 MPa√m respectively. All specimens fractured in the ductile dimple mode. If the toughness values reported here prove to be representative, it seems likely that further process modifications or the use of a different composition filler wire will enable the alloy to meet the U.S. requirements in the welded condition.

CONCLUSIONS

The 4.2 K yield strength of the 18Mn-5Ni-16Cr-0.22N alloy after autogenous pulsed current GTAW using a 75% He/25% Ar shield gas is approximately 100 MPa below the strength of the base metal. The addition of nitrogen to the shield gas causes an increase in the nitrogen concentration in the weld fusion zone and a concommitant increase in the 4.2 K yield strength of the alloy. The yield strength of the weld and base metal are comparable when a 6 vol% addition of nitrogen is made to the shield gas.

Although the fracture toughness data are still preliminary, autogenous welds made on this alloy approach the U.S. requirements for 4.2 K yield strength and fracture toughness.

ACKNOWLEDGEMENTS

The alloys studied here were supplied by Kobe Steel, Ltd., Kobe, Japan. This research was supported by the Director, Office of Energy Research, Office of Development and Technology, Magnetic Systems Division of the U.S. Department of Energy under contract #DE-AC03-76SF00098.

REFERENCES

1. K. Yoshida, K. Koizumi, H. Nakajima, M. Shimada, Y. Sanada, Y. Takahashi, E. Tada, H. Tsuji and S. Shimamoto, in: "Proceedings of the ICMC, Kobe, Japan", Butterworths, Guildford, (1982), pp. 417-420.

2. E. N. C. Dalder, Lawrence Livermore National Laboratory, personal communication.

3. R. Ogawa and J.W. Morris, Jr., in: "Advances in Cryogenic Engineering", vol. 30, Plenum Press, New York (1983), pp. 177-184.

4. R. Ogawa and J.W. Morris, Jr., in: "Proc. of the International Cryogenic Materials Conference, Kobe, Japan", Butterworths, Guildford, (1982), pp. 124-128.

5. T.A. Whipple, H.I. McHenry, and D.T. Read, Weld. J. Res. Suppl., 60:72-75 (1981).

6. H.I. McHenry, J.W. Elmer and T. Inoue, in: "Proc. of the International Cryogenic Materials Conference, Kobe, Japan", Butterworths, Guildford, (1982), pp. 413-416.

IMPROVEMENTS IN THE WELDABILITY OF A SUPERCONDUCTOR SHEATH MATERIAL

L. T. Summers

Materials Fabrication Division
Lawrence Livermore National Laboratory
Livermore, California

J. W. Morris, Jr.

Department of Materials Science and Mineral Engineering
Lawrence Berkeley Laboratory
Berkeley, California

ABSTRACT

This paper investigates the effects of chemistry and heat treatment variation on the 4-K tensile properties of A-286, a candidate sheath material for force-cooled superconductors. Currently, full use of A-286 and similar superalloys is limited by the observed low yield and ultimate tensile strengths in the welded and aged condition. The low strength is shown to be associated with the formation of precipitate-free-zones as a result of alloying-element segregation during weld pool solidification. It has been determined that minor modifications of the weld-metal chemistry by the addition of Ti reduce precipitate-free-zone formation, resulting in matching weld-metal and base-plate strengths at 4 K. Furthermore, nucleation of the γ' hardening phase has been found to be a strong function of temperature and composition. Modified heat-treatment schedules have been determined that are amenable to superconductor fabrication and that resulted in increased weld hardening and improved 4-K tensile properties.

INTRODUCTION

Advanced superconducting magnets will incorporate unalloyed or alloyed Nb_3Sn superconductors, surrounded by a thin-walled pressure vessel (sheath), and cooled by the forced flow of liquid helium inside the pressure vessel.[1,2,3] This is the so-called "internally cooled coiled superconductor" (ICCS). This report details an investigation to improve the cryogenic mechanical properties of gas tungsten arc (GTA) welds of A-286, a candidate for use as the conductor pressure vessel.

The high strain sensitivity of bronze-processed Nb_3Sn superconductors requires that the conductor experience strains of less than 0.5% following the high-temperature reaction heat treatment used to precipitate the superconducting phase.[4] An accepted method of manufacturing ICCS conductors is to perform most fabrication steps, including wrapping and seam welding of the sheath, prior to the reaction heat treatment. Thus the conductor sheath is exposed to reaction heat treatments that are typically more than 30 hours in length and at temperatures of 700 °C or higher.

A previous investigation has shown JBK-75, an iron-base superalloy similar to A-286, to be acceptable for use as the sheath material in applications such as the Westinghouse magnet for the Oak Ridge National Laboratory Large Coil Project.[5] However, this investigation determined that heat-treated GTA welds of JBK-75 have yield and ultimate tensile strengths that are 7 to 8% lower than similarly aged base metal.

It has been shown that the low weld strength is due to the formation of precipitate-free-zones resulting from the segregation during weld pool solidification.[6,7] Dendrite cores are consequently too lean in Ti to form the γ' hardening phase. The low weld strength has raised concern over the future use of this alloy, particularly in magnets that will encounter higher operational loads.

In the present investigation, methods of increasing the weld strength to levels approaching those of the base metal have been evaluated. This has included the use of Nb_3Sn-compatible, multistep heat treatments and weld pool chemistry control.

PROCEDURE

The starting material consisted of commercial round stock of A-286, a variant of JBK-75 that has been determined to have a similar aging response. This material was hot rolled to a thickness of 3.0 mm, sealed in stainless steel bags, and solution heat treated at 1100 °C for one hour with a final water quench. The composition of A-286 is given in Table 1.

Welding was full penetration bead on plate. All welds used a torch velocity of 2.5 mm/sec, 110-A current, and 18-V straight polarity. The weld bead width at the top surface was approximately 6 mm. The heat input is estimated to be 735 J/mm. Modifications of the weld pool chemistry were made by buttering the surface of the plate with 0.4-mm-diameter, high-purity Ti wires. Two weld compositions were produced, 3.06 wt% Ti and 3.8 wt% Ti.

Isothermally aged specimens were heat treated for 48 hours at 725 °C. Multistep heat treatments were selected on the basis of a review of the literature that indicated low temperature steps early in the aging sequence would increase the driving force for nucleation of the hardening phase.[8,9] Multistep aged specimens were heat treated for 4 hours at 725 °C, plus 5 hours at 650 °C, plus 44 hours at 725 °C. Temperature was controlled to $\pm$ 2 °C for all heat treatment. These heat treatments are similar to those used to produce bronze-processed Nb_3Sn superconductors.

Table 1. Composition of A-286 Stainless Steel

C	Mn	Si	Cr	Ni	Ti	Al	Mo	Fe
0.05	1.3	0.6	15.0	25.0	2.15	0.15	1.0	BAL

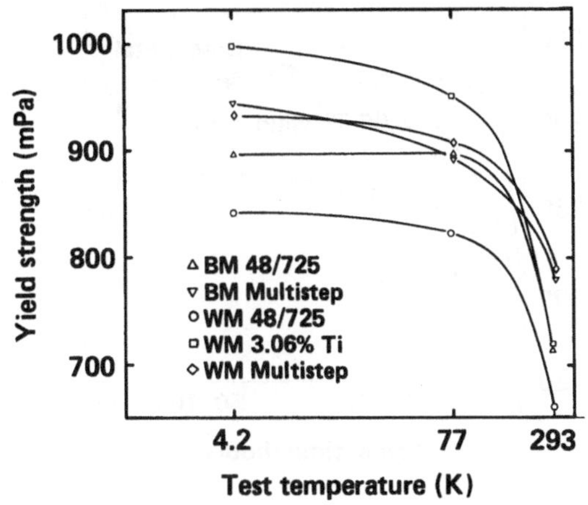

Figure 1. Yield strength vs temperature for weld and base metal.

Tensile tests were performed using a machine equipped with a compression-tube load train and a liquid helium cryostat. Tests were conducted using a strain rate of 3.3×10^{-4}/s. Test specimens were flat tensiles with a 25.4-mm gauge length. Elongation during liquid helium tests was monitored using a clip gauge.

The clip gauge, which was constructed for cryogenic service, employed four SK strain gauges arranged in a Wheatstone-bridge configuration. The use of low excitation voltage avoided local boiling of helium.

Aging response was determined using microhardness measurements. All values reported are the average of at least five indents. Weld indents were made at the weld centerline.

RESULTS

The yield strengths versus temperature for commercial-weld and base-metal, modified-chemistry and multistep heat-treated weld specimens are shown in Figure 1. The commercial-composition weld metal aged for 48 hours at 725 °C (WM 48/725) shows a lower yield strength throughout the temperature range than similarly heat-treated base plate. Multistep heat-treated weld metal shows a close matching of yield strength with isothermally aged base plate at room temperature and 77 K; however, the 4-K yield strength is higher. It should be noted that multistep aging influences the yield strength of both the weld and base plate. The multistep aged base plate yield strength has increased to that of comparably aged weld metal. Increasing the weld Ti content to 3.06 wt% followed by isothermal aging at 725 °C increases the yield strength, particularly at low temperatures where it exceeds that of the base plate.

Failure in the isothermal and modified-chemistry-weld specimens is by ductile rupture at the center of the weld zone. Failure in the multistep aged weld occurred by intergranular fracture in the heat-affected zone (HAZ). The failure of the base metal specimens also occurred by intergranular fracture.

Elongation in the composite weld-metal specimens was largely confined to the weld-metal section of the gauge length. A typical weld-metal elongation was 30% with a base-metal section elongation of 7% at 4 K.

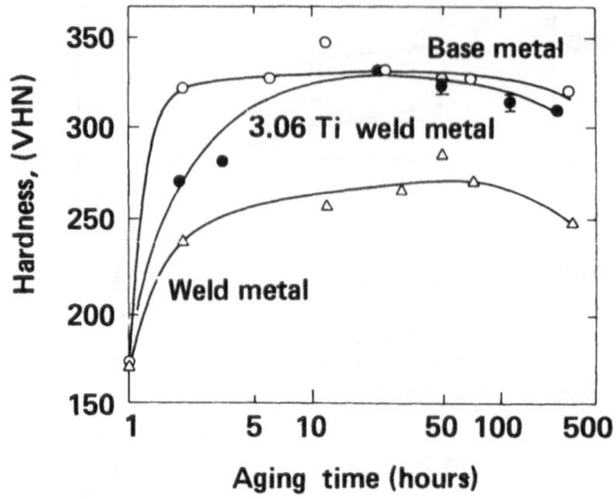

Figure 2. Isothermal aging response of base metal, commercial-, and modified-chemistry weld metal.

Figure 2 shows the isothermal aging response of the base metal, commercial-chemistry, and modified-chemistry (3.06 wt% Ti) welds heat treated at 725 °C. The commercial-chemistry weld shows a slower aging response and lower peak hardness than does the base metal. The addition of Ti to the weld pool increases the aging rate and gives a peak hardness that is similar to the base plate. The hardness peak occurs at times used for Nb_3Sn superconductor processing.

Figure 3 shows the aging response of isothermal, and multistep heat-treated weld and base metal. The inclusion of a low temperature step (5 hours/650 °C) raises the weld and base metal hardness slightly. Returning to 725 °C aging results in no significant change in the base metal hardness. The weld, however, shows a continuing hardening trend. The peak hardness is close to that of the base metal and occurs near 40 hours of aging.

Characterization of the weld microstructure shows the isothermally aged weld metal to have a precipitate-free-zone volume fraction of approximately 40% (Figure 4a). The 3.06% Ti weld is approximately 20% precipitate free (Figure 4b). The multistep aged welds show no evidence of distinct precipitate-free-zone formation (Figure 4c). However, TEM

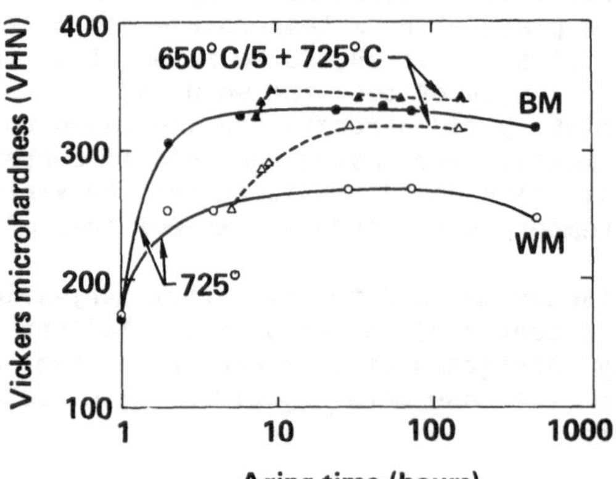

Figure 3. Aging response of multistep heat-treated weld and base metal.

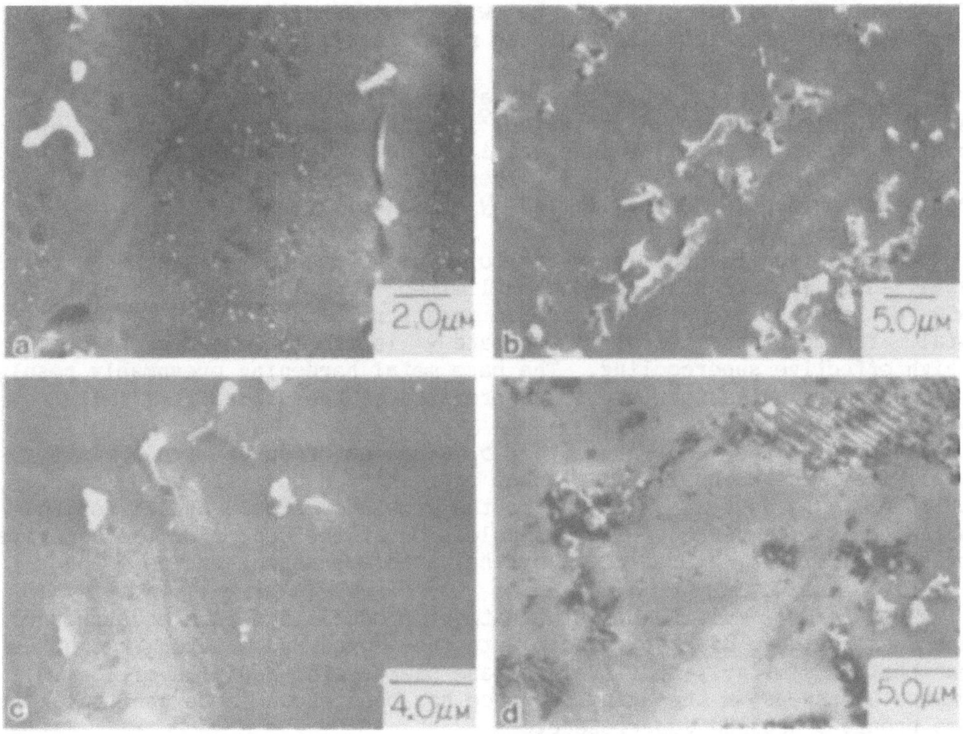

Figure 4. Weld microstructure: (a) commercial chemistry (48/725 °C), (b) 3.06% Ti (48/725 °C), (c) multistep aging, and (d) 3.81% Ti (48/725 °C).

analysis shows considerable coarsening of the γ' hardening phase at the dendrite cores. Isothermally aged weld containing 3.80 wt% Ti shows no evidence of precipitate-free-zones, but rapid formation of the cellular η phase is seen (Figure 4d).

DISCUSSION

The low yield strengths of commercial-chemistry, isothermally heat-treated welds of A-286 and JBK-75 are directly attributable to the formation of precipitate-free-zones. As shown previously, the precipitate-free-zones result from the segregation of Ti during weld pool solidification.[6,7] As determined by STEM microanalysis, the minimum Ti concentration for the precipitation of the hardening phase during 725 °C aging is approximately 1.8 wt%.

Additions of Ti to the weld pool are effective in decreasing the precipitate-free-zone size. High local volume fractions of γ' are seen in the outer regions of the dendrites and contribute to strengthening.[6] Formation of the precipitate-free-zones in the modified-chemistry welds was observed to occur at Ti contents of 1.8 wt%.

The formation of the η phase in the weld containing 3.8 wt% Ti is not unexpected. High levels of Ti have long been known to cause early overaging in Fe-base superalloys.[10,11] Therefore, Ti levels should be maintained well below this level. Ti levels of 3.06 wt% have not been observed to result in the extensive formation of η phase. Close control of the weld Ti content at levels below 3.06 wt% should result in matching yield strengths for the weld and base metal at 4 K.

The work of Irvine et al.[8] indicates that the precipitation of γ' is also temperature dependent. Hughes[9] has shown that, at 800 °C, 25-Ni 15-Cr superalloys should precipitate the γ' phase at Ti concentrations as low as 1.0 wt%. Therefore, in A-286 and JBK-75 welds, precipitation should not be limited by Ti solubility in the matrix.

The observed response of A-286 to multistep heat treatments can be explained if γ' growth is nucleation limited. The low temperature step (650 °C) results in a slight increase in strength in the weld and base metal. The weld dendrite cores nucleate γ' in the Ti-lean regions due to a high effective supercooling. The base metal hardening presumably results from secondary nucleation due to suppression of the solubility limit.

Increasing the temperature to 725 °C has no effect on the base metal, i.e., the solubility limit has been shifted up. The weld, however, continues to harden. This is due to the growth of stable nuclei formed during the low temperature step. Nucleation at the center of the dendrite cores is difficult, as reflected by the rapid coarsening of the γ'.[12]

The use of a multistep heat treatment produces a weld microstructure that does not have distinct precipitate-free-zones and a strength that is near that of similarly heat-treated base plate. The addition of low temperature steps in the aging sequence has been shown to be beneficial in some bronze-processed superconductors.[13,14]

The intergranular failure observed in the base plate regardless of aging conditions results from the high solution temperatures employed.[15,16] These high temperatures were required to yield large grain sizes in the base plate so as to eliminate Hall-Petch effects when comparing base-metal and coarse-grained-weld microstructures.

Localized elongation in the composite weld-metal specimens may lead to an overestimate of the yield strength of the weld. However the error is considered to be small.

CONCLUSIONS

Isothermal aging at temperatures of 725 °C and higher results in the formation of precipitate-free-zones in A-286 weldments. The precipitate-free-zones persist at long aging times and are responsible for the lower yield strength of the weld as compared to similarly aged base metal.

Small additions of Ti to the weld reduce the extent of the precipitate-free-zones and increase the low temperature yield strength. Weld Ti contents of 3.06 wt% raise the strength above that of the base plate at 4 K without significant formation of η phase at long aging times. Ti levels of 3.80 wt% result in the formation of η phase at short aging times.

Precipitation of the hardening phase in A-286 is nucleation limited in the Ti-lean dendrite cores. Multistep heat treatments are effective in promoting nucleation of the hardening phase at the dendrite cores, and increase the weld yield strength. Matching yield strengths at 4 K are achievable between weld and base metal using multistep heat treatments. JBK-75, due to its similar chemistry, is expected to behave similarly to heat treatment and chemistry modification.

ACKNOWLEDGMENTS

This work was supported by the Director, Office of Energy Research, Office of Basic Energy Science, Materials Science Division, of the U.S. Department of Energy under contract No. DE-AC-03-76SS00098.

Work was performed under the auspices of the U.S. Department of Energy by the Lawrence Livermore National Laboratory under contract number W-7405-ENG-48.

REFERENCES

1. J. R. Miller, "Design of Aggressive Superconducting TFCX Magnet Systems," Lawrence Livermore National Laboratory, Livermore, California, UCRL-92371 (1985).
2. B. Blackwell, M. Greenwald, D. Gwinn, B. Lipschultz, E. Marmar, B. Montgomery, R. Parker, P. Politzer, M. Porkolab, J. Schultz, S. Wolfe, "Alcator DCT Proposal," Plasma Fusion Center, Massachusetts Institute of Technology, Sept. (1983).
3. K. Yoshida, H. Nakajima, K. Koizumi, M. Shimada, Y. Sanada, Y. Takahashi, E. Tada, H. Tsuji, S. Shimamoto, Development of Cryogenic Structural Materials for Tokamak Reactors, in: "Austenitic Stainless Steels at Low Temperatures," R. Reed and T. Horiuchi, eds., Plenum Press, N.Y. (1983), pp. 29-39.
4. J. W. Ekin, in "Filamentary A-15 Superconductors," M. Suenaga and A. F. Clark, eds., Plenum Press, N.Y. (1980).
5. M. J. Strum, M.S. Thesis, University of California at Berkeley (1982).
6. L. T. Summers, Ph.D. Thesis, University of California at Berkeley (1984).
7. L. T. Summers and J. W. Morris, Jr., Adv. Cryog. Eng. 30:231 (1983).
8. K. J. Irvine, D. J. Llewellyn, and F. B. Pickering, JISI 199:153 (1961).
9. H. Hughes, JISI 203:1019 (1965).
10. J. R. Mihalisin and R. F. Decker, Trans. AIME 218:507 (1960).
11. F. G. Wilson and F. B. Pickering, JISI 204:628 (1966).
12. B. F. Clark and F. B. Pickering, JISI 205:70 (1967).
13. I. W. Wu, D. R. Dietderich, W. V. Hassenzahl, and J. W. Morris, Jr., "The Influence of Microstructure on the Properties of Bronze-Processed Multifilamentary Wire," presented at the United States-Japan Workshop on High Field Superconducting Materials for Fusion, San Diego, CA, May 23-25 (1983).
14. I. W. Wu, D. R. Dietderich, J. T. Holthuis, M. Hong, W. V. Hassenzahl, and J. W. Morris, Jr., J. Appl. Phys. 54(12):7139 (1983).
15. A. Ma, M.S. Thesis, University of California at Berkeley (1978).
17. K. M. Chang, Ph.D. Thesis, University of California at Berkeley (1979).

WELDABILITY AND MECHANICAL PROPERTIES OF AGE-HARDENED Fe-Ni-Cr-Mn-Ti AUSTENITIC ALLOY FOR CRYOGENIC USE

K. Hiraga, K. Nagai, T. Ogata, Y. Nakasone,
T. Yuri, and K. Ishikawa

National Research Institute for Metals
Tsukuba Laboratories
Sakura-mura, Niihari-gun, Ibaraki, Japan

ABSTRACT

The electron beam weldability and the low temperature mechanical and magnetic properties of Mn-modified iron-base superalloys have been investigated. Mn addition to an Fe-30Ni-13Cr-2.4Ti alloy suppressed the occurrence of ferromagnetism at 4.2K and did not cause any deleterious effects on the mechanical properties of the alloy and its weldability, i.e., the alloys containing from 3 to 9 mass% Mn were welded without any fusion zone hot cracking and HAZ micro-fissuring. The post-weld heat-treatment which consists of solutionizing followed by aging was very effective to diminish the strength mismatch between base metal and weld metal regions. The absorbed energy of impact specimens of weldments for both the as-welded and heat-treated conditions showed higher values than those of the base materials.

INTRODUCTION

In previous works[1-3], Mn-modified iron-base superalloys were proposed as high strength and nonmagnetic structural materials for cryogenic use, and compositional criteria for the alloys were clarified to obtain the $\gamma + \gamma'$ structure which is indispensable for excellent mechanical properties at low temperatures. Recently, the weldability and low temperature mechanical properties of the weldments of iron-base superalloys have been regarded as important matter in cryogenic use of the alloys such as rotor components of electricity generator and superconductor sheath[4]. Since Mn has been regarded as rather a undesirable element for weldability, the compositional modification in which Mn content is much lowered than in A-286 is seen for a weldable alloy (JBK-75)[5-7]. But there have been no explicit studies on the influence of Mn upon the weldability of iron-base superalloys with a wide range of Mn content. Therefore, the purpose of this study was to investigate the effects of Mn on the weldability and low temperature mechanical properties of the weldments of an 30% Ni iron-base superalloy.

EXPERIMENTAL METHODS

A series of alloys listed in Table 1 was melted by a 20kg induction furnace under an Ar atmosphere. The chemical compositions of Alloys 2

Table 1. Chemical Analysis of Used Alloys

Alloy	C	Si	Mn	Ni	Cr (mass%)	Mo	Ti	Al	Others
1	0.060	0.48	1.52	26.1	15.3	1.20	2.60	0.20	V:0.23, B:0.005
2	0.012	0.05	0.00	29.5	13.0	1.18	2.43	0.26	----------
3	0.007	0.05	2.99	29.2	13.0	1.13	2.40	0.22	----------
4	0.007	0.03	5.67	29.4	13.1	1.15	2.41	0.23	----------
5	0.007	0.03	8.54	29.5	13.2	1.14	2.44	0.23	----------
	Alloys 1 to 5: P 0.002, S 0.004, and Fe:bal.								

Alloy 1 and 2 correspond to A-286 and the base composition of a weldable alloy(JBK-75), respectively.

to 5 were chosen such that the average electron concentration of the alloys is higher than 8.05 so as to perfectly avoid the occurrence of δ-ferrite, χ, and σ phases in weld metal regions. Ingots were hot-rolled at 1373K to 18mm thick and 60mm wide plates, then solution-treated at 1373K for 3.6ks followed by water quenching. The plates were mechanically ground to 15mm thick and aged at 973K for 144ks in a vacuum furnace followed by cooling to 600K by Ar gas and subsequent water cooling to 300K. These heat-treatments gave average grain sizes of 44μm in dia. for Alloy 1 and about 78μm in dia. for Alloy 2 to 5, respectively. This pre-weld condition of the base material, i.e., fully hardened condition with relatively coarsened grains due to the higher solutionizing temperature than usual, is a much severer one for welding[8] than as-solutionized and fine-grained conditions.

Bead on plate welds with welding parameters of 50kV, 170mA, and travel speed in 125cm/min were made along the hot-rolling direction on the center of each plate. Charpy 2mm-V-notched specimens and smooth tensile specimens with 20mm gage length and 3.5mm diameter were cut from the base materials and the welded plates. The specimens were oriented perpendicular to the hot-rolling direction, and the welded region was centered on the gage length in tensile specimens and on the notched region of impact specimens. The same heat-treatment as that for the base materials was perfomed as a post-weld heat-treatment for the half of the welded specimens.

Tensile and Charpy impact tests at 300, 77, and 4.2K and metallographic examination were carried out on the base and welded materials. Magnetization curves of the base materials were measured on a vibrating sample magnetometer at 4.2K in fields up to 6MA/m. Magnetic measurements at 300K for the fusion zone of the weldments were also made to examine the formation of δ-ferrite during welding.

RESULTS AND DISCUSSION

Mechanical and Magnetic Properties of Base Materials

Mechanical properties of the base materials are listed in table 2. Though Mn addition caused a slight decrease in ductility at 300K and in impact toughness, the Mn-modified alloys showed excellent strength and higher ductility and toughness at 4K than those of Alloy 1(A-286).

Figures 1 and 2 are the magnetization curves of the base materials and Arrott plots of the curves for 4.2K, respectively. Magnetization plots at 300K are linear, indicative of paramagnetic behavior of the alloys at this temperature. At temperature down to 4.2K, however, the magnetization curves shift towards higher M direction and show negative

Table 2. Mechanical Properties of Base Materials

Alloy	T(K)	$\sigma_{0.2}$(MPa)	σ_u(MPa)	ε(%)	φ(%)	vE(J)
1	300	756	1063	20.1	41.5	48
	77	974	1394	26.2	27.5	47
	4.2	1061	1548	25.4	24.0	46
2	300	780	1098	25.5	45.9	86
	77	922	1390	35.2	39.5	75
	4.2	1004	1544	34.2	34.4	84
3	300	730	1054	18.6	55.8	73
	4.2	1001	1500	28.9	31.2	86
4	300	819	1082	17.3	25.0	71
	4.2	987	1560	35.4	40.1	76
5	300	812	1076	17.8	22.2	71
	77	898	1363	34.1	34.8	73
	4.2	954	1511	33.1	30.3	76

T:temperature, $\sigma_{0.2}$:proof Strength, σ_u:Tensile strength
ε:total elongation, φ:reduction of area, vE:absorbed energy

curvature, suggesting ferro- or superparamagnetic behavior of the alloys. For the Mn free alloy(Alloy 2), the magnetization curve shows slight hysteresis and its Arrott plot clearly makes a positive intercept with the M^2 axis, which indicate that spontaneous magnetization occurred in the alloy, i.e., the alloy is ferromagnetic at 4.2K. Since the intercepts for Alloys 1 and 3 are about 0, these alloys are on the transition point from ferromagnetic alloys to strong superparamagnetic ones at 4.2K.

From Figs.1 and 2, the beneficial effects of Mn addition to Alloy 2 are very clear: Mn addition suppresses the occurrence of ferromagnetism in

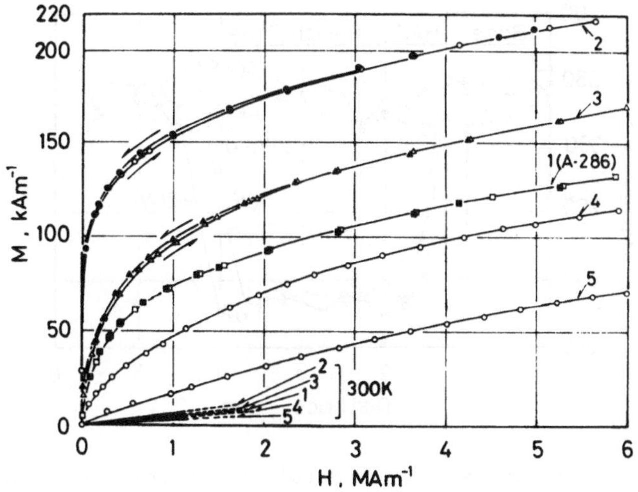

Fig. 1. Magnetization curves at 4.2K and 300K for base materials.

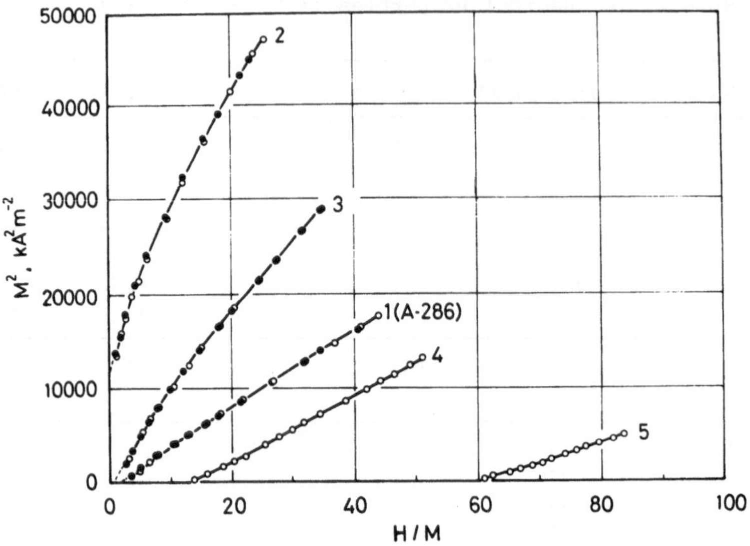

Fig. 2. Arrott plots of magnetization curves at 4.2K.

the Mn free alloy at 4.2K; and remarkably decreases the initial slopes of the magnetization curves and also decreases the magnetization at respective applied fields.

Electron Beam Weldability

As shown in Photo.1, the alloys containing up to 9% Mn were welded without any fusion zone hot cracking and HAZ micro-fissuring which markedly occurred in Alloy 1(A-286). Magnetization measurements at 300K revealed that no δ-ferrite was formed during solidification of the weld metal. Therefore, it is concluded that Mn does not cause any detrimental effects on the weldability of iron-base superalloys. Photograph 1(c) represents the microstructure of the weld metal region after the post-weld treatment. There is no reheat cracking in the fusion zone and HAZ. The fusion zone is found to be recrystalized by the solution treatment and thus an equi-axial structure same as those of the base materials are obtained.

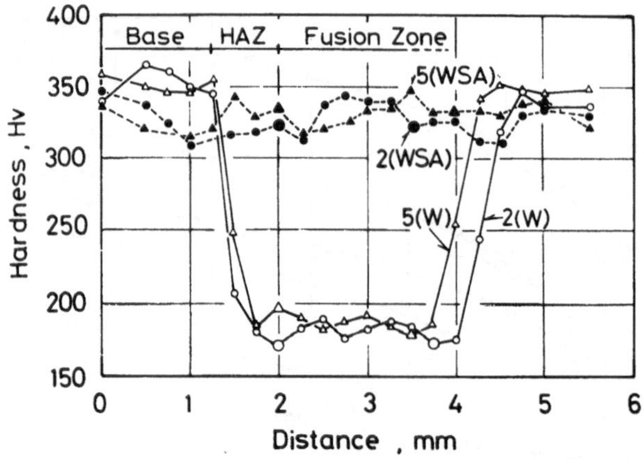

Fig. 3. Microhardness of weld metal region for as-welded(W) and heat-treated(WSA) conditions.

It is found in Fig. 3 that the weldments have the soft zone, which consists of about 1mm wide HAZ and about 2mm wide fusion zone. Such hardness gaps between the base metal and the weld metal region can be almost removed by the post-weld heat treatment. There are no differences between the Mn-free alloy and the 9% Mn alloy in the hardness profiles described above.

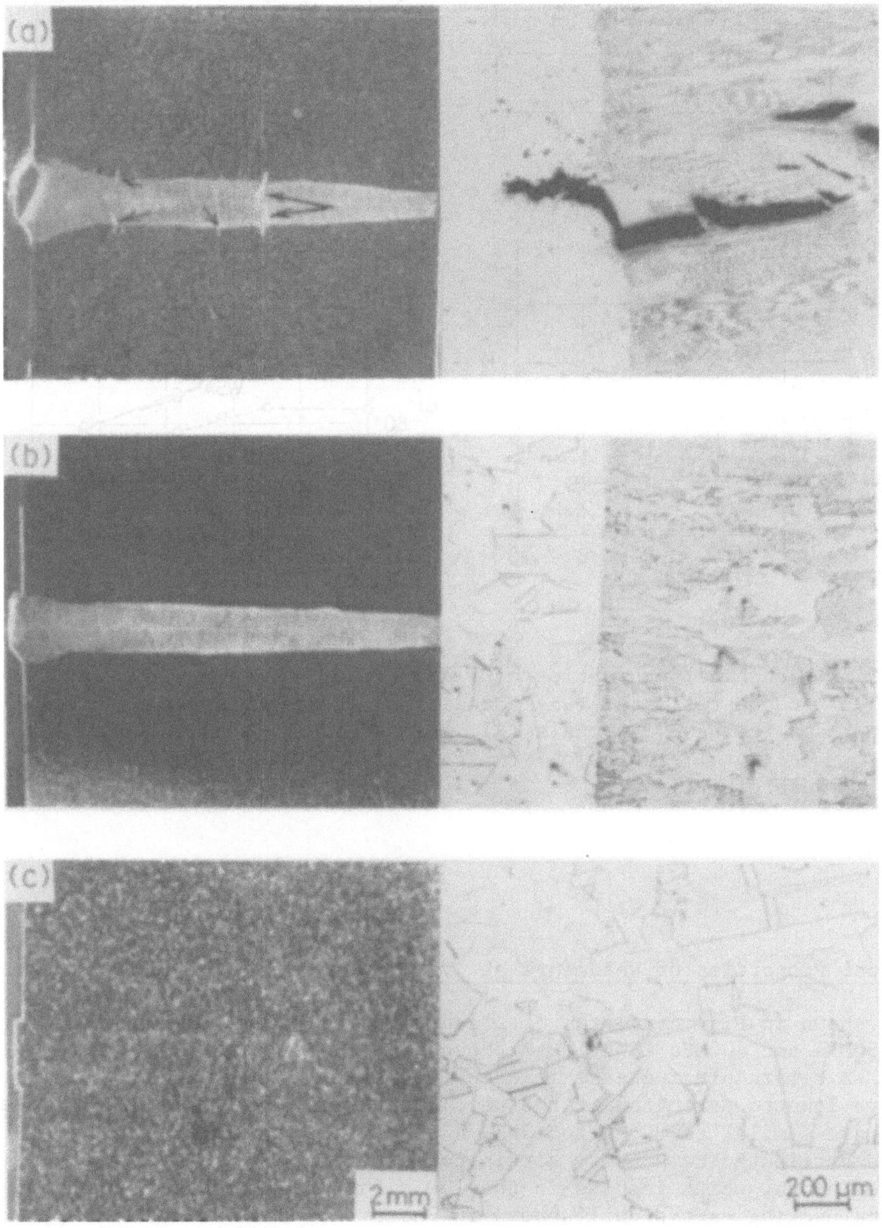

Photo. 1. Microstructures of weld metal region. (a) as-welded condition for Alloy 1(A-286), showing marked weld hot cracking indicated by arrows; (b) as-welded condition for Alloy 5, and (c) heat-treated condition for Alloy 5.

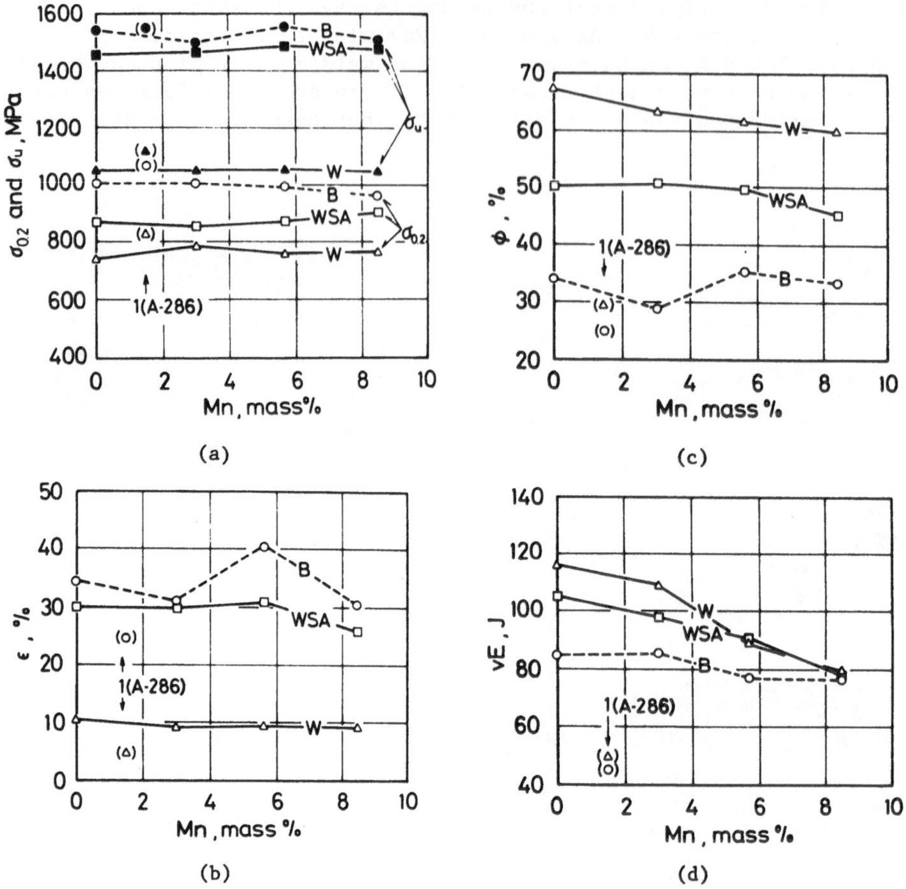

Fig. 4. Mechanical properties of weldments at 4.2K for as-welded (W) and heat-treated (WSA) conditions in comparison with those of the base materials (B). (a) yield and tensile strengths; (b) total elongation; (c) reduction of area, and (d) absorbed energy (vE).

Mechanical Properties of Weldments at 4.2K

As shown in Fig. 4(a), the $\sigma_{0.2}$ and σ_u of the as-welded materials are about 250MPa and 400MPa lower than those of the base materials, respectively. A remarkable increase in ϕ and decrease in ε are accompanied by the above lowered strengths of the weldments [Figs. 4(B) and (c)]. These results are similar to those for JBK-75 reported by Strum et al.[7], and they are attributed to a heavy strain concentration on the soft zone described above, which results in the low ε for the 20mm gage length and the large necking as typically shown in Photo. 2. The absorbed energy of the weldments shown in Fig. 4(d) is higher than that of the base materials and it decreases gradually with an increase in Mn content. The low ductility and absorbed energy for the weldment of Alloy 1 are due to the presence of weld hot cracking shown in Photo. 1.

The post-weld heat treatment is very effective in diminishing the strength mismatch between the base metal and welded regions. Especially,

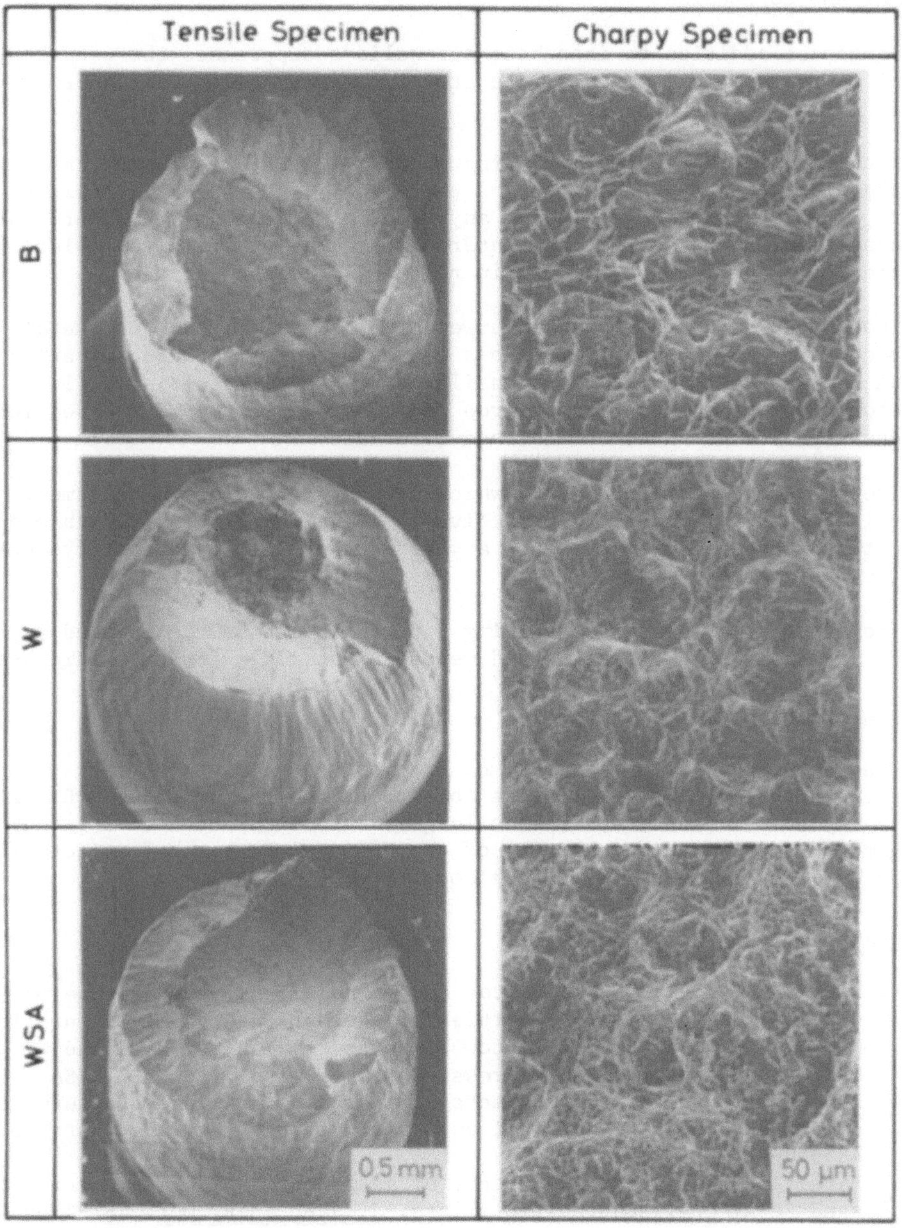

Photo. 2. Fractographs of tensile and Charpy impact specimens tested at 4.2K for base(B), welded(W), and post-weld heat-treated (WSA) materials of Alloy 5.

this treatment results in marked recovery of the tensile strength of the weldments without any harmful influence on ductility and toughness at 4.2K.

Photograph 2 is typical fractographs of tensile and Charpy impact specimens of Mn-modified alloys tested at 4.2K. Fractographs for the base materials, the weldments, and the heat-treated weldments of the Mn-modified alloys revealed transgranular fracture surfaces covered with ductile dimples.

SUMMARY AND CONCLUSIONS

1. Manganese addition to iron-base superalloys suppresses the occurrence of ferromagnetism in the alloys at 4.2K and causes a remarkable decrease in initial susceptibility.

2. Manganese addition does not bring about any deleterious influences on the electron beam weldability of the alloys. The alloys containing 3 to 9% Mn can be welded without any weld hot cracking even under a severe condition such that the base materials are fully aged after solution treatment at 1373K, resulting in relatively coarse γ grains.

3. Though the strength mismatch between the base metal and weld metal regions of the Mn-modified alloys causes heavy strain concentration on the latter region during tensile deformation, resulting in low total elongation, the reduction of area and absorbed energy of the weldments are higher than those of the base materials.

4. The post-weld heat treatment, which consists of solution treatment followed by aging is very effective in diminishing the strength mismatch in the weldments without any harmful effects on ductility and toughness at 4.2K.

5. Iron-base super alloys containing 6 to 9% Mn can be recommended as nonmagnetic, high-strength, and weldable materials for cryogenic use.

ACKNOWLEDGEMENTS

The authors wish to thank Dr. H. Irie for conducting the electron beam welding. They also would like to express their gratitude to Dr. K. Tachikawa, Dr. H. Maeda, and the colleagues in the Superconducting and Cryogenic Materials Research Group of the National Research Institute for Metals for their many useful suggestions.

REFERENCES

1. K. Hiraga, K. Ishikawa, T. Ogata, and K. Nagai, Low temperature mechanical and physical properties of age-hardened Fe-Ni-Cr-Mn alloys, in: "Austenitic Steels at Low Temperatures," R. P. Reed and T. Horiuchi, eds., Plenum Press, New York (1983), pp. 277-286.
2. K. Hiraga and K. Ishikawa, Effects of manganese on microstructures of solution-treated Fe-Ni-Cr-Ti austenitic alloys, J. Jap. Inst. Met. 48:950-957 (1984).
3. K. Hiraga and K. Ishikawa, Effects of manganese on microstructures of aged Fe-Ni-Cr-Ti austenitic alloys, J. Jap. Inst. Met. 48:957-964 (1984).
4. W. A. Logsdon, P, K. Liaw, and M. H. Attar, Cryogenic fatigue crack growth rate properties of JBK-75 base and autogeneous tungsten arc weld metal, in: "Advances in Cryogenic Engineering--Materials," vol. 30, Plenum Press, New York (1984), pp. 883-892.
5. J. A. Brooks, Progress toward a more weldable A-286, Weld. J. 53:242s-245s (1974).
6. J. A. Brooks, Effect of alloy modifications on HAZ cracking of A-286 stainless steel, Weld. J. 53:517s-523s (1974).
7. M. J. Strum, L. T. Summers, and J. W. Morris, Jr., The aging response of a welded iron-based superalloy, Weld. J. 62:235s-242s (1983).
8. R. Thamburaj, W. Wallace, and J. A. Goldak, Post-weld heat-treatment cracking in superalloys, Int. Met. Rev. 28:1-22.

CRYOGENIC COMPRESSIVE PROPERTIES OF BASIC EPOXY RESIN SYSTEMS

F.W. Markley, J.A. Hoffman, D.P. Muniz

Fermi National Accelerator Laboratory

Batavia, Illinois

ABSTRACT

The compressive properties of short cylindrical samples of many different epoxy resin systems have been measured at ambient temperature and at 77 degrees Kelvin. These are pure resin systems of known chemistry, without the inorganic fillers or fibrous reinforcements needed in final cryogenic systems. Of course, chemically incorporated modifiers, such as flexibilizing resins, have been included. These data should make possible inferences about cryogenic properties from molecular structures and provide specific data useful to formulators and end users.

Measurements on some other plastics such as polytetrafluoroethylene (PTFE), polyimides, and ultra-high molecular weight polyethylene (UHMWPE) have been made for comparison purposes.

INTRODUCTION

Epoxy resin systems have been widely used in the construction of superconducting coils and other cryogenic apparatus. Fiber reinforced laminates such as NEMA grades G-10 and G-11, heavily filled room cured putties, and "B-staged" adhesives have been popular. Studies of the mechanical cryogenic properties of many useful systems have been presented at these conferences.[1-7]

The practical focus of much testing has been on proprietary, commercial systems. There has been a need for additional data on simple matrix systems of known constituents.

EXPERIMENTAL APPARATUS AND PROCEDURE

Figure 1 shows the overall arrangement of the apparatus in the Instron testing machine. Everything is supported through an insulating rod of G-10 by the load cell on the stationary frame of the machine. The moveable crosshead of the machine applies the load through another insulating rod of G-10 to the sample clamped between steel plates. The surrounding foam insulated steel container can be filled with liquid nitrogen. Two glass tubes transmit relative motions to the linear variable differential transformer located above the crosshead. The inner tube rests on the lower steel compression member, and the outer tube rests on a thin steel washer on top of the sample. Figure 2 shows the details of the sample and its surroundings. The iron core of the LVDT may be adjusted for zero output with the plastic extension and the threaded attachment of the core to the inner glass tube. The output of the load cell amplifier and the LVDT read-out are connected to an analogue X-Y recorder. Data have been manually digitized and entered into the central computer for further manipulations and analysis. All stress-strain curves have a manually fitted slope and a subsequent shift of strain values to give a consistent 0,0 intercept. This procedure eliminates small effects of sample end deviation from a true flat and parallel condition.

The sample is loaded into the apparatus, and a room temperature stress-strain curve is taken with care not to exceed the yield point. The load is then reduced to zero and the container filled with liquid nitrogen. A shrinkage measurement is taken before reloading. The springs between the steel plates ensure contact with the sample during cooldown. These shrinkage measurements are reported here, but should be used with caution since the thermal-mechanical shock of filling can cause the glass tubes to move on their support areas, resulting in errors.

When rapid boiling subsides in the liquid nitrogen, the sample is reloaded - this time it is taken to ultimate failure. The failure is sudden, and accompanied by a loud report. Failure often causes breaking of the glass tubes which is why quartz was not used. During loading, cracking noises are often heard and are accompanied by discontinuities in the stress measurement. The modulus measurement is not disturbed by these events, but they are probably responsible for the large variation in ultimate stress and strain values. They are most serious for the weak and brittle samples. Very careful machining of the samples may reduce their occurrence.

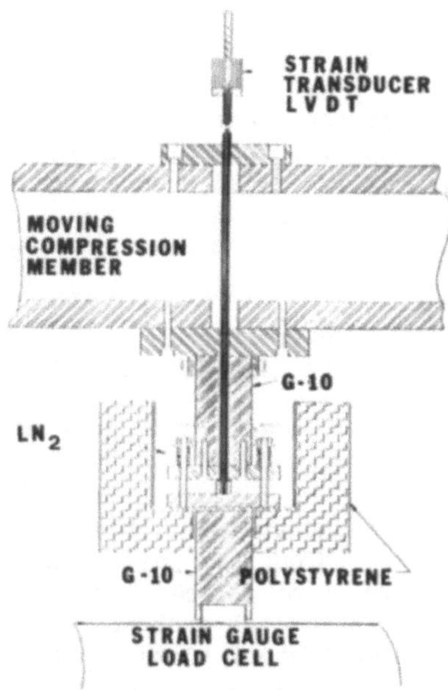

Fig. 1 Test apparatus.

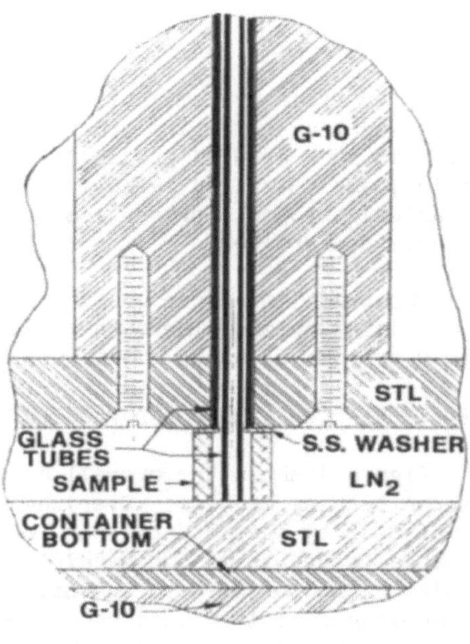

Fig. 2. Sample detail.

MATERIALS

The materials used in this study have been the standard commercial epoxy resin (DGEBA) and its minor variations, combined with many different curing agents. Each major chemical category of curing agent has been represented. The materials are identified in Table I by chemical name, by structural formula, and by a shorthand code, often an acronym. Only the hydrogen atoms active in the curing reactions are shown in the structural formulas. Epoxy resin and flexibilizing agent structural formulas are also shown. Inclusion of this detail should encourage analysis of the data in terms of chemical structure by the readers.

Figure 3 is a graph of the difference between the 77 K modulus and the 300 K modulus versus the 300 K modulus. It shows a general trend for the change in the modulus to be inversely proportional to room temperature modulus. The 77 K modulus tends to be constant and independent of the 300 K modulus. Flexibilizing agents that greatly reduce the warm modulus do not affect this cold modulus. In fact, they usually cause a slight increase in the cold modulus. Even the variations in warm modulus within a system are "frozen out" at low temperatures. Specifically DER 732 increases the cold modulus of a TETA cured system; castor oil increases the cold modulus of an EM 308 cured system; and EMPOL 1040 increases the modulus of a DDSA cured system. This is consistent with the observation reported by Ekin, et al.[7] in 1983 at these conferences: "The addition of flexibilizers to the epoxy systems also produced no improvement in 4 K properties . . . "

RESULTS AND CONCLUSIONS

Figures 4, 5, and 6 show the stress-strain curves grouped into general areas. Real differences between systems are apparent. The aromatic amine (Shell Z) gives greater elongations than the group of aliphatic amines. The aliphatic amine TETA gives a higher modulus and greater elongation than all the rest of the aliphatic amines, including the structurally similar DETA. Table 2 gives the actual measured data in detail for design use by the cryogenic materials engineer.

The catalytic curing agents give lower moduli and greater elongations than the aliphatic amines. BDMA is even better than DMP-30 in this regard. In fact, it gives the greatest elongations of any epoxy system. The catalyst BF_3-400 showed a lot of scatter in modulus and elongation, but its room temperature latency can be so valuable as to commend it for further study. The catalyst Shell D should be similar to DMP-30. The amido-polyamines of low molecular weight behaved like the better aliphatic amines, while higher molecular weights gave lower moduli, but disappointingly smaller elongations.

The anhydrides have smaller elongations than the BDMA system, but also show a trend toward smaller moduli. The modulus of the HHPA, PA, and MTHPA systems equals that of the BDMA, but the NMA systems moduli are lower, and the DDSA gives the lowest modulus of any epoxy system. The anhydrides show more scatter than other systems. Perhaps this is due to the complexity of their chemical reactions. They should be sensitive to epoxy/anhydride ratio, type of catalyst, presence of -OH initiators including water, and cure schedule. However, they show a lot of promise should these variables be carefully controlled and studied.

Selected thermoplastics showed much lower moduli and very much greater elongations, but the rigid brittle nature of polymethylmethacrylate (PMMA) warns against any hasty conclusion that eliminating cross-links will give a flexible low temperature material.

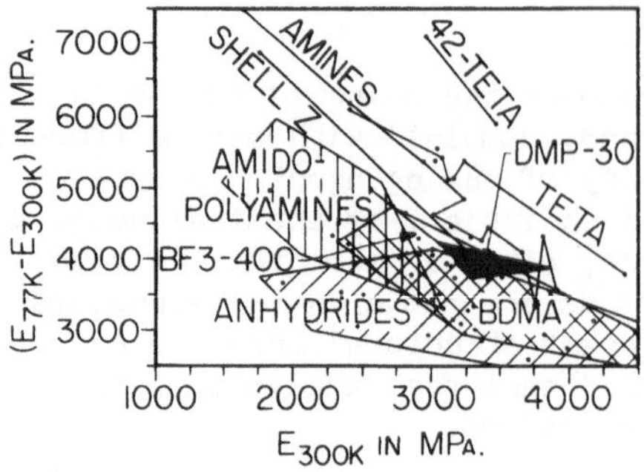

Fig. 3. Modulus change

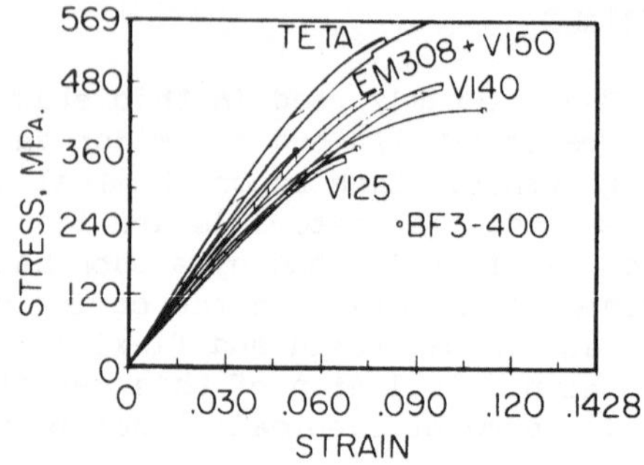

Fig. 4. TETA and amido amines

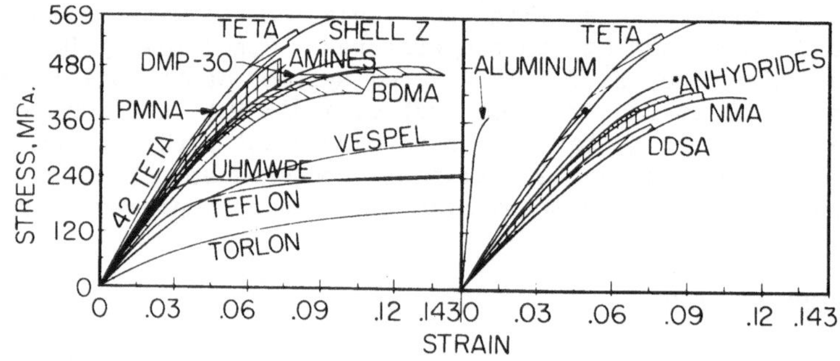

Fig. 5. TETA and miscellaneous systems Fig. 6. TETA and anhydrides

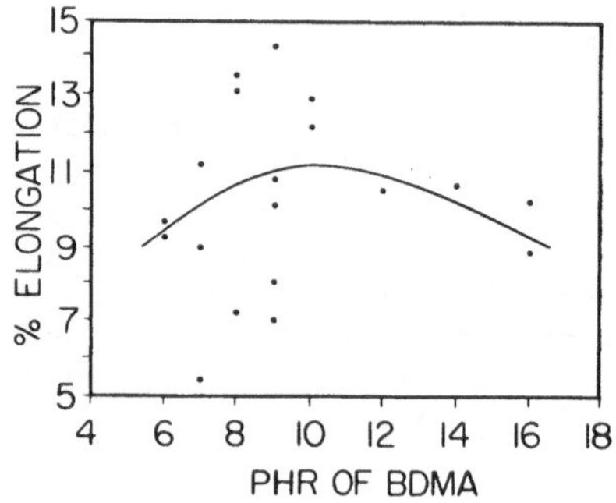

Fig. 7. Maximizing elongation

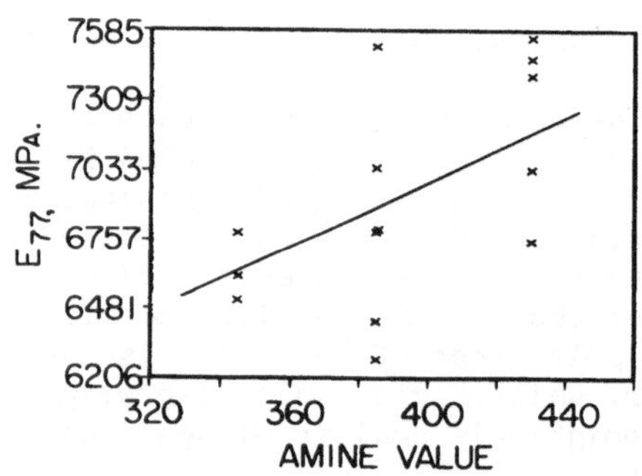

Fig. 8. Molecular weight influence

In fact, the two samples of 42 Phr TETA tested can be expected to give very low cross-link densities since the primary amine hydrogens are much more reactive than the secondary amine hydrogens. Their low elongations are probably only indicative of low molecular weight. However, their high moduli (like PMMA) again shows low cross-link density does not necessarily give high elongation. It is evident that the broad category of thermoplastics must be studied with reference to polarity, chain length, chain branching, chain structure and rigidity, crystallinity, and molecular weight. UHMWPE, polytetrafluoroethylene, and polyimides should be very useful, but they are difficult to use with fillers, or to mold, cast, use as adhesives, etc.

The aluminum curve is given for reference and was not taken to destruction. Figure 7 shows the variations of elongation with curing agent concentration for the catalyst BDMA. It has a broad curve with a peak around 10 Phr. Figure 8 gives the cold elastic modulus of the amido-polyamine systems versus amine numbers, which are inversely related to curing agent molecular weight. A reasonable correlation is shown. However, the high molecular weight resin 1001, when cured with PA, was not appreciably different from the low molecular weight resins cured with HHPA or MTHPA. In spite of their scatter the shrinkage measurements show a definite inverse relationship between shrinkage and 300 K elastic modulus.

SUMMARY

1. Adding 3 common flexibilizing agents actually causes slight increases in 77 K modulus.

2. Epoxy systems ranked in order of decreasing modulus at 77 K are TETA > aliphatic amines and aromatic amines > DMP-30 > BDMA > most anhydrides > DDSA > selected thermoplastics.

3. Epoxy systems ranked in order of increasing elongation at 77 K are aliphatic amines < all others < BDMA < selected thermoplastics.

4. Improvements in cryogenic properties of epoxy resins can be achieved and additional research is encouraged.

Table 1. Structural Formulas

DETA	TETA	AEP	DEAPA	BAC
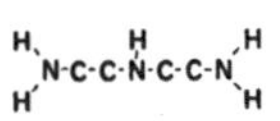	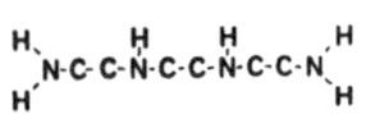	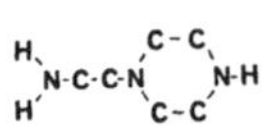	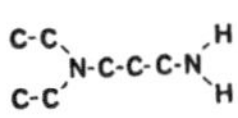	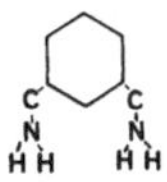
Diethylene triamine	Triethylene tetramine	Aminoethyl piperazine	Diethyl aminopropyl amine	1,3-Bis (aminomethyl) cyclohexane

POLYCAT - 352	D230	MXDA	BF_3-400
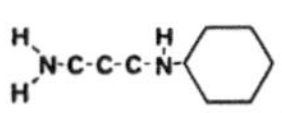	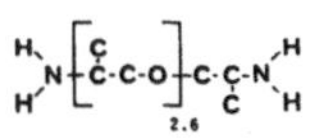	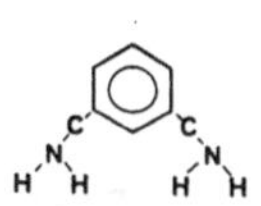	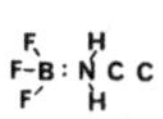
N (3 aminopropyl) cyclohexylamine	Polyoxypropylene diamine	m-Xylylene diamine	Boron trifluoride-monoethylamine

<u>SHELL Z</u>

MPDA 70%	MDA 30%	BDMA	DMP 30	DICY
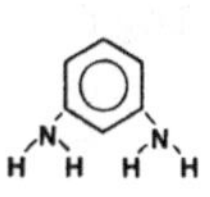	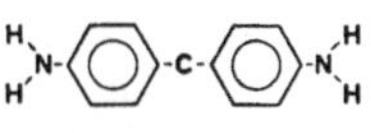	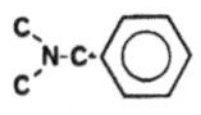	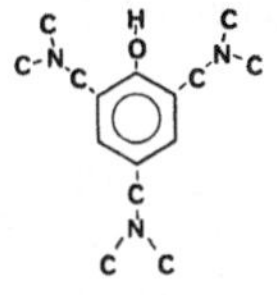	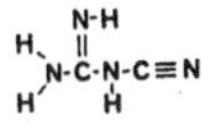
m-Phenylene diamine	4,4'-Methylene dianaline	Benzyl dimethyl-amine	Tris (dimethyl-aminomethyl)-phenol	Dicyan-diamide

NMA	MTHPA	HHPA	DDSA	PA
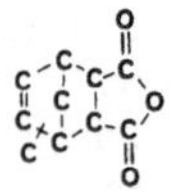	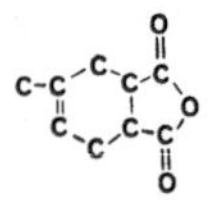	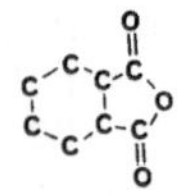	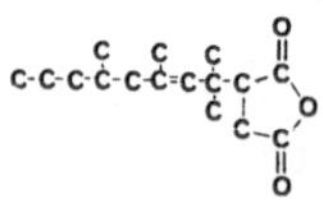	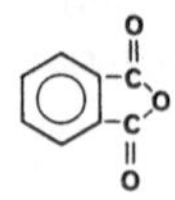
Nadic methyl anhydride	Methyl-tetrahydro-phthalic anhydride	Hexahydro-phthalic anhydride	Dodecenyl-succinic anhydride	Phthalic anhydride

AMIDOPOLYAMINE V150
V140
V125

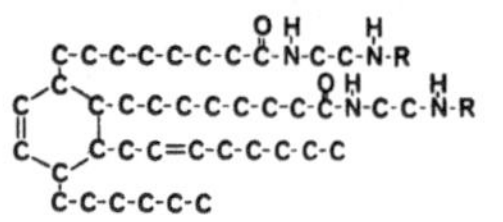

R=H or another dimer

Dimerized linoleic acid & ethylenetriamine

AMIDOPOLYAMINE EM 308

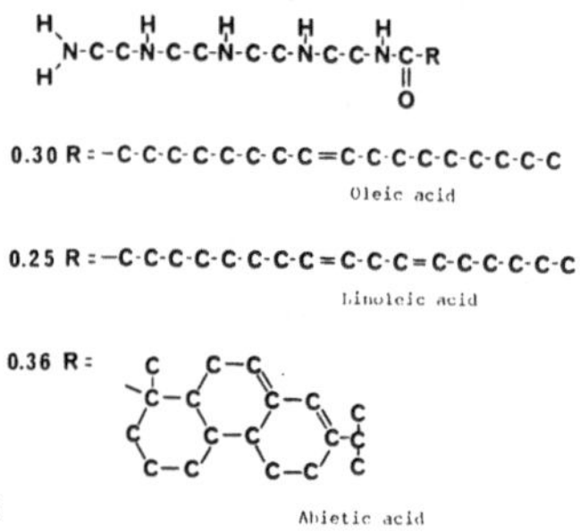

Tall oil fatty acid with tetraethylenepentamine

SHELL D

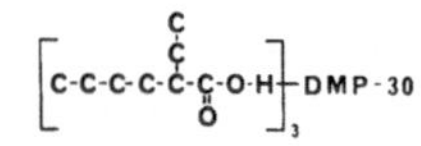

Tri(2 ethylhexoate) salt of tri[dimethyl(amino methyl)]phenol

CASTOR OIL

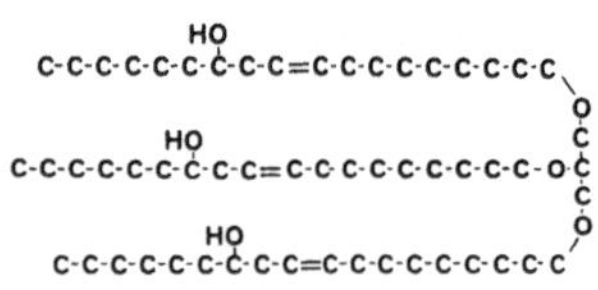

Triglyceride of ricinoleic acid

DER 732

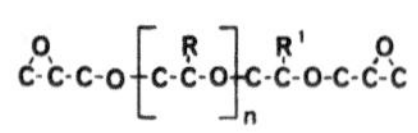

R and R[1] may be methyl
n may be about 9

Diglycidyl ether of polypropylene glycol

DER 331, 332 and Shell 826, 828, 1001, and BIS-F

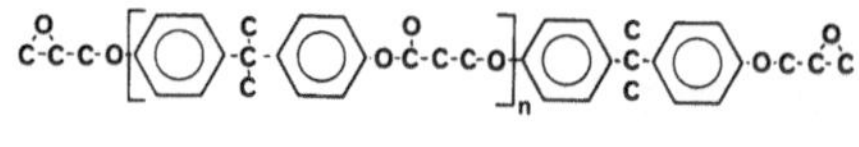

Diglycidylether of bisphenol A and higher homologues

For 332 and 826 n ≅ 0
For 331 and 828 n ≅ 0.13
For 1001 n ≅ 2.5

For BIS-F n ≅ 0
and the 3 carbon chain is reduced to 1

Table 2. Compressive Properties of Epoxy Resins

Cure Agent/ Addit.	PHR Agent/ Addit.	300 K Mod MPa	77 K Mod MPa	σ Ult 77 K MPa	ε ULT 77 K %	Λ L %	Resin/ Sample/ Addit.	°C-hr* /post Cure	∫σdε MPa
Z	20	3320	7380	523	10.2	0.73	332-1	/65-2	32.4
Z	20		7580	519	10.6	0.75	332-2	/65-2	34.2
Z	20	1740	8620	516	7.4	0.70	826	/65-2	20.2
AEP	24.7	2890	7510	416	6.6	0.79	332	/65-16	14.0
BAC	16.0	3890	7450	415	6.6	0.77	332-1	/65-21	14.9
BAC	16.0	3740	7100	407	6.0	0.75	332-2	/65-21	13.2
DEAPA	6.0	3360	7580	425	7.1	0.86	332	72/65-24	17.0
DETA	12.1	3400	7860	361	4.9	0.90	332	*	9.0
D230	30.0	3680	7790	459	7.4	0.84	332	/65-4	18.7
MXDA	15.0	3810	8130	478	7.0	0.77	332	/65-21	18.5
POLYCAT	30.0	2920	8270	275	4.0	0.75	332	16/100-3	5.8
U	25.0	2390	8550	563	10.3	0.77	332-1	/65-16	36.3
U	25.0	3230	8070	426	6.1	1.76	331-2	/65-16	13.2
U/732	33/33	1630	9100	385	5.4	1.50	331-732	/65-16	10.7
TETA	14.0	3061	8480	576	10.2	0.87	332-1	6/65-15	36.4
TETA	14.0	2410	8550	456	6.0	0.62	332-2	*	14.2
TETA	13.0	3100	8340	411	5.6		331	*	13.6
TETA	15.4	3150	8410	492	2.7	0.74	BIS-F	*	21.5
TETA732	15/33	3230	8620	540	7.8	0.76	331-732	*	4.6
TETACGE	17/33	4390	8200	428	5.8		331	*	14.4
TETA	42.0	2910	10000	254	2.5	0.72	332-1	*	3.2
TETA	42.0	3440	9310	256	2.5	0.80	332-2	*	3.3
BDMA	6.0	3130	7030	443	9.3	0.84	332-1	64/65-4	25.2
BDMA	6.0	2760	6750	449	9.6	0.79	332-2	64/65-4	26.5
BDMA	7.0	4480	7100	345	5.4	1.59	332-1	4/65-4	10.3
BDMA	7.0	3990	7450	445	9.0	0.65	332-2	4/65-4	22.4
BDMA	7.0	2870	6960	443	11.3	0.64	332-3	4/65-4	30.0
BDMA	8.0	4180	7310	397	7.2	0.63	828	16/65-4	15.9
BDMA	8.0	3300	6300	464	13.1	0.81	332-1	16/65-4	41.1
BDMA	8.0	3040	6530	465	13.5	0.82	332-2	16/65-4	41.5
BDMA	9.0	3810	6740	376	7.0	0.79	332-1	/65-4	14.8
BDMA	9.0	3100	7030	444	10.2	0.85	332-2	/65-4	28.8
BDMA	9.0	3090	7170	450	14.2	0.67	332-3	/65-4	45.0
BDMA	9.0	2390	6700	399	8.0	0.65	332-4	/65-4	16.6
BDMA	9.0	2900	7170	447	10.8	0.62	332-5	/65-4	30.3
BDMA	10.0	3030	8550	461	12.2	0.85	332-1	16/65-4	41.5
BDMA	10.0	3080	6960	440	12.9	0.80	332-2	16/65-4	37.3
BDMA	12.0	3400	7100	458	10.6	0.77	332	16/65-4	29.6
BDMA	14.0	2900	6960	439	10.6	0.88	332	/80-1	30.1
BDMA	16.0	2800	7450	439	8.9	0.86	332-1	16/65-4	24.2
BDMA	16.0	2850	6630	410	10.3	0.89	332-2	16/65-4	28.11
DMP-30	6	3170	7240	429	8.2	0.81	332	/80-1	21.0
DMP-30	8	3300	7030	450	10.3	0.69	332	/80-1	28.7
DMP-30	10	3040	7240	458	11.4	0.74	332	/80-1	33.9
DMP-30	12	3840	7720	448	9.5	0.70	332	/65-1	29.6
DMP-30	14	3160	7310	463	11.0	0.68	332-1	/65-4	32.6
DMP-30	14	3430	7520	465	10.5	0.84	332-2	/65-1	33.4
Shell-D	12	3120	7580	417	6.9	0.78	332	/150-2	15.9
BF3-400	3	2550	7520	425	11.2	0.70	332-1	120-3	30.7
BF3-400	3	3280	7580	356	5.5	0.76	332-2		9.6
BF3-400	3	3240	7100	336	6.6	0.64	332-3	/200-1	11.4
V 150	83.6	1810	7520	449	7.5	1.3	332	12/65-4	18.9
V 140	50.0		6280	270	4.6	5.9	331	24	5.8
V 140	83.6	2990	6430	467	9.8	1.1	332-1	12/65-4	26.5
V 140	83.6	2320	6810	401	8.1	1.7	332-2	24	18.8

V 140	100.0		7030	349	5.9	0.5	331	24	9.6
V 140	150.0		6800	348	6.1		331	24	11.3
V 125	150.0	1830	6790	352	6.3		331	24	11.7
V 125	93.3	1610	6520	271	4.8	1.2	332	48	6.7
V 125	100.0	2280	6610	290	4.8	2.4	331	24	7.4
Em308	80.0	2520	7380	460	6.6	1.3	332	24	20.1
Em308	180.0						826	24	
Em308C	80/20		7580	352	5.3	1.4	826-1-C	24	9.4
Em308	100.0		6760	420	4.7		826	24	6.7
NMA 50-1.5DMP		3710	6300	384	7.4	0.90	332	143-24	15.2
NMA 70-1.5DMP		3350	6220	404	8.9	1.03	332	143-24	21.6
NMA 92-1.5DMP		2970	7030	449	8.6	1.04	332	143-26	21.9
NMA 102-1.5DMP		2970	5910	410	11.3	1.09	332	143-24	29.9
NMA 90-2BDMA		4000	6750	419	9.4	0.80	332-1	150-24	23.6
NMA 90-2BDMA		4910	8270	371	4.8	0.96	332-2	150-1	9.0
NMA 90-5DBVIII		200	6290	404	9.8	0.90	332-1	143-24	23.4
NMA 90-5DBVIII		3500	7240	380	6.9	0.82	332-2	143-24	13.9
HHPA 80-1.5DMP		3170	6900	412	7.9	1.04	332	148-2	18.7
MTHPA	84	3290	6630	411	9.5	0.78	332	79-4	19.9
NMA DBVIII							826	150-24	10.9
DDSA 138-1.5 DMP		2470	5530	381	9.2	1.5	332		20.2
DDSA 116-201BDM		2340	5860	351	7.6	1.0	332-1-EM	100-1	14.7
DDSA 116-201BDM		1930	5590	346	8.5	1.3	828-1-EM	/150-2	14.6
DDSA 116-201BDM		2340	5820	333	7.0	1.3	828-2-EM		13.5
PA	30	3540	6830	363	6.3	0.78	1001	120-4	12.1
TEFLON®			6420	233	16.4				31.7
TEFLON®			5830	169	6.0				5.0
TORLON®		1570	5050	379	29.9	1.2			29.9
VESPEL®			5110	322	17.9	0.70			42.3
PMMACRYLATE		3020	8410	367	4.7	0.88			8.2
UHMWPE			8210	325	18.6				39.7
TETA[1]	14/90	2490	5940	202	3.8	0.78	332	24	3.6
TETA[1]	14/125	3130	6270	307	5.9	0.67	332	24	9.5
TORLON[2]		7990	12760	507	4.7	0.29			12.5
G-10CR[3]		15000	32200			0.26			
ALUM		69600	89600						

* - Initial cure room temperature - 4 hours. Exceptions noted; [1] - TETA/Torlon; [2] - Torlon - 40% Glass; [3] - G-10 CR Rod

ACKNOWLEDGEMENTS

The work as presented was performed at Fermi National Accelerator Laboratory, which is operated by Universities Research Association, Inc., under contract with the U.S. Department of Energy.

REFERENCES

1. M. B. Kasen, G. R. MacDonald, D. H. Beekman, Jr., and R. E. Schramm, Adv. Cryog. Eng.--Mater. 26:235 (1980).
2. K. Dahlerup-Peterson, Adv. Cryog. Eng.--Mater. 26:268 (1980).
3. D. Evans and J. T. Morgan, Adv. Cryog. Eng.--Mater. 28:147 (1982).
4. G. Hartwig, Adv. Cryog. Eng.--Mater. 28:179 (1982).
5. J. V. Gauchel, J. L. Olinger, and D. C. Lupton, Adv. Cryog. Eng.--Mater. 28:211 (1982).
6. G. Hartwig and G. Schwarz, Adv. Cryog. Eng.--Mater. 30:61 (1984).
7. J. W. Ekin, E. S. Pittman, R. B. Goldfarb, M. J. Superczynski, and D. J. Waltman, Adv. Cryog. Eng.--Mater. 30:977 (1984).

CRYOGEN CONTAINMENT IN COMPOSITE VESSELS

D. Evans and J. T. Morgan

Rutherford Appleton Laboratory
Chilton, Didcot, Oxfordshire, United Kingdom

ABSTRACT

The paper draws attention to the need for designers to consider the long term ageing (thermal cycling) characteristics of composite materials rather than merely short term mechanical properties. The anisotropic nature of fibre reinforced composites is highlighted and the potential problems arising from this are discussed. Detailed consideration is given to outgassing and permeability, including gaseous evolution resulting from ionising radiation, and time constants for the permeation of helium gas through epoxide resins. Where the composite vessel is not exposed to low temperature, a reduced specification is considered in order to provide a vacuum container at an economic price.

INTRODUCTION

Much of the existing data on composite materials at low temperatures is based on short term mechanical tests and on this basis, many readily available materials exhibit properties that make them attractive to cryogenic engineers. It is made clear by Gauchel[1] that such initial properties do not guarantee long term success. With composite materials, long term properties may be adversely affected by degradation of the matrix/reinforcement bond. In the context of containment vessels for cryogens, internal damage of this nature could not only lead to a reduction in mechanical integrity but more seriously, to an increased permeability and therefore decreased vacuum characteristics and an associated increased heat leak.

MORE ON COMPOSITE MATERIALS

Glass, carbon, Kevlar and boron fibres may be used to reinforce resins. They may be unidirectional, woven into a fabric using any one of a variety of weaves (except for boron), wound onto a former with a predetermined winding angle (filament winding), or mixed in several different ways. The surface treatment associated with the fibre is an important variable which can often have a greater influence on the long term stability of the composite than it does on the absolute value of the short term properties.[2]

Kasen[3] has adequately summarised the short term cryogenic mechanical properties of a range of composites, showing the effect of fibre type and directionality. It should be noted that only with unidirectional materials can the very high moduli often claimed for "advanced composites" be realised. The low interlaminar tensile strength, together with the much reduced modulus at 90° to the principal axis of reinforcement, renders such unidirectional composites relatively unattractive as engineering materials. Cylindrical vessels made by filament winding have a predetermined winding angle to enhance properties in the direction of the principal stresses.

When considering cryogen storage, the usual requirement is for vacuum insulation. Frequently therefore, the containment vessel has to operate with an external positive pressure relative to the vacuum on the inside. This is the reverse of the pressure vessel situation in which the great strength and stiffness, in tension, of unidirectional fibre reinforced composites may be realistically used. Several papers[4,5,6] have presented details on cryogenic pressure vessels and further consideration will not be given here.

WHY COMPOSITES?

There are situations where a nonmetallic cryostat may be the only acceptable specification. Other situations may not demand nonmetallics specifically, but weight saving may be an important goal, as may cost savings, if the design and manufacturing possibilities of composites offer a more economical route than is available with established metallic containers. Composite necks on Dewars will minimise heat leaks and are already widely used. In vacuum based insulation, support for the atmospheric/vacuum load may be necessary and this simple requirement can best be met by insulating plastics materials.

Assume a sufficient thickness of composite to support an applied vacuum load, and compare the relative weights of different constructional materials.

Table 1. Relative Thickness and Weight for Equal Bending Stiffness

Composite	Modulus, GPa	Density, g/ml	Relative Thickness	Relative Weight
Woven Glass Fabric	18	1.7	2.2	0.48
Filament Wound Glass at 20°	21	1.8	2.1	0.50
Woven Carbon Fabric	49	1.45	1.6	0.29
Filament Wound Carbon at 20°	29	1.53	1.9	0.37
Filament Wound Kevlar	25	1.33	2.0	0.34
Aluminium	70	2.7	1.4	0.49
Stainless Steel	190	7.8	1	1
Reduced Specification	9	1.36	2.8	0.48

Based on conservative figures for modulus (Table 1), it is clear that weight savings are possible, a modest glass reinforced structure rivaling the weight of say an all aluminium construction.

Consider the following reduced specification:

Reinforcement: Heavyweight woven roving (glass) with alternate layers of glass chopped strand mat. Estimated weight of two layers of reinforcement approximately 1 kg/m^2, giving composite density of about 1.3 to 1.4 at 30% w/w fibre. Matrix material to be polyester. The low modulus of this composite means an increased thickness to maintain the equal bending stiffness requirement but the relatively low density (arising from the low glass content) compensates in overall weight terms. However, this "reduced specification" could be considered only where the composite material is performing at room temperature. For low temperature applications the high resin content may be unacceptable.

Where there is a firm technical requirement to use a composite storage vessel, cost may not be a major consideration but where the scientific aims can be met by metallic vessels, then the cost of a composite structure needs careful consideration.

OUTGASSING, PERMEABILITY AND DIFFUSION

Outgassing

This is generally taken to mean the loss of volatiles loosely bonded to the surface of a material exposed to a reduced pressure. For metals the rate of outgassing is approximately inversely proportional to the pumping (exposure) time, i.e.,

$$\text{RATE} \propto 1/t \qquad \text{where,} \quad t = \text{time}$$

For many plastics materials this simple relationship is not followed and it is found that the outgassing rate is more accurately described by:

$$\text{RATE} \propto 1/t^{1/2}$$

These relationships, illustrated in figure 1, arise almost exclusively from the loss of water vapour. Note the change in slope after a short pumping time for the resin previously vacuum baked and allowed to equilibriate in air for 24 h prior to test. It can be argued that water vapour within the structure gradually diffuses to the surface and this gives rise to the $1/t^{1/2}$ relationship. The use of composite vessels could conceivably lead to longer pumping times prior to starting a cooling cycle, but in many cases the presence of super-insulation will offer a much larger surface area for outgassing, relegating the effects from the vessel surface to second place.

Diffusion and Permeability

All polymeric materials show some measurable permeability to gases and vapours and in general, the smaller the gas molecule (permeant) the greater is the permeability through the polymeric barrier. The mechanism by which the permeant is transported through the barrier is a twofold process:

(i) interstitially through pores

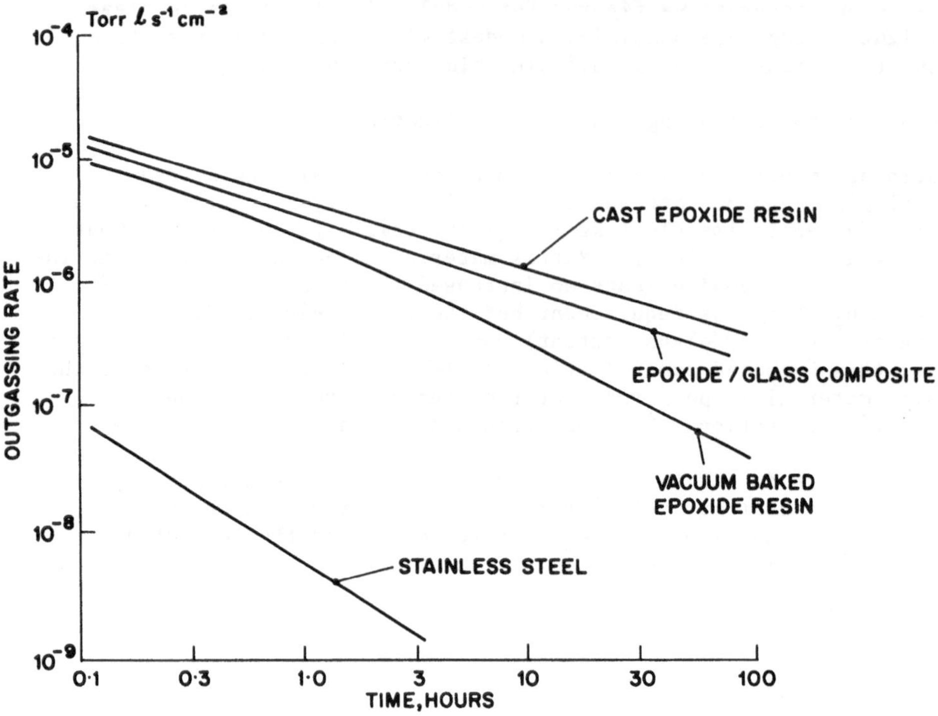

Fig. 1. Outgassing rates for various surfaces.

(ii) by a process of sorption or dissolution on the high pressure side of the barrier, subsequent diffusion through it followed by desorption on the low pressure side.

It has been recorded[7] that the quantity of gas (Q) passing through a barrier of cross section (a) and thickness (dx) in a time (t) with a pressure difference (dp) is given by:

$$Q = P\ a\ t\ dp/dx \qquad (1)$$

where P is the permeability constant.

Data from reference 8 have been used to derive the following relationships for the permeability constants for hydrogen and helium gases:

For hydrogen:

$$\text{Log}_{10}P = -1948/K - 4.3 \qquad K = \text{absolute temperature} \qquad (2)$$

For helium

$$\text{Log}_{10}P = -1793/K - 4.65 \qquad P = \text{mole m}^{-2}\ \text{s}^{-1}\ \text{bar}^{-1}\ \text{m} \qquad (3)$$

From these data, the permeability constant may be calculated at any temperature, accepting that the original data from which the above equations were derived only spanned the temperature range from 298 K to 328 K. From equation (1) it is apparent that if P is constant, the rate of gas flow (at constant pressure difference) must be inversely proportional to the barrier thickness. Data from Fujioka et al.[9] have been processed and found to follow this relationship. Equation (3) enables the rate of pressure increase in any defined geometry to be readily calculated for helium gas.

Under conditions such that the amount of permeant entering the barrier is equal to that leaving it, then the permeability and solubility of the permeant are related by the expression:

$$P = DS \quad \text{where } D = \text{diffusion constant}, \; S = \text{solubility} \qquad (4)$$

The concept of "lag time," introduced in reference 8, is the time taken for gas to appear on the other side of the barrier, indicated by a pressure rise. For the thin (about 0.5 mm) specimens considered in that work, lag times of only several minutes were recorded. Using the simple relationship:

$$D = d^2/6L \quad \text{where } D = \text{diffusion constant}, \; d = \text{thickness}, \; L = \text{lag time} \qquad (5)$$

Permeability, solubility, diffusion and lag time are all temperature dependant.

For helium at 298 K, a permeability constant for epoxide resin of 2.2×10^{-11} mole m^{-2} s^{-1} bar^{-1} m may be obtained from equation (3); using this value and a typical solubility of helium in epoxide resin of 0.14 mole m^{-3} bar^{-1}, then using equations (4) and (5) a lag time of approximately 3×10^4 seconds may be derived for epoxide resin of 5 mm thickness.

The transmission of helium gas through a composite involves little or no interaction between the gas and barrier. Vapours condensing readily (such as water), interact with most barrier materials, including epoxide and polyester resins. If the barrier has functional groups or chain segments chemically similar to the vapour, dissolution results, leading to a high permeability. It precludes the possibility of predicting the permeability of a barrier material to water vapour from data on any other gas or vapour. Suffice it to say that significant permeability of composites to water vapour should be expected but there is a desperate shortage of information on the subject. It has been argued that water vapour permeability is unimportant because of the low temperatures involved. However, there are many designs of vacuum insulated cryogen container where large surface areas of composite are exposed to a vacuum environment but not at very low temperature. Metal foil barriers have been considered in several designs of cryostat, not always with success.

When testing for permeability or leaks, the "lag time" must be taken into consideration. A long time constant for examination is required to determine permeability and a short term response should be taken to indicate penetration of gas by an easier route than interstitial permeation, ie. via voids or cracks in the material. Earlier work by the authors[2] indicated the growth of interlaminar damage, assessed by N.D.T. methods, when thermally cycling a glass/epoxide resin composite. Such damage, present as cracks in the matrix or debonding at the resin/fibre interface, will change any permeability constants for the material that have been measured or calculated. Gauchel,[1] illustrated this point, with fully restrained and thermally cycled specimens (298 - 77 k) which were helium leak tested at room temperature every 100 cycles. The criteria selected by Gauchel was a 'pass' for the material if helium flow through the sample was not evident after 5 minutes on the 3mm thick specimens. Substituting this information in equation (5), the permeability constant may be calculated to be (7×10^{-10} mole m^{-2} s^{-1} bar^{-1} m). This is some 30 times greater than that previously calculated for unfilled epoxide resin, indicating that significant damage may have been present in specimens where the time for passage of gas is 5 minutes.

If the environment in which the cryogen is being stored involves exposure to ionising radiation, the outgassing from this source cannot be ignored.[10] There are two possible sources of gas contamination:-

(a) Evolution of gaseous by-products from the polymer.

(b) Cracking, crazing etc. of the material by radiation, leading to high permeability factors.

The first factor is somewhat easier to quantify than the second and therefore the assumption is made that total integrated radiation doses are not at a level likely to cause mechanical damage. Data from reference 10 has been used to provide the information in table 2 on the total gas evolved per square metre of surface exposed to ionising radiation on one face with vacuum on the other.

If half the total gas is lost through the vacuum side of the composite, some 7.6 litres of gas/square metre of exposed surface will be lost into the vacuum for each 10^7 Gy of radiation absorbed by the composite. Factors such as the rate of deposition of energy (radiation dose rate) will control the rate at which gas is evolved. The situation is further complicated by temperature effects, especially temperatures in the cryogenic range. A more detailed discussion on this topic is contained in references 11 and 12.

THERMAL CONTRACTION

As with the mechanical properties, the thermal characteristics of composite materials are anisotropic. This may create difficulties that can present serious engineering problems.

Large changes in thermal contraction values may result from changes in fibre type, fibre angle, and content. A reinforced composite cylinder, which is a likely form for a vacuum insulated cryogen container, will exhibit different values for circumferential and radial contraction. The circumferential contraction will vary with fibre type, direction, concentration etc., but thermal strain in the radial direction will vary with thickness and diameter in addition to all other factors. Reference 2 contains a discussion on these aspects of a glass reinforced composite and indicates a limiting thickness to diameter ratio. Whilst the point has not been investigated, it is likely that for carbon and Kevlar fibre composites, since the radial/circumferential contraction difference is larger, the induced thermal strain is greater and therefore the limiting thickness to diameter ratio will be smaller (figure 2).

Table 2. Radiation Induced Gas Evolution from Epoxide Composite

Filler Content, Vol. %	Composite Weight, kg/m^2	Thickness, mm	Resin Weight, kg/m^2	Gas Evolved ml/g of resin per Gy	Total Gas, $ml/m^2/10^7$ Gy
65	19.3	10.7	3.8	0.04	15,200

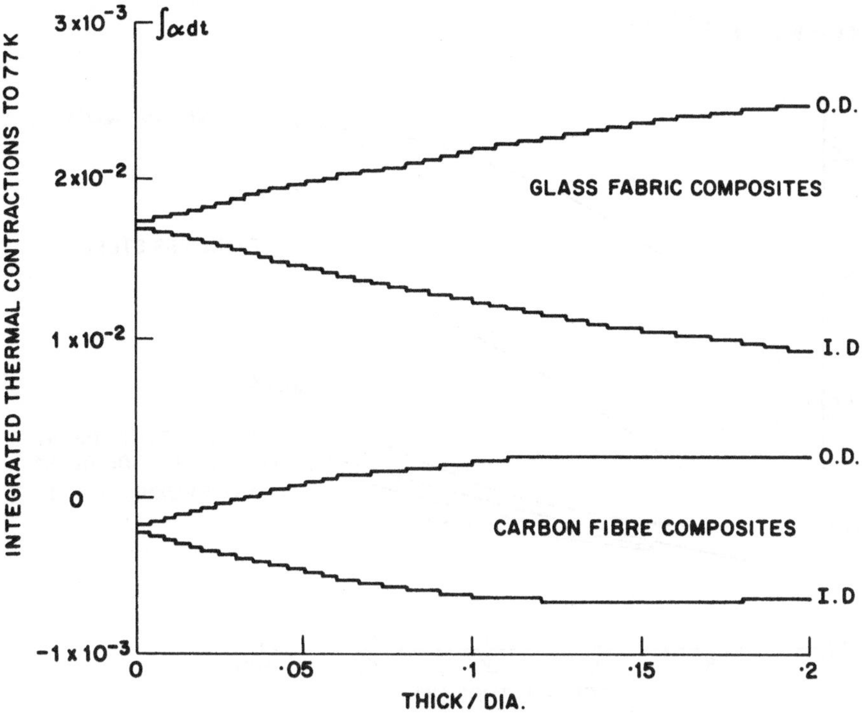

Fig. 2. Diametral contraction of composite rings.

The anisotropic thermal behaviour of composites may give rise to problems where composites are bonded to themselves or other materials. For example, the outer diameter of a composite tube normally exhibits a higher thermal contraction than the inner diameter. When bonding tubes together, the joint is best effected with an outer tube having a slightly higher thermal contraction than the inner, so that the bond is not subjected to tensile forces on cooling. For composite tubes, an outer wrap of a similar composite does not achieve this because of its anisotropy.

Composites produced from blending particulate fillers with resins usually exhibit isotropic properties, and with careful formulation, the thermal contraction may be engineered to match other materials of construction. The strength and modulus relative to fibre reinforced materials is reduced but in designs where weight saving is not the prime objective, "cast" composite Dewars are possible. Table 3 lists some of the important characteristics of a filled resin system suitable for casting.

Table 3. Properties of a Filled Resin System

Modulus, GPa	Density, g/ml	Thermal Contraction 77 K	4.2 K	Relative Thickness*	Relative Weight*
9	1.7	0.003	0.0035	2.8	0.6

* For comparison with Table 1

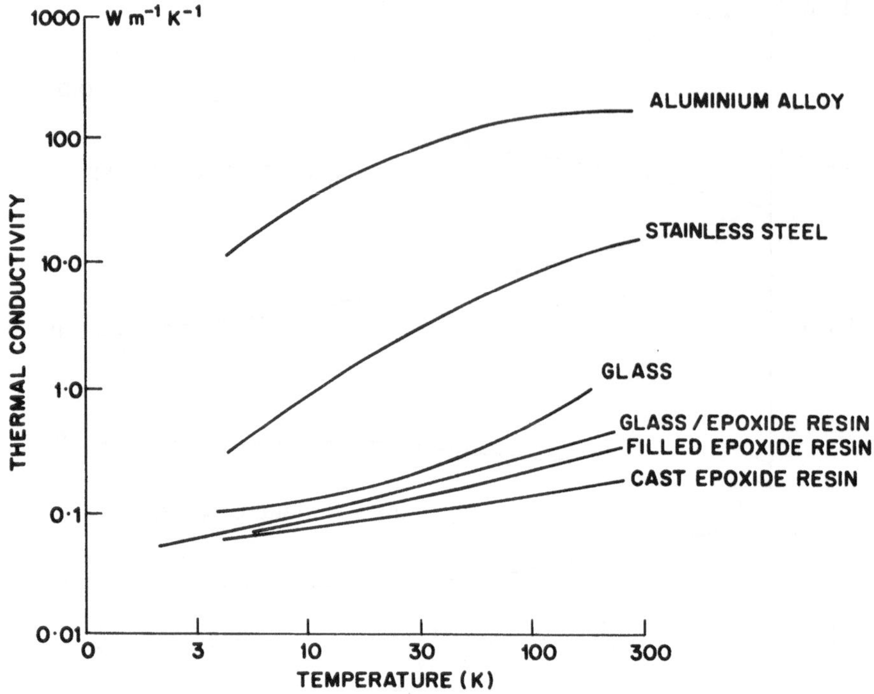

Fig. 3. Thermal conductivity as a function of temperature.

THERMAL CONDUCTIVITY

Apart from specific areas such as the neck, thermal conductivity of materials used in Dewar construction is not a major consideration. In general, composites have lower thermal conductivities than metals but note (figure 3) that in the fibre direction unidirectional carbon composites and stainless steel have similar conductivities. At room temperature these materials have a thermal conductivities at least an order of magnitude higher than that of a comparable glass reinforced composite, while at 4.2 K the difference is reduced to approximately a factor of two. However at the temperature of liquid helium, the thermal conductivity of a composite is dependent more on the fibre direction and is less sensitive to other variables, such as fibre type and concentration. Information on the thermal conductivity of a filled casting system is also presented in figure 3.

SPECIFIC HEAT AND ENTHALPY

Specific heat is normally expressed per unit mass of material. The values in $Jg^{-1}K^{-1}$ for various composite materials together with those for aluminium alloy and stainless steel are shown in figure 4. In practical terms, it may be better to compare the specific heats of materials on a volume basis, i.e., $Jcm^{-3}K^{-1}$. Specific heat varies with temperature, becoming zero at zero Kelvin, so, for cryostat construction the heat content (or enthalpy) of a material at a given temperature is a more useful parameter as this gives a direct comparison of the amount of cryogen lost in cooling the cryostat. Whilst this is relatively unimportant for long term storage of cryogens, it is more important where a cryostat is frequently subjected to large temperature changes.

The histograms (figures 5 and 6) show the relative enthalpy values for various materials together with the specific heat requirement in

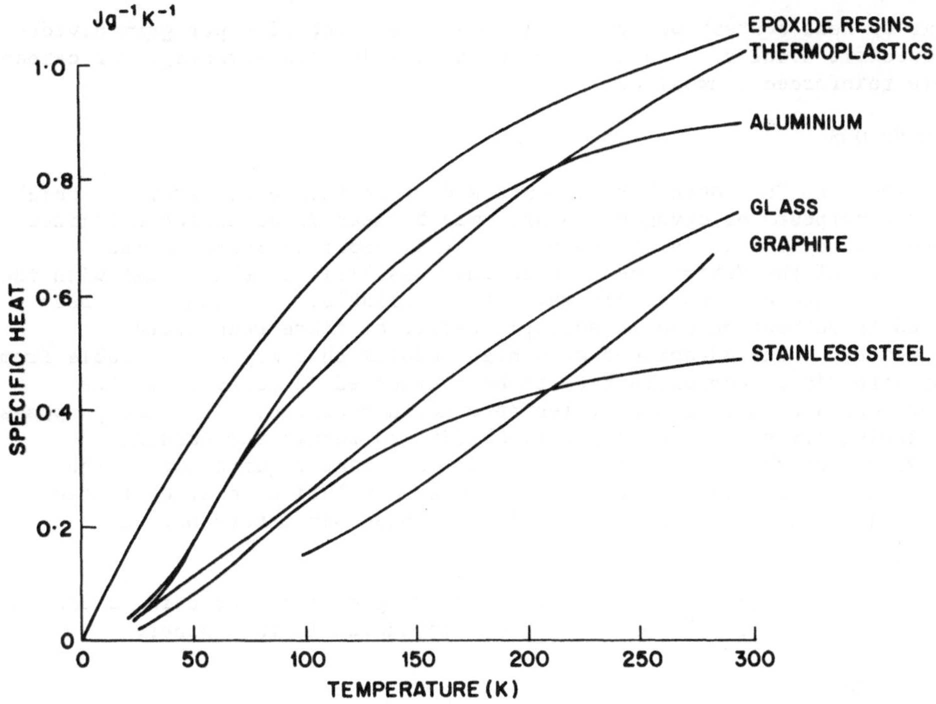

Fig. 4. Specific heat as a function of temperature.

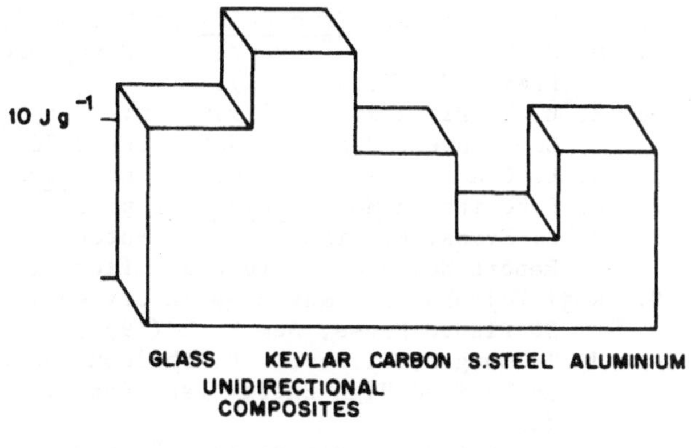

Fig. 5. Enthalpy at 80 K.

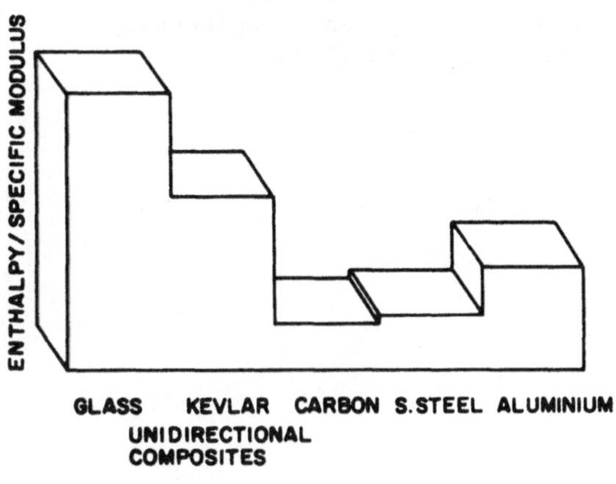

Fig. 6. Cooldown requirement from 80 K (relative).

terms of heat content per unit stiffness i.e., enthalpy per gram divided by specific modulus. This indicates a considerable advantage for carbon fibre reinforced composites.

CONCLUSIONS

The term "composite" represents a diverse family of materials which, for the purposes of cryogen storage, may be used in an almost unlimited number of ways. It is the purpose of this paper to increase the awareness of the design and formulating possibilities that exist with the exciting range of nonmetallic materials available. Emphasis has been placed throughout on the anisotropic nature of fibre reinforced composites. The advantages of a high modulus that may be available from a unidirectional composite have to be considered in relation to the mechanical and thermal properties in other directions. For the purposes of storing cryogens, the long term ability to retain the vacuum, necessary for thermal insulation, is an important requirement. The outgassing (including radiation induced gas evolution) characteristics are of importance, as are less well documented parameters such as diffusion and permeability.

The possibility of manufacturing a cryogen container using a casting process and a "filled" resin has been considered in this paper.

REFERENCES

1. J. V. Gauchel, J. L. Olinger and D. C. Lupton in: "Advances in Cryogenic Engineering." Vol. 28. Plenum Press, New York (1982).
2. D. Evans & J. T. Morgan in: "Nonmetallic Materials & Composites at Low Temperatures 2". Plenum Publishing Corporation (1982).
3. M. B. Kasen in: Cryogenics 15 No. 6 p327-349 (1975).
4. M. P. Hanson in: "Advances in Cryogenic Engineering." Plenum Press, New York (1971).
5. E. E. Morris & R. J. Alfring in: "Cryogenic Boron Filament Wound Containment Vessels" NASA, CR-72330 (1967).
6. R. L. Conder & N. L. Newhouse in: Cryogenics 20 No. 12 p697-701.
7. A. Lebovits in Modern Plastics p139-213 (1966).
8. F. P. Evans, R. Gibson, C. G. Hutcheson and P. M. S. Jones: AWRE Report No. 0-45/63 (unclassified) (1963).
9. Koji Fujioka, in "Advances in Cryogenic Engineering" vol. 30 Plenum Press, New York (1984).
10. J. T. Morgan, G. Scott, R. Sheldon and G. B. Stapleton: RHEL/R196 Rutherford High Energy Lab. Report (1970).
11. D. Evans & J. T. Morgan in: "Advances in Cryogenic Engineering (Materials)" vol. 28 Plenum Press, New York (1982).
12. D. Evans & J. T. Morgan in: "Advances in Cryogenic Engineering (Materials)" vol. 30 Plenum Press, New York (1984).

INFLUENCE OF DAMAGE ON MECHANICAL PERFORMANCE OF WOVEN LAMINATES AT LOW TEMPERATURES

Ronald D. Kriz

Fracture and Deformation Division
National Bureau of Standards
Boulder, Colorado

Walter J. Muster

Metals Division
Swiss Federal Laboratories for
Materials Testing and Research (EMPA)
Dubendorf, Switzerland

ABSTRACT

Large quantities of nonmetallic woven composites will be used in magnetic fusion energy structures at low temperatures. We predicted and measured the influence of crack formation on the mechanical performance of standard glass/epoxy laminates (G-10CR, G-11CR) at low temperatures. From experiments with tension loads, we studied the formation of damage as a collection of fiber breaks, fiber bundle cracks, and delaminations between adjacent fiber bundles. We measured fiber bundle cracks in the laminate interior and individual fiber fracture at the laminate edges. We discovered that the sequence and type of damage control the discontinuities ("knees") in the load-deformation (stress-strain) diagrams. We found that G-11CR has two knees and three distinct moduli, whereas G-10CR has only two moduli and a single knee at a lower strain than G-11CR. Decrease in moduli measured near the knees compared well with predictions from a finite element model.

INTRODUCTION

Magnetic-fusion-energy (MFE) power plants may use large quantities of nonmetallic composites.[1] There are a variety of applications where woven glass/epoxy laminates can provide thermal insulation, electrical insulation, and structural support.[2] In the structural design of superconducting magnets at 4 K, nonmetallics are used primarily as electrical insulation where only small mechanical loads occur. However, when cooled to liquid helium temperature (4 K), large internal stresses can occur within these structures because of large differential thermal contractions between metals and nonmetals. When these thermally induced stresses are combined with stresses induced by small mechanical loads (magnetic loads), microcracks may initiate at applied loads well below the yield strength. The formation of large numbers of microcracks could limit the electrical, thermal, and mechanical usefulness of these materials for MFE applications. The coalescence of

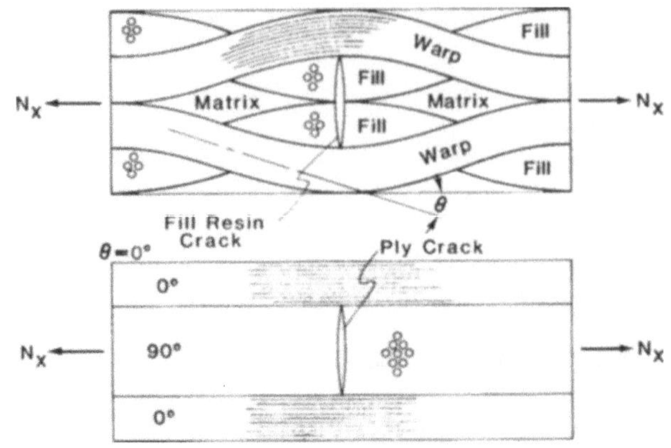

Fig. 1. Repeating pattern for a two-layer woven laminate with an interior fill resin crack and a 0°/90° laminate (θ = 0°) with 90° ply crack.

microcracks and voids could also limit their usefulness as barriers for containment of cryogenic liquids and gas. In this study, we measured the variability in mechanical performance caused by the formation of damage in a unit cell of plain weave.

THEORY

The unit cell of a woven structure was defined by Ishikawa[3] as the smallest repeating structural unit. The four sides or cross sections (see Fig. 1) of a plain-weave unit cell are identical.[4] Hence, from a thin slice of the common cross section we approximate the three-dimensional stress state by a finite element model (FEM)[4] that assumes generalized plane strain. Details of this analysis are given in ref. 4.

Damage in this model is limited to transverse cracks in the resin of the internal fill fiber bundles (see Fig. 1). Hence, only 50 percent of all possible fill resin cracks were modeled. To study the significance of fill resin cracks with weave geometry, we varied θ (Fig. 1) from 0° to 45°. The condition of θ = 0° in Fig. 1 represents a special limiting case for the woven laminate, where the fill resin crack becomes a ply crack in a 0°/90° laminate.

From experiment, the laminate load, N_x = 175 N/mm, near the knee was used to calculate the strains ε_x = 0.011 and ε_y = 0.0017. Young's moduli in the damaged and undamaged state were calculated using these strains. A worst-case thermal load of ΔT = -318 K was chosen for the operating temperature of 76 K, and a strain-free state was assumed at 395 K. Under these load conditions, we predicted the influence of weave curvature and fill-crack damage on the elastic behavior of a unit cell at 76 K. Results of these calculations are compared with experimental results in the following section.

EXPERIMENT

Ten G-10CR and ten G-11CR tensile specimens were cut by shear from panels of cryogenic grade laminates.[2] For all specimens, the warp fiber bundles were aligned with the load axis. Both panels were 0.5-mm thick with three layers of a plain weave. All specimens were 5.0-cm long and 1.77-cm wide. The edges of four specimens of each material were polished so that edge damage could be recorded by the replication technique described in ref. 5. There are four test groups: equal numbers of specimens of G-10CR and G-11CR were tested at 295 K and 76 K.

All specimens were quasi-statistically loaded in tension at 20 N/s. The first three specimens from each group were loaded in tension to ultimate fracture strength. Two specimens with polished edges were loaded in increments of 200 N. Damage at the edge of these specimens was recorded by replication[5] at room temperature after each load increment. For 76-K tests, a warm-up period was required before replication of edge damage at room temperature. Densities of fill cracks were determined by observation through the thicknesses of the semitransparent thin laminates.

RESULTS AND DISCUSSION

A summary of experimental results is given in Table 1. Our results compare well with the previous studies.[2,6] We report two additional Young's moduli (E_2, E_3) and a strain (ε_T^1) that locates the first knee in the stress-strain diagram. A second knee was observed for G-11CR, but is only reported here as a change in modulus between E_2^{11} and E_3^{11}. From experimental observation we correlate individual damage events to nonlinear behavior in the stress-strain diagrams (see Figs. 2-7). New parameters for these nonlinearities are defined geometrically at 76 K in Fig. 4 for G-10CR and in Fig. 7 for G-11CR. At room temperature (295 K), bimodulus behavior does not exist (see Fig. 2). Other studies[7,8] reported similar bimodulus results at low temperatures, but no explanations are given for this behavior. In this study, we examine relationships between damage accumulation and these nonlinearities.

The stress-strain response of a bisphenol A epoxy resin, representative of resin used in G-10CR and G-11CR,[4] is included in Figs. 2, 4, 6, and 7 for comparison. The failure strain of the resin at 295 K is larger than those of G-10CR and G-11CR. At 76 K, the opposite is true: we observed that the resin failure strain, ε_{resin}, was much lower than laminate fracture strains. We also observed that ε_T^1 is slightly lower than ε_{resin}. This observation implies that the first knee at 76 K could be related to a resin failure in the laminate. Because the resin is constrained in the fiber-reinforced laminate, we predict that the resin will fail at a lower strain ($\varepsilon_T^1 < \varepsilon_{resin}$). At 76 K we observed several epoxy failures in the fill fiber bundles of G-10CR

Table 1. Summary of experimental results.

	G-10CR		G-11CR	
Moduli	295 K	76 K	295 K	76 K
E_1 (GPa)	26.9	32.4	29.9	34.3
E_2 (GPa)		19.4		17.7
E_3 (GPa)				20.4
Strains				
ε_T^1 (%)		1.15		0.91
ε^{UTS} (%)	1.73	3.33	2.10	3.95
Strength				
UTS (MPa)	417	809	478	918

UTS indicates ultimate strength.
Subscripted values defined in Figs. 4 and 7.

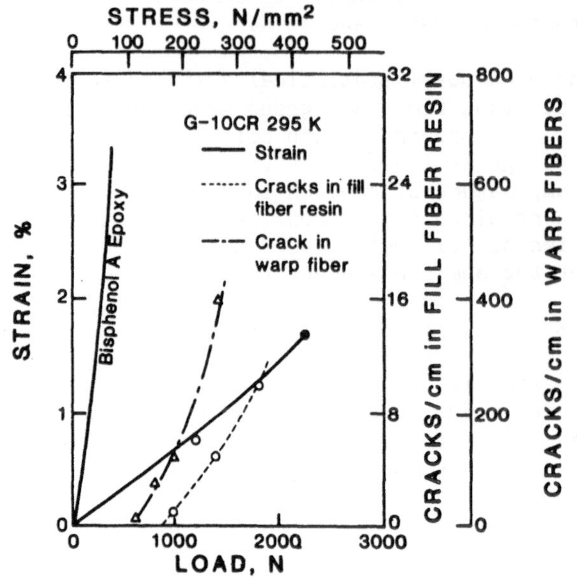

Fig. 2. Correlation between formation and stress-strain response of G-10CR in tension at 295 K. Note: stress-strain curve is inverted.

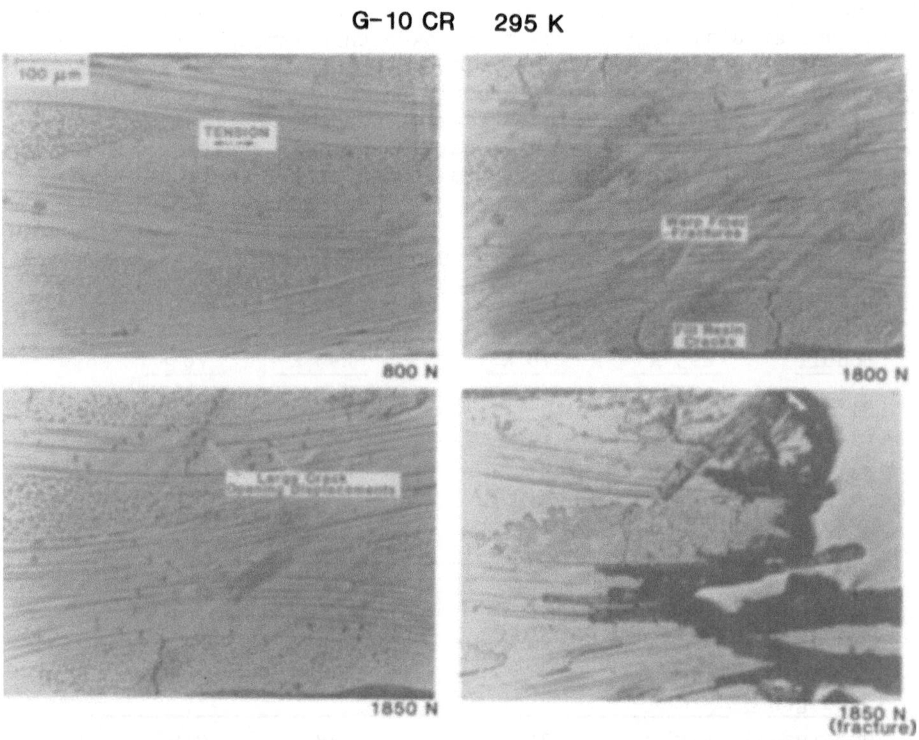

Fig. 3. Edge damage corresponding to loads of Fig. 2. Note: Large crack opening displacements between broken fiber ends at 1850 N in the region of final failure.

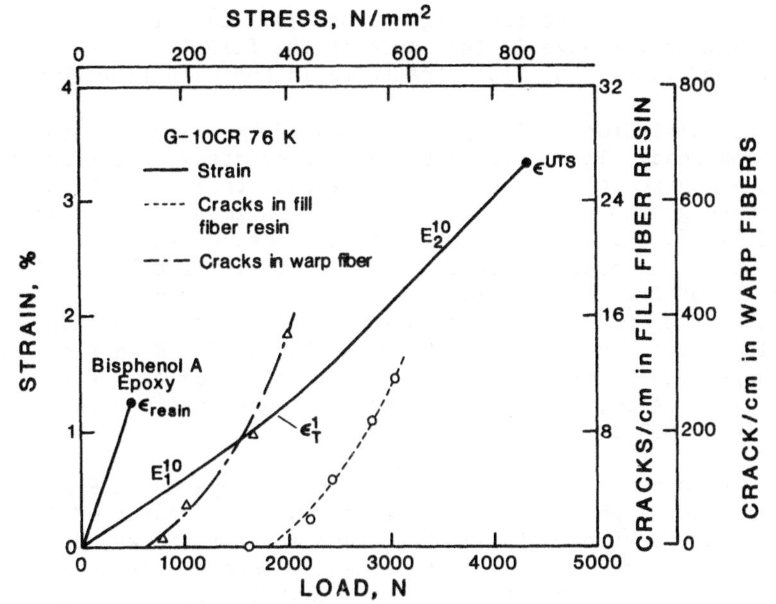

Fig. 4. Same as Fig. 2 except at 76 K with geometric parameters defined.

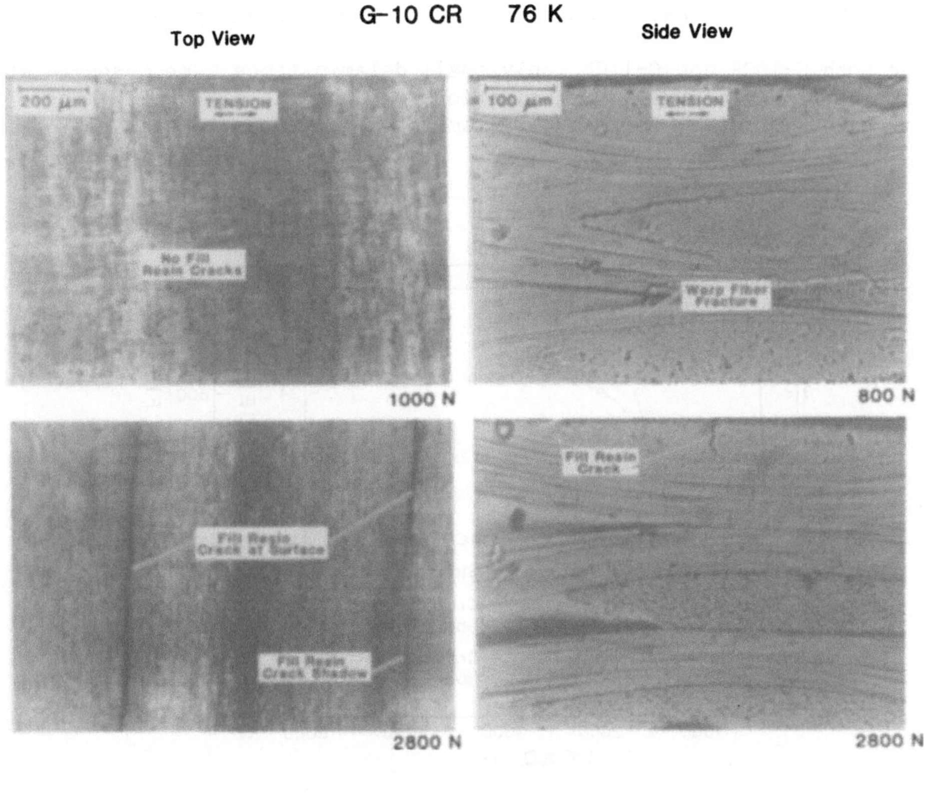

Fig. 5. Top and edge views of damage corresponding to loads of Fig. 4. Damage in top view is fill resin cracks often observed as shadows below the surface.

and G-11CR prior to the knee (see Figs. 4 and 7). In all cases before fracture we observed only cracks in the resin of the fill fiber bundles (fill resin cracks) and individual fiber fractures in the warp fiber bundles.

After the initiation of damage, we measured the growth of this damage with increasing load. To correlate the damage events with stress-strain nonlinearities, we inverted the conventional stress-strain format and superposed the damage growth on the same diagrams with load as the common abscissa (see Figs. 2, 4, 6, 7). In all cases damage increased exponentially with increasing load. The nonlinearities shown in Figs. 2 and 4 correlated well with the growth of fill resin cracks in G-10CR at 76 K and 295 K, but the growth of warp fiber fractures at much lower loads had no influence on these nonlinearities. The edge damage state of G-10CR at 295 K in Fig. 3 revealed unusually large displacements between broken fiber ends only within the fractured region. Also, fewer fiber fractures were observed on the edge of G-11CR than in G-10CR at 76 K.

The bimodulus behavior of G-10CR at 76 K is compared in Fig. 8 with the change in modulus predicted by FEM where only 50 percent of fill fiber bundles were modeled with resin cracks. From experiment, we observed resin cracks in all fill fiber bundles (100 percent). Hence, the decrease in modulus for G-10CR in Fig. 8 is larger than that predicted by the FEM. For G-11CR the increase in modulus at the second knee in Fig. 7 could be caused by the straightening of warp fiber bundles after multiple resin cracks in the fill fiber bundles. Unfortunately, there is no evidence of warp bundle straightening at 76 K because replicas were taken at reduced loads after warm-up to prevent formation of new damage. The two-dimensional FEM[4] approximation appears to be a serious oversimplification, which models only a portion of the woven structure. A complete three-dimensional model that includes fiber straightening would provide a more accurate prediction.

For both G-10CR and G-11CR, only small delaminations were observed after fracture. Before fracture no delaminations occurred at either 295 K or 76 K. Previous studies[9] for nonwoven laminates showed that delaminations

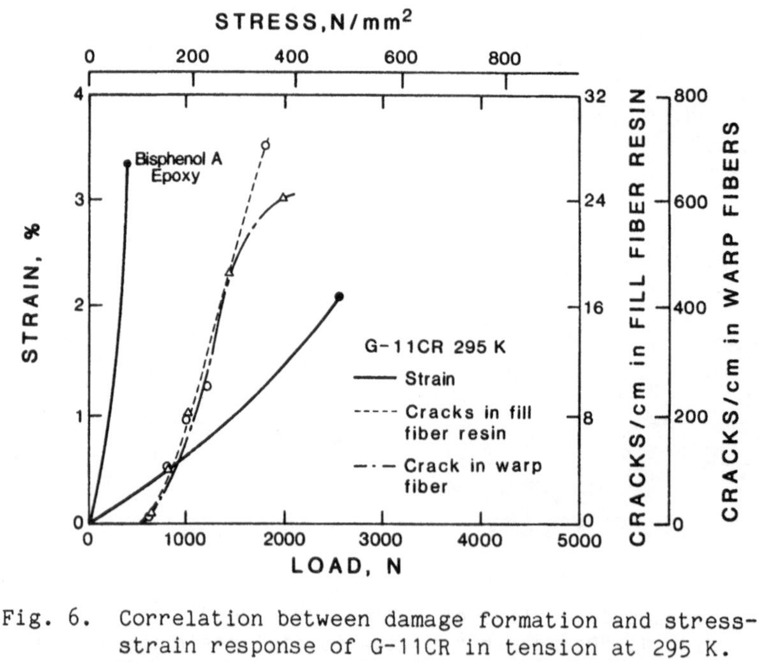

Fig. 6. Correlation between damage formation and stress-strain response of G-11CR in tension at 295 K. Note: stress-strain curve is inverted.

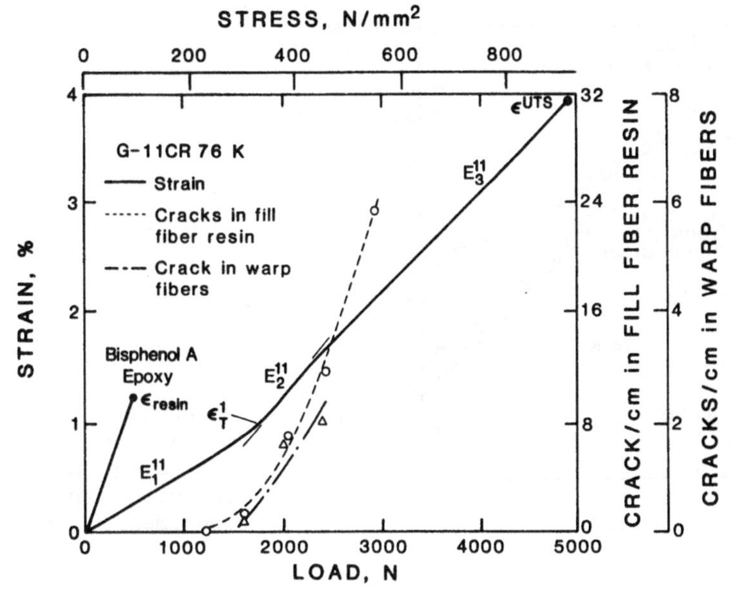

Fig. 7. Same as Fig. 6 except at 76 K with geometric parameters defined.

can originate from the tips of transverse ply cracks at the interlayer interface. For woven laminates with θ = 45°, we predicted by the FEM[4] a decrease in delamination stresses when the fill resin crack occurs and when the temperature decreases to 76 K. Hence, in this study we verified that delaminations are prevented for moderate to large weave curvatures by the formation of fill resin cracks at low temperatures.

SUMMARY

From experiments at 76 K and 295 K we observed that damage in specimens of G-10CR and G-11CR accumulated as resin cracks in fill fiber bundles and individual fiber fractures in warp fiber bundles. Only small regions of delamination were observed between warp and fill fiber bundles after laminate fracture. For both materials, larger loads were required to initiate damage at 76 K.

Fiber fractures had no influence on the appearance of the stress-strain diagrams. For G-10CR at 295 K, large numbers of fiber fractures occurred randomly and large crack opening displacements were observed in the region of laminate fracture. At a lower temperature we observed fewer fiber fractures and measured a higher laminate fracture strength. From theory we predicted that a high density of fill resin cracks would result in a reduced modulus, or a knee, in the stress-strain diagram. In G-10CR and G-11CR at 76 K, a high density of fill resin cracks was observed at a strain similar to the failure strain of the epoxy resin. At 76 K we observed two knees and three moduli for G-11CR where the third modulus increased in value. For G-10CR at 76 K, we observed only one knee at a strain lower than that of the first knee of G-11CR. Most of the damage and nonlinearities in the stress-strain diagrams occurred at strains well below fracture. This is an important consideration for design applications where these materials are only loaded to small strains.

From both theory and experiment, we conclude that the weave geometry is beneficial at low temperatures: when fill resin cracks occur delaminations are prevented at the fill-warp interface.

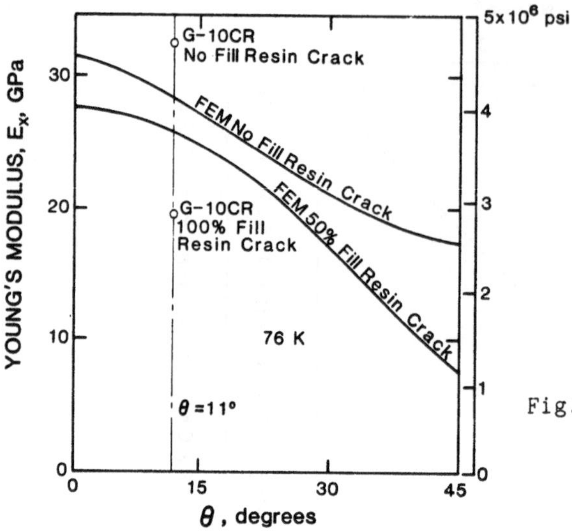

Fig. 8. Influence of internal fill cracks on Young's modulus of a woven unit cell: a comparison between FEM and experiment.

ACKNOWLEDGMENTS

This study was sponsored by the U.S. Department of Energy, Office of Fusion Energy. Material was supplied by Spaulding Fiber Company. The cryogenic tension fixture was machined by Mr. Dale Thoel at NBS.

REFERENCES

1. M. B. Kasen, Composite laminate applications in magnetic fusion energy superconducting magnet systems, in: "Proceedings of the 1978 International Conference on Composite Materials," AIME, New York (1978), pp. 1493-1507.
2. M. B. Kasen, G. R. MacDonald, D. H. Beekman Jr., and R. E. Schramm, Mechanical, electrical and thermal characterization of G-10CR and G-11CR glass/epoxy laminates between room temperature and 4 K, in: "Advances in Cryogenic Engineering--Materials," vol. 26, Plenum Press, New York (1980), pp. 235-244.
3. T. Ishikawa, Anti-symmetric elastic properties of composite plates of satin weave cloth, Fiber Sci. Technol. 15:127-145 (1981).
4. R. D. Kriz, Mechanical-damage effects in woven laminates at low temperatures, in: "Materials Studies for Magnetic Fusion Energy Applications at Low Temperatures--VIII," R. P. Reed, ed., NBSIR 85-3025, National Bureau of Standards, Boulder, Colorado (1985), pp. 49-86.
5. D. O. Stalnaker and W. W. Stinchcomb, in: "Composite Materials: Testing and Design (Fifth Conference)," ASTM STP 674, American Society for Testing and Materials, Philadelphia (1979),pp. 620-641.
6. H. M. Ledbetter, Dynamic elastic modulus and internal friction in G-10CR and G-11CR fiberglass-cloth-epoxy composites, Cryogenics, 20:637-640 (1980).
7. H. Benz, I. Horvath, K. Kwasnitza, R. K. Maix, and G. Meyer, Worldwide cryogenics--Switzerland cryogenics at BBC Brown, Boveri & Co., Ltd., Cryogenics 19:3-15 (1979).
8. M. B. Kasen, Mechanical and thermal properties of filamentary-reinforced structural composites at cryogenic temperatures 1: Glass-reinforced composites, Cryogenics, 15:327-349 (1975).
9. A. L. Highsmith, W. W. Stinchcomb, and K. L. Reifsnider, Effect of fatigue-induced defects on the residual response of composite laminates, in: "Effects of Defects in Composite Materials," ASTM STP 836, American Society for Testing and Materials, Philadelphia, (1984), pp. 194-216.

RADIATION DAMAGE OF COMPOSITE MATERIAL--

METHOD AND EVALUATION

T. Okada, S. Nishijima
ISIR Osaka University
Ibaraki, Osaka Japan

and

H. Yamaoka
KURRI, Kyoto University
Kumatori, Osaka, Japan

ABSTRACT

The fracture mode of irradiated composite materials was analyzed in flexural and tensile tests. The fracture mode varies from tensile to shear with increasing the radiation dose because interlaminar strength is thought to be decreased by irradiation. The change of fracture mode of composite materials should therefore be considered when the radiation damage of composite materials is studied. The change of interlaminar shear strength of irradiated composite materials was studied to establish a method of evaluating radiation damage in composite materials. Gamma irradiation was performed on a commercially available composite material. It is concluded that the change of the fracture mode after irradiation is caused by a decrease in interlaminar strength.

INTRODUCTION

It is important to estimate the degradation caused by neutron irradiation of the superconducting magnets that are employed as field generators in fusion reactors. Among the components of the superconducting magnets, organic insulators are thought to be most sensitive to irradiation, and hence, the evaluation of radiation damage in organic insulators is essential.

In this study, the degradation of mechanical properties in organic composite materials was investigated. The mechanical properties are thought to be most essential and important because the degradation of electrical properties is not introduced until the mechanical defects are induced.[1]

ANALYSIS

Problems in Evaluation of Radiation Damage

The mechanical strength of composite materials should be determined by the strength of their reinforcement, because that of the matrix is considerably lower than that of the reinforcements. The degradation of reinforcements caused by irradiation is more difficult to induce than that of organic materials. This leads to the conclusion that the organic composite materials should not show degradation. On the contrary, the experimental data demonstrate the degradation of composite materials. A possible explanation is that the ability of the organic matrix to maintain the shape of composite materials is lost by irradiation, that is, the fracture mode of composite materials has been changed by irradiation. The change of fracture mode should be taken into account when the radiation effects on organic composite materials are studied.

There are other problems to be solved: (1) the scatter of literature data[2] and (2) the threshold dose in mechanical degradation.[3] The threshold dose in the radiation-induced degradation is also frequently obtained. The origin of the phenomena should be understood sufficiently in order to grasp the essentials of degradation induced in composite materials by irradiation.

Change of Fracture Mode in Flexural Tests

Flexural tests are discussed first because the tests have been often performed in the study of radiation damage in composite materials.

Figure 1 shows the distributions of bending and shear stresses in a specimen during flexural tests. To simplify the discussion the simple beam theory is adopted here, although the stress distribution is more complicated due to the stress concentration at the loading tip.[4] Bending fracture occurs when the bending stress reaches the bending strength, σ. Shear fracture takes place when the shear stress reaches the shear strength, τ. The shape of the specimen, span length, and the magnitude of σ and τ determine the fracture mode. The applied load, which causes the bending fracture, P_σ, and the shear fracture, P_τ, are defined as follows:

$$P_\sigma = 2bh^2\sigma/3L$$

$$P_\tau = 4bh\tau/3$$

Where b is the width, h is the height, and L is the span length, as presented in Fig. 1. When P_τ is larger than P_σ, bending fracture will occur and vice versa. The change of P_σ and P_τ with L/h is also shown in Fig. 1.

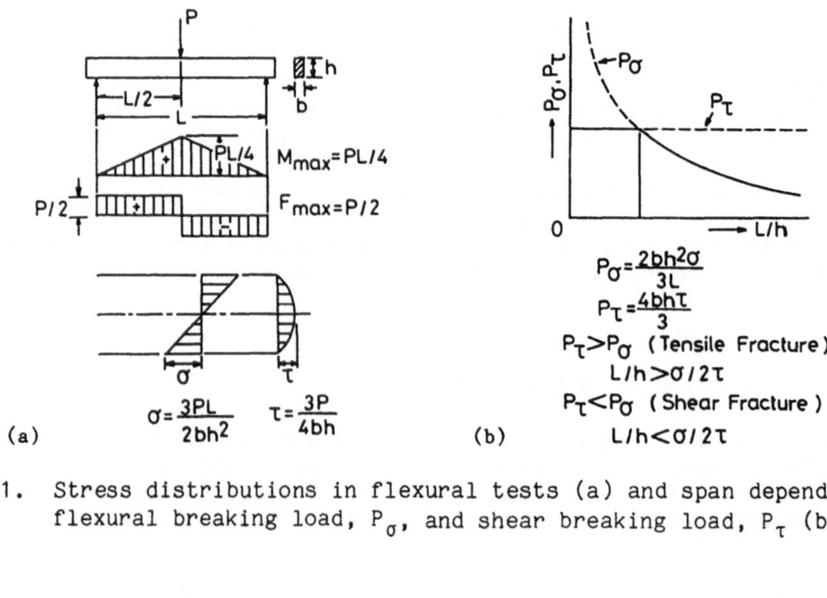

Fig. 1. Stress distributions in flexural tests (a) and span dependence of flexural breaking load, P_σ, and shear breaking load, P_τ (b).

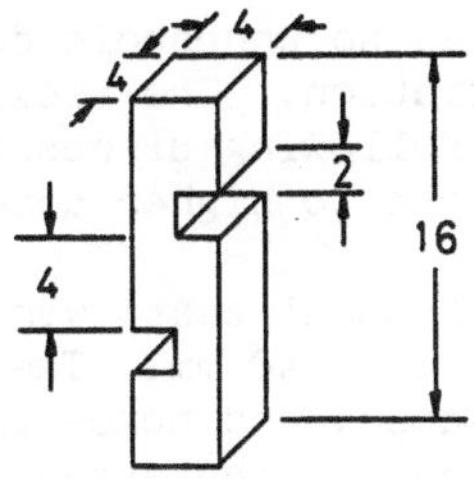

Fig. 2. Shape and size of the specimen. Dimensions are in mm.

When L/h is small enough, shear fracture occurs, and bending fracture takes place when the L/h is large. The short-beam shear test is performed at small L/h.

The degradation of σ and τ induced by irradiation is considered here. The σ is determined by the reinforcement, and hence the change of σ is thought to be small. On the other hand, the τ is thought to be degraded by irradiation because it reflects the characteristics of the matrix. The flexural tests ought to be performed at the area where the bending fracture occurs in order to obtain σ. The irradiation, however, would cause a decrease in τ, which results in shear fracture even at the area where the bending fracture occurred before the irradiation. The change of fracture mode should be taken into account in estimating the radiation damage.

An experiment was performed to estimate the degradation of the shear strength, τ. The shape and the size of the specimen are presented in Fig. 2. This size was determined considering the following:

(i) small enough to make the radiation experiment easy
(ii) large enough to get significant data
(iii) simplicity of handling the irradiated material

The shear strength was derived by compressing the specimen to induce shear fracture between the notches. The mean shear strength was used to make the discussion clear. The stress concentration is discussed later. The specimen was glass cloth reinforced epoxy. Lamiverre-A supplied by Nitto Electric Industry Co., Ltd. The weight fraction of reinforcement was approximately 62%. The irradiation was performed by Co-60 at RT. The interlaminar shear tests were also made at RT.

Figure 3 shows the degradation of interlaminar shear strength induced by the gamma irradiation. Five specimens were tested at each dose. The

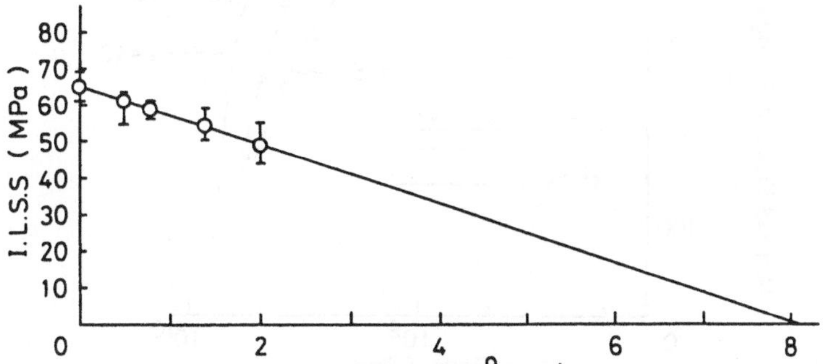

Fig. 3. Degradation of interlaminar shear strength induced by gamma irradiation.

shear strength decreased with dose, and there was no threshold dose. The 25% degradation occurred at 2.1 x 10^9 rad irradiation. The flexural test was discussed on the basis of these data. The following discussion uses the extrapolated change of interlaminar shear strength to higher dose.

The dimensions of specimens subjected to flexural tests were 4 mm x 4 mm x 40 mm. The span length was varied from 15 to 40 mm. The interlaminar shear strength, τ, decreased with each dose, and hence the fracture mode changed from bending to shear fracture. The bending strength calculated from the simple beam theory, in which the change of fracture mode was not considered, is presented in Fig. 4. The tensile strength (or bending strength), σ, was obtained from tensile tests. The nominal flexural strength demonstrated a threshold, as shown in Fig. 4. At the threshold dose, the change of fracture mode occurs. The threshold dose varied with the span length. The threshold and the scatter of literature data could be explained by the change of fracture model.

Change of Fracture Mode in Tensile Tests

The shear stress would not be present in the specimen if the specimen has identical longitudinal and transverse elastic moduli and constant width and thickness throughout the specimen. The specimen, however, almost always has a cut to define the initiation area of fracture. Because of this, shear fracture is induced at a high dose in the same manner as in the flexural test.

Figure 5 shows the shape of the specimens used in this study. The arced cuts are in the thickness direction. The shear stress in the specimen was calculated by the finite element method.[4] Shear fracture should occur when the maximum shear stress reaches the shear strength. Tensile fracture should take place when the tensile stress reaches the tensile strength.

The change of shear fracture load with the radius of arced cut is demonstrated in Fig. 6. An increase of radius causes tensile fracture and a decrease causes shear fracture. In the same manner as in a flexural test, the degradation of the interlaminar shear strength causes the change of fracture mode from tensile to shear.

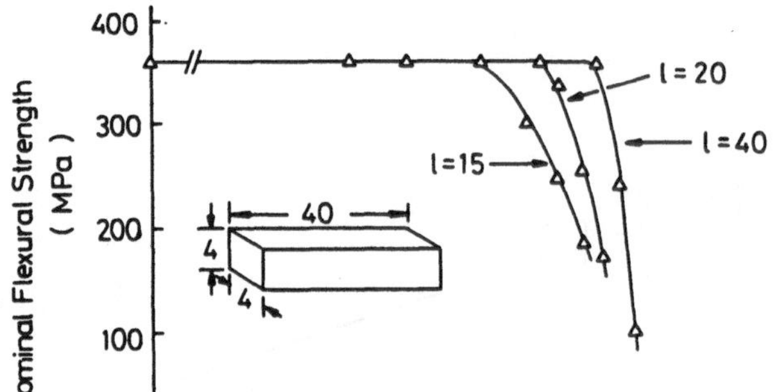

Fig. 4. Change of nominal flexural strength induced by gamma irradiation.

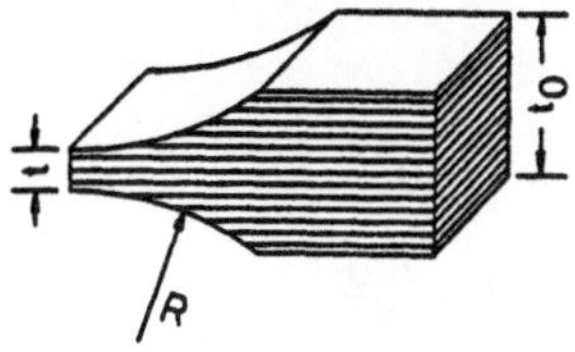

Fig. 5. Shape of the tensile specimen.

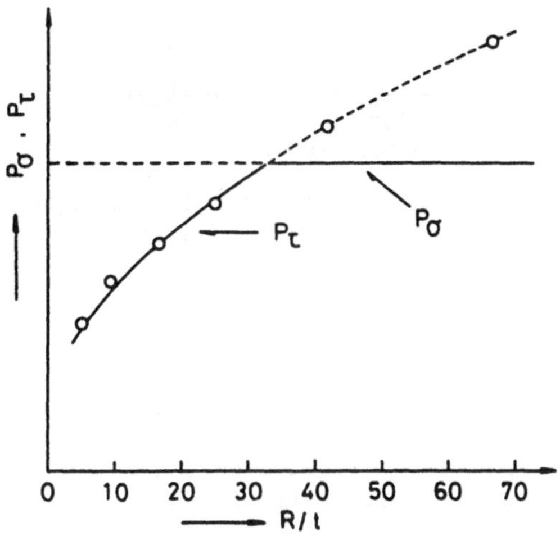

Fig. 6. Notch radius dependence of tensile fracture load, P_σ, and shear fracture load, P_τ.

Using Fig. 3, in which the change of fracture mode is not seen, the nominal tensile strength was evaluated in Fig. 7. The nominal tensile strength also shows the threshold dose and varies with the radius of the arced cut. Visual observation of the tensile specimen indicated that the irradiation increased the pull-out length of the fibers.[5,6] This coincides well with the conclusions of this study.

Interlaminar Shear Strength

The radiation induced degradation in flexural and tensile tests is essentially caused by the degradation of the interlaminar shear strength. The dose at which the change of fracture mode occurs can be estimated when the change of interlaminar shear strength and the system are given. Above this dose the nominal flexural and tensile strengths begin to decrease, and hence the dose can be considered the upper dose limit in the sense that the fracture mode remains the same under fixed experimental conditions.

Consequently, estimation of the interlaminar shear strength is important. There are various methods to evaluate the shear strength, and each method has its own problems.

The problems of the method employed in this study are pointed out; they are subjects for a future study.

The interlaminar shear strength obtained in this study was the mean shear strength without considering the stress distributions between the notches. The stress distribution between the notches is not uniform, as shown in Fig. 8. In this figure, the stress distributions were obtained by the finite element method. At the bottom of the notch, stress approximately 1.7 times the magnitude of the mean shear stress is induced. This stress is

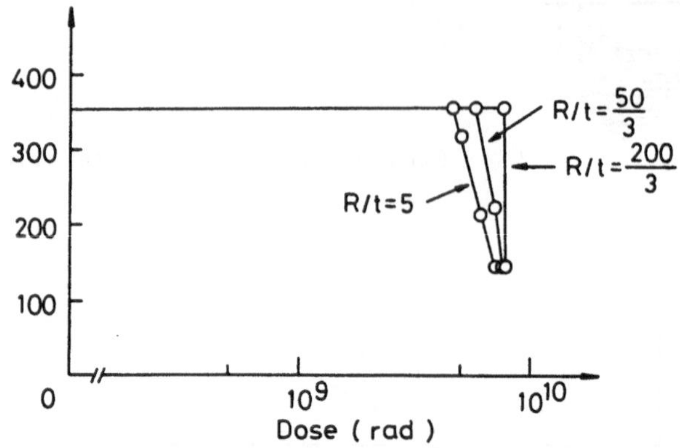

Fig. 7. Change of nominal tensile strength induced by gamma irradiation.

much larger than that estimated from shear lag theory.[7,8] The estimated interlaminar shear strengths in this study are not accurate values in this sense. Therefore, it is important to establish the methodology to estimate the interlaminar shear strength accurately with small specimens.

CONCLUSIONS

The degradation in flexural and tensile strength of organic composite materials is thought to be caused by the change of fracture mode from bending and tensile to shear fracture. This conclusion explains the experimental data and coincides with the visual observations.

The essential property that contributes to the degradation is thought to be the interlaminar shear strength. A method to estimate the interlaminar shear strength should be established.

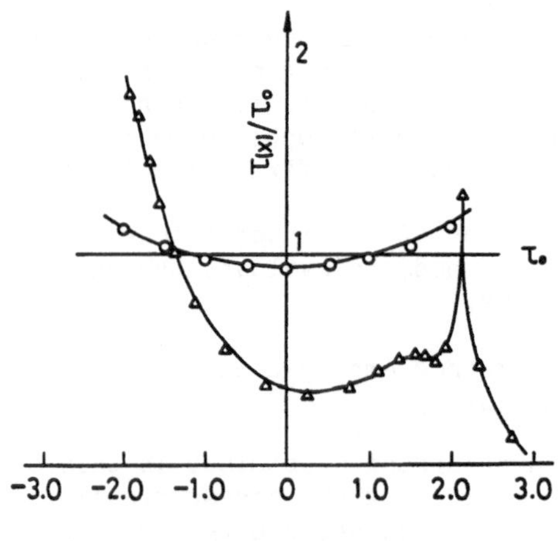

Fig. 8. Shear stress distribution in the specimen calculated by the finite element method and shear lag theory.

ACKNOWLEDGMENT

The authors would like to thank Prof. S. Namba at the University of Osaka Profecture for suggestions and discussions. They would like to thank Mr. N. Ikeda at ISIR for gamma irradiation. This work is partly supported by Grant in Aid for Scientific Research No. 60050044, Ministry of Education in Japan.

REFERENCES

1. R. H. Kernohan, C. J. Long, and R. R. Coltman, Jr., Cryogenic radiation effects on electric insulator, J. Nucl. Mater. 85&86:379-383 (1979).
2. S. Nishijima and T. Okada, Low temperature irradiation effects on mechanical properties of epoxy used in superconducting magnets, Cryogenics 18:215 (1978).
3. R. R. Coltman, Jr. and C. E. Klabunde, The strength of G-10CR and G-11CR epoxies after irradiation at 5K by gamma rays, J. Nucl. Mater. 113: 268-272 (1981).
4. M. Uemura, Problems and design standard for mechanical test of FRP, J. Japan Soc. Compos. Mater. 7:154-161 (1981) (in Japanese).
5. H. W. Weber, E. Kubasta, W. Steiner, H. Benz, and K. Nylund, Low temperature neutron and gamma irradiation of glass fiber reinforced epoxies, J. Nucl. Mater. 115:11-15 (1983).
6. S. Takamura and T. Kato, Mechanical properties of organic insulators for superconducting magnets after low temperature irradiation, in: "Advances in Cryogenic Engineering - Materials," vol. 30, Plenum Press, New York (1981), p. 41.
7. H. Becker, Problems of cryogenic interlaminar shear strength, in: "Advances in Cryogenic Engineering - Materials," vol. 30, Plenum Press, New York (1984), p. 33.
8. M. F. Markhan and D. Dawson, Composites 6:173-176 (1975).

SCREENING THE PERFORMANCE OF ORGANIC INSULATORS

UNDER CRYOGENIC NEUTRON IRRADIATION*

Maurice B. Kasen and Robert B. Stoddard

Fracture and Deformation Division
National Bureau of Standards
Boulder, Colorado

ABSTRACT

Specimens and test procedures are being developed for determining the significant parameters influencing resistance of organic insulators to neutron irradiation at 4 K. The specimens are 3.2-mm-diameter rods of exceptionally high quality produced by a method allowing a large number of experimental variables to be evaluated. Flexural and torsional shear tests perfomed with these specimens indicate that such tests will be useful in studying cryogenic neutron irradiation damage to the fiber-matrix interface. Results of 76-K tests on unreinforced and glass-fiber reinforced epoxy and polyimide materials are presented.

INTRODUCTION

A serious material shortcoming in building superconducting magnets for magnetic fusion energy (MFE) systems is believed to be the inability of conventional organic insulators to withstand the level of neutron irradiation to which they will be exposed in a functioning reactor. It is estimated that portions of the magnets must be exposed to a fluence level equivalent to 550 to 1000 MGy (MGy = 10^8 rads) if excessive shielding costs are to be avoided.[1] Although it is technically feasible to fabricate magnets with much more radiation resistant inorganic materials, the cost may be prohibitive.

A facility providing irradiation conditions approximating those in an MFE reactor has thus far been unavailable. Fortunately, this will be rectified with the coming on line of the National Low Temperature Neutron Irradiation Facility (NLTNIF) currently under construction at the Oak Ridge National Laboratory. This facility will provide the desired radiation spectrum at fluence levels that will allow the required experiments to be performed in a relatively short period of time. The comparatively large irradiation volume and associated test chamber will permit simultaneous irradiation of a large number of specimens as well as in-situ postirradiation testing at 4.2 K without warmup.

*Publication of NBS, not subject to copyright.

In the United States, the main research effort involves cooperative research between individuals at the National Bureau of Standards, the Oak Ridge National Laboratory, the Los Alamos National Laboratory, and the Massachusetts Institute of Technology. Within this group, NBS has the primary responsibility for providing well-characterized research materials, conducting pre- and postirradiation mechanical test programs, and determining the influence of irradiation on the material failure mode. Informal but active cooperation in this research has been established with individuals at Osaka University in Japan and at the Rutherford Appleton Laboratories in England.

EXPERIMENTAL PROGRAM

Our primary objective is to determine the significant parameters influencing the degradation of glass-fiber reinforced polymer insulators during neutron irradiation at 4.2 K. Since many parameters must be evaluated, it is necessary to establish an efficient, systematic screening program.

Specimen Selection and Preparation

Differential radiation heating of the specimen cross section, the need to minimize residual radioactivity, and the many variables that must be studied dictate the need for small, high-quality, well-characterized specimens. Surface defects introduced by machining should be eliminated. The same specimen configuration must be available in a variety of unreinforced (neat) resin specimens and as specimens reinforced with any desired type and amount of glass fiber. The specimen should lend itself to testing in a variety of modes. There is no commercial source of materials meeting these requirements. We have therefore developed in-house capability of producing the required specimens.

We have selected the rod-shaped specimens illustrated in Fig. 1(a) as best meeting these requirements. Glass-fiber bundles are vacuum impregnated with well-degassed resin, pulled into a 300-mm long, 3.2-mm i.d. smooth-bore TFE mold and cured under approximately 420-kPa nitrogen gas pressure. Neat

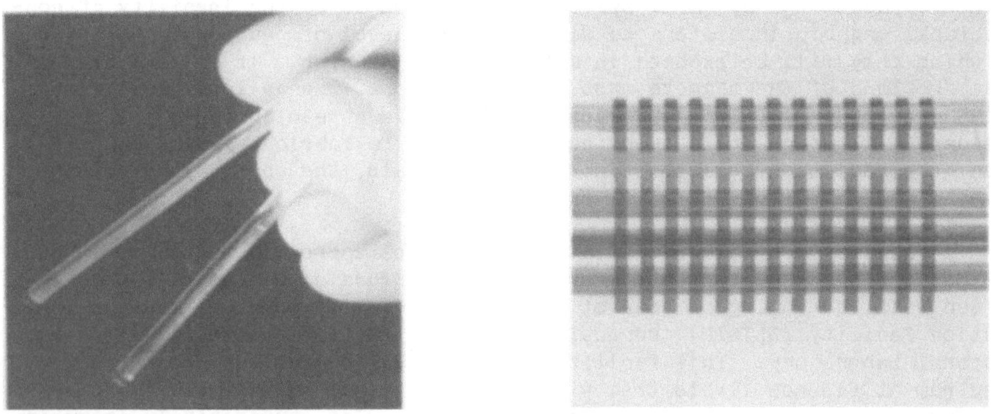

(a) (b)

Fig. 1. (a) Typical 3.2-mm-diameter rod specimens. (b) An indication of the rod quality as revealed by its transparency. Top to bottom in (b) are neat polyimide, reinforced G-11CR, neat G-11CR, reinforced Rutherford resin, and neat Rutherford resin. Reinforced specimens have 58 v/o glass content.

Table 1. Experimental Resin Materials

1)	Standard G-11CR epoxy system
2)	Standard G-11CR epoxy without BF_3MEA accelerator
3)	Rutherford bisphenol A epoxy impregnating resin
4)	Proprietary bismaleimide polyimide
5)	Bisphenol A epoxy with polyamide cure

resin specimens are produced by vacuum drawing degassed resin into the same molds. As illustrated in Fig. 1(b), specimen quality is very high. Fiber volume fractions can be calculated from knowledge of the tube dimensions and glass density. Resin chemistry, cure state, fiber chemistry, and fiber finish can be altered as desired for a given experiment. The specimen surface finish is excellent, and the coefficient of variability in diameter among rods produced with a given mold is on the order of 0.001 to 0.003, indicating excellent dimensional reproducibility. These specimens can be cut to length and directly tested.

Some polymer systems of interest require blending of a liquid resin with a solid cure agent. In industrial practice, this is facilitated by the addition of solvents, but this cannot be done with the NBS production method because evolution of the solvents would cause porosity during curing in the closed mold. We avoid this problem by the ultrasonic blending of such systems.[2]

This method was used to produce unreinforced and type E glass-fiber inforced specimens with the resin systems listed in Table 1. Materials 1 through 4 will be used in the initial radiation program to be conducted in the NLTNIF reactor at Oak Ridge. The G-11CR system (material 1) provides a baseline of material performance. However, this basic system contains BF_3MEA as an accelerator to shorten the cure cycle. Since boron has a high neutron capture cross section, this addition may be contributing to excessive degradation of the resin. We therefore produced a variant of the G-11CR formulation without this addition to evaluate this parameter (material 2). An impregnation resin (material 3) was included in response to suggestions made at the 1984 NBS/DOE workshop that such a system would be of value to magnet fabricators. The proprietary bismaleimide polyimide system (material 4) represents the state-of-the-art in advanced resin design for this type of material. The bisphenol A/polyamide system (material 5) is a simple system used during the development stage of the program and will not be included in the radiation study.

Test Method Development

Evidence suggests that the primary cause of degradation in the mechanical properties of reinforced insulators will be radiation-induced damage to the interface between the fiber and the polymer matrix.[3,4] We therefore concentrate on developing test methods that are sensitive to such damage. Primary candidates are short-beam shear and torsional shear tests.

The fixture used for short-beam shear tests is similar to a conventional three-point fixture for apparent interlaminar shear strength tests, differing primarily in contouring the supports and the loading nose to accommodate round specimens. At the proper span/depth ratio, this test produces an essentially mode-II type loading in which the shear stress is resolved parallel to the fiber direction. Required specimen length is only 15 mm for a span/depth ratio of 3.0, making it very economical of material. Typical load-time curves produced at 76 K and 295 K by this test method are shown in Fig. 2. The load drops appearing on the 76-K curve reflect the onset of consecutive shear failures with increasing stress. Continuous bending over of the 295 K curve is attributed to the high room temperature ductility of the bisphenol A/polyimide resin system.

A typical 76-K shear failure is illustrated in Fig. 3. Narrow shear bands initially progressed into the specimen in an inverted V configuration from the point of contact with the loading nose. These bands subsequently acted as nuclei for massive shear failures that propagated along the specimen length, giving rise to the observed load drops.

Although the flexural shear test appears potentially useful, the change in specimen shape during deformation introduces failure modes which are a combination of tension, compression and shear. This makes it difficult to quantify the results. An alternative is to conduct tests in torsion. Such a test in a uniaxially reinforced composite also produces a combined shear, tension and compression loading on the interface, but in this case the symmetry of deformation simplifies calculation of the various stress gradients and permits the shear component of the stress to be isolated.

The appearance of 76-K failures produced by applying dead-weight torsional loading to unreinforced and to glass-fiber reinforced specimens bonded into aluminum end caps is illustrated in Fig. 4(a-b). The unreinforced specimens developed typical 45° shear failures. In most cases, seven to ten longitudinal cracks were observed in the reinforced specimens. But some specimens burst open, revealing individual fiber bundles, indicating a lower integrity of the fiber-matrix interface. A typical load-deflection curve obtained in torsion at 76 K is shown in Fig. 5. Under dead weight loading, cracks are manifested as sudden jumps in displacement.

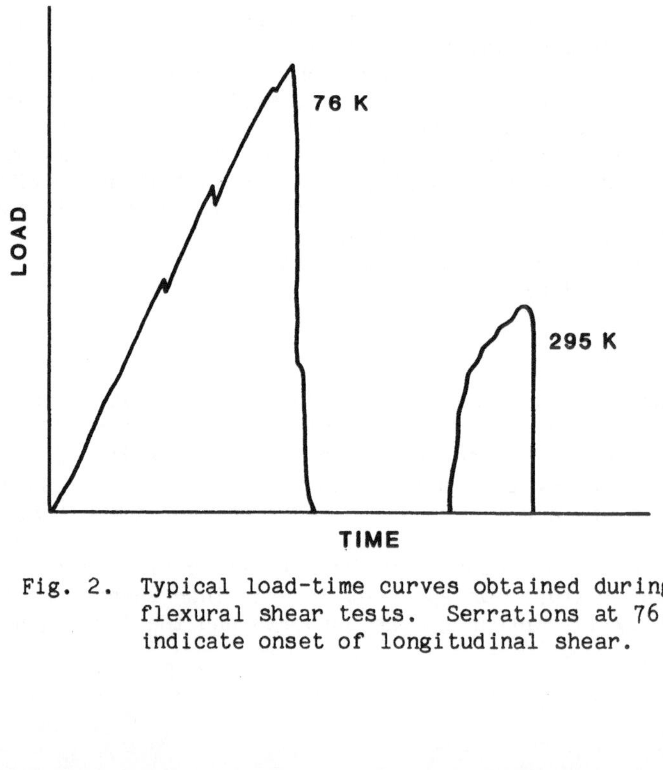

Fig. 2. Typical load-time curves obtained during flexural shear tests. Serrations at 76 K indicate onset of longitudinal shear.

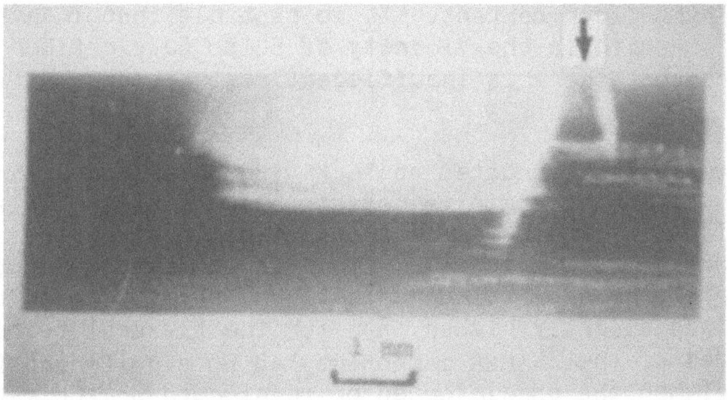

Fig. 3. Typical longitudinal shear failure at 76 K. Arrow indicates the point of contact of the loading nose. Shear bands have propagated into the specimen, forming an inverted V and acting as the nucleating point for massive longitudinal shear failure.

We investigated the feasibility of freezing the specimens into the end caps as an alternative to potting with epoxy. This would permit recovery of the undamaged parts of the specimens for additional testing by other methods. It would also allow specimens to be periodically removed from the end caps during a test run for quantitative analysis of damage accumulation by dynamic modulus and damping techniques. Although this approach was found to be feasible using several types of aqueous slurries, it was also found that specimens prepared in this way might develop progressive cracking of the bond between the specimen and the end cap from the gage section into the gripped portion as the torsional load increased. Some of the displacement jumps in Fig. 5 may reflect such cracking. This approach to the test method is undergoing further development.

Preliminary Test Results

A series of flexural shear tests was conducted with the bisphenol A/polyamide material (material 5) at 76 K. Five specimens each from two rods containing either 48 or 58 v/o type E glass were tested at each temperature at a span/depth ratio of 3.0. The data, summarized in Table 2, reflect the stress required to initiate the first massive longitudinal shear failure (first load drop). The low coefficients of variability indicate good consistency from specimen to specimen and between specimens from different rods. No significant difference in shear strength was noted between

(a) 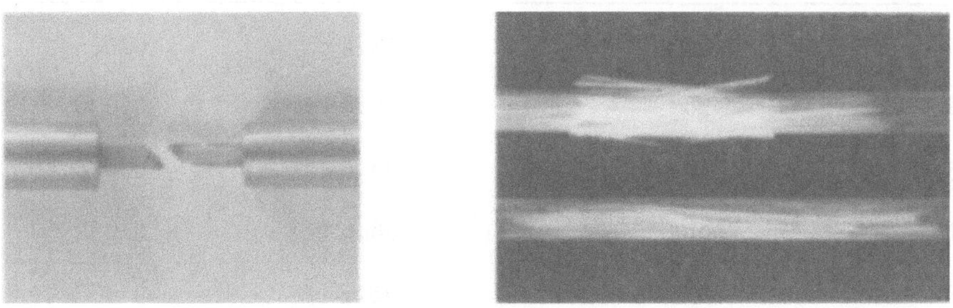 (b)

Fig. 4. (a) Typical 76-K torsional failure in an unreinforced resin specimen illustrating the method of holding in aluminum end caps. (b) Typical 76-K torsional failures in glass-reinforced Rutherford resin system (material 3, Table 1).

the rods of differing fiber content. It is probable that this reflects a maximum in shear strength in the vicinity of 50 to 60 v/o fiber content. At very high fiber content there is insufficient resin to ensure fiber separation.

Torsional tests were conducted at 76 K with materials 1 through 4 (Table 1). The ultimate strength values are summarized in Table 3. A somewhat greater variability was found in the neat resin strength as compared with that of the reinforced materials. This is expected owing to the low strain to failure of the embrittled resins. Variability among the reinforced materials was slightly higher than for the flexural tests, but still reasonable. Removal of the BF_3MEA component had no significant influence on the strength of either the unreinforced or reinforced G-11CR materials. The Rutherford resin developed the highest intrinsic strength at 76 K, but the lower strength of the reinforced product made with this resin, as compared with that of G-11CR, indicates a lower fiber-matrix bond integrity. The strength of the polyimide resin was about that of G-11CR.

These tests appear to be useful in characterizing cryogenic neutron irradiation damage to the various types of materials that will be evaluated during this program.

Table 2. Flexural Shear Data, Material 5, at 76 K (10 Specimens of Each Condition)

V/O Glass	Rod No.	Ultimate Strength (MPa)	CV
48	1	23.6	0.042
	2	23.0	0.087
		23.3	0.064
58	1	23.3	0.047
	2	23.6	0.013
		23.5	0.034

Table 3. Torsional Shear Data, 76 K (5 Specimens of Each Condition)

Material No.	V/O Glass	Ultimate Strength (MPa)	CV
1	58	158.4	0.086
2	58	160.8	0.036
3	58	148.6	0.078
1	none	96.6	0.099
2	none	93.3	0.253
3	none	120.8	0.099
4	none	94.7	0.190

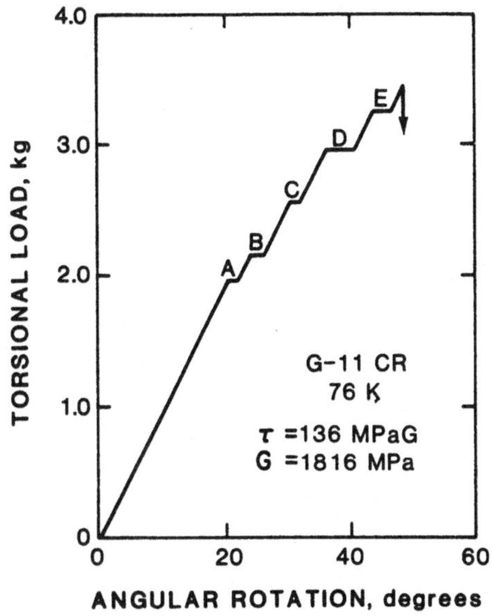

Fig. 5. Torsional load-deflection curve obtained at 76 K illustrating sequential interfacial crack development (A-E).

ACKNOWLEDGMENTS

This work is supported by the Office of Fusion Energy, U.S. Department of Energy, Dr. Marvin Cohen, Project Manager.

REFERENCES

1. G. F. Hurley and R. R. Coltman, Organic materials for fusion reactor applications, J. Nucl. Mater. 123 & 124:1327-1337 (1984).
2. W. T. Hodges and T. L. St. Clair, "Ultrasonic Mixing of Epoxy Curing Agents," Technical Memorandum 85643, NASA Langley Research Center, Hampton, Virginia (1983).
3. S. Egusa, M. A. Kirk, R. C. Birtcher, M. Hagiwara, and S. Kawanishi, Irradiation effects on the mechanical properties of composite organic insulators, Nucl. Phys. Methods Phys. B229:610-616 (1984).
4. A. Udagawa, S. Kawanishi, S. Egusa, and M. Hagiwara, Radiation induced debonding of matrix-interface in organic composite materials, J. Mat. Sci. Lett. 3:68-70 (1984).

EFFECT OF CRYOGENIC IRRADIATION ON THE MECHANICAL PROPERTIES OF ORGANIC INSULATOR FILMS

Hitoshi Yamaoka and Kiyomi Miyata

Research Reactor Institute
Kyoto University
Kumatori, Osaka, Japan

ABSTRACT

The object of this study is to acquire data on radiation damage of organic insulator films at low temperature. The specimens used were thin films of polyphenylene sulfide (PPS), polyarylether etherketone (PEEK), polyethylene terephthalate (PET), and polypyromellitimide (PPMI, Kapton H). Experiments were mainly done with PPS, since at the present there is little information on the cryogenic properties and radiation effects for this material. Irradiation of the films was performed at 20 K in the low-temperature irradiation facility of the Kyoto University Reactor and the mechanical properties of the films were measured at 77 K. In tensile tests of PPS films, both the unirradiated and irradiated specimens showed a gradual increase in strength beyond the elastic limit at all test temperatures. The tensile strength and the ultimate elongation of PPS films were almost independent of irradiation dose up to 8 MGy at 20 K. For comparison, similar mechanical tests were carried out with irradiated films of PEEK, PET, and PPMI. As far as the present experiments are concerned, the radiation tolerance of PPS films has been proved to be on the same level as that of PEEK and PPMI films.

INTRODUCTION

Fusion reactors based on magnetic confinement, such as Tokamak and mirror machines, employ a variety of superconducting magnets which are operated at extremely low temperatures. Recent studies on radiation effects in fusion magnet materials have shown that the maximum radiation level to the magnet will be on the order of 10^8 Gy at the end of a plant life of 20 or 30 years.[1,2]

A number of studies have been carried out in recent years on radiation damage of filamentary-reinforced resins at low temperature. However, few reports on the cryogenic irradiation of polymer films have been published. This is probably due to the fact that most of conventional polymers can hardly withstand the severe radiation environment for a long operation time of fusion reactors.

Recently several aromatic polymers have been developed as heat-resistant materials in the field of space engineering. Judging from the chemical structures, these polymers are expected to be highly stable to ionizing radiation. Polyphenylene sulfide (PPS) and polyarylether etherketone (PEEK), newly developed engineering plastics, are also known as radiation-resistant polymers at ambient temperature, but no experimental results on the radiation tolerance at low temperature have been reported so far.

For the purpose of searching the suitable polymers for organic insulators of the fusion magnet, the cryogenic properties of PPS film and the effect of low temperature irradiation on the mechanical properties of PPS and PEEK films have been studied and compared with the results obtained for irradiated polyethylene terephthalate (PET) and polypyromellitimide (PPMI, Kapton H) films.

EXPERIMENT

The specimens were commercially available polymer films of PPS, PEEK, PET, and PPMI. The chemical structures of these polymers are shown in Fig. 1. Sizes of the specimens in mm were 5 x 90, and thickness of the films was 100 μm for PEEK and 50 μm for PET, PPS, and PPMI.

The tensile properties of unirradiated PPS film were measured at 300, 173, and 77 K with a Shimadzu Autograph test machine, Model AG-500, using 5 mm-wide strips and 50 mm in gauge length. The crosshead speed was varied in the range from 0.02 to 0.6 mm/mm·min.

Irradiation of the films was performed in the low-temperature irradiation facility of Kyoto University Reactor, the average temperature of which was about 20 K under reactor operation at 5 MW. The highest γ-ray dose rate, fast ($E > 0.1$ MeV) and thermal meutron fluences of the facility are 1.2×10^5 Gy/h, 2.5×10^{15} n/m^2/s and 2.3×10^{16} n/m^2/s, respectively. After irradiation, the specimens were gradually warmed up to room temperature and stored in a dry atmosphere for a few weeks.

The tensile tests of the irradiated films were carried out at 77 K with a constant crosshead speed of 0.05 mm/mm·min. Average values of the tensile properties for the unirradiated and the irradiated films were based on tests of 5 specimens.

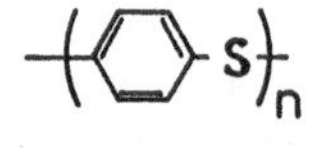

Polyphenylene sulfide (PPS)

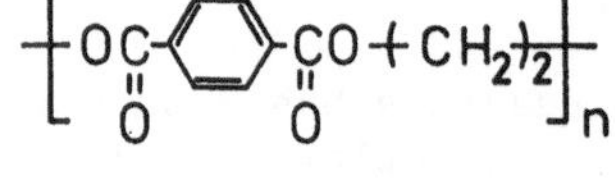

Polyethylene terephthalate (PET)

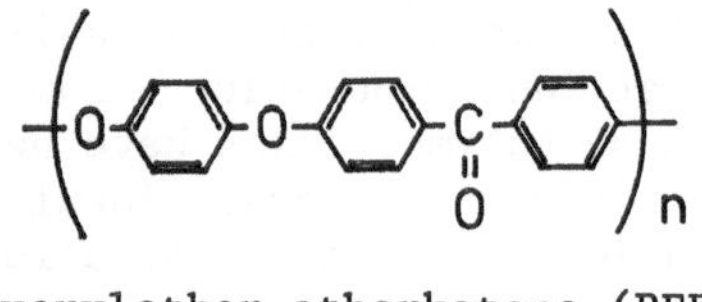

Polyarylether etherketone (PEEK)

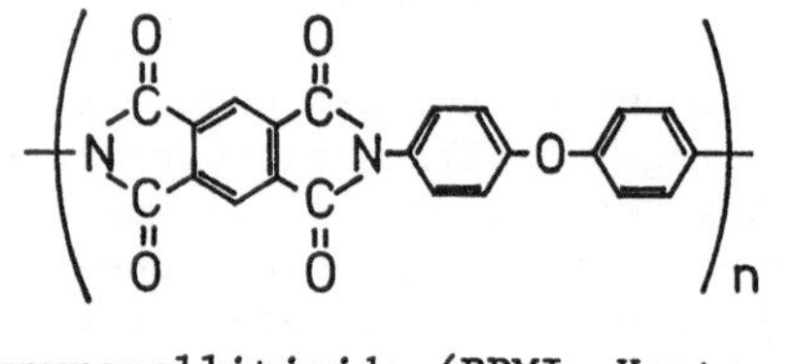

Polypyromellitimide (PPMI, Kapton H)

Fig. 1. Chemical structures of the polymer films.

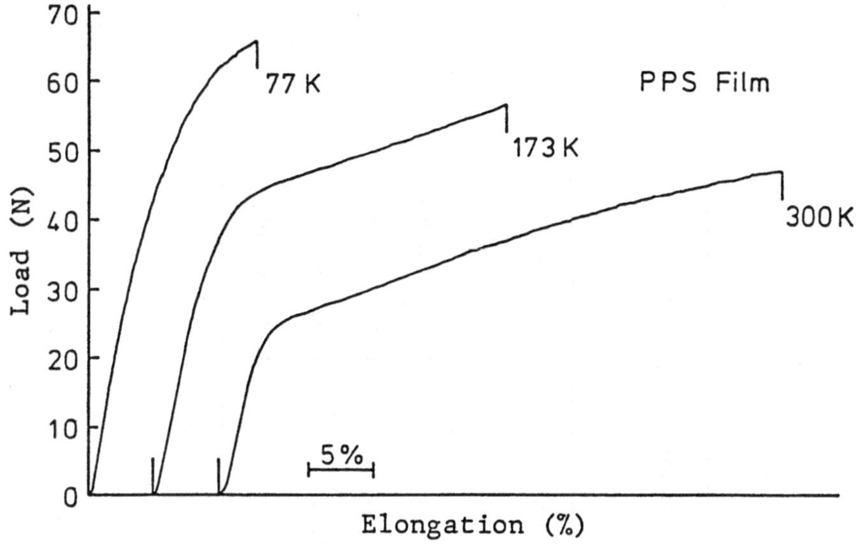

Fig. 2. Stress-strain curves of PPS film at different temperatures.

RESULTS AND DISCUSSION

Mechanical Properties of Unirradiated PPS Film

The tensile experiments on unirradiated PPS film were first carried out because there was little information about the cryogenic properties of this material.

Fig. 2 shows the stress-strain curves of PPS film at different temperatures as measured at a constant rate of strain (0.05 mm/mm·min). The curve at 300 K is typical of hard ductile polymers and shows a gradual increase in strength beyond the elastic limit. This feature remained unchanged with decreasing temperature, indicating that no transition region from ductile to brittle exists between 300 K and 77 K. Table 1 shows the

Table 1. Mechanical Properties of PPS Films at Different Temperatures

Temperature (K)	Tensile Modulus (GPa)	Tensile Strength (MPa)	Ultimate Elongation (%)	Energy (J)
300	3.80 (0.06)	188 (3)	38.2 (2.5)	0.67 (0.05)
173	4.21 (0.17)	215 (12)	22.8 (2.5)	0.47 (0.08)
77	4.82 (0.14)	260 (4)	12.2 (0.7)	0.27 (0.02)

( ) : Standard Deviation

Table 2. Strain Rate Dependence of Mechanical Properties of PPS Films in Tensile Test at 77 K

Strain Rate (mm/mm min)	Tensile Modulus (GPa)	Tensile Strength (MPa)	Ultimate Elongation (%)	Energy (J)
0.02	4.77 (0.20)	252 (5)	10.8 (0.6)	0.23 (0.02)
0.05	4.82 (0.14)	260 (4)	12.2 (0.7)	0.27 (0.02)
0.1	4.87 (0.12)	264 (6)	12.4 (1.9)	0.29 (0.06)
0.2	4.90 (0.17)	256 (19)	11.8 (3.5)	0.27 (0.12)
0.4	4.86 (0.25)	259 (17)	11.9 (2.7)	0.27 (0.10)
0.6	4.80 (0.15)	262 (11)	11.8 (1.1)	0.26 (0.04)

( ) : Standard Deviation

Strain Rate (mm/mm min) = Test Speed (mm/min) / Gauge Length (mm)

mechanical properties of PPS film at three different temperatures. Both the tensile modulus and the ultimate tensile strength gradually increase with a decrease in temperature. The ultimate elongation remarkably decreases with decreasing temperature, but it appears that PPS retains some ductility even at 77 K.

The effects of strain rate on the mechanical properties of PPS film measured at 77 K are shown in Table 2. The tensile strength is almost constant despite the large variation of strain rate. Other properties such as the tensile modulus and the ultimate elongation are also practically independent of the variation in strain rate. This means that PPS film is capable of deforming in a predictable manner at 77 K under high mechanical loads.

These results indicate that PPS film possesses excellent mechanical properties at low temperature.

Cryogenic Irradiation of Organic Films

PET film is often used at cryogenic temperature for the electrical insulation or for thermal superinsulation in a metallized form. However, the radiation tolerance of PET at low temperature is qualitatively known to be rather poor. Takamura and Kato[3] reported that PET film was too brittle to handle after irradiation of 6.2×10^6 Gy at 5 K. In Table 3 is summarized the results of mechanical tests on the irradiated PET films obtained in the present study, where the absorbed doses due to fast neutrons were estimated under the assumption of 1 Gy = 10^{15} n/m^2 (E > 0.1 MeV). Although the tensile modulus was little affected by a dose of 8×10^6 Gy, the tensile strength decreased with increasing absorbed dose above 2×10^6 Gy and was reduced by 50% after irradiation to about 6.8×10^6 Gy. The ultimate elongation also remarkably decreased with increasing dose and fell to half of the original elongation at approximately 2.7×10^6 Gy. These results mean that the use of PET film at low temperature should be rather limited in radiation environments, confirming the qualitative reports in previous papers.[3-5]

Table 3. Effects of Reactor Irradiation at 20 K on the Mechanical Properties of PET Films

Dose (MGy)	Tensile Modulus GPa)	Tensile Strength (MPa)	Ultimate Elongation (%)
0	8.29 (0.25)	285 (5)	8.6 (1.8)
1.8	8.53 (0.29)	284 (4)	4.9 (0.3)
4.9	8.65 (0.23)	197 (16)	2.6 (0.4)
5.5	8.07 (0.20)	185 (38)	2.5 (0.6)
8.0	8.16 (0.13)	107 (13)	1.4 (0.2)

( ) : Standard Deviation

As to the effects of cryogenic irradiation on PPMI film, Takamura and Kato[3] reported that the breaking stress of PPMI did not appreciably change after irradiation to 1.2×10^7 Gy at 5 K. Coltman and his co-workers[6,7] also confirmed that PPMI film was still useful for cryogenic systems at least up to a dose of 1×10^8 Gy. In the present study, the tensile modulus, the tensile strength, and the ultimate elongation of irradiated PPMI films were not significantly altered by irradiation to 8×10^6 Gy at 20 K, indicating that the radiation stability of PPMI film is extremely excellent at cryogenic conditions.

The results of tensile tests on irradiated PPS and PEEK films are shown in Fig. 3. Both the tensile strength and the ultimate elongation of PPS were essentially independent of the absorbed dose up to 8×10^6 Gy. The tensile modulus of PPS film was found to be about 4.8 GPa and also to remain unchanged during the course of irradiation up to 8×10^6 Gy. The high radiation tolerance, the sufficient strength, and the adequate elongation of this film at low temperature suggest that PPS is useful as an insulator for superconducting magnets in fusion reactors. Similar results were also obtained in the case of PEEK film. The tensile strength and the ultimate elongation of PEEK were constant up to a maximum dose at 8×10^6 Gy as seen in Fig. 3. The tensile modulus of this film was found to be about 3.5 GPa and to be constant during the irradiation. However, the film of PEEK at low temperature completely lost its flexibity even in an unirradiated state and crushed into pieces during tensile testing at 77 K. This means that PEEK film has almost no utility as an insulator under cryogenic conditions.

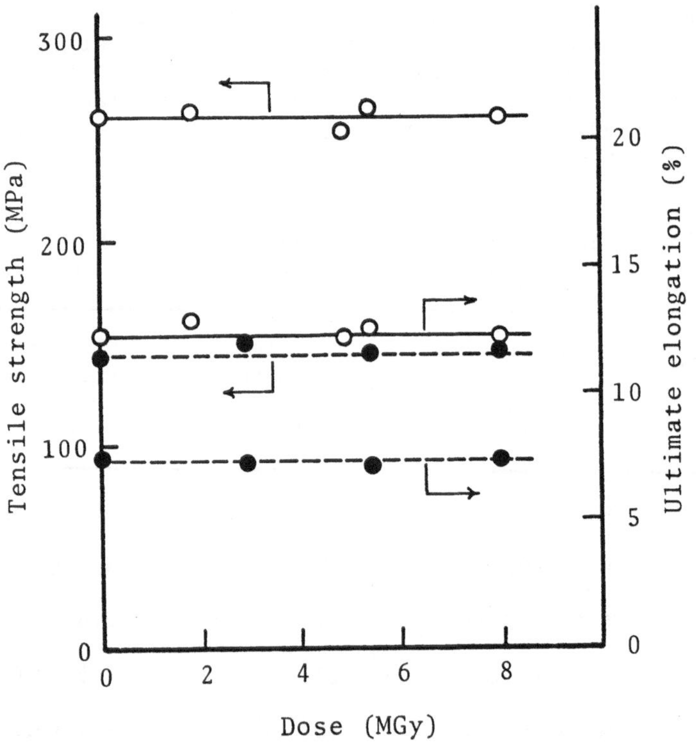

Fig. 3. Effect of reactor irradiation at 20 K on mechanical properties of PPS (O) and PEEK (●) films. (Test at 77 K after warm up to room temperature.)

CONCLUSION

In tensile tests of unirradiated PPS film at 77 K, a stress-strain curve typical of hard, ductile polymers was observed, and several mechanical properties were found to be almost constant despite a large variation in strain rate. These results suggest that PPS film is a useful material for cryogenic applications.

During reactor irradiation at 20 K, PET film showed embrittlement with an increase in absorbed dose, indicating a limited utility in radiation environments at low temperature. On the contrary, PPS and PEEK films were found to be highly radiation resistant. No significant change in mechanical properties of either film was observed during the course of irradiation up to 8 x 10^6 Gy at 20 K. In the case of PEEK film, however, its practical use at cryogenic conditions would be extremely restricted because of its brittleness at low temperature.

The present study shows that the radiation tolerance and the mechanical properties of PPS film in cryogenic environments are on the same level as those of PPMI film. This means that PPS is one of the candidate materials as an insulator for superconducting magnets in fusion reactors.

ACKNOWLEDGEMENTS

The authors express their thanks to Toray Industries, Inc. and to Mitsui Tohatsu Chemical Co., Ltd. for supplying the sample films. This work was partly supported by Grant in Aid for Scientific Research No. 59050035, Ministry of Education in Japan.

REFERENCES

1. M. A. Abdou, Radiation considerations for superconducting fusion magnets, J. Nucl. Mater. 72: 147-167 (1978).
2. B. S. Brown, Radiation effects in superconducting fusion magnet materials, J. Nucl. Mater. 97: 1-14 (1981).
3. S. Takamura and T. Kato, Effect of low temperature irradiation on insulators and other materials for superconducting magnets, J. Nucl. Mater. 103/104: 729-734 (1981).
4. M. H. Van de Voorde, Radiation resistance of organic materials at cryo-temperatures, IEEE Trans. Nucl. Sci. 20: 784-785 (1973).
5. D. Evans and J. T. Morgan, A review of the effects of ionizing radiation on plastic materials at low temperatures, in: "Advances in Cryogenic Engineering - Materials," vol. 28, Plenum Press, New York (1982), pp. 147-164.
6. R. R. Coltman, Jr., C. E. Klabunde, R. H. Kernohan, and C. J. Long, "Radiation Effects on Organic Insulators for Superconducting Magnets," ORNL/TM-7077, Oak Ridge National Laboratory, Oak Ridge, Tennessee (1979).
7. R. R. Coltman, Jr. and C. E. Klabunde, Mechanical strength of low-temperature-irradiated polyimides: A five-to-tenfold improvement in dose-resistance over epoxies, J. Nucl. Mater. 103/104: 717-722 (1981).

FRACTURE PROPERTIES OF POLYMERS AND COMPOSITES AT CRYOGENIC TEMPERATURES

G. Hartwig, B. Kneifel and K. Pöhlmann

Kernforschungszentrum Karlsruhe
Institute of Material and Solid State Research, IMF IV
Karlsruhe, West Germany

ABSTRACT

At low temperatures many properties are rather independent of polymer structure. This is true for the static fracture strength σ_{UT}, but not for the fracture toughness K_{Ic}, which is directly correlated to σ_{UT}. The reason is that dynamic processes with adiabatic heating and plastification of the crack tips (arrest lines) are involved in the measurement of K_{Ic} and the energy of fracture γ. The amount of plastification therefore depends on properties at higher temperatures which are different for most polymers. Consequently the question arises whether or not adiabatic heating occurs in a similar manner for σ_{UT} measured at high deformation rates. It has been found that polymers become more ductile if a certain deformation rate is exceeded. Thus, brittleness of polymers is not increased under impact loading.

For composites one important type of load is bending. The great influence of matrix properties on the interlaminar shear strength has been demonstrated for a brittle epoxy resin and polycarbonate PC, which is ductile even at 4.2K. For PC no interlaminar shear failure occured until fibre fracture by bending.

INTRODUCTION

Polymeric fibre composites are attractive alternatives or necessary supplements to metals. Their electrical insulation capability is necessary for avoiding eddy currents in pulsed superconducting magnet systems, e.g., in superconducting generators or in Tokamak fusion reactors. Their further advantage lies in the high fatigue endurance limit, expecially for carbon-fibre composites. For many low cryogenic applications their high thermal insulation capability is necessary.[1]

The disadvantage of fibre composites is the polymeric matrix, which results in low transverse and shear strengths. At low temperatures most polymers are very brittle, leading to specific fracture modes or degradation of properties. A few thermoplastic polymers show some cryogenic ductility, which is attributed to the high toughness of crazes. Matrices with cryogenic ductility are able to balance (thermal) stress concentrations in composites under load without matrix failure prior to fibre breakage.

Brittleness of polymers is usually increased by fast deformation rates or impact load. However, there are processes that reduce cryogenic brittleness even under fast deformations.

A further problem arises from the fibre-matrix bond, especially for carbon- and Kevlar fibre composites. Many fracture modes under static and fatigue loading or thermal cycling arise from debonding or delamination. Improvements of these interfacial properties are necessary for increasing transverse and shear strengths.

These investigations are intended to improve and enable more efficient use of the expensive composite materials and to reduce failure modes arising from matrix and fibre-matrix bonds.

CRYOGENIC FRACTURE MODES OF POLYMERS

Ductility

Most polymers are brittle at very low temperatures and show a linear elastic stress-strain behavior. The fracture strain at 4.2 K is $\varepsilon_F \approx 2\%$ or lower. There are a few thermoplastic polymers, such as polycarbonate (PC), polysulfone (PSU) or polyethersulfone (PES), which possess some cryogenic ductility: the static fracture strain at 4.2 K is between 3 and 3.5% or 5 and 9% at 77 K (see Table 1). The stress-strain diagram is nonlinear (see Fig. 1). It is very probable that one reason for this behavior arises from crazes existing in those polymers. The morphology of crazes is shown in Fig. 2. The tips of crazes are rather tough because of fibrils, which are loosely packed and aligned almost in stress direction.

The flexibility of those isolated fibrils is large because of less steric hindrance. It was shown in ref. 2 that the extensibility of craze-fibrils up to fracture is 100% or more; the concentration of crazes is

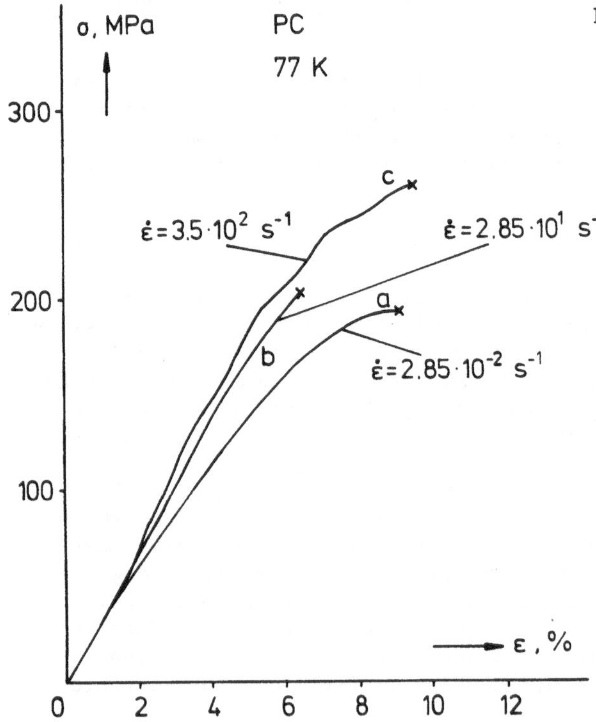

Fig. 1. Stress-strain diagrams at various strain rates $\dot{\varepsilon}$
a) quasi static deformation
b) more brittle behavior at medium $\dot{\varepsilon}$
c) ductile behavior at high $\dot{\varepsilon}$ (vibrations from the impact bar are superimposed)

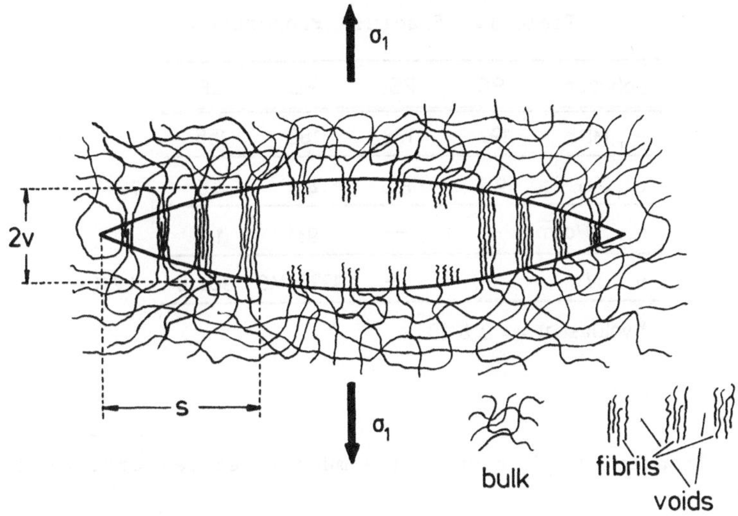

Fig. 2 Morphology of crazes. $\rho_{craze} \approx 0.6\ \rho_{bulk}$

about 5 vol %. Those polymers used as a matrix for fibre composites offer the great advantage of balancing residual and thermal residual strains when cooled.

Fracture properties

The static fracture strength of polymers at cryogenic temperatures is within a range of 100 to 200 MPa. Two important parameters for describing fracture properties are the stress intensity factor, K_{Ic} (fracture toughness) and the fracture energy γ.

Measurement of K_{Ic} and γ on HDPE and EP at 4.2 K and 77 K indicated that those parameters are not restricted to a small range but are very different for different polymers.[3,4]

Values of K_{Ic} and γ at 4.2 K for HDPE are higher by a factor of 4 and 9, respectively than those for EP (see Table 1). This is surprising since the fracture strengths of HDPE and EP are nearly equal.

The reason for different K_{Ic} or γ values is different initial conditions in measuring crack propagation. K_{Ic} was measured on a CT-specimen with an initial crack, a_o, which was generated at the temperature of measurement. At low temperatures cracks propagate uncontrolled with a velocity about 1/3 that of sound. At those velocities heating at crack tips occurs nearly adiabatically, which leads to a temperature rise, ΔT.

Now, the question arises whether or not ΔT is large enough to drive the material in the tip zones into a temperature range where the polymer is plastic. If this is the case, then the stress concentrations are reduced to some extent by relaxation processes in the plastic tip zone. If the crack stops at some remaining stress, this favorable condition is frozen, thus creating arrest lines which are the starting position for measuring K_{Ic}. If the arrest lines have a large radius, then a large value of K_{Ic} will result. As seen from Fig. 3 the arrest lines of HDPE are much larger than those of EP; this is reflected in the values of K_{Ic}.

A brittle-plastic transition is indicated by maxima of relaxation spectra (damping spectra). These measurements[5] show that a partial main

Table 1. Fracture Properties

polymer	PC	PSU	PE	EP	
σ_{UT}, MPa *)	192	178	160	156	77 K
ε_{UT}, % *)	9.2	8.5	4.0	4.5	
K_{Ic}, MPa$\sqrt{m}$	—	—	9±1	1.9	
γ, J/m^2	—	—	1370±140	90±10	4.2 K

*) values at $\dot{\varepsilon} = 2.8 \cdot 10^{-2}\,s^{-1}$

chain flexibilisation of PE occurs at a much lower temperature than that of EP.

For EP, temperature at the crack tip was not large enough to reach this transition temperature, and thus the arrest line remained small. A similar situation holds for the fracture energy, γ, which is consumed by creating a new fracture area, dA, and γ can be expressed as follows:

$$\gamma = \frac{dE_o}{dA} + \frac{dE_{pl}}{dA} + \frac{dE_{kin}}{dA} \tag{1}$$

E_o = surface energy

E_{kin} = kinetic energy

E_{pl} = plastification energy

For polymers whose temperature at the propagating crack tip is above the brittle-plastic transition temperature, the plastification energy is rather large, resulting in a large γ. The occurance of an adiabatic transition depends on the crack velocity, v, and the thermal diffusivity, a.

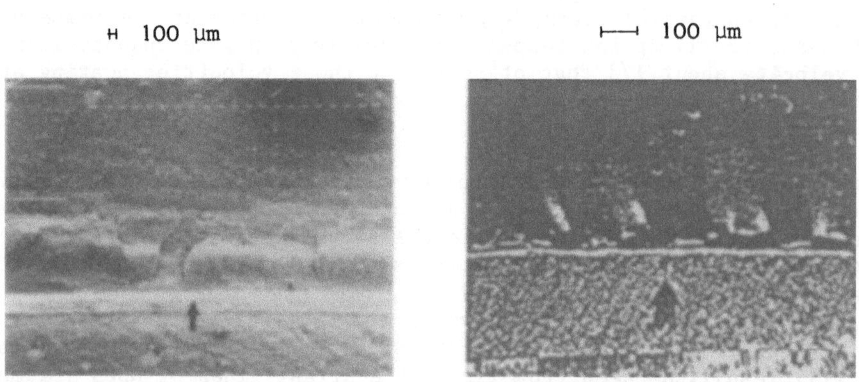

(a) (b)

Fig. 3. Arrest lines of a) PE and b) EP at 77 K.

$$a = \frac{\lambda}{\rho \cdot c}$$

λ : thermal conductivity
ρ : density
c : specific heat

The temperature rise is a function of c, v, and γ, where γ is a measure of the heat source: $\Delta T = \Delta T(a, c, v, \gamma)$. The brittle-plastic transition is a function of ΔT and T_i, where T_i is the temperature of relaxation maximum which indicates flexibility of main chain segments at a frequency related to the time scale of v.

K_{Ic} and γ are determined by the processes discussed above. The static value of the tensile fracture strength is controlled by different types of processes. Microcracks, crazes or voids exist in every polymeric specimen; they are the precursors of macroscopic cracks. Under static loading at cryogenic temperatures no plastic zones are developed, which could be the reason for the different behavior at fracture. But if the previous conclusions are correct, then at high tensile deformation rates, the crack tips should heat up in a similar manner. Different temperature rises and specific material brittle-plastic transition temperatures should change the tensile fracture behavior differently for different polymers. At very low deformation rates some relaxation may occur at existing microcracks. At larger $\dot{\varepsilon}$, the deformation time is too short for relaxation and the material becomes more brittle.

At a faster deformation, crack tips are heated and at very high more heat is generated locally than is diffused, and an adiabatic transition occurs. When $\dot{\varepsilon}$ increases to very high values, this leads to a brittle-plastic transition at the tips, depending on the ΔT compared to T_i.

The material becomes more ductile when the tip becomes plastic. This is shown in Figs. 1 and 4 for various materials with different T_i. PE has the lowest T_i and a rise in fracture strain and stress starts at a lower $\dot{\varepsilon}$ than for EP, which has a higher T_i. The experimentally attainable tensile deformation rates are not high enough in comparison to the velocities acting when measuring K_{Ic} and γ. But the tendency is evident, and by extrapolation one can speculate that at comparable deformation velocities the fracture stresses of PE and EP would differ by a similar factor, as do K_{Ic} and γ. This result indicates that at cryogenic temperatures the brittleness of polymers is not increased much by faster deformations and that at very high deformation rates (impact) the polymers become tougher.

THEORY

The temperature rise, ΔT, is a function of the thermal conductivity, λ; the specific heat, c; the deformation rate, $\dot{\varepsilon}$, and the source of heat generation at the crack tip.

The heat source might be the internal friction, which can be expressed at low strains, ε_T, by the mechanical loss factor, $\tan\delta$.

For the adiabatic case it holds:

$$\Delta T = 0.54 \tan\delta \cdot E' \cdot \varepsilon_T^2 (c \cdot \rho)^{-1} \qquad (2)$$

ε_T: Strain at the crack tip
E': Young's modulus
ρ : Density

This yields a lower limit, since one is confronted with large tensile deformations up to the fracture strain at the tip.

For a more realistic estimation, the fracture energy, γ (energy release rate), is taken as heat source. γ includes all terms of energy necessary for the processes before and during crack formation up to macroscopic fracture. The approximation is made that even if the crack is not yet propagating, about 60 % of γ is needed for plastification of the crack tip and thus converted into heat;[6] this yields an upper limit for ΔT.

Under these assumptions one gets for the adiabatic case:

$$\Delta T_{ad} \approx 0.6 \cdot \frac{\gamma}{\rho \cdot c \cdot d} = 0.6 \, \frac{K_{Ic}^2}{\rho \cdot c \cdot d \cdot E'} \qquad (3)$$

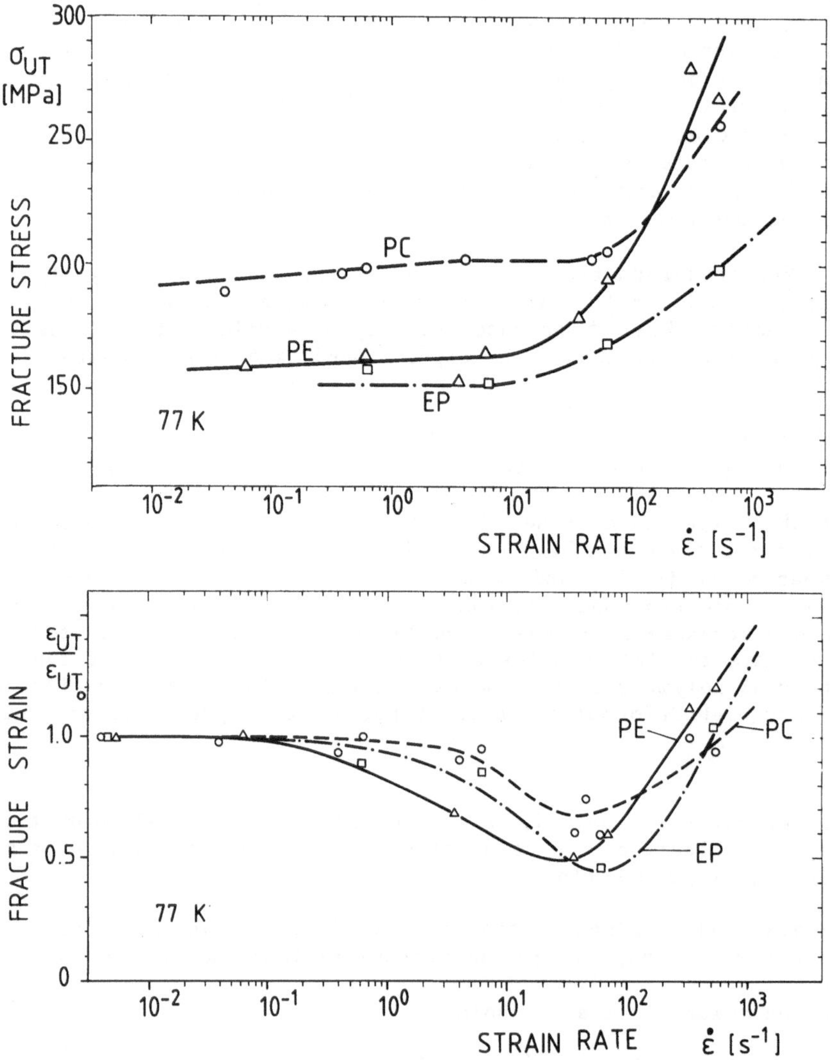

Fig. 4 Fracture stress and strain versus strain rate, $\dot{\varepsilon}$.
The fracture strain, ε_{UT}, is normalized to the static value, ε_{UT_0}.

d = diameter of the crack tip zone, calculated by the Dugdale- or Irwin model[7,8]

Assuming a tip diameter calculated by the Dugdale model, one gets for PE and EP at 77 K:

$$\Delta T_{ad} \approx 210 \text{ K}.$$

(γ and d are counter-acting parameters)

At the strain rates of this investigation, no real adiabatic case is established. Detailed calculations yield:

$$\Delta T \approx 0.6 \text{ to } 0.8 \cdot \Delta T_{ad} \quad \text{for } \dot{\varepsilon} = 0.35 \cdot 10^3 \text{sec}^{-1}$$

This is the order of magnitude for driving the material of the crack tip near or above the brittle-plastic transition temperature, T_i ($T_i \approx 170$ K for PE and $T_i \approx 250$ K for EP).

EXPERIMENTAL METHODS

K_{Ic} and γ were measured with a tensile machine, which was made very stiff by means of a lever arm. Thus, no energy stored in the machine falsifies the energy balance of γ. The specimens used were compact tension specimens.

The tensile fracture properties at high strain rates were measured with a machine having a crosshead speed of up to 10 m/s. Having reached the desired speed, an impact bar was connected with one clamp of the specimen by means of a push rod. All parts were made very stiff in order to minimize vibrations induced by the impact bar. The stress was measured with a very stiff quartz transducer near the clamps. The deformation was detected by a high-speed camera which took 32 000 pictures per second of a fine-structured grating on the specimen. In addition, an inductive strain gage at the impact bar served as a monitor. A transient recorder collected the signals of the strain gages and of the quartz load cells. The specimens were cooled in LN_2 and quickly put into the clamps. No remarkable temperature rise was measured during the short time experiment. At LHe temperatures a special minicryostat was used (not reported in this paper).

FRACTURE MODES OF FIBRE COMPOSITES; THERMAL RESIDUAL STRAIN OF FIBRE COMPOSITES

The low fracture strain of most polymers is reduced by thermal residual strain arising from different thermal contractions of the fibres and the matrix. Assuming low contractive fibres, it amounts to 1 to 1.4% during cooling to LHe. The remaining free strain is thus 0.6 to 1%, which is available for external loading. For angle plies thermal residual shear strain are superimposed, in addition. Under fatigue load, the situation of carbon and Kevlar fibre composites is more stressed since they possess a much higher fatigue endurance limit than the pure epoxy resin.[9] Thus, epoxy matrices fracture earlier than the fibres, and these composites cannot be used in an optimum manner. Thermoplastic matrices, such as PC, PSU, PES with a fracture strain at 4.2 K of 3 to 3.5% are able to compensate residual strains and therefore are on the safe side. A brittle matrix even if not load bearing, can initiate special fracture modes that reduce the strength of the composite. In crossplies cracks in layers perpendicular to the stress direction can propagate and destroy the load-bearing layers. This fracture mode was shown for carbon fibre epoxy crossplies.[10]

Carbon and Kevlar fibres are weaker in the transverse direction than tough, isotropic glass fibres. These can act as crack stoppers, which was shown in ref. 10.

One important fracture mode is debonding and delamination at the fibre matrix interface. These fracture modes occur predominantly under fatigue loading and thermal cycling. The matrix fibre bond is very important for increasing transverse and shear strengths of composites.

The interlaminar shear strength tested in short beam tests is determined by the weakest link, which could be:

- the matrix strength
- the matrix-fibre bond
- the fibre shear strength.

The interlaminar shear fracture at cryogenic temperatures occurs by failure of:

- matrix for fibre glass-composites
  $\tau_{ILS} \approx$ 150-170 MPa (4.2 K)
- matrix or bond for carbon fibre composites
  $\tau_{ILS} \approx$ 80-120 MPa (4.2 K)
- bond or fibre for Kevlar fibre composites
  $\tau_{ILS} \leq$ 40 MPa.

Epoxy resins could be covalently bonded to surface treated (oxydized) carbon fibres; however, thermoplastic polymers, such as PC, PSU and PES are only bonded by Van der Waals forces. It is interesting to note that τ_{ILS} values for composites with both types of matrices are nearly the same. Thus, probably the dipole-dipole forces between matrix and oxydized fibre surface dominate. Another interesting result is derived from the load-deflection curves when τ_{ILS} is measured in short-beam bending tests on carbon fibre composites having a rather high bond strength. A clear failure occurs when a brittle epoxy resin matrix is used. For more ductile PC and PSU matrices there is no real interlaminar shear failure until fibre breakage occurs (see Fig. 5). Only an upper limit of τ_{ILS} exists, but no interlaminar shear failure. This favourable fracture behavior is a consequence of cryogenic ductile matrices.

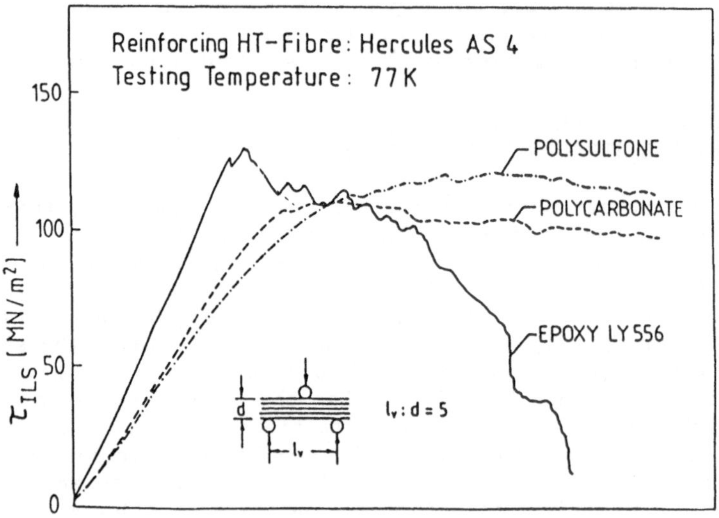

Fig. 5 Load-deflection diagram of HERCULES AS4 reinforced with epoxy LY 556, PC, and PSU at 77 K. The load is converted into τ_{ILS}.

CONCLUSIONS

- Several cryogenic fracture modes of polymers and composites can be avoided by low temperature ductile polymers.
- Polymers, usually more brittle at higher deformation rates, become more ductile at very high deformation rates (impact) when adiabatic heating and brittle-plastic transition take place.
- For carbon and Kevlar fibre composites the matrix-fibre bond properties need further investigation, especially the importance of Van der Waals forces (dipole-dipole forces) for interfacial properties need further studies. Sputtering of monomers with high dipole moments onto the fibre surface may be one possibility for improving fibre composites.

ACKNOWLEDGMENT

The assistance of Dr. Schmitt and A. Geissler of the Fraunhofer-Gesellschaft, ICT, Berghausen, FRG, is greatly appreciated.

REFERENCES

1. G. Hartwig, Nonmetallic materials in present and future low temperature technology, in: "Proceedings of the International Cryogenic Materials Conference", K. Tachikawa and A. F. Clark, eds., Butterworths, London (1982), pp. 495-502.
2. W. Döll, Optical interference measurement and fracture mechanics analysis, in: Advances in Polymer Science 52/53, H. H. Kausch, ed., Springer (1983), pp. 105-168.
3. B. Kneifel, in: "Nonmetallic Materials and Composites at Low Temperatures", A. F. Clark, R. P. Reed, and G. Hartwig, eds., vol. 1, Plenum Press, New York (1978), pp. 123-130.
4. B. Kneifel in: "Nonmetallic Materials and Composites at Low Temperatures", G. Hartwig and D. Evans, eds., vol. 2, Plenum Press, New York (1982), pp. 125-138.
5. G. Hartwig and G. Schwarz in: "Advances in Cryogenic Engineering-Materials," vol. 30, Plenum Press, New York (1984), pp. 61-70.
6. R. Weichert and K. Schönert, Heat generation at the tip of a moving crack, J. Mech. Phys. Solids 26:151-161 (1978).
7. D. S. Dugdale, Yielding of steel sheets containing slits, J. Mech. Phys. Solids 8:100-108 (1960).
8. G. R. Irwin, Plastic zone near a crack and fracture toughness, Proc. 7th Sagamore Conf., p IV-63 (1960).
9. G. Hartwig and S. Knaak, in: "Cryogenics," vol. 24, Butterworths, London (1984), p. 639.
10. G. Hartwig, in: "Advances in Cryogenic Engineering-Materials," vol. 28, Plenum Press, New York (1981), p. 179.
11. E. Fitzer and R. Weiss, in: "Processing and Uses of Carbon Fibre Reinforced Plastics," VDI-Verlag, Düsseldorf, FRG (1981), p. 45.
12. G. Hartwig, H. Jaeger, and S. Knaak in: "Nonmetallic Materials and Composites at Low Temperatures," vol. III, Plenum Press, New York (1985), in press.

MECHANICAL BEHAVIOR AND FATIGUE IN POLYMERIC COMPOSITES AT LOW TEMPERATURES

Y. Katz, A. Bussiba and H. Mathias

Nuclear Research Center-Negev
Beer-Sheva, Israel

ABSTRACT

Advanced fiber reinforced polymeric composite materials are often suggested as structural materials at low temperature. In this study, graphite epoxy and Kevlar-49/epoxy systems were investigated. Fatigue behavior was emphasized after establishing the standard monotonic mechanical properties, including fracture resistance parameters at 77, 190, and 296 K. Tension-tension fatigue crack propagation testing was carried out at nominal constant stress intensity amplitudes using precracked compact tensile specimens. The crack tip damage zone was measured and tracked by an electro-potential device, opening displacement gage, microscopic observation, and acoustic emission activity recording. Fractographic and metallographic studies were performed with emphasis on fracture morphology and modes, failure processes, and description of sequential events. On the basis of these experimental results, the problem of fatigue resistance, including low temperature effects, is analyzed and discussed. The fundamental concepts of fatigue in composites are assessed, particularly in terms of fracture mechanics methods.

INTRODUCTION

Extensive research activity has been devoted to fatigue resistance studies in composite materials.[1-5] These anisotropic and inhomogeneous systems introduce specific aspects regarding the monotonic and the cyclic load behavior. Generally subcritical or critical faults, which are nucleated or propagated by a mechanical driving force, are the result of a balanced equation between the mechanical force and the localized or global material resistance term. In composites, the resistance term combines two major components: (1) an intrinsic component, which is basically related to the composite constituents, including their mutual interactions and (2) a more extrinsic resistance component influenced by the typical sequence of events that accompanies the damage process. The component is dependent on the material superstructure dictated by the composite design. Here, orientation, lay-up influences and geometrical factors dominate, resulting in branching, debonding, or complex spatial interconnected cracking with typical configuration that occurs even during the extension of a relatively large and sharp major crack. The occurrence of such phenomena affects the nature of the damage zone and the global material resistance.

This study is based on an experimental program emphasizing fatigue behavior at low temperatures. This topic was prompted by the realization of the hidden potential in graphite or aramid fiber-reinforced polymeric composites for low temperature applications. The prospective combination of physical and mechanical properties in terms of density, modulus, strength, toughness, magnetism and thermal conductivity, demands additional information for load-bearing structural purposes and design data. Consequently, the experimental plan intends to elaborate on the micromechanisms involved in fatigue crack extension with attention to the thermal influences on the damage affected zone. Fracture mode studies are included that consider fiber breakage, resin cracking, interfacial debonding and delamination between plies and laminates.

Chang et al.[6] have described the damage growth in G/E laminates and the role of crack tip geometry. The damage morphology in composites has also been addressed by Ueng et al.[7] and Yeow et al.[8] Considering the sequential events and the typical alternative modes, the current work attempts to evaluate fatigue behavior in composites by considering the general fatigue concepts gathered so far in metallic systems.

EXPERIMENTAL PROCEDURE

Two advanced reinforced polymeric materials were selected. The first was graphite/epoxy (G/E) 80-ply plate, 12 mm in thickness with a lay-up orientation of 0°/90°. This composite was processed from magnamite graphite prepreg tape AS4/3502. The composite constituents were manufactured by Hercules according to the following description: The continuous fibers were high strength filaments, surface treated, designated AS4, and the resin was amine-cured epoxy, designated 3502. Generally, the surface treatment was intended to improve the composite shear and transverse tensile strength. The second material selected was Kevlar-49 (Kev) plate, 4 mm in thickness with 12 plies. Basically the lay-up of the woven reinforcement was along the same principal direction, processed from prepreg organic fiber produced by DuPont and following the reinforcement style of 285 and Hexcel epoxy system, designated F155. The curing cycles for each resin were performed according to the commercially recommended procedure.

Uniaxial tensile properties were obtained by using uniform specimens with a 25-mm gage length and 6-mm width. A constant strain rate of $1 \times 10^{-3}\ s^{-1}$ was applied utilizing proper devices at low temperatures. Three-point bending tests were also performed on the G/E composite to provide flexural and shear strength data. Uniform bend specimens with rectangular cross sections of 6 mm x 12 mm and a 40-mm span were used.

For both composites, fracture toughness and tension-tension fatigue studies were carried out by using a compact tensile specimen (CTS) after crack sharpening. The tests were performed by an electro-hydraulic closed-loop machine following general calculation procedures according to ASTM E-399. In the case of fatigue, a frequency of 15 Hz was applied with load ratio of $R \cong 0$. The crack length or the crack tip damage zone was monitored by a crack opening displacement gage (COD), electropotential device, and microscopic observations. All mechanical tests were performed at temperatures of 77, 190 and 296 K and tracked continuously by stress wave emission activity.

Table 1. Mechanical Properties for the G/E and the Kevlar Composites

Test Temp. (K)	σ_{UTS} (MPa) G/E	σ_{UTS} (MPa) Kev.	e (%) G/E	e (%) Kev.	τ (MPa) G/E	σ_f (MPa) G/E	$\sigma_{f/\tau}$ G/E	K_{crit} (MPa $\sqrt{m}$) G/E	K_{crit} (MPa $\sqrt{m}$) Kev.
296	530	460	3.0	3.5	68	520	7.7	44	27
190	430	360	→ 0	1	90	620	6.7	20	
77	350	340	→ 0	1	74	445	5.9	14.5	23

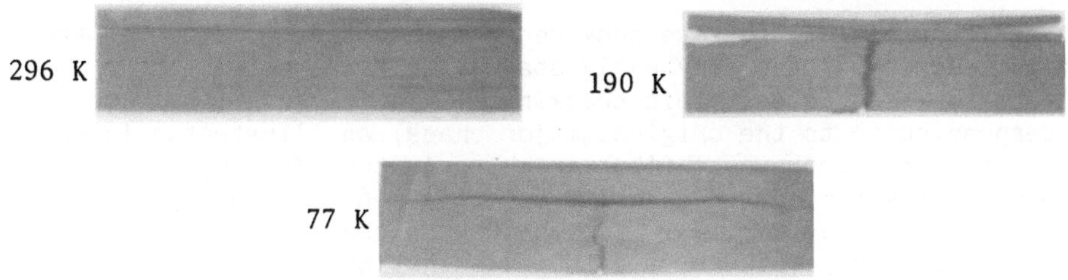

Fig. 1. Macroscopic fracture appearance of the G/E composite at various temperatures.

Fractographic and metallographic studies were included in the experimental scheme. Particular attention was paid to the damage zone characterization and the sequential events by light and SEM observations. Fracture surface modes and the typical failure micromechanisms, including their thermal dependency, were investigated as well.

EXPERIMENTAL RESULTS

The mechanical properties at different temperatures, including the tensile strength (σ_{UTS}), elongation (e), flexural and shear strength (σ_f and τ, respectively), and the critical fracture toughness (K_{crit}), are summarized in Table 1. For the G/E, τ and σ_f were calculated according to the following relationships:

$$\tau = (3/4)\ [P_{max}/(B \cdot W)] \qquad (1)$$

$$\sigma_f = (3/2)\ [P_{max} \cdot L/(B \cdot W)] \qquad (2)$$

where L is the span, B is the thickness, and W is the width.

As shown in Table 1 σ_{UTS} and K_{crit} values increase with the temperature. Compared to that of the G/E, the thermal dependency of the Kev was quite moderate, as indicated by the critical stress intensity factor values. For the G/E, the σ_f/τ ratio reflects the bend specimen behavior during monotonic loading at ambient and low temperatures, as illustrated in Fig. 1. A pure delamination mode was obtained at 296 K, whereas a mode transition occurred at low temperatures where transverse fibers crackings were dominating.

Figure 2 shows a typical load-displacement curve in the G/E composite tracked by acoustic emission activity at 296 K. The load drop due to delamination cracking is indicated by the stress wave rate activity. The stress-strain response for the G/E was almost linear at all temperatures, whereas the Kev was characterized by nonlinear mechanical behavior, even at 77 K. During

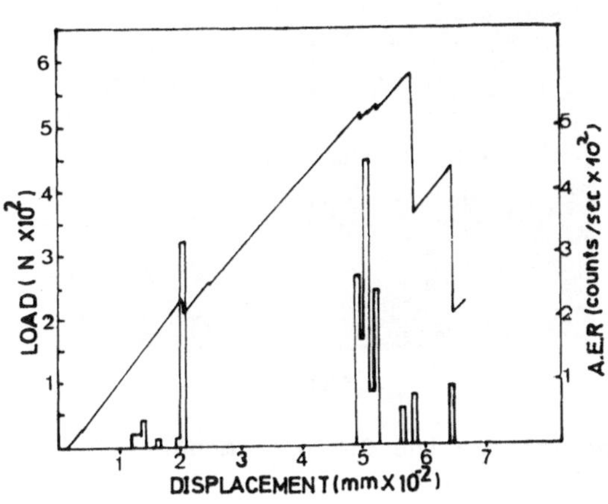

Fig. 2. Load and stress wave emission rate vs. displacement for the G/E bend specimen at 296 K.

fatigue tests, the resistance gage provided complementary evidence of the delamination cracking at the very early stages; this was confirmed by the liquid penetrator techniques. This cracking, 10 to 15 mm in length, developed at 296 K perpendicular to the original major crack, as illustrated in Fig. 3.

On the contrary, transverse fiber cracking in the G/E composite was the dominant mode at low temperatures, while delamination cracking was limited to the crack tip region (see Fig. 4). For the Kev no change in the cracking mode was observed at low temperatures, and the delamination cracking mode was readily observed on fatigue fractured specimens, as illustrated in Fig. 5.

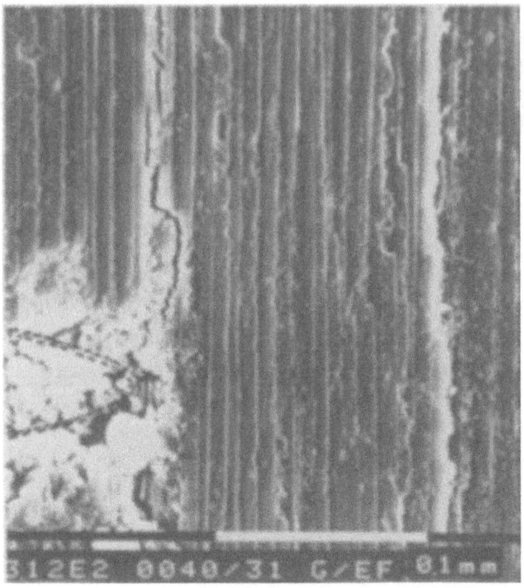

Fig. 4. Transverse cracking developed after 3 x 10^5 cycles at 77 K. The nominal ΔK was 4-14 MPa$\sqrt{m}$.

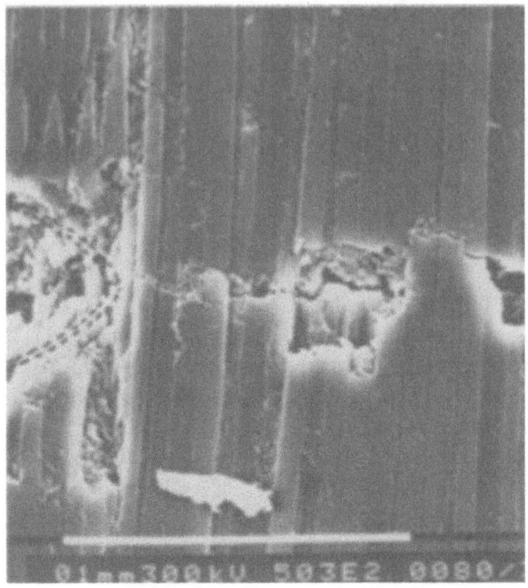

Fig. 3. Delamination cracking developed after 2.5 x 10^5 cycles at 296 K. The nominal ΔK was 7-14 MPa$\sqrt{m}$.

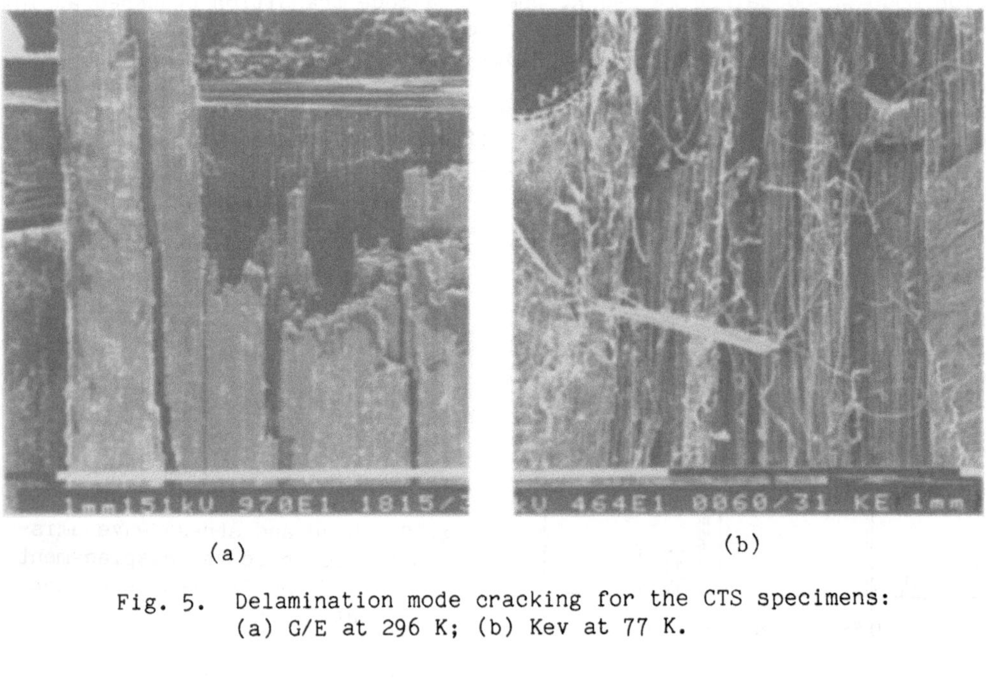

(a) (b)

Fig. 5. Delamination mode cracking for the CTS specimens: (a) G/E at 296 K; (b) Kev at 77 K.

Crack opening displacement measurements provided the following information for the G/E composite: (a) The ΔCOD values increase monotonically with the nominal cyclic stress intensity factor amplitude (ΔK) at the early stage of the crack extension. (b) Higher ΔCOD values were obtained at 296 K than at 77 K for equal ΔK. (c) The increasing ΔCOD values at 296 K and cyclic condition that corresponded to $\Delta K = 12\ MPa\sqrt{m}$ after 5×10^4 cycles were discontinuous in nature. Under these conditions, no ΔCOD change was recorded at 77 K. The typical fracture mode transition at low temperatures enabled us to measure steady state crack extension rate in the G/E composite. For example:

$$\Delta K = 9\text{-}12\ MPa\sqrt{m} \quad - \quad da/dN = 3 \times 10^{-7}\ mm/cycle$$

$$\Delta K = 13.5\ MPa\sqrt{m} \quad - \quad da/dN = 1.1 \times 10^{-4}\ mm/cycle$$

The CTS fatigue specimens failed critically at a nominal $\Delta K \cong \Delta K_{crit}$ value. This behavior occurred in the G/E and the Kev composites at all investigated temperatures.

With respect to the uniform tensile specimens, Fig. 6 illustrates the fracture mode in G/E where the macromode appeared flat, even at 296 K. Typical differences were observed on a more refined scale (Fig. 7). At 296 K broken 90°-ply fibers (Fig. 7a) were completely different from those at 77 K (Fig. 7b).

The macrofracture of the G/E bend specimens at low temperatures included zones in which the bundle of fibers were torn out by a mixed mode of delamination and transverse cracking (Figs. 1 and 8). The main findings of fatigue tests of the G/E were that bundles of fibers projected from the fracture surface mainly at 296 K temperature (see Fig. 9) and the fracture mode of the fibers perpendicular to the load direction were not affected by the temperature.

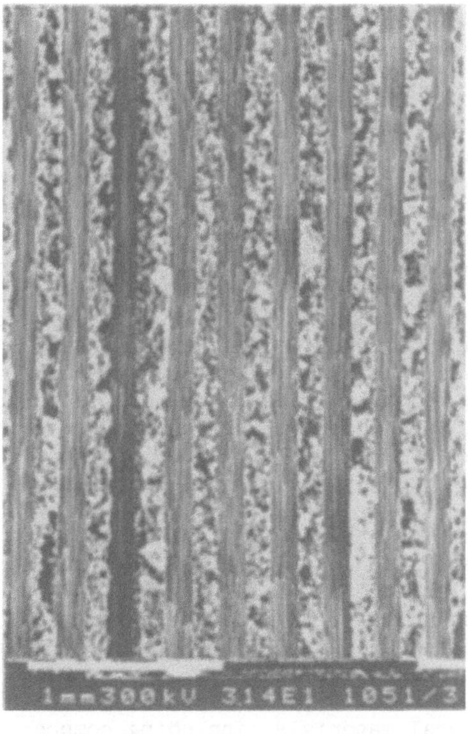

Fig. 6. Macrofracture of G/E tensile specimen at 296 K.

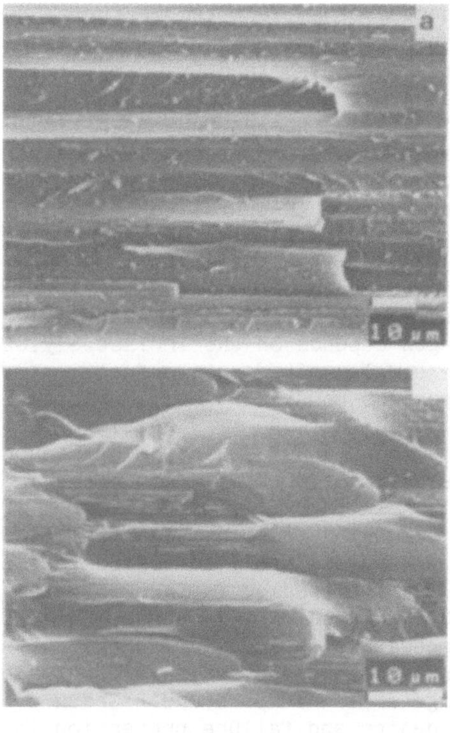

Fig. 7. The G/E fracture appearance of the 90° ply (a) at 296 K; (b) at 77 K.

Fig. 8. SEM fractograph of the G/E bend specimen at 190 K.

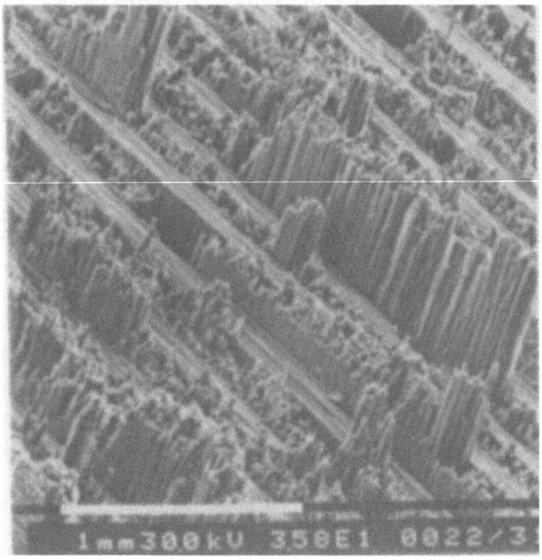

Fig. 9. SEM fractograph of the G/E specimen at the damage zone at 296 K.

DISCUSSION

The progressive damage in the crack tip area during cyclic loading was attributed to delamination processes. This typical behavior was particularly observed at ambient temperatures, causing dramatic variations of the effective mechanical environment. This evidence confirms the fatigue damage characterization, as described by Chang et al.[6], where a modified x-ray nondestructive evaluation technique has been utilized in the G/E laminated composite. The main issue concerns the subcritical cracking morphology, since the extension stage is not dominated by a single crack and is not confined to a single dominating fracture mode. Gaggar and Broutman[9] have suggested the resistance curve method (R-curve) to overcome some of the difficulties. Clearly the use of an R-curve method for a composite introduces a certain conceptual modification, since the material resistance increases with the damage growth owing to crack tip geometrical factors. This is a completely different origin of resistance than that of metals, where the monotonic increase of R with the crack length depends on the inelastic yielding and hardening at the process zone. The complex mixed active fracture mode classification in composites remains an important research task. For example, Fig. 10 illustrates two kinds of faults in the G/E at 296 K. Additional information regarding classification of fracture modes and description of sequential events during delayed failure processes might provide a sound background for design criteria and composite assessment.

Tension-tension axial fatigue has been investigated by Tobler and Read[10] in an S-glass/epoxy composite in the temperature range 4 and 295 K. Toth et al.[11] have studied fatigue in glass/epoxy laminate composite systems at 20 and 300 K. Based on S-N curves, these studies indicated fatigue resistance at low temperature is superior to that at ambient temperature. However, the current work does not support this conclusion with respect to the crack propagation stage at low temperature. Crack propagation behavior is an important factor in design and failure prevention in structural materials, including composites. This investigation demonstrates that fracture mechanics methods can be applied at low temperatures (particularly in the G/E composite) owing to the thermal effects on the fatigue fracture mode. In fact, an extremely

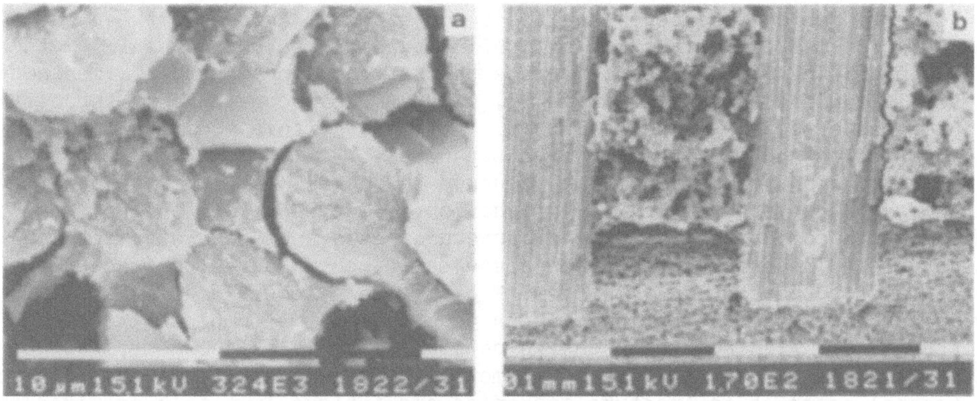

Fig. 10. G/E interfacial debonding (a) and delamination cracking (b) during fatigue at room temperature.

strong dependency was obtained between the crack extension rate and the mechanical driving force. Similar to fatigue in metals, a proper comparison (which should be based on a global view) between ambient and low temperature properties is still a major issue in evaluating low temperature design procedures that are based on ambient data.

With respect to the fatigue crack propagation rate, Wachnicki and Radon[5] have also indicated the strong dependency between da/dN and ΔK in CSM composites at 293 K. These findings could be expected from fatigue studies in metals, where alternative fracture modes dominate. For example, a drastic change in the sigmoidal-shaped da/dN vs. ΔK fatigue curves in metals has been shown and discussed for the ductile-brittle transition.[12]

The detection and classification of fracture processes in composites at low temperatures by acoustic emission have been suggested by Nishijima and Okada.[13] In the current experimental plan, stress wave emission was included as a complementary method. In G/E, damage initiation at the crack tip ($\Delta K = 7\ MPa\sqrt{m}$) was detected by the stress wave emission technique utilizing the energy mode. Further cycling of about 10^5 cycles enhanced the number of events, as illustrated in Fig. 11. Increasing the stress intensity factor to $10\ MPa\sqrt{m}$ caused cracking delamination at the 90° ply, which extended significantly. This progressive damage was indicated by acoustic emission and crack opening displacement data.

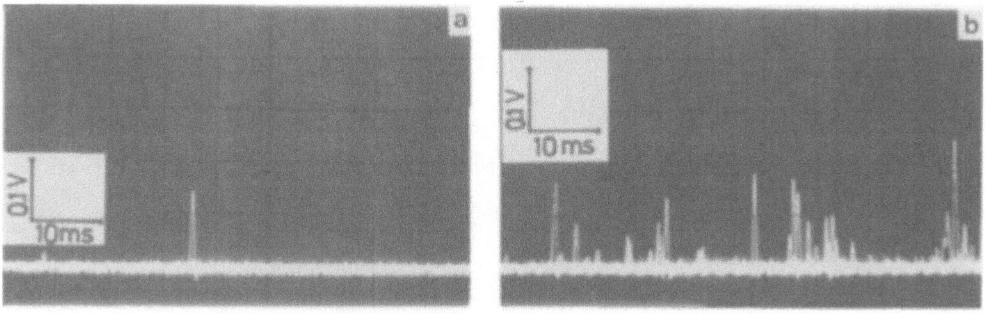

Fig. 11. Acoustic emission signals in the G/E composite during fatigue at room temperature; $\Delta K = 7\ MPa\sqrt{m}$
(a) 2×10^3 cycles; (b) 1×10^5 cycles.

The orientation dependence and the crack rate differences between the plies call for multiaxial testing and refinement of fracture mechanics methods at low temperature. An extension of the theoretical approach to a mixed mode analysis, including the role of mode II, might improve considerably the definition of the effective mechanical term. Attempts at this modification are being carried on and will be addressed separately.

In conclusion, mode transition occurred in the G/E composite at low temperatures, contrary to the Kevlar-49 behavior. A degradation tendency was observed in both composites in terms of fracture resistance parameters at low temperatures. The crack propagation stage could be evaluated by fracture mechanics methods, particularly in the G/E composite. Acoustic emission was very useful as complementary information to classify the complex mixed fracture processes. With regard to design data and material assessment, fatigue concepts that have been established in metallic systems should not be ignored.

General progress in fracture mechanics and fatigue in metals provided the drive for composite development. Consequently the high toughness and fatigue resistance of composites fulfill the designed and expected mechanical properties. Modification of methods and classification of fractures in composites are essential to design and mechanical stability.

REFERENCES

1. R. S. Williams and K. L. Reifsnider, J. Compos. Mater. 8:340-355 (1975).
2. R. Papirno, J. Compos. Mater. 10:41-50 (1977).
3. H. C. Kim and L. J. Ebert, J. Compos. Mater. 12:139-152 (1978).
4. S. S. Wang, E. S. M. Chim, and N. M. Zahlan, J. Compos. Mater. 17:250-266 (1983).
5. C. R. Wachnicki and J. C. Radon, Composites 15:211-216 (1984).
6. F. H. Chang, D. E. Gordon, B. T. Rodini, and R. H. McDaniel, J. Compos. Mater. 10:182-192 (1976).
7. C. E. S. Ueng, J. A. Aberson, and B. A. Lafitle, J. Compos. Mater. 11:222-234 (1977).
8. Y. T. Yeow, D. H. Morris, and H. F. Brinson, Exper. Mech. 1-8 (1979).
9. S. G. Gaggar and L. J. Broutman, J. Compos. Mater. 9:216-227 (1975).
10. R. L. Tobler and D. T. Read, J. Compos. Mat. 10:32-43 (1976).
11. J. M. Toth, Jr., W. J. Bailey, and D. A. Boyce, in: "Fatigue at Low Temperatures," R. L. Stephens, ed., STP 857, American Society for Testing and Materials, Philadelphia (1983), pp. 163-172.
12. Y. Katz, A. Bussiba, and H. Mathias, in: "Advances in Cryogenic Engineering - Materials," vol. 30, Plenum Press, New York (1984) pp. 339-347.
13. S. Nishijima, T. Okada, and S. Namba, in: "Advances in Cryogenic Engineering - Materials," vol. 30, Plenum Press, New York (1984), pp. 25-32.

THERMO-STIMULATED CURRENT AND DIELECTRIC LOSS IN COMPOSITE MATERIALS

S. Nishijima and T. Okada
ISIR Osaka University
Ibaraki, Osaka, Japan

and

T. Hagihara
Department of Physics
Osaka Kyoiku University
Tennoji, Osaka, Japan

ABSTRACT

Thermo-stimulated current and dielectric loss measurements have been performed on five kinds of commercially available composite materials in order to study the electric properties of composite materials at low temperatures. Thermo-stimulated current measurements have been made on the composite materials in which the matrix quality was changed intentionally. The changes in the matrices were introduced by gamma irradiation or different curing conditions. Thermo-stimulated current and dielectric loss measurements revealed the number and the molecular weight of dipolar molecules. The different features of thermo-stimulated currents and dielectric losses were determined for different composite materials. The gamma irradiation and the curing conditions especially affect the thermo-stimulated current features. The changes in macroscopic mechanical properties reflect those of thermo-stimulated current. It was found that the change in quality and/or degradation of the composite materials could be detected by means of thermo-stimulated current and/or dielectric loss measurements.

INTRODUCTION

Organic composite materials have frequently been employed as insulating and/or structural materials in superconducting magnets. The organic matrices probably cause overall degradation in properties of composite materials. The changes and degradation in quality of matrices should be understood sufficiently and their effect on macroscopic properties of composite materials must be estimated. Especially, the insulating materials used in the fusion energy magnet should be investigated thoroughly in advance.

In this work, thermo-stimulated current (TSC) and dielectric loss (DL) measurements were performed to detect the changes and/or degradation of the matrices. The effects of curing conditions and gamma irradiation were investigated. Dynamic Young's modulus was measured,[1] and interlaminar shear tests[2,3] were performed in order to provide a correlation between TSC (or DL) and the macroscopic mechanical properties.[4-6]

EXPERIMENT

Samples

The five kinds of commercially available samples chosen were G-10CR®, G-11CR®, Spaulrad®, Lamiverre-A® and Hoxan®. Except for Spaulrad, the matrices of the composite materials are epoxy. The curing time for Hoxan was changed. The curing times were 2, 4, 8 and 16 h. The curing temperature was kept at 423 K. The gamma irradiation was performed on Lamiverre-A. The irradiation temperature was room temperature (RT). The absorbed dose rate was 2.4×10^6 rad/h, and the absorbed doses were 8.0×10^8, 1.4×10^9, and 2.1×10^9 rad.

The DL and TSC of the materials were measured. The dynamic Young's modulus was also measured on Hoxan. The interlaminar shear tests were performed on the gamma-irradiated Lamiverre-A.

The specimens used for TSC and DL measurements were cut from the sheets 1 mm in thickness. Electrodes were made with silver paste on both sides of the specimen. The electrodes used in DL measurements were circular and 10 mm in diameter. Those for TSC measurements were rectangular, 5 mm x 10 mm.

Dielectric Loss Measurements

Since the dipolar molecules in composite materials can be detected in DL measurement, certain qualities of the composites could be evaluated, such as degree of curing, degradation, and hygroscopicity. Consequently, these measurements are expected to be valuable for quality control and irradiation studies.

Measurements were performed with a Schering bridge. The frequency and the applied voltage was 1 kHz and 14 V, respectively. The temperature range of measurement was 150 to 300 K. Measurements were also made at 60 and 1 MHz in order to compare the result with the handbook values.

Thermo-stimulated Current Measurements

This measurement provides information about mobility and the quantity of the dipolar molecules. At RT or higher temperatures, an electric field (400 V/mm) was applied to the specimen for 5 min so as to arrange the dipoles in the external field direction. Then the specimen was rapidly cooled down to approximately 100 K. The oriented molecules were frozen and still oriented when the external field was removed at this temperature. An increase in temperature enables the frozen dipoles to rearrange. The induced electric current due to dipole rearrangement was measured while the temperature was raised at a constant rate (2 K/min). This method is suitable for detection of the dielectric relaxation at very low frequencies. The dipolar molecules in the matrices, whose relaxation times are relatively long, could be detected.

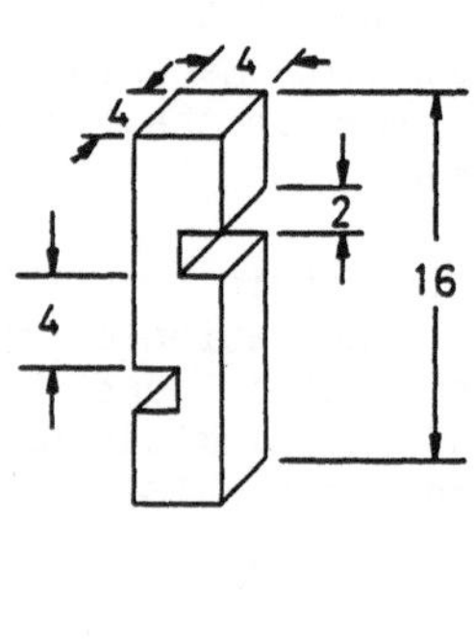

Fig. 1. Shape and dimensions of the specimen for the interlaminar shear test.

Table 1. Dielectric Constants and Losses

Material		Handbook Values ε	tan σ (10^{-2})	This Work ε	tan σ (10^{-2})
Spaulrad	(at 1 MHz)	4.80	1.2	5.09	0.85
G-10CR	(at 1 MHz)	4.90-5.10	2.0-2.2	5.08	2.2
G-11CR	(at 1 MHz)	5.10-5.30	1.6-1.8	5.27	1.6
Lamiverre-A	(at 60 Hz)	5.19	0.50	5.12	0.46

Interlaminar Shear Tests

Interlaminar shear tests were made on the irradiated Lamiverre-A at a deformation rate of 1 mm/min at RT. The shape and dimensions are presented in Fig. 1. Five samples were tested at the same absorbed dose, and their mean shear strengths were evaluated.

Dynamic Young's Modulus and Internal Friction Measurements

Dynamic Young's modulus and internal friction were measured on Hoxan. The flexural vibration and the free-decay method were used for Young's modulus and internal friction, respectively. The measurements were performed down to 130 K at a strain amplitude of 7×10^{-7}.

RESULTS AND DISCUSSION

In Table 1 the measured dielectric constants and losses and the handbook values are presented. Although the specimens tested in this work were quite small, the data obtained were significant.

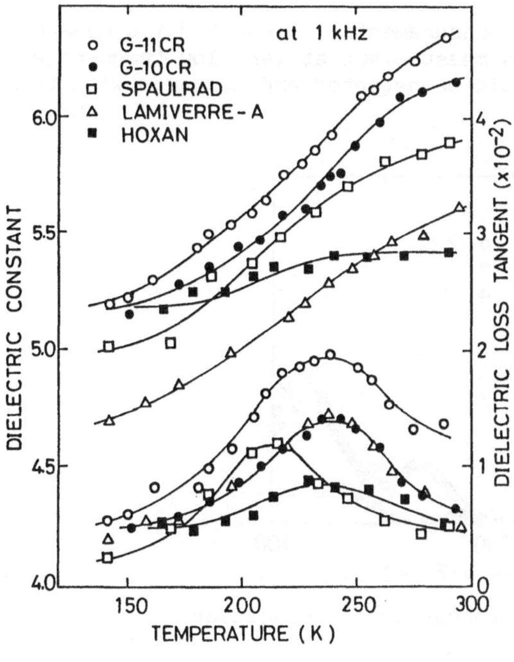

Fig. 2. Temperature dependence of dielectric loss and constant.

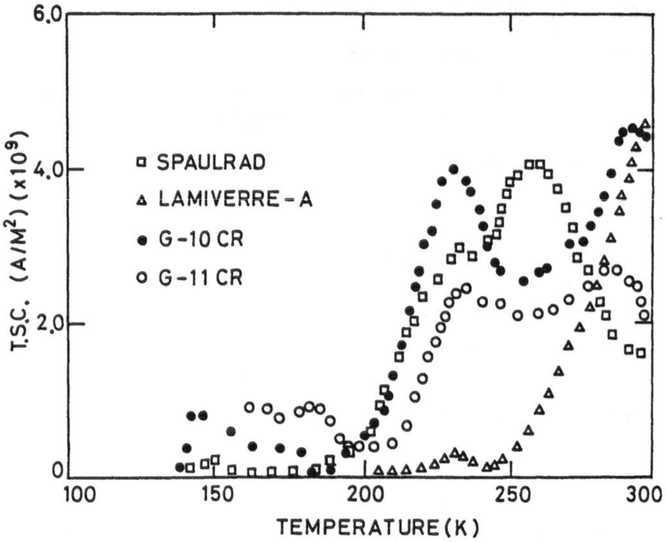

Fig. 3. TSC in various specimens.

Figure 2 shows the temperature dependence of the dielectric constant and loss of each specimen. Since the matrices of most specimens are epoxies (except for Spaulrad), the dielectric losses showed peaks at approximately 240 K. Spaulrad, whose matrix is polyimide, presented a peak at about 210 K. These results imply that the sample with a different matrix could be distinguished. The composites, whose dielectric losses showed peaks at the same frequency and temperature, had the same relaxation time. The polyimide and epoxy showed peaks in a different temperature range, and hence their relaxation times were different. The lower temperature range of polyimide means a higher mobility of polyimide molecules. The magnitude of the loss peak shows the relative number of dipolar molecules. The G-11CR has a relatively large number of dipolar molecules, but direct comparison was not possible because the volume fraction of the matrices varies from sample to sample.

Figure 3 shows the results of TSC measurements. Since the TSC measurement corresponds to the dielectric loss measurement at very low frequencies, the motions of the larger molecules could be detected and compared with those

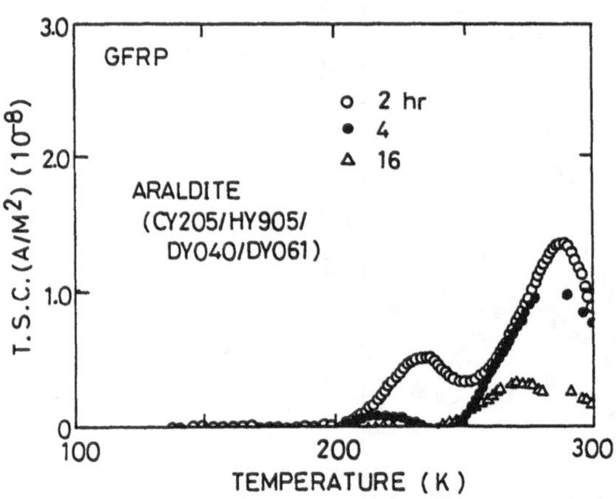

Fig. 4. TSC of Hoxan with different curing conditions.

Table 2. Dynamic Young's Modulus and Internal Friction of Hoxan at RT

Curing Time	Dynamic Young's Modulus (GPa)	Internal Friction (10^{-2})
2	21.4	1.08
4	22.4	0.84
8	23.3	0.79
16	23.7	0.69

in the dielectric loss measurements. The composites having epoxy matrices presented two peaks around 230 K and 290 K; polyimide had peaks at 230 K and 260 K. These results suggest that the measurements enable us to detect the characteristics of the matrix.

It is difficult to compare the obtained values directly in these commercial composites because the volume fraction of each matrix is different. Consequently, the composites were made with identical matrices and volume fractions, but with different curing conditions. Figure 4 shows the results of TSC measurements of Hoxan whose curing conditions are different. The Hoxan shows the two TSC peaks, which suggest the existence of two types of dipolar molecules. As the curing proceeds, the magnitude of peaks decreases, and finally the peak at the lower temperature range almost disappears. This process was understood to be the formation of three-dimensional networks of epoxy resin. By changing the TSC features, the curing process could be detected.

The dynamic Young's modulus and internal friction of Hoxan are presented in Table 2. Enhanced TSC corresponded with a lower Young's modulus and higher internal friction.

The TSC of gamma-irradiated Lamiverre-A was also measured. Figure 5 presents the results. The gamma irradiation induced an increase in TSC, which suggests the creation of dipolar molecules.

To study the change in TSC closely, the change of TSC with temperature was studied and is shown in Fig. 6. The onset temperature of the increment decreases with absorbed dose. It means that heavy irradiation results in lower weight dipolar molecules.

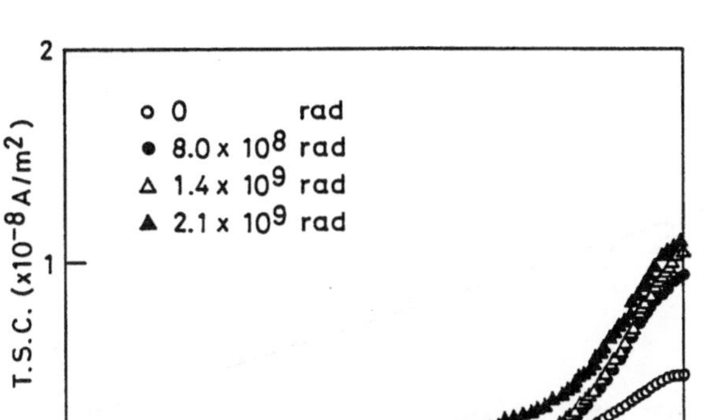

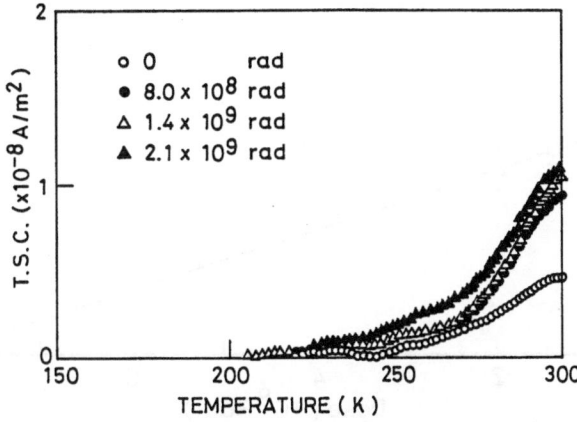

Fig. 5. TSC of gamma-irradiated Lamiverre-A.

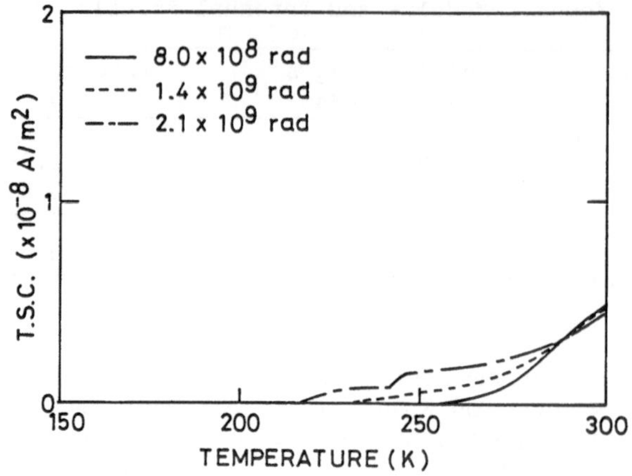

Fig. 6. Increment of TSC induced by gamma irradiation.

To correlate the change of TSC with the macroscopic mechanical properties, the interlaminar shear strength of gamma-irradiated specimens[7] was measured, as shown in Fig. 7. The 25% decrement of interlaminar shear strength was induced by 2.1 x 10^9 rad, which corresponds to the degradation of epoxy, as expected from the TSC measurements. The change of shear strength coincides with the increment of TSC.

CONCLUSIONS

The TSC and DL measurements were performed on commercially available composite materials and the following conclusions were drawn:

1. The different matrices could be distinguished by means of DL or TSC measurements.
2. With TSC measurements, the process of forming three-dimensional networks in epoxy resin with curing could be traced.
3. The creation of lower weight dipolar molecules induced by gamma irradiation could be detected, which coincided with the degradation of interlaminar shear strength.
4. It is concluded that the change and degradation of matrix was monitored by means of TSC and DL measurements.

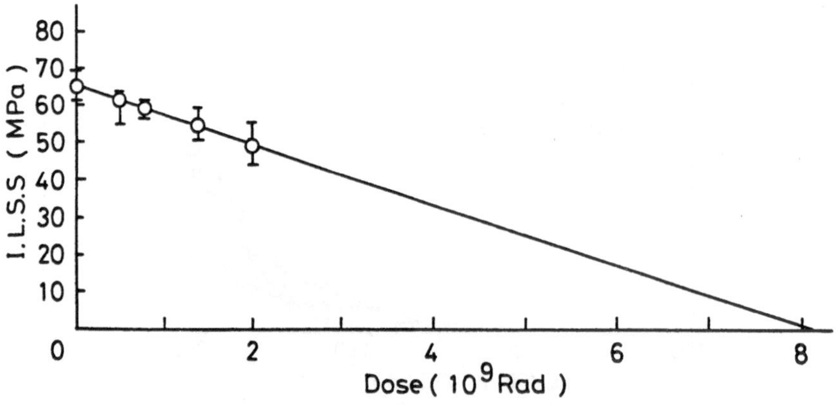

Fig. 7. Change of interlaminar shear strength (ILSS) induced by gamma irradiation.

ACKNOWLEDGMENT

The authors would like to thank Prof. T. Okamoto and Mr. K. Mastushita at ISIR for the dynamic Young's modulus measurements. They would like to thank Mr. N. Ikeda at ISIR for gamma irradiation. This work is partly supported by Grant in Aid for Scientific Research No. 60050044, Ministry of Education in Japan.

REFERENCES

1. G. Hartwig and G. Schwarz, Correlation of dielectric and mechanical damping at low temperatures, in: "Advances in Cryogenic Engineering - Materials," vol. 30, Plenum Press, New York (1984) p. 61.
2. H. Becker and E. A. Erez, A study of interlaminar shear strength at cryogenic temperatures, in: "Advances in Cryogenic Engineering - Materials," vol. 24, Plenum Press, New York (1980), p. 259.
3. H. Becker, Problems of cryogenic interlaminar shear strength testing, in: "Advances in Cryogenic Engineering - Materials," vol. 30, Plenum Press, New York (1984), p. 33.
4. J. V. Gauchel, J. L. Olinger, and D. C. Lupton, Characterization of glass-reinforced composites, in: "Advances in Cryogenic Engineering - Materials," vol. 28, Plenum Press, New York (1982), p. 211.
5. J. R. Benzinger, Manufacturing capabilities of CR-grade laminates, in: "Advances in Cryogenic Engineering - Materials," vol. 24, Plenum Press, New York (1980), p. 252.
6. J. R. Benzinger, The manufacture and properties of radiation-resistant laminates, in: "Advances in Cryogenic Engineering - Materials," vol. 26, Plenum Press, New York (1982), p. 231.
7. P. E. Schmunk, G. R. Imel, and Y. D. Harker, Irradiation studies of magnet insulator materials, in: "Proc. Second Topical Meeting of Fusion Reactor Materials" Seattle (1981) 3D-3.

MECHANICAL PROPERTIES OF CARBON-GLASS HYBRID COMPOSITE MATERIALS AT LOW TEMPERATURE

S. Nishijima, T. Okada
ISIR Osaka University
Ibaraki, Osaka, Japan

K. Fujioka, Y. Kuraoka
Cryogenic Technology Development Center
Hoxan Corporation
Tsukuba, Ibaragi, Japan

and

S. Namba
University of Osaka Prefecture
Sakai, Osaka, Japan

ABSTRACT

Mechanical properties of carbon-glass hybrid composite materials have been studied in order to check the applicability of hybrid composite materials. Two types of interlaminate hybrid composites which were reinforced by carbon and glass cloth were prepared. Their volume fractions of reinforcement were identical, but the geometrical arrangement of the reinforcement was different. Tensile tests were performed and acoustic emission was monitored during the mechanical tests. Dynamic Young's moduli were also measured by the flexural vibration method. Mechanical properties were found to vary with the arrangement of the reinforcements. The acoustic emission characteristics were different from those of GFRP. The applicability of hybrid composite materials is also discussed. The mechanical properties of hybrid materials could be calculated by the law of mixture or the composite beam approximation.

INTRODUCTION

Organic composite materials have been used as insulating and structural materials at cryogenic temperatures. Their advantages are mainly attributed to their good electrical or thermal insulating properties and high specific strength at low temperatures. The specific moduli, however, are relatively low compared with those of metallic materials, and hence it has been difficult to construct large-scale structures with the usual GFRP (glass fiber reinforced plastics).[1] The development of carbon fibers, has enabled the production of CFRP (carbon fiber reinforced plastics) with high moduli.[2,3]

Composite materials have been developed aiming at (i) high specific modulus, (ii) high specific strength, and (iii) low cost. Besides, these the hybrid materials have shown resistance to fatigue, creep, and wear.[4]

Even though hybrid materials seem promising, their characteristics at low temperatures are not yet sufficiently clarified. The hybrid materials in this study were developed for use in support members of cryogenic apparatus. The mechanical properties of these hybrid materials at low temperatures have been studied and the results are reported.

EXPERIMENTAL METHODS

Samples

The samples made were interlaminated hybrid material reinforced by carbon and glass cloths. The volume faction of each reinforcement was approximately 24%, that is, the total volume fraction of reinforcement was 50%. Two types of hybrid materials were made, as illustrated in Fig. 1. The order of lamination was different, that is, carbon, glass, carbon (called CGC) or glass, carbon, glass (GCG). The thickness of the hybrid materials was approximately 3.2 mm.

The glass cloths used were plain woven E-glass cloths, WE35D® supplied by Nitto Boseki Co., Ltd. The carbon cloths were also plain woven cloths fabricated from high modulus carbon fiber M-40® supplied by Toray Co., Ltd. Both cloths have similar density of mesh. The specifications of each cloth are presented in Table 1. The matrix of the hybrid materials was Araldite® (CY205/HY905/DY040/DY061). The specimens were cut along the warp direction.

Dynamic Flexural Modulus Measurement

Dynamic flexural modulus of 10-mm x 100-mm x 3.2-mm specimens was measured by the flexural vibration method[5] at temperatures ranging from 150 to 300 K. The rate of temperature change was 3 K/min. The measurements were performed with a strain amplitude of 7×10^{-7}. The dynamic flexural modulus (GPa) was calculated from the resonant frequency using the following formula:

$$E = 9.465 \times 10^{-10} l^3 w f^2 / a^3 b,$$

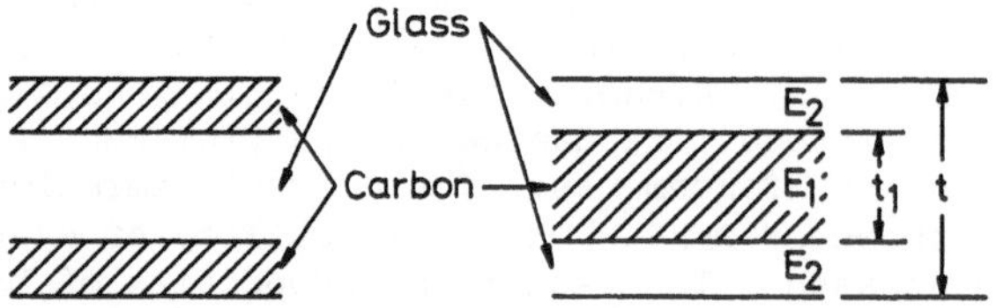

Fig. 1. Schematic illustration of CGC and GCG hybrid material.

Table 1. Specifications of Reinforcements

	Type	Density of Mesh (/cm)	Density (g/m^2)	Thickness (mm)
Glass cloth	E-glass (WE35D)	6.3 x 5.9	335	0.30
Carbon cloth	high modulus (M-40)	5.1 x 5.1	180	0.22

where l is the length of the specimen (mm); w, the weight of the specimen (g): f, frequency; d, thickness (mm); and b, width (mm).

Tensile Modulus Measurement

Tensile modulus was measured by attaching a strain gauge on the surface of the specimen, whose dimensions were 10 mm x 100 mm x 3.2 mm. The measurements were performed at room (RT) and liquid nitrogen temperature (LNT).

Tensile Strength and AE Measurement

Tensile tests were made with Universal Testing Machine at RT and LNT. The testing speed was 1.0 mm/min. The shape and the dimensions of the specimens are shown in Fig. 2. An AE sensor (PZT transducer) was attached to the surface of the specimen with vacuum grease.[6] The resonant frequency of the sensor was 140 kHz.

RESULTS AND DISCUSSION

Table 2 shows the tensile moduli obtained in the tensile test. The CFRP shows the largest tensile modulus, and the GFRP, the lowest. The calculated moduli of hybrid materials based on the law of mixture (parallel model) are also tabulated. The 5% and 13% differences between obtained and calculated values were found in CGC and GCG, respectively. Considering the misalignment of the cloth reinforcement, close agreement between the experimental and calculated data could be obtained. It is concluded that the tensile modulus of hybrid material could be calculated from the law of mixture.

The temperature dependence of dynamic flexural modulus is shown in Fig. 3. The flexural modulus of CFRP is the highest and that of GFRP is the lowest at every temperature. Concerning the hybrid materials, CGC presents a larger modulus than GCG.

One of the main purposes of hybridization is the improvement of the specific modulus. Especially, the flexural modulus can be improved by arranging the high modulus materials outside. The flexural modulus of hybrid material E_H is derived from the following formula:

$$E_H=\{E_1 t_1^3+E_2(t^3-t_1^3)\}/t^3,$$

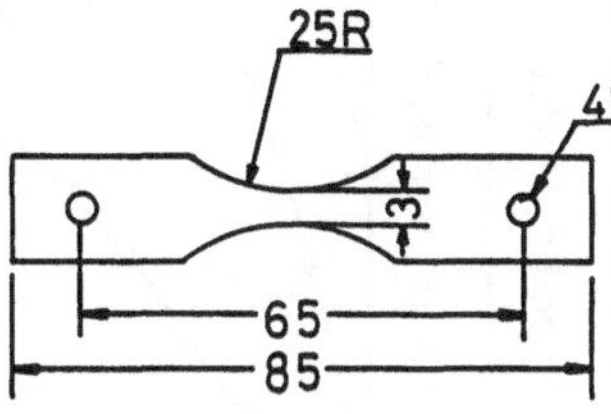

Fig. 2. Shape and size of the specimen in the tensile test.

Table 2. Tensile Modulus of Each Composite (GPa).

	CFRP	GFRP	GCG	CGC	THEORY*
RT	84.4	22.8	60.9	52.3	53.6
LNT	87.0	29.0	65.8	55.0	58.0

*Calculated by the law of mixture

where E is the Young's modulus of material 1; E_2, the Young's modulus of material 2; t, thickness of the hybrid material; t_1, thickness of material 1 (see Fig. 1).

Using the Young's modulus of CFRP, E_C and GFRP, E_G in Table 2, the flexural moduli of hybrid materials were calculated. The calculated flexural moduli of CGC, E_{CGC} and GCG, E_{GCG} were 76.7 and 30.4 GPa, respectively. Good agreement was found between calculated and experimental data (Fig. 3).

The increased rate of flexural modulus, R, was calculated as follows when the carbon cloths are arranged outside:

$$R=(t_1/t)^3+(E_C/E_G)\{1-(t_1/t)^3\}$$

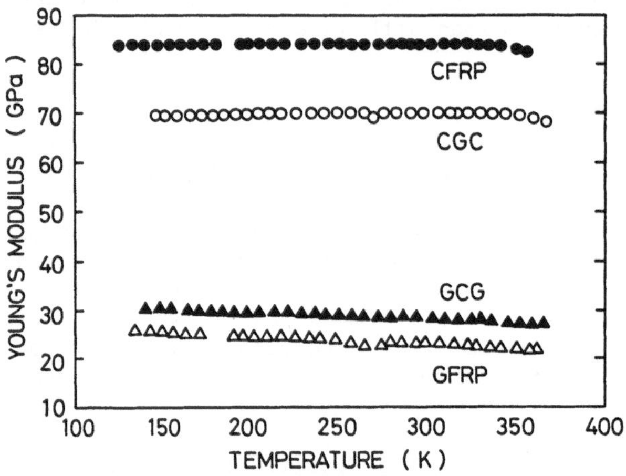

Fig. 3. Temperature dependence of dynamic flexural modulus in each material.

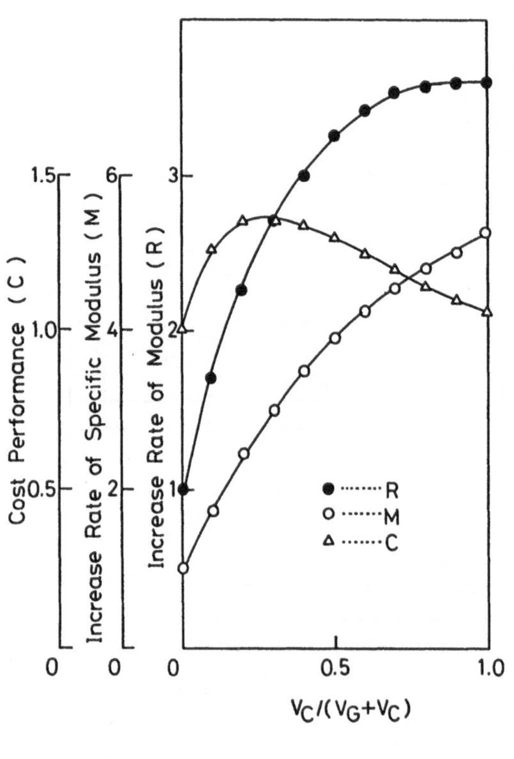

Fig. 4. Carbon fraction dependence of increased rate of modulus (R), increased rate of specific modulus (M), and cost performance (C).

Here, let the total sum of glass fraction, V_G, and carbon fraction, V_C equal 50% and also E_C/E_G equal 3.6. The carbon fraction dependence of R is calculated as in Fig. 4. The validity of hybridization can be understood. The upper limit of R is E_C/E_G, that is, 3.6, and hence the excessive increase of carbon is not advantageous. The change of specific flexural modulus, M, is also presented in this figure. The increase of carbon fraction causes the decrease of density, and hence the specific flexural modulus increases with carbon fraction. Since the other objective of hybridization is lower cost, the cost performance, C, was calculated and is also shown in this figure. The cost performance was calculated as follows: The prices of hybrid materials are normalized by that of GFRP and then the M is divided by the normalized price: The C means the improvement of M per unit cost. The hybrid materials containing 10 to 20% carbon reinforcement showed the highest cost performance.

Figure 5 shows the load-displacement curves of each material obtained at RT and LNT. The tensile strength of CFRP was smaller than that of GFRP at both temperatures. The hybrid materials showed almost the same tensile strength between CGC and GCG. In the hybrid materials the carbon and the glass fibers broke simultaneously at RT. On the contrary, at LNT the load-displacement curves of hybrid materials presented characteristic behavior, that is, they showed a sudden decrease of load followed by the lower slope of the curve to fracture.

The strength of hybrid materials is analyzed in Fig. 6. In this figure, the total sum of V_G and V_C remains 50%. Since the fracture displacement of the carbon fiber should be lower than that of the glass fiber, the carbon fibers break earlier than glass fibers in the hybrid materials. Just before the fracture of the carbon fibers, the mean stress in hybrid materials is presented as line BD in this figure. The point B means the stress in GFRP corresponding to the strain of the carbon fracture. This point can be obtained from Fig. 5 experimentally.

The point D means the tensile strength of CFRP. After the fracture of carbon fibers, the stress would be supported by glass fibers. The durable stress in glass fibers is presented as line AE. The point A corresponds to the tensile strength of GFRP. The point E is the contribution of matrix, which is quite small in comparison with the tensile strength of hybrid materials. At the lower carbon fraction, the stress can be supported by

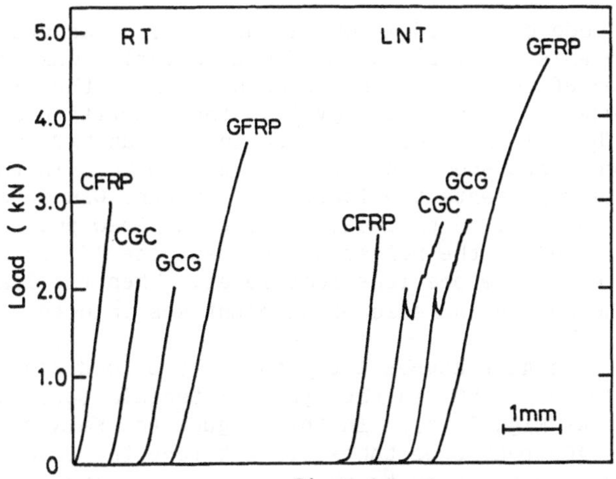

Fig. 5. Load-displacement curves obtained at RT and LNT for each material.

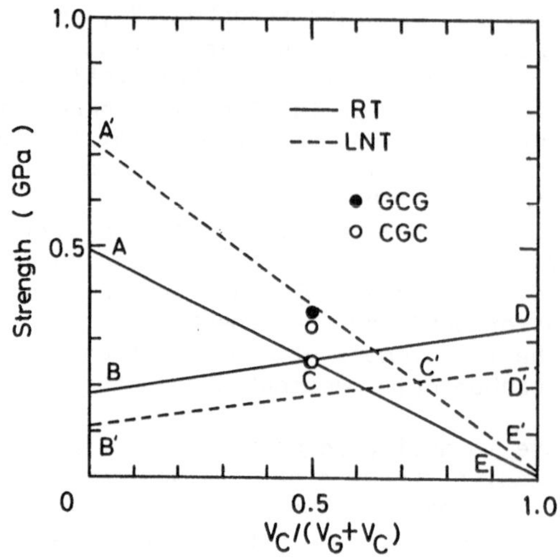

Fig. 6. Law of mixture of tensile strength in hybrid materials.

glass fibers even after the fracture of carbon fibers, as is shown by line AC. When the carbon fraction is high, the fracture of carbon fibers induces fracture of the hybrid materials and then the strength is presented as line CD. The volume fractions of the carbon and the glass fibers are both 25% in these hybrid materials, and hence the strength of the hybrid materials is presented as C at RT. This figure predicts the simultaneous fracture of carbon and glass fibers. The experimental data are also shown in this figure and close agreement was obtained.

At LNT the different behavior of strength can also be derived from Fig. 5. The diagram is demonstrated as broken lines and dashed points. At LNT simultaneous fracture will occur at point C'. In the LNT experiment the strength was determined only by the strength of the glass fibers, and there was no contribution of the carbon fibers. This prediction can be confirmed experimentally, as given in Fig. 5. The experimental data are also presented as circle markings in Fig. 6. It is concluded that the tensile strength of hybrid materials can be calculated by the law of mixture even at low temperature.

The fracture mode of hybrid materials at LNT is characteristic, and hence the acoustic emission (AE) during the deformation was monitored. There is the Kaiser effect on the AE features in which the AE is not emitted until the load exceeds the previous level in the repetition of the load-unload process. The Kaiser effect obtained in GFRP at LNT is presented in Fig. 7 (a). The ordinate shows the cumulative AE counts in the ring-down mode and the abscissa presents the load. This figure was obtained in the repetition of the load-unload process. In the unloaded process there is no occurrence of AE. Even in the reloading process, the AE was not emitted until the load exceeded the previous load level. When the load exceeded the previous level, the sudden increase of AE count was induced.

In Fig. 7 (b) the AE features accompanied by a load drop (corresponding to Fig. 5) obtained at LNT in the hybrid materials (GCG) are presented in the same manner as Fig. 7 (a). In this figure the Kaiser effect does not exist, and the AE occurrence takes place before the load reaches the previous level. This is attributed to the fact that the stress rearrangement in the glass fibers (or GFRP) is caused after the fracture of the carbon fibers. The stress rearranging process could be detected by means of AE techniques.

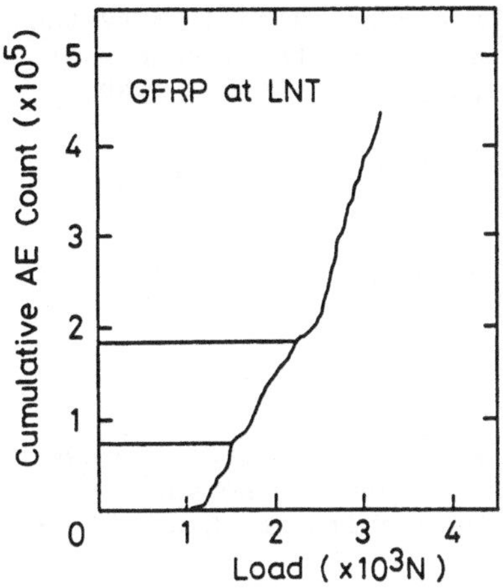

(a)

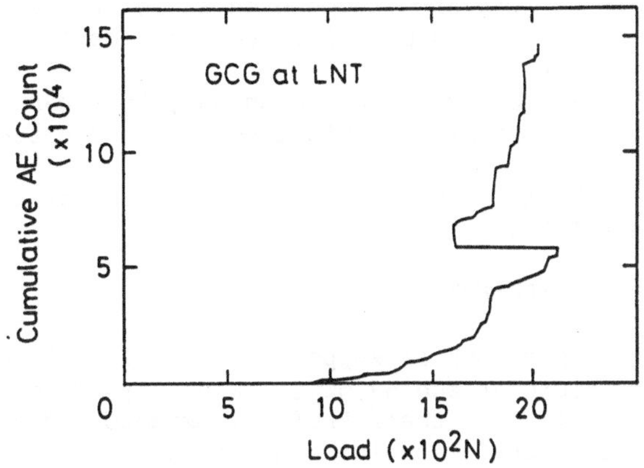

(b)

Fig. 7. (a) Kaiser effect in GFRP at LNT.
(b) AE accompanied by a load drop in hybrid material (GCG) at LNT.

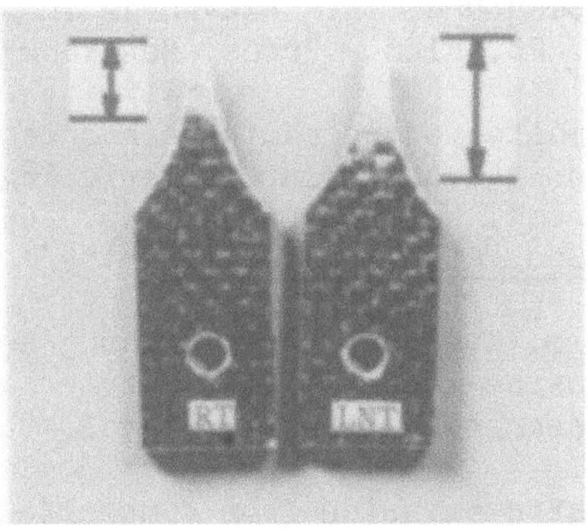

Fig. 8. Hybrid materials (GCG) near the areas of fractures that occurred at RT and LNT. The extent of the whitened area is shown by arrows.

Figure 8 shows the photographs of the fractured area of GCG hybrid material broken at both RT and LNT. The deformation of hybrid materials is still restricted by carbon fibers even after the fracture of carbon fibers. The excess deformation causes either the carbon fracture at another area (multiple fracture) or interlaminar failure between CFRP and GFRP. In this case the latter phenomenon takes place. Since this material is the GCG hybrid, the area of interlaminar failure could be distinguished as the whitened area. The area in the specimen tested at RT is smaller than that at LNT. This means that the glass fibers show a larger fracture displacement at LNT compared with that at RT, and hence the degree of mismatching of fracture strain between the carbon and the glass fibers at LNT is enhanced compared with that at RT.

CONCLUSIONS

The hybrid composite materials were made and were tested at RT and LNT. The following conclusions were drawn:

1. The tensile moduli of the hybrid materials can be calculated by the law of mixture. The flexural modulus is also predicted by the composite beam approximation.

2. The tensile strength of hybrid materials can be calculated by the law of mixture even at low temperature and hence the strength of hybrid materials could be derived form those of GFRP and CFRP.

3. Using the AE technique, the characteristics of fracture process in hybrid materials could be detected.

We have been designing a FRP cryostat having thinner walls with hybrid materials.

ACKNOWLEDGMENT

This work is supported in part by Grant in Aid for Scientific Research No. 59050044 and 59750053, Ministry of Education in Japan. The authors would like to thank Prof. T. Okamoto and Mr. K. Matsushita at ISIR for dynamic Young's modulus measurements.

REFERENCES

1. M. Takeno, S. Nishijima, T. Okada, K. Fujioka, and Y. Kuraoka, Thermal and mechanical properties of advanced composites materials at low temperatures, in: "Advances in Cryogenic Engineering - Materials," vol. 32, Plenum Press, New York (1986).
2. G. Hartwig, Reinforced polymers at low temperature, in: "Advances in Cryogenic Engineering - Materials," vol. 28, Plenum Press, New York (1982), p. 179.
3. M. B. Kasen, Mechanical performance of graphite- and ceramic-reinforced composites at cryogenic temperatures, in: "Advances in Cryogenic Engineering - Materials," vol. 28, Plenum Press, New York (1982), p. 165.
4. H. Fukuda, Mechanical of Hybrid, J. Japan Soc. Comp. Mat. 9: 153-159 (1983).
5. T. Okada, S. Nishijima, K. Matsushita, and Y. Kuraoka, Dynamic Young's modulus and internal friction in a composite material at low temperatures, in: "Advances Cryogenic Engineering - Materials," vol. 30, Plenum Press, New York (1984), p. 9.
6. S. Nishijima, T. Okada, and S. Namba, Acoustic emission from composite materials at low temperatures, in: "Advances Cryogenic Engineering - Materials," vol. 30, Plenum Press, New York (1984), p. 25.

MECHANICAL PROPERTIES OF UNIDIRECTIONALLY REINFORCED COMPOSITE MATERIALS

T. Okada and S. Nishijima
ISIR, Osaka University
Ibaraki, Osaka, Japan

H. Yamaoka and K. Miyata
KURRI, Kyoto University
Kumatori, Osaka, Japan

K. Fujioka and Y. Kuraoka
Cryogenic Technology Development Center
Hoxan Corporation
Tsukuba, Ibaragi, Japan

S. Namba
University of Osaka Prefecture
Sakai, Osaka, Japan

ABSTRACT

Mechanical properties of unidirectional glass fiber reinforced composite materials with systematic variations of the matrix have been studied aiming at standardization of materials and testing methods. Mechanical properties of identical materials were measured independently at three different laboratories, and the results were compared. The deviation of data caused by different experimental conditions is discussed. The desirable testing conditions and the possibility of standardization are also discussed.

INTRODUCTION

The mechanical properties of composite materials are scattered even for identical materials, depending on the size and/or shape of the specimen and the manner and/or direction of the applied force. Furthermore, there are a number of possible reinforcements and matrices and many ways of constructing composites.[1] Consequently, in this field of study, standard methods for accumulation of various data and the establishment of standard testing methods are both desirable in the development of new materials. This is particularly true at low temperatures because of the difficulty in performing the experiments.

In Japan, three separate groups have made flexural tests on standard unidirectionally reinforced specimens aiming at establishing a reliable testing method.[2] The results were compared to establish optimal conditions desired for such tests. This series of experiments is directed toward the development of radiation-resistant composite materials.

Table 1. Experimental Conditions in Each Group

Type of Bending	Three Point			Four Point
Laboratory	ISIR	KUR	HOXAN	ISIR
Span (mm)	60	40	60	54
Loading tip radius (mm)	2	5	5	5
Supporting tip radius (mm)	0.5	2.5	2	5
Deformation speed (mm/min)	1.6	1	5	1.6

EXPERIMENT

Sample

The samples are unidirectional glass fiber reinforced plastics made by pultrusion in vacuum.[3] The specimen has a cylindrical shape, 100 mm in length and 3 mm (average: 3.025 mm) or 4 mm (average: 4.015 mm) in diameter. The volume fraction of the reinforcement is approximately 50%. The matrix is Araldite® (CY205/HY905/DY040/DY061) supplied by the Ciba Geigy Company. Two types of epoxy matrix are made by changing the amount of the diluent (DY040), that is matrix A (100/100/15/1) and matrix B (100/100/60/1) of weight fraction. The reinforcement is E-glass fiber with silane finish fiber supplied by the Nitto Boseki Company.

Flexural Test

Flexural tests were performed at room (RT) and liquid nitrogen temperature (LNT). Both three- and four-point test methods were studied. The effects of span length and loading tip radius on the mechanical behavior were investigated for the three-point method. In Table 1 the conditions of each test are presented. In the three-point bending test, comparison of the data obtained by Hoxan and ISIR reveals the effects of loading tip radius, and that between Hoxan and KUR clarifies the effects of span length. The difference between three- and four-point bending was revealed by comparing the data obtained at Hoxan with that obtained at ISIR.

RESULTS

The Young's moduli and the breaking stresses were calculated by simple beam theory. At RT the specimen containing matrix B did not fracture because of the ductility of the matrix. Table 2 shows the results obtained by each group. The number of specimens tested was three. The values in brackets give the standard deviation.

The results of tests on the specimen with matrix A are plotted against span/diameter ratio in Fig. 1. The open and solid markings present the results obtained by 5-mm and 2-mm tip radius in three-point bending, respectively. The symbols with a dot inside are results obtained in the four-point bending test of 5-mm tip radius. Concerning the flexural strength, the span/diameter ratio did not affect it much either at RT or LNT in the present experiments. Young's modulus, however, increased with the increase in span/diameter ratio at both RT and LNT. Data of both Young's modulus and the breaking stress obtained with four-point bending were larger values than those of three-point bending. When the span/diameter ratio was small, Young's modulus and the breaking stress obtained by the 2-mm tip radius were smaller than those by the 5-mm tip radius. For large ratios, the effect of the tip radius was minor.

Table 2. Results Obtained in Both Three- and Four-Point Bending

		Three-Point Bending				Four-Point Bending
		Matrix B		Matrix A		Matrix A
		3mmΦ	4mmΦ	3mmΦ	4mmΦ	4mmΦ
R T	ISIR	20.6	19.9	40.2	33.6	36.5/2/
		(1.7)	(0.26)	(3.2)	(0.33)	(0.35)
Young's modulus	KUR	7.54	5.18	33.0	27.7	-------
		(0.34)	(0.35)	(0.42)	(0.37)	
(GPa)	HOXAN	29.7/1/	27.6	37.5	34.6/2/	-------
		(0.0)	(0.32)	(0.42)	(0.44)	
	ISIR	-----	-----	0.940	0.773	1.07/2/
				(0.017)	(0.012)	(0.02)
Breaking stress	KUR	-----	-----	0.909	0.836	-------
				(0.054)	(0.017)	
(GPa)	HOXAN	-----	-----	1.19	0.956/2/	-------
				(0.041)	(0.0034)	
L N T	ISIR	38.2	35.3	40.1	37.4	40.1/2/
		(1.7)	(1.0)	(1.4)	(1.2)	(0.75)
Young's modulus	KUR	36.0	34.3	36.7	34.1	-------
		(0.13)	(0.42)	(0.16)	(0.49)	
(GPa)	HOXAN	38.6/2/	38.8	40.5	38.6	-------
		(0.44)	(0.98)	(0.88)	(0.24)	
	ISIR	2.12	1.88	1.95	1.75	2.28/2/
		(0.16)	(0.053)	(0.070)	(0.066)	(0.045)
Breaking stress	KUR	2.51	2.17	2.49	2.10	-------
		(0.18)	(0.021)	(0.087)	(0.023)	
(GPa)	HOXAN	1.93/2/	2.19	>2.09	2.10	-------
		(0.15)	(0.0056)	(0.034)	(0.068)	

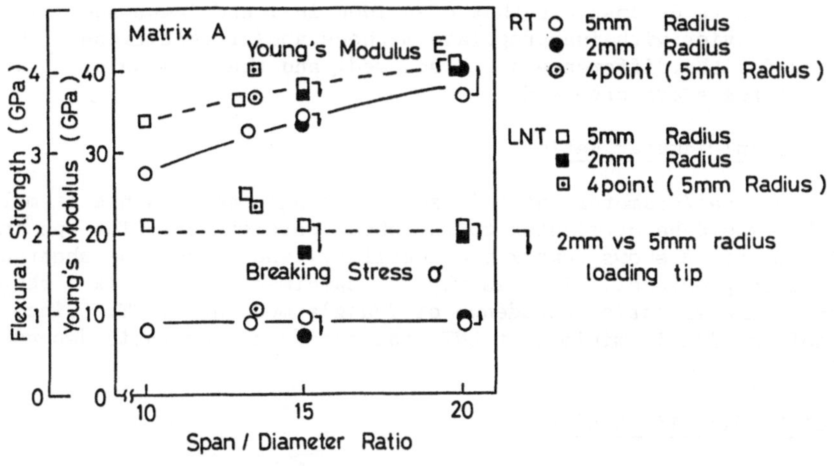

Fig. 1. Span/diameter ratio dependence of breaking stress and Young's modulus with matrix A. Downward arrows show that the results obtained by 2-mm radius are smaller than those by 5-mm radius.

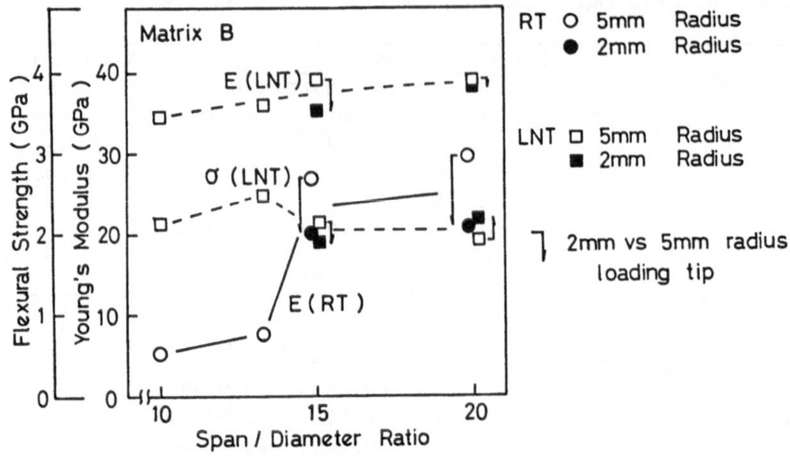

Fig. 2. Span/diameter ratio dependence of breaking stress and Young's modulus with matrix B. Upward arrows indicate where the results obtained by 2-mm radius were larger than those by 5-mm radius.

In Fig. 2, the results obtained for matrix B are shown. The Young's modulus obtained at RT depends remarkably on the span/diameter ratio. Loading tip radius had an effect with the smaller span/diameter ratios.

It could be summarized:

1. Young's modulus depended on span/diameter ratio and increased with the ratio.

2. At smaller span/diameter ratios, the effect of loading tip radius was notable.

3. The data obtained in four-point bending were larger than those in three-point bending.

4. The difference of matrix became negligible at low temperature.

DISCUSSION

These results demonstrate that even identical specimens show different mechanical behavior with inappropriate methods and/or conditions. The The reasons for the difference are discussed, and the desirable conditions for flexural tests are proposed.

Effects of Span/Diameter Ratio

When the span/diameter ratio is small, Young's modulus has a smaller value. This would be attributed to the shear deformation. The specimen containing matrix B shows remarkable ductility, and hence, the shear deformation becomes prominent. This is why the specimen with matrix B shows a large span/diameter ratio dependence of Young's modulus at RT. Since plastic deformation is minimal at LNT, the span/diameter ratio dependence is small.

Effects of Loading Tip Radius

The smaller tip radius resulted in a smaller breaking stress at a small span/diameter ratio. This could be explained by the stress concentration. The finite element method was used to analyze this. The calculation was made by assuming the elastic constants as follows: E_L = 3600 kg/mm^2,

E_T = 970 kg/mm^2, G_{LT} = 360 kg/mm^2, and υ = 0.27, where the subscripts L and T refer to the longitudinal and transverse directions, respectively. The shear stress distributions in the vicinity of the loading tip caused by the 2-mm and 5-mm tip are presented in Fig. 3. The deviation from the simple beam theory is relatively large. The 2-mm and 5-mm loading tips bring 2.5 and 1.5 times the stress concentration just under the loading tip, respectively.

In the three-point bending test, the large concentrated stress appears just under the loading tip, and it could initiate the fracture. The magnitude of the concentrated stress caused by the 2-mm tip radius is larger than that caused by the 5-mm tip radius, and hence, the breaking stress obtained for the 2-mm tip radius should be smaller than that for the 5-mm tip radius. Under the same loading, the displacement in the 2-mm tip radius is approximately 2% larger than that in the 5-mm tip radius, which leads to a decrease in Young's modulus.

When the span/diameter ratio is large enough, the stress concentration becomes small.[4] This is why the tip radius effect is minor for large span/diameter ratios.

Comparison between Three- and Four-Point Bending

In three-point bending, even when the tip radius is 5 mm, the stress just under the tip is larger than that expected from simple beam theory, and hence, the fracture starts just under the tip. This is confirmed by visual observation. The stress concentration in four-point bending was also calculated by the finite element method. The stress concentration is considerably smaller than that in three-point bending, and the stress distribution agrees well with the calculation by simple beam theory. This suggests that the data obtained in four-point bending are closer to the actual values than those in three-point bending, and therefore four-point

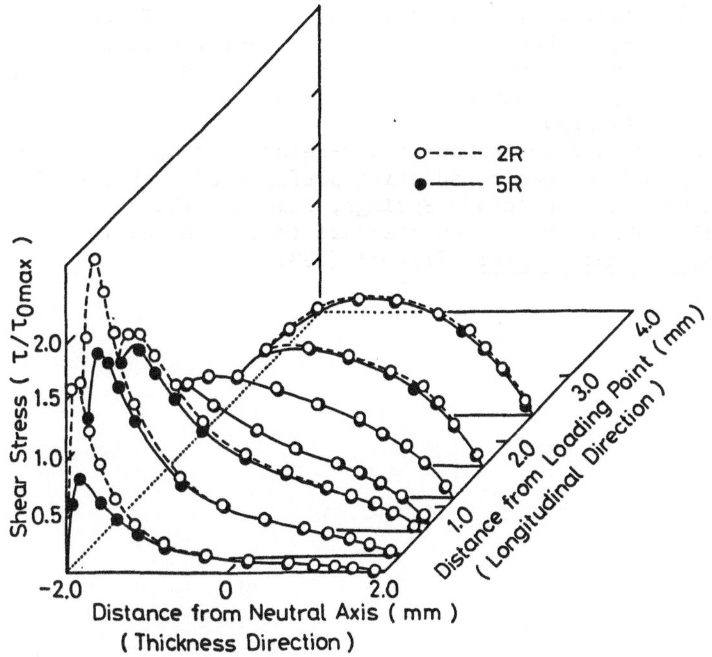

Fig. 3. Shear stress distributions in the vicinity of the loading point.

ending should be preferable. This is why data obtained in four-point bending are larger than those in three-point bending.

CONCLUSIONS

Flexural tests were performed under several conditions for the purpose of establishing a reliable testing method at cryogenic temperatures. Results were compared, and the following conclusions were drawn:

1. When the span/diameter ratio is too small, the true bending value can not be obtained because of the shear deformation and/or stress concentration. The span/diameter ratio should be larger than 20 in order to reduce the effects of both shear fracture and loading tip radius.

2. The tip radius must be larger than 5 mm, especially when the span/diameter ratio is smaller than 15. Even the 5-mm tip radius brings the stress concentration just under the tip, which should initiate fracture.

3. Four-point bending is preferable as a flexural test from the viewpoint of stress concentration

ACKNOWLEDGMENT

This work is partly supported by Grant in Aid for Scientific Research No. 60050044, Ministry of Education in Japan. This investigation originated from a suggestion by M. B. Kasen of NBS concerning the international joint research program on the standardization of testing methods for composite materials.

REFERENCES

1. M. B. Kasen, Standardizing nonmetallic composite materials for cryogenic applications, in: "Nonmetallic Materials and Composites at Low Temperatures - 2," G. Hartwig and D. Evans, eds., Plenum Press, New York (1982), pp. 232-337.
2. T. Okada, S. Nishijima, H. Yamaoka, K. Miyata, Y. Tsuchida, K. Kuraoka, and S. Namba, Mechanical properties of unidirectionally reinforced materials, in: "Nonmetallic Materials and Composites at Low Temperatures - 3," G. Hartwig and D. Evans, eds., Plenum Press, New York. In press.
3. M. B. Kasen, Method for making fiber-reinforced specimens for use in screening of cryogenic radiation performance, paper CP-13 presented at the CEC-ICMC, Colorado Springs, Colorado (1983).
4. M. Uemura, Problem and design standard in mechanical test of FRP, <u>Jap. Soc. Compos. Mater</u>. 7:74-81 (1981).

TWO-DIMENSIONAL THERMAL CONTRACTION OF COMPOSITES

S. Nakahara, T. Fujita, K. Sugihara
Department of Mechanical Engineering
Kansai University
Suita, Osaka, Japan

and

S. Nishijima, M. Takeno, T. Okada
ISIR, Osaka University
Ibaraki, Osaka, Japan

ABSTRACT

The thermal contraction of composite materials has been studied by changing the off-axis angle and the content of the glass fibers in order to analyze the thermal strain of anisotropic materials. The distributions of thermal contraction in a composite plate have also been measured down to liquid nitrogen temperature using laser speckle photography. Square-shaped composite plates were given the thermal gradient by cooling the center of the specimen. In-plane displacements were measured by speckle photography. The calculations of the strain distribution were made by a finite element method based on the off-axis angle dependence of the thermal contraction. Good agreement was found between calculated and experimental results.

INTRODUCTION

The thermal behavior of composite materials, either thermal contraction or thermal conductivity, is one of the most important properties when the composite materials are employed as a structural material for cryogenic use. The thermal contraction down to cryogenic temperatures is considerably greater than that of metals. Especially when the composite materials are used together with metals, the thermal stress due to differences of thermal contraction can be large. Since the thermal conductivity of the composite materials is also relatively small, the thermal shock associated with the cooling process could become a serious problem. It is, therefore, important to grasp the anisotropic characteristics of the thermal contraction in composite materials.

In order to examine the anisotropy of GFRP (glass fiber reinforced plastics), the off-axis angle dependence of thermal contraction was measured down to liquid nitrogen temperature (LNT). The glass content dependence was also obtained. The two-dimensional thermal contraction of GFRP plate was measured by the laser speckle photography technique.[1-3] The obtained results were compared with the results calculated by a finite element method (FEM).

The applicability of laser speckle photography to cryogenic engineering[3] will also be discussed.

EXPERIMENT

Figure 1 shows the configuration of the apparatus for measurement of the thermal contraction. The change of size induced by cooling was measured by the differential transducer equipped in room temperature space. The measurement was performed down to LNT. The temperature control was performed by changing the liquid nitrogen level outside of the quartz tube. As shown in Fig. 2, the cylindrical shapes of the specimens were cut from a GFRP plate. The size of the specimen is 35 mm in length and 10 mm in diameter. The matrix of the GFRP is epoxy and is reinforced by plane woven E-glass cloths. The GFRP was produced by a vacuum impregnation method. The glass content is approximately 73% by weight.

The glass content dependence of thermal contraction was also measured with respect to the fiber and the thickness direction. In this case, two types of production methods were used, that is, hand lay up and vacuum impregnation methods.

The optical system for laser speckle photography technique (LSP) is shown in Fig. 3. The cryostat gives the temperature gradient in the specimen and has a observation window made of an acrylic acid resin for LSP. The temperature gradient is given by cooling the center of the specimen with thermal conduction. The material tested is commercially available glass cloth reinforced epoxy, G-10CR.[4] The glass content is approximately 66% by weight.

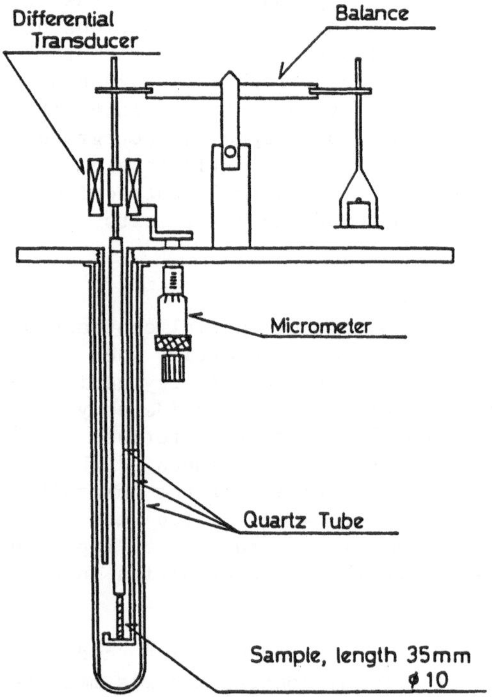

Fig. 1. Apparatus for thermal contraction measurement.

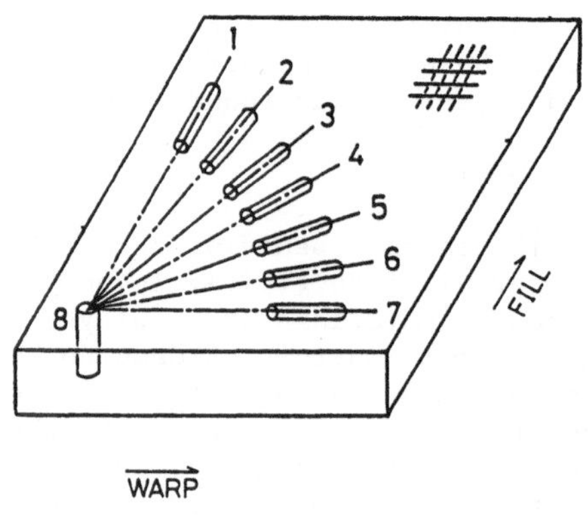

Fig. 2. Cutting out configuration of the specimen from GFRP plate. Specimens were cut out every 15° from the fiber direction.

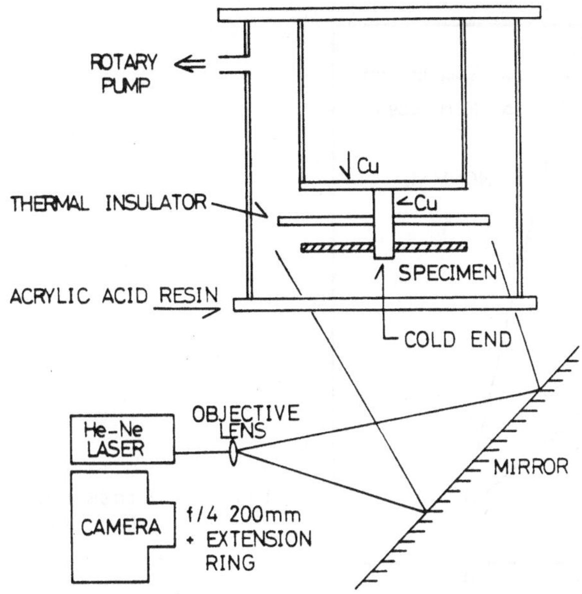

Fig. 3. Cryostat and optical system for laser speckle photography.

The square-shaped GFRP plate, with dimensions of 85x85x1 mm, was attached to the cold end by screws. The camera system used to record the speckle photographs was made up of an f/4 Nikkor lens of 200-mm focal length and an extension ring (Nikon PK11). The object-camera separation was chosen so that the image substantially filled the 35-mm photographic film (minicopy film HRII, Fuji Photo Film Co., 200 lines/mm), giving a demagnification factor of about 5.3. The sample was illuminated by light from a helium-neon laser (about 2 mW), extended by a microscope objective lens, and reflected by the mirror. The temperature distribution of the specimen was measured by thermocouples.

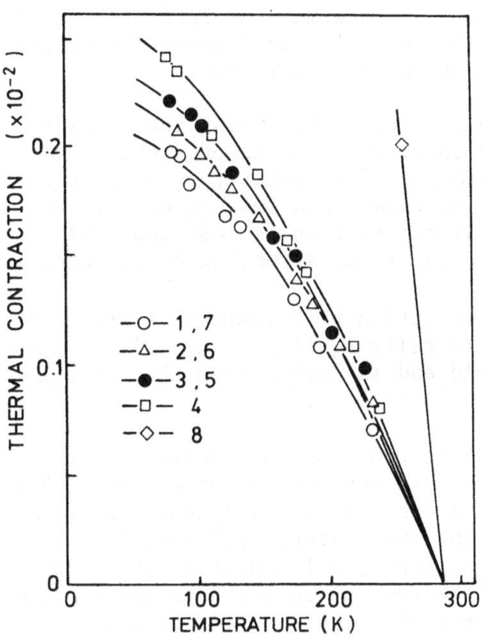

Fig. 4. Off-axis angle dependence of thermal contraction. The numbers in the figure correspond to those in Fig. 2.

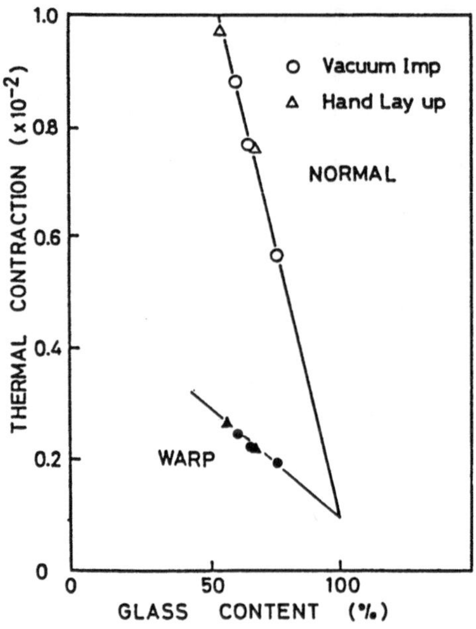

Fig. 5. Glass content dependence of thermal contraction.

RESULTS AND DISCUSSION

Figure 4 shows the off-axis angle dependence of the thermal contraction. The thermal contraction was defined as the ratio between the decrement of the size and the initial size. It is seen that the thermal contraction changes symmetrically with respect to 45° orientation and is minimized with respect to 0° and 90°. The maximum value of thermal contraction was obtained in the thickness direction.

The glass content dependence of the thermal contraction is shown in Fig. 5. The measured values can be approximated by two straight lines. The two straight lines intersect at glass content of 100%. The thermal contraction at the point corresponds well to that of glass fibers. It seems, however, that with respect to GFRP containing less than 40% glass content, the law of mixture does not hold. The thermal contraction of epoxy resin was found to be approximately 1%. There is no significant difference in thermal contraction in spite of different production methods.

Figure 6 shows the temperature change of GFRP plate during the cooling down process at the various points of the specimen. It takes about 40 minutes for the sample to reach the steady state. The temperature difference between the observed surface and the back surface was a few degrees at the most. Then, the difference of the thermal contraction is less than 5×10^{-5}. The large out-of-plane deformations in the thickness direction do not occur.

Figure 7 presents the temperature distribution with respect to the distance from the cold end after the system is fully equilibrated. It is seen that the temperature gradient near the cold end is larger than that at distant region.

In Fig. 8, the two-dimensional thermal contractions are shown. The magnitude and direction of the displacement are represented as arrows. The direction of fiber in the specimen tested is not uniform and hence the distribution is complicated in comparison with ideal materials.[5] Young's fringe on the center of the sample represents a uniform displacement of the optical system. The net displacement induced by the thermal contraction was obtained by subtracting the uniform displacement from the experimental values.

The displacement caused by the thermal contraction is derived from the Young's diffraction fringes that are obtained by optical Fourier transformation of the double exposed films. From the spacing and the orientation of these fringes, the magnitude and the direction of the speckle displacement were derived.

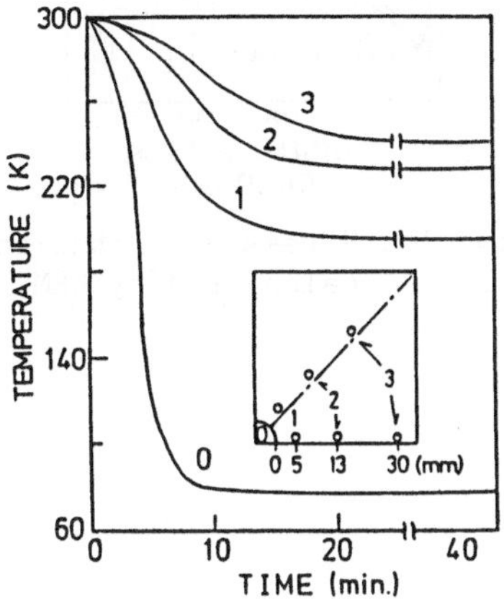

Fig. 6. Temperature change of GFRP plate during cooling test at various points.

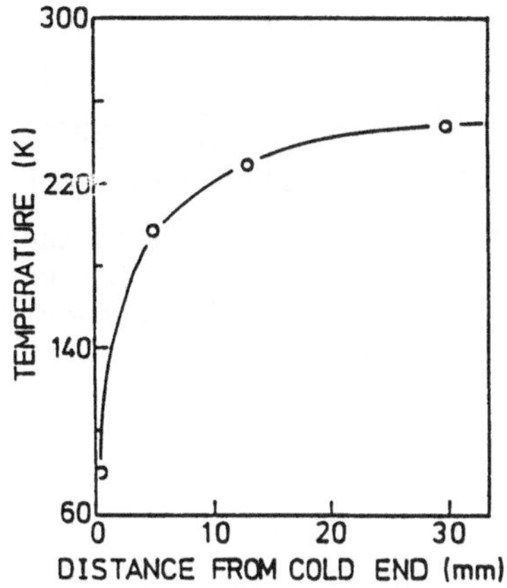

Fig. 7. Temperature distribution on the GFRP plate in steady state.

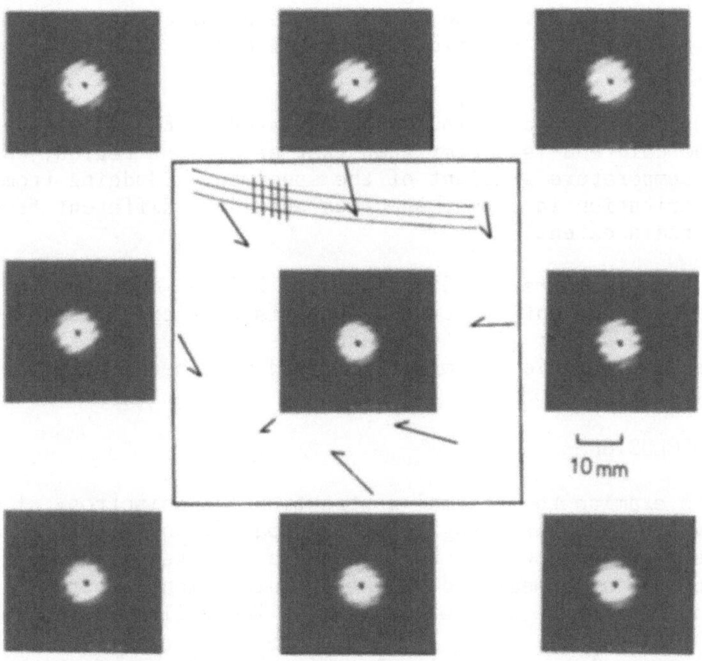

Fig. 8. An example of the fringe patterns and the two-dimensional thermal contraction on the square-shaped GFRP plate. Fiber orientations, which are not always parallel to edges, are also shown with thin solid lines.

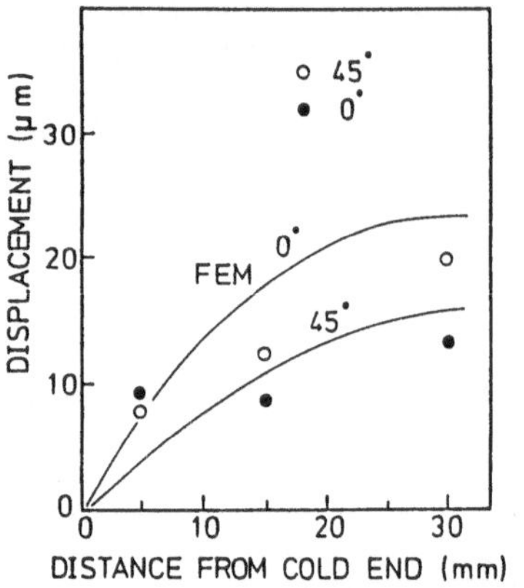

Fig. 9. Correlation between thermal contraction vs. distance from cold end together with calculated value by FEM.

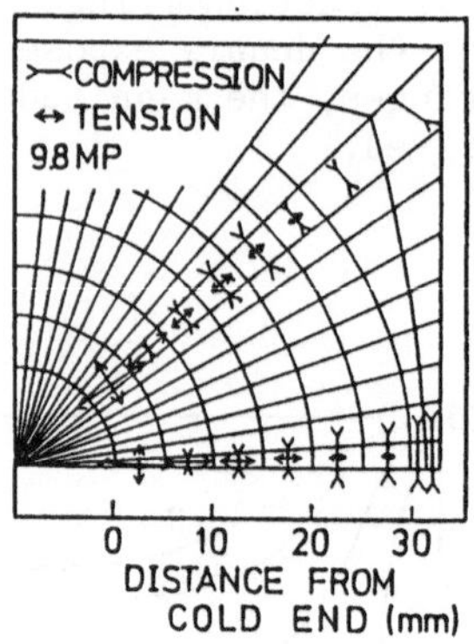

Fig. 10. Stress distribution calculated by FEM.

Figure 9 shows the magnitude of displacement with respect to the distance from the cold end. The solid lines are calculated values of the thermal contraction by FEM. Good agreement between calculated and experimental values were obtained. The direction of the contraction, however, does not point toward the cold end exactly, as expected in Fig. 8. The elastic modulus of the specimen varies with the off-axis angle and hence the out-plane displacement could be induced. The misalignment of the reinforcement would be another reason. As shown in Fig. 8 the orientation of the reinforcement is disordered considerably. The out-of-plane displacement and the misalignment of reinforcement made the two dimensional contraction more complicated than expected.

In Fig. 10 the stress distribution calculated by FEM is presented. The stress near the cold end is larger than that at distant region, which corresponds to the temperature gradient of the specimen. Judging from Fig. 8, the stress distribution in a real specimen should be different from the calculation to certain extent.

In spite of the temperature distribution, LSP can measure the displacement, while the measurement by strain gauges is difficult because of the change of the gauge factor with temperature. The LSP is not affected by the temperature, electric field and magnetic field and, hence, has the possibility of cryogenic applications.

SUMMARY AND CONCLUSIONS

In order to examine the thermal contraction and anisotropy of GFRP, the samples with various off-axis angles were prepared, and their thermal contractions were measured down to LNT. The glass content dependence of thermal contraction was also measured. The two-dimensional thermal contraction was measured on GFRP plate.by the LSP method. The following conclusions were drawn:

(1) The off-axis angle and the glass content dependence of thermal contraction was clarified.

(2) With respect to the two dimensional in-plane thermal contraction of GFRP, the magnitude of the displacement measured by LSP coincided with the calculation by FEM. Concerning the direction of the thermal contraction,

the calculation did not always agree with the experimental results. This is thought to be caused by the out-of-plane deformation and/or misalignment of the reinforcements.
(3) By using LSP, the behavior of the two-dimensional thermal contraction of GFRP plate was measured. The application of LSP to cryogenic displacement measurement was confirmed. This noncontacting and temperature-independent method is an advantage at low temperatures.

ACKNOWLEDGMENT

This work is supported in part by Foundation for In-Service Training and Welfare of the Private School Personnel in Japan. The authors would like to thank Dr. Y. Kuraoka and Mr. Y. Tsuchida at Hoxan Corp. for making the specimen.

REFERENCES

1. A. E. Ennos and M. S. Virdee, Laser speckle photography as a practical alternative to holographic interferometry for measuring plate deformation, Opt. Eng. 21: 478-482 (1982).
2. I. Yamaguchi, Fring formation in speckle photography, J. Opt. Soc. Am. A1: 81-86 (1984)
3. E. A. Fuchs and R. E. Rowlands, Photomechanical stress analysis under cryogenic environments, in: "Advances in Cryogenic Engineering--Materials," vol. 30, R. P. Reed and A. F. Clark, eds., Plenum Press, New York (1984), pp. 111-117.
4. M. B. Kasen, G. R. MacDonald, D. H. Beekman, Jr., and R. E. Schramm, Mechanical, electrical, and thermal characterization of G-10CR and G-11CR glass-cloth/epoxy laminates between room temperature and 4 K, in "Advances in Cryogenic Engineering--Materials," vol. 26, A. F. Clark and R. P. Reed, eds., Plenum Press, New York (1980), pp. 235-244.
5. S. P. Timosenko and J. N. Goodier, Thermal stress, in: "Theory of Elasticity," third edition, McGraw-Hill (1970), pp. 441-443.

THERMAL AND MECHANICAL PROPERTIES OF ADVANCED COMPOSITE MATERIALS AT LOW TEMPERATURES

M. Takeno, S. Nishijima, and T. Okada
ISIR Osaka University
Ibaraki, Osaka, Japan

K. Fujioka, Y. Tsuchida, and Y. Kuraoka
Cryogenic Technology Development Center
Hoxan Corporation
Tsukuba, Ibaragi, Japan

ABSTRACT

Thermal and mechanical properties of advanced composite materials have been studied in order to examine their applicability to cryogenic use. Carbon, silicon carbide and alumina fiber unidirectionally reinforced epoxies were prepared as advanced composite materials and were compared with glass fiber reinforced epoxy. Thermal conductivity and thermal contraction have been measured in both directions parallel and perpendicular to fibers down to liquid helium and liquid nitrogen temperatures, respectively. To investigate mechanical properties, a four-point flexural test was also performed at room, liquid nitrogen and liquid helium temperatures, and Young's moduli and breaking stress were calculated. Considering the specific modulus and the ratio of breaking stress to thermal conductivity, the advanced composite materials are found to be suitable for cryogenic structural support members.

INTRODUCTION

Composite materials reinforced by glass fiber (GFRP) have been widely used as insulating and/or structural materials for cryogenic apparatus.[1,2] They show not only good thermal and electrical insulating properties but also nonmagnetism, which is desirable for cryostats with SQUID, NMR and pulsed magnets. The specific modulus of the glass fiber reinforced epoxy is, however, quite low compared with those of metals. Composite materials which are reinforced by carbon (CFRP), silicon carbide (SFRP) and alumina (ALFRP) fiber could be desirable for the cryogenic structural materials because they have high specific strength and modulus. Some of them show good thermal and electrical properties at room temperature (RT).

When considering the advanced composite materials for cryogenic use, it is necessary to investigate their thermal and mechanical properties at cryogenic temperatures. In this work, thermal conductivity and thermal contraction were measured down to liquid helium (LHeT) and liquid nitrogen temperature (LNT), respectively. The four-point flexural test was performed at cryogenic temperatures and the applicability of advanced composite materials is also discussed.

Table 1. Properties and Size of Each Reinforcement from Manufacturer

Fiber	Young's Modulus (GPa)	Tensile Strength (GPa)	Density (g/cc)	Fiber Diameter (μm)
Glass	75.5	2.45	2.5	9
Carbon	235	2.74	1.7	8
Silicon carbide	176	2.45	2.5	10
Alumina	206	1.76	3.2	17

EXPERIMENTS

Samples

The samples were unidirectionally reinforced plastics made by the pultrusion method. The matrix is ARALDITE® (CY205/HY905/DY040/DY061) supplied by CIBA GEIGY Co. Their proportion was (100/100/15/1) by weight, respectively. Four kinds of reinforcement were chosen: E-glass, carbon (high strength), silicon carbide (SiC) and alumina (γ-$A\ell_2O_3$) fiber. The E-glass fiber was supplied by Nitto Boseki Co., Carbon fiber, T-300® by Toray Co., Silicon carbide fiber, Nicalon® by Nippon Carbon Co., and Alumina fiber, Sumica Alumina® by Sumitomo Chemical Co. The properties and specifications of each reinforcement are shown in Table 1. The fiber contents of samples were approximately 50 percent in volume.

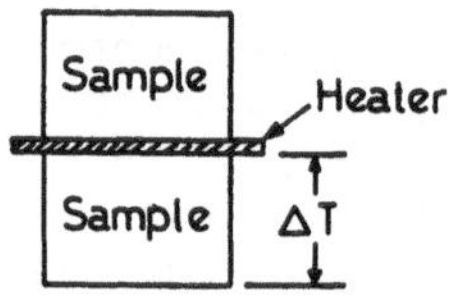

Fig. 1. Sample setup in thermal conductivity measurement.

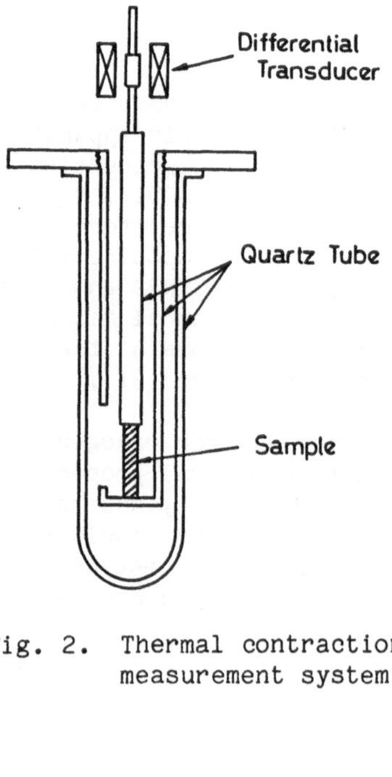

Fig. 2. Thermal contraction measurement system.

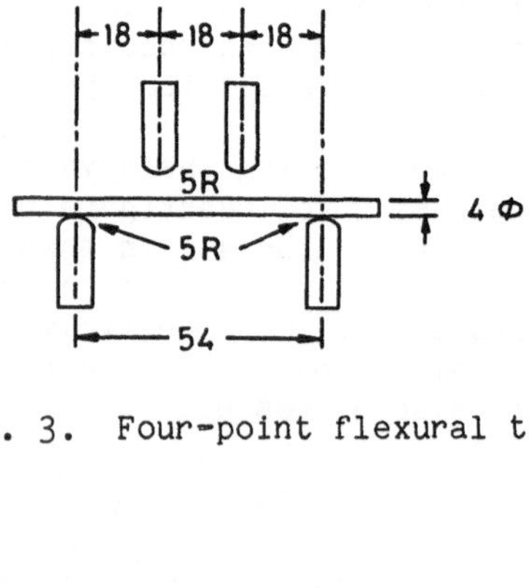

Fig. 3. Four-point flexural test.

Thermal Conductivity

The thermal conductivity, which is important in thermal design, has been measured from LHeT to RT in directions parallel and perpendicular to the fibers with a steady state technique.[6] The specimen is basically a disc shape, 10 mm in height and 16 mm in diameter. The cold end temperature was allowed to vary from LHeT to RT while maintaining a temperature gradient between both ends of approximately 1 K. The schematic setup of the specimen is shown in Fig. 1. The thermal conductivity, κ, was obtained from following formula:

$$\kappa = PH/\Delta TS$$

where P is thermal input per unit time, H is the height of specimen, ΔT is the temperature difference between the both ends of the specimen, and S is the cross section.

Thermal Contraction

The thermal contractions of the samples were also measured in directions parallel and perpendicular to fibers down to LNT. The specimen was of cylindrical shape, 35 mm in height and 10 mm in diameter for measurement in the longitudinal direction. In the case of the transverse direction, the discs were 10 mm in height and 16 mm in diameter. A schematic illustration of the thermal contraction measurement system is shown in Fig. 2. The temperature was controlled by changing the liquid nitrogen level in which the end of quartz tube was immersed. The displacement of the specimen was measured by a differential transducer attached to the other end of the quartz tube.

Flexural Test

To investigate mechanical properties, four-point flexural tests were performed at RT, LNT, and LHeT. The cylindrical specimen was 100 mm in length and 4 mm in diameter. A schematic illustration of four-point flexural test is shown in Fig. 3. The Young's modulus and the breaking stress were calculated by simple beam theory.

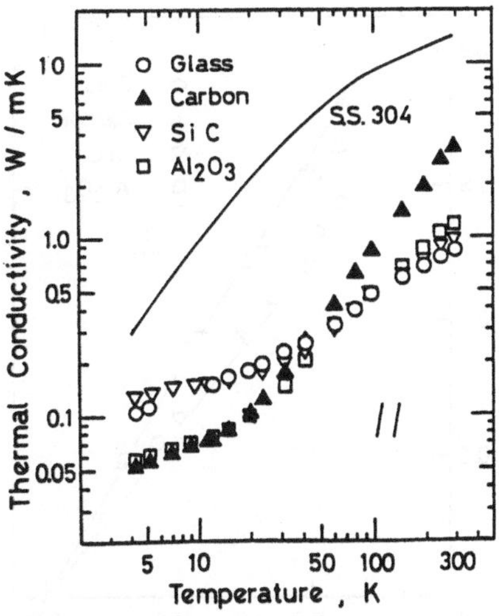

Fig. 4. Temperature dependence of the thermal conductivity in the fiber direction.

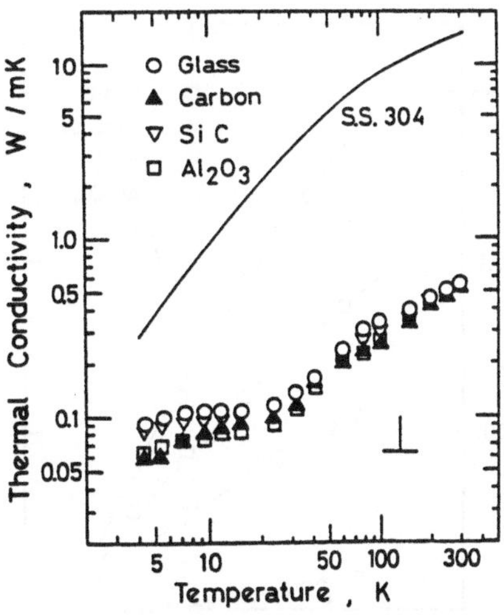

Fig. 5. Temperature dependence of the thermal conductivity in the transverse direction.

RESULTS AND DISCUSSION

Figure 4 shows the temperature dependence of the thermal conductivity in the fiber direction for the composite specimens and for 304 stainless steel from 300 K to 4.2 K. The thermal conductivity of advanced composite materials is lower than that of 304 stainless steel throughout the temperature range. It is obvious that the thermal insulating ability of composites is superior to that of 304 stainless steel even though stainless steel has a relatively low thermal conductivity compared with that of other metals. It is seen in Fig. 4 that the thermal conductivity in the fiber direction depends on that of the fibers. The thermal conductivity of CFRP is about 4 times as much as that of GFRP at RT; it is lower than that of GFRP below 40 K. The thermal conductivity of SFRP and ALFRP is almost equal to GFRP and CFRP respectively below 40 K.

Figure 5 shows the temperature dependence of the thermal conductivity in the transverse direction (perpendicular to fibers). It is found that the thermal conductivity in this direction is lower than that in the fiber direction. Each specimen demonstrates almost the same thermal conductivity throughout the temperature range, and hence, the effects of fibers on thermal conductivity are minor. It is indicated that the thermal conductivity in this direction is dominated by the epoxy matrix. Experimental results closely agree with the theory.[7]

Figure 6 shows the thermal contraction of each sample in the fiber direction. The thermal contraction of GFRP from RT down to LNT is approximately 0.15 percent, and those of ALFRP, SFRP, and CFRP are approximately 0.09, 0.06, and 0.015 percent, respectively. Figure 7 shows the thermal contraction in the transverse direction. In this direction, the thermal contraction, which is larger than that in the fiber direction, ranges from 0.5 to 0.7 percent down to LNT and does not strongly depend on fiber type.

Figure 8 shows the load-displacement curves obtained in a four-point flexural test at LHeT. It is noticeable that the Young's modulus of GFRP and the breaking stress of CFRP are smaller than those of other composites.

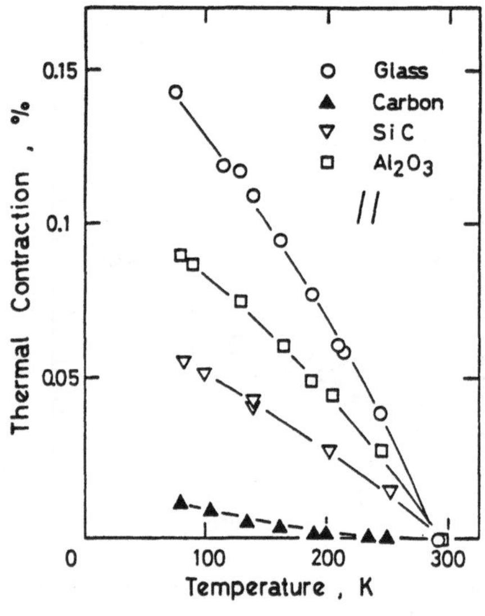

Fig. 6. Thermal contraction in the fiber direction.

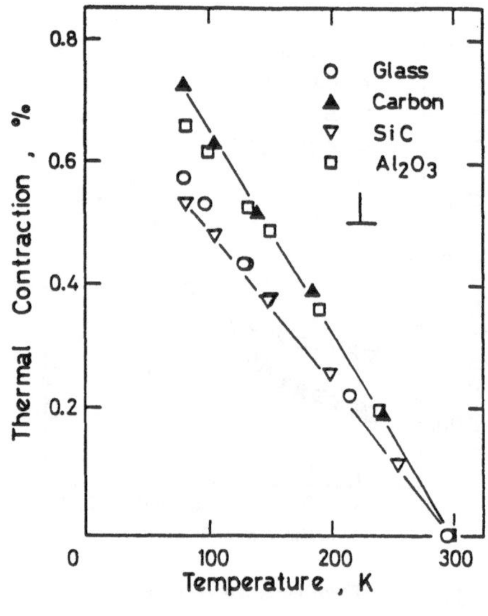

Fig. 7. Thermal contraction in the transverse direction.

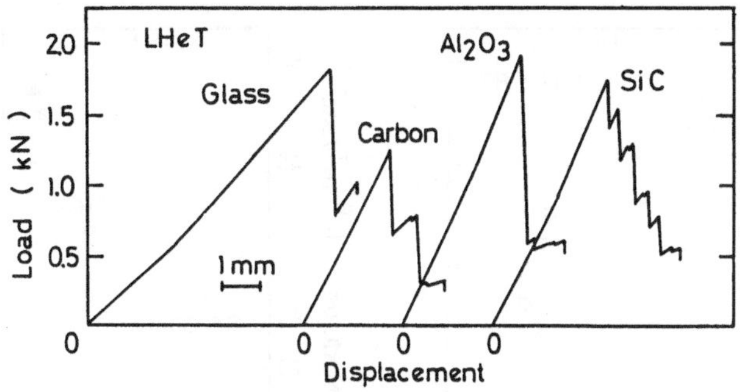

Fig. 8. Load-displacement curves obtained in four-point flexural test at LHeT.

Table 2. Young's Modulus and Breaking Stress Obtained in four-point flexural test at RT, LNT, and LHeT.

	T (K)	GFRP	CFRP	SFRP	ALFRP
Young's Modulus (GPa)	300	36.5	79.5	79.6	82.6
	77	40.1	79.8	81.3	86.6
	4.2	42.2	81.2	84.9	95.6
Breaking Stress (GPa)	300	1.07	1.14	1.33	1.30
	77	2.28	1.68	2.38	2.23
	4.2	2.32	1.64	2.21	2.37

By comparison, SFRP and ALFRP have good mechanical properties at low temperatures. The results of the four-point flexural test shown in Table 2 are the means of two or three specimens. It is seen in Table 2 that the Young's modulus and the breaking stress of advanced composite materials increase with decreasing temperature. Since there is no remarkable difference between LNT and LHeT, it is possible to make use of the data at LNT in place of that at LHeT when the experiment is difficult to perform at LHeT.

Application Note

Based on the experimental results, an evaluation was made as to the performance of each material from the viewpoint of the structural members.

The ordinate of Fig. 9 shows the specific modulus, and the abscissa demonstrates the ratio of average breaking stress to average thermal conductivity in the fiber direction, σ/κ, between LNT and RT. A high σ/κ means durability against load with unit thermal penetration through the material.[8] Consequently, larger values in both axes are desirable for support members. It is found that all the present advanced composite materials are better structural support materials than the 304 stainless steel. The GFRP has high value of σ/κ and low specific modulus. The CFRP has high specific modulus and small σ/κ conversely. It appears that SFRP and ALFRP are well-balanced materials, and should be considered for the structural support materials in this temperature range.

Figure 10 shows the performance of each material between LHeT and LNT. In this temperature range, SFRP and ALFRP also appear to be superior to GFRP or 304 stainless steel.

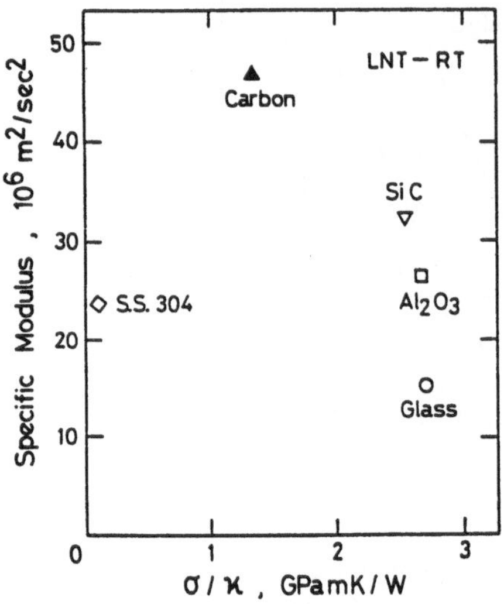

Fig. 9. Specific modulus vs. σ/κ between LNT and RT.

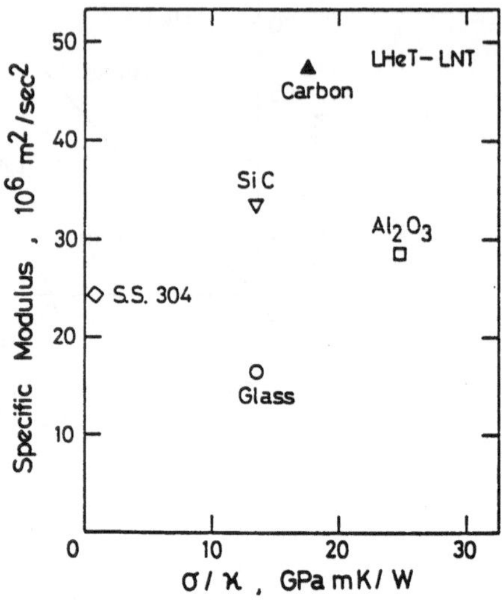

Fig. 10. Specific modulus vs. σ/κ between LHeT and LNT.

On the basis of experimental results, the application of advanced composite materials was examined in a practical cryostat system for the NMR imaging magnet. The design study of such a cryostat using the present advanced composite materials has been performed, and the schematic illustration is shown in Fig. 11. The diameter of the supporting rod is based on the mechanical strength at room temperature. The helium vessel is suspended with eight support rods from the nitrogen vessel. The heat penetration to the helium vessel is mainly by thermal radiation from the nitrogen vessel and thermal conductivity through the support materials. The heat penetration is estimated in Table 3 when the advanced composite materials are used for support materials. It is found that the penetrating heat through ALFRP is about half that through GFRP, and hence the ALFRP should be an excellent choice for support materials.

CONCLUSIONS

Thermal conductivity and thermal contraction measurements and four-point flexural tests of advanced composite materials were performed at low

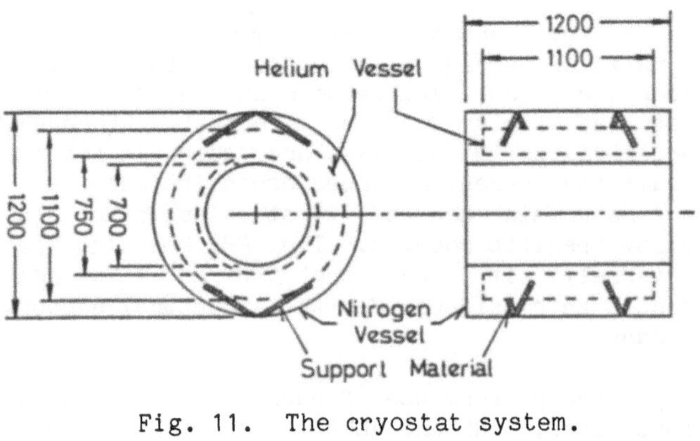

Fig. 11. The cryostat system.

Table 3. Heat Penetration to Helium Vessel

	Through Supporting Rod (mW)	Radiation (mW)	Total Heat Penetration (mW)
GFRP	219	26	245
CFRP	144	26	170
SFRP	181	26	207
ALFRP	98	26	124

temperatures; the following conclusions were drawn:

(1) The thermal conductivity in the fiber direction depends on the kind of fibers. The thermal conductivity of SFRP and ALFRP is almost the same as that of GFRP and CFRP below 40 K, respectively. ALFRP provides the lowest integrated thermal conductivity from RT to LHeT. In the transverse direction, the thermal conductivity is determined mainly by the epoxy matrix, and hence the influence of the fibers on thermal conductivity is minor.

(2) The thermal contraction of GFRP in the fiber direction is approximately 0.15 percent down to LNT, and those of ALFRP, SFRP and CFRP are approximately 0.09, 0.06, and 0.015 percent, respectively. The thermal contraction in the transverse direction ranges from 0.5 to 0.7 percent down to LNT and the effects of fibers are thought to be minor.

(3) It is found from the results of four-point flexural tests that the Young's modulus and the breaking stress of advanced composite materials increase with decreasing temperature.

(4) The SFRP and ALFRP materials have good thermal and mechanical properties, and hence they should be desirable for structural support members.

ACKNOWLEDGMENT

The authors are grateful to all members of the Low Temperature Center for the liquid helium used in the experiment. This work is supported in part by Grant in Aid for Scientific Research No. 60050044, Ministry of Education in Japan.

REFERENCES

1. M. Nagai, K. Kadodani, E. Toda, M. Nishi, and S. Shimamoto, Mechanical properties of a insulator for the Japanese LCT Coil, in: "Advances in Cryogenic Engineering - Materials," vol. 28, Plenum Press, New York (1982), pp. 223-230.
2. A. Khalil and K. S. Han, Mechanical and thermal properties of glass-fiber-reinforced composites at cryogenic temperatures, in: "Advances in Cryogenic Engineering - Materials," vol. 28, Plenum Press, New York (1982), pp. 243-252.

3. M. B. Kasen, G. R. MacDonald, D. H. Beakman, and R. E. Schramm, Mechanical, electrical, and thermal characterization of G-10CR and G-11CR glass-cloth/epoxy laminates between room temperature and 4 K in: "Advances in Cryogenic Engineering - Materials," vol. 26, Plenum Press, New York (1980), pp. 235-244.
4. M. B. Kasen, Cryogenic properties of filamentary-reinforced composites: an update, Cryogenics 221:323-340 (1981).
5. S. Nishijima, M. Takeno, and T. Okada, Impact tests of reinforced plastics at low temperatures, in: "Advances in Cryogenic Engineering - Materials," vol. 28, Plenum Press, New York (1982), pp. 261-270.
6. K. Fujioka, Y. Kuraoka, S. Nishijima, and T. Okada, Automatic measuring system of thermal constants for nonmetallic materials, in: Proceedings of ICEC 10, Helsinki, Finland," Butterworths, UK (1984), pp. 429-432.
7. D. Radcliffe and H. M. Rosenberg, The thermal conductivity of glass fiber and carbon fiber/epoxy composites from 2 to 80 K, Cryogenics 22:245-249 (1982).
8. Y. Kawate, T. Holiuchi, H. Sonoi, and T. Oui, Thermal and mechanical properties of structural support materials at cryogenic temperatures, Cryog. Eng. Jap. 14(4):164-177 (1979).

INORGANIC DIELECTRIC INSULATION FOR SUPERCONDUCTING Nb_3Sn WIRE

T. K. Gupta*

Alcoa Laboratories
Alcoa Center, Pennsylvania

W. N. Lawless

CeramPhysics
Westerville, Ohio

C. E. Oberly

Wright-Patterson Air Force Base
Dayton, Ohio

G. J. Bich and G. R. Wagner

Westinghouse R&D Center
Pittsburgh, Pennsylvania

D. S. Holmes**

University of Wisconsin
Madison, Wisconsin

ABSTRACT

The need for high current, high field, low loss, stable superconductors has led to the development of multifilamentary Nb_3Sn as the most promising candidate for use in superconducting machines. However, the brittle nature of Nb_3Sn and the high reaction temperature (~700°C) required to form it preclude the use of standard organic insulation systems. A recently developed class of high temperature dielectric materials, which are characterized by unusually large specific heats and thermal conductivities at cryogenic temperatures, offers the opportunity of providing increased enthalpy stabilization in a superconducting winding, as well as the required dielectric strength. In recent years the Air Force has supported a series of programs to develop a film-type, dielectric insulation system for superconductors using these materials incorporated in a glassy matrix.[1] The development of this composite insulation system is described in this paper.

*The author was with Westinghouse R&D Center, Pittsburgh, Pennsylvania, at the time of doing this work.
**This author was with Wright-Patterson AFB, Dayton, Ohio, at the time of doing this work.

INTRODUCTION

Fast pulse superconducting magnets for use in airborne ac generators, MHD generators, and energy storage devices require high current, high field, low loss, stable superconductors and associated insulation systems. The multifilamentary Nb_3Sn superconductor has emerged as the most promising conductor for use in these machines. A major concern in developing an insulation system for Nb_3Sn is the high formation temperature (700-800°C) of Nb_3Sn and the fact that it is brittle after its formation and cannot be deformed. This requires that the insulation must withstand this high formation temperature and yet be electrically satisfactory at low use temperature (4-8 K). A further requirement of the insulation is that it must absorb the energy dissipated during fast charge and discharge of the magnet. The energy must be absorbed without allowing the temperature of the conductor to rise high enough to quench the magnet.

A unique opportunity to achieve these goals is offered by the possible application of a new class of dielectric material as insulation coating for the Nb_3Sn wire. These materials[1] are characterized by high specific heats and thermal conductivities at low temperature and thus can provide large enthalpy for thermal stabilization of superconducting wires. Recent advances in the development of thin dielectric coatings have proved beyond doubt that these enthalpy stabilizing materials can be successfully applied very close to the potential heat source, i.e., directly on the superconductor as an insulation coating.

With this objective in mind, the Air Force has supported a series of programs for developing an enthalpy stabilizing dielectric film-type coating for superconducting Nb_3Sn wire. The broad picture is to incorporate the enthalpy stabilizing dielectric powders into appropriate glasses for the purpose of coating Nb_3Sn wire, thereby providing not only the electrical insulation, but also a significant enthalpy stabilization to the superconductor. This paper will present the state-of-the-art characteristic of this insulation system, with special emphasis on properties and compatibility between the insulation and the wire.

CRITICAL CONSIDERATIONS FOR DEVELOPING ENTHALPY STABILIZED DIELECTRIC COATING

Three critical considerations are required for designing the dielectric coating: 1) enthalpy stabilization of the coating, 2) high formation temperature (700-800°C) of Nb_3Sn, and 3) compatibility between coating and Nb_3Sn. These considerations are briefly described in the following:

1) Enthalpy stabilization refers to the ability of the coating to absorb thermal energy produced within the magnet winding. The coating acts as a buffer; the thermal energy is eventually conducted to the helium coolant. The time scale of the disturbance and the geometry of the winding determine which thermal properties of the coating material are most important. The spatial distribution of the disturbance does not affect which thermal properties are most important for the coating, but only the relative importance of the thermal properties of the coating and the wire. Three cases can be identified. The thermal relaxation time, τ, of the coating is needed to identify these cases and is given by

$$\tau = d^2/\alpha \tag{1}$$

where d is the coating thickness, and α is the thermal diffusion coefficient. For disturbances shorter than the thermal relaxation time, the thermal energy absorbed by the coating is proportional to the thermal property group[2]

$$\eta = \sqrt{K\rho C} = K/\sqrt{\alpha} \qquad (2)$$

where K is the thermal conductivity, ρ is the density, and C is the gravimetric specific heat. The second case occurs when the disturbance time is longer than τ and the coating is not in direct contact with helium or any other thermal energy sink (e.g., in a fully potted winding). The thermal energy absorbed by the coating is then proportional to the volumetric heat capacity, $C = \rho c$. The third case occurs when the disturbance time is longer than τ and the coating is in direct contact with a thermal energy sink, such as a helium bath. The thermal energy absorbed by the coating can then be passed on to the heat sink, and the energy absorbed becomes proportional to the thermal conductivity, K, of the coating. To summarize, either η, K, or C can be the most important thermal property of the coating, depending upon the temporal length of the disturbance and the winding geometry.

2) The second constraint comes from the high formation temperature of Nb_3Sn and the poor strain tolerance of the formed Nb_3Sn. The Nb_3Sn is generally formed by heat treating at temperature between 700-800°C for many hours. Once it is formed, it is brittle, and straining the wire more than 0.07% degrades the critical current. This requires that the winding be carried out prior to formation reaction. Consequently, the insulation that needs to be developed should be capable of being applied before the reaction, be able to withstand the time-temperature excursion during the reaction, and should exhibit no adverse effect on electrical or mechanical properties of the wire. Because of the high reaction temperature, the organics are thus ruled out, and inorganics are prescribed.

3) The third constraint is imposed by the compatibility requirement between the insulation and the superconductor. The insulation must be compatible with electrical, mechanical, thermal, and chemical properties of the Nb_3Sn superconductor. It must not degrade the superconducting current while exhibiting high electrical resistance, high breakdown voltage, and satisfactory enthalpy stabilization in zero and intense magnetic field. It is desired that the formation temperature of the coating and the reaction temperature of Nb_3Sn be the same, and no adverse chemical reaction occurs between the insulation and the superconductor. To achieve this goal, it was necessary to use a protective layer of nickel on the outer copper substrate of the Nb_3Sn wire, although other metals were also found to be satisfactory.

EXPERIMENTAL RESULTS

A large number of materials and combinations thereof were studied, all of which will be impossible to describe within the scope of this paper. We will therefore present a few examples of the coating composition and its characteristics from a broad range of studies undertaken in the present program.

Two high specific heat ceramics reported here are: SC1C, a chromite-spinel ceramic, and SC1A, a niobate-columbite ceramic. The SC1C has an enormous specific heat maxima at 8 K of 2J $cm^{-3}K^{-1}$, and SC1A has a specific heat maximum at 4.2 K of 0.3 J $cm^{-3}K^{-1}$. Of several glasses investigated in this study, the one that will be discussed here is labeled 3072, developed specifically for this study. In a previous study,[3] two Corning glasses, 7570 and 7052 were investigated in making composites with SC1C and SC1B, a ceramic analogous to SC1A. Adhesion of the Corning glasses to the wire proved unsatisfactory, so new glasses were developed which showed much improved adhesion. Pore-free bulk samples of cylindrical shape were prepared of various compositions by hot pressing at 800°C. The composite compositions that will be reported in this paper for bulk samples are: 50/50 weight ratios of SC1A/3072, SC1C/3072, and a ternary composite 32/18/50 SC1C/SC1A/3072. Simultaneous studies were conducted on composite compositions for coating and the one that will be described here is 40/60 SC1C/3072. The reason for higher glass content was to ensure better adhesion of the coating to the wire. This reduced the thermal characteristic only very slightly. The coating was applied by first preparing a slurry of a mixture of glass, ceramic, organic binder, and solvent, and then drawing the wire through the slurry using the conventional wire enameling equipment. The coating thus developed is called the "green coating," which is then heat treated at 700-800°C to obtain what is known as "fired" or "vitrified" coating. After heating, the glass has melted with the ceramic embedded in the glass and adhered strongly to the wire. The green coating is flexible enough to allow twisting, bending, and coiling. The test samples are prepared from the green coating in the form of a twisted pair, standard in organic enamel industry (according to IEEE specification), and heat treated at various temperatures to obtain the vitrified test samples.

Bulk Samples

Specific heat and thermal conductivity data (zero field) were measured in the adiabatic calorimeter described elsewhere.[4] Specific heat measurements in intense magnetic fields were performed at the National Magnet Laboratory (MIT) by a drift technique using capacitance thermometry.[5]

The specific heat data on pore-free SC1A/3072, SC1C/3072, and SC1A/SC1C/3072 composites are shown in Fig. 1. The sharp peak for SC1A and the broad-based peak for SC1C bulk ceramics at appropriate

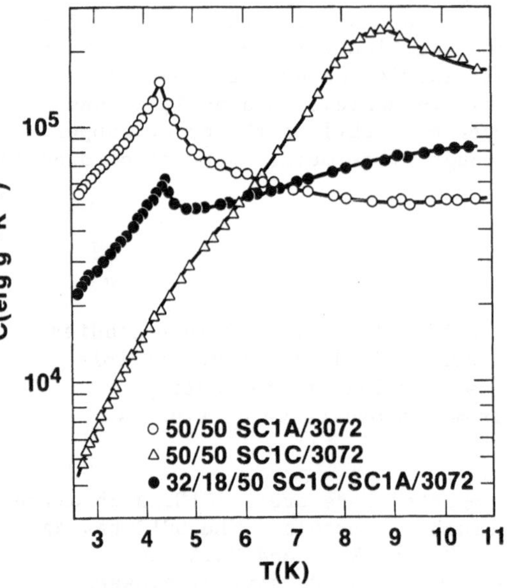

Fig. 1. Specific heat of ceramic/glass composites.

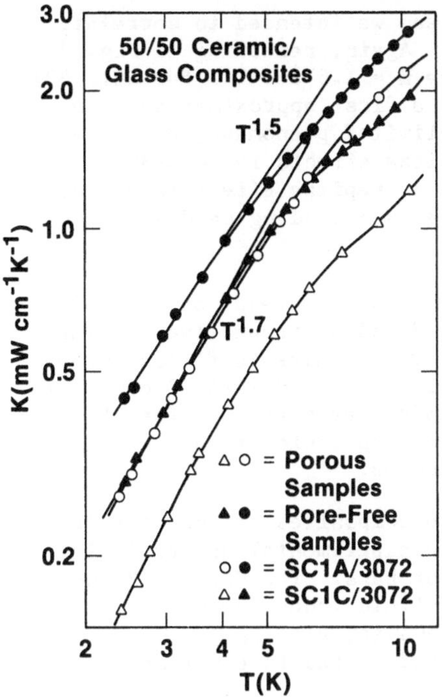

Fig. 2. Thermal conductivities of dense and porous ceramic/glass composites.

temperatures are maintained in the composites, indicating the absence of any adverse reaction between the ceramic and the glass. The broad maximum centered at approximately 8 K for SC1C/3072 composite may be due to correlation effects in the fine grain (~1μm) SC1C powder as also observed previously for SC1C/7570 composite.[3]

Thermal conductivity data measured on the 50/50 ceramic/glass composites are shown in Fig. 2. Also shown in this figure are the results on samples which had approximately 10% porosity. Porosity decreases the SC1C/3072 thermal conductivity by approximately 41% and the SC1A/3072 thermal conductivity by approximately 31%. These results are important

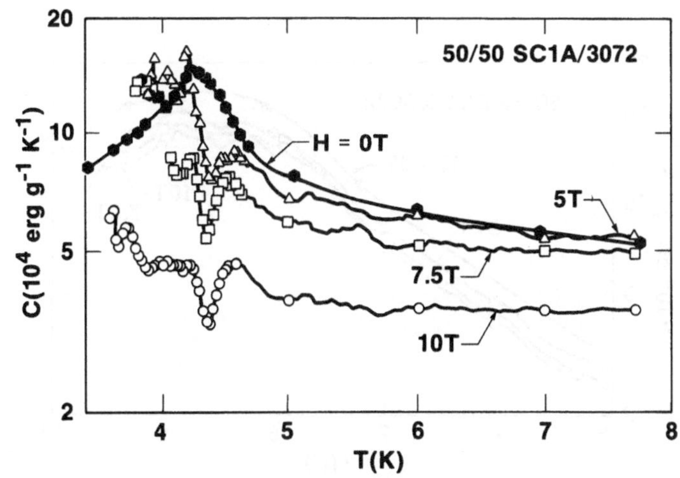

Fig. 3. Magnetic-field dependence of the specific heat of the SC1A/3072 composite.

because the coating was nearly 100% dense, and we intended to correlate the bulk thermal data to those of coatings. Again, referring to Fig. 2, the thermal conductivities follow $K \alpha T^m$ where m = 1.5 and 1.7 for SC1A/3072 and SC1C/3072 composites, respectively. To a first approximation, one would associate these exponents to Kapitza-limited phenomena, but other studies have indicated that there is no Kapitza effects in SC1C/3072 composite. We remark here that the absence of Kapitza effect is one of the reasons for choosing SC1C/3072 composite as a leading candidate for wire coating.

Specific heat data measured in intense magnetic fields on SC1A/3072 and SC1C/3072 composites are shown in Figs. 3 and 4, respectively. The two composites behave very differently in intense magnetic fields. The specific heat maximum is depressed with field for the SC1A/3072 composite, whereas the broad-based maxima is still retained even at 10 T for the SC1C/3072 composite. The reason for the peak depression is not readily understood and would require fundamental studies.

The thermal conductivity data for these composites, along with the ternary composite, are shown in Fig. 5. The experimental uncertainties for these measurements are shown by the error bars. The SC1C/3072 and SC1A/SC1C/3072 composites have apparently H-field independent thermal conductivities. The SC1A/3072 composite, however, appears to have a slight H-field dependence, which cannot be attributed to experimental uncertainties.

Coated Wire Samples

The coating studies can be broadly divided into two categories: those conducted on Cu wire and those on Nb_3Sn wire. The bulk of the coating evaluation was done on samples made from Cu wire. This is due to the fact that most Nb_3Sn wire has an exposed surface of Cu, and, as such, it was argued that an insulation system developed for Cu will apply equally well to Nb_3Sn. The other reason for testing Cu wire is the high cost and scarcity of Nb_3Sn wire. Both wires had Ni-plating on the outer surface. Tests were conducted on both "green" and "vitrified" twisted pairs. The thickness of the green coating was typically 5.5×10^{-3} - 7.5×10^{-3} cm (2-3 mil) and that of the vitrified coating was typically 2.5×10^{-3} - 3.8×10^{-3} cm (1-1.5 mil).

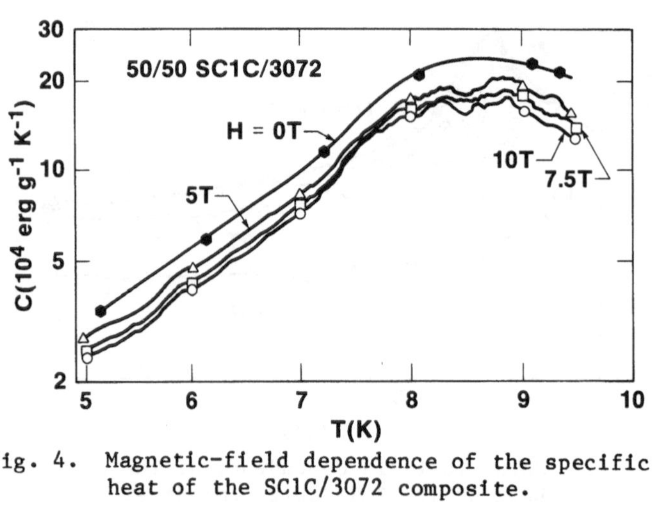

Fig. 4. Magnetic-field dependence of the specific heat of the SC1C/3072 composite.

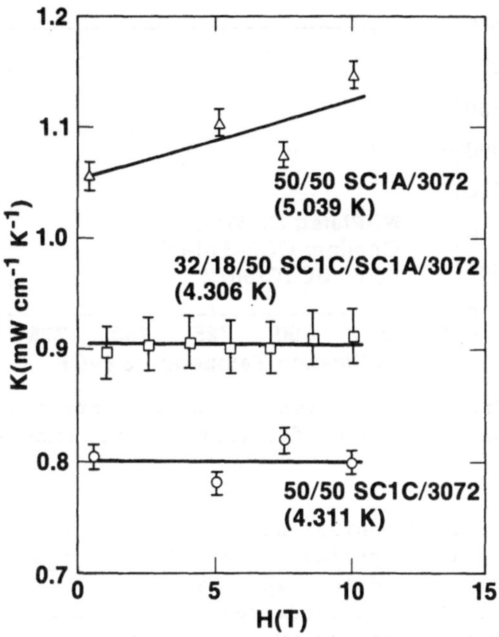

Fig. 5. Magnetic-field dependence of the thermal conductivities of ceramic/glass composites.

Table 1. "Green" Coating Characteristics

Coating Identification C03-310-00
Coating Composition 40/60 SC1C/3072

Parameters	Ni-Plated Cu Wire	Ni-Plated Nb_3Sn Wire
Wire Size	#20 AWG	#20 AWG
Coated Surface	Fairly Smooth	Very Smooth
Build	5.0×10^{-3} cm (2 mil)	$5.0\text{-}5.5 \times 10^{-3}$ cm (2.0-2.2 mil)
Flexibility:		
1. Quick Snap	passes	passes
2. Elongation + 5X Mandrel Wrap	20%	<5%
3. e bend	passes	passes
Scrape Abrasion (No. of cycles @ 350 g lead)	33, 48, 56 52, 63, 17	12, 34, 24 36, 22, 18
Electric Breakdown (volts)	1000, 960, 820 790, 920, 990	1700, 1600, 1250 1200, 1100, 1500
Insulation Resistance MΩ @ 50 V (on twisted pair)	>20,000	>20,000

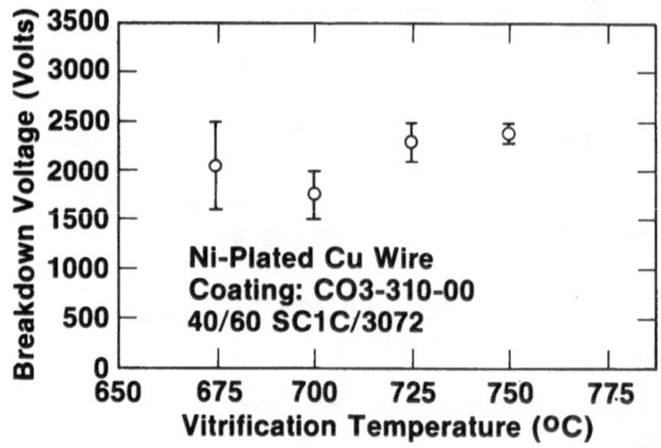

Fig. 6. Breakdown voltage of the dielectric coating as a function of vitrification temperature.

Typical examples of mechanical and electrical characteristics of the "green" dielectric coatings applied on both Cu and Nb_3Sn wires are shown in Table 1. Since no standard test specifications exist for inorganic dielectric insulation, the test parameters were generated along the line of organic enamel coating modified by the brittle nature of the ceramic/glass coating. The test data presented in Table 1, when compared with the characteristics of a typical organic insulation, can be considered quite satisfactory. The coating exhibits high flexibility, high resistance to scrape and abrasion, and high electrical strength. The breakdown strength of the vitrified coating as a function of vitrified temperature is shown in Fig. 6 for Ni-plated Cu wire. These data represent one of the best breakdown voltages obtained in this program and are substantially in excess of the 500 V targeted for this program. These coatings also showed excellent adhesion to the wire after vitrification, and in some cases sand blasting was required to remove the coatings from the wire.

DISCUSSION

What we have tried to demonstrate in this paper is a novel method of enthalpy stabilization of superconducting magnets through the development of inorganic film-type coatings on Nb_3Sn wire, the coating having the characteristic of high thermal capacity combined with high electrical strength. Taking the case of enthalpy stabilization, it is instructive to evaluate the η-parameters for the composites from the thermal data of the bulk samples. The procedure followed was to compute η for $H = 0$ and $H = 7.5$ T using the appropriate specific heat, thermal conductivity, and density data. The η-parameter data at temperatures 5-8 K are shown in Fig. 7 where the band of η-values ($H = 0$, 7.5 T) is shown for each of the three composites discussed in this paper. Also shown for comparison are the η-data for glasses, which are the only other potential dielectric insulation for Nb_3Sn. The composites represent an improvement in the η-parameter by a factor of about 20.

One can conclude that for 50/50 ceramic/glass composites, the SC1A/3072 composite is superior up to 6.5 K; the SC1C/3072 composite is superior above 6.5 K. The ternary SC1C/SC1A/3072 composite appears to offer no particular advantage. For all composites, the η-parameters are decreased by the field ($H = 7.5$ T), but this suppression is relatively minor. This is one of the major findings in this program, namely, that the thermal properties of these insulation coatings are not greatly affected by intense magnetic fields.

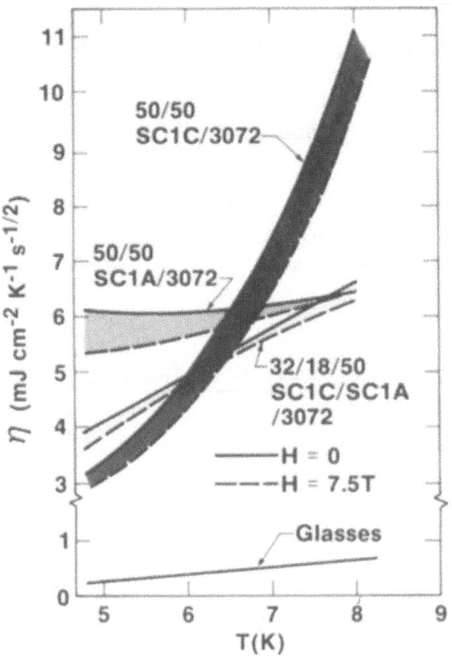

Fig. 7. Transient thermal parameter, η, as a function of temperature and field.

Fig. 8. Green-coated Nb_3Sn wire in the form of a coil.

Turning to the dielectric coating, it can be stated that the inorganic film-type coating consisting of a ceramic and a glass offers a very satisfactory solution to the conflicting requirements of a high reaction temperature of Nb_3Sn and a low use temperature (4-8 K) of a superconducting environment. The additional attractive feature is that the coating formation (vitrification) and Nb_3Sn reaction can be accomplished simultaneously. The vitrified coating shows a high degree of scrape and abrasion resistance combined with resistance to thermal shock upon immersion to liquid nitrogen or helium bath. The vitrified coating maintains a high insulation resistance with resistivity values measured at $\rho \sim 10^{11-12}$ ohm-cm and a high breakdown voltage with values in excess of 500 V, both in liquid nitrogen. Furthermore, the residual resistance ratio (RRR), measured for coated and uncoated wires which were heat treated similarly, showed no significant difference in values. The coating is very compatible with the Nb_3Sn wire.

Finally, the green coating is highly conformable, i.e., it is extremely amenable to coil formation which can be vitrified to give final coil configuration. A photograph of a green-coated unreacted wire wrapped on a ~6.4 mm (1/4 in.) mandrel is shown in Fig. 8, illustrating the flaw-free green coating. This then readily conforms to the "react-after-winding" (RAW) scheme that is now being pursued in a separate Air Force program for making small coils of Nb_3Sn wire.

ACKNOWLEDGMENT

This research was sponsored by the U.S. Air Force.

REFERENCES

1. W. N. Lawless, C. F. Clark, and R. W. Arenz, Final Report: WPAFB Contract 33615-80-C-2022 (1982).

2. M. Jakob and G. Hopkins, "Elements of Heat Transfer," Wiley, New York (1957), p. 65.

3. W. N. Lawless, C. F. Clark, and T. K. Gupta, in: "Advances in Cryogenic Engineering - Materials," vol. 30, A. F. Clark and R. P. Reed, eds., Plenum Press, New York (1984), pp. 433-440.

4. W. N. Lawless, Phys. Rev. B 14:134 (1976).

5. W. N. Lawless, C. F. Clark, and R. W. Arnez, Rev. Sci. Instrum. 53:1647 (1982).

MEASUREMENT OF THERMAL CONDUCTIVITY OF INSULATING CRYOGENIC STRUCTURAL MATERIALS

J. Waynert

Applied Superconductivity Center
University of Wisconsin-Madison
Madison, Wisconsin

ABSTRACT

This paper describes an apparatus for measuring thermal conductivities of less than 1 $Wm^{-1}K^{-1}$ using a longitudinal heat flow method. Two methods are discussed; one is a relative measurement method and the other is an absolute method. Contact resistance effects are discussed. Results are presented for measurements on Pyrex 7740, G-10CR, and a new material from Du Pont, FP/epoxy, a continuous, uniaxial alumina fiber filled epoxy.

INTRODUCTION

The goal was to build an apparatus to measure thermal conductivity of insulating materials from 300 K to 4.2 K with an accuracy of at least ± 10%.

The thermal conductivity is approximated from the equation

$$Q = kA\ \Delta T/\Delta X \qquad (1)$$

where k is the thermal conductivity in $Wm^{-1}K^{-1}$, A is the cross-sectional area of the sample in square meters, ΔT is the temperature difference across the sample in Kelvins, ΔX is the sample length in meters, and Q is the heat flux in watts. The longitudinal heat flow method assumes a temperature gradient is established in the axial direction of a sample in rod form. The temperature difference across a known length of sample is then used in eq. (1) to determine k.

Diode thermometers were selected for their ability to cover the entire temperature range desired. A continuous flow cryostat which carries liquid helium or liquid nitrogen through a transfer line from a storage dewar to a cold finger sample mount, permits stable operation at low temperatures for extended periods. The apparatus was computer controlled so that temperatures could be recorded as a function of time. A temperature controller monitored one diode at the cold end. Using a feedback mechanism between a heater and the cryogen, the controller could maintain the cold finger - the temperature regulated block - at any preselected temperature higher than the cryogen temperature. Once the system had stabilized and the data, temperatures and the lower heater

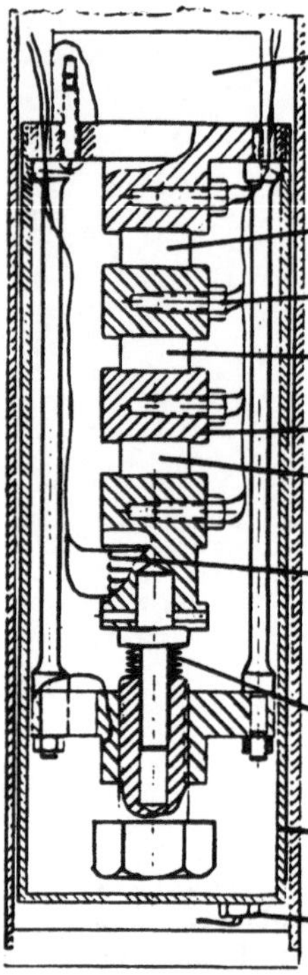

Fig. 1. Original sample holder. All hatched regions are copper. Super insulation is not shown.

current and voltage (see Fig. 1) were recorded, the temperature of the cold finger could be increased or decreased to the next desired temperature. The power to the lower heater which established the temperature gradient in the sample was provided by a constant current source. The voltage across this heater was monitored using a four-probe technique.

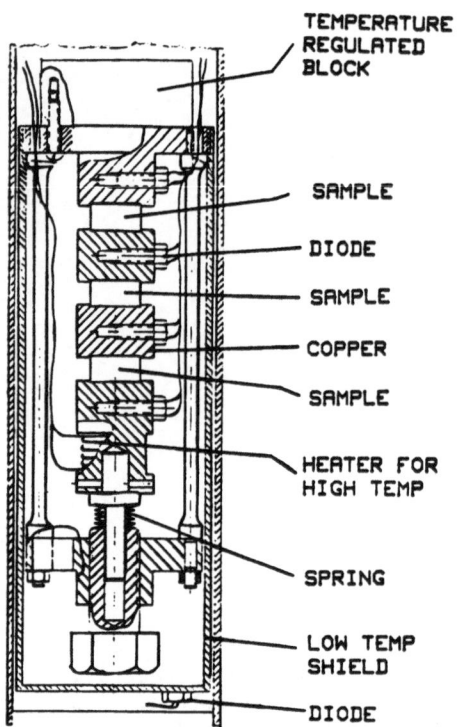

All the thermometers used are silicon diode thermometers purchased from Lake Shore Cryotronics. Two were purchased calibrated, while the rest were calibrated in the lab. All diodes were connected for four-wire, forward bias voltage measurements with a 10 μA excitation current. The leads were of 16.4 Ω/m manganin wire, thermally anchored at the cold end of the cryostat. Two additional manganin leads for voltage and two 34 gauge copper current leads were used for the heater to generate the temperature gradient. Two more manganin current leads controlled the heater on the temperature regulated block, the upper heater of Fig. 1. The total calculated heat conduction down the leads was 0.22 mW, when operating at 4.2 K.

A Janis Research continuous flow cryostat was used to obtain temperatures ranging from 4.2 to 300 K. Liquid nitrogen was used for measurements down to 80 K and liquid helium for measurements below 80 K.

ORIGINAL SAMPLE HOLDER

The sample rig, as initially designed, is shown in Fig. 1. The original idea was to be able to check several operational parameters by having three samples. At first, the top and bottom samples were to be a standard material, and the middle one, the sample to be measured. Pyrex 7740 was the standard chosen. Any difference in heat flux through the two standards could then be attributed to radiative losses. Once radiation losses were determined, the rig could be adjusted to accommodate one sample and the standard or two samples and one standard. In any case, the rig provided a wide range of freedom.

The initial experiments to measure radiation losses by the decrease of heat flux indicated that these losses were small and less than the precision of the measurements. Later studies in which a thermometer was added to the inner-most radiation shield, the lowest diode in Fig. 1, indicated that the gradient in the shield was similar to that in the sample. With super insulation on the inside of this shield, radiation losses are calculated to be less than 1% of the total heat flux at any temperature.

Three samples of Pyrex 7740 were used in the first experiments. The temperature difference across one of the samples was used to determine the heat flux through all the samples. The results agreed favorably with the published data;[1] see Fig. 2. One difficulty which arose, was the time required to reach thermal equilibrium increased with decreasing temperature. Measurements that take only 5 hours at room temperature, required over 60 hours at 16 K. Previous results on amorphous material[2] indicated that the diffusivity should be a monotonically increasing function of decreasing temperature. This implies that the stabilizing time, hereafter the time constant, should have decreased instead of increased. Several possible causes of this problem were explored.

It was thought that one cause could have been a contact resistance problem at the interfaces. CRY CON grease, silver print paint and thin sheets of indium were tried as interface materials. There was a slight improvement in reducing the time constant in the order given, with indium being the best material with a time constant about 60% of that for CRY CON. The CRY CON was also found to lose some of its high conductance properties after extended periods under vacuum.

The effect of contact pressure was also considered. Time constants were found to vary inversely with pressure. Preliminary measurements showed the contact resistance to decrease by a factor of two as the pressure was increased from about 0.5 MPa to 1.4 MPa. At higher pressures the resistance was almost independent of the pressure.

Using the high contact pressure and the indium contacts, the apparatus still required unacceptably long time constants. In order to obtain data in reasonable periods of time and at reasonable cryogen costs, the system was altered.

PRESENT SAMPLE HOLDER

The old rig was modified by removing the nylon support rods and gluing a single sample to the temperature regulated block, while the heater with a thermometer is glued to the other end. Several different glues were tried with the idea of making sample removal easy but most failed on cool down. Finally Stycast 2850 was chosen. The difficulty with Stycast is that the sample and holder must be heated to high temperature to separate them; we used a torch. This damages the sample and prevents any further testing with it. Furthermore, too rapid a cool down now resulted in fractured samples. It was difficult to understand this behavior since dunking the glued sample and rig in the liquid nitrogen and then in boiling water had no such effect. The fractured surface of the pyrex left a smooth annular region on the outside and a rougher bubble in the center. This would seem to indicate a radial temperature or stress gradient. Yet it does not seem reasonable to develop any sizeable temperature gradients in the attached copper for the cool down rates used. To avoid the problem, cool down rates were adjusted so that longitudinal temperature differences across the sample never exceeded 15 K.

The present sample holder uses an absolute measurement whereas the original holder was a relative measurement. Heat flux is presently measured by monitoring the heater current and voltage by the four probe technique.

RESULTS ON PYREX 7740, FP/EPOXY, AND G-10CR

Pyrex 7740

Pyrex 7740 was chosen for the standard material since it is fairly well characterized, easily obtained, and relatively accurate data can be found for its thermal conductivity. Typical sample dimensions were 1.0 cm diameter and 0.6 cm length. Fig. 2 shows the results of the comparison of UW data, X's, and the collected data obtained by NBS. The data at 4.7, 10, and 20 K were obtained with the single sample method. The higher temperature data are from the original sample rig.

The higher temperature data, which required over one hour to stabilize were handled in a unique way. Once the constant temperature copper block had reached the selected temperature, the diode voltages, which were computer monitored, were recorded as a function of time. Since the copper blocks separating samples have much shorter thermal time constants than the pyrex, the diode temperatures responded as if to a simple diffusion process,

$$T - T_o = (T_f - T_o)\,(1 - e^{-t/\tau}) \qquad (2)$$

where T_o is the temperature at which data recording started, T_f is the final stable temperature and τ is the thermal time constant. For small

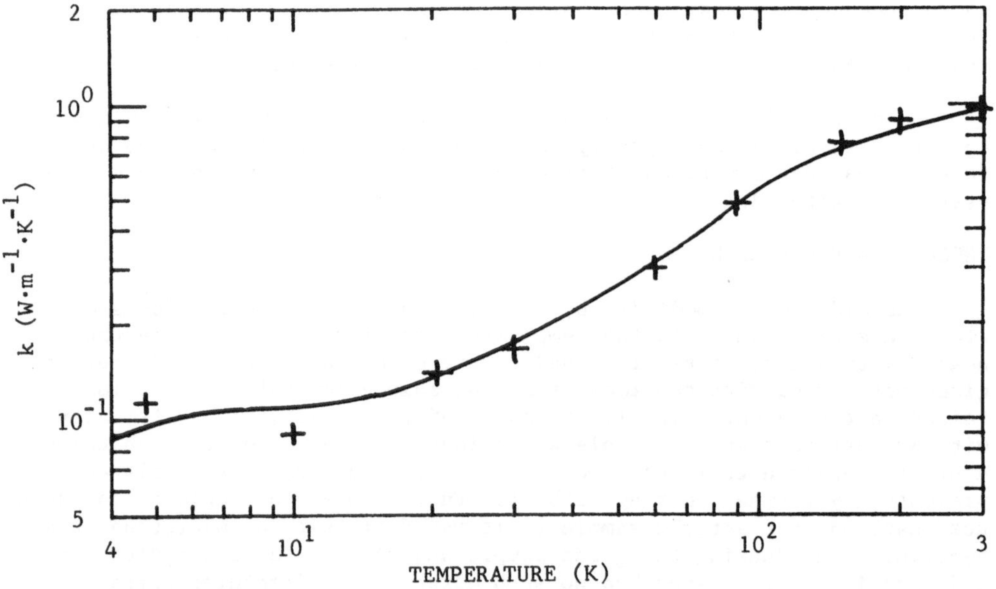

Fig. 2. NBS collected data on pyrex from NBS Monograph 131. UW data are marked by X's. Data above 20 K were taken in original sample rig. Lower temperature data were taken using single sample method.

changes in T, about 5 K at temperatures above 80 K, τ is assumed independent of temperature. Eq. (2) is rewritten as

$$\ln \{1 - (T-T_o)/(T_F-T_o)\} = -t/\tau \qquad (3)$$

Diode data were recorded as a function of time. The diode voltages are converted to the calibrated diode voltage and then to temperature via a cubic spline fitting. Then a linear regression was performed for an initial guess of T_f when using Eq. (3). An iterative procedure obtained the best value of T_f which minimizes the square of the residuals. An added benefit of this method is that not only can k be calculated from the temperature difference, but the diffusivity may be gotten from the slope. Knowing the diffusivity and the thermal conductivity, the volumetric specific heat may be calculated.

This procedure worked well for temperatures above about 60 K. For lower temperatures, the diffusivity changes too rapidly and an assumed temperature dependence for the diffusivity would be required to solve the diffusion equation. Time constraints prevented studying this approach in detail.

Fiber FP in Epoxy

Fiber FP is a continuous alumina fiber manufactured by Du Pont. The excellent compressive strength of 2300 MPa (335 ksi) at 50% volume fraction of fiber makes the material a good candidate for cryogenic structural supports. Data were taken using the single sample technique with the results shown in Fig. 3. The method of taking data as a function of time was not used. Our sample is 0.625 cm long and 0.64 cm in diameter. The initial procedure was to leave the heater on at all times and record temperatures when the sample stabilized. Even at room temperature, stability was reached in about 30 - 40 minutes. It was discovered though that with the heater off, the heated end seldom achieved the same temperature as the cool end, sometimes differing by as much as 0.4 K. Both the

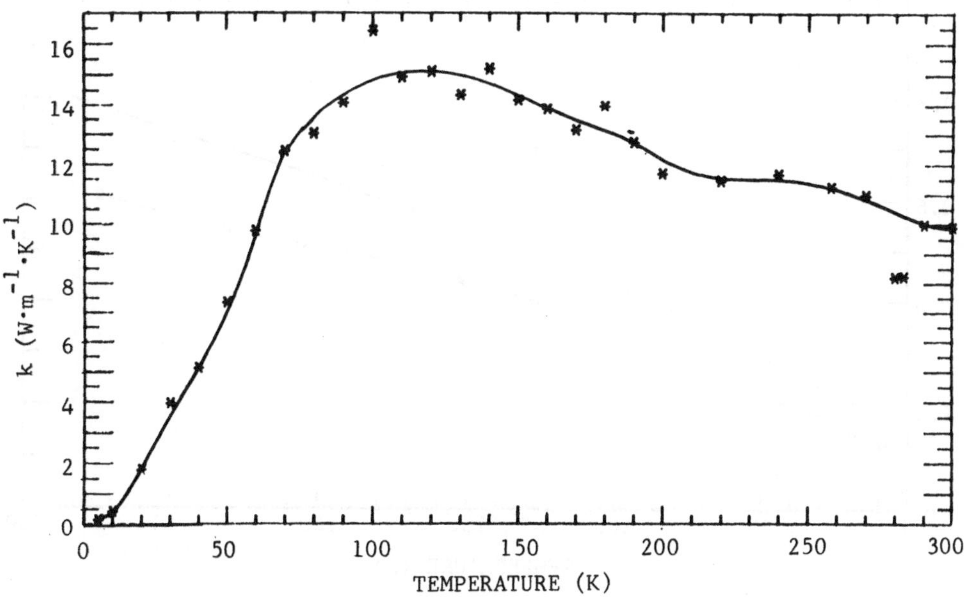

Fig. 3. Thermal conductivity results on FP/epoxy. Curve is for clarity only.

diodes being used were ones calibrated by Lake Shore Cryotronics with an alleged accuracy of ± 50 mK. The two diodes were eventually found to differ by 145 mK at room temperature and by 220 mK when in a liquid nitrogen bath. To minimize this effect, the temperature difference across the sample was determined from the lower, hot-end, sample diode with the heater on and with the heater off. The cold end was monitored to be sure it had not drifted.

The longitudinal thermal conductivity of fiber FP/epoxy obeys a simple rule of mixtures. The relatively large room temperature value, 10 $Wm^{-1}K^{-1}$, and the peak near liquid nitrogen temperature are a manifestation of the properties of alumina.[1] Using chopped fibers would reduce the thermal conduction. It remains to be seen if further configurational changes can make FP-epoxy a good load bearing thermal insulator.

G-10CR

The thermal conductivity of G-10CR in the warp direction was measured as a further check of the equipment. The sample used was 0.503 cm long with a radius of 0.631 cm, purchased in rod form.

NBS has published results[3,4] on measurements of sheet form G-10CR. Their results are compared to those of UW on Fig. 4. The UW results never differ by more than 23% from those of NBS.

This difference arises from the different characteristics, both thermal and mechanical, arising from the methods of making rods versus sheets. Sheets have parallel layers of reinforcement, while the rods have roughly circumferential layers. The rod formation results in a lower fiber volume fraction at the axis of the cylinder giving a lower thermal conductivity.

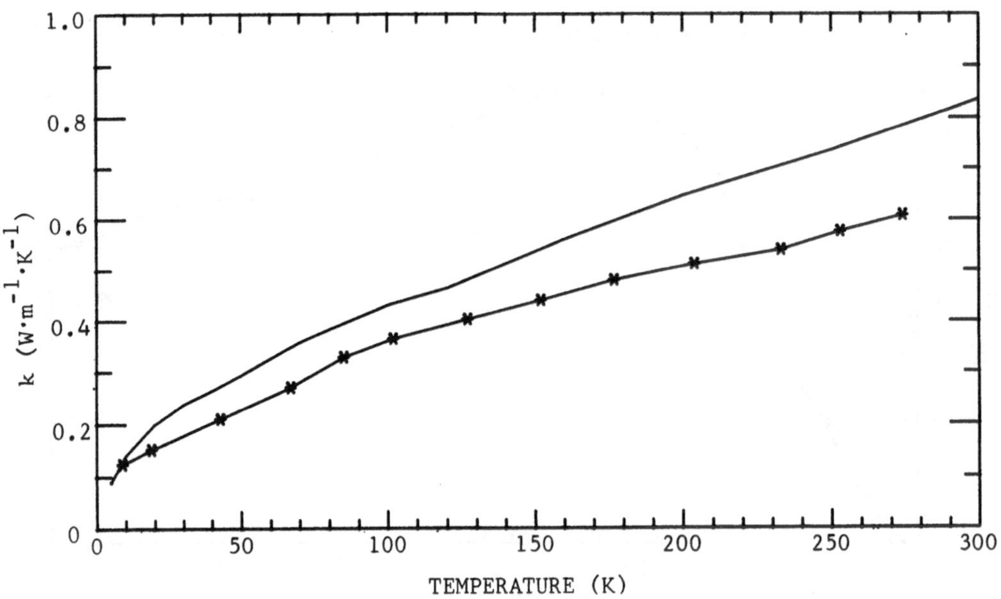

Fig. 4. NBS and UW data on G-10CR. NBS is for sheet sample. UW measurements were on a rod. NBS is the upper curve; UW is the lower.

ERROR ANALYSIS

The uncertainty in length and cross-sectional area is less than 0.5% in each. Using the absolute measurement method, the uncertainty in the heat flux applied by the heater is 0.1%, but at the lowest temperatures, the calculated contribution of the leads may be as much as 5%. The greatest uncertainty seems to be in the temperature measurement. Lake Shore estimates the repeatability of the diodes to be 50 mK. In some cases, the temperature difference maintained across a sample was as small as 1.0 K. This represents a 5% uncertainty. The total accuracy is believed to be within ± 10%.

DISCUSSION

The goal was to design a device which could measure thermal conductivity from 300 K to 4.2 K for low conductivity samples with an accuracy of ± 10%. The goal has been achieved, but many interesting questions have arisen. Some further studies on contact resistance are planned. Measurements on a promising new compressive structural material, FP-epoxy, show that although the room temperature compressive strength may be as much as six times that of G-10, the thermal conductivity is about ten times larger also. Thus, the ratio of strength to thermal conductivity implies that G-10 is the better cryogenic structural material unless some configurational changes in FP/epoxy can lower its conductivity without affecting its strength too much.

ACKNOWLEDGMENTS

Thanks are due to D. McCabe and D. Iuliano of Du Pont for providing the FP-epoxy samples and extensive advice on both thermal and mechanical aspects of the experiments.

This work was supported by Los Alamos National Laboratory subcontract 9-X44-J8677-1 and the Wisconsin Electric Utilities Research Foundation.

REFERENCES

1. G. E. Childs, L. J. Erichs, R. L. Powell, "Thermal Conductivity of Solids at Room Temperature and Below," NBS Monograph 131, 1973.
2. A. Khalil, Thermal conductivity and diffusivity measurements at cryogenic temperatures by double specimen technique, Cryogenics, 22:310-312 (1982).
3. J. G. Hust, "Materials Studies for Magnetic Fusion Energy," NBSIR 82-1667, National Bureau of Standards, Boulder, Colorado (1982).
4. M. B. Kasen et al., Mechanical, electrical and thermal characterization of G-10CR, in "Adv. Cry. Eng.-Materials", vol. 26, Plenum Press, New York (1980), pp. 235-244.

COMPRESSIVE PROPERTIES OF SILICA AEROGEL AT 295, 76, AND 20 K*

J. M. Arvidson and L. L. Scull

Fracture and Deformation Division
National Bureau of Standards
Boulder, Colorado

ABSTRACT

Specimens of silica aerogel were tested in compression at 295, 76, and 20 K in a helium gas environment. The properties reported include Young's modulus, the proportional limit, and yield strength. Compressive stress-versus-strain curves at these temperatures are also given.

A test apparatus was developed specifically to determine the compressive properties of low strength materials. To measure specimen strain a concentric, overlapping-cylinder, capacitance extensometer was developed. This frictionless device has the capability to conduct variable temperature tests at any temperature from 1.8 to 295 K.

Results from the compression tests indicate that at low temperatures the material is not only stronger, but tougher. During 295-K compression tests, the samples fractured and, in some cases, crumbled. After 76- or 20-K compression tests, the specimens remained intact.

INTRODUCTION

Foam materials have many important applications in energy-related fields. They are utilized not only for insulation, but also for load-bearing, structural applications. For these applications, it is necessary to know the mechanical properties.

The compressive properties of a silica aerogel foam were determined at 295, 76, and 20 K in a gaseous helium environment. The properties reported include Young's modulus, proportional limit, yield strength (at 0.2% offset), and strain (elastic, permanent, and total).

MATERIAL CHARACTERIZATION

The silica aerogel foam was produced using a supercritical solvent extraction process from a colloidal silica gel in an autoclave.[1] Its cell size

*Work sponsored by the Fusion Target Fabrication Group at the Lawrence Livermore National Laboratory, Livermore, California. Contribution of the National Bureau of Standards, not subject to copyright.

is 60 nm; its cell structure is open and interconnected. The density of the silica aerogel tested was nominally 0.10 g/cm^3. The volume percents of silica and air were calculated to be approximately 4% silica and 96% air, using a solid silica density of 2.660 g/cm^3 and air density of 1.204 mg/cm^3.[2]

SPECIMENS

The compression specimens were right circular cylinders with a height of 19 mm (0.75 in) and a diameter of 19 mm. Because the extremely small cells of the specimens adsorb moisture readily, the specimens were stored in a desiccator for at least 30 days prior to testing. All specimens were inspected for defects before they were loaded into the test apparatus. The defects were recorded for possible correlation with data scatter, since flaws may alter the strength of the material.

To prevent damage during installation into the compression fixture, special tongs were constructed from styrofoam, with a low-density flexible foam insert, to handle the delicate aerogel specimens.

TEST APPARATUS

The apparatus was specifically designed to test delicate materials such as silica aerogel.[3-6] The same apparatus has the capability to perform tests in either tension or compression, by changing the specimen grips and pull-rod fixtures. It is fitted with a cryostat for low temperature testing and sealed such that the test environment can be varied.

The specimen displacement is measured using a concentric, overlapping-cylinder, capacitive strain extensometer. Since the extensometer system is not attached directly to the specimen, it does not influence the test results. Because it is unnecessary to attach a strain gauge or other device, this system is ideally suited for testing delicate or viscoelastic materials at cryogenic to ambient temperatures. Another benefit is that the capacitive strain extensometer is extremely sensitive and linear for relatively large specimen displacements.[6-8] For the specific geometry of the extensometer used (4.5-cm diameter x 9-cm height), the range of linearity is >25 mm (1.0 in) with a resolution of <25 μm (0.001 in). The extensometer geometry can be easily optimized to the specific specimen geometry required.[9]

When the extensometer is situated in a stable fluid or gas (i.e., single phase), an extremely stable output signal may be obtained. The relative noise level with this extensometer is on the order of 10^{-6}.

This apparatus also has the capability for variable temperature tests at any temperature from 1.8 to 295 K, with an accuracy of ±0.1 K. This is accomplished using a resistive heater embedded in the inner capacitance cylinder, with a Type-E thermocouple, for temperature monitoring and control. Long-term tests, 3 h and more, have been conducted with a maximum temperature variation of <0.25 K during the entire test. The environmental pressure in the test apparatus can also be adjusted in the range 0.1 to 200 kPa (0.015 to 30 psia).

TEST PROCEDURE

After the specimen had been loaded into the apparatus, the system was sealed and purged with dry helium gas at ambient temperature for 5 h. This purge was done to remove any atmospheric contaminants that could solidify inside the aerogel as the liquid helium was introduced into the apparatus. A single specimen (#12), at ambient temperature, was pumped in a vacuum and slowly backfilled with helium gas three times prior to cooling. This specimen exhibited a stress-strain curve that fell within the minimum and maximum curves obtained from other tests, indicating that the 5-h helium purge technique prior to testing was sufficient.

After the purging operation, the room temperature modulus was determined for all specimens by loading the specimen to not more than 50% of the proportional limit and then unloading to zero compressive force. The average of two loadings was taken as the room temperature Young's modulus.

For the 20- and 76-K tests, the liquid helium was introduced slowly (less than 0.5 L/min) such that the specimen would not be thermally shocked. When the liquid helium level in the apparatus was sufficient, power was applied to the heating element to warm the specimen to 20 or 76 K. The desired test temperature was maintained at ±0.25 K for 30 min prior to testing such that the specimen would achieve thermal equilibrium.

All tests were conducted on a commercial, universal tension/compression testing machine. An initial strain rate of 5×10^{-3} min^{-1} was chosen to obtain the Young's modulus, proportional limit, and yield strength. The test was continued at that rate until a strain of approximately 0.2 was achieved. The strain rate was then increased to 5×10^{-2} min^{-1} to reduce the duration of the test. Previous experiments[3-6] indicate that there are no differences in the mechanical properties when using these two strain rates. Compressive tests were continued to the capacity of the load frame corresponding to a stress of 15.6 MPa (2270 psi). The displacement at this point was used to calculate the total strain in the specimen. After testing, the specimens were unloaded, removed from the apparatus, and warmed to ambient temperature. The final lengths measured using an optical comparator were used to calculate the permanent strain. The "elastic" strain was calculated as the difference between the total and permanent strains. All specimens tested were retained for subsequent examination.

RESULTS

The results are presented in Table 1 and Figs. 1 through 5. Figure 1 shows the average stress-versus-strain curves for silica aerogel at 20, 76, and 295 K. This material exhibits increased rigidity as the test temperature is reduced. Figures 2 through 4 show that Young's modulus, proportional limit, and yield strength (at 0.2% offset) are inversely proportional to temperature. The stress-versus-strain curves approach each other at the higher stress levels above 10 MPa. Therefore, since all tests were conducted to a stress of 15.6 MPa (2270 psi), the total strain was essentially constant with respect to test temperature (see Fig. 5). The "elastic" strain increases with decreasing test temperature and conversely, the permanent strain decreases with decreasing test temperature.

DISCUSSION OF RESULTS

The Young's modulus, proportional limit, and yield strength (at 0.2% offset) all show a temperature dependence typical of engineering materials in general; that is, they increase with decreasing temperature.

When the strain data are examined, behavior atypical of most materials is observed in the elastic component of strain. This behavior can be explained in terms of material properties: The silica aerogel is composed of glass fibers, 4 to 10 nm in diameter, linked together to form the lattice structure. At room temperature, glass is a brittle solid. When cooled, the strength of the glass increases significantly without a corresponding change in the modulus. Thus, the strain required to cause failure must increase as the temperature decreases. This behavior is exhibited by glass-reinforced composites.[10]

Table 1. Compression Test Results for Silica Aerogel (Nominal Density = 0.1 g/cm^3)

Specimen Number	Test Temp., K	Young's Modulus, MPa	Young's Modulus, psi	Proportional Limit, MPa	Proportional Limit, psi	Yield Strength (0.2% offset), MPa	Yield Strength (0.2% offset), psi	Strain, %** Elastic	Strain, %** Permanent	Strain, %** Total
1	295	1.55	225	0.116	16.8	0.145	21.0	3.6	78.5	82.1
2		2.71	393	0.119	17.3	0.150	21.8	2.8	79.2	82.0
3		1.85	268	0.085	12.3	0.122	17.7	0.9	79.4	80.3
4		1.96	284	0.110	15.9	0.130	18.9	2.1	79.5	81.6
7		3.31	480	-	-	-	-	-	-	-
8		2.16	313	-	-	-	-	-	-	-
9		1.99	289	-	-	-	-	-	-	-
10		1.87	271	-	-	-	-	-	-	-
11		2.29	332	-	-	-	-	-	-	-
12*		2.76	400	-	-	-	-	-	-	-
	$\bar{x}$ =	2.25	326	0.108	15.6	0.137	19.9	2.4	79.1	81.5
10	76	2.44	354	0.102	14.8	0.136	19.4	16.7	61.5	78.2
11		2.73	396	-	-	0.227	33.0	29.1	51.7	80.8
	$\bar{x}$ =	2.59	375	0.102	14.8	0.182	26.2	22.9	56.6	79.5
7	20	3.84	556	0.164	23.8	0.246	35.5	28.7	46.7	75.4
8		2.91	422	0.143	20.7	0.175	25.4	27.4	56.1	83.5
9		2.93	425	0.146	21.1	0.187	27.1	28.7	52.4	81.1
12*		3.64	528	0.177	25.7	0.232	33.6	27.6	51.3	78.9
	$\bar{x}$ =	3.33	483	0.158	22.8	0.210	30.4	28.1	51.6	79.7

* Vacuum/purged specimen. All other specimens were helium purged only.

** Total strain was measured at end of test at 15.6 MPa (2270 psi). Permanent strain was measured at end of test at zero stress. Elastic strain is the difference of the total and permanent strains (see Fig. 1).

Without a corresponding loss in ductility, the effect of the strength increase on the specimen integrity is dramatic. The room temperature tests typically resulted in catastrophic specimen failure, fracturing and sometimes pulverizing the specimens. This is due to the lower strength of the glass fibers at room temperature, resulting in easier cell failure and collapse. Fiber failure causes most of the strain to be permanent. When cooled, each fiber is greatly strengthened without a substantial loss of ductility, and the specimen can support larger loads before cellular failure occurs, giving more elastic strain at lower temperatures. After testing at 20 and 76 K, the specimens were usually intact and remained cylindrical in shape. Ashby provides a mathematical model to illustrate strengthening mechanisms for open cell foams, using beams to model the open cell structure.[11]

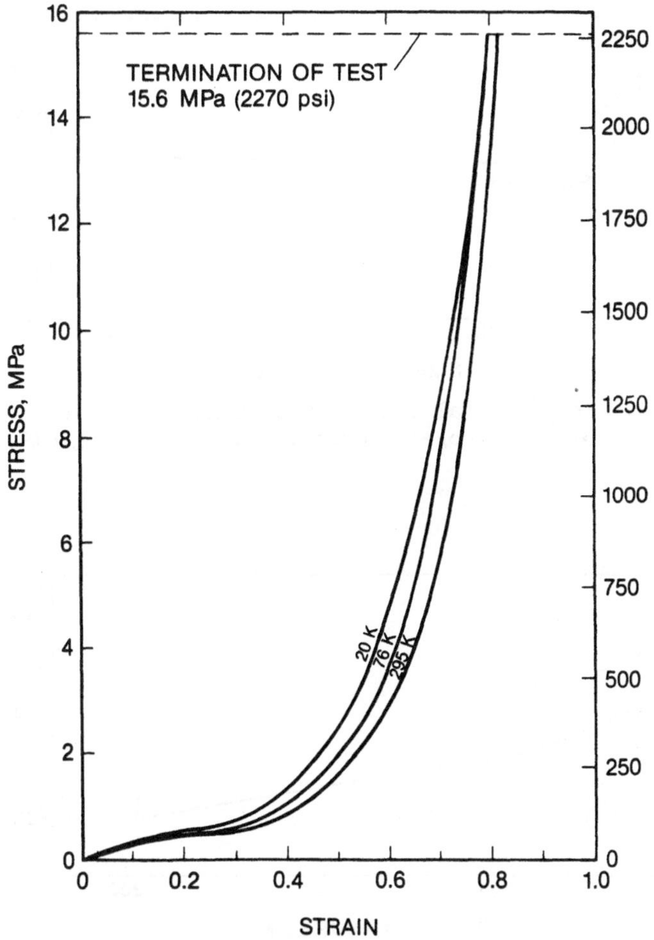

Fig. 1. Stress-versus-strain behavior for silica aerogel at 20, ,76, and 295 K (nominal density = 0.1 g/cm^3).

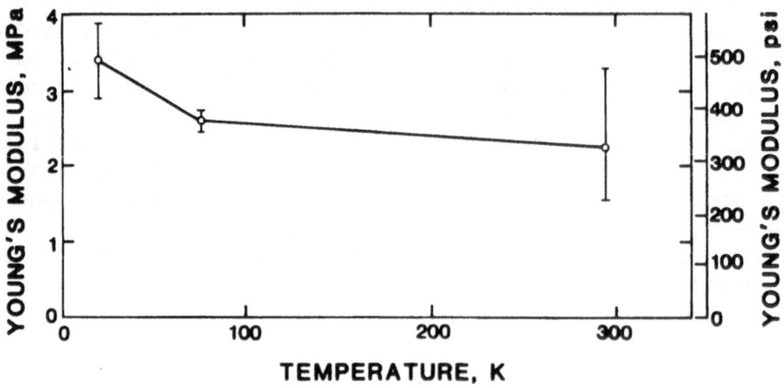

Fig. 2. Young's modulus versus temperature for silica aerogel (nominal density = 0.1 g/cm^3).

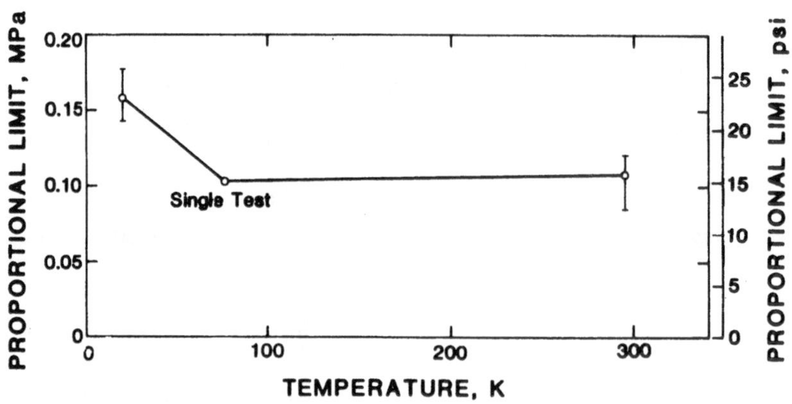

Fig. 3. Proportional limit versus temperature for silica aerogel (nominal density - 0.1 g/cm^3).

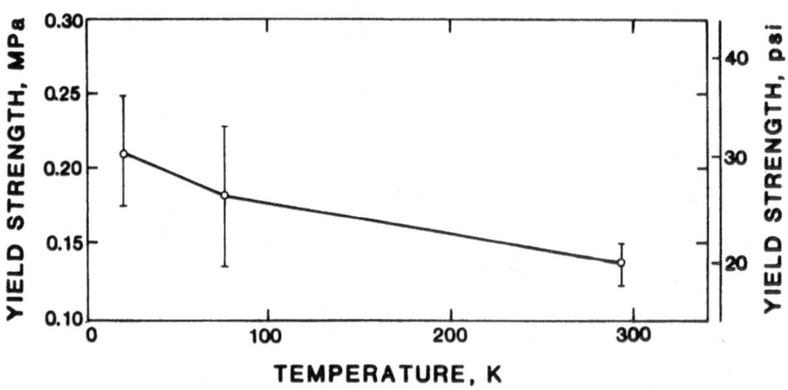

Fig. 4. Yield strength (at 0.2% offset) versus temperature for silica aerogel (nominal density = 0.1 g/cm^3).

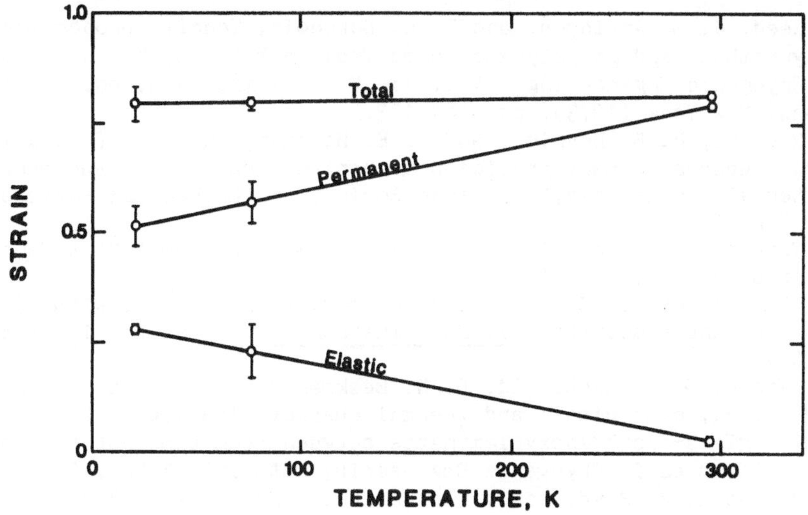

Fig. 5. Strain (elastic, permanent, and total) versus temperature for silica aerogel (nominal density = 0.1 g/cm^3).

CONCLUSIONS

Silica aerogel is an extremely delicate material with very low strength. Its behavior, with respect to temperature, is unlike polymeric foams. This is due to the large increase in strength of the glass fibers that form the aerogel, with no significant loss in ductility as the material is cooled.

ACKNOWLEDGMENTS

This research was sponsored and the silica aerogel foam was supplied by the Fusion Target Fabrication Group at the Lawrence Livermore National Laboratory, Livermore, California. The authors are indebted to Drs. Jack Campbell, Dale Darling, and Irving Stowers for their support and the many helpful discussions concerning this project.

REFERENCES

1. W. J. Schmitt, R. A. Greiger-Block, and T. W. Chapman, The preparation of acid-catalyzed silica aerogel, in: "Chemical Engineering at Supercritical Fluid Conditions," M. E. Paulaitis, J. M. L. Penninger, R. D. Gray Jr., and P. Davidson, eds., Ann Arbor Science, Ann Arbor, Michigan (1983), pp. 445-460.
2. "Handbook of Physics and Chemistry," R. C. Weast, ed., Chemical Rubber Company, Boca Raton, Florida (1982).
3. J. M. Arvidson and L. L. Sparks, "Low Temperature Mechanical Properties of a Polyurethane Foam," NBSIR 81-1654, National Bureau of Standards, Boulder, Colorado (1981).
4. J. M. Arvidson, L. L. Sparks, and G. Chen, "Tensile, Compressive, and Shear Properties of a 64 kg/m^3 Polyurethane Foam at Low Temperatures," NBSIR 83-1684, National Bureau of Standards, Boulder, Colorado (1983).
5. J. M. Arvidson, R. S. Bell, L. L. Sparks, and G. Chen, "Tensile Compressive, and Shear Properties of a 96 kg/m^3 Polyurethane Foam at Low Temperatures," NBSIR 83-1696, National Bureau of Standards, Boulder, Colorado (1983).

6. R. P. Reed, J. M. Arvidson, and R. L. Durcholz, Tensile properties of polyurethane and polystyrene foams from 76 K to 300 K, in: "Advances in Cryogenic Engineering," Vol. 18, K. D. Timmerhaus, ed., Plenum Press, New York (1973), pp. 184-193.
7. J. M. Roberts, R. B. Herring, and D. E. Hartman, The use of capacitance gauge sensors to make precision mechanical property measurements, in: "Materials Technology," American Society for Mechanical Engineers, New York (1968).
8. "High Temperature Capacitive Strain Measurement System," NASA Tech. Brief B75-10069, NASA (1975).
9. P. C. F. Wolfendale, Capacitive displacement transducers with high accuracy and resolution, J. Sci. Instrum. (J. Phys. E.) 1:817-818 (1968).
10. M. B. Kasen, G. R. MacDonald, D. H. Beekman Jr., and R. E. Schramm, Mechanical, electrical, and thermal characterization of G-10CR and G-11CR glass-cloth/epoxy laminates between room temperature and 4 K, in: "Advances in Cryogenic Engineering Materials," Vol. 26, A. F. Clark and R. P. Reed, eds., Plenum Press, New York (1980), pp. 235-244.
11. M. F. Ashby, The mechanical properties of cellular solids, Metall. Trans. A, 14A:1755-1769 (1983).

LH_2 ON-ORBIT STORAGE TANK SUPPORT TRUNNION DESIGN AND VERIFICATION*

William J. Bailey, Dale A. Fester, and Joseph M. Toth, Jr.

Martin Marietta Denver Aerospace
Denver, Colorado

ABSTRACT

A subcritical liquid hydrogen orbital storage and transfer experiment is being designed for flight in the shuttle cargo bay. The Cryogenic Fluid Management Facility (CFMF) includes a liquid hydrogen storage tank supported in a vacuum jacket by two fiberglass epoxy composite trunnion mounts. The capability of the CFMF to meet a seven mission requirement is extremely sensitive to the fatigue life of the composite trunnions at cryogenic temperatures. An E-glass/S-glass epoxy composite material was selected for the trunnions since it provided desirable strength, weight and thermal characteristics. Because of the limited extent of analytical or experimental treatment of the fatigue life of this composite at cryogenic temperature, an experimental program was conducted to provide verification of the trunnion design and performance capability at ambient and liquid hydrogen temperatures. Basic material fatigue property data were obtained for the laminate of interest using specifically prepared test specimens. Full-scale trunnions were manufactured and subjected to cyclic load testing to verify fatigue life. An analytical evaluation of the thermal performance of the trunnions was conducted, and a test setup is being manufactured to correlate analytical predictions with test results.

INTRODUCTION

The Cryogenic Fluid Management Facility (CFMF) (shown in Fig. 1) is a reusable Shuttle payload bay test designed to investigate systems and technologies required to manage cryogens in space efficiently and effectively. In particular, it is the first flight system designed to provide low-g verification of fluid and thermal models of cryogenic fluid storage, expulsion, transfer and resupply, and to characterize overall performance of these systems in a space environment. Significant design data and criteria for future subcritical cryogenic fluid storage and transfer systems will be obtained. (See Ref. 1 for a more detailed description of CFMF).

---

*This work was performed for the NASA Lewis Research Center under contracts NAS3-23355[2] and NAS3-23245.

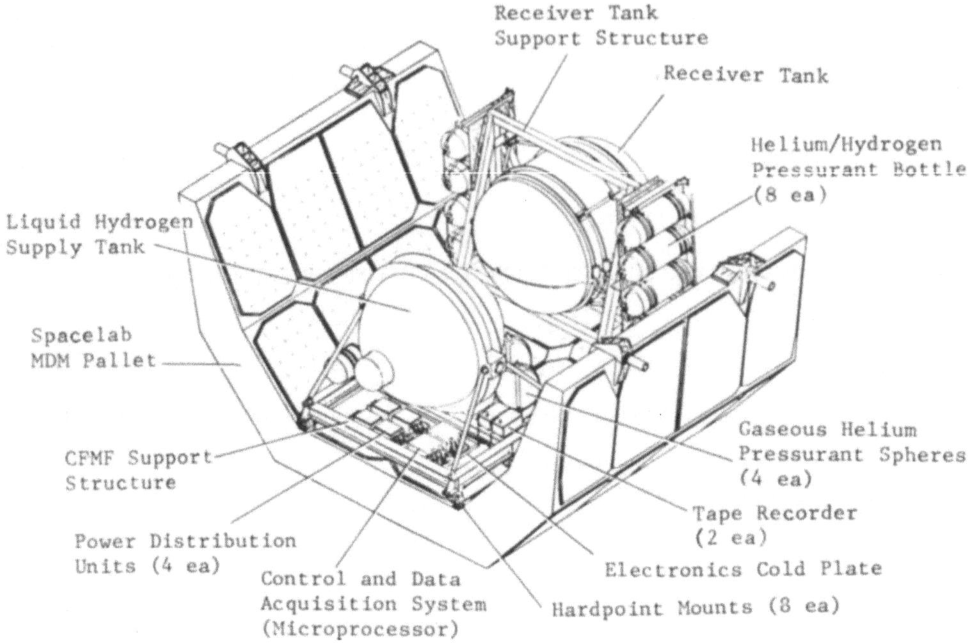

Fig. 1. Cryogenic Fluid Management Facility (CFMF)

Future space missions will require the use of cryogenic liquids for various applications. Storing and supplying subcritical, rather than supercritical, cryogens offers size and weight advantages with the lower pressures associated with subcritical storage. Trade offs between structural design margins and thermal heat leak considerations (which multiply as support cross-sectional areas increase) must be made to achieve an optimal design. A low heat leak thermal requirement led to the selection of fiberglass epoxy trunnions as structural supports for the pressure vessel within the vacuum jacket.

Two trunnion configurations are used for the CFMF supply tank assembly, one defined as "fixed" (shown in Fig. 2), and the other referred to as "floating". The floating trunnion has a tubular open end and is free to move to allow for thermal contraction as the tank is loaded with liquid hydrogen. The threaded fitting in the fixed trunnion allows for proper initial positioning of the inner vessel within the vacuum jacket so that it will be centered after contraction due to cryogenic liquid loading.

A detailed fatigue analysis was conducted as part of the design activity to assess capability to meet the seven mission life requirement.[3] All areas were found to have sufficient fatigue life with the possible exception of the trunnions. Only limited tension-compression fatigue data were available for this material and no data were available for the actual trunnion configuration.

An experimental program was conducted to evaluate fatigue life of the composite material and to provide verification of the trunnion design.[2] Basic material property data were obtained from laminate test material E-glass/S-glass specimens at both ambient and liquid hydrogen temperatures. These data supported the adequacy of the trunnion design for seven-mission life. Three trunnions were then manufactured and subjected to cyclic load testing to verify fatigue life of the design. They were then loaded to ultimate failure for margin assessment.

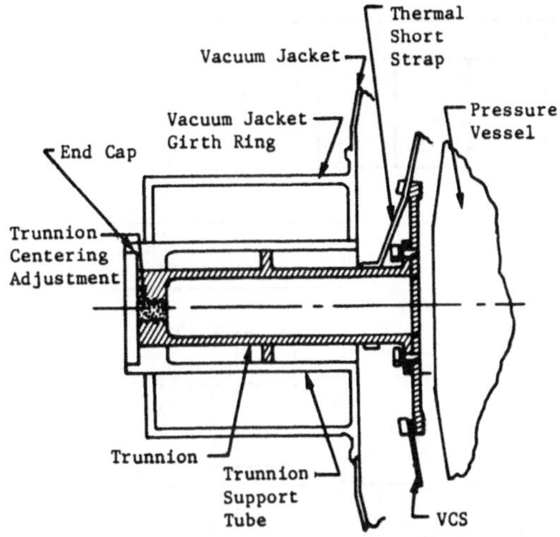

Fig. 2. CFMF fixed trunnion-vacuum jacket-pressure vessel interface.

MATERIAL PROPERTIES DETERMINATION

The original trunnion analysis was conducted using best available material property data which suggested that the design was marginal with a damage factor of 0.813.[2] This identified the trunnions as the most critical item of the design and became the factor which initiated this investigation. The first objective was to experimentally determine the material properties of the CFMF trunnion laminate material. Since fatigue criteria drove the original analysis, fatigue data became the prime property data desired.

Thermal, stiffness, and strength requirements drove the design and resulted in a layup of $(-45°/0°_3/+45°/0°_3/+45°/-45°/0°_3/+45°/0°_3/-45°)$, where the 45° material was style No. 1581 E-glass fabric at a cured thickness of 0.023 cm (0.009 in.)/ply and the 0° material was S2 unidirectional glass fibers at a cured thickness of 0.0305 cm (0.012 in.)/ply. The epoxy matrix was Fiberite Corporation's 934 resin. The laminate was balanced to avoid in- plane coupling effects. The laminate was not symmetrical, because the 1581 fabric was only slightly unbalanced with respect to warp and fill modulus (i.e., $E_{fill\ direction}$ = 95% of $E_{warp\ direction}$). The unsymmetrical plies were grouped about the centerline to minimize the coupling between in-plane and bending response.

Material fatigue test specimen design in a flat configuration became a complex issue which underwent several iterations before an acceptable configuration was established as having the greatest chance of success. Little information was available in the literature on fiberglass epoxy composite fatigue specimen designs suitable for use at both ambient and liquid hydrogen temperatures under the fully reversed (R = -1.0) tension-compression loading.

Specimen designs evolved from the following considerations:

- Specimen/test machine interface must be compatible to minimize buckling on the compression part of the cycle.
- Specimen gage section cross-sectional area must be compatible with existing LH_2 cryostat and fatigue fixture capacities.
- Specimen end configuration and machine attachment must minimize alignment complexities.

Specimen Layup									
-45	0°_{3}	+45°	0°_{3}	+45°	-45°	0°_{3}	+45°	0°_{3}	-45°
.023 (.009	.091 (.036)	.023 (.009)	.091 (.036)	.023 (.009)	.023 (.009)	.091 (.036)	.023 (.009)	.091 (.036)	.023 (.009)

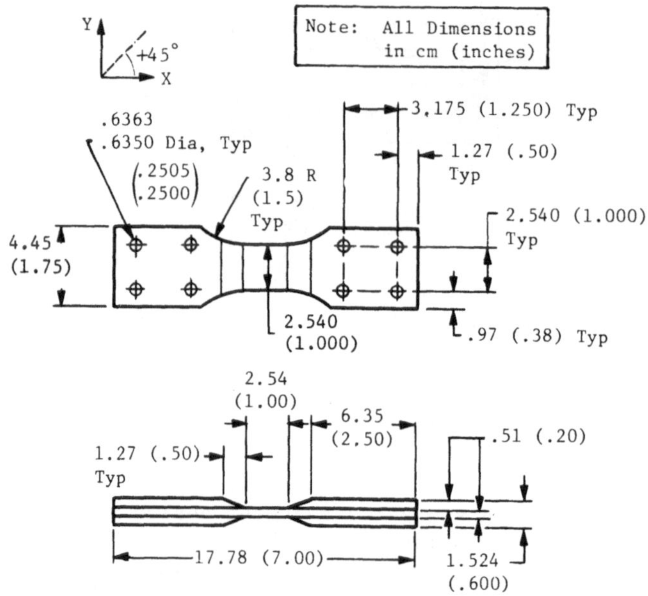

Fig. 3. Laminate test specimen configuration.

The resulting configuration is shown in Fig. 3. The S-glass roving was aligned with the longitudinal specimen axis. This layup was selected for its stiffness characteristics in the longitudinal direction. Total laminate thickness was approximately 0.51 cm (0.20 in.). End tabs were bonded at the specimen ends using a high-strain capacity, low temperature adhesive and beveled along one edge at approximately 20° to provide for a gradual transfer of the load from the end tabs to the gage section. The overall specimen length and gage section were minimized to reduce buckling during compression. A four bolt attachment interfaced with the test fixture and aligned the center of the coupon with the load line of the test machine.

Installation of the test specimen in the test fixture, which was configured as an integral part of the lid of a LH_2 cryostat, is shown in Fig. 4. For ambient temperature tests, the cryostat reservoir was not installed. Tests were performed at both ambient conditions and with the test specimen completely immersed in LH_2. Loads were transferred through a loading rod attached to the top of the specimen by a mating clevis. The rod was flange-mounted to a load cell which was attached to a hydraulic actuator mounted on top of a MTS Testing Machine. The other end of the specimen was rigidly fixed to a similar mating clevis bolted to the test fixture bottom.

Tension-compression loads were applied to each specimen in a sinusoidal manner. The loading conditions were selected based on a measured ultimate strength in tension of $1.323 X 10^6$ kN/m^2 (192 ksi) at 300 K (80°F). Four specimens were tested in fatigue at LH_2 temperature 20 K (-423°F). The ambient temperature results and LH_2 temperature results are shown for comparison purposes in Fig. 5. As expected, the fatigue resistance of the laminate increases at cryogenic temperature.

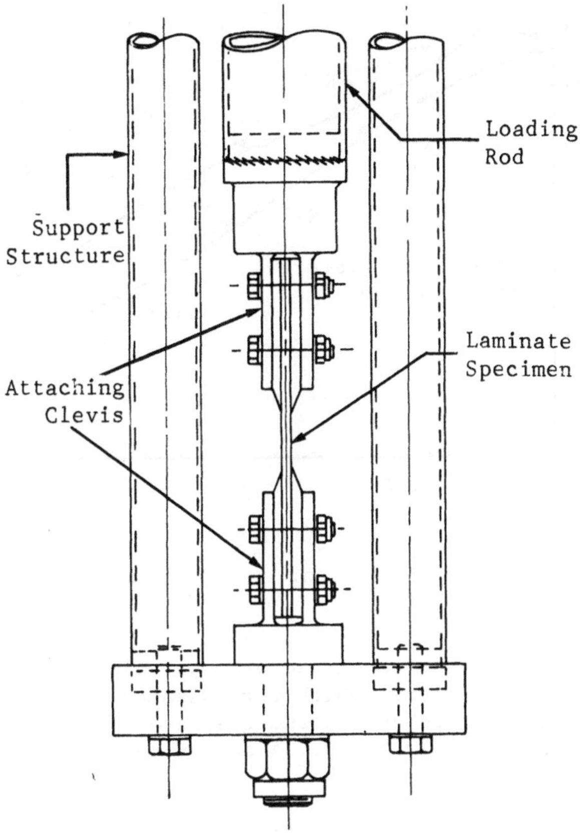

Fig. 4. Specimen—test fixture interface.

The static strength of filament-wound fiberglass (0/90) laminate composites at 20 K (-423°F) has been shown to be approximately 20-30% higher than that at 300K(80°F). The limited fatigue test data generated here indicates an increase in stress level at a given cyclic level from 10% to greater than 50% since specimens H-1, H-3, and H-4 were not tested to failure.

The test data showed that the initial analysis had been overly conservative and attested to the adequacy of the proposed design. A new damage factor was calculated using the experimental ambient S/N curve, resulting in a number less than 0.1 (the original damage factor was 0.813).

EVALUATION OF TRUNNION STRUCTURAL INTEGRITY

Data collected on the laminate material pertained to the longitudinal direction only. Since the trunnion experiences combined loading, testing the actual part was needed to resolve the remaining uncertainties.

A stress and dynamic fatigue analysis was conducted to verify design load factors based on Spacelab Payload Accommodation Handbook (SLP/2104) requirements and a preliminary random vibration evaluation. The loading spectrum for the trunnion fatigue tests was established. Areas of highest stress were identified as being next to the trunnion support ring, which is shown in the simplified trunnion configuration of Fig. 6. This figure also shows load application locations and instrumentation locations.

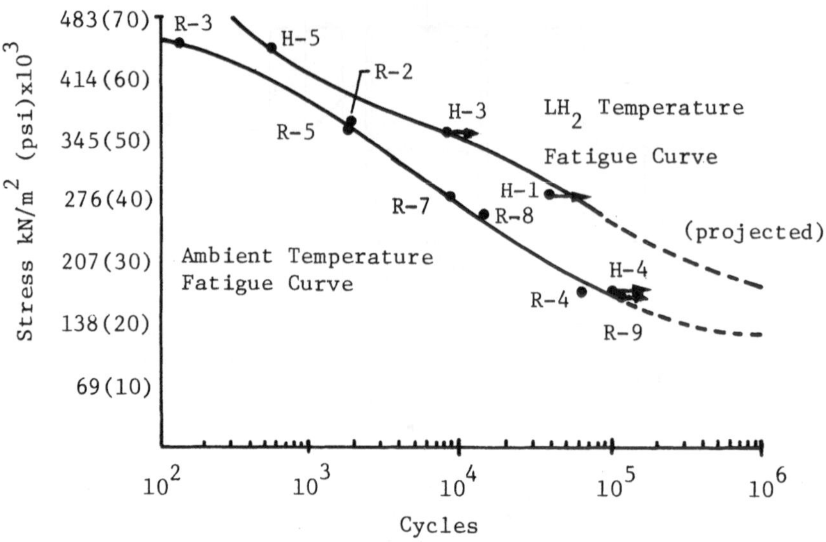

Fig. 5. S/N curve for ambient and LH_2 temperature data.

The loading conditions consisted of mean value fixed quasi-static loads and random vibration loads which were assumed to vary over a Rayleigh distribution. Fatigue cycles over a shuttle seven-mission life were determined to be approximately 5,000. A safety factor of 4 resulted in testing to a 20,000 cycle level.

A test system (Fig. 7) was constructed to apply fatigue loading at the trunnion end, support ring, and threaded end fitting (for the fixed trunnion only). Actuators P1 and P2 mated with the trunnion through rings to distribute the load uniformly around the trunnion cylinder/support ring circumference. The axial actuator had a threaded end to mate with the titanium insert on the fixed trunnion.

The primary objective was to verify the capability of the flight trunnion to survive the fatigue cycle environment over a distribution up to and including limit load. Additional testing above limit load established the degree of design margin. One fixed and two floating trunnions were

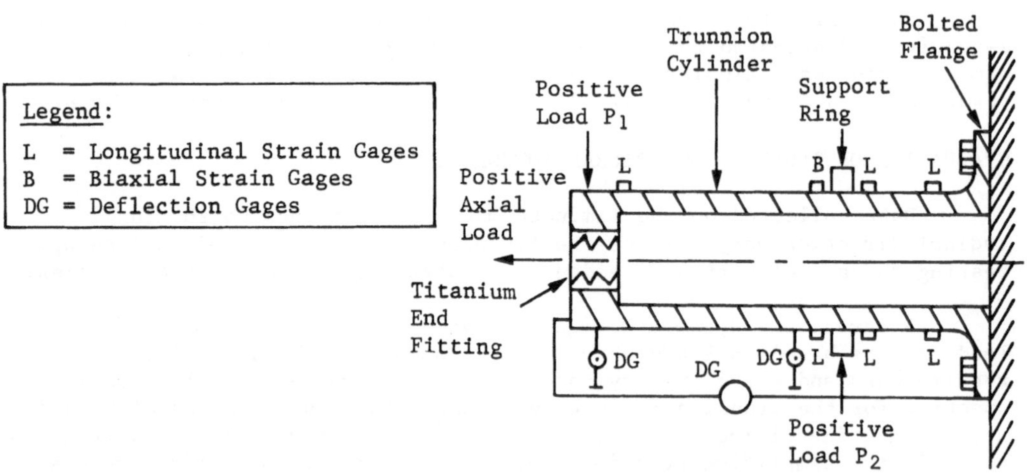

Fig. 6. Trunnion test configuration.

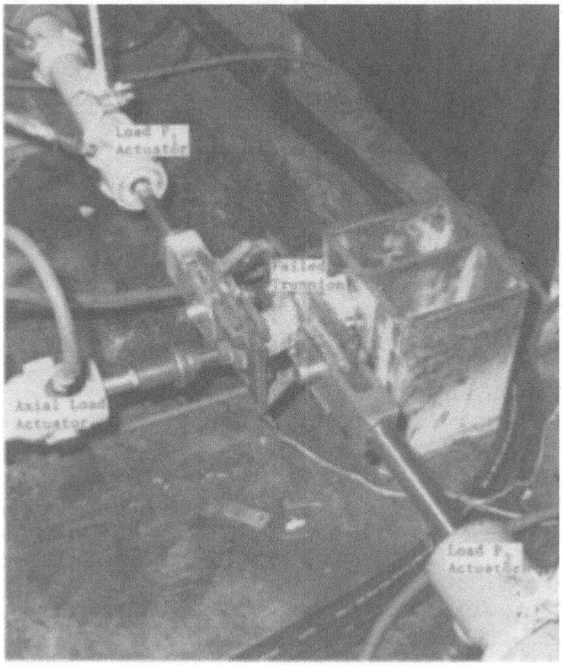

Fig. 7. Trunnion fatigue test system.

fabricated and fatigue tested for 20,000 cycles with the load distributions shown in Tables 1 and 2. One fixed and one floating trunnion were then subjected to margin testing for an additional 2,540 cycles up to and including the ultimate loading condition of the design without failure. Following cycling, all three articles were loaded to failure.

All of the trunnions exceeded the design load-carrying requirement, with the fixed trunnion (S/N 001) failing at 190% of limit load (127% of design ultimate) after margin testing. The floating trunnion (S/N 001) subjected to margin testing failed at 210% of limit (140% of design ultimate) while the floating trunnion (S/N 002) not subjected to margin testing failed at 230% of limit (153% of design ultimate). In each instance, primary failure occurred adjacent to the support ring as predicted. The results clearly show the adequacy of the trunnion supports to meet the required CFMF seven-mission design life.

TRUNNION THERMAL ANALYSIS

The thermal performance for both the fixed and floating configurations depends on two factors; (1) the low thermal conductivity of the S-glass/E-glass composite epoxy material at cryogenic temperature, and (2) the contact resistance of the sliding surfaces at the spacer location and at the hot-end sleeve. Only limited information is available on the thermal properties of S-glass and E-glass materials and effects of the fiber orientations on these properties. Thermal performance analysis is extremely sensitive to the assumptions made on contact conductivity between the trunnion and the sleeve so a conservative number of $1.14 kW/m^2$-K(200 BTU/hr-ft^2-°R) was selected. This represents a reasonable value for illustrative purposes. The area of the spacer in contact with the support sleeve was taken to be 5% of the rib area at its outer edge. The remaining

Table 1 20,000 Cycle Fatigue Loading (0 to 2σ)

Cyclic Axial Load, N (lbs)	Cyclic P_1 Load, N (lbs)	Cyclic P_2 Load, N (lbs)	σ	Cycles 0 to 2σ = 20,000
3649/-534 (820/-120)	9372/5794 (2106/1302)	17275/10680 93882/2400	0.4	1780
6786/-3671 (1525/-825)	12055/3110 2709/699)	22223/5732 (4994/1288)	1.0	7320
9924/-6809 (2230/-1530)	14738/427 (3312/96)	27167/788 (6105/177)	1.6	7600
10969/-7854 (2465/-1765)	15633/-467 (3513/-105)	28818/863 (6476/-194)	1.8	1860
12015/8900 (2700/-2000)	16527/-1362 (3714/-306)	30465/-2510 (6846/-564)	2.0	1440

Table 2 2,540 Cycle Fatigue Loading Margin Testing (2σ to 3σ)

Cyclic Axial Load, N (lbs)	Cyclic P Load, N (lbs)	Cyclic P_2 Load, N (lbs)	σ	Cycles 2σ to 3σ = 2,540
15152/-12037 (3405/-2705)	19211/-4045 (4317/-909)	34413/-7458 (7958/-1676)	2.6	2060
17244/-14129 (3875/-3175)	21000/-5834 (4719/-1311)	38711/-10756 (8699/-2417)	3.0	480

95% was connected to the sleeve via radiation with a view factor of 1.0 and an emissivity of 0.8. Conduction along the length of the trunnion is divided such that 70% of the cross-sectional area represented axial fibers of S-glass unidirectional material and 30% represents fibers of E-glass cloth woven into a 45 degree orientation to the axis. A thermal model for the trunnion was developed to evaluate the heat inputs to the storage tank through the trunnions. The results of this analysis are presented in Fig. 8. There is essentially no difference in the heat leak of the fixed and floating trunnion configurations.

TRUNNION THERMAL PERFORMANCE CHARACTERISTICS

A test system was designed (Fig. 9) to measure the trunnion thermal conductance for correlation with analytical predictions. This test setup, consisting of a liquid nitrogen reservoir to which the test trunnion is attached, simulates the actual trunnion/LH_2 tank connection. A trunnion housing simulates the actual vacuum jacket support tube configuration. An electrical heating element is attached to the housing and the trunnion is thoroughly instrumented with temperature sensors. The entire test item (trunnion and housing) is wrapped with multi-layer insulation (MLI). A separate MLI blanket insulates the LN_2 reservoir. The entire test setup is then installed within a vacuum chamber. Trunnion thermal conductance will be calculated by measuring the temperature distribution along the trunnion with heaters off and then again while varying the input power level to the heater element. The liquid nitrogen reservoir will act as a

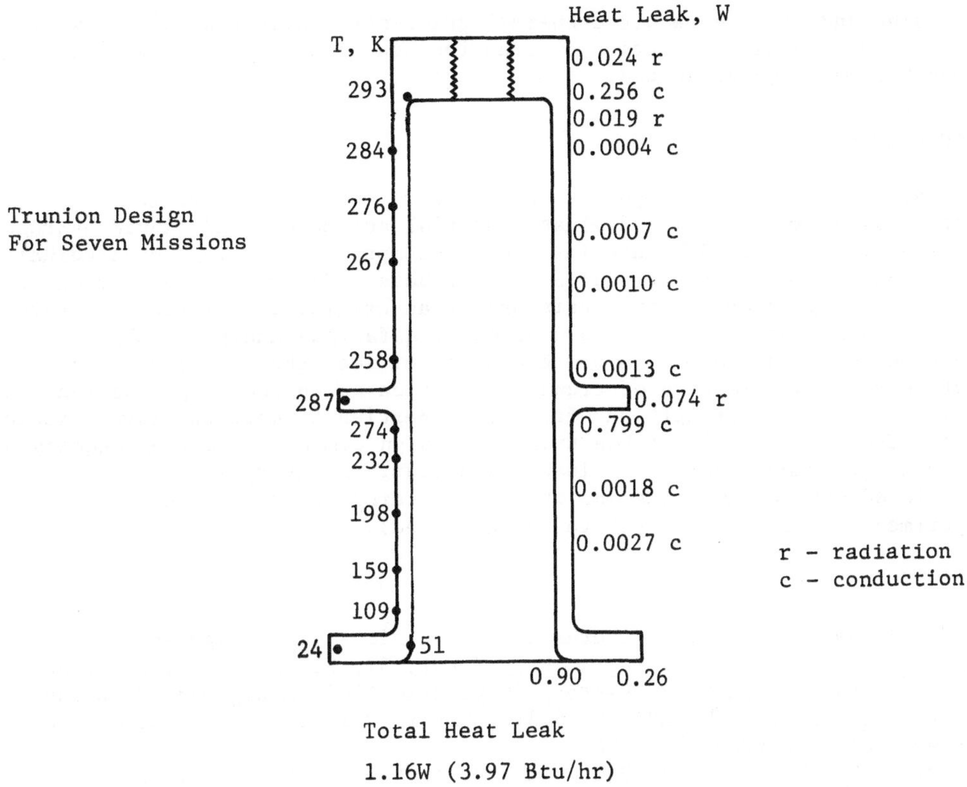

Fig. 8. Trunnion analytical heat leaks and temperature distributions.

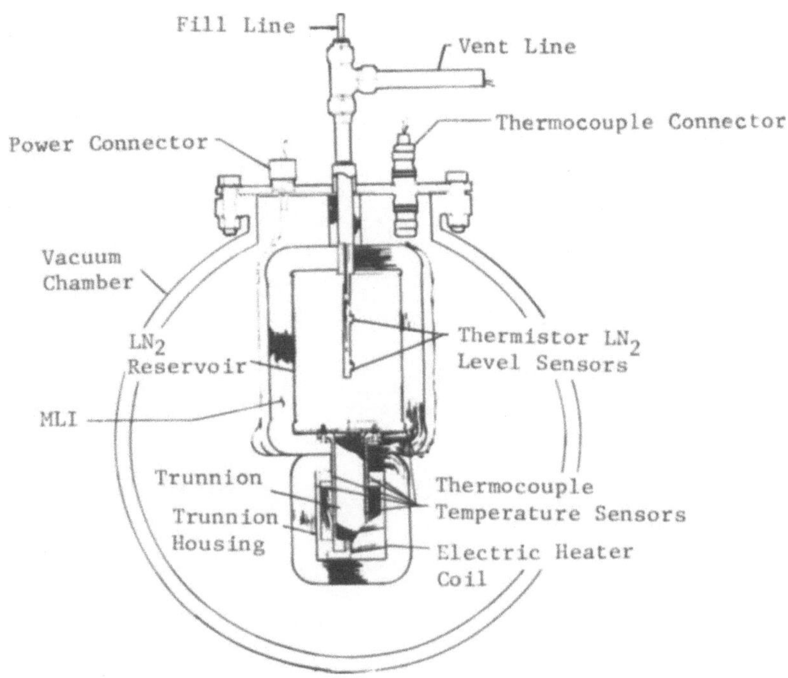

Fig. 9. Trunnion thermal test system.

heat sink and provide required thermal gradients. Nitrogen boiloff will provide additional information on trunnion heat leak. Testing with in-house funds has not been conducted.

CONCLUSIONS

Ultimate strength and fatigue properties were established for the fiberglass epoxy composite laminate material at ambient and liquid hydrogen temperatures. These data are for tension-compression in the longitudinal direction of the unidirectional S-glass fibers. The material was found to provide much greater fatigue resistance than originally derived from very limited data. Based on the measured data, a fatigue analysis for the trunnion supports produced a damage factor of less than 0.1 for seven shuttle flights. Testing of trunnions fabricated to the flight design verified adequate strength and fatigue properties to meet the seven mission life. Thermal analysis of the trunnion design demonstrated the adequacy of the configuration from a heat leak standpoint. However, sensitivities associated with assumptions and other uncertainties created the impetus for experimental verfication (yet to be performed).

REFERENCES

1. W. J. Bailey and R. N. Eberhardt, Cryogenic Fluid Management Facility, in: "Advance in Cryogenic Engineering," Vol. 31, Plenum, New York (1986).
2. W. J. Bailey and D. A. Fester, "Cryogenic Fluid Management Trunnion Verification Testing," NASA-CR-168310, Martin Marietta Denver Aerospace, Denver, Colorado (December 1983).
3. R. N. Eberhardt, W. J. Bailey and D. A. Fester, "Cryogenic Fluid Management Experiment," NASA-CR-165495, Martin Marietta Denver Aerospace, Denver, Colorado (October 1981).
4. J. M. Toth, Jr. and D. J. Soltysiak, "Investigation of Smooth-Bonded Metal Liners for Glass Fiber Filament-Would Pressure Vessels, Final Report," Contract NAS3-6293, McDonnell Douglas Corporation (May 1967).

RECENT INVESTIGATIONS ON REFRIGERANTS FOR MAGNETIC REFRIGERATORS

T. Hashimoto

Department of Applied Physics
Tokyo Institute of Technology
Oh-okayama, Meguro, Tokyo

ABSTRACT

In development of the magnetic refrigerator, an important problem is selection of magnetic materials as refrigerants. The main purpose of the present paper is to discuss the magnetic and thermal properties necessary for these refrigerants and to report recent investigations. Magnetic refrigerants can be expediently divided into two groups, one for the Carnot-type magnetic refrigerator below 20 K and the other for the Ericsson-type refrigerator. The required physical properties of refrigerants in each type of the magnetic refrigerator are first discussed. And then, the results of recent investigations on the magnetic, thermal and magnetocaloric characters of several promising magnetic refrigerants are shown. Finally, a brief prospect of the magnetic refrigerants and refrigerators is given.

INTRODUCTION

After the notable success achieved with the magnetic refrigerator near room temperature by Brown,[1] many investigators began work on the magnetic refrigerator because of its novel and excellent character. Recently, investigations in France[2] and Japan[3] resulted in development of prototype magnetic refrigerators operating in the low temperature range below 20 K. Its considerable high efficiency in comparison with a gas refrigerator was verified.

An important reason for this success in spite of the failure of earlier investigations[4] is thought to utilize the $Gd_3Ga_5O_{12}$ (GGG) as the refrigerant. This material has excellent properties of magnetocaloric effect and thermal conductivity, and this demonstrates that the refrigerant plays a very important role in the performance of the magnetic refrigerator.

GGG, however, is not always a perfect refrigerant, even if the temperature range of operation of the magnetic refrigerator is limited to below 20 K. For instance, the He liquefaction magnetic refrigerator developed by a Japanese group[5] is only usable below ~16 K, and this limitation of temperature range is due to the considerable decrement of the magnetic entropy change caused by the external magnetic field and the increment of the lattice entropy.[5]

In the present paper, for selection of the magnetic materials usable for a refrigerant, the physical properties necessary will be discussed first. Here, we will deal with only the physical properties with emphasis on the magnetocaloric and thermal conductive characteristics. The electric conductivity, chemical stability and so on are not covered. Recent topical progress in investigations of several promising magnetic refrigerants will then be reviewed and, finally, the current status of investigations on the magnetic refrigerator and refrigerants will be briefly discussed.

PROPERTIES NECESSARY FOR A MAGNETIC REFRIGERANT

The thermal agitation energy of the spin system increases with temperature T as ~kT, where k is the Boltzmann constant, and in magnetic refrigeration only the magnetic field below 6 T is available for the external magnetic field to produce the magnetic entropy change ΔS_J. Different kinds of magnetic substances must, therefore, be used for refrigerants in the temperature ranges above and below ~20 K.

On the other hand, the lattice entropy, S_L, which cannot be controlled by the magnetic field, increases with temperature as shown in Fig. 1.[6] In the temperature range above ~20 K, where S_L is not negligible in comparison with the magnetic entropy, we must select the Ericsson cycle. In this case, the characteristics necessary for a magnetic refrigerant are slightly different from those in the Carnot cycle. Therefore, the magnetic refrigerants will be discussed separately from the viewpoints of temperature range and refrigeration cycle.

Magnetic Refrigerant in the Low Temperature Range

In the temperature range below ~20 K, the Carnot-type magnetic refrigerator was selected for its simplicity of mechanism and operation; it has been studied by many investigators. Here, only the rule governing selection of a refrigerant for the Carnot-type will be considered.

Below ~20 K, as the thermal agitation of the spin system is not strong, paramagnetic materials, in which the magnetic exchange interactions among spins are very weak, have been used for the refrigerants. In Fig. 2 the Carnot cycle is shown as A B C D in the entropy vs. temperature plane of a typical paramagnetic material, $Gd_3Ga_5O_{12}$ (GGG); this has been used

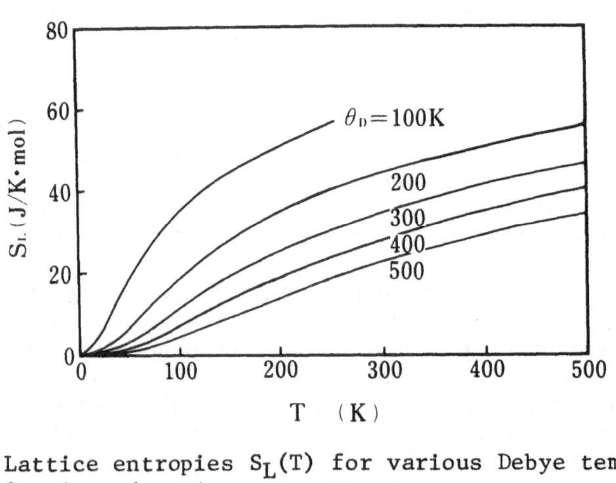

Fig. 1. Lattice entropies $S_L(T)$ for various Debye temperature θ_D plotted against temperature.

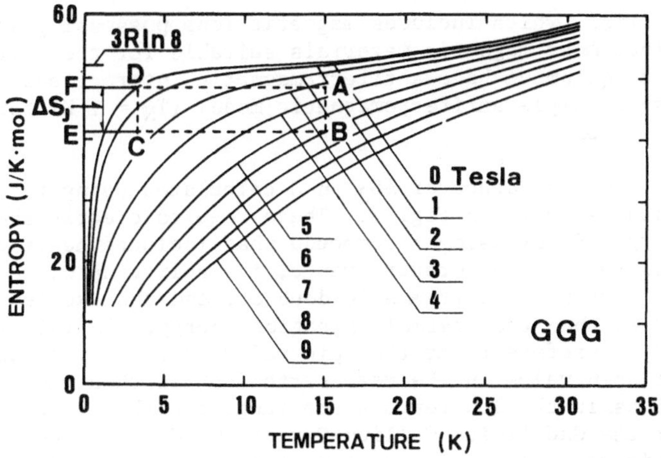

Fig. 2. Entropy of GGG in various magnetic fields plotted against temperature. Rectangle ABCD show the Carnot cycle.

widely for the magnetic refrigerator in the low temperature range. From this figure, we can conclude the following three restrictions are necessary in selecting a magnetic refrigerant:

(1) The refrigeration power per cycle is defined as the area of the rectangle E C D F. Therefore, in order to obtain the large amount of refrigeration power per cycle, the magnetic entropy change, ΔS_J, produced by the external magnetic field must be large.

(2) The magnetic entropy, S_J, decreases gradually with decreasing temperature in the range lower than about twice the phase transition temperature, T_t, and abruptly near T_t. These reductions of the magnetic entropy result in the reduction of area of the rectangle E C D F, that is, the amount of refrigeration power per cycle. Therefore, it is desirable that T_t is lower than 2 K.

(3) The refrigeration power also depends on the operating frequency of the cycle. In order to shorten the time necessary for the isothermal processes A → B and C → D, it is desirable that the thermal conductivity of the refrigerant is comparable to that of Cu, at least.

First, let us consider the magnetic entropy change, ΔS_J. According to the thermodynamic theory, ΔS_J is given by the following formula,

$$\Delta S_J = \int_0^B \left(\frac{\partial M}{\partial T}\right)_B dB. \tag{1}$$

In the paramagnetic region, using the simplest approximation, the Curie-Weiss law, we can obtain ΔS_J as

$$\Delta S_J = -\lambda B^2/2(T - \theta)^2$$

and

$$\lambda = N\mu_0 g^2 \mu_B^2 J(J + 1)/3k, \tag{2}$$

where λ is the Curie constant; B, the magnetic field; N, the number of spins; μ_0, the permeability in vacuum; g, the g-factor of electron spin; μ_B, the Bohr magneton number; J, the quantum number of the total angular momentum; k, the Boltzmann constant; and θ, the paramagnetic Curie temperature, which takes a value close to T_t in the simplest model. Among these, the characteristic physical parameters depending on the materials are g, J and θ. Therefore, in order to obtain a large ΔS_J, we must select a ferromagnetic material, which includes magnetic ions possessing large J and g values. Since most magnetic materials suitable for the refrigerant in the Carnot-type equipment have antiferromagnetic interactions, in this section I will use a simple antiferromagnetic model (T_N = 1 K) for the theoretical calculation.

For example, the dependence of ΔS_J on J calculated using the molecular field approximation is shown in Fig. 3. These results clearly show that we can get the large ΔS_J by using a compound that includes heavy rare earth ions. However, in the low temperature range, the splitting width between the lowest state, that is, the Krammer's doublet, and the excited state in the Dy ion is larger than the thermal agitation energy. In this temperature range we must therefore treat the spin value in these ions as if J = 1/2. The largest spin value in the rare earth ions is thus thought to be J = 7/2 of Gd^{3+} ions in the low temperature range and this is one of the main reasons that the Gadolinium-Gallium-Garnet, $Gd_3Ga_5O_{12}$ (GGG) has been used for the refrigerant of the Carnot-type magnetic refrigerator. The experimental results of GGG shown in Fig. 2 clearly verify the above conclusion below ~15 K. In the temperature range above ~15 K, meanwhile, since there is considerable decrement of ΔS_J and obvious increment of the lattice entropy S_L, we cannot get enough ΔS_J to refrigerate using the magnetic field of 6 T.

Equation (2) is thought to express one of the promising ways to use a compound that includes magnetic ions of large g-value. Therefore, the g-factor dependence of S_J in J = 1/2 was calculated theoretically as a function of the temperature using the molecular field approximation (Fig. 4). For comparison, the calculated result of ΔS_J in GGG is also shown. The large g-factor clearly gives a large ΔS_J, more than the ΔS_J of GGG, especially at higher temperatures. Thus, when we desire to make the Carnot cycle of a large temperature span, for instance from 20 K to 4.2 K, we have to use compounds that include magnetic ions having a large g-factor rather than a large J-value.

Let us consider next the thermal conductivity and indicate values necessary for making the refrigeration cycle. At the CEC in 1983, Numazawa

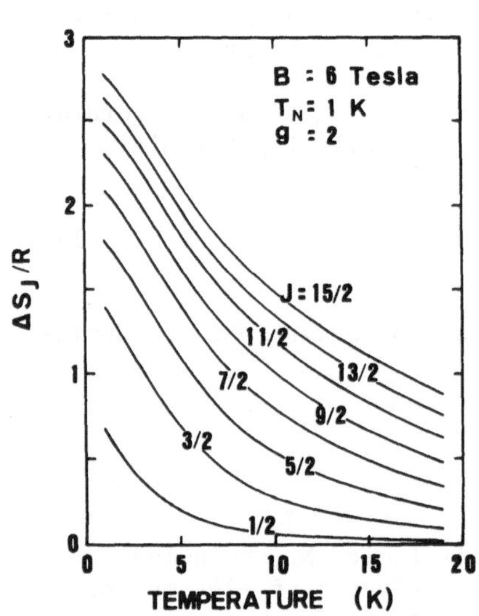

Fig. 3. Dependence of magnetic entropy change, ΔS_J, on J-value (assuming g=2).

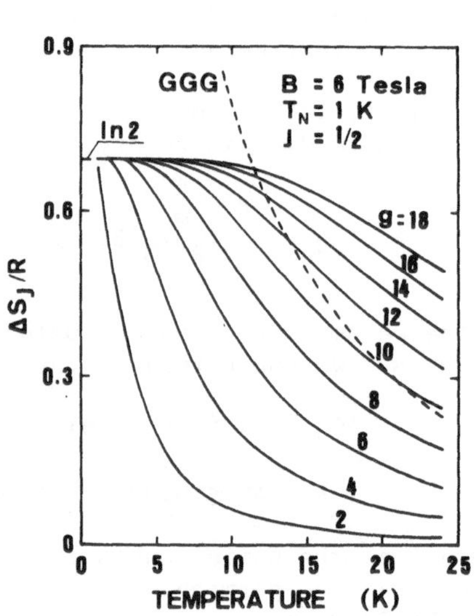

Fig. 4. Dependence of magnetic entropy change, ΔS_J, on g-factor (assuming J=1/2).

et al.[7] reported experimental results on the liquefaction character in a simple Carnot-type refrigerator. In this apparatus a heat pipe operating at 4.2 K was used as the heat switch of the low temperature side and a single crystal of GGG 50 mm in diameter and 100 mm in length was used for the refrigerant. The heat transfer coefficient of the heat pipe from the low temperature heat bath, that is, the He gas to GGG in the operating stage at 4.2 K, is larger than that of pure Cu metal. From this experiment they concluded that about 5 s is required to accomplish the isothermal demagnetization process in the Carnot cycle and that it is possible to make liquid He of 2.5 l/h from gas He at 17 K. These results are thought to indicate that the thermal conductivity, κ, of the refrigerant must be nearly the same as that of GGG to result in a highly efficient magnetic refrigerator.

On the other hand, according to our previous paper[8] the value of κ in GGG is about one third that of κ in pure Cu, though the ratio of κ's in those two compounds depends slightly on temperature. From those two experimental results it may be concluded that the value of κ of the refrigerant must be the same order of magnitude as that of Cu. From the viewpoint of thermal conductivity, it may also be concluded that paramagnetic oxide compounds of garnet structure are among the most promising magnetic refrigerants in the low temperature range, since all have very high conductivity, comparable to that of GGG.

Magnetic Refrigerant above ~20 K

In the temperature range above ~20 K, since the thermal agitation energy ~kT of spins becomes larger than the Zeeman splitting energy, $g\mu_B B$, obtained using the magnetic field of 6 T, the usual paramagnetic refrigerants in the low temperature range are not suitable. Therefore, we have to use ferromagnetics, which themselves have an internal magnetic field that can arrange the spins parallel.

This expedient method is also introduced theoretically.[6] According to eq.(2), if the temperature T is close to θ, ΔS_J increases greatly. In an ideal 3-dimensional ferromagnet, the value of θ may be almost the same as the Curie temperature, T_c. Therefore, even near room temperature, using ferromagnetic materials whose T_c values are near this temperature, we can obtain large ΔS_J values. As an example, the calculated results of ΔS_J in EuS using the molecular field approximation are shown in Fig. 5(a). The temperature dependence of ΔS_J clearly bears out the above theory.

On the other hand, above ~20 K the lattice entropy, S_L, which plays the role only of thermal load in magnetic refrigeration, becomes comparable to or more than the magnetic entropy S_J [Fig. 5(b)].[6] If we use the Carnot cycle, therefore, we cannot obtain a large temperature span of refrigeration and large refrigeration power. To achieve high power and high efficiency refrigeration, the Ericsson or the Brayton cycle, shown as ABCD (Ericsson) or A'B'C'D' (Brayton) in the entropy plane of EuS in Fig. 5(b), must be selected as the refrigeration cycle.

It is thus concluded that the following characteristics are required of a refrigerant for the Ericsson type refrigerator:
(1) To obtain large ΔS_J, the refrigerant must be ferromagnetic with magnetic ions whose g-factor and J-value are large.
(2) Their Curie points must be within the temperature span of the magnetic refrigeration cycle.
(3) It is desirable that the Curie temperatures of the compounds can be changed in the temperature range by substitution of the magnetic ions or by making solid solution of these compounds.

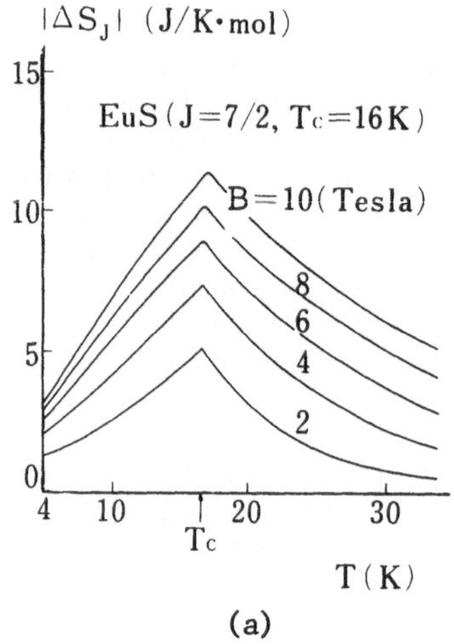

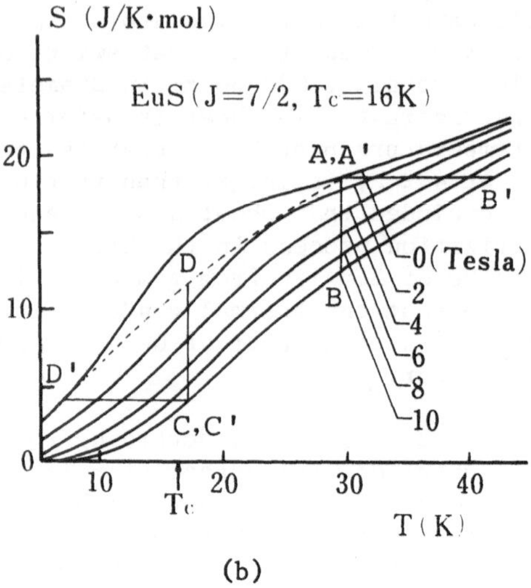

Fig. 5. (a) ΔS_J of EuS in various fields. (b) Entropy S of EuS in various fields. ABCD and A'B'C'D' show the Ericsson and Brayton cycle.

A more precise discussion on the last characteristic (3) can be found in our future manuscript and will be referred to again in the next section. The thermal properties of these refrigerants are not discussed here because investigation on Ericsson-type magnetic refrigeration is still in the early stage.

RECENT INVESTIGATIONS OF MAGNETIC REFRIGERANTS

For refrigerants used for the Carnot-type magnetic refrigerator in the low temperature range, two approaches are being investigated, improvements of the ΔS_J character near ~20 K by external magnetic field and improvement of the heat transfer characteristics.

As mentioned previously, in most commonly used GGG, the value of ΔS_J decreases considerably as the temperature approaches 20 K. GGG was selected with the expectation of getting the largest ΔS_J, because Gd ions included in GGG are thought to have the largest J-value in the low temperature region. It may therefore be concluded that a paramagnetic material superior to GGG is probably not available as far as we consider the J-value.

Recently, investigators in the Tokyo Institute of Technology studied the g-factor dependence of ΔS_J theoretically and believe that the dysprosium compounds are the most promising refrigerant for the Carnot refrigerator having a large temperature span. They therefore investigated the entropy of Dysprosium-Aluminum-Garnet $Dy_3Al_5O_{12}$ (DAG) experimentally.[9] In order to clarify qualitatively the difference between DAG and GGG, a comparison of the temperature dependences of ΔS_J at several magnetic fields (4 T, 6 T and 8 T) is shown in Fig. 6. This clearly indicates the superiority of DAG to GGG above ~10 K, depending slightly on the strength of the applied magnetic field. It is therefore concluded that the character of DAG for the refrigerant in the Carnot cycle for a temperature from above ~15 K to 4.2 K is far superior to that of GGG.

By increasing the operational frequency of the refrigeration cycle, on the other hand, it is also possible to increase the refrigeration power of the refrigerator. In this case, the refrigerant must have a thermal

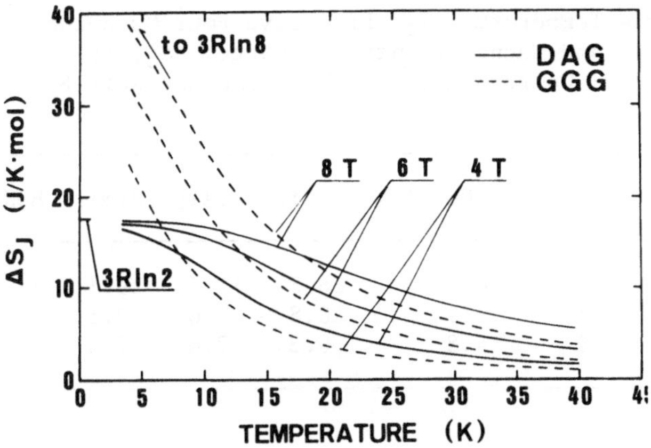

Fig. 6. Comparison of temperature dependences of ΔS_J in DAG and GGG at several magnetic fields (4 T, 6 T and 8 T).

conductivity superior to that of GGG. One of the candidates fulfilling this requirement is $Gd_3Al_5O_{12}$ (GAG), whose thermal conductivity is about twice that in GGG. Workers in the National Research Institute for Metals (Japan) investigated the solid solution of GAG and GGG, $Gd_3(Ga_{1-x}Al_x)_5O_{12}$ where $x\leq0.4$, because from a phase diagram of GAG it would be difficult to make a single crystal of GAG as large as that of GGG. As an example of their results, the entropy of $Gd_3(Ga_{0.8}Al_{0.2})_5O_{12}$ is shown as a function of temperature and magnetic field in Fig. 7.[10] If the perfect order between Ga ions and Al ions can be realized, it is expected that the thermal conductivity of $Gd_3(Ga_{0.6}Al_{0.4})_5O_{12}$ can be considerably improved.

As a refrigerant for the Ericsson cycle in the low temperature range, the workers in the Grenoble group (C.E.A/C.E.N-G) reported a promising solid solution between $DyVO_4$ and $Gd_3Ga_5O_{12}$.[11] Recently, the investigators at the Tokyo Institute of Technology and Kyushu University studied the magnetocaloric effect of the dysprosium gallium garnet $Dy_3Ga_5O_{12}$ (DGG). From the viewpoint of heat transfer character, there is no matter regarding DGG because DGG has a high thermal conductivity comparable to the other

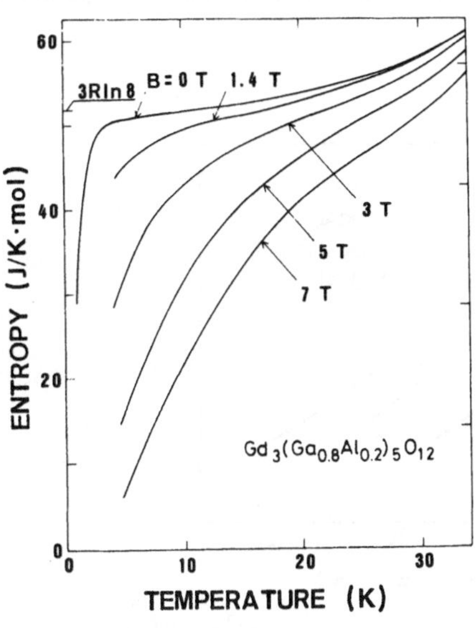

Fig. 7. Temperature dependence of S of $Gd_3(Ga_{0.8}Al_{0.2})_5O_{12}$ in various magnetic field.

Table I. Curie Temperatue, T_c, Effective Bohr Magneton Number, μ_{eff}, and Magnetic Entropy Change, ΔS_J, at T_c of Several Compounds in the RAl_2 and the RNi_2 System at 5 T.

Compound	$DyAl_2$	$Dy_{0.5}Ho_{0.5}Al_2$	$HoAl_2$	$ErAl_2$	$DyNi_2$	$HoNi_2$	$ErNi_2$
T_c	55.9	46.0	26.8	11.7	19.3	12.3	6.7
μ_{eff}	11.1	11.7	10.8	9.6	10.5	10.1	9.6
ΔS_J(J/mol·K)	4.1	4.8	6.3	7.6	7.0	6.3	6.7

compounds of garnet structure. They had made clear that DGG has to be used in the Ericsson cycle.[12] Presently, Sakaguchi and his colleagues in Shin-etsu Chemical Co. Ltd. succeeded in industrially making large single crystals of DAG and DGG; for instance, the size of DAG is 50 mm in diameter and 120 mm in length. This will greatly facilitate our use of these superior refrigerants.

The results of investigations on refrigerants for the Ericsson cycle above 10 K to 77 K show that because of the limitation of magnetic field (less than 6 T), only ferromagnetic substances can be used as refrigerants. Moreover, for the purpose of satisfying the condition to construct a high efficiency cycle, in the constant magnetic field process of the Ericsson cycle, a mixed compound (not solid solution) of an appropriate ratio of ferromagnetics must be used, whose Curie temperatures are within the temperature span of the Ericsson cycle. Therefore, a series of compounds whose Curie temperature can be changed by the substitution of magnetic ions is necessary for the refrigerant. Investigators in the Tokyo Institute of Technology and Kyushu University have studied a series of two kinds of compounds, RAl_2 and RNi_2, where R is the rare earth ion. Their Curie temperatures T_c and effective Bohr magnetons are shown in Table I. Since these compounds have an identical crystal structure, T_c of the solid solution can be selected arbitrarily by changing the ratio of the two mixed compounds.

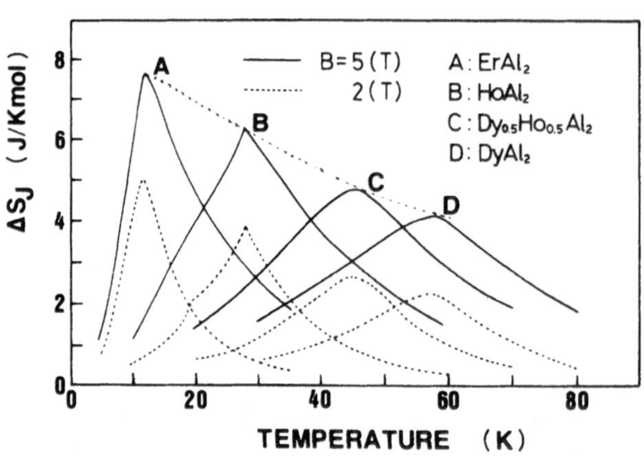

Fig. 8. Temperature dependence of ΔS_J in the RAl_2 series (R : rare earth ions).

For example, the temperature dependence of ΔS_J of RAl_2 for several R ions obtained by the external magnetic field at 5 T (tesla) is shown in Fig. 8. The limiting condition for the refrigerant of the Ericsson is sufficiently satisfied by the RAl_2 series. Results of the RNi_2 series similar to those of RAl_2 have also been obtained. More precise experimental detail and discussion is given in the report of ICMC '85.[13,14]

BRIEF DISCUSSION AND CONCLUSION

As expected, the magnetic refrigerator, especially in the low temperature range, has been verified to possess very high efficiency, for instance, ~45 % in the superfluid He refrigerator and 40 % in the He liquefaction refrigerator. These values indicate great promise for the future of the magnetic refrigerator; however, it is also clear that the refrigeration performance depends largely on the character of refrigerant used, especially in a machine with a large temperature span of, for instance, ~20 K to 4.2 K. On the other hand, an envisioned usage of the magnetic refrigerator below ~20 K is the cascade refrigerator connected with a gas refrigerator above ~20 K. To realize these machines, development of a refrigerant other than GGG is urgently needed.

DAG is refrigerant candidate, however, since three kinds of principal (z) axes, <100>, <010> and <011>, of the g-factor for Dy ions exist in DAG, the effective g-factor, g_e, is equal to 10.5 in spite of $g_z = 18.1$ and the anisotropic effect on the g-factor cannot be sufficiently utilized. For the Carnot-type refrigerant with a temperature span from 20 ~ 23 K to 4.2 K, it is concluded that the dysprosium paramagnetic compounds, which have only one principal axis for g-factor, are the most promising magnetic refrigerants.

Since investigations on the Ericsson-type refrigerator are not so advanced, the precise characteristics required of its refrigerant cannot be discussed here. However, from the theoretical condition to make two kinds of constant magnetic field processes satisfying the ideal condition in the Ericsson cycle (B - C and D - A process shown in Fig. 5), the ferromagnetic series is needed to make this refrigerant. More detailed theoretical and experimental results on this subject will be published elsewhere. There is also another way to make the refrigerant for the Ericsson machine. The phase transition in the amorphous magnetic compound is not as sharp as that in the regular lattice compounds. Accordingly, it can be expected that the temperature dependence of ΔS_J in the amorphous compounds becomes flat. However, our investigations show that the absolute value of ΔS_J is smaller than ΔS_J in the mixed compound of the regular lattice and, moreover, the thermal conductive character is generally inferior to that of the regular lattice. Therefore, the amorphous materials are not always superior to the regular lattice materials.

Finally, magnetic refrigeration is one of the most promising methods for the future. As the example of the He liquefaction refrigerator shows, investigations on magnetic refrigerants will become increasingly important with the development of the magnetic refrigerator.

ACKNOWLEDGMENTS

This study was performed through Special Coordination Funds for Promoting Science and Technology of Science and Technology Agency of the Japanese Government.

REFERENCES

1. G. V. Brown, Magnetic heat pumping near room temperature, J. Appl. Phys. 47:3673 (1976).
2. For instance, A. F. Lacaze, R. Béranger, G. Bon Mardion, G. Claudet and A. A. Lacaze, Double acting reciprocating magnetic refrigerator: Recent improvements, in: "Advances in Cryogenic Engineering," vol. 29, Plenum Press, New York (1983), pp. 573-579.
3. For instance, H. Nakagome, N. Tanji, O. Horigami, H. Ogiwara, T. Numazawa Y. Watanabe and T. Hashimoto, The helium magnetic refrigerator I: Development and experimental results, in "Advances in Cryogenic Engineering", vol. 29, Plenum Press, New York (1983), pp. 581-587.
4. For instance, W. A. Steyert, Rotating Carnot-cycle magnetic refrigerator for use near 2 K, J. Appl. Phys. 49: 1227 (1978).
5. For instance, T. Fujita, T. Yazawa, T. Hashimoto, H. Nakagome and T. Kuriyama, Analysis of liquefaction process and losses of a reciprocating magnetic refrigerator for helium liquefaction, in: "Advances in Cryogenic Engineering," vol. 31, Plenum, New York (1986).
6. T. Hashimoto, T. Numazawa, M. Shiino and T. Okada, Magnetic refrigeration in the temperature range from 10 K to room temperature: the ferromagnetic refrigerants, Cryogenics 21: 647-653 (1981).
7. T. Numazawa, T. Hashimoto, H. Nakagome, N. Tanji and O. Horigami, The helium magnetic refrigerator II: Liquefaction process and efficiency, in: "Advances in Cryogenic Engineering" vol. 29 Plenum Press, New York (1984). pp. 589-596.
8. T. Numazawa, Y. Watanabe, T. Hashimoto, A. Sato, H. Nakagome, O. Horigami, S. Takayama and M. Watanabe, The magnetic refrigeration characteristics of several magnetic refrigerants below 20 K: II Thermal properties, in: "Proceedings of ICEC 9", Butterworths, UK (1983), pp. 30-33.
9. R. Li, T. Numazawa, T. Hashimoto, A. Tomokiyo, T. Goto and S. Todo, Magnetic and thermal properties of $Dy_3Al_5O_{12}$ as a magnetic refrigerant, in: "Advances in Cryogenic Engineering-Materials," vol. 32, Plenum New York (1986).
10. H. Maeda et al. Private communication.
11. B. Daudin, A. A. Lacaze and B. Salce, $DyVO_4$-$Gd_3Ga_5O_{12}$: A composite material to achieve magnetic refrigeration using a cycle with internal heat transfer, Cryogenics 22: 439-440 (1982).
12. A. Tomokiyo, H. Yayama, T. Hoshimoto, T. Aomine, M. Nishida and S. Sakaguchi, Specific heat and entropy of dysprosium gallium garnet in magnetic fields, Cryogenics 25: 271-274 (1985).
13. T. Hashimoto, K. Matsumoto, T. Numazawa, T. Kurihara, A. Tomokiyo, T. Goto and M. Sahashi, Investigation on possibility of the RAl_2 system as a refrigerant in a Ericsson-type magnetic refrigerator, in: "Advances in Cryogenic Engineering-Materials," vol. 32, Plenum, New York (1986).
14. A. Tomokiyo, H. Yayama, H. Wakabayashi, T. Kuzuhara, T. Hashimoto, M. Sahashi and K. Inomata, Specific heat and entropy RNi_2 in magnetic field, in: "Advances in Cryogenic Engineering-Materials," vol. 32, Plenum, New York (1986).

HIGH ENTHALPY MATERIALS FOR USE IN SUPERCONDUCTOR STABILIZATION AND IN LOW TEMPERATURE CRYOCOOLER REGENERATORS

B. Barbisch, J.L. Olsen

Solid State Physics
Swiss Federal Institute of Technology
Zürich, Switzerland

and

K. Kwasnitza

Swiss Federal Institute for Nuclear Research
Villigen, Switzerland

ABSTRACT

At helium temperatures the heat capacity of most materials becomes very small. Materials which undergo phase transitions at low temperature may, however, show anomalies in the specific heat making it one or two orders of magnitude higher than in normal metals. Such materials are of technical interest for superconductor stabiliztion or regenerative refrigeration. We present some metallic materials with very high specific heat anomalies due to various kinds of phase transitions and Schottky effects. A magnetic field up to 7 Tesla does not influence their heat capacity significantly. We report different possibilities of using these substances for superconductor stabilization, and show our first results. Regenerative gas-cycle refrigerators with usual regenerators cease operation below about 15 K because the heat capacity of the regenerator material becomes smaller than that of the helium working fluid. New materials with very high specific heats in this low temperature range are proposed.

INTRODUCTION

At low temperature the specific heats of solids are very small, typically 10^{-3} of their room-temperature values, and a rapidly decreasing function of the temperature. In most materials, only the lattice vibrations (phonons) and the conduction electrons make a contribution to the specific heat. The first goes approximately as T^3 and the second is linear in T. Figure 1 shows the specific heat of some of the materials most commonly used at low temperatures. These are copper, aluminium, lead and the two superconducting materials Nb_3Sn and NbTi. The specific heat of lead is about one order of magnitude larger than that of copper or aluminium due to its very low Debye temperature. Lead was therefore proposed very early as an enthalpy stabilizer for superconducting magnets[1] and is also used in refrigerators as the regenerating material at low temperature[2].

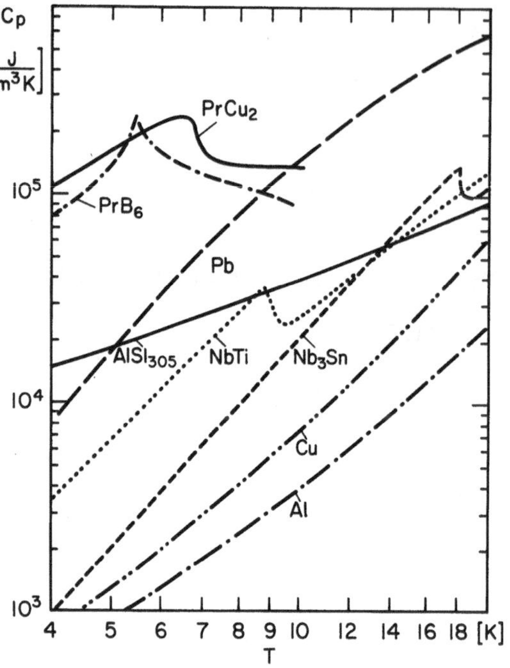

Fig. 1 Specific heat per unit volume of some materials commonly used at low temperatures. Data for NbTi, Nb_3Sn, copper and aluminium is from Wilson[5]. Data for lead is from Johnson[6] data for stainless steel, AISI 305 is from Touloukian[7], and data for $PrCu_2$ and PrB_6 is from Barbisch[8].

Phase transitions or Schottky anomalies can yield a large contribution to the specific heat. Schottky anomalies show a broad peak in specific heat. Its maximum is of the order of 0.5 R (R is the gas constant). This leads to specific heat values of 4 kJ/kmol K, about 1200 times the value of copper at 4.2 K. It has to be noted the specific heat per volume unit is much less increased because substances showing Schottky anomalies are normally compounds with a much larger molecular weight than copper. Phase transitions, such as magnetic ordering, are accompanied by a peak of smaller area showing a very high narrow maximum. B-phase (monoclinic) Gadoliniumoxide Gd_2O_3, proposed as enthalpy stabilizer by Rosenblum[3] has a maximum specific heat of 65 J/kg K at 3.8 K.

For applications as enthalpy stabilizers, the thermal conductivity λ plays an important role while the high specific heat c_p reduces the thermal diffusivity a, defined by

$$a = \frac{\lambda}{\rho c_p} \quad (1)$$

where ρ is the density of the material. It is well known[4] that a constant flux F_0 of heat into a semi-infinite solid incrases the temperature at the surface by

$$\Delta T = \frac{2F_0}{\sqrt{\pi}} \left\{\frac{t}{\lambda \rho c_p}\right\}^{1/2} \quad (2)$$

This shows the importance of both properties, specific heat per volume and thermal conductivity for a good enthalpy stabilizer. Here we concentrate our attention on metallic materials undergoing phase transitions at low temperature. Metallic because this promises a good thermal conductivity due to the heat transport by electrons.

Our preliminary results in enthalpy stabilization of technical superconductors have been published elsewhere[8]. The materials we investigated were $PrCu_2$ and PrB_6. Our results for their heat capacity in zero magnetic field are shown in Fig. 1. We also measured the specific heat in applied magnetic fields up to 7 Tesla; only a weak magnetic field dependence of the specific heat was found[8]. The specific heat anomaly in $PrCu_2$ can be explained as a Jahn-Teller effect (structural phase transition) which is superimposed on a Schottky anomaly with a maximum at 12 K to 13 K [9]. $PrCu_2$ shows a metallic lustre, but it oxidizes slowly in air. In PrB_6, the specific heat anomaly seems to be due to antiferromagnetic ordering[10]. We found that the PrB_6 compound is stable at room temperature and no significant oxidation was observed in air.

The two materials mentioned above are hard, nonductile, intermetallic compounds. They cannot be formed to wires by pressing or drawing. Therefore some attempts were made to bring them in a form useful for technical application. The easiest possibility is to use them in the form of a powder which can be incorporated in the epoxy resin impregnating the coil of a superconducting magnet or in the solder of a superconducting cable. In both cases there will be an increase of the specific heat, and a stabilizing effect can be expected. To show such an effect the following experiments were carried out.

We ground some $PrCu_2$ to a grain size of about 50 µm. This powder was covered with silver by evaporation and then mixed with pure liquid indium. The silver covering helped to make a bond between powder and solder. On measuring the specific heat, no increase or anomaly was observed. We suppose that the $PrCu_2$ powder was oxidized in the air before it was covered with silver and protected, but the possibility that $PrCu_2$ reacts with indium cannot be excluded.

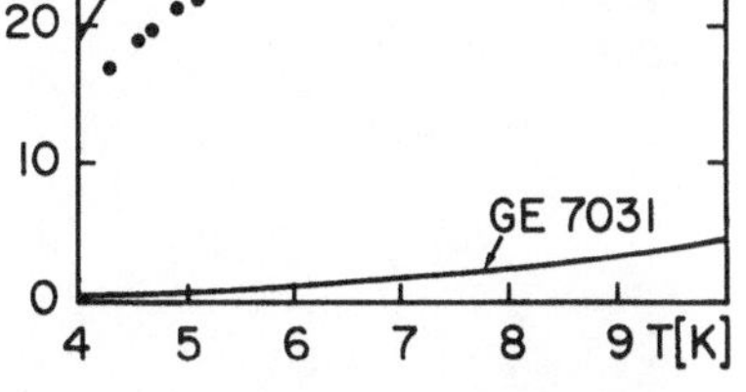

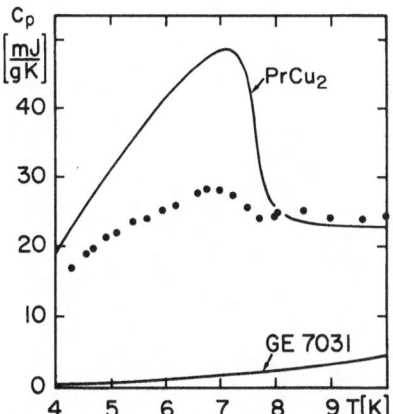

Fig. 2 - Specific heat measurements of $PrCu_2$ powder in GE 7031 varnish (open circles). Solid lines represent the expected contributions of $PrCu_2$ and varnish.

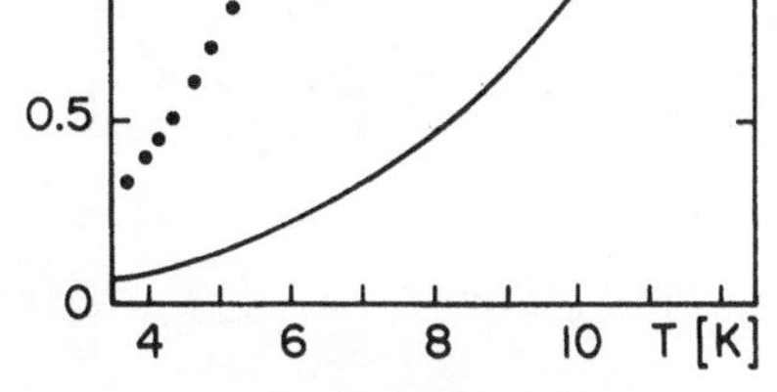

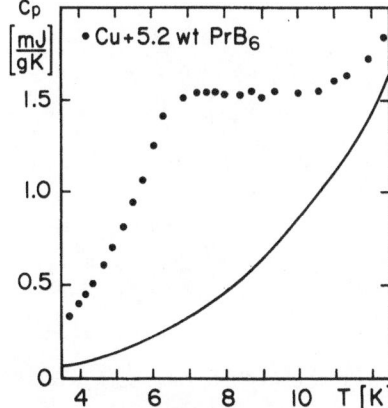

Fig. 3 - The measured specific heat of the sinter wire. The specific heat of pure copper is displayed by the solid line.

A second experiment was to place the powder in a varnish. Fig. 2 shows the measured specific heat of a mixture of GE 7031 varnish and $PrCu_2$ powder. The specific heat is not raised as much as one could expect from the heat capacities of the two parts. The characteristic peak of $PrCu_2$ in the temperature range between 4 K to 8 K is rather suppressed but not the enhancement above 8 K. It seems that the Jahn Teller effect disappeared, and that the level scheme of the crystal field producing the Schottky anomaly remains more or less unchanged. The specific heat values are still high enough to be useful, and we believe that $PrCu_2$ powder would be advantageous as a filler in the epoxy used for coil impregnation.

In collaboration with the company Metallwerke Plansee (Austria) we produced a sinter wire containing PrB_6. PrB_6 was ground and mixed with copper powder, filled in a copper tube, heat treated under pressure, and then drawn to a wire. The length of the wire was about 40 m and the diameter was 0.65 mm. For the first attempt we used not more than 5.2 wt% PrB_6 (about 10 % in volume). Fig. 4 shows an electron microscope photograph and in Fig. 3 the specific heat of this wire is reported.

At 5 to 7 K the specific heat of the wire is increased by a factor 5 above that of copper. The anomaly due to the PrB_6 grains is not as sharp as shown in Fig. 3 but the specific heat is still enhanced. The somewhat broadened peak is not disadvantageous. The electrical resistivity of this wire was 2.4 μΩ cm at 0°C and the residual reistance ratio could be raised by heat treatment up to 200. To check the stabilizing effect of this sinter wire, we investigated the decay of normal zones (quenches) in superconducting cables. The apparatus used for this experiment, the analysis and the results are described and discussed in detail elsewhere[11]. We found that the thermal time constant for heat transfer between the PrB_6 grain and the copper was unexpectedly high in the order of one second. This anomalously poor heat transfer can be explained by assuming that there is no metallic binding between the two components. The grains lie practically free in the copper matrix as shown in the electron microscopic picture in Fig. 4.

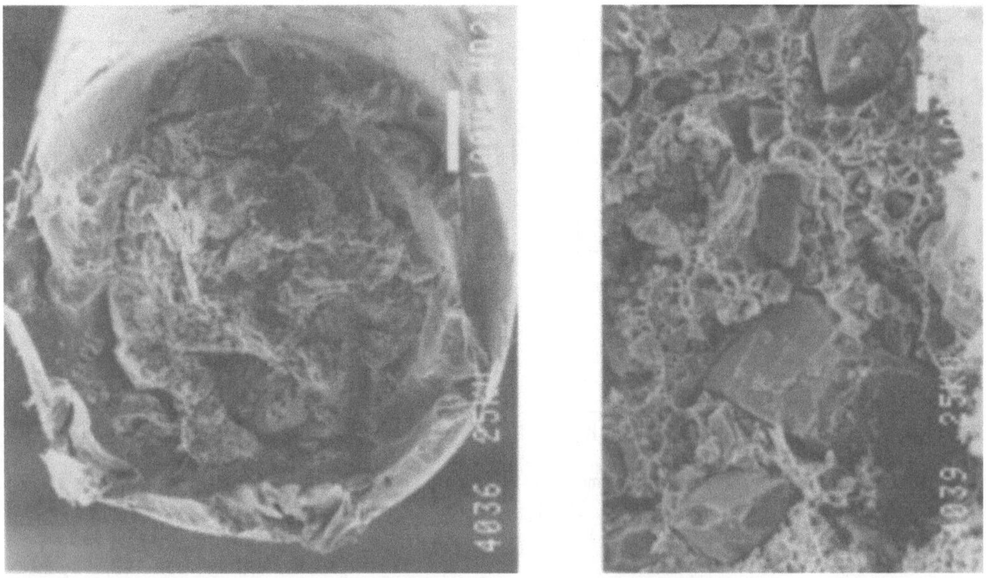

Fig. 4 - Electron microscope photographs of the sinter wire. Note the different scale on the right side of the two pictures.

RARE EARTH COMPOUNDS AS REGENERATOR MATERIALS IN STIRLING-CYCLE CRYOCOOLERS

Several authors[12,13] have mentioned the need for new materials with high c_p values in the temperature range between about 4 and 15 K. Lead as regenerator material is not very efficient at these low temperatures as its c_p drops strongly and becomes even smaller than the pressurized He gas. Therefore in 1971 prototype tests were made by Daniels and du Pré[14] with a regenerator made from EuS. It has a large peak in c_p at 16 K due to ferromagnetic ordering. For $T < 16$ K, c_p decreases sharply. The result was somewhat disappointing as only 7.8 K as minimum temperature were reached. In our opinion the negative aspects of EuS are that it is an insulator and that the c_p-peak temperature of 16 K seems to be somewhat too high if it is not used together with other materials. Recently Nakashima et al[15] based on the specific heat measurements of Buschow et al[16] used the compounds GdRh and GdErRh with great success in powdered form for their Stirling cycle regenerator. Depending on the composition $Gd_xEr_{1-x}Rh$ has high peaks in c_p between about 3 and 20 K. The disadvantage of this material might be the high price of Rhodium as is also the case for elementary Europium.

In this paper we propose alternative rare earth compounds with large heat capacity in the low temperature range between 2 and 15 K due mainly to magnetic phase transitions and contributions to c_p from Schottky anomalies.

The materials were selected by the following criteria:

- very high c_p values not only per mol but also per unit volume, partly with a broad maximum,
- good heat conductivity due to their metallic character,
- relatively low price of those rare earth elements which are used, or
- possibility to fabricate the compounds starting from RE-oxydes
- proven or probable stability against oxidation to allow also the use as coarse-grained powders in the regenerator,
- c_p maxima of the different compounds dispersed over the whole interesting temperature range. Therefore the regenerator may be optimized by the use of several compounds with successively lower ordering temperatures.

The materials we propose are mainly some rare earth hexaborides and rare earth monosulfides. Their data are given in Table 1 and their c_p-

Table 1. High c_p rare earth compounds

Compound	Ordering temperature Kelvin	References
CeB_6	2.1, 3.3	19, 20
PrB_6	6.9	8, 10, 17
NdB_6	7.5	18
$EuB_{5.97}C_{0.03}$	5 - 6	22
GdB_6	16	29
CeS	8.3	27
NdS	8.4, 9.3, 9.7, 11.5	26

values have been compiled from literature. In Fig. 5 the c_p values of CeB_6, PrB_6, NdB_6 and $EuB_{5.97}C_{0.03}$ are presented together with those of Pb and He gas of 2 atm. All substances order antiferromagnetically and are metallic. PrB_6 does not loose its high c_p-values[10,17] by powdering[8,11] and has no tendency for oxidation. Electrical resistivity $\rho \simeq 1$ μΩ cm at 4 K and heat conductivity $\lambda \simeq 88$ mW/cm K. CeB_6 shows, beside the antiferromagnetic phase transition[19] at 2.1 K, a second ordering[20] at 3.3 K.

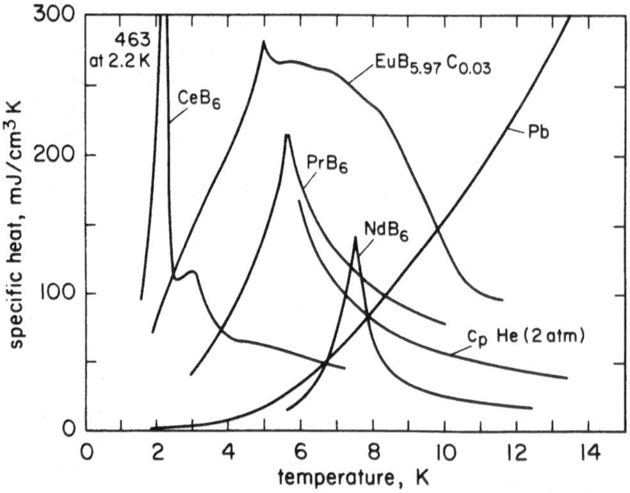

Fig. 5. Specific heat values of CeB_6 measured by Fuyita et al[19], c_p of PrB_6 from Barbisch and Kwasnitza[8], c_p of NdB_6 from Westrum et al[18] and c_p curve of $EuB_{5.97}C_{0.03}$ from Fuyita et al[22]. The c_p data of lead and He gas of 2 atm are added for comparison.

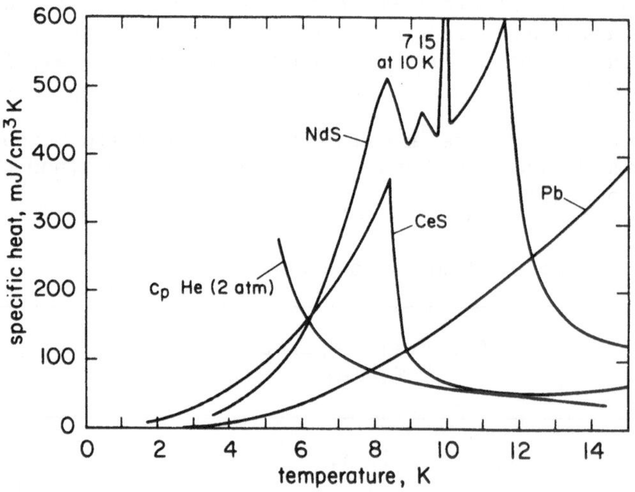

Fig. 6. Specific heat curves of NdS, measured by Vasilev et al[26] and of CeS from Ott et al[27]. The c_p data of lead and He of 2 atm are again included.

Electrical resistivity ρ ≃ 75 μΩ cm at 4.2 K. While the purely stoichiometric EuB_6 is an insulator with a ferromagnetic transition[21,22,23] at about 8.6 K, $EuB_{5.97}C_{0.03}$ orders antiferromagnetically[22] at 5 - 6 K as C acts as electron donor. Therefore we expect in this case a reasonable heat conductivity. Because of their metallic character and their low antiferromagnetic ordering temperatures (in brackets) also the two rare earth tetraborides[24] ErB_4(13 K) and HoB_4(9 K) and the rare earth diborodicarbides[25] CeB_2C_2(7 K), HoB_2C_2(7 K) and NdB_2C_2 (9 K) deserve attention.

In Fig. 6 the c_p values of NdS and CeS are given[26,27]. NdS has rather high c_p-values over the broad temperature range between about 5 and 12 K due to four antiferromagnetic transitions. At room temperature the electrical resistivity ρ is $3.4 \cdot 10^{-5}$ Ω cm. $(Gd_{0.8}Nd_{0.2})_2S_3$ (not shown) orders[28] at 9.7 K and has also rather high c_p-values but possibly a bad heat conductivity. Finally also the initially mentioned $PrCu_2$ could be used if oxydation can be prevented by any surface coating. We have not made a systematic search for all known high c_p-materials between 4 and 15 K. Therefore other candidates might well be found.

REFERENCES

1. R. Hancox, Enthalpy Stabilized Superconducting Magnets, in "IEEE Transactions on Magnetics", vol. MAG-4, IEEE Inc., New York (1968), pp. 486-488.
2. W. E. Gifford and T.E. Hoffmann, A new refrigeration system for 4.2°K, in "Advances in Cryogenic Engineering", vol. 6, Plenum Press, New York (1961) pp. 82-94.
3. S. S. Rosenblum, M. Sheinberg and W. A. Steyert, High specific heat metals for use in superconducting composites, in "IEEE Transactions on Magnetics", vol. MAG-13, IEEE Inc., New York (1977), pp. 834-835.
4. M. S. Carslaw and J.C. Jaeger, "Conduction of heat in solids", 2nd ed., Clarendon Press, Oxford (1959) pp. 75-77.
5. M. V. Wilson "Superconducting Magnets", Monographs on Cryogenics, Clarendon Press, Oxford (1983) p. 72.
6. V. J. Johnson, "Properties of Materials at low temperature (Phase 1)", Pergamon Press, Oxford (1961) p. 4.142-3.
7. V. S. Touloukian and E. H. Buyco, "Specific Heat, Metallic elements and alloys" in Thermophysical Properties of Matter, vol. 4, Plenum New York (1979) p. 702.
8. B. Barbisch and K. Kwasnitza, Two new metallic materials with high specific heat for superconductor stabilization, in "Journal de Physique, colloque C1", vol. 45, Les Editions de Physique, Les Ulis, France (1984), pp. 561-565.
9. M. Wum and N. E. Phillips, Low temperature heat capacity of $PrCu_2$, Phys. Letters 50A, 195-196 (1974).
10. K. N. Lee, R. Bachmann, T. H. Geballe, and J. P. Maita, Magnetic Ordering in PrB_6, Phys. Rev. B2, 4380-4584 (1970).
11. B. Barbisch and K. Kwasnitza, Experiments on Enthalpy Stabilization of Technical Superconductors, in "Proc. of 10th Int. Cryogenic Eng. Conf. ICEC10. IPC Press, Guildford, GB (1984) pp. 689-692.
12. G. Walker, "Cryocoolers", part 2, p. 33, Plenum Press, New York (1984).
13. J. L. Smith, Jr., G. Y. Robinson, Jr., and Y. Iwasa, Survey of the State of the Art of Miniature Cryocoolers for Superconductive Devices, Superconductive Technology in Review, No. 84-3, 1 (1972).
14. A. Daniels and F. K. du Pré, Adv. Cryog. Eng., 16:178 (1971).
15. H. Nakashima, K. Ishibashi, and Y. Ishizaki, Stirling Cycle Refrigerator Test Result Below 5 K, in "Proc. ICEC10", Helsinki (1984), pp. 54-57.

16. H. J. Buschow, J. F. Olijhoek, and A. R. Miedema, Extremely Large Heat Capacities Between 4 and 10 K, Cryogenics 15: 26-264 (1975).
17. C. M. Mc Carthy, C. W. Tompson, R. J. Graves, H. W. White, Z. Fisk, and J. R. Ott, Low Temperature Phase Transitions and Magnetic Structure of PrB_6, Solid State Communn. 36:861-868 (1980).
18. E. F. Westrum, Jr., H. L. Clever, J. T. S. Andrews, and G. Feick, Thermodynamics of Lanthanum and Neodymium Hexaborides, in "Proc. Conf. Rare Earth Res. 4th", Phoenix, Arizona (1984) pp. 597-605.
19. T. Fujita, M. Suzuki, T. Komatsubara, S. Kunu, T. Kasuya, and T. Ontsuka, Anomalous Specific Heat of CeB_6, Solid State Commun., 35:569-572 (1980).
20. B. Lüthi, S. Blumenröder, B. Hillebrands, E, Zirngiebl, G. Güntherodt and K. Winzer, Elastic and Magnetoelastic Effects in CeB_6, Z. Phys. B.-Condensed Matter 58:31-38 (1984).
21. B. T. Matthias, T. H. Geballe, K. Andres, E. Corenzwit, G. W. Hull, and J. P. Maita, Magnetic Ordering in the Rare-Earth Hexaborides, Science 160:1443-1444 (1968).
22. T. Fujita, M. Suzuki, and Y. Isokawa, Specific Heat of EuB_6, Solid State Commun. 33: 947-950 (1980).
23. Y. Isikawa, M. M. Bajaj, and M. Kasaya, on The Correlations between Magnetic Structure and their Electrical Properties of EuB_6, Solid State Commun. 22: 573-576 (1977).
24. J. Etourneau, J. P. Mercurio, A. Berrada, P. Hagenmüller, R. Georges, R. Bourezg, and J.C. Gianduzzo, The Magnetic and Electrical Properties of Some Rare Earth Tetraborides, J. Less Comm. Met. 67: 531-539 (1979).
25. T. Sakai, G. Adachi, and J. Shiokawa, Solid State Commun. 40: 445-449 (1981).
26. L. N. Vasilev, V. M. Grabov, A. V Golubkov, A. G. Gorobets, V.S. Oskotskii, I. A. Smirnov, and V.V. Tikhonov, Physical Properties Phase Transitions of the Rare Earth Monosilfides in the Homogeneity Range, phys. stat. sol. 80: 237-244 (1983).
27. F. Hulliger, B. Natterer, H. R. Ott, Magnetic Properties of the Cerium Monochalcogenides, J. of Magn. and Magn. Mat. 8: 87-98 (1978).
28. S. M. A. Taher, J. B. Gruber, J. C. Ho, and D. C. Yeh, Low Temperature Heat Capacities of Rare Earth Sesquisulfides, in "The Rare Earth in Modern Science and Technology", G. J. Mc Carthy and J. J. Rhyne, eds. (1978), pp. 359-365.
29. E. F. Westrum, Jr., J. T. S. Andrews, and G. A. Clay, Thermal Anomalies and Thermodynamic Properties of Gadolinium Hexaboride, U.S. Dept. Com., Clearing House, Sci. Tech. Inform. AD 627224, 9 (1965).

INVESTIGATIONS ON THE POSSIBILITY OF THE RAl_2 SYSTEM AS

A REFRIGERANT IN AN ERICSSON TYPE MAGNETIC REFRIGERATOR

T. Hashimoto, K. Matsumoto, T. Kurihara and T. Numazawa

Department of Applied Physics
Tokyo Institute of Technology
Ohokayama, Meguroku, Tokyo

A. Tomokiyo and H. Yayama

College of General Education
Kyushu University
Chuohku, Fukuoka

T. Goto and S. Todo

Institute for Solid State Physics
University of Tokyo
Minatoku, Tokyo

M. Sahashi

Toshiba Research and Development Center
Saiwaiku, Kawasaki, Kanagawa

ABSTRACT

We investigated the Ericsson type magnetic refrigerators in the range below 77 K. This is the first report of experimental results of the refrigeration character, especially the magnetocaloric character of RAl_2, where R is a rare earth atom.

The specific heat and magnetization were measured precisely as a function of temperature and magnetic field. From analyses of these results the total entropies S and the magnetic entropy changes ΔS_J of the RAl_2 series were determined.

It is made clear that a large enough value of ΔS_J to make refrigeration can be induced by the external magnetic field near T_c, and also that the temperature corresponding to the maximum value of ΔS_J is adjustable by making a solid solution between the compounds in the RAl_2 series. It is therefore concluded that the RAl_2 series is a promising candidate for refrigerant at above 20 K.

INTRODUCTION

After the work on the magnetic refrigerator near room temperature by Brown,[1] several groups[2,3,4,5] began to study the magnetic refrigerator, particularly because of its high efficiency potential. However, these studies focussed on the Carnot type refrigerator in the low temperature range, attracted by its simplicity of structure and operation. We recently investigated the Ericsson type magnetic refrigerator in the temperature range from above ~20 K down to 4.2 K. Our first report of this investigation will cover information gained on the magnetic refrigerant used for the Ericsson type.

In the temperature range above ~20 K, the thermal agitation energy of the spin system becomes larger than the Zeeman energy produced by the external magnetic field at ~6 T (tesla). As far as using the magnetic field below 6 T, we can get a magnetic entropy change ΔS_J sufficient to refrigerate only by using ferromagnetic materials. Moreover, in this range, since the lattice entropy is comparable to or larger than S_J, we must adopt the Ericsson cycle instead of the Carnot cycle.

To assure a high efficiency Ericsson cycle over a wide temperature span requires as refrigerant a series of ferromagnetic materials whose magnetic moments are large and whose Curie temperatures are easily adjustable to a desired level within the temperature range of the refrigeration cycle.

One of the most promising candidates to satisfy these conditions is thought to be the RAl_2 system, where R is the rare earth atom, because the system has a large magnetic moment of about 10 μ_B per rare earth ion and the Curie temperature can be changed from 10 K for $ErAl_2$ to 165 K for $GdAl_2$ by making a solid solutions between these compounds.

We measured the magnetization and specific heat of the polycrystalline RAl_2 compounds as a function of temperature at various strengths of the applied magnetic field and obtained the total entropies and magnetic entropy changes.

EXPERIMENT

All compounds were prepared by arc melting in a water-cooled copper boat in an atmosphere of pure argon. The ingots were crushed and remelted several times for homogenization. From the polycrystalline ingots, cylindrical samples were cut and polished. The size of samples for specific heat measurements was ~8 ϕmm × ~20 mm, and that for magnetization measurements was 2 ϕmm × 10 mm. Before the measurement, samples were annealed at 1000°C for 70 hours to make them homogeneous and to eliminate internal strain.

The samples were analyzed by powder X-ray diffraction technique. The diffraction patterns showed that the samples had the Laves phase structure and no extra peaks corresponding to other structural phases were observed.

The heat capacities were measured in an adiabatic calorimeter above 4.2 K. The temperature of the sample was determined by a carbon glass resistance thermometer. Under conditions, where constant heat was supplied to the sample by a heater, the temperature variation of the sample against time was obtained.

The magnetization was measured by the fluxmetric method in the applied magnetic field up to 9 T. In a constant external magnetic field, the sample was suddenly removed from the pick-up coils and the induced voltage in the coils proportional to the change of magnetic flux was simultaneously integrated by a computer after conversion to a digital code.

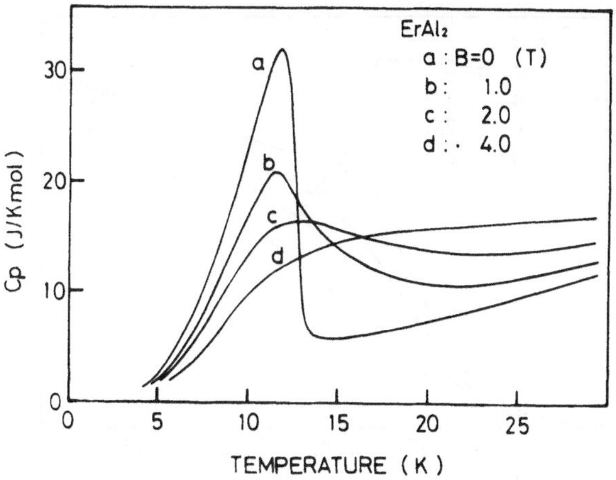

Fig. 1. Temperature dependence of the specific heat C_p of $ErAl_2$ in various magnetic fields.

RESULTS AND DISCUSSION

As an example, the specific heat of $ErAl_2$ between 4.2 K and 30 K in various applied magnetic fields is shown in Fig. 1. The zero field specific heat exhibits a steep peak at 12 K. This anomalous peak is attributed to the entropy change with the phase transition from the ferromagnetic to the paramagnetic state.

In the temperature dependence of the specific heat of $HoAl_2$ in the zero magnetic field, two kinds of peaks were observed as shown in Fig. 2. The sharp and large anomalous peak at 28 K is due to the usual order-disorder phase transition. The broad one at 13 K may be attributed to complicated change of its magnetic structure.[6] However, since most of the magnetic entropy change ΔS_J occurs at T_c and the latter peak is noticeably smaller than that at T_c, the existence of the latter peak is not considered to affect adversely the refrigeration character.

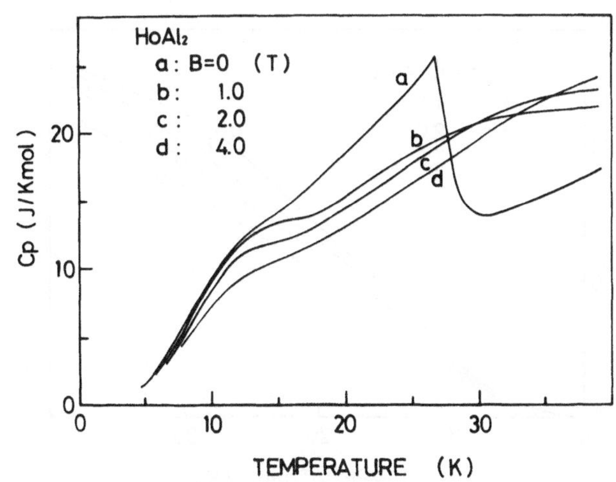

Fig. 2. Temperature dependence of the specific heat C_p of $HoAl_2$ in various magnetic fields.

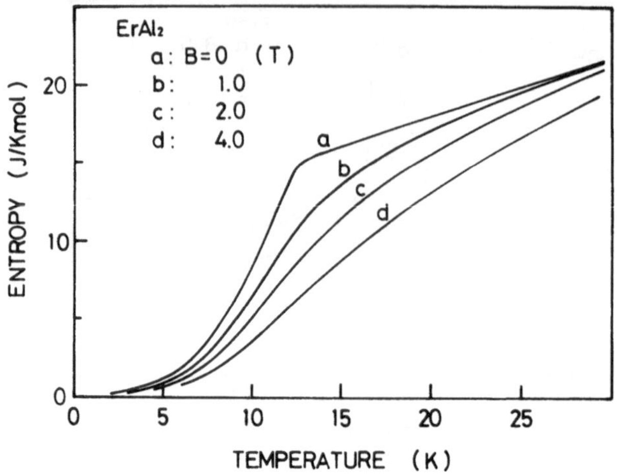

Fig. 3. Temperature dependence of the entropy of $ErAl_2$ obtained from specific heat in various magnetic fields.

The temperature variation of the entropy in the magnetic field can be evaluated from the results of the specific heat using the following relation

$$S(T,B) = \int_0^T \frac{C(T,B)}{T}\, dT, \quad (1)$$

where S(T,B) is the entropy as a function of temperature, T, and magnetic field, B, and C(T,B) is the specific heat in various fields.

As the specific heat measurements were performed in the range above 4.2 K, the experimental results of C(T,0) by Wallace[7] were used for the calculation of entropy in zero field S(T,0) below 4.2 K. Because no report exists on C(T,B) below 4.2 K, the values of C(T,B) smoothly extrapolated to 0 at 0 K were used for the estimation of S(T,B) in this temperature range.

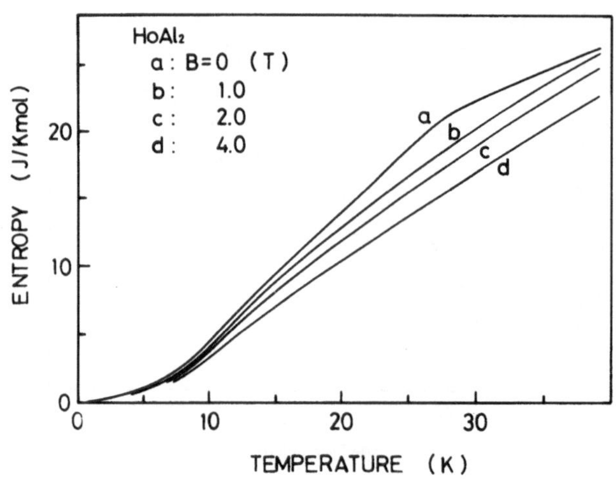

Fig. 4. Temperature dependence of the entropy of $HoAl_2$ obtained from specific heat in various magnetic fields.

The variation of the entropies of $ErAl_2$ and $HoAl_2$ with temperature in various magnetic fields is shown in Fig. 3 and Fig. 4, respectively. These results confirm that the magnetic entropy change, ΔS_J, caused by the external magnetic field takes a sufficiently large value to refrigerate near the Curie temperature, T_c.

The magnetic entropy change ΔS_J was determined from measurement of the magnetization M(T,B). In the temperature range near and below T_c, since permeability of a ferromagnetic material increases considerably, the magnetic field actually applied to the spin in the sample, that is, the effective magnetic field B_{eff} is reduced by the demagnetization field. In this case, we have to decide the effective field using the following relation,

$$B_{eff} = B_{app} - 4\pi N_d M, \tag{2}$$

where B_{app} is the applied magnetic field, N_d the demagnetization factor and M the magnetization which is observed in the applied magnetic field. The demagnetization factor is determined by the aspect ratio of the sample. Because the samples which we measured are cylindrical in shape, we approximate that the demagnetization factor is 0.04 which is the factor of an ellipsoid having the same aspect ratio. The results of magnetic susceptibility have verified that rare earth ions (Er^{3+}, Ho^{3+} and Dy^{3+}) have large magnetic moments which are almost the same value as free ions.

The total entropy of magnetic material S is thought to be composed of four kinds of entropies, that is, the magnetic entropy S_J, the lattice entropy S_L due to the lattice vibration, the entropy due to the free electrons and the other entropies, such as that of the nuclear spin system. However, in the range above 10 K, because S_J and S_L are dominant in comparison with the other kinds of entropies, we can write this approximately as

$$S(T,B) = S_J(T,0) + \Delta S_J(T,B) + S_L(T). \tag{3}$$

On the other hand, from the magnetization M(T,B), the entropy change induced by the external magnetic field $\Delta S_J(T,B)$ is obtained from the thermodynamical relation as

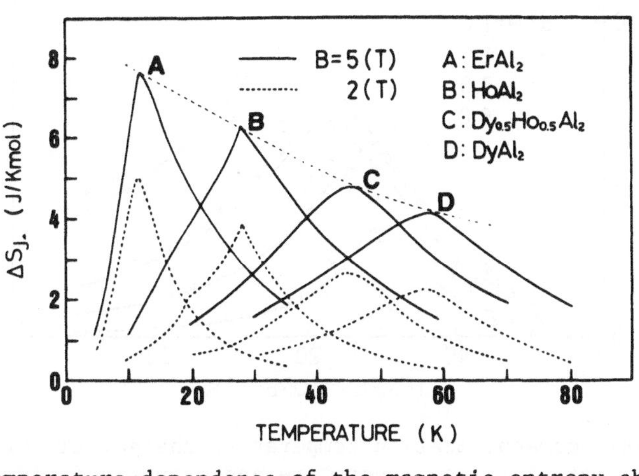

Fig. 5. Temperature dependence of the magnetic entropy changes, ΔS_J, obtained from magnetization in the RAl_2 system.

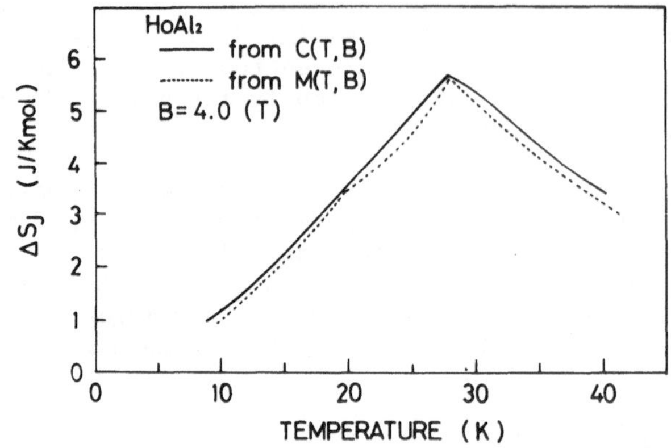

Fig. 6 Magnetic entropy change ΔS_J of $HoAl_2$ obtained from magnetization is compared with that obtained from specific heat. In spite of no correction for demagnetization field in the specific heat measurement, ΔS_J from specific heat is comparatively coincident with ΔS_J from magnetization.

$$\Delta S_J(T,B) = \int_0^B \left(\frac{\partial M}{\partial T}\right)_B dB. \tag{4}$$

We have measured the magnetization of $ErAl_2$, $HoAl_2$, $Ho_{0.5}Dy_{0.5}Al_2$ and $DyAl_2$ in various magnetic fields as a function of temperature. From these results, ΔS_J's were calculated by the relation (4) and they are shown in Fig. 5 in the effective magnetic field of 2 and 5 T. The temperature dependence curves of those ΔS_J's have peaks at $\sim T_c$.

In order to show clearly the difference between ΔS_J's obtained from C(T,B) and M(T,B), the entropy change of $HoAl_2$ obtained from both are shown in Fig. 6 and are seen to coincide reasonably well.

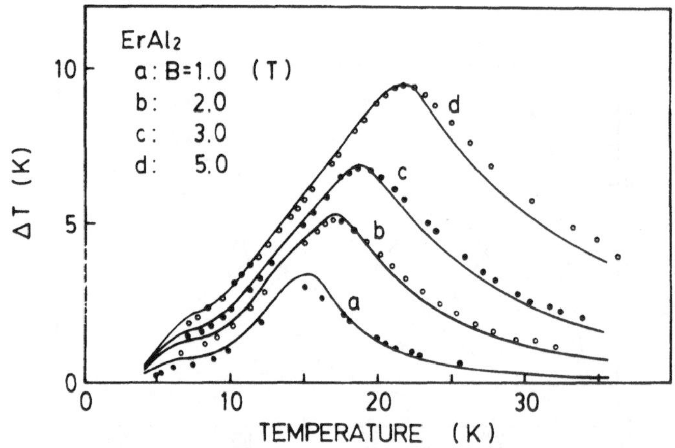

Fig. 7. Adiabatic demagnetization temperature change, ΔT, of $ErAl_2$ as a function of starting temperature and magnetic field. Curves are calculated from our magnetization measurement, and the circles are experimental data of adiabatic demagnetization.

Table 1. Curie Temperature, T_c, Effective Bohr Magneton Number (μ_{eff}) and Magnetic Entropy Change, ΔS_J at T_c.

	T_c(K)	μ_{eff}	$\Delta S_{J(max)}$ (J/K·mol)	
$ErAl_2$	11.7	9.6	8.3(4T)[(a)*] 7.0(4T)[(b)*]	7.6(5T)[(b)]
$HoAl_2$	26.8	10.8	5.7(4T)[(a)] 5.7(4T)[(b)]	6.3(5T)[(b)]
$Dy_{0.5}Ho_{0.5}Al_2$	46.0	11.7	4.2(4T)[(b)]	4.8(5T)[(b)]
$DyAl_2$	55.9	11.7	3.6(4T)[(b)]	4.1(5T)[(b)]

(a) $\Delta S_{J(max)}$ from Specific heat
(b) $\Delta S_{J(max)}$ from Magnetization

* These differences between ΔS_J's from specific heat and from magnetization are attributed to the difference of distribution on the crystallographic orientation of the single domains in the samples.

In Fig. 7, we plotted the adiabatic demagnetization cooling temperature span ΔT of $ErAl_2$ with various starting temperatures and magnetic fields. As an example, adiabatic demagnetization starting at 21.7 K and 50.0 T and reduction of the field to 0 T results in cooling the $ErAl_2$ by 9.5 K. The curves in Fig. 7 represent calculated adiabatic temperature changes of $ErAl_2$ that are obtained from the ΔS_J in Fig. 5 and the specific heat in zero field. It is confirmed from the coincidence of those results that the entropies obtained from our experiment in the RAl_2 series are correct.

CONCLUSIONS

The important magnetic properties and the magnetic entropy changes of the RAl_2 series are shown in Table 1. Study results (in Table 1) show that the RAl_2 system has a large magnetic moment (about 10 μ_B) and thus that a value of ΔS_J sufficient to make a refrigerant can be obtained by an external field of ~6 T. Therefore, we conclude that the compounds in the RAl_2 series have promise as refrigerants.

In this temperature range, however, as the lattice entropy, S_L, is comparable to the magnetic entropy, S_J, we have to utilize the Erisccon cycle for the magnetic refrigeration. To create the ideal Ericsson cycle, a series of magnetic compounds, whose Curie temperatures are arbitrarily adjustable in the temperature span of the Ericsson cycle, are necessary as refrigerants. More precise investigative results about the above argument will be reported in the near future.

The results in Table 1 show that the Curie temperatures of the RAl_2 series are changeable from 10 K or $ErAl_2$ up to 62 K of $DyAl_2$ and, therefore, from the viewpoint of its magnetic properties, this series may be considered to be a promising refrigerant for the Ericsson cycle refrigerator.

ACKNOWLEDGMENTS

This study was performed through Special Coordination Funds for Promoting Science and Technology of the Science and Technology Agency of the Japanese Government.

REFERENCES

1. G. V. Brown, Magnetic heat pumping near room temperature, J. Appl. Phys. 47: 3673 (1976).
2. For instance, A. F. Lacaze, R. Béranger, G. Bon Mardion, G. Claudet and A.A. Lacaze, Double acting reciprocating magnetic refrigerator: Recent improvements, in: "Advances in Cryogenic Engineering", vol. 29, Plenum Press, New York (1983), pp. 573-579.
3. For instance, H. Nakagome, N. Tanji, O. Horigami, H. Ogiwara, T. Numazawa, Y. Watanabe and T. Hashimoto, The helium magnetic refrigerator I: Development and experimental results, in: "Advances in Cryogenic Engineering", vol. 29, Plenum Press, New York (1983), pp. 581-587.
4. For instance, W. A. Steyert, Rotating Carnot-cycle magnetic refrigerator for use near 2 K, J. Appl. Phys. 49: 1227 (1978).
5. For instance, T. Fujita, T. Yazawa, T. Hashimoto, H. Nakagome and T. Kuriyama, Analysis of liquefaction process and losses of a reciprocating magnetic refrigerator for helium liquefaction, in: "Proceedings of CEC '85", DD-2.
6. T. W. Hill, W. E. Wallace, R. S. Craig and I. Inoue, Low temperature heat capacities and related thermal properties of $TbAl_2$ and $HoAl_2$, J. Solid State Chem. 8: 364 (1973).
7. W. E. Wallace, in: "Rare Earth Intermetallic Compounds". Academic Press, New York and London (1973), pp. 112-126.

MAGNETIC AND THERMAL PROPERTIES OF $Dy_3Al_5O_{12}$ AS A MAGNETIC REFRIGERANT

R. Li, T. Numazawa and T. Hashimoto

Department of Applied Physics
Tokyo Institute of Technology
Oh-okayama, Meguro, Tokyo

A. Tomokiyo

College of General Education
Kyushu University
Chuoku, Fukuoka

T. Goto and S. Todo

Institute for Solid State Physics
University of Tokyo
Minatoku, Tokyo

ABSTRACT

We investigated the magnetic entropy of $Dy_3Al_5O_{12}$ (DAG), which is one of the most promising refrigerants for the Carnot-type magnetic refrigerator. In the present investigation we measured the specific heat in zero magnetic field and the magnetization as a function of temperature and magnetic field, and from the analysis of these experimental results, the magnetic entropy change ΔS_J and entropy S are obtained. The values of ΔS_J and S of DAG were compared with those of $Gd_3Ga_5O_{12}$(GGG), which is frequently used as a refrigerant for the Carnot-type magnetic refrigerator. The g-factor of the magnetic ion in DAG was shown to play a more important role in determining ΔS_J near 20 K than the J-value. It is therefore clear that DAG is a more useful refrigerant than GGG for the Carnot-type refrigerator having a broad temperature span from ~20 K to 4.2 K.

INTRODUCTION

Recently, several kinds of magnetic refrigerators have been developed.[1,2,3] One of the most promising of those is the Carnot-type magnetic refrigerator for He liquefaction from ~15 K, which has very high efficiency. One anticipated use of this refrigerator is as a cascade refrigerator connected with a gas refrigerator at ~20 K. The latter has comparatively high efficiency above ~20 K in contrast with its low efficiency in the low temperature range below ~20 K.

Because of its high heat transfer character[4] and magnetocaloric effect,[5] a single crystal of industrially produced gadolinium-gallium-garnet $Gd_3Ga_5O_{12}$

(GGG) (T_N=0.85 K) is now commonly used as refrigerant in the Carnot-type magnetic refrigerator. In GGG, a magnetic entropy change ΔS_J sufficient for refrigeration can be obtained by using an external magnetic field of ~6 T (tesla) below ~15 K. However, near ~20 K, the usefulness of ΔS_J for refrigeration considerably decreases because of the decline of the magnetocaloric character and increase of the lattice entropy. In this paper, we report on investigations on a new refrigerant.

It is well known that the determining parameters of ΔS_J in the paramagnetic region are the J-value and the g-factor.[6] Because the Gd ions in GGG have the largest J-value (J=7/2) in the low temperature range a single crystal of GGG has been used as the refrigerant. However, as mentioned previously, GGG has a serious flaw above ~17 K as a Carnot-type refrigerant. Our investigation therefore focused on the g-factor dependence of the entropy instead of the J-value.

Many investigators have reported that Dy ions in anisotropic crystalline fields have large g-values, for instance, g_z=19 in $DyPO_4$.[7] Since the refrigerant must also possess high thermal conductivity, we here selected the dysprosium-aluminum-garnet $Dy_3Al_5O_{12}$ (DAG) (T_N=2.53 K), which has a thermal conductive character equal to that of GGG.[8]

The adiabatic demagnetization experiments of the polycrystalline DAG and GGG have been made by Hashimoto et al.[5] and it is made clear that the magnetocaloric character of the polycrystalline DAG is almost equivalent to that of GGG. This result is not coincident with the conclusion from the theoretical consideration, and therefore, we investigated the entropy of the single crystal DAG. In the present paper, the entropy of DAG is determined from specific heat measurements in zero magnetic field and measurements of the magnetization.

EXPERIMENTS

A single crystal of $Dy_3Al_5O_{12}$ (DAG) made by the Bridgman method was supplied by Dr. S. Sakaguchi of Shin-etsu Chemical Co., Ltd. Its size was ~5 cm in diameter and ~12 cm in length as shown in Fig. 1. The specific heat of DAG in zero magnetic field, C(T,0), where T is temperature, was

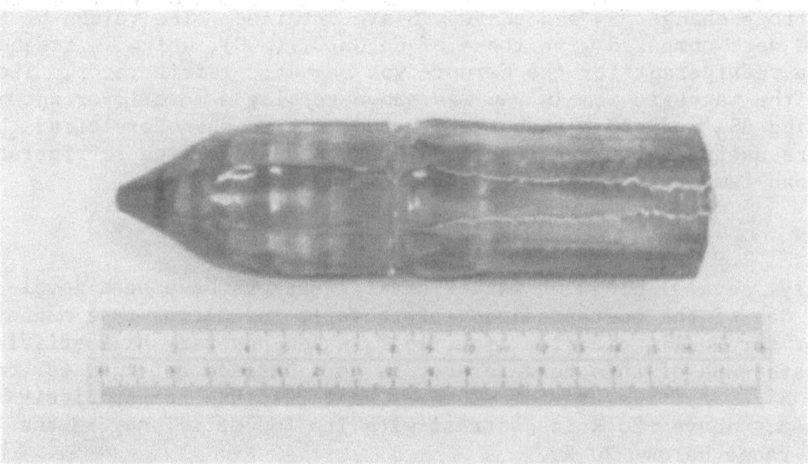

Fig. 1. Photograph of a single crystal of DAG produced by Dr. S. Sakaguchi et al. of Shin-etsu Chemical Co., Ltd.

measured using the adiabatic calorimeter in the temperature range from 30 K to 1.5 K. The entropy in the zero field, S(T,0), is obtained from the analysis of C(T,0) by the following formula,[9]

$$S(T,0) = \int_0^T \frac{C(T,0)}{T}\, dT. \quad (1)$$

In the low temperature range, when the physical properties concerned with ferromagnetic ordering are observed in strong magnetic fields, the applied field must be corrected by the demagnetization field to obtain the effective field in the sample. This correction for the magnetization is easier and more accurate than that for specific heat in a magnetic field. In observing the magnetization, we can easily utilize a particular sample shape with known demagnetization factor. The magnetic field B dependence of entropy change $\Delta S_J(T,B)$ caused by the external magnetic field was determined from an analysis of the magnetization as a function of T and B, M(T,B), using the following formula,[9]

$$\Delta S_J(T,B) = \int_0^B \left\{\frac{\partial M(T,B)}{\partial T}\right\}_B dB. \quad (2)$$

The total entropy S(T,B) is given by

$$S(T,B) = S(T,0) + \Delta S_J(T,B). \quad (3)$$

In the observation of the magnetization to determine the anisotropy of $\Delta S_J(T,B)$, we used two kinds of cylindrical single crystals whose axes were parallel to the <111> and <110> directions, where the <111> direction is the easy axis.

The difference between the cylindrical axis of the sample and the specific crystallographic axis (<111> or <110>) was within 5 degrees. Experimental errors in the measurement of temperature and magnetic field strength were less than 1 percent.

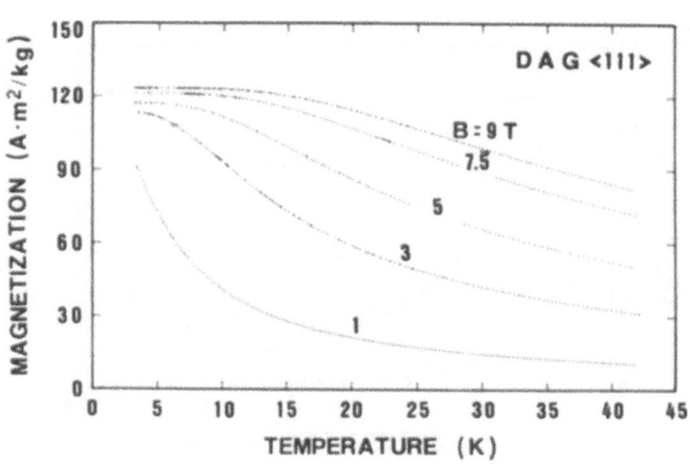

Fig. 2. The magnetization plotted against temperature for various magnetic field strengths applied parallel to the <111> direction.

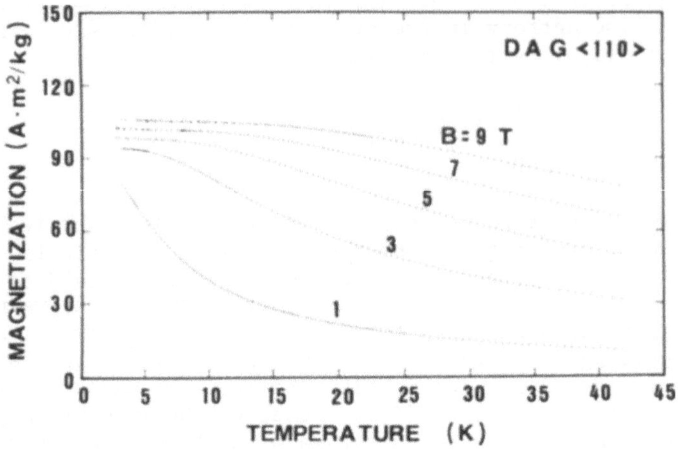

Fig. 3. The magnetization plotted against temperature for various magnetic field strengths applied parallel to the <110> direction.

RESULTS

The magnetizations of the two kinds of cylindrical samples I and II, whose axes are parallel to the <111> and the <110> directions, respectively, are given as a function of temperature and magnetic field (Figs. 2 and 3).

Near 4.2 K the magnetizations of both samples are almost saturated by the magnetic field above 6 T (tesla) and their anisotropy is clearly observed. The ratio of the saturation value of the magnetization in sample I to that in sample II is 1.21, and this result is compared with the theoretical result in the next section.

The entropy change ΔS_J of samples I and II is shown in Figs. 4 and 5, respectively. The value of ΔS_J near 4.2 K in the strong magnetic field at

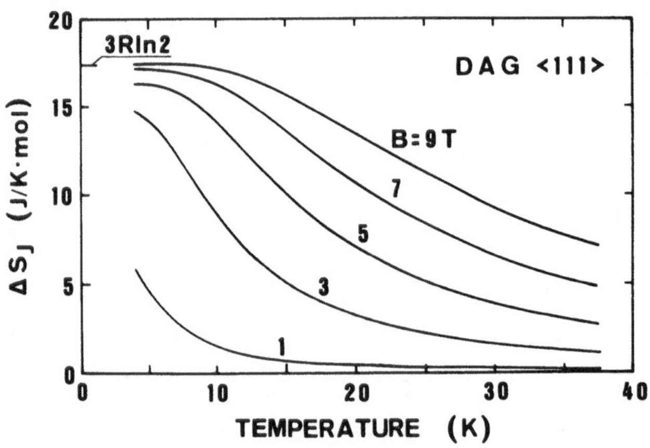

Fig. 4. The magnetic entropy change ΔS_J plotted against temperature, caused by various magnetic fields applied parallel to the <111> direction.

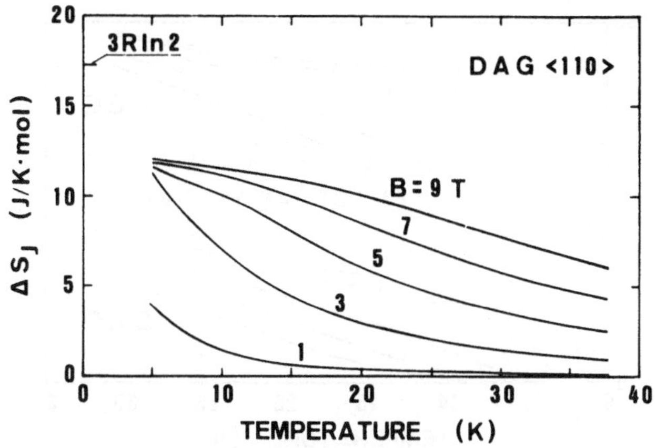

Fig. 5. The magnetic entropy change ΔS_J plotted aginst temperature, caused by various magnetic fields applied parallel to the <110> direction.

9 T is almost coincident with 3Rln2, where R is the gas constant 8.3 J/mol·K, and this fact clearly verifies that the assumption of J = 1/2 for the spin value of the Dy ion is a good approximation in the low temperature range.

The total entropy of $Dy_3Al_5O_{12}$ (DAG), applied magnetic field parallel to the <111> direction, is shown in Fig. 6. For easy comparison, the total entropy of $Gd_3Ga_5O_{12}$ (GGG) is also shown. It is clear that in the temperature range below ~25 K the absolute value of the entropy in DAG is almost one third of that in GGG. The most noteworthy aspect is that the value of ΔS_J in DAG is noticeably larger than that in GGG, especially in the temperature range above ~15 K.

DISCUSSION

First, we consider the origin of the anisotropy of the magnetization and the entropy in DAG. The g-factor of the Dy ion in DAG was obtained from observation of the ESR spectrum[10] as g_z = 18.1 and $g_x = g_y$ = 0.5, where the z-axis is the principal axis of the crystalline field. In the unit cell of DAG there are three kinds of Dy ions, whose principal axes of the g-factor are parallel to the <100>, <010> and <001> directions. The effective value of the g-factor, g_e, in DAG thus depends on the crystallographic axis as shown in Table I.

In general, the saturation magnetization in each crystallographic direction is described by ~$g_e \mu_B J$. Therefore, the ratio of the magnetization of sample I to sample II, M_{111}/M_{110} is thought to be proportional to the ratio of g_{111}/g_{110}. As the experimental result of (M_{111}/M_{110}) = 1.21 is in good agreement with the theoretical ratio of (g_{111}/g_{110}) = 1.21, the anisotropy of the magnetization is believed to reflect that of the g-factor.

Table I. Effective g-factors, g_e, in three crystallographic axis directions.

	<111>	<110>	<100>
g_e value	10.5	8.7	6.4

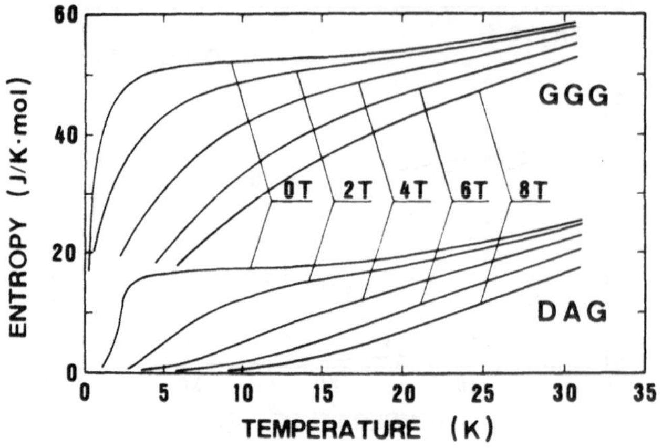

Fig. 6. The entropy of $Dy_3Al_5O_{12}$(DAG) and $Gd_3Ga_5O_{12}$(GGG) plotted against temperature in various fields.

The entropy change ΔS_J in the paramagnetic region far from T_c is approximately given by

$$\Delta S_J = -\{N\mu_0 g^2 \mu_B{}^2 J(J+1)/3k\}B^2/2(T-\theta)^2, \qquad (4)$$

where N is the population of spins; μ_B, the Bohr magneton; k, the Boltzmann constant, and θ is the paramagnetic Curie temperature. From Eq.(4), the anisotropy of ΔS_J is considered to reflect that of the g-factor. Therefore, in reference to ΔS and S, we will show only the results of sample I in which the magnetic field is applied in the direction of the <111> axis, because in this case the values of ΔS_J and S are maximum.

The magnetic entropy changes of ΔS_J in DAG and GGG in magnetic fields of 4, 6 and 8 T (tesla) are shown in Fig. 7. The maximum value of the magnetic entropy S_J is determined by the J-value as N k ln (2J + 1). In the low temperature range, especially below ~10 K, since the thermal energy ~kT is less than or comparable to the Zeeman energy $g\mu_B B$ at 6 T, and most

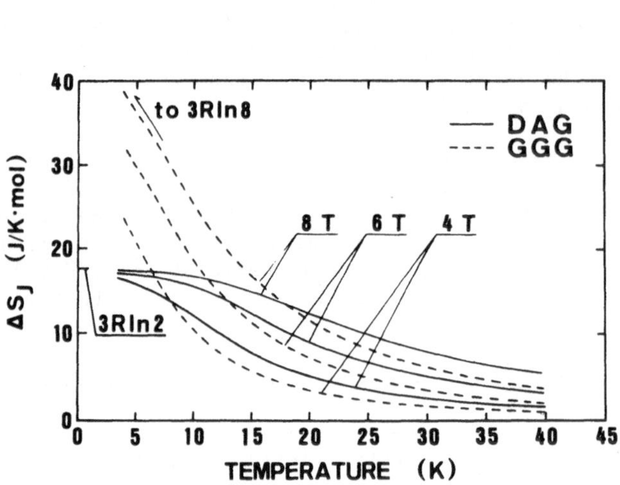

Fig. 7. The magnetic entropy changes of S_J in DAG and GGG in magnetic fields of 8 T, 6 T and 4 T.

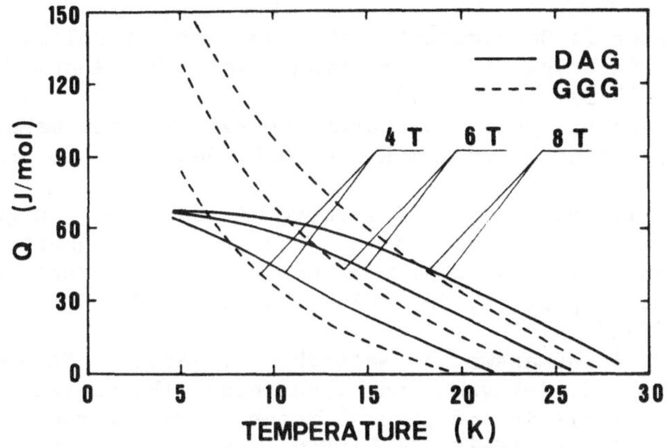

Fig. 8. The quantity of heat Q of $Dy_3Al_5O_{12}$ and $Gd_3Ga_5O_{12}$ absorbed in the isothermal process of the Carnot cycle. The horizontal axis shows the starting temperature of the Carnot cycle down to 4.2 K.

of the spins occupy the ground state, ΔS_J is mainly controlled by the J-value. Therefore, the value of ΔS_J in GGG is considerably larger than in DAG (Figs. 6 and 7).

In contrast, in the temperature range above ~15 K, since kT is larger than $g\mu_B B$, the entropy change ΔS is decided by the difference between the population of spins in the excited state and the ground state. The value of ΔS_J is then mainly controlled by the factor of $g_e^2 J(J + 1)$, which is important in the determination of the Boltzmann distribution. The value of ΔS_J in DAG is thus seen to become larger than that in GGG as it approaches ~20 K because the factors $g_e^2 J(J + 1)$ are 82 in DAG and 63 in GGG.

The heat absorption and release characteristics of the refrigerant in the isothermal processes of the refrigeration cycle are affected not only by the value of ΔS_J, but also by the temperature dependence of S_L. To clarify this refrigeration character, the quantity of heat absorbed or released in the isothermal process, Q, was estimated by drawing the ideal Carnot cycle on the entropy vs. temperature planes in DAG and GGG. A comparison of the Q of DAG and GGG obtained in magnetic fields of 4, 6 and 8 T is shown in Fig. 8. Considering the maximum value of the usable magnetic field (~6 T), these results indicate that DAG is a more appropriate refrigerant than GGG for the Carnot-type magnetic refrigerator, especially one having a large temperature span from 20 K to 4.2 K.

ACKNOWLEDGMENTS

This study was performed through Special Coordination Funds of the Science and Technology Agency of the Japanese Government.

REFERENCES

1. For instance, A. F. Lacaze, R. Béranger, G. Bon Mardion, G. Claudet and A. A. Lacaze, Double acting reciprocating magnetic refrigerator: recent improvements, in: "Advances in Cryogenic Engineering", vol. 29, Plenum Press, New York (1983), pp. 573-579.
2. For instance, H. Nakagome, N. Tanji, O. Horigami, H. Ogiwara, T. Numazawa, Y. Watanabe and T. Hashimoto, The helium magnetic

refrigerator I: Development and experimental results, in: "Advances in Cryogenic Engineering", vol. 29, Plenum Press, New York (1983), pp. 581-587.
3. Y. Hakuraku and H. Ogata, Conceptual design of a new magnetic refrigerator operating between 4 K and 20 K, Submitted to Jan. J. Appl. Phys.
4. T. Numazawa, Y. Watanabe, T. Hashimoto, A. Sato, H. Nakagome, O. Horigami, S. Takayama and M. Watanabe, The magnetic refrigeration characteristics of several magnetic refrigerants below 20 K: II Thermal properties, in "ICEC9, Kobe, Japan, 1982", Butterworth, UK, pp. 30-33.
5. T. Hashimoto, T. Numazawa, Y. Watanabe, A. Sato, H. Nakagome, O. Horigami, S. Takayama and M. Watanabe, The magnetic refrigeration characteristics of several magnetic refrigerants below 20 K: I Magnetocaloric effect, in "ICEC9, Kobe, Japan, 1982", Butterworth, UK, pp. 26-29.
6. T. Numazawa, M. Ikisawa, T. Hashimoto, A. Tomokiyo, The magnetocaloric and thermal properties of garnet type magnetic refrigerants between 4 K and 30 K, to be published in ICMC. 1985.
7. J. A. Barclay, A 4 K to 20 K rotational-cooling magnetic refrigerator capable of 1 mW to > 1 W operation, Cryogenics, 20:467-471 (1980).
8. G. A. Slack and D. W. Oliver, Thermal conductivity of garnets and phonon scattering by rare-earth ions, Phys. Rev. B. 4:592-609 (1971).
9. T. Hashimoto, T. Numazawa, M. Shiino and T. Okada, Magnetic refrigeration in the temperature range from 10 K to room temperature: the ferromagnetic refrigerants, Cryogenics, 21:647-653 (1981).
10. W. P. Wolf, B. Schneider, D. P. Landau and B. E. Keen, Magnetic and thermal properties of dysprosium aluminum garnet II. Characteristic parameters of an ising anti-ferromagnet, Phys. Rev. B. 5:4472-4496 (1972).

SPECIFIC HEAT AND ENTROPY OF RNi_2 (R: RARE EARTH HEAVY METALS) IN MAGNETIC FIELD

A. Tomokiyo and H. Yayama
Department of Physics, College of General Education
Kyushu University, Chuoku, Fukuoka, Japan

H. Wakabayashi, T. Kuzuhara and T. Hashimoto
Department of Applied Physics, Faculty of Science, Tokyo
Institute of Technology, Ohokayama, Meguroku, Tokyo, Japan

M. Sahashi and K. Inomata
Toshiba Research and Development Center
Saiwaiku, Kawasaki, Kanagawa, Japan

ABSTRACT

The system RNi_2 (R=Dy, Ho and Er) is verified to be promising as magnetic refrigerant in the wide temperature range below 35 K. We prepared these compounds and measured the temperature variation of the specific heat and the magnetization in magnetic field up to 5 T. From these results we determined the total entropy and the magnetic entropy change by a magnetic field. It is made clear that these compounds are applicable to the magnetic refrigerants using an Ericsson-type cooling cycle below 35 K.

INTRODUCTION

Recently, though many kinds of investigation of the magnetic refrigeration have been made, the large parts of these studies have focussed on the production of liquid helium[1] from the gas phase near 15 K and superfluid helium[2-4] below 2 K. In this temperature region, a paramagnetic material such as gadolinium sulfate or gadolinium gallium garnet is used as a magnetic refrigerant. For those paramagnetic materials, a sufficiently large entropy change is obtained in a magnetic field of several Tesla and the lattice entropy is negligibly small compared with the magnetic one. Therefore, a Carnot-type cooling cycle can be used with a suitable temperature span of magnetic cycle and without a regenerator, and a refrigerator of a simple structure is possible. Some of the above results show that the magnetic refrigeration is a considerably efficient and hopeful refrigeration method.

We are very interested in a magnetic refrigeration method operating in a temperature region higher than 15 K. This kind of investigation has not been tried yet except the one around room temperature by Brown[5]. In magnetic refrigeration above 15 K, one of the most important problems is the choice of suitable working materials. In this temperature region, the entropy change of paramagnetic material decreases with an increase in temperature; besides the lattice part of entropy increases and cannot be neglected. In this case, we have to employ an Ericsson-type cooling cycle instead of Carnot cycle and use a ferromagnetic material as a refrigerant. Since the maximum entropy change under the applied magnetic field is obtained at around the Curie temperature, T_c, the cooling cycle may efficiently operate at around T_c.

The magnetic measurements for the system RNi_2 (R=Gd, Tb, Dy, Ho and Er) by Wallace et al.[6] show that these are ferromagnetic and their T_c's decrease monotonically along the series from 85 K for $GdNi_2$ to 21 K for $ErNi_2$. It should be noted that those T_c's were obtained in a magnetic field of 2 T. The striking feature is that nickel in these compounds appears to be nonmagnetic. Neither T_c in zero magnetic field nor magnetocaloric properties of these compounds are known. So we prepared the compounds RNi_2 (R=Dy, Ho and Er) and measured the temperature variation of the specific heat (for $DyNi_2$ and $ErNi_2$) and the magnetization (for $HoNi_2$) in a magnetic field up to 5 T. From the results of the total entropy and the entropy change of these compounds which were derived from the specific heat and the magnetization measurements, we discuss their refrigeration characteristics and possibilities as magnetic refrigerants for the Ericsson-type cooling cycle in the wide temperature range.

SAMPLE PREPARATION AND EXPERIMENT

Constituent amounts of each element were arc melted together in argon gas and were cooled down. The compounds were turned over after cooling and were melted again. This procedure was repeated several times to get homogeneous quality. Then they were formed into columnar shape, 8 mm in diameter and 20 mm in length for the specific heat measurements, 2 mm in diameter and 10 mm in length for the magnetization measurements. These samples were covered with tantalum foil to avoid oxidation and were annealed in vacuum for 2 days at 900 ^{o}C.

For the specific heat measurements of $DyNi_2$ and $ErNi_2$, the continuous heating method was employed. To supply the heater power to the sample, a manganine wire, 70 μm in diameter, was wound noninductively around the sample, which was suspended in a stainless steel cylindrical chamber with four monofilament nylon lines, 90 μm in diameter, to maintain sufficient themal isolation. The sample chamber was evacuated to the pressure 2.7×10^{-4} Pa. The temperature of the sample was determined with a carbon glass resistor thermometer that was sufficiently insensitive to magnetic fields.

When heater power was supplied in a constant magnetic field, the temperature of the sample rose gradually. From the gradient of the measured temperature-time curves, dT/dt, wattage of the supplied power, P, the sample weight, m, and the molecular weight of the sample, M, the specific heat, C(T,B), could be evaluated by the relation

$$C(T,B) = \frac{M\,P}{m}\left(\frac{dT}{dt}\right)^{-1}, \qquad (1)$$

where T is absolute temperature and B the applied magnetic field.

Magnetization of $HoNi_2$ was measured by the conventional vibrational method as a function of temperature and magnetic field.

From the measured specific heat C(T,B), we can determine the total entropy S(T,B) using the relation,

$$S(T,B) = \int_0^T \frac{C(T,B)}{T}\, dT. \tag{2}$$

The total entropy is written as,

$$S(T,B) = S_J(T,B) + S_L(T) + S_E(T), \tag{3}$$

where $S_J(T,B)$, $S_L(T)$ and $S_E(T)$ are magnetic, lattice and electronic parts of entropy, respectively. In the temperature range of our interest, the electronic part is much smaller than the others. Only $S_J(T,B)$ changes with an applied magnetic field. An entropy change ΔS_J in an applied field can be determined from measured magnetization M(T,B) using the equation,

$$\Delta S_J = S_J(T,B) - S_J(T,0) \tag{4}$$

$$= \int_0^B \left\{ \frac{\partial M(T,B)}{\partial T} \right\}_B dB. \tag{5}$$

The magnetic entropy changes ΔS_J's of $DyNi_2$ and $ErNi_2$ were evaluated directly from the results of total entropy using the eqs. (3) and (4), while the ΔS_J of $HoNi_2$ was derived from the measured value of M(T,B) using eq.(5).

RESULTS AND DISCUSSION

The temperature variation of the specific heat of $DyNi_2$ in magnetic field up to 5 T is shown in Fig. 1 (a). The peak of the zero field specific heat corresponds to the Curie temperature, T_c=20 K. When the applied magnetic field is increased, the peak is broadened and shifts to higher temperature.

Figure 1 (b) shows the specific heat of $ErNi_2$ in various magnetic fields. The sharp peak of the zero field curve demonstrates a typical λ-type thermal anomaly indicating T_c=6.6 K. The temperature and magnetic field dependence of the specific heat shows the behavior of usual ferromagnetic substances.

The total entropies of $DyNi_2$ and $ErNi_2$ were evaluated from the values of specific heat using eq.(2) and are shown in Fig. 2 (a) and (b), respectively. To discuss the lattice entropy of the system RNi_2, the entropy derived from specific heat of $CeNi_2$[6], which behaves as a nonmagnetic compound, is shown with a dashed curve in Fig. 2 (a). The entropy of $CeNi_2$ may be a estimate of the lattice entropy of the series of compound RNi_2. It is found that the lattice part of the entropy of $DyNi_2$ is not negligible at temperatures above 15 K, and that, near T_c, more than half of the magnetic entropy is removed at 5 T.

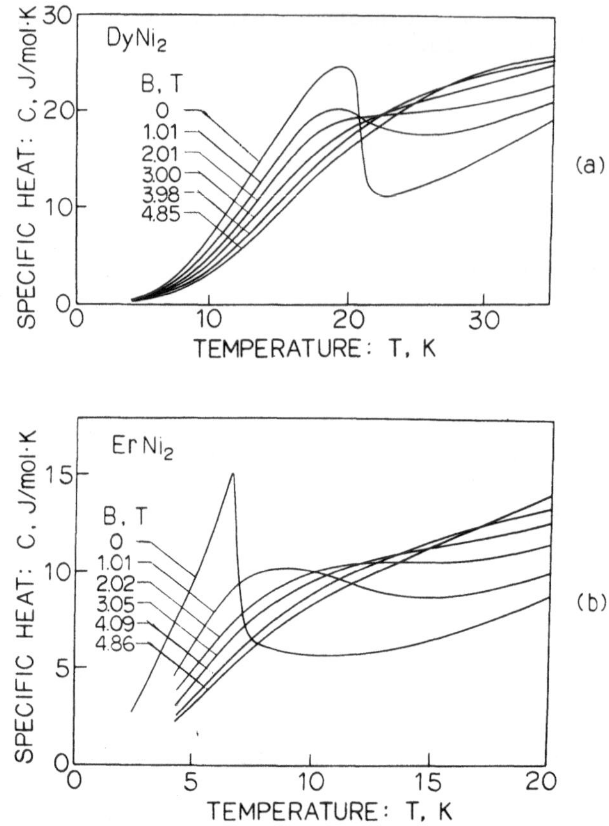

Fig. 1. Temperature dependence of the specific heat of $DyNi_2$ and $ErNi_2$ in a magnetic field.

As predicted theoretically, these entropy curves indicate that the entropy reduction caused by an applied magnetic field gives maximum at around T_c of the compound. The amounts of the entropy change, ΔS_J, in various magnetic fields are shown in Fig. 3. The values of $DyNi_2$ and $ErNi_2$ were derived directly from the total entropy in Fig. 2 (a) and (b), and that of $HoNi_2$ was obtained from the measured magnetization data. Curie temperature, T_c, effective Bohr magneton number, μ_B[6] and the maximum value of the entropy change, ΔS_J(max), in a 5 T field are listed in the Table 1. We believe that our values of T_c are correct, because we determined them in zero magnetic field, while those by Wallace et al.[6] were determined in a field of 2 T; therefore, they were considerably larger than ours. The ΔS_J exhibits a maximum value of about 7 J/mol K at near the T_c of each compound at the field 5 T. As far as the amount of the entropy change is concerned, these values of ΔS_J are large enough to use as a magnetic refrigerant.

In Fig. 4, two types of thermodynamic cooling cycle, the Carnot and the Ericsson cycles, are illustrated on the entropy-temperature plane of $DyNi_2$ as an example. Apparently, a large temperature span of the cycle cannot be obtained using a Carnot cycle like ABCD. While the Ericsson-type cooling cycle, A'B'C'D', gives a sufficient temperature span and appreciable cooling power, as estimated below.

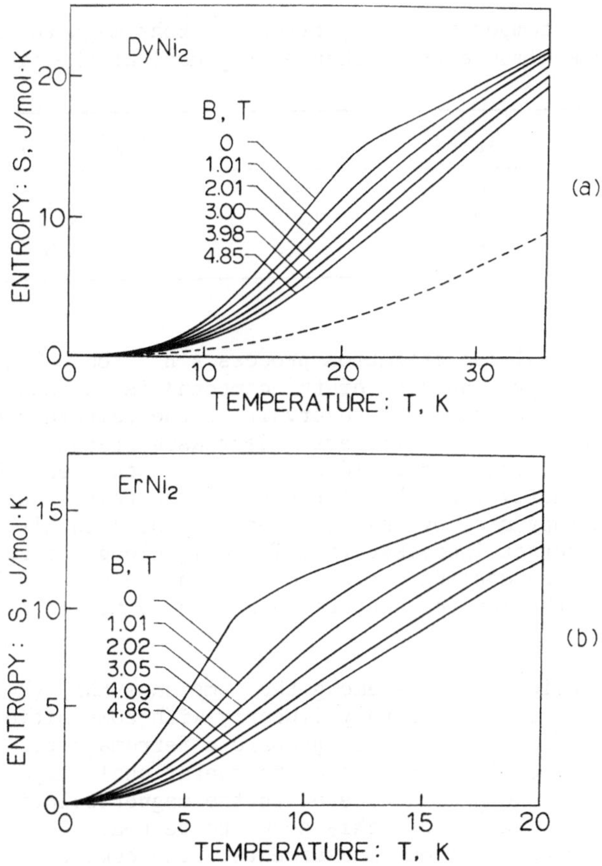

Fig. 2. Temperature dependence of the total entropy of $DyNi_2$ and $ErNi_2$ in a magnetic field. Dashed curve is the entropy of $CeNi_2$.

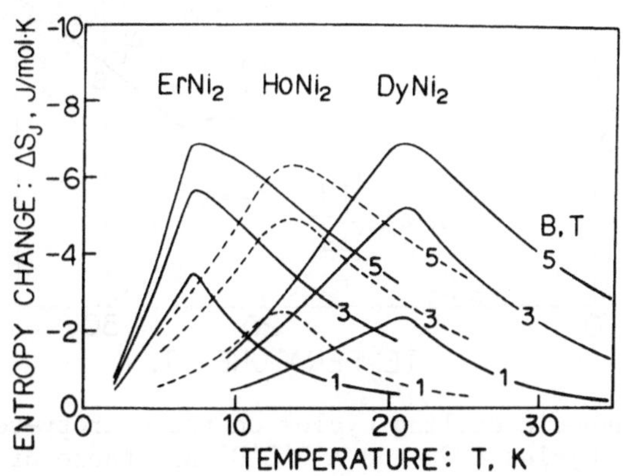

Fig. 3. Temperature dependence of the entropy change of $DyNi_2$, $HoNi_2$ and $ErNi_2$ in a magnetic field.

Table 1. Curie temperature, T_c, effective Bohr magneton number, μ_B,[6] and maximum entropy change, $\Delta S_J(max)$ at the field 5 T.

Compound	T_c(K)	μ_B	ΔS_J(max) (J/mol K)
$DyNi_2$	20	10.4	6.9
$HoNi_2$	13	10.5	6.3
$ErNi_2$	6.6	9.4	6.9

In the isothermal magnetization process A'B' from 0 T to 5 T as shown in Fig. 4, the entropy reduction of the compound is ΔS_2 and the heat Q_2 ($=T_2\Delta S_2$) is released to the heat reservoir of the temperature T_2 (about 30 K). The temperature of the refrigerant decreases gradually along the path B'C' in the constant magnetic field of 5 T. Then the heat, Q_1 ($=T_1\Delta S_1$), is removed isothermally from the heat bath of the temperature T_1 (about 15 K) due to the entropy rise by the demagnetization. Finally, the temperature of the refrigerant increases from T_1 to T_2 along the zero magnetic segment D'A', and the cooling cycle closes. In the case of $DyNi_2$, the heat removed from the low temperature bath is 60 J/mol in a single cycle operation.

Continuous solid solutions among the compounds RNi_2 (R=Dy, Ho and Er) can be obtained and their T_c's vary linearly with composition of the rare earth elements.[6] Therefore, we can prepare a ferromagnetic material that has a desired value of T_c. To extend the operational temperature region to higher than 35 K, we are interested in the magnetocaloric properties of the compound $GdNi_2$. The T_c of $GdNi_2$ seems to be near 40 K from extrapolation of the values of T_c of our measurments. The compound is a promising magnetic refrigerant that works well above at 35 K.

We propose a magnetic refrigerator system operating in a wide temperature range. This is constructed with three or four refrigerators, which are connected in a cascade, and each of them operates with Ericsson-type cooling cycle in a separate temperature range.

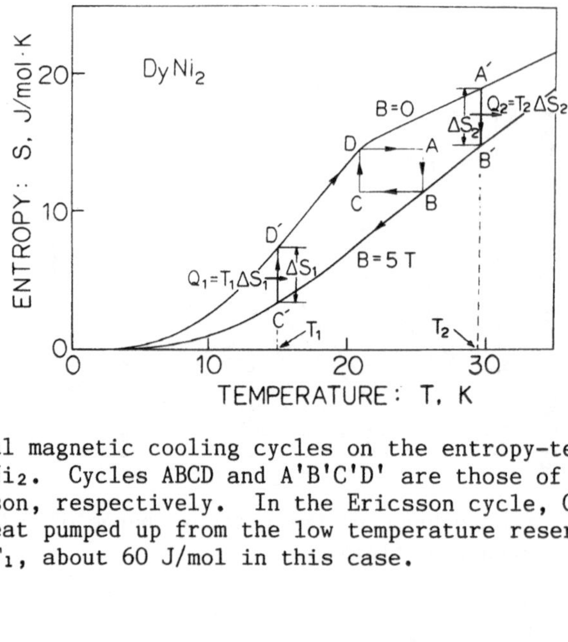

Fig. 4. Typical magnetic cooling cycles on the entropy-temperature plane of $DyNi_2$. Cycles ABCD and A'B'C'D' are those of Carnot and Ericsson, respectively. In the Ericsson cycle, Q_1 is equal to the heat pumped up from the low temperature reservoir at temperature T_1, about 60 J/mol in this case.

CONCLUSION

From our investigation on the system RNi_2, it was made clear that the entropy change ΔS_J of the system was very large at around T_c and they were good magnetic refrigerants.

ACKNOWLEDGMENT

This study was performed through Special Coordination Funds for promoting Science and Technology of the Science and Technology Agency of the Japanese Government.

REFERENCES

1. T. Numazawa, T. Hashimoto, H. Nakagome, N. Tanji and O. Horigami, The helium magnetic refrigerator II: Liquefaction process and efficiency, in: "Advances in Cryogenic Engineering" vol. 29, Plenum Press, New York (1984), pp. 589-596.
2. W. P. Pratt, S. S. Rosenblum, W. A. Steyert and J. A. Barcley, A continuous demagnetization refrigerator operating near 2 K and a study of magnetic refrigerants, Cryogenics 17: 689-693 (1977).
3. W. A. Steyert, Rotating Carnot-cycle magnetic refrigerator for use near 2 K, J. Appl. Phys. 49: 1227-1231 (1978).
4. A. F. Lacaze, A. A. Lacaze, R. Beranger and G. Bon Mardion, Thermodynamical analysis of a double acting reciprocating magnetic refrigerator, "Proceedings of ICEC", Butterworths, England (1982).
5. G. V. Brown, Magnetic heat pumping near room temperature, J. Appl. Phys. 47: 3673-3690 (1976).
6. W. E. Wallace, "Rare Earth Intermetallics", Academic Press, Inc., New York and London (1973), pp. 111-128.

LOW TEMPERATURE DEFORMATION OF COPPER AND AN AUSTENITIC STAINLESS STEEL*

R. P. Reed and R. P. Walsh

Fracture and Deformation Division
National Bureau of Standards
Boulder, Colorado

ABSTRACT

The tensile-deformation characteristics and effect of strain rate were studied on relatively pure CDA 102 Cu and solid-solution-strengthened AISI 310. Tensile strain rate was varied between two orders of magnitude (2×10^{-3}; 2×10^{-5} s^{-1}) at temperatures ranging from 4 to 295 K. Tensile stress-strain and strain-hardening curves were determined for these temperatures. The effect of strain-rate changes on tensile flow strength was measured from strains near 0.002 (yield strength) to over 0.300. The data reflect three distinct ranges of face-centered cubic, polycrystalline plastic deformation, which have different characteristics depending on solute content.

INTRODUCTION

Superconducting magnet structural materials deform at greatly different rates, varying from relatively slow stress increases due to rising magnetic fields to quite high strain rates due to quenching effects. Controlled strain-rate experiments not only provide data on the effects of strain rate on tensile properties, but they also lead to a better understanding of deformation mechanisms. Guntner and Reed[1] varied the strain rate from 0.0008 s^{-1} to 0.017 s^{-1} in separate tests to assess the low temperature (4-295 K) tensile properties of various bar, sheet, and plate forms of alloy 304. Variations of yield strength up to 10% and of tensile strength up to 25% were observed from increasing the strain rate for this metastable austenitic stainless steel.

The dependence of the strain to initiate load drops on the strain rate during tensile testing of alloy 304L was measured by Read and Reed[2] at 4 K. As the strain rate was increased from about 0.003 s^{-1} to 0.1 s^{-1}, the strain prior to the first load drop decreased from about 0.05 to 0.01%. From thermocouples embedded within tensile specimens, temperature rises up to 52 K were detected during load drops. Load drops are associated with macroscopic slipbands, closely resembling the Lüders bands that appear during ambient temperature tensile tests. The temperature rise is associated with the very

*Partially supported by the Office of Fusion Energy, U.S. Department of Energy. Contribution of NBS, not subject to copyright.

low specific heat and low thermal conductivity of metals at very low temperatures, which result in nearly adiabatic localized conditions. Using fine superconducting wire bonded within a slot on the surface of tensile specimens, Read and Reed did not detect temperature rises up to 10 K (T_c of the NbTi wire) during plastic deformation prior to load drops. They concluded that the triggering event for discontinuous yielding in a macroscopic sense was essentially mechanical in nature and that only during the dislocation avalanche, associated with the load drops, was the temperature adiabatically driven up to much higher temperatures. The microscopic triggering event for the macroscopic slipband (dislocation avalanche) has not been adequately identified.

Conrad[3] reviewed the low temperature deformation of face-centered cubic metals. He analyzed existing strain-rate-dependent data of Al, Ag, Cu, and Ni, plotting a strain-rate sensitivity parameter ($\Delta \ell n\sigma/\Delta \ell n\dot{\varepsilon}$) versus temperature over the range 4 to 300 K. He illustrated that $\Delta \ell n\sigma/\Delta \ell n\dot{\varepsilon}$ gradually decreases at lower temperatures, falling to zero at 0 K for these metals. Conrad showed that the activation volume (reciprocal of stress sensitivity) decreases as a function of stress (and strain) at constant temperature and that the activation volume of Cu is greater than that of Al.

Tien and Yen[4] summarized work on low temperature creep. The review includes some of the tensile measurements on Cu that use strain-rate-change experiments to measure activation volume. The apparent disparity of experimental results on this subject is compounded by the dependency of activation volume on flow stress (σ), strain (ε), and temperature (T). There have been no low temperature strain-rate-change measurements on austenitic stainless steels; however, Warren and Reed[5] published tensile stress-strain curves at low temperatures on alloy 310.

Kocks et al.[6-10] and Mulford[11] presented a clearer representation of strain hardening, strain-rate sensitivity, and dynamic aging for polycrystalline metals and alloys. They argued that the flow strength at low temperatures is composed of two additive contributions--from dislocation-solute interactions and dislocation-dislocation interactions. Their approach is discussed in more detail later and used to describe some of our data.

This paper presents stress-strain, strain-hardening, and strain-rate sensitivity data at temperatures from 4 to 295 K for relatively pure Cu and for a solid-solution strengthened austenitic stainless steel.

EXPERIMENTAL PROCEDURE

Specimens of relatively pure Cu (oxygen-free, high-conductivity CDA 102) and a common solid-solution-strengthened austenitic stainless steel (AISI 310) were obtained in 1.9-cm (0.75-in) diameter bar stock and tested in the

Table 1. Material Characterization

Alloy	Chemical Composition (wt.%)	Hardness (R_B)	Grain Size (μm)
CDA 102	99.99 Cu + Ag	22	36
310	24.8 Cr, 20.8 Ni, 1.7 Mn, 0.093 C, 0.031 N, 0.02 P, 0.02 P, 0.02 S, 0.7 Si, 0.1 Mo, 0.1 Cu	72	161

annealed condition. Alloy 310 was annealed at 2050°C for 0.5 h after machining and CDA 102 Cu, at 650°C for 1 h. Round tensile specimens were machined to a 6.35-mm diameter along a gage length of 41.9 mm, increasing with a radius of 9.52 mm to 12.7-mm-diameter threaded ends. The entire specimen length was 71.4 mm. The hardness, grain size, and chemical composition of both alloys are listed in Table 1.

Two loading frames were used. At room temperature the loading compliance of the screw-driven assembly was measured as 8×10^{-3} cm/kN (8×10^{-7} cm/kgf). The compliance of the servohydraulic load train varied from 11×10^{-3} cm/kN at loads less than about 2 kN to 6×10^{-3} cm/kN at larger loads. The Cu specimens were measured at the lower compliance levels of the load train; the austenitic stainless steels, at the higher compliance levels.

The actuator velocity and crosshead movement of the machines were independently measured and found to deviate from expected rates. The "low" crosshead rate of the screw-driven machine was 0.0047 cm/min (stated as 0.005 cm/min) and the "high" crosshead rate for this machine was 0.455 cm/min (stated as 0.5 cm/min). The servohydraulic "high" actuator speed was 0.540 cm/min (stated as 0.5 cm/min), significantly higher than that of the other machine. The "low" actuator speed of the servohydraulic machine was 0.005 cm/min.

In all tests, specimens were initially loaded at the low strain rate (2×10^{-5} s^{-1}) at constant temperature. After sufficient strain to permit measurement of the yield strength (0.002 plastic strain), the strain rate ($\dot{\varepsilon}$) was incrementally increased (to 2×10^{-3} s^{-1}) and held long enough for good characterization of the stress-strain curve. The strain rate was subsequently reduced, and the specimen was strained to the next measurement range. The difference between the flow strengths at equivalent strain-hardening rates was considered the stress change ($\Delta\sigma$) associated with the strain-rate change. The exception to this was measurement of alloy 310 at higher levels of strain, where apparent "yield points" were observed immediately following changes of strain rate. In these cases we used the maximum value of $\Delta\sigma$. Use of extrapolated $\Delta\sigma$ to the onset of the high strain-rate range usually produced a negative dependence (but always positive values) of $\Delta\sigma$ on ε or σ.

Temperature measurement and control, measurement equipment variables that affected the accuracy of $\Delta\sigma$ data from strain-rate changes, strain measurement, and other procedures are discussed in Ref. 12. Reference 12 also contains all data from which the strain-rate sensitivity figures of this paper were derived.

At low strains (~<0.04) the attainment of an equivalent strain-hardening rate on changing to the higher strain rate usually consumes more strain. Therefore, interpretation of $\Delta\sigma$ at lower σ or ε values is more difficult. The strain range over which the strain-hardening rates tended to be slower to change was always less than 0.01. This early portion of the σ-ε curve naturally exhibited larger strain-hardening rates and, when reliable measurements could be obtained, lower values of $\Delta\sigma$. No data in this low strain range were included unless both σ-ε curves (high ε and low ε) assumed equivalent strain-hardening rates.

Discontinuous yielding at temperatures below about 20 K was observed in both materials. The strain at which load drops began was much less in the austenitic steel than in Cu. This is attributed to their differences of low temperature thermal conductivity. The thermal conductivity of alloy 310 at 4 K is about 0.5 $W \cdot m^{-1} \cdot K^{-1}$, and that of Cu is 667 $W \cdot m^{-1} \cdot K^{-1}$. The specific heat of the austenitic steel is actually somewhat higher (~ 1.8 $J \cdot kg^{-1} \cdot K^{-1}$) than that of Cu (0.10 $J \cdot kg^{-1} \cdot K^{-1}$). The occurrence of discontinuous yielding precluded assessment of strain-rate effects and strain-hardening rates on flow stress. Changes to the higher strain rate immediately triggered load drops and more erratic stress-strain behavior. At the higher strain rate it is possible that the flow strengths at 4 K reflect thermal instability, even at lower strains prior to the onset of discontinuous yielding.

RESULTS AND DISCUSSION

Copper

The tensile stress-strain curves of CDA 102 Cu are shown in Fig. 1. At low temperatures there was a decided increase in strain hardening but little increase in yield strength. The tensile yield strength increased from about 27 MPa at room temperature to 32 MPa at 4 K. Only very modest discontinuous yielding occurred at 4 K at large strains.

The data depicting the dependence of $\Delta\sigma$ on σ for Cu are summarized in Fig. 2. At 4 K there is a linear dependence of $\Delta\sigma$ on σ. At higher temperatures the dependence of $\Delta\sigma$ on σ is linear at low strain and slightly concave upwards at higher strain. The $\Delta\sigma$ decreases with decreasing temperature and extrapolates to zero at zero stress.

Alloy 310

Alloy 310, a stable austenitic stainless steel, has completely different stress-strain and strain-rate sensitivity traits. After deformation neither martensitic product (hexagonal close-packed ε; body-centered cubic α') was detected at any temperature. The stress-strain curves from 4 to 295 K are shown in Fig. 3. Alloy 310 shows considerably more strength in the early region of the stress-strain curve than relatively pure Cu. Indeed, the tensile yield strength increased from about 215 MPa at 295 K to 730 MPa at 4 K. The rate of strain-hardening of alloy 310 at high strain was rather independent of temperature, unlike that of Cu.

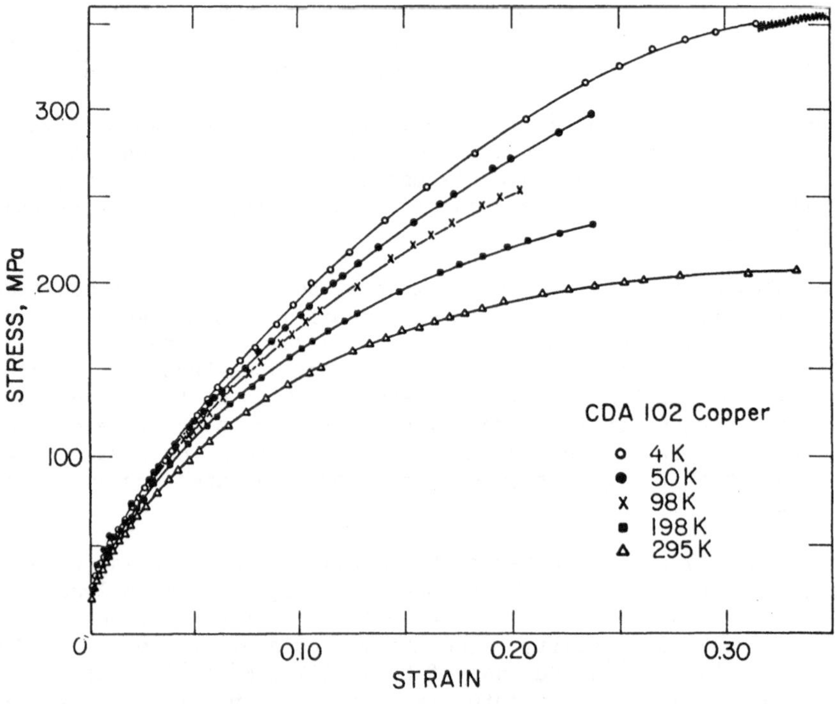

Fig. 1. The tensile stress-strain curves of CDA 102 Cu at various temperatures.

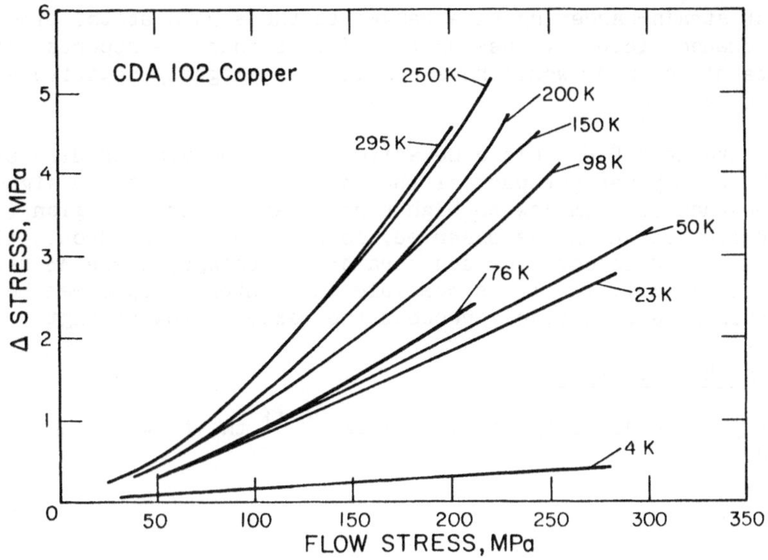

Fig. 2. General characterization for CDA 102 Cu of the effects of strain-rate change (2×10^{-5} to 2×10^{-3} s^{-1}) on the stress change as a function of flow stress at various test temperatures.

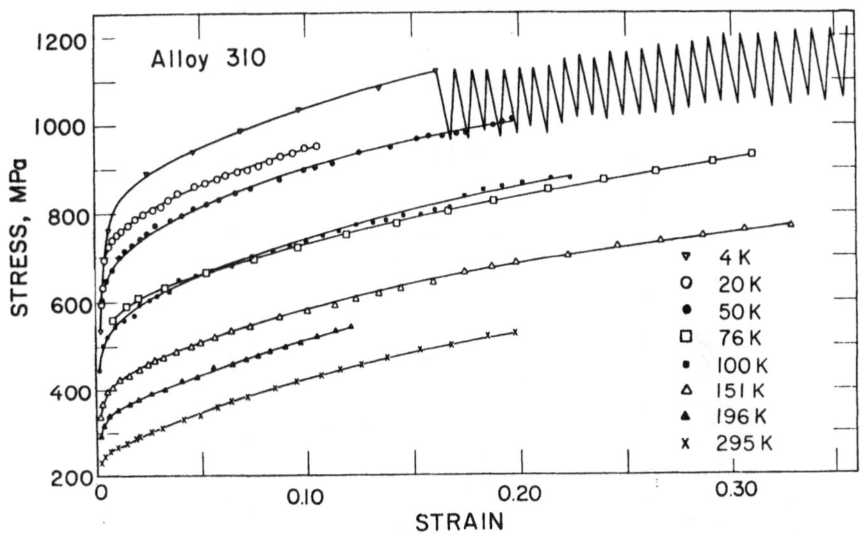

Fig. 3. The tensile stress-strain curves of alloy 310 at various temperatures.

Data on $\Delta\sigma/\Delta\ln\dot{\varepsilon}$ versus σ for alloy 310 are summarized in Fig. 4. Notice the differences first between Cu (Fig. 2) and alloy 310. The $\Delta\sigma$ values of alloy 310 were nearly constant, independent of stress or strain and, extrapolated to $\sigma = 0$, always are positive. The maximum of $\Delta\sigma$ as a function of temperature occurred in the range 76 to 100 K. From 76 to 4 K, there was a sharp decrease of strain-rate sensitivity.

At strains in excess of about 0.02 at temperatures near 76 K and strains of about 0.05 at temperatures near 295 K, very small yield points began to occur in alloy 310 following strain-rate changes. Again, the maximum $\Delta\sigma$ values on changing from a low to high strain rate were plotted. Extrapolation of equivalent strain-hardening rate ranges to the strain at the time of strain-rate change yielded values up to 15% less than the reported data. Use of these data of lower $\Delta\sigma$ would have resulted in slightly negative slopes of $\Delta\sigma$ versus σ or ε.

The $\Delta\sigma$ data at 4 K exhibit a negative slope; however, as discussed earlier, these data may reflect partial thermal instability at the higher strain rates. After changes from low to high strain rate, a brief region (~0.0003ε) of increased flow strength was observed, followed by a load drop. Although the maximum $\Delta\sigma$ was recorded for each strain-rate change, these $\Delta\sigma$ data should be regarded as minimal; for it is possible that internal specimen heating at the high strain rate effectively reduced the maximum flow strength.

<u>Stress-Strain Characteristics</u>

Following the general approach of Kocks,[6-11] the flow strength (σ) can be expressed as:

$$\sigma = \sigma_f + \sigma_d \qquad (1)$$

where σ_f is the frictional strength contribution dependent upon dislocation-solute interactions, and σ_d is the dislocation-dislocation interaction contribution dependent primarily on number and structure of dislocations. The frictional strength can be regarded as independent of strain; the yield strength is a good measure. The term σ_d is strain dependent; tensile strain-hardening characteristics are reflective of its contribution. Typically,

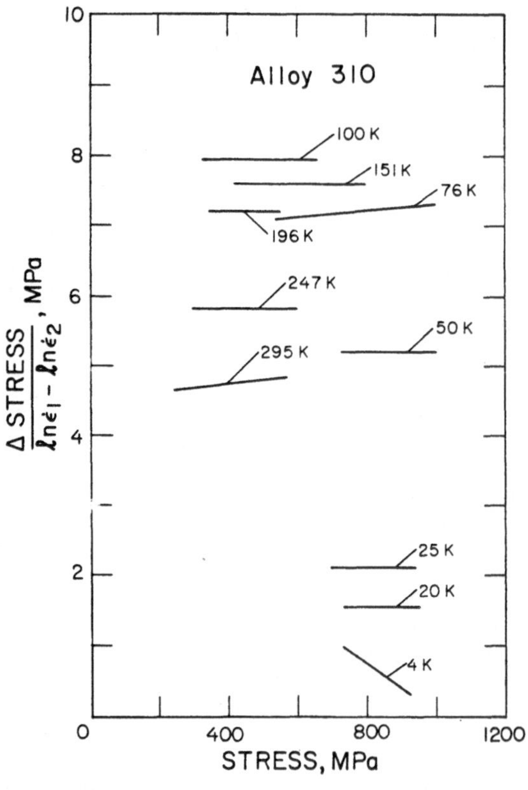

Fig. 4. General characterization for alloy 310 of the effects of strain-rate change (2×10^{-5} to 2×10^{-3} s^{-1}) on the stress change as a function of stress at various test temperatures.

$$\sigma_d = \alpha\mu b\sqrt{p} \tag{2}$$

where μ is the shear modulus; b, the Burgers vector; p, the active dislocation density; and α, a constant, which may be dependent on the solute content, c, and is slightly less than unity. In face-centered cubic alloys, the strain-hardening rate ($d\sigma/d\varepsilon$) is dependent on solute content. To account for this, Bloom et al.[9] suggest

$$\sigma_d = [1+k(c)]\ \alpha\mu b\sqrt{p} \tag{3}$$

with k(c) expressing the effect of solute. The strain-hardening rate is expressed as

$$\Theta = \frac{d\sigma}{d\varepsilon} = \frac{\alpha\mu b}{2}\ p^{-1/2}\ \frac{dp}{d\varepsilon} \tag{4}$$

The change of dislocation density with strain is considered to have two components: one concerns the "storage rate" and the other, the dynamic recovery rate. The rate of dislocation accumulation is athermal in nature; the recovery rate is obviously affected by thermal activation. Thus, with these considerations, the strain-hardening rate should have two major terms:

$$\Theta = d\sigma/d\varepsilon = \Theta_h(c) - \Theta_r(\dot{\varepsilon},T,\sigma,c) \tag{5}$$

where Θ_h is the athermal hardening rate and Θ_r is the dynamic recovery rate contributing negatively to Θ.

The strain-hardening curves derived from the stress-strain curves for Cu are shown in Fig. 5. They are plotted as $\sigma\ d\sigma/d\varepsilon$ versus σ, following Mecking and Kocks.[7] There is a linear region with positive slope at low stress followed by deviation to lower strain-hardening rates at higher stress. The linear region is constant at all temperatures and deviations from linearity occur at higher stresses with lower temperatures.

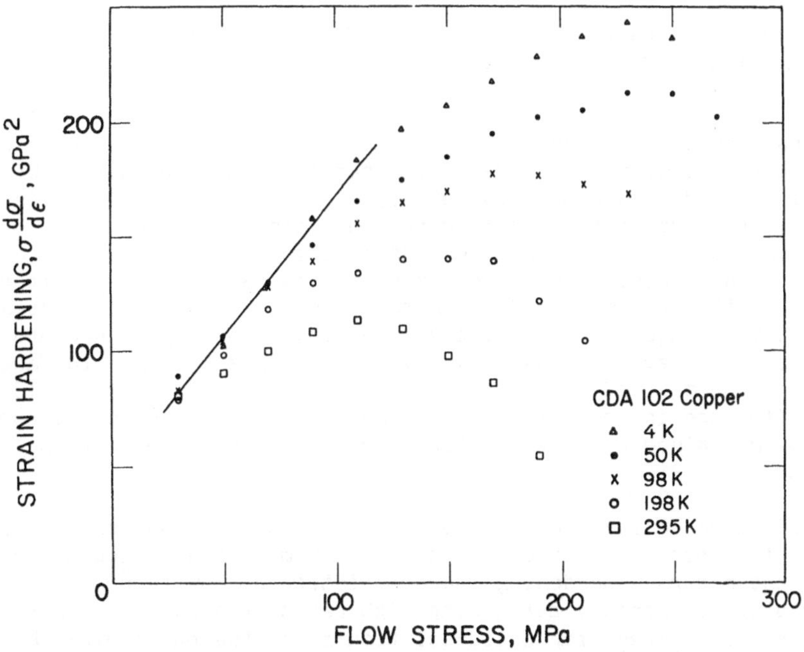

Fig. 5. Strain-hardening curves for CDA 102 Cu derived from stress-strain curves of Fig. 1.

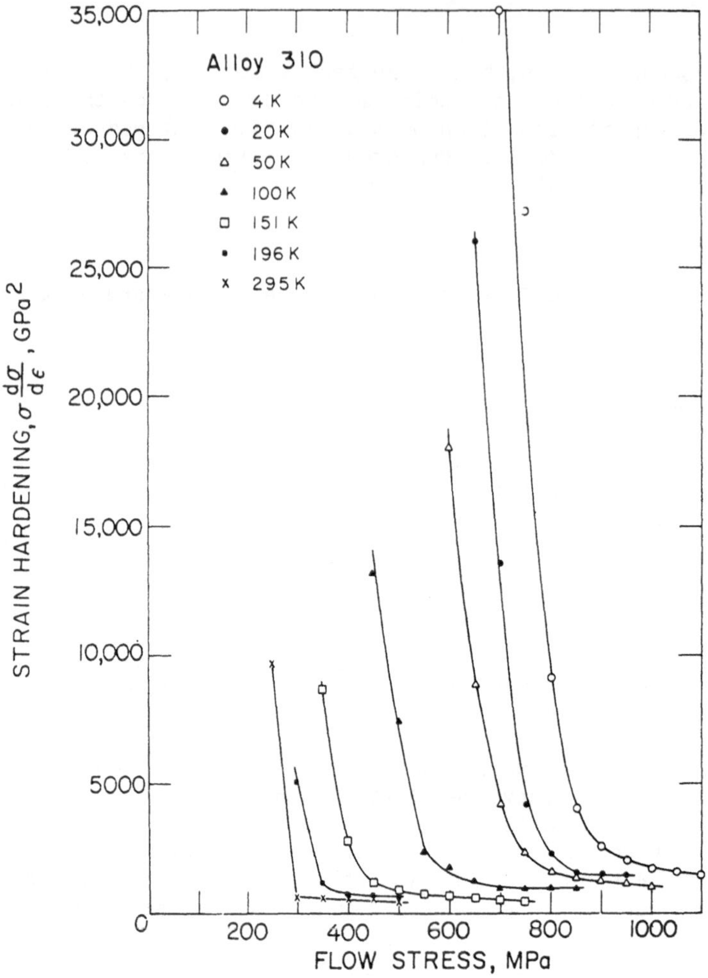

Fig. 6. Strain-hardening curves for alloy 310 derived from stress-strain curves of Fig. 3.

The strain-hardening curves (Fig. 6) at various low temperatures of alloy 310 indicate two distinct stages. The strain at the transition of the two stages increased from about 0.02 at 295 K to 0.04 at 4 K. This is, approximately, the restricted strain range that separates the early (difficult-to-measure) stage of strain-rate sensitivity from the higher strain region. The strain-hardening in the early strain range was very sensitive to temperature, similar to the yield strength. The strain-hardening rates in the higher strain range were relatively insensitive to temperature and assumed the same slope, which was negative, but only slightly. In fact, if the curves were normalized by dividing the ordinate by μ^2, there would be even less difference in their absolute magnitude. There was no evidence of dynamic recovery, since a recovery stage should be sensitive to temperature and nonlinear.

Thus, if Cu and alloy 310 are representative of pure and solid-solution-strengthened face-centered cubic metals, one may make the following conclusions about their strain-hardening characteristics at low temperatures: There is an early stage of rapidly decreasing high rates of strain-hardening in the region of the yield stress for solid-solution-strengthened alloys; this region is strongly temperature dependent. The next stage (first stage for pure metals) exhibits a linear dependence of $\sigma\theta$ on σ and, for Cu, is athermal. The slopes of this stage depend on solute content (solute effects on dislocation interactions); in pure metals the strain-hardening dependence extrapolates to zero (indicating only dislocation-dislocation interactions). The final stage is a region of dynamic recovery, not necessarily present in solid-solution-strengthened alloys. This stage typically exhibits a nonlinear dependence of $\sigma\theta$ on σ.

Strain-Rate Effects

Alloy 310, with much larger flow strength than Cu, exhibited strength changes of the order of 20 to 40 MPa with each change of strain rate; Cu had strength changes of less than 5 MPa. In both alloys, the stress change approached zero at very low temperatures. However, if the data were normalized by dividing the stress change by the yield strength at temperature, all stress changes would be found to be less than 15% of the flow strength. All percentages decreased to zero at low temperatures. There is a distinction between measurements of strain-rate effects from experiments that maintain a given strain rate and those that change strain rate during straining. During continuous straining, higher strain rates induce internal specimen heating more rapidly. This heating cannot be quickly removed at the high strain rate; thus the specimen probably remains permanently at a higher temperature during the test. This results in lower flow strengths and failure strengths. On the other hand, the strain-rate change experiments provide time at the low strain rate for the specimen to return to thermal equilibrium and, thus, to return to the higher flow strength at 4 K.

Mulford and Kocks[10] have shown that, provided the sum of the contributions of solution strengthening (σ_f) and work hardening (σ_d) are equivalent to the flow strength, the strain-rate sensitivity has two components:

$$\left(\frac{\partial\sigma}{\partial \ln\dot{\varepsilon}}\right)_T = \left(\frac{\partial\sigma_f}{\partial \ln\dot{\varepsilon}}\right)_T + \left(\frac{\partial\sigma_d}{\partial \ln\dot{\varepsilon}}\right)_T = \sigma_f\,(m_f - m_d) + \sigma\, m_d \tag{6}$$

where $m_f = \partial \ln\sigma_f/\partial \ln\dot{\varepsilon}$ and $m_d = \partial \ln\sigma_d/\partial \ln\dot{\varepsilon}$ at constant temperature. A plot of $\partial\sigma/\partial \ln\dot{\varepsilon}$ versus σ, called the Haasen plot, should have a slope of m_d at stresses somewhat above σ_y and extrapolate to $\sigma_f(m_f - m_d)$ at zero stress. If two contributions are present, they should be reflected in a change of slope at rather low stress: at low stress the contribution from dislocations must be small and, therefore, the contribution from solute strengthening dominates. The Cu data of Fig. 2 extrapolate to zero as flow stress approaches zero. This indicates the dominance of dislocation-dislocation interactions and the absence of solute strengthening ($\sigma_f = 0$). For alloy 310, the range of constant $\Delta\sigma/\Delta \ln\dot{\varepsilon}$ values corresponds to the second stage of $\sigma\Theta$ versus σ (Fig. 3), where $\sigma\Theta$ is relatively insensitive to σ.

Alloy 310 exhibited considerable solute strengthening, and extrapolation of the $\Delta\sigma/\Delta \ln\dot{\varepsilon}$ data to zero stress yielded large values of frictional stress (σ_f is positive), dependent on temperature. Unlike Cu, the slope of $\Delta\sigma/\Delta \ln\dot{\varepsilon}$ was practically independent of flow stress. Therefore, the solute-strengthening mechanism operative in this second stage (related to strain-hardening behavior) is very rate sensitive and temperature dependent. Both strain-hardening and strain-rate data confirm the absence of dynamic recovery in alloy 310 at the measured σ levels. There is a distinction between the rate-limiting process that controls the strain-rate response at low strain levels and the process that influences the response at higher strain, and it appears that both are influenced strongly by solutes. At low strains, perhaps the interaction of individual dislocations with discrete solutes is rate controlling and short range. At high strains, perhaps dislocation cell wall interaction with solutes or dislocation multiple solute interactions are rate controlling.

The temperature dependences of both strain-hardening and strain-rate sensitivity showed no unexpected transitions; therefore it appears that the same rate-controlling deformation mechanism that operates at room temperature remains operative at very low temperatures in both metals.

SUMMARY

The stress-strain curves, strain-hardening curves, and strain-rate sensitivities have been determined for Cu and alloy 310 austenitic stainless steel at temperatures from 4 to 295 K. The results suggest three regions of plastic deformation of polycrystalline face-centered cubic metals and alloys:

1. For solute-strengthened alloys at low strain, a region of rapidly decreasing strain-hardening rate and rapidly rising strain-rate sensitivity is strongly dependent on temperature and solute content.

2. At low strain for pure metals and at intermediate strains for solute-strengthened alloys, a strain-hardening region where the slope of $\sigma\Theta$ versus σ is constant and strongly dependent on solute content. In this region, the strain-rate sensitivity plotted against σ has a positive or zero slope, dependent also on solute content.

3. At high strain, a region of dynamic recovery in metals and some alloys. Here, the $\sigma\Theta$-versus-σ trend shows an ever-increasing negative slope, and the plot of strain-rate sensitivity versus stress shows an ever-increasing positive slope.

Finally, there is less strain-rate sensitivity at very low temperatures, although specimen adiabatic heating at higher strain rates does result in the lowering of the flow stress.

REFERENCES

1. C. J. Guntner and R. P. Reed, ASM Trans. Q. 55:399-419 (1962).
2. D. T. Read and R. P. Reed, Adv. Cryo. Eng.--Mater. 26:91-101 (1980).
3. H. Conrad, in "High Strength Materials," J. Wiley & Sons, New York (1965), pp. 436-505.
4. J. K. Tien and C.-T. Yen, Adv. Cryo. Eng.--Mater. 30:319-338 (1984).
5. K. A. Warren and R. P. Reed, "Tensile and Impact Properties of Selected Materials from 20 to 300 K." NBS Monograph 63, U.S. Government Printing Office, Washington, D.C. (1963).
6. U. F. Kocks, R. E. Cook, and R. A. Mulford, Acta Metall. 33:623-638 (1985).
7. H. Mecking and U. F. Kocks, Acta Metall. 29:1865-1875 (1981).
8. R. A. Mulford and U. F. Kocks, Acta Metall. 27:1125-1134 (1979).
9. T. A. Bloom, U. F. Kocks, and P. Nash, Acta Metall. 33:265-272 (1985).
10. R. A. Mulford and U. F. Kocks, Scripta Metall. 13:729-732 (1979).
11. R. A. Mulford, Acta Metall. 27:1115-1124 (1979).
12. R. P. Reed and R. P. Walsh, Low temperature strain hardening and strain rate effects of copper and aluminum, to be published in "Materials Studies for Magnetic Fusion Energy Applications at Low Temperatures--IX," NBSIR, National Bureau of Standards, Boulder, Colorado.

EFFECTS OF GAGE DIAMETER AND STRAIN RATE ON TENSILE

DEFORMATION BEHAVIOR OF 32Mn-7Cr STEEL AT 4 K

Kotobu Nagai, Tetsumi Yuri, Yuji Nakasone, Toshio Ogata, and Keisuke Ishikawa

Tsukuba Labs, National Research Institute for Metals
Sakura-mura, Niihari-gun, Ibaraki, Japan

ABSTRACT

The cryogenic deformation behavior under uniaxial tensile stress was investigated for 32Mn-7Cr high manganese steel using a diametral extensometer. The tensile deformation and serration behavior were observed for varying gage diameters and strain rates ($< 5 \times 10^{-4} s^{-1}$). A significant effect of gage diameter and strain rate on strengths and total elongation was not seen at 4 K. As the gage diameter became smaller or the strain rate lower, the strain to initiate serration decreased. The geometrical changes in the specimen during serrated yielding are also discussed in relation to the internal specimen heating.

INTRODUCTION

With advances in the applied technology of low temperature physics, the necessity of precise determination of the mechanical properties for cryogenic structural materials has increased. However, a reliable method has not been established, even for measurement of displacement or strain at liquid helium temperature (4 K). The occurrence of serration at cryogenic temperatures causes additional difficulties in taking an accurate measurement of displacement. Mechanical tests at cryogenic temperature are generally done in an insulated chamber, which creates many problems in measuring the displacement precisely. The adjustable diametral extensometer is available equipment that can be used to measure the diametral displacement of cylindrical or hourglass-type specimens for tensile or fatigue tests in the cryostat. In the present study, therefore, the cryogenic deformation behavior under uniaxial tensile stress was investigated for a high manganese steel using the diametral extensometer in order to know the basis of strain measurement at cryogenic temperatures.

EXPERIMENT

Diametral Extensometer

A strain gage type, immersible extensometer (MTS 632.19C-21) is provided for the cryogenic tests. However, the strain gage type extensometer has a zero drift and gage factor deviation with the change of operating temperature. In advance, the zero drift and the gage factor deviation were evaluated at 77 and 4 K. It is very difficult to obtain an accurate value of the diameter at cryogenic temperatures. As an alternative method, the relative change in diameter was measured as follows: First the output was calibrated at room temperature and the zero balance was set at the testing temperature. The output was corrected by the gage factor deviation and reduced to displacement data. This method is valid when the linearity of the output is guaranteed at the operating temperature. At 77 and 4 K, the linearity was within 0.25% of full scale and the gage factor was 104.0% that at room temperature. Thus, the extensometer gives high-quality measurements of displacement at cryogenic temperatures. The diameter measured with a micrometer at room temperature was adopted as the initial diameter at the testing temperature.

A personal computer was utilized for digital acquisition and storage of load, displacement, and temperature signals. The signals were amplified so that the full scale was ±10 V converted to 12-bit resolution data by A/D converters, and stored on a diskette. The fastest acquisition rate was 155 data per second per channel.

Material and Tensile Test

Hot-rolled and solution-treated 32Mn-7Cr steel plates were provided by Japan Steel Works. The 30-mm-thick plates were hot-rolled again to 15 mm at 1423 K. Cylindrical tensile test pieces were machined from the rerolled plates. The longitudinal direction was perpendicular to the rolling direction. The total length was 80 mm and the gage length, 30 mm. The gage diameter ranged between 3.5 and 6.5 mm. The tensile test machine was an Instron. The crosshead speed was chosen between 0.03 and 1 mm/min so that the maximum strain rate was less than $5 \times 10^{-4} s^{-1}$, since higher strain rates significantly affect the deformation behavior owing to specimen heating.[1,2]

RESULTS AND DISCUSSION

Stress-Strain Curve

Figure 1 represents the stress-diametral strain curve for a 4.5ϕ specimen at 4 K. After uniform deformation of a few percent elongation, serration yielding begins and occurs repeatedly until rupture. The diametral strain changes irregularly in the serrated region. The stress and the diametral strain are described as a function of time in Fig. 2. The upper curve of Fig. 2 is similar to the stress-axial strain curve usually observed in a constant crosshead speed test. In the lower curve, there are some steps that correspond to an abrupt shrinkage of the point at which the extensometer is attached. The characteristics of the stress-strain curve are discussed later.

Tensile Properties

Yield strengths (0.2% offset stress) are listed in Table 1. Ultimate tensile strength (UTS), yield strength, and total elongation are shown as function of diameter for the specimen tested at a crosshead speed of 1 mm/min in Fig. 3. There is no large scatter in properties over the range of diameter or strain rate investigated. Only the 3.5ϕ specimen has a little higher strength among the specimens tested at the same strain rate.

The strain to initiate serrated yielding was highly dependent on strain rate and diameter (Fig. 4). As the strain rate decreases, the strain to initiate serrated yielding increases. At higher strain rates, the elongation at the start of serrated yielding is more dependent on the diameter. Although the strain to initiate serrated yielding increases with a decrease in diameter when the diameter is larger than 4 mm, there is a significant

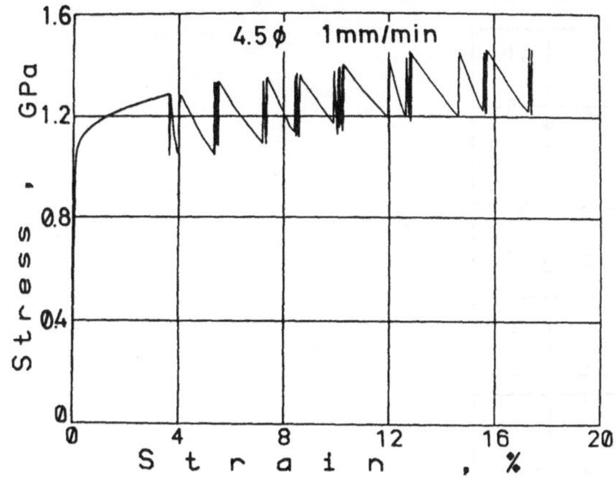

Fig. 1. Stress-diametral strain curve for a 4.5-mm diameter specimen tested at a crosshead speed of 1 mm/min.

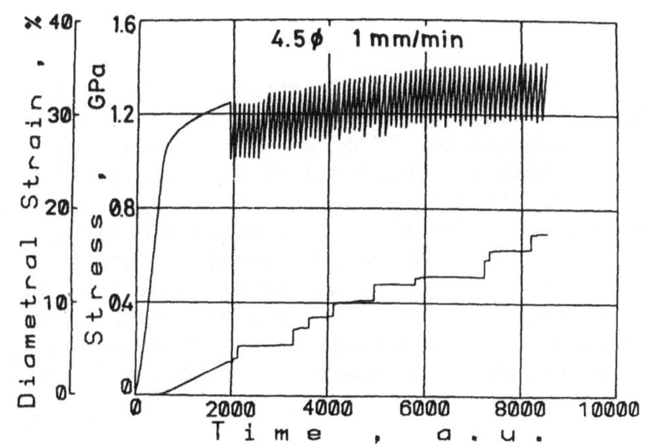

Fig. 2. Changes of stress (upper) and diametral strain (lower) as a function of time for a 4.5-mm diameter specimen tested at a crosshead speed of 1 mm/min.

Table 1. Yield Strength* (0.2% offset stress) at 4 K for Specimens of Various Diameters Tested at Four Crosshead Speeds

Diameter (mm)	Crosshead Speed (mm/min) 0.03	0.1	0.3	1.0
3.5	--	1151	1171	1139
4.0	--	1119	1115	1101
4.5	1109	1128	1152	1101
5.0	--	--	--	1096
5.5	--	--	1126	1098
6.0	--	1109	1119	1087
6.5	--	--	--	1128

*in MPa

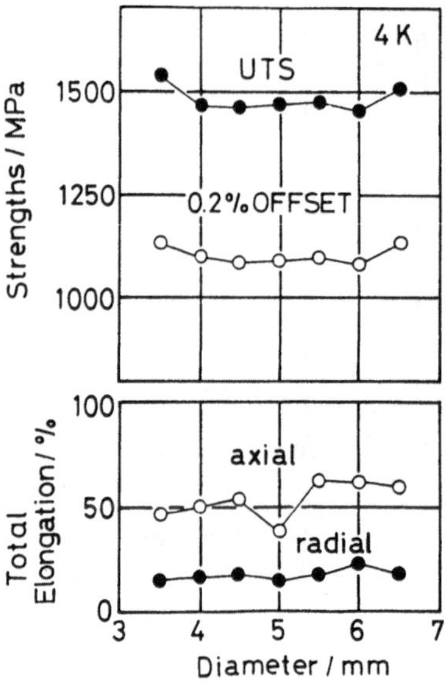

Fig. 3. Effects of diameter on strength and total elongation. Crosshead speed was 1 mm/min.

drop in the elongation at the start of serrated yielding for the 3.5-mm-diameter specimen. The asterisked data in Fig. 4 are those of the 3.4-mm-diameter specimen, which was made from a 4-mm-diameter specimen by removing the surface layer chemically. These values are nearly equal to that of the 4-mm-diameter specimen. It means that the hardening due to mechanical machining may promote the initiation of serrated yielding. Then, 5-mm-diameter specimens prestrained at room temperature were prepared and the strain to initiate serrated yielding was obtained at 4 K (Table 2). The elongation at the initiation of serrated yielding decreases obviously with an increase in prestrain. Accordingly, it can be concluded that the strain to initiate serrated yielding increases and reaches a constant value as the diameter decreases.

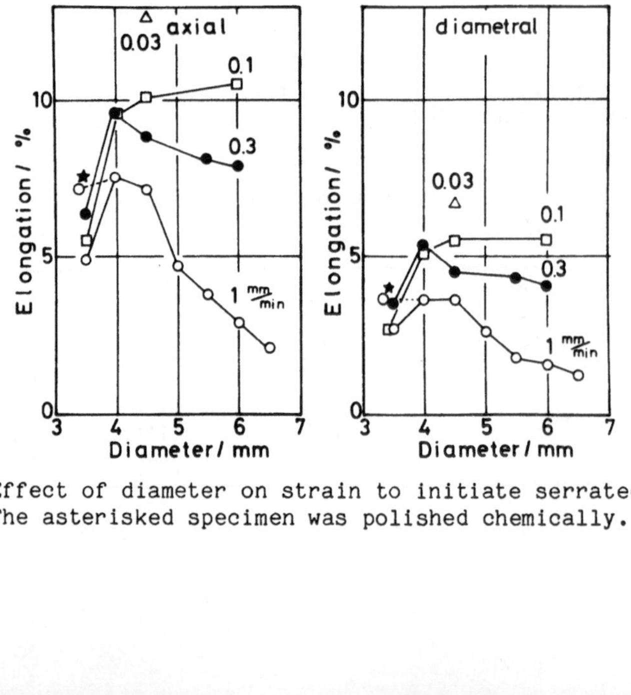

Fig. 4. Effect of diameter on strain to initiate serrated yielding. The asterisked specimen was polished chemically.

Table 2. Effect of Prestrain at Room Temperature on Serration Beginning Elongation for a 5-mm-diameter Specimen

Prestrain at Room Temperature (%)	Serration Beginning Elongation (%)
--	4.70
0.37	3.27
0.87	3.83
2.00	2.61

NOTE: Crosshead speed was 1 mm/min.

Size of Deformation Zone By Serration

There are some regularities in the serration phenomena. Table 3 shows the maximum deflection during one serration for various diameters and the probability that one serration occurred including the measuring point within the deformation zone of one serration. The maximum deflection is about 2% in diameter and the probability about 16%, independent of the diameter. Provided that each serration is accompanied by the same scale deformation zone, the deformation zone has an axial width of about 5 mm, i.e., 30 mm (gage length) multiplied by 0.16 (probability). Figure 5 represents the variation in diameter measured by an optical microscope over the whole range of gage length after the specimen was deformed until one or five serrations occurred. In Fig. 6 one serration is enlarged. These results demonstrate that every serration has the nearly same size, namely an axial length of 5 to 6 mm and a diametral shrinkage of about 2%. In certain cases, the serrations occurred at different points as if each serration selected a softer part of the specimen. As represented in Fig. 2, there is a plateau in the stress-time curve during several early serrations. This plateau demonstrates that each serration takes place in a softer part; finally serrations cover the entire gage length.

Specimen Heating During Serration

A marked temperature increase during each serration was reported.[1] Figure 6 shows the temperature increase calculated on the assumption that all plastic work done is converted into heating the deformation zone caused by serrated yielding. The maximum temperature is about 70 K at the center of the deformation zone.

Table 3. Characteristics of Serrated Yield

Diameter (mm)	Number of Serrations	Number of Steps	Maximum Deflection (%)	Probability (%)
3.5	73	10	2.3	14
4.0	68	9	1.9	13
4.5	63	10	1.8	16
5.0	44	8	2.0	18
5.5	66	12	2.1	18
6.0	62	12	2.1	19
6.5	55	9	2.5	16

NOTE: Crosshead speed was 1 mm/min.

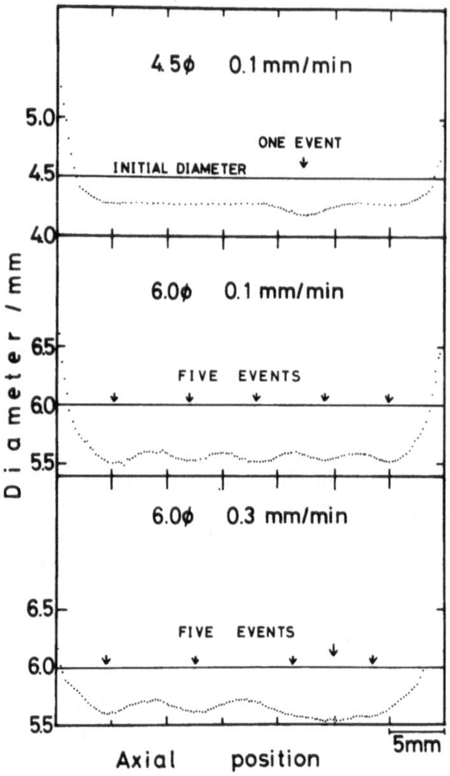

Fig. 5. Diameter distribution of specimen deformed until one or five serrations occurred. Diameter was measured by an optical microscope.

The internal specimen temperature can be measured if a center-hole specimen is used.[1] However, the manganese steel could not be drilled. Therefore, a temperature increase was observed in 310S stainless steel, since the steel also has uniform plastic deformation at an earlier stage of the stress-strain curve and a comparatively large load drop during serrated yielding. The 2-mmφ center-hole specimen, whose gage diameter was 6.25 mm, was utilized for the three-channel (load, diametral strain, and temperature) measurement. A thermocouple was inserted into the hole and attached near the center of the gage length and sealed to liquid helium by stopping up the hole with varnish. Figure 7 shows examples of the three-channel charts, and Fig. 8 is shown in the magnified scale of time. The specimen

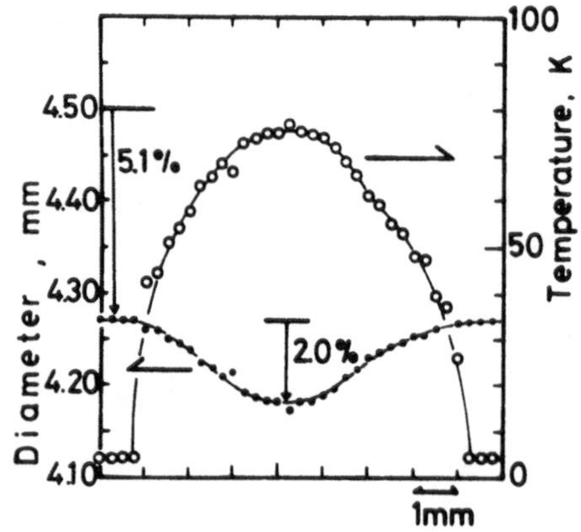

*specific heat, c, J/kg·K

$$c = 0.3837 \times T^{1.115} \quad (T \leq 20\ K)$$

$$= 1.325 \times 10^{-23} \times T^{2.906}$$

$$(20\ K < T < 200\ K)$$

SUS316L (from NBS data)

Fig. 6. Distribution of diameter measured and temperature calculated. Calculation was performed on the assumption that plastic work done was converted into heating the deformed portion.

temperature is at the base level, i.e., 4 K during the elastic deformation and increases slightly by 1 to 2 K when the specimen is deformed plastically. Serrated yielding occurs suddenly accompanied by an abrupt load drop and sharp diametral shrinkage and is followed by a large temperature increase. The total time for the load drop or the diametral shrinkage is about 100 ms and that for the increase and decrease of specimen temperature is nearly 1 s. The time of peak temperature does not coincide with the time when the load is minimum, probably because the temperature monitoring point is not always equal to the very center of the deformation zone of serration and there is a delay due to thermal conduction. In Fig. 9 the maximum temperatures are plotted on a logarithmic scale as a function of time from the start of load drop to the start of temperature increase. The plotted data are quite linear. This indicates that the maximum temperature at the center of the deformation zone of each serration is constant and estimated to be approximately 50 K. The similarity of the maximum temperature corresponds with the constant size of the deformation zone of each serration.

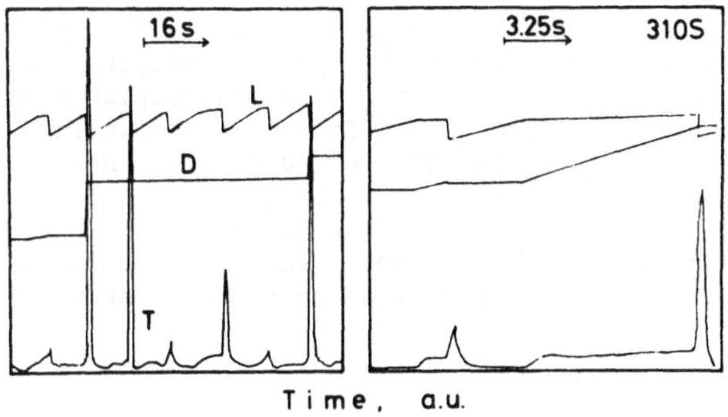

Fig. 7. Examples of three-channel charts of load, diametral strain, and temperature. The material tested was 310S.

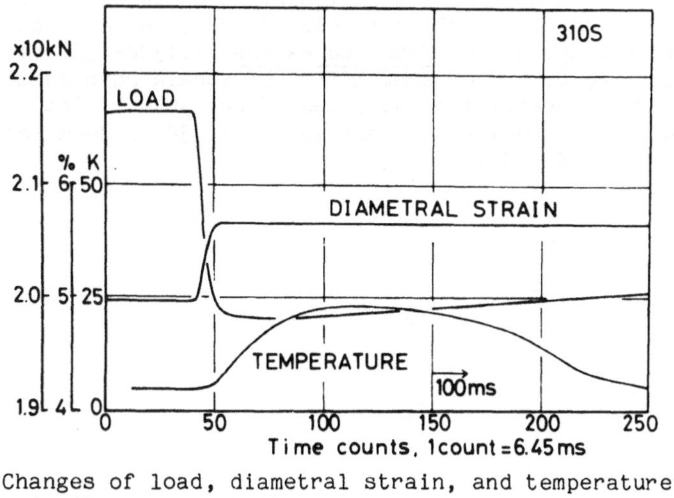

Fig. 8. Changes of load, diametral strain, and temperature before and after one serration.

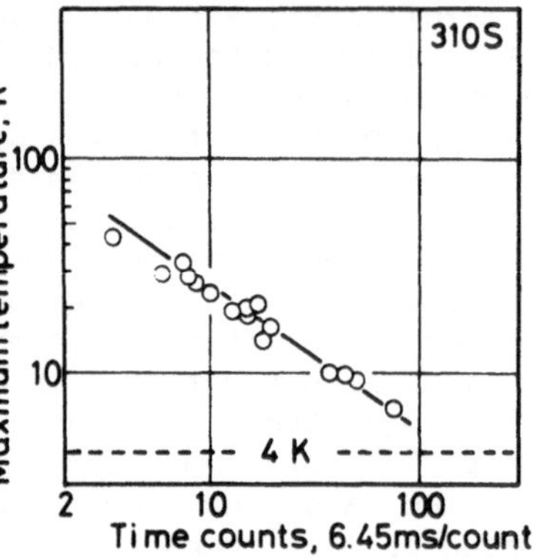

Fig. 9. Maximum temperature measured as a function of time taken from the start of load drop to the start of temperature increase. Time is described in terms of time count, and one count is 6.45 ms.

SUMMARY

The extensometer using a strain gage is available equipment that can be used to measure the displacement at cryogenic temperature. The linearity of the output is very good, and the relative displacement is obtained by a linear correction of the output. There is no significant effect of strain rate and gage diameter on tensile properties. However, the strain to initiate serrated yielding is dependent on strain rate and gage diameter. As strain rate becomes lower or gage diameter smaller, the strain to initiate serrated yielding usually increases. Each serration has nearly the same deformation zone size and is accompanied by about the same amount of internal heating.

ACKNOWLEDGMENTS

The authors thank Mr. R. Miura, JSW for offering the testing material. This study was performed through Special Coordination of Science and Technology Agency of the Japanese Government.

REFERENCES

1. T. Ogata, K. Ishikawa,and K. Nagai, Effects of strain rate on the tensile behavior of stainless steels, copper, and an aluminum alloy at cryogenic temperature, Tetsu-to-Hagane 71:1390-1397 (1985).
2. W. G. Dobson and D. L. Johnson, Effect of strain rate on measured mechanical properties of stainless steel at 4 K, in: "Advances in Cryogenic Engineering - Materials," vol. 30, Plenum Press, New York (1984), pp. 185-192.

NEAR-THRESHOLD FATIGUE CRACK GROWTH BEHAVIOR

OF AISI 316 STAINLESS STEEL†

R. L. Tobler

Fracture and Deformation Division
Center for Materials Science and Engineering
National Bureau of Standards
Boulder, Colorado

ABSTRACT

The near-threshold fatigue behavior of an AISI 316 alloy was characterized using a newly developed, fully automatic fatigue test apparatus. Significant differences in the near-threshold behavior at temperatures of 295 and 4 K are observed. At 295 K, where the operationally defined threshold at 10^{-10} m/cycle is insensitive to stress ratio and strongly affected by crack closure, the effective threshold stress intensity factor $(\Delta K_{Th})_{eff}$ is about 4.65 $MPa{\cdot}m^{1/2}$ at R = 0.1 and R = 0.3. At 4 K, the threshold is higher, crack closure is less pronounced, and there is a stress ratio dependency: $(\Delta K_{Th})_{eff}$ is 5.1 $MPa{\cdot}m^{1/2}$ at R = 0.3 and 6.1 $MPa{\cdot}m^{1/2}$ at R = 0.1. There is also a significant difference in the form of the da/dN-versus-ΔK curves on log-log coordinates: at 4 K the curve has the expected sigmoidal shape, but at 295 K the trend is linear over the region of da/dN from 10^{-6} to 10^{-10} m/cycle. Other results suggest that the near-threshold measurements of a 6.4-mm-thick specimen of this alloy are insensitive to cyclic test frequency below 40 Hz.

INTRODUCTION

Near-threshold fatigue crack growth rates (da/dN-versus-ΔK curves) and threshold stress intensity factors (ΔK_{Th}) are relevant mechanical parameters for superconducting magnet design since they lead to the prediction of safe operating conditions below which failures from fatigue crack growth can not occur. Owing to measurement difficulties and costs, however, the existing data base at 4 K is still quite limited. Liaw and coworkers[1-3] designed a fully automatic fatigue test apparatus and reported threshold data for precipitation hardenable Fe-29Ni-14Cr-1.2Mo austenitic stainless steel and its welds at 297, 77, and 4 K. Tobler and Cheng designed a similar apparatus and presented data for AISI 304L, 304LN, and 316 stainless steels at 4 K.[4] The present paper considers the AISI 316 alloy in greater detail, contrasting the behaviors at 295 and 4 K and exploring the effects of some mechanical test variables.

---

† Contribution of NBS, not subject to copyright in the U.S.

MATERIAL

The tested material was a 50.8-mm-thick plate of AISI 316 austenitic stainless steel in the annealed condition. The chemical composition in wt.% is Fe-17.25Cr-13.48Ni-0.057C-1.86Mn-2.34Mo-0.024P-0.019S-0.58Si-0.030N. Tensile, fracture toughness, and midrange fatigue crack growth data were reported previously.[5] Selected tensile properties are listed in Table 1, along with Young's moduli for a similar grade of steel.[6]

EXPERIMENTAL PROCEDURE

The threshold fatigue test apparatus used in this study includes a 100-kN servohydraulic test machine and cryostat with minicomputer, programmable arbitrary waveform generator, programmable digital oscilloscope, and liquid helium refill system. The procedure is fully automated, allowing continuous data collection and load adjustments to be performed in the absence of a machine operator. Tests were conducted at 40 Hz and lasted about 4 days with continuous 24-h operation. Typically, 13 million load cycles are needed to complete one threshold measurement and to accumulate rate measurements ranging from 10^{-8} to 10^{-10} m/cycle.

Table 1. Mechanical properties and grain size of the test material.

Test Temperature (K)	Yield Strength, σ_y (MPa)	Ultimate Strength, σ_u (MPa)	Elongation (%)	Red. of Area (%)	Young's Modulus, E (GPa)	ASTM Grain Size No.
295	228	576	56	73	195	5.0
4	711	1301	48	57	208	--

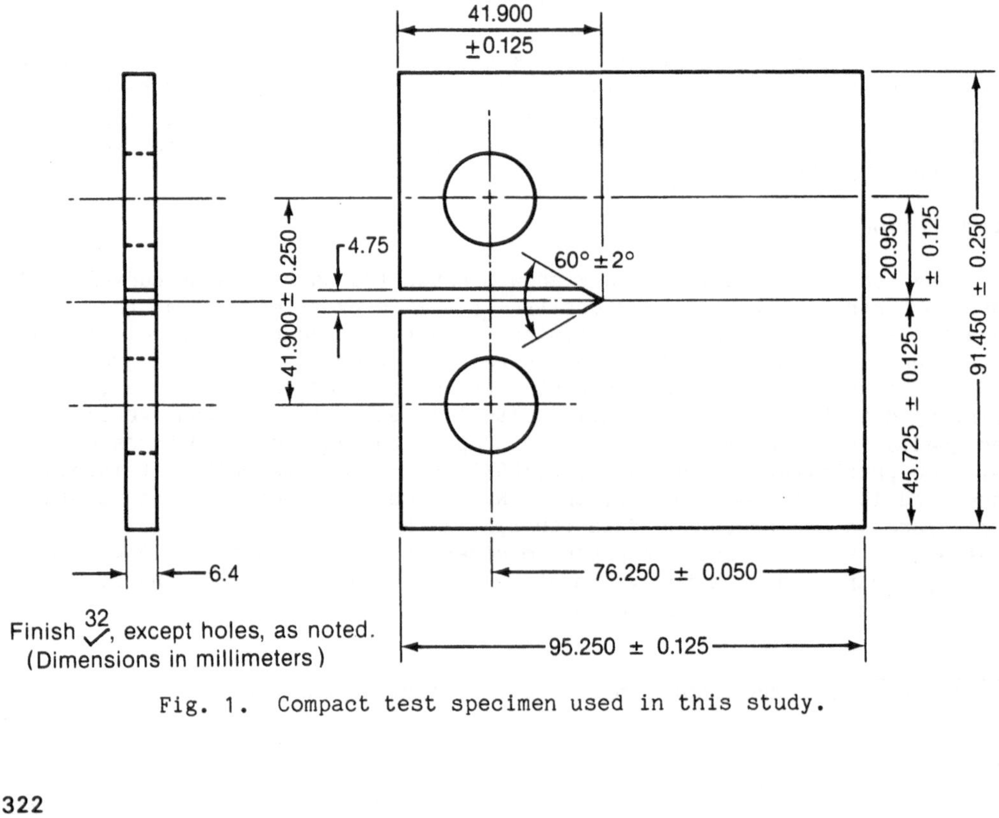

Fig. 1. Compact test specimen used in this study.

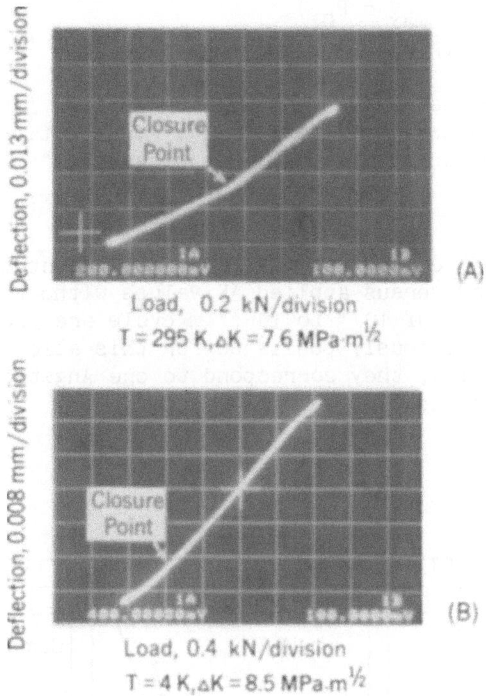

Fig. 2. Deflection-versus-load curves for AISI 316 test specimens at the termination of threshold fatigue tests: (A) T = 295 K; (B) T = 4 K.

The 6.4-mm-thick compact specimen used in this study is shown in Fig. 1. A ring gage was attached to the specimen edge, and the load cell and gage outputs were interfaced to the computer. The deflection-versus-load curves provided the basis for crack growth measurements; they were continuously displayed on the oscilloscope. Representative oscilloscope traces for tests at 295 and 4 K are shown in Fig. 2.

As crack growth proceeded, ΔK was systematically decreased according to equation (1):

$$\Delta K = \Delta K_o \exp[C(a - a_o)] \qquad (1)$$

Here ΔK_o and a_o are the initial stress intensity factor and crack length, respectively, a is the current crack length, and C is a constant chosen to be -0.08 mm^{-1}.[4] The da/dN data were plotted versus the nominal ΔK, and ΔK_{Th} was operationally defined as the ΔK value corresponding to a growth rate of 10^{-10} m/cycle.[7]

Following the determination of ΔK_{Th}, a correction for crack closure was applied. Crack closure causes the cracks to remain closed for a portion of the load cycle; this portion of the load is, therefore, ineffective in contributing to the driving force for crack growth.[8] In Fig. 2, the lower segments of the bilinear curves are associated with closure and must be disregarded in the calculation of the effective ΔK. Correction is made as follows:

$$(\Delta K_{Th})_{eff} = (\Delta K)_{Th} \frac{(P_{max} - P_{Cl})}{(P_{max} - P_{min})} \tag{2}$$

Here P_{max} and P_{min} are the maximum and minimum applied fatigue loads, and P_{Cl} is the closure load as identified in Fig. 2. Additional details of the test apparatus and procedures are published elsewhere.[4]

RESULTS

The da/dN-versus-ΔK curves for AISI 316 are presented in Fig. 3. The da/dN values are plotted versus applied ΔK values without correction for crack closure. The rates of 10^{-8} to 10^{-10} m/cycle are two orders of magnitude lower than rates previously published for this alloy, and at the threshold level (10^{-10} m/cycle), they correspond to one Angstrom per load cycle or less than one lattice spacing per load cycle.

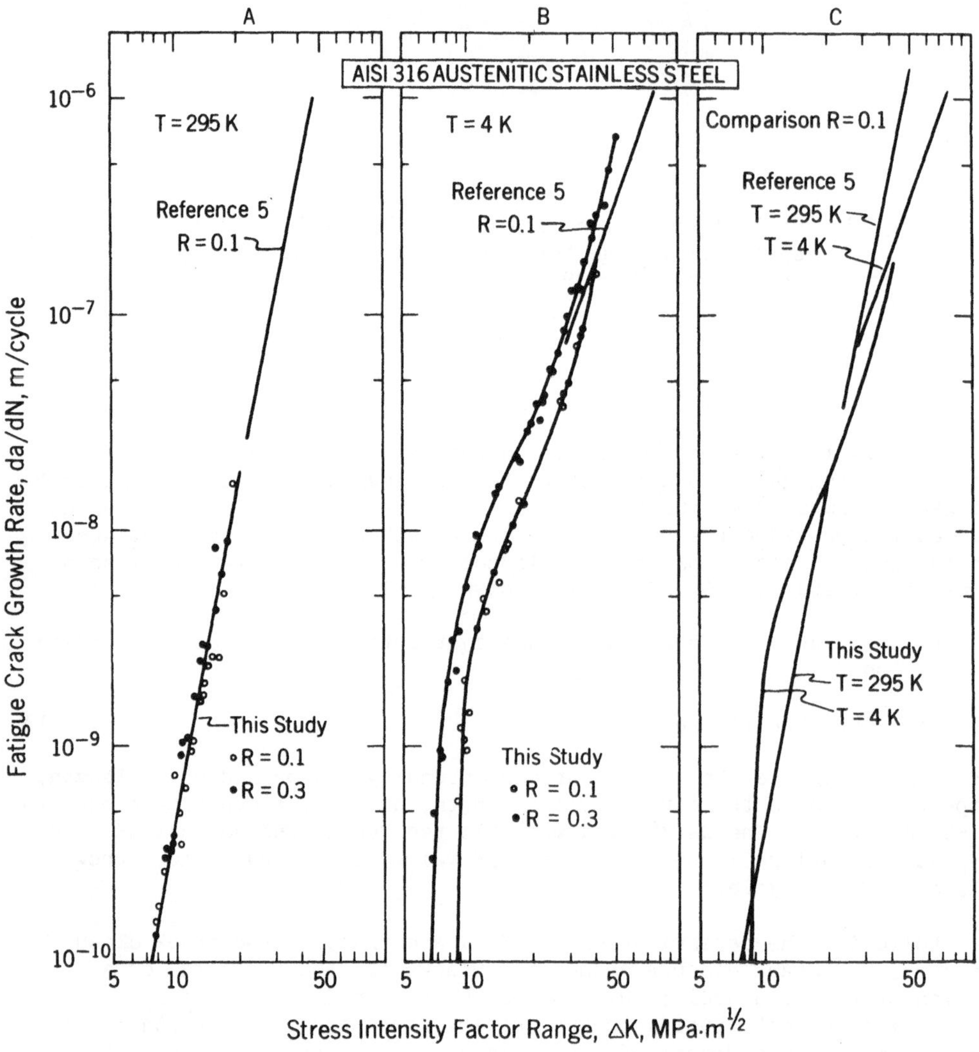

Fig. 3. Fatigue crack growth rate data for AISI 316 stainless steel: (A) at 295 K, (B) at 4 K, and (C) comparison at R = 0.1.

Table 2. Nominal and effective threshold stress intensity factors for AISI 316 stainless steel.

Test Temperature (K)	Stress Ratio, R	ΔK_{Th} (MPa•m$^{1/2}$)	$(\Delta K_{Th})_{eff}$ (MPa•m$^{1/2}$)
295	0.1	7.6	4.6
	0.3	7.6	4.7
4	0.1	8.5	6.1
	0.3	6.4	5.1

At 295 K (Fig. 3A), the rates at R = 0.1 and R = 0.3 are nearly indistinguishable, showing no effect of stress ratio. The data trend is linear on log-log coordinates, and the rates do not exhibit a knee or approach an asymptote at the near-threshold level. Read and Reed[5] tested the same heat of AISI 316 at ΔK from 25 to 50 MPa•m$^{1/2}$. Their data, representing 25.4-mm-thick specimens at 20 Hz, are in excellent agreement with and complement the present results.

The results at 4 K (Fig. 3B) differ from those at 295 K in that there is a significant effect of R ratio. The rates at R = 0.3 are higher than at R = 0.1, and the threshold stress intensity factor is larger at R = 0.1. At this temperature, the da/dN curves do exhibit the expected sigmoidal shape, and there is approximate agreement with the results for 25.4-mm-thick specimens.[5]

The effect of test temperature at R = 0.1 is also shown in Fig. 3C. The linear and sigmoidal trends at 295 and 4 K are distinct and cross twice. The ΔK_{Th} value at 4 K is slightly higher than that at 295 K, and when crack closure corrections are applied, the temperature effect is more pronounced. This can be seen by comparison of the $(\Delta K_{Th})_{eff}$ values in Table 2. It is evident from these data that reducing the temperature from 295 to 4 K increases the threshold fatigue resistance.

Figure 4 presents the effect of frequency variations at 4 K on the crack growth rate at ΔK of 20 and 42 MPa•m$^{1/2}$. For these tests the fatigue apparatus was programmed for constant ΔK, and replicate crack growth rate measurements were performed. Frequency was varied by two orders of magnitude from 0.4 to 40 Hz. When the range of experimental error and some effects of crack closure and variability are taken into account, these data show that there is no measurable effect of test frequency on the crack growth rates.

DISCUSSION

At 295 K, the fatigue crack growth rates for AISI 316 appear to follow a linear trend for at least four orders of magnitude in da/dN, and no asymptote is approached at the threshold level of 10^{-10} m/cycle. This has several implications. First, it implies that no true threshold is exhibited in this range. Hence, the operational definition of ΔK_{Th} at 10^{-10} m/cycle is inadequate for this alloy/temperature combination, and a new definition must be sought. Second, it points to the necessity of performing additional measurements at still lower growth rates. Note that an average growth rate of 10^{-10} m/cycle corresponds to 1 Å/cycle, while the lattice spacing for austenite is about 3.6 Å. Rates of crack growth less than a lattice parameter per cycle are possible if the crack growth occurs nonuniformly, that is, in spurts or in highly localized areas along the crack front.

In these tests, crack closure became more significant as the cracks propagated and the loads were reduced. As shown in Fig. 2, the crack closure was much more significant at 295 K than at 4 K. The plasticity-induced closure mechanism can be used to explain this observation. According to this mechanism, closure is attributed to residual stresses left in the wake of the advancing crack tip plastic zone.[8] This closure mechanism would contribute greater closure effects at 295 K where the yield strength is lower. The fact that the threshold ΔK values at $R = 0.1$ and $R = 0.3$ are larger at 4 K than at 295 K can be explained as an intrinsic effect of the increased lattice resistance to dislocation movement at cryogenic temperatures.

Frequency effects on da/dN have been observed in AISI 316 at elevated test temperatures because of creep interactions with fatigue[9]. At low temperatures creep is less consequential and cyclic frequency is expected to have little influence on fatigue crack growth rates. At $T = 4$ K, however, an exception is conceivable, since local heating can occur in the crack-tip plastic zone due to plastic strain energy dissipation. The amount of heat generated must decrease at lower cycle frequencies and at lower ΔK. Since Fig. 4 shows no measurable variation in rates at $\Delta K = 20$ or 42 $MPa \cdot m^{1/2}$ between 0.4 and 40 Hz, and since the plastic zone size is significantly smaller at lower ΔK, we tentatively conclude that the ΔK_{Th} and $(\Delta K_{Th})_{eff}$ parameters for AISI 316 are not frequency dependent below 40 Hz.

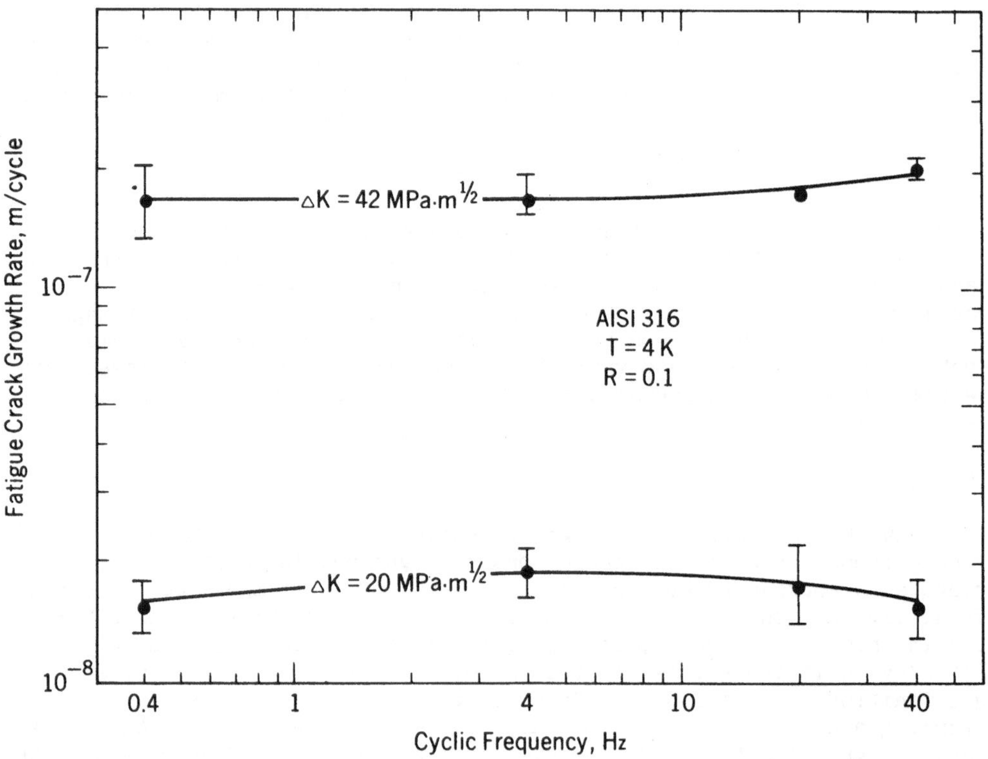

Fig. 4. Effect of cyclic test frequency on 4-K fatigue crack growth rates at two intermediate ΔK values.

SUMMARY AND CONCLUSIONS

This study of the near-threshold fatigue crack growth rate behavior of annealed AISI 316 stainless steel at 295 and 4 K gives the following results:

1. At 295 K, the da/dN-versus-ΔK curves on log-log coordinates follow a linear trend for four orders of magnitude. The fact that the data do not approach an asymptote at 10^{-10} m/cycle leads to the conclusion that still lower rate measurements are necessary to reach a true fatigue threshold, and that the conventional operational definition of ΔK_{Th} is not satisfactory for this material and temperature combination.

2. At 4 K, the da/dN-versus-ΔK curve is sigmoidal, and the operational definition of ΔK_{Th} is satisfactory at this temperature.

3. The near-threshold behavior of AISI 316 is stress ratio dependent at 4 K, but not at 295 K, for R ratios between 0.1 and 0.3.

4. The threshold stress intensity factor for this material is temperature dependent, increasing as temperature is reduced from 295 to 4 K.

ACKNOWLEDGMENTS

This project was supported by the Office of Fusion Energy (DOE), Dr. V. Der, project monitor.

REFERENCES

1. P. K. Liaw, W. A. Logsdon, and M. H. Attaar, in: "Austenitic Steels at Low Temperatures," R. P. Reed and T. Horiuchi, Eds., Plenum Press, New York (1983), pp. 171-185.
2. R. S. Williams, P. K. Liaw, M. G. Peck, and T. R. Leax, Eng. Fract. Mech. 18:953-964 (1983).
3. P. K. Liaw, W. A. Logsdon, and M. H. Attaar, in: "Fatigue at Low Temperatures," ASTM STP 859, R. I. Stephens, Ed., American Society for Testing and Materials, Philadelphia, Pennsylvania (1985), pp. 173-190.
4. R. L. Tobler and Y. W. Cheng, Automatic near-threshold fatigue crack growth rate measurements at liquid helium temperature, Int. J. Fat. 7:191-197 (1985).
5. D. T. Read and R. P. Reed, in: "Materials Studies for Magnetic Fusion Energy Applications at Low Temperatures--II," NBSIR 79-1609, National Bureau of Standards, Boulder, Colorado (1979), pp. 81-122.
6. H. M. Ledbetter, J. Appl. Phys. 52:1587-1589 (1981).
7. R. J. Bucci, in: "Fatigue Crack Growth Measurement and Data Analysis," ASTM STP 738, S. J. Hudak, Jr. and R. J. Bucci, Eds., American Society for Testing and Materials, Philadelphia, Pennsylvania (1981), pp. 5-28.
8. S. Suresh and R. O. Ritchie, in: "Fatigue Crack Growth Thresholds Concepts," D. Davidson and S. Suresh, Eds., Met. Soc. AIME (1984), pp. 227-262.
9. L. A. James, At. Energy Rev. 14:37-85 (1976).

FATIGUE TESTING AT 4 K WITH A HELIUM RECONDENSATION SYSTEM

K. Nagai, T. Ogata, T. Yuri, and K. Ishikawa

National Research Institute for Metals
Tsukuba Laboratories
Sakura-mura, Niihari-gun, Ibaraki, Japan

ABSTRACT

A liquid helium temperature fatigue testing system was developed at the National Research Institute for Metals, Tsukuba Laboratories. The system is equipped with a recondenser installed in the test machine cryostat. Helium mist is transferred to the recondenser from the refrigerator, and the evaporated helium gas in the cryostat is recondensed into liquid helium. Thus, the liquid helium level in the cryostat is kept constant without the addition of liquid helium during testing. Continuous operation of about 500 h has been achieved with this fatigue testing system.

INTRODUCTION

Fatigue testing systems for liquid helium temperature were developed at the National Research Institute for Metals, Tsukuba Laboratories and have been used to evaluate the fatigue life of structural materials at cryogenic temperatures. The progress in superconducting technology increasingly demands highly reliable structural materials. However, few data about fatigue at liquid helium temperature have been accumulated, except for those obtained by NBS (the National Bureau of Standards in the U.S.A.).[1] This is mainly ascribed to difficulties in conducting fatigue tests at liquid helium temperature. For example, fatigue tests take a long time because about 10^6 fatigue cycles are needed to estimate the fatigue life of certain materials. Moreover, fatigue tests cannot be accelerated because the specimen temperature is raised considerably at a high frequencies at liquid helium temperature.

There are two cooling methods for fatigue tests at liquid helium temperature. One requires replenishment of the liquid helium evaporated during testing; the other reliquefies the evaporated helium gas. The first method is the simplest, but it consumes a large amount of expensive liquid helium during long-term tests. The second method reduces the consumption of liquid helium and eliminates its troublesome replenishment, but then the contamination by moisture or air components is problematic. It is also not easy to carry out long-term operation of the liquefier accompanied with the purifier. Consequently, we adopted the recondensation closed-loop system for the long-term fatigue tests and have achieved a continuous operation of

Table 1. Specifications of Apparatus

Refrigeration system;	
Refrigeration cycle	Claude cycle with recondenser
Refrigeration capacity	minimum 20 W
Delivery of coolant	helium mist at 4.4 K, 0.117 MPa (0.2 atm)
Compressor outlet condition	300 K, 1.7 MPa (16 atm)
Fatigue test machine;	
Type of load	5 ton, maximum tensile and compressive loading
Frequency	0.01 to 50 Hz

about 500 h.[2] Besides, the system has a great advantage because the duration of a fatigue test cannot be estimated in advance.

DETAILS OF THE SYSTEM

The specifications of this system are summarized in Table 1.[3] Figure 1 shows the principle of a recondensation system. A cryostat with liquid helium is completely closed. Evaporation of liquid helium is caused by the heat invading from the surroundings and from the heating of the specimen tested. On the other hand, helium mist is produced at the Joule-Thomson (JT) valve in the refrigerator, and the mist is transferred to the recondenser in the cryostat. The evaporated gas in the cryostat is recondensed and reliquefied using the latent heat of the mist at the recondenser and returned to the liquid helium bath. Accordingly, the liquid helium level in the cryostat is kept constant. The mist in the recondenser is evaporated with the heat flux from the outside of the recondenser. The refrigeration system can work for hundreds of hours, since there is no problem of contamination.

Flow Diagram

Figure 2 is a helium gas flow diagram of this system. The system is composed of a compressor, a gas holder, a purifier, a refrigerator, and a fatigue test machine. The compressor is fully sealed to avoid the contamination of the helium gas. The maximum flow rate is 300 N/m^3, and inlet gas pressure of the expansion engine and the JT valve in the refrigerator is increased to about 1.63 MPa (15 atm). The purifier contains a molecular sheaves adsorber for the removal of moisture and a charcoal adsorber

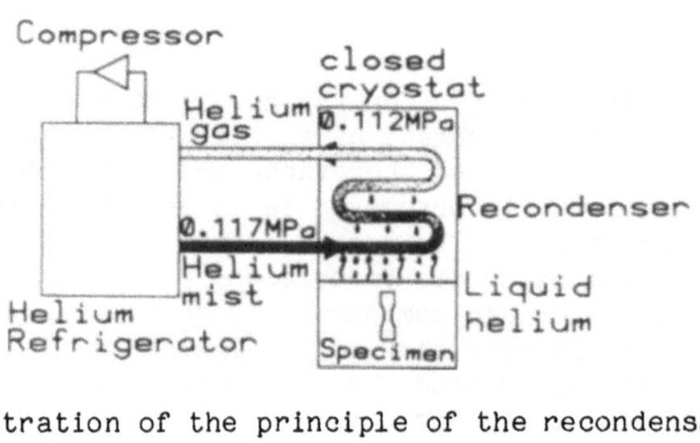

Fig. 1. Illustration of the principle of the recondensation system.

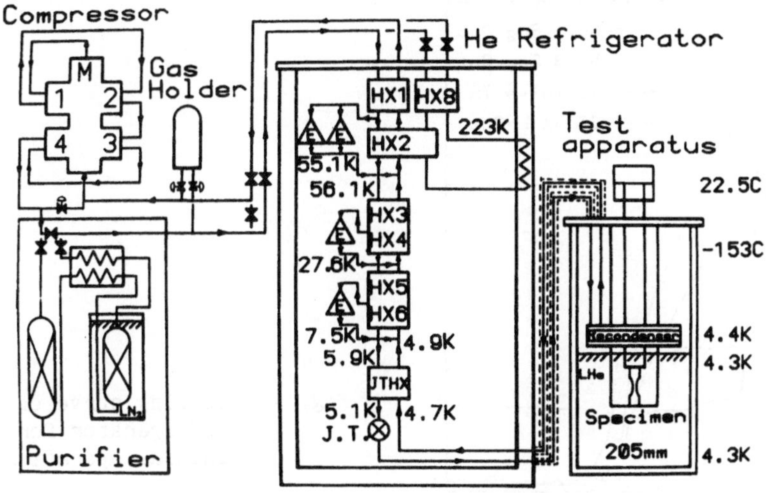

Fig. 2. Helium gas flow diagram of the fatigue test system for liquid helium temperature.

(cooled to 77 K) for the elimination of oxygen and nitrogen. The purifier is used to reduce the density of moisture or oxygen and nitrogen to less than 10 ppm prior to long-term operation.

The refrigerator has three-stage Claude cycle engines and can provide 20 W at 4.4 K. The helium gas that enters into expansion engines is cooled to about 70 K at the first-stage engine, to about 30 K at the second-stage engine, and to about 9 K at the third-stage engine. The high-pressure gas is cooled to about 5 K by heat exchangers Nos. 1 through 6 and the JT heat exchanger. The helium mist is generated at the JT valve by adiabatic free expansion to 0.11 MPa and is transferred through the vacuum-insulated mist transfer tube to the recondenser in the test machine cryostat, where the mist almost becomes gaseous owing to heat exchange. Then it returns to the refrigerator. The five major heat exchangers are the aluminum plate fin type to reduce the weight, improve the heat efficiency, and prevent blocking of the passage. The JT heat exchanger is composed of a dual tube of OFHC copper.

Fig. 3. Overall view of the system installed at NRIM, Tsukuba.

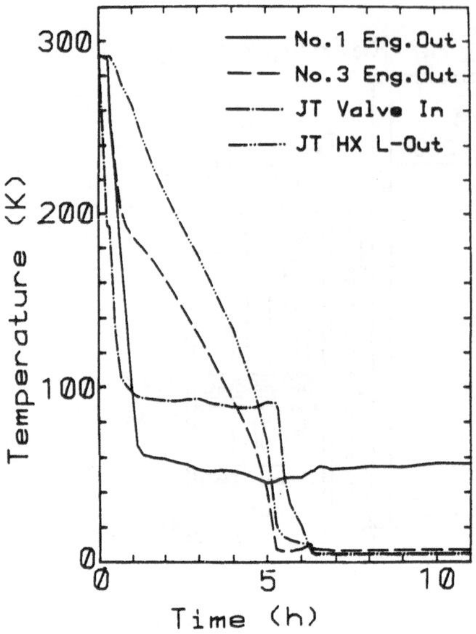

Fig. 4. Cooling curve of the refrigerator for initial operation.

The temperature values in the figure are typical of various parts of the refrigerator in a steady state; they are indicated by a temperature monitoring system assisted by a microcomputer real-time data acquisition system. The fatigue test machine is a hydraulic servo-pulse type. The dynamic maximum load is ±5 tons and the static maximum load is ±7.5 tons. The test frequency range covers 0.01 to 50 Hz. Figure 3 shows a view of the entire system. Compressors are installed in the next room.

Cooling Curve

Since the heat exchangers have large latent heat capacities, it takes several hours for each part of the system to reach a steady-state temperature. Figure 4 shows the cooling behavior of each part of the refrigerator. The cryostat can be closed within only six hours. The amounts of liquid nitrogen for precooling and liquid helium to fill the cryostat are about 300 kg and about 50 liters, respectively.

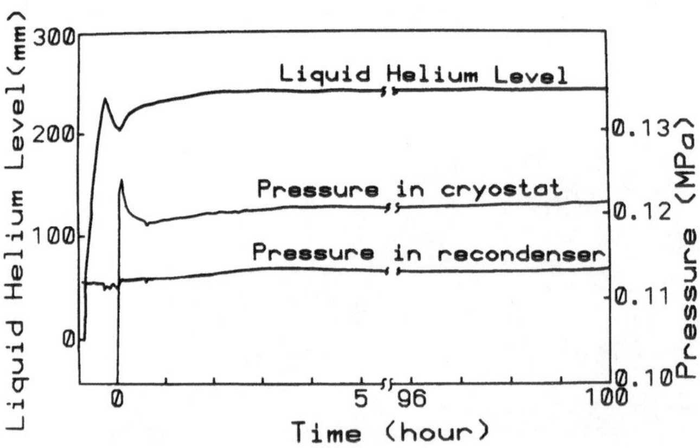

Fig. 5. Pressure of gaseous helium and level of liquid helium in the cryostat after closing.

Pressure Behavior after Closing

After the refrigerator has been sufficiently cooled by pouring liquid helium into the cryostat, the cryostat is closed. Figure 5 shows the liquid helium levels and the pressures in the cryostat and in the recondenser after the cryostat was completely closed by shutting the blow-off valve. Once the cryostat was closed, the pressure in the cryostat rapidly increased and the liquid helium level, which was reduced due to initial vaporization, began to recover gradually. There was little change in the pressure in the recondenser. The liquid helium level and pressure difference between the cryostat and the recondenser were kept constant in the steady-state condition.

Specimen Exchange

When the refrigerator is stopped in order to replace the specimen after failure, the liquid helium in the cryostat completely evaporates, and the temperature inside the cryostat raises to room temperature. It takes a lot of time for the system to return to the steady state. So, the specimen replacement was carried out without stopping the refrigeration.

Figure 6 gives the results. The recondenser was exposed to the atmosphere during replacement of the specimens. There was no problem in this procedure. The temperature of each part in the refrigerator was raised somewhat but was not remarkably disturbed, and the system returned to the steady state within about two hours. No extra heat invasion to the recondenser was observed during the procedure. The liquid helium level was lowered very slowly after opening the cryostat. The temperatures of the mist transfer line and the inlet of the JT valve increased to a maximum of 20 K, but returned soon once liquid helium was added. On the contrary, the temperature of the No. 2 expansion engine outlet decreased, which is ascribed to the partial closing of the JT valve and the increase of the flow rate to the engine. Liquid helium in the cryostat finally decreased to half the initial amount.

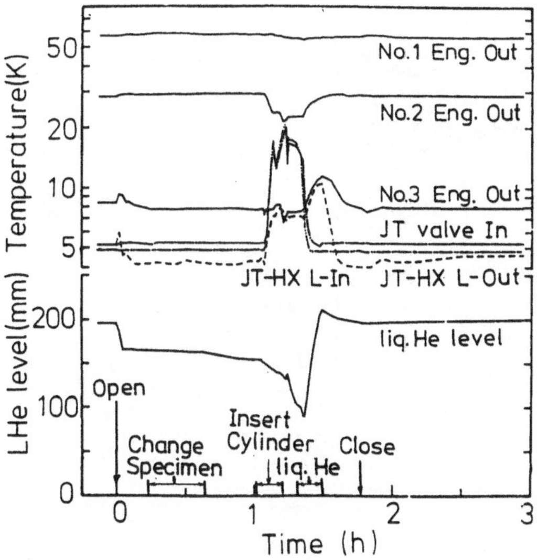

Fig. 6. Temperature of the refrigerator and level of liquid helium during specimen replacement.

Table 2. Chemical Composition of Base Material and Weldment (wt.%).

	C	Fe	N	O	H	Al	Sn	Ti
base	0.012	0.19	0.0024	0.057	0.0058	5.15	2.66	bal.
weldment	0.008	0.21	-	0.102	0.0040	5.13	2.64	bal.

Table 3. Tensile Properties of Ti-5Al-2.5Sn ELI Alloys at Cryogenic Temperatures.

	base material			weldment
testing temperature (K)	293	77	4	4
yield strength (MPa)	705	1209	1405	1368
tensile strength (MPa)	772	1277	1483	1459
fracture strain (%)	14.6	15.3	9.9	6.7

TESTING RESULTS

Fatigue tests have been carried out on titanium alloys, stainless steels and other materials with this system. The material used here was a Ti-5Al-2.5Sn ELI alloy. Table 2 shows the chemical composition of the base material and weldment; their tensile properties are given in Table 3. Figure 7 illustrates a cross-sectional view of the test machine cryostat. The specimen is gripped by the dual cylinder, whose specific frequency is above 100 Hz to avoid sympathetic vibration with the machine.

Figure 8 gives the dimensions of the specimens. The internal temperature of specimens during fatigue tests was measured. A thermocouple was used to measure the difference in the temperature between the surrounding liquid helium and the specimen.[4]

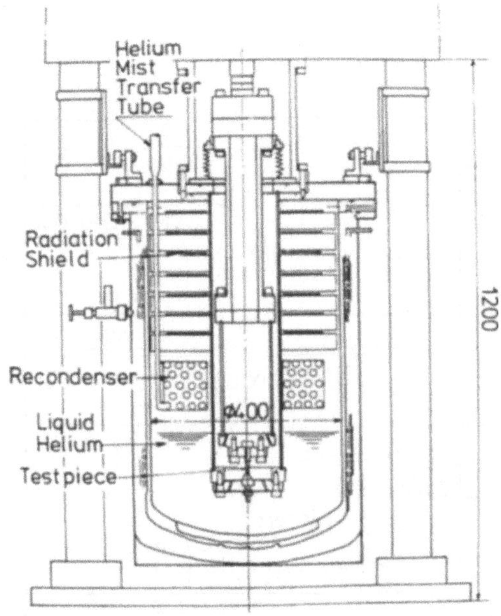

Fig. 7. Cross-sectional view of the cryostat for the fatigue test apparatus.

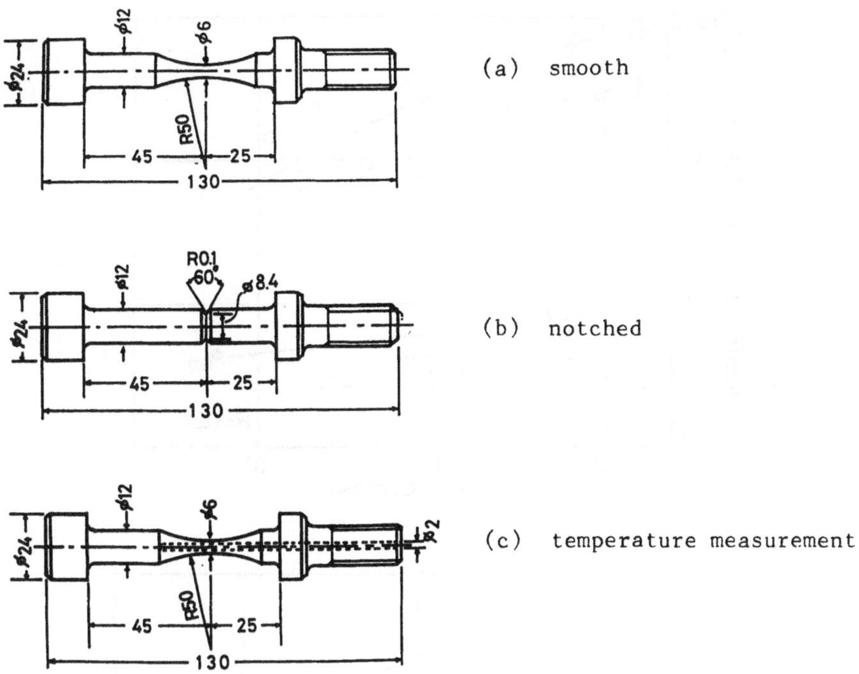

Fig. 8. Dimensions of specimens used.

Internal Specimen Heating

As shown in Figure 9, the temperature rise of the specimen was kept at less than 1 K, at a stress of about half the yield stress. At a stress of four-fifths of the yield stress, the temperature rose more than 1 K when the frequency exceeded 5 Hz. However, there was no temperature rise at the surface of the specimen at any frequency. Fatigue tests were carried out at a frequency of 4 Hz for titanium alloys.

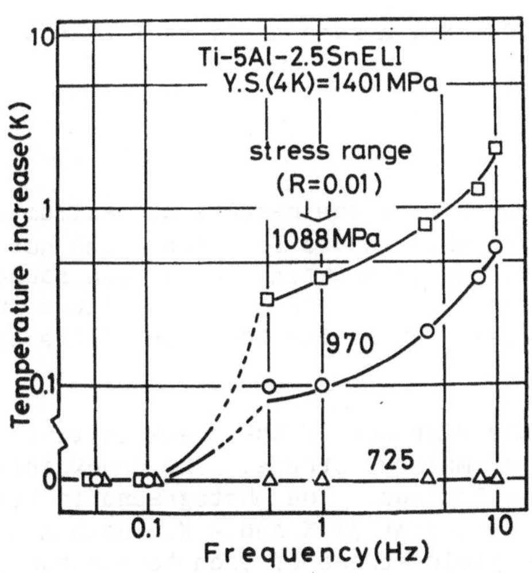

Fig. 9. Dependence of temperature increase on frequency in the fatigue test.

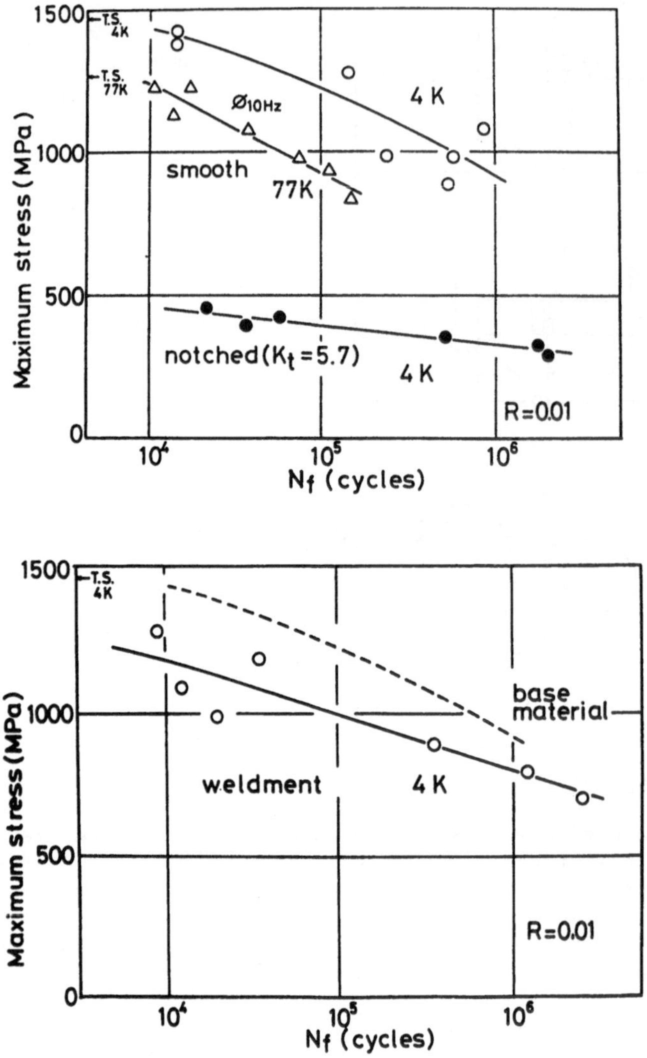

(a)

(b)

Fig. 10. S-N curves of Ti-5Al-2.5Sn alloy.

S-N Curves

In Figure 10 a) and b) are the results of maximum stress vs. the number of cycles to failure (N_f) curves for smooth and notched specimens, base material and weldments. An interesting result was found for crack initiation of smooth specimens at 4 K: The crack initiated at the surface of the specimen at higher stress, but at the interior of the specimen at lower stress.

Figure 11 shows the distance of the crack initiation site from the surface as a function of maximum stress. The crack initiation site moves inside as the stress is lowered. The photographs in Figure 12 show the crack initiation site tested at 77 K and 4 K. Both maximum stresses are at a similar ratio to the yield stress of each temperature. The crack initiation site at 77 K is always located at the surface. The site at 4 K moves inside with an increase in the stress.

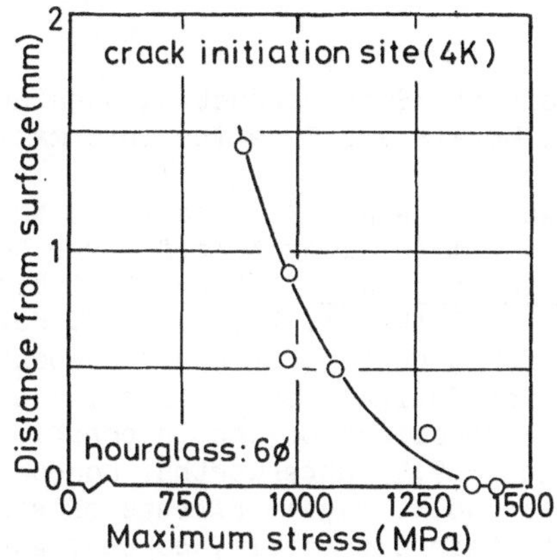

Fig. 11. Crack initiation.

(a) 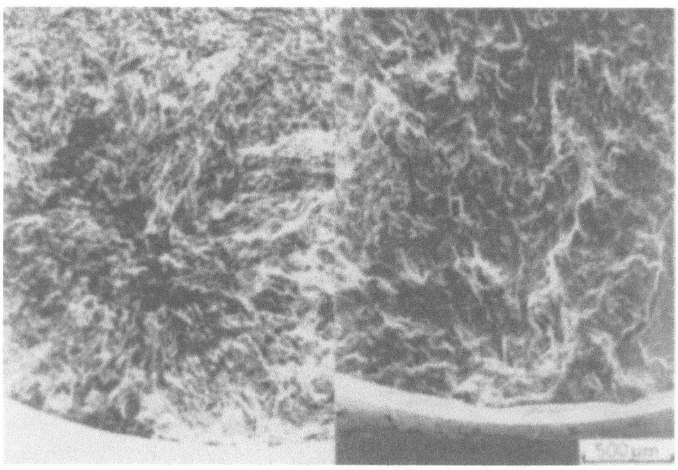 (b)

Fig. 12. Crack initiation in fracture surface
a) at 4 K, 980 MPa;
b) at 77 K, 833 MPa.

CONCLUSIONS

The present fatigue test machine with its closed loop recondensation system enables long-term fatigue tests at liquid helium temperature without an additional supply of liquid helium. A fatigue test of more than a week (2 x 10^6 cycles) could be carried out and continued for a longer period by an efficient exchange of specimens. A total continuous operation time of about 500 h has been achieved.

ACKNOWLEDGMENT

The authors wish to thank Kobe Steel Company for collaboration in developing and supporting the operation of this system and in supplying the specimens.

REFERENCES

1. "Handbook on Materials for Superconducting Machinery," Sheet No. 8.1.3-4 (11/76), Metals and Ceramics Information Center, Columbus, Ohio (1977).
2. T. Ogata, K. Ishikawa, K. Nagai, K. Hiraga, Y. Nakasone, and T. Yuri, Helium cooling system for long-term fatigue test at liquid helium temperature, Tetsu-to-Hagane 71:236-241 (1985), in Japanese.
3. T. Ohtani, Y. Ohtsu, N. Shiki, Y. Tomisaka, Y. Kawate, Y. Monju, and T. Horiuchi, Performance test of new recondensing type cooling system designed for fatigue testing machine at liquid helium temperature, in: "Proceedings of the International Cryogenic Engineering Conference - 9," Butterworths, London (1982), pp. 604-607.
4. T. Ogata, K. Ishikawa and K. Nagai, Effects of strain rate on the tensile behavior of stainless steels, copper, and an aluminum alloy at cryogenic temperatures, Tetsu-to-Hagane 71:1390-1397 (1985), in Japanese.

A MODEL FOR FATIGUE STRENGTH DEGRADATION INCLUDING TEMPERATURE EFFECT

M. Hamdi AbdelMohsen, M. K. Abdelsalam and R. E. Rowlands

Applied Superconductivity Center
University of Wisconsin-Madison
Madison, Wisconsin

ABSTRACT

A general statistical model that includes the interaction of applied stress, number of fatigue cycles and temperature is developed to analyze the fatigue degradation of composites. It is verified experimentally using fatigue data at room and liquid nitrogen temperatures for three different composites, e.g. cloth glass/epoxy, unidirectional glass/polyester and glass/polyester laminates. This model is useful in calculating the cyclic strength degradation curve at any desired temperature from static strength test results at this temperature and both static and fatigue results at room temperature.

INTRODUCTION

Fiber reinforced composites are prospective materials for cryogenic temperature applications in which high strength-to-density and high strength-to-thermal conductivity ratios are required. A superconductive energy storage magnet requiring a supporting structure to carry the compressive radial loads from 1.8 K at the magnet surface to 300 K at the bedrock is an example of such applications. The Applied Superconductivity Center at the University of Wisconsin-Madison has undertaken a program to generate cryogenic engineering data on commercial composites emphasizing the strut requirements for superconductive magnetic energy storage systems (SMES).[1-5]

The objective of this paper is to present a model that can be used to examine and compare the stiffness degradation and associated damage mechanisms resulting from compression-compression fatigue cycling. The literature contains an enormous number of fatigue degradation models for composites.[6-10] In this paper, a statistical model based on rate-type equation is used. The model considers the interaction of the applied stress, the temperature, and the number of cycles on the fatigue degradation phenomena. Fatigue data on three different composites; fiberglass cloth reinforced epoxy (G-10CR), laminated glass fiber reinforced polyester (Extren) and unidirectional glass fiber reinforced polyester (HIR) are analyzed using this model.

FATIGUE DEGRADATION MODEL

Assume that the rate of change in residual strength during cycling can be described by a power-law equation

$$\frac{dF}{dn} = - AF^{-m} \tag{1}$$

where F is the residual strength at any temperature T after n number of cycles, m is an exponent characteristic of material behavior and is independent of F, n or T. A depends in general on the characteristics of fatigue loading, such as stress amplitude, stress ratio, frequency and temperature. To simulate the loading pattern of a SMES structure both the frequency and the minimum applied stress were held constant throughout the tests. Hence the only independent variables remaining in the parameter A are the maximum applied stress σ_m and the temperature T. Integrating equation (1) for $m > 0$ between $n = 0$ and n yields

$$F(T,n)^{m+1} = F(T,0)^{m+1} - (m+1)\, A(\sigma_m, T)\, n \tag{2}$$

where F(T,0) is the static strength at any temperature T. It is reasonable to assume that the static strength F(T,0) follows a two-parameter Weibull distribution[2-4]

$$P_s(F) = P[F(T,0) \leq F] = 1 - \exp[-(F/\beta(T))^{\alpha(T)}] \tag{3}$$

where $P_s(F)$ is the probability that the static strength F(T,0) is less than or equal to F, and $\alpha(T)$ and $\beta(T)$ are the shape and the scale parameters at a certain temperature T. The values of these parameters for each of the three composites are listed in table 1. The distribution of the residual strength of the samples that have a strength greater than the applied stress σ_m based on the samples that already survived the σ_m stress level is

$$P_s(F) = 1 - \exp\{-[(F/\beta(T)]^{\alpha(T)} + [(\sigma_m/\beta(T)]^{\alpha(T)}\} \tag{4}$$

Table 1. Compression Strength, Scale Factor, Shape Factor of the Tested Materials at 300 K and 77 K, and Parameters of Eq. (8)

Material	σ_u ksi		β		α		a	$B_o \times 10^{-6}$
	300 K	77 K	300 K	77 K	300 K	77 K		
G-10	64.2	135.5	66.3	130	12.1	6.0	.073	2.4
Extren	20.7	37.6	21.0	39.0	23.5	7.7	.085	3.0
HIR	110	200	115.7	205	8.6	7.0	.107	6.8

Substituting for $F = F(T,0)$ in equation (3) and assuming that $A(\sigma_m,T) = A_1(\sigma_m)\ A_2(T)$ we get the residual strength distribution $P_r(F)$ at any number of cycles n

$$P_r(F) = 1 - \exp\left\{-\left[\frac{F(T,n)^{m+1}+(m+1)A_1(\sigma_m)A_2(T)n}{\beta(T)^{m+1}}\right]^{\frac{\alpha(T)}{m+1}} + \left(\frac{\sigma_m}{\beta(T)}\right)^{\alpha(T)}\right\} \quad (5)$$

where A_1 and A_2 are separate functions of the maximum fatigue stress and temperature. Equation (5) shows that the residual strength is a function of the number of cycles, the maximum applied stress and the temperature. One can choose reasonable forms for the functions $A_1(\sigma_m)$ and $A_2(T)$ and use optimization techniques to fit these functions to the available experimental data. A particular case for $A_2(T) = 1$, and $A_1(\sigma_m)$ = constant, corresponds to the theory proposed in (6), where temperature effects were ignored and only one stress level was used. At the instance of failure n = N where N is the number of cycles to failure, $F(T,n) = F(T,N) = \sigma_m$ and the fatigue life distribution $P_N(N)$ will be equal to the residual strength distribution $P_r(\sigma_m)$

$$P_N(N) = 1 - \exp\left\{-\left[\left(\frac{\sigma_m}{\beta(T)}\right)^{m+1} + A_1(\sigma_m)\ K(T)\ N\right]^{\frac{\alpha(T)}{m+1}} + \left[\frac{\sigma_m}{\beta(T)}\right]^{\alpha(T)}\right\} \quad (6)$$

$$\text{where}\quad K(T) = \frac{(m+1)\ A_2(T)}{\beta(T)^{m+1}} \quad (7)$$

The only unknowns in equation (6) are $A(\sigma_m)$, $K(T)$ and m. $\alpha(T)$ and $\beta(T)$ are known a priori from the ultimate static strength results at different temperatures, as shown in table 1.

RESULTS AND DISCUSSION

Figure 1 is a sketch of the variation of the calculated parameter $B = K(T)*A_1(\sigma_m)$ with the applied stress σ_m. The same parameter is plotted in figure 2 as a function of the maximum applied stress σ_m normalized with respect to the average ultimate compressive strength σ_u. From this plot one can see that the parameter B is a function of the normalized applied stress only. For each material, experimental results appear to be scattered around a straight line that can be represented by

$$B = B_o\ e^{q/a} \quad (8)$$

where $q = \sigma_m/\sigma_u$, B_o and a are constants. The optimum values of these constants are listed in Table 1 for each of the three materials studied.

The predicted fatigue life distributions of G-10CR along with the corresponding experimental data are plotted in Figs. 3 and 4 for different stress ratios and different temperatures. The same data for Extren and HIR at room temperature are plotted in Figs. 5 and 6. These figures show a reasonable agreement between the model and the experimental data.

The above results indicate that degradation mechanism by fatigue cycling at any two temperatures is the same provided that it is created by the same normalized applied stress. The residual strength (Eq. 2) of

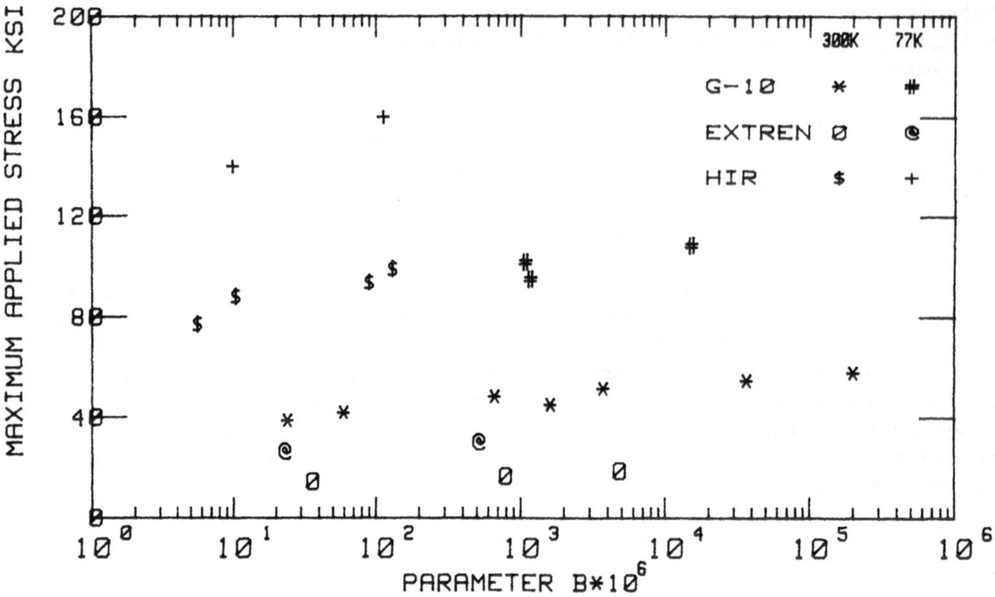

Fig. 1. Correlation between the parameter B and the maximum applied stress σ_m for G-10CR, Extren and HIR.

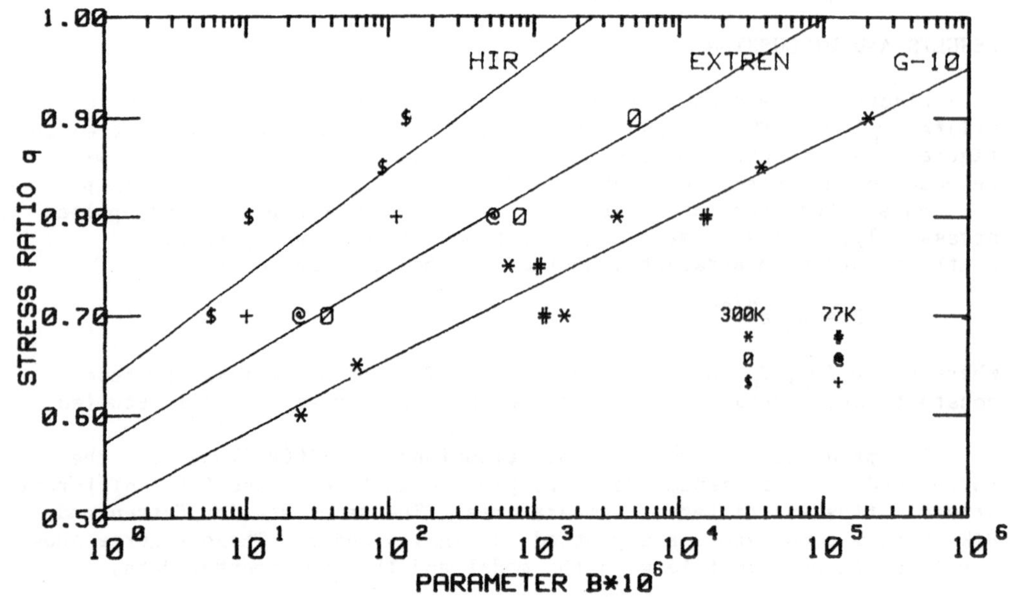

Fig. 2. Correlation between the parameter B and the maximum applied stress σ_m normalized with respect to the average ultimate compressive strength, ($q = \sigma_m/\sigma_u$) for G-10CR, Extren and HIR.

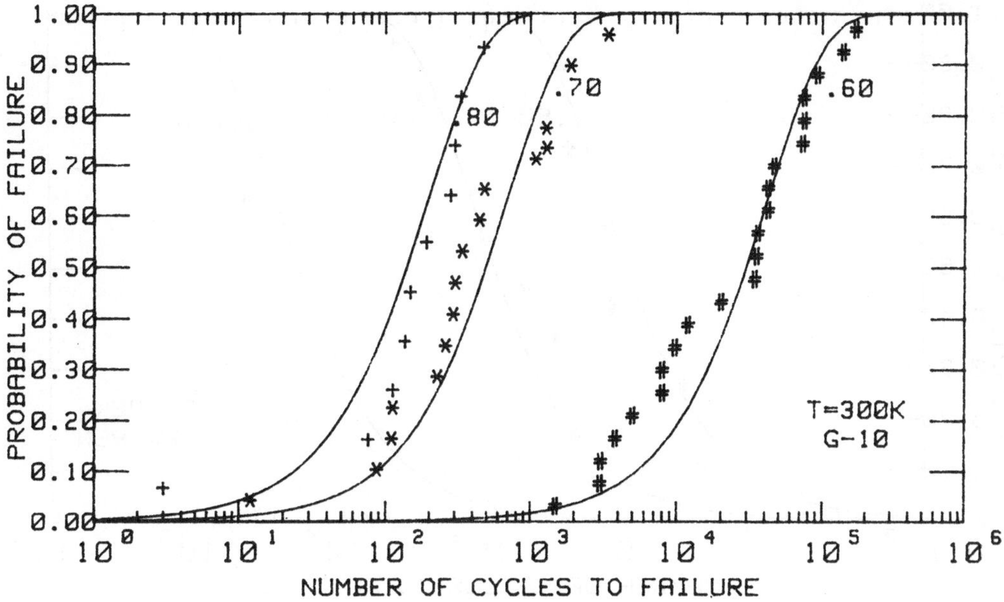

Fig. 3. Predicted and experimental room temperature compression-compression fatigue life distribution for G-10CR at maximum stress levels of 60, 70 and 80% of the ultimate compressive strength.

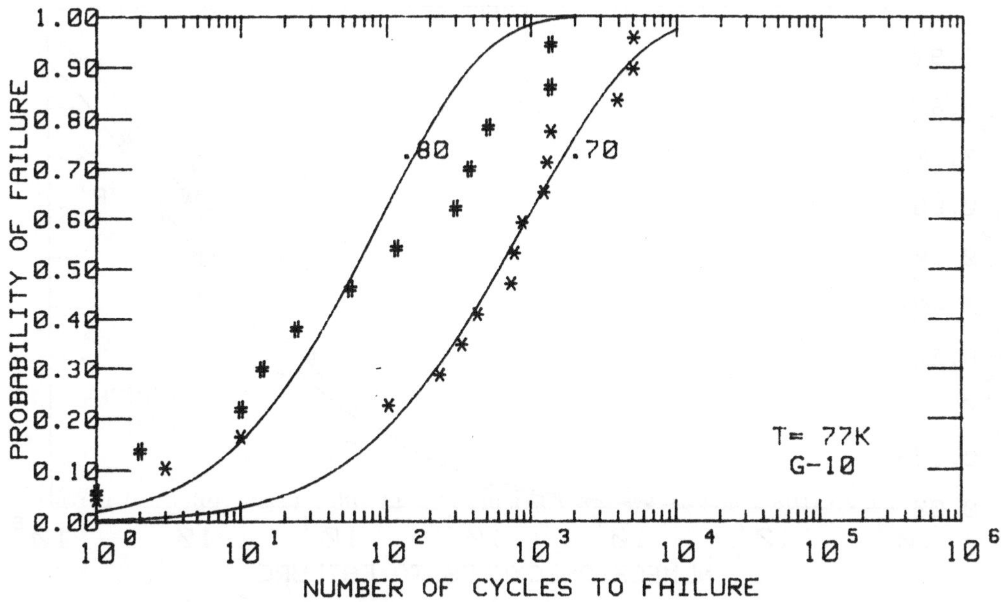

Fig. 4. Predicted and experimental liquid nitrogen temperature compression-compression fatigue life distribution for G-10CR at maximum stress levels of 70 and 80% of the ultimate compressive strength.

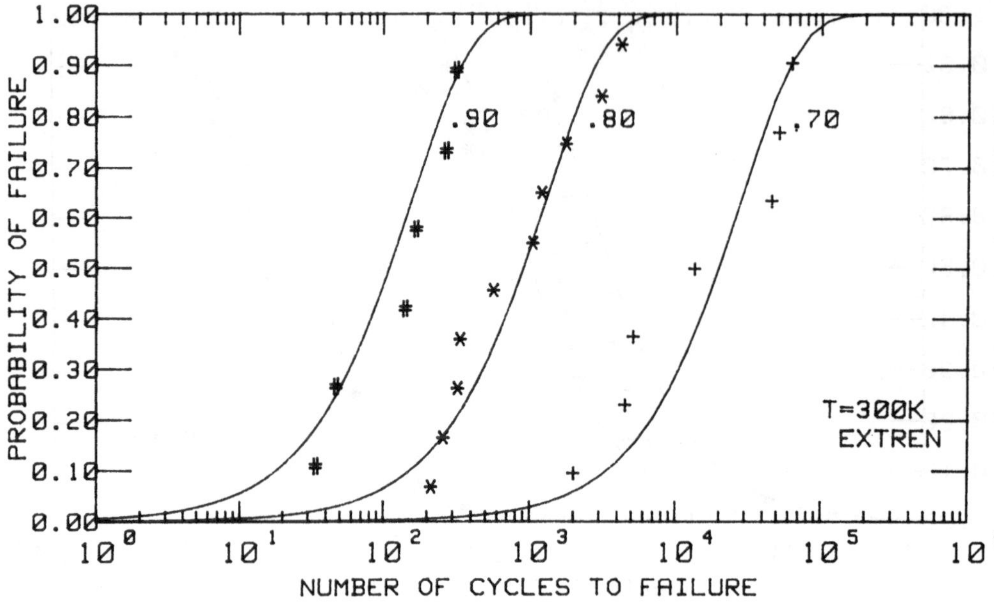

Fig. 5. Predicted and experimental room temperature compression-compression fatigue life distribution for laminated fiberglass reinforced polyester (Extren) at maximum stress levels of 70, 80 and 90% of the ultimate compressive strength.

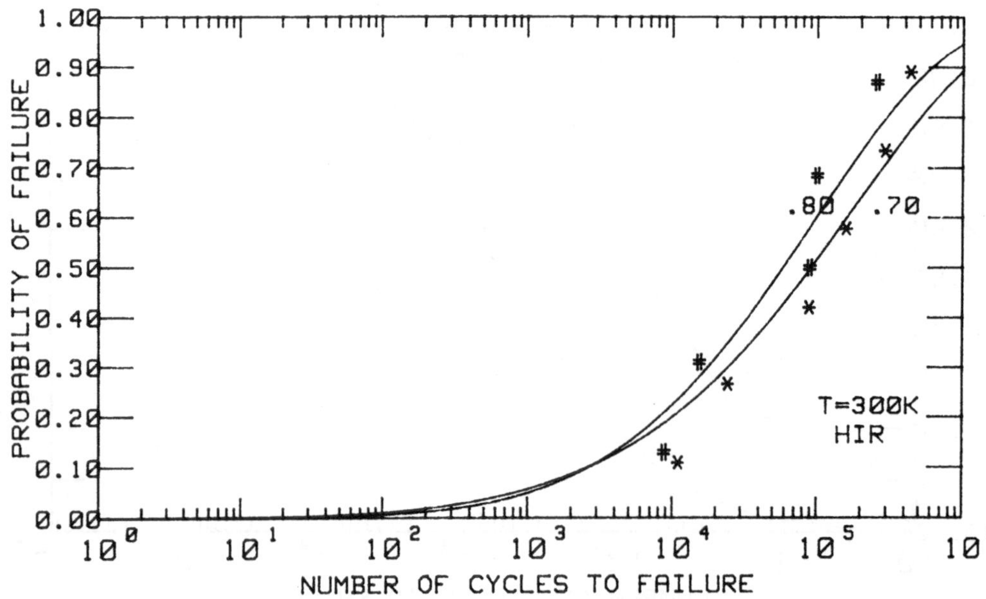

Fig. 6. Predicted and experimental room temperature compression-compression fatigue life distribution for unidirectional glass fiber reinforced polyester (HIR) at maximum stress levels of 70 and 80% of the ultimate compressive strength.

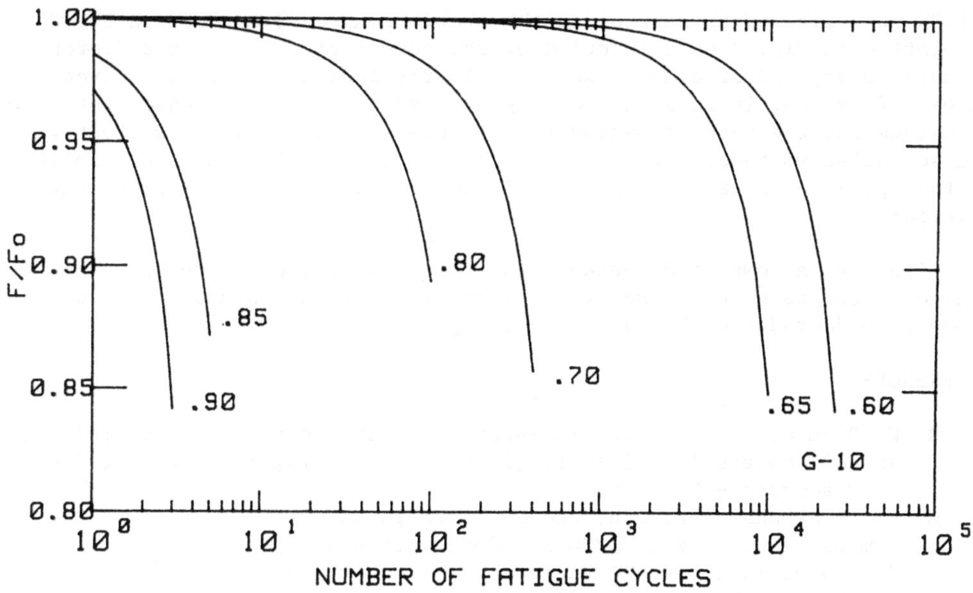

Fig. 7. Normalized residual strength (with respect to static compressive strength) versus number of cycles for G-10CR. Maximum fatigue stress is 60, 65, 70, 80, 85 and 90% of ultimate compressive strength.

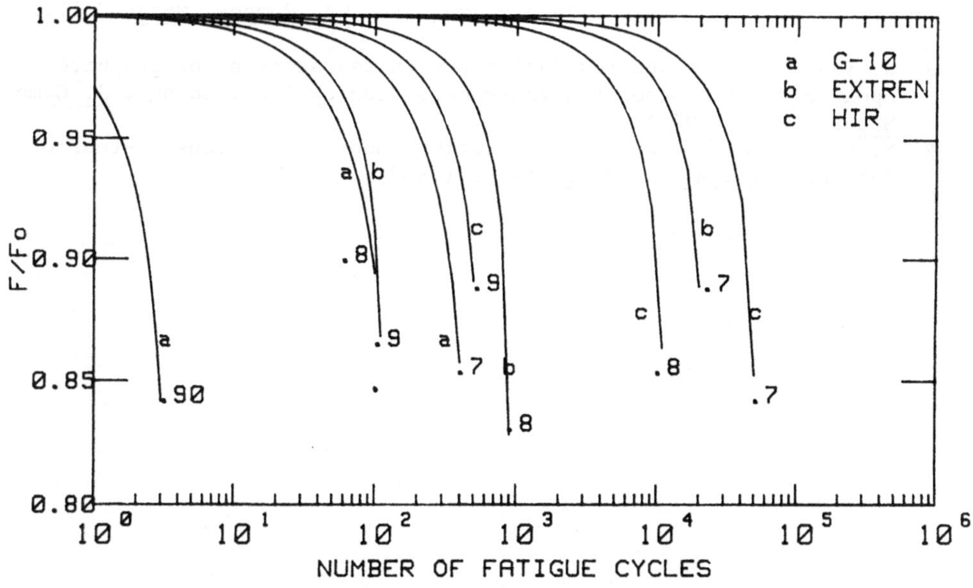

Fig. 8. Comparison between fatigue degradation (in terms of normalized residual strength versus number of cycles) for G-10CR, Extren and HIR at stress levels of 70, 80 and 90% of ultimate compressive strength.

G-10CR normalized with respect to the static compressive strength F(T,0) is plotted in Fig. 7 as a function of the number of cycles for different normalized applied stresses. Results clearly indicate that at a given number of cycles, increasing the range of cyclic stress increases both the magnitude and the rate of degradation. This phenomenon is consistent with the so-called wear-out process in which damage accumulation occurs uniformly in the specimen so as to progressively deteriorate the stiffness of the material.

Finally, a comparison between the three composites is demonstrated by Fig. 8. Results show the superiority of HIR over Extren and G-10CR with respect to degradation by fatigue cycling.

REFERENCES

1. R. W. Boom et al., "Wisconsin Superconductive Energy Storage Project," Annual Reports I (1974), II (1976), III (1977), IV (1981), University of Wisconsin - Madison.
2. M. H. AbdelMohsen, K. Han, and R. E. Rowlands, Fatigue of glass-epoxy composite at 77 K and 300 K: Observation and prediction, in: "Advances in Cryogenic Engineering - Materials," vol. 30, Plenum, New York (1984), pp. 17-25.
3. S. Han and M. Hamdi AbdelMohsen, "Fatigue Life Scattering of RP/C," The Society of the Plastic Industry (Feb. 1983).
4. K. Han, Composite fatigue behavior of glass fiber polyester composite composites at 300 K and 77 K, J. Compos. Mater. (1982).
5. E. L. Stone, L. O. El-Marazki, and W. C. Young, Compressive fatigue tests on unidirectional glass/polyester composite at cryogenic temperatures, in: "Nonmetallic Materials and Composites at Low Temperatures," Plenum (1978), p. 283.
6. H. T. Hahn and R. Y. Kim, Proof testing of composite materials, J. Compos. Mater. 9:297 (1975).
7. J. N. Yang, R. K. Miller, and C. T. Sun, Statistical fatigue of unnotched graphite/epoxy laminate, J. Compos. Mater. 14:82 (1980).
8. J. N. Yang, Residual strength degradation model and theory of periodic proof tests for graphite/epoxy laminates, J. Compos. Mater. 11:176 (1976).
9. J. N. Yang, Fatigue and residual strength degradation for graphite/epoxy composite under tension-compression cyclic loadings, J. Compos. Mater. 12:19 (1978).
10. R. Y. Kim and W. J. Park, Proof testing under cyclic tension-tension fatigue, J. Compos. Mater. 14:69 (1980).

FRACTURE TOUGHNESS OF NEWLY DEVELOPED STRUCTURAL MATERIALS

FOR SUPERCONDUCTING COILS OF FUSION EXPERIMENTAL REACTOR

H. Nakajima, K. Yoshida, K. Okuno, M. Oshikiri,
E. Tada, and S. Shimamoto
Japan Atomic Energy Research Institute
Naka-Machi, Naka-gun, Ibaraki-Ken, Japan

R. Miura
The Japan Steel Works, Ltd.
Hokkaido, Japan

M. Shimada and S. Tone
Kobe Steel, Ltd.
Hyogo, Japan

K. Suemune and T. Sakamoto
Nippon Steel Corporation
Tokyo, Japan

K. Nohara
Kawasaki Steel Corporation
Chiba, Japan

ABSTRACT

There are three main problems--electrical, thermal, and mechanical--in the design of large, high magnetic field coils. It has been recognized that there is not enough margin in mechanical design because large superconducting coils for future fusion reactors are subject to high electromagnetic force. The safety factor of mechanical design is smaller than that of electrical design; the mechanical problems must be solved immediately. Development of high strength and toughness materials is one way to solve this problem. For three years, the Japan Atomic Energy Research Institute (JAERI), in collaboration with four Japanese steel industries, has been developing new cryogenic structural materials that have yield strengths of more than 1,200 MPa and plain-strain fracture toughnesses of 200 MPa$\sqrt{m}$. This paper describes the procedures and results of fracture toughness tests performed at JAERI on newly developed structural materials and presents data to clarify the relationship between Charpy absorbed energy and plain-strain fracture toughness.

INTRODUCTION

JAERI identified targets on engineering properties of structural materials for superconducting coils of Fusion Experimental Reactor (FER) in 1982.[1] As these targets required much higher properties than those of ordinary cryogenic structural materials at that time, it seemed that it was difficult to develop steels that satisfied these requirements. However, JAERI and Japanese steel industries have successfully developed new cryogenic materials that have high yield strength and toughness. Some results of tensile tests, Charpy impact tests, and fracture toughness steels were presented in 1983.[2] JAERI selected five high manganese austenitic stainless steels and five austenitic stainless steels at that time. One of the high manganese austenitic stainless steels had a yield strength of more than 1,200 MPa and a fracture toughness of more than 200 MPa$\sqrt{m}$. An elastic-plastic fracture toughness, J_{Ic}, measurement system was installed at the beginning of 1984 at JAERI.

JAERI has been evaluating mechanical properties of steels to select candidate materials for superconducting coils of FER. At present, JAERI has been conducting fracture toughness tests on newly developed steels. The major concern is whether the fracture toughness of selected materials exceeds 200 MPa$\sqrt{m}$. There are six steels that satisfy the requirements. Weld materials for these steels have been developed in parallel with this work.

CANDIDATE MATERIALS

JAERI has completed tensile and Charpy impact tests on eighty-one kinds of steels. Figure 1 shows the characteristics of candidate steels in terms of yield strength and Charpy absorbed energy. JAERI's targets require higher mechanical properties than those of 304LN or 316LN stainless steels. In spite of the high level requirements, steels have been developed that satisfy these requirements.

JAERI selected candidate materials, which consisted of five high manganese austenitic stainless steels (steels A1, A2, B2, C1, and C2) and seven austenitic stainless steels (D3, D4,D5, D6, D7, E1 and F1). The selected steels have yield strengths over 1,200 MPa and Charpy absorbed energy over 80 J. The number of candidate materials increased from the previous report[2] because they include some steels with improved toughness and some newly developed materials. Steels B2 and D3 were improved with regard to toughness from steels B1 and D1, D2, respectively, as shown in Fig. 1.

ELASTIC-PLASTIC FRACTURE TOUGHNESS

Large specimens and large capacity machines are necessary to measure valid plane-strain fracture toughness, K_{Ic}, of high fracture toughness material, according to ASTM E399. Plane-strain fracture toughness, K_{Ic},

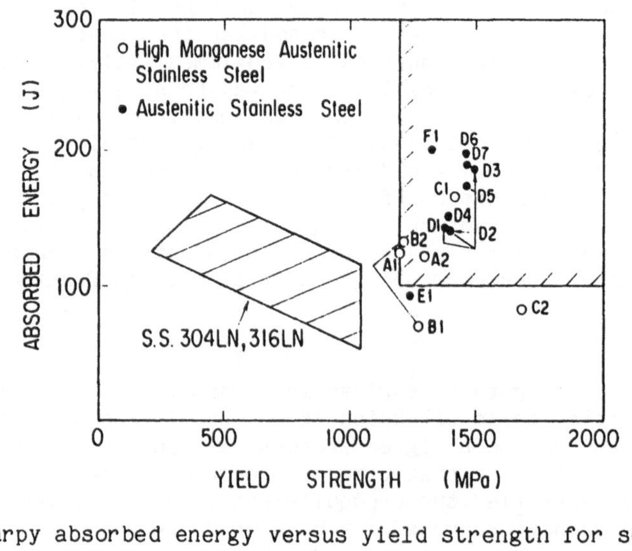

Fig. 1. Charpy absorbed energy versus yield strength for selected steels.

corresponds to elastic-plastic fracture toughness, J_{Ic}. Accordingly, the K_{Ic} value can be estimated from the J_{Ic} value. Therefore, the J_{Ic} test, in which small specimens can be used, is usually conducted instead of the K_{Ic} test. There are several methods for J_{Ic} determination, and these are divided into two methods: the multiple-specimen method and the single-specimen method.

The key factors for measuring the mechanical properties of structural materials at 4.2 K are:

1. Efficient use of liquid helium
2. Simple test procedure

The single-specimen method is the most suitable one at liquid helium temperature owing to its small liquid helium consumption. There are several single-specimen methods, such as unloading compliance, electrical potential, and supersonic and acoustic emission methods. JAERI has adopted a computerized J_{Ic} test method using the unloading compliance technique because this method satisfied the above requirements. This method is also effective from an engineering point of view without microscopic observation.

TEST FACILITY AND PROCEDURE

The test facility to measure fracture toughness is composed of a cryogenic test machine, which has a capacity of 10 tons, control unit, J_{Ic} interface unit, and a computer system. Specimen configuration and dimensions are in accordance with ASTM E813. Compact-tension specimens (25.4-mm thickness, T-L orientation) are used in fracture toughness tests. In the future, JAERI will also use this facility to measure fatigue characteristics of newly developed materials.

The data acquisition and control are performed by a minicomputer, PDP-11/70. The J_{Ic} test is conducted automatically and the J value is plotted against crack extension in real time during a test. The computer output (R-Curve) to a plotter is shown in Fig. 2. Measurements of crack extension in the computerized J_{Ic} test are based on the unloading compliance method,[3] and the test procedure is according to ASTM E813. The fracture toughness test is conducted by displacement control. Control by computer is not influenced when serration occurs. The measured data are stored on magnetic tape so that the data can be dealt with at any time.

JAERI verified the reliability of the unloading compliance method by comparing this method and the standard method according to ASTM E813 at 4.2 K. The test material was 25Cr-13Ni austenitic stainless steel, supplied from 70-ton production heats. This steel, which has yield strength of 1450 MPa, is one of the newly developed materials. The results of the verification test are shown in Fig. 3. The R-curve obtained by the unloading compliance method corresponds to that measured by the standard method. The K_{Ic} values measured by both methods are 190 and 182 MPa$\sqrt{m}$, respectively.

RESULTS AND DISCUSSION

Yield Strength and Fracture Toughness

Table 1 shows the nominal chemical composition of selected steels. Up to now, JAERI has completed fracture toughness tests of three high manganese austenitic stainless steels (A1, A2, and B2) and six austenitic stainless steels (D3, D4, D5, D6, D7, and E1) but not three steels (C1, C2, and F1). Fracture toughness tests were conducted not only on selected materials, but also on other materials, such as higher yield strength and lower fracture toughness material (D8), lower yield strength and higher fracture toughness materials (A3, A4,and B3) in order to study the relationship between Charpy absorbed energy and fracture toughness.

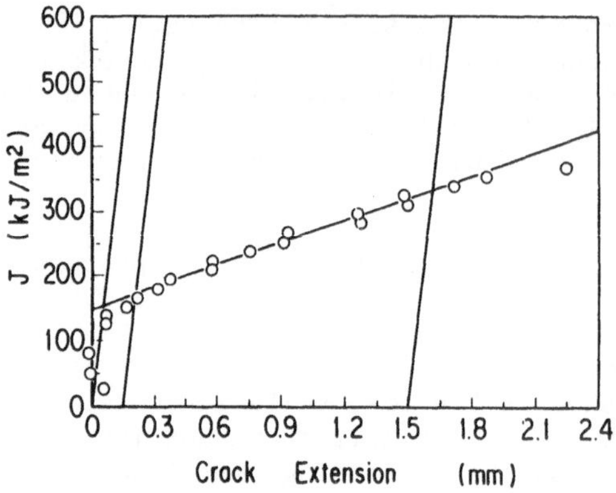

Fig. 2. R-curve measured by unloading compliance method.

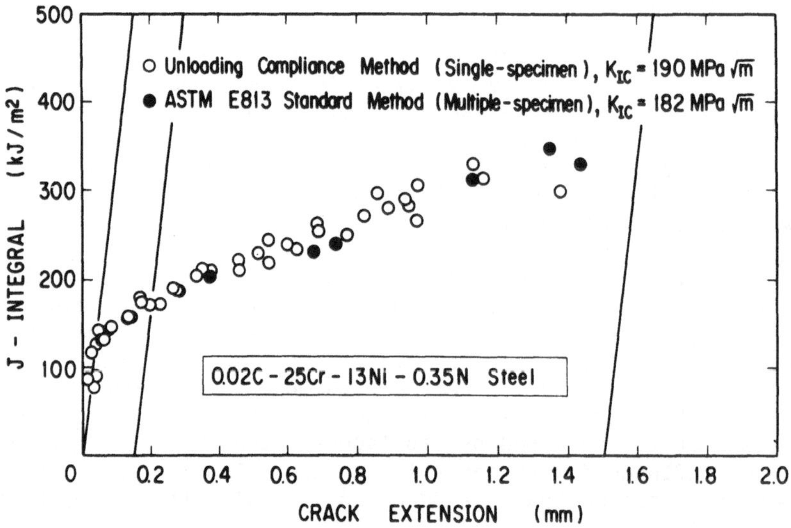

Fig. 3. Comparison between unloading compliance method and ASTM E813 standard method.

Figure 4 shows the fracture toughness and yield strength characteristics of steels tested at 4.2 K. The trend line presented by the National Bureau of Standards, which is the data base of ordinary austenitic stainless steels at 4.2 K, is drawn on this figure.[4] The mechanical properties of the Japanese LCT coil structure material (304 LN austenitic stainless steel) are also plotted on this line. JAERI's target, called the JAERI Box, is located over the NBS trend line. Both higher yield strength and higher fracture toughness are required than those of common austenitic stainless steels.

The fracture toughnesses of tested materials were greatly improved by changing chemical composition, production processes, and quality control. This improvement is especially remarkable in austenitic stainless steels D3-D8. Six steels, which consist of one high manganese austenitic stainless steel (A1) and five austenitic stainless steels (D3, D4, D6, D7, and E1), entered the JAERI Box.

Table 1. Nominal Chemical Composition

High Manganese Austenitic Stainless Steels	
A1 - A4	22Mn - 13Cr - 5Ni - 0.2N
B1 - B3	25Mn - 15Cr - 1Ni - 1Cu - Nb - 0.2N
C1,C2	9Mn - 21Cr - 7Ni - 0.3N
Austenitic Stainless Steels	
D1 - D8	25Cr - 13Ni - Mn - 0.35N
E1	12Cr - 12Ni - 10Mn - 5Mo - 0.2N
F1	20Cr - 14Ni - 1Mn - Mo - V - 0.3N

The fracture toughnesses of the tested materials are 180 to 240 MPa$\sqrt{m}$, independent of the type of steel. The yield strengths divide the tested materials into two groups: The first group consists of steels A1-A4, B2, B3, E1; the second group consists of steels D3-D8. The differences of yield strength are due to the effect of nitrogen content on cryogenic strength. Figure 5 shows plots of yield strength and fracture toughness against nitrogen content. The nitrogen contents of the first and second group of steels are about 0.2% and 0.35%, respectively. The yield strength dependence on nitrogen content is well-known. However, the fracture toughness is independent of nitrogen content.

Mechanical properties of selected steels are summarized in Table 2. At present, steels A1-A4 and D5 are supplied from 15- and 70-ton production heats, respectively. Steel A1 was selected from the four steels A1-A4.

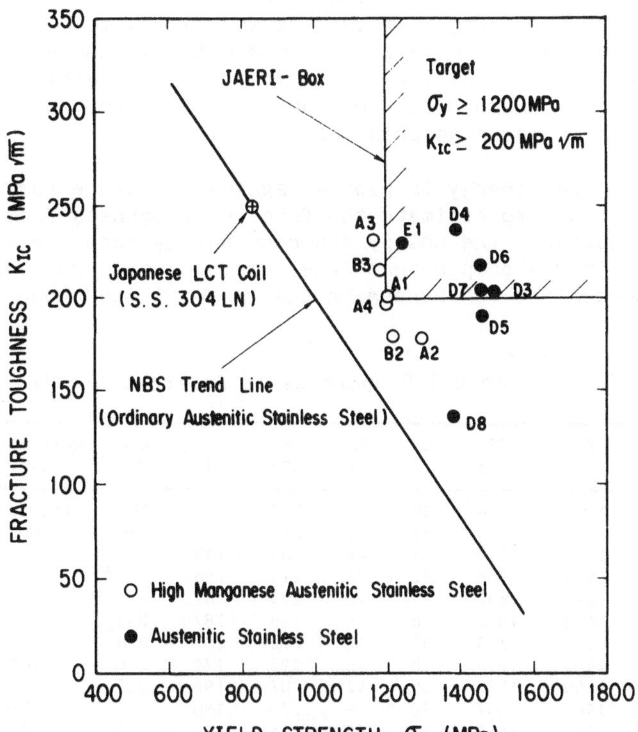

Fig. 4. Fracture toughness versus yield strength for selected steels.

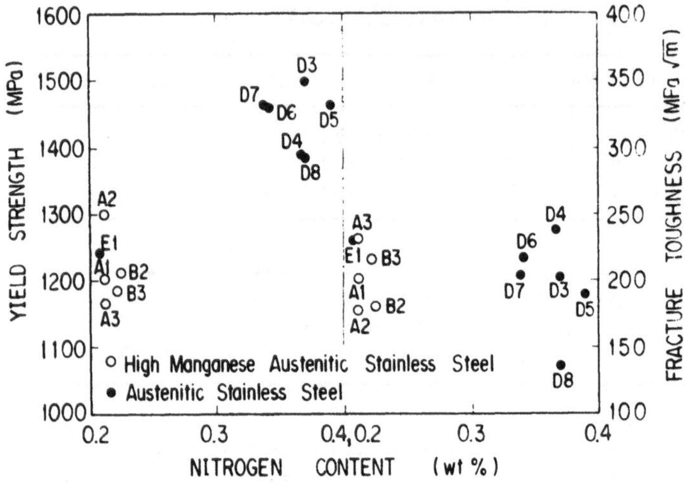

Fig. 5. Yield strength and fracture toughness versus nitrogen content.

Steel D5 did not achieve fracture toughness of more than 200 MPa$\sqrt{m}$. But very soon a new steel, which belongs to the D3-D7 class, will be produced in industrial quantities in view of its excellent test results. Steel B1 will be supplied in the form of production heats, too. One of these steels is used as reinforced material of the Poloidal Unit Pancake being developed at JAERI and will be applied to sheath material of the Demonstration Poloidal Coil, which will also be developed at JAERI.

Charpy Absorbed Energy and Fracture Toughness

There are two main reasons for conducting Charpy impact tests on newly developed steels: 1) screening of materials and 2) estimation of fracture toughness. Charpy absorbed energy itself does not give useful data for structural design of the superconducting coil. If it were possible to estimate fracture toughness from Charpy absorbed energy, the Charpy impact test would be practical from an engineering point of view.

The Charpy absorbed energy is plotted against fracture toughness in Fig. 6. It is difficult to estimate the fracture toughness from Charpy absorbed energy. However, the Charpy absorbed energy data are useful to screen materials. It was proper that JAERI employed a Charpy absorbed energy of 100 J as the borderline for screening the materials as shown in this figure.

Table 2. Mechanical Properties of Selected Steels

Steels	YS MPa	TS MPa	EL %	RA %	E GPa	CVN J	K_{IC} MPa$\sqrt{m}$	Melting kg
A1	1204	1624	39	52	217	124	202	15,000
A2	1300	1690	37	48	217	123	178	15,000
B2	1214	1568	35	44	208	133	180	300
C1	1415	1706	21	23	211	166	---[a]	100
C2	1677	1867	15	22	205	82	---[a]	100
D3	1498	1852	26	--	213	187	203	150
D4	1390	1793	32	42	216	151	238	150
D5	1465	1918	28	40	222	174	190	70,000
D6	1460	1832	29	42	207	198	217	150
D7	1465	1857	28	--	220	190	204	150
E1	1242	1615	41	49	214	91	230	50
F1	1322	1603	17	27	186	201	---[a]	100

[a] To be measured before the end of August

If the data are examined closely, there is a relationship between Charpy absorbed energy and fracture toughness in the case of high manganese austenitic stainless steels (A1, A2, A3, B2, B3) and in the case of austenitic stainless steels (D3, D5, D6, D7), respectively, except for three steels (D4, D8, E1). However, this relationship is not true for all types of steel.

Steel E1 has a high fracture toughness of 230 MPa$\sqrt{m}$ in spite of its low Charpy absorbed energy of 90 J. Thus, fracture toughness is not always low even if Charpy absorbed energy is low. However, it may be considered that this is a special characteristic for a newly developed material. A fracture toughness test of steel D8 was conducted to confirm this characteristic. The chemical composition of steel D8 is the same as that of steel D5. Charpy absorbed energy of steel D8 is less than that of steel D5, despite similar yield strength, because it contains many more inclusions than steel D5. The fracture toughness of steel D8 is less than that of steel D5, too. Inclusions contained in materials affect both the Charpy absorbed energy and the fracture toughness. Low charpy absorbed energy generally gives low fracture toughness.

We tried to plot the Charpy absorbed energy considering the yield strength $CVN/\sigma_y^{1.5}$ value against the fracture toughness K_{Ic}^2/E value of newly developed materials. This result is shown in Fig. 7. There is some connection between both values that is independent of the type of steel; of course, there are some materials for which this relationship does not apply, such as steels D4 and E1. As the fracture toughness of steel D5 is improved by quality control of steel D8, this relationship may apply to steel D8. It will become clear if data for the fracture toughness, K_{Ic}^2/E, from 100 kJ/m^2 to 150 kJ/m^2, are generated.

Thus, there is some relationship between Charpy absorbed energy considering yield strength and fracture toughness. However, the fracture toughness can not be estimated from Charpy absorbed energy accurately.

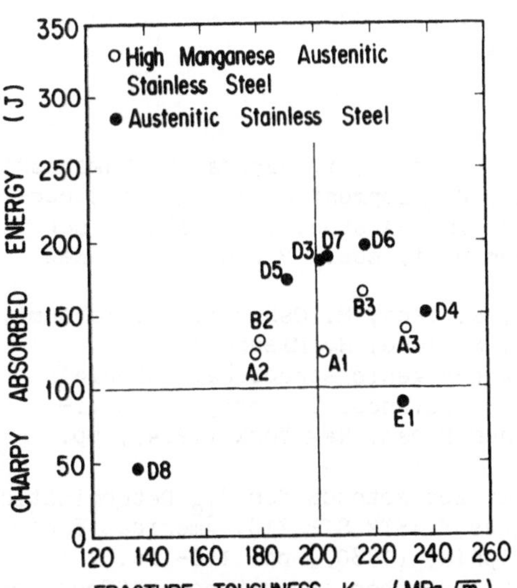

Fig. 6. Charpy absorbed energy versus fracture toughness.

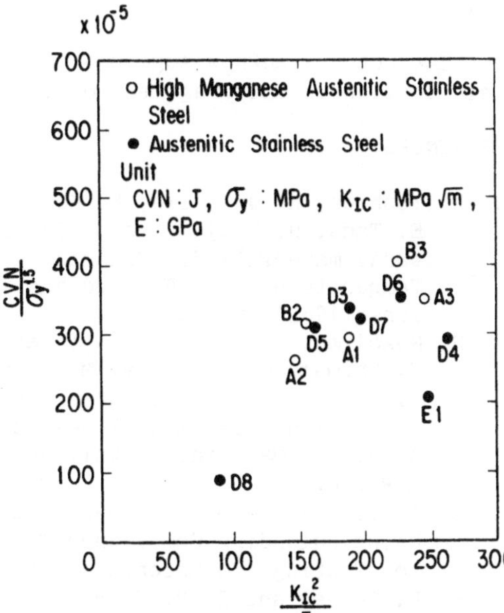

Fig. 7. Charpy absorbed energy considering yield strength versus fracture toughness.

Weldments

Weldments have been developed in parallel with the base metals. Tensile and fracture toughness tests on welded specimens of steel A1 were conducted. The steel was welded using an electron beam welding process. Joint efficiency was more than 97%. The fracture toughness of the welded joint was higher than that of the base metal. Details of the weldments will be presented in the next conference.

CONCLUSIONS

The Japan Atomic Energy Research Institute has been conducting elastic-plastic fracture toughness, J_{Ic}, tests on newly developed structural materials for superconducting coils of the Fusion Experiment Reactor. The main conclusions are:

1. The fracture toughness measured by a computerized J_{Ic} test using the unloading compliance technique corresponds to that measured by the standard method according to ASTM E813. The unloading compliance method is the most suitable one at liquid helium temperature.

2. Fracture toughnesses of newly developed structural materials for superconducting coils of FER are greatly improved by changing chemical composition, production processes, and quality control. There are six steels inside the JAERI Box, and the development of base metals has almost been completed.

3. There is some relationship between the Charpy absorbed energy considering the yield strength and the fracture toughness of newly developed materials (except for three steels). However, it is difficult to estimate the fracture toughness from the Charpy absorbed energy accurately. The Charpy impact test can be applied to the screening of materials, but right now accurate fracture toughness measurements should be performed using the standard fracture toughness tests.

ACKNOWLEDGMENTS

The authors acknowledge Drs. S. Mori, A. Tomabechi, and M. Tanaka for their continuous encouragement of this work.

REFERENCES

1. K. Yoshida, H. Nakajima, K. Koizumi, M. Shimada, Y. Sanada, Y. Takahashi, E. Tada, H. Tsuji, and S. Shimamoto, Development of cryogenic structural materials for Tokamak reactor, in: "Austenitic Steels at Low Temperatures," R. P. Reed and T. Horiuchi, eds., Plenum Press, New York (1983), pp. 29-39.
2. H. Nakajima, K. Yoshida, Y. Takahashi, E. Tada, M. Oshikiri, K. Koizumi, S. Shimamoto, R. Miura, M. Shimada, S. Tone, H. Masumoto, and T. Sakamoto, Development of the new cryogenic structural materials for Fusion Experimental Reactor, in: "Advances in Cryogenic Engineering - Materials," Vol. 30, Plenum Press, New York (1984), pp. 219-226.
3. G. A. Clarke and G. M. Brown, "Computerized Methods for J_{Ic} Determination Using Unloading Compliance Techniques," ASTM STP 710, American Society for Testing and Materials, Philadelphia, (1980), pp. 110-126.
4. R. L. Tobler and R. P. Reed, Interstitial carbon and nitrogen effects on the tensile and fracture parameters of AISI 304 stainless steels, in: "Materials Studies for Magnetic Fusion Energy Applications at Low Temperatures - III," NBSIR 80-1627, National Bureau of Standards, Boulder, Colorado (1980), pp. 15-48.

TENSILE, FRACTURE TOUGHNESS AND FATIGUE

CRACK GROWTH RATE PROPERTIES OF HP 9-4-30

W. A. Logsdon

Westinghouse R & D Center
Pittsburgh, Pennsylvania

ABSTRACT

Tensile, fracture toughness and fatigue crack growth rate (FCGR) properties were developed at room and cryogenic temperatures on HP 9-4-30, a 9Ni-4Co-0.30C content high strength martensitic steel. HP 9-4-30 exhibited excellent room temperature strength, ductility and fracture toughness as well as good FCGR properties. Because the fracture toughness of HP 9-4-30 decreased significantly with temperature over the temperature range 297 to 77 K (75 to -320°F), indicating a ductile-to-brittle transition, combined with the fact that the growth rate of fatigue cracks in HP 9-4-30 increased with decreasing temperature, HP 9-4-30 is not recommended for extreme low temperature structural applications.

INTRODUCTION

Westinghouse Electric Corporation and the Electric Power Research Institute (EPRI) jointly funded the design and development of a 300 MVA, 3600 rpm superconducting generator. HP-9-4-30, a low alloy, high strength martensitic steel, was proposed for the thermal contraction joint laminations and inner rotor end plate of the superconducting generator. Under normal conditions, both the thermal contraction joint laminations and the inner rotor end plate operate at room temperature. In the unlikely event of a gross helium leak, however, the remote possibility exists that the HP 9-4-30 thermal contraction joint laminations or inner rotor end plate will be required to sustain operational loads while exposed to extremely low temperatures. Consequently, the tensile, fracture toughness and FCGR properties of HP 9-4-30 were developed at both room and cryogenic temperatures.

MATERIAL

HP-9Ni-4Co-0.30C (designated HP 9-4-30) is a low alloy, high strength steel with high nickel and cobalt contents to promote toughness. This quenched and tempered martensitic steel was produced by the vacuum consumable electrode process. The alloy was designed to have high hardenability, good weldability in the heat-treated condition and excellent fracture toughness. Typical applications include aircraft structural components, rocket motor cases, armor plate, hydrofoil assemblies and underwater pressure vessels.

Table 1. Chemical Composition and Heat Treatment of HP 9-4-30

Chemical Composition, Wt. Percent

C	Mn	P	S	Si	Ni	Cr	Mo	V	Co	Cu
0.32	0.25	0.005	0.005	0.16	7.22	1.06	0.99	0.09	4.34	0.09

Heat Treatment

Double Anneal	704°C (1300°F), hold 2 h, air cool
	649°C (1200°F), hold 4 h, air cool
Normalize	899°C (1650°F), hold 3 h, air cool
Austenitize	816°C (1500°F), hold 3 h, water quench
Refrigerate	-73°C (-100°F), hold 1 h
Double Temper	538°C (1000°F), hold 2 h, air cool
	538°C (1000°F), hold 2 h, air cool

The subject HP 9-4-30 was received as a 6.4 cm (2.5 in.) diameter by 45.7 cm (18.0 in.) long bar. The chemical composition and heat treatment of this HP 9-4-30 bar are listed in Table 1. The HP 9-4-30 satisfied the nominal chemical requirements typically specified for this alloy.[1] Hardness in the as-received condition, i.e., following the double anneal, equaled 343 to 358 Brinell (approximately 38.5 R_c). Hardness in the final heat treated condition averaged 45.7 R_c.

EXPERIMENTAL PROCEDURES

Tensile tests were conducted on longitudinal and radial button head 0.51 cm (0.20 in.) diameter by 2.54 cm (1.0 in.) long gage length specimens. All tests were performed according to the American Society for Testing and Materials (ASTM) Standard Test Methods of Tension Testing of Metallic Materials (ASTM E8).

Fracture toughness and FCGR tests were performed on 5.1 cm (2.0 in.) wide (W) by 6.1 cm (2.4 in.) high compact type specimens machined so that the specimen crack opening displacement could be measured at the centerline of loading. Compact specimens for fracture toughness testing were 2.54 cm (1.0 in.) thick (B) while the specimens for FCGR testing were 0.64 cm (0.25 in.) thick. All of the compact specimens were oriented with their notch directions parallel to the longitudinal bar dimension or R-L per the ASTM Standard Test Method for Plane-Strain Fracture Toughness of Metallic Materials (ASTM E399).

The fracture toughness of HP 9-4-30 at 297 K (75°F) was determined via the ASTM Standard Test Method for J_{Ic}, A Measure of Fracture Toughness (ASTM E813), utilizing the unloading compliance test technique. The J resistance curve thus obtained is illustrated in Fig. 1. Unfortunately, because the specimen experienced a rather large and unstable load drop immediately following the maximum load point, sufficient data points were not available within the boundary limits specified in ASTM E813. The J_{Ic} value was obtained in two ways; these are explained in detail in the Results and Discussion section of this paper.

At 77 K (-320°F), the fracture toughness test on HP 9-4-30 produced a linear load-deflection record to failure. As such, the compact specimen loaded at this low temperature experienced essentially zero plasticity and the ASTM E399 plane strain fracture toughness standard was appropriate for evaluating fracture toughness. The 2.54 cm (1.0 in.) thick compact specimens tested at 297 or 77 K (75 or -320°F) both were of adequate size to satisfy the respective ASTM E813 or ASTM E399 dimensional requirements necessary for a valid test.

The FCGR tests on HP 9-4-30 were conducted according to the ASTM Standard Test Method for Constant-Load-Amplitude Fatigue Crack Growth Rates Above 10^{-8} m/cycle (ASTM E647). A sinusoidal waveform was utilized at a frequency of 10 Hz. The load ratio ($R = P_{min}/P_{max}$ where P_{min} and P_{max} are the applied minimum and maximum loads, respectively) equaled 0.10.

Crack length was determined by the elastic compliance method. The stress intensity factor range (ΔK) expression developed for the compact specimen by Saxena and Hudak was utilized in this investigation.[2] The seven-point incremental polynominal method was employed to convert crack length (a) versus elapsed cycles (N) to crack growth rate (da/dN) for the 297 and 77 K (75 and -320°F) FCGR tests. At 4 K (-452°F), the fracture toughness of HP 9-4-30 was rather low; as such, the specimen cycled in fatigue failed after generating only three crack length versus elapsed cycle data points. The three-point incremental polynominal method was utilized to salvage a single da/dN versus ΔK data point from this test.

RESULTS AND DISCUSSION

Tensile

The tensile properties of HP 9-4-30 are illustrated in Fig. 2. Test specimens were loaded either longitudinally or radially. Duplicate specimens were tested only at liquid helium-temperature.

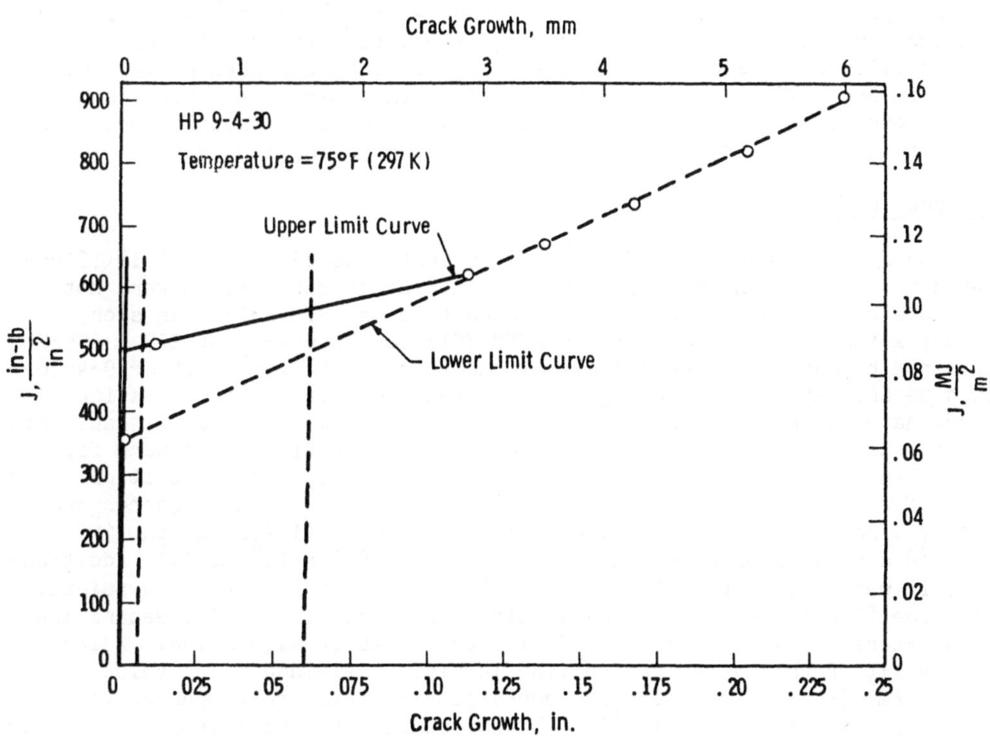

Fig. 1. J resistance curve for HP 9-4-30 at 297 K (75°F).

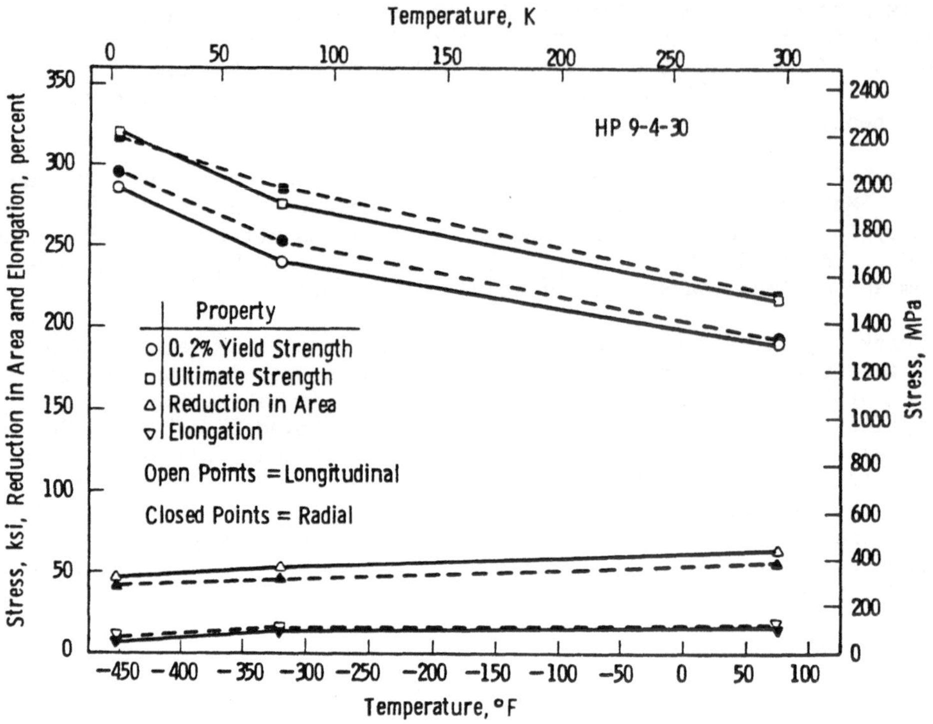

Fig. 2. Tensile properties of HP 9-4-30.

The yield and ultimate strengths of HP 9-4-30 both increased with decreasing temperature while the ductility (reduction in area and elongation) of this alloy decreased moderately with decreasing temperature over the temperature range 297 to 4.2 K (75 to -452°F). In general, the yield and ultimate strengths of HP 9-4-30 were superior when measured radially while the ductility of this alloy was superior when measured longitudinally.

Fracture Toughness

The J resistance curve for HP 9-4-30 tested at 297 K (75°F) is illustrated in Fig. 1. As previously noted, sufficient data points were not available within the boundary limits specified in ASTM E813. As such, a valid fracture toughness value per ASTM E813 for HP 9-4-30 at 297 K (75°F) could not be obtained. Nevertheless, the fracture toughness of HP 9-4-30 could be bracketed by employing two least-squares fits of the existing J versus Δa data points. Initially, only the first two J versus Δa data points (apart from the initial point which fell on the blunting line) were fit (solid line in Fig. 1). This produced an upper limit for the critical value of J (J_{Ic}) equal to 0.087 MJ/m^2 (495 in·lb/in.^{2}), which corresponds to a K_{Ic} fracture toughness value of 131.7 MPa·$m^{1/2}$ (119.0 ksi·in.$^{1/2}$) based on a modulus of elasticity of 0.197 GPa (28.6 x 10^6 psi).[1] Additionally, it was obvious from Fig. 1 that all of the J versus Δa data points, with exception of the initial data point where actual material separation crack extension was experienced, fell into a neat straight line. A least-squares fit of this data (broken line in Fig. 1) yielded a J_{Ic} value of 0.062 MJ/m^2 (356 in.·lb/in.^{2}) and corresponding fracture toughness value of 111.7 MPa·$m^{1/2}$ (100.9 ksi·in.$^{1/2}$). Consequently, the room temperature fracture toughness of the subject HP 9-4-30 fell within the range of 111.7 to 131.7 MPa·$m^{1/2}$ (100.9 to 119.0 ksi·in.$^{1/2}$).

Furthermore, a material's resistance to tearing instability is related to the slope of the J-integral resistance curve through the tearing modulus, T.[3] The higher the tearing modulus value, the greater the ability of a material to absorb energy upon additional crack extension. Materials which have a tearing modulus over 100 (nondimensional) typically demonstrate a high degree of stability against tearing mechanisms for all crack configurations. A tearing modulus below 10, however, virtually ensures tearing instability in some crack configurations as soon as J_{Ic} and limit load are reached.

The tearing modulus for the subject HP 9-4-30 at 297 K (75°F) equaled 0.75 to 1.56, depending on whether the two-point or five-point J-resistance curve was considered. Consequently, HP 9-4-30 will be rather unstable in most or all crack configurations as soon as J_{Ic} or limit load is reached. This crack instability behavior was substantiated by the fact that the HP 9-4-30 compact fracture toughness specimen tested at room temperature experienced a rather large and unstable load drop immediately following the maximum load point, as previously noted.

The fracture toughness of HP 9-4-30 tested at 77 K (-320°F) equaled 44.6 MPa·m$^{1/2}$ (40.3 ksi·in.$^{1/2}$), which was less than half of the fracture toughness demonstrated by this material at room temperature. Keep in mind that the load-deflection record for HP 9-4-30 at liquid nitrogen temperature was linear to the point of failure and that this fracture toughness value was calculated according to the ASTM E399 plane-strain fracture toughness standard. Consequently, it appears that this high strength, low alloy steel experienced a ductile-to-brittle transition at some intermediate temperature between 297 and 77 K (75 and -320°F).

Fatigue Crack Growth Rate

The FCGR properties of HP 9-4-30 are illustrated in Fig. 3. All three of the FCGR tests were conducted at a load ratio of 0.10 and a frequency of 10 Hz using a sinusoidal waveform. Obviously, the growth rate of fatigue

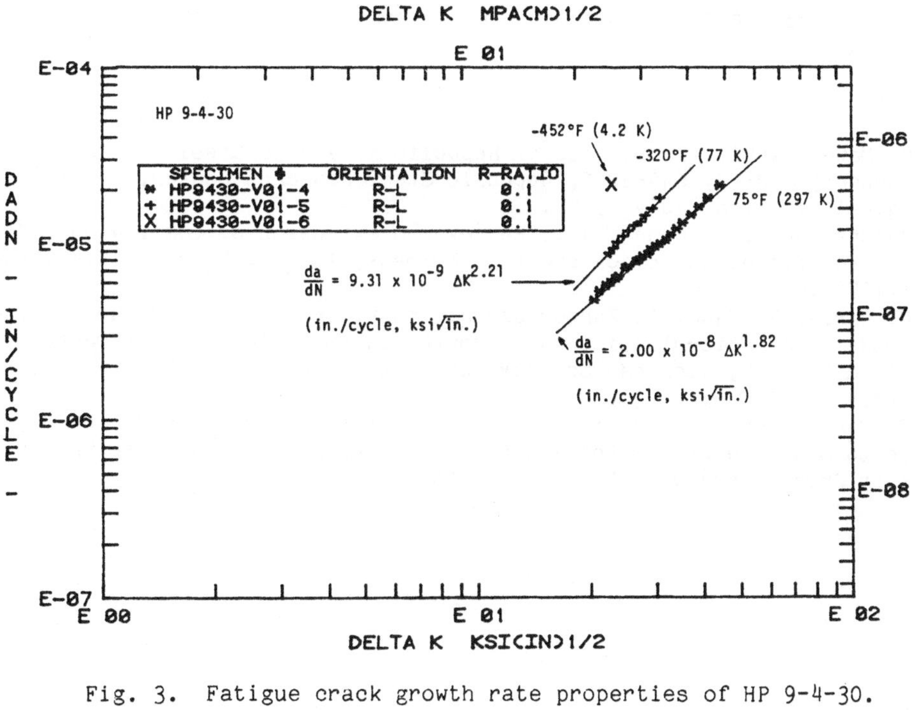

Fig. 3. Fatigue crack growth rate properties of HP 9-4-30.

cracks in HP 9-4-30 increased with decreasing temperature. This behavior was similar to that previously observed on other steels, i.e., those materials that are martensitic at room temperature, such as HP 9-4-30, or those that transform from austenite to martensite when strained at low temperatures. On the other hand, structurally stable austenitic stainless steels and nickel-base superalloys, materials typically utilized in cryogenic structural applications, demonstrate the opposite behavior, i.e., the growth rate of fatigue cracks decreases with decreasing temperature.

A nearly linear relationship between log (da/dN) and log (ΔK) for the Stage II region of fatigue crack growth was obtained for HP 9-4-30 at 297 and 77 K (75 and -320°F). This behavior is typical of most FCGR data. Since this linear relationship existed, the crack growth rate data was expressed in terms of the generalized FCGR law developed by Paris.[4] The appropriate expressions for HP 9-4-30 are included in Fig. 3. The slope of the log-log plot for HP 9-4-30, in particular at room temperature, was surprisingly shallow.

Recall, the compact specimen tested in fatigue at liquid helium temperature failed prematurely such that only a single da/dN versus ΔK data point was obtained. Consequently, it was impossible to develop an expression to describe the growth rate of fatigue cracks in HP 9-4-30 at 4.2 K (-452°F). Nevertheless, the single point that was obtained for HP 9-4-30 at liquid helium temperature permitted an extrapolation of the effect of temperature on FCGR for this material from room temperature all the way down to 4.2 K (-452°F).

CONCLUSIONS

HP 9-4-30 exhibited excellent room temperature strength, ductility and fracture toughness as well as good FCGR properties. In addition, the fracture toughness of HP 9-4-30 decreased substantially over the temperature range 297 to 77 K (75 to -320°F), indicating a ductile-to-brittle transition. Moreover, the growth rate of fatigue cracks in HP 9-4-30 increased with decreasing temperature over the temperature range 297 to 4.2 K (75 to -452°F). Finally, HP 9-4-30 is not recommended for extreme low temperature structural applications.

REFERENCES

1. J. P. Materkowski, "Properties of Republic HP 9-4-30 Steel: A Summary Report," PR-12, 049-82-2, Republic Steel Research Center, Metallurgical Division (July 1982).
2. A. Saxena and S. J. Hudak, Jr., Review and extension of compliance information for common crack growth specimens, Int. J. Fract., 14:453-468 (1978).
3. P. C. Paris, H. Tada, A. Zahoor and H. Ernst, Instability of the tearing mode of elastic-plastic crack growth, in "Elastic-Plastic Fracture," ASTM STP 668, American Society for Testing and Materials, Philadelphia (1979). pp. 5-36.
4. P. C. Paris, The fracture mechanics approach to fatigue, in "Proceedings Tenth Sagamore Army Materials Research Conference," Syracuse University Press, Syracuse, New York (1964).

NOTCH TENSILE MEASUREMENTS AND FRACTURE TOUGHNESS CORRELATIONS FOR AUSTENITIC STAINLESS STEELS

R. P. Reed, D. T. Read, and R. L. Tobler

Fracture and Deformation Division
National Bureau of Standards
Boulder, Colorado

ABSTRACT

Thirty-two alloys were included in a study of notch tensile testing as a method of fracture toughness characterization for austenitic stainless steels at 4 K. For the same austenitic stainless steels, tensile and J-integral fracture toughness [$K_{Ic}(J)$] measurements have been conducted. The notch tensile strength (σ_{NTS}) generally increases with yield strength (σ_y), and the σ_{NTS}/σ_y ratios are typically much greater than 1.0. Correlations between σ_{NTS}, $K_{Ic}(J)$, and σ_y were assessed. The best data fit was found between the ratio $\sigma_{NTS}/K_{Ic}(J)$ and the toughness, $K_{Ic}(J)$. Unfortunately, from this relation there is not uniqueness of K_{Ic} from σ_{NTS}. There are three regions in plots of J-integral fracture toughness vs. cylindrical bar notch-tensile measurements: (1) linear elastic [σ_{NTS} increases as $K_{Ic}(J)$ increases]; (2) elastic-plastic [σ_{NTS} is essentially independent of $K_{Ic}(J)$]; (3) plastic [σ_{NTS} decreases as $K_{Ic}(J)$ increases]. The elastic-plastic region is associated with a plastic zone that extends completely through the notched cross-sectional area.

INTRODUCTION

For austenitic stainless steels used in structural components of superconducting magnets, elastic-plastic fracture toughness evaluations at 4 K are usually required for qualification of base metals and welds. Two acceptable methods of quantitative fracture toughness measurement under elastic-plastic conditions are the J-integral and the crack-tip-opening displacement (CTOD) tests; both require large specimens with fatigue precracking to provide a sharp crack front. The complexity of measurement methodologies and difficulties in interpretation of test data prevent many laboratories from adopting these tests.

Impact tests are less expensive. Unfortunately, they provide only a qualitative measure of toughness and are unsuitable at 4 K because adiabatic heating occurs during straining prior to fracture. Adiabatic temperature rises of up to 150 K have been reported[1-3] and are possible because of the very low specific heat of metals at 4 K and the high strain rate from impact loading.

This paper assesses the use of the notch tensile test to estimate the elastic-plastic fracture toughness of austenitic stainless steels. Notch tensile tests were performed on austenitic steels at low temperatures[4-6] long before quantitative elastic-plastic fracture toughness methods became available. Subsequently, they have been used for direct measurement of the plane strain fracture toughness of aluminum and steel alloys.[7-11] Specimen machining costs less than compact tensile specimens and the test methodology is simple, requiring no strain gages or displacement measurements.

The ratio of notch tensile strength (σ_{NTS}) to unnotched tensile yield strength (σ_y) is accepted as a qualitative indicator of fracture toughness: ratios greater than 1.0 indicate tough behavior (ability to deform plastically in the presence of stress concentration). A simplified stress concentration factor, K_t, is given by

$$K_t = (d/2r)^{1/2} \quad (1)$$

where d is the width between the notches and r is the notch radius. Typical K_t values used with cylindrical or flat specimens range from 4 to 7 for steels[4] and from 16 to 23 for aluminum alloys.[12]

Kaufman et al.[12] summarized an extensive room-temperature notch-tensile measurement program on aluminum cylindrical specimens. The alloys (age-hardened 2024, 2124, 7075, 7475) generally exhibited linear-elastic fracture behavior. The plane-strain fracture toughness (K_{Ic}) initially increased with increasing σ_{NTS}. However, at higher toughness levels, K_{Ic} tended to be less dependent on σ_{NTS}. They found that the ratio σ_{NTS}/σ_y was approximately linearly dependent on K_{Ic} at low ratios (~0.85 to 1.25). At higher values (~1.25 to 1.45), the ratio tends toward less dependency on K_{Ic}.

All previous notch tensile tests of alloys exhibiting elastic-plastic fracture behavior, such as austenitic steels at 4 K, have been used for qualitative toughness evaluation, and until now there appears to have been no attempt to correlate notch tensile measurements with K_{Ic} measurements. In this study, thirty-two austenitic stainless steels were tested at 4 K, and the relationships between σ_{NTS}, σ_y, and K_{Ic} were determined from the previously published 4-K tensile and J-integral fracture toughness data for these alloys.[13]

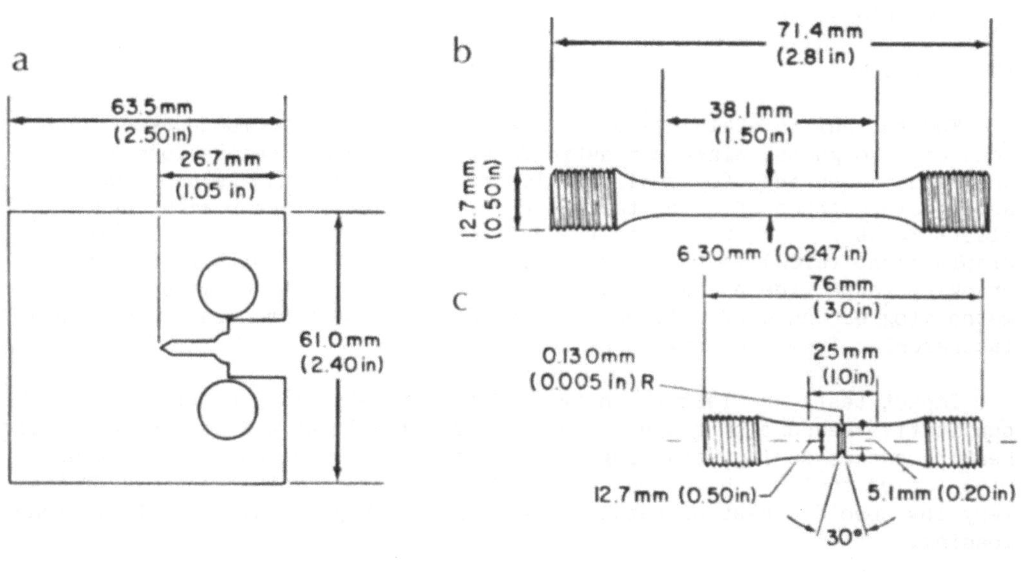

Fig. 1. Compact tension (a), tensile (b), and notch tensile (c specimens used in mechanical property measurements.

EXPERIMENTAL PROCEDURES

The compositions and metallurgical variables for the alloys have been previously summarized.[13] All alloys were completely austenitic (prior to testing) and were tested in the annealed condition. The tensile, notch tensile, and compact-tension (fracture-toughness) specimen geometries are shown in Fig. 1.

Notch tensile tests were conducted at 4 K using a crosshead rate of 8.3 x 10^{-4} mm/s on two specimens per alloy. A 0.127-mm notch-tip radius was specified for each specimen. Notch radii, measured after grinding, varied from 0.125 to 0.135 mm. The stress concentration factor, K_t, from Eq. 1 was 4.3 to 4.5. The σ_{NTS} was calculated from the maximum load divided by the original notched cross-sectional area.

The tensile properties used in this paper for analysis were previously measured at a plastic strain rate of approximately 1.5 x 10^{-5}. At 4 K the yield and ultimate strength inaccuracies are about ± 2%, strain measurement inaccuracies are about ± 3%, and estimates of material variability for yield and ultimate strengths are about ± 5%. Notch tensile strength measurements reflect inaccuracies and material variabilities similar to those for the tensile strength measurements.

In most cases, the 4-K fracture-toughness data referred to in this paper represent K_{Ic} estimates based on J-integral tests. For several high-strength steels (N-50, 18/18, N-40, N-60, and AISI 216), the data represent direct measurements of K_{Ic}.

RESULTS AND DISCUSSION

The mechanical property data for each austenitic stainless steel are listed in Table 1. All data represent averages of at least two measurements taken at 4 K. All steels tested in this study have a σ_{NTS}/σ_y ratio greater than 1.0, ranging from 1.07 to 4.15. Therefore, all the annealed austenitic stainless steels are relatively tough at 4 K by the conventional notch tensile test criterion. Usually as the σ_y of austenitic stainless steels increases, their σ_{NTS} also increases, as shown in Fig. 2. This contrasts with the accepted inverse relation between $K_{Ic}(J)$ and σ_y, shown in Fig. 3 for the austenitic steels included in this study. The correlation between the σ_{NTS} and $K_{Ic}(J)$ at 4 K is not obvious (see Fig. 4). We found excellent correlation between the log of the ratio $\sigma_{NTS}/K_{Ic}(J)$ and $K_{Ic}(J)$ (see Fig. 5). The best fit equation of data in Fig. 5 is

$$\sigma_{NTS} = 44K_{Ic}(J) \exp -6.54 \times 10^{-3}K_{Ic}(J) \qquad (2)$$

where σ_{NTS} has units of MPa and $K_{Ic}(J)$ has units of MPa$\sqrt{m}$. Equation 2, also plotted in Fig. 4, adequately depicts most of the data. Unfortunately, $K_{Ic}(J)$ is not unique for given σ_{NTS} values in this formulation. The ratio σ_{NTS}/σ_y increases with increasing $K_{Ic}(J)$, as shown in Fig. 6. Here Eq. 2 is also included and adequately represents the data.

Because multiple values of $K_{Ic}(J)$ correspond to single σ_{NTS} values (Fig. 4) and because the ratio of σ_{NTS} to σ_y is rather insensitive to the value of $K_{Ic}(J)$ (Fig. 6), measurement of σ_y does not provide unique quantitative fracture toughness information.

The results plotted in Figs. 2 through 6 are interpreted further: From Fig. 4 it is evident that, under linear-elastic fracture mechanics conditions, there is an approximate correlation between σ_{NTS} and $K_{Ic}(J)$ following Eq. 2, when $K_{Ic}(J) \leq 100$ MPa$\sqrt{m}$. This same trend of direct proportionality between K_{Ic} and σ_{NTS} was demonstrated by linear-elastic notch tensile

Table 1. Mechanical Properties of Austenitic Stainless Steels at 4 K.

Specimen Number	σ_{NTS} (MPa)	σ_u (MPa)	σ_y (MPa)	σ_{NTS}/σ_y	$K_{Ic}(J)$ (MPa$\sqrt{m}$)
739	1910	1480	460	4.2	273
740	2190	1370	720	3.0	244
741	2340	1420	870	2.7	254
414	2538	1652	949	2.7	143
520	1377	1220	961	1.4	87
413	2920	1580	1020	2.9	204
415	2040	1390	1040	2.0	75
416	2690	1630	1160	2.3	238
2	1700	1550	440	3.9	--
3	1870	1530	530	3.5	337
4	2030	1570	740	2.7	266
5	2120	1700	880	2.4	230
6	2230	1630	900	2.5	235
7	2460	1640	1190	2.1	124
8	2480	1740	1180	2.1	137
9	2390	1700	1290	1.9	124
70	2060	1620	660	3.1	192
71	2030	1580	720	2.8	215
72	2180	1510	760	2.9	243
73	2170	1810	880	2.5	117
74	2300	1520	960	2.4	121
75	2340	1670	990	2.4	149
76	2340	1580	1050	2.2	162
77	2250	1650	860	2.6	206
78	2400	1570	900	2.7	205
304L	1880	1680	610	3.1	310
304LN	2080	1660	790	2.6	214
316	1880	1300	710	2.6	268
N-50	2330	1890	1450	1.6	106
310	1880	1260	640	2.9	285
18/18	1910	2050	1780	1.1	75
N-40	2520	1660	1200	2.1	136
N-60	1540	1350	990	1.6	91
216	2280	1820	1400	1.6	107

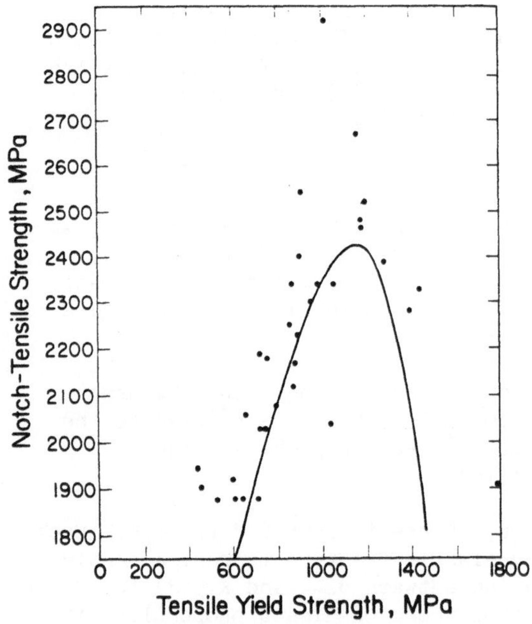

Fig. 2. Notch tensile strength versus tensile yield strength of annealed austenitic stainless steels at 4 K.

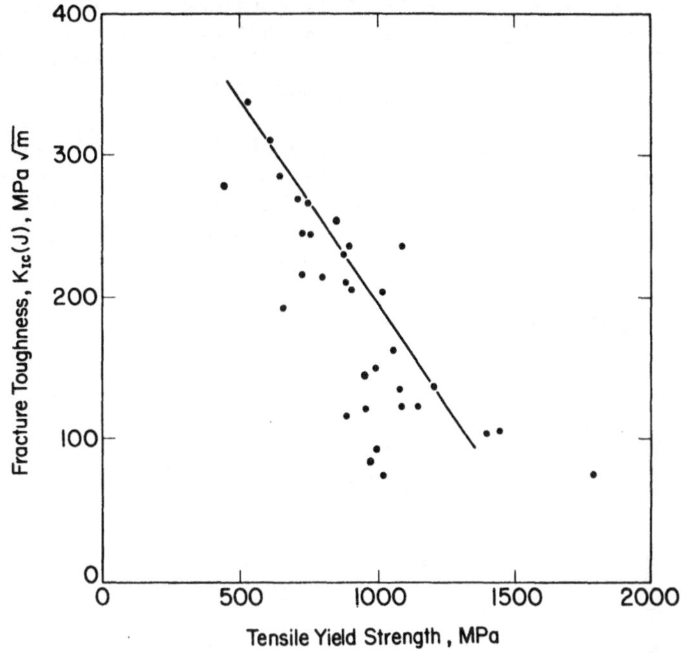

Fig. 3. Tensile yield strength versus fracture toughness of austenitic stainless steels at 4 K. Solid line is the trend line reported earlier for austenitic stainless steels at 4 K.[7-11]

measurements of aluminum alloys by Kaufman et al.[12] At high toughness levels [$K_{Ic}(J) \geq 200$ MPa$\sqrt{m}$, corresponding to lower yield-strength steels], σ_{NTS} decreases with increasing $K_{Ic}(J)$. The lower yield-strength alloys are ductile, and under the notch tensile loading conditions of our specimens, they experienced significant plastic deformation prior to fracture. In the intermediate toughness range of austenitic stainless steels at 4 K [100 MPa$\sqrt{m} \leq K_{Ic}(J) \leq 200$ MPa$\sqrt{m}$], σ_{NTS} is approximately independent of $K_{Ic}(J)$, and there is transition from linear-elastic to fully plastic fracture behavior. The tendency toward transition and independence of σ_{NTS} on K_{Ic} was also demonstrated by Kaufman et al.[12] for aluminum alloys at room temperature. Ductile alloys were not included in their study; therefore, they did not detect the entire transition range.

One can approximate the transition by considering the plastic zone size (r_p) under plane stress conditions:[14]

$$r_p = (1/\pi)(K/\sigma_f)^2 \tag{3}$$

where K is the stress intensity factor, r_p is the distance ahead of the crack tip to the edge of the plastic zone, and σ_f is the flow strength, $(\sigma_y + \sigma_u)/2$. The flow strength in Eq. 3 represents the effective local yield strength for the range of plastic strain considered. To achieve a plastic zone throughout the cross section, $2r_p$ must be at least equivalent to the diameter of the notched specimen (5.1 mm). Then

$$K/\sigma_f = [(5.1 \text{ mm}/2)(\pi)]^{1/2} \tag{4}$$

Assuming that such conditions are reached when the stress intensity factor reaches $K_{Ic}(J)$, then from Eq. 4:

$$\sigma_f \cong 11K_{Ic}(J) \qquad \text{(metric units)} \tag{5}$$

The ratio of σ_f to $K_{Ic}(J)$ in Eq. 5 is obtained when $K_{Ic}(J)$ is about 135 MPa$\sqrt{m}$. When $K_{Ic}(J)$ is 135 MPa$\sqrt{m}$, σ_{NTS} is 2430 MPa, the approximate midpoint of the transition from linear-elastic to fully plastic behavior (Fig. 4). A flow strength of about 1445 MPa corresponds to these associated σ_{NTS} and $K_{Ic}(J)$ values (Figs. 2 and 3; Table 1). Or, σ_f = 10.7 $K_{Ic}(J)$, close to the prediction of Eq. 5. The close similarity of the ratios implies that plane stress conditions begin when the fracture toughness reaches about 135 MPa$\sqrt{m}$ and dominate at higher toughness values. The existence of plane stress conditions at high $K_{Ic}(J)$ causes a close correlation with σ_y and, therefore, an inverse correlation of σ_{NTS} with $K_{Ic}(J)$.

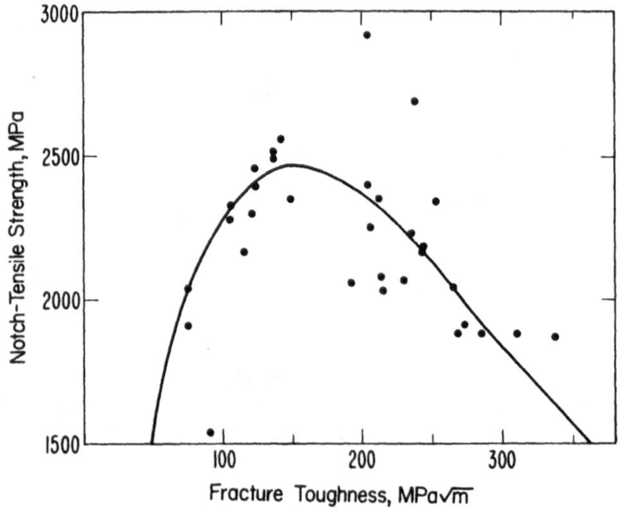

Fig. 4. Notch tensile strength versus fracture toughness of austenitic stainless steels at 4 K. Solid line is Eq. 2.

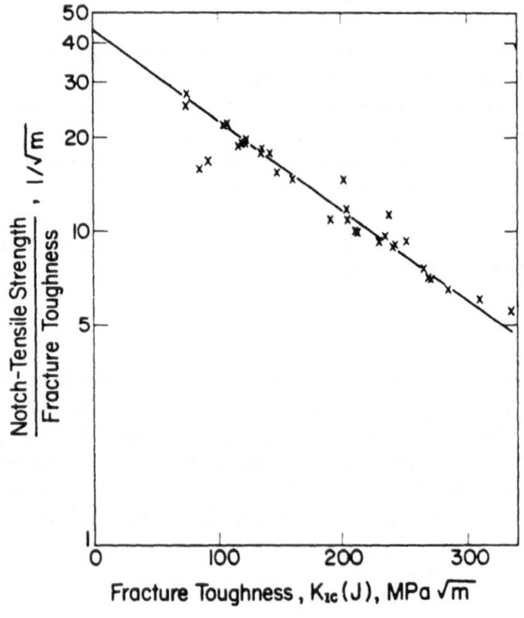

Fig. 5. Ratio of notch tensile strength to fracture toughness plotted against fracture toughness of austenitic stainless steels at 4 K. Solid line is Eq. 2.

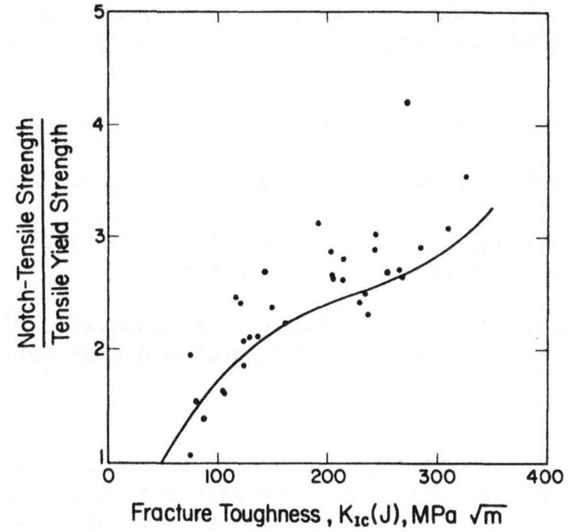

Fig. 6. Ratio of notch tensile strength to tensile yield strength versus fracture toughness of austenitic stainless steels at 4 K.

To extend the range of one-to-one correlation of σ_{NTS} with K_{Ic} or $K_{Ic}(J)$, the stress concentration factor at the notch tip must be increased, but decreasing the notch radius requires costly machining or grinding. Increasing specimen diameter precludes the use of the test for plate thicknesses of 12 mm or less and also increases the required cryostat load capacity above that which is normally available.

The notch tensile strength may be described as proportional to the square of the plane strain fracture toughness (K_{Ic}).[15] The J-integral (J) is related to δ, the crack-tip-opening displacement,[15] by

$$J = m \, \sigma_f \, \delta \tag{6}$$

where m is a constraint factor and, for notch tensile measurements, may be equated to σ_{NTS}/σ_f. The J-integral is related to K_{Ic} under plane strain conditions[15] by

$$J = [K_{Ic}(J)]^2 \, (1 - \nu^2)/E \tag{7}$$

where ν is Poisson's ratio and E is Young's modulus. Combining Eqs. 6 and 7 and substituting for m yields

$$\sigma_{NTS}/K_{Ic}(J) = K_{Ic}(J) \, (1 - \nu^2)/\delta E \tag{8}$$

Equation 8, of the same form as Eq. 2, is the function found to have the best fit when σ_{NTS} is plotted against $K_{Ic}(J)$ (Fig. 4). The slope of Fig. 4 (described in Eq. 2) should reflect the dependence of the CTOD on the ratio $\sigma_{NTS}/K_{Ic}(J)$, as both E and ν may be considered constant. We are now measuring the notch-opening displacement directly with clamp-on strain-gage extensometers. Further analysis of notch tensile CTOD measurements awaits new data from this approach.

CONCLUSIONS

At present there is no advantage in using notch tensile measurements of austenitic stainless steels to obtain quantitative fracture toughness information.

J-integral fracture toughness (compact-tension specimens) and notch tensile (cylindrical-bar specimens) measurements have been compared. There are three regions:

1. Linear elastic, where σ_{NTS} increases as $K_{Ic}(J)$ increases

2. Elastic-plastic, where σ_{NTS} is essentially independent of $K_{Ic}(J)$

3. Plastic, where σ_{NTS} is directly related to σ_y and inversely related to $K_{Ic}(J)$ [since $K_{Ic}(J)$ is inversely related to σ_y]

The transition region (elastic-plastic) is associated with a plastic zone that extends completely through the notched cross-sectional area.

ACKNOWLEDGMENTS

This research was sponsored by the Department of Energy, Office of Fusion Energy, Dr. V. Der, Project Monitor. The test materials include developmental heats prepared by the research laboratories of U.S. Steel (alloys 2-9), Armco Steel (alloys 70-78), and E. O. Paton Institute of Electrowelding (alloys 739-416). The careful work of J. Gerlitz and R. Walsh of NBS in performing the notch tensile tests is very much appreciated.

REFERENCES

1. W. G. Dobson and D. L. Johnson, in: "Advances in Cryogenic Engineering--Materials," vol. 30, A. F. Clark and R. P. Reed, eds. Plenum Press, New York (1984), pp. 185-192.
2. D. T. Read and R. P. Reed, in: "Advances in Cryogenic Engineering--Materials," vol. 26, Plenum Press, New York (1980), pp. 91-101.
3. K. Ishikawa, K. Hiraga, T. Ogata, and K. Nagai, Cryog. Eng. (Tokyo) 18(1):2-9(1983).
4. C. J. Guntner and R. P. Reed, Trans. Am. Soc. Met. 55:399-419 (1962).
5. A. Randak, W. Wessling, H. E. Bock, H. Steinmaurer, and L. Faust, Stahl Eisen, 91:1255-1269 (1971).
6. A. Hurlich and W. G. Scheck, in: "Advances in the Technology of Stainless Steels and Related Alloys," ASTM STP No. 369, American Society for Testing and Materials, Philadelphia (1963), pp. 127-143.
7. G. R. Irwin, Report 4763, Naval Research Laboratories, Washington, D.C. (1956).
8. J. G. Kaufman and E. W. Johnson, in: "Advances in Cryogenic Engineering," vol. 6, Plenum Press, New York (1960), pp. 637-649.
9. C. M. Carmen, D. F. Armiento, and H. Markus, J. Basic Eng. 87:904-916 (1965).
10. R. P. Wei and F. J. Lauta, Mater. Res. Stand. 5:305-310 (1965).
11. W. F. Brown, Jr. and J. E. Srawley, "Plane Strain Crack Toughness Testing of High Strength Metallic Materials," ASTM STP 410, American Society for Testing and Materials, Philadelphia (1966).
12. J. G. Kaufman, G.-T. Sha, R. F. Kohm, and R. J. Bucci, in: "Cracks and Fatigue," ASTM STP 601, American Society for Testing and Materials, Philadelphia (1976), pp. 169-190.
13. R. P. Reed, R. L. Tobler, and D. T. Read, in: Materials Studies for Magnetic Fusion Energy Applications at Low Temperatures--VIII," NBSIR 85-3025, National Bureau of Standards, Boulder, Colorado (1985), pp. 181-192.
14. J. W. Hutchinson, Trans. ASME 50:1042:1051 (1983).
15. J. H. Giovanola and I. Finnie, Solid Mech. Arch., 9:227-257 (1984).

CREEP OF 304 LN AND 316 L STAINLESS STEELS AT CRYOGENIC TEMPERATURES

L. D. Roth and A. E. Manhardt

Westinghouse R&D Center
Pittsburgh, Pennsylvania

E. N. C. Dalder and R. P. Kershaw, Jr.

Lawrence Livermore National Laboratory
Livermore, California

ABSTRACT

Creep behavior of Type 304 LN plate and 316 L shielded-metal-arc (SMA) deposited stainless weld metal was investigated at 4 K. Testing was performed at constant load in a creep machine with a cryostat designed for stability. Both transient and steady-state creep were observed during tests lasting over 200 h. Steady-state creep rates were much greater than expected from extrapolations of 300 K creep data. Creep rates on the order of 10^{-10} s^{-1} were observed at stresses around the yield stress for both materials. The stress exponent under these conditions is ~2.3. Possible creep mechanisms at this temperature and the impact of these results on the design of engineering structures for long-term structural stability at cryogenic temperatures are discussed.

INTRODUCTION

Lawrence Livermore National Laboratory (LLNL) is the site for the Mirror Fusion Test Facility (MFTF-B), a tandem-mirror fusion experiment that depends on high-field superconducting magnets for plasma containment. Magnet cases are made from 304 LN stainless-steel plate that is joined with 316 L stainless-steel filler metal. The magnet cases are at 4 K during operation.

The design lifetime of the MFTF-B is 10^3 full-power cycles over ten years. Design stresses are 67% of the 4-K 0.2% yield strength (σ_y) of the material. Time-dependent deformation that produces plastic strains above 0.2% during the lifetime of MFTF-B might cause difficulties.

Time-dependent plastic deformation, or creep, is thought to be an elevated-temperature problem, becoming a factor at temperatures greater than 0.4 T_m, where T_m is the melting temperature of the material in K. Extrapolating elevated temperature steady-state creep rates to cryogenic temperatures, using the activation energy determined at elevated temperatures, yields the conclusion that creep is a negligible factor in design at 4 K. Recent studies of long-term creep of copper and its alloys[1] show that creep occurs at temperatures as low as 4 K, and is classical in its behavior and time evolution. Thus, the activation energy must decrease for creep to occur at cryogenic temperatures, which has been confirmed for copper and its alloys.[1]

Primary and steady-state creep are observed at 77 K for copper and its alloys, and both increase with increasing stress. The steady-state creep rates are small ($\sim 10^{-10}$ s^{-1}), but are orders of magnitude greater than that predicted by extrapolation of high-temperature data to 77 K. The activation energy measured for steady-state creep of copper at 77 K is about 0.02 eV. The activation energy for steady-state creep of copper by self-diffusion at elevated temperatures is 2.0 eV. Activation-energy changes indicate mechanism changes for creep, so creep at cryogenic temperatures cannot be predicted by extrapolation from higher temperature creep data. Creep rates similar to those measured at cryogenic temperatures could pose an operational problem for some structures. Tien et al.[2] reviewed published cryogenic creep data and described creep behavior by alloy system. Besides copper, austenitic stainless steels creep at cryogenic temperatures as low as 20 K for tests up to 100 h.[3]

To evaluate the creep resistance of the structural materials used in MFTF-B magnet cases at cryogenic temperatures, the nature and extent of creep observed in these materials must be known. This report summarizes the results of creep tests performed on 304 LN plate and 316 L weld metal. The objective of this program was to evaluate whether creep is a significant design factor for magnet-case structural materials at 4 K.

EXPERIMENT

Specimens

Threaded creep test specimens with a 31.8 mm length and a 6.4 mm gage diameter were made from materials used in the MFTF-B magnet system. The composition of the materials in weight percent is given in Table 1. Weld and base plate 4 K properties are presented in Table 2. Tensile properties of the 304 LN materials were higher than expected.

Weld specimens were taken from a shielded-metal-arc (SMA) deposited weld joint in 75 mm thick 304 LN plate (Weld 82). The joint was welded in the horizontal position, using 3.2 mm diameter E316L-15 electrodes. No post weld heat treatement was used. Radiography and metallographic examination of the weld revealed a few microfissures. Magne-gage measurements on each of the weld specimens before and after creep testing showed less than 0.5 FN (Ferrite Number).

The creep test specimens are identified as follows. Weld specimens are identified with the prefix "W82". Longitudinal (L) specimens were

Table 1. Composition of the Materials Tested in This Program (Weight Percent)

Alloy	C	Mn	P	S	Si	Cr	Ni	Mo	N_2
304 LN	0.03	1.7	0.015	0.015	0.35	19.0	9.0	---	0.15 (nominal)
316 L	0.042	2.6	0.024	0.009	0.35	17.69	13.58	2.22	0.025

Table 2. Tensile Properties of 304 LN Plate and 316 L Weld Metal at 4 K

Material	Specimen Orientation	0.2% Yield Stress (MPa)	Ultimate Tensile Strength (MPa)	Elongation (%)	Reduction in Area, RA (%)	Modulus of Elasticity (GPa)
316 L	L	672	1200	53.8	43.7	138
316 L	TL	635	1104	36.2	21.7	229.1
316 L	TM	634	1096	21.6	20.9	216.7
316 L	TB	667	1446	38.6	37.9	199.4
304 LN	L	827	1599	31.7	NA	NA
304 LN	T	949	1838	43.8	NA	NA

taken near the top of the weld. Long-transverse specimens were taken from near-top, midwall, and near-bottom locations and are designated TL,TM, and TB, respectively. Specimen gage lengths are all weld material. Plate material is identified as "BM". Longitudinal (L) and transverse (T) specimens were taken at midwall from 19 mm thick 304 LN plate.

Method

Dead load creep testing was performed using methods described in Ref 4. A schematic diagram of the test facility is shown in Fig. 1. The dead load creep machine was configured with a 20:1 load amplifying lever arm and a Janis 10DT vacuum insulated cryostat with a 7 liter liquid helium (LHe) reservoir. The cryostat was fitted with a tensile testing insert to apply the creep load to the specimen. Specimens were fixtured in K-Monel grips. Prior to loading the test chamber was purged, precooled and then flooded with LHe. The system was allowed to come to thermal equilibrium at 4 K for 24 hours prior to loading. A nitrogen jacket surrounding the LHe chamber is automatically refilled on a 30 minute schedule to maintain thermal stability of the pull rod assembly. The LHe level in the test chamber was maintained above the top of the upper specimen grip.

Temperature was measured with Teflon-sheaved copper-constantan thermocouples on the upper and lower grips. A liquid nitrogen reference junction was used. During testing, no temperature gradient was observed across the specimen. Temperature measurements are accurate to 1 K.

Creep strain was measured using two noncontacting capacitance displacement gages mounted 180° apart on the test specimen. The gages are model 2500 capacitance strain-gage system, made by Mechanical Testing Instrument Company; they were mounted on the upper grip. Polished targets for the gages were mounted on the lower shank of the specimen. Gage displacement was calibrated in LHe and 300-K air. This gage system has a linear range of 1.25 mm and a resolution of 0.127 μm in air. Differences in displacement gages on loading indicated that specimens were subject to bending. This condition was detected and avoided. Creep strain is calculated from the averaged reading of the two gages.

Eight 200 h cryogenic creep tests at 4 K were performed in this program; four on each of the materials. Tests were performed at

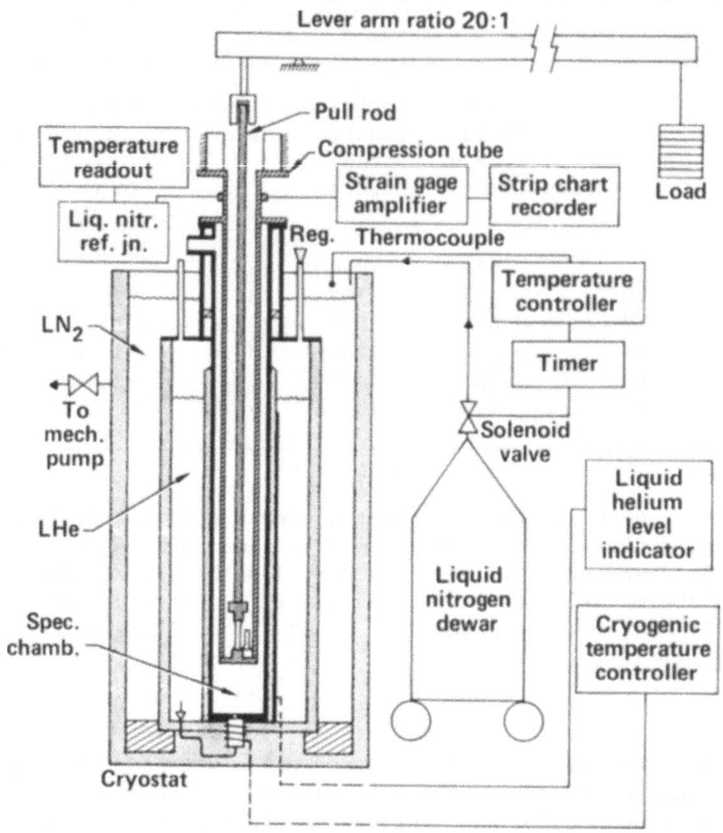

Figure 1. Facility for creep testing at cryogenic temperatures.

selected stress levels; nominally 50%, 67%, 100% and 120% of the yield strength at 4 K. The nominal design stress for the magnet materials is 67% of the 4 K σ_y.

RESULTS AND DISCUSSION

The results of the 4 K creep tests for 304 LN and 316 L materials are presented in Table 3. The strain-versus-time curves are presented in Fig 2a and b, respectively.

Many lessons were learned in performing the first few creep tests on the weld metal at 4-K and the resultant data is noisy. Some general trends can be observed. The amount of creep and creep rate increases with increasing stress. At 0.67 σ_y, creep was below the resolution of the test equipment. Creep rates as high as 8.2 x 10^{-11} s^{-1} were measured. The weld metal was not affected by work hardening during loading above σ_y. The amount of creep and its rate were larger than when stressing below $_y$. This test, W82L2, exceeded the range of the capacitance gage and was terminated early. The gage was incorrectly positioned for 5% full range. This was avoided in subsequent tests.

Only 3 of 4 creep curves for the weld material are shown in Fig. 2b. An anomolous strain excursion occurred for weld specimen W82TM2 which made it impractical to plot it with the other data. No temperature excursions were observed to account for the strain excursion. Weld

Table 3. 4K Creep Results for Types 304 LN Plate and 316 L Weld Metal

Specimen	Applied Stress, σ (MPa)	$\frac{\sigma}{\sigma_y}$	Strain on Loading ($x10^6$)	Strain on Unloading ($x10^6$)	Test Time (h)
BML1	555	0.67	4880	4710	193.1
BML2	828	1.0	11322	7826	215.6
BML3	414	0.5	5218	4116	216.7
BML4	936	1.13	27250	9034	239.6
W82L2	828	1.2	2800	---	191.5
W82TM2	607	1.0	7436	6140	215.0
W82TL2	425	0.67	5925	4620	142.6
W82TB2	697	1.0	8054	6793	190.5

Speci- men	Primary Creep Strain ($x10^6$)	Duration of Primary Creep Strain (h)	Secondary Creep Amount ($x10^6$)	Secondary Creep Rate (s^{-1})	Duration of Secondary Creep (h)	Total Creep Strain ($x10^6$)
BML1	350	20	46	7.13×10^{-11}	180	396
BML2	500	60	71	1.98×10^{-10}	155	571
BML3	140	2	---	$<1.0 \times 10^{-11}$	---	141
BML4	100	100	31	1.0 x 10-10	89	132
W82L2	>100	---	50	8.2×10^{-11}	>100	50
W82TM2	450	35	17,800	$0.001 - 1.0 \times 10^{-8}$	160	18,290
W82TL2	< 10	---	< 10	$<1.0 \times 10^{-11}$	---	< 10
W82TB2	20	50	40	8.0×10^{-11}	140	60

specimen W82TM2 showed periods with steady-state creep rates as high as 1×10^{-8} s^{-1} between large strain jumps. Total creep strain for test W82TM2 was 1.8% in 200 h (Table 3). Severe slip patterns were observed on the surface of the test specimen after unloading. This was the only specimen that this behavior was observed.

Microcracking, microstructural inhomogenieties, such as ferrite at dendrite boundaries, inclusions, slag, and porosity, are present in weld 82 and might contribute to the observed strain behavior. While the cause of the anomolous strain behavior is not understood, it is known that the lack of change in the Ferrite Number after creep testing indicates that no martensitic transformation occurred as a result of the creep test.

Cryogenic creep tests on 304 LN proved to be better controlled than the tests on the weld material. Significant primary and steady-state creep were seen for each 304 LN specimen (see Table 3). Primary creep lasted from 2 to 100 h, producing primary creep strains of 0.01 to 0.05%. Steady-state creep rates up to 1.98×10^{-10} s^{-1} were seen. Creep magnitude and rate varied as a function of applied stress and material strength. Applied stresses below 0.5 σ_y resulted in creep below the resolution of the test equipment. Applied stresses between 0.5 σ_y and σ_y produced enough creep to be considered in design.

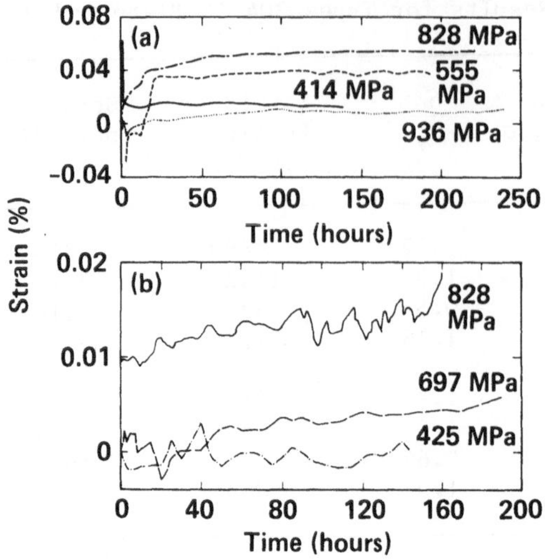

Figure 2. Strain vs time at 4 K for (a) 304 LN plate and (b) 316 L weld metal.

The stress dependence of the steady-state creep rate for 304 LN tests is shown in Fig. 3. The two specimens tested at 0.67 σ_y and at σ_y yield a plot of steady-state creep rate versus stress with a slope of 2.3. Specimen BML4, which work-hardened during loading, showed a lower steady-state creep rate and a smaller amount of primary creep than other 304 LN specimens. Suppression of primary creep by prior plastic deformation is known at elevated temperatures and applies for cryogenic creep of 304 LN as well.

The creep magnitude, steady-state creep rates, and the value of the stress dependence of steady-state creep are consistent with those observed for copper and copper alloys at 77 K,[1] suggesting that mechanisms for steady-state creep of copper might apply to 304 LN. Determination of the activation energy for steady-state creep and activation volume for primary creep is required to verify this hypothesis.

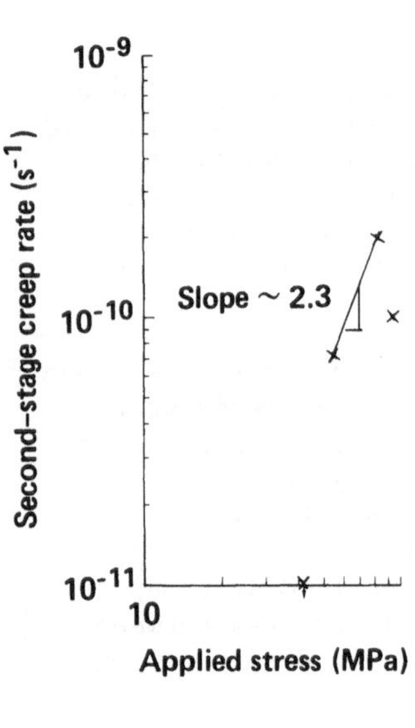

Figure 3. Stress dependence of second-stage creep rate for 304 LN plate at 4 K.

Table 4. Estimated Creep Effect for MFTF-B and MARs

Machine	Planned Number of Full-Power Years of Operation	Maximum Stress in Magnet Case	Estimated Strain at End of Life
MFTF-B	~ 0.0001	462 MPa	0.095%
MARS	24	462 MPa	3.5%

Two additional comments about the 304 LN test data are necessary. In test BML1 (555 MPa) a negative creep strain is manifest in the first few hours of the test. This is believed to be experimental artifact. A specimen under tensile load cannot contract. A martensitic transformation can be ruled out as well, since a volume expansion is expected. Most likely it is caused by deformation induced boiling of the LHe in the vicinity of the gap between capacitance probe and target. Similar but much smaller evidence of this was observed on other specimens. This will be alleviated in the future with a clip gage. Second, in specimen BML2, several temperature excursions occurred during testing. A broken helium gas-pressure release valve caused this problem until 112 hours into the test. After the temperature was restabilized at 4 K steady-state creep proceeded smoothly. The temperature induced strain excursions were subtracted out of the test data prior to plotting the curve in Fig. 2a. Such excursions raise the question of possible creep of magnet case materials during transient warm-up conditions under load.

Table 4 compares the effect of creep in 304 LN plate on the planned operation of two, tandem mirror fusion experiments; MFTF-B and MARS (Mirror Advanced Research Study).[5] For the planned operation scenarios, strain accumulation by creep will not be a problem in MFTF-B magnet cases, but will definitely be a problem in a power reactor such as proposed for MARS.

CONCLUSIONS

1. Significant primary and steady-state creep is seen for 304 LN plate and SMA-deposited 3-116 L weld metal at 4 K. Steady-state creep rates about 1×10^{-10} s^{-1} were measured at applied stresses close to σ_y at 4 K.

2. From an engineering standpoint, accumulation of creep damage in 304 LN plate will not be a problem in MFTF-B (0.095% strain), but will be a problem in power-producing fusion plants, such as MARS (3.5% strain).

3. Appreciable creep strains (on the order of 2% plastic strain) and creep rates were observed in 316 L weld material in one case. The microstructural features causing the large creep rate are possibly associated with segregation and small cracks (i.e., microfissures) in the weld metal.

ACKNOWLEDGMENTS

The authors would like to thank Mr. Chuck Fox and Tom Caulfied for their help, patience, and attention to detail in performing these creep tests. The work was performed under the auspices of the U.S. Department of Energy by the Lawrence Livermore National Laboratory under contract number W-7405-ENG-48.

REFERENCES

1. C. T. Yen, Doctoral Thesis, Columbia University, New York (1983).

2. J. K. Tien and Chin-Tang Yen, Cryogenic Creep of Metals, Adv. Cryog. Eng. 30:319 (1984).

3. R. Voyer and L. Weil, Creep Properties of 18-10 and 25-20 Stainless Steels at Cryogenic Temperatures, Adv. Cryog. Eng. 10:110 (1965).

4. C. Yen, L. Roth, J. Wells, and J. Tien, Equipment for Long-Time Creep Testing at Cryogenic Temperatures, Cryogenics 24(8):410 (1984).

5. C. D. Henning et al., "Mirror Advanced Reactor Study (MARS) Final Summary Report," Lawrence Livermore National Laboratory, California, UCRL-89370, presented at 10th Symp. on Fusion Eng., IEEE, Philadelphia, December 1983.

MAGNETO-MECHANICAL EFFECTS IN TWO STEELS WITH METASTABLE AUSTENITE

B. Fultz*, G. O. Fior, G. M. Chang, R. Kopa and J. W. Morris, Jr.

Dept. of Materials Science and Mineral Engineering, University of California, and the Lawrence Berkeley Laboratory, Berkeley, California

* Present Address: Keck Laboratory of Engineering Materials, California Institute of Technology, Pasadena, California

ABSTRACT

Magneto-mechanical effects, or effects of magnetic fields on mechanical properties of materials, are reported for two steels containing austenite that undergoes an fcc→bcc martensitic transformation during plastic deformation. Stress-strain curves from tensile tests of AISI 304 stainless steels and 9Ni steels were measured in magnetic fields as large as 18 T at temperatures of 4 K, 77 K and room temperature. Even in 18 T magnetic fields at cryogenic temperatures the magneto-mechanical effects were small, but they were reproducible and scaled with the strength of the magnetic field and the amount of transformation. Magneto-mechanical effects in steels with metastable austenite provide a unique means of determining how martensitic transformations affect mechanical behavior. The fcc→bcc transformation makes an important contribution to the work hardening of both AISI 304 and 9Ni steel, so the more rapid transformation during magnetic exposure results in a higher strength and a reduced elongation of tensile specimens. In AISI 304 stainless steel a reduced flow stress in the magnetic field was found at small plastic strains.

INTRODUCTION

AISI 304 stainless steel and 9Ni steel are representative of two classes of cryogenic structural steels which contain metastable austenite. Hot rolled AISI 304 stainless steel is nearly 100 % fcc austenite. Commercial 9Ni steel is typically prepared with about 10 % precipitated austenite particles in the bcc martensitic matrix. The austenite phase in both these steels is thermodynamically unstable at low temperatures, but the fcc→bcc martensitic transformation occurs only after some plastic deformation of these materials[1,2]. It is known that as the fcc→bcc martensitic transformation occurs during plastic deformation, the flow stress is changed. It is also known that high magnetic fields promote the fcc→bcc martensitic transformation of metastable austenite[3-5]. It therefore seems reasonable that additional martensitic transformation will occur with magnetic exposure during deformation and this additional transformation will change the flow stress. However, mechanical effects of magnetically induced martensitic transformations have only recently received attention. Magneto-mechanical effects may be of interest to fusion energy engineering because the operating environment for structural steels in large superconducting magnets includes high magnetic fields together with high stresses at low temperatures.

Recently we induced additional martensitic transformation in AISI 304 stainless and 9Ni steels by exposing them to steady 8 T and pulsed 16 T magnetic fields during plastic deformation. Small changes in deformation behavior were found during magnetic exposure[2,6,7]. In particular, the flow stress of AISI 304 stainless steels in tensile tests was found to increase during exposure to steady 8 T magnetic fields. These increases in flow stress were small, but reproducible. We have now extended these measurements of stress-strain curves of AISI 304 and 9Ni steels to steady fields of 12.5 T and 18.1 T. We have also performed a transmission electron microscopic (TEM) study of these materials, including observations of the fcc→bcc martensitic transformation. In addition to establishing the existence of magneto-mechanical effects, this recent work helps to demonstrate the microstructural mechanisms through which fcc→bcc martensitic transformations affect the flow stress of AISI 304 and 9Ni steels.

EXPERIMENT

The AISI 304L and AISI 304LN stainless steels had net concentrations of carbon plus nitrogen of 0.098 and 0.160 wt.%, respectively. The N.K.K. 9Ni steel was tested in both the as-received "QT" condition with relatively stable precipitated austenite, and after the following overtempering heat treatment to produce unstable austenite: (1000°C 3 hrs / WQ) , (800°C 1.5 hrs / WQ) , (600°C 300 hrs / WQ) . Specimens were machined with their tensile axes parallel to the rolling direction of the plate. Gauge sections of the specimens were 38x4.2x3.2 mm for the 12.5 T experiments, and 25x3.0x2.0 mm for the 18.1 T experiments. All tensile tests were performed in pairs; one specimen was exposed to the magnetic field and a control specimen was not. Care was taken to prepare both specimens with the same machining procedures and from the same region of the plate.

Experiments at 12.5 T were performed at 4 K with the specimen between a pair of superconducting solenoids that provided a magnetic field perpendicular to the tensile axis. Experiments at 18.1 T were performed at 4 K, 77 K and 290 K within the bore of a water cooled solenoid that provided a magnetic field parallel to the tensile axis. All tests were performed under "stroke control". For the experiments at 18.1 T the feedback control loop included an active correction of the "stroke signal" for the elastic deformation of the load frame itself, so the specimens were pulled at an approximately constant strain rate. Strain rates for all tests were 2.5×10^{-4}/sec. Interactions between the high magnetic field gradients and the paramagnetic components of the load frame developed forces as large as 200 Nt at 4 K, but these forces were measured and later subtracted from the data.

TEM studies employed a Phillips 301 transmission electron microscope operated at 100 kV. Specimens were prepared from bulk material by cutting 400 μm sections with a slow speed diamond wafering saw under flood cooling. After further chemical thinning and mechanical grinding on SiC papers, 3 mm disks were thinned to perforation in a twin-jet electropolishing apparatus. The solution for the 9Ni steel foils was 400 ml glacial acetic acid, 75 g chromium trioxide, and 21 ml distilled water. Final preparation of foils of AISI 304 stainless steel used cooled solutions of 10 % nitric acid and 90 % methanol.

RESULTS and DISCUSSION

AISI 304 Stainless Steels

Engineering stress-strain curves of AISI 304L and 304LN stainless steels at 4 K and 77 K are shown in Figs. 1 and 2. Three effects of magnetic exposure during testing are seen*. 1) The flow stress is suppressed at small plastic strains (this is not seen in Fig. 2b, however). 2) At intermediate strains the work-hardening rate is greater for the specimen exposed to the magnetic field, and its flow stress becomes greater than that of the control specimen. 3) The total elongation of the specimens tested in the magnetic field was less than that of the control specimens. These three effects of high magnetic fields on the cryogenic stress-strain curves of AISI 304 stainless steels are consistent with an increased amount of fcc→bcc martensitic transformation during magnetic exposure.

The effects of 18.1 T magnetic fields on stress-strain curves of both AISI 304L and 304LN stainless steels at room temperature were small, and perhaps experimentally insignificant. This is consistent with the fact that little martensitic transformation occurs during tensile tests of either material at room temperature[2].

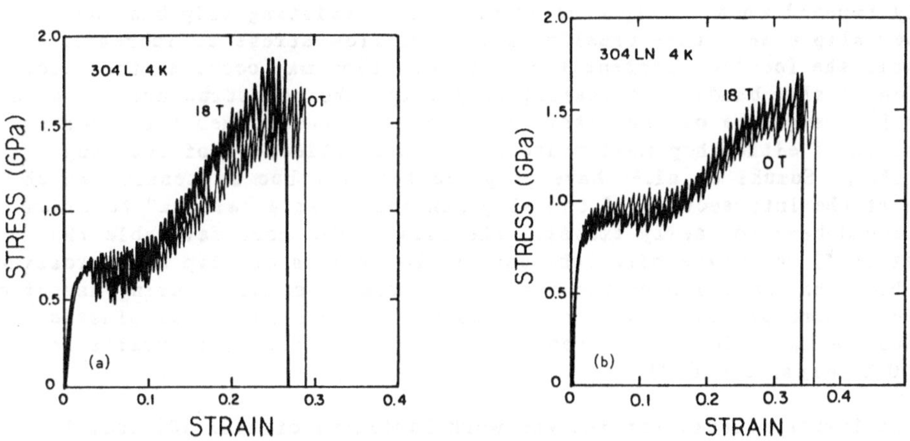

Fig. 1. Engineering stress-strain curves at 4 K.
a. AISI 304L b. AISI 304LN

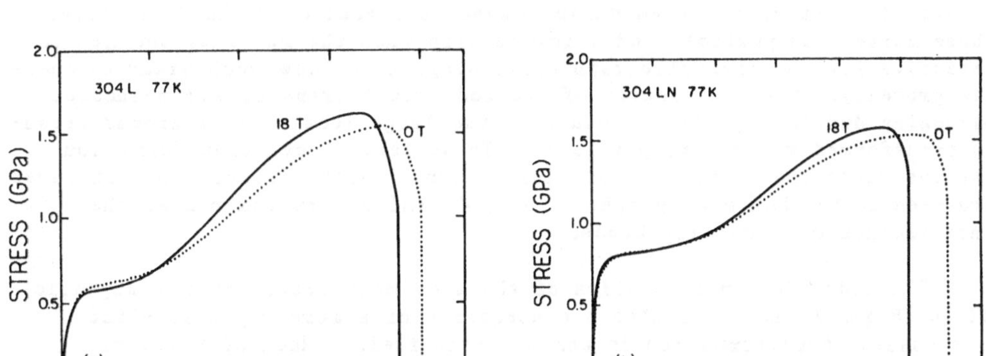

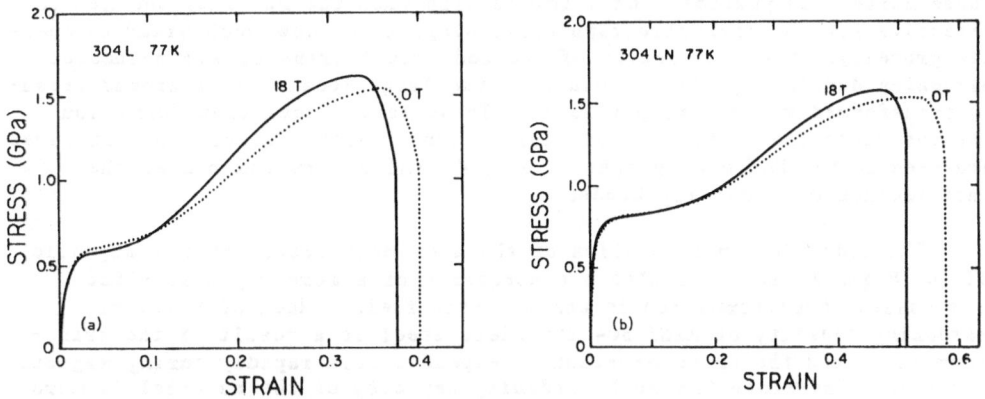

Fig. 2. Engineering stress-strain curves at 77 K.
a. AISI 304L b. AISI 304LN

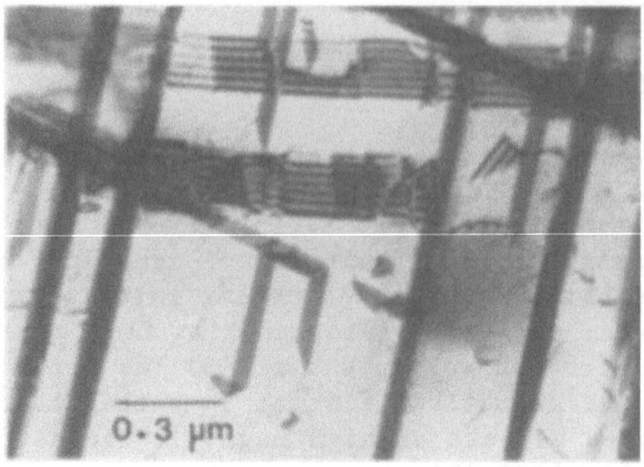

Fig. 3. TEM bright field micrograph of a region with low strain in AISI 304L strained at 77 K. Stacking faults on {111}fcc planes are seen with (000)+(002)fcc two-beam condition and [013]fcc zone axis.

At small plastic strains, slip in AISI 304 stainless steel occurs in well-defined bands. These bands are shown in Fig. 3 to begin as stacking faults on {111} planes. Some of the vertical bands in Fig. 3 are composed of layered stacking faults. The process of plastic deformation is generally impeded when a slip band encounters an existing slip band on another slip plane; this usually causes the flow stress to increase. However, the fcc→bcc martensitic transformation may occur at the intersection of slip bands. Crystallographically, the bcc structure can be a natural consequence of the intersection of slip bands when the bands are comprised of either hcp martensite or a dense collection of stacking faults[8,9]. Suzuki et al.[10] have proposed that the bcc martensite which forms at the intersection of two slip bands acts as a "window" to allow the second band to easily traverse the first. The more favorable the formation of bcc martensite, the easier the process of slip band crossing will be. The promotion of the martensitic transformation during magnetic exposure therefore results in a reduced flow stress for small plastic strains. A magnetically suppresed flow stress is seen for strains less than 0.1 (Figs. 1 and 2).

At larger plastic strains the work hardening of AISI 304 stainless steel is enhanced in high magnetic fields. The flow stress of the specimen tested in the magnetic field becomes greater than that of the control specimen (Figs. 1 and 2). In part this occurs because the additional bcc martensite particles formed during magnetic exposure are hard particles. These martensite particles will impede slip when the material becomes too defective for the crystallographically elegant "window mechanism" to operate properly. A second source of the additional transformation-induced hardening during magnetic exposure is the large local strain around freshly transformed martensite particles. These large local transformation strains cause the generation of dislocations. Such dislocation structures are seen in the TEM micrographs of Fig. 4, which were taken near the intersection of wide slip bands.

The reduction in elongation of the specimens tested in the magnetic field (Figs. 1 and 2) is also a consequence of a more rapid fcc→bcc martensitic transformation in the magnetic field. Much of the work hardening capacity of AISI 304 stainless steel is a result of the transformation, and the transformation is expended more rapidly during magnetic exposure. Therefore the work hardening capacity of the material is more rapidly exhausted during magnetic exposure, and the Considerè criterion for necking occurs at a lower strain, reducing the elongation of the specimen.

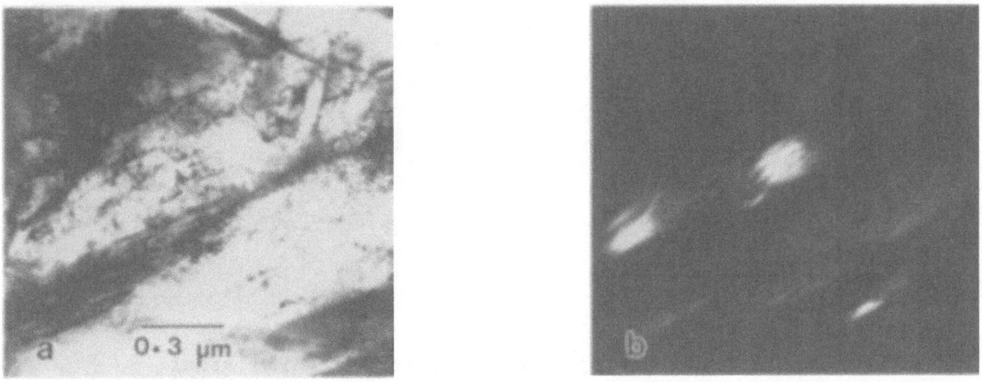

Fig. 4. a. TEM bright field micrograph of a region with high strain in AISI 304L strained at 77 K. b. complementary dark field micrograph of (002)bcc diffraction with [012]bcc zone axis.

<u>9Ni Steel</u>

Engineering stress-strain curves of commercial "QT" and overtempered 9Ni steel at 77 K and room temperature are shown in Fig. 5. The "window mechanism" for supppression of flow stress in AISI 304 stainless steel cannot operate in 9Ni steel because its microstructure primarily consists of bcc martensite particles. On the other hand, work-hardening effects due to the fcc→bcc martensitic transformation are expected for 9Ni steel. The TEM micrographs of overtempered material (Fig. 6) show particles of austenite which have partially transformed. Dislocation structures similar to those previously reported[1,11] are seen near the transformed particles, and some dislocations extend deeply into the tempered bcc martensite matrix. As in AISI 304 stainless steel, both the hard, fresh martensite particles and the dislocation structures around them will harden the material. The more rapid martensitic transformation during magnetic exposure causes an increased flow stress in both commercial "QT" and overtempered 9Ni steels during magnetic exposure. The austenite in the

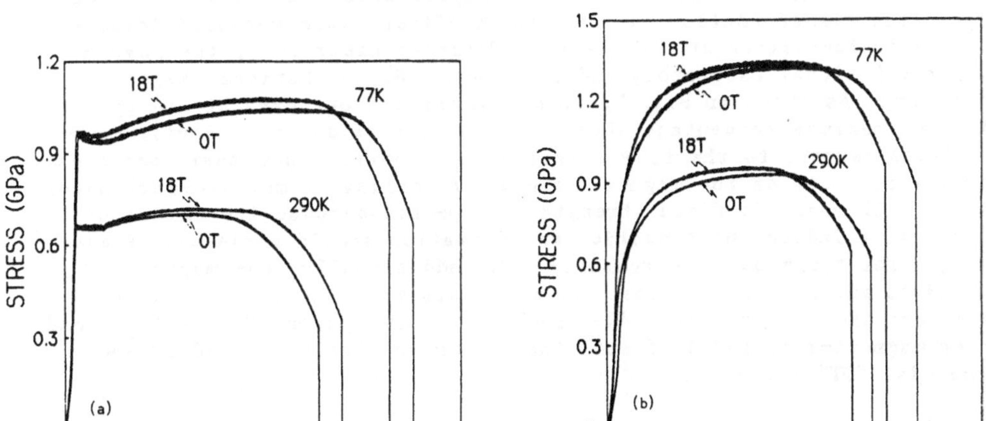

Fig. 5. Engineering stress-strain curves of 9Ni steel at 77 K (LNT) and room temperature (RT) in 18 T field and without field.
a. commercial "QT" treated b. overtempered

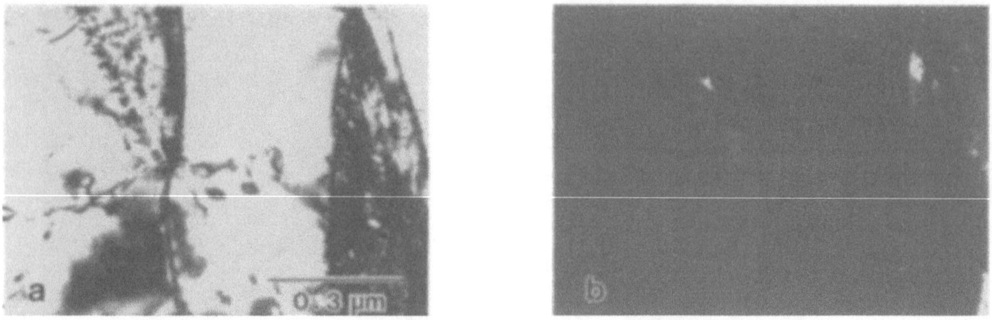

Fig. 6. a. TEM bright field micrograph of overtempered 9Ni steel
b. complementary dark field micrograph of (002)fcc diffraction with [111]bcc ∥ [110]fcc zone axes.

overtempered material is significantly less stable than in commercial "QT" material[1], and it tends to transform at smaller strains. Consequently the magnetically induced increase in flow stress tends to occur at smaller strains for the overtempered material than for the commercial "QT" material, as can be seen by comparing Figs. 5a and 5b. The mechanically induced transformation of precipitated austenite in 9Ni steel evidently makes an important contribution to work-hardening as it does for AISI 304 stainless steel. Because this transformation is expended more rapidly during magnetic exposure, necking occurs at smaller strains, and the elongation of the specimen is reduced. An exception to this occured for commercial "QT" 9Ni steel at room temperature (Fig. 5a), in which the martensitic transformation may not go to completion.

<u>Magneto-Mechanical Effects</u>

Although the phenomena studied in this research are probably of little engineering concern for the present designs of superconducting magnet structures, our experimental data overwhelmingly support the existence of magneto-mechanical effects in steels with metastable austenite. We found additional work-hardening during magnetic exposure for every one of the 25 pairs of specimens tested at temperatures where the fcc→ bcc martensitic transformation occurs. These effects were measured independently with facilities at the Lawrence Berkeley Laboratory, the Lawrence Livermore National Laboratory and the Francis Bitter National Magnet Laboratory. As shown in Fig. 7, which summarizes our present and previous data, the maximum percentage change of flow stress during magnetic exposure (with respect to the flow stress of the control specimen) increases with the strength of the magnetic field. A scaling of magneto-mechanical effects with magnetic field strength is expected because the amount of magnetically induced martensitic transformation should scale with field strength for thermodynamic reasons[2,3,7]. Additionally, the magneto-mechanical effects were largest for materials with the most unstable austenite; the more unstable AISI 304L and overtempered 9Ni steels showed larger magneto-mechanical effects than the more stable AISI 304LN and commercial "QT" 9Ni steels.

CONCLUSIONS

Magneto-mechanical effects exist for steels containing metastable austenite. The effects scale with the strength of the magnetic field, and are larger when more fcc→bcc martensitic transformation occurs during

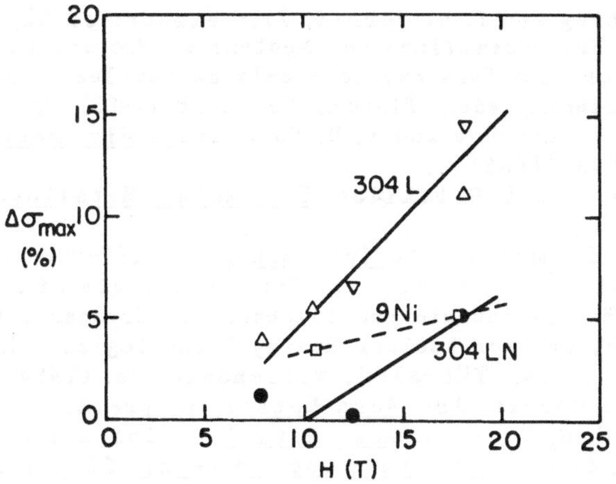

Fig. 7. Maximum percentage change of flow stress in the magnetic field versus magnetic field. point up triangles: AISI 304L at 4 K, point down triangles: AISI 304L at 77 K, dots: AISI 304LN at 4 K and 77 K, squares: overtempered 9Ni steel at 77 K.

plastic deformation. Increasing the amount of fcc→bcc martensitic transformation with high magnetic fields offers a direct technique for determining how the transformation affects plastic deformation. We found three effects of high magnetic fields on the flow stress in tensile tests. 1.) In AISI 304 stainless steels at 4 K and 77 K, high magnetic fields cause a reduction in flow stress at small plastic strains. Together with TEM observations these results are consistent with a "window mechanism" for work softening. 2.) Both AISI 304 and 9Ni steels exhibit increased work hardening during magnetic exposure, resulting in an increased flow stress at large strains. This increased work hardening in the magnetic field results from additional particles of hard bcc martensite and the dislocation structures around them. The fcc→bcc martensitic transformation is an important mechanism of work hardening in both AISI 304 and 9Ni steels. 3.) The more rapid martensitic transformation during magnetic exposure causes necking at smaller strains and thereby reduces the elongation of specimens in tensile tests.

ACKNOWLEDGEMENTS

The authors acknowledge the help of M. Ledezma of the Lawrence Berkeley Laboratory, S. Foner and L. G. Rubin of the Francis Bitter National Magnet Laboratory, and E. N. C. Dalder and R. W. Hoard of the Lawrence Livermore National Laboratory.

The 12.5 T experiments were performed at the Lawrence Livermore National Laboratory, which is supported by the U. S. Dept. of Energy under contract #W-7405-ENG-26.

The 18 T experiments were performed at the Francis Bitter National Magnet Laboratory, MIT, operated for the National Science Foundation.

This work was supported by the Director, Office of Energy Research, Office of Development and Technology, Magnetic Systems Division of the U. S. Dept. of Energy under contract #DE-AC03-76SF00098.

REFERENCES

1. B. Fultz, J. I. Kim, Y. H. Kim, H. J. Kim, G. O. Fior and J. W. Morris, Jr., Metall. Trans. A in press.

2. B. Fultz, G. M. Chang and J. W. Morris, Jr.: Effects of Magnetic Fields on Martensitic Transformations and Mechanical Properties of Steels at Low Temperatures, in: "Austenitic Steels at Low Temperatures", R. P. Reed and T. Horiuchi, eds., Plenum, New York (1983) pp. 199-209.
3. Ye. A. Fokina, L. V. Smirnov and V. D. Sadovskiy: Fiz. Metal. Metalloved. 19:592-595 (1965).
4. L. V. Voronchikhin and I. G. Fakidov: Fiz. Metal. Metalloved. 21:436-441 (1966).
5. T. Kakeshita and K. Shimizu: Scripta Metall. 17:897-900 (1983).
6. G. O. Fior, B. Fultz and J. W. Morris, Jr: The Effects of High Magnetic Fields on the Microstructure and Toughness of Cryogenic 9Ni Steel, in: "Ferritic Alloys for Nuclear Energy Technologies", J. W. Davis and D. J. Michel, eds., TMS-AIME, Warrendale, Pa. (1984) pp. 543-548.
7. B. Fultz and J. W. Morris, Jr.: Acta Metall. in press.
8. A. J. Bogers and W. G. Burgers: Acta Metall. 12:255-261 (1964).
9. G. B. Olson and M. Cohen: Jour. Less Common Metals 28:107-118 (1972).
10. T. Suzuki et al.: Scripta Metall. 10:353 (1976).
11. B. Fultz and J. W. Morris, Jr: Metall. Trans. A in press.

IN SITU OBSERVATION OF MARTENSITIC TRANSFORMATION DURING TENSILE DEFORMATION OF AN Fe-Mn-C ALLOY

Li Yi-yi and Zhang Xiu-mu

Institute of Metal Research
Academia Sinica
Shenyang, China

ABSTRACT

A scanning electron microscope was used to observe martensitic transformations in a Fe-14Mn-0.4 alloy. The morphology and types of thermal- and strain-induced martensite are identified and discussed.

INTRODUCTION

Most studies of deformation-induced martensitic transformation have dealt with Fe-Ni-C alloys and austenitic stainless steels.[1-3] A few studies have been done on Fe-Mn-C alloys.[4-6] Martensitic transformations in Fe-Mn-C alloys are induced by deformation or cooling. Two types of martensite are formed: ε martensite, with a hexagonal close-packed (hcp) structure in the shape of thin laths, and α' martensite, with a body-centered cubic (bcc) structure in the shape of lenticular plates or needles. It is not clear what happens when both thermally induced martensite and austenite are subjected to tensile stress.

In the study of an Fe-14Mn-0.4C alloy, both types of transformations were observed with a scanning electron microscope (SEM). At the same time, we examined crack nucleation and propagation.

EXPERIMENTAL PROCEDURE

The Fe-14Mn-0.4C alloy was prepared by induction melting and homogenization at 1523 K for 24 h. The 25-kg ingots were forged into 14 mm x 14 mm bars. Specimens were machined and cut for tensile testing and microstructural studies, as shown in Fig. 1. The specimens were solution-treated for 1 h at 1323 K, followed by water quenching to room temperature, grinding, polishing, and then cooling at 77 K for 1 h. They were examined by the Nomarski differential interference contrast metallographic method. Then the samples were strained in tension and observed on an SEM as they deformed.

RESULTS AND DISCUSSION

For the Fe-14Mn-0.4C alloy, the temperature at which thermally induced α' martensitic transformation begins, M_s, is 233 K; the temperature at which the deformation-induced transformation begins, M_d, is 300 K.[7] The transformation was continuously monitored by x-ray diffraction measurements. After

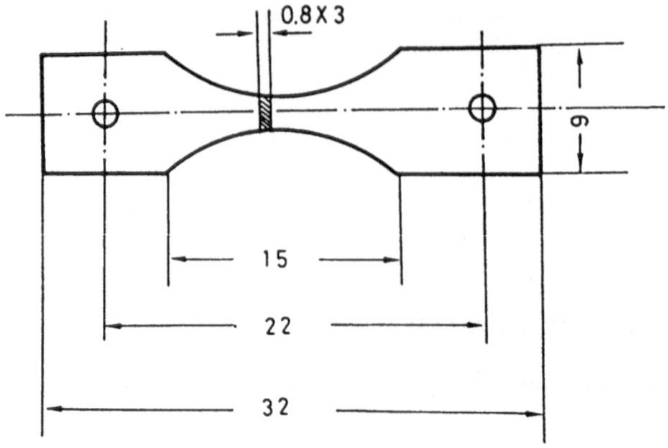

Fig. 1. Specimen configuration; dimensions are in millimeters.

cooling to 77 K, the sample was brought to room temperature, and its microstructure was examined by the Nomarski differential interference contrast metallographic method, as shown in Fig. 2.

The ε martensitic forms with a thin lath morphology, and the α' martensite forms as lenticular plates with paired reliefs. Most of the paired plates of martensite run across the grains; just a few formed along the slip bands. The α' paired plates of martensite are shown in SEM micrograph of Fig. 3a. The morphology of the paired martensite can be regarded as two lenticular plates shaped like wings and meeting edge-on along a junction plane. This angle-profile martensite is usually found in Fe-Ni and Fe-Ni-C alloys.[8,9] The remarkable features of this morphology in the Fe-Mn-C alloy are that the pairs form in the same direction and at an acute angle, and the "wing" interiors contain no midribs. In situ observation by SEM during the initial step of tensile deformation showed that there were many slip bands in the deformed austenite. The austenite was gradually strengthened by successive formation of strain-induced ε and α' martensites. In Fig. 3b, some strain-induced α' martensite appeared, and the arrow in Fig. 3c indicates an area of their growth. In the upper region, ε martensite, formed by

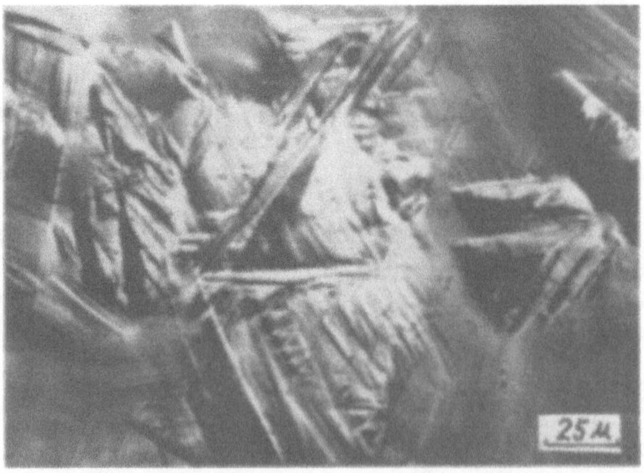

Fig. 2. Microstructure of the Fe-14Mn-0.4C alloy revealed by Nomarski interference contrast metallography.

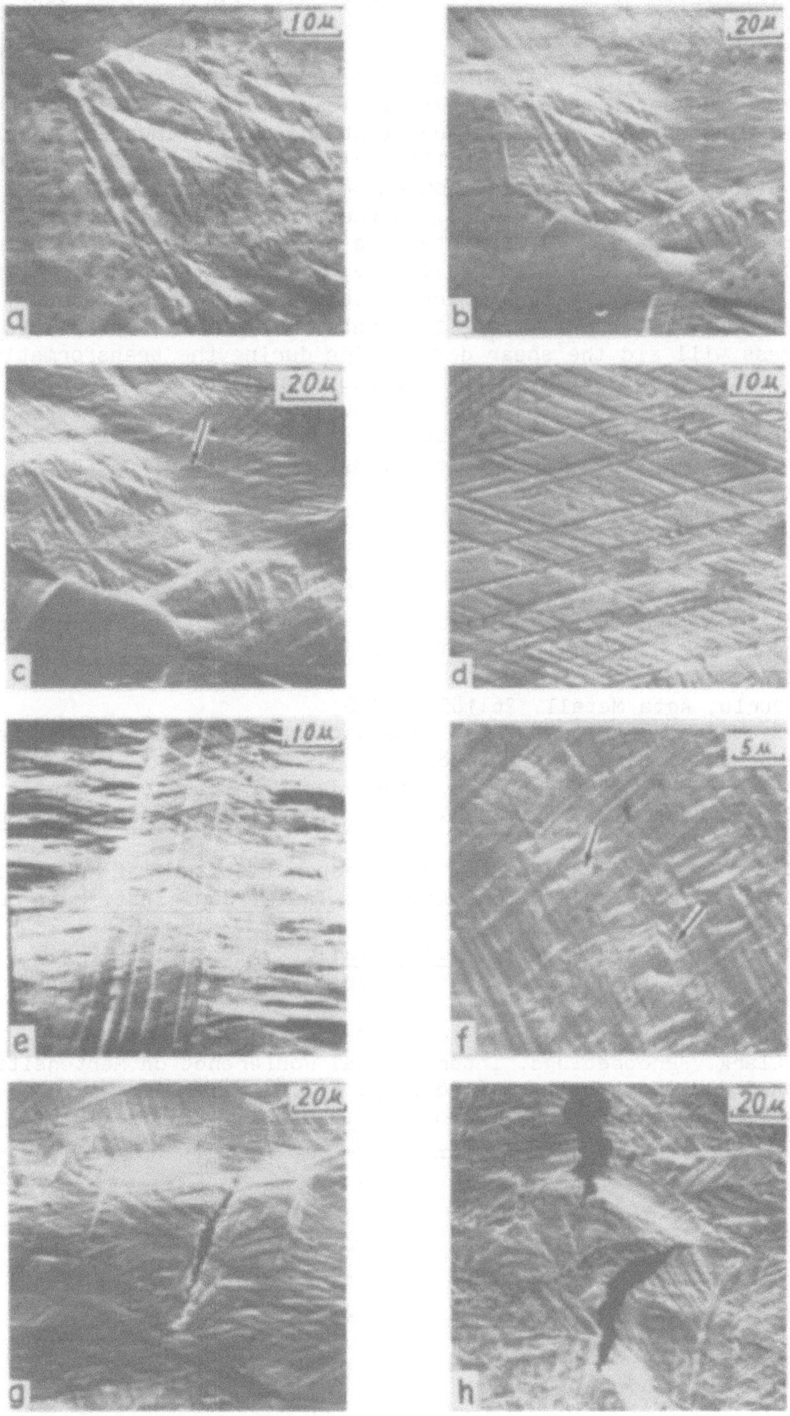

Fig. 3. SEM micrographs showing the microstructural characteristics during the transformation process. (a) initial deformation, (b) during deformation, (c) growth of α' martensite, (d) ε martensite, (e,f) α' martensite within a deformation twin, (g) interface of α' martensite and matrix, (h) interface of ε martensite and matrix.

simple cooling, was thickened by the applied stress. In contrast with the α' martensite, strain-induced ε martensite formed with a volume contraction along the $\{111\}_\gamma$ slip planes (see the micrograph Fig. 3d). Figure 3e shows the strain-induced α' and deformation twins in the lower region. From this micrograph, we found that strain-induced, paired martensite increases as the applied stress increases, but the angle is less acute than that of the thermally formed martensite. We also found that the strain-induced α' martensite forms along slip bands of the $\{111\}_\gamma$ planes. Figure 3f is a micrograph of a deformation twin, and strain-induced, paired martensite has formed along the primary and critical slip planes, as the arrows indicate. Since the martensitic transformation is characterized by cooperative shear movement of atoms to change the lattice from fcc to bcc, according to Y. Higo[10] and M. Kato,[11] if the austenite-to-α' martensite transformation occurs by a combination of shears along the $(111)_\gamma\langle 211\rangle_\gamma$ system in Fe-14Mn-0.4C alloys, the applied stress will aid the shear deformation during the transformation. During the latter stages of deformation, the microprocess of fracture was examined. We observed that the crack was nucleated preferentially in two ways: at the α' martensite-matrix and the ε martensite-matrix interfaces. The cracking phenomena are shown in Figs. 3g and 3h. The fracture is ductile, and the crack propagation cuts through the shear deformation bands.

REFERENCES

1. M. Tokizane, "Proceedings, First Japan Institute of Metals International Symposium," H. Suzuki, ed., Trans. Jap. Inst. Met. Suppl. 17:345 (1976).
2. T. N. Durlu, Acta Metall. 26:1855 (1978).
3. F. D. S. Margues, "Proceedings, International Conference on Martensitic Transformation," Belgium (1982), p. 569.
4. H. Onodera et al., "Proceedings, First Japan Institute of Metals International Symposium," H. Suzuki, ed., Trans. Jap. Inst. Met. Suppl. 17:327 (1976).
5. H. Fujita et al., "Proceedings, First Japan Institute of Metals International Symposium," H. Suzuki, ed., Trans. Jap. Inst. Met. Suppl. 17:351 (1976).
6. E. Gartstein and A. Rabinkin, Acta Metall. 27:1053 (1979).
7. Y.-Y. Li and F. Zhang, Martensitic transformation in Fe-Mn-C, to be published.
8. C. M. Wayman, Metallography 8:105 (1975).
9. H. R. Clark, "Proceedings, International Conference on Martensitic Transformation," Belgium (1982), p. 567.
10. Y. Higo, F. Lecroisey, and T. Mori, Acta Metall. 22:313 (1975).
11. M. Kato and T. Mori, Acta Metall. 24:853 (1976).

FRACTURE TOUGHNESS OF MODERN 9% NICKEL CRYOGENIC STEELS

Robert D. Stout

Lehigh University
Bethlehem, Pennsylvania

Steve J. Wiersma

Gas Research Institute
Chicago, Illinois

ABSTRACT

Storage tanks for liquefied natural gas must contain a highly flammable gas at 111K with an ample margin of safety. The 9% nickel steels have served in more than 50 LNG tanks without failure for over a decade. The Gas Research Institute has sponsored an international cooperative program to demonstrate the capabilities of 9% Ni steels to resist fracture initiation and arrest propagation. One project has considered the enhancement of properties attained in the past 15 years by lowering impurity levels. A statistical study has shown the large gains in the notch toughness of production heats made by reducing P and S contents below 0.005%. In contrast to the ASTM minimum requirements of 27J Charpy at 77K, levels over 100J have readily been attained in recent production. Other phases of the GRI program have studied the crack arrest capacity of 9% Ni heats ranging from 35 to 200J Charpy at 77K, including the effects of welding. At 40J toughness level, 25mm thick plates can arrest cracks 100mm long, but at 250J Charpy, 25mm plates could arrest cracks over 750mm long. An important ingredient of crack-arrest is the large energy absorption by the surface layers of these relatively thin storage tank gages. While weld planar flaws and property alteration in the weld heat-affected zone may have adverse effects on welded joints the excellent fracture toughness of the austenitic weld metal used in tanks compensates for these effects.

INTRODUCTION

Storage tanks for liquefied natural gas (LNG) must be designed to contain a highly flammable substance at low temperature (-110K) with an ample margin of safety against the initiation or propagation of brittle fracture. To meet these requirements, the containment material must possess a high notch toughness at operating temperatures as well as weldability characteristics to resist hot and cold cracking and embrittlement of the heat-affected zone. Strength requirements are not a problem at 110K but must be adequate to withstand the stresses of construction and hydrotesting. ASTM specifications for 9% nickel steel (A553 Class I and A353 Class I) address tensile strength and ductility and Charpy test notch toughness, e.g., 27J in the transverse-to-rolling direction at 77K. The quality of construction of LNG storage tanks is governed by the API Code Standard 620 Appendix Q or its equivalent.

The adequacy of 9% nickel steel fabricated by Code procedures for LNG service is attested by the absence of tank failures in over 50 storage tanks that have been put into service worldwide. The Gas Research Institute considered it prudent to confirm the safety features of 9% nickel steel in LNG tanks by instituting an international cooperative research program following the suggestions proposed in an interpretive report[1] prepared under its support. Projects were undertaken in England, Japan, and the United States on four topics: (1) the significance of the weld heat-affected zone to crack arrest, (2) the correlation of the several crack initiation and crack arrest tests, (3) the influence of fabrication flaws on fracture toughness, and (4) the characterization of strength and notch toughness of present-day production heats of 9% nickel steels. This paper concerns primarily the 4th project listed, but some pertinent results from the other projects are also included.

CHARACTERISTICS OF PRODUCTION HEATS

A statistical study was conducted of 9% nickel steels produced during the past two decades to determine (1) the trend of mechanical properties with the year of manufacture, and (2) the effect of compositional variables on the mechanical properties of production heats. For these purposes heat sheet data were assembled from sources in Europe, Japan, and the United States. Over 600 heats were represented in the data bank. A versatile computer package was chosen for the regression analysis of the data. The package permits forward or backward stepwise examination of the variables, prints histograms, identifies outlier data points, and analyzes the data to suggest possible data transformations.

Figure 1 shows the relation of the average Charpy notch toughness of 9% nickel steels to the year of manufacture. A three-year weighted running average was used to smooth out the ups and downs of the numbers of heats reported for each year. The improvement in notch toughness over fifteen years is striking, with a discontinuous upward jump in the last 5 years.

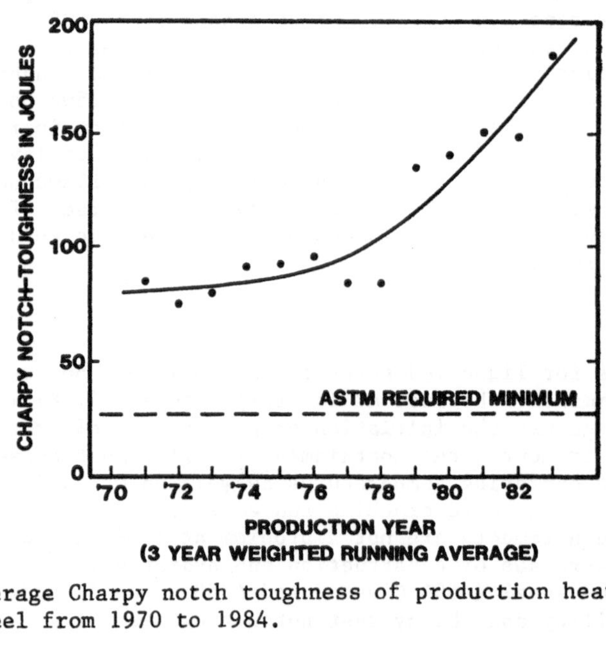

Fig. 1. Average Charpy notch toughness of production heats of 9% nickel steel from 1970 to 1984.

The same trend was exhibited by the heat data obtained from all of the sources. It is obvious that modern 9% nickel steels provide a toughness level far above the minimum level specified by the ASTM A553 Class I grade. As will be seen later from the regression analyses, the single most important factor responsible for the improvement is the minimization of sulfur content.

As an addendum it should be mentioned that the data from all of the sources displayed a similar improvement over the fifteen year period.

The regression analyses considered nine independent variables: gage, C, Mn, P, S, Si, Ni, yield strength, and tensile strength. The latter two were included in an attempt to introduce an indirect index of heat treatment procedures which were not supplied to the data bank. The multiple regression obtained with Charpy test Joules at 77K as the dependent variable was as follows:

$$Cv=96-0.16t-600C+157Mn-190P-4300S+470Si+39Ni-0.34Y.S.-0.31T.S.$$

The R^2 value for this equation was 56% and the standard error of estimate was 40. The computer program results suggested that an improved and simplified correlation could be obtained by choosing interactive predictors, such as Mn/C and Mn/S. When this transformation was applied, R^2 rose to 64%, and the equation became

$$C_v=335+293Si+0.14Mn/S+5.25Mn/C-0.50\ (\text{Tens. Strength})$$

It is interesting to note that a large fraction of the variance was accounted for by relatively few of the variables. For example, C, Mn, S, Si, and T.S. produced and R^2 of 53% in the first equation and 64% in the second one. Thus it is the control of these elements which leads to most of the improved toughness of more recent heats. It is known that grain size, retained austenite, and tempering temperature are important and interrelated factors, but data were lacking to allow their quantitative evaluation.

Fracture Toughness of 9% Nickel Steels

The Gas Research Institute international cooperative program is concerned with the fracture toughness of 9% nickel steel used in LNG storage tanks with particular emphasis on its ability to arrest cracks in the presence of weld flaws, as influenced by the heat-affected zone and weld metal properties. For the study heats were obtained comprising two thicknesses and three Charpy-notch toughness levels as shown in Fig. 2. The

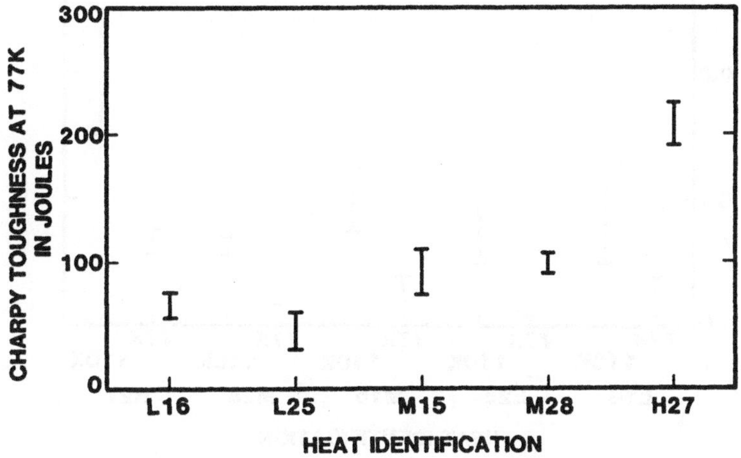

Fig. 2. Charpy notch toughness of steels.

Table 1. Composition and Tensile Properties of the 9% Nickel Heats

A. Chemical Composition, in %

Heat	C	Mn	P	S	Si	Ni	Cr	Mo	Cu	Al
L16	0.07	0.48	0.006	0.006	0.23	9.03	0.08	0.05	0.04	0.03
L25	0.04	0.68	0.019	0.10	0.27	8.70	0.04	0.01	0.02	0.04
M14	0.06	0.62	0.011	0.004	0.28	8.80	0.12	0.04	0.13	0.05
M28	0.05	0.55	-0.005	-0.005	0.18	9.20	0.04	0.04	0.14	0.03
H17	0.06	0.59	0.010	0.005	0.27	9.00	0.02	-0.01	-	-

B. Tensile Properties

Heat	Yield Strength(MPa)		Tensile Strength(MPa)		Elongation,%		Red.Area,%	
	295K	110K	295K	110K	295K	110K	295K	110K
L16	710	790	870	1130	24	23	72	68
L25	715	935	740	970	24-31	28	66-80	64-74
M14	695	950	745	1055	23	24	75	70
M28	665	850	840	1150	27	25	77	73
H27	680	940	730	990	30	27	78	74

heats were coded L, M or H for low, medium or high toughness and with numbers indicating the plate thickness in mm. Chemical compositions and tensile properties are listed in Table 1. The heats were characterized by CTOD tests, compact crack arrest tests and wide plate tests in which welding effects were incorporated. The temperatures of test were 111K, the B.P. of LNG, and 77K, the B.P. of nitrogen.

The results of the tests on the various base metals are presented in Figs. 3 to 5. The trend of the CTOD tests (Fig. 3) is broadly parallel to the Charpy tests but the effects of the plate thickness and test temperature are marked. The CCA tests (Fig. 4) reveal the remarkable ability of 9% nickel steel to arrest a running crack. In these tests the influence of

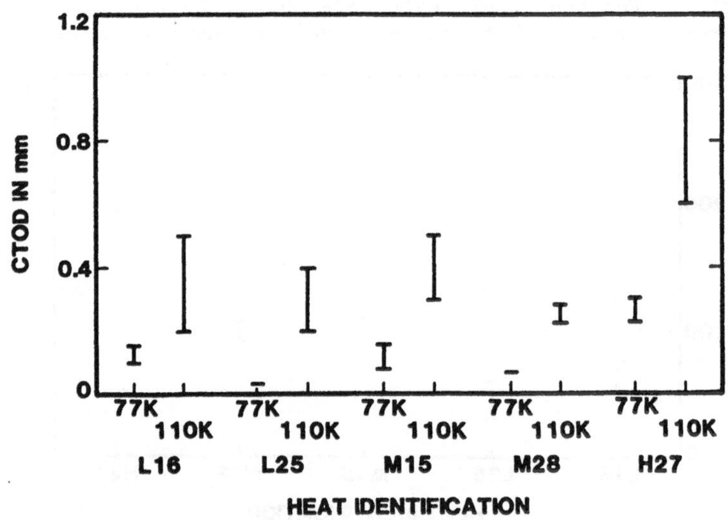

Fig. 3. CTOD test results on steels.

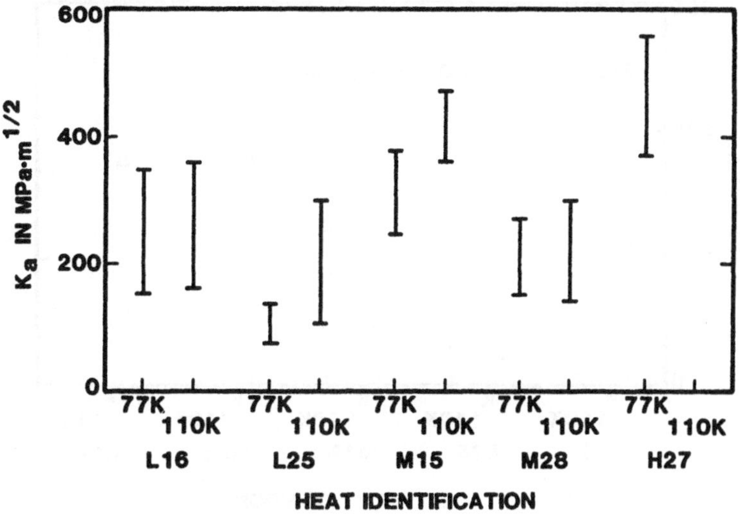

Fig. 4. Compact crack arrest test results.

thickness is apparent but little change in K_a occurs with test temperature. Generally the wide plate tests (Fig. 5) support the small-scale CCA tests, but more of these expensive tests will be needed to confirm the correlation.

The toughnesses of the base-metal heat-affected zones and of the weld metals produced by the shielded metal-arc process are shown in Figs. 6 and 7. Data are shown from weldments made at a high heat input, 3.6 kJ/mm, and thus are the least favorable values. Except for Heat L25, the borderline-toughness heat, the heat-affected zones (Fig. 6) are as tough or tougher than the parent metals. The weld metal toughnesses (Fig. 7) indicate a clear advantage for Incoweld A, but this weld metal is appreciably lower in tensile strength and thereby limits design stress levels in LNG storage tanks.

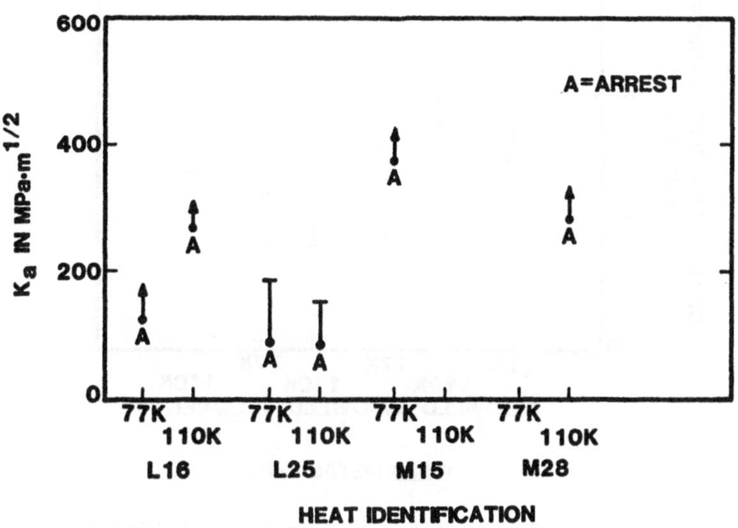

Fig. 5. Wide-plate test results.

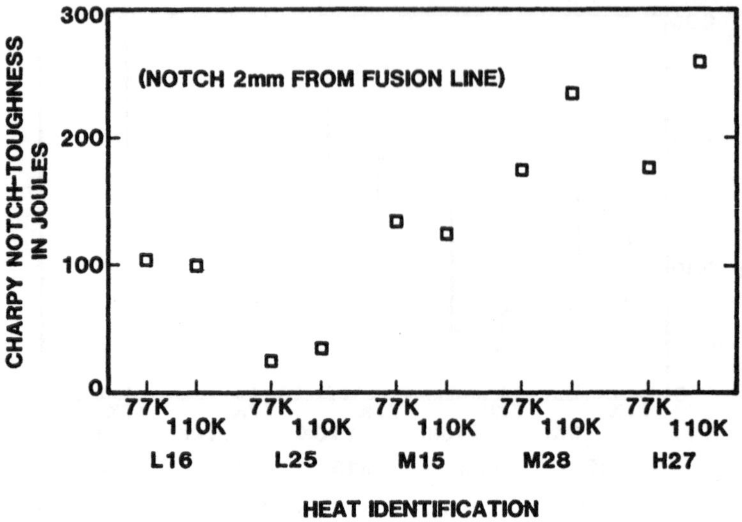

Fig. 6. Heat-affected zone Charpy notch toughness (shielded metal arc welds, 3.6 kJ/mm heat input).

SUMMARY DISCUSSION

It is apparent from the data bank assembled that the notch-toughness properties of production heats of 9% nickel steel have been notably improved in the past fifteen years. It is arguable that the ASTM specifications for A553 Class I steels could be revised to take advantage of this improvement and thus ensure an even higher level of safety in cryogenic structures without hardship to steel producers.

The regression analysis demonstrated clearly that much of the enhancement of notch toughness in the 9% nickel steels has sprung from the reduction of impurities content, such as sulfur, and the control of carbon and alloy

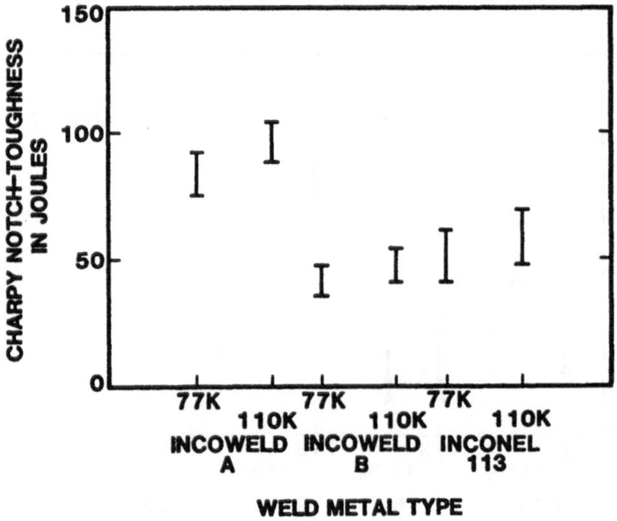

Fig. 7. Weld metal Charpy notch toughness (shielded metal-arc welds, 3.6 kJ/mm heat input).

levels. While no data were provided to permit an analysis of heat treatment effects, it is known from laboratory studies that the temperatures and times of heat treatment can be optimized to obtain superior toughness.

As an outcome of the GRI cooperative program the remarkable ability of the 9% nickel steels to arrest running cracks has been quantitatively demonstrated. A favorable factor operating in LNG storage tanks is the relatively limited thickness of the material used, typically 12 to 30 mm. In these gages the surface layers provide very high energy dissipation during crack growth, so that the crack front is strongly impeded by surface ligaments that lag considerably behind the internal crack tip. As an example, for Heat M28 operating at a stress of 200 MPa at 110K, a crack triggered by some local condition or accident would not run in the base plate unless it were over 2000 mm in length. Since a crack this long would lead to excessive leakage and other problems, the program has operated under the assumption that no cracks above 150 mm are of practical moment.

ACKNOWLEDGMENT

The information on which this paper is based has been made available by the courtesy of the Gas Research Institute and of the participating organizations connected with the cooperative program on 9% nickel cryogenic steels. Further technical information is available from the progress reports submitted to GRI by the participating laboratories. Any inquiries should be made to GRI.

The authors take sole responsibility for the contents of this paper and caution against the use of any of the data for design or other engineering purposes.

REFERENCE

1. R. D. Stout and A. W. Pense, "Crack Arrest Properties of 9% Nickel Cryogenic Steels," Final Report GRI 80/0037 May 1, 1981, The Gas Research Institute, Chicago, Illinois.

CRYOGENIC MECHANICAL PROPERTIES OF Al-Cu-Li-Zr ALLOY 2090

J. Glazer, S. L. Verzasconi, E. N. C. Dalder,* W. Yu, R. A. Emigh, R. O. Ritchie, and J. W. Morris, Jr.

Materials and Molecular Research Division,
Lawrence Berkeley Laboratory, and
Department of Materials Science and Mineral Engineering
University of California, Berkeley, California

*Lawrence Livermore National Laboratory
University of California, Livermore, California

ABSTRACT

The mechanical properties of aluminum-lithium alloy 2090-T8E41 were evaluated at 298 K, 77 K, and 4 K. Previously reported tensile and fracture toughness properties at room temperature were confirmed. This alloy exhibits substantially improved properties at cryogenic temperatures; the strength, elongation, fracture toughness and fatigue crack growth resistance all improve simultaneously as the testing temperature decreases. This alloy has cryogenic properties superior to those of aluminum alloys currently used for cryogenic applications.

INTRODUCTION

The dual objectives of minimizing operating costs and maximizing performance of aircraft and aerospace systems provide a powerful incentive to reduce aircraft empty weight. Recent design studies indicate that structural weight is more effectively lowered by reducing the density of structural materials than by improving their mechanical properties.[1] This conclusion has provided the rationale for the development and application of resin composites. However, the highly anisotropic properties of composites make their application difficult, and it seems likely that at least commercial aircraft will remain primarily aluminum. As a consequence, there is a strong impetus to develop advanced high strength aluminum alloys. This challenge is responsible for a renewed interest in producing low density aluminum alloys to replace current alloys. A number of promising alloys have been developed that contain additions of lithium to reduce their density.

Intensive research and development in the last several years have led to the registration of several aluminum-lithium alloys intended to replace at lower density standard commercial aluminum alloys such as 2024, an Al-Cu-Mg-Mn alloy, and 7075, an Al-Zn-Mg alloy. One of the new alloys is 2090, designed to have properties similar to those of 7075-T651. In addition to having a significantly lower density, 2090 is superior to 7075 in many respects; in fact, although low toughness has been a problem of aluminum-

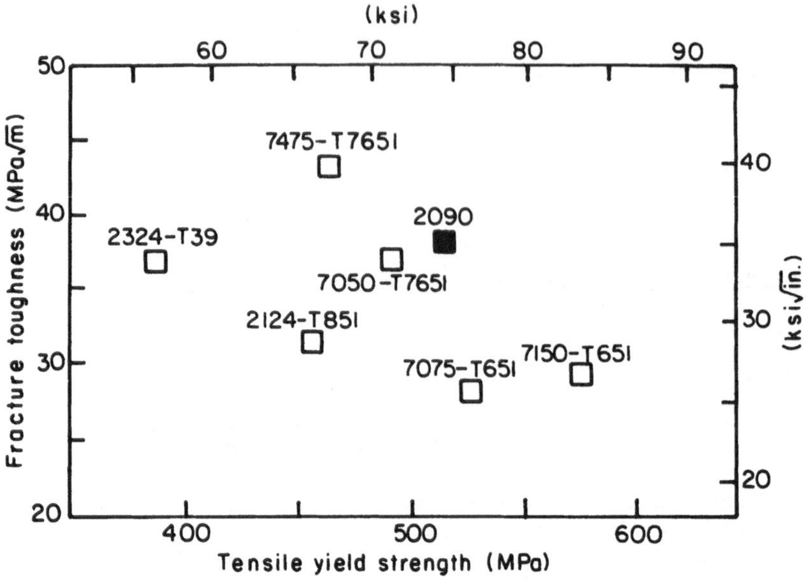

Fig. 1. Comparison of the longitudinal properties in plate material of alloy 2090-T8 to standard aerospace aluminum alloys.

lithium alloys in the past, the room temperature strength-toughness relationship of 2090 is better than that of any other standard aerospace alloy[2] (see Fig. 1). It appears likely that aluminum-lithium alloys will see service within the next decade.

Although aluminum-lithium alloys have been developed with aircraft structures in mind, there are a number of potential cryogenic applications for a low density, high strength, aluminum alloy. These applications include space systems, cryogenic tankage, and high-field magnet cases. Preliminary investigations by other workers have suggested that aluminum-lithium alloys may be particularly promising cryogenic alloys because their toughness tends to improve at low temperatures.[3,4] In addition, preliminary results indicate that 2090 is sufficiently weldable for cryogenic applications.

This paper is both a characterization of the room temperature mechanical properties of this new alloy, 2090, and a study of the alloy's low temperature properties.

EXPERIMENTAL PROCEDURE

All specimens were taken from a 12.7 mm (0.5 in) plate of 2090 supplied by the Alcoa Technical Center. The plate was received in the T8E41 condition, where E41 is a proprietary thermomechanical processing treatment. The nominal composition of 2090 is Al-2.7Cu-2.2Li-0.12Zr. Registered composition limits are given in Table 1. The actual composition of the plate is still under investigation. Chemical analysis of Al-Li alloys is hampered by a lack of standards. Several analyses were performed and although all composition determinations were within the specified composition limits, they scattered widely. The alloy is hardened in the T8 condition by a combination of the coherent ordered precipitate phases δ', T_1', and T_2'.[5]

Table 1. Registered Composition Limits for 2090 in Weight Percent

Al	Cu	Li	Zr	Fe	Si	Mg	Mn	Ti
bal.	2.4–3.0	1.9–2.6	0.08–0.15	0.10	0.12	0.25	0.05	0.15

Elastic-plastic fracture toughness tests and tensile tests were conducted at room (298 K), liquid nitrogen (77 K), and liquid helium (4 K) temperatures. Round tensile specimens with a 2.54 cm (1.0 in) gauge length were tested according to ASTM standard E8-82 in both longitudinal (L) and long-transverse (LT) orientations. The 0.2% offset yield stress, ultimate tensile strength, and total elongation were determined for pairs of specimens. Compact tension specimens were tested according to ASTM standard E813-81 in both LT and TL orientations to determine the crack initiation toughness, J_{Ic}. The measured J_{Ic} data were used to calcualte the plane strain fracture toughness K_{Ic} values. The room temperature elastic modulus of 79 GPa was used for all calculations. Since the elastic modulus of most aluminum alloys increases with decreasing temperature,[6] the low temperature K_{Ic} values represent a lower limit on the actual K_{Ic} values. Fracture modes were characterized using a scanning electron microscope.

Constant amplitude fatigue crack growth rate behavior of 2090 was determined over a range of ΔK values using manual load shedding techniques on 6.4 mm thick single-edge-notched four-point-bend specimens. Tests were conducted at room temperature in room air (relative humidity, 25%) and at 77 K (submerged in liquid nitrogen) at a load ratio $R = K_{min}/K_{max}$ of 0.1. The test frequency was 50 Hz at fatigue crack propagation rates below 10^{-5} mm/cycle and 20 Hz at higher crack growth rates. Crack growth was monitored using dc electrical potential methods. The fatigue threshold was defined as the value of ΔK at which the fatigue crack growth rate decreased below 10^{-8} mm/cycle.

Fig. 2. Grain structure of 12.7 mm (0.5 in) plate of 2090.

Table 2. Strength, Elongation, and Fracture Toughness of 2090-T8E41 at Room, Liquid Nitrogen, and Liquid Helium Temperatures.

Test Temperature (K)	Yield Strength (MPa)		U.T.S. (MPa)		Elongation (in 25.4 mm) (%)		K_{Ic} (MPa$\sqrt{m}$)	
	L	LT	L	LT	L	LT	LT	TL
298	535	535	565	565	11.0	5.5	34	25
77	600	625	715	695	13.5	5.5	52	34
4	615	705	820	815	17.5	6.5	65	39

Specimens for metallography were polished to 0.05 μm and then etched with Keller's etch—2.5% HNO_3, 1.5% HCl, 0.5% HF, balance H_2O—for 15 to 30 s. The grain structure of the 12.7 mm plate of 2090 is shown in Fig. 2. Large constituent particles were not observed.

RESULTS AND DISCUSSION

The tensile and fracture toughness properties of 2090-T8 are shown in Table 2. The data are displayed graphically in Fig. 3. The room temperature data differ only slightly from those published by Alcoa.[2] Significantly, strength, elongation, and toughness in both orientations increase simultaneously with decreasing temperature. The data are compared with the properties of standard aluminum alloys at room temperature in Fig. 4.

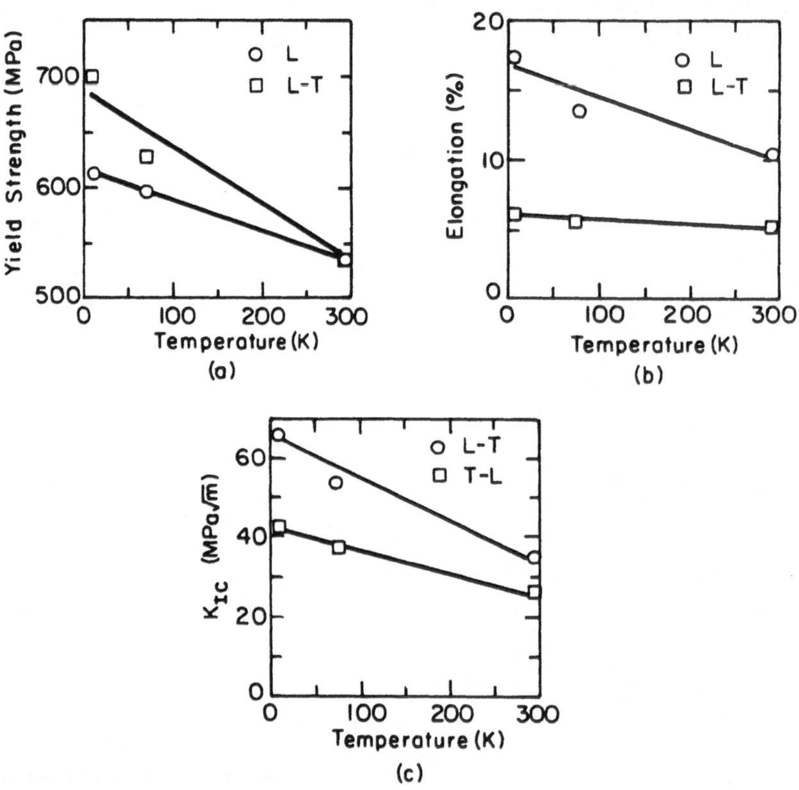

Fig. 3. Mechanical properties of 2090-T8E41 as a function of temperature.

The yield strength increases with decreasing temperature in both the L and LT directions. The yield strengths are identical in the two orientations at room temperature; however, at lower temperatures, the strength is higher in the LT direction. The increase in yield strength is fairly typical of precipitation-hardened aluminum alloys.

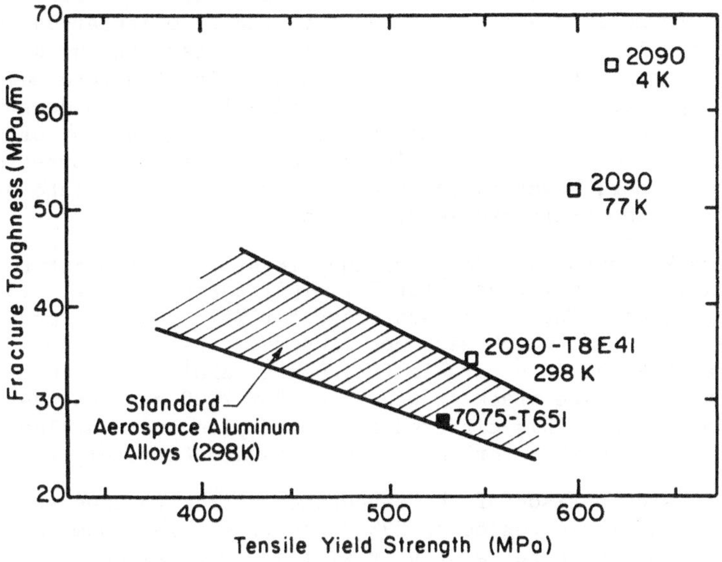

Fig. 4. Comparison of longitudinal cryogenic properties of 2090-T8E41 to standard aerospace alloys.

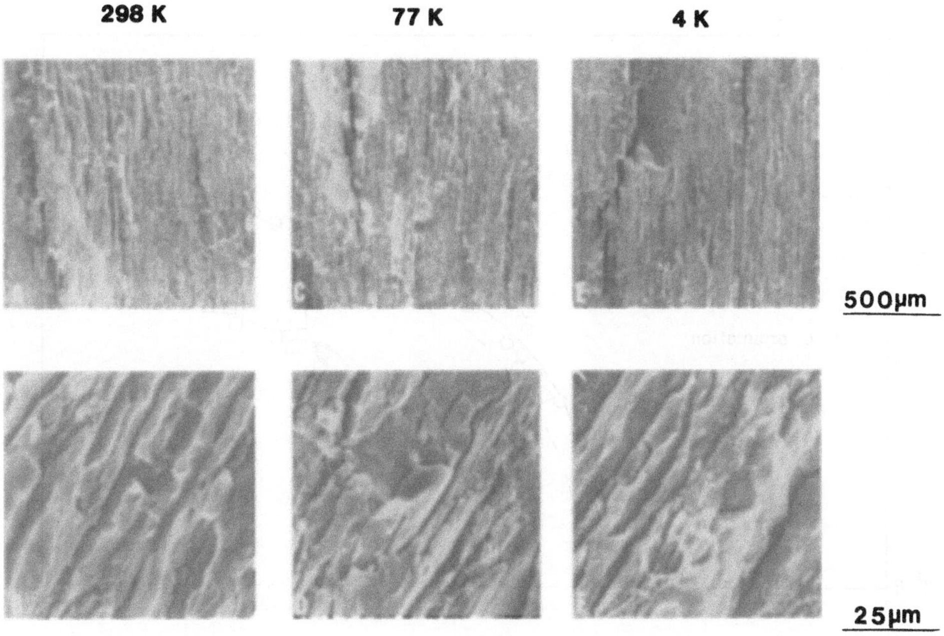

Fig. 5. Comparison of fracture surfaces of J_{Ic} specimens in the LT orientation broken at 298 K (A,B), 77 K (C,D) and 4 K (E,F).

The fracture toughness in the LT orientation is consistently higher than in the TL direction. Similarly, the tensile elongation in the LT direction is considerably lower than in the L direction. Although a small difference in properties parallel to and perpendicular to the rolling direction is not unusual, the separation seen here is atypical.

It might be expected that the large increase in fracture toughness at low temperatures would be reflected in the observed fracture mode. However, the J_{Ic} fracture surfaces at all three test temperatures are strikingly similar. The fracture mode at all temperatures in the LT orientation is illustrated in Fig. 5. It appears to be controlled by intergranular delamination that is linked by regions of transgranular shear. The fracture mode in the TL orientation appears to be similar mechanistically and also does not display an obvious change with test temperature. The large amount of intergranular cracking observed in this alloy suggests that its short-transverse properties may be poor; however, the small thickness of the plate precluded any testing in the short-transverse direction.

Fatigue crack growth rate data for 2090-T8 at room temperature and 77 K are shown in Fig. 6. Similar data for 7475-T651 taken from ref. 10 are included for comparison. The 123 K data shown for 7475-T651 are probably quite similar to the 77 K data, which were not measured. As can be seen in the figure, 7475 and 2090 have comparable fatigue thresholds in both temperature regimes. However, fatigue crack growth rates for 2090 are considerably lower at higher values of ΔK. This difference is most pronounced at low temperatures above $\Delta K \sim 8$ MPa$\sqrt{m}$ where crack growth rates are extremely rapid for 7475.

Figure 7 is a comparison of the appearance of the fatigue fracture surface in the near threshold regime at 298 K and 77 K. The room temperature fatigue surface appears to be relatively similar to those observed in other aluminum-lithium alloys; relatively brittle, crystallographic crack growth features are observed. The 77 K fatigue crack surface contains unusual ductile regions. The crack path is highly irregular and branched, a typical morphology in aluminum-lithium alloys.[11,12]

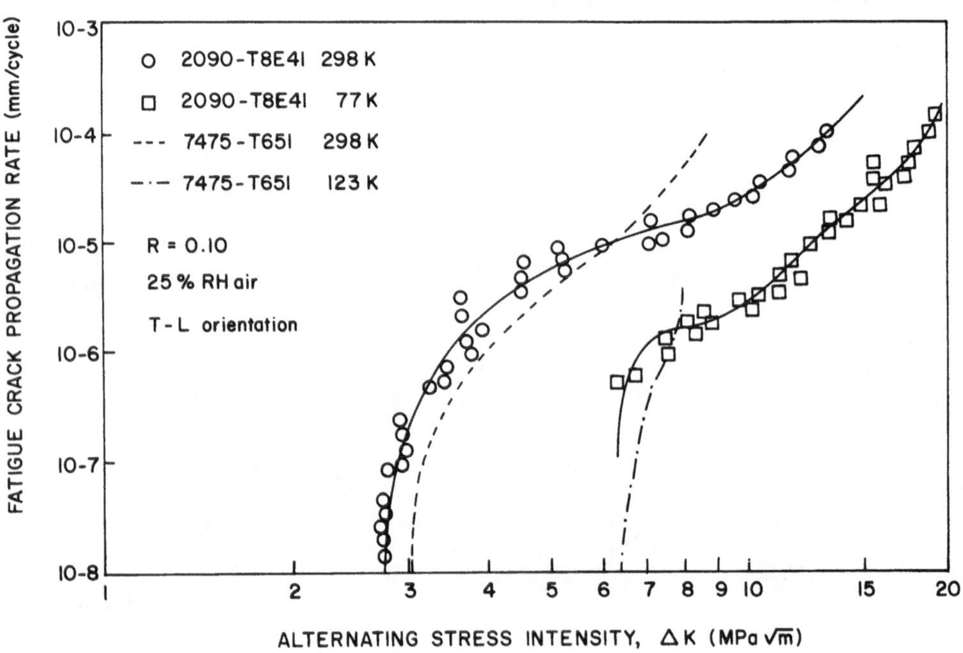

Fig. 6. Effect of temperature on the fatigue crack propagation rates in 2090-T8 and 7475-T651.

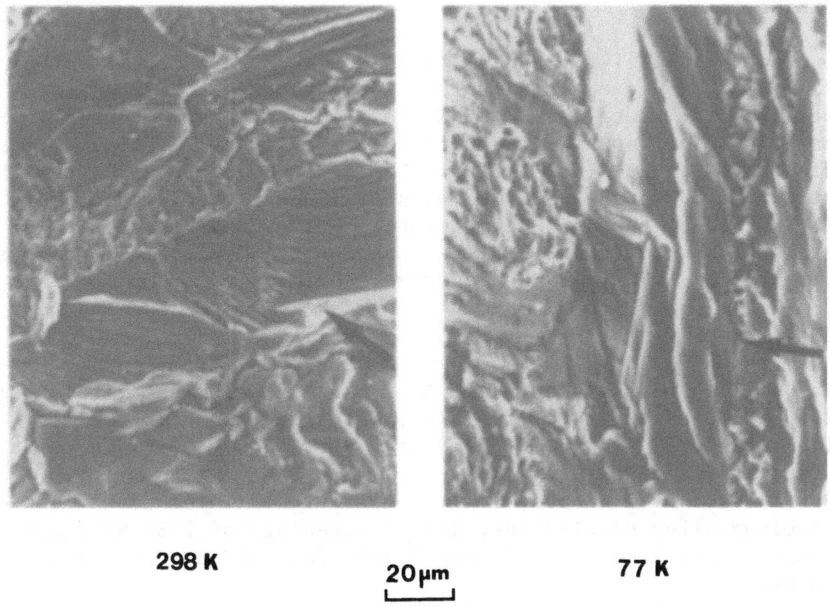

Fig. 7. Fatigue crack surface appearance in the near-threshold region of specimens. A) $\Delta K = 2.6$ MPa$\sqrt{m}$ at 298 K; B) $\Delta K = 6.3$ MPa$\sqrt{m}$ at 77 K. Arrow indicates general direction of crack growth.

Many aluminum alloys have improved strength-toughness relationships at low temperatures. The reasons behind this phenomenon are not understood. The much greater improvement in yield strength and fracture toughness observed here is consistent with data obtained by Webster on binary Al-Li alloys[3] and on the Al-Li-Cu-Mg alloy 8090,[4] but is also unexplained. The properties of 2090-T8E41 are compared in Table 3 with those of 2219-T87, a standard high-strength cryogenic alloy. The 77 K and 4 K properties of 2090-T8 appear to surpass those of any currently available aluminum alloy.

CONCLUSIONS

Alloy 2090-T8E41 has a better combination of strength and toughness at room temperature than the standard aerospace alloys. Its cryogenic mechanical properties—strenth, elongation, fracture toughness, and fatigue crack growth resistance—all improve dramatically with decreasing temperature. The mechanism behind these improvements is not well understood at this time. The cryogenic properties of 2090 appear to be superior to existing high-strength cryogenic aluminum alloys.

Table 3. Comparison of Mechanical Properties of 2090-T8E41 with 2219-T87

Alloy	Yield Strength (MPa)			Elongation (%)*			K_{Ic} (MPa$\sqrt{m}$)		
	298 K	77 K	4 K	298 K	77 K	4 K	298 K	77 K	4 K
2090-T8E41	535	600	615	11.0	13.5	17.5	34	52	65
2219-T87†	386	461	512	11.8	14.0	-	36	43	48

*2090: elongation in 25.4 mm (1.0 in); 2219: elongation in 50.8 mm (2.0 in)
†2219 data are from refs. 7 (298 K, 77 K tensile properties), 8 (298 K, 77 K fracture toughness), and 9 (4 K).

ACKNOWLEDGMENT

The authors would like to thank the Alcoa Technical Center for supplying the material used in these experiments. The authors are also grateful to R. R. Sawtell, P. E. Bretz and R. J. Rioja, Alcoa Technical Center, for helpful discussions. This work was supported by the Director, Office of Basic Energy Sciences, Materials Science Division of the U.S. Department of Energy under Contract No. DE-AC03-76SF00098 with Lawrence Berkeley Laboratory. It was also supported by the U.S. Department of Energy under Contract No. W-7405-ENG-48 with Lawrence Livermore National Laboratory. One of the authors, J.G., was supported by an AT&T Bell Laboratories scholarship.

REFERENCES

1. W. E. Quist, G. H. Narayanan and A. L Wingert, Aluminum-lithium alloys for aircraft structure—overview, in: "Aluminum-Lithium Alloys, II," T. H. Sanders, Jr. and E. A. Starke, Jr., eds., AIME, Warrendale, Pennsylvania (1981), p. 313.
2. R. R. Sawtell, P. E. Bretz, J. I. Petit and A. K. Vasudévan, Low density aluminum alloy development, in: "Proceedings of 1984 SAE/Aerospace Congress and Exposition," Long Beach, California, Oct. 15-18, 1984, in press.
3. D. Webster, Aluminum-lithium alloys, Met. Prog. 125:33 (1984).
4. D. Webster, Temperature dependence of toughness in various Al-Li alloys, in: "Proceedings of the 3rd International Aluminium-Lithium Conference," Oxford, England, July, 1985, Pergamon, in press.
5. R. J. Rioja and E. A. Ludwiczak, Identification of metastable phases in an Al-Cu-Li-Zr alloy (2090), in: "Proceedings of the 3rd International Aluminium-Lithium Conference," Oxford, England, July, 1985, Pergamon, in press.
6. J. G. Kaufman and E. W. Johnson, New data on aluminum alloys for cryogenic applications, Adv. Cryog. Eng. 6:637 (1960).
7. J. G. Kaufman, F. G. Nelson, and E. W. Johnson, The properties of aluminum alloy 2219 sheet, plate, and welded joints at low temperatures, Adv. Cryog. Eng. 8:661 (1963).
8. J. G. Kaufman and M. Holt, Evaluation of fracture characteristics of aluminum alloys at cryogenic temperatures, Adv. Cryog. Eng. 10:77 (1965).
9. C. Fiftal, "NBS/DOE Report on Cryogenic Structural Materials for Superconducting Magnets," vol. 1 (1978).
10. W. Yu, Thermally activated processes associated with fatigue thresholds in Fe-Si-binary and aluminum alloys, Ph.D. Thesis, University of Minnesota (1983).
11. S. Suresh and A. K. Vasudévan, Influence of composition and microstructure on constant and variable amplitude fatigue crack growth in aluminum-lithium alloys, in: "Proceedings of the 3rd International Aluminium-Lithium Conference," Oxford, England, July, 1985, Pergamon, in press.
12. P. E. Bretz, L. N. Mueller, and A. K. Vasudévan, Fatigue properties of 2020-T651 aluminum alloy, in: "Aluminum-Lithium Alloys II," T. H. Sanders, Jr. and E. A. Starke, Jr., eds., AIME, Warrendale, Pennsylvania (1981), p. 543.

STRAIN RESISTIVITY AT 4.2 K IN PURE ALUMINUM

K. T. Hartwig, G. S. Yuan* and P. Lehmann

Applied Superconductivity Center
University of Wisconsin-Madison
Madison, Wisconsin

ABSTRACT

Strain induced resistivity increases are examined in pure aluminum under conditions of uniaxial tensile and compressive stress at 4.2 K. Metal studied has an annealed residual resistivity ratio (RRR = $\rho_{273K}/\rho_{4.2K}$) covering the range 150 - 7100, is in some cases work hardened prior to testing, and is strained to levels below 10%. Resistivity is monitored during strain by eddy current coils located directly over strain gages bonded to centimeter diameter specimens. We find strain-resistivity to be slightly dependent on purity and similar for tension and compression. For all cases the empirical expression $\Delta\rho = A\ (\varepsilon_p)^B$ holds for plastic strains (ε_p) above about 0.1% where A and B are constants. Values of B group around 1.5 for annealed metal and decrease to near 1.15 for the work hardened cases examined.

INTRODUCTION

A major consideration in the design of high field magnets is the high mechanical stress on the current windings and the corresponding strain in the conductor. Such strain will cause an increase in the resistivity of the conductor metal and may affect magnet performance. It is the objective of this work to investigate strain induced resistivity in aluminum over a wide range of purities and in the region of low strains at low temperature so that designers can set limits on magnet operating conditions.

The application of stress to metals always causes electrical resistivity to increase.[1] For stress levels in the elastic range, the increase is small and reversible. For plastic strains however, lattice defects are formed which cause an irreversible increase in resistivity if the defects remain stable. Arp, Kasen and Reed[2] found that for annealed pure aluminum the resistivity caused by plastic strains at 4 K follows the empirical expression:

---

*On leave from the General Research Institute for Non-Ferrous Metals, Beijing, China.

$$\Delta\rho \text{ (n}\Omega\text{ cm)} = 110\ (\varepsilon_p)^{1.19} \quad . \qquad (1)$$

for ε_ρ from about 0.2 to 3 percent. In this study initial residual resistivity, yield strength and average grain size were ~ 1.3 nΩ cm, ~ 25 N/mm^2 and ~ 0.2 mm respectively. Agreement of Eq. (1) with earlier work (see Ref. 2) suggests that this expression is valid over a wider strain range and is rather independent of temperature, grain size and orientation effects.

The present work confirms that tension and compression strains give a similar resistivity increase and that the strain-resistivity relationship at 4.2 K is exponential. Our findings indicate that strain-resistivity is quite insensitive to purity and depends to a greater extent on level of cold work and/or original defect concentration. The results for annealed metal generally agree with the earlier work of Arp et al. for plastic strains below about 3%.

MATERIAL AND SPECIMEN PREPARATION

The specifications and manufacturers of the aluminum used for this investigation are listed in Table 1.

Specimens are prepared as follows:

1. Swage round stock to specified beginning diameter and chemically clean surface.
2. Anneal in air at 500°C for one hour.
3. Swage or draw to 16.3 mm diameter.
4. Machine to test specimen dimensions and chemically clean surface.
5. Anneal in air at 500°C for one hour.
6. Chemically polish surface and remove oxide with 7 H_3PO_4:2.5 H_2SO_4:1 HNO_3 at T > 50°C.
7. Mount two strain gages by bonding in air at 160°C for two hours.

The above sequence is modified for the cold worked (prestrained) specimens by eliminating step 5 (annealing) and changing the strain gage bonding heat treatment (step 7) to 100°C for 4 hours or 50°C for 6

Table 1. Aluminum Specifications

Manufacturer	Grade	Form	Purity[a] (% Al)	RRR[a] $(\rho_{273K}/\rho_{4.2K})^b$
Alcoa	--	multi-ton casting	99.98	150
Sumitomo	RNS	2.5 cm dia. extruded rod	> 99.9966	1000
Griffiths Metals	--	1 kg notch bar ingot	99.999	5000
VAW	Kryal-O	13 cm dia. casting	> 99.9995	> 9170

a) nominal
b) ρ_{273K} is taken as 2.43 μΩ cm.

hours. Specimen gage length dimensions are: diameter 10 mm and length 80 mm. Refer to past work for more details.[3,4]

EXPERIMENTAL PROCEDURES

All test specimens are held in special collet grips fastened to a cryogenic load frame and submerged in liquid helium at 4.2 K for stress and resistivity measurements.[5] Samples are mounted and cooled to 4.2 K without applying significant stress by operating the hydraulic actuator-control system in the zero-load-control mode. Tensile and compressive loads are limited to 9800 N (2200 lb) applied in the strain-or-stroke-control mode. Strain rate is estimated to be near 10^{-4}/s. Measurement of both plastic strain and resistivity are taken after extension (or compression) and return to zero load.

All specimens have two Micromeasurement WK-13-125BT-350 strain gages mounted opposite to one another at the sample midsection. A block of aluminum with two similar gages is located near the specimen for temperature compensation and bridge completion. Relative strain measurements are accurate to within 0.005% ε up to about one percent strain. Above this level, the strain gages fail and strain is deduced from stroke measurements and assumed accurate to better than 0.1% ε.

Resistivity at the strain gage location is determined by the Eddy Current Decay Method[6] which for the present case gives resistivity perpendicular to the direction of applied stress. These measurements are corrected for the size change caused by plastic deformation and assumed to be accurate to within one percent for strains below one percent. At higher strains, accuracy decreases because of dimension uncertainties.

Table 2. Sample Characteristics

ID	Purity[a]	Grain Size Trans.	Grain Size Long.	Yield Strength[b]	Initial Residual Resistivity	Initial RRR
	(% Al)	(mm)	(mm)	(N/mm^2)	(nΩ cm)	
			(Tension/Annealed)			
84	3N8	0.5	0.6	17.7	15.8	153.8
87	4N7	1.4	1.4	20.6	2.38	1024
48	--	--	--	12.9	0.705	3450
89	5N7	0.8	1.4	16.8	0.394	6170
			(Tension/Cold Worked)			
91	4N7	0.5	3.3	117	6.81	357
90	4N7	0.6	0.6	84.1	3.55	685
93	5N7	--	--	86.4	1.84	1318
			(Compression/Annealed)			
94	3N8	0.5	0.6	16.0	15.75	154.3
88	4N7	1.4	1.4	18.0	2.40	1013
49	--	--	--	14.1	0.730	3340
86	5N7	1.1	2.0	12.6	0.342	7105

a) A nines code is used here, i.e. 99.98 = 3N8.
b) Yield strength is stress at 0.2% offset strain. 1 N/mm^2 = 0.145 ksi.

RESULTS

Specimens with purities ranging from 99.98 to 99.9997% aluminum and corresponding to 150 < RRR < 7100 were successfully tested. Strains were limited to below 6% because of the actuator force maximum of 9800 N. Most of the samples were in the fully annealed condition with an average grain diameter of one-half to two millimeters. The work hardened samples had a grain size similar to or smaller than the annealed material,

The characterisitics of all specimens are listed in Table 2. Results are presented for tension and compression in annealed metal as well as for tension in cold worked metal. Grain size measurements are accurate to ± 0.2 mm. Initial residual resistivity and RRR are values for starting material before any strain.

Figure 1 presents tensile engineering stress-strain curves for the range of purities and cold work studied. Compression results are similar.

Figure 2 shows the resistivity increase ($\Delta\rho$) with increased plastic strain at 4.2 K for typical samples in tension and compression. These samples cover the range of purities studied for annealed metal and show the response of a cold worked sample.

DISCUSSION

Initial residual resistivity for the samples listed in Table 2 reflects metal purity and cold work effects consistent with Matthiessen's rule:[1]

$$\rho_{tot} = \rho_o + \Sigma\rho_i + \Sigma\rho_d \tag{2}$$

where ρ_{tot} is total resistivity and ρ_o, ρ_i, and ρ_d are the intrinsic (phonon), impurity and defect contributions, respectively. Simple linear relationships between initial resistivity and impurity level or level of cold work are not present (Table 2 data) because each sample type contains a different mix of impurity elements and defect densities and structures.

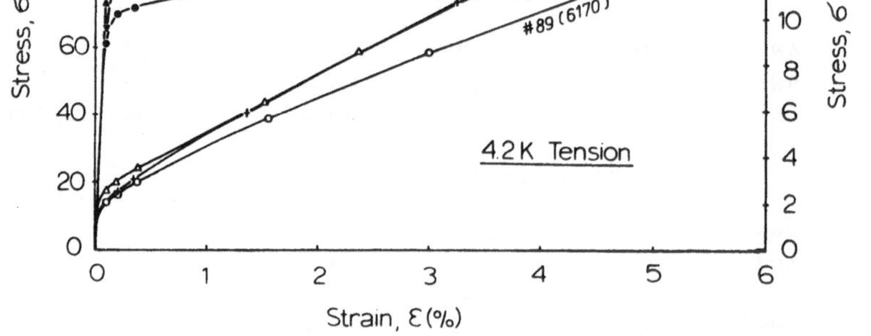

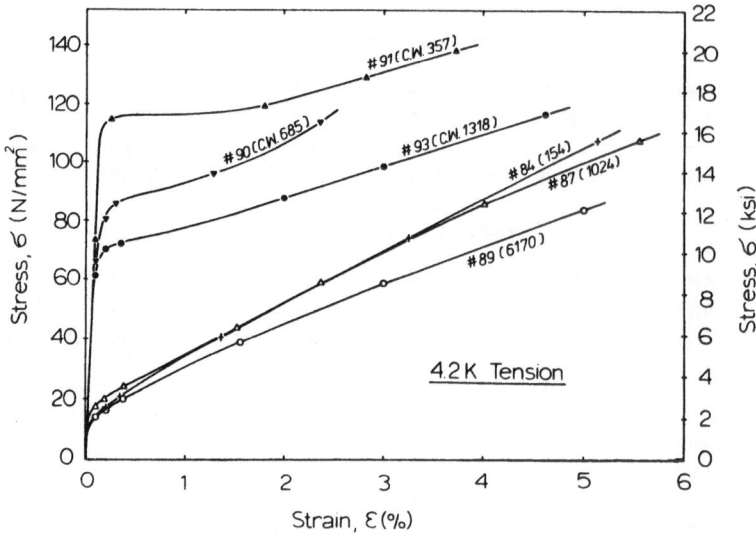

Fig. 1. Stress versus strain at 4.2 K for pure aluminum. Sample ID numbers are shown with initial RRR in parentheses.

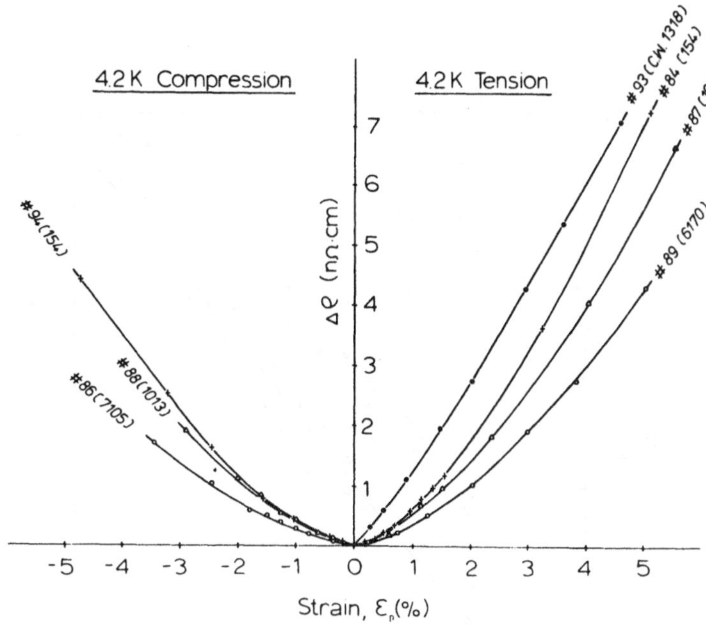

Fig. 2. Resistivity increase versus plastic strain at 4.2 K for pure aluminum. Sample ID numbers are shown with initial RRR in parentheses.

The yield strength in tension or compression for the annealed samples is between 12.6 and 20.6 N/mm^2 (1.8 - 3.0 ksi) and insensitive to purity level. The rate of strain hardening as seen in the curves of Fig. 1 for the annealed case is however slightly dependent on purity as expected. The work hardened samples are all at least four times as strong as those annealed despite a similar grain size for all samples. The added strength of the hardened aluminum does not correlate well with grain size and must come from sub-boundaries and/or other defect structures present but not observed.

Strain induced resistivity is dependent on purity and level of cold work as shown in Fig. 2. For the annealed case, tension and compression behave similarly with $\Delta\rho \propto |\varepsilon_p|$ and $\Delta\rho$ slightly dependent on purity. As the impurity level increases so does $\Delta\rho$ for a given strain. The fact that $\Delta\rho$ increases with an increase in ε_p, and is larger in more impure metal for a given ε_p is because matrix defects (whether they are impurities or lattice imperfections) give rise themselves to more defects during plastic strain. The trend is clear as one moves along any given $\Delta\rho$ vs. ε_p curve: the $\Delta\rho$ increment increases as $\Delta\varepsilon_p$ increases.

For all cases we find that the strain-resistivity data is fit by the empirical expression:

$$\Delta\rho = A\,(\varepsilon_p)^B \qquad (3)$$

where A and B are constants. Figs. 3 and 4 present strain-resistivity results plotted as ln ($\Delta\rho$) vs. ln (ε_p) and show that a good fit to Eq. (3) is possible for strains above about ε_p = 0.1%. Values of A and B for each sample are listed in Table 3 and are determined by a least squares curve fitting method with points below good exponential behavior neglected. The derived expressions underestimate $\Delta\rho$ for strain below 0.1 to 1% ε_p.

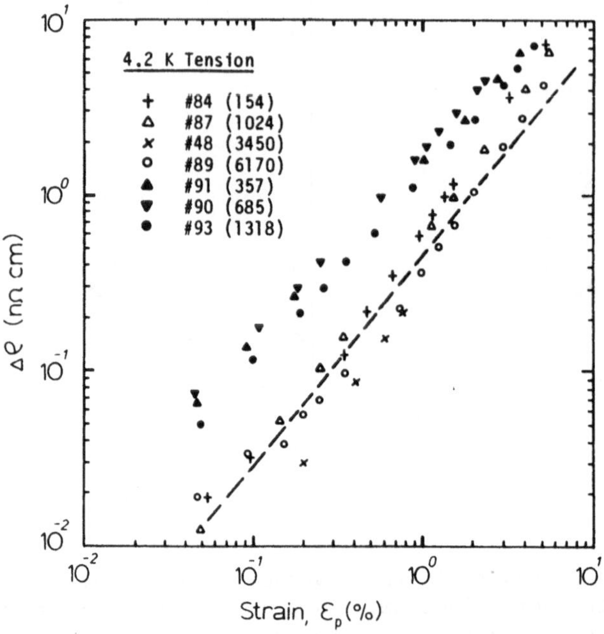

Fig. 3. ln (Δρ) versus ln (ε_p) for pure aluminum in tension at 4.2 K. Dashed line is Eq. (1).

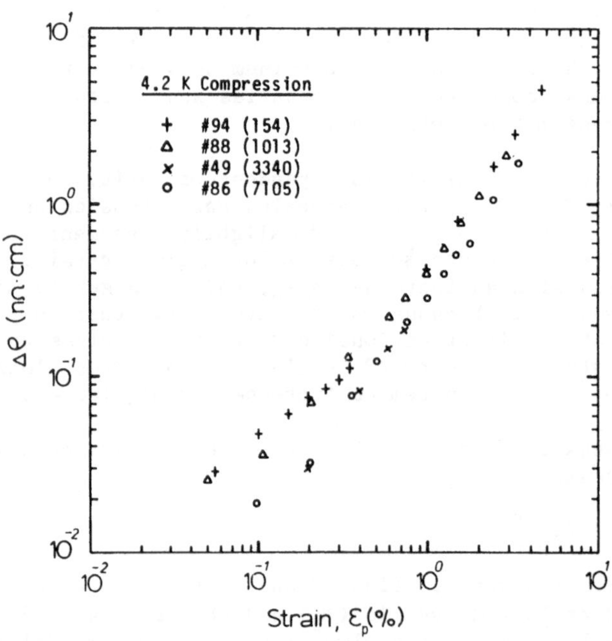

Fig. 4. ln (Δρ) versus ln (ε_p) for pure aluminum in compression at 4.2 K.

Table 3. The Constants A and B for Strain Induced Resistivity[a]

ID	Strain Range[b] (%)	A	B
84	0.2 - 5.2	630	1.51
87	1.1 - 5.6	520	1.51
48	0.2 - 0.8	210	1.41
89	0.7 - 5.1	430	1.55
91	1.0 - 3.8	330	1.19
90	0.18 - 2.4	280	1.11
93	0.05 - 4.7	230	1.13
94	0.2 - 3.5	470	1.53
88	0.7 - 2.9	340	1.47
49	0.2 - 0.8	120	1.32
86	1.0 - 4.8	250	1.48

a) Where $\Delta\rho$ (nΩ cm) = $A(\varepsilon_p)^B$.
b) Data below this range is neglected in fitting calculations except for the case of zero strain.

The values of B for tension and compression in annealed metal (Table 3) are similar, independent of purity and group around 1.5. B for the work hardened case is near 1.15 and agrees with earlier work[2] where a value of 1.19 was found. The Arp et al. data overlie the present data for the annealed case (see dashed line in Fig. 3) but with a different slope, and generally agree over the range $0.2\% < \varepsilon_p < 3\%$. It is not clear how or why B and yield strength (or g.s. for the earlier work) correlate other than to say that B probably decreases as initial matrix defects increase. The values of A, on the other hand, seem to be independent of strength and generally decrease as metal purity increases. A curve fit to equation (3) with A = 500 ± 100 and B = 1.5 nicely models the tension annealed strain-resistivity results for $0.2\% < \varepsilon_p < 5\%$.

CONCLUSIONS

1. Resistivity increases caused by plastic strain are similar for tension and compression.

2. Strain induced resistivity is slightly dependent on metal purity. An increase in impurity concentration (ppm) by 20 X causes $\Delta\rho$ to increase by only about 80% for plastic strains to 5%.

3. Strain induced resistivity is dependent on the level of prior cold work. Prestrain causes significant increases in the $\Delta\rho$ arising from additional plastic strain at 4.2 K.

4. Strain resistivity in annealed aluminum with a purity in the range 3N8 to 5N7 fits the empirical expression:

$$\Delta\rho\ (\mathrm{n\Omega\ cm}) = (500 \pm 100)\ (\varepsilon_p)^{1.5}$$

for tensile strains from 0.2 to 5 percent.

ACKNOWLEDGMENT

The authors thank Alcoa and Sumitomo Corp. for donating aluminum and Mr. H. Lau for assistance with metallography and grain size measurements. Financial support from EPRI and the Wisconsin Electric Ultilities Research Foundation is gratefully acknowledged.

REFERENCES

1. G. T. Meaden, "Electrical Resistance of Metals," Plenum Press, New York (1965).
2. V. D. Arp, M. B. Kasen, and R. P. Reed, "Magnetic Energy Storage and Cryogenic Aluminum Magnets," Technical Report AFAPL-TR-68-87 (Feb. 1969).
3. K. T. Hartwig, G. S. Yuan, and P. Lehmann, The effects of low temperature fatigue on the RRR and strength of pure aluminum, IEEE Trans. Magn. MAG-21(2):161-164 (1985).
4. "Superconducting Magnetic Energy Storage: Basic R&D," EPRI Research Project RP2572-2 Annual Report (Jan. 1985).
5. P. Lehmann, G. S. Yuan, and K. T. Hartwig, A grip for fatigue testing pure aluminum, Cryogenics 25:164-165 (1985).
6. A. F. Clark, V. A. Deason, J. G. Hust, and R. L. Powell, "Standard Reference Materials: The Eddy Current Decay Method for Resistivity Characterization of High Purity Metals," NBS Special Publication 260-39, National Bureau of Standards, Boulder, Colorado (1972).

THE EFFECT OF PRESTRAIN ON LOW TEMPERATURE FATIGUE INDUCED RESISTIVITY IN PURE ALUMINUM

G. S. Yuan,[†] P. Lehmann and K. T. Hartwig

Applied Superconductivity Center
University of Wisconsin-Madison
Madison, Wisconsin

ABSTRACT

Residual resistivity increases are examined in pure aluminum work hardened at room temperature and then cyclically strained at 4.2 K. The metal studied has an annealed residual resistivity ratio (RRR = $\rho_{273K}/\rho_{4.2K}$) of 1000, is swagged conventionally to area reductions of between 8 and 82%, and is taken through 3000 cycles of 0.1, 0.2 or 0.3% strain. Fatigue induced resistivity ($\Delta\rho$) is found to be dramatically reduced in prestrained metal when compared to the increase seen in annealed metal. Even though the initial residual resistivity of the work hardened aluminum is higher than that for the annealed case, the final resistivity after cycling is lower for the prestrained metal. The results of this study show that prestrain may provide a way to avoid much of the serious resistivity increase in the stabilizer aluminum used in composite superconductors and cryoconductors subjected to cyclic strain during operation.

INTRODUCTION

The composite superconductors proposed for large energy storage magnets will be subjected to cyclic diurnal stress at liquid helium temperatures. This situation could lead to a serious increase in the residual resistivity of the annealed pure aluminum if used as the stabilizer in such devices. The objective of this investigation is to find a way to inhibit low temperature fatigue degradation in pure aluminum.

Past work[1] on fully annealed aluminum has shown that strength and residual resistivity both rapidly increase during the first 100 strain cycles at 4.2 K and approach saturation after about 1000 cycles. The stress-strain curve shows predominantly elastic behavior at saturation where the peak stress (σ_{max}) is found to be:

$$\sigma_{max} = \frac{\varepsilon E}{2} \qquad (N/mm^2) \qquad (1)$$

[†] On leave from the General Research Institute for Non-Ferrous Metals, Beijing, China.

where ε is the cyclic strain range and E the elastic modulus. Resistivity at saturation (ρ) increases exponentially with strain. For the case of 99.997% aluminum (RRR = 1000), resistivity is related to strain as follows:

$$\rho = \rho_o + 2.2 \times 10^6 \varepsilon^2 \quad (\text{n}\Omega \text{ cm}) \qquad (2)$$

where ρ_o is the initial residual resistivity of annealed material and ε the cyclic strain range.[2]

We have found that work hardened pure aluminum undergoes much less of a resistivity increase due to cycle strain at 4.2 K than does annealed aluminum. The increase in resistivity due to prestrain (cold work or work hardening) plus the increase due to cyclic strain at 4.2 K in work hardened material is far below the increase in resistivity seen in fully annealed aluminum taken through the same cyclic strain history. As an example, consider grade 1000 RRR material strained from 0 to 0.2% at 4.2 K through 3000 cycles. For fully annealed material the residual resistivity increases from 2.41 (RRR_i = 1008) to 11.21 nΩ cm (RRR_f = 217). In 40 percent work hardened aluminum, residual resistivity increases from 3.64 (668 RRR_i) to 5.61 nΩ cm (433 RRR_f). Prestrain is a successful method for reducing resistivity increases caused by cyclic strain at 4.2 K.

EXPERIMENTAL PROCEDURES

Preparation of Specimens

The aluminum used in this study is supplied by Sumitomo Corporation as grade RNS with a purity of 99.997% and has a nominal RRR of 1000. Test specimens are prepared by carrying annealed rod through the following sequence:

1) form to intermediate diameter (15.0, 15.6, 16.6, 18.5, 22.7 and 34.1 mm) by swagging
2) anneal at 400°C for 3 hours in air
3) swage to 14.4 mm diameter.

The result is work hardened metal with percent area reductions

$$(\text{PAR} = \frac{\text{initial area-final area}}{\text{initial area}}) \text{ of 8, 15, 25, 40, 60 and 82\%.}$$

Specimens are machined to standard dimensions with a gage diameter of one centimeter as previously reported and instrumented with two strain gages.[1] Strain gage mounting includes a heat treatment at 160°C for two hours in air.

Test Apparatus and Strain Schedule

Test specimens are held in place and subjected to cyclic strain at 4.2 K by special grips and load frame system.[3] Tensile and compressive loads are applied by a hydraulic actuator which is driven by an MTS control system in the strain control mode. Resistivity at the strain gage location is measured throughout the process of fatigue by an eddy current decay method. Specimens are carried through 3000 cycles for strain ranges of 0 - 0.1, 0 - 0.2 or 0 - 0.3% ε. The strain rate schedule is: 0.01 Hz for cycles 1 to 10, 0.1 Hz for cycles 11 - 100, 0.5 Hz for cycles 101 - 200 and 1.0 Hz for cycles 201 and above. Strain rate effects are not seen for frequencies at 1.0 Hz and below.

Table 1. Pure Aluminum Characteristics

Specimen	Cyclic Strain Range	Percent Area Reduction	Maximum Tensile Stress on First Cycle (σ_1)	Residual Resistivity ($n\Omega$ cm)		Residual Resistivity Ratio* ($RRR = \rho_{273K}/\rho_{4.2K}$)	
No.	ε (%)	(%)	(N/mm^2)	Initial ρ_i	Final ρ_f	Initial RRR_i	Final** RRR_f
A	0 - 0.1	0	14.8	2.41	4.61	1008	527
68	"	8	44.2	2.75	2.86	885	848
69	"	15	48.5	2.92	2.99	833	813
B	0 - 0.2	0	17.6	2.41	11.2	1008	217
59	"	8	43.1	2.68	7.01	906	346
57	"	15	52.4	2.85	6.00	852	405
58	"	25	61.3	3.17	5.70	768	426
66	"	40	71.0	3.64	5.61	668	433
67	"	60	84.1	4.63	6.00	525	405
78	"	82	103.4	6.18	7.33	393	332
C	0 - 0.3	0	19.6	2.41	22.2	1008	109
70	"	25	65.0	3.23	12.9	752	189
71	"	40	68.4	3.52	12.8	690	190
72	"	60	83.4	4.45	11.8	547	206
79	"	82	118.5	6.65	11.6	365	209

*ρ_{273K} is 2430 $n\Omega$ cm.
**Initial and Final represent before and after fatigue respectively.

RESULTS

Specimen characteristics are listed in Table 1. The samples identified as A, B and C represent fully annealed aluminum; resistivity data for these cases are calculated using the empirical expression (2).

The maximum tensile deformation strength for cycle number one (σ_1), reflects material initial strength and is shown in column 4 of Table 1. In looking down column 4, note that increased cold work (percent area reduction) causes an increase in first cycle maximum stress as expected. After saturation at 3000 strain cycles, all samples are fully hardened and found to obey the relationship:

$$\sigma_t + |\sigma_c| = \varepsilon E \qquad (N/mm^2) \tag{3}$$

where σ_t and σ_c are the maximum tensile and compressive stress respectively.

Residual resistivity versus percent area reduction (PAR) for all specimens before and after cyclic stress is plotted in Fig. 1. The resistivities at zero PAR are for fully annealed material. Points A, B and C are for fully annealed aluminum which has undergone cyclic strain to saturation at strain levels of 0.1, 0.2 and 0.3% respectively. The dashed line beginning with point I is residual resistivity versus PAR for initial state material prior to any cyclic strain. The solid lines show resistivity after 3000 strain cycles versus PAR for the 0.1, 0.2 and 0.3% strain cases.

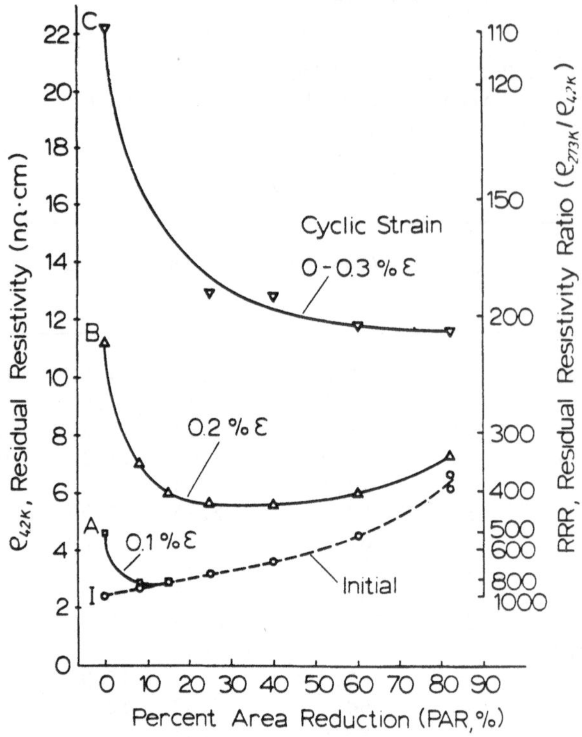

Figure 1. Residual resistivity versus percent area reduction (PAR) or degree of cold work for 1000 RRR pure aluminum. The dashed curve labeled "initial" is the resistivity before fatigue versus PAR. The solid curves are for saturation resistivity at 3000 strain cycles versus PAR.

From the solid curves of Fig. 1 it is clear that the saturation resistivity of prestrained material is below that for material initially in the annealed condition. This is true despite the fact that initial resistivity is higher in prestrained than in annealed material.

Fig. 2 shows resistivity degradation versus cycle number for an annealed and a 40% work hardened specimen. It is obvious that the resistivity of the prestrained specimen reaches saturation earlier as stress cycles build and has a lower final resistivity than the fully annealed specimen, even though the initial resistivity is higher for the prestrained aluminum.

Fig. 1 also shows that a wide range exists for choosing an amount of prestrain that gives the minimum resistivity for a particular level of cyclic strain. Consider for example the 0.2% cyclic strain case. Here, the optimum level of prestrain is 40% where the fatigue induced resistivity increase is 1.97 nΩ cm. For initially soft aluminum, the resistivity increase associated with 0.2% strain is 8.79 nΩ cm. The prestrained material shows a degradation four times smaller than that for the annealed material. Another example is shown by the 0.1% strain curve. Here, the resistivity increase can be kept to practically zero by imparting only a modest amount of prestrain (12%).

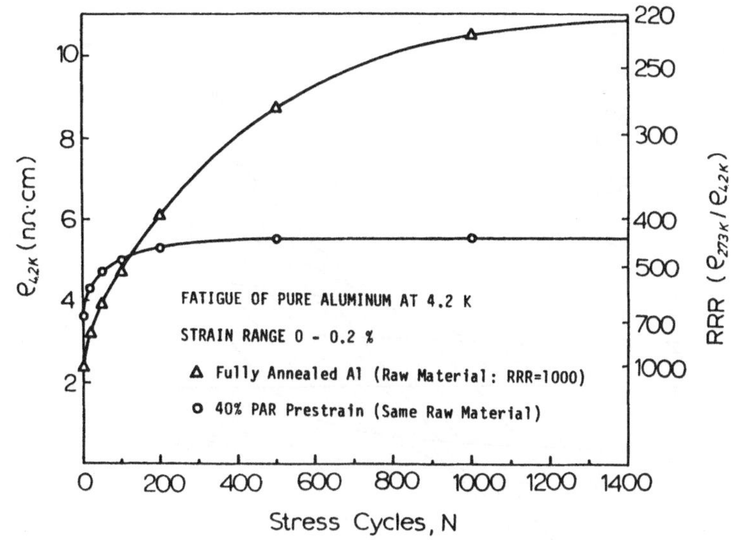

Figure 2. Residual resistivity ($\rho_{4.2\ K}$) and residual resistivity ratio (RRR) versus number of stress cycles for annealed and work hardened pure aluminum.

DISCUSSION

As we indicated, all of the aluminum specimens become fully hardened after fatigue and the deformation strength reaches $\sigma_t + |\sigma_c| = \varepsilon E$. This process is however quite different between fully annealed and prestrain specimens. Figures 3 and 4 show typical stress-strain charts for these two cases. As shown in Fig. 4, mechanical equilibrium (stable hysteresis) occurs after just about 200 cycles for the cold worked specimen. In general, equilibrium is approached after about 1000 cycles for fully annealed aluminum as shown in Fig. 3. Since the cold worked specimen has a higher initial strength, it reaches mechanical equilibrium much faster than the fully annealed sample.

Our results show that the cumulative value of cyclic strain damage as measured by resistivity is much less for the cold worked specimen than for the annealed specimen. Two points are relevant. First, since the work hardened metal is strong to start with, fewer added defects are needed to bring the strength up to the point where elastic behavior occurs (saturation). We have measured that the sum of hysteresis loop widths (plastic strain component) out to a given cycle number for the work hardened metal is less than that for the annealed case. The cumulative value of plastic deformation reflects cyclic strain damage and the resistivity degradation. Therefore the resistivity increment out to a given cycle number should be less for the work hardened metal, just as we find.

The second point is that the increase in $\rho_{4.2}$ as prestrain increases is related to the buildup of crystal defects with increased work hardening.[4] We find that the resistivity increase caused by room temperature prestrain is not as large as that caused by cryogenic fatigue even though the resulting strengths at 4.2 K are the same. One of the reasons is that the point defects which are a major contributor to the residual resistivity are stable at 4.2 K and will be annealed out at room temperature.

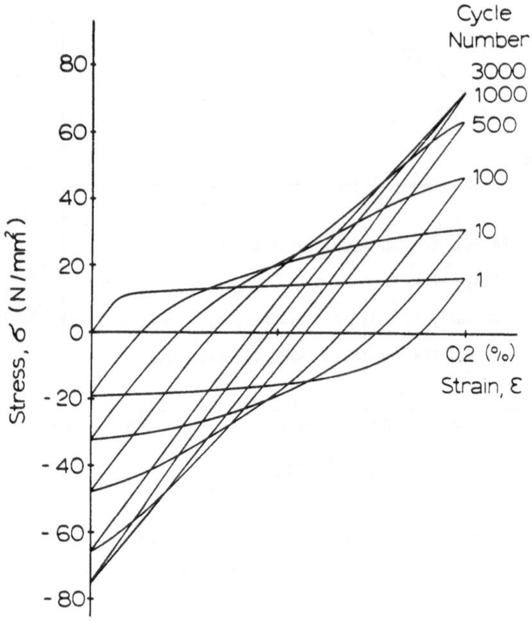

Figure 3. Stress versus strain hysteresis curves for fully annealed pure aluminum at 4.2 K.

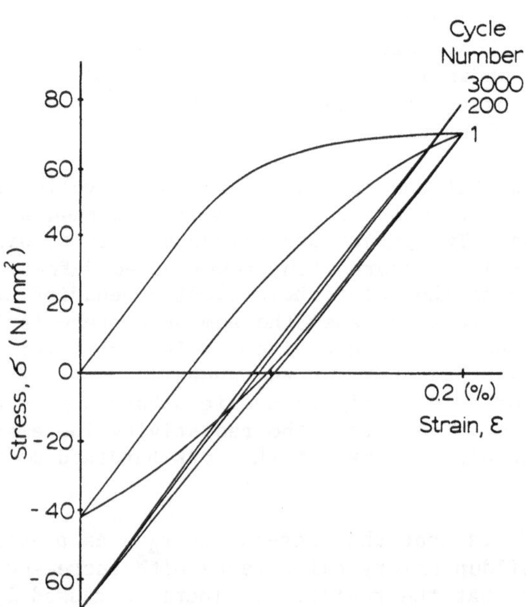

Figure 4. Stress versus strain hysteresis curves for 40% PAR cold worked pure aluminum at 4.2 K.

Inspection of Fig. 1 leads to two main conclusions:

1. Prestrain inhibits resistivity degradation due to cyclic strain at 4.2 K,
2. The best level of prestrain depends on the cyclic strain environment. The best level of prestrain for the 0 - 0.1% and 0 - 0.2% cyclic strain cases is the range of 8 - 15% and 20 - 40% area reduction (PAR) respectively.

ACKNOWLEDGMENTS

This work was sponsored by the Wisconsin Electric Utilities Research Foundation.

REFERENCES

1. K. T. Hartwig, G. S. Yuan and P. Lehmann, The Effect of Low Temperature Fatigue on the RRR and Strength of Pure Aluminum, in: "IEEE Trans. Magn.", Vol. MAG-21, No. 2, pp. 161-164 (1985).
2. "Superconducting Magnetic Energy Storage: Basic R & D", EPRI Research Project RP2572-2 Annual Report, January 1985.
3. P. Lehmann, G. S. Yuan and K. T. Hartwig, Grip for Fatigue Testing Pure Aluminum, Cryogenics, 3:164-165 (1985).
4. K. T. Hartwig and P. G. Appius, Strength and Electrical Resistance of Cold Drawn Al and Cu, Proceedings of the 8th ICEC, IPC Science and Technology Press, pp. 720-725 (1980).

RELATIONSHIPS BETWEEN MECHANICAL AND MAGNETOELECTRIC PROPERTIES OF OXYGEN-FREE COPPER AT 4 K*

F. R. Fickett and T. E. Capobianco

Electromagnetic Technology Division
National Bureau of Standards
Boulder, Colorado

ABSTRACT

Commercially pure, oxygen-free copper is the material of choice for nearly all superconductor stabilization. Straining relatively pure copper at 4 K can result in significant increases in the residual resistivity and, thus, a decreased ability of the copper to stabilize the superconductor. In this paper we quantify the effect of strain on the resistivity and magnetoresistivity of a number of oxygen-free coppers from various sources and in various tempers. In addition, the low temperature stress-strain behavior of these materials and its correlation with room temperature data and the residual resistivity ratio (RRR) prior to straining is discussed. An apparatus developed for testing of mechanical properties of relatively small wire samples at low temperatures is described.

INTRODUCTION

In this report we present data from our continuing program to develop a data base on the properties of oxygen-free copper at low temperatures. We chose to concentrate on this material because it is the sole type of copper used in the stabilization of commercial superconductors. However, the fact that the ore comes from many sources and is subject to different processing results in quite a wide range of properties in low temperature applications. Several other reports describe earlier work on this project[1-3].

In this phase of the investigation, we concentrated on the mechanical properties of oxygen-free copper. We wanted to know how strain affects the residual resistance ratio (RRR) and the magnetoresistance. We also hoped that there might be a correlation between some easily measured electrical parameter, such as RRR, and the strength of the copper at 4 K. This would be especially valuable because the strength of oxygen-free copper in the annealed (soft temper) condition is highly variable, as seen in Fig. 1, and strength is a property that is reasonably difficult to measure in the average laboratory.

Copper wires meeting the oxygen-free specification were acquired from six processors and drawn to four tempers either in the laboratory or by the

* Work supported by the International Copper Research Association
Contribution of NBS, not subject to copyright

supplier. The final diameter of all wires was the same (1.0 mm), and thus we could use resistance rather than resistivity in many of the data plots that compare samples. We measured the magnetoresistivity and the stress-strain properties of the wires at 4 K in a number of different apparatus. Most of the data presented here were measured with one of two cryostats: a strain-magnetoresistance device in which a long (30 cm) sample, instrumented with electrical leads, is strained while in a magnetic field that may be as high as 10 T[2]; and a device that allows rapid measurement of the stress-strain curve of the wire without either magnetic field or electrical instrumentation--more on this below. The first cryostat measures strain by monitoring crosshead displacement; the second uses a full-bridge extensometer. Agreement of stress-strain data measured by these two devices on the same sample material was excellent, usually within 2%. This is a result of the care taken to avoid annealing the sample while soldering voltage taps and current connections in the magnetoresistance cryostat.

EFFECT OF DEFORMATION ON RESISTANCE AND MAGNETORESISTANCE

In magnet applications, there are two major ways in which the resistivity of copper is altered from that of the annealed state: (1) strain induced by energizing of the magnet and (2) deformation intentionally induced to increase strength, usually by drawing of the wire. In addition, strain due to winding of the wire on the coil form can degrade the RRR, but this is not usually a major contributor in magnets of reasonable size (bore larger than about 5 cm). In special applications, radiation damage can degrade the stabilizer, and there are some data to indicate that the effect of that sort of damage on the magnetoresistance may be significantly different from the mechanisms treated here[4].

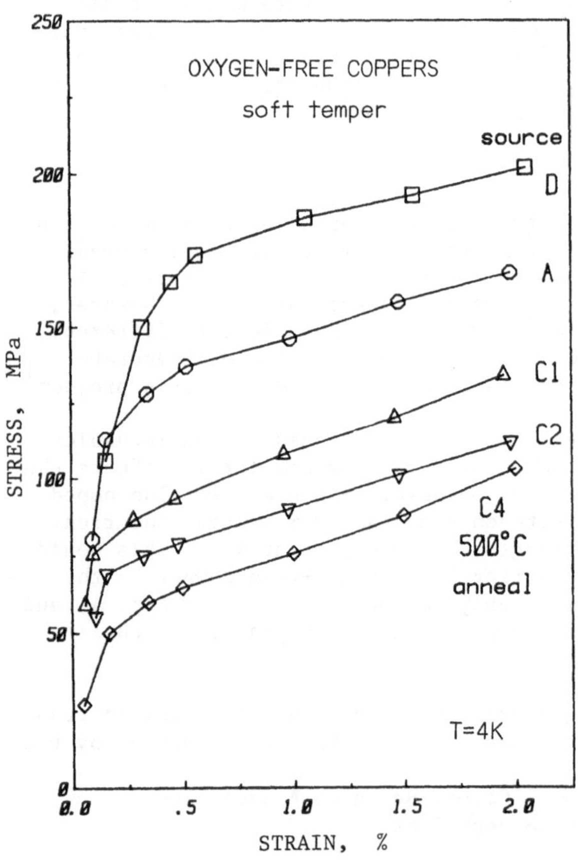

Fig. 1. Variation of stress-strain data among the coppers in their as-received condition. All are annealed to full soft by the supplier, except C4, which was annealed in the laboratory.

Strain degradation of the RRR for a variety of soft temper coppers is shown in Fig. 2. Notice the wide range of RRR values in the as-received copper (zero strain). Similar behavior is seen when other tempers are strained, but the percentage change is smaller. Some rough rules of thumb upon straining to 2% at 4 K: soft copper RRR degrades by 10 to 20% with the larger value more likely for the higher RRR materials; quarter hard copper (21% reduction in area) shows 5 to 10% reduction; half hard copper (37% reduction in area), 3 to 6%; and full hard copper (60% reduction in area), 1 to 3% decrease in RRR. These values are about the same whether measured under strain or after the strain is released, since the strain limit is well into the plastic region. In fact, in the tensile testing work described below, we measured the RRR only after straining and removing from the grips.

The effect of strain and temper on the magnetoresistance is illustrated in Fig. 3. The curves are for copper from a single source, but the behavior is similar for all of our materials. The curve bounding the lower edge of the shaded area for each temper is the magnetoresistance as measured with no applied strain. The zero field resistance increases with temper, as one would expect, but the magnetoresistance in the high field ($\Delta R/\Delta H$) region is not changed greatly from one temper to the next. Furthermore, a 2% applied strain (upper curve of each set) only changes the zero field resistance, not the magnetoresistance. The preceding statements are made in the spirit of engineering applications. In fact, the slopes of the curves are somewhat different. However, it is also true that a Kohler plot of all the data[5] shows that all samples of all materials follow a single curve at high fields. More precise information can be obtained from such a plot if it is required.

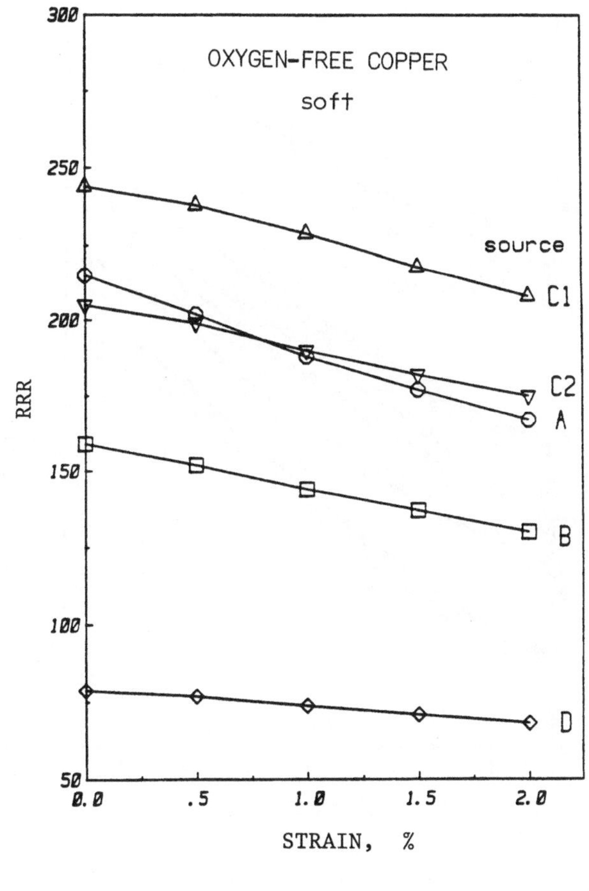

Fig. 2. Effect of tensile strain at 4 K on the RRR of as-received, soft-annealed oxygen-free coppers.

TENSILE TESTING OF COPPER WIRES

Our investigation of the possibility of a relationship between yield strength and RRR for oxygen-free coppers required that we perform a large number of tensile tests on the wire samples in liquid helium. Development of an apparatus to perform these tests reliably and reproducibly was a major undertaking. The most basic criterion for a good test is that, when pulled to failure, the wire should not break at the grips and, if an extensometer is used, it should break between the extensometer attachment points.

Several gripping fixtures were developed to hold the wire during test. The first was a set of sliding serrated jaw grips based on the grip design of a wire drawing bench. Later we added universal joint mounts for the grips. This system was abandoned when it was determined that the jaws introduced significant stress risers where the wire left the jaws, causing sample breakage at that point, before yielding occurred in the gauge length.

The final grip design was a pair of spools (Fig. 4). The stationary aluminum spools are mounted on an aluminum bracket that connects the fixtures to a conventional tensile test machine. Three turns of the sample wire are wrapped around a machined groove in the spool with the loose end held by the set screw on the bracket. The active length of the sample is aligned along the axis of the pull rod of the test machine by being constrained in the machined groove of the spool. The other end of the sample is held in a grip which is a mirror image of the first. We also tried knurled, hardened steel spools without grooves (as are commonly used in the

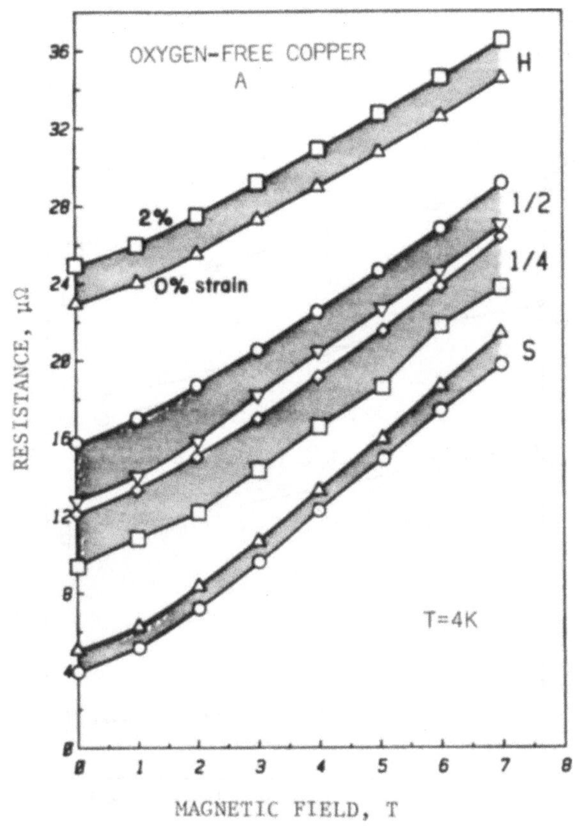

Fig. 3. Effect of strain and temper on the magnetoresistance of oxygen-free copper measured at 4 K.

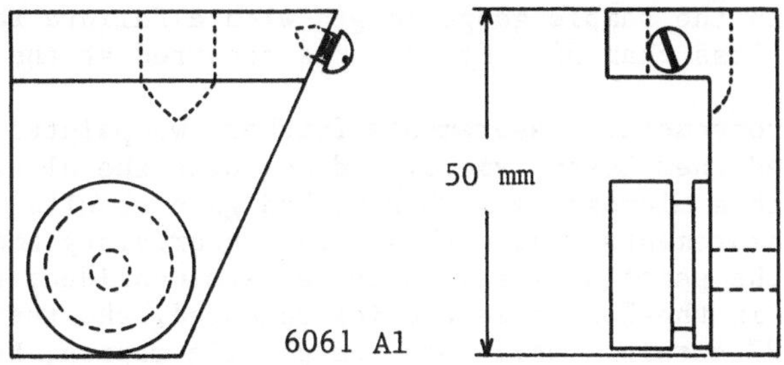

Fig. 4. One of the grips used for wire tensile properties measurements.

copper wire industry) but they were considerably more difficult to use and did not produce any better results.

With the aluminum spool fixtures, sample breakage at the grip was infrequent (about 5% of the samples). When an extensometer* was placed on the wire to measure elongation, the samples began to break consistently at either the upper or lower grip of the extensometer as a result of stress introduced by the grip mechanism. After several modifications, the final extensometer grip configuration was a spring loaded post with a flat head and a groove cut into the head (Fig. 5). The wire runs in this groove and is gently pushed against a knife edge cut into the body of the extensometer. Unfortunately, a sharp-edged grip is necessary to define the gauge length adequately and we still observed frequent breakage at this point.

Next, we decided to electropolish the sample for a length equivalent to the gauge length of the extensometer (about 12 mm). This operation removed only 0.025 to 0.035 mm from the diameter of the wire. We were able to control the length of this region so that the extensometer grips mounted just at the edges of the modified length. The samples were checked for irregularities in diameter along the length of the electropolished area. The variations were consistently less than 0.01 mm. This combination of spool grips, modified extensometer, and electropolished samples resulted in

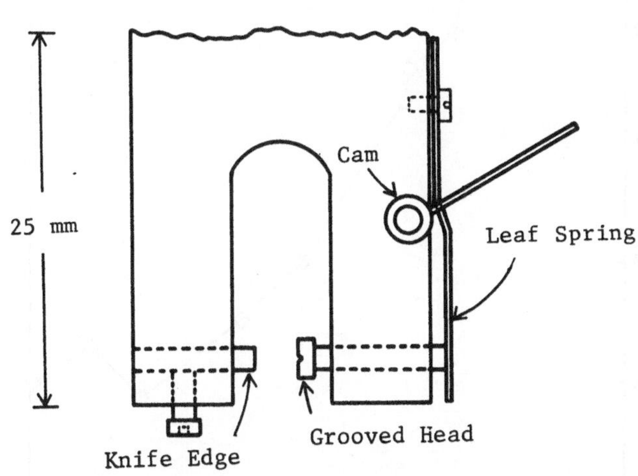

Fig. 5. Detail of one grip of the extensometer. The wire under test would be directed into the page in this view.

* A full-bridge extensometer custom-built for use at 4K by J. Shepic, Lakewood, Colorado

consistent yielding in the sample gauge length with a failure rate at the extensometer grip of less than 5%. No failures occurred at the spool grips.

To check our extensometer measurements further, we painted fine lines on the samples, pulled them beyond yield, and measured the elongation between the lines with a microscope. This method gave results very close to the electrical measurements, but with greater uncertainty because of the finite dimension of the paint stripes. Once we were confident that the strain was occurring in the length we had instrumented, the test runs were made only to around 2% strain - well into the plastic region, but far from failure. The extensometer itself was further checked for calibration at liquid helium temperature with a specially constructed micrometer calibrator.

STRENGTH AND ELECTRICAL PROPERTIES

Using the apparatus described above, we made stress-strain measurements at room temperature and at 4 K on more than 50 samples from our stock of materials. The 4 K results are summarized in Fig. 6 in which we plot the measured yield strength (at 0.2% offset) as a function of the RRR prior to straining. A very similar curve is obtained for the room temperature data except that the overall yield strength values are somewhat lower. The precision of the measurement is about 2%. The sample-to-sample variation for a given material is also of this order, 2-5% for repeated measurements

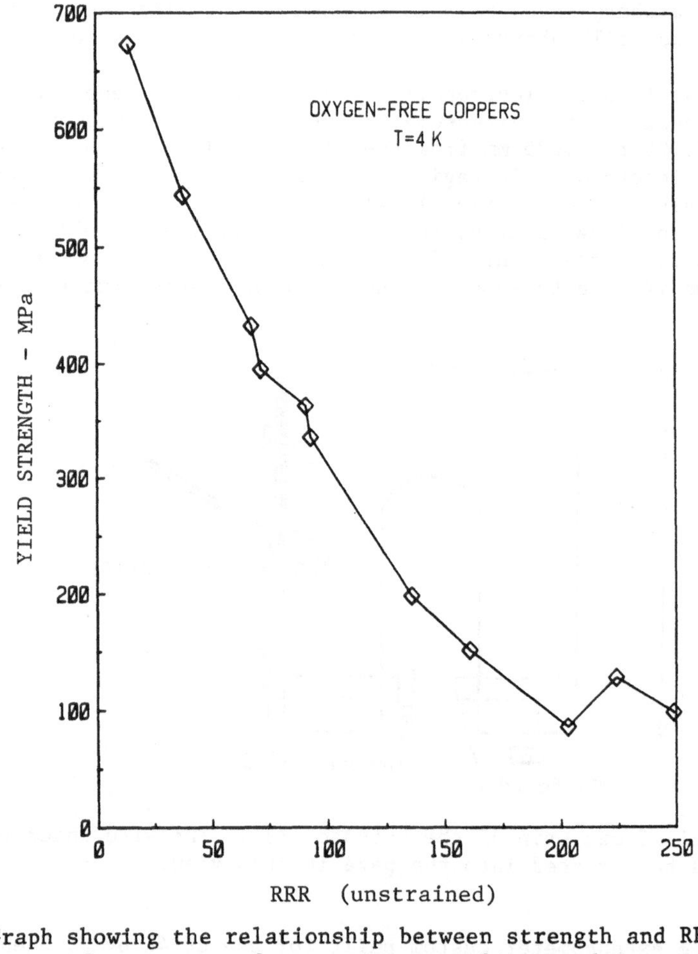

Fig. 6. Graph showing the relationship between strength and RRR for oxygen-free coppers from various suppliers and in several tempers.

of the same material. It is important to note that the data shown are only for oxygen-free coppers, although the highest yield strength material is a dispersion hardened version. Other coppers, such as ETP, do not appear to fall on this curve. The variation at the high RRR end of the curve is essentially that from one material to another in the soft anneal state. This implies that there may be a reasonably large spread in this region. Fortunately, at this strength level, the copper is seldom being used as a structural member. A further caution should be observed in applying these data to copper pieces of significantly different cross section than our 1.0 mm wires. Drawing of large diameter rods distributes the coldwork across the diameter in a manner different from drawing of relatively small wires. Thus, it is not certain that the curve shown would give correct yield stress data for much larger drawn samples.

In spite of the few caveats given above, the relationship between RRR and yield strength is very good over much of the range, and our conclusion is that this curve provides a relatively easy way to estimate the strength at 4 K of drawn oxygen-free copper in a variety of applications and even at different points along a conductor where variations in strength may occur due to local thermal treatment (soldering) or localized stress due to bending.

CONCLUSIONS

The oxygen-free coppers show a wide variation in properties depending on their source. However, the effect of applied strain and magnetic field is similar for all of them. It appears that it is possible to predict the yield strength of oxygen-free copper wires to within about 5% from a measurement of the residual resistance ratio.

REFERENCES

1. F. R. Fickett, Oxygen-free copper at 4 K: resistance and magnetoresistance, IEEE Trans. Magn. MAG-19:228-231 (1983).
2. F. R. Fickett, "Conductors for Advanced Energy Systems," INCRA 321A, International Copper Research Association, New York (1983).
3. F. R. Fickett, The effect of mill temper on the mechanical and magnetoresistive properties of oxygen-free copper in liquid helium, in: "Advances in Cryogenic Engineering - Materials," vol. 30, Plenum Press, New York (1984), pp. 453-460.
4. M. W. Guinan and R. A. Van Konynenburg, Fusion neutron effects on magnetoresistivity of copper stabilizer materials, J. Nucl. Mater. 122 & 123:1365-1370 (1984).
5. F. R. Fickett, "Conductors for Advanced Energy Systems," INCRA 321B, International Copper Research Association, New York (1985). To be published.

ANOMALOUS DECREASE OF RESISTANCE AT 250 K IN ULTRATHIN Au-Nb FILM ON SINGLE-CRYSTAL SILICON

Hiroshi Yamamoto, Toshitaka Kawashima, and Masaichi Tanaka

College of Science and Technology
Nihon University
Funabashi-shi, Chiba, Japan

ABSTRACT

Ultrathin Au-Nb films as thin as 0.2 ~ 10 nm were deposited on clean surfaces of single-crystal silicon in order to investigate interfacial excitonic superconductivity. The samples were classified into two types, Nb-Au/Si and Au-Nb-Au/Si. In the latter case, the secondary Au film was deposited on the former sample cooled by liquid nitrogen. In the Nb-Au/Si type of sample, a sheet resistance, R_s at room temperature abruptly increased from 10^3 Ωsq^{-1} order to about 10^5 Ωsq^{-1} in several days a few months after the sample preparation. Then the sample showed an anomalous decrease of R_s at about 250 K and an approximately null resistance at lower temperatures. This phenomenon was not so stable and was observed only for a few days. The Au-Nb-Au/Si type of sample showed low R_s (10^2~10^3 Ωsq^{-1}) at room temperature. A decrease and disappearance of R_s were also observed at about 240 K in the sample with comparatively good reproducibility. These phenomena are discussed qualitatively, based on the excitonic superconductive model for an interface of metal/semiconductor by Allender, Bray, and Bardeen.

INTRODUCTION

There have been many theoretical and experimental efforts to find material systems of high temperature superconductivity. The microscopic theory for superconductivity was established by BCS.[1] The BCS theory gives us the guiding principle that if electrons attract each other in a certain material system through an elementary excitation with high characteristic energy (it is given a general name "exciton" for convenience), the system reveals high superconductive critical temperature, T_c. Little[2] and Ginzburg[3] already suggested the typical model for one-dimensional and two-dimensional excitonic superconductivity, respectively. After their exciting suggestions, various models[4] have been investigated for the excitonic superconductivity. Allender, Bray and Bardeen (ABB)[5] discussed also the theoretical model expected of the interface of metal/semiconductor in which electrons tunneled into the semiconductor gap interact with virtual excitons and net attractive interactions can be formed in that electron system. Based on the ABB theory, Miller et al.[6] tried to synthesize the interfacial excitonic superconductor and reported that no excitonic enhancement of T_c was observed in epitaxial Pb films as thin as 1.0 ~ 1.5 nm deposited on PbTe crystals.

The purposes of this work are to synthesize a quasi two-dimensional interface of metal/semiconductor by characteristic low temperature reactions at Si-metal contacts and to investigate the conduction of the sample relevant to excitonic superconductivity. The sample is prepared by deposition of ultrathin Au-Nb film on a very clean surface of Si in ultrahigh vacuum. The sheet resistance of the sample changes as the time proceeds. Then the sample shows an anomalous temperature dependence of sheet resistance for only a few days, accompanied by a rapid increase of sheet resistance at room temperature. Typically the sample shows abrupt decrease of sheet resistance at about 250 K and approximately null resistance at lower temperatures. Though the decisive interpretation for this anomaly is not obtained, a structure for the sample is suggested, and the possibility of excitonic superconductivity based on the ABB theory is investigated for this material system.

PREPARATION AND MEASUREMENT OF SAMPLE

The process of sample preparation is shown schematically in Fig. 1. The samples prepared are classified into three types in accordance with their structual differences. The substrate is a (100) plane of n-Si wafer with sheet resistance of $12 \sim 75\ \Omega sq^{-1}$. The substrate ($7x20x0.4mm^3$) was mounted in a Mo holder with electrodes for thermal flash and set up in a vacuum chamber. The chamber was evacuated to about 0.1 μPa by an ion pump. Before a film deposition, the surface of the substrate was cleaned by a thermal flash for a few seconds in an ultrahigh vacuum.

For type A and B samples, Au film was deposited on an ambient substrate by thermal evaporation from W wire. Thickness of the film was monitored in situ by an oscillating quartz thickness meter. Deposition rate was typically about 6.7 nm/h. In the case of a type B sample, Au deposition was done in the following two steps: in the first step, Au film of about 5 nm thick was deposited on a hot substrate of about $700 \sim 800\,°C$, and in the second step, Au film of about 2 nm thick was over-deposited on an ambient substrate. Furthermore, ultrathin Nb film was successively deposited by evaporation using a work-accelerating type of electron bombardment. Thickness of the Nb film can not be monitored accurately in situ and was estimated after the deposition from the X-ray intensity of Nb in electron probe microanalysis. A type C sample was prepared by the secondary Au coating on a type B sample which was reset after exposing an atmosphere in the other holder cooled by liquid nitrogen. The Au coating was done under the same conditions as the other samples, and thickness of the Au film was varied in the range between 1 nm and 5 nm.

Sheet resistance, R_s, was measured by the dc four-probe technique. Electrical contacts were W needles for room temperature measurement and silver paste for cryogenic measurement. When a large voltage was induced by electrode contacts, R_s was determined by increasing the current.

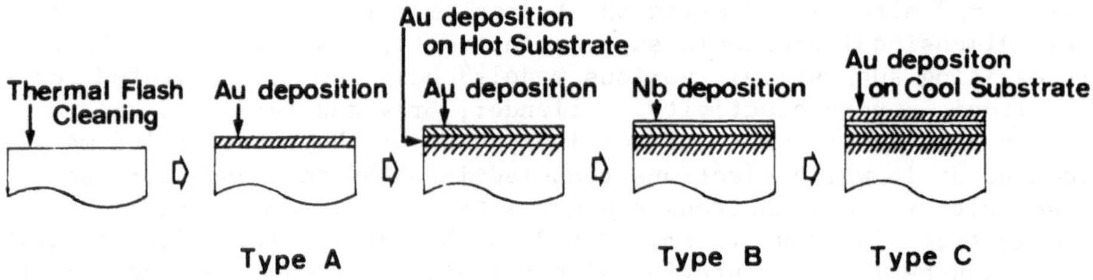

Fig. 1. Typical process of sample preparation and classification of the samples.

Electric current was supplied by a constant current source in the range between 0.01 mA and 100 mA with an accuracy of 0.05%. Temperature was measured by Au-0.07%Fe/chromel thermocouple within an error of 1 K. Since the sample is not homogeneous in the direction to the film thickness, the sheet resistance observed should be recognized as an apparent value and a resistivity of a metallic layer can not be obtained. Changes of R_s at room temperature are monitored in the atmosphere.

RESULTS

As shown in Fig. 2, R_s of type A sample depended on the thickness of the Au film, t_{Au}. In the range of t_{Au} less than about 10 nm, R_s was found to increase with increasing t_{Au} and to attain up to about 10^5 Ωsq^{-1}, in contrast to the case of Au film deposited on a nonflashed Si. A comparatively thick Au coating resulted in low R_s of about 10^{-1} Ωsq^{-1}, which agrees with the value estimated from the resistivity of Au. It should be mentioned that R_s of the sample changed with time. For example, R_s increased by two orders of magnitude in several days in the sample coated by Au film of about 10 nm thick. At the same time, the color of the sample changed from uniform gold to an interference color pattern. Here the data shown in Fig. 2 are the values observed in a comparatively steady state. The temperature coefficient of R_s in those samples is very small, similar to that observed in amorphous metal or alloys.

In type B sample, R_s changed very slowly, depending on time. As shown in Fig. 3, R_s typically increased from about $2x10^2$ Ωsq^{-1} to about 10^3 Ωsq^{-1} in a few months. After the moderate increase of R_s, it rapidly increased and attained up to about 10^5 Ωsq^{-1} over a period of several days. When the abrupt increase of R_s was observed, an anomalous temperature dependence of R_s was also observed. The typical data of the **a** and **b** states in the type B sample are shown in Fig. 4. Though each value of R_s at room temperature is quite different, R_s showed an abrupt decrease by two orders of magnitude at about 230 K in the **a** state and at about 250 K in the **b** state, respectively. Then R_s was almost zero in the range of temperatures lower than 200 K. When the measuring current was large enough, about 100 mA, R_s returned to almost the same value observed at

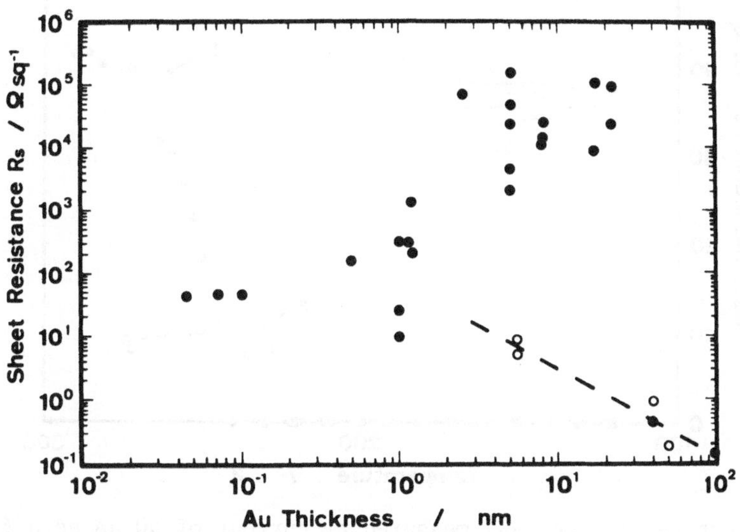

Fig. 2. Sheet resistance, R_s, as a function of thickness of Au film, where solid circles represent the data observed in type A samples, open circles in Au films deposited on a nonflashed Si.

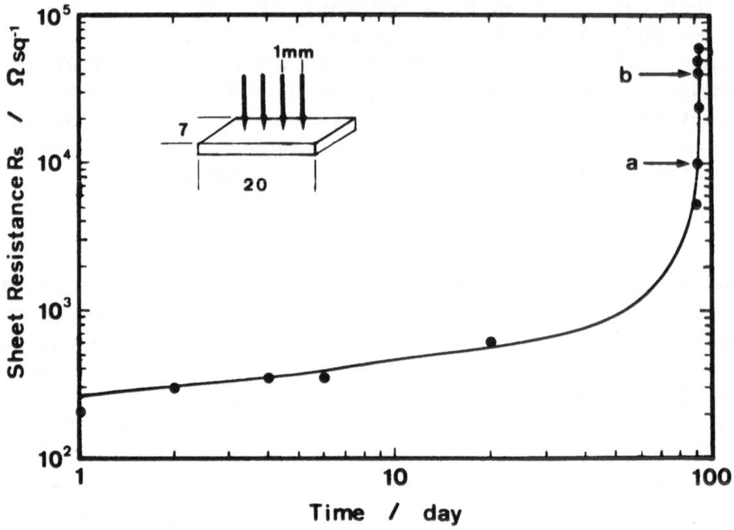

Fig. 3. Typical sheet resistance, R_s, as a function of time in the type B sample, where **a** and **b** represent the data observed 92 and 93 days after the sample preparation, respectively.

room temperature, even at low temperatures. This resistance anomaly can be observed only for a few days, accompanied by an abrupt increase of R_s at room temperature. After R_s attained its saturation value, about 10^5 Ωsq^{-1}, R_s did not show any anomaly and it depended monotonically on temperature, similar to the case of the type A sample.

It was found that the secondary Au coating in type C sample gave more stable observation and better reproducibility of the resistance anomaly. Figure 5 shows the typical result of the type C sample with the secondary coating of Au film of about 1 nm thick. The anomalous decrease

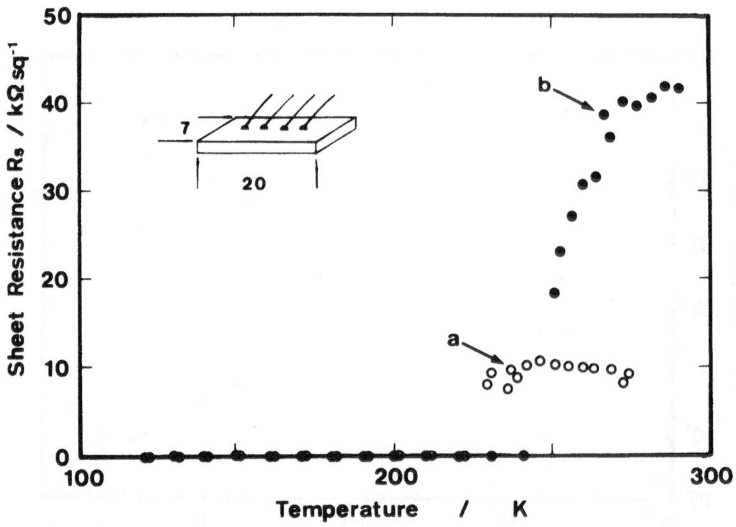

Fig. 4. Sheet resistance, R_s, measured by current of 50 μA as a function of temperature in **a** and **b** state of the type B sample shown in Fig. 3.

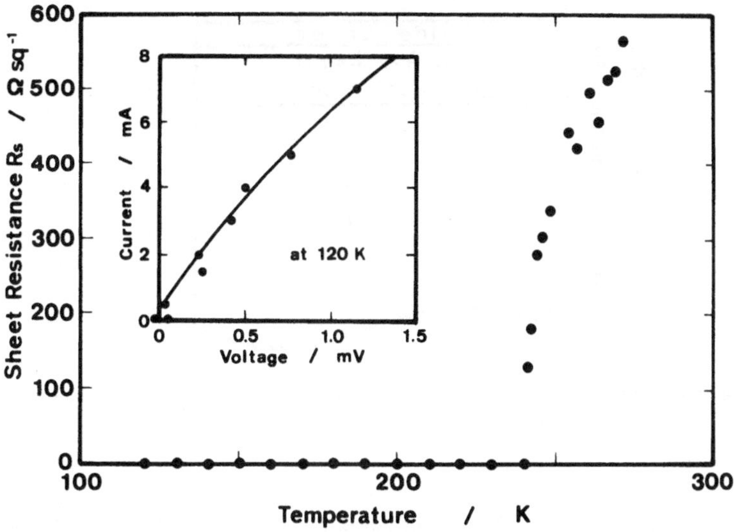

Fig. 5. Sheet resistance, R_s, measured by current of 10 μA as a function of temperature in type C sample 74 days after the sample preparation. The inset is a I-V characteristic observed at 120 K.

of R_s was observed even in the range of comparatively low R_s. Then R_s decreased with decreasing temperature and almost disappeared at 240 K in the same manner as that of the type B sample. The inset shows a I-V characteristic at 120 K. When the measuring current was of the order of mA, R_s at low temperature was three orders of magnitude less than that at room temperature. It was also noticed that R_s was almost zero in the range of measuring current less than 500 μA. Furthermore, R_s increased to the room temperature value under irradiation of He-Ne laser with a power density of about 5 kWm^{-2} even in the null resistance region and R_s also fell into the null state without irradiation. This response of R_s for an irradiation of light was observed with good reproducibility.

DISCUSSION

Since the surface layer was too thin and complex to analyze the structure of the sample, it is very difficult to discuss the resistance anomaly observed and a decisive interpretation for the anomaly was not obtained. In the following, the material structure of the sample is proposed qualitatively. In such a material system, an excitonic superconductivity was also investigated with relation to the ABB theory.

It is known that Au atoms deposited on very clean Si readily react or interdiffuse at low temperature (≲200 °C) as long as the density of Au is large enough, above 2×10^{19} atoms/m^2.[7] Hiraki[8] has proposed a model for the characteristic phenomenon, postulating electronic screening of the Coulomb interaction responsible for the covalent bonding due to mobile free electrons in the metal films on a covalent semiconductor, like Si. The low temperature reactions have been investigated in various semiconductor-metal (Au, Pd, Cu etc.) systems. According to the phenomenon, Au atoms react with Si even at room temperature to induce peculiar effects, such as thick SiO_2 formation in the atmosphere. The alloyed Si/Au interface region grows with the thickening the SiO_2 layer until the Au layer is completely consumed in the Si/Au alloy formation. As a result of the formation of the SiO_2 layer, the color of the sample changes from original gold color to interference color, like the result

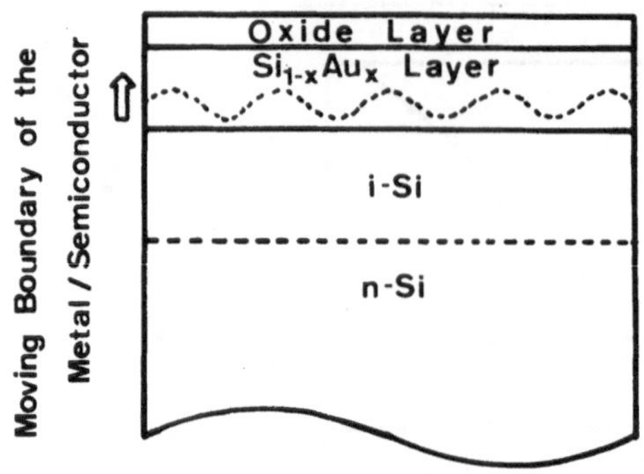

Fig. 6. Material structure of the typical sample formed by low temperature reactions between Si and Au film.

observed in type A samples. This phenomenon is the most important mechanism in understanding the structure of the sample prepared.

In the type B sample, Au atoms deposited on a hot substrate can be diffused into the substrate thermally. Then the interface of Au/n-Si is changed into the Au heavy-doped i-Si layer, of which the resistivity is two orders of magnitude larger than that of the original n-Si. After the Au coating, the ultrathin Nb film is deposited in order to suppress the intermixing reactions and/or to harden the surface of the sample. In the case of type C sample, the secondary Au coating supplies additional metal atoms on the sample surface. After reactions have occurred between Au and Si, the surface of the sample is expected to have the heterogeneous multilayered structure, as schematically illustrated in Fig. 6. Then the Nb layer is neglected because it plays a very small role in the material effect. For an amorphous $Si_{1-x}Au_x$ alloy system, Nishida et al.[9] reported the metal-insulator transition which is observed in the range of x of about 0.14. The boundary of the phase transition is expressed by the wavy line in Fig. 6, since the intermixing reaction depth may fluctuate in the direction of the film thickness. After the reactions proceed thoroughly, the surface layer changes to a highly resistive layer. When an ultrathin Au film as thin as 1 nm is deposited, R_s in the type A sample is primarily the value of the Si substrate, about $10^2\ \Omega sq^{-1}$. In the range of t_{Au} between 1 nm and 10 nm, R_s values observed almost agree

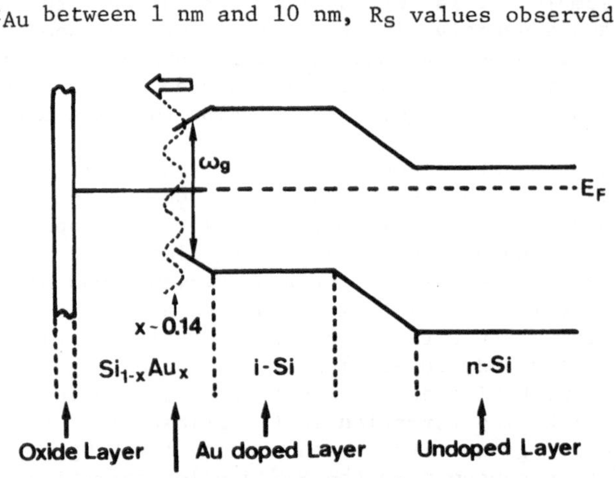

Fig. 7. Imaginary energy band corresponding to the material structure of the sample shown in Fig. 6.

with the values calculated from the resistivity of amorphous $Si_{1-x}Au_x$, $10^{-3} \sim 10^{-4}$ $℧m^{-1}$. According to these results, it is thought that the value of R_s measured is not definitely influenced by the comparatively thin oxide layer formed in the surface of the sample.

Ginzburg[3] investigated the excitonic superconductivity expected in the interface of metal/insulator (semiconductor) and proposed practical parameter values of the materials used in his model: a relative dielectric constant less than 30, energy gap less than 3 eV for an insulative layer, and Fermi energy between $1 \sim 10$ eV for a metallic layer; especially the thickness of each layer should be in order of nm. These parameter values are satisfied in the samples investigated here. Furthermore, ABB[5] suggested the interface of metal/semiconductor in which electrons tunneled into the gap of the semiconductor interact with the virtual excitons and transit into superconductive states. Their model really corresponds to the material system expected in the samples. Then the ABB theory can be applied to the imaginary band structure illustrated in Fig. 7. In the ideal interface of metal/semiconductor, T_c can be estimated by the following approximate equation without solving a gap equation.

$$T_c = \theta_o \exp(-1/g_{eff}), \qquad (1)$$

$$g_{eff} = \lambda_{ph}^* + (\lambda_{ex}^* - \mu')/[1-(\lambda_{ex}^* - \mu')\ln(\omega_g/\omega_{po})],$$

$$\mu' = \mu/[1+\mu\ln(\omega_F/\omega_g)], \quad \theta_o = \omega_{po}/k_B.$$

Where ω_F is the Fermi energy of metal; ω_g, the gap energy of semiconductor; ω_{po}, the center of phonon spectrum; and μ, the density of states times an average of the screened Coulomb interaction. The λ^* parameter is defined as follows: $\lambda^* = \lambda/(1+\lambda)$, where λ_{ph} and λ_{ex} represent the degree of the attractive interaction of electrons through phonon and exciton mechanism, respectively. Paying attention to the boundary of metal/semiconductor in $Si_{1-x}Au_x$ layer, T_c values were calculated as shown in Fig. 8 from Eq. 1. Then it is assumed that the gap energy of $Si_{1-x}Au_x$ is proportional to the concentration of Si. The results calculated are available only in the range of x less than about 0.14 from the consideration of the metal-insulator transition. When λ_{ex} is about 0.6, high T_c

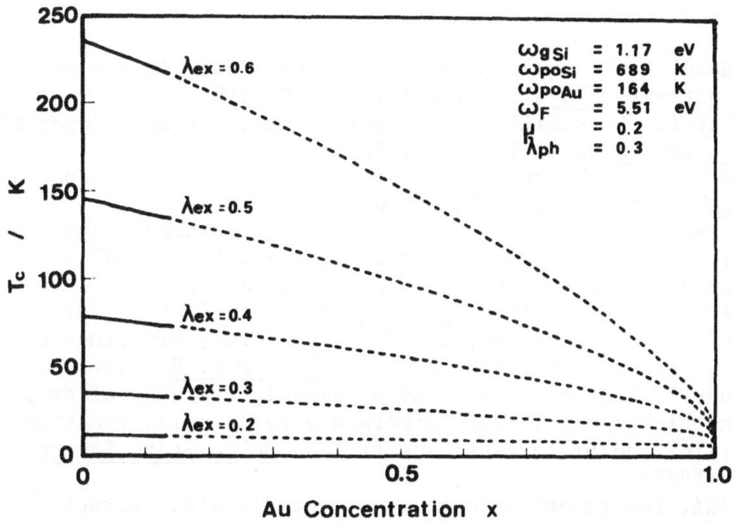

Fig. 8. Approximately calculated T_c values using the indicated parameter values as a function of Au concentration for the interface of metallic Au layer and semiconductive layer in $Si_{1-x}Au_x$ system. Solid lines mean available calculations in this system.

values, above 200 K, can be expected from this model. It can be also understood that the resistance anomaly strongly depends on the time because the important interface of metal/semiconductor grows and diffuses as the time proceeds, consistent with the instability of the resistance anomaly observed in type B and C samples. As a conclusion, these qualitative investigations agree well with the phenomena observed in this work.

Miller et al.[6] succeeded in synthesis of the ideal interface of Pb/PbTe. However, they failed to observe excitonic enhancement of T_c partially because electrons may be trapped in an inversion layer of conduction band states rather than the gap of the semiconductor in their samples. On the contrary, it is the feature in the sample of this work that the diffused region of metal-insulator transition has a quasi two-dimensional extent and it provides opportunity for many excitonic interactions between conductive electrons.

SUMMARY

A new type of an ultrathin film sample has been prepared by deposition of ultrathin Au-Nb film on a single-crystal Si in order to investigate the Ginzburg or ABB model of an excitonic superconductivity. The sample prepared revealed a typically anomalous decrease of sheet resistance, R_s at about 250 K for only a few days, accompanied by an abrupt increase of R_s as the time proceeds. The resistance anomaly is discussed qualitatively in relation to the ABB theory. Since the structure of the sample has not been understood quantitatively, the resistance anomaly may be interpreted by other mechanisms. Though the phenomenon observed is really unstable, the preparation process of the sample is comparatively simple with good reproducibility. Therefore, it can be hoped that detailed analyses of the sample are continued and that the mechanism of the anomalous conduction is clarified in near future.

ACKNOWLEDGMENT

The authors would like to thank Mr. A. Watanabe for his assistance in doing experiments.

REFERENCES

1. J. Bardeen, L. N. Cooper, and J. R. Schrieffer, Theory of superconductivity, Phys. Rev. 108:1175-1204 (1957).
2. W. A. Little, Possibility of synthesizing an organic superconductor, Phys. Rev. A 134:1416-1424 (1964).
3. V. L. Ginzburg, The problem of high temperature superconductivity, Comtemp. Phys. 9:355-374 (1968).
4. "High-Temperature Superconductivity," V. L. Ginzburg and D. A. Kirzhnits, eds., Consultants Bureau, New York (1982).
5. D. Allender, J. Bray, and J. Bardeen, Model for an exciton mechanism of superconductivity, Phys. Rev. B 7:1020-1029 (1973).
6. D. L. Miller, M. Strongin, and O. F. Kammerer, Experimental search for excitonic superconductivity, Phys. Rev. B 13:4834-4844 (1976).
7. K. Okuno, T. Ito, M. Iwami, and A. Hiraki, Presence of critical Au-film thickness for room temperature interfacial reaction between Au(film) and Si (crystal substrate), Solid State Commun. 34:493-497 (1980).
8. A. Hiraki, Low temperature reactions at Si-metal contacts--from SiO_2 growth due to Si-Au reaction to the mechanism of silicide formation, Jpn. J. Appl. Phys. 22:549-562 (1983).
9. N. Nishida, M. Yamaguchi, T. Furubayashi, K. Morigaki, H. Ishimoto, and K. Ono, Superconductivity and metal-insulator transition in amorphous $Si_{1-x}Au_x$ system, Solid State Commun. 44:305-309 (1982).

A NEW ALUMINUM-BASE ALLOY WITH POTENTIAL CRYOGENIC APPLICATIONS

J. C. Ho
Wichita State University
Wichita, Kansas

C. E. Oberly
Air Force Aero Propulsion Laboratory
Wright-Patterson Air Force Base, Ohio

H. L. Gegel, W. M. Griffith, J. T. Morgan,
W. T. O'Hara, and Y. V. R. K. Prasad
Air Force Materials Laboratory
Wright-Patterson Air Force Base, Ohio

ABSTRACT

An Al-Fe-Ce alloy has been recently demonstrated to be suitable as a matrix material for a composite conductor with high-purity aluminum filaments which can be used in devices operating at liquid hydrogen temperatures. The alloy is lightweight and has high strength, but just as importantly, its major alloying elements (Fe and Ce) are practically diffusionless in Al. The material was initially synthesized by ALCOA using powder metallurgy. Dynamic recrystallization processes were further developed at the Air Force Materials Laboratory, resulting in a more homogeneous microstructure. This paper describes mainly the results of microstructural, electrical resistivity, and low temperature heat capacity measurements on samples before and after dynamic recrystallization.

INTRODUCTION

Potential applications of aluminum-conductor devices operating at cryogenic temperatures were considered many years ago.[1-3] Research and development activities gradually gave way in the early 1970s to the then successful development of multifilament superconductors. However, it is generally recognized that, in special circumstances where liquid hydrogen is available, aluminum conductors may still be attractive. This is particularly true if the overall system weight, including that associated with conductors as well as cryogen liquefaction and storage, is a critical design factor. For practical Al conductors under consideration, the electrical resistivity at 20 K is only about twice as high as that at 4.2 K. For superconductors, on the other hand, liquid hydrogen does not yield a low enough temperature, and liquid helium is definitely required.

Earlier studies[2-7] provided a broad range of data in terms of purity, magnetic field, size, strain, and fatigue effect on low temperature resistivity of high-purity aluminum. As a result of these studies, one would easily identify aluminum as a better cryoconductor than copper for applications where weight and magnetoresistance are among the main concerns. The only major disadvantage for aluminum is its extremely low strength. This fact is even more serious for such applications as pulsed power devices where, in order to allow full current penetration during a short pulse, the conductors must be in the form of fine filaments. One could conceivably braid the filaments with structural materials for mechanical support. However, a better approach would be to embed Al filaments in an appropriate matrix. Manufacturing technology similar to that for producing multifilament superconductors could also be employed if suitable matrix materials can be found. This report describes a new Al-Fe-Ce alloy that can be used effectively for this purpose.

MATERIAL DESCRIPTION

The matrix material should satisfy several basic requirements: (I) light weight, (II) high strength, (III) good thermal conductivity, (IV) reasonably high electrical resistivity to minimize eddy current loss and to enhance electromagnetic diffusion rate, (V) workability compatible with that of high-purity aluminum, and (VI) diffusionless alloying elements, if present. The last requirement is most essential to warrant the high purity of aluminum filaments in the final product. It is indeed this requirement that eliminates all commercially available aluminum-based alloys from being considered here.

The new Al-Fe-Ce alloy,[8,9] containing 8.4 wt.% Fe and 3.6 wt.% Ce, has a low density of 2.95 g/cm^3. Primarily developed for high temperature aerospace applications, it derives its strength from densely dispersed fine intermetallic particles yet to be identified. On the other hand, this material has been successfully used as the matrix in a feasibility study[10] for fabricating multifilament aluminum conductors. Specifically, it was coprocessed with high-purity aluminum inserts by multistep extrusions with very uniform deformation -- a clear indication of compatible workability. More importantly, significant diffusion of Fe or Ce from the matrix to the aluminum filaments was not detected through residual resistivity ratio (RRR) measurements.

A complete documentation of physical and mechanical properties of this alloy will be undoubtedly needed for further development. Of particular interest is that such properties can be greatly affected in processing. Described below are results of some comparative experiments on the material before and after dynamic recrystallization.

EXPERIMENTAL RESULTS AND DISCUSSION

Microstructures and Mechanical Properties

To successfully fabricate a multifilament aluminum conductor by extrusion, swaging, and/or wire drawing, the matrix material must have suitable microstructure and mechanical properties. This will ensure the integrity of the filaments in the product and eliminate the possiblity of crack formation. The starting Al-Fe-Ce material used in this work was prepared at ALCOA from powders through cold compaction followed by vacuum hot pressing (VHP). It was further recrystallized based on dynamic process modeling developed at the Air Force Materials Laboratory.[11] The remarkably different microstructures before and after the dynamic recrystallization are shown in Fig. 1: Elimination of prior powder particle boundaries in the VHP material yields a more homogeneous distribution of intermetallic particles in the recrystallized condition. As expected, this process results in some softening of the material. However, even with the associated decrease in

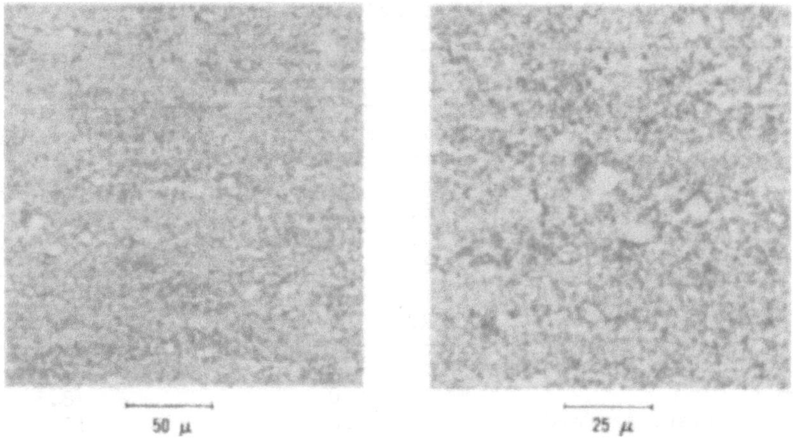

Vacuum Hot Pressed

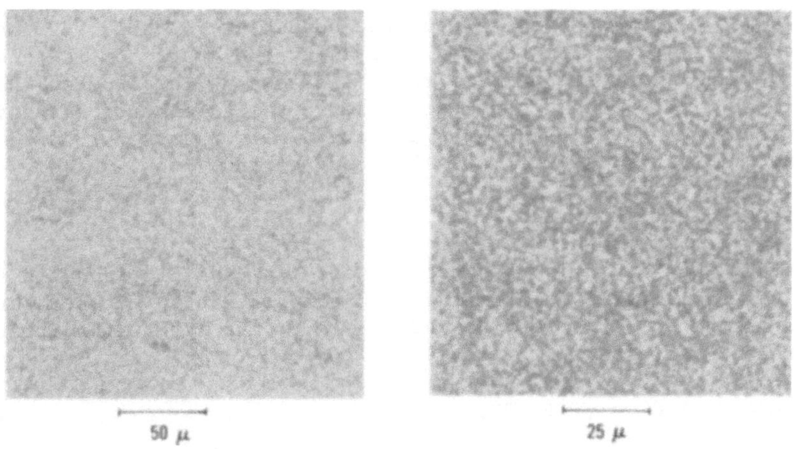

Dynamically Recrystallized

Fig. 1. Micrographs of Al-Fe-Ce alloy before and after dynamic recrystallization.

room temperature yield strength (from above 410 MPa to 170 MPa) and ultimate tensile strength (from 480 MPa to 250 MPa), as well as an increase in elongation (from 8-10% to 15%), the alloy can still be considered as a good structural matrix material for the multifilament aluminum conductors. It should be pointed out that, once the alloy is in the soft, recrystallized condition, it can be coprocessed with high-purity aluminum filaments at much reduced temperatures. This should, in turn, minimize any possible impurity diffusion from the matrix to the filaments.

Electrical Resistivity

Using the standard four-probe method, electrical resistivities and, consequently, RRR values of both VHP and dynamically recrystallized samples were determined as given in Table 1 (1100 Al and 6061-T6 Al alloy data[12] are also included for comparison). The heavily alloyed Al-Fe-Ce in the dynamically recrystallized condition exhibits behavior similar to 1100 Al insofar as electrical conductivity is concerned. In practice, it is anticipated that, as the matrix material in a composite conductor, this alloy will have

Table 1. Electrical Resistivities and RRR Values

	ρ (μΩ·cm)		
	Room Temp.	Liq. Nitrogen	Liq. Helium
6061-T6 Al alloy	3.9	1.9	1.4
VHP AL-Fe-Ce	5.5	1.1	0.66
Dynamically recryst. Al-Fe-Ce	4.6	0.66	0.27
1100 Al	3.0	0.46	0.20
	ρ(RT)/ρ(L. Nitrogen)	ρ(RT)/ρ(L. Helium)	
6061-T6 Al alloy	2.1	2.8	
VHP Al-Fe-Ce	5.0	8.3	
Dynamically recryst. Al-Fe-Ce	7.0	17	
1100 Al	6.5	15	

certain degrees of work-hardening when the conductor reaches its final wire form. The effect may not be removed even after the conductor is annealed (at relatively low temperatures) to relieve stress in the aluminum filaments. As a result, the alloy could have the reasonably high electrical resistance at cryogenic temperatures needed for a short time constant for current penetration into filaments, as well as a low eddy current loss.

Low Temperature Heat Capacity

This study was for the purpose of identifying possible magnetic (Fe,Ce) clusters in the alloy. Typical metals and alloys have low temperature heat capacities as the sum of a lattice and an electronic term with different temperature dependencies:

$$C = C_e + C_\ell = \gamma T + \beta T^3, \quad (1)$$

or, $$C/T = \gamma + \beta T^2. \quad (2)$$

This is true for our calorimetric data between 2 and 7 K for high-purity Al, as indicated in Fig. 2 by a linear fit of C/T vs. T^2. However, for the VHP Al-Fe-Ce alloy, the observed nonlinearity in the same plot requires a third term to fit the data:

$$C = A + \gamma T + \beta T^3, \quad (3)$$

or, $$C/T = A/T + \gamma + \beta T^2. \quad (4)$$

The temperature-independent term, A, responsible for the upturn in C/T as temperature decreases, has been previously observed in other nonferromagnetic alloys containing one or more of the elements Fe, Co, and Ni. It is generally recognized as a magnetic clusters contribution:[13] basically, ferromagnetic clusters in a nonferromagnetic matrix oscillate at low temperatures within directions given by the energy of crystal anisotropy. These extra degrees of freedom lead to a constant heat capacity contibution if the thermal energy is much greater than the energy associated with interaction between the magnetic moments of the clusters and the crystal field. It is of interest in this regard to note that this effect was much reduced in the heat capacity data obtained from the dynmaically recrystallized Al-Fe-Ce sample, again an indication of a more homogeneous microstructure.

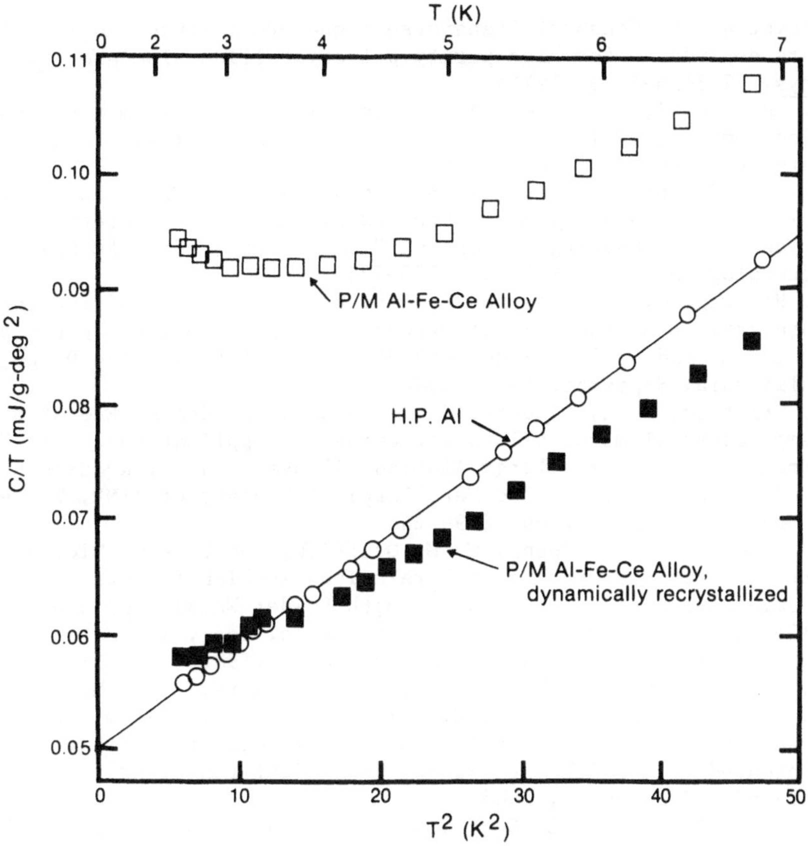

Fig. 2. Low temperature heat capacities of high-purity Al and Al-Fe-Ce alloy before and after dynamic recrystallization.

CONCLUSION

This work has demonstrated that newly developed high temperature Al-Fe-Ce alloy can be further processed to optimize its mechanical and physical properties towards cryogenic applications.

ACKNOWLEDGMENTS

We are grateful for Dr. Alan Male and Mr. Natraj Iyer of Westinghouse R & D Center for helping us carry out several extrusion experiments. We also acknowledge Mr. Don Linder of the University of Dayton for his expert technical assistance in preparing the extrusion billets.

REFERENCES

1. R. R. Barthelemy and C. E. Oberly, Optimum magnets for MHD generators, presented at the International Symposium on MHD Electrical Power Generation, Salzburg, Austria, July 4-8, 1966.
2. V. D. Arp, M. B. Kasen, and R. P. Reed, "Magnetic Energy Storage and Cryogenic Aluminum Magnets," AFAPL-TR-68-87, Air Force Aero Propulsion Laboratory, Wright-Patterson Air Force Base, Ohio (1969).
3. W. Schauer, Conductor materials for cryomagnets, in: "Proceedings of the 5th International Conference on Magnet Technology," N. Sachetti, M. Spadoni, and S. Stipcich, eds., Laboratori Nazionali del CNEN, Frascati, Italy (1975), pp. 629-643.

4. B. Krevet and W. Schauer, Transverse magnetoresistance and its temperature dependence for high-purity polycrystalline aluminum, J. Appl. Phys. 47:3656-3669 (1976).
5. R. J. Corruccini, "The Electrical Properties of Aluminum for Cryogenic Electromagnets," Technical Note No. 218, National Bureau of Standards, Boulder, Colorado (1964).
6. K. T. Hartwig and P. G. Appius, Strength and electrical resistance of cold drawn Al and Cu, in: "Proceedings of the 8th International Cryogenic Engineering Conference," C. Rizzuto, ed., IPC Science and Technology Press (1980), pp. 720-725.
7. K. T. Hartwig, G. S. Yuan, and P. Lehmann, The effects of low temperature fatigue on the RRR and strength of pure aluminum, paper presented at the Applied Superconductivity Conference, San Diego, California, September 9-13, 1984.
8. W. M. Griffith, R. E. Sanders, Jr., and G. J. Hildeman, Elevated temperature aluminum alloys for aerospace applications, in: "High Strength Powder Metallurgy Aluminum Alloys," J. J. Koczak and G. J. Hildeman, eds., The Metallurgical Society of AIME, Warrendale, Pennsylvania (1982), pp. 209-224.
9. S. D. Kirchoff, R. H. Young, W. M. Griffith, and Y. Kim, Microstructure/strength/fatigue crack growth relations in high temperature P/M aluminum alloys, in: "High Strength Powder Metallurgy Aluminum Alloys," M. J. Koczak and G. J. Hildeman, eds., The Metallurgical Society of AIME, Warrendale, Pennsylvania (1982), pp. 237-248.
10. J. C. Ho, C. E. Oberly, H. L. Gegel, W. T. O'Hara, J. T. Morgan, Y. V. R. K. Prasad, and W. M. Griffith, Composite aluminum conductors for pulsed power applications at hydrogen temperatures, paper presented at the 5th IEEE Pulsed Power Conference, Arlington, Virginia, June 10-12, 1985.
11. Y. V. R. K. Prasad, H. L. Gegel, S. M. Doraivelu, J. T. Morgan, K. A. Lark, and D. R. Barker, Modeling of dynamic material behavior in hot deformation: Forging of Ti-6242, Metal. Trans. 15A:1883-1892 (1984).
12. "Handbook on Materials for Superconducting Machinery, Battelle Memorial Institute, Columbus, Ohio (1974).
13. K. Schroder and C. H. Cheng, J. Appl. Phys. 31:2154-2155 (1960).

PRODUCTION OF HIGH-CONDUCTIVITY, HIGH-STRENGTH IN-SITU Cu-Nb MULTIFILAMENTARY COMPOSITE WIRE AND STRIP

C.V. Renaud, E. Gregory, and J. Wong

Supercon, Inc.
Shrewsbury, Massachusetts

ABSTRACT

Production of high-conductivity, high-strength wire and strip in useful sizes from in-situ CuNb multifilamentary composite material is described. Variation of mechanical properties and electrical conductivity vs. processing parameters are discussed. The material's thermomechanical properties, i.e., the relatively large resistance ratio and mechanical strength, suggest that this material will have applications where it is superior to copper and other generally used copper alloys, such as CuBe.

INTRODUCTION

The purpose of this on-going research is to investigate the technical feasibility of producing high-strength, high-conductivity material using in-situ CuNb composite. This material is of particular interest due to tensile values that are considerably higher than those predicted by the "Rule of Mixtures" based on the highest strengths of copper and niobium. In-situ CuNb composite was extensively investigated on a laboratory scale, in the literature,[1] where tensile values of 2.230 GPa (323 ksi) at 300 K (80°F) and 2.850 GPa (413 ksi) at 77 K (-321°F) were reported. While this material exhibited some excellent properties, including excellent strength properties at elevated temperatures,[2] it was of little practical value, as only small diameter conductors, 0.024 to 0.507 mm (0.001-0.020 in.) were studied.

The aim of our research is to determine if CuNb composite material can be mechanically processed by standard industry procedures, such as extrusion, swaging, drawing, and rolling, and whether the excellent mechanical properties reported in the literature could be obtained in a conductor of 2.54 to 6.35 mm (0.100-0.250 in.) diameter. Various billet designs, processing schedules, area reductions and heat treatments are being used to control material properties. Parameters for operations, such as extrusion and hot isostatic pressing (HIP), are being adjusted in order to process multifilament billets at the lowest possible temperatures that will provide good bonding and also minimize material degradation and particle agglomeration.

The potential commercial applications of CuNb in-situ composite encompass all areas in which high-strength, high-conductivity materials are presently used. However, the unusually high-strength and conductivity of in-situ material introduce a new standard for such materials. This material could potentially extend the life of existing electrical parts, such as contacts and relays, and it could result in overall weight and size reductions of items containing in-situ components. Its use in magnet design would extend the present limit of maximum field strength.[3] Other applications include rotating electrical machinery, electron tubes, connectors, springs, diaphragms, bellows, brake pad linings, and interior walls of fusion reactors.[4]

FABRICATION PROCEDURES

CuNb in-situ composites of 12, 18, and 24 wt %, obtained from the Materials Preparation Center of Ames Laboratory at Iowa State University, were the starting materials. Cast to 76.2 mm (3 in.) diameter by the consumable electrode arc technique,[5] the as-received materials were in the form of a 25.4 mm (1 in.) diameter extruded rod, with an in-situ core and a 0.635 mm (0.025 in.) thick copper jacket. Extrusion was conducted at 873 K (112°F). The thickness of the niobium filaments ranged from 5 to 12 μm (50,000-120,000 Å). Figure 1 shows an optical photomicrograph of a transverse section of 12 wt % Nb at the as-received size. Figure 2 shows a scanning electron micrograph of the same material after a 10-min. deep etch in a solution of 55% phosphoric acid, 25% acetic acid, 20% nitric acid, which preferentially removed the copper phase. Note the globular appearance of the Nb particles. During processing these dendritic particles will be transformed into randomly aligned discontinuous filaments embedded in the in-situ matrix. The processing of the in-situ into a large diameter conductor with fine Nb filaments is similar to that employed in the fabrication of superconductors. The process involves swaging, drawing, billet fabrication, and extrusion, repeating these steps until the desired conductor diameter and filament sizes are obtained.

Extrusion parameters for multifilament billets were varied to determine an optium extrusion temperature and reduction ratio that would provide good bonding of the in-situ subelements. Extrusions were performed at temperatures ranging from 723 K (842°F) to 1033 K (1400°F), and area reduction ratios from 10:1 up to 16:1. Low temperature extrusions resulted in

Fig. 1. Optical photomicrograph of 12 wt % Nb-Cu in-situ composite in the as received condition.

100 μm

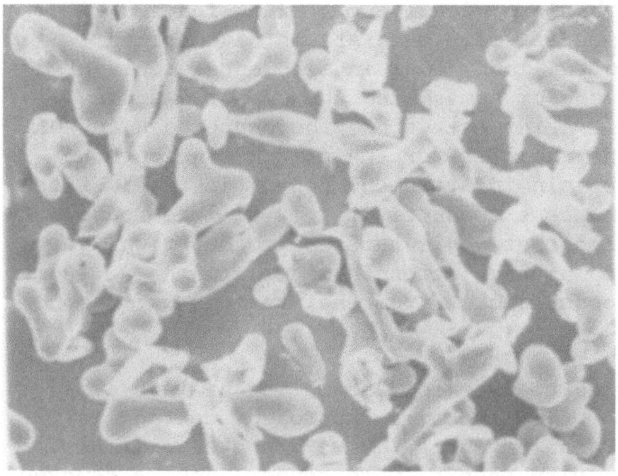

Fig. 2. SEM micrograph of 12 wt % Nb-Cu in-situ composite in the as received condition.

10 μm

little or no subelement bonding regardless of the reduction ratio. Medium temperature extrusions provided enough bonding so that the material could be cold worked but not without occasional breakages. High temperature extrusions with a better than 15:1 reduction ratio provided excellent bonding, but the high temperature contributed to material degradation by causing the fine Nb filaments to agglomerate into larger particles; thereby removing a good deal of the cold work previously applied. To counter this problem, hot isostatic pressing (HIP) has recently been employed to consolidate the billet prior to extrusion. Test billets were HIP processed at a temperature of 1033 K (1400°F) and a pressure of 103.4 MPa (15 ksi) for one hour. Figure 3 shows a photomicrograph of a HIP processed billet exhibiting grain growth across subelement boundaries. The billets were then extruded at 1033 K and cold worked extensively with no bonding problems. However, due to the high temperatures employed, particle agglomeration occurred. It is believed that by doubling the pressure, lowering the temperature (approx. 810 K), and increasing the time at cycle, the billet will be preconsolidated well enough to allow for lower temperature extrusions and thereby avoid material degradation each time a billet is repacked and extruded. We are presently assembling billets to be processed in this manner.

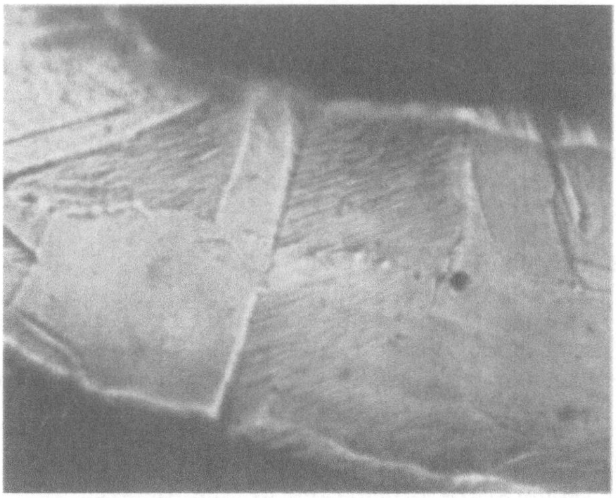

Fig. 3. Micrograph of HIP processed rectangular billet showing evidence of grain growth across subunit boundary.

Many variations of the same basic process, e.g., extrusion, swaging, drawing, repacking, have been investigated. Some attempts contained varying amounts of copper between the individual in-situ subelements of the multielement conductor. The copper comes from the copper can used during the extrusion process. If, after extrusion, the copper can is not removed prior to assembling the next billet for the next extrusion, the subsequently extruded rod will contain interelement copper in the in-situ matrix. While this copper may increase conductivity, it will decrease overall strength. Studies revealed that any interelement copper was very detrimental with regard to developing high strength material. As a result, all subsequent billets had the copper jacket from the previous extrusion removed during processing.

Two variations have been tried to date for assembling billets aimed at making strip as a final product. In the first approach, rectangular billet cans were packed with rectangular shaped, wire-drawn subelements. In the second approach, rolled in-situ sheets, with copper jackets removed, were also assembled into rectangular billet cans. Due to bonding problems and defects associated with rolling, we have as yet been unable to produce any data on strip material produced by either of the above two methods.

RESULTS

The results demonstrate the feasibility of extruding multielement in-situ CuNb material. A strong metallurgical bond was formed between in-situ material. The integrity of the bond was evident from the fact that extruded rod could be drawn after its copper jacket had been removed. In general, the multielement in-situ extrusions drew well from 12.7 mm (0.5 in.) to 2.54 mm (0.1 in.) in diameter. Further reductions usually required annealing heat treatments.

The effects that extruding and annealing had on the mechanical, electrical, and metallurgical properties of the in-situ material constituted the bulk of our research effort. The room temperature mechanical and electrical properties of unannealed multielement 18 wt % CuNb in-situ material are given in Table 1. The conductor with the highest UTS, 1.792 GPa (260 ksi), is the as-received single element drawn to 0.152 mm (0.006 in.) diameter, with an area reduction of 99.996%. The small diameter of this wire precludes its usefulness. Of greater importance, the conductor with the second highest UTS, 1.325 GPa (192 ksi), had a diameter of 2.311 mm (0.091 in.), and an area reduction of 99.993% after the first extrusion and subsequent drawing.

The UTS of the extruded conductor was low in comparison with the UTS of a single element conductor having a similar area reduction. This effect is even more pronounced after the second extrusion. An explanation of the effects of extrusion on the mechanical properties was revealed by SEM analysis. The loss in strength after each extrusion is due to the annealing of the copper in the in-situ matrix coupled with agglomeration of the fine Nb filaments back into globules. This is brought about mainly by the high extrusion temperatures mentioned previously.

It is significant to point out that the multielement conductor with the highest UTS also had an electrical conductivity of 64% IACS. These values exceed those of commercial high-strength, high-conductivity copper-beryllium alloys.

In an effort to increase electrical conductivity and study the effects of high temperatures on the room temperature properties of the in-situ composites, a few samples were annealed for various lengths of time at

Table 1: Mechanical & Electrical Properties of Unannealed 18 Wt% In-Situ

SIZE (mm)	0.2% YS (MPa)	UTS (MPa)	ELONGATION TO FAILURE	% REDUCTION	CONDUCTIVITY (% IACS)
AS RECEIVED					
3.1242 Hex				98.33	
2.5908	488.8	630.	9.3	98.96	78.7
0.5080				99.96	66
0.2540		1702.9		99.99	52
0.1524		1792.5		99.996	
1st EXTRUSION					
6.0960				99.95	69.5
4.386	832.8	1128.6	4.6	99.98	
3.556 Hex				99.98	63.9
2.794	1052.1	1138.9	1.5	99.99	
2.311	974.8	1325.1	3.1	99.993	64
2nd EXTRUSION					
6.121	737.7	854.9	8.5	99.9995	
2.616	748.0	1096.2	4.6	99.9999	61.8
2.032	910.1	1206.5	4.6	99.99994	

Table 2: Mechanical & Electrical Properties of Annealed 18 Wt% In-Situ

SIZE (mm)	0.2% YS (MPa)	UTS (MPa)	ELONGATION TO FAILURE	% REDUCTION	CONDUCTIVITY (% IACS)
FIRST EXTRUSION 1 h @ 923 K @ 2.794 mm diameter and subsequently drawn					
2.066	719.8	777.7	7.8	99.994	
1.524		930.7		99.997	
2 h @ 923 K @ 2.794 mm diameter and subsequently drawn					
2.006	700.5	807.3	7.8	99.994	
1.524	772.2	930.7	5.4	99.997	
1.143		1096.2	4.6	99.998	
0.812		1068.6		99.999	
3 h @ 673 K @ 2.311 mm diameter					
2.311	868.7	1148.6	6.2	99.993	67.3
20 h @ 673 K @ 2.31 mm diameter and subsequently drawn					
2.311	848.0	1123.8	6.2	99.993	69
1.626	985.9	1103.1	3.1	99.997	
1.143	1082.4	1447.8	3.1	99.998	66.6
SECOND EXTRUSION 3 h @ 673 K @ 2.616 mm diameter					
2.616	670.1	1009.3	9.3	99.9999	71.5
20 h @ 673 K @ 2.616 mm and subsequently drawn					
2.616	653.5	925.9	7.8	99.999	
1.625	987.9	1167.9	7.8	99.99996	65.3

temperatures of 673 K (752°F) and 923 K (1200°F), Table 2. The 923 K heat treatments were generally deleterious to the material properties. In contrast, the 673 K heat treatments had several positive effects on the in-situ composites, % IACS increased by 13% while UTS decreased by only 8% and elongation to failure nearly doubled. Longer time at temperature did not significantly alter the properties which may suggest the use of this material at elevated temperatures.

The most significant data was obtained from the conductor in Table 2, first extrusion, after a 20 h heat treatment at 673 K, at 2.311 mm (0.091 in.) and subsequent drawing to 1.143 mm (0.045 in.). This material exhibited a UTS of 1.447 GPa (210 ksi) and electrical conductivity of 66.6% IACS. To the best of our knowledge, this combination of properties exceeds those of most other copper based materials at this size.

In addition to producing wire, several samples were rolled into strip. Figure 4 shows a scanning electron micrograph of in-situ material rolled to a thickness of 0.457 mm (0.018 in.). Notice that unlike wire, where the dendritic particles elongate into discontinuous filaments, here the Nb globules flatten into layers of thin sheet oriented along the rolling axis. This material exhibited a UTS of 1.08 GPa (158 ksi) and an electrical conductivity of 55% IACS.

SUMMARY

While a considerable amount of work remains to be done, it is reasonable to assume from the results presented, a UTS of 1.447 GPa (210 ksi) and electrical conductivity of 66.6% IACS in a 1.143 mm diameter conductor, that a multielement high-strength, high-conductivity CuNb in-situ composite is technically feasible, and that its material properties show considerable promise. CuNb in-situ conductors have been fabricated, whose properties exceed the combined mechanical and electrical properties of most other copper based alloys. In addition, investigations show that although as yet not as well developed as in wire products, outstanding properties can also be fabricated in the form of in-situ strip material.

Fig. 4. SEM micrograph of 12 wt % Nb-Cu rolled to 0.457.

ACKNOWLEDGEMENTS

This research is being funded by a grant of the Small Business Innovative Research Program and is administered by the Basic Energy Sciences Division of the Department of Energy. We would also like to thank the members of the Materials Preparation Center at Ames Laboratory for their helpful assistance on numerous occasions during this work.

REFERENCES

1. J. Bevk, J. P. Harbison, and J. L. Bell, Anomalous increase in strength of in-situ formed Cu-Nb multifilamentary composites, J. Appl. Phys. 49:6031-6038 (1978).
2. K. Karasek and J. Bevk, High temperature strength of in-situ formed Cu-Nb multifilamentary composite, Scripta Met. 13:259-262 (1979).
3. D. K. Finnemore, J. E. Ostenson, J. D. Verhoeven, and E. D. Gibson, Use of in-situ wire in small magnets, IEEE Trans. Magn. MAG-17: 255-256 (1981).
4. O. K. Harling, G. P. Yu, N. J. Grant, and J. E. Meyer, Application of high strength copper alloys for a fusion reactor first wall, J. Nucl. Mater. 103:127-132 (1981).
5. J. D. Verhoeven, E. D. Gibson, F. A. Schmidt, and D. K. Finnemore, Preparation of Cu-Nb alloys for multifilamentary in-situ superconducting wire, J. Mater. Sci. 15:1449-1455 (1980).

DATA ON A MAGNETIC SHIELD ALLOY AT LOW MAGNETIC FIELDS FROM 300 KELVIN TO 1.91 KELVIN, AND 0.08 HZ TO 10KHZ

R.F. Arentz and M.H. Johnson

Ball Aerospace Systems Division
Boulder, Colorado

L. Dant

AD-VANCE Magnetics, Inc.
Rochester, Indiana

ABSTRACT

We tested two samples of a commercial shielding alloy at 300, 77, 4 and 1.91 Kelvin, using field strengths of 4 and 50 gauss rms and frequencies from 0.08 Hz to 10 KHz. Both samples were effective shields, even at 1.91 Kelvin. One of the samples, which had received a special treatment, shielded even more effectively at 4 Kelvin than at 300 Kelvin, particularly at higher frequencies.

INTRODUCTION

In a design for a helium cooled SQUID-based satellite we needed a 2 Kelvin magnetic shield. A material called "Cryoperm-10"[1] was called to our attention by the Electromagnetic Technology Division of the National Bureau of Standards in Boulder, Colorado. This material had been measured at 4.4 Kelvin to have an initial permeability of 8.81×10^{-2} $H.m^{-1}$ $(70{,}000\mu_o)$, and a maximum permeability of 3.14×10^{-1} $H.m^{-1}$ $(250{,}000\mu_o)$.[2] This material had very good magnetic properties but is hard to get.

A literature search[3-9] produced only a few references[3,4,5,6] dealing directly with the measurements of ferromagnetic alloys at low field strengths (0.6×10^{-4} T), low temperatures (1.91 to 4 Kelvin) and low frequencies (0.01 to 10 kHz), and only one reference[7] dealing with all of these. We agree completely with Suzuki et al.,[3] "that the performance of ferromagnetic shields immersed directly in liquid helium has rarely been reported."

We were quite pleased therefore when the opportunity arose to test two likely samples for our application. The results of those tests are the subject of this paper.

MEASUREMENT APPROACH

We chose to measure the shielding factor of an enclosed right circular cylinder that was 17.8 cm high, 7.62 cm wide (OD) and with a wall thickness of 1 mm. The material we tested is an alloy called AD-MU78.[11]

AD-MU78 is a high permeability, ferromagnetically soft alloy with a nominal composition by weight percentage of Ni(75-77), Fe(12-15), Cu(4-6), Cr(3), Mn(1.8), Si(0.5), P(0.02), S(0.02) and C(0.05). We received two cans with end caps for study. Although the two specimens were made from the same ingot of material, they had undergone separate and distinct processing. Can No. 1 (henceforth called AD-MU78) underwent the standard processing used by the manufacturer. Can No. 2 (henceforth called CPEXP-1184) underwent a proprietary process in addition to the standard process. Both cans were formed by a heli-arced, butt-weld seam up their side and around their bottom. The end caps slipped tightly over the outer walls of the cylinder with a 2 cm overlap. In the center of each cap there was a 0.64 cm diameter hole through which leads for the pickup coil could be routed, and through which cryogenic fluids could fill the interior.

All in all we found the alloy to be very tractable from all points of view. It forms and welds easily. It does not corrode in water as far as we could see. It can be purchased in sheets up to 40.6 cm wide and in lengths up to 3 meters long. It comes in a wide variety of thicknesses. Its alloy composition is well documented and it can be purchased to MIL-SPEC MIL-N-14411 Composition 2. It is made in the U.S. and interacting with the manufacturer is very easy. Its price is competitive, and its magnetic properties are quite good to 1.91K and it has a 300 Kelvin permeability in the range of $\simeq 9.0 \times 10^{-2}$ to $\simeq 3.5 \times 10^{-1}$ $H.m^{-1}$ depending on the exact alloy type and its processing.

EXPERIMENTAL PROCEDURE

In order to measure the shielding factor we used a setup that was quite simple. We drove a sinusoidal current into a pair of coils hooked in series-aiding and measured the induced, "open-circuit" output voltages of a third coil, located between them.

The geometry of the three coils was always fixed, and the coupling between the coils was measured first without and then with a can inserted between the coils. The shielding factor is then simply the ratio of the two signal levels. We varied the drive frequency over the range of 0.008 Hz to 12 kHz, and selected two drive currents of 20mA and 160mA rms. These currents gave us fields whose rms values were 4×10^{-4} T and 50×10^{-4} T (4 and 50 gauss) at the contact interface between a drive coil and the wall of the can. These field levels were measured with a thin Hall-effect probe inserted between the coil and the can.

We measured the input current on a Fluke 8502A digital multimeter; and measured the output voltage on a Hewlett-Packard 3582A spectrum analyzer. The entire system was dc coupled. Our test converted an input rms current into an output rms voltage, so our transfer function is dimensionally transconductance. We have inverted our numbers into ohms to display the data.

All the displayed data is referred to the input. The noise floor of the system, checked with the spectrum analyzer, was just 6.75 nV/√ Hz. The worst case signal to noise ratio occurred at 20mA of drive current, T = 300K and f = 0.008 Hz, when either of the cans was present. At that limit the SNR was 4 to 1. In all other cases the SNR was greater than 10, and frequently greater than 1,000.

The coils were hand wound on cardboard tubes for mandrels. Once wound, each coil was trimmed for an inductance of 11 mH ±10%. In the final configuration there was a low-Q resonance (caused by all the associated stray capacitances) at 32 KHz.

The coils had IDs of 4.5 cm and ODs of 5.5 cm. The internal pickup coil was left on its cardboard mandrel, and the mandrel was cut and shaped to fit transversly across the inside of the can. The two drive coils were removed from their mandrels and tied as slightly bent circular coils on the outside of the can. They had a winding thickness of 0.68 cm. The drive coils were connected in series-aiding and were coaxial with the pick up coil.

The coupling between the coils (curve D in our Figures 1 and 2) was measured with the coils mounted on a cardboard model of a can. This kept the geometry of the three coils the same both with and without the cans being present. The coils were mounted in the bottom third of a can to get them away from any field leaking out of the small hole in the top of the can, or fringing from the small gap between the lid and the body. Both sets of coils were connected to the warm electronics through No. 28 AWG twisted shielded-pair cable, and the shields were all connected to single-point ground.

The cans were supported by a stainless steel tube with a thin wall. The tube carried the interconnect cables. It was sealed at the top to allow for a superfluid pump down. At the bottom of the tube, just above its entrance into the can, we cut a small notch into the tube through which the drive-coil cable came out, and through which the cryogenic fluid could get into the interior of the can. The entire assembly was strapped together with 3M Fiberglass strapping tape (which worked surprisingly well at 4 K).

Data were taken inside our aluminum-walled, fiberglass-necked helium dewar. Three data sets were taken. First with the cardboard mock-up; second, with the AD-MU78 canister; and then third, with the CPEXP-1184 canister. The temperature order in all three sets was 300, 77, 4 and 1.91 Kelvin. At all times the coils were 11.5 cm from the nearest conductive wall, which we believe keeps any eddy currents in the dewar from affecting the data.

We were also suspicious enough of the deep cusp at f = 1 kHz to retest the coils and cans with the unit suspended in a large styrofoam cooler which we then filled with liquid nitrogen. During these tests all extraneous conductive materials were at least 1.5 m from the coils. The cusp remained unchanged during these retests.

RESULTS

Figures 1 and 2 are plots of the raw data. Figures 3 and 4 show the actual shielding factor ratios of both materials for all of the experimental parameters. Below 10 Hz the two materials are similar in their shielding factors for all parameters, with the CPEXP-1184 being slightly better in all cases. Between 10 Hz and 1 kHz, both materials show a marked increase in their effectiveness, peaking near a value of 30,000. Also in this frequency domain there is a temperature-dependent shift in peak efficiency upwards in frequency by about a factor of 10.

Above about 1 kHz the two materials begin to show their greatest differences. The CPEXP-1184 is continuing to <u>increase</u> in its efficiency, while the AD-MU78 is rapidly <u>decreasing</u>. Please keep in mind, however, that the worst case degradation of about 100 for the AD-MU78 material still leaves you with a good shield, particularly when it is contrasted with the large degradations described in the references.

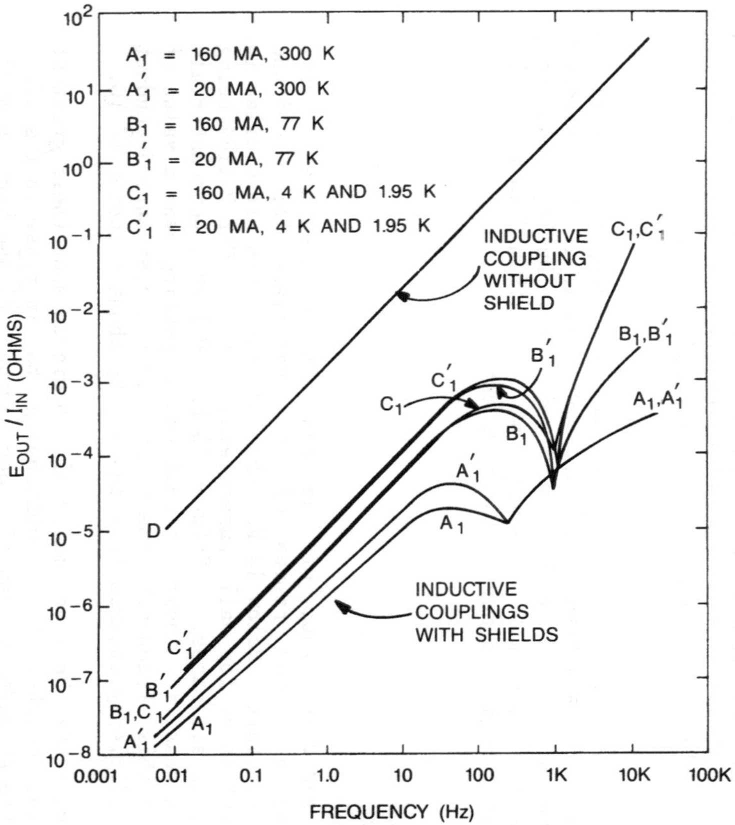

Figure 1. AD-MU78 - raw data.

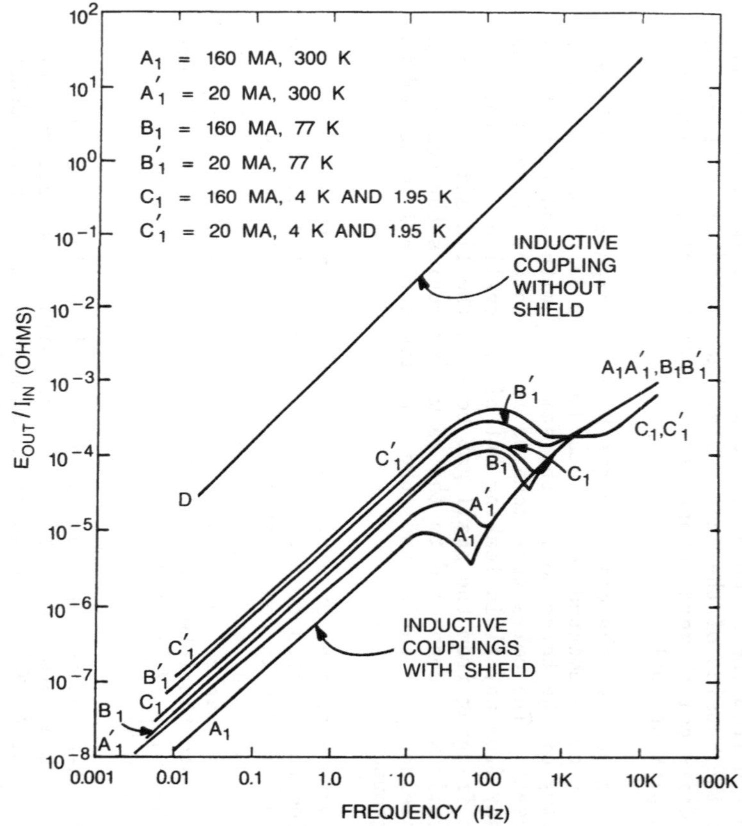

Figure 2. CPEXP-1184 - raw data.

Curves A_1 thru C_1' are raw data plots of Eout/Iin as a function of frequency with excitation currents and temperatures as shown. Curve D is the inductive coupling between the input and output coils as a function of frequency with no shielding between them. Curve D is identical for all currents and temperatures within the domain of interest.

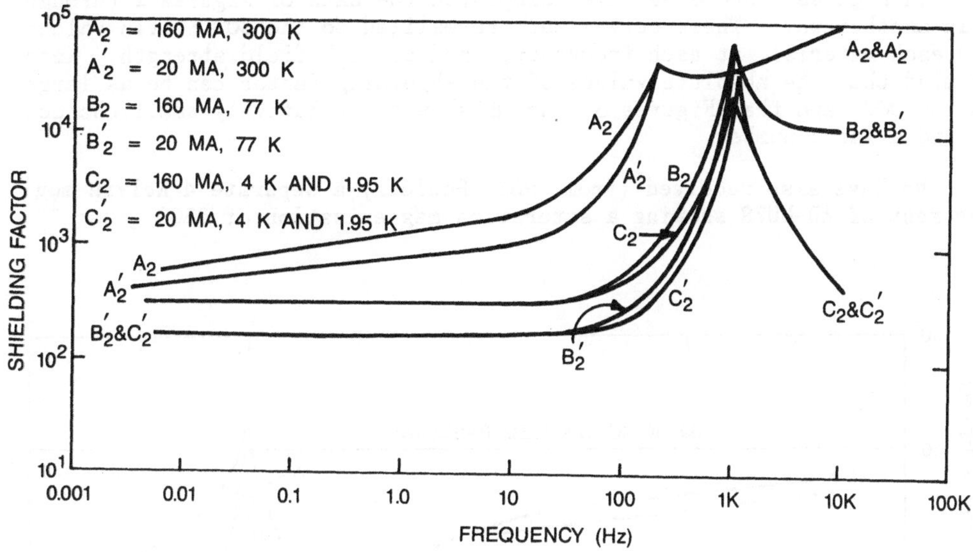

Figure 3. AD-MU78 - shielding factor vs. frequency, temperature, and field.

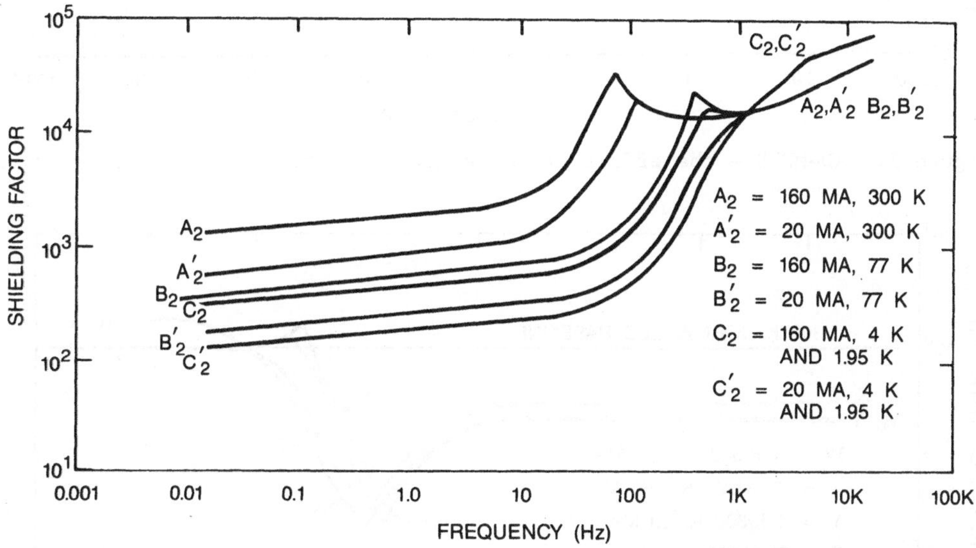

Figure 4. CPEXP-1184 - shielding factor vs. frequency, temperature, and field.

Figures 3 and 4 are plots of the shielding factor as a function of frequency. Curves A_2 thru C_2' were obtained by dividing curve D in Figures 1 and 2 by the raw data curves A_1 thru C_1'. The resulting set of curves represents the ratio of the inductive coupling between the input and output coils without shielding to the inductive coupling between the coils with shielding as a function of frequency, temperature, and excitation field strength.

In Figures 5 and 6 we have displayed the data of Figures 1 through 4 in another way. These curves are normalized to the 300 Kelvin values for each material, at each frequency, and at each field strength. Keep in mind that the absolute values of the shielding factor can be as large as 100,000, and that Figures 5 and 6 show the relatively small changes around a large number.

We have also received (from NBS Boulder) a separate 4 Kelvin measurement[10] of AD-MU78 showing a saturation magnetization of 6300 gauss.

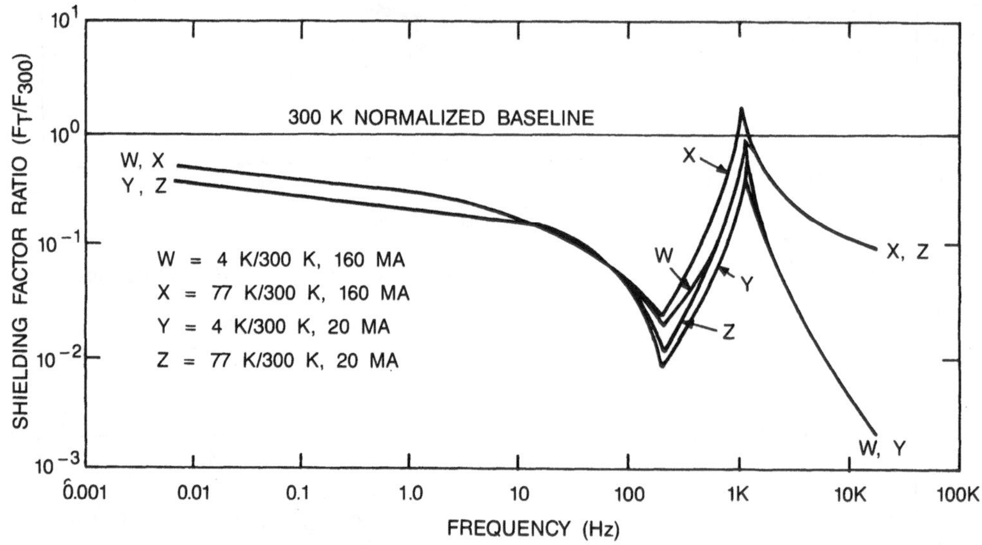

Figure 5. AD-MU78 - the effect of temperature on the shielding factor ratio.

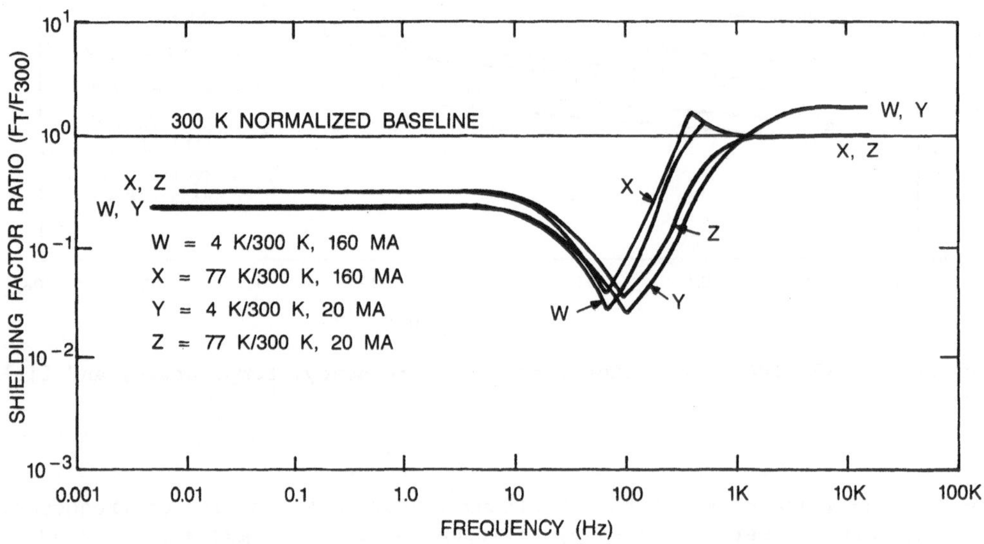

Figure 6. CPEXP-1184 - the effect of temperature on the shielding factor ratio.

Figures 5 and 6 represent the effect of temperature on the shielding factor ration as a function of frequency, normalized to a 300°K baseline.

SUMMARY

These two samples are good cryogenic magnetic shield materials. They are of the same chemical composition and differ only by their processing. CPEXP-1184 is clearly superior for frequencies between 1 and 10 kHz. Below 1 kHz both materials show good magnetic rejection with the CPEXP-1184 being consistently better by about a factor of 5.

Because of its relatively constant behavior from 300 to 1.91K CPEXP-1184 would be quite useful for localized, cryogenically cooled magnetic shielding.

REFERENCES

1. Cryo-Perm is manufactured by Vacuum Schmelze GMBH, Gruner WEG37, D-6450, Hanau #1, Germany.

2. F. R. Fickett, Private Communications, Electro-Magnetic Technology Division, National Bureau of Standards, Boulder, CO.

3. Y. Suzuki, E. Horikoshi and K. Niwa, "Magnetic Properties and Ferromagnetic Shielding of Ni-Fe-Mo Alloys at Cryogenic Temperatures," FUJITSU Science and Technology Journal, Vol. 20, No. 2, pp. 167-177, June 1984.

4. D. L. Martin, R. L. Snowdon, "Effect of Temperature on Magnetic Shielding with Co-netic," Rev. of Sci. Instrum., Vol. 46, No. 5, May 1975.

5. F. W. Ackerman, W. A. Klawitter, J. J. Drautman, "Magnetic Properties of Commercial Soft Magnetic Allows at Cryogenic Temperatures," Proceedings of the Cryogenic Engineering Conference, Vol. B-3, pp. 46-50, June 1970.

6. J. J. Gniewek, E. Ploge, "Cryogenic Behavior of Selected Magnetic Materials," Journal of Research of the National Bureau of Standards-Engineering and Instrumentation, Vol. 69C, No. 3, pp. 225-236, July-September 1965.

7. J. F. Dufresne, I. G. Ritchie, P. Moser, "On the Question of Low-Temperature Magnetic Relaxation in Pure Iron," Phys. Stat. Sol., Vol. 42, No. 2, pp 751-756, August 1977.

8. F. Brailsford, "Physical Principles of Magnetism," Appendix 2, p. 259, D. Van Nostrand Company, Ltd., London, 1966.

9. R. Bozworth, "Ferromagnetism," D. Van Nostrand Company, Inc., New Jersey.

10. R. Goldfarb, Private Communication, Electromagnetics Technology Division, National Bureau of Standards, Boulder, CO.

11. AD-MU78 and CPEXP-1184 are manufactured by AD-VANCE Magnetics, Inc., Rochester, Indiana, 46975.

THE CHARACTERIZATION OF CRYOGENIC MATERIALS BY PHOTOACOUSTIC SPECTROSCOPY

Barrie S.H. Royce

Applied Physics and Materials Laboratory
Princeton University
Princeton, New Jersey

ABSTRACT

Photoacoustic spectroscopy provides a method for measuring optical and thermal properties of materials in a variety of physical forms. Measurement of the sample response to modulated radiation can employ pressure transducers, sensitive temperature probes, laser beam deflection, surface displacement, and piezoelectric detectors. This range of detectors permits samples to be studied in liquid, gaseous, or high vacuum environments over a wide range of temperatures. This paper reviews the generation of a PAS signal and the experimental configurations used for various modes of signal detection. Attention is paid to the special features associated with low temperature measurements and the relatively sparse literature extant in this area is briefly discussed.

INTRODUCTION

Photoacoustic spectroscopy (PAS) is a well established technique for making absorption measurements over a spectral range that extends from the microwave to the ultraviolet. The method measures the sample heating that results from the absorption of modulated radiation and the nonradiative return of the system to a ground state. PAS is, therefore, complementary to fluorescence measurements and provides essentially the same optical information as conventional absorption spectroscopy. Because PAS involves periodic sample heating, it can provide information about the thermal properties of materials. In the limit of a thermally thick sample, the PAS signal depends upon the product of the optical absorption coefficient and the thermal diffusion length of the absorbing material, and will, therefore, depend upon the thermal conductivity, heat capacity, and density of the sample. If the optical properties are known, the method may be used to measure the thermal diffusivity of the sample. The PAS technique has been discussed in several books[1,2] and review papers.[3,4,5] This paper looks at some features of the method that make it of interest for low temperature studies and briefly reviews some contributions in the literature.

SIGNAL GENERATION

The primary step in photoacoustic signal generation involves the absorption of modulated electromagnetic radiation by the sample and the nonradiative conversion of part of this energy into heat. Figure 1, taken from the review of Tam,[3] illustrates the signal path and some of the characteristic lengths associated with signal generation from a homogeneous solid. Sample heating due to the absorption of the modulated radiation within an optical absorption length of the surface produces an acoustic wave which causes the surface to oscillate and give rise to pressure fluctuations in the coupling gas within the PAS cell. In addition, the periodic heat released within a thermal diffusion length of the sample interface causes periodic heating of this coupling gas within a thermal diffusion length of the sample surface. This temperature change may be detected in a number of different ways, and for most systems is the major source of the PAS signal.

The first PAS measurements detected the pressure fluctuations in the gas contacting the sample in a closed volume. This signal has a periodic component at the frequency of the modulated radiation and may be shifted in phase with respect to the stimulating radiation. Except under special circumstances, the sound propagation in the gas due to the oscillating interface makes a negligible contribution to the signal. In general, the photoacoustic signal amplitude depends upon the product of the sample's optical absorption coefficient, α, and the thermal diffusion length, μ, at the acoustic excitation frequency. The expression for the signal is complex, except in certain limiting cases which, fortunately, correspond

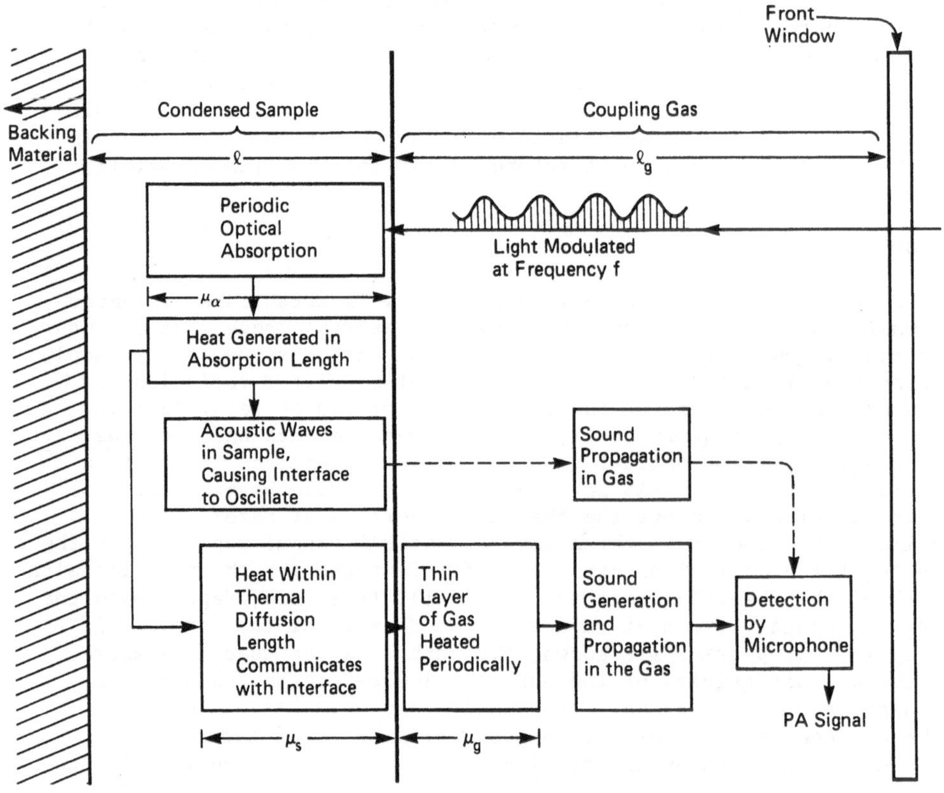

Fig. 1. Schematic diagram of signal paths in condensed matter PAS.

to experimentally realizable configurations. In order for the signal to be independent of the properties of the sample support, a thermally thick specimen should be chosen, i.e., the sample thickness should be greater than its thermal diffusion length. To avoid phase contributions from the coupling gas, this should also be longer than its thermal diffusion length.

For spectroscopic measurements, the pressure signal requires normalization for the photon flux arriving in any wavelength interval and this is most easily measured by using a reference sample with a high and uniform absorption coefficient over the spectral range of interest. For samples of this type, the energy deposition occurs in a distance much less than a thermal diffusion length in the sample and the signal depends only upon the incident radiation intensity. Such a sample is said to be in photoacoustic saturation. The choice of a suitable "black" reference sample presents some problems, particularly in the infra-red spectral range where carbon samples may show absorption features due to residual hydrocarbons. Graphite powders have been found to be satisfactory over the range from 200 nm to 25 microns, although they are grey bodies rather than true black absorbers.

The spectral data that can be obtained from PAS provide accurate wavelength information for absorption features; however, the signal amplitude has a nonlinear dependence on the optical absorption coefficent at high values of this parameter. This nonlinearity gives the method a high dynamic range, a feature that may be of considerable value if both surface species and bulk lattice modes are to be studied on the same sample. For homogeneous samples, the use of both amplitude and phase data permits the absolute optical absorption coefficient to be determined,[6] but this procedure is not valid for powdered samples. For these systems the pressure signal is modified by contributions from the interstitial gas[7] to which rapid heat transfer occurs. This pressure contribution is the equivalent of the acoustic wave generated in a homogeneous sample, but, due to the larger thermal expansion coefficient of the gas, it may make a significant contribution at low values of the bulk optical absorption coefficient. Since it is transmitted to the transducer gas at a velocity of sound, it alters both the amplitude and phase of the measured signal. The pressure contribution is greatest for a weak absorber of high porosity and is, therefore, important when surface species are being studied on samples of high specific surface area. Figure 2 shows the effect of this term on the total amplitude and phase of the PAS response of a high porosity powder.

Heat transfer between a sample and the transducer fluid in contact with it produces a refractive index gradient near the sample surface that may be detected directly. Photothermal deflection spectroscopy (PDS)[8,9] uses a probe laser beam to measure this gradient. For solid samples in a liquid or gaseous ambient, the probe beam is passed tangentially with respect to the solid surface within one thermal diffusion length of the fluid at the chosen modulation frequency. Probe beam deflection is measured using sensitive position detectors and lock-in techniques. This method is of comparable sensitivity to PAS and for a powdered sample will only measure the thermal signal as the pressure contribution to the refractive index of the gas is negligible. If the method is to be used over a wide range of temperature and pressure, contributions to the signal arising from changes in the thermal diffusion length, heat capacity, and refractive index of the transducer fluid must be taken into account.[10]

At cryogenic temperatures the heat transferred from a solid sample may be measured directly by employing sensitive thermometric methods.[11] Smith and Laguna[12] were the first to report the use of second sound in

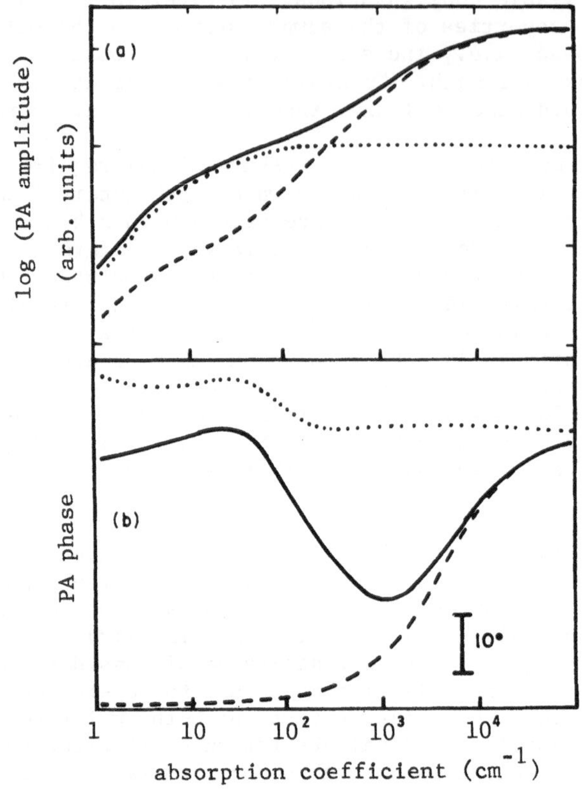

Fig. 2. Calculated PAS response for a high porosity powder. a) Magnitude, b) phase. Total PAS signal (Solid line), pressure term (dotted line), and the thermal term (dashed line).

liquid helium II as a suitable detector for the PAS response of a solid sample. Main and McCann[13] have also used this technique for ESR spectroscopy. Superconducting bolometers may be employed as second sound detectors, however, magnetic field variations in ESR experiments make other thermometers more desirable. The major advantage of this detection mode is the freedom it provides from the constraint of the thermal diffusion length in the liquid that is present in normal periodic thermal conduction measurements.

If radiation is absorbed in a local region of a solid sample, the surface of the sample is deflected due to local thermal expansion. Amer and Olmstead[14] have extended the laser beam deflection technique to measure this surface displacement. The method is of considerable interest since it can be applied in an ultra-high vacuum environment with the sample temperature varying over a wide range. The deflection is measured using a sensitive position detector in conjunction with a lock-in amplifier so that both phase and amplitude data can be recorded. The authors claim a sensitivity for this method of $\alpha \ell = 10^{-6}$. The signal magnitude depends upon the ratio of the thermal expansion coefficient of the sample to its heat capacity, a ratio that should be approximately temperature independent. The method should, therefore, function well under cryogenic conditions.

The expansion of a solid or liquid sample may also be detected using piezoelectric transducers in direct contact with the sample or with a liquid or solid transmitter between the sample and the detector. Most piezoelectric materials are also light sensitive and care must be taken to prevent scattered radiation from reaching them. Patel and Tam[15] have performed a wide range of pulsed photoacoustic measurements on condensed matter using this method in conjunction with laser light sources. With a high intensity source, very low optical absorption coefficients can be measured, and Nelson and Patel,[16] who measured the absorption spectra of liquid ethylene at 113 K between 0.7 and 1.6 microns, were able to see features with absorption coefficients less than 5×10^{-5} cm^{-1}. Ikari et al.[17] have employed ZnO piezoelectric transducers to measure the absorption spectrum of n-InSe as a function of temperature using the data to provide information about recombination processes in the semiconductor, and Etxebarria et al.[18] used this detection method to study first and second order phase transitions in solid samples.

METHODS FOR LOW TEMPERATURE MEASUREMENTS

The earliest PAS measurements at cryogenic temperatures were made using gas-microphone cells, frequently in a Helmholtz resonator configuration.[19,20,21,22,23] A typical cell[23] is shown in Figure 3. With an

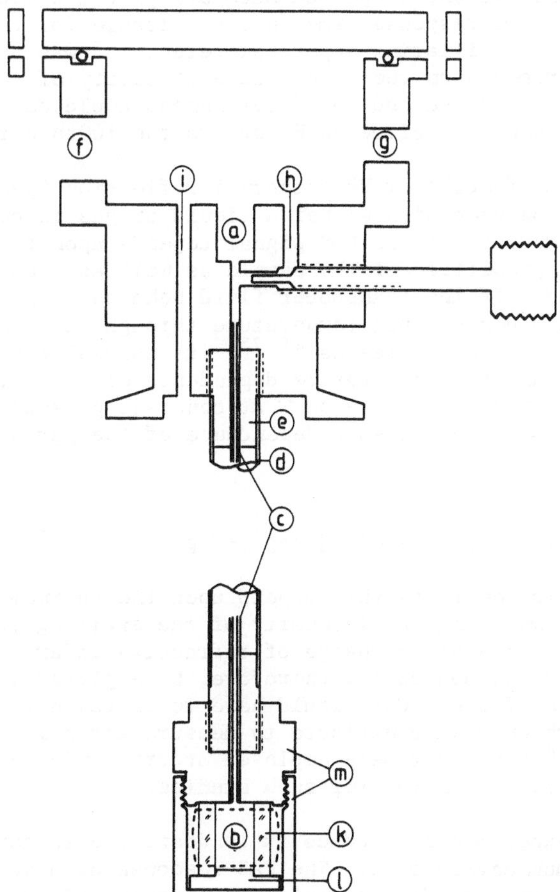

Fig. 3. Schematic diagram of a variable temperature PAS cell for the 4.2 to 300 K range. a) Microphone, b) sample cell, c) resonance tube, d) support, e) brass fitting, f) hermetic BNC connector, g) dry He supply, h) bypass valve, i) thermocouple feedthrough, k) infralux tube, l) base of sample carrier, m) sample carrier holder.

arrangement of this type, the microphone and preamplifier can be at a different temperature than the sample. Two problems are encountered with this design: acoustic noise due to the boiling of the cryogenic fluid and a restriction on the temperature range due to the need for a transducer gas in thermal contact with the sample. These limitations may be overcome by using an exchange gas to cool the sample cold-finger and by employing other detectors such as piezoelectrics, laser beams, or second sound in the helium refrigerant.

Smith and Laguna[12] used second sound detection in liquid helium II to study the PAS response of a doped selenium edge-filter between 400 and 700 nm. The second sound detector exploited the amplification in signal provided by the Q-factor of the cavity (which was 100 for the conditions employed). The authors suggest that, with proper cavity design and lock-in detection, fluctuating second sound amplitudes as small as 3×10^{-9} K should be detectable. Recent papers by Main and McCann[13] and Thijssen et al.[24] also describe the use of second sound in helium to detect the heat generated by nonradiative de-excitation processes. In the ESR measurements, the second sound detector could not be a superconductor because of a varying magnetic field in the region of the detector. A carbon resistance film was field insensitive and responded adequately at low modulation frequencies but, due to its 1 ms response time, was unsuitable for the upper audio frequency range. An oscillating superleak detector[25] was tried but placed stringent requirements on the temperature stability of the detector; however, with this device second sound resonances could be followed from the fundamental frequency of circa 70 Hz out to the 100th harmonic.

The "mirage effect"[8] or Photothermal Deflection Spectroscopy, PDS,[9,26,27] may be used with either a liquid or gas in contact with the sample. The magnitude of the PDS signal depends upon the product of thermal and optical properties of the sample as well as properties of the transducer fluid. If the transducer fluid behaves as an ideal gas, the PDS signal should depend upon temperature through the thermal diffusion length of the gas which varies as $T^{0.75}$, its thermal diffusivity which varies as $T^{1.5}$, and the temperature dependence of the refractive index which has an inverse dependence on T at constant pressure. These terms yield a temperature and pressure dependence of the periodic PDS signal that has the form:[10]

$$\text{Signal} = B[p^{1.5}\ T^{-2.75}]\ \exp\{-C\ p^{0.5}\ T^{-0.75}\}$$

where B and C are constants that depend upon the thermal and optical properties of the sample and the intensity of the exciting light. Cryogenic liquids will also exhibit a change of refractive index as their temperature changes. The PDS method could, therefore, be employed with these materials as the transducer fluid. Care would have to be taken to minimize convection currents which will contribute to measurement noise. To the author's knowledge, PDS has not yet been employed at cryogenic temperatures but would seem to have potential for such studies.

For many experimental studies it is desirable to have the sample in an ultra high vacuum environment. The PAS response of a solid sample can be measured under these conditions using a piezoelectric detector[3] or by directly measuring the displacement of the sample surfaces with the laser beam deflection method developed by Olmstead et al.[14] When piezoelectric detection is used, powder samples may be clamped between transparent plates of quartz or sapphire which will permit the sample to be cooled to low temperatures[28] and also serve as the acoustic transfer path to the piezoelectric detector.

MEASUREMENT OF THERMAL PROPERTIES

The absorption of radiation in the PAS process produces a distributed heat source in the sample that is periodic in time. The thermal properties of the sample are, therefore, involved in the resulting PAS signal and may be measured provided that the optical parameters are known. Physical processes that change the thermal properties of a sample will also give rise to changes in the amplitude and phase of the photoacoustic response under constant light amplitude and modulation frequency conditions. Changes in sample dimension that are due to phase transitions or thermal expansion may also contribute to the photoacoustic signal.

The use of modulated heat flow methods to measure the thermal properties of materials was initially developed by Angstrom.[29] Optically heating a sample in PAS measurements is a straightforward way of providing the required periodic temperature fluctuations and has been exploited by several groups to obtain the thermal parameters of thin film samples.

Thermal diffusivity measurements on thin metallic films using either microphone or mirage effect detectors have been reported by Charpientier et al.[30] and Rousset and Lepoutre.[31] The two sets of measurements were made at room temperature and recorded both the amplitude and phase of the signal that resulted when the samples, which were either self-supporting or in good thermal contact with a rigid backing, were illuminated with the modulated radiation from a 450W Xenon arc lamp. For the unsupported films a contribution to the total signal arose from their thermal expansion which caused the membrane to oscillate coherently with the excitation. A similar phenomenon has been observed to arise in gas-microphone cells when scattered light falls on the microphone diaphragm. These thermal expansion related mechanical vibrations caused about a 6% overestimate of the samples thermal diffusivity which was computed from a theoretical fit to the frequency dependence of the data. The effect of sample dimension changes had a negligible effect on the mirage detector which measured the near surface temperature gradient from the amplitude of the PDS signal as a function of position above the sample surface at several modulation frequencies. The thermal diffusivity of the zinc sample was obtained by fitting these data to theoretical expressions for the signal amplitude from a one-dimensional thermal analysis. Recently, Lachaine[32] has reported measurements of the thermal conductivity and specific heat of supported samples made using the frequency dependence of the photoacoustic phase and samples of different thickness. The thermal parameters were, once again, extracted from the data by fitting to the theoretical model of Rosencwaig and Gersho[33] and found to be in good agreement with published values. Although these measurements were all made at room temperature, the methods and analytical techniques discussed are applicable at cryogenic temperatures.

The measurement of first and second order phase transitions has received considerable photoacoustic attention. Pichon et al.[19] used a gas-microphone cell and thermally thick samples to measure the specific heat anomaly at the magnetic phase transition transition temperatures of MnF_2 and $CrCl_3$ which occur at about 68 K and 16 K, respectively, The measured temperature dependence of the PAS signals is shown in Figure 4 which also displays the independently measured value of the specific heat[34] or the value computed from Rosencwaig-Gersho theory.[33] The reciprocal of the PAS signal is proportional to the specific heat at constant pressure and is seen to exhibit the lambda-type anomaly expected at the transition temperature. Agreement between the PAS and direct measurements is also seen to be good with the small differences being attributed to sample inhomogeity. Another cryogenic measurement has been reported by Kuhnert and Helbig,[23] who investigated the structural phase transition occurring at circa 22 K in $NaC_2H_3O_2$. Only the

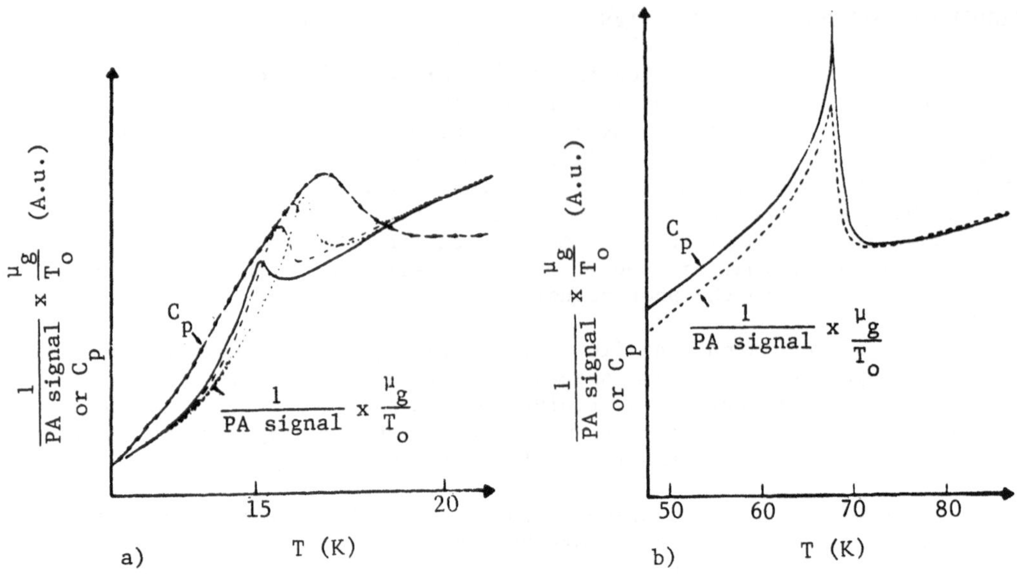

Fig. 4. Temperature dependence of the PAS response at a magnetic phase transition. a) CrCl sample at various laser powers. Solid line: 20mW, dot-dash line: 10mW, dotted line: 5mW, dashed line: 2.5mW.

PAS signal, rather than its reciprocal, is presented but this shows amplitude changes in the region of the transition. Measurements were made with the sample being heated and cooled through the transition temperature. The authors suggested that the difference between these two sets of data results from the heat of transition being provided by the light absorption during the heating cycle, whereas only the difference between the specific heats of the two phases alters the signal during the cooling process.

Korpiun and Tilgner[35] have extended the one-dimensional thermal analysis of Rosencwaig and Gersho to take into account first order phase transitions occurring as a result of the optical excitation process. To simplify the thermal analysis, they assume a thermally thick sample that has a high optical absorption coefficient so that the absorbed radiation gives rise to a plane heat source. The phase boundary separating the two different thermal states of the sample occurs at the isotherm corresponding to the transition temperature and its location depends upon the ambient temperature and the optical energy input. A periodic variation in the sample surface temperature will result in the periodic motion of this boundary, the magnitude of which depends upon the ratio of the heat of transformation to the specific heat. When the ambient temperature is close to the phase transition temperature, the heat of transformation associated with the optically stimulated phase change causes a change in the amplitude and phase of the PAS signal.

Florian et al.[36] measured the melting transition in gallium at circa 300 K prior to the development of the above model. They observed an apparent assymetry in signal behavior upon heating or cooling through the melting transition and attributed this to the effect of the heat of transformation. When the sample is heated, the optical energy input provides the heat of transformation required for the melting process, thus reducing the heat transferred to the transducer gas. Upon cooling, the amplitude signal only changes due to the difference between the specific heats and thermal conductivities of the two phases. These authors also pointed out that phase transitions in transparent materials could be monitored if they contained inert particles, such as graphite, that could absorb the modulated radiation. The periodic temperature fluctuations due to the heat transfer between these particles and the transforming material would be modified as the material is heated through the transition temperature.

Korpiun and co-workers[37] repeated the measurements on the melting transition of gallium and also studied the structural phase transitions in VO_2 (340 K) and $BaTiO_3$/0.45 weight percent Fe_2O_3 (393 K). They interpreted their results in terms of their theoretical model and were able to observe the expected amplitude and phase shifts for both sample heating and cooling. Experimentally, it was necessary to use a low heating rate and avoid supercooling of the sample when the temperature was decreased. They attribute the difference between their results and the earlier data of Florian et al. to the heating rate used and lack of thermal homogeneity. Bechtold et al.[38] have applied similar methods to measuring first and second order phase transitions in metal-hydrogen alloys between 300 and 500 K; Louis et al.[39] have detected transitions in 2 mg samples of liquid crystals that occurred between 300 and 320 K; and Bibi and Jenkins[40] have measured a first order transition in cobalt hexaquo-hexafluorosilicate at 250 K. All of these measurements were made using variable temperature gas-microphone PAS cells.

Recently, Etxebarria et al.[41] used piezoelectric detection to measure transitions in $CsCuCl_3$ and $(N(CH_3)_4)_2CoCl_4$. They extended the theory of Korpiun et al. to the piezoelectric case, but in order to simplify the model, did not take the temperature dependence of the elastic constants or the sample thermal expansion coefficient into account. The form of the amplitude and phase signals is predicted for both first and second order phase transitions and used to interpret the data. Since the ambient temperature range over which the phase transition occurs is determined by the thermal input from the modulated radiation, control of the radiation intensity permits this to be adjusted to a convenient value. $CsCuCl_3$ has a well defined first order transition at 423 K at which the PAS amplitude showed the expected drop and return to its prior value. The $(N(CH_3)_4)_2CoCl_4$ sample has five transitions between 100 and 500 K and each of these was clearly resolved from the amplitude of the signal. The authors point out that the temperature dependence of the sample's thermal expansion coefficient must be taken into account in interpreting the data since this acts as a scaling factor in the expressions for the signal. In comparing their results with the predictions of their extended theory and also with the Rosencwaig and Gersho model, the authors note that the two models give essentially the same amplitude prediction, but only their extended model describes the observed phase behavior. These authors emphasize the importance of the thermal expansion term in a subsequent letter[42] concerning the observed PAS response at a ferro-paraelastic phase transition.

CONCLUSIONS

The major emphasis in the use of PAS at low temperatures has so far been the study of the optical properties of materials. The technique has been used at cryogenic temperatures to detect phase transitions, but the majority of other thermal measurements have been conducted at room temperature or above. Since the PAS method for measuring thermal diffusivity is a modification of Angstrom's technique, a suitable choice of sample supporting material should permit the method to be exploited at cryogenic temperatures Both the piezoelectric and gas-microphone techniques for phase transition detection have a preliminary theoretical foundation and the directions required for their extension are clear.

One of the important spectroscopic advantages of PAS is its ability to provide optical absorption data from powdered samples without special sample preparation. Unfortunately, the complex heat transfer paths involved in signal generation from samples of this type would make them unsuitable for thermal diffusivity measurements. However, if only the temperature at which thermal parameter changes occur is of interest, then powdered samples could be employed. This could be an advantage in exploratory studies on new materials.

ACKNOWLEDGMENTS

This work was partially supported by NSF Grant #CPE-8217364. Thanks are due to Drs. Tam, Helbig, and Boccara for permission to use figures from their papers.

REFERENCES

1. "Optoacoustic Spectroscopy and Detection," Yo-Han Pao, editor. Academic Press, New York (1977).
2. A. Rosencwaig, "Photoacoustics and Photoacoustic Spectroscopy," John Wiley, New York (1980).
3. A.C. Tam, in "Ultrasensitive Spectroscopic Techniques," D. Klinger, editor, Academic Press, New York (1982).
4. J.B. Kinney and R.H. Staley, "Annual Reviews of Materials Science," 12 (1982), p. 295.
5. D.W. Vidrine, in "Fourier Transform Infra-Red and Raman Spectroscopy," J.R. Ferraro and L.J. Basile, editors, Academic Press, New York (1982).
6. Y.C. Teng and B.S.H. Royce, J. Opt. Soc. Am., 70 (1980), pp. 557.
7. S.J. McGovern, B.S.H. Royce, and J.B. Benziger, J. Appl. Phys., 57 (1985), pp. 1710.
8. A.C. Boccara, D. Fournier, and J. Bardoz, Appl. Phys. Lett., 36 (1980), pp. 130.
9. J.C. Murphy and L.C. Aamodt, J. Appl. Phys., 51 (1980), pp. 4580.
10. N. Moore, J.D. Spear, J.B. Benziger, and B.S.H. Royce, Proc. 4th International Topical Meeting on Photoacoustic, Thermal, and Related Sciences, Montreal (1985).
11. M.B. Robin and N.A. Kuebler, J. Chem. Phys., 66 (1977), pp. 169.
12. J.B. Smith and G.A. Laguna, Phys. Lett., 56A (1976), pp. 223.
13. P.C. Main and J.P.J. McCann, J. Phys. D: Appl. Phys., 17 (1984), pp. 523.
14. M.A. Olmstead and N.M. Amer, J. Vac. Sci. Technol., B1 (1983), pp. 751; and Surface Science, 132 (1983), pp. 68.
15. C.K.N. Patel and A.C. Tam, Reviews of Modern Physics, 53 (1981), pp. 517.
16. E.T. Nelson and C.K.N. Patel, Appl. Phys. Lett., 39 (1981), p. 537.
17. T. Ikari, S. Shigetomi, Y. Koga, and S. Shigetomi, J. Phys. C: Solid State Phys., 17 (1984), pp. 435.
18. J. Etxebarria, S. Uriarte, J. Fernandez, M.J. Tello, and A. Gomez-Cuevas, J. Phys. C: Solid State Phys., 17 (1984), pp. 6601.
19. C. Pichon, M. LeLiboux, D. Fournier, and A.C. Boccara, Appl. Phys. Lett., 35 (1979), pp. 435.
20. P.S. Bechtold, M. Campagna, and J. Chatzipetros, Optics Comm., 36 (1981), pp. 369.
21. O.A. Cleves Nunes, A.M.M. Monteiro, and K. Skeff Neto, Rev. Sci. Instrum., 52 (1981), pp. 1413.
22. J. Pelzel, K. Klein, and O. Nordhaus, Applied Optics, 21 (1982), pp. 94.
23. R. Kuhnert and R. Helbig, Applied Optics, 20 (1981), pp. 4191.
24. H.P.H. Thijssen, R. van den Berg, S. Volker, J.H. van der Waals, and L.P.J. Husson, Chemical Physics Letters, 111 (1984), pp. 11.

25. R.A. Sherlock and D.O. Edwards, Rev. Sci. Inst., 44 (1970, pp. 1603.
26. W.D. Jackson, N.M. Amer, A.C. Boccara, and D. Fournier, Appl. Opt., 20 (1981), pp. 1333.
27. B.S.H. Royce, F. Sanchez-Sinencio, R. Goldstein, R. Muratore, R. Williams, and W.M. Yim, J. Electrochem. Soc., 129 (1982), pp. 2393.
28. R.W. Shaw and H.E. Howell, Applied Optics, 21 (1982), pp. 100.
29. A.J. Angstrom, Ann. Physik, 114 (1961), pp. 513; Phil. Mag. 25 (1963), p. 130; and Phil. Mag., 26 (1963), pp. 161.
30. P. Charpentier, F. Lepoutre, and L. Bertrand, J. Appl. Phys., 53 (1982), p. 608.
31. G. Rousset and F. Lepoutre, Revue Phys. Appl., 17 (1982), pp. 201.
32. A. Lachaine, J. Appl. Phys., 57 (1985), pp. 5075.
33. A. Rosencwaig and A. Gersho, J. Appl. Phys., 47 (1976), pp. 64.
34. "Properties of Materials at Low Temperatures," V.J. Johnson, editor, Pergamon, New York (1961).
35. P. Korpiun and R. Tilgner, J. Appl. Phys., 51 (1980), pp. 6115.
36. R. Florian, J. Pelzel, M. Rosenberg, H. Vargas, and R. Wernhardt, Phys. Stat. Sol., (a)48 (1978), pp. K35.
37. P. Korpiun, J. Baumann, E. Luscher, E. Papamokos, and R. Tilgner, Phys. Stat. Sol., (a)58 (1980), pp. K13.
38. P.S. Bechtold, M. Campagna, and T. Schober, Solid State Comm., 36 (1980), p. 225.
39. G. Louis, P. Peretti, and J. Billard, C.R. Acad. Sci. Paris, 298 (1984), pp. 435.
40. I. Bibi and T.E. Jenkins, J. Phys. C: Solid State Phys., 16 (1983), pp. L57.
41. J. Etxebarria, S. Uriarte, J. Fernandez, M.J. Tello, and A. Gomez-Cuevas, J. Phys. C: Solid State Phys., 17 (1984), p. 6601.
42. J. Etxebarria, J. Fernandez, M.A. Arriandiaga, and M.J. Tello, J. Phys. C: Solid State Phys., 18 (1985), pp. L13.

THE CHARACTERIZATION OF CRYOGENIC MATERIALS BY X-RAY ABSORPTION METHODS

S. M. Heald and J. M. Tranquada

Brookhaven National Laboratory
Upton, New York

ABSTRACT

X-ray absorption techniques have in recent years been developed into powerful probes of the electronic and structural properties of materials difficult to study by other techniques. In particular, the extended x-ray absorption fine structure (EXAFS) technique can be applied to a variety of cryogenic materials. Three examples will be used to demonstrate the power of the technique. The first is the determination of the lattice location of dilute alloying additions such as Ta and Zr in Nb_3Sn. The Ta additions are shown to reside predominately in Nb lattice sites, while Zr is not uniquely located at either Nb or Sn sites. In addition to structural information, temperature dependent EXAFS studies can be used to determine the rms deviations of atomic bond lengths, providing information about the temperature dependence of interatomic force constants. For Nb_3Sn deviations are found from simple harmonic behavior at low temperatures which indicate a softening of the Nb-Sn bond strength. The final example is the study of interfacial properties in thin film systems. This is accomplished by making x-ray absorption measurements under conditions of total external reflection of the incident x-rays. As some examples will show this technique has great potential for studying interfacial reactions, a process used in the fabrication of many superconducting materials.

INTRODUCTION

Applications of Extended X-ray Absorption Fine Structure (EXAFS) measurements have expanded rapidly in recent years.[1-3] This has come about as a result of an improved understanding of the theoretical foundations of EXAFS and the availability of intense tunable x-rays from synchrotron radiation sources.[4] As a result of these developments the EXAFS technique can be fruitfully applied to a number of cryogenic materials. In particular EXAFS measurements can provide local structural and dynamic information in a variety of systems including highly dilute or disordered systems for which traditional diffraction techniques have difficulties.

After a brief discussion of the origins of EXAFS, three different experiments are used to show the range of EXAFS measurements. The problems to be discussed are the site determination of dilute Ta additions to Nb_3Sn, a study of the temperature dependence of the interatomic forces in Nb_3Sn using the EXAFS Debye-Waller factor, and the use of total external reflection techniques to measure EXAFS of interfacial reactions.

BASIC PRINCIPLES OF EXAFS

EXAFS is the modulation in the x-ray absorption coefficient found above x-ray absorption edges for most materials. Figure 1a shows an example for the Nb K absorption edge in Nb_3Sn. The absorption increase or edge is a result of the opening up of a new channel for the x-ray absorption process. In the case of the K-edge it is the energy at which the 1 s electrons can begin to be ejected. When this occurs the electrons are emitted as photoelectrons which can be described as an outgoing wave as shown in Fig. 2. This outgoing wave will scatter off the surrounding atoms resulting in interference. The interference can be

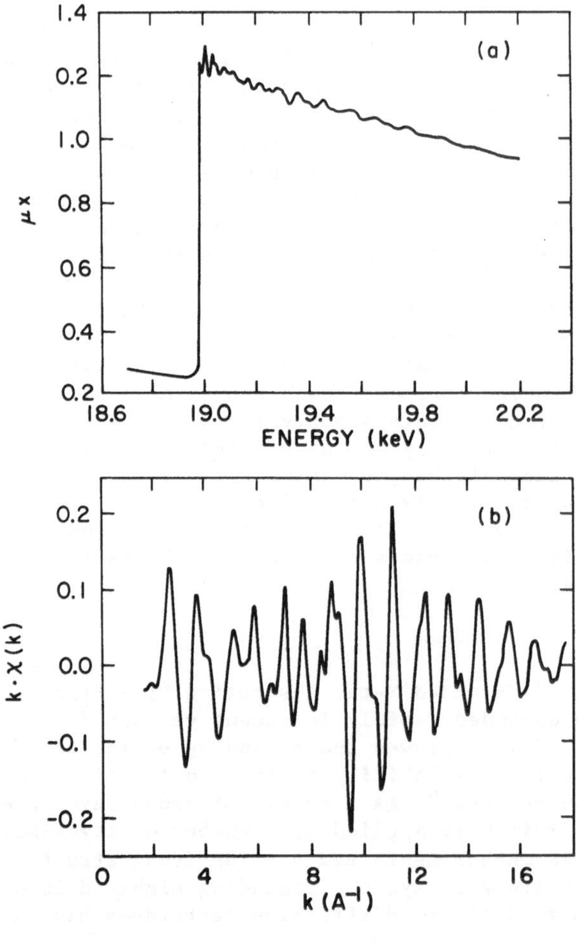

Fig. 1. a) X-ray absorption at the Nb K edge in Nb_3Sn. b) The EXAFS oscillations, $\chi(k)$, in (a) multiplied by the wave vector k. The smoothly varying background absorption has been subtracted and the data normalized to the edge jump.

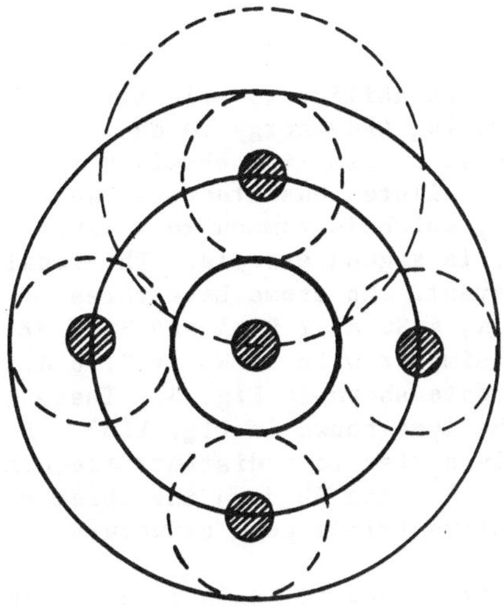

Fig. 2. Origin of the EXAFS oscillations. The spherical wave of the outgoing photoelectron scatters off the surrounding atoms. The interference between the outgoing and scattered waves modulates the X-ray absorption probability.

constructive or destructive depending on the relative phase of the outgoing and scattered wave. This gives rise to the oscillations in the absorption probability shown in Fig. 1. Thus, the EXAFS is a local probe providing information about the first few shells of surrounding atoms. In addition, since each element has its own distinctive absorption edge energy, the local environment of each component in a system can be probed by tuning the x-ray energy to the appropriate value.

For discussion purposes a simple independent particle model can be used. Putting the above qualitative discussion into mathematical form gives[1]:

$$\chi(k) = \frac{\mu-\mu_o}{\mu_o}$$

$$= \frac{m}{4\pi h^2 k} \sum_j \frac{N_j}{R^2_j} t_j(2k)\exp(-2R_j/\lambda)Q_j(k) \sin[2kR_j+\delta_j(k)+\phi_j(k)] \qquad (1)$$

where k is the photoelectron wave vector, and μ_o is the smoothly varying background absorption. The sum is over the neighboring coordination shells which contain N_j neighbors at distances R_j. λ is the mean free path of the photoelectron and the functions $Q_j(k)$ and $\phi_j(k)$ take account of any static or dynamic disorder in the jth shell of atoms from their average distance R_j. The backscattering amplitude, denoted $t_j(2k)$, is a function of the type of backscattering atom, and $\delta_j(k)$ is the phase shift due to scattering from both the neighboring and central atoms. Typically $t(2k)$ and $\delta(k)$ are calibrated by measuring standard materials for which the structure is known.

As can be seen each shell of atoms contributes an oscillatory component to the EXAFS which has a frequency determined mainly by the interatomic distance, and an amplitude factor related to the types and number of atoms in a given shell along with any disorder present. Since each shell contributes a separate frequency, Fourier analysis is often used to analyze the data. Details of data analysis procedures are given elsewhere.[1,2,5]

A strength of the EXAFS technique is the ability to selectively study a dilute component in a system by tuning the energy to one of its absorption edges. From the local structural information obtained from the EXAFS it is often possible to locate a dilute constituent at an unique lattice position. The A15 lattice, which is common to a large class of superconductors including Nb_3Sn, is a good example. The local environments of Nb and Sn are quite different. Nb atoms have three nearby shells consisting of 2 Nb at 2.64 A, 4 Sn at 2.96 A and 8 Nb at 3.23 A, while the Sn environment is much simpler with 12 Nb at 2.96 A. These differences are easily seen in the data shown in Fig. 3. These are Fourier transforms of $\chi(k)$ data of the type shown in Fig. 1(b). As mentioned, each near neighbor distance gives rise to a distinct frequency which results in a peak in the transform. For the Nb data the three near neighbor distances give rise to a distinctive triple peak structure.

Alloying additions such as Ta or Zr are added to Nb_3Sn in an attempt to improve superconducting properties such as H_{c2}.[6,7] Significant improvements in H_{c2} are found for Ta additions while adding Zr gives little

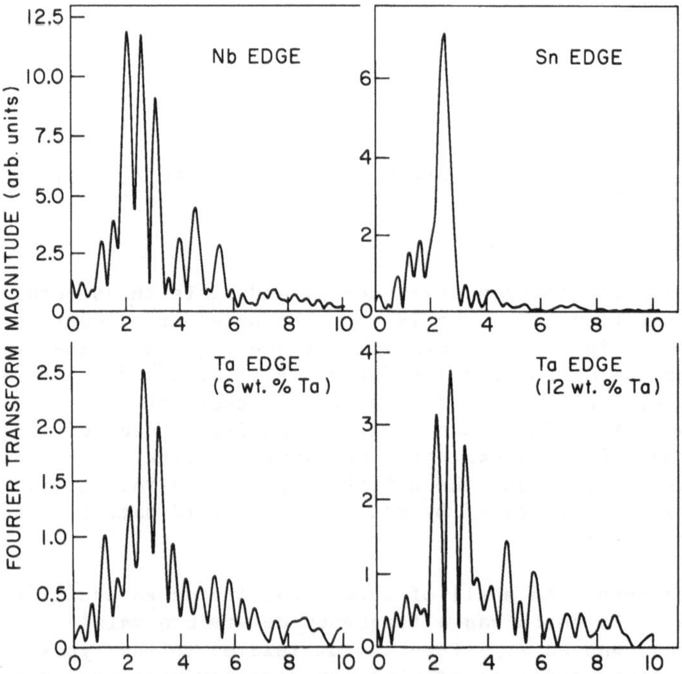

Fig. 3. Fourier transforms of $k \cdot \chi(k)$ for four cases. The Nb and Sn edges are for pure Nb_3Sn while the Ta edges are for Nb_3Sn samples which contain the indicated amounts of Ta.

change. Figure 3 shows transforms of EXAFS data taken for the Ta site in Ta-containing Nb_3Sn formed by the bronze process. It is clear from the similarities in the transforms that at higher concentrations the Ta atoms are sitting in Nb sites. This is in agreement with recent electron channeling results.[8] The situation is not as clear for the lower concentration, but it still appears that most of the Ta atoms are on Nb sites. There is evidence from increased reaction rates of Ta-alloyed Nb_3Sn that some of the Ta resides in the grain boundaries.[9] This would presumably involve a larger fraction of Ta atoms when the Ta concentration is low and could explain the observed differences.

For the EXAFS data obtained at the Zr edge the situation is different. The Zr transforms do not resemble either the Nb or Sn transforms. Further work is required to quantitatively determine the local Zr environment which will be the subject of a later publication.

INTERATOMIC FORCES IN Nb_3Sn

The parameters $Q(k)$ and $\phi(k)$ take account of any disorder present in the bond length distribution of a particular coordination shell. The disorder can be due to static variations in bond lengths related to the inherent structure of the material, or a dynamic effect arising from the interatomic vibrations of the atoms and the fact that the x-rays take an instantaneous snapshot of the distribution. If there is no static disorder and a harmonic distribution is sufficient to describe the thermal vibrations then $\phi(k)\sim 0$ and $Q(k)=\exp(-2k^2\sigma^2)$ where σ^2 is the mean square variation of bond length r. This EXAFS Debye-Waller factor is similar to the x-ray Debye-Waller factor except that it measures the variation in the bond lengths rather than deviations from the ideal lattice position. Therefore, the EXAFS σ^2 is a sensitive probe of interatomic force constants. Also, anharmonicity reveals itself by giving a nonzero $\phi(k)$ and higher order terms in $Q(k)$.

The anomalous lattice dynamical behavior in high T_c A15 superconductors is well established. For Nb_3Sn, measurements of phonon dispersion,[10] phonon density of states,[11] and elastic constants[12] all indicate significant softening of phonon modes as the temperature is lowered towards the Martensitic transition temperature ($T_M\sim 45$ K) where the cubic phase undergoes a small distortion resulting in a tetragonal structure. Questions remain concerning the connection between the phonon softening, the electronic structure, and the Martensitic phase transition. Some theoretical work[13] has suggested that the Nb atoms see an anharmonic effective potential, and an x-ray diffraction study[14] of V_3Si indicates that the V atoms vibrate anharmonically at low temperature.

EXAFS measurements of σ^2 for the Nb-Sn bond are shown in Fig. 4. Our results for the mean square fluctuations in the Nb-Sn nearest-neighbor distance indicate that as the temperature decreases, σ^2 decreases much less than would be expected from a harmonic Einstein model which fits most ordinary solids.[1] The extra contribution to σ^2 at low temperatures is most likely due to weakening of the effective Nb-Sn bond strength. Increasing static disorder could also explain the results, but there is no evidence for such disorder from other experiments. Anharmonic contributions to the EXAFS Debye-Waller factor are found to be negligible, so that at a given temperature the motion along the Nb-Sn bonds is harmonic. Further analysis should provide similar information about the motions of Nb-Nb pairs.

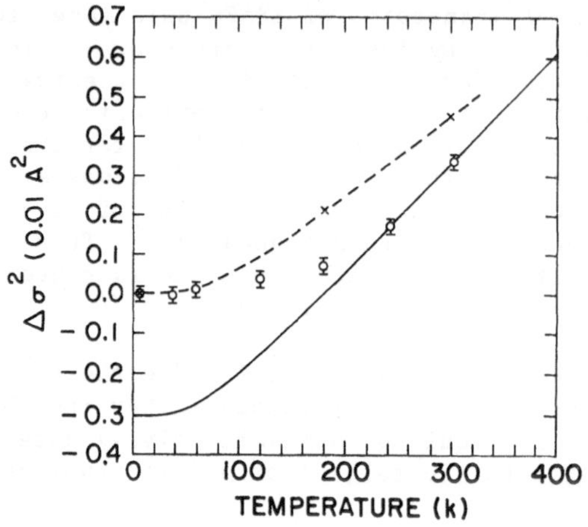

Fig. 4. Change in the EXAFS Debye-Waller parameter, σ^2, as a function of temperature for the Sn edge in Nb_3Sn. The solid line is an Einstein model fit to the two highest temperature points and the deviation of the data indicates a softening of the Nb-Sn bond as the temperature is lowered. The x's are data for pure Nb which is fit well by an Einstein model (dashed line).

INTERFACE EXAFS

The index of refraction for x-rays is slightly less than one. If the incidence angle is small enough total external reflection can occur, in which case the penetration of the x-rays into the surface is very small. For a material such as Cu the penetration is 20-30 A. Similarly a reflection can be made to occur at the interface between an overlayer and substrate if the electron density of the substrate is larger. Again the penetration is only 20-30 A, so that if a signal specific to the substrate is detected it will be due only to atoms located close to the interface. Thus, high sensitivity to interfacial regions can be obtained. One such element specific signal is the x-ray fluorescence excited by the incoming x-rays. The fluorescence intensity is directly proportional to the probability of absorption and, therefore, can be used to obtain the EXAFS spectrum.

An Al-Cu bilayer consisting of 1000 A of Al on 1000 A of Cu deposited on float glass was used to test the above ideas. For incident angles above the Al critical angle and below the Cu critical angle a strong reflection occurs from the interface. The EXAFS from the interfacial Cu atoms was then obtained by monitoring the Cu fluorescence as the incident x-ray energy was varied. The results are shown in Fig. 5 for various annealing temperatures. For temperatures below 140°C the spectra show the Cu to be in its normal metallic structure. At 160°C, however, an interfacial reaction forms a thin layer of $CuAl_2$,[15,16] and the resulting spectra are a combination of those for Cu and $CuAl_2$. At higher temperatures the $CuAl_2$ becomes thicker and the underlying Cu atoms are no longer reached by the reflecting x-rays; thus, the spectra show only $CuAl_2$ features.

Many materials, of which bronze process Nb_3Sn is a good example, are formed by solid state reactions. These glancing angle EXAFS techniques

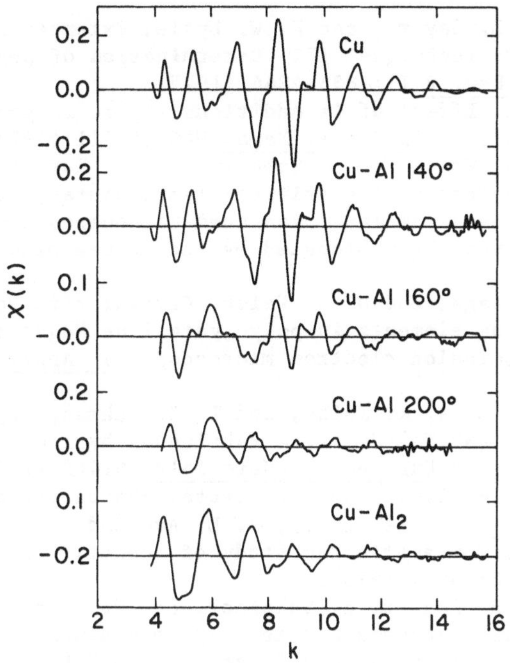

Fig. 5. EXAFS signal for a Cu-Al interface after annealing at 140, 160 and 200°C obtained using fluorescence detection and an incident angle of 4.5 mrad. For comparison the spectra for pure Cu (top) and pure $CuAl_2$ (bottom) are also shown.

should find wide applications in the study of the microscopic structure of interfacial compounds. The high sensitivity to very thin interface regions make glancing angle EXAFS techniques especially sensitive to the initial stages of compound formation where nucleation problems are important.

ACKNOWLEDGMENTS

This work was supported in part by the U.S. Department of Energy under Contract Nos. DE-AS05-80-ER10742 and DE-AC02-76CH00016.

REFERENCES

1. E. A. Stern and S. M. Heald, Basic principles and applications of EXAFS, in: "Handbook on Synchrotron Radiation," E. E. Koch, ed., North Holland, Amsterdam (1983), pp. 955-1014.
2. "EXAFS Spectroscopy," B. K. Teo and D. C. Joy, eds., Plenum, New York (1981).
3. "EXAFS and Near Edge Structure," A. Bianconi, L. Incoccia, and S. Stipcich, eds., Springer-Verlag, Berlin (1983).
4. "Synchrotron Radiation Research," H. Winick and S. Doniach, eds., Plenum, New York (1980).

5. E. A. Stern, D. E. Sayers, and F. W. Lytle, Extended x-ray absorption fine-structure technique. III. Determination of physical parameters, Phys. Rev. B 11:4836-4846 (1975).
6. J. D. Livingston, Effect of Ta additions to bronze-processed Nb_3Sn superconductors, IEEE Trans. Magn. MAG-14:611 (1978).
7. M. Suenaga, D. O. Welch, R. L. Sabatini, O. F. Kammerer, and S. Okuda, Superconducting critical temperatures, critical magnetic fields, lattice parameters, and chemical compositions of "bulk" pure and alloyed Nb_3Sn produced by the bronze process, to be published.
8. J. Tafto, M. Suenaga, and D. O. Welch, Crystal site determination of dilute alloying elements in polycrystalline Nb_3Sn superconductors using a transmission electron microscope, J. Appl. Phys. 55:4330-4333 (1984).
9. M. Suenaga, K. Aihara, K. Kaiho, and T. S. Luhman, Superconducting properties of $(Nb,Ta)_3Sn$ wires fabricated by the bronze process, Adv. in Cryogenic Engineering-Materials 26:442 (1980).
10. L. Pintschovius, H. Takei, and N. Toyota, Phonon anomalies in Nb_3Sn Phys. Rev. Lett. 54:12 (1985); J. D. Axe and G. Shirane, Inelastic-neutron-scattering study of acoustic phonons in Nb_3Sn, Phys. Rev. B 8:1965 (1973).
11. B. P. Schweiss, B. Renker, E. Scheider, and W. Reichardt, Phonon spectra of A15 compounds and ternary molybdenum chalcogenides, in: "Superconductivity in d- and f-band Metals," D. H. Douglass, ed., Plenum Press, New York (1976), pp. 189-208.
12. K. R. Keller and J. J. Hanak, Ultrasonic measurements in single-crystal Nb_3Sn Phys. Rev. 154:629 (1967).
13. C. C. Yu and P. W. Anderson, Local-phonon model of strong electron-phonon interactions in A15 compounds, Phys. Rev. B 29:6165 (1984); L. R. Testardi, Structural instability, anharmonicity, and high temperature superconductivity in A15-structure compounds, Phys. Rev. B 5:4342 (1972).
14. J.-L. Staudenmann and L. R. Testardi, X-ray determination of anharmonicity in V_3Si, Phys. Rev. Lett. 43:40 (1979).
15. R. A. Hamm and J. M. Vandenberg, A study of the initial growth kinetics of the copper-aluminum thin-film interface reaction by in situ x-ray diffraction and Rutherford backscattering analysis, J. Appl. Phys. 56:293 (1984).
16. H. T. G. Hentzell, R. D. Thompson, and K. N. Tu, Interdiffusion in copper-aluminum thin film bilayers. I. Structure and kinetics of sequential compound formation, J. Appl. Phys. 54:6923 (1983).

NONDESTRUCTIVE CHARACTERIZATION OF CRYOGENIC

MATERIALS FOR QUALITY AND PROCESS CONTROL

Jean F. Bussière

Industrial Materials Research Institute
National Research Council Canada
Boucherville, Québec, Canada

ABSTRACT

Nondestructive characterization techniques offer not only the possibility of testing total production or completed structures but also of monitoring microstructures while they evolve during fabrication. Amid the wide range of techniques presently available or under development, a few are described which relate to cryogenic structural materials: nondestructive measurements of grain size and mechanical properties such as yield strength and fracture toughness, the development of noncontact ultrasonic transducers and cure monitoring of fibre-reinforced polymers.

INTRODUCTION

Present trends in nondestructive evaluation (NDE) include the development of techniques to characterize materials nondestructively[1]. This relatively new development is complementary to traditional NDE in assessing the reliability of completed structures or their degradation from exposure to service environments and also opens the possibility of enhancing both productivity and quality in manufacturing through better process control[2-4]. Regardless of whether we are considering traditional materials such as steel or aluminum or newer materials such as polymers, composites or structural ceramics, the lack of adequate materials property sensors is presently a major limitation for the direct, intelligent or adaptive control of desired properties during fabrication. Present steel and aluminum mill control systems, in all their sophistication, only control stability and dimensional characteristics. They do not control material properties such as strength, fracture toughness, formability, coating adhesion, etc. Similarly, the control systems for polymerization reactors and autoclaves used in the fabrication of fiber reinforced composites, control temperatures, pressures, flow, etc, without feedback relating to changes of the desired properties such as viscosity, degree of cure or polymerization, molecular weight, branching, etc.

Nondestructive techniques can in principle be used to characterize a broad range of material characteristics such as mechanical properties, porosity, state of fatigue, residual stress, moisture content, degree of cure, etc., the techniques ranging from eddy currents and ultrasonics to positron annihilation and NMR. In this paper only a restricted number of examples are presented which are thought relevant to cryogenic materials such as ferritic and austenitic steels and fibre reinforced polymer matrix composites. These include assessments of grain size, mechanical properties such as yield strength, tensile strength and fracture toughness in steels using magnetic and ultrasonic techniques, the recent development of laser based noncontact ultrasonic transducers for remote monitoring of these properties during processing and applications of ultrasonic and dielectric measurements to monitor the cure of epoxies in composites during fabrication. Most of the technology presented is still in the development or research stage, but in some cases is already commercialized or has been integrated as part of manufacturing processes.

GRAIN SIZE

It is well known that the grain size of metals is an important factor in controlling strength, ductility, fracture toughness, formability and final appearance of metallic products. Several physical mechanisms can be used as a basis for grain size determination. Among these are magnetic hysteresis and Barkhausen noise (in magnetic steels), X-ray diffraction, and ultrasonic scattering. Although all the above techniques may be promising and should probably be further developed, (see ref. 4), measurements of grain size based on ultrasonic techniques appear to have the greatest potential and to be applicable to a much broader range of materials.

Ultrasonic attenuation, α, is a measure of the exponential decay of the amplitude A(x) of ultrasonic waves as they propagate say along the x direction in a medium,

$$A(x) = A_o e^{-\alpha x} \qquad (1)$$

where A_o is a constant. The total attenuation, α, generally results from the combined effects of scattering, α_s and absorption, α_a.

$$\alpha = \alpha_s + \alpha_a \qquad (2)$$

Scattering is the reorientation or mode conversion of acoustic energy due to interactions with material defects generally comparable to a wavelength in size such as grain boundaries, second phase particles or porosity, whereas absorption can be linked to inhomogeneities on a much finer scale such as dislocations and interstitials as well as magnetic domain wall motion.

The general problem of attenuation from scattering and velocity dispersion of ultrasound in grainy anisotropic solids has recently been solved theoretically[5]. The formalism is general enough to include effects of the size and shape distribution of the grains, preferred crystallographic orientation, elastic anisotropy and frequency. However, because of the complexity of the problem, only numerical solutions can be obtained and for interpretation of experimental data, one often uses simpler models which are valid under restrictive assumptions such as 1) no preferred orientation, 2) spherical grains and 3) large or small wavelength with respect to grain diameter. A particularly useful approximation is the so-called long wavelength or Rayleigh scattering regime which requires $\lambda > 2\pi d$. In this case we have:

$$\alpha_{sc} = S\, d^3 f^4 \qquad (3)$$

where S is a scattering coefficient (which depends on elastic anisotropy, density and both shear and longitudinal velocities), d is the average grain size and f is frequency. Values of S applicable to shear waves are listed in table 1 for a number of materials of cryogenic importance[6].

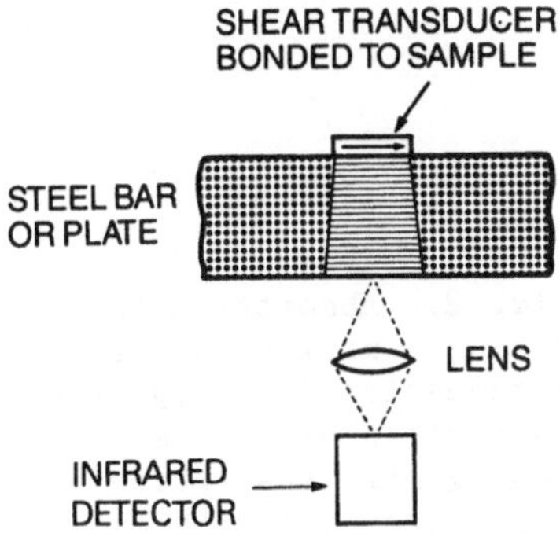

Fig. 1. Schematic of an experimental setup for measuring ultrasonic absorption. The infrared detector monitors the surface temperature modulation due to a chopped ultrasonic beam.

Experimentally, attenuation is usually measured by comparing the amplitude of several echoes obtained in samples with smooth parallel surfaces. Alternatively, it is possible to obtain similar information by analysing the noise backscattered into the transducer using traditional pulse-echo setups without being restricted to normal incidence. This has the advantage of not requiring a second reflecting parallel surface and also offering the possibility of probing different layers by time gating. However, these techniques only yield total attenuation values and the grain scattering contribution must be deduced by making simplifying assumptions on the absorption contribution α_a in eq. 2. These assumptions vary from complete neglect to linear and square dependences on ultrasonic frequency.

Using a recently developed technique[9,10] shown schematically in figure 1 for normal ultrasonic incidence, it is possible to measure α_a directly and thus assess its importance relative to scattering and determine its frequency dependence. The technique is based on the detection of the heat produced by the absorption of ultrasound. The ultrasonic beam is chopped at low frequency ($\approx$ 1 Hz) and the resulting temperature modulation at the surface observed with an infrared detector. Results from this type of measurement, combined with pulse-echo in a type A36 ferritic steel plate are shown in figure 2. Total attenuation is shown to be due to three separate contributions: an f^4 term from grain scattering, an f term from magnetic domain absorption and an f^2 contribution probably due to dislocation damping. The magnetic contribution was obtained by making measurements with and without a saturating magnetic field. Of particular interest is the fact that at the low frequencies commonly used in nondestructive evaluation (< 5 MHz), magnetic domains are the dominant source of attenuation (absorption). Also of interest is the existence of a measurable contribution tentatively associated with dislocation interactions and increasing as f^2. Measurements of this contribution may eventually help in the nondestructive determination of the state of fatigue of metals or other strength related properties. However, for the present, results such

Table 1. Shear Wave Scattering Coefficients, S_s, for a Few Materials of Cryogenic Interest[6]

Material	Aluminum	Copper	Iron	Niobium	316 SS
S_s [dB/cm $(MHz)^4$ cm^3]	284	24,600	3,260	6,990	787

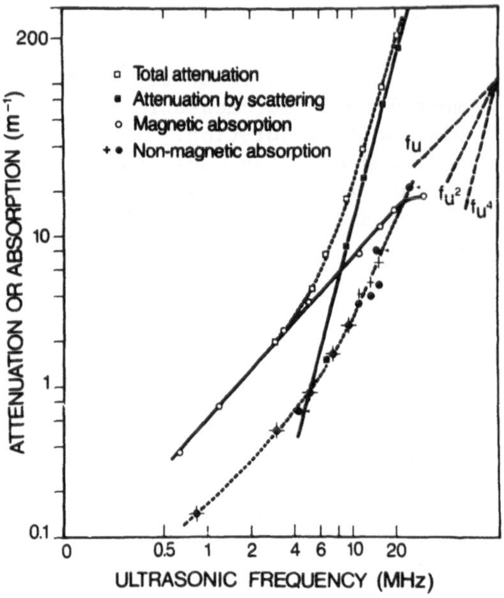

Fig. 2. Absorption and scattering contributions to total attenuation in a type A36 ferritic/pearlitic steel. The nonmagnetic absorption was deduced from infrared measurements. Scattering was obtained by subtracting absorption from total attenuation.

as shown in figure 2 allow to clearly deduce contributions to attenuation associated with grain scattering even at low frequencies. It may therefore be hoped that this new technique will both be useful in itself as a microstructural characterization tool and also indirectly by elucidating the various sources of attenuation in materials, in particular clearly delineating contributions due to scattering.

Attenuation due to grain scattering has been the subject of numerous experimental and theoretical studies (see for instance refs. 5, 7, 11 to 13 and references therein). The difficulties associated with the inverse problem of deducing grain size from attenuation have been recently described by Papadakis[14]. These include the contribution of absorption to attenuation, as mentioned above, varying grain size distributions, and the presence of grain substructures. Nevertheless, a number of encouraging results have been reported in the literature for copper[15], austenitic stainless steels[16] and as described in more detail below for steels of varied structure.

By assuming a linear frequency dependence for absorption and using a backscattering technique at two different frequencies, Goebbels and co-workers[13] have been able to measure grain size in the range 8-250 µm for steels of ferritic, austenitic and mixed structures such as ferrite-pearlite and ferrite-bainite. Results are shown in figure 3, where shear waves were used and a constant value of scattering coefficient (S) and absorption coefficient assumed for all samples. The excellent agreement may be somewhat surprising in view of recent measurements[17] which show that the original austenite grain size strongly affects scattering, indicating that the previous austenite grains (grains prior to low temperature transformations) continue to be effective scattering centers after they have transformed to ferritic, pearlitic, bainitic or martensitic microstructures after cooling or heat treatment. It was also shown that attenuation from ferritic-pearlitic two phase steels strongly depends on the percentage of each phase[18], with a maximum scattering occuring for ≈ 60% pearlite. It should be noted, however, that because of the cubic dependence of attenuation on grain size in the Rayleigh regime, an order of magnitude error in attenuation only leads to a factor of 2 or so error in grain size. At present, ultrasonic instruments to

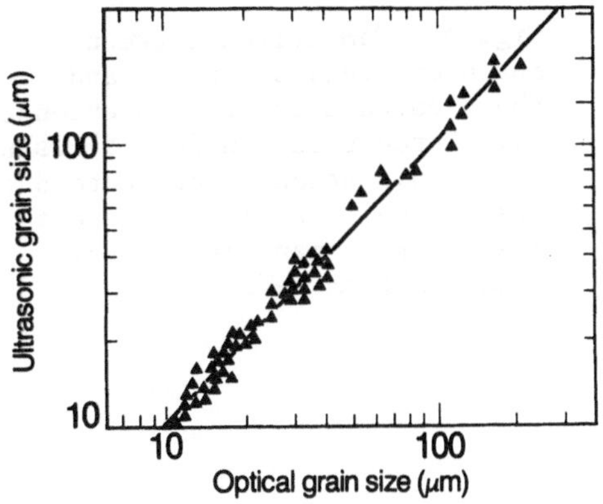

Fig. 3: Comparison between grain size obtained ultrasonically by backscattering and metallographically for steel of various microstructures (adapted from ref 13).

measure grain size have been installed on a steel cold strip line by Krupp in Germany[19], and are presently being developed in England[20].

YIELD STRENGTH, TOUGHNESS, HARDNESS

Tensile strength, yield strength and fracture toughness are key mechanical properties which often determine the suitability of structural materials for large cryogenic structures such as nuclear fusion reactors, MHD power generators, magnetically levitated trains, transportation of LNG.

Knowledge of these properties presently requires lengthy and costly destructive measurements which are performed on coupons which only represent a small fraction of production. Nondestructive techniques offer the possibility of monitoring 100% of production during or after fabrication and of monitoring actual structures. In addition, variations which occur in welds or in heat affected zones can also be monitored. This is particularly important since variations in grain size and fracture properties can often occur in certain areas, eg. in welds of cryogenic ferritic steels such as 12 Ni and 9 Ni[21].

One approach to nondestructive mechanical property determination is to take advantage of the close relation which often exists between mechanical properties and grain size as expressed in the well-known Hall-Petch relation:

$$M = M_0 + M_1 d^{-\frac{1}{2}} \qquad (4)$$

where M is a mechanical property, M_0 and M_1 are constants and d is grain size. Recent studies at Bethlehem Steel[22,23] on plain carbon-manganese steels have shown that it is possible to accurately predict mechanical properties and ductile to brittle transition temperatures by replacing the last term in equation (4) with ultrasonic attenuation, provided that chemical composition is taken into account. The following expression was used for yield strength, Y.

$$Y\ (\text{MPa}) = 166 + 317C + 78Mn + 130Si - 1.40\alpha_{10} \qquad (5)$$

where C, Mn and Si represent the wt% of these elements and α_{10} is ultrasonic attenuation at 10 MHz in dB/cm. This produced a standard error of 19.6 MPa as compared to 15.4 using metallographic grain size and 27.4 MPa using chemical composition only. These results are illustrated in figure 4. It should be noted that a small number of specimens were found to have significantly higher attenuation for a given grain size (4 out of 58

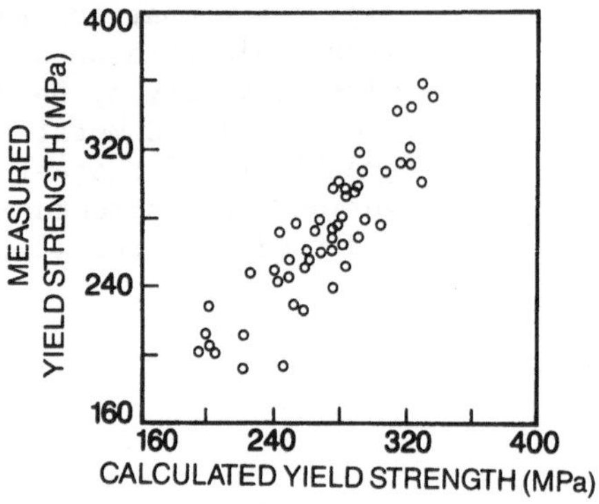

Fig. 4. Comparison between measured yield strengths and those calculated from an extended Hall-Petch relation (eq. 5) using ultrasonic attenuation instead of grain size. The materials are plain carbon-manganese steels (adapted from ref. 23).

specimens) but did not have unusual strength for their composition. No good explanation was found for these unusually high attenuations, although the possibility of inclusions was considered. In view of our previous discussion on absorption and figure 2, it is possible that absorption played a dominant role for some of the specimens studied leading to significant errors in the estimate of grain size effects from attenuation.

Another approach to ultrasonic mechanical property measurements was taken by A. Vary[24], who suggested a general empirical relation between fracture toughness, K_{1c}, yield strength, and ultrasonic attenuation. Using attenuation measurements in the 10 to 50 MHz range, good results were obtained for two maraging steels and a titanium alloy aged at various temperatures. Other studies, however, indicated the doubtful applicability of the general relation for Fe-C alloys[25] and for a 403 stainless steel whose K_{1c} was varied by changing the temperature, thus keeping grain size and microstructure constant[26].

Mechanical properties can also be deduced nondestructively from magnetic measurements because of the frequently observed correlations between hysteresis related parameters such as coercivity and yield strength, tensile strength, or hardness[27,28]. This is used as a basis for sorting small objects such as fasteners for improper heat treatment. In this case the entire object is initially magnetized by passing through a solenoid and external magnetic fields associated with remnant magnetization measured as the objects are moved on a conveyor belt. In the case of larger objects such as steel plates or bars which cannot be placed inside a solenoid, the technique used is to magnetize the structure locally and then measure the external field in the vicinity of the previously magnetized area. The field intensity is closely related to coercivity and can in principle be used to determine hardness or yield strength after calibration. For instance this technique has been used for the on-line monitoring of the degree of anneal of steel at the exit of continuous annealing lines[28].

One major problem with magnetic techniques is the lack of a unique relationship between magnetic and mechanical properties. This situation can be significantly improved by taking into account the simultaneous variation of several parameters such as chemical composition[30] and texture[31]. Composition is usually known before or during processing. Texture can be partially monitored with ultrasonic measurements. Hence it may be possible in the future to increase reliability by combining several techniques (eg magnetic and ultrasonic) to improve the overall reliability of such techniques.

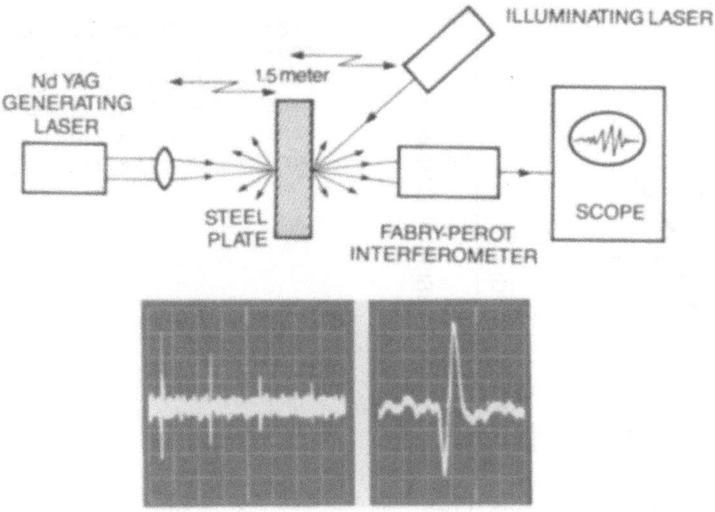

Fig. 5. Schematic of a laser ultrasonic system allowing remote ultrasonic measurements at distances of ≈ 1.5 m. Echoes at bottom left are for a 12 mm thick steel plate; bottom right is detail of one echo (from ref. 29). Such a system can be used in principle for the remote monitoring of microstructures during processing.

NONCONTACT ULTRASONICS

The recent development of noncontact ultrasonic transducers offers the possibility of in-situ monitoring of material properties during fabrication steps which would not normally be accessible with piezoelectric transducers. Two types are presently available or under development: 1) electromagnetic acoustic transducers (EMAT)[32] and 2) pulsed lasers for generation and optical interferometry for detection[33]. EMAT devices have been developed longer and are presently more widespread than laser based systems which mostly exist in research laboratories. Laser systems offer the possibility of more remote inspection ($\approx$1 meter compared to $\approx$1 mm with EMAT's) and could find useful applications in harsh environments. One system which has been developed in our laboratory allows detection of ultrasound with good sensitivity at distances of 1 m or greater. Figure 5 shows a schematic of the system including a Nd:YAG laser for generation and a Fabry-Perot interferometer for detection[34]. Also shown are ultrasonic signals observed for a 12 mm thick steel plate placed at 1.5 m from both generating and detecting systems. With such systems it appears possible in the future to monitor grain size, microstructure, degree of anneal in hot metals during processing. Of particular interest is the monitoring of weld pool dimension during welding and of microstructure in the weld and heat-affected zones.

CURE MONITORING IN COMPOSITES

The mechanical properties of fiber reinforced polymer composites depend on a number of process-related parameters. One of these is the details of the cure cycle based on a combination of time, temperature and pressure. Usually a standard procedure is used which does not take into account the molecular condition of the resin prior to the cure. In some cases the resin may be partially cured while being transported, stored or processed. The degree of cure of an area in a particular structure will also depend on its local environment in the autoclave, such as the local thickness or the presence of heat sinks. An important parameter is to

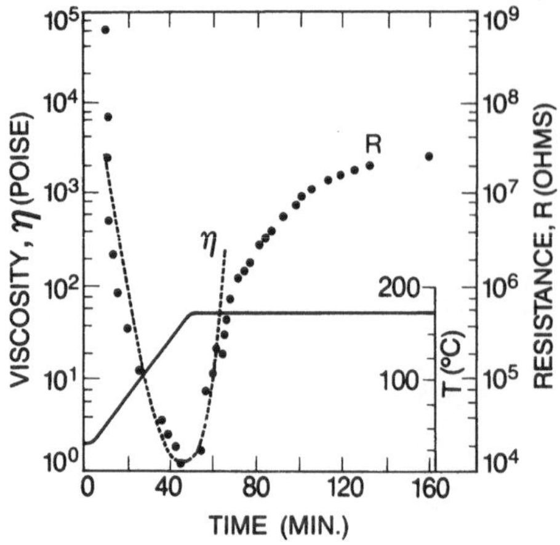

Fig. 6. Change in electrical resistance, R, and viscosity, η, during curing of an epoxy laminate. The close correlation between η and R can be used for cure monitoring (ref. 35).

apply pressure at the correct time, which should occur when the resin reaches the gelation or "gel" stage. At that point flow occurs under pressure and a void-free structure results. If pressure is applied too early, resin flow is excessive (low viscosity), and the composite may be resin starved. If pressure is applied too late, the resin will be too hard for pressure to suppress voids. Because of this, composite structures may be rejected due to improper cure and nondestructive techniques for cure monitoring and measurements of resin viscosity are of great importance.

Several promising techniques are presently under development based on dielectric and ultrasonic properties. Measurements of dc electrical resistivity, dielectric constant and loss tangent, tan δ, have been shown to correlate with viscosity during curing and to be a useful tool for determining the time for applying pressure[35,36]. Figure 6 shows how electrical resistivity closely follows viscosity during a cure cycle, where the temperature is initially increased linearly and then held constant at 175°C [35]. Deviations however do occur; these are particularly severe at high viscosity. Similar results are obtained by measuring the loss tangent or the phase angle. Recently a miniature dielectric cure monitor probe based on integrated circuit technology and which can be embedded in a composite has been developed at MIT[36] and is presently being commercially developed[37]. Work at Westinghouse has concentrated on full-scale applications in an autoclave where curing is monitored with measurements of the dissipation factor - indicating that process control is feasible, especially prior to gelation[38]. Problems remain, however, in accurately characterizing the cured state.

Ultrasonic wave velocity and attenuation have also been shown to be very sensitive to the degree of cure and appear to have potential for cure monitoring but have not yet been applied under industrial conditions[39,40].

CONCLUSION

The availability of material property sensors is an important key to the development of intelligent adaptive materials processing systems which will insure the production of high quality materials for cryogenic and other structures. The development of such sensors will depend on an improved understanding of the complex interactions between materials and the fields which can be used to nondestructively probe them. In addition, new transducer technologies to surmount the limited access restrictions or harsh environments often encountered on the production floor, must be developed. A limited number of examples of laboratory research and evolving industrial technology allowing the nondestructive measurement of mechanical properties and grain size in metals, and cure monitoring of fibre reinforced polymer composites were presented, indicating that great progress has already been achieved in this area and that considerable potential exists for future development.

ACKNOWLEDGEMENT

The author is grateful to his colleagues P. Cielo, J.-P. Monchalin, L. Piché and C. K. Jen for stimulating discussions and to M. Lord for technical help.

REFERENCES

1. "Nondestructive Methods for Material Property Determination," C. O. Ruud and R. E. Green, eds., Plenum, New York (1984).
2. N. M. Tallan, NDE, a key to enhanced productivity, in: "Progress in Quantitative Nondestructive Evaluation," vol. 1, Plenum (1982), pp. 1-5.
3. R. Mehrabian and H. Wadley, Needs for process control in advanced materials processing, J. Met. 37(2):51-58 (1985).
4. J. F. Bussière, Nondestructive materials characterization - an important tool in manufacturing, Eleventh World Conference on NDT, Las Vegas, Nov. 3-8, 1985, to be published.
5. F. E. stanke and G. S. Kino, A unified theory for elastic wave propagation in polycrystalline materials, J. Acoust. Soc. Am. 75(3): 665-681 (1984).
6. Except for 316SS, data are from ref. 7; for 316SS elastic constants and velocities in ref. 8 were used to calculate S.
7. E. P. Papadakis, Scattering in polycrystalline media, in: "Methods in Experimental Physics," vol. 19, Academic, New York (1981), p. 237.
8. H. M. Ledbetter, Monocrystal elastic constants in the ultrasonic study of welds, Ultrasonics 23:9-13 (1985).
9. J.-P. Monchalin and J. F. Bussière, Measurement of ultrasonic absorption by thermo-emissivity, ref. 1, pp. 298-298.
10. J.-P. Monchalin and J. F. Bussière, Infrared detection of ultrasonic absorption and application to the determination of absorption in steel, in: "Review of Progress in Quantitative Nondestructive Evaluation," vol. 4B, Plenum, New York (1985), pp. 965-973.
11. K. Goebbels, Structure analysis by scattered ultrasonic radiation, ch. 4 in: "Research Techniques in Nondestructive Testing," vol. II, R. S. Sharpe, ed., Academic, New York (1980). p. 87.
12. S. Serabian, Frequency and grain size dependency of ultrasonic attenuation in polycrystalline materials, Brit. J. NDT March:69 (1980).
13. H. Willems and K. Goebbels, Characterization of microstructure by backscattered ultrasonic waves, Met. Sci. 15:549-553 (1981).
14. E. P. Papadakis, The inverse problem in materials characterization through ultrasonic attenuation and velcoity measurements, in: Ref. 1, pp. 151-160.
15. N. Grayeli, F. Stanke, and J. C. Shyne, Prediction of grain size in copper using acoustic attenuation measurements, in: "1982 IEEE Ultrasonics Symposium," vol. 2, p 954.
16. A. Hecht, R. Thiel, E. Neumann, and E. Mundry, Nondestructive determination of grain size in austenitic sheet by ultrasonic backscattering, Mater. Eval. 39:934-938 (1981).
17. N. Grayeli and J. C. Shyne, Effect of microstructure and prior austenite grain size on acoustic velocity and attenuation in steel, in: "Review of Progress in Nondestructive Evaluation," vol. 4B, Plenum, New York (1985), pp. 927-937.
18. N. Grayeli and J. C. Shyne, Acoustic attenuation in two-phase materials, in: "Review of Progress in Quantitative Nondestructive Evaluation," vol. 3B, Plenum, New York (1984), p. 1107.

19. A. Randak, On-line inspection of semi-finished and finished steel products for surface defects and internal quality, in: "On-line Inspection of Steel Products," International Iron and Steel Institute, Committee on Technology, Brussels (1983), pp. 1-20.
20. On-line monitor to check grain size of alloy strip, Met. Prog. March:18 (1985).
21. H. J. Kim, C. K. Syn and J. W. Morris, Jr., Ferritic weldment of grain-refined ferritic steels for cryogenic service, in: "Advances in Cryogenic Engineering--Materials," vol. 30, Plenum, New York (1982), pp. 873-882.
22. R. Klinman, C. R. Webster, F. J. Marsh, and E. T. Stephenson, Ultrasonic prediction of grain size, strength and toughness in plain carbon steel, Mater. Eval. 39:26-32 (1980).
23. R. Klinman and E. T. Stephenson, Ultrasonic predition of grain size and mechanical properties of plain carbon steel, Mater. Eval. 39:1116-1120 (1981).
24. A. Vary, Ultrasonic measurement of material properties, ch. 5 in: "Research Techniques in Nondestructive Testing," vol. II, R. S. Sharpe, ed., Academic, New York (1980), pp. 159-204.
25. R. S. Smith, F. L. Rusbridge, W. N. Reynolds, and B. Hudson, Ultrasonic attenuation, macrostructure and ductile to brittle transition temperature in Fe-C alloys, Mater. Eval. 41(2):219 (1983).
26. F. Nadeau, J. F. Bussière and G. Van Drunen, On the relation between ultrasonic attenuation and fracture toughness in type 403 stainless steel, Mater. Eval. 43:101-104 (1985).
27. F. Forster and W. Stumm, Application of magnetic and electromagnetic nondestructive test methods for measuring physical and technological material values, Mater. Eval. 33(1):5-11, 14-15 (1975).
28. J. F. Bussière, On-line measurement of the microstructure and mechanical properties of steel, Mater. Eval. (in press).
29. G. Tomilov, Sov. J. NDT (translated from Defktoskopia) 4:70 (1966).
30. L. Devyatchenko and L. Valyaeva, Improving the accuracy of magnetic inspection of the mechanical properties of high-quality steel, Sov. J. NDT (translated from Defektoskopia) 2:52-57 (1983).
31. N. N. Timoshenko, G. A. Pyatunin and V. I. Slavov, Method of magnetic inspection of the mechanical properties of rolled sheet, Ind. Lab. 47(2):184-186 (1981).
32. R. B. Thompson, Noncontact transducers, in: "1977 Ultrasonics Symposium," IEEE cat. #77 CHI264-1SU, pp. 74-83.
33. C. Birnbaum and G. S. White, Laser techniques in NDE, in: "Research Techniques in Nondestructive Testing, vol. 7, R. S. Sharpe, ed., Academic, New York (1984), pp. 259-365.
34. J.-P. Monchalin, Optical detection of ultrasound at a distance using a Confocal Fabry-Perot Interferometer, Appl. Phys. Lett. 47:14 (1985).
35. A. T. Tajima, Monitoring cure viscosity of epoxy composite, Polym. Compos. 3(3):162-169 (1982).
36. N. Sheppard, S. Garverich, D. Day and S. Senturia, Micro-dielectrometry: a new method for in-situ cure monitoring, ibid, pp. 65-76.
37. Micromet Instruments, 21 Erie St., Cambridge, MA 02139.
38. J. Chottiner, Z. N. Sanjana, M. R. Kodami, K. W. Lengel, and G. B. Rosenblatt, Monitoring cure of autoclave-molded parts by dielectric analysis, in: "Proceedings of the 26th National SAMPE Symposium," April 28-30, 1981, pp. 77-88.
39. D. L. Hunston, Cure monitoring of thermosetting polymers by an ultrasonic technique, in "Progress in Quantitative Nondestructive Evaluation," vol. 2B, Plenum, New York (1983), pp. 1711-1729.
40. T. H. Hahn, Application of ultrasonic technique to cure characterization of epoxies, in: ref. 1, pp. 315-326.

CHARACTERIZATION OF TUNNEL BARRIERS BY FLICKER NOISE SPECTROSCOPY

C. T. Rogers and R. A. Buhrman

School of Applied and Engineering Physics
Cornell University
Ithaca, New York

ABSTRACT

This paper reviews recent results concerning the nature of excess low frequency noise in tunnel junctions. Through the study of small area junctions, the individual components of this noise have been directly observed; it has been found that the noise originates with the slow filling and emptying of trap states in the barrier. The rate limiting step in this process appears to be a reversible change in the configuration of the ions constituting the trap site. The density and strength of these trap states in different tunnel junctions is discussed.

INTRODUCTION

The study and minimization of the excess low frequency noise (1/f or flicker noise), in Josephson tunnel junctions is a matter of considerable practical concern since such noise can directly limit the low frequency sensitivity of high performance SQUID magnetometers and gradiometers.[1,2] On the fundamental side, the investigation of this noise in small area tunnel junctions is rapidly developing into a very effective means for addressing some of the fundamental questions concerning the origin and nature of 1/f noise in electronic systems. Moveover, since current studies are clearly demonstrating relationships between the noise magnitude and junction materials and quality, measurements of this phenomena show considerable promise for developing into a new and effective means of characterizing tunnel barriers.

The key to the realization of such possibilities has been the recent discovery that, by carefully studying the resistance fluctuations of small tunnel junctions at low temperatures, it is straightforward to resolve the individual noise sources that combine to yield 1/f noise in large devices.[2,3] It has been this identification and subsequent study of the individual building blocks of 1/f noise that has led to new insights into both the origin of tunnel junction 1/f noise and the nature of actual tunnel barriers[4]. It is the purpose of this paper to briefly review recent work in this area and to discuss some of its implications for tunnel junction development efforts.

Measurements of low frequency noise in tunnel junctions are straightforward. The tunnel junction is current biased at some d.c. level at fixed T(2K < T < 300K) and the voltage fluctuation across the junction are amplified by a low noise preamplifier using a four point probe. The amplified signal is then studied through both spectrum analysis, resulting in a voltage noise power spectral $S_V(f)$, and through the time record of the fluctuations. $S_V(f)$ is typically measured in the bandwidth 1Hz to 25kHz to an accuracy of ≈ 5%.

The type of tunnel junctions that[2] have been studied most extensively are small area (.05 μm^2 < A < 5 μm^2) $Nb\text{-}Nb_2O_5\text{-}PbBi$ edge junctions with barriers formed by reactive ion beam oxidation. Studies have also been made on NbN-PbBi edge junctions,[5] Nb-αSi-Nb[6] junctions and on very high quality $Nb\text{-}Nb_2O_5\text{-}PbInAu$[7] junctions where the barriers have been formed by reactive plasma oxidation.

For large area junctions and/or at higher temperature, the spectral measurements yield $S_V(f)$ which scales very accurately with V^2 (for junctions with linear I-V characteristics), proving that the origin of this noise is barrier resistance fluctuations; the noise amplitude also scales inversely with the junction area.[2] The spectral variation of S_V consistently trends about 1/f, with some systems exhibiting classic behavior over many decades of frequency. This S_V behavior is illustrated in Fig. 1 for a 6 μm x 7 μm Nb-αSi-Nb tunnel junction at T = 38K and V = 10 mV.

For small junctions, at low temperatures strong departures from 1/f behavior are observed. In the frequency domain the result is that S_V is clearly dominated by a few, 1-3, distinct Lorentzian spectra each of the form:

$$S_\ell \frac{\tau_{eff}}{1 + \omega^2 \tau_{eff}^2} \qquad (1)$$

with each Lorentzian characterized by its own S_ℓ and $1/\tau_{eff}$. In the time domain, the junction resistance switches randomly between well defined resistance levels: If only one strong Lorentzian is observed in S_V, then random switching of the resistance between two distinct levels is observed. This switching is characterized by the mean time τ_1 spent in the high resistance state, the mean time τ_2 spent in the low resistance state, and the resistance jump between the levels δR. If two

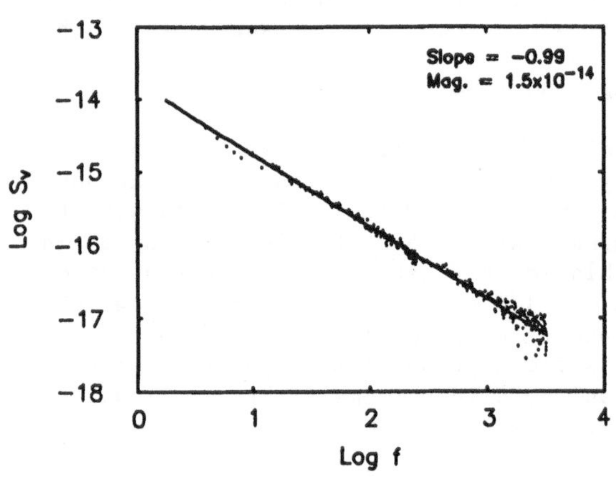

Fig. 1. Classic 1/f noise for a Nb-αSi-Nb tunnel junction of A = 6x7 μm^2 at T = 38K and V = 10 mV.

Lorentzians are present, then a second pair of discrete resistance levels appears allowing transitions to four distinct levels. Each additional Lorentzian in S_V accounts for another pair of discrete levels in the resistance fluctuations.

Machlup[8] has analyzed this type of "telegraph noise" and has shown that each two level resistance switching process produces a Lorentzian noise spectrum of the form eq. (1) with:

$$\frac{1}{\tau_{eff}} = \frac{1}{\tau_1} + \frac{1}{\tau_2}$$

$$S_\ell = (\delta R)^2 \frac{\tau_{eff}}{\tau_1 + \tau_2} \qquad (2)$$

By direct measurements of τ_1, τ_2 and δR from the resistance time record, we have verified that these equations accurately describe the Lorentzians visible in $S_V(f)$. Examples of this phenomena are shown in Fig. 2. Figure 2a is a digitized voltage time record for a tunnel junction biased at the point where two sets of discrete two level transitions dominate the junction noise. Fig. 2b shows the spectral power density for this junction with the same bias conditions. The two Lorentzian noise components are observed as the strong peaks in the quantity $f \cdot S_V(f)$ at 10 Hz and 10 kHz.

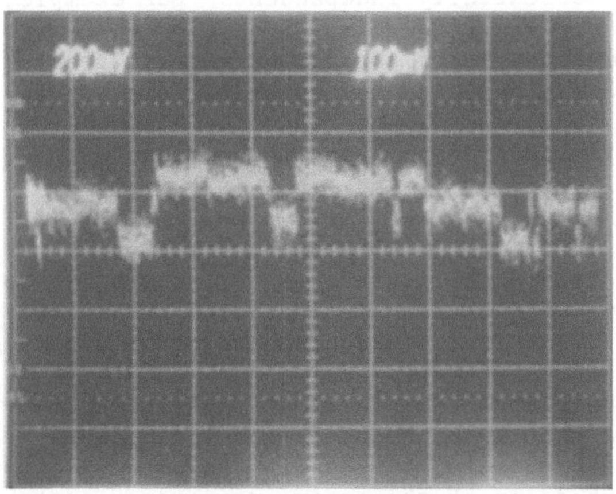

Fig. 2.a) Digitized time record of junction resistance showing two sets of discrete two level transitions. Total time length ≈ 205 msec.

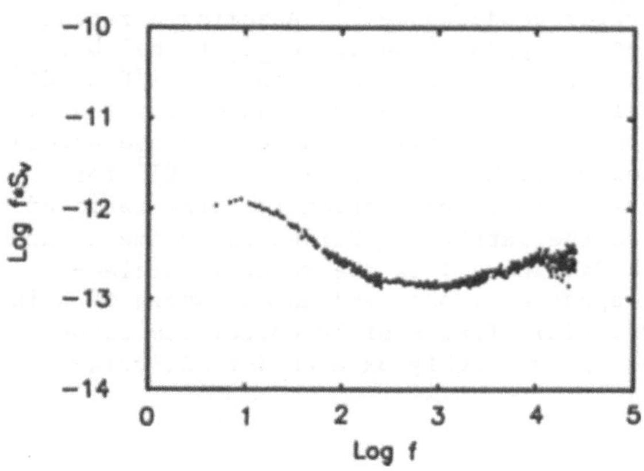

Fig. 2.b) $f \cdot S_V(f)$. The fast square spikes in the trace above yield the Lorentzian of $1/\tau_{eff} \approx$ 10 kHz while the slow switching gives the low frequency Lorentzian of $1/\tau_{eff} \approx$ 10 Hz.

The manner in which these distinct Lorentzian spectra superpose to yield 1/f noise is clearly observed by studying both large A and T. The number of two level states, or equivalently the number of Lorentzian noise oscillators, that are observed in the measurement bandwidth varies with bias voltage and temperature. But in general smaller junctions of a given type have a lower average number of distinct oscillators and greater variation in the amplitude of S_V than is observed in large junctions. As junction size is increased, the Lorentzian oscillators in S_V begin to overlap: Only the strongest can then be observed directly; the rest add together to form what appears to be a roughly 1/f background component. In such junctions the discrete steps in the time record are obscured, but S_V can still reveal the behavior of τ_1 and τ_2 for the largest Lorentzians.

When the temperature of the tunnel junctions is increased the number of Lorentzian oscillators in the measurement bandwidth eventually increases and they start to overlap. Two different types of behavior have been observed: In the first case, for Nb_2O_5 barriers the overall noise level increases significantly, apparently as more strong two level states become active. In the other case, αSi barriers, the noise spectrum varies smoothly over to a featureless 1/f behavior with no significant increase in overall amplitude; the high temperature active states in this system are characterized not only by an increased density but also by a broad range of δR roughly bounded from above by δR for the strongest low T Lorentzians.

Returning to the single Lorentzian behavior, it has been found that these distinct noise sources are essentially independent. For example, the effective lifetime of each oscillator in a given junction varies both with bias voltage and temperature, each at its own particular rate. Thus the evidence seems clear that the source of the excess low frequency noise in these tunnel junctions is the random combination of distinct, localized resistance fluctuations of tunnel junctions.

LOCALIZED ELECTRON TRAPS

The origin of the discrete resistance fluctuation is attributed to the slow trapping and slow escape of electrons at localized trap sites in the barrier. The effect of this charging and discharging of the trap is a local raising and lowering of the tunnel barrier for those electrons tunneling through the barrier in the vicinity of the trap site.[9] This picture of local barrier height fluctuations is consistent both with the scaling of the Lorentzian noise amplitude with inverse junction area and with the apparent independence of the active noise components in a given junction. This picture has been further supported by model calculations based on the work of Schmidlin concerning the effect of ionized impurities on tunnel barrier conductance.[10] Assuming a rectangular barrier and barrier parameters appropriate for Nb_2O_5 tunnel barrier, Schmidlin's work indicates a resistance fluctuation of $\delta R/R \approx 10^{-5}$ for the presence of a single ionized impurity in the middle of a tunnel barrier of 10^{-9} cm^2 cross sectional area. This is in quite close accord with typical observed fluctuation amplitude of $\delta R/R \approx 2.5 \times 10^{-5}$ for such junctions.[2] In this picture, $\delta R/R$ is proportional to the ratio of the electrostatic energy by which the barrier is raised to the mean barrier height E_B, i.e. $\delta R/R \propto 1/(\varepsilon E_B)$ where ε is the relative dielectric constant. This scaling is apparently observed in αSi where $\delta R/R$ is typically 10x that seen in Nb_2O_5. This factor of 10 correlates rather well with the reduction of ε and E_B by roughly 3x each for αSi relative to Nb_2O_5.

With the hypothesis above, we see that the observed discrete resistance jumps provide a direct measure of the electron population kinetics of single traps: $1/\tau_1$ and $1/\tau_2$ are the electron emission and capture rates of a single trap site. By carefully measuring the changes in the Lorentzian contributions to S_V over a wide range of voltage and temperature it has proven possible to gain considerable insight into the nature of these trap sites for various barrier materials. These measurements are made by determining τ_{eff} and S_ℓ for each oscillator as a function of V and T and using eqs. (2) to derive $1/\tau_1$ and $1/\tau_2$. For Nb_2O_5 barriers, which have been the most extensively studied, τ_{eff}^{-1} generally increases with T in a thermally activated manner above ≈ 15K. The noise power S_ℓ varies much more slowly with T, and as a result, tends to exhibit more clearly the effect of coulombic interactions between nearby trap sites. However, from such measurements and inspection of eqs. (2) it is clear that a thermally activated τ_{eff} and a slowly varying S_ℓ require that both the charge capture and charge emission are roughly thermally activated processes, or that

$$1/\tau_i = 1/\tau_{0i}\, e^{-E_{Bi}/kT} \qquad (3)$$

To illustrate this, Fig. 3 shows τ_{eff}^{-1} and S_ℓ for a Lorentzian oscillator that is free from significant interaction effects over a wide temperature range. The solid line through the data is a fit obtained using eqs. (2) and (3).

This fit yields five quantities: 1) The two activation energies for $1/\tau_1$ and $1/\tau_2$, 2) The two attempt rates, 3) The total squared resistance change δR^2. It is usually found that the two attempt rates are the same within a factor of three although they can differ by a factor of ten; they lie in the range $10^9 < 1\tau_0 < 10^{13}$ sec^{-1}. The two barrier

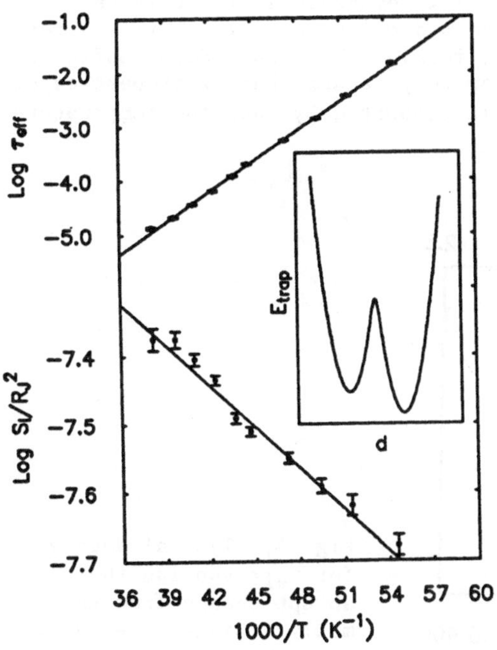

Fig. 3. Plot of log τ_{eff} and log S_l vs. 1/T showing thermally activated behavior. Solid lines are fits assuming eqs. (2) and (3) in the text. Inset shows a two well model which adequately describes the rate limiting kinetics.

heights usually differ by about 20% with no apparent correlation with the difference in $1/\tau_0$. E_B varies from roughly 10 meV at the lowest T to about 150 meV at 100K at which point single Lorentzians can no longer be resolved.

From data such as in Fig. 3 it is concluded that the population kinetics of the individual traps are governed by a two well potential energy picture shown schematically in the Fig. 3 inset. $1/\tau_1$ and $1/\tau_2$ then represent the transition rates between the wells. This double well picture is similar to the two level ionic reconfiguration model used to describe the low temperature properties of glasses.[11,12,13] In the present system, the energy barrier represents the rate limiting step in both the charge capture and emission processes which proceed via thermally activated transitions over the energy barrier separating the two nearly energy equivalent, but charge inequivalent wells.

Although the observed range of $1/\tau_0$ extends to rather low frequencies, the upper end of the range is characteristic of ionic vibrational frequencies and the relatively low energy scale of E_B is typical of that assumed to explain low T properties of bulk amorphous materials. It appears then that the charge trapping and escape process at high T involves reversible trapping of the ion or ions forming the trap between two nearly energy equivalent configurations. The charge emission and capture mainly then occur either during this transition or immediately before and after the transition.

The nature of this two well trap system is further clarified by examining the low temperature behavior of the Lorentzians. For $T < \approx 15K$ there are striking deviations from the thermally activated kinetics described above. Figure 4 is a plot of all the τ_{eff} at 10 mV bias for a typical device as a function of 1/T. Above ~ 15K, all the τ_{eff} show the thermal activation behavior described above. However, below this temperature, there is a rather abrupt change in behavior leading to a very weakly T dependent region at the lowest temperatures.

The proposed explanation of this low temperature effect is that the rate limiting step in electron capture and emission at low T is a direct configurational tunneling of the ions forming the trap. Such configurational tunneling also explains the low temperature thermal properties of amorphous systems. This conclusion is supported by both the temperature

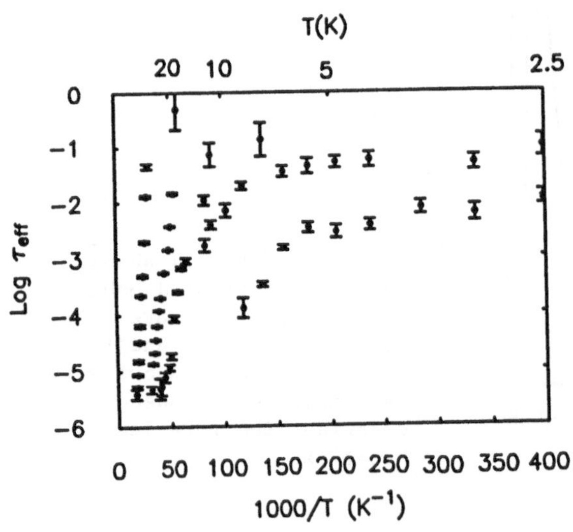

Fig. 4. Typical data set for τ_{eff} showing the abrupt change from thermally activated behavior above to non-activated behavior below T ≈ 15K.

and voltage dependence of τ_{eff} at low T. Figure 5 shows Log τ_{eff} vs. T at four different bias voltages for one particular defect state along with a thermally activated state at higher temperatures shown for comparison. Each low T trace shows the transition from activated to non-activated behavior. Below ~15K, $1/\tau_{eff}$ has an exponential voltage dependence and is nearly independent of T, while at high T the voltage effect is quite weak. The temperature independent exponential voltage dependence as well as the weak temperature dependence at fixed V is common in tunneling systems. Thus the data clearly point to configuration tunneling as the rate limiting step in the trap population kinetics at low temperatures.

Further support for this idea is found by making comparisons between the low temperature transition rates and the rates calculated from a simple WKB ionic tunneling model. Using barrier heights and attempt frequencies determined in the high temperature, thermally activated regime, it has been found that the low temperature transitions are quantitatively consistent with the tunneling of an ion through the barrier for a distance of a few angstroms. The solid lines through the data in Fig. 5 are the result of a non-linear fit of such a model.[4]

Thus, while much more work needs to be done, the ability to directly observe individual trap states in tunnel barriers is clearly providing new insights into their microscopic nature. The results summarized here clearly indicate that the rate limiting factor in the trapping process is a reversible displacement or reconfiguration of the ions forming the trap. This process proceeds by thermal activation at high T and by ionic tunneling at low T. While data described here was obtained from Nb_2O_5 barriers, qualitatively similar results have recently been found in αSi tunnel barriers,[6] this suggests that the above is a fairly general phenomena.

EFFECTS OF TUNNEL BARRIER QUALITY AND TYPE ON NOISE

If the low frequency individual noise sources in junctions are slow trap states in the barrier, then the noise magnitude should reflect the density of such states in a given junction. It might then be reasonable to expect a correlation between noise magnitude and other measurements of junction quality. For Josephson junctions the most common method of characterizing the junction is through some measurement of the non-ideal leakage current below the gap voltage. While there are several possible causes of leakage currents it certainly can depend on the purity and quality of the barrier.

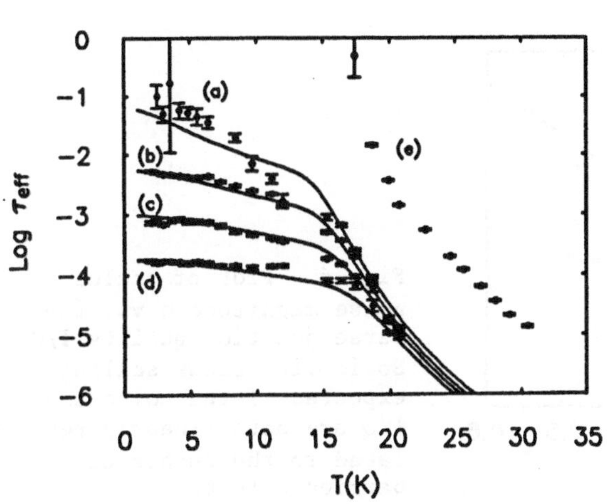

Fig. 5. Plot of τ_{eff} vs. T for four different voltages: a) 8 meV, b) 15 meV, c) 20 meV, d) 25 meV. Solid lines are fits which explain the deviation from thermal activation as due to a parallel tunneling process. e) Shows thermally activated data for comparison.

An apparent correlation between low frequency noise and subgap leakage has in fact been observed.[2] This is illustrated in Fig. 6. The data shown there is obtained by approximating the measured $S_V(f) = I^2 R_j^2 \eta/Af$. Thus the individual Lorentzian structure has been averaged and a characteristic mean scaled amplitude η has been obtained for each junction. This averaging is a reasonable approximation provided the density of strong Lorentzians is of the order of 1 per decade of frequency, or greater. Figure 6 shows a plot of this quantity η vs. $1/Q$ where Q is the junction quality as measured by the ratio $R_j(2\ mV)/R_j(15\ mV)$. Junctions with low subgap leakage currents clearly have less noise than junctions with high leakage. We can make a tentative connection between the measured conductance fluctuations and the number of active defects in the tunnel barriers: Halbritter[14] has shown that the subgap leakage current due to resonant tunneling through localized states in the barrier increases linearly with the number of states when such states are assumed to be independent. Thus, we expect the subgap conductance to have a term for the ideal barrier case and a second term proportional to the number of defect states N_d. At 4.2K, the ideal subgap conductance is small so that we approximately have $R_j(2\ mV) \propto 1/N_d$. For $Q \gg 1$, the supergap conductance is nearly independent of the resonant tunneling path so that $1/Q$ gives a nearly direct measurement of the number of defect states. If the noise should depend linearly on the number of defects, we would expect that $\eta \propto 1/Q$. The solid line in Fig. 6 shows that this relation seems to hold fairly well. Obviously, other mechanisms which cause subgap leakage such as the presence of normal metal shorts will modify the argument. Moreover, it is obvious that the number of active trap states in our bandwidth is only a very small fraction of those states responsible for the leakage current.

Once the leakage current of a tunnel junction is sufficiently high to reduce Q to order unity, the $Q \propto 1/N_d$ scaling should break down: A continued increase in N_d will have no effect on Q since resonant tunneling will then dominate at all voltages. However, since pair tunneling cannot proceed by resonant tunneling, the defect number can be estimated from the reduced I_cR_j product of the junction. While we have not yet examined this strongly self shunted regime, Fig. 6 suggests that a junction with a greatly reduced I_cR_j would have a very high 1/f noise level.

When the junction quality improves to the point where the number of noise sources decreases below 1 per frequency decade, characterization of the noise level by an average value η becomes inappropriate.

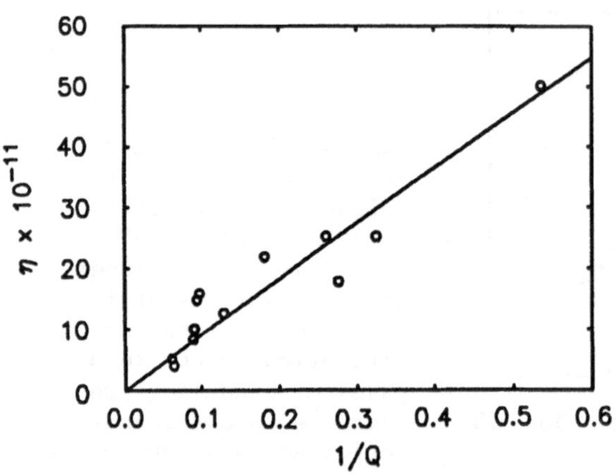

Fig. 6. Plot of scaled noise magnitude η vs. inverse junction quality $1/Q$ Solid line shows scaling expected if the noise and $1/Q$ are both linearly related to the number of barrier defects.

Then one must discuss the average behavior in terms of average Lorentzian power and the typical spacing between Lorentzians. This situation has been reached most clearly in some recent measurements on high quality $Nb-Nb_2O_5$-PbInAu edge junctions.[7] There in 25 μm wide x 0.3 μm high junctions the observed density of strong Lorentzian noise oscillators was no greater than 1 per 3 decades of frequency. The amplitude of these discrete Lorentzians, appropriately scaled for the junction area, was quite similar to that found in less high quality Nb_2O_5 junctions, with $S_\ell/R_j^2 \approx 10^{-10}$ for a 1 μm^2 tunnel junction. This indicates a strong similarity in the behavior of individual traps in all Nb_2O_5 junctions. The key difference between higher and lower quality junctions was simply that the trap density was much lower in the best junctions.

Apart from these scattered Lorentzians, whose location varied of course with V and T in the typical manner, the remaining low frequency noise in these high quality junctions was +5 dB or more below the peak Lorentzian value. Assuming that the number of noise oscillators scales with the junction area, smaller edge junctions of this quality would be expected to have essentially no Lorentzian noise oscillators visible over wide ranges of frequency. This is in fact the likely explanation of the greatly reduced or non-existent 1/f noise levels found in dc SQUIDs fabricated with such high quality junctions.[15]

Finally, we note that no consistent difference was found between αSi barriers and barriers consisting of a layer of hydrogenated αSi sandwiched between two αSi buffer layers. This is not necessarily surprising since the effect of the hydrogenated layer was to reduce the subgap leakage by about a factor of two to three.[16,17] While this is a very significant effect with regard to I-V quality, more extensive noise measurements are needed before the associated reduction in average trap state density by a factor of three can be clearly resolved.

Barrier Uniformity, Shape, and Noise

The magnitude of δR associated with a particular trap depends on the shape and height of the tunnel barrier only at or near the trap site. Thus, for non-uniform barriers, we expect greatly increased resistance fluctuations for those traps located near high transmission regions of the barrier as compared to those located in uniform regions. Thus, the resistance fluctuations show promise as local probes of barrier uniformity.

This point is illustrated most clearly by a noise measurement on early effort NbN-NbN edge junctions. After initial barrier oxidation, a discontinuous Au film was deposited, the edge was reoxidized to completion and the NbN counter electrode was deposited. Although these devices showed gap structure, they were very "leaky". Low frequency noise measurements on this junction revealed a few two level states with a general behavior consistent with that described previously but with an amplitude two orders of magnitude larger. Thus in a junction of a nominal cross sectional area of 10^{-9} cm^2 discrete resistance fluctuations as large as 1% were seen with these very large fluctuations apparently due due to individual trap states. It seems safe to conclude that the effective area of this device was much smaller than 10^{-9} cm^2.

A similar effect is expected for barriers which deviate strongly from a rectangular shape, e.g. Schottky barriers. In this case, traps strategically located under high regions of the barrier could have a much stronger effect on barrier transmission than would be expected for a rectangular barrier with the same total resistance. Since αSi barriers are most likely back to back Schottky barriers, it seems likely that the observed range in δR described above is caused by this effect.

SUMMARY

The ability to directly observe the individual Lorentzian components of the excess low frequency noise of Josephson tunnel junctions has resulted in an significant advance in the understanding of the origin of this noise and of the nature of the tunnel barriers. Detailed studies of these noise components have shown that they are due to slow trap states in the tunnel barrier. The data indicate that the rate limiting step in trapping and escape processes at these localized trap states is a reversible motion or reconfiguration of the ion or ions constituting the trap state. At low temperature this reconfiguration proceeds by ionic tunneling. Measurements of the density and amplitude of the individual noise components shows promise as developing into a new means of characterizing tunnel barriers.

ACKNOWLEDGMENTS

We wish to acknowledge the essential contributions of H. Kroger, L. N. Smith, W. J. Gallagher and S. I. Raider to various aspects of the work summarized here. This research was supported in part by the Office of Naval Resarch.

REFERENCES

1. C.T. Rogers and R.A. Buhrman, IEEE Trans. on Magn. MAG-19:453 (1983).
2. C.T. Rogers and R.A. Buhrman, IEEE Trans. on Magn. MAG-21:126 (1985).
3. C.T. Rogers and R.A. Buhrman, Phys. Rev. Lett. 53:1272 (1984).
4. C.T. Rogers and R.A. Buhrman, Phys. Rev. Lett. 55 (1985) in press.
5. H. Hallen, C.T. Rogers, and R.A. Buhrman, unpublished.
6. C.T. Rogers, R.A. Buhrman, H. Kroger, L. Smith, to be published.
7. C.T. Rogers, R.A. Buhrman, W.J. Gallagher, and S.I. Raider, to be published.
8. S. Machlup, J. Appl. Phys. 25:341 (1954).
9. M. Celasco, A. Masoero, P. Mazzetti, and A. Stepanescu, Phys. Rev. B. 17:2553 (1978).
10. F.W. Schmidlin, J. Appl. Phys. 37:2823 (1966).
11. P.W. Anderson, B.R. Halperin, C.M. Varna, Phil. Mag. 25:1 (1972).
12. W.A. Phillips, Phil. Mag. 34:983 (1976).
13. R.O. Pohl, in: "Phonon Scattering in Solids", L.J. Callis et.al., eds., Plenum Press, N.Y. (1976). p. 107.
14. J. Hallbritter, Surf. Sci. 122:80 (1982).
15. C.D. Tesche, IEEE Trans. on Magn. MAG-21:1032 (1985).
16. H. Kroger et.al., IEEE Trans. on Magn. MAG-19:783 (1983).
17. H. Kroger et.al., IEEE Trans. on Magn. MAG-21:870 (1985).

ROLE OF LOCALIZED STATES IN AMORPHOUS SILICON TUNNEL BARRIERS

S. Bending, R. Brynsvold, and M. R. Beasley

Department of Applied Physics
Stanford University
Stanford, California

ABSTRACT

Results of an experimental study of the role of localized states in amorphous silicon tunneling barriers are reported. Composite barriers formed of multilayers of SiOx and a-Si, the latter containing a known density of localized states, have been developed in which the dominant tunneling mechanism can be made to be either direct or resonant tunneling. The effect of these localized states on the behavior of the conductance with barrier thickness, and on the superconducting tunneling current-voltage characteristics, is described.

INTRODUCTION

The recent progress in superconducting electronic materials has been remarkable. Prototype all-high-T_c Josephson junctions have been demonstrated,[1,2] and even simple circuits fabricated in some cases.[3,4] All-refractory integrated circuit processes are increasingly the norm for systems work. As part of these developments has come the appreciation that so-called artificial tunnel barriers are quite practical and even superior to native oxide barriers in many cases.[5] At the same time, it has become apparent that our understanding of the physics and materials science of tunnel barriers is primitive compared with our understanding of the superconducting electrodes. Among the various outstanding questions that have been raised, one of the most persistent is the role of localized states in the properties of tunnel barriers. Localized states have been implicated in a wide variety of the nonidealities commonly seen in superconductive junctions, for example conduction below the energy gap and as the origin of 1/f noise. In particular, Halbritter and coworkers have catalogued the many ways that localized states might affect tunneling behavior.[6,7]

Clear-cut experimental evidence has been hard to come by, however. Unfortunately, the best one can usually do is to conjecture that one or another nonideality is due to localized states, based on the available theoretical calculations. In a few cases much more specific evidence has been found; for example, the recent work of the Cornell group reported at this meeting.[8,9] Nevertheless, in general clear-cut experimental tests have been hard to come by. The problem has been the lack of a good experimental system for study in which the existence, nature and number of localized states are known a priori and can be systematically varied. As we have recently reported, multilayered SiOx/amorphous silicon (a-Si) tunnel barriers appear to have many of the qualities needed for a good test system.[10,11] Amorphous silicon is readily deposited and is known to have a relatively large density of localized states. Amorphous silicon has the added interest that it is also the basis of several artificial barriers of considerable practical importance.[12,13] Finally, it is possible that from the physics point of view a-Si can serve as an archtype for the various other amorphous barrier materials that have been demonstrated to have practical utility.[2,14] In this paper we summarize the initial results of our study of the role of localized states on the properties of tunnel barriers and try to relate our emerging understanding to the behavior of practical superconducting tunnel junctions.

Effects of Localized States

Both to motivate the particular tunnel barrier structures we have studied, and to provide a basis for the interpretation of our results, it is helpful first to review qualitatively the existing understanding of the role of localized states in tunnel barriers and the tunneling process itself.[6,15-18] Figure 1 illustrates schematically a tunnel barrier containing two representative localized states. The effect of a particular localized state is a strong function of its position in the barrier. Those near the "electric center" of the barrier play the dominant role in actual transport through the barrier since they are in the resonant condition.[6,15] (By electrical center we mean that location at which the tunneling probabilities to the base and counter-electrode are the same.) Those near the edges, on the other hand, hybridize more strongly with the electrodes, and therefore play a larger role in any changes in the electronic properties of the electron states in the metal that may arise.[6]

Also shown in Fig. 1 are the various transport processes that are believed to take place. These are direct tunneling, resonant tunneling, and hopping. Both resonant tunneling and hopping occur via a localized state. They differ, however, in that in the case of resonant tunneling the process is quantum mechanically coherent (and therefore elastic) all the way across the barrier. In the case of hopping, transport occurs via two quantum mechanically incoherent tunneling processes in which phase coherence across the barrier is not maintained, i.e., it involves an inelastic process in one of the hops to or from the localized state. Also, the occupancy (charging) of a localized state, whether the electron goes back to the original electrode or all the way across the barrier, will

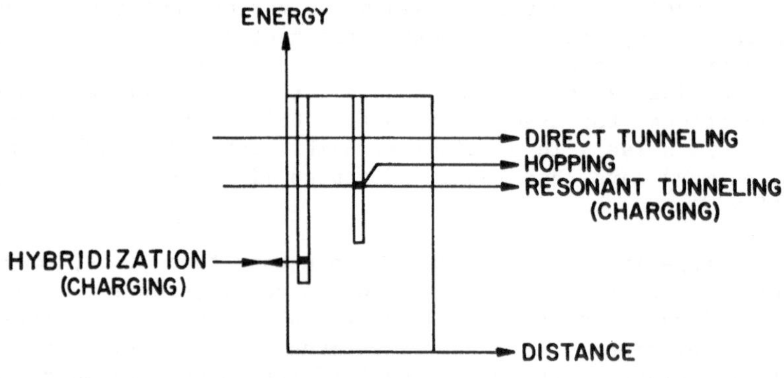

Fig. 1. Schematic tunnel barrier showing two representative localized states and the various possible tunneling processes.

also affect the direct tunneling process in the immediate vicinity of the localized state by virtue of changes in the local barrier potential. The net effect is conductance fluctuations, and it is just these conductance fluctuations that are believed to be the origin of 1/f noise in tunnel junctions.

Setting aside hopping for the moment, as discussed by various authors the total conductance of a tunnel junction is composed of two channels, the direct channel, for which the conductance $G_D \propto \exp(-2ad)$, and the resonant channel for which the conductance $G_R \propto \exp(-ad)$, where a^{-1} is the decay length of the electronic wave functions in the barrier, and d is the thickness of the barrier.[6,15] G_R is also proportional to the density of localized states near the electrical center of the junction. It is important to note that these two channels depend differently on the thickness of the barrier. The exponential decrease of conductance with thickness is slower for the resonant channel than for the direct (1/a versus 1/2a). Hence resonant tunneling should dominate in a sufficiently thick barrier, the exact crossover point depending on the density of localized states. It is this difference in the decay length of these two channels that has permitted us to distinguish between them in our a-Si tunnel barriers. Moreover, having distinguished between them, it is permitting us to go on and investigate the dependence of the tunnel junction properties on whether the conduction process is dominated by direct or resonant tunneling. These results are described below.

EXPERIMENTAL RESULTS

The two barrier configurations we have been studying are showm schematically in Fig. 2.[11] The first is an asymmetric bilayer structure of a-Si and SiOx. The second is a symmetric trilayer structure consisting of an a-Si layer sandwiched between two SiOx layers. The bilayers were formed by depositing successively thicker layers of a-Si that were then oxidized. The thicker the original a-Si layer, the thicker the resultant a-Si layer in the final structure. For the thinnest a-Si depositions the barrier is completely oxidized. Only as the thickness of the original deposited a-Si layer is increased does an unoxidized a-Si layer exist. The trilayers were formed by first depositing an a-Si layer of fixed thickness (~20Å) that was subsequently oxidized and then followed by a second a-Si deposition of successively increasing thicknesses that was also oxidized, resulting in a trilayer structure with successively thicker a-Si layers between the two SiOx layers. Since resonant tunneling requires equal tunneling probabilities to both electrodes, clearly the trilayer structure is more favorable for resonant tunneling than the bilayer, at least for thin bilayers.

Figure 2 shows how the conductances of these two types of barriers depends on the thickness of the barrier. Note that the x-axis in this figure is the original thickness of the a-Si layer before oxidation in the case of the bilayers, and the original thickness of the second a-Si deposition in the case of the trilayers. Thus in both cases the actual thickness of the a-Si in the final structure must be corrected for the thickness of the oxide. Incremental thickness changes are accurately reflected, however. Consider the case of the bilayer with a Nb base electrode. The conductance is seen first to decrease rapidly and then more slowly for thickness greater than about 25Å. The initial rapid fall off in G reflects the increasing oxide thickness in the thinner barriers, which, as we have noted, are oxidized all the way through. Beyond 25Å a layer of unoxidized a-Si begins to build up and the fall off of the conductance is slower, reflecting the longer tunneling length of the unoxidized a-Si. Similar behavior is seen for the trilayer structure with the important difference that the delay length through the unoxidized

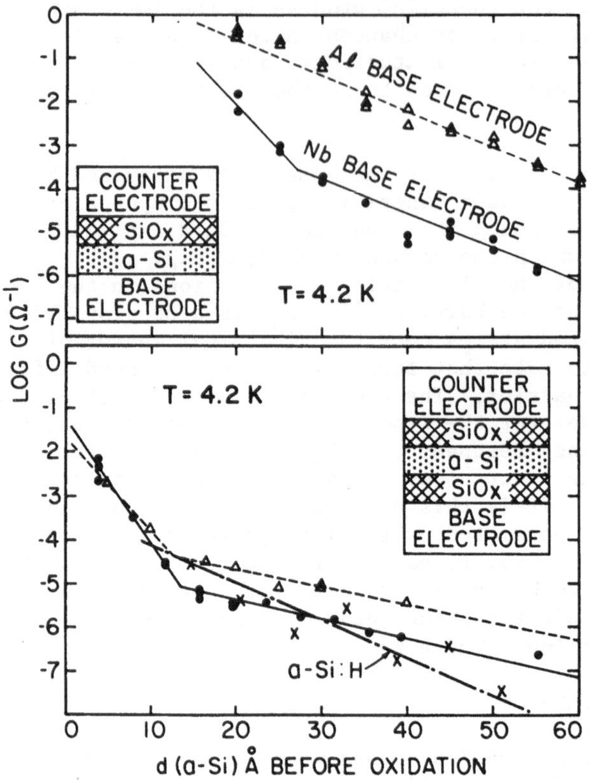

Fig. 2. Tunneling conductance as a function of the a-Si layer thickness before oxidation (see text) for the bilayer (top), trilayer (bottom), and hydrogenated trilayer (bottom) barriers at 4.2 K.

a-Si is larger by a factor of 2, suggesting that in this case the tunneling is resonant, consistent with its electrically more symmetric structure. This interpretation is further supported by the observation that when the a-Si layer in the trilayer is hydrogenated, so as to reduce the density of localized states, the decay length reverts to the value seen in the bilayers. Thus almost certainly the conduction process is by direct tunneling in the bilayers, and by resonant tunneling in the unhydrogenated trilayer structures. Note further that this behavior is independent of whether the base electrode was Nb or Al. The tunneling length a^{-1} of the a-Si inferred from these data is 11Å. Other interesting aspects of the physics of these tunneling barriers, including direct observation of the individual resonant tunneling channels, have been discussed elsewhere.[11,19]

Having established the tunneling mechanism in these various structures, we can now inquire as to the differences in their behavior as superconducting tunnel junctions. The I-V curves of a sequence of bilayers, trilayers, and hydrogenated trilayers, as a function of increasing barrier thickness, are shown in Fig. 3. Recall that the thicknesses quoted are for the original thickness of the a-Si layer. Physically the curves are more appropriately distinguished by the characteristic regimes revealed in Fig. 2. To do this, we have indicated those barriers in the initial oxide barrier regime by a solid line, those in the transition region in which some a-Si first appears by a dashed line, and those for which an a-Si layer was present by dotted and dash-dotted lines.

As seen from the figure, each of the three types of junctions behaves differently as a function of the barrier thickness. The general picture that emerges is as follows. In each case for a fully oxidized barrier good tunneling I-V curves are obtained. (Data not shown for the hydrogenated trilayer.) In fact, the normalized I-V curves are essentially identical in this case for each type of junction, as one would expect. However, as the unoxidized a-Si layer starts to form, differences in the I-V curves begin to appear. For the bilayer the initial change is small, possibly a slight reduction in the gap structure. For the trilayer the I-V begins to deteriorate seriously as soon as some unoxidized a-Si is present, and the deterioration progressively worsens as the thickness of this layer increases. For the thicker barriers, the I-V curves

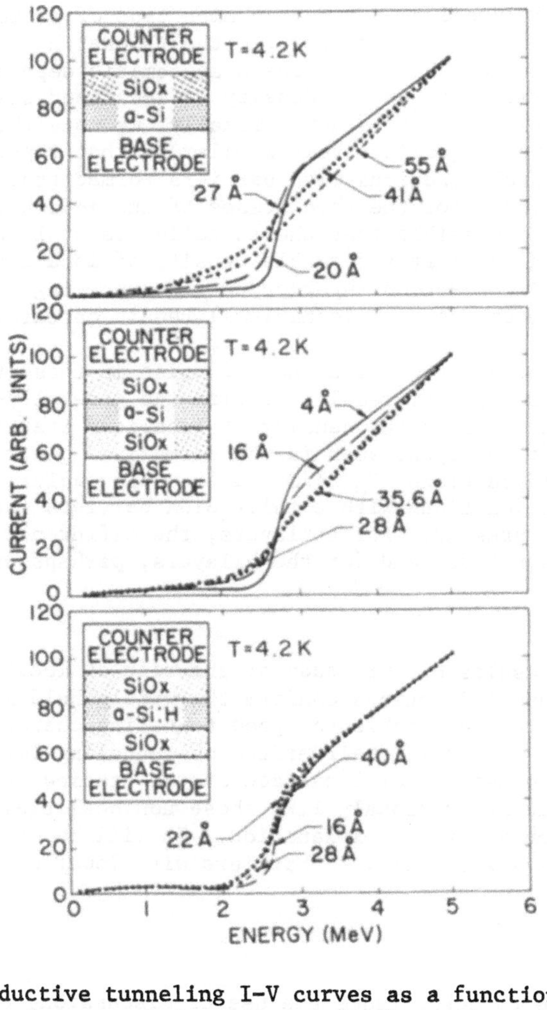

Fig. 3. Superconductive tunneling I-V curves as a function of barrier thickness (see text) for bilayer (top), trilayer (middle), and hydrogenated trilayer (bottom) barriers.

are very similar to those seen for the bilayers. By contrast, for the hydrogenated trilayers, there is no tendency of the I-V curve to wash out and at worst there is some increase in the conduction below the gap as the a-Si:H layer thickness increases.

Clearly the existence of a large number of localized states in the barrier has a substantial detrimental effect on the tunneling I-V curve. Only the hydrogenated trilayer has reasonable superconducting tunneling I-V's to very large barrier thicknesses, and even here there is some deterioration at the larger thicknesses. Beyond this observation, the meaning of these data is not so clear. In both the bilayers and the trilayers the ultimate effect of a thick a-Si region is to wash out the I-V curve. Given the conclusions obtained from Fig. 2, however, there appears to be no correlation with whether the dominant tunneling mechanism is direct or resonant tunneling. Similarly, there appears to be no correlation with the degree of hybridization with the electrodes, which is much smaller in the trilayer. While no definite evidence is available, one wonders if two mechanisms of deterioration may not be acting which only coincidently lead to the same end result. If not, some new idea seems to be required to provide a universal explanation for the deterioration. We also note that the nature of the deterioration--a washing out of the gap structure--is not one of the more common nonidealities seen in conventional tunnel junctions. A more common type of nonidentity is conducting below the gap, like that seen for the thicker hydrogenated trilayer barriers. Whether this indicates a systematic dependence of the nature of the deterioration on the density of localized states is not yet clear and will require further study. Finally, we note that this washing out of the I-V curve is qualitatively similar to that reported by Brooks and Meservey in tunnel junctions with barriers formed from a single layer of a-Si.[19] (Note that, for the thicknesses of the a-Si barriers reported in this paper, it is possible that the tunneling is still direct, not resonant.) We conclude that a sizeable density of localized states is clearly detrimental for a good superconducting tunneling I-V curve, but that the specific mechanism (or mechanisms) remain obscure.

We have also begun to measure the 1/f noise in these tunneling structures. Our preliminary results confirm the generally-held belief that 1/f noise increases as the density of localized states increases. Very much lower 1/f noise was seen in the normalized power spectrum of the conductance fluctuations $S_G(f)/G^2$ at low bias voltages (approximately 10 meV) for junctions with totally SiOx barriers than for those with an a-Si layer present. For trilayers, the difference was at least three orders of magnitude, and for the bilayers, perhaps over six.

CONCLUSIONS

These early results of our study of the role of localized states of superconductive tunnel junctions confirm the widely held suspicion that localized states are undesirable for good tunneling characteristics and for low 1/f noise. However, their effect on tunneling I-V curves, at least at the high densities of localized states studied here ($\sim 10^{18}$/ev cm^3), is not simple nor obviously like those nonidealities for which such states are a commonly invoked explanation. It will be of interest to extend this work systematically to barriers with lower concentrations of localized states.

ACKNOWLEDGEMENTS

We would like to acknowledge the useful discussions of the role of localized states in tunneling with J. Halbritter, H. Kroger, and P.A. Lee. This work is supported by the U.S. Office of Naval Research.

REFERENCES

1. A. de Lozanne, M. S. DiIorio, and M. R. Beasley, Fabrication and Josephson Behavior of High-T_c Superconductor-Normal-Superconductor Microbridges, Appl. Phys. Lett. 42:541-543 (1983).

2. A. K. Shoji, M. Aoyagi, S. Kosaka, F. Shinoki, and H. Hayakawa, Niobium Nitride Josephson Tunnel Junctions with Magnesium Oxide Barriers, Appl. Phys. Lett. 46:1098-1100 (1985).

3. M. S. DiIorio and M. R. Beasley, High-T_c Superconducting Integrated Circuit: A dc SQUID with Input Coil, IEEE Trans MAG-21:532-535 (1985).

4. H. Rogalla, B. David, M. Mück, and Y. Kato, Study of Preparation Techniques for a Practical Microbridge dc-SQUID Structure Fabricated from Nb_3Ge, IEEE Trans. MAG-21:536-538 (1985).

5. For a discussion of these various barriers and a listing of the original references, see M. R. Beasley, in "Percolation Localization and Superconducting," A. M. Goldman and S. Wolf, eds., NATO ASI Series, Series B: Physics, Vol. 109, Plenum (1984), p. 133.

6. J. Halbritter, On Resonant Tunneling, Surf. Sci. 122:80-98 (1982).

7. V. Kresin and J. Halbritter, Superconducting Junction with Resonant Tunneling, IEEE Trans. MAG-21:862-865 (1985).

8. C.T. Rogers and R. A. Buhrman, Discrete Lorentzian Structure in Low Frequency Voltage Noise Spectra of Very Small Area Josephson Tunnel Junctions, IEEE Trans. MAG-21:126-129 (1985).

9. C. T. Rogers and R. A. Buhrman, Composition of 1/f Noise in Metal-Insulator-Metal Tunnel Junctions, Phys. Rev. Lett. 53:1272-1275 (1984).

10. M. R. Beasley, S. Bending, and J. Graybeal, in "Localization, Interaction and Transport Phenomena," B. Kramer, G. Bergmann, and Y. Bruynseraeda, eds., Solid State Sciences Series No. 61, Springer-Verlag (1984), p. 138.

11. S. Bending and M. R. Beasley, Transport Processes via Localized States in Thin a-Si Tunnel Barriers, Phys. Rev. Lett. 55:324-327 (1985).

12. H. Kroger, C. N. Potter, and D. W. Jillie, Niobium Josephson Junctions with Doped Amorphous Silicon Barriers, IEEE Trans. MAG-15:488-489 (1979).

13. D. A. Rudman and M. R. Beasley, Oxidized Amorphous-Silicon Superconducting Tunnel Junction Barriers, Appl. Phys. Lett. 36:1010-1013 (1980).

14. H. Asano. K. Tanabe, O. Michikami, M. Igarashi, and M. R. Beasley, Fluoride Barriers in Nb/Pb Josephson Junctions, Jap. J. Appl. Phys. 24:289-292 (1985).

15. B. Ricco, M. Ya. Azbel, and M. H. Brodsky, Novel Mechanism for Tunneling and Breakdown of Thin SiO_2 Films, Phys. Rev. Lett. 51:1795-1798 (1983).

16. M. Ya. Azbel, A. Hartstein, and D. P. DiVincenzo, T Dependence of the Conductance in Quasi One-Dimensional Systems, Phys. Rev. Lett. 52:1641-1644 (1984).

17. A.P. Stone and P. A. Lee, Private Communication.

18. Y. Gefen and Gerd Schön, Effect of Inelastic Processes on Localization in One Dimension, Phys. Rev. B. 30:7323-7327 (1984).

19. R. Meservey, P. M. Tedrow, and J. S. Brooks, Tunneling Characteristics of Amorphous Si Barriers, J. Appl. Phys. 53:1563-1570 (1982).

ALL REFRACTORY JOSEPHSON INTEGRATION PROCESS WITH NbN JUNCTIONS

Shin Kosaka

Electrotechnical Laboratory
Umezono, Sakura-mura, Niihari-gun
Ibaraki, Japan

ABSTRACT

An integration process for the fabrication of an all refractory Josephson circuit using niobium nitride / oxide / niobium nitride junctions is described. Materials and processing technologies have been selected to meet the requirement for integrating Josephson circuits with LSI level complexity. In order to overcome the problems peculiar to the use of refractory materials for Josephson LSI, highly anisotropic and selective reactive ion etching processes and planarization technologies have been developed. Details of the key fabrication steps in the integration process developed for improving the chip yield are also described.

INTRODUCTION

In order to get stable and high quality Josephson tunnel junctions applicable to the high performance digital and analog circuits, the research on tunnel junctions made of refractory electrode materials have received much attention. Recently, several innovations have been done in fabricating Josephson tunnel junctions with refractory electrode materials and significant progress has been made in obtaining good tunneling characteristics.[1-4] Owing to these innovations, qualities of the all refractory junctions have reached the level for the practical use in several application fields. Inherent mechanical stability, accurate fine pattern manufacturability, and uniformity of the characteristics are the substantial advantages of the all refractory junctions for making integrated circuits. Several papers on fabricating logic[5-7] and memory[8,9] circuits using all refractory junctions have been reported so far.

In this paper, the development of the all refractory Josephson integration process using NbN junctions is described. In particular, attention is focused on the development of reliable and reproducible processings for integrating Josephson circuits with LSI complexity.

PROCESS CONSIDERATIONS

Various combinations of refractory electrode materials (Nb, NbN, etc.) and tunneling barrier materials (oxide, Al, MgO, Si, etc.) have been investigated.[10] We have chosen an NbN/oxide/NbN junction structure for the attempt to fabricate the all refractory Josephson integrated circuits with LSI level complexity. Since NbN is less chemically reactive than Nb, NbN junctions are kept stable during complicated integration process. This chemical stability of NbN junctions makes it possible to increase the process temperature up to 310 °C,[3] allowing us to employ various fabrication technologies developed in the field of semiconductor electronics. The freedom to introduce various elaborated fabrication technologies to the circuit integration process has significant advantages for fabricating Josephson LSI circuits with high reliability.

During the deposition of the refractory electrode materials, the substrate subjects to a substantial temperature rise, even if it is not heated intentionally. So, the photoresist lift-off technology, used in the Pb-alloy junction processes, is difficult to be applied for the all refractory junctions. On the contrary, an etching technology is widely applicable for the patterning of the refractory materials. In particular, a reactive ion etching (RIE) technique is preferable because of its anisotropic and selective etching nature. The anisotropy in etching is needed to realize accurate pattern transfer between photoresist masks and etched materials. The selectivity with respect to resist mask also affects the accuracy in pattern transfer. High selectivity with respect to the underlying material is also needed in order to etch the film having non uniform thickness distribution with appropriate processing margins.

The edge profiles of the electrode etched by highly anisotropic RIE shows steep vertical step structures. These surface topography significantly degrades the step coverage of overlying insulation layers. Minute cracks in the insulation layer generally develop at the step edges due to the shadowing effects during film growth, which often cause shorts in insulating films. In addition to this, the insulating films deposited on a step grow in a concave shape. This causes a fatal circuit failure in metal layer deposited on it, especially when patterning the overlying metal layer by anisotropic RIE. Metal films deposited inside the concave region of the insulating films cannot be etched by anisotropic RIE and a residue is remained in it even if the overetching is increased, which results in shorts through the metal residue along the steps. In order to prevent these circuit failures, special attention must be paid to control the edge profile of electrode so as to improve the step coverages of the overlying insulating films.

KEY STEPS IN DEVICE FORMATION

Outline of the process

Schematic cross section of the all refractory Josephson integrated circuit is shown in Fig. 1. The process flow for fabricating the circuit is summarized in Table 1. Nb films are used for a ground plane, base electrodes and wirings. NbN films are used for making Josephson junctions. The tunneling barrier is formed by oxidizing the surface of the base NbN film in 10%O_2-Ar plasma before the deposition of the counter NbN film.[2] A molybdenum (Mo) film is used for resistors. The Mo film has 100 nm thickness and its sheet resistance is 2Ω/□. SiO films are used for insulating and planarizing layers. Spin-on-glass (SOG) films are used for smoothing the edge profile of the electrodes. Deposition conditions and growth rates of metal and insulation layers used for this process are summarized in Table 2. An SEM photograph of a circuit fabricated by this process is shown in Fig. 2.

Reactive ion etching

Highly anisotropic and selective RIE processes have been developed for patterning various films used in the all refractory Josephson

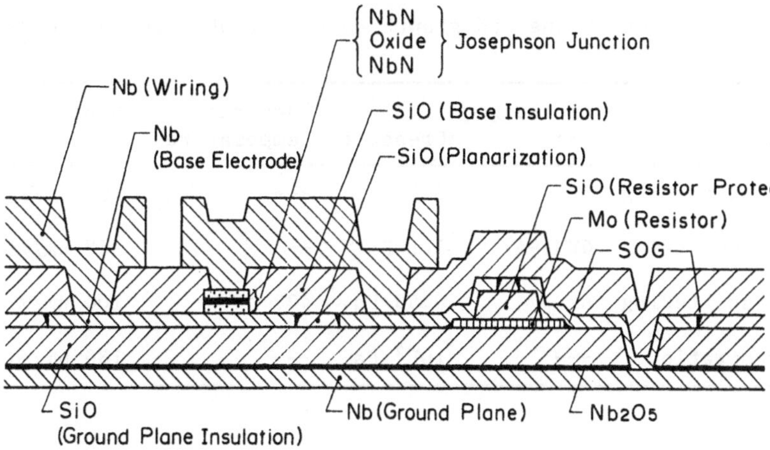

Fig. 1. Schematic cross section of the all refractory Josephson logic integrated circuit.

integrated circuits.[7] The RIE conditions and the etching characteristics for each layers are summarized in Table 3. In the Nb and NbN etchings, an accurate pattern transfer is realized due to the high anisotropy and high selectivity. In the Mo etching, a small amount of decrease in pattern width is caused by the relatively small selectivity against the mask, but it is small (about 100 nm) due to the thin Mo thickness. In the SiO etching, etching profile is slightly tapered (about 75 deg.) due to the small selectivity against the mask. However, this profile is preferable to improve the coverage of the overlying metal film. In each etching steps, considerable processing margins are obtained due to the high selectivity against the underlying materials.

Table 1. Process outline

1.	Ground plane layer deposition, Nb sputtering, 250 nm.
2.	Ground plane etch, CCl_2F_2+Ne RIE.
3.	Planarization, SAILOP.*
4.	Anodization, wet (ammonium pentaborate+ethylene glycohl), 20 nm.
5.	Ground plane insulation layer deposition, SiO, 400 nm.
6.	Ground contact hole etch, CF_4 RIE.
7.	Resistor layer deposition, Mo rf sputtering, Ts=400 °C, 100 nm.
8.	Resistor etch, $CCl_2F_2+O_2$ RIE.
9.	Edge profile control, SOG etchback.
10.	Resistor protect layer deposition, SiO, 320 nm.
11.	Resistor protect etch, CF_4 RIE.
12.	Base/JJ multilayer deposition, Nb/NbN/oxide/NbN, sputtering.
13.	Base electrode etch, CCl_2F_2+Ne RIE.
14.	Planarization, SAILOP.*
15.	JJ etch, CF_4 RIE.
16.	JJ periphery isolation, O_2 RIE.
17.	Base insulation layer deposition, SiO, 480 nm.
18.	JJ contact open, SNIP.**
19.	Base contact hole etch, CF_4 RIE.
20.	Wiring layer deposition, Nb sputtering, 750 nm.
21.	Wiring etch, CCl_2F_2+Ne RIE.

* self-aligned insulation lift-off planarization.[7]
** self-aligned niobium nitride isolation process.[3]

Table 2. Deposition conditions and growth rates for Nb, NbN, Mo and SiO films

Material	Method	Discharge gas	Pressure	Substrate temperature	Deposition rate
Nb	dc sputtering	Ar	1.1 Pa	---	100 nm/min
NbN	rf sputtering	6%N_2-Ar	1.1 Pa	---	45 nm/min
Mo	rf sputtering	Ar	0.65 Pa	400 °C	100 nm/min
SiO	evaporation	---	10^{-4} Pa	---	130 nm/min

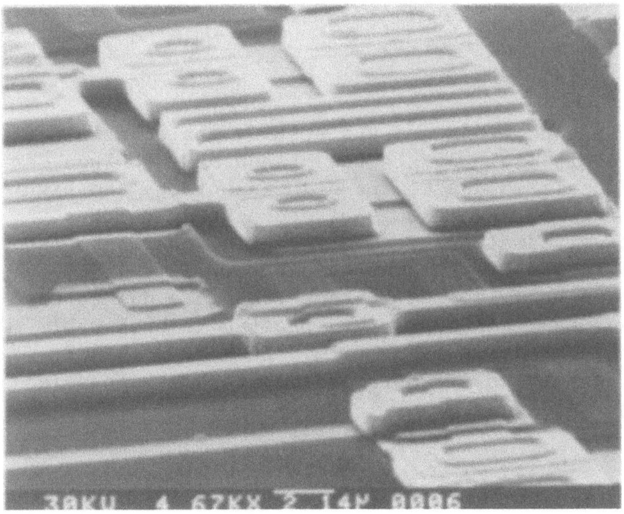

Fig. 2.
SEM photograph of an all refractory Josephson logic integrated circuit with 2.5 μm minimum feature.

Table 3. RIE conditions for the paterning of Nb, Mo, NbN and SiO layers

Layer	Gas	Pres-sure (Pa)	Rf power (W/cm^2)	Etch rate (nm/m.)	Selectivity to SiO	Nb	AZ-1450J	Etching profile
Nb	65%Ne-CCl_2F_2	13	0.16	160	6.9	--	4.0	anisotropic
Mo	75%O_2-CCl_2F_2	13	0.16	105	6.5	--	1.3	anisotropic
NbN	CF_4	27	0.16	42	1.9	--	2.6	anisotropic
SiO	CF_4	2.7	0.16	60	--	4.8	1.3	tapered

Planarization technology

The existence of steep step structure degrades the step coverages of overlying films, resulting in substantial circuit failures. It is essential to smooth the edge profile of the electrode fabricated by anisotropic RIE in order to eliminate these failures and to obtain the expected chip yield. Various planarization techniques, such as P-glass flow[11], rf bias sputtering[12], sacrificial resist etching[13], and lift-off[14] have been developed for realizing VLSI multilevel interconnect structures in semiconductor fields. A planarization technology which is applicable to the all refractory Josephson integration process has been developed, in which SiO lift-off and spin-on-glass (SOG) etch back methods are used.[7] We call this planarization process SAILOP (self-aligned insulation lift-off planarization).

The fundamental process steps for SAILOP are shown in Fig. 3, and explained as follows.

(a) Nb layer is etched with a resist mask using a CCl_2F_2+Ne anisotropic RIE.
(b) Nb layer is overetched using a CF_4+O_2 isotropic RIE, resulting in an about 150 nm undercut in the Nb layer from the mask edge.
(c) SiO film for planarization layer is evaporated.
(d) SiO film is lifted-off by removing resist in acetone. Due to the overhang of the resist mask, high lift-off yield is obtained in a good reproducibility. Crevasses are formed along the edge of the Nb electrode after the lift-off.

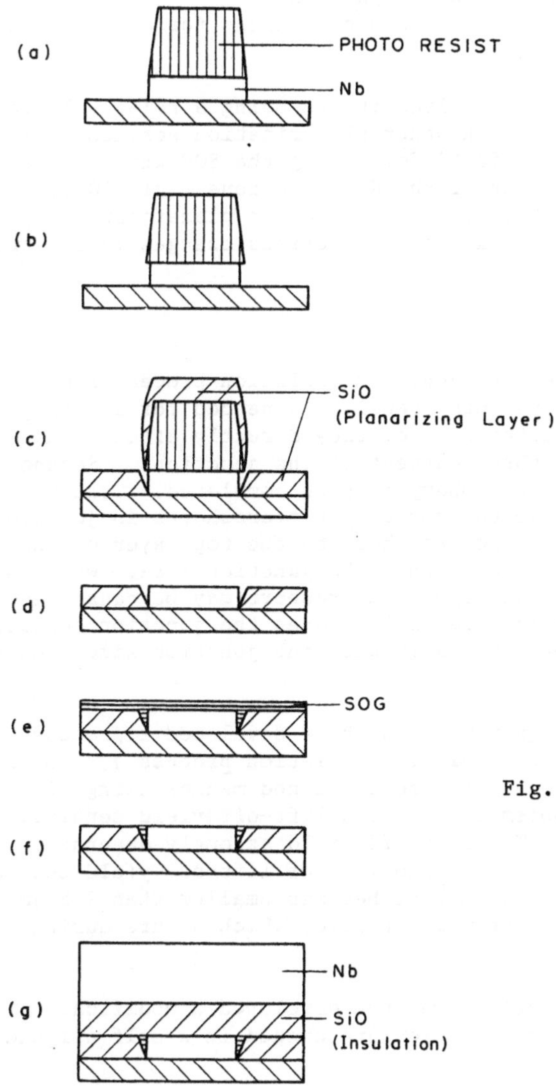

Fig. 3. SAILOP process flow :
(a) Nb anisotropic RIE.
(b) Nb isotropic RIE.
(c) SiO (planarization layer) deposition.
(d) SiO lift-off.
(e) SOG coat.
(g) SOG RIE.
(f) SiO (insulation) deposition and Nb deposition.

Table 4. Detailed process conditions for SAILOP

Step (b)	Step (e)	Step (f)
Isotropic RIE of Nb	SOG coating	RIE of SOG
Gas : 5%O_2-CF_4	Spin coat : 3500 rpm	Gas : 21%H_2-CF_4
Pressure : 27 Pa	20 sec	Pressure : 2.7 Pa
Rf power : 0.16W/cm^2	Cure : 250 °C	Rf power : 0.16W/cm^2
Etch rate : 100 nm/m.	30 min	Etch rate : 80 nm/m.
	Thickness : 120 nm	Selectivity
		to Nb : 16.0
		to NbN : 8.0

(e) SOG is spin coated and baked. Due to the flow characteristics of the SOG film, the crevasses are completely filled by the SOG.
(f) SOG is reactively ion etched to the level of the Nb surface, resulting in a plane surface topography.
(g) SiO layer for the insulation is evaporated, on which an Nb layer for the wiring is deposited and patterned.

Conditions for the isotropic RIE of Nb, and conditions for coating and etching SOG are summarized in Table 4.

An attractive feature of SAILOP lies in its being a relatively low temperature technique comparing with other planarization schemes.[11-14] The highest temperature used is 250 °C for baking the SOG which is well below the stable temperature limit of the NbN junctions (310 °C). Another merit of SAILOP is that it is applicable for planarizing steps with various heights without introducing any serious changes in the process conditions.

Junction isolation process

The basic requirements for the junction isolation process are as follows. First, the accurate pattern transfer is needed for defining the junction area, because this patterning accuracy directly affects the variations in the maximum Josephson current of the junctions. Second, the insulation around the junction periphery must be completed. Otherwise, the junction area will give rise to problems of microshorts in junction I-V characteristics. Third, the contact hole to the top layer of the junction must be completely aligned within the junction area. When the contact hole touches the junction edge, a microshort may be caused at the junction edge. Therefore, it is essential to open the junction contact holes in a self-aligned manner especially when the junction size becomes small.

In order to meet these requirements we have developed a process called SNIP (self-aligned niobium nitride isolation process),[3] in which junction contact holes are opened in a self-aligned manner using lift-off of SiO films. However, a problem lies in its lift-off yield sensitivity on the processing conditions. The lift-off yield strongly depends on the edge acuity of the photoresist mask. The loss of lift-off yield becomes serious especially when the junction size becomes smaller than 2.5 μm due to the deterioration of the resist wall profile, which occurs during lithography and RIE steps.

In order to avoid this problem, we have developed a modification of the SNIP, in which junction contact holes are opened in a self-aligned

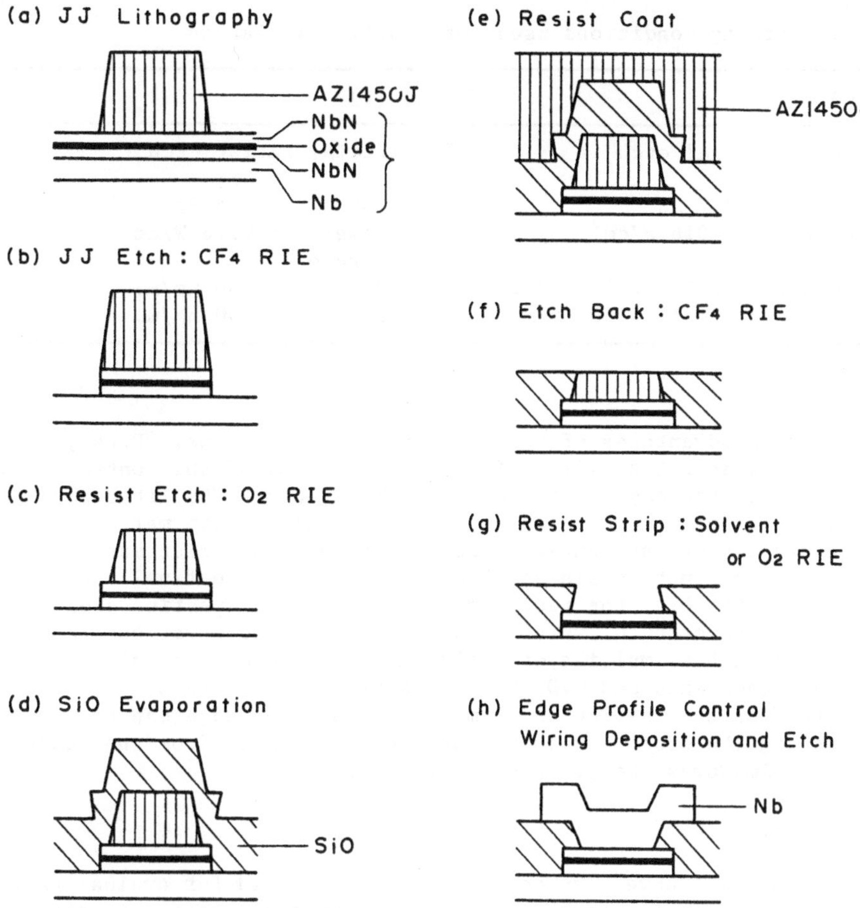

Fig. 4. Fabrication steps for opening junction contact hole.

manner using a sacrificial resist etch back technology. The process steps are shown in Fig. 4 and are explained as follows.

(a) Junction area is defined by photoresist mask.

(b) Junction multilayer is etched to the top of the base Nb layer using RIE in CF_4.

(c) Resist mask is eroded laterally from the junction edge using RIE in O_2.

(d) SiO film is evaporated on the junction and the resist mask. Due to the lateral erosion of the resist mask, junction periphery can be covered completely by the SiO film.

(e) Sacrificial resist is coated to planarize the surface.

(f) Resist and SiO film are etched using RIE in CF_4 until the sacrificial resist is completely etched off. In this step, etching conditions are chosen such that the etching rate of the resist is same as that of the SiO film. At the end point, SiO film on the top of the junction area is completely opened.

(g) Resist mask is removed by dipping in acetone or by using RIE in O_2. Contact hole is opened on the top of the junction in a self-aligned manner.

(h) Overhang in the contact hole is round etched by RIE in CF_4+H_2 after SOG coating. Then an Nb layer for wiring is deposited and patterned, completing the junction.

Table 5. Etching conditions used for junction isolation

Step (c)		Step (f)	
Resist mask lateral etch		Sacrificial resist etch	
Gas	: O_2	Gas	: CF_4
Pressure	: 2.7 Pa	Pressure	: 6.5 Pa
Rf power	: 0.016 W/cm^2	Rf power	: 0.16 W/cm^2
Etch rate		Etch rate	
AZ1450J(lateral)	: 30 nm/m.	AZ1450J	: 50 nm/m.
(vertical)	: 50 nm/m.	SiO	: 50 nm/m.

The major advantages of this process are as follows. First, because this process does not use lift-off method, opening of the contact hole is not affected by the edge acuity of the resist mask. The lithographic and the junction etching conditions have considerable safety margin to insure complete opening of the contact holes. Second, the resist mask defining the junction need not remain soluble to accomplish the final contact hole opening. Due to this, the resist mask defining the junction area can be used in high temperature, which allows us to introduce various fabrication technologies such as multilayer resist methods, deposition of insulation layers by plasma enhanced CVD or rf bias sputtering, and so on. Third, the coverage of the insulation film on the junction edge can be completed by controling the erosion distance of the resist mask from the junction edge, which improves the junction reliability.

CIRCUIT FABRICATION

Test circuits have been designed and fabricated for evaluating the material and processing capability when the present process is applied to the LSI level integration. The largest circuit tested was a 4x4 bit parallel multiplier, the features and performances of which are summarized in Table 6. The multiplier has been successfully operated with the power margin of $\pm$10%. Decrease in the observed power margin and multiply time from the designed value can be attributed mainly to the variation in the maximum Josephson current of the junctions, which is estimated to be about $\pm$10% from the test element group measurements. The variations in the resistance of Mo resistors are also responsible for the decrease in the power margin.

The chip yield for the 4x4 bit parallel multiplier by the present integration process is not high and cannot be evaluated quantitatively at the present stage. The qualitative nature of the yield limiting factors of the present process will be discussed in the followings.

Table 6. Features and performance summary of the 4x4 bit multiplier

Features		Performance summary		
Minimum width	: 2.5 μm	Item	Measured	Designed
Minimum junction size	: 2.5x2.5 μm^2	Multiply time (min.)	1.0 ns	0.5 ns
Number of gates	: 652 (4JL)			
Number of junctions	: 2968	Power dissipation (max.)	0.21 mW	0.86 mW
Number of resistors	: 2035			
Chip size	: 3.5x3.5 mm^2	Power margin	$\pm$10%	$\pm$30%

First, most of the circuit failures were resulted from macroscopic pattern failures, mainly due to particulates which were introduced during the film deposition steps. Adhesion of the lifted-off SiO films on the substrate was also responsible for the problem. These problems may be solved provided that the process technology matures further.

Second, the microscopic circuit reliability, such as the integrity of the interconnects or the integrity of the overcrossings against open lines or shorts, was very high in the present fabrication process. In fact, the chips passed the initial screening by the optical microscope inspection were operated with a high provability.

Third, though high uniformity developed in SNIP has realized the wide operating margin of the experimental circuit, the fabrication variations are still the major part of the margin loss. This problem may become serious when the circuit complexity increases further. In order to preserve wide operating margin of the circuits, two points are required from the view point of material and processing technology. One is to improve the junction I-V characteristics. The reduced subgap conductance and the steeper current rise at the gap voltage in the I-V characteristics increase the fan-out capability of the logic gates and decrease the sensitivity of the gates against the variations of circuit parameters. The recent development of NbN/a-MgO/NbN junctions reported by Shoji et al.[15] offers an attractive candidate for junctions of Josephson LSI circuits. Another is to improve the uniformity of the maximum Josephson currents. Recently, Aoyagi et al. developed a bi-layer resist processing technology applicable to the SNIP process, in which significant decrease in the maximum Josephson current variations has been demonstrated.[16] It may be expected that the introduction of these advanced processing technologies will improve the operating margin and the chip yield of the Josephson LSI circuit.

CONCLUSION

An integration process for the fabrication of an all refractory Josephson integrated circuit with NbN junctions has been described. An attractive feature of the NbN junctions lies in its high temperature stability, which provides freedom to employ various processing technologies for fabricating circuits. An integration process is designed to meet the requirements for fabricating circuits with complexities in the realm of LSI. Highly anisotropic and selective RIE and planarization technology have been developed in order to attain appropriate processing margins and chip yields. Application of this process for fabricating logic circuits with near LSI level complexity has been successfully performed.

Efforts to improve junction characteristics and to develop processing technologies for decreasing fabrication variations are now being consistently continued. These combined efforts will open up the Josephson LSI technology in near future.

ACKNOWLEDGEMENT

The author wishes to thank Dr. H. Hayakawa for his valuable suggestions and discussions, Dr. F. Shinoki, Mr. A. Shoji and Mr. M. Aoyagi for their cooperation and discussions, and other members of " Special section for Josephson computer technology " in ETL for helpful comments.

REFERENCES

1. H. Kroger, L. N. Smith, and D.W. Jillie, Selective niobium

anodization process for fabricating Josephson tunnel junctions, Appl. Phys. Lett. 39:280-282 (1981).
2. A. Shoji, F. Shinoki, S. Kosaka, M. Aoyagi, and H. Hayakawa, New fabrication process for Josephson tunnel junctions with (niobium nitride,niobium) double-layered electrodes, Appl. Phys. Lett. 41:1097-1099 (1982).
3. A. Shoji, NbN based Josephson junctions, 3rd international conference on superconducting quantum devices, Berlin (1985).
4. M. Gurvitch, M. A. Washington, and H. A. Huggins, High quality refractory Josephson tunnel junctions utilizing thin aluminum layers, Appl. Phys. Lett. 42:472-474 (1983).
5. S. Kosaka, A. Shoji, M. Aoyagi, F. Shinoki, H. Nakagawa, S. Takada, and H. Hayakawa, High speed logic operations of all refractory Josephson integrated circuits, Appl. Phys. Lett. 43:213-215 (1983).
6. D. Jillie, L. N. Smith, H. Kroger, L. W. Currier, R. L. Payer, C. Potter, and D. M. Shaw, All refractory Josephson logic circuits, IEEE J. Solid-State Circuits, SC-18:173-180 (1983).
7. S. Kosaka, A. Shoji, M. Aoyagi, F. Shinoki, S. Tahara, H. Ohigashi, H. Nakagawa, S. Takada, and H. Hayakawa, An integration of all refractory Josephson logic LSI circuit, IEEE Trans. Magn. MAG-21: 102-109 (1985).
8. A. Shoji, S. Kosaka, F. Shinoki, M. Aoyagi, and H. Hayakawa, All refractory Josephson tunnel junctions fabricated by reactive ion etching, IEEE Trans. Magn. MAG-19:827-830 (1983).
9. S. Tahara, S. Kosaka, A. Shoji, M. Aoyagi, F. Shinoki, and H. Hayakawa, Fabrication and performance of all refractory Josephson logic circuits for 1 Kbit SFQ memory, IEEE Trans. Magn. MAG-21: 733-736 (1985).
10. S. I. Raider, Josephson tunnel junctions with refractory electrodes, IEEE Trans. Magn. MAG-21:110-117 (1985).
11. A. C. Adams, and C. D. Capio, Planarization of phosphorus-doped silicondioxide, J. Electrochem. Soc. 128:423-429 (1981).
12. C. Y. Ting, V. J. Vivalda, and H. G. Schaefer,Study of planarized sputter-deposited SiO_2, J. Vac. Sci. Technol. 15:1105-1112 (1978).
13. Y. Homma, and S. Harada, LSI surface leveling by rf sputter etching, J. Electrochem. Soc. 126:1531-1533 (1979).
14. K. Ehara, T. Morimoto, S. Muramoto, and S. Matsuo, Planar interconnection technology for LSI fabrication utilizing lift-off process, J. Electrochem. Soc. 131:419-424 (1984).
15. A. Shoji, M. Aoyagi, S. Kosaka, F. Shinoki, and H. Hayakawa, Niobium nitride Josephson tunnel junctions with magnesium oxide barriers, Appl. Phys. Lett. 46:1098-1100 (1985).
16. M. Aoyagi, A. Shoji, S. Kosaka, F. Shinoki, and H. Hayakawa, A 1 μm cross line junction process, 2nd ICMC Symposium on materials and processing for superconducting electronics, Cambridge (1985).

THIN FILM PREPARATION AND DEVICE APPLICATION OF LOW CARRIER DENSITY SUPERCONDUCTOR $BaPb_{1-x}Bi_xO_3$

Minoru Suzuki

NTT Ibaraki Electrical Communication Laboratories

Tokai, Ibaraki, Japan

ABSTRACT

$BaPb_{1-x}Bi_xO_3$ (BPB), a low carrier density oxide superconductor, offers a wealth of potential applications to new cryoelectronic devices because of such semiconductor-like properties as structure sensitiveness, low-carrier density and optical transmittance. BPB thin films are prepared by rf (conventional or magnetron) sputtering. Because deposited thin films tend to lack Pb, Bi and oxygen, particular attention is given to controlling the composition. To compensate for the Pb and Bi deficiency, sputtering targets are enriched (x1.5) with Pb and Bi. To reduce oxygen vacancies incorporated during deposition, sputtering is performed under relatively high oxygen partial pressure. The BPB films deposited on sapphire substrates are polycrystalline and need annealing at 550-600°C to render them superconductive. A lead oxide atmosphere during annealing is effective in suppressing Pb and Bi vaporization. Single crystal BPB films are also grown on $SrTiO_3$ at substrate temperatures higher than 700°C. The grain boundaries of polycrystalline BPB thin films function as Josephson junctions. These films are applied as two-dimensional arrays to switching devices, electromagnetic coupling devices, and optical detection devices.

I. INTRODUCTION

The low carrier density superconductor $BaPb_{1-x}Bi_xO_3$ (BPB)[1] is a solid solution between semimetallic $BaPbO_3$[2] and semiconducting $BaBiO_3$[1], both of which have the same perovskite crystal structure with a slight distortion from cubic symmetry. This oxide system exhibits superconductivity over the compositional range of $x<0.35$,[3,4] and undergoes a metal-semiconductor transition near $x=0.35$. The highest superconducting transition temperature T_c reaches 13 K near $x=0.3$.

The discovery of BPB and its superconductivity by Sleight et al.[1] in 1975 was followed by extensive detailed experiments by Kahn et al.[5] and also by Thanh et al.[3] These pioneering works revealed unusual properties in this system, some of which needed subsequent experimental and theoretical investigations for sufficient interpretation.

The most important feature of BPB is that the carrier density ($2x10^{21}$ cm^{-3})[6] is more than one order of magnitude smaller than those for usual metal or for intermetallic alloy superconductors, and that, nonetheless, T_c is extraordinarily high. This property, reflecting the existence of a conspicuously strong electron-phonon coupling,[7] characterizes this material as an unusual low carrier density superconductor. It may be due to these properties that the earliest experimental results misleadingly made the room for a superconductivity mechanism other than Bardeen-Cooper-Schrieffer (BCS) theory.[8] It was not until BPB single crystals were grown[9,10] that BPB was characterized as an extreme type II, strong coupling, BCS superconductor.[6,11]

The resistivity of BPB is more than two orders of magnitude larger than those of usual metals, showing semiconductor-like temperature dependence for $X>0.25$. Although both the real-space pairing model by Rice and Sneddon,[12] and the electronic band calculation by Mattheiss and Hamann[13] shed some light on the superconductivity mechanism and the electronic conduction mechanism, there still remains some unclear areas associated with the extreme strong-coupling superconductivity.

On the other hand, owing to these unique superconducting properties, BPB seems to be favored with a wealth of novel cryoelectronic applications in addition to conventional applications. First, the structure sensitiveness like semiconductors leads to the formation of grain boundary Josephson junctions in polycrystals.[14,15] Second, because the density of states at Fermi level is exceptionally small, a remarkable effect due to nonequilibrium superconductivity is expected, making it possible to use this material in a highly sensitive optical detector.[16] All these, as well as future applications can be attained by means of thin film preparation techniques.[17-19,22]

This paper reviews the thin film preparation of BPB using the conventional and magnetron rf sputtering method. As an extended consequence of the sputtering technique thus developed for BPB thin films, the epitaxial growth of single crystal thin film is also described.[19]

The latter half of this paper deals with the grain boundary Josephson junctions in polycrystalline BPB thin films and their applications to novel kinds of cryoelectronic devices.

II. THIN FILM PREPARATION BY CONVENTIONAL SPUTTERING

A. Sputtering Targets

The essential point in preparing BPB thin films is to attain the stoichiometry including oxygen. In the thin film deposition process, the main problem arises from the evaporation of Pb and Bi during deposition as well as from the oxygen deficiency resulting in vacancies in BPB. Both, as will be stated, work to depress the T_c of BPB. Fortunately, a slight deviation from the stoichiometry to a Pb rich composition was found not to depress T_c seriously. Furthermore, this change in the composition even works to improve crystal quality. This made it possible to use Pb and Bi enriched targets to compensate for the Pb and Bi deficiency, without any serious problem due to a possible overcompensation.

Another problem is oxygen deficiency, which is inevitably involved in the oxide thin film preparation process. Since the oxygen 2p bands are responsible for BPB electronic conduction, the amount of oxygen deficiency must be suppressed to as low as possible. This requirement restricts the range of sputtering conditions that lead to excellent electrical and superconducting properties.

The sputtered BPB thin films strongly tend to lack Pb and Bi.[18] Therefore, the targets must be enriched with these elements in advance to compensate for these deficiencies. To obtain excellent electrical properties and crystal qualities, as well as for reliability and efficiency, it is not advantageous to utilize a powder target consisting of BPB and PbO (or PbO_2, Pb_3O_4) and Bi_2O_3 powders for compensation, even though it certainly gives superconducting BPB thin films with limited properties.[21] The targets used are sintered ceramics with a nominal composition $Ba(Pb_{1-x}Bi_x)_{1.5}O_4$. To fabricate these targets, properly weighed and mixed powders consisting of $BaCO_3$, PbO_2 and Bi_2O_3 are reacted at 720°C for 4 hr and sintered after being pressed into a target shape at 800°C for 2 hr in an oxygen atmosphere

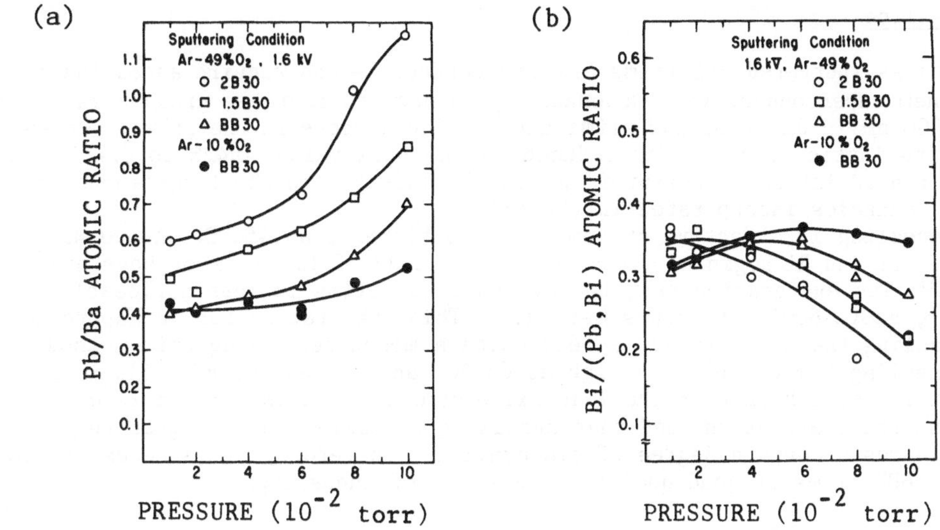

Fig.1 Pb/Ba and Bi/(Pb,Bi) atomic ratio for as-deposited BPB films sputtered under various pressures.

(x=0.3). These temperatures must be changed in accordance with composition x. Since the fabrication process is accompanied by the evaporation of Pb and Bi, the (Pb,Bi)/Ba ratio of the target obtained is reduced to 1.4 or less. The compensated targets are actually composed of BPB and lead-bismuth oxides.

B. Deposition Rate, Film Composition, and Crystal Quality

A typical deposition rate is about 4 nm/min when the films are deposited by conventional rf sputtering under optimum conditions, which will be described later. Changing the sputtering pressure or applied rf voltage to increase the deposition rate results in degradation of the electrical properties and crystal qualities.

The compositions of as-sputtered BPB films are plotted in Fig.1 as a function of the sputtering pressure. When a stoichiometric target (B30) is used, the deposited BPB films have Pb and Bi deficiencies of 30% or more. On the other hand, The (Pb,Bi) enriched targets (1.5B30 is 1.5 times enriched and 2B30 2 times) compensate for these deficiencies, and the film composition most closely approaches stoichiometry when a 1.5B30 target is used. The 2B30 target overcompensates for the deficiencies since sputtering must be performed under a relatively high pressure to minimize oxygen deficiency. Therefore, the 1.530 target was selected as the optimum target. When this target is used, Bi/(Pb,Bi) is about 0.3 or less, the x value that results in the highest T_c.

BPB polycrystalline thin films are usually deposited on sapphire substrates. Fused quartz or microsheets can be used for substrates, though it is difficult to obtain film thicknesses of less than a few 100 nm because of a slight chemical reaction between BPB and the substrate glass when annealing is carried out above 530°C.

As-sputtered BPB films are crystallized at substrate temperatures higher than 200 C. The films are composed of small crystallites about 50 nm in diameter and thin amorphous layers surrounding them. The lattice constant of the as-sputtered films is larger than the bulk value by about 3%. BPB films tend to grow with a preferred crystal orientation along [100]. By increasing oxygen partial pressure, this tendency is enhanced and crystal quality is improved. Too high a sputtering pressure, however, results in amorphous films.

C. Annealing

The as-sputtered BPB films are semiconductive and require annealing to make them superconductive. An annealing is used to increase grain size up to about 200 nm in diameter and eliminate lattice defects in the films. Moreover, the lattice constant is reduced to the bulk value (0.428 to 0.429 nm). The increased lattice constant is primarily ascribed to the large amount of oxygen vacancies incorporated in the films.

Annealing is effective at temperatures higher than 450 °C. Above 450 °C, however, Pb and Bi begin to evaporate from the BPB films if amorphous areas exist.[20] The evaporation from the crystallized films is less noticeable; however, above 600°C it occurs markedly. This evaporation can be suppressed by annealing the BPB films in a lead oxide atmosphere. Using this method, the annealing temperatures as high as 650°C can be used for BPB films.

It is noted here that the lead oxide atmosphere does not completely suppress the evaporation, and that annealing at temperatures higher than 600 °C causes a certain degree of evaporation. Therefore, it is necessary to prepare BPB films of good quality in advance of annealing.

D. Electrical and Superconducting Properties

The resistivity of as-sputtered BPB films has a remarkable dependence on sputtering pressure, as shown in Fig.2. The resistivity decreases drastically when oxygen partial pressure is increased. This is because the electronic conduction originates in the oxygen octahedrons surrounding Pb and Bi, and because the oxygen vacancies reduce the electrical conductivity. All of these oxygen vacancies cannot be recovered by annealing, since annealing temperatures must be held below 600 °C in order to suppress Pb and Bi evaporation to a minimum. Therefore, it is desirable that the amount of oxygen vacancies be reduced as small as possible at the film preparation stage.

Figure 3 shows the temperature dependence of resistivities for annealed BPB thin films sputtered under various sputtering pressures. The resistivities and T_c clearly improve as the pressure is increased within the range in which crystalline BPB films can be obtained. The films obtained under the optimum sputtering conditions (1.4kV applied rf voltage and $8x10^{-2}$ Torr Ar-50%O_2 pressure) are characterized by a T_c of higher than 9 K, a transition width of less than 0.4 K, and a resistivity of $3x10^{-3}$ ohm-cm. Film thick-

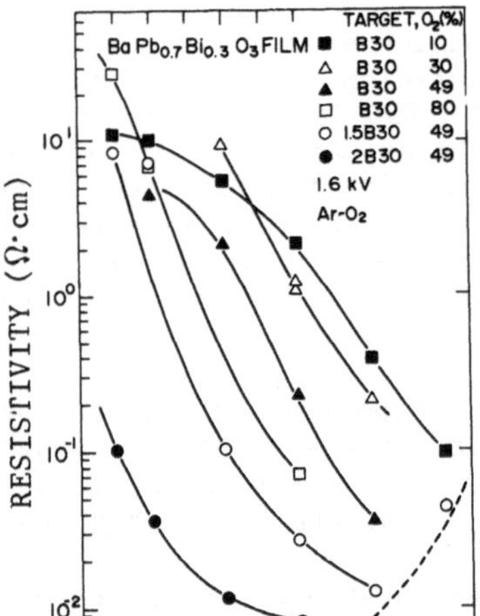

Fig.2 Resistivities for as-deposited BPB films sputtered under various pressures.

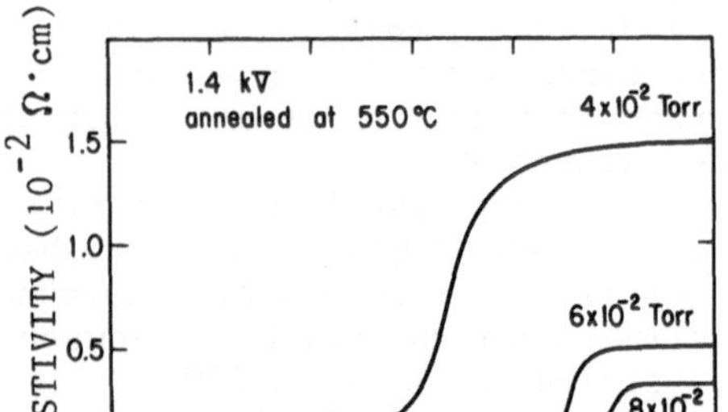

Fig.3 ρ(T) for annealed BPB films sputtered from a 1.5B30 target under various pressures.

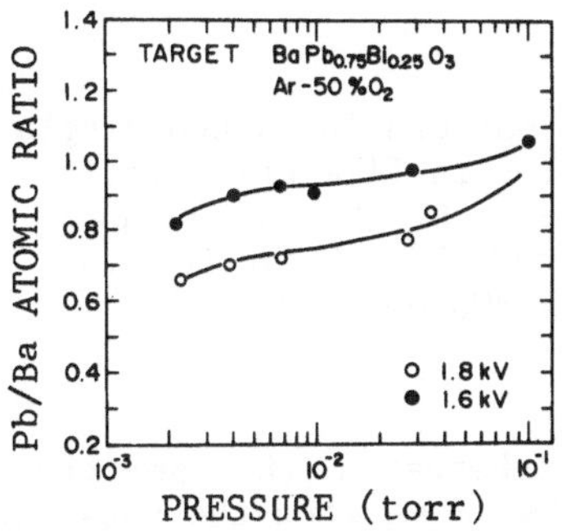

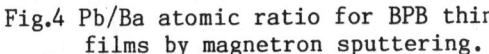

Fig.4 Pb/Ba atomic ratio for BPB thin films by magnetron sputtering.

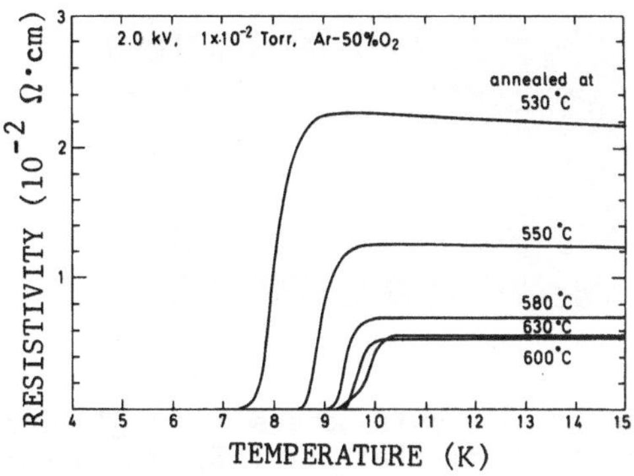

Fig.5 ρ(T) for BPB films by magnetron sputtering from 1.5B30.

nesses for attaining these values are 150 nm or more. Further reductions in film thickness result in depression in T_c and conductivity.[20,22]

The annealing temperature is usually set at 550°C. Annealing at higher than 600°C temperatures also causes a depression in T_c, and sometimes even deterioration of the films. These all result from the Pb and Bi evaporation.

III. THIN FILM PREPARATION BY MAGNETRON SPUTTERING

A. Deposition Rate, Film Composition, and Crystal Quality

Using the magnetron rf sputtering method for BPB thin film preparation[22] improves the following three points: the deposition rate (40 nm/min), crystal quality and dependences of the BPB film composition on sputtering conditions.

The most remarkable change brought about by this method is the dependence of the film composition on the sputtering conditions. Figure 4 shows the Pb/Ba atomic ratio for BPB films deposited under various sputtering pressures. The Pb/Ba ratio is almost independent of the pressure over a considerably wide pressure range of around 10^{-2} Torr. This implies that a small change in pressure during deposition does not seriously affect the film composition The improvement in the reproducibility and reliability obtained by this method can partly be ascribed to this factor, and partly to the improved crystal quality.

In contrast to the conventional method, the BPB films by the magnetron method tends to grow with a preferred orientation along [110].

Although using the B30 target attains BPB film stoichiometry under the limited sputtering condition range, T_c, ΔT_c, conductivity, and surface texture do not exceed those of the films sputtered using a 1.5B30 target. Therefore, the 1.5B30 target is preferable.

B. Superconductivity

The BPB films deposited at a substrate temperature of about 260°C are semiconductive and becomes superconductive by annealing at 530°C or higher. Figure 5 shows the temperature dependence of resistivities for BPB thin films annealed at various temperatures, indicating that annealing at 600°C is optimum. Annealing at 630°C, in some instances, results in films having a slightly higher resistivity and a somewhat lower T_c. This corresponds to the fact that Pb and Bi begin to evaporate appreciably above 630°C.

C. Epitaxial Growth

Among the several advantages resulting from the combination of magnetron sputtering and (Pb,Bi)-compensated targets, the most remarkable one is that it enables the epitaxial growth of BPB single crystal thin films.[19] The

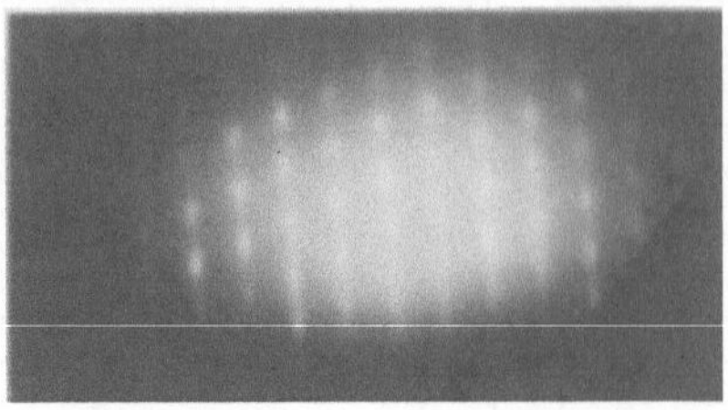

Fig.6 An RED pattern for a BPB single crystal thin film epitaxially grown on a $SrTiO_3$ (100) substrateat700°C. Film Thickness is 390 nm.

epitaxial growth occurs on single crystal $SrTiO_3$ substrates (the same crystal structure as that for BPB) and sapphire (01$\bar{1}$2) substrates when the substrate temperatures are held above about 500°C. Lattice mismatch is 8.8%, which is rather large considering the normal epitaxial growth technique. A typical RED pattern is shown in Fig.6, indicating that the film is single crystalline. The condition for epitaxial growth is mainly limited owing to the large lattice mismatch. Therefore, the sputtering conditions must be suitably selected in order that the epitaxial plane coincides with the preferred growth plane - (110) plane for a moderately high deposition rate (more than 30 nm/min), and (100) plane for a low deposition rate.

$BaPb_{0.7}Bi_{0.3}O_3$ single crystal thin films epitaxially grown on (100) $SrTiO_3$ have the resistivity of $8x10^{-4}$ ohm-cm, a Hall mobility of 3 cm^2/Vsec (Ref.23), a T_c of 10 K and the transition width of 0.2 K. The critical currents at 4.2 K for these films are $2.5x10^5$, which is one order of magnitude larger than those for polycrystalline BPB thin films.

IV. GRAIN BOUNDARY JOSEPHSON JUNCTIONS

A. Grain Boundaries of BPB Thin Films

It is well known that in polycrystals there always exist grain boundary potential barriers due to the chemical potential shift in these regions from the bulk value. The width of this barrier is primarily determined by the carrier density through Poisson's equation.

In the case of semimetals or degenerate semiconductors (an intermediate case between the two cases of metals and semiconductors) the barrier width is several nm, and electron tunneling through the barrier dominates the electronic transport. This electronic transport characterizes the electrical properties of BPB grain boundaries.

In such materials, whose polycrystalline grain boundary potential makes a tunneling barrier, the boundaries function as Josephson junctions when the material is kept in the superconducting state. BPB is probably the first material to be studied that provides the phenomena associated with grain boundariy Josephson junctions.

The Josephson current and the tunneling I-V characteristics are usually observed in polycrystalline BPB thin films prepared by sputtering.[14,15] Recently, tunneling Josephson I-V charactreristics were also observed by Stepankin et al.[24] for a BPB bicrystal.

Figures 7 (a) and (b) respectively show TEM photographs of the grain structure of BPB thin films by sputtering and the lattice image of a grain boundary. The atomic ordering is regular over the range except for in the grain bounadry region, indicating the absence of second phase segregation at the boundaries. This confirms that BPB grain boundaries form potential barriers for tunnel Josephson junctions.

B. Characteristics of Boundary Josephson Junctions (BJJ)

Since a polycrystalline film consists of a two-dimensional network of grain boundaries, the I-V characteristics observed for a BPB thin film are those for series-connected tunneling Josephson junctions. Figure 8 (a)-(c)

(a) (b)

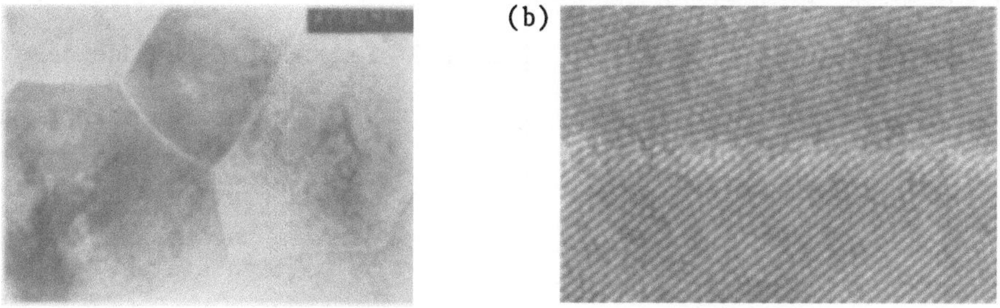

Fig.7 (a) A TEM photograph (x 3.2×10^4) for a polycrystalline BPB thin film. (b) An enlargement of a grain boundary part.

show I-V characteristics for BPB films at x=0.3, 0.25 and 0.2. The step of the voltage state curves, which roughly equals the gap voltage, corresponds to $2\Delta/e$ for each film, where Δ is the superconducting energy gap.

The hysteresis of the I-V curves tend to decrease for a smaller x. The gap voltage decreases systematically with the decrease in T_c when x is varied. With increasing temperature, the hysteresis gradually decrease, disappearing near T_c. The temperature dependence of the gap voltage corresponds quite closely to that for the energy gap in the BCS theory. Its magnitude, however, exceeds the value expected from the theory for x=0.3, while it becomes considerably smaller than the theoretical value for x<0.25. The former fact can be attributed to the strong-couplig nature of BPB superconductivity. The latter reflects the existing proximity effect in BJJ barriers.

The maximum Josephson current, I_J, has its largest value (2×10^4 A/cm^2) at x=0.25. For $x < 0.25$, I_J decreases mainly due to the reduction in Δ, while when $x > 0.25$, I_J decreases due to the increase in the normal tunneling resistance, which tends to increase with x. The I_J for BPB films at x=0.3 (usually used for device application) ranges from 3×10^2 to 3×10^3 A/cm^2.

Temperature dependence of I_J at x=0.3 is shown in Fig.9. Although the overall behavior resembles the theoretical curve by Ambegaokar and Baratoff, the local behavior in the temperature range near T_c contrasts with the theoretical curve, i.e., the experimental plots become convex in this region. This tendency becomes more conspicuous the smaller x becomes. This behavior can be explained by the proximity effect associated with the spatial variation in the BJJ potential barriers.

Magnetic field dependence of I_J for BJJ does not exhibit the Fraunhofer interference pattern, probably because the film forms a two-dimensional Josephson array. When the applied magnetic field is increased, I_J decreases first rapidly, then rather gradually. With a further increase in the magnetic field above 200 Oe, the Josephosn current in the I-V characteristics

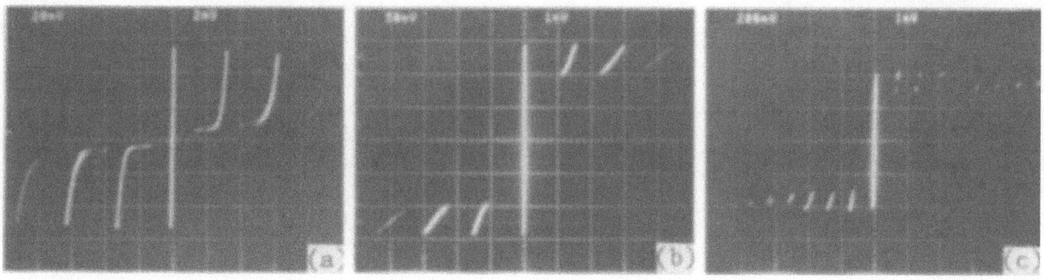

Fig.8 I-V characteristics of BJJs for BPB thin films with 50 μm constriction. (a) x=0.3. T=4.49K. Y:20μA/div, X:2mV/div. (b) x=0.25. T=4.48K. Y:1mA/div, X:0.5mV/div. (c) x=0.2. T=2.41K. Y:0.2mA/div, X:1mV/div.

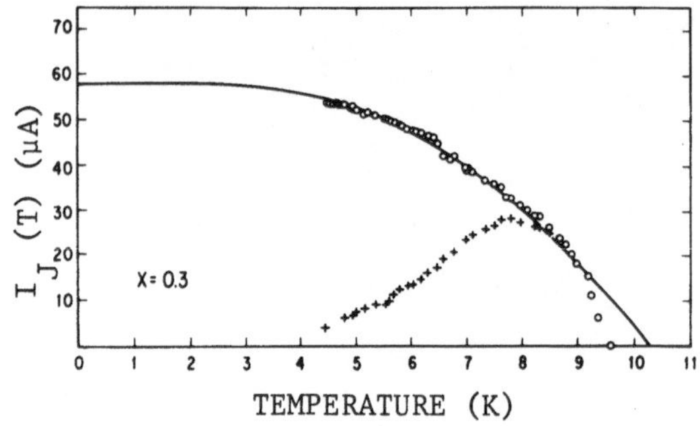

Fig. 9 Temperature dependence of the maximum Josephson current of BJJs for BPB thin films (x=0.3). The solid line is a theoretical curve by Ambegaokar and Baratoff. + denotes the minimum current at voltage states.

begins to incline appreciably. This is caused by the onset of a vortex flow along the grain boundaries.

In the case of films with a larger I_J, it behaves similarly but depends weakly on the magnetic field. Decreasing x has a similar effect to the case of a large I_J.

V. DEVICE APPLICATION

A. Phenomena in the BJJ Array

A polycrystalline BPB films is electrically equivalent to a two-dimensional tunneling Josephson junction array. The junction barriers stand vertically against the substrate and the spacing between the junctions is about 200 nm. In such a film, some interesting phenomena are manifested as a result of the mutual interactions among BJJs.

A method of observing these phenomena[25] is described below. First, a cross-type pattern is made of a BPB film by etching. The line width, W_x and W_y for each direction, are 10 to 50 μm. While observing the I_y-V_y characteristics along one direction(y), a dc current is injected into the other direction(x). For $W_x > W_y$, when the critical injection current is exceeded, the I_y-V_y curves (typical BJJ characteristics) are observed to shift toward the voltage direction by a unit of gap-voltage. With a further increse in the injection current, discrete shifts with the same step increment occur. Changing the direction of injection current results in a voltage shift to the opposite direction.

For a second case in which $W_x \leq W_y$, no voltage shift is obseved. When I_x exceeds the critical value, the I_y-V_y curves change to those shown in Fig.10. In this situation the current is found to flow in both the normal and reverse direction at a finite voltage. This is similar to the first case in that the injection current generates a definite voltage in the lateral direction.

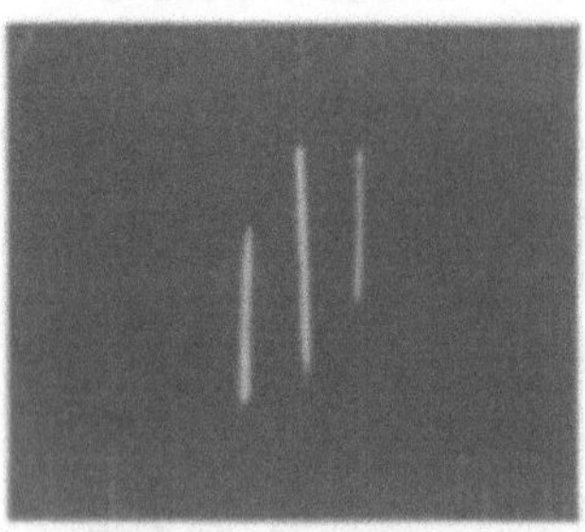

Fig.10 I_y-V_y characteristics for a sample of W_x=W_y =50 μ m. Y:100 μ A/div., X:2mV/div., I_{inj}=520 μA.

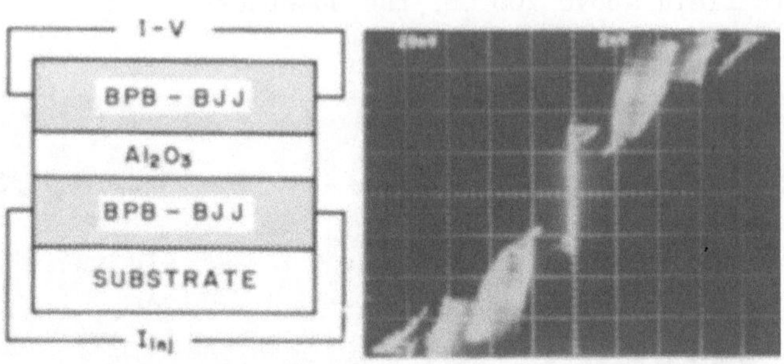

Fig.11 Device configuration (Left) and I-V characteristics showing a spike-like voltage generation. X:2mV/div., Y:20μA/div., I_{inj}=1.6 mA.

The mechanisms for the phenomena in these two cases have not been clarified yet. However, the essential point is that a chain of voltage state BJJs in the cross section plays a main role in these phenomena. Moreover, the electromagnetic radiation from the voltage state BJJs effectively assist the tunneling of quasiparticles and Cooper pairs. Niemeyer et al.[26] have recently observed similar I-V curves in 1474 series Josephson array under a microwave field.

Another novel phenomenon[27] is obseved in the configuration described in Fig.11. Two BPB films with 20 μm line widths are separated by an insulating Al_2O_3 layer (200 nm thick). When a dc current above a threshould level is injected into one of the electrodes (BPB-BJJ), a spike-like voltage generation occurs at the other electrode. The duration time of this voltage generation is about 20 msec. This voltage generation is found to occur easily (at a small injection current) when BJJs on the other electrode are in the voltage state.

This phenomenon can be explained by means of the inverse ac Josephson effect. With a sufficient injection current, many BJJs enter the voltage states, emitting electromagnetic radiation. When all of the voltage state BJJs are phase-locked, the electrode becomes superradiant. This superradiance induces voltage generation in the BJJs at the other electrode. A short duration time probably implies that the superradiance is not sufficiently stable in this configuration due to the geometrical irregularity of BJJs in BPB films.

In addition to these phenomena, a discrete current phenomenon has also been obseved.[28]

B. Optical Detector

The absorption coefficient is on the order of 10^4 cm^{-1} at 570 nm. The plasma wavelength of BPB is located in the near infrared region, where the reflectance becomes very small. These optical properties make the interaction with light quite effective in BPB thin films.

When light is irradiated on a superconductor, the Cooper pairs are destroyed and yield quasiparticles. This excessive number of quasiparticles induces nonequilibrium superconductivity, reducing the superconducting energy gap. We can detect the light as a result of this voltage change. This is the mechanism for detecting light using Josephson junctions. The change in energy gap, $\delta\Delta$, is determined by the relation such that $\delta\Delta=-n_{qp}/2N(0)$, where n_{qp} the induced quasiparticle density, and N(0) the density of states at the Fermi level. It should be noted that N(0) of BPB is more than one order of magnitude smaller than that of usual metals. Moreover, by biasing the higher order voltage state, the output voltage is multiplied by the number of voltage steps. Taking advantage of these properties, BPB-BJJ can be used for a highly sensitive infrared optical detector.

The detector[16] utilizes a polycrystalline BPB film with a 10 μm wide constriction to which a light is irradiated through an optical fiber. The BJJs are appropriately current-biased. The light signal is measured by a differential amplifier. In this configuration, a sensitivity of 10^{-10}W can be attained in the more than 1 μm wavelength range. Responsivity is about 10^4 V/W with a constant frequency response up to 500 MHz (the instrumental limit in the experiment).

VI. CONCLUSION

Thin film preparation and device application of a low carrier density superconductor $BaPb_{1-x}Bi_xO_3$ are reviewed. The essential point for BPB thin film preparation is to attain the stoichiometry including oxygen. It is shown that the (Pb,Bi)-compensated sputtering targets are quite effective for obtaining high quality BPB films (even single crystalline thin films). The grain boundaries of polycrystalline BPB films function as tunnel Josephson junctions, having the proximity sandwich feature. These films, as a two-dimensional Josephson array, give rise to a voltage-switching phenomena and voltage generation. The films are also applied to a new type of highly sensitive optical detector that utilizes the enhanced nonequilibrium superconductivity effect resulting from the low carrier density properties of this oxide system.

ACKNOWLEDGEMENTS

The author would like to express his gratitude to Drs. T. Murakami, Y. Enomoto, Y. Hidaka, and K. Moriwaki, without whose experience and skill this work would not have been possible. He would also like to thank Drs. H. Iwasaki, Y. Katayama, T. Manabe, and S. Takahashi for their continuous encouragement.

REFERENCES

1. A.W.Sleight, J.L.Gillson, and P.E.Bierstedt; Solid State Commun. 17 27 (1975).
2. P.D.Shannon and P.E.Bierstedt; J.Am.Ceramic Soc. 53 635 (1970).
3. T.D.Thanh, A.Koma, and S.Tanaka; Appl. Phys. 22 205 (1980).
4. V.V.Bogatko and Yu.N.Venevtsev; Soviet Phys. Solid State 22 705 (1980).
5. Y.Kahn, K.Nahm, M.Rosenberg, and H.Willner; Physica Status Solidi A39 79 (1977).
6. B.Btlogg; Physica 126B 275 (1984).
7. T.M.Rice; Superconductivity in Magnetic and Exotic Materials, T.Matsubara and A.Kotani, ed., (Springer Verlag, Berlin, 1984) p178.
8. C.Methfessel and S.Methfessel, in Superconductivity in d- and f-band Metals, W.Buckel and W.Weber, ed., (Kernforschungszentrum, Karlsruhe, FRG, 1982) p.393.
9. A.Katsui and M.Suzuki; Jpn.J.Appl.Phys. 21 L157 (1982).
10. J.P.Remeika, B.Batlogg, F.DiSalvo, A.S.Cooper, and H.Barz; Bull. Am. Phys. Soc. 27 318 (1982).
11. B.Batlogg, J.P.Remeika, R.C.Dynes, H.Barz, A.S.Cooper, and J.P.Garno; in Superconductivity in d- and f-band Metals, W.Buckel and W.Weber, ed., (Kernforschungszentrum, Karlsruhe, FRG, 1982) p.401.
12. T.M.Rice and L.Sneddon; Phys. Rev. Lett. 47 689 (1981).
13. L.F.Mattheiss and D.R.Hamann; Phys. Rev. B26 2686 (1982); ibid B28 4227 (1983).
14. Y.Enomoto, M.Suzuki, T.Murakami, T.Inukai, and T.Inamura; Jpn. J. Appl. Phys. 20 L661 (1981).
15. M.Suzuki, Y.Enomoto, and T.Murakami; J. Appl. Phys. 56 2083 (1984).
16. Y.Enomoto, M.Suzuki, and T.Murakami; Jpn. J. Appl. Phys. 23 L333 (1984).
17. M.Suzuki, T.Murakami, and T.Inamura; Jpn. J. Appl. Phys. 19 L231 (1981).
18. M.Suzuki, Y.Enomoto, T.Murakami, and T.Inamura; J. Appl. Phys. 53 1622 (1982).
19. M.Suzuki and T.Murakami; J. Appl. Phys. 56 2330 (1984).
20. Y.Hidaka, M.Suzuki, T.Murakami, and T.Inamura; Thin Solid Films 106 311 (1983).
21. M.Suzuki, T.Murakami, and T.Inamura; J. Vac. Soc. Jpn. 24 19 (1981) (in Japanese).
22. M.Suzuki and T.Murakami; Jpn. J. Appl. Phys. 22 1794 (1983).
23. M.Suzuki and T.Murakami; Solid State Commun. 53 691 (1985).
24. V.N.Stepankin, E.A.Protasov, A.V.Kuznetsov, and S.V.Zaitsev-Zotov; JETP Lett. 41 27 (1985).
25. Y.Enomoto, M.Suzuki, T.Murakami, and T.Inamura; Jpn. J. Appl. Phys. 21 L384 (1982).
26. J.Niemeyer, J.H.Hinken and R.L.Kautz, Appl. Phys. Lett. 45 478 (1984).
27. K.Moriwaki, M.Suzuki, and T.Murakami; Jpn. J. Appl. Phys. 23 L181 (1984).
28. K.Moriwaki, M.Suzuki, and T.Murakami; Jpn. J. Appl. Phys. 23 L115 (1984).

UHV DEPOSITION AND IN-SITU ANALYSIS OF THIN-FILM SUPERCONDUCTORS

J. Talvacchio, M. A. Janocko, J. R. Gavaler,
and A. I. Braginski

Westinghouse R&D Center
Pittsburgh, Pennsylvania

ABSTRACT

The application of a UHV deposition and surface analysis facility to the fabrication of high-transition-temperature superconducting films for studies of epitaxial growth and development of refractory tunnel junctions is discussed. A description is given of the vacuum system with some detail of the chambers used for co-evaporation and reactive dc magnetron sputtering. Specific examples presented to illustrate the effect of MBE-like deposition conditions, and the role of RHEED, XPS, and, to a lesser extent, other surface-sensitive probes, are: (a) the preparation of clean and damage-free sapphire and Nb_3Ir substrates, (b) the epitaxial relationships in a number of superconductor-insulator systems such as sapphire and NbN, Nb_3Sn, and Mo-Re, (c) the development of CaF_2 and ion-beam oxidized Al and Mg tunnel barriers, and (d) the deposition of refractory counterelectrodes. Correlations are made with tunneling characteristics for appropriate examples.

INTRODUCTION

The fundamental aspects of Molecular Beam Epitaxy (MBE) have been established in the study of semiconductor thin films. We have attempted to prepare and characterize superconductor films in an analogous fashion. Some similarities to the growth of semiconductor films by MBE which are desirable are:

1. The use of single-crystal substrates with surfaces that are characterized with respect to structure (periodicity), composition, and contaminants.

2. Low ($<$ 0.1 nm/sec) deposition rates without having film properties dominated by background impurities which are incorporated into the films.

3. Precise control of composition by achieving stable deposition rates.

4. A UHV environment which allows sufficient time for characterization before surfaces are contaminated.

The realization of these growth conditions for high-T_c superconductors is different from semiconductors not just due to the different growth properties of metals and semiconductors. The high-T_c superconductors are all composed of at least one refractory element. Whereas effusion cells can be used to evaporate most of the elements needed for semiconductors, refractory metals must be deposited by electron-beam evaporation or by sputtering. Vacuum chambers must be designed so that there is no compromise in the level of contaminants due to the presence of a plasma for sputtering, or due to a greater heat load from higher source and substrate temperatures. Evaporation rates from electron-gun sources are inherently less stable than from effusion cells. Rate fluctuations are particularly significant for compounds with structural or superconducting properties which are sensitive to composition such as the A15 compounds.[1]

This paper is organized as follows: The capabilities of our vacuum system will be described with an emphasis on the conditions necessary for MBE-type growth and analysis of metallic films. Examples will be given of the manner in which surface analytical tools are used to help prepare and to investigate epitaxial high-T_c superconducting films and refractory tunnel junctions. The examples are in the areas of substrate preparation, niobium nitride, niobium-tin, and molybdenum-rhenium.

AN INTEGRATED DEPOSITION AND ANALYSIS FACILITY

The four chambers of the UHV deposition and analysis facility at Westinghouse are shown schematically in Figure 1. Three of the chambers were designed and fabricated by ISA/Riber. The prototype design was based, in part, on specifications and concepts developed by future users. The magnetron sputtering chamber was designed and assembled by Westinghouse. Samples as large as 2-inch wafers are clamped to molybdenum blocks before insertion to the system. The molybdenum blocks can be transferred between chambers, without exposure to atmosphere, through gate valves using transfer rods which are magnetically coupled to the exterior of the system.

Each chamber has two positions on its manipulator which hold the blocks - one heated stage and one unheated stage. The blocks have a small cavity on the back into which a fixed W-Re thermocouple is inserted as the block is mounted on a heated stage. A larger cavity in the molybdenum block surrounds the fixed heater filament. The transfer mechanism and heater configurations are standard Riber equipment although the 1250°C heaters were designed for this system.

Evaporation Chamber

Figures 2a and 2b show vertical and horizontal cross sections, drawn to scale, of the evaporation chamber. The levels labeled in Figure 2a are the transfer level, substrate and RHEED (Reflection High-Energy Electron Diffraction) level, rate monitor sensor and main shutter level, and effusion cell and electron-beam source level. The walls of the chamber are largely shielded from radiation from the electron guns by liquid-nitrogen-filled cryopanels (stippled areas). Figure 2b shows the arrangement of evaporation sources at the level of the electron-beam guns. Guns #1 and #2 have 40 cc hearths and are mounted on the same flange. Gun #3 has 4 hearths in-line, each with a 7 cc capacity. Gun #4 has 3 small in-line hearths. The position available for effusion cell #1 currently has a standard evaporation boat mounted in place of a third effusion cell.

Shuttered vibrating-crystal rate monitors are mounted above guns #1, #2, and #3. A fourth crystal monitor is positioned as close to the sample

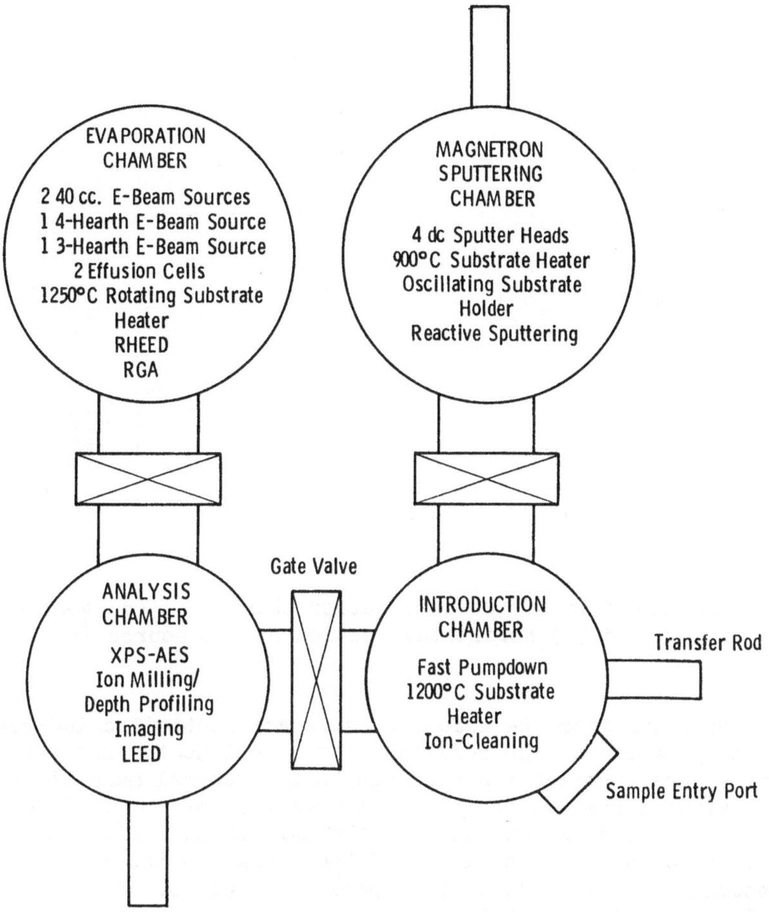

Fig. 1. Schematic of the Westinghouse deposition and analysis facility.

block as possible on the substrate level. Rate information from the crystal monitors is use in a feedback loop to control the filament current for guns #3 and #4. "Sentinel" rate monitors made by Leybold-Heraeus/Inficon, which use Electron Impact Emission Spectroscopy (EIES) to obtain a photon count proportional to the flux of evaporant, are positioned above guns #1 and #2. The vibrating-crystal rate monitors are used to calibrate the Sentinel sensors before each evaporation. The beam position on the hearth of a 40 cc gun is swept both laterally and longitudinally. Following Hammond,[1] the signal from the rate monitor can be used to control the amplitude of the sweep in one direction for a faster response to fluctuations than can be achieved by changes to the filament current alone. The response time is limited to 100 msec, the period at which the Sentinel updates its output signal. One reason that the Sentinel rate monitors are used in place of crystal monitors over the large sources is that there is no accumulated deposit to require frequent periodic maintenance. Our experience indicates that replacing crystals, removing flakes and debris, and replenishing sources has to be performed every 10 to 14 weeks.

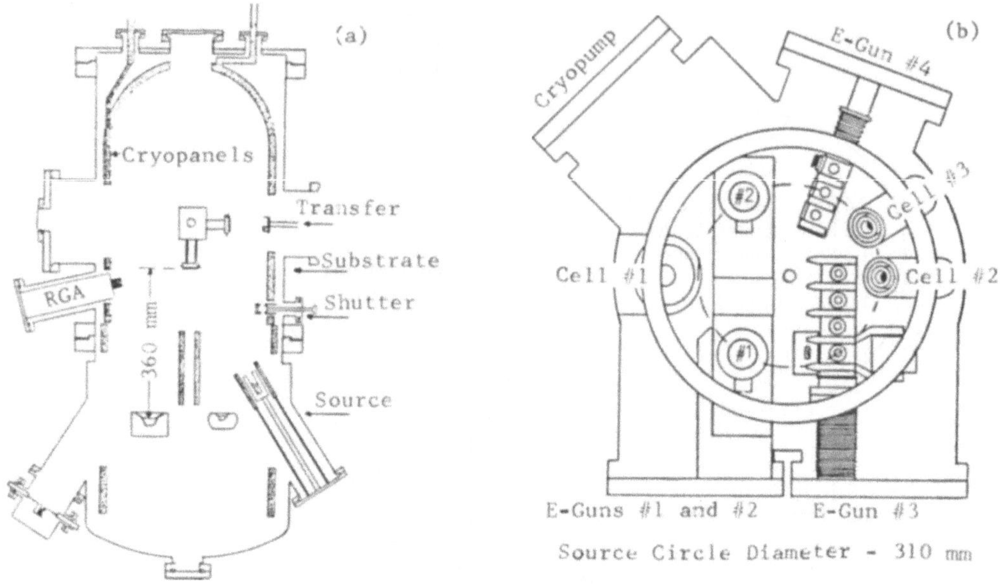

Fig. 2. Scale drawings of the evaporation chamber in (a) vertical section and (b) horizontal section at the source level.

Evaporation rates from the effusion cells are controlled indirectly by using the temperature of the cell to control cell heater current. The short-term rates are inherently stable due to the thermal mass of the cell. Several minutes are needed for the temperature of the cell to equilibrate after its shutter is opened, but the main shutter remains closed during that time. Changes in the level of the charge in the cell have not been great enough in typical 10-14 week periods to affect the reproducibility of rates versus temperature.

The evaporation chamber is pumped by a 400 liter/sec ion pump and a titanium sublimation pump. The normal base pressure (cryopanels cold) is < 3×10^{-9} Pa. Typical background pressure during evaporation of a film using a 40 cc electron-beam source is 2×10^{-8} Pa. The background pressure is somewhat lower if only the small guns or effusion cells are used. Substrates can be rotated during deposition and heated up to 1250°C.

Sputtering Chamber

A horizontal cross-section of the sputtering chamber at the level of the substrates and sputtering targets is shown in Figure 3. There are four sputter guns which accommodate 2-inch targets. Three of the guns are dc magnetron sources and the fourth, an rf magnetron gun, has been added recently. The substrates can be rotated to face any source, oscillated ±30° in front of a single source to promote uniform coverage, or oscillated ±60° between two guns for co-depositions.

A base pressure in the mid 10^{-7} Pa range is reached with a cryopump, titanium sublimation pump, and a 100 liter/sec ion pump. With the cryopump pumping through a nearly-closed throttle valve, just before adding the sputtering gas, the pressure is about 1×10^{-6} Pa. The level of impurities due to the background pressure is lower than the level introduced by research-grade argon with a pressure of 1 Pa. However, clean surfaces become contaminated with about a monolayer of oxygen in 10-30 minutes after deposition. The presence of a plasma apparently increases the out-

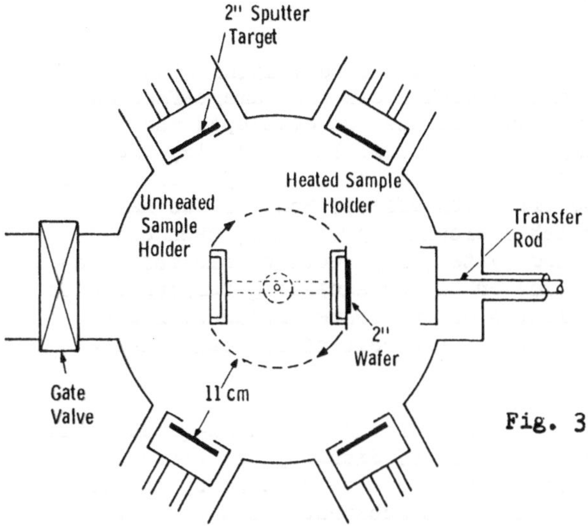

Fig. 3. Horizontal cross section of the magnetron sputtering chamber at the level of the targets and substrates.

gassing rate from the chamber walls. To qualify as an analogue to MBE systems, this chamber should have a liquid-nitrogen-filled cryopanel similar to the evaporation chamber or the sputtering system described in reference 2.

Analysis Chamber

The most important of the surface analysis techniques available in the analysis chamber has been X-ray Photoelectron Spectroscopy (XPS). The primary advantage of Auger Electron Spectroscopy (AES) over XPS is the possibility of focusing the beam to analyze a small area. However, for analysis of freshly-deposited films which cover large areas, XPS is preferred because the chemical shifts are more easily interpreted. AES is more frequently used in conjunction with ion milling to obtain depth profiles so that only a small area needs to be milled. However, with deposition and analysis capabilities both present in UHV, the uncertainties of ion-induced chemical change and uneven milling rates can be avoided by obtaining "depth profile" information as a multilayer structure is being formed.

The analysis chamber is pumped by a 200 liter/sec ion pump and a titanium sublimator. Typical pressure is 5 x 10^{-9} Pa, without cooling the sublimator cryoshroud. The electron spectrometer is a Riber MAC-1.

Introduction Chamber

The introduction chamber is the only one which is not baked each time it is vented. It has one 6" port, normally sealed with a viton o-ring, which can be quickly opened and closed for loading samples into the system. A rough vacuum is obtained with sorption pumps, and the chamber is pumped with a cryopump. The pressure can be reduced from 1 atm to < 5 x 10^{-6} Pa in 20 min. Then the sample block can be transferred to the sputtering or analysis chamber without deteriorating the vacuum level in those chambers. Pumping overnight reduces the pressure to about 1 x 10^{-7} Pa.

The introduction chamber is used for operations which would compromise the cleanliness of the other chambers such as an initial degassing of sample blocks, thermal oxidation in oxygen pressures up to 10 Pa, low-energy ion beam cleaning, and ion-beam oxidation in an argon/oxygen background.

SUBSTRATE PREPARATION

An essential preliminary step to the growth of epitaxial films is the preparation of single-crystal substrates with clean and damage-free surfaces. A procedure for preparing the surface must be developed and evaluated without removing the sample from vacuum. In principle, the procedure may be developed in one vacuum system and then repeated in the deposition system without characterizing its effect. However, we have found that the application of published cleaning processes, even one developed by us in a different vacuum system, may require some modification such as a longer ion-milling time or higher annealing temperature to obtain equivalent results. Some examples of substrate preparation are presented below with an emphasis placed on the role of in-situ characterization.

Sapphire (α-Al_2O_3) surfaces with a commercial epitaxial-grade polish have a layer of adsorbed carbon on the surface. Sinharoy has shown that the carbon AES signal from the (0001) surface can be reduced by thermal desorption up to 1000°C.[3] Therefore, sapphire substrates are routinely pre-heated to 900-1000°C even for low-temperature depositions and then brought to the deposition temperature. No carbon has been detected by in-situ XPS after heating. RHEED patterns show that the surface is smooth and crystalline in contrast to sapphire with a fine, non-epitaxial polish. The samples with the lower-grade polish exhibit a RHEED pattern characteristic of an amorphous surface. Single-crystal and polycrystalline Nb films have been grown side-by-side on sapphires with the different polishes for diffusion-couple experiments investigating the formation of A15 phase Nb_3Sn and Nb_3Al.[4]

Single crystals of A15 Nb_3Ir, grown by E. Walker at the Universite de Geneve, have been used for the epitaxial growth of Nb_3Ge. The cleaning procedure, developed in a separate surface analysis system, used an ion heat treatment, that is, a 500 eV argon ion milling while the sample was maintained at a temperature of > 700°C.[5] This combination was more effective than heating alone or ion milling followed by heating, in removing oxygen from the surface. Oxygen-free (100) and (111) surfaces exhibited unreconstructed (1x1) Low-Energy Electron Diffraction (LEED) patterns. Only the (110) surface exhibited a reconstructed surface when there was no oxygen contamination.

The prescribed ion heat treatment for Nb_3Ir was repeated in the deposition and analysis system. Although there were some differences in the apparatus compared with the vacuum system where the recipe was developed (such as a 45° incidence for the ion beam instead of normal incidence), it was surprising to find that two cycles of the ion heat treatment at 800°C followed by annealing at 1000°C were needed to remove all oxygen from the surface as determined by XPS. XPS and AES sampled comparable distances of about 1.5 nm into the film in this case because the O(1s) photoelectrons and O(KVV) Auger electrons have comparable kinetic energies. This experience emphasized the necessity of in-situ characterization of the actual substrate to be used for film growth.

Single crystals of MgO were polished with Syton (0.035 micron silica particles in a basic solution), degreased, and heated to 700°C to remove adsorbed gases. Without additional treatment, LEED patterns were observed and the surface was free of carbon.[6]

NIOBIUM NITRIDE

All NbN films have been prepared by reactive dc magnetron sputtering. Epitaxial, single-crystal films have been grown on several orientations of sapphire and MgO substrates,[7] following Noskov et al. and Oya and Onodera.[8,9] Details of the deposition parameters and the role of epitaxy in stabilizing the high-T_c composition of NbN are reported in reference 7.

The epitaxial relationship between the substrate and film can be easily observed with RHEED. Figure 4a is a RHEED pattern of a (100) MgO crystal. The azimuthal angle of the electron beam with respect to the crystal was adjusted by rotating the sample block until a high-symmetry pattern was observed. In this case, the beam is parallel to the (010) direction. After a 100 nm thick NbN film was deposited at 700°C in the sputtering chamber, the block was transferred back to the evaporation chamber and the azimuthal angle was set to the value used for the substrate (± 1°). The pattern in Figure 4b was observed. The presence of streaks instead of spots, Kikuchi lines, and the ring of lines at the bottom of the photograph, indicates that the film is a smooth single crystal.[10]

The RHEED patterns are more difficult to interpret for epitaxial relationships between crystals with different structures such as NbN grown on sapphire, although x-ray diffraction can be used for thick enough films. Figures 4c and 4d show RHEED patterns, at a fixed azimuthal angle, of an (0001) sapphire substrate and a 100 nm thick NbN film deposited at 700°. In this case, the growth direction was easily identified by comparison with the RHEED pattern of a (111) NbN film (Figure 4f) grown at

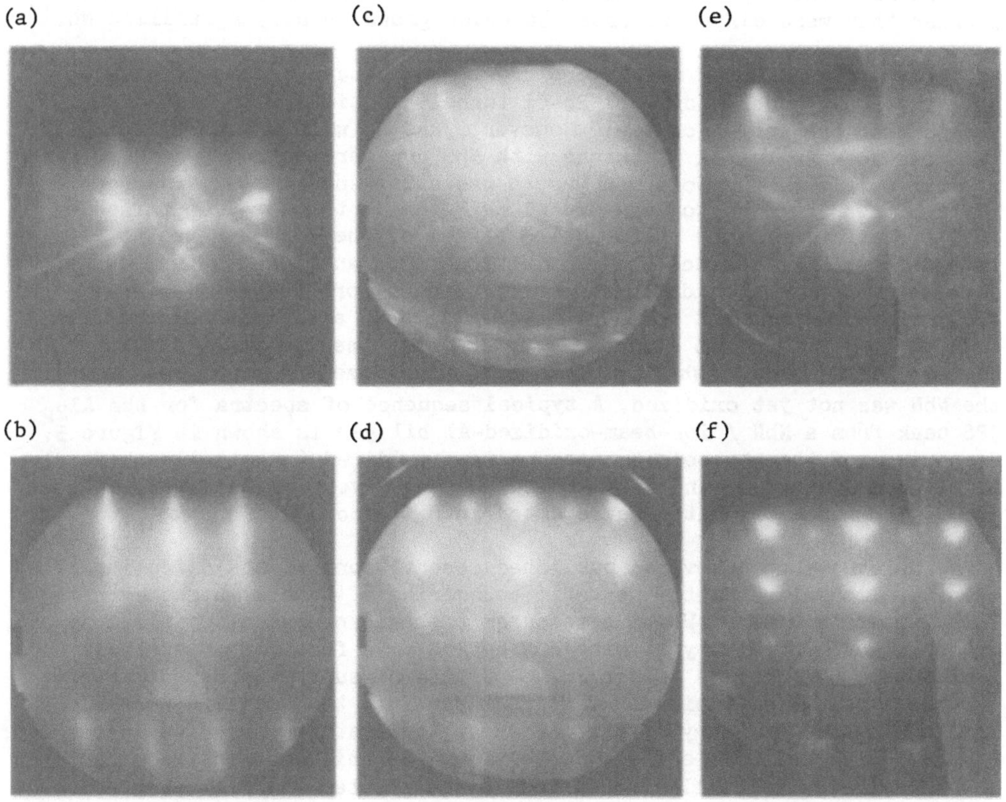

Fig. 4. Pairs of RHEED patterns for substrate and film with a fixed azimuthal angle: (a) (100) MgO substrate and (b) (100) NbN film deposited at 700°C; (c) (0001) sapphire substrate and (d) (111) NbN film deposited at 700°C; (e) (111) MgO substrate and (f) (111) NbN film deposited at 300°C.

300°C on a (111) MgO substrate (Figure 4e). For the (111) NbN films, the RHEED patterns exhibited "arrowhead"-shaped spots rather than streaks. The spot pattern indicated that the electron beam was diffracted by a three-dimensional crystal caused by the roughness of the surface of the film. The shape of the spots was a signature of a faceted surface.[11]

These samples provide an opportunity to study fundamental superconducting properties of NbN that may be anisotropic, such as tunneling $\alpha^2F(\omega)$ and H_{c2}. We have used XPS to measure the anisotropic growth of the native oxide of (100), (111), and randomly-oriented polycrystalline surfaces at room temperature.[12] After three days exposure to air, the respective thicknesses were 1.5 nm, 1.9 nm, and 1.8 nm based on assumptions of a uniform thickness for the oxide layer and a photoelectron escape depth of 2.0 nm. A number of groups have successfully used the thermally-grown native oxide for tunnel barriers.[13-16] We have also made low-leakage junctions with the native oxide of polycrystalline films using Pb-Bi counterelectrodes. The ratio of the current at 5 mV (above the gap) to the current below the gap, i(5)/i(2.5), was as high as 110 at 4.2K. However, the oxide that formed on the (100) and (111) surfaces had much different barrier properties. The relatively thin oxide of the (100) surface formed a low-leakage barrier that had a resistance 2 orders of magnitude higher than the oxide of the polycrystalline film. The thicker oxide of the (111) surface had a similarly high resistance but the tunnel junctions had high leakage currents, perhaps related to the faceted surface observed with RHEED. The native oxide grown thermally on single-crystal Nb films has also been reported to have properties as a tunnel barrier that were different from the oxide grown on polycrystalline Nb.[17]

The oxides of thin overlayers of Al and Mg have also been used to make low-leakage NbN / oxide / Pb-Bi tunnel junctions with higher resistances than the native oxide.[18] However, just as with the native oxide, attempts to make tunnel junctions with NbN counterelectrodes resulted in shorted barriers. One solution was to use a low-energy (300 V was optimum) argon-oxygen ion beam to make the Al or Mg oxide thicker (determined by XPS) and, perhaps, more uniform. The effect of the ion beam was to remove material from the surface while oxidizing in an analogue to the Greiner process.[19] Ion-beam oxidation has been used to form the native oxide of Nb films,[20-22] but has not been used before with an artificial barrier. The role of in-situ XPS was crucial in determining the end-point of the process, the point at which a minimum of unoxidized Al or Mg remained and the NbN was not yet oxidized. A typical sequence of spectra for the Al_{2p} XPS peak from a NbN / ion-beam-oxidized-Al bilayer is shown in Figure 5. Comparison of the photoelectron count from oxidized (chemically-shifted) Al with unoxidized Al and with Nb (not shown) gives the thickness of the Al_2O_3. The tunneling data are reported in reference 18.

There have been several published observations of the diffusion of a thin metal overlayer into the grain boundaries of a metal base, even at room temperature.[23-25] We have measured the thicknesses of oxidized Al and Mg overlayers on polycrystalline NbN by XPS as a function of analysis temperature (Figures 6a and 6b). The purpose of such measurements was to observe possible diffusion of the overlayer into the base and possible changes in the chemistry of the oxides at temperatures which would be desirable for counterelectrode deposition. The thickness of the unoxidized metal layer was initially two to four times greater than for samples used for tunnel junctions so that diffusion would be easier to observe. The samples were maintained at each temperature for 15 minutes before collecting the XPS spectra during an additional hour. In the case of the oxidized Al overlayer, there was no observable change in the oxide thickness or chemical shift up to 800°C. The thickness of the unoxidized Al layer decreased to only half of its room temperature value at 800°C, above the

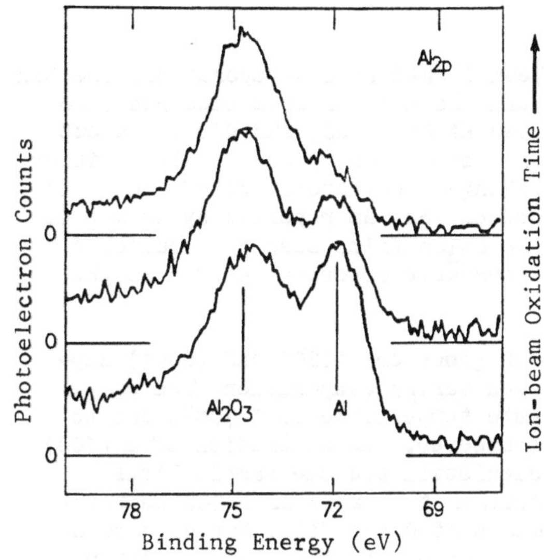

Fig. 5. XPS spectra for Al_{2p} photoelectrons used for monitoring the effect of ion-beam oxidation. The first spectrum was recorded after an initial thermal oxidation at room temperature for 15 min in 13 Pa oxygen.

melting point of Al. However, for the oxidized Mg overlayer, the unoxidized Mg thickness started to decrease at 300°C and disappeared by 500°C. No changes were seen in the MgO.

The results suggest that high temperatures might be used for counterelectrode growth, particularly for oxidized Al barriers. We have deposited NbN counterelectrodes on NbN / oxidized Al bilayers at temperatures up to 300°C. The leakage currents for junctions formed at 300°C were comparable to junctions with NbN counterelectrodes deposited at room temperature. Tunnel junctions with oxidized Mg barriers had a lower V_m compared to junctions with oxidized Al barriers due to smearing of the NbN gap rather than higher leakage at zero bias.[18] We attribute the smearing to the diffusion of Mg into the NbN base electrode. Since the ionic radius of Mg is larger than that of Al, the diffusion of Mg may be accompanied by a reaction forming magnesium nitride which does not occur in the case of Al. We have not been able to determine with XPS whether there was a reaction.

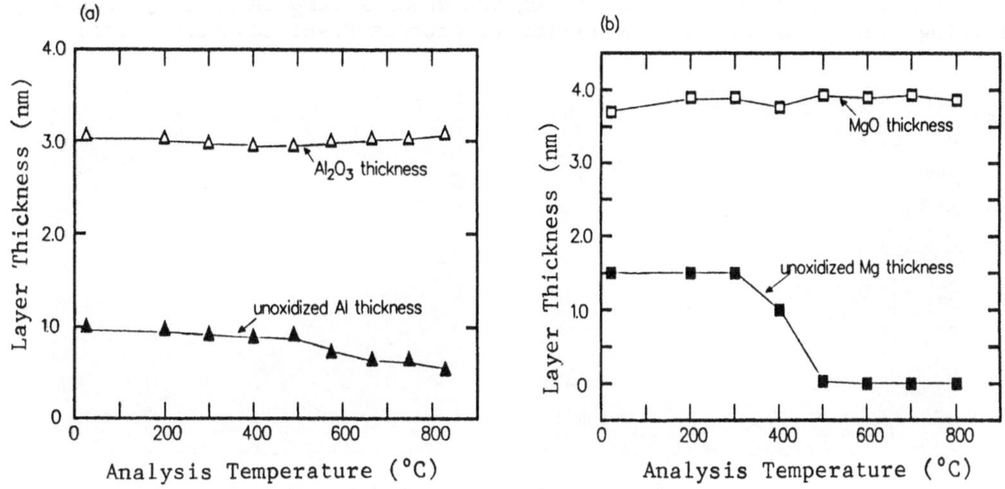

Fig. 6. The thicknesses of (a) Al_2O_3 and Al and, (b) MgO and Mg as a function of XPS analysis temperature for thin oxidized overlayers of Al and Mg on polycystalline NbN.

A15-structure Nb_3Sn films have been formed by co-evaporation. The tin was evaporated from effusion cell #2 and the niobium from electron beam gun #1. A series of films were deposited at 850, 900, and 950°C without rotating the sample block. Plots of Sn composition versus sample positior on the block showed a "composition locking" (re-evaporation of excess Sn from the Nb_6Sn_5 phase) at 25 atomic percent Sn, as reported by Rudman et al.,[26] for all but the 850°C films. The Nb_3Sn films used for studies of epitaxial growth and for tunnel junctions were deposited with a slight excess of Sn and at > 900°C.

Epitaxial films of Nb_3Sn have been grown on (11$\bar{2}$0) and (0001) sapphire. RHEED patterns cannot be observed during evaporation from an electron-beam source or if the substrate temperature is > 700°C due to background light on the RHEED screen. However, the deposition of a (100) film was interrupted after 6.0 nm accumulation and the sample block cooled. We observed a RHEED pattern similar to that from a 200 nm thick single-crystal (x-ray rocking curve width of 0.4°) film, but with more diffuse spots.[27] There was no evidence that crystallites with another orientation were competing with the single crystal matrix as reported by Marshall et al. for a film grown on (1$\bar{1}$02) sapphire.[28]

Figure 7 contains RHEED patterns of a 200 nm thick Nb_3Sn film and an 8.0 nm thick overlayer of CaF_2 observed from the same azimuthal angle. The CaF_2 was evaporated from a standard evaporation boat mounted in the position for effusion cell #1 in the evaporation chamber while the substrates were maintained at room temperature. The Nb_3Sn film was a smooth single crystal with an (100) growth direction. The RHEED pattern of the CaF_2 overlayer contains both rings and a pattern of focussed spots, suggesting that there was a mixture of randomly-oriented grains and an epitaxial matrix. Although the existence of an epitaxial relation between these structures has not been firmly established, both Nb_3Sn and CaF_2 have cubic structures and a lattice mismatch of only 3% (0.529 and 0.546 nm). Growth of CaF_2 at higher temperatures may provide an epitaxial insulator-superconductor system analogous to MgO-NbN which could promote the formation of a high-T_c layer in tunnel junction counterelectrodes within a coherence length of the barrier. The best epitaxial layers of CaF_2 grown on Si have been formed at 600±25°C.[29]

Tunnel junctions were made using the Nb_3Sn / CaF_2 samples by transferring them to another vacuum system to deposit Pb-Bi counterelectrodes.

(a) (b)

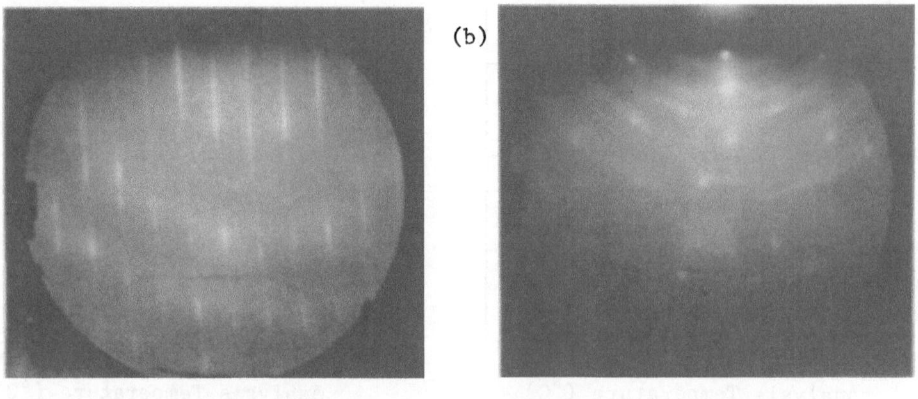

Fig. 7. RHEED patterns of (a) a 200 nm thick film of Nb_3Sn grown at 900°C, and (b) an 8.0 nm thick overlayer of CaF_2 deposited at 20°C.

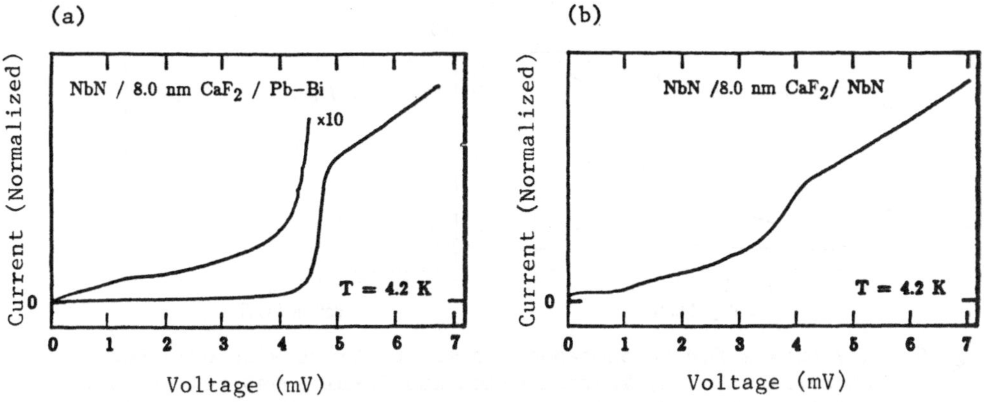

Fig. 8. Quasiparticle I-V curves for (a) NbN / CaF_2 / Pb-Bi and (b) NbN / CaF_2 / NbN tunnel junctions. The surface of the base electrode was oxidized prior to the CaF_2 deposition.

We measured samples with a range of thicknesses of the CaF_2 layer from 6.0 to 10.0 nm (based on evaporation rates and confirmed by XPS). All junctions had I-V curves characteristic of superconducting microbridges even though one might expect that the exposure to air just prior to depositing the counterelectrode would oxidize any area of the base electrode that was left exposed by pinholes in the CaF_2.[30] Similar I-V characteristics were observed for NbN / CaF_2 / Pb-Bi tunnel junctions.

In a second experiment with NbN base electrodes, a native oxide barrier was formed, prior to the deposition of 8.0 nm of CaF_2, by bleeding oxygen into the introduction chamber. The quasiparticle I-V curves for the latter set of samples, shown in Figure 8a for a Pb-Bi counterelectrode and Figure 8b for a NbN counterelectrode, indicated that the pinholes were sealed. The resistances of the junctions with Pb-Bi counterelectrodes were approximately 5×10^{-3} ohm-cm^2, two orders of magnitude higher than we typically obtained for the native oxide alone. The higher leakage at zero voltage for the junctions with NbN counterelectrodes was consistent with the fact, stated earlier, that the thermally-grown native oxide barrier was always shorted during the deposition of an NbN overlayer. The CaF_2 deposited on NbN was polycrystalline with randomly-oriented grains. Asano et al. have reported the successful use of amorphous ZrF_4 and AlF_3 tunnel barriers with Nb / Pb junctions.[31] They also had a low-resistance barrier, formed by cleaning in an Ar/CF_4 rf plasma, underneath the artificial fluoride barrier. Unresolved issues are whether the composite barrier structure is necessary for all of the fluoride compounds and whether the presence of pinholes is related to the evaporation of such compounds as undissociated molecular units.[32]

Figure 9a shows the I-V curve of a typical junction formed on a Nb_3Sn base with a thermally-oxidized Al barrier and a Pb-Bi counterelectrode. The tunnel junctions with thermally-oxidized barriers established that the top layer of the Nb_3Sn film was homogeneous with a high T_c,[27] but the thermal-oxide barrier was shorted by the deposition of refractory counterelectrodes. Figure 9b shows the I-V curve of a tunnel junction of the form, Nb_3Sn / Al_2O_3 / NbN, which was fabricated with the ion-beam oxidized Al tunnel barriers discussed in the previous section and in reference 18. The subgap conductance was primarily due to damage to the Nb_3Sn by the 300 V Ar-O_2 ion beam as indicated by the rise in conductance at the NbN gap.

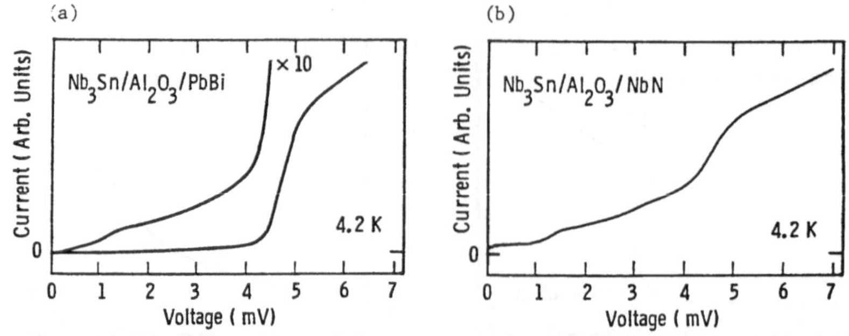

Fig. 9. I-V curves for Nb_3Sn-based tunnel junctions with oxidized Al barriers. (a) Thermal oxide and Pb-Bi counterelectrode. (b) Oxide formed by ion-beam treatment and NbN counterelectrode deposited at room temperature. The rise in conductance at the NbN gap, 1.2 mV, was from damage to the Nb_3Sn surface caused by the ion milling.

MOLYBDENUM-RHENIUM

We have studied the Mo-Re system in the range of 30 to 40 atomic percent Re for the following reasons:

1. A15 structure Mo_3Re provides an opportunity to study the stabilization of an A15 compound which does not exist in equilibrium at any composition.[33]
2. With a T_c of 12K for the bcc alloy and up to 15K for the A15 structure,[34] Mo-Re provides an alternative high-T_c material for refractory tunnel junctions.
3. In either the bcc or A15 structure, the T_c of Mo-Re is relatively insensitive to disorder. The coherence length, 20 nm, estimated from normal-state and superconducting properties in the literature,[35,36] is 4 to 5 times longer than for other high-T_c superconductors. Both are desirable properties for obtaining the T_c of the bulk material within a coherence length of a tunnel barrier / counterelectrode interface, particularly at low temperatures.

Single crystal films of bcc (α-Mo) structure $Mo_{65}Re_{35}$ with a (110) orientation and an x-ray rocking curve width of 0.3° have been grown by evaporation at 800°C on (11$\bar{2}$0) sapphire. The Mo source was electron-beam gun #2 and the Re source was gun #3. Although gun #3 is not equipped for high-frequency rate control, tunnel junctions made on these films with

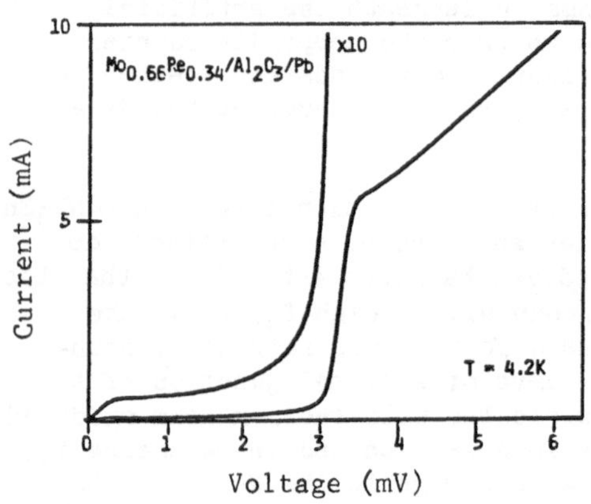

Fig. 10. Tunneling into a 20.0 nm thick $Mo_{65}Re_{35}$ film with a polycrystalline bcc structure. The width and value of the gap voltage showed that the film was homogeneous with T_c = 11.5K.

oxidized Al barriers and Pb counterelectrodes showed that the samples were homogeneous with a gap voltage only 0.2 meV wide. Films grown at 100°C had a polycrystalline bcc structure. The T_c's of the films agreed with values reported for bulk samples (12K) independent of substrate temperature in the range of 100 - 1000°C.[33] The T_c decreased by 0.5K for films grown both at low temperature (100°C) and only 20.0 nm thick, and were still homogeneous based on tunneling measurements (Figure 10).

A set of Mo-Re films were deposited at 1000°C in a background pressure of 6×10^{-7} Pa, consisting of mostly methane and nitrogen, and typical for the first set of films made after the evaporation chamber has been opened for maintenance, pumped out, and baked. Although other samples grown at the same temperature in a background of $< 2 \times 10^{-8}$ Pa had a single-crystal bcc structure, x-ray diffraction showed that the films grew with an A15 structure in a (100) orientation. The RHEED pattern showed a times-3 surface reconstruction that had not been seen for other Mo-Re films. A study of another single-crystal transition metal A15 compound, Nb_3Ir, found that the (100) surface also exhibited a (3x3) reconstruction when contaminated by oxygen and carbon.[5]

The formation of the A15 structure by evaporation has been reported by Postnikov et al. for comparable substrate temperatures but in a background pressure of about 10^{-4} Pa.[37] In contrast to the earlier A15 Mo-Re films formed by evaporation or sputtering, the single-crystal films had T_c values of approximately 12K. The onset of the transition for one film was 12.7K, 0.5K higher than found in any of the bcc samples. These results suggest that the impurities are needed not just to stabilize the A15 phase, but can also affect the T_c.

CONCLUSIONS

The growth of films of high-T_c superconductor materials in UHV has produced, with the appropriate choice and preparation of substrates, single-crystal samples for all superconducting materials that have been tested. For stable superconductors which have been grown as bulk single crystals, the existence of thin-film single crystals expands the number of characterization techniques which can be used to investigate the mechanisms of superconductivity, to include, for example, tunneling. Epitaxial thin-film growth is the only way to obtain single crystals of metastable compounds such as A15 Mo-Re and stoichiometric Nb_3Ge.

The combination of UHV film growth with in-situ surface treatment and analysis makes feasible the development of all-epitaxial tunnel junctions for high-operating-temperature circuit applications with low-loss transmission lines. At the least, clean, well-characterized interfaces and tunnel barriers used with homogeneous superconducting electrodes can serve as model systems for the study of barrier physics, metal-surface physics, nucleation and growth in metal-insulator epitaxial systems, and superlattices.

ACKNOWLEDGMENTS

This work was supported in part by AFOSR Contract No. F49620-78-C-0031 and ONR Contract No. N00014-82-C-0617. The authors would like to acknowledge ISA/Riber and especially M. Picault for their role in building the vacuum apparatus for this work. J. Bevk of Bell Laboratories and C. M. Falco and I. K. Schuller of Argonne National Laboratory contributed to the design of the evaporation chamber. H. Pohl and A. L. Foley have contributed to the installation and operation of the facility.

REFERENCES

1. R. H. Hammond, Electron beam evaporation synthesis of A15 superconducting compounds: Accomplishments and prospects, IEEE Trans. Magn. MAG-11(2):201 (1975).
2. J. M. Lumley, R. E. Somekh, J. E. Evetts, and J. H. James, High quality all refractory tunnel junctions for SQUID applications, IEEE Trans. Magn. MAG-21(2):539 (1985).
3. S. Sinharoy, LEED and Auger studies of the structure and composition of the (0001) surface of sapphire, submitted to Surface Science.
4. A. I. Braginski, J. R. Gavaler, and K. Schulze, Formation of A15 phase in epitaxial and polycrystalline Nb-Sn and Nb-Al diffusion couples, presented at the International Cryogenic Materials Conference, Boston, 1985.
5. S. Sinharoy, A. I. Braginski, J. Talvacchio, and E. Walker, A LEED, AES, and XPS study of single-crystal Nb_3Ir surfaces, submitted to Surf. Sci..
6. S. Sinharoy, private communication
7. J. R. Gavaler, J. Talvacchio, and A. I. Braginski, Epitaxial growth of NbN films, presented at the International Cryogenics Materials Conference, Boston, 1985.
8. V. L. Noskov, Y. V. Titenko, F. I. Korzhinskii, R. L. Zelenkevich, and V. A. Komashko, Heteroepitaxial layers of niobium nitride on sapphire, Sov. Phys. Crystallogr. 25(4):504 (1980).
9. G. Oya and Y. Onodera, Transition temperatures and crystal structures of single-crystal and polycrystalline NbN films, J. Appl. Phys. 45(3):1389 (1974).
10. M. G. Lagally, Diffraction techniques, in: "Methods of Experimental Physics, Vol.22," R. L. Park and M. G. Lagally, eds., Academic Press, New York (1983), pp. 237-298.
11. B. A. Joyce, J. H. Neave, P. J. Dobson, and P. K. Larsen, Analysis of reflection high-energy electron-diffraction data from reconstructed semiconductor surfaces, Phys. Rev. B 29(2):814 (1984).
12. J. R. Gavaler, A. I. Braginski, M. A. Janocko, and J. Talvacchio, Epitaxial growth of high-T_c superconducting films, presented at the International Conference on the Materials and Mechanisms of Superconductivity, Ames, Iowa, May, 1985.
13. J. C. Villegier, L. Vieux-Rochaz, M. Goniche, P. Renard, and M. Vabre, NbN tunnel junctions, IEEE Trans. Magn. MAG-21(2):498 (1985).
14. R. B. van Dover, D. D. Bacon, and W. R. Sinclair, Superconductive tunneling into NbN deposited near room temperature, Appl. Phys. Lett. 41(8):764 (1982).
15. V. M. Pan, V. P. Gorishnyak, E. M. Rudenko, V. E. Shaternik, M. V. Belous, S. A. Koziychuk, and F. I. Korzhinsky, Investigation of the properties of niobium nitride films, Cryogenics 23(5):258 (1983).
16. M. Igarashi, M. Hikita, and K. Takei, Barrier/electrode interface structure and I-V characteristics of NbN Josephson junctions, in: "Advances in Cryogenic Engineering - Materials," vol. 30, Plenum Press, New York (1984), p. 535.
17. S. Celaschi, T. H. Geballe, and W. P. Lowe, Tunneling properties of single crystal Nb / Nb_2O_5 / Pb Josephson junctions, Appl. Phys. Lett. 43(8):794 (1983).
18. J. Talvacchio, J. R. Gavaler, A. I. Braginski, and M. A. Janocko, Artificial oxide barriers for NbN tunnel junctions, to be published in J. Appl. Phys.
19. J. H. Greiner, Oxidation of Pb films by rf sputter etching in an oxygen plasma, J. Appl. Phys. 45(1):32 (1974).
20. R. Herwig, Ion-beam oxidation of Nb-based Josephson junctions, Electron. Lett. 16:850 (1980).

21. A. W. Kleinsasser and R. A. Buhrman, High-quality submicron Nb tunnel junctions with reactive ion-beam oxidation, Appl. Phys. Lett. 37(9):841 (1980).
22. S. S. Pei and R. B. van Dover, Ion beam oxidation for Josephson circuit applications, Appl. Phys. Lett. 44(7):703 (1984).
23. J. Kwo, G. K. Wertheim, M. Gurvitch, and D. N. E. Buchanan, X-ray photoelectron study of surface oxidation of Nb/Al overlayer structures, Appl. Phys. Lett. 40(8):675 (1982).
24. J. Kwo, G. K. Wertheim, M. Gurvitch, and D. N. E. Buchanan, XPS and tunneling study of air-oxidized overlayer structures of Nb with thin Mg, Y, and Er, IEEE Trans. Magn. MAG-19(3):795 (1983).
25. A. I. Braginski, J. R. Gavaler, M. A. Janocko, and J. Talvacchio, New materials for refractory tunnel junctions: Fundamental aspects, presented at the Third International Conference of Superconducting Quantum Devices, Berlin, June, 1985.
26. D. A. Rudman, F. Hellman, R. H. Hammond, and M. R. Beasley, A15 Nb-Sn Tunnel Junction Fabrication and Properties, J. Appl. Phys. 55(10):3544-3553 (1984).
27. J. Talvacchio, A. I. Braginski, M. A. Janocko, and S. J. Bending, Tunneling and interface structure of oxidized metal barriers on A15 superconductors, IEEE Trans. Magn. MAG-21(2):521 (1985).
28. A. F. Marshall, F. Hellman, and B. Oh, Epitaxy of Nb_3Sn films on sapphire, in: "Layered structures, Epitaxy, and Interfaces," Materials Research Society, Pittsburgh (1985), p. 517.
29. J. M. Phillips, Recent progress in epitaxial fluoride growth on semiconductors, in: "Layered structures, Epitaxy, and Interfaces," Materials Research Society, Pittsburgh (1985), p. 143.
30. D. A. Rudman and M. R. Beasley, Oxidized amorphous-silicon superconductung tunnel junction barriers, Appl. Phys. Lett. 36(12):1010-1013 (1980).
31. H. Asano, K. Tanabe, O. Michikami, M. Igarashi, and M. Beasley, Fluoride barriers in Nb/Pb tunnel junctions, Jap. J. Appl. Phys. 24:289 (1985).
32. R. F. C. Farrow, P. W. Sullivan, G. M. Williams, G. R. Jones, and D. C. Cameron, MBE-grown fluoride films: A new class of epitaxial dielectrics, J. Vac. Sci. Technol. 19(3):415 (1981).
33. R. D. Blaugher, A. Taylor, and J. K. Hulm, The superconductivity of some intermetallic compounds, IBM J. of Res. and Dev. 6(1):116 (1962).
34. J. R. Gavaler, M. A. Janocko, and C. K. Jones, A15 Structure Mo-Re superconductor, Appl. Phys. Lett. 21(4):179 (1972).
35. F. J. Morin and J. P. Maita, Specific heats of transition metal superconductors, Phys. Rev. 129(3):1115 (1963).
36. A. Echarri, M. J. Witcomb, D. Dew-Hughes, and A. V. Narlikar, Dependence of the lower critical field on normal state resistivity in superconducting alloys, Philos. Mag. 18:1089 (1968).
37. V. S. Postnikov, V. V. Postnikov, and V. S. Zheleznyi, Superconductivity in Mo-Re system alloy films produced by electron beam evaporation in high vacuum, Phys. Status Solidi A 39:K21 (1977)

A MULTILAYER TECHNOLOGY WITH HIGH T_c Nb_3Ge FILMS

B.David, M.Mück and H.Rogalla

Institut für Angewandte Physik
Justus-Liebig-Universität Giessen
Giessen, Germany

ABSTRACT

For practical high-T_c Josephson devices like Nb_3Ge dc-SQUIDs, a multilayer technology is indispensable. A consequence is that interconnections between different layers are needed. This requires high-T_c superconducting contacts from one Nb_3Ge layer to the other. We therefore developed a versatile Nb_3Ge multilayer technology based on SiO_2 as insulating material. Bottom- as well as top Nb_3Ge layers have a T_c of nearly 21 K. Interconnections between two Nb_3Ge films have T_c's up to 20.5 K. As a first application for this described multilayer technology, a Nb_3Ge dc-SQUID of the microbridge type with integrated input coil was prepared. The whole device is working up to 19 K.

INTRODUCTION

The fabrication technique for superconducting electronics has developed towards a multiple-layer thin-film technology that involves the successive deposition of various superconducters, insulaters, and normal metals. By contrast to the common Pb- and Nb-technology, application of the high-T_c superconducting materials is much more difficult. In case of the superconductor with the highest known T_c near 23 K, Nb_3Ge, substrate temperatures during preparation of 800°C-900°C are needed. Patterning by standard lift-off processes is not possible. For practical devices it is furthermore indispensable to use insulating layers between the different superconducting films. For the case of Nb_3Ge this requires an insulating material of high thermal and mechanical stability. Also the superconducting properties of the Nb_3Ge films should not be affected by deposition of the insulating layer.

Therefore we studied in more detail different insulating materials as well as various preparation techniques to develop a versatile Nb_3Ge integration process which yields real high-T_c superconducting electronic circuits.

PREPARATION OF THE INSULATING LAYER

The first step towards this aim is the choice of an insulating material that withstands the elevated substrate temperatures necessary to deposit a high quality superconducting Nb_3Ge film.

SiO is normally used as insulator in the Pb-In- and Nb technology. It offers good insulating properties but cannot be used here, because it decomposes into silicon and silica at temperatures above 600°C [1]

Si_3N_4 is of great practical importance as an insulating material in semiconductor electronics. Therefore we tried to use this material for our purposes. We prepared silicon nitride films by reactive sputtering from a silicon target in a pure nitrogen atmosphere. Sputtering at ambient temperatures results in soft and porous films. The deposition at substrate temperatures up to 800°C leads to hard and pinhole free films. Unfortunately the Nb_3Ge films covered with Si_3N_4 suffered a drop in T_c of about 4 K. We suppose that the nitrogen reacts with the superconductor at these high temperatures resulting in decomposition of the A15 material.

SiO_2 and Al_2O_3 have excellent insulating and thermal properties. For both materials deposition is done by rf-magnetron sputtering in an argon atmosphere. SiO_2 is sputtered from a fused quartz target and a sintered Al_2O_3 plate is used for the deposition of aluminum oxide films. Using a magnetron cathode deposition rates of 12 nm/min for SiO_2 and 20 nm/min for Al_2O_3 were realized. To remove water from the surface of the substrate before deposition and to minimize stress during deposition the substrate is held at a temperature of 600°C. Films of 200 nm thickness (SiO_2) or 150 nm (Al_2O_3) are very hard, pinhole free, and stable against thermal cycling. For these reasons both films can successfully act as insulating layers in a Nb_3Ge technology.

A very important step in the fabrication of superconducting electronic circuits is the patterning of the deposited films. In principal there are two major ways: liftoff and etching.

The liftoff-process cannot easily be used with high quality Al_2O_3- and SiO_2- films since there are no organic resists available which can withstand the deposition temperatures of 600°C. Therefore the second way was chosen which implies that the insulating layer has to be structured by etching. For this purpose one needs a selective etchant of sufficient etch rate which should not degrade the properties of the superconducting film.

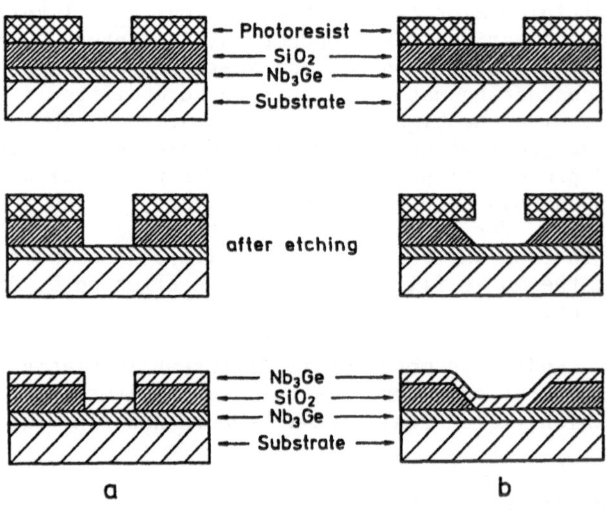

Fig. 1. Schematic view of the fabrication process when using
a) reactive ion etching
b) wet chemical etching

The etching of Al_2O_3 is difficult. To our knowledge there exists no suitable wet chemical etchant with sufficient etch rate and selectivity. One acceptable method to remove aluminium oxide films is reactive ion etching using BCl_3 or CCl_4. To structure a 100 nm film with an etch rate of about 10 nm/min one needs a total of 10min, which is the maximum time a Shipley AZ-1450J resist mask can withstand these etching conditions.

The etching of SiO_2 is less difficult. Reactive ion etching with SF_6 results in etch rates of about 300 nm/min leading to a high selectivity with the resist mask. A problem occurs if a SiO_2 film is deposited on top of a Nb_3Ge film since the etch rates for both materials are nearly the same so that etching will not stop at the films boundary. This difficulty can be overcome by a thin 2-3 nm Al_2O_3 layer deposited between the Nb_3Ge- and SiO_2-film acting as etch stopper during the reactive ion etching. The Al_2O_3 layer can be removed with BCl_3 or CCl_4 as described above.

Wet chemical etching with a mixture of 10 vol. conc. HF + 90 vol. NH_4F is even an easier way to structure SiO_2. This method has two additional advantages. First one needs no etch stopper on top of the Nb_3Ge since the etch rate for Nb_3Ge is very small. Secondly a slight undercutting under the photoresist takes place. This results in oblique etching edges which improve the covering by a possible toplayer as shown in Fig.1.

In conclusion, SiO_2 seems to be a very suitable material. It's good insulating and thermal properties as well as it's simple patterning were the reasons for us to use SiO_2 in a high-T_c multilayer technology.

PROPERTIES OF THE Nb_3Ge FILMS

The Nb_3Ge films used in this study were prepared by dc-sputtering in an Ar atmosphere. The vacuum chamber is turbopumped and baked prior to use. A detailed description of the preparation of the Nb_3Ge films is given elsewhere [2]. The properties of the films can be characterized by their T_c and their crystal structure. As first step in the multilayer process thin Nb_3Ge films are deposited onto heated sapphire substrates (T=900°C). These films have a T_c of 21 K, containing only the A15-phase and having a lattice constant of 5.14 Å as measured by x-ray diffraction. The grain size is of the same magnitude as the film's thickness (50 nm - 150 nm).

Table 1: Deposition method, etching conditions, and thermal stability range for some insulating materials

material	formation	etching	thermal stability
SiO	evaporation	reactive ion etching with SF_6	up to 600°C
SiO_2	rf-magnetron sputtering	HF+NH_4 wet etching, reactive ion etching with SF_6	>1000°C
Si_3N_4	reactive sputtering	conc.HF wet etching	900°C
Al_2O_3	rf-magnetron sputtering	reactive ion etching with BCl_3,CCl_4	>1000°C

Important for the successful use of a multilayer process is that the properties of the individual superconducting layers are not affected by the different process steps like sputtering, etching, or thermal cycling. Especially the A15-structure of Nb_3Ge is very sensitive to damage. We therefore covered a Nb_3Ge layer with SiO_2 and a second Nb_3Ge layer under usual deposition conditions and measured the Tc of the bottom layer before and after the deposition. Only a small drop of 0.2 K in T_c was observed.

In order to optimize the T_c of a second Nb_3Ge layer, an optimization process was performed similiar to that described earlier [2]. One of the most important parameters in the preparation of high-T_c Nb_3Ge films is the substrate heater temperature T_H. For this purpose SiO_2 covered Nb_3Ge films were covered by a second Nb_3Ge layer in one vacuum run under different substrate heater temperatures. Fig.2 shows the dependence of T_c on T_H. A broad maximum in T_c can be observed at a substrate temperature of about 900°C. This is nearly the same optimum temperature as for a single Nb_3Ge film sputtered on a sapphire substrate.

SUPERCONDUCTING Nb_3Ge CONTACTS

Another important step in the multilayer technology is the preparation of superconducting contacts between the superconducting films at appropriate locations. The T_c of the link should be nearly the same as the T_c of the superconducting layers and the critical current density j_c has to be high enough for the contact to remain superconducting under all working conditions.

For measurements of j_c and T_c of the contacts the pattern of Fig.3 was used. It additionally allows the determination of the superconducting properties of the bottom film A and the top film B. The fabrication process begins with the deposition of a 100 nm Nb_3Ge film. Next this layer is structured by means of photolithography and reactive ion etching. Then a 200 nm SiO_2 layer is sputter deposited on top of film A at a substrate temperature of 600°C. In the next step the contact window is etched into the SiO_2 film via photolithography and wet chemical etching as described above. As last step the top layer B is sputtered and structured under the same conditions as film A. In Fig.4 an enlarged view of a superconducting interconnection is illustrated.

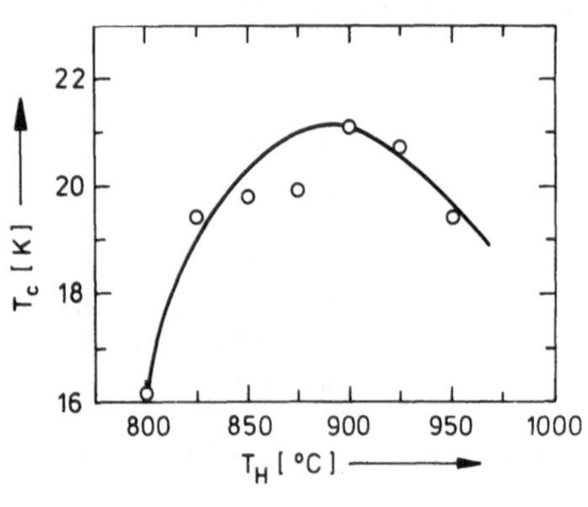

Fig. 2. Dependence of zero resistance temperature T_c on heater temperature T_H for Nb_3Ge films sputtered on SiO_2-covered Nb_3Ge films.

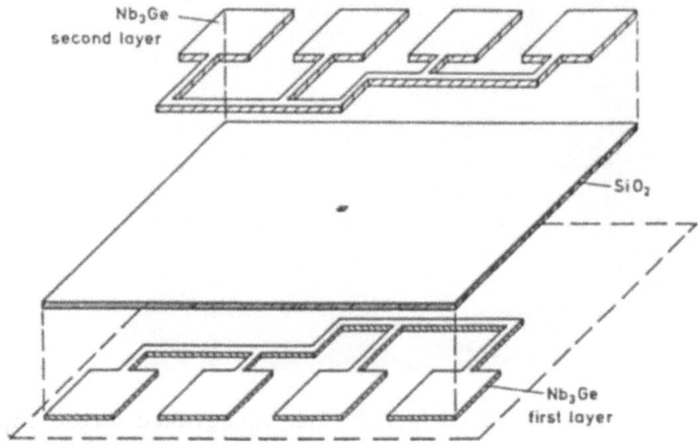

Fig. 3. Exploded view of the test pattern used to measure the superconducting link as well as the T_c of the bottom and the top Nb_3Ge film.

The critical current and its temperature dependence were measured by a four point method with a voltage criterium of 2 μV. The temperature was determined by a germanium resistor. Data for a typical contact are plotted in Fig.5. The contact is fully superconducting up to 20.5 K and has a critical current density j_c of 60 A/cm^2 at a temperature of 20 K. These values show that the interface between layer A and B inside the contact region is fully superconducting. This leads to the assumption that the top layer B grows quasi homoepitaxial on top of the bottom layer. Possible damage in the interface region due to the previous etching process seems to be annealed by the high substrate temperature of about 900°C during the deposition of the counter electrode.

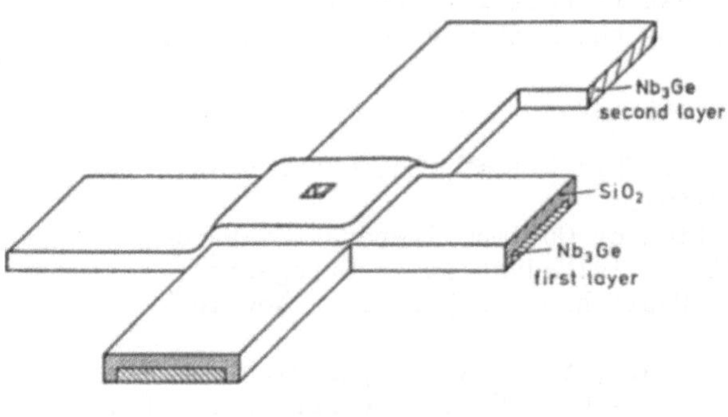

Fig. 4. Enlarged view of the test pattern showing the superconducting interconnection between two Nb_3Ge layers.

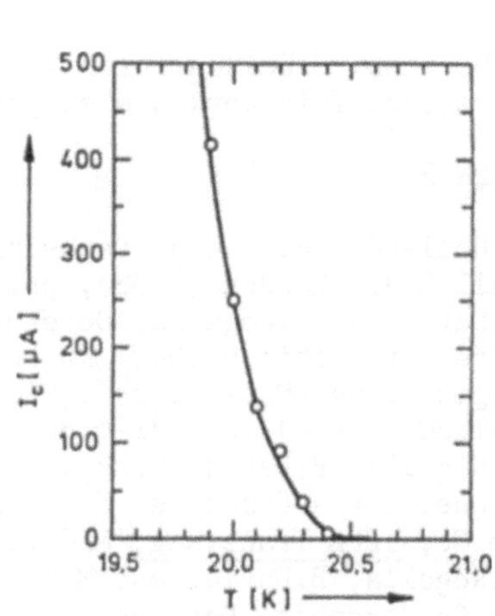

Fig. 5. Dependence of the critical current I_c on temperature for a superconducting Nb_3Ge interconnection (cf. fig.4). The contact area is 400 μm^2.

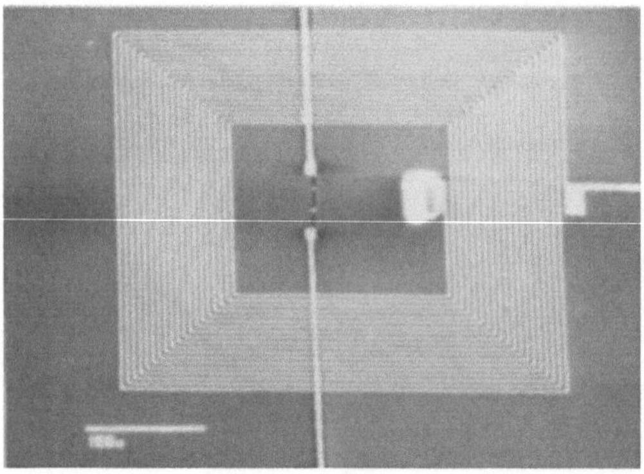

Fig. 6. SEM picture of a Nb_3Ge dc-SQUID with integrated Nb_3Ge input coil

MULTILAYER Nb_3Ge dc-SQUID

As a first application for the described multilayer technology a Nb_3Ge dc-SQUID with integrated input coil was prepared. Details on the preparation as well as on the performance are described elsewhere [3] (see also [4]). The dc-SQUID using two microbridges as weak links was structured into the bottom film by electron beam lithography and reactive ion etching. This dc-SQUID has similiar properties as those prepared earlier [5]. Next an insulating SiO_2 layer was deposited. Since the inner electrode of the input coil has to be connected to the outside a window was etched into the SiO_2 layer prior to the deposition of the second Nb_3Ge film. This film then was structered into a 20 turns planar coil. One electrode of the dc-SQUID additionally serves as one coil port. The whole dc-SQUID operates up to 19 K. The input coil has a mutual inductance to the SQUID of about 200 pH. The coupling coefficient was estimated to be 0.16. A SEM picture of this device is shown in Fig.6 .Thus it has been demonstrated that practical high-T_c dc-SQUIDs can be made of Nb_3Sn [6] and Nb_3Ge films.

ACKNOWLEDGEMENTS

We thank C.Heiden for his interest in this work and many useful discussions. This work was supported by the Stiftung Volkswagenwerk.

REFERENCES

1. L.Holland, in:"Vacuum Deposition of Thin Films", ed. Chapman and Hall Ltd., London (1958), pp.483-491.
2. B.David and H.Rogalla, Determination of Optimum Preparation Parameters for Thin high-T_c Nb_3Ge Films, in:"Advances in Cryogenic Engineering (Materials)", Plenum Publishing Corporation, New York, vol.30, (1984), pp.631-638.
3. H.Rogalla, B.David, M.Mück, and Y.Kato, Study of Preparation Techniques for a Practical Microbridge dc-SQUID Structure Fabricated from Nb_3Ge, IEEE Trans.Mag. MAG 21, vol.2, (1985), pp. 536-538.
4. H.Rogalla, B.David, and M.Mück, Fabrication Techniques for Thin Film Nb_3Ge Josephson Devices, to be published in: "SQUID '85, Superconducting Quantum Interference Devices and their applications" vol.3, Walter de Gruyter, Berlin - New York, (1985).
5. M.Mück, H.Rogalla, and B.David, DC-SQUID Operation above 20 K, Phys.Stat.Sol.(a) 87:K105-K107 (1985).
6. M.S.DiIorio and M.R.Beasley, High-T_c Superconducting Integrated Circuit: A dc-SQUID with Input coil, IEEE Trans. Mag. MAG 21, vol.2 (1985), pp. 532-535.

JOSEPHSON TUNNELING JUNCTIONS WITH HIGH Tc $Nb_3(Ge,Si)$ ELECTRODE

M. Naoe*, N. Terada** and Y. Hoshi***

* Tokyo Institute of Technology, Oh-okayama, Meguro-ku, Tokyo, Japan
** Electrotechnical Laboratory, Sakura-mura, Niihari-gun, Ibaraki, Japan
***Tokyo Institute of Polytechnics, Iiyama, Atsugi-shi, Kanagawa, Japan

ABSTRACT

Three kinds of Josephson tunneling junctions have been fabricated and their electrical characteristics and stability have been investigated. Lower electrode of all junctions is composed of high Tc $Nb_3(Ge, Si)$ of A15 type of crystal structure. Pb, Nb and $Nb_3(Ge,Si)$ have been used as upper electrode of them. Patterning of Nb and $Nb_3(Ge,Si)$ films was successfully performed by means of lift-off method with ZnO stencil. Oxidized amorphous Si thin film was used as barrier layer;oxidization of it was performed in two ways, one is natural oxidization and another is thermal oxidization in deposition chamber just after the deposition of this film. Application of thermal oxidization results in good reproducibility of electrical properties of the junctions. By adjusting thickness of the barrier layer in the range from 20 to 25 Å and oxidizing it thermally, the junctions with Pb upper electrode have good electrical properties as follows; gap voltage Δ_{sum}/e about 4.7 mV, product of critical current and resistance in normal state $Ic \cdot R_{NN}$ of 4 mV and good hysteresis in I-V characteristics. However, thermal stability of this junction is not so good. Then the junctions of which all electrode are composed of refractory materials were fabricated. The exchange of upper electrode to Nb and $Nb_3(Ge,Si)$ results in the further increase of electrical properties of junctions; for gap voltage, that of junctions with Nb and $Nb_3(Ge,Si)$ upper electrode are 4.85 mV and 6.0 mV, respectively. Besides, electrical properties of them show no change even by application of thermal cycles more than 10 times.

INTRODUCTION

Tunneling type of Josephson junctions of which electrodes are composed of materials with high superconducting critical temperature Tc are promising application of superconducting devices because of their large energy gap, potential of high temperature operation and thermal-mechanical stability. These characters should improve the ratio of signal to noise and reliability of superconducting circuits[1]. In order to realize it, it is necessary to develope the high Tc film in which good superconducting properties can be obtained reproducibly even on the condition of moderate deposition temperature and small thickness. Besides, processing techniques suitable for such materials have to be developed. $Nb_3(Ge,Si)$ of A15 type of crystal structure is one of the compounds which have the highest Tc about 23 K. And this film can satisfy the requests for fabrication process of junction described above, by optimizing film composition. In this study, the authors have attmpted to fabricate tunneling type of Josephson junctions with high Tc $Nb_3(Ge,Si)$ electrode, In all types of junctions, oxidized amorphous Si thin films were used as barrier layer because it is suitable for growth of upper electrode. At first, the junction with Pb upper electrode has been fabricated in order to investigate basic properties of $Nb_3(Ge,Si)$ film and to optimize the process. Then, the junctions with the refractory upper electrodes with high Tc, that is Nb and $Nb_3(Ge,Si)$ have been fabricated for obtaining junctions with excellent properties. This paper concerns fabrication process of these junctions and the dependence of their electrical properties and stability on fablication conditions.

FABRICATION PROCESS OF JUNCTIONS

Figure 1 shows the fabrication process of $Nb_3(Ge,Si)$ lower electrode and barrier layer of all types of junctions. At first, ZnO films were deposited as a lift-off stencil by means of Targets Facing type of sputtering. By adjusting oxygen gas pressure during deposition of it in the low range, ZnO films with uniform structure and thermal resistivity were obtained. The films were patterned chemically. Then, $Nb_3(Ge,Si)$ films were deposited by means of DC magnetron sputtering on the conditions listed in Table I. Composition of the films was fixed at $Nb_{75}(Ge_{0.88}Si_{0.12})_{25}$. On these conditions, high Tc above 21 K ($\Delta Tc<1$ K) is obtained reproducibly even in the early stage of film growth. Just after the deposition of $Nb_3(Ge,Si)$ film, amorphous Si films with various thickness were deposited on it without braking vacuum by means of DC diode sputtering. In this deposition process, Argon gas pressure P_{Ar} and distance between the target and the substrate D were set at 100 mTorr and 10 cm, respectively, in order to suppress the degradation in superconductivity of the $Nb_3(Ge,Si)$ lower electrode which would result from the bombardment of high energy particles in post deposition process. Oxidization of a-Si film was performed in two ways; (i) samples were exposed in air at room temperature for 10

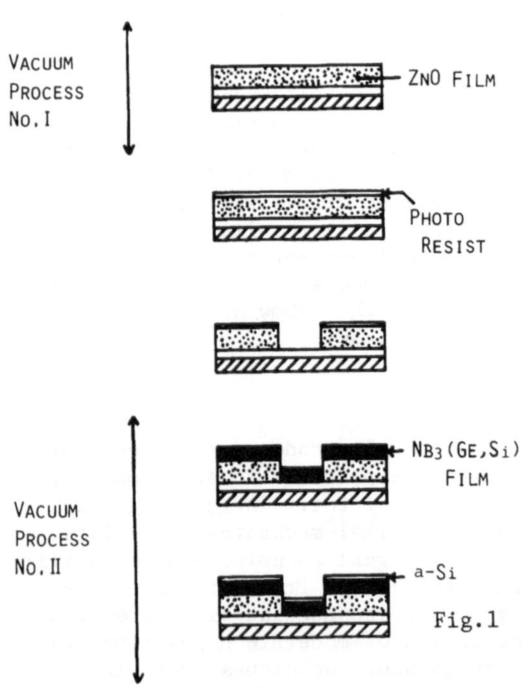

Fig.1 Fabrication process of $Nb_3(Ge,Si)$ lower electrode and a-Si film.

Table I. Preparation Conditions of $Nb_3(Ge,Si)$

Depo. Rate	80	(Å/min)
P_{Ar}	50	(mTorr)
D	10	(cm)
Ts	450	(°C)

Table II. Change of Gap Voltage with Barrier Layer Thickness, d

d	(Å)	10	15	20	25	50
Δ_{sum}/e	(mV)	4.30	4.60	4.75	4.70	4.45

hours (oxidization in air), (ii) samples were oxidized in pure oxygen gas for four hours, which was introduced into the deposition chamber just after the deposition of a-Si film, under the condition of substrate temperature of 300°C (oxidization in chamber). Pb films for upper electrode were deposited by evaporation and were patterned by using photoresist as lift-off stencil. On the other hand, the deposition temperature for obtaining high Tc Nb and $Nb_3(Ge,Si)$ films exceeds the deformation temperature of photo resist. Therefore, these films were patterned in the same way as that used in patterning of $Nb_3(Ge,Si)$ lower electroden. Tc of of upper electrode is about 22 K on thr case of $Nb_3(Ge,Si)$ and 9 K on that of Nb film. All of junctions fabricated in this study have cross line type of shape. And the area of weak link was fixed at 20x20 μm^2.

CHARACTERISTICS OF THE JUNCTIONS

Characteristics of the junctions such as gap voltage and leak current depend mainly on the thickness of the barrierlayer and kind of oxidization methods for it. Table II shows the dependence of gap voltage $\Delta sum/e$ of the junctions with Pb upper electrode on barrier layer thickness d. Gap voltage of the junction with thermal oxidized barrier layer takes the maximun about 4.7 mV in the range of barrier layer thickness d for 20 to 25 Å. Energy gap of $Nb_3(Ge,Si)$ lower electrode calculated on the assumption that energy gap of Pb at 4.2 K is 1.2 meV reaches high value about 3.5 meV. Basing on this value, $2\Delta_{Nb3(Ge,Si)}/k\cdot Tc$ is rather high about 4.1. This means that the surface layer of $Nb_3(Ge,Si)$ lower electrode has good superconductivity. It should be noted that there is no application of cleaning process to the surface layer. The application of cleaning to the fabrication process of the junction should result in further increase of $\Delta_{Nb3(Ge,Si)}$[4]. Figure 2 shows the dependence of leak current $I_{sub\ gap}$ on barrier layer thickness d. For the junctions of which barrier layer is oxidized in air, the value of $I_{sub\ gap}$ is rather high and it shows rather wide dispersion. On the other hand, $I_{sub\ gap}$ of the junctions with the barrier layer which is oxidized in chamber is low and it shows little dispersion for the junctions with a same barrier layer thickness. For the latter junction, the relation between logarithm of $I_{sub\ gap}$ and barrier layer thickness has a linearlity which is similor th the ideal properties for tunneling-conduction. There results means that the oxidization in chamber results in an inhomogeneity in insulation properties of barrier layer; uncertainties in this oxidization condition such as adhesion of water vapor onto surface of barrier layer may cause the inhomogeneity. And oxidization in chamber is an useful technique for obtaining the barrier layer with good insulation properties and uniformity.

Other electrical properties of the junction are also improved by the application of the oxidization in chamber. Figure 3 shows the dependence of ratio of sub gap resistance to that in normal conducting state $R_{sub\ gap}/R_{NN}$ on barrier layer thickness d. R_{subgap} was calculated by using the value of

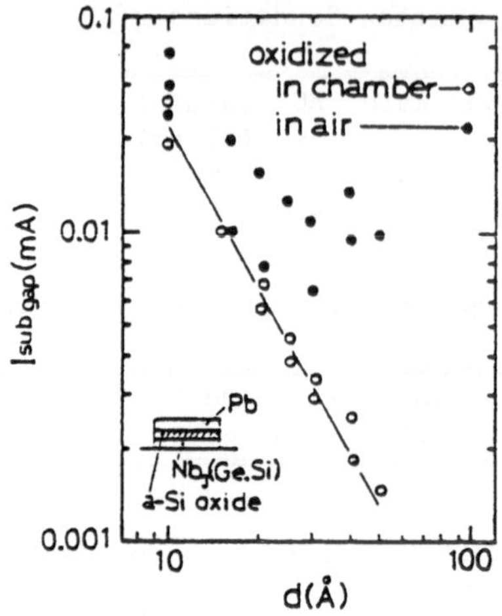

Fig. 2. Dependence of subgap current I_{subgap} on barrier layer thickness d.

I_{subgap} at 3.5 mV. The ratio of the junctions with the barrier layer oxidized in chamber and in air take the maxima in the same range of d from 15 to 30 Å. However, the value of the ratio of the junctions with barrier layer oxidized in air is about half as large as that of another junction. This difference results from an increase of R_{subgap} by the application of oxidization process in chamber. The maximum of the ratio reaches about 18, which is one of the highests of the resistance ratio in the tunneling type of junctions with high Tc A15 compound lower electrodes[4]. Therefore, the junctions with good hysteresis in I-V characteristics are reproducibly obtained by adjusting d and application of oxidization process in chamber.

Figure 4 shows the dependence of the product of critical current and resistance in normal conducting state $Ic{\cdot}R_{NN}$ on barrier layer thickness d. Similar to the resistance ratio, $Ic{\cdot}R_{NN}$ of the junctions with barrier layer oxidized in air is not so high, 3.5 mV even for its maximum, and its dispersion is large, which due mainly to small value and large dispersion of R_{NN} in such junction. On the other hand, the high value in the product and little dispersion of it are achieved in the junctions with barrier layer oxidized in chamber. $Ic{\cdot}R_{NN}$ of the latter junction rakes the maximum of 4

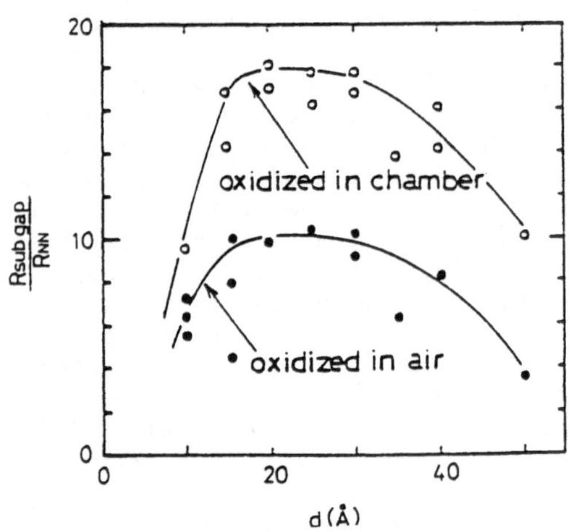

Fig. 3. Dependence of resistance ratio R_{subgap} on d.

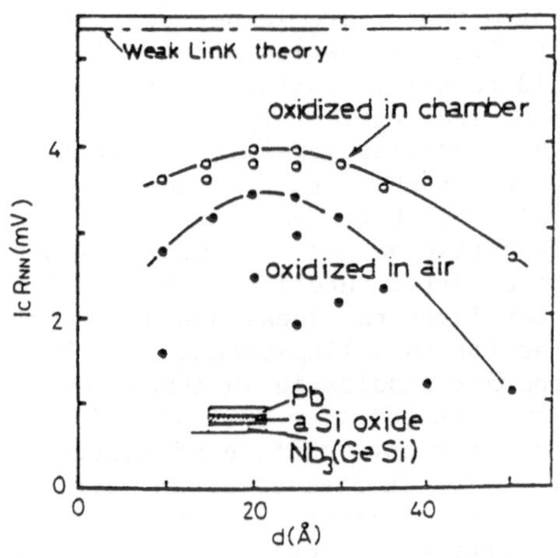

Fig. 4. Dependence of $I_c{\cdot}R_{NN}$ on d.

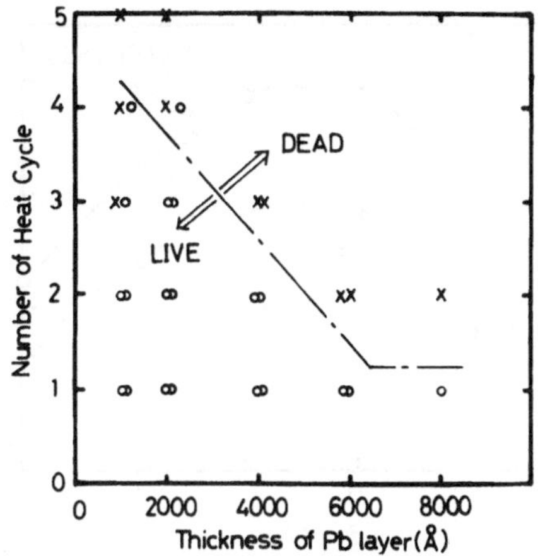

Fig. 5. Relationship between thermal resistance of $Nb_3(Ge,Si)/Pb$ junctions and thickness of Pb upper electrode.

mV in the range of d from 20 to 25 Å. From the theory in which revision for strong couple supercondutivity is neglected, the value of $Ic \cdot R_{NN}$ of $Nb_3(Ge,Si)/Pb$ junction at 4.2 K should rise up to 5.20 mV. The experimental value is about 65 % of the theoretical value. So, a certain strong coupling revision should be necessary for a design of electrical properties of this junction. These results concerning I-V characteristics mean that the optimum thickness of barrier layer is in the range from 20 to 25 Å.

Figure 5 shows the relationship between the ability for operation of the junctions and number of applied thermal cycles N, thickness of Pb upper electrode d. Barrier layer of all junctions in this figure were fabricated by oxidization in chamber and thickness of it was set in the optimized range. Simbol of circle means that junctions have operationability and another simbol means that they do not have it. The junctions with Pb upper electrode thicker than 6000 A break down by the application of only two thermal cycle. Electron micrographs of the breaked junctions show that break of Pb upper electrode at the edge of linking area causes the breakdown of them. On the other hand, $Nb_3(Ge,Si)$ lower electrode shows no change in its morphology and supercoductivity even by application of numbers of thermal cycles. A decrease in thickness of Pb layer results in a increase of N within which the junctionsare able to be operated. These results indicate that thermal stability of $Nb_3(Ge,Si)$/a-Si-oxide/Pb junction is limited only by that of Pb upper electrode which is easy to break by an accumulation of thermal stress, and $Nb_3(Ge,Si)$ electrode is rather stable against thermal cycles.

From the results concering the junctions with Pb upper electrode, it is confirmed that $Nb_3(Ge,Si)$ film has good superconductivity and thermal mechanical stability. Then, in order to improve not only the stability but also electrical properties, substitution of Nb or $Nb_3(Ge,Si)$ for Pb films as upper electrodes have been attempted. For all of the junctions described below, their barrier layer was fabricated by oxidization in chamber.

Figure 6 shows the dependence of gap voltage Δ_{sum}/e of the junctions with Nb and $Nb_3(Ge,Si)$ upper electrode on barrier layer thickness d. As well as in the junctions with Pb upper electrode, gap voltage of both types of junctions takes the maximum in the range of d from 20 to 25 Å. Therefore, d of these junctions was set at this range.

Figure 7 shows the temperature dependence of gap voltage Δ_{sum}/e of three types of the junctions fabricated in this study and that of conventional junction with Pb alloy electrode. Gap voltage is far advanced by

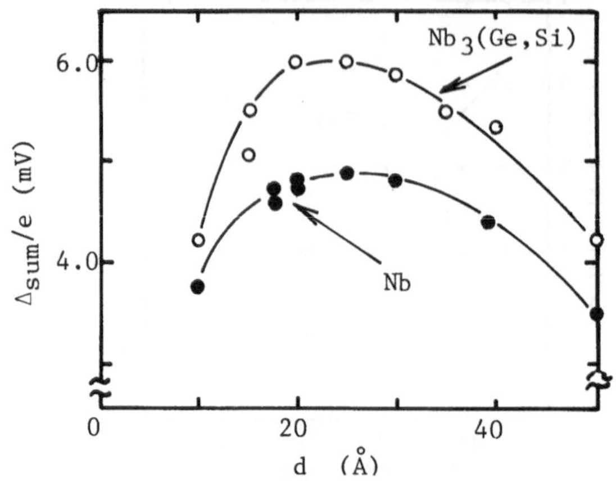

Fig. 6. Dependence of Δ_{sum}/e of the junctions with Nb and $Nb_3(Ge,Si)$ upper electrode on barrier layer thickness d.

using $Nb_3(Ge,Si)$ films only as the lower electrode; the gap voltage of $Nb_3(Ge,Si)$/a-Si-oxide/Pb junction at 4.2 K is more than 1.8 times as large as that of conventional junction. And the voltage shows further increase by application of high Tc materials to upper electrode. The voltage of the junctions with Nb or $Nb_3(Ge,Si)$ upper electrode are as high as 4.8 mV and 6.0 mV, respectively. Therefore, the application of high Tc materials, especially of $Nb_3(Ge,Si)$, to electrodes is useful to improve resistivity of junctions against electrical noise.

Figure 8 shows the temperature dependence of $Ic\cdot R_{NN}$ of the junctions fabricated in this study. $Ic\cdot R_{NN}$ also shows a significant increase by the application of $Nb_3(Ge,Si)$ film only to lower electrode. And use of Nb and $Nb_3(Ge,Si)$ films as the upper electrode causes a further increase of it. Especially, in the $Nb_3(Ge,Si)$/a-Si-oxide/$Nb_3(Ge,Si)$ junction, $Ic\cdot R_{NN}$ keeps the value higher than that of conventional junction at 4.2 K even in the high temperature region. This result indicates that tha application of high Tc $Nb_3(Ge,Si)$ is effective not only for the increase of gap voltage but also for extending operational margin of junctions.

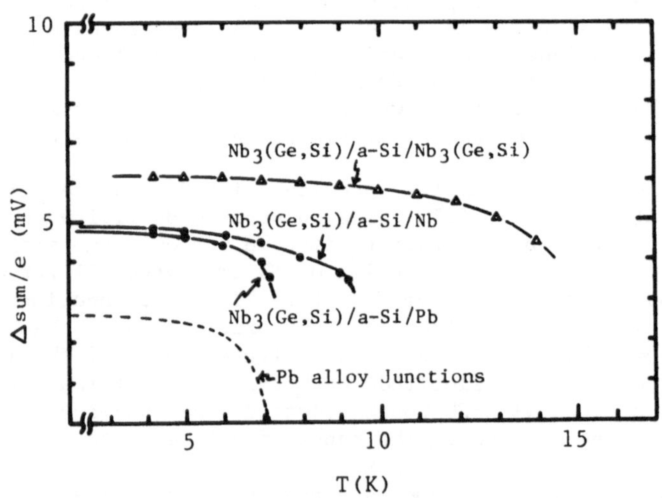

Fig. 7. Temperature dependence of Δ_{sum}/e of the junctions fabricated in this study.

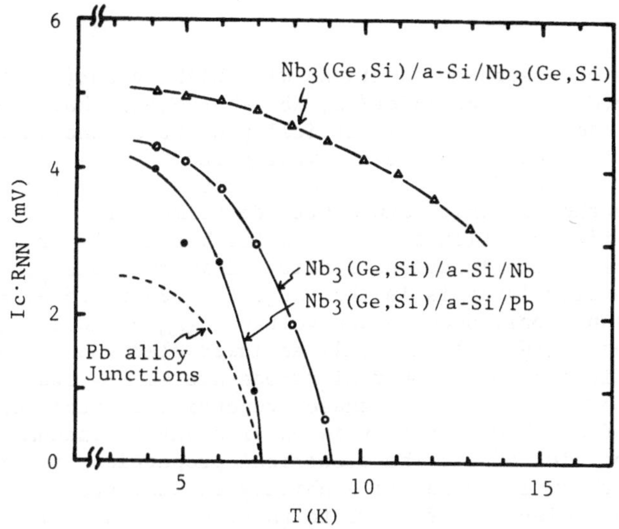

Fig. 8. Temperature dependence of $I_c \cdot R_{NN}$ of the junctions fabricated in this study.

Figure 9 shows the relationship between $I_c \cdot R_{NN}$ and the number of thermal cycles applied to the junctions. $I_c \cdot R_{NN}$ of the junctions with Nb or $Nb_3(Ge,Si)$ upper electrode shows no change even after application of thermal cycles more than 10 times. It was expected that the junctions of which all electrodes are composed of refractory superconductors should have excellent thermal-mechanical stability, and it is experimentally confirmed. The application of $Nb_3(Ge,Si)$ film to electrode of Josephson tunneling junction is useful to achieve not only good electrical performance but also excellent stability.

The merits obtained in this study due mainly to the development of A15 $Nb_3(Ge,Si)$ film which is suitable for electrode of junctions and to that of fabrication process with thermal resistivity, fitness for this film.

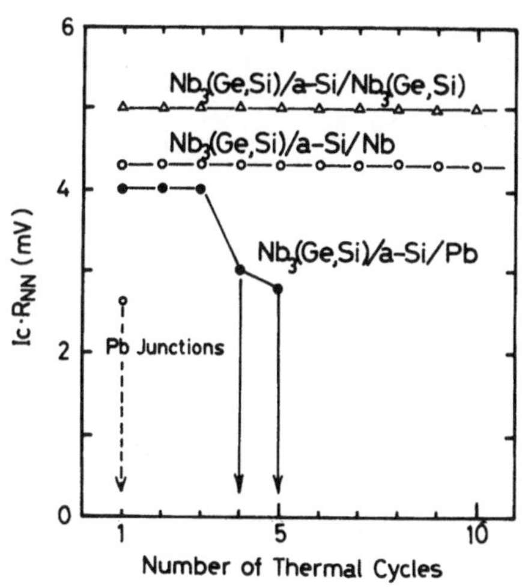

Fig. 9. Thermal resistance of the junctions fabricated in this study.

CONCLUSION

Tunneling type of Josephson junctions with high Tc $Nb_3(Ge,Si)$ electrodes, that is $Nb_3(Ge,Si)$/a-Si-oxide/Pb, Nb, $Nb_3(Ge,Si)$, have been fabricated and the dependence of their electrical properties and thermal stability on fabrication process has been investigated.

Patterning of $Nb_3(Ge,Si)$ electrode was successfully performed by means of lifi-off method in which ZnO sputtered films were used as stencil. And amorphous Si oxide thin film is suitable for barrier layer of the junctions. For all types of the junctions fabricated in this study, the optimum thickness of the barrier layer is in the range from 20 to 25 Å. Electrical properties of the junctions, such as gap voltage Δ_{sum}/e and $Ic \cdot R_{NN}$ rise drastically by using $Nb_3(Ge,Si)$ film only as lower electrode; Δ_{sum}/e and $Ic \cdot R_{NN}$ of the junctions with Pb upper electrode are 4.7 mV and 4.0 mV, respectively. Good reproducibility for these properties is achieved by using the barrier layer of which oxidization is carried out in vacuum chamber just after the deposition of it. The electrical properties are improved more and excellent thermal-mechanical stability is achieved by exchanging the material of upper electrode from Pb to Nb or $Nb_3(Ge,Si)$; for example, Δ_{sum}/e of the junctions with Nb and $Nb_3(Ge,Si)$ upper electrodes are 4.8 mV and 6.0 mV, respectively. Especially, the junctions of which all electrodes are composed of high Tc $Nb_3(Ge,Si)$ film have the high temperature operativeness with electrical performance superior to that of conventional Pb alloy junctions. High Tc $Nb_3(Ge,Si)$ is an useful material for electrode of Josephson junctions with excellent performance and the junctions fabricated in this study are able to be offered as elements for superconducting integrated circuits.

REFERENCES

1. R. F. Broom, R. B. Laibowotz, Th. O Moohr and W. Walter, IBM J. Res. Develop., 212 (March 1980).
2. N. Terada, Y. Hoshi, M. Naoe and S. Yamanaka, IEEE Trans. Magn., MAG-18, 1580 (1982).
3. N. Terada, Y. Hoshi, M. Naoe and S. Yamanaka, Advances in Cryogenic Engng., 30, 559 (1984).
4. O. Michigami, Y. Katoh and S. Yoshii, Jpn. J. Appl. Phys., 22, L91 (1983).

A 1 μm CROSS-LINE JUNCTION PROCESS

Masahiro Aoyagi, Akira Shoji, Shin Kosaka,
Fujitoshi Shinoki, and Hisao Hayakawa

Electrotechnical Laboratory,
Sakura-mura, Niihari-gun, Ibaraki, Japan

ABSTRACT

A new fabrication process for all refractory Josephson tunnel junctions with dimensions less than 2.5 μm-square is proposed, in which junctions are formed by reactively ion etching a full wafer junction sandwich using a cross-line patterning method. As a mask of the junction patterning process, a two-layer resist system has been employed for improving the patterning accuracy. The cross-line patterning method has made it possible to fabricate small area junction with small spreads of the maximum critical currents. All niobium nitride (NbN) Josephson junctions with dimensions from 1.25 μm-square to 0.5 μm-square have been integrated on a chip by this method. The standard deviation of the maximum critical currents for 1024 1 μm-square junctions has been measured to be about 4%.

INTRODUCTION

In Josephson integrated circuit technology, it is important to fabricate Josephson tunnel junctions with small dimensions and small spread of the maximum critical current I_c. The small dimension of junctions makes the switching speed of junction high. The small spread of I_c makes the operating margin of circuit large.

So far, we have fabricated Josephson tunnel junctions with niobium nitride (NbN) electrodes using a process called SNIP (self-aligned niobium nitride isolation process)[1-4]. In SNIP, junctions are defined from a full wafer junction sandwich by reactive ion etching and are isolated by successive deposition of an insulation film. This fabrication procedure of SNIP has allowed us to obtain a small spread of I_c for a large number of junctions integrated on a chip (σ = 5% for 1024 junctions whose dimensions are 2.5 μm-square). A 4×4 bit multiplier which contains 2800 junctions has been fabricated by SNIP and successfully operated with a multiply time of 1 ns.[5] However, when we fabricate more largely integrated circuits and operate them with a higher operating speed, smaller spread of I_c and smaller dimension of junctions are necessary.

From this point of view, we have advanced SNIP at two points. One is the use of a two-layer resist system in the junction patterning process.[6,7] The two layer resist system has several advantages, for example, (1) improvement of resolution, (2) enlargement of aspect ratio, and (3) elimination of line width variation. The other is the use of cross-line patterning (CLIP) method in which two orthogonally crossing resist lines are used for patterning junctions. The CLIP method allows us to fabricate junctions with dimensions less than 1.5 μm-square.

In this paper, we report the fabrication of all NbN Josephson tunnel junctions by the two-layer resist system and the CLIP method. The standard deviation of the maximum critical current for fabricated junctions is discussed concerned with the junction dimension.

TWO-LAYER RESIST

There are four requirements for the resist which is used in the junction patterning process of SNIP. The first is high resolution; the second is high dry etching durability; the third is large aspect ratio of the resist image; and the fourth is good removability in solvents for the lift-off process. When we reduce the resist thickness in order to improve its resolution, the aspect ratio of the resist image becomes small. And when we bake a resist at higher temperature in order to improve its dry etching durability, its removability in solvents becomes worse. Therefore, it is difficult to satisfy all these four requirements by a single-layer resist.

To satisfy these four requirements, we have enployed a two-layer resist system consisting of AZ-1400 and PMMA layers. AZ-1400 has been employed as the top layer because it has high resolution to UV exposure and a low etch rate for a CF_4 plasma (15 nm/min, at 27 Pa, 0.16 W/cm^2) which is half as much as that of PMMA. On the other hand, PMMA has been employed as the bottom layer because it has good removability even when it is baked at high temperature.

Figure 1 shows the lithographic process for the two-layer resist system and fundamental process steps are listed as follows.

(a) Spin-coating of 500 nm-thick PMMA (OEBR-1000, 200 cp, Tokyo Oka).
Baking for 60 min at 120 °C.
Spin-coating of 500 nm-thick PMMA.
Baking for 60 min at 150 °C.
Spin-coating of 400 nm-thick AZ-1400 (MP-1400, 17 cp, Shipley).
Baking for 20 min at 90 °C.

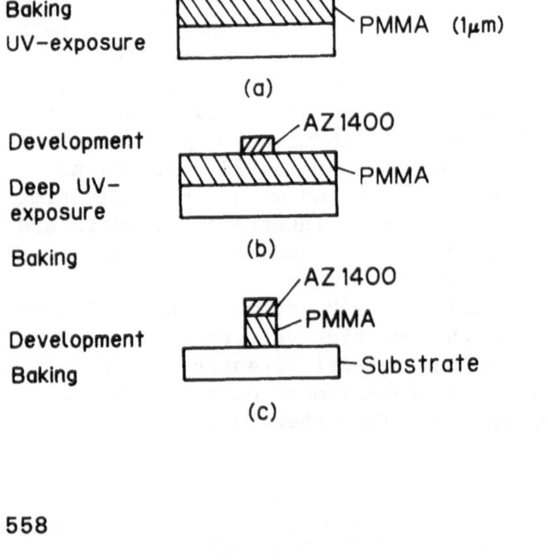

Fig. 1. Schematic diagram of two-layer resist system.

(b) UV exposure of the top AZ-1400 layer by a 4:1 stepper (Cannon FPA-141) using both **g** (436 nm) and **h** (405 nm) lines.
Development of AZ-1400 in AZ developer.

(c) Deep-UV (200-260 nm) blanket exposure by a proximity printer (Cannon PLA-521F) with a 500 Watt Hg-Xe lamp. In this Deep-UV exposure, AZ-1400 works as the mask.[8]
Baking for 40 min at 120 °C. This baking makes AZ-1400 insoluble for methyl isobutyl ketone (MIBK).
Development of PMMA in MIBK.
Baking for 20 min at 150 °C.

Figure 2 (a) shows a scanning electron microscope(SEM) photograph of 1 μm-width two-layer resist lines formed on a niobium (Nb) film. The resist line has a stencil of an AZ-1400 top layer. The aspect ratio of the resist line is about 1.5. Figure 2 (b) shows an SEM photograph of the resist lines after the dry etching of the Nb film was performed in a CF_4 plasma at 27 Pa, 0.16 W/cm^2. Although the stencil of the AZ-1400 top layer was reduced by the dry etching, it well protected the PMMA bottom layer during the dry etching, resulting in small shrinkage of the width of the PMMA bottom layer. The width of the lines transferred on the Nb film was the same as that of the PMMA bottom layer. As shown in Fig. 2(b), the aspect ratio of the two-layer resist was held after the dry etching of Nb. The minimum width of two-layer resist line obtained in this work was 0.5 μm which is nearly equal to the resolution limit of UV photolithography.

Figure 3 shows the standard deviations of I_c for 1024 all NbN junctions fabricated using the two-layer resist system as a function of the junction size L, in comparison with those for junctions fabricated using single-layer resist system (AZ-1400). As shown in Fig. 3, standard deviations were much decreased to about 1/3 using the two-layer resist system. The minimum standard deviations obtained for 2.5 μm-square, 5 μm-square, and 10 μm-square junctions were 1.6%, 0.8%, and 0.4%, respectively. These results show that the use of the two-layer resist system in SNIP is quite useful for obtaining small spreads of I_c for junctions with dimensions larger than 2.5 μm-square.

However, when the dimension of junctions becomes smaller than 2 μm, standard deviation steeply increases as shown in Fig. 3. Furthermore, we could not fabricate junctions with dimensions less than 1.5 μm-square by SNIP even when two-layer resist system was used. This is due to the reduction of the aspect ratio for small square resist masks (<1.5 μm-square).

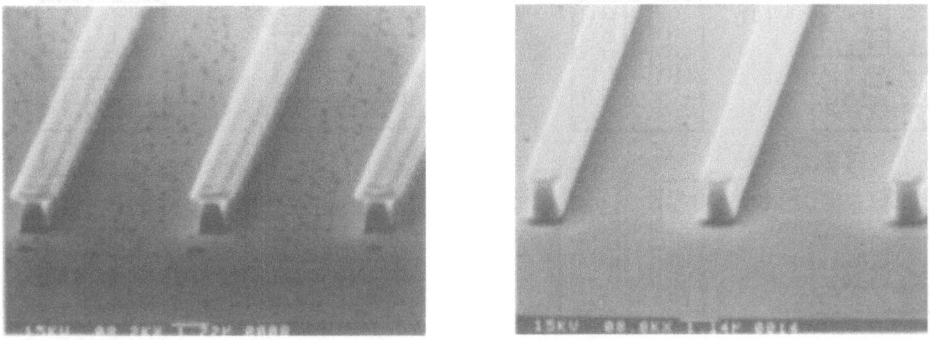

(a) (b)

Fig. 2. SEM photographs of (a) 1 μm two-layer resist lines on a Nb film and (b) 1 μm two-layer resist lines after dry etching of Nb.

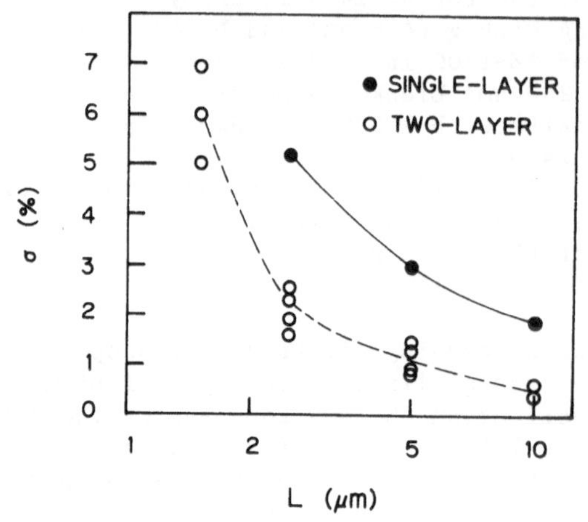

Fig. 3. The standard deviation σ of I_c versus the junction size L for 1024 all NbN junctions fabricated using two-layer resist (circle) and for junctions fabricated using single-layer resist (dot).

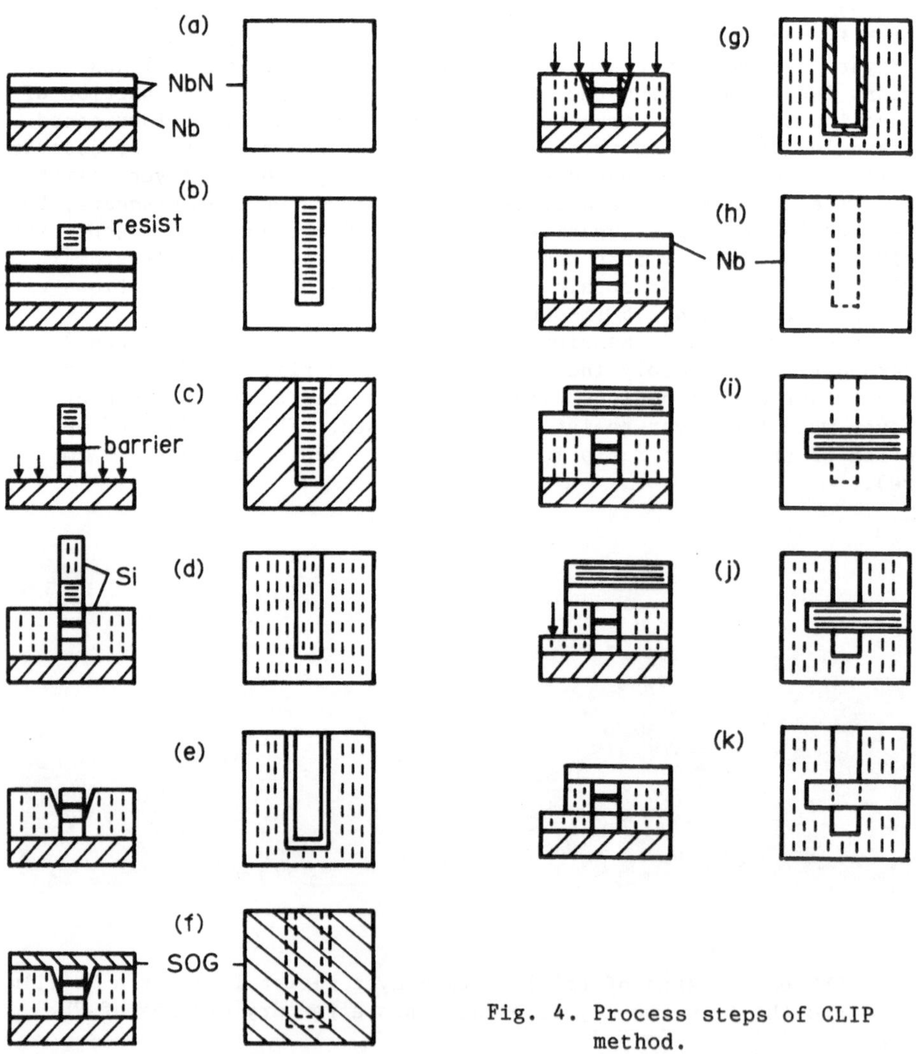

Fig. 4. Process steps of CLIP method.

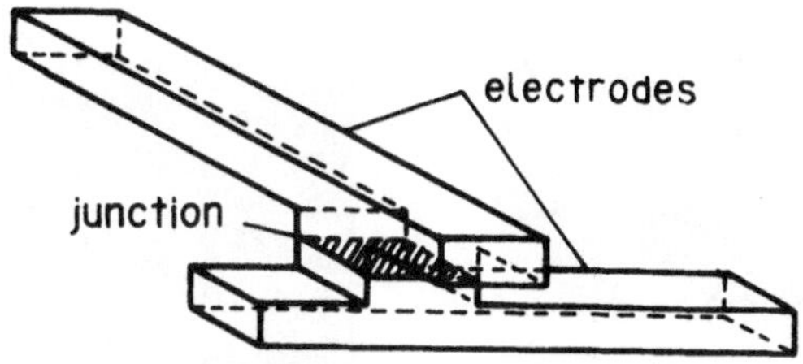

Fig. 5. Cross sectional view of a junction fabricated by CLIP method.

CROSS-LINE PATTERNING METHOD

In order to overcome this problem, we have developed a cross line patterning (CLIP) method in which linear resist masks are used instead of square resist masks. Larger aspect ratio can be obtained for linear resist masks compared with square resist masks. Figure 4 shows the process steps of the CLIP method.

(a) A multilayered junction sandwich of Nb-NbN-barrier-NbN films is formed on a whole silicon wafer.
(b) A two-layer resist line is formed on the junction sandwich.
(c) The junction sandwich is patterned by dry etching using the two-layer resist line as a mask in CF_4 plasma.
(d) An insulation layer (SiO or Si) is deposited by evaporation.
(e) The insulation layer on the resist mask is lifted off in acetone.
(f) An SOG (spin on glass) layer of 80 nm-thick is coated and baked for 30 min at 250 °C to fill the dip between the insulation layer and the junction periphery.
(g) Then the SOG layer is etched back up to the counter electrode, resulting in almost flat surface around the junction.[5]
(h) After cleaning the surface of the counter electrode by Ar sputter etching, an Nb film is deposited by RF sputtering.
(i) Another two-layer resist line which is crossed orthogonally with the first resist line is formed on the Nb film.
(j) Nb and NbN films are patterned by dry etching using the second two-layer resist line as a mask.
(k) The second resist line is removed in acetone.

Figure 5 schematically shows a cross sectional view of a completed junction. The junction area is placed at the cross section of two linear electrodes. The use of CLIP method has made it possible to fabricate junctions with dimensions less than 1.5 μm-square. The minimum junction area obtained in this study was 0.5 μm-square.

Figure 6 (a) shows an SEM photograph of a part of an array of series-connected 1024 1 μm-square all NbN junctions fabricated by CLIP method. Figure 6 (b) shows the I-V characteristic of the array. The tunnel barrier of the NbN junction was native oxide and the critical current density of the junction was 15 kA/cm^2. The area occupied by this array was 160×160 μm^2. From Fig. 6 (b), the spread and the standard deviation of I_c were measured to be ±8.9 % and 3.5%, respectively.

In Fig.7, the standard deviations of I_c for the junctions fabricated by CLIP method are plotted as a function of junction size L, in comparison with those for the junctions fabricated using the square two-layer resist masks. The standard deviations for cross-line type and square type junctions with dimensions larger than 2.5 μm-square are nearly equal. On the other hand, the standard deviations with dimensions less

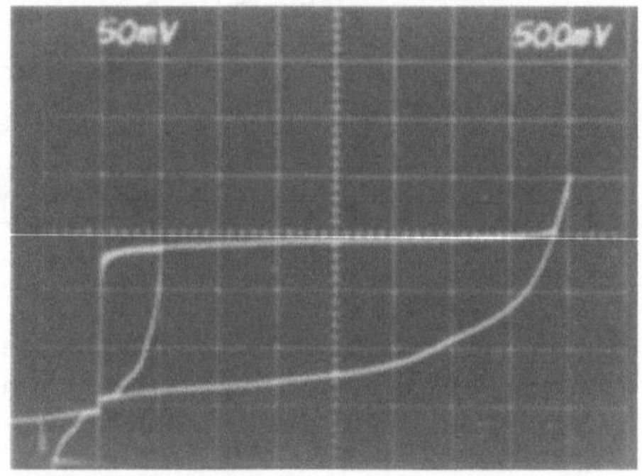

(a) (b)

Fig. 6. (a) SEM photograph of series-connected 1 μm-square all NbN junctions, (b) I-V characteristic for an array of series-connected 1024 1 μm-square all NbN junctions; I: 50 μA/div, V: 0.5 V/div

than 2.5 μm-square are much improved for cross-line type junctions. The standard deviation of 3.5% obtained for 1 μm-square junctions is encouraging for fabricating Josephson large scale integration with 1 μm design rule.

SUMMARY

We have advanced the fabrication process for all NbN Josephson junctions (SNIP) by introducing a two-layer resist system and a cross-line patterning (CLIP) method. The use of the two-layer resist system has much improved the spread of I_c for junctions integrated on a chip. A standard deviation of 1.6% has been obtained for 2.5 μm-square 1024 junctions fabricated by the two-layer system. On the other hand, the use of CLIP method has allowed us to fabricate junctions whose dimensions are less than 1.5 μm-square. By the conjunction of the two-layer resist system and the CLIP method, we have obtained a standard deviation of 3.5% for 1024 1 μm-square junctions.

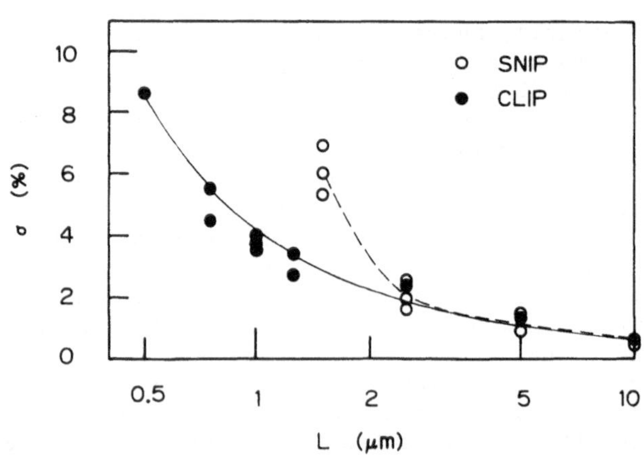

Fig. 7. The standard deviation σ of I_c versus the junction size L for 1024 all NbN junctions fabricated by SNIP using square resist mask (circle) and for junctions fabricated by CLIP method (dot).

REFERENCES

1. A. Shoji, F. Shinoki, S. Kosaka, M. Aoyagi and H. Hayakawa, New Fabrication Process for Josephson Tunnel Junctions with (Niobium Nitride, Niobium) Double-Layered Electrodes, Appl. Phys. Lett. 41:1097-1099(1982)
2. A. Shoji, F. Shinoki, S. Kosaka, M. Aoyagi and H. Hayakawa, All Refractory Josephson Tunnel Junctions Fabricated by Reactive Ion Etching, IEEE Trans. Magn. MAG-19:827-830(1983)
3. A. Shoji, M. Aoyagi, S. Kosaka, F. Shinoki, and H. Hayakawa, Niobium Nitride Josephson Tunnel Junctions with Magnesium Oxide Barriers, Appl. Phys. Lett. 46:1098-1100(1985).
4. A. Shoji, NbN based Josephson Junctions, Proc. of IC SQUID '85, Berlin (1985) to be published.
5. S. Kosaka, A. Shoji, M. Aoyagi, F. Shinoki, S. Takada, H. Ohhigashi, H. Nakagawa, S. Takada and H. Hayakawa, An Integration of All Refractory Josephson LSI Circuit, IEEE Trans. Magn. MAG-21:102-109(1985).
6. B. J. Lin, Multilayer Resist Systems and Processing, Solid State Technol. 26:105-112(1983).
7. J. Nakano, Y. Mimura, K. Nagata, Y. Hasumi and T. Waho, RF CO_2 Oxidation and Double-Layer Resist Technique Realizing Josephson Junctions Very Small I_j-Spread, Extended Abstracts of the 16th Conference on Solid State Devices and Materials, Kobe(1985), pp. 635-638
8. B. J. Lin, AZ1350J as a Deep-UV Mask Material, J. Electrochem. Soc. 127:202-205(1980).

SELECTIVE TRILAYER ION-BEAM ETCHING PROCESS FOR FABRICATING Nb/Nb OXIDE/Pb-ALLOY TUNNEL JUNCTIONS

H. Tsuge

Microelectronics Research Laboratories
NEC Corporation
Miyamae-ku, Kawasaki, Japan

ABSTRACT

A new process, suitable for integrated circuits, for fabricating Nb/Nb oxide/Pb-alloy tunnel junctions is described. The process includes the following key steps: Nb/Nb oxide/Pb-alloy trilayer formation, involving in-situ thermal oxidation; junction area delineation by deep UV photolithography; highly selective Pb-alloy layer patterning by ion-beam etching; lift-off of an evaporated insulation layer. The fabricated junctions show good junction characteristics and excellent uniformity in critical currents. High-quality junctions with critical currents of about 80% of the theoretical values and without any knee structure have been obtained. Standard deviations in critical currents for 100 series-connected junctions with 4 x 4 and 2 x 2 μm areas are 0.8% and 2.4%, respectively. The junctions with PbIn and PbAuIn counterelectrodes exhibit sufficient thermal stability for this process.

INTRODUCTION

Primary requirements for tunnel junctions in integrated circuits include thermal stability, junction quality, critical current uniformity and reproducibility. To date, stable and high-quality junctions have been achieved with Nb/Nb oxide/Pb-alloy[1,2] or Nb/artificial barrier/Nb structures.[3] Recently, several investigators[4-6] have proposed novel fabrication techniques, in which the junction trilayer, involving a base electrode, tunnel barrier and counter electrode, has been fabricated on an entire substrate without breaking vacuum before any patterning steps. These techniques also have the individual advantage of reproducibility because the trilayer interfaces are not contaminated by exposure to the atmosphere. However, critical current uniformity achieved has not been good enough for integrated circuit applications, presumably because of the difficulty in accurate definition of small junction areas.

This paper describes a new process for fabricating Nb/Nb oxide/Pb-alloy junctions that satisfy the requirement for uniformity, which is of primary importance, especially for practical integrated circuits. The distinct feature of this process is that ion-beam etching is used for defining the junction areas after the trilayer formation.[7] Ion-beam etching allows highly accurate, spatially uniform patterning, which leads to good critical current uniformity for small area junctions.

Figure 1 shows the selective trilayer ion-beam etching process (STIEP). A junction trilayer of Nb/Nb oxide/Pb or Pb-alloy films is formed on the entire substrate (Fig.1(a)). The Nb base electrode film, which has 9.2K Tc and 9.3-9.7 resistance ratios, R(296K)/R(10K), is deposited by rf magnetron sputtering without intentional substrate heating. The Nb film surface is then thermally oxidized in-situ in oxygen atmosphere for 30 minutes to form a tunnel barrier. The barrier thickness is controlled by changing oxygen pressure. After the Nb film oxidation, the Pb or Pb-alloy counter electrode film is deposited at room temperature by E-gun evaporation. The Pb alloys used are Pb(95wt%)-In(5wt%) and Pb(84wt%)-Au(4wt%)-In(12wt%), evaporated sequentially in the order given, and Pb(71wt%)-Bi(29wt%) evaporated from the alloy. Typical thicknesses for the Nb film and Pb or Pb-alloy film are 200 nm and 150 nm, respectively.

The trilayer is patterned by ion-beam etching the Pb or Pb-alloy film and by reactive ion etching the Nb film sequentially with a photoresist mask to form base electrodes and interconnection lines (Fig. 1(b)). Junction areas are then defined by selectively ion-beam etching the Pb or Pb-alloy film, as discussed below, with ODUR1013 photoresist mask (Fig. 1(c)). Particularly in this step, deep UV lithography was used because of its greater resolution over UV lithography. Junction areas are 4 x 4 μm and 2 x 2 μm.

While the photoresist mask is retained, a 300 nm thick SiO film for base electrode insulation is evaporated and then patterned by lifting off the SiO film on the photoresist mask. After sputter etching counter electrodes to remove any surface contaminants resulting from exposure to the atmosphere, a 600 nm thick Pb or Pb-alloy film for forming interconnecting lines is evaporated and then patterned by ion-beam etching (Fig. 1(d)).

A feature of STIEP is that ion-beam etching is used to define junction areas. The ion-beam etching was performed using a system with a 3-inch Kaufman ion source and a rotating substrate holder. The following etching conditions were used: 2.7×10^{-2} Pa Ar or Ar/O_2 pressure; 0.8 mA/cm^2 current density; and 500 V acceleration voltage.

The etch rates for Pb, Nb, Nb_2O_5, SiO_2 and ODUR1013, and the etch rate ratios for Pb to the other materials in Ar ion-beam etching are given in Table 1. Nb, Nb_2O_5 and SiO_2 are underlayer materials and ODUR1013 is a masking material, used during Pb etching for achieving junction definition. The high etch rate ratios for Pb to Nb, Nb_2O_5 and SiO_2 indicate that Nb, Nb_2O_5 and SiO_2 films act as an effective stopper

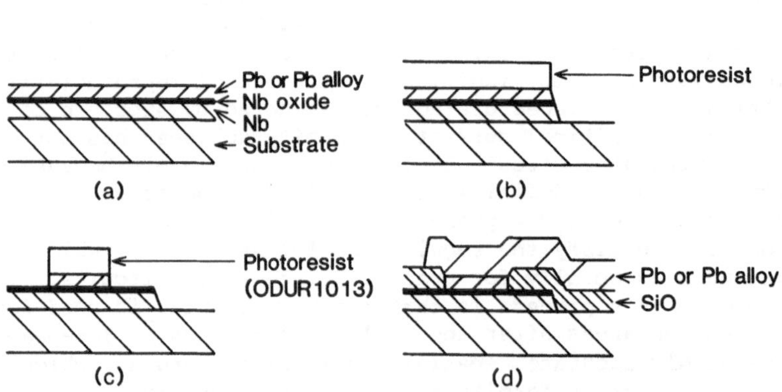

Fig. 1. Fabrication steps for STIEP junctions.

Table 1. Etch rates for Pb, Nb, Nb_2O_5, SiO_2 and ODUR1013, and etch rate ratios for Pb to other materials in Ar ion-beam etching

Material	Pb	Nb	Nb_2O_5	SiO_2	ODUR1013
Etch Rate (nm/min)	245	13	27	33	27
Etch Rate Ratio Pb/M	-	19	9.1	7.4	9.1

for the Pb etching. Moreover, the high etch rate ratio for Pb to ODUR1013 means that ODUR1013 resist can be used as a masking material.

The effect of oxygen partial pressure on the etch rates was studied to further improve the etching selectivity of Pb to Nb and Nb_2O_5, because Nb films redeposited on junction pattern sidewalls, caused by unavoidable overetching, affect the junction quality. The etch rate dependence on oxygen partial pressure, P_{O_2}, is shown in Fig. 2. The Pb etch rate is almost constant in the measured P_{O_2} range, while the etch rates for Nb, Nb_2O_5 and SiO_2 decrease and that for ODUR1013 increases, particulary from 10^{-3} Pa with increasing P_{O_2}. Consequently, by using an Ar/O_2 mixture, the etching selectivity of Pb to Nb, Nb_2O_5 and SiO_2 is extremely improved, while still retaining the etching selectivity of Pb to ODUR1013 within a reasonable value. The etch rate ratios for Pb to Nb, Nb_2O_5, SiO_2 and ODUR1013 at $P_{O_2}=4 \times 10^{-3}$ Pa are 51, 19, 14 and 4.3, respectively. The etch rates for Pb alloys are listed in Table 2, in comparison with that for Pb. Although the etch rates for Pb alloys are lower than that for Pb, the etch rates are still high enough for forming their counter electrodes.

The patterning accuracy and spatial pattern uniformity in the ion-beam etching were evaluated, because they were important factors for obtaining good critical current uniformity. It was found that the pattern width gain caused by redeposition effects, was about 80% of the

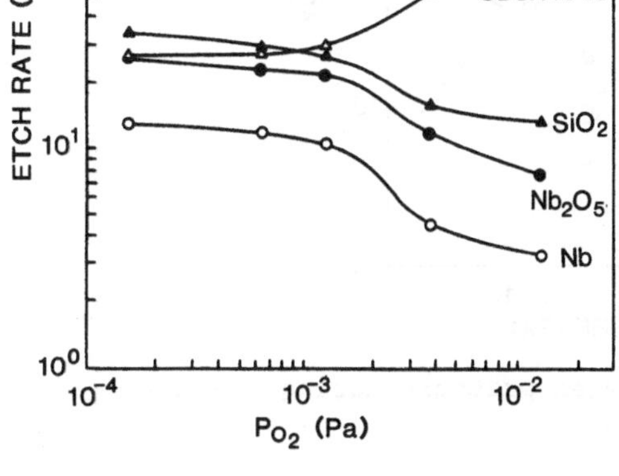

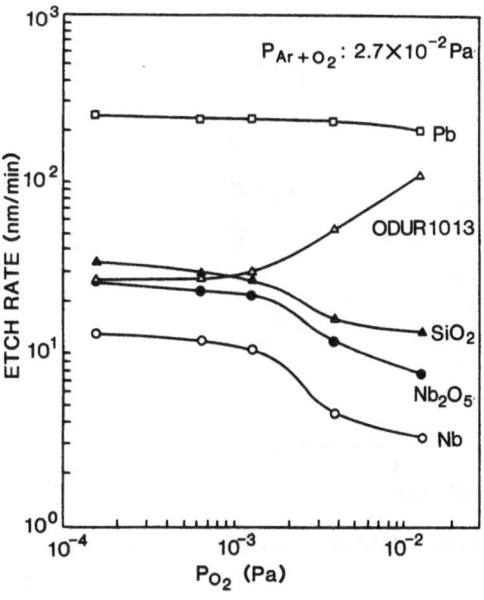

Fig. 2. Etch rates as a function of oxygen partial pressure in ion-beam etching.

Table 2. Etch rates for Pb and Pb alloys at 4 x 10^{-3} Pa oxygen partial pressure

Material	Pb	PbIn	PbAuIn	PbBi
Etch Rate (nm/min)	225	198	152	160

thickness of etched Pb or Pb alloy films. The variation in the pattern width gain over a 2-inch wafer was within measurement errors.

JUNCTION CHARACTERISTICS

Junction characteristics are affected by parameters, such as oxygen pressure, oxidation time and temperature during thermal oxidation. In this work, only the oxygen pressure was selected as an oxidation parameter. The oxidation time and temperature are fixed at 30 minutes and room temperature. Figure 3 shows the critical current density, Jc, and normalized normal tunneling resistance, ARn, dependence on oxygen pressure for Nb/Nb oxide/Pb junctions. Measurements were carried out at 4.2K. Both Jc and ARn exhibit a power-law dependence on oxygen pressure with -0.58 and 0.56 exponents, respectively. These exponents indicate that Jc is reciprocally nearly propotional to Rn, as expected from Ambegaokar-Baratoff's equation.[8] The above Jc exponent is much smaller than that obtained for rf plasma oxidation.[1] This means that thermal oxidation is more controllable than rf plasma oxidation. Jc can be controlled effectively over two orders of magnitude by simply changing oxygen pressure. The Jc deviation from the power law in higher oxygen pressure may be due to noise effects during the measurement.

Figure 4 shows the current-voltage (I-V) characteristics for Nb/Nb oxide/Pb junctions with different Jc values. The junction area is

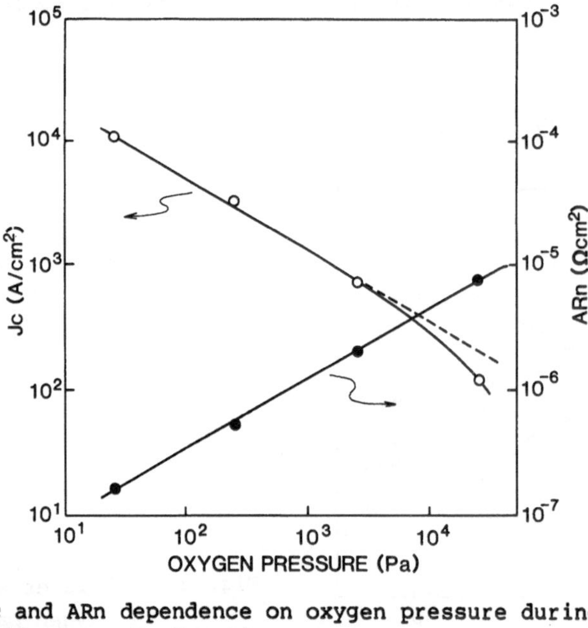

Fig. 3. Jc and ARn dependence on oxygen pressure during thermal oxidation. A: Junction area.

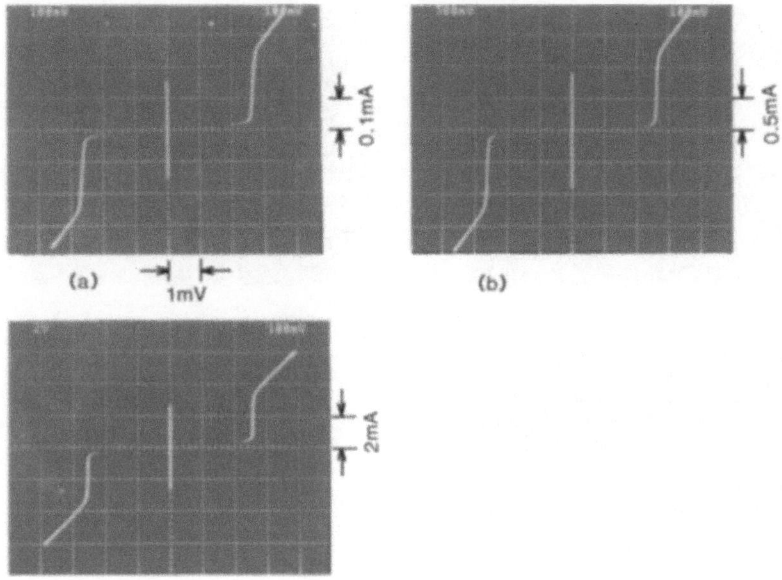

(c)

Fig. 4. I-V characteristics for Nb/Nb oxide/Pb junctions with different Jc values: (a) 970 A/cm^2, (b) 5600 A/cm^2, (c) 17000 A/cm^2.

4 x 4 μm. A "knee" structure, which is known to result from a proximity layer,[9-11] was not observed at a voltage just above the gap voltage for any junctions measured. The result is attributed to the fact that the base electrode surfaces are neither contaminated by the atmosphere nor ion-damaged by sputter cleaning or rf plasma oxidation, since, in this process, tunnel barriers are formed by in-situ thermal oxidation. The quality parameters for the junctions shown in Fig. 4 are listed in Table 3. The junctions are high-quality, with Vm=21-29 mV and IcRn=1.5-1.7 mV over a wide Jc range. The cause of Vm dependence on Jc has not been revealed yet. The Ic(experiment)/Ic(BCS) values indicate that the junctions have critical currents near the BCS value.[8] The smaller Vg value for Jc=17000 A/cm^2 is due to the appearance of a negative resistance, presumably caused by heating effects.

Representative I-V characteristcs for Nb junctions with three different counter electrodes are shown in Fig. 5. Critical current density is 2500 to 8500 A/cm^2. Also in the case of Pb-alloy counter electrodes, no knee structure is observed. The quality parameters for the junctions

Table 3. Quality parameters for Nb/Nb oxide/Pb junctions with different Jc values

Jc(A/cm^2)	Vg(mV)	Vm(mV)	IcRn(mV)	Ic(expt.)/Ic(BCS)
970	2.66	21	1.5	0.71
5600	2.65	29	1.7	0.81
17000	2.59	24	1.7	0.85

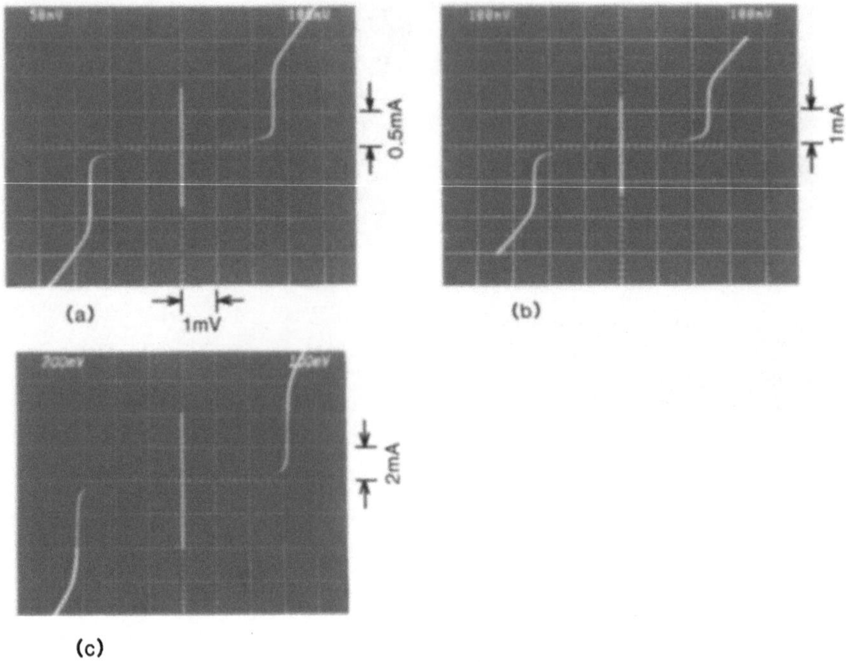

Fig. 5. Representative I-V characteristics for Nb junctions with (a) PbIn, (b) PbAuIn and (c) PbBi counter electrodes.

shown in Fig. 5 are listed in Table 4. The quality parameters for 2 x 2 μm junctions are almost the same as for 4 x 4 μm junctions. The critical currents for Pb-alloy counter electrodes are also reasonablely like BCS limits. However, for PbIn and PbAuIn, in particular for PbAuIn, Vm is smaller than for Pb. The reduction in Vm is attributed to the increase in the subgap current, which may result from a reduced-gap or normal material caused by the diffusion of In or Au into the tunnel barrier vicinity.

CRITICAL CURRENT UNIFORMITY

The critical current uniformity is of prime importance for practical large scale application. Figure 6 shows the I-V characteristics of 100 series-connected Nb/Nb oxide/PbAuIn junctions. Standard deviation, σ, in critical currents is 0.8% for 4 x 4 μm junctions, and 2.4% for 2 x 2 μm junctions. These values demonstrate excellent uniformity. For PbIn counter electrodes, almost the same σ values as for PbAuIn counter elec-

Table 4. Quality parameters for Nb/Nb oxide/Pb-alloy junctions

Counter electrode	Vg (mV)	Vm (mV)	IcRn (mV)	Ic(expt.)/Ic(BCS)
PbIn	2.60	18	1.6	0.79
PbAuIn	2.44	13	1.5	0.80
PbBi	3.11	39	2.0	0.82

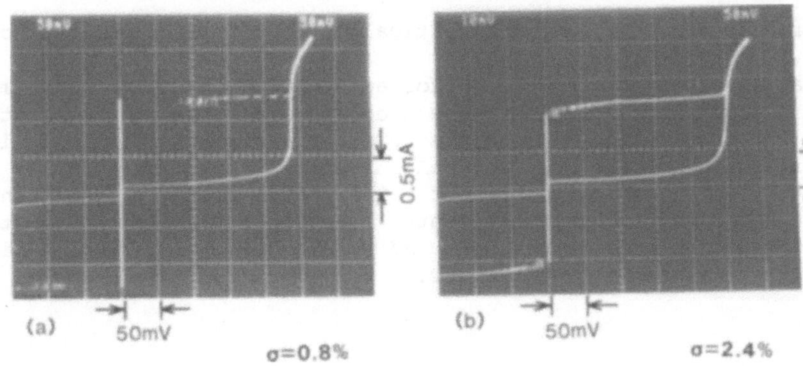

Fig. 6. I-V characteristics for 100 series-connected Nb/Nb oxide/PbAuIn junctions. Junction area: (a) 4 x 4 μm, (b) 2 x 2 μm.

trodes are obtained. However, the σ values for Pb are approximately two times as large as for PbAuIn. The values for PbBi are an order of magnitude larger than for PbAuIn. The larger values for Pb are due to the deterioration and/or lateral erosion around the counter electrode pattern periphery exposed to the air, after defining junction areas by ion-beam etching. The much larger values for PbBi are mainly attributable to the remarkable grain growth during its evaporation. The grain size for a 150 nm thick PbBi film, used for counter electrode forming is about 1 μm, which is comparable with the junction size.

Table 7 summarizes the maximum to minimum spreads and standard deviations in mean critical currents in 100 series-connected junctions distributed over a 3 x 3 cm area in a 2-inch wafer. The data shows, especially for PbIn and PbAuIn counter electrodes, that better uniformity is obtained for both 4 x 4 μm and 2 x 2 μm junctions.

The good critical current uniformity, in particular for junctions with PbIn or PbAuIn counter electrodes, represents a advantage of STIEP in regard to achieving processing uniformity.

JUNCTION STABILITY

Thermal annealing tests were performed to determine the thermal conditions allowed in the fabrication steps after junction formation. Junctions used in the tests were passivated with E-gun evaporated SiO films to eliminate the air effect on junction characteristics. The

Table 5. Maximum to minimum spreads and standard deviations in mean Ic for 100 series-connected junctions distributed over a 3 x 3 cm area

Counter electrode	4 x 4 μm junction max-min spread(%)	σ (%)	2 x 2 μm junction max-min spread(%)	σ (%)
Pb	9.3	3.8	30	9.0
PbIn	11	4.1	23	9.0
PbAuIn	9.9	3.6	21	8.4
PbBi	15	7.8	-	-

junction area is 4 x 4 μm, and the critical current density is 2500 to 8500 A/cm^2.

Thermal annealing tests were performed at 100°C in a nitrogen atmosphere. This annealing temperature was used since the maximum temperature throughout the fabrication process was 90°C. The reduction in Ic is less than 10% after 5 hours for PbIn and PbAuIn, while it is more than 10% for Pb and about 20% for PbBi after 1 hour. The change in Vm is less than 10% after 5 hours for each of the counter electrodes measured. From the results, it has been proved that PbIn and PbAuIn counter electrodes have sufficient thermal stability for STIEP.

CONCLUSIONS

A new process (STIEP) has been developed to prepare Nb/Nb oxide/Pb-alloy tunnel junctions that satisfy the principal requirements for integrated circuits. The distinct feature of this process is that ion-beam etching is used for defining the junction areas after the junction tri-layer formation. The fabricated junctions exhibit good junction characteristics and excellent uniformity in critical currents. High-quality junctions with critical currents near the BCS values and without any knee structure have been obtained for all the counter electrodes employed. Standard deviations in critical currents for the junctions with a PbIn or PbAuIn counter electrode represent the reported best results. These junctions also exhibit sufficient thermal stability for STIEP. It has been found that STIEP offers excellent potential for Josephson integrated circuit applications.

ACKNOWLEDGEMENT

The author would like to thank Y. Takayama, H. Abe, Y. Wada and T. Yoshida for helpful discussions and encouragement. The present research effort is part of the National Research and Development Program on "Scientific Computing System", conducted under a program set by the Agency of Industrial Science and Technology, Ministry of International Trade and Industry.

REFERENCES

1. R. F. Broom, S. I. Raider, A. Oosenburg, R. F. Drake, and W. Walter, IEEE Trans. on electron Dev. ED-27:1998 (1980).
2. S. I. Raider and R. E. Drake, IEEE Trans. Magn. MAG-17:299 (1981).
3. M. Gurvitch and J. Kwo, Adv. Cryogenic Eng. 30:509 (1984).
4. H. Kroger, L. N. Smith, and D. W. Jillie, Appl. Phys. Lett. 39:280 (1981).
5. A. Shoji, F. Shinoki, S. Kosaka, M. Aoyagi, and H. Hayakawa, Appl. Phys. Lett. 41:1097 (1982).
6. M. Gurvitch, M. A. Washington, and H. A. Huggins, Appl. Phys. Lett. 42:472 (1983).
7. H. Tsuge, T. Yoshida, and H. Abe, 45th Conv. Japan Society of Applied Physics, Okayama (1984).
8. V. Ambegaoker and A. Baratoff, Phys. Rev. Lett. 10:486 (1963).
9. W. L. McMillan, Phys. Rev. 175:537 (1968).
10. P. W. Wyatt, R. C.Barker, and A. Yelon, Phys. Rev. B6:4169 (1972).
11. V. Keith and J. D. Leslie, Phys. Rev. B18:4739 (1978).

A NOVEL DESIGN OF SUBMICRON THIN FILM POINT CONTACTS

H. Koch

Physikalisch-Technische Bundesanstalt
Institut Berlin
Federal Republic of Germany

ABSTRACT

A thin film point contact design applicable to SIS-, SNS-, and microbridge-type Josephson junctions is presented, which offers potentially advanced junction characteristics (low capacitance, low stray inductance, increased quasi-particle resistance). The design philosophy is based on the fact that a point contact results if two planes having a common symmetry axis but oriented perpendicular to each other are brought into contact with each other. For the case of thin films, instead of two-dimensional planes, the cross section of the resulting "point"-contact is defined by the thicknesses of the two thin films. Film thicknesses can be controlled much more precisely than lateral dimensions created by lithography. Hence, submicron junction geometries can be achieved using only conventional fabrication techniques. Following this idea, Josephson weak links of the ultrashort microbridge-type have been fabricated by an all-Nb technique having a 0.3-μm x 0.2-μm cross section with a R_qI_c product (R_q = quasi-particle resistance, I_c = critical current) of more than 20 mV.

INTRODUCTION

Recent papers on superconducting microstructures for DC-SQUIDs, SIS mixers,[1] MQT, and MQC experiments[2,3] (MQT = macroscopic quantum tunneling, MQC = macroscopic quantum coherence) argue that Josephson junctions should contain even smaller contact areas than the present state of the art (10^{-10} cm^2).[4] This goal requires new approaches in submicron thin film fabrication. In this paper a novel junction design is proposed. The essential fabrication steps are described, and some experimental results are presented, i.e., I-V characteristics of ultrashort variable thickness bridges produced by the described microfabrication technique. The addition of only a minor fabrication step should yield either SNS- or SIS-tunnel junctions with advanced properties.

DESIGN PHILOSOPHY

To explain the idea behind the new concept, Fig. 1 shows schematically the minimum achievable contact area of three different types of tunnel junctions. The minimum thin film thicknesses practically applicable and/or controllable are denoted by "a," and the minimum lateral width maintainable by lithography, by "b." Whatever lithographic method is used (e.g., UV or e-beam lithography), it is always possible to keep "a" considerably smaller than "b." Thus, if the contact area is solely defined by lithographic resolution limits (i.e., window junctions), the minimum contact area is given by "b x b." With edge junctions,[4,5] one dimension of the contact area between the two superconducting thin film banks is defined by the film thickness "a," but the other dimension still depends on the resolution limit "b" of the lithographic process; hence, the minimum contact area scales with "a x b." Junctions fabricated with the overlap technique[6,7] yield contact areas somewhat between these two cases.

The design proposed in this paper relies solely on the easily controllable thin film thickness, since the superconducting banks are oriented perpendicular to each other, as illustrated in Fig. 1. (The crosswise orientation of the thin film planes led to the proposed name x-junction.) Thus, lithographic resolution limits no longer play a role, and the contact area scales with "a x a." Another advantage of this design is that the constriction is limited to the contact area only. The cross section of the current path increases in the proximity of the contact, and thus the current density decreases very rapidly with respect to the distance to the contact zone. This feature is very favourable for reducing heating effects and for improved heat removal. If junctions of the microbridge type are formed by this technique, ultrashort weak links result.

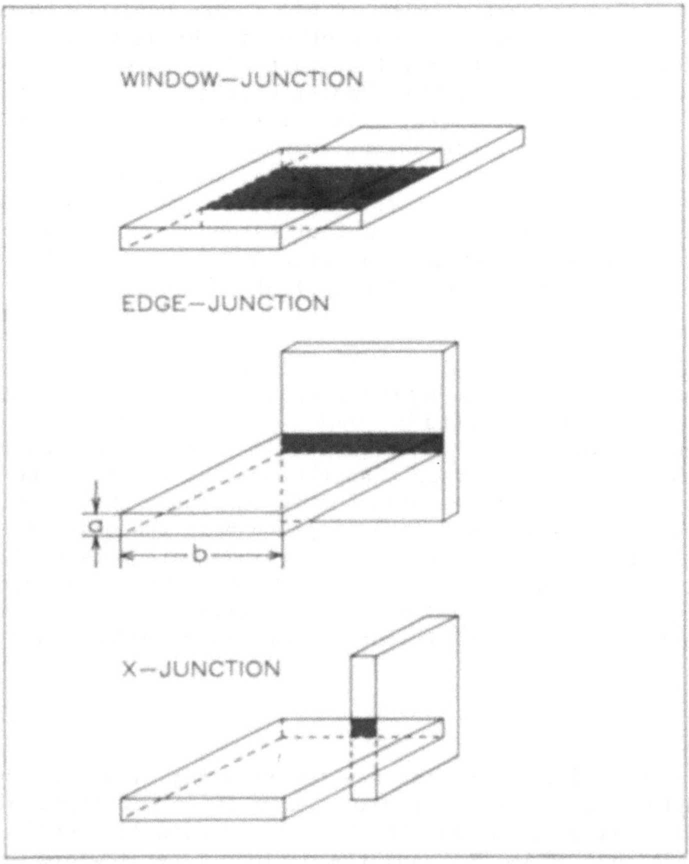

Fig. 1. Contact areas for three different junction designs.

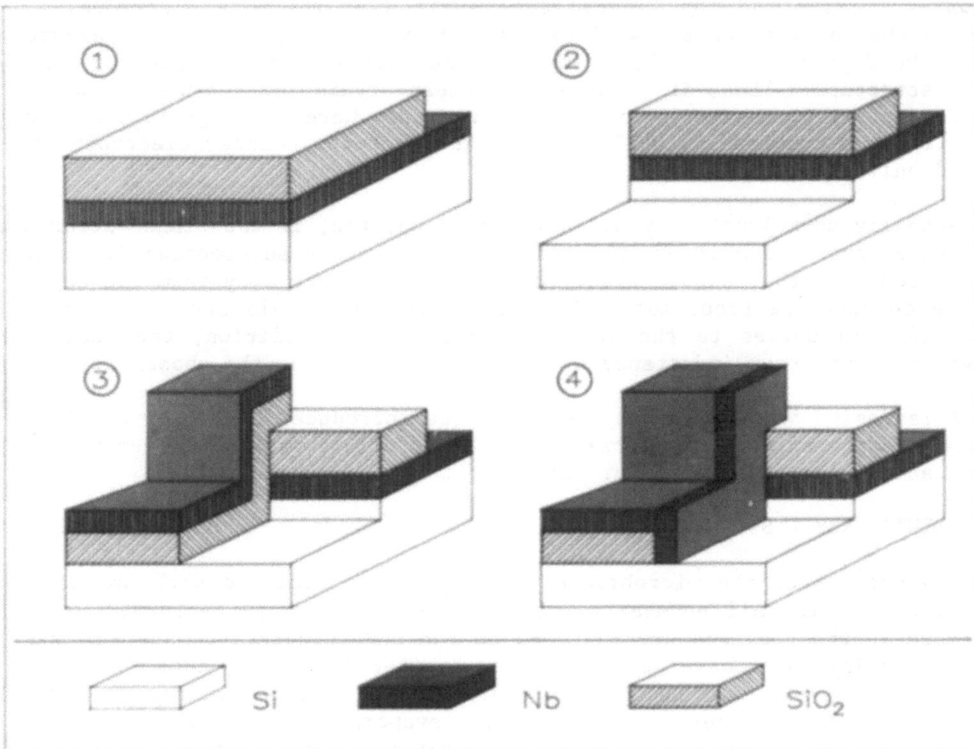

Fig. 2. Microfabrication steps for a x-junction.

MICROFABRICATION PROCESS STEPS

Figure 2 illustrates different stages of the practical realization of x-junctions. The first three stages resemble the process steps necessary to fabricate edge junctions:

1. At the location where the junction is to be prepared, the deposited thin film layer of superconductive material (e.g., Nb) is covered with an insulating material (e.g., SiO_2). (For conductive substrates like Si wafers, a passivation, for instance by oxidation prior to the Nb deposition, may be advantageous.)

2. A well-defined faceted edge structure is formed by etching away part of the insulator-superconductor-substrate sandwich, as shown. Depending on the desired junction type at this stage, either the faceted superconductor edge may be anodized or covered with an artificial barrier leading later to SIS-junctions, or when covered with a thin normal layer, to SNS-junctions. Alternatively, the edge can be left as it is for a later ultrashort microbridge-type junction.

3. By means of the lift-off technique, an overlap "finger" structure can be attached. If this finger is entirely of superconductive material, a conventional edge junction results. The design presented here differs from that technique in such a way that the superconductive finger is provided with a thick insulating underlayer. Thus, until this stage, no electrical connection between the base electrode and the counter electrode is maintained.

4. The process sequence is completed when the final layer for the counter electrode is evaporated under rather oblique angles with respect to

the wafer surface and the faceted edge surface of the base electrode. Although the whole structure will be covered by the superconductive material due to the deposition, the layer will be the thickest at the "luv" of the finger structure. Thus, if a subsequent clean-etching of the whole structure is performed, the very thin layers in areas where coverage is unwanted can be removed completely, while the thickness of the counter electrode is reduced only negligibly.

Actually an x-junction would have shown up, too, if the finger-structure in stage 3 consisted only of insulating material and a superconductive bank were attached to the counter electrode a short distance away from the edge. But the counter electrode with its small dimensions would act as a long microbridge in series to the junction contact. In addition, the heat removal and the stray inductance are greatly improved by the chosen design.

This design also resembles the technique introduced by Prober et al.,[8] but that design allows only variable thickness bridges with a longer bridge length and is not applicable for SIS- or SNS-type junctions.

EXPERIMENTAL RESULTS

X-junctions of the microbridge type have been produced with sputter equipment by which all necessary fabrication steps can be performed, as described in the previous section: Nb as the superconductive material and SiO_2 as the insulator have been magnetron sputter deposited; edge formation occurred via Ar-plasma etching. Of course, edge profile definition and atom beam collimation during oblique angle evaporation are rather poor in quality with the techniques involved. Therefore, considerable deviations from the theoretically possible contact geometries (0.2 μm x 0.3 μm) may have developed. E-beam evaporation deposition and ion beam milling techniques could also have been employed and most probably with better results. But the feasibility of the design can be demonstrated by the I-V characteristics presented in Fig. 3 a-e. Quite different I-V characteristics can be obtained according to the process conditions; Fig. 3 a-e displays a choice of examples.

Sample 3a shows hysteresis (whether thermally or capacatively in origin cannot be decided at present). If a McCumber parameter of $\beta_c = 2\pi I_c R_n^2 / \phi_o \sim 1$ is supposed, then the junction capacity should be in any case lower than 10^{-16}F as the critical current yields I_c = 0.8 mA and the quasi-particle resistance R_n = 65 Ω (ϕ_o being the flux quantum, $\phi_o = 2 \times 10^{-15}$ Wb). Such a capacitance value is even lower than the estimation of de Waal et al.[9] for the parasitic capacitances due to the leads of the overlap junctions employed in their DC-SQUID.

With sample 3b, a considerably lower value for I_c has been obtained (I_c = 200 μA) that is more favourable for applications, for instance, in DC-SQUIDs, especially if the large value for the quasi-particle resistance R_n = 57 Ω is considered. Here the shoulder-feature common to I-V characteristics of microbridges is quite apparent.

Unlike the samples described previously, the following samples were treated by a current-pulse-trimming technique.[10,11,12] Sample 3c exhibits no hysteresis after pulse trimming, although I_c = 3.3 mA and R = 8.5 Ω are still very large values. It seems as if the junction could be modeled as a number of junctions with differing parameters connected in series.

Sample 3d (3e is of the same sample but the voltage scale is expanded) features parameters suitable for DC-SQUID applications: I_c = 60 μA, R_n = 7 Ω (differential resistance for DC-SQUID bias conditions).

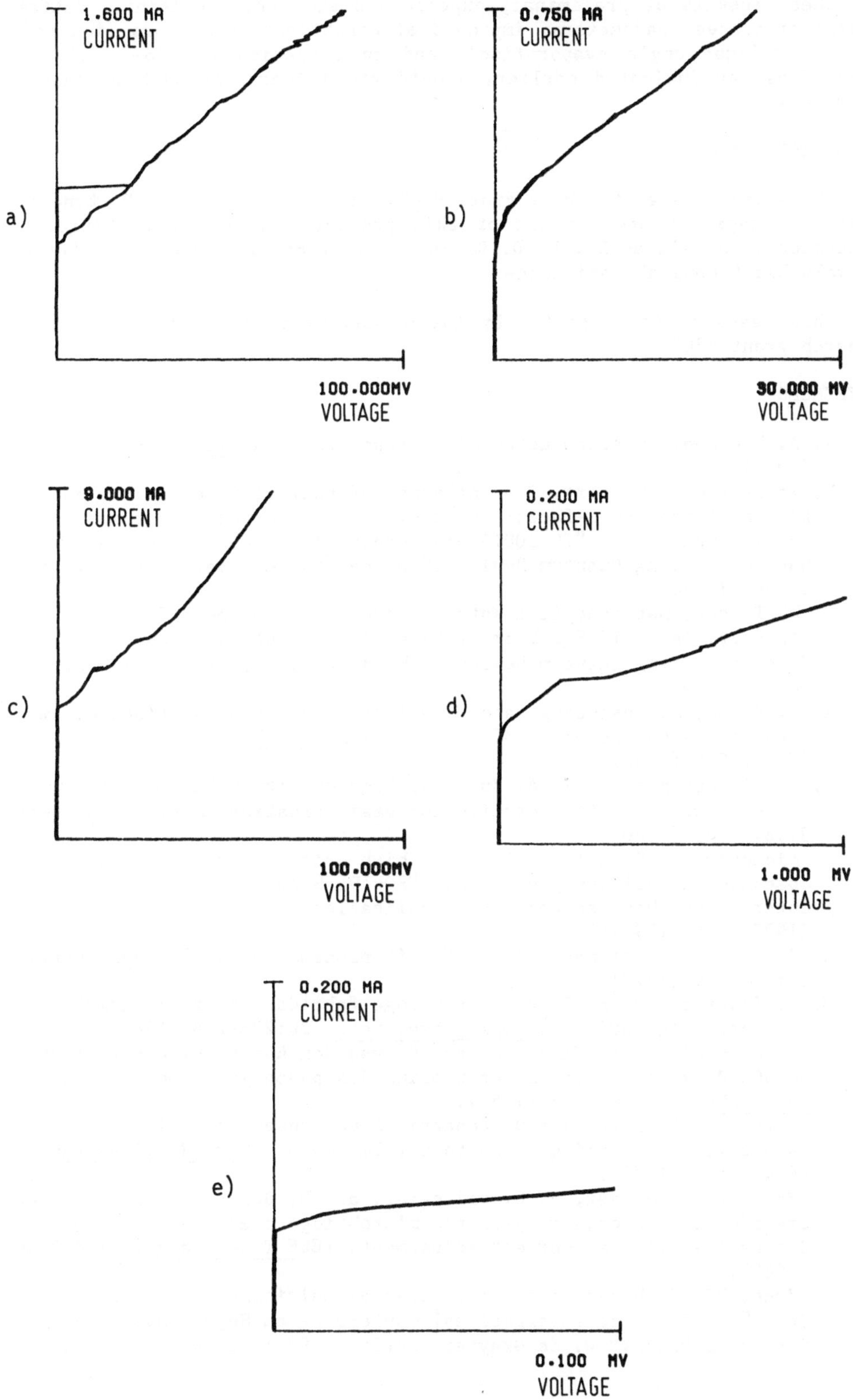

Fig. 3 (a-e). I-V characteristics of four x-junction samples.

These results of preliminary experiments are very promising. Better control of process parameters, improved microfabrication methods (ion beam milling, oblique angle evaporation), and an extension to SIS- and/or SNS-designs, as indicated earlier, should yield junctions with advanced parameters.

ACKNOWLEDGEMENTS

The author wishes to thank Prof. H.-D. Hahlbohm and Dr. S. N. Erné for their encouragement and support of this project and for many helpful discussions, as well as Dr. K. D. Klein, Mrs. M. Peters, and Mr. H. Luther for valuable technical assistance.

This research is supported by the German Ministry of Economics by research grant 9303.

REFERENCES

1. R. A. Buhrman, Superconductor microstructures, Physics 126B:62-69 (1984).
2. U. Weiss and H. Grabert, The influence of thermal fluctuations and biasing forces on Q-factor and amplitude in macroscopic quantum oscillations, in: "IC SQUID 3rd International Conference on Superconducting Quantum Devices," W. de Gruyter, Berlin (1985), to be published.
3. C. D. Tesche, Macroscopic quantum coherence: an experimental strategy, in: "IC SQUID 3rd International Conference on Superconducting Quantum Devices," W. de Gruyter, Berlin (1985), to be published.
4. R. F. Broom, A. Dosenbrug, and W. Walter, Josephson junctions of small area formed on the edges of niobium films, Appl. Phys. Lett. 37:237-239 (1980).
5. A. W. Kleinsasser and R. A. Buhrman, High quality submicron niobium tunnel junctions with reactive ion beam oxidation, Appl. Phys. Lett. 37:841-843 (1980).
6. J. Niemeyer and V. Kose, Effects in high critical current density Josephson tunnel junctions, in: "SQUID, Superconducting Quantum Interference Devices and Their Applications," W. de Gruyter, Berlin (1977), pp. 179-191.
7. G. J. Dolan, Offset masks for lift-off processing, Appl. Phys. Lett. 31:337-339 (1977).
8. M. D. Feuer and D. E. Prober, Step-edge fabrication of ultrasmall Josephson microbridges, Appl. Phys. Lett. 36:226-228 (1980).
9. V. J. de Waal, T. M. Klapwijk, and P. van den Hamer, High performance DC-SQUIDs with submicrometer niobium Josephson junctions, J. Low Temp. Phys. 53:287-312 (1983).
10. D. Duret, P. Bernard, and D. Zenatti, A uhf superconducting magnetometer utilizing a new thin film sensor, Rev. Sci. Instrum. 46:474-480 (1975).
11. Y. Monfort, D. Bloyet, J. C. Villegier, and D. Duret, Nb metallurgical transformations occurring in the microbridge area of an RF-SQUID during its critical current adjustment, IEEE Trans. Magn. 21:866-869 (1985).
12. H. Koch, DC-SQUID design for a 25 channel multisensor test circuit, in: "IC SQUID 3rd International Conference on Superconducting Quantum Devices," W. de Gruyter, Berlin (1985), to be published.

SUPERCONDUCTING WIRES FOR ELECTRONIC APPLICATIONS

V. LeTourneau, J.H. Claassen, S.A. Wolf, D.U. Gubser,
T.L. Francavilla

Naval Research Laboratory
Washington, DC

R.A. Hein

Catholic University
Washington, DC

For more than a decade, Superconducting QUantum Interference Device (SQUID) magnetometer and gradiometer systems have been used for high sensitivity magnetic field measurements in the laboratory. More recently, attempts to utilize these systems in an ambient field, mobile environment have uncovered hitherto ignored sources of noise which severely limit the ultimate sensitivity of these circuits. We have already learned that a significant source of noise arises from the effect of thermal fluctuations on magnetic field penetration in the superconducting shield surrounding the SQUID and coupling transformer.[1,2] Measurements of different shielding materials and configurations have indicated that shields made from Nb_3Sn are significantly better than the conventional Pb and Nb shields due primarily to its higher transition temperature. These measurements also found that surface condition plays a tremendous role in minimizing the depth of penetration of magnetic fields into a superconductor.[3]

We have more recently realized that the superconducting wires which form the various flux transformer assemblies can also contribute greatly to the thermal fluctuation and thermal gradient fluctuation noise of a SQUID circuit. For this reason we have undertaken a survey of a series of niobium and niobium-titanium wires that have been traditionally used to form the various coils and transformers used in magnetometers and gradiometers. We have used a series of wires obtained from BTI[4] (formerly SHE) manufactured by Magnet Wire Supply, California Fine Wire, KBI, and Supercon as well as wires obtained indepenently from these and other sources. In addition we obtained samples of annealed Nb wire from California Fine Wire.

Based on our experience with shields and SQUIDs and somewhat on intuition we believe that the best superconducting wire for this application would have a high T_c, yet would be closest to "ideal" in its narrow ΔT_c and more nearly reversible magnetization curve indicating low J_c and low flux trapping, and high residual resistivity ratio (RRR) indicating high purity materials [for the Nb-Ti the RRR is not indicative of anything significant].

Table 1. Summary of Superconducing and Normal State Properties of Various Wires

SPECIMEN	O.D. (M)	T_c^m (K)	ΔT_c (K)	ρ(300)	RR(77)	RR(T_o)	T_c (K)	#T_c (K)
		RESISTIVE					MAGNETIC	
Niobium								
:KBI*-Uninsul	2.54×10^{-4}	9.11	0.10	15.7	3.8	12.2	8.86	0.03
:CFW**-inwul	2.54×10^{-4}	9.21	0.06	15.2	4.6	22.1	9.12	0.05
:CFW-insuf	7.62×10^{-5}	8.70	0.09	-	-	5.3	-	-
:CFW-uninsul	7.62×10^{-5}	9.22	0.09	-	-	13.0	-	-
:MWS-insul	1.27×10^{-4}	9.45	0.10	-	4.5	19.7	-	-
" "		9.51	0.10	13.5	4.6	20.1	9.16	0.05
:MWS[+]-insul	1.27×10^{-4}	9.09	0.06	18.8	4.1	11.6	9.04	(Tail)
:MWS-insul	1.27×10^{-4}	9.09	0.12	15.5	3.9	9.6	8.86	0.10
:MWS-insul	1.27×10^{-4}	9.22	0.09	-	-	-	-	-
" "	" "	9.29	0.08	15.3	3.8	11.4	-	-
" "	" "	9.08	0.04	17.4	3.7	10.9	8.99	0.14
:(Spool)	5.08×10^{-5}	-	-	-	-	-	8.94	0.36
Niobium-Titanium								
:Supercon[++]MFC	2.08×10^{-4}	9.52	0.35	-	6.9	74.0	-	-
" " "	"	9.32	0.07	-	6.0	74.0	-	-
" " "	"	9.67	0.19	-	7.7	93.6	8.87	0.10
" " "	"	-	-	-	-	-	8.90	0.07
:Supercon	1.27×10^{-4}	9.31	0.04	70.4	1.15	1.36	9.15	0.05
" "MFC	1.27×10^{-4}	9.23	0.10	-	6.34	49.6	8.44	(Tail)
" "	1.14×10^{-4}	9.20	0.04	-	1.15	1.53	-	-
" "	7.62×10^{-5}	9.08	0.03	-	1.13	1.38	8.92	0.13
:(CuNi)MF	2.54×10^{-4}	-	-	-	-	-	9.17	0.38
:MonoFil	1.14×10^{-4}	-	-	-	-	-	8.82	0.09

MF - Multifilament C = Copper Clad

[*]KBI, Div. of Cabot Corp., P.O. Box 1462, Reading, PA 19603.

[**]California Fine Wire, 338-40 South Fourth Street, P.O. Box 446, Grover City, CA 93433.

[+]Magnet Wire Supply, 31200 Cedar Valley Dr., Thousand Oaks, CA 91362.

[++]Supercon, 830-T Boston Turnpike, Shrewsburg, MA 01545.

Table I summarizes measurement of the midpoint of the superconducting transition T_c in both resistive and in ac susceptibility, the width of the transition [90% to 10%] ΔT_c, the resistivity at 300°K, the resistivity ratio at 77K (RR77) and the resistivity ratio just above the superconducting transition RR(T_o). It is very difficult to pick out any totally inambigious trends. However, for the Nb wires, lower resistivities, higher resistivity ratios generally yield higher T_c materials.

To gain more insight into differences between various Nb wires as well as differences between Nb and Nb-Ti we measured the magnetization curves of two Nb wires drawn from the same starting material by California Fine Wire. One was unannealed and uninsulated [hard] and the other was given a final strand anneal at 1400'F in argon at 25 to 30 ft/min and insulated [soft]. In addition, we measured one unclad NbTi and one copper clad NbTi wire. From the magnetization data we deduced the J_c vs field above H_{c1} and extrapolated to estimate $J_c(o)$.

The magnetization data was obtained with a 900 Series BTI susceptometer[4] with a 90 K gauss superconducting magnet. Figure 1a shows the magnetization data for the hard and soft Nb wires at 4.2K. Note that the soft Nb has a more reversible, hysteresis loop with a significantly smaller trapped moment corresponding to a smaller trapped flux. This may be significant for SQUID applications because trapped flux is a serious suspect for contributing to thermal noise. Unfortunately that final annealing process, through reducing the hysteresis degrades both the T_c and the RR in this material as is seen in Table I. A cleaner anneal should definitely be used.

Figure 1b shows similar data at 4.2K for the two NbTi wires. Noting the size of the trapped moments on this figure, the value for the unclad NbTi is comparable to the hard Nb while the clad NbTi is much worse.

To demonstrate that trapped flux can contribute to thermal noise we took each wire through its hysteresis loop and then monitored the moment as the wire was warmed from 4.2 up to its transition temperature. This data is plotted in Fig. 2. Note that the residual moment decays nearly linearly with temperature with a slope proportional to the 4.2 value. Thus, a larger trapped flux decays with temperature even around 4.2K. This decay although non-reversible might be an ultimate source of SQUID circuit noise.

Figure 3 shows the values of J_c(H) and the extrapolated J_c(o) for the 4 wires obtained using the model of Fietz and Webb.[5] This data completes our characterization of these wires. Notice that the trends reflected by the <u>trapped</u> flux are also reflected in the J_c values.

In another series of measurements using a different technique flux motion in Nb wires was studied in low magnetic fields (a few gauss). Here it was found that the response to temperature changes was entirely reversible. Furthermore, the response to small field excursions ($\pm$ 1G) was non-hysteretic to within the accuracy of the measurements.

In conclusion, we have characterized a large number of Nb and NbTi wires in an attempt to determine characteristics that would be desirable for use in SQUID circuits. At this point we believe that of the wires we tested the cleanest Nb samples with the highest T_c and lowest trapped flux would be the most suitable for SQUID applications. We intend to test this hypothesis by replacing any NbTi in present generation SQUID's with high quality Nb wire and determining if their noise performance is improved.

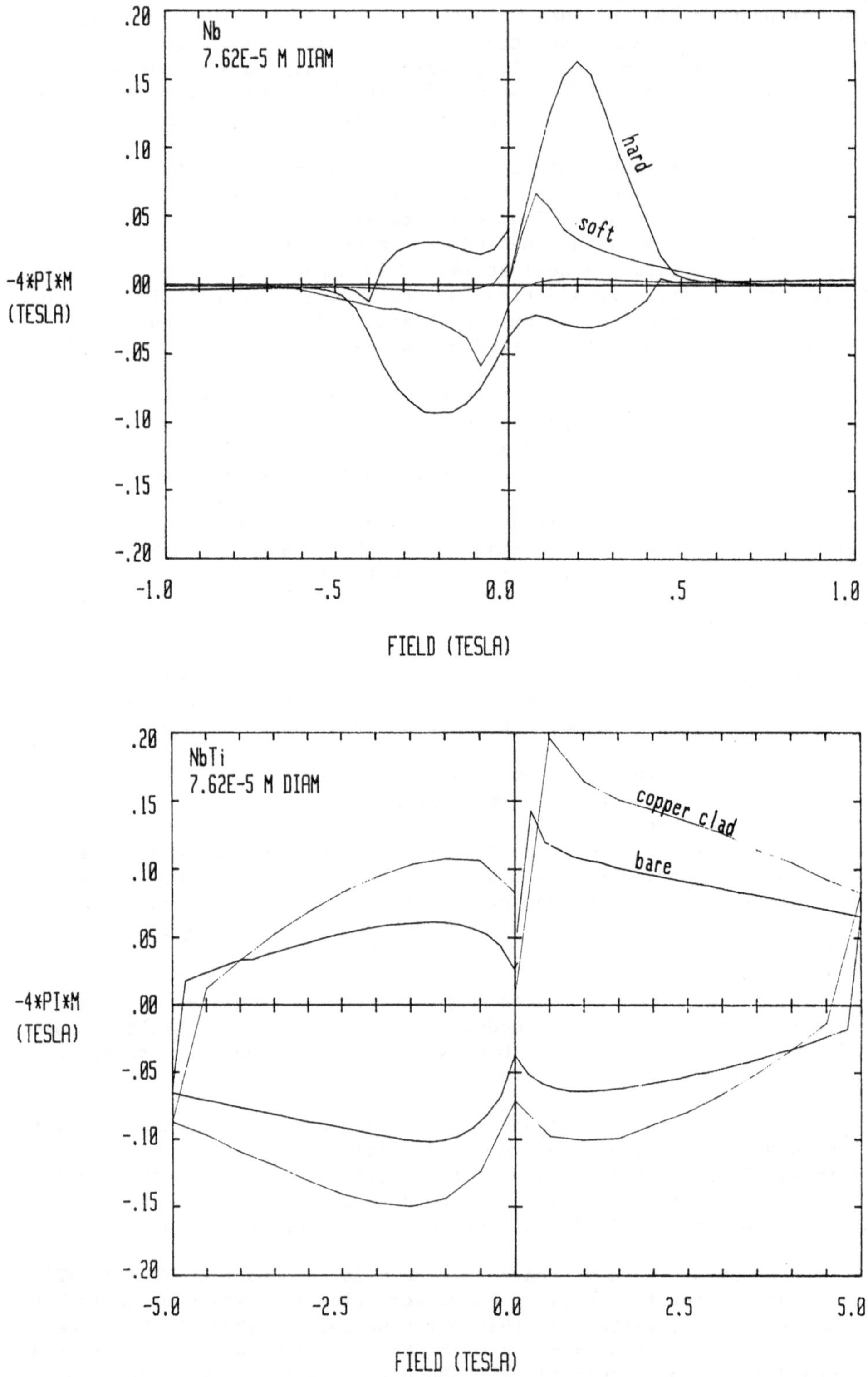

Fig. 1. Magnetization data for Nb and NbTi wires at 4.2 K.

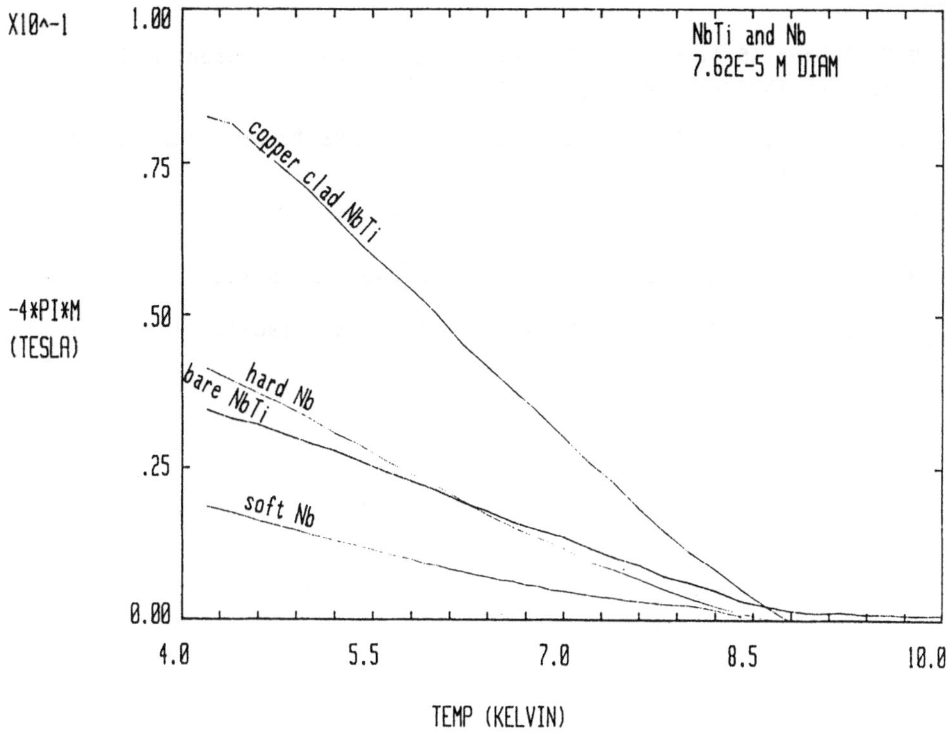

Fig. 2. Trapped flux decay as a function of temperature.

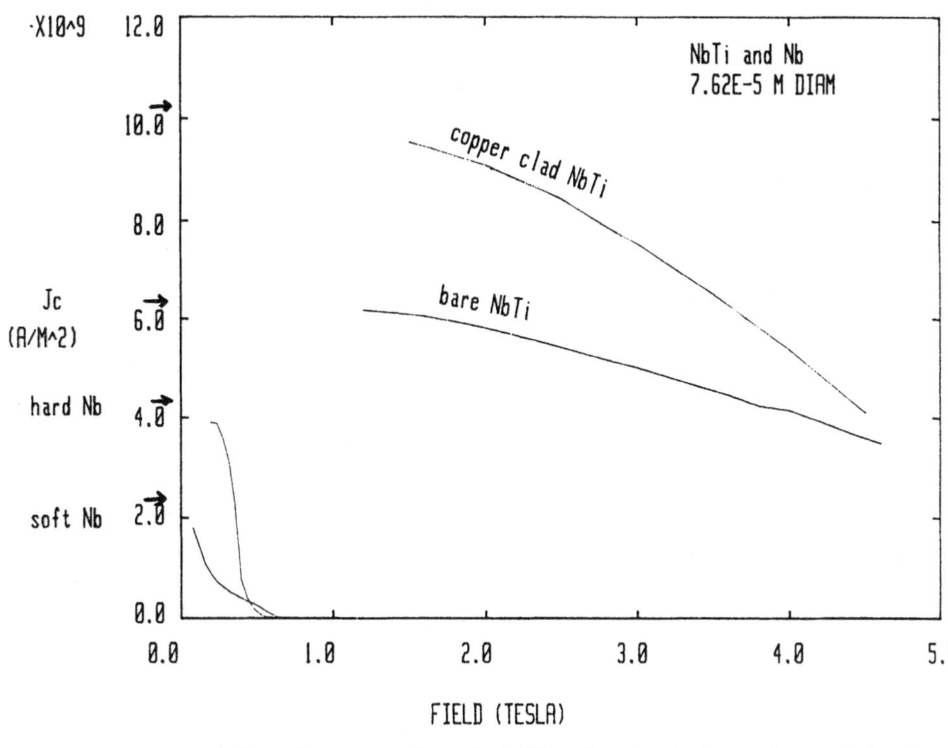

Fig. 3. $J_c(H)$ and extrapolated $J_c(0)$ obtained from the model of Fietz and Webb.

REFERENCES

1. D.U. Gubser, S.A. Wolf, T.L. Francavilla, J.H. Claassen, and B.N. Das, IEEE Trans. Magn. MAG-21, 320 (1985).

2. J. Claassen, S.A. Wolf, and D.U. Gubser, IEEE Trans Magn. MAG-21, 434 (1 985).

3. T. Clem, IEEE Trans. Magn. MAG-21, 606 (1985).

4. BTI, 4174 Sorrento Valley Blvd., San Diego, CA 92121.

5. W.A. Fietz and W.W. Webb, Phys. Rev. 178, 657 (1969).

FORMATION OF A15 PHASE IN EPITAXIAL AND POLYCRYSTALLINE Nb-Sn DIFFUSION COUPLES*

A. I. Braginski and J. R. Gavaler

Westinghouse R&D Center, Pittsburgh, PA

K. Schulze

Max Planck Institut für Metallforschung, Stuttgart, FRG

ABSTRACT

We investigated the effect of oxygen or oxide impurities and grain boundaries in binary diffusion reactions to form the A15 structure Nb_3Sn. Single-crystal and polycrystalline thin film Nb-Sn diffusion couples were fabricated and annealed in ultrahigh vacuum using very pure and ultrapure materials. Up to 700C, grain boundary diffusion produced A15 structure Nb_3Sn while bulk diffusion produced Nb_6Sn_5. At 850C the A15 formed also by bulk diffusion. Between 650 and 850C the oxygen or oxide impurity enhanced the formation of an A15 phase closer to 3/1 stoichiometry.

INTRODUCTION

The purpose of this study was to investigate the effect of oxygen or oxide impurities and grain boundaries in binary diffusion reactions between elements A and B to form a stable A_3B phase crystallizing in the A15 structure. It is known that the prototype β-tungsten metastable A15 phase can best be formed by hydrogen reduction of WO_3, and was in the past observed as a product of other reactions where oxygen impurities were present.[1] We and other authors have shown that oxygen presence is necessary to synthesize the metastable Nb_3Ge.[2-4] Later, we reported that several A15 compounds, some of them stable, can be formed by equilibrium diffusion reactions between the A metal and B oxide or vice-versa.[5] Also, nearly stoichiometric A15 phases were seen to coexist with B-deficient ones in films sputtered from targets having $A/B >> 3$ when oxygen impurities were present. All these results lead us to suggest that oxygen or oxides contribute to, and may be required for, the nucleation of both metastable and stable A15's.[5] In some instances only minute amounts of oxygen could be present in diffusion and non-equilibrium low-temperature reactions to form stable A15's. Also, high-temperature A15 phases obtained by quenching in some binary systems are not known to contain any appreciable amount of oxygen. We thus invoked the possibility of a catalytic effect due to oxygen's presence.[6] In polycrystals, and at lower temperatures, the dissolved oxygen usually segregates to grain

*Supported in part by AFOSR Contract No. 49620-C-85-0043.

boundaries. The existence of grain boundaries in A- and B-reactants was thus expected to not only increase the reaction rate but, perhaps, to also influence its outcome.

In this paper we report on results of an attempted direct test of our ideas concerning the formation of stable A15 phases. The approach was to employ the A-B binary diffusion couple technique, in ultrahigh vacuum (UHV) and with ultrapure materials, to compare reactions between single crystal and polycrystalline layers and to observe the effect of added oxygen or oxide on the phase composition of reaction products. We investigated the formation of Nb_3Sn, a stable A15 compound having a high superconducting transition temperature, T_c = 18K.

EXPERIMENTAL

Diffusion couples were prepared by sequential deposition of elements, and subsequent in-situ annealing in an UHV deposition and surface analytical system.[7] A film of niobium 200 to 300 nm thick was first e-beam evaporated on various precleaned sapphire substrates. On top of Nb, approximately 50 nm of tin (average thickness inferred from the deposited mass) was evaporated from an effusion cell. The sample holder with sapphire substrates was rotated during deposition at 20 rpm. The purity of source materials is given in Table 1. The background pressure was not exceeding 7×10^{-9} Pa during the evaporation of ultrapure Nb, Sn and during annealing. The residual gas analysis (RGA) indicated, in the decreasing order of partial pressure, the presence of CH_4, H_2, N_2 or CO as main gas impurities. Free oxygen was not detected. The MARZ grade Nb was being evaporated in 1.5 to 3×10^{-8} Pa. The sapphire substrates were 0.64 x 0.64 x 0.05 cm. To obtain single crystal Nb, epitaxial-grade-polished sapphire was used as a substrate, and the substrate temperature during deposition was typically 800C, although even at room temperature single crystal films were produced. To deposit polycrystalline Nb, non-epitaxial or ground sapphire substrates were kept at a temperature not exceeding 100C. Tin was deposited at either 300C or below 100C.

The surface of Nb was oxidized in some experiments prior to the deposition of the subsequent Sn layer. Oxidation was performed in about 13 Pa of dry oxygen, for one hour at room temperature. Annealing temperatures were T_a = 650, 700 and 850C, for 0.5, 5 or 50 hours at the peak temperature. The W-Re thermocouple was calibrated by direct, visual observation of melting points of Ge and Al. The substrates were always clamped to a rotatable molybdenum support block (sample holder) to insure a good thermal contact,

Table 1. Source Materials

(Contaminants' content in at. ppm, or purity)

Ultrapure-Nb (Max-Planck)	Pure-Nb MARZ (MRC)	Sn MARZ (MRC)
Ta: 1.2 W : 0.3 O : 5-10	99.995% — nominal; Ta: 200, W: 0.1 O + N + C > 50 Other met.: > 11	99.995%

and the thermocouple was positioned in the block's cavity shaped for this purpose. The reproducibility of clamping and thermocouple positioning was tested by a direct pyrometric observation of annealed samples. The temperature variation between neighboring samples was ± 10C. The measurement error was estimated at ± 20C.

The crystallinity of deposits was determined in-situ by reflection high energy electron diffraction (RHEED). The in-situ surface analysis, phase determination from chemical shift, and measurement of Nb oxide thickness was performed using the X-ray photoelectron spectroscopy (XPS). The ex-situ characterization included phase analysis by X-ray diffractometer, determination of composition depth profiles by scanning Auger electron spectroscopy (SAES), and resistive measurements of T_c by the four-point van der Pauw method and Ge-thermometry. The resistivity ratio, RR = $\rho(300K)/\rho(10K)$ was measured in reacted couples since it provides a sensitive indirect probe of a compound formation and impurity content. The RR was also measured in separate Nb films to obtain an indication of their purity. In the ultra-pure Nb the conductivity was thickness-limited, as shown in Figure 1, and thus determined by surface scattering rather than by impurities.

RESULTS AND DISCUSSION

The crystallinity of Nb and Sn films was determined from RHEED patterns such as shown in Figure 2. On epitaxial (11$\bar{2}$0) sapphire, single-crystal (110) Nb was formed with the X-ray rocking curve width of 0.1 degree. Tin grew epitaxially on Nb. On nonepitaxial sapphire polycrystalline Nb was obtained. In this case no electron diffraction pattern of Sn overlayer was observed. The tin deposits grew with a 3D-habit to form islands on the Nb surface. This was evidenced by XPS detecting Nb 3d spectra before annealing, by SAES and by scanning electron microscopy. Chemical shift in the XPS spectra indicated that oxidation of Nb surface produced Nb_2O_5 which was polycrystalline, according to RHEED. Upon heating to 850C in UHV the oxide disappeared completely, presumably by dissolving in Nb, and a clean Nb surface was again observed by XPS upon cooling to the ambient. Consequently, it is not known whether the oxide was present at the Nb/Sn interface at the inception of diffusion.

The interdiffusion resulting from annealing was monitored ex-situ by measuring SAES depth profiles of tin. Examples of such profiles are shown in Figure 3. The average profiling (sputtering) rate was determined for unreacted and reacted Nb-Sn specimens of different thicknesses which were

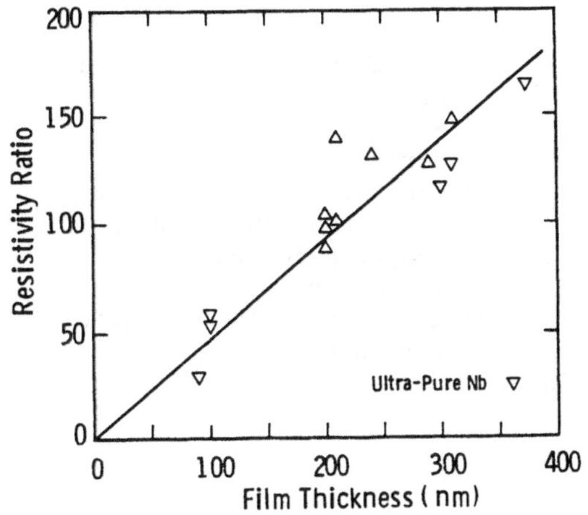

Fig. 1. Niobium resistivity ratio vs. thickness for films deposited from ultrapure Nb.

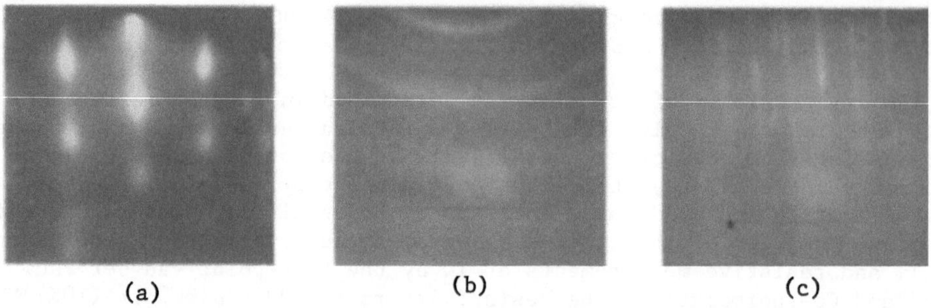

Fig. 2. RHEED patterns: a) Nb (110) on (11$\bar{2}$0) sapphire; b) Nb on non-epitaxially polished sapphire; c) Sn on Nb (110).

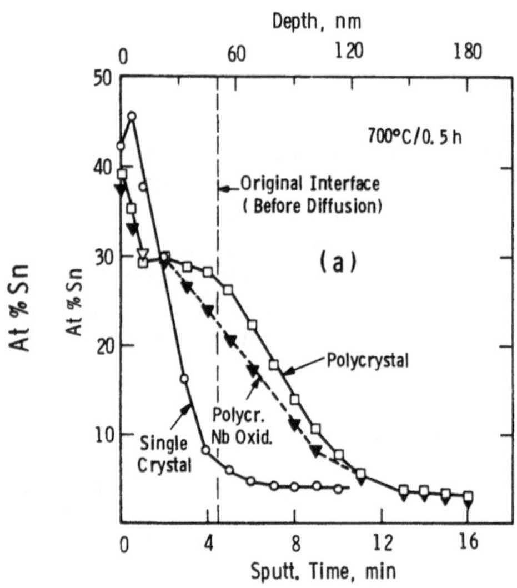

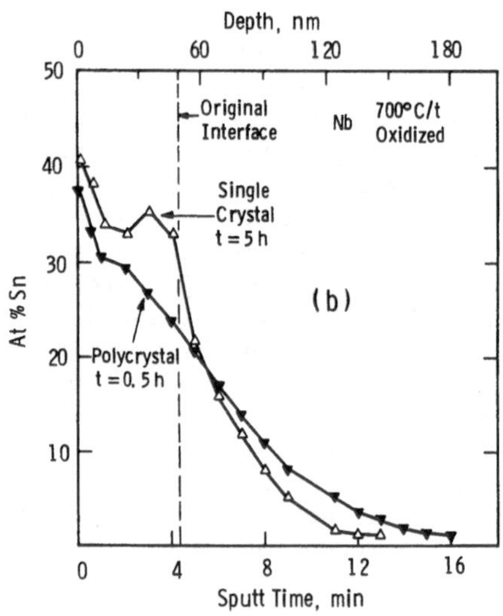

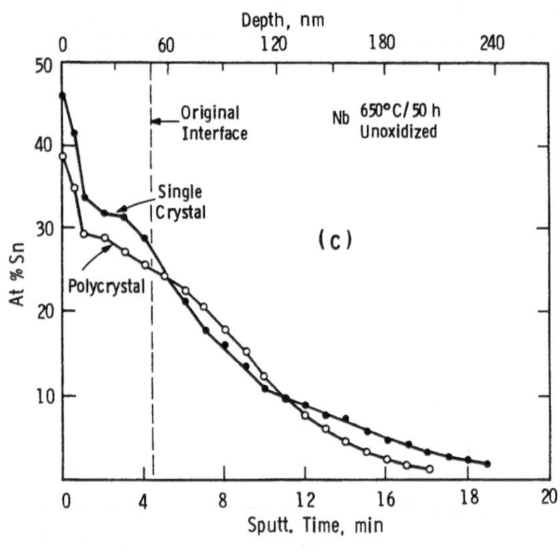

Fig. 3.

Tin depth profiles obtained by AES:
a) After annealing at 700°C for 0.5 hour
b) Comparison with the profile in a single-crystal couple annealed at 700°C for 5 hours
c) After annealing at 650°C for 50 hours.

independently measured by a profilometer. The rate was defined by the thickness and the sputtering time required to detect the substrate aluminum and oxygen Auger signals. The obtained value of 11.2 nm/min. was within 1%, independent of the thickness and degree of interdiffusion. It represented, however, an average of rates of sputtering of the several species present (Sn, Nb and Nb-Sn reaction products) which were not determined independently. Consequently, the depth scale in Figure 3 is only approximate. The composition (atomic % tin) indicated in Figure 3 is semiquantitative, since it was obtained from elemental Auger peak heights without calibrating by Nb-Sn compound standards. Such a calibration was not performed since the non-uniform coverage of Nb by Sn would make a quantitative analysis unreliable.

Figure 3a shows tin profiles after 0.5 hour annealing at 700C. The profile in the Nb-Sn single crystal couple was the same as in an unreacted sample and defined by the non-uniform coverage of the Nb surface by Sn. In contrast, tin visibly diffused into polycrystalline Nb. The comparison with the tin profile in a single crystal couple after 5 hours at the same temperature (Figure 3b) indicated that bulk diffusion into the Nb-crystal was taking place, although at a rate over an order of magnitude slower than the diffusion into polycrystalline Nb which we assumed to occur along the grain boundaries. After 50 hours at 650C, however, very similar profiles were obtained for both types of couples (Figure 3c) thus suggesting an approach to the equilibrium distribution. We believe that after the very long annealing time the difference in diffusion rates between polycrystalline and single crystal couples was no longer easily observable since the free tin was largely consumed to form the Nb-Sn compound(s). Surface oxidation of polycrystalline Nb reduced the diffusion rate (Figure 3a). Qualitatively, the effect of the grain boundaries and of the oxygen impurity appeared to us as consistent with established models of diffusion.[8] The microstructure (grain size) of Nb films was not determined, so that a quantitative analysis of diffusion rates was not attempted.

The critical temperatures and RR values in couples formed on sapphire and reacted at 700C for 0.5 hour depended upon the crystallinity and oxidation, as shown in Table 2. The T_c (1 to 99% transition) and RR data given represent averages of 4 to 6 specimens, to minimize a possible effect of variation in thermal anchoring of specimens clamped on the Mo-block. In single crystalline, unoxidized couples with ultrapure Nb the T_c = 9.30 to 9.25K did not exceed that of very clean Nb, and the RR remained high, although reduced from the original value of 140 to 150 determined for a corresponding Nb-film alone. This reduction indicated that tin diffusion had progressed somewhat, even if not detected by SAES. In couples with oxidized Nb the T_c = 7.3 to 7.2K was reduced, and the RR exhibited a drastic drop to values less than three, both typical of dirty Nb with impurity-dominated

Table 2. Nb-Sn Diffusion Couples Annealed at 700°C/30 min ($p < 6.7 \times 10^{-9}$ Pa)

RR of Nb Film Prior to Anneal.		140 to 150		>20	
Diff. Couples		T_c, K (1% to 99%)	RR	T_c, K (1% to 99%)	RR
Single Crystal	No Oxid.	9.3 - 9.25	46	9.1 - 9.0	15.3
	Oxid.	7.3 - 7.2	2.8	--	--
Polycrystalline	No Oxid.	12.8 - 11.0	5.1	15.4 - 14.8	7.3
	Oxid.	15.9 - 14.7	2.7	--	--

Table 3. Nb-Sn Diffusion Couples Annealed at 850°C/30 min ($p < 6.7 \times 10^{-9}$ Pa)

RR of Nb Film Prior to Anneal.	140-150		100-120		100-120	
Diff. Couples	T_c, K (1% to 99%)	RR	T_c, K (1% to 99%)	RR	T_c, K (1% to 99%)	RR
Single Crystal	14.9 - 12.7 (Sapphire subst.)	22	17.7 - 17.3 (Sapphire subst.)	25	--	--
Polycrystalline	--	--	17.7 - 17.5 (Sapphire subst.)	9.4	16.5 - 10 (Niobium subst.)	7.6

scattering. In polycrystalline, unoxidized couples with ultrapure Nb the superconductive transition between 12.8 and 11.0K and the low RR = 5.1 proved that an off-stoichiometric A15 phase was formed, and that the diffusion proceeded far deeper into the Nb-film, presumably along its grain boundaries. Couples with oxidized Nb exhibited an even higher T_c = 15.9 to 14.7K, indicative of an A15 phase closer to stoichiometry. The RR was low, as in oxidized single-crystal couples.

Single-crystal, unoxidized couples fabricated from lower purity, MARZ grade Nb (RR > 20) showed lower T_c = 9.1 to 9.0K and RR than those of ultrapure material, consistent with a higher impurity content in Nb. In contrast, corresponding polycrystalline couples exhibited a T_c = 15.4 to 14.8K, higher than that of their ultrapure counterpart, and almost identical to that of ultrapure but intentionally oxidized couples. The X-ray phase identification of all these samples was not attempted because of the thinness of the diffusion layers (Figure 3a). The data of Table 2 suggested, nonetheless, that grain boundaries and the presence of oxygen in Nb enhance the formation of stoichiometric Nb_3Sn, in line with our earlier speculations.

The results obtained for a subsequent series of couples annealed at much higher temperature, T_c = 850C for 0.5 hour, are shown in Table 3. These results indicate that the A15 phase had formed even in the cleanest attained conditions, albeit with a T_c somewhat lower than that of analogous couples fabricated from MARZ-Nb of somewhat lesser purity (RR = 100 to 120 vs 140 to 150). This Nb source was gradually purified by remelting in consecutive depositions, so that the contrast with ultrapure Nb was reduced. To verify whether the sapphire (Al_2O_3) substrate could stimulate the formation of A15, we also formed a couple on polycrystalline, low-purity Nb foil, with identical results (Table 3). Until the deposition of Sn on ultrapure, bulk single crystal Nb (RR = 10^4) is attempted, we are left with results that are suggestive of some effect of oxygen, but clearly not definitive. It is possible that the effect of oxygen is to increase the rate of A15 formation at lower temperatures, 700C and below. Alternatively, very minute amounts of oxygen or oxide are required for the A15 phase nucleation by an unidentified mechanism.

The X-ray phase analysis was performed to determine the products of interdiffusion at lower temperatures, where no evidence of A15 formation was obtained from the T_c of single-crystal couples. The results of phase analysis and T_c measurement are summarized in Table 4 for the couples annealed at 650C for approximately 50 hours which have similar tin diffusion profiles as shown in Figure 3c. Corresponding diffractometer traces are reproduced in Figure 4. These traces show that the reaction products were strongly textured. Unfortunately, this made the interpretation less definitive. In Figure 4a the Nb_6Sn_5 is seen to form in the absence of grain boun-

Table 4. Nb-Sn Diffusion Couples Annealed at 650°C for 50 Hours, $p = 4 \times 10^{-9}$ Pa

Diffusion Couples	T_c (K) (1% to 99%)	Phases Present (from X-Ray Diffractometer)
Single Crystal	9.3 - 9.1	Nb_6Sn_5 - Major, Nb
Polycrystalline	16.9 - 16.4	Nb_3Sn - Major, Nb

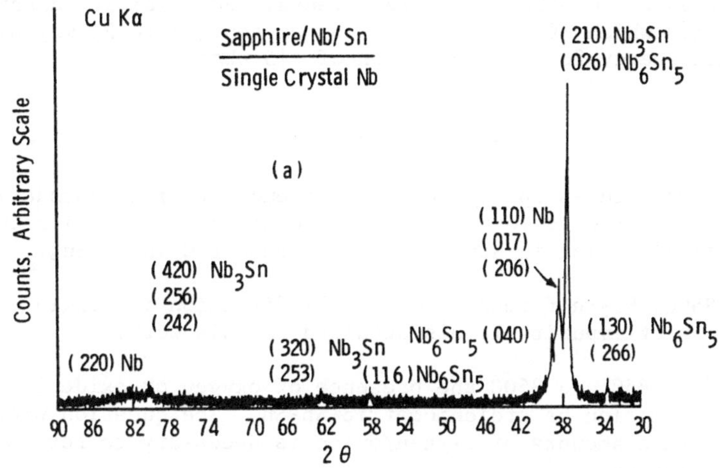

Fig. 4(a). Diffractometer trace for a single crystal couple annealed at 650C for 50 hours.

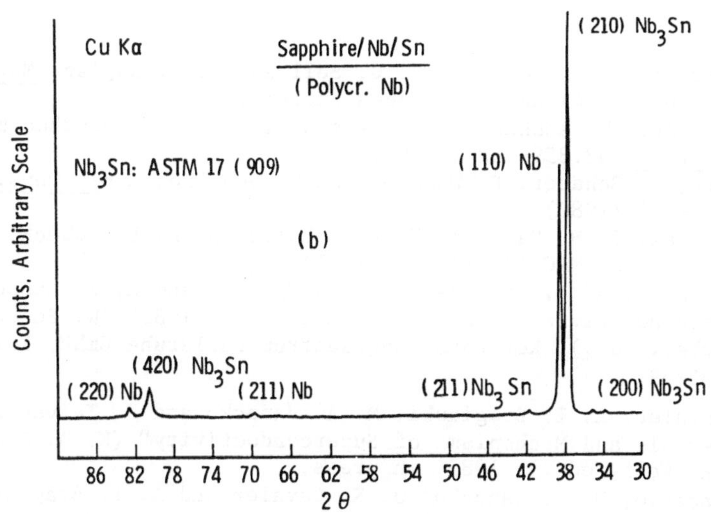

Fig. 4(b). Diffractometer trace for a polycrystalline couple annealed at 650C for 50 hours.

daries. Figures 3b and 3c show that at sputtering times up to five minutes single crystal couples exhibited tin plateaus much higher than in polycrystalline samples. These higher plateaus are consistent with the Nb_6Sn_5 presence. Unfortunately, the co-presence of the A15 phase cannot be excluded based on the X-ray data alone since the A15 reflection peaks overlap those of the orthorombic Nb_6Sn_5. The absence of A15 is, however corroborated by the electrical properties. The T_c given in Table 4 is typical of the very pure Nb still present. In contrast, only A15 was obtained by diffusion reaction in polycrystalline couples, as evidenced by Figure 4b and corroborated by the high T_c in Table 4.

At this point we do not know whether the difference between the reaction products in single crystal and polycrystalline couples is due to oxygen at grain boundaries or to other possible effects. For example, disorder at grain boundaries which is only enhanced by oxygen (Table 2) may simply reduce the activation energy for the formation of A15 phase.

CONCLUSIONS

1. At temperatures up to 700C the presence of grain boundaries in high-purity Nb was required to form the A15 phase Nb_3Sn by diffusion in UHV environment. Bulk diffusion in single crystals resulted in Nb_6Sn_5.

2. At 850C, Nb_3Sn formed also by bulk diffusion in single crystals, even at the lowest impurity levels attained in this work.

3. Between 650 and 850C the presence of oxygen or oxide impurity enhanced the formation of stoichiometric Nb_3Sn. The question whether the presence of minute amounts of oxygen/oxide is necessary to form Nb_3Sn by diffusion remains open.

ACKNOWLEDGEMENT

We acknowledge the important contribution of R. Kuznicki who performed the X-ray diffractometer analysis.

REFERENCES

1. W. R. Morcom, W. L. Worrell, H. G. Sell and H. I. Kaplan, Met. Trans. 5:155-161 (1974) and references therein.
2. J. R. Gavaler, M. Ashkin, A. I. Braginski and A. T. Santhanam, Appl. Phys. Lett. 33:359-362 (1978).
3. B. Krevet, W. Schauer, F. Wüchner and K. Schulze, Appl. Phys. Lett. 36:704-706 (1980).
4. A. B. Hallak, R. H. Hammond, T. H. Geballe and R. B. Zubeck, IEEE Trans. on Magnetics, MAG-13:311- (1977).
5. J. R. Gavaler, A. I. Braginski, M. A. Janocko and A. S. Manocha, in: "Superconductivity in d- and f-Band Metals 1982" (W. Buckel and W. Weber, eds.), Kernforschungszentrum Karlsruhe GmbH, Karlsruhe 1982, pp. 35-39.
6. J. R. Gavaler, A. I. Braginski, M. A. Janocko and J. Talvacchio, in: "Materials and Mechanisms of Superconductivity" (K. A. Gschneider Jr. and E. Wolf, eds.), 1985, in press.
7. J. Talvacchio, M. A. Janocko, J. R. Gavaler and A. I. Braginski, this volume.
8. P. Haasen, "Physical Metallurgy," Cambridge Univ. Press, London, 1978.

A NEW LOOK AT THE GROWTH OF THIN FILMS OF Nb-Sn

Frances Hellman,[†] J. Talvacchio[*] and T. H. Geballe[**]

Department of Applied Physics
Stanford University

and

A. F. Marshall

Center for Materials Research
Stanford University

ABSTRACT

A15 films grown by vapor deposition onto heated substrates have been generally observed to form in the equilibrium phase. However, vapor deposition is not in fact an equilibrium process and effects of the layer-by-layer growth may be seen. Two such effects will be described here. Both are related to the presence of grains in the growing film, itself a non-equilibrium effect. The first effect to be described is a grain overgrowth process resulting in cone-shaped rather than columnar grains in polycrystalline films grown at 800°C, and single-crystal films grown from a polycrystalline region near the substrate at higher deposition temperatures. The second is a segregation effect believed to be previously unobserved. It is related to surface segregation but dependent on the low bulk mobility and high surface mobility characteristic of materials grown below half their melting temperature. This effect we believe causes certain polycrystalline films to become compositionally inhomogeneous.

INTRODUCTION

The films analyzed in this work are 1-2 μm thick, far thicker than any relevant length scale, and hence are essentially bulk samples. The question arises as to how these samples differ from ideal bulk samples, and how the layer-by-layer growth affects the final material.

We will first discuss how the films are grown and how they are characterized. Transmission electron microscopy (TEM) and low temperature specific heat were the primary measurements utilized. The overgrowth phenomenon and the inhomogeneity will be discussed in turn. A brief description of the model for the segregation process causing the inhomogeneity will then be presented. Finally, practical suggestions for reducing the inhomogeneity will be given.

FILM PREPARATION AND CHARACTERIZATION

We will primarily discuss A15 Nb-Sn samples grown by electron beam coevaporation under different deposition conditions and at different compositions. The electron beam evaporation system was built by Hammond.[1,2] Three sources, arranged colinearly in order to eliminate any geometry-related phase spread, are heated by electron beams. The rates are monitored and controlled by ionization-gauge-type rate monitors; control is believed better than 1%. Substrates, typically (1$\bar{1}$02) nominally epitaxially polished sapphire, are clamped to a substrate holder which is mounted on a furnace (see reference 3 for more detail). For the work relevant to this paper, substrate temperatures from 200-1000°C and rates from 0.2-10 nm/sec were used. The so-called "standard" deposition parameters of 800°C and 2 nm/sec had been previously optimized;[4] 800°C is essentially the hottest deposition temperature possible with no worry of Sn reevaporation from the A15, and 2 nm/sec yields optimal rate control. Typical background pressure is 1.3×10^{-5} Pa during the evaporation.

Two methods of characterization will be discussed in this paper: transmission electron microscopy (TEM) and low temperature specific heat. The microscopy was done in a Phillips 400 microscope. Samples were prepared by cutting a 3 mm disk from the standard 1/4" × 1/4" sapphire substrate. The disk was ground to ≈ 100 μm thickness. The thinned disks were then ion-milled through from the back to perforation. This allows the top (vacuum interface) surface to be examined by TEM. In order to examine the bottom (sapphire interface) surface, the samples were subsequently carefully ion-milled from the top. Selected area diffraction (SAD) and convergent beam diffraction were used to characterize the orientation of the films.

Low temperature specific heat was measured in an apparatus designed by S. R. Early for the measurement of small (< 1 mg) samples.[5,6] Samples are grown directly on the back of silicon-on-sapphire chips. These chips consist of two thermometers and a heater made from doped Si grown on (1$\bar{1}$02) sapphire. The back of this sapphire is polished to the same nominal epitaxial finish as the other substrates. They are thinner (125-175 μm thick) than the other substrates, but should otherwise be identical.

OVERGROWTH

It is observed that randomly-oriented grains are overgrown by the apparently dominant (100) grains. This overgrowth manifests itself somewhat differently in the single crystal samples grown at 900°C and above than it does in the polycrystalline samples grown at 800°C.

Shown in Fig. 1 are TEM images of the top and bottom surfaces of a Nb-Sn sample grown at 900°C. The top surface shows the contrast effects of a single crystal sample, which SAD pictures show it to be. The characteristic defects are discussed elsewhere.[7] The bottom surface shows large areas of the epitaxially growing (100) orientation, but also shows polycrystalline regions. These polycrystalline regions are gone by approximately 1000 Å into the film and the remaining 2.4 μm appears as in Fig. 1a. Small rotated (100) inclusions (see reference 7 for more detail) persist through more of the film but also are finally overgrown.

Fig. 2 shows TEM images taken at equally spaced depths through a 2.5 μm thick A15 Nb-Sn film grown at 800°C. The images are all at the same magnification; they clearly show the enormous change in average grain diameter through the film thickness. The change, which is approximately an order of magnitude, occurs continuously through the film. The grains, rather than columnar as usually supposed, are actually more like cones. No voids are seen between the grains at any depth.

The overgrowth phenomenon in the polycrystalline films is more complicated than simply the (100) grains overgrowing all others. Firstly, there are still grains of other orientations and they too are larger than in the underlying layers, indicating that orientations other than the (100) may overgrow other orientations. Secondly, recent results indicate that polycrystalline A15 Nb-Sn grown on (1$\bar{1}$02) sapphire, while predominantly (100), has a sizable fraction of (311) grains, the remaining grains being seemingly randomly oriented. Preliminary results also suggest that when grown on

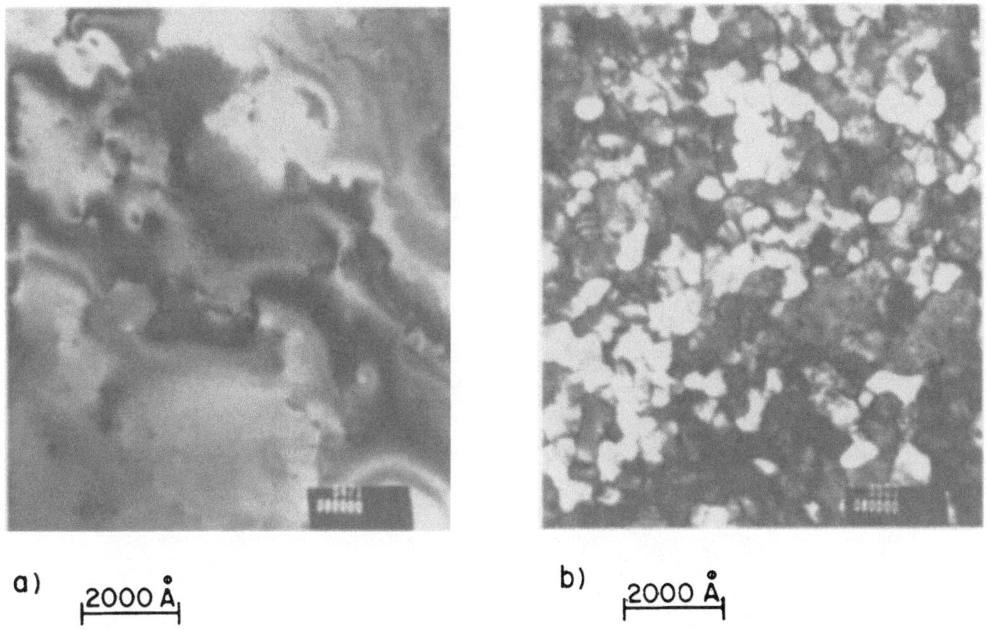

Fig. 1 TEM images of A15 Nb-Sn sample grown at 900°C.
a) Single crystal top surface of film.
b) Bottom surface of same sample as shown in a). Note mixed large epitaxial (100) regions and polycrystalline regions.

(0001) sapphire, a high fraction of (210) orientation is found. Polycrystalline A15 V-Ga grown at 750°C on 1102 sapphire shows a double texturing of (210) and (100) orientation.[6] Epitaxial relations with the sapphire substrate must be affecting the nucleation rates of the different orientations. The factors controlling the subsequent rate of overgrowth of the different orientations are not clear.

SURFACE SEGREGATION AND INHOMOGENEITY

The second growth effect to be discussed here is linked to the different orientations present at the surface of the polycrystalline samples. We have found what we believe to be a previously unobserved process in film growth. This process will cause certain of the polycrystalline films to become compositionally inhomogeneous, despite both a homogeneous deposition and the fact that a homogeneous film is the equilibrium configuration in the materials to be considered. The inhomogeneity we believe occurs while the atoms are in the surface layer, before they are buried by the next layer of material.

As will be shown below, the inhomogeneity is only $\pm 1-1.5$ at.% at its most extreme and hence is difficult to measure directly. Instead, we make use of the fact that in the A15's, the superconducting properties, especially the transition temperature T_c, are a strong function of composition. An inhomogeneity therefore manifests itself as a spread in T_c, as long as the length scale of the inhomogeneity is greater than a superconducting coherence length ξ, which is equal to approximately 50Å at $T = 0$ in the A15's. The width of T_c may be quite large, up to 6 K, and is thus easily measured by low temperature specific heat. The spread in T_c can of course be caused by an inhomogeneity in material properties other than composition. Results documented elsewhere prove fairly conclusively that composition is in fact the origin.[8]

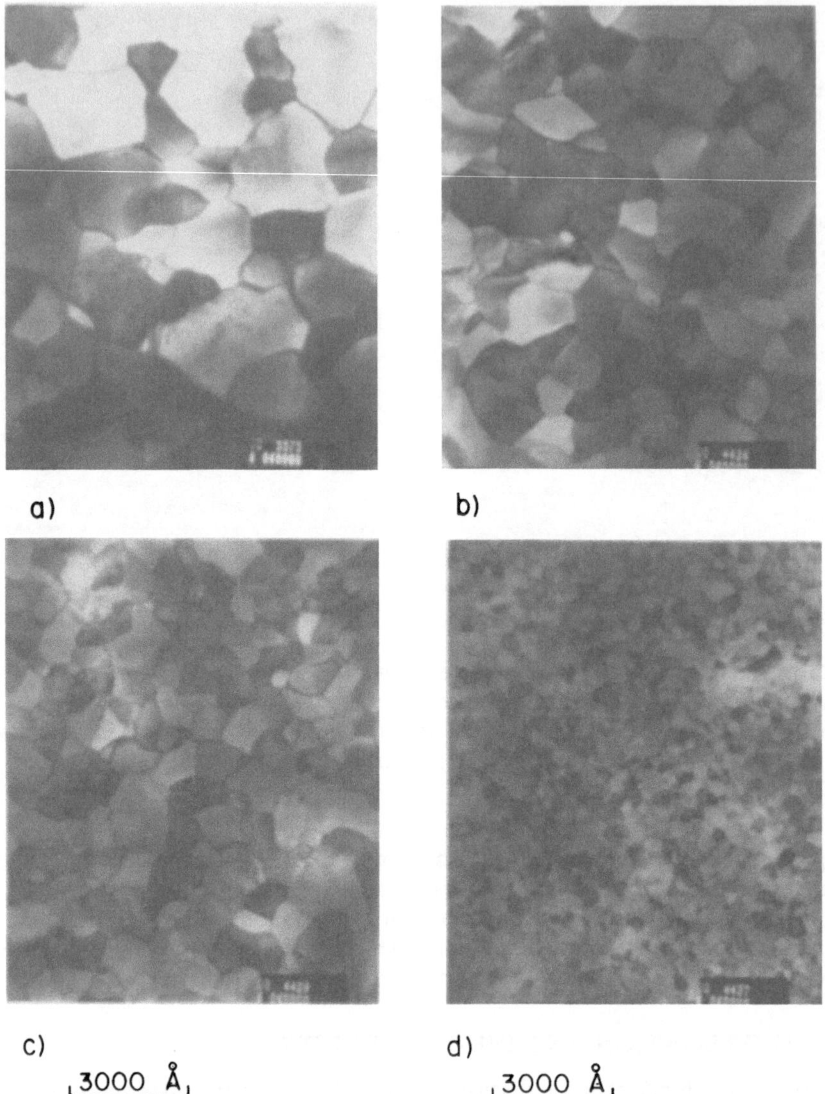

Fig. 2 TEM images of polycrystalline A15 Nb-Sn film grown at 800°C at equally spaced intervals through film thickness: a) top surface, b) 1.4 μm, c) 0.7 μm and d) bottom surface.

Figure 3 shows the specific heat from 1.5-28 K for six Nb-Sn samples spanning the accepted equilibrium phase field. These samples were grown at 800°C at 20 Å/sec. Compositions were determined by electron microprobe. After the initial measurement, two of the samples shown were annealed at 800°C for 24 hours and remeasured. There was no significant change in the specific heat; the T_c of one shifted upward approximately 0.5 K, but the shape of the curve was unchanged.

The data for the 25 at.% Sn sample are well within experimental error (estimated to be under 10%) of the literature values for the specific heat of bulk Nb_3Sn. The transition is quite sharp; the width is only 0.3 K. This width, however, increases immediately upon decreasing the Sn content; at 23.5 at.% Sn, it is approximately 4 K and at 22 at.% Sn, it is approximately 6 K. It then decreases towards the 18 at.% Sn phase boundary. The width of the transition has here been specified as follows. The onset is taken as the temperature at which the data break from a polynomial describing

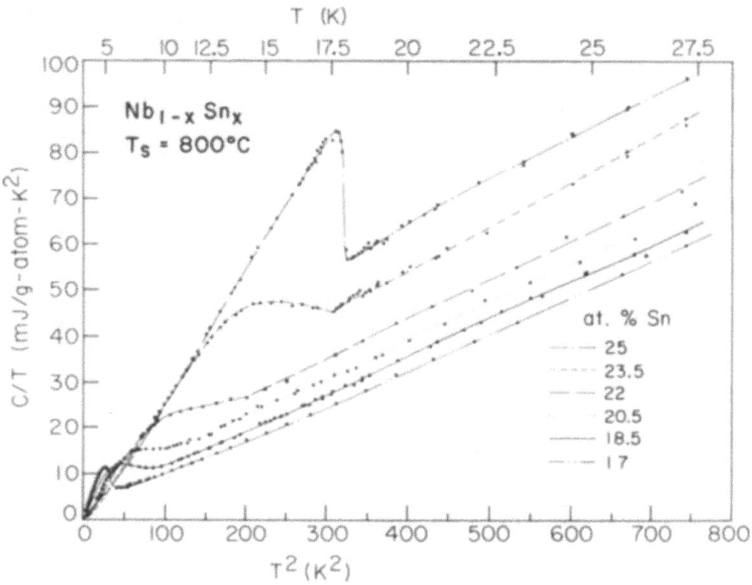

Fig. 3 Specific heat for "standard" samples: T_s = 770-830°C, deposition rate = 1-2 nm/sec, $P < 2.5\times10^{-5}$ Pa. Compositions shown: 17, 18.5, 20.5, 22, 23.5 and 25 at.% Sn. Lines are drawn through data to guide the eye.

the normal state. The lower end of the transition is conservatively estimated as the temperature at which the data clearly deviate from BCS or strong-coupled behavior.

Figure 4 shows the transition temperature T_c versus composition for A15 Nb-Sn grown at 800°C. Both the inductively measured T_c and the T_c from specific heat are shown. This plot re-emphasizes the sharpness of T_c in the 17 at.% Sn and, especially, the 25 at.% Sn, and the breadth at 22 at.% Sn. Figure 4 also emphasizes the disparity between the width measured by specific heat and by the inductive method. Even at 22 at.% Sn, the inductive T_c 5%−95% points are only 0.7 K apart. The resistive transition is even sharper. This fact explains why the inhomogeneity was so long unobserved in otherwise well-characterized samples similar to these.

Figure 5 shows the upper critical field Hc_2 and critical current density J_c as a function of temperature for the 25 at.% Sn and 23.5 at.% Sn samples whose specific heat was shown in Fig. 3. Hc_2 shows no sign of the inhomogeneity; it is measured by determining the transition temperature either inductively or resistively as a function of field. J_c on the other hand is a bulk measurement like the specific heat, although somewhat more geometry dependent and hence is sensitive to the inhomogeneity. There is no intrinsic reason for J_c to extrapolate linearly to T_c. Empirically, it has been observed that in materials such as the A15's, J_c shows a curved portion near T_c, followed by a linear region at lower temperature which extrapolates to T_c^*.[9] In the stoichiometric 25 at.% Sn film, T_c^* is 0.8 K less than T_c. However, in the 23.5 at.% Sn film, T_c^* is 2 K less than T_c. Presuming the pinning mechanism in off-stoichiometric Nb-Sn to be the same as in the stoichiometric material, T_c^* for a homogeneous 23.5 at.% Sn film would be expected to be 0.8 K less than T_c; the increased curvature is a direct reflection of the inhomogeneity.

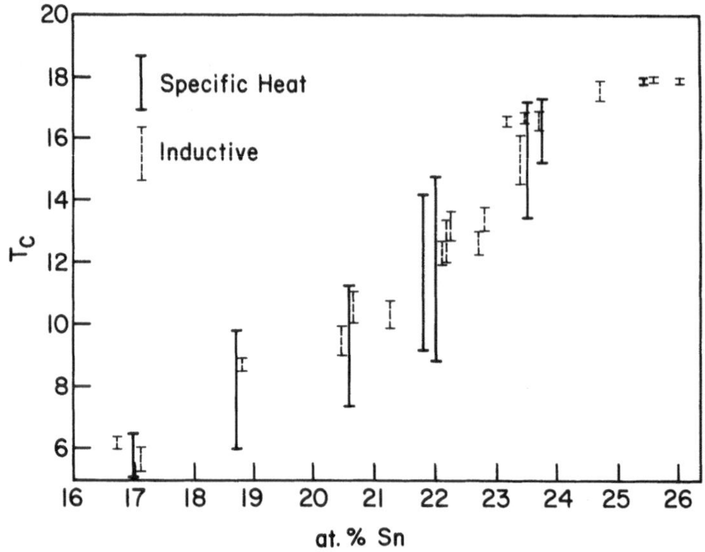

Fig. 4 Composition dependence of T_c for Nb-Sn grown at 800°C determined inductively and by specific heat.

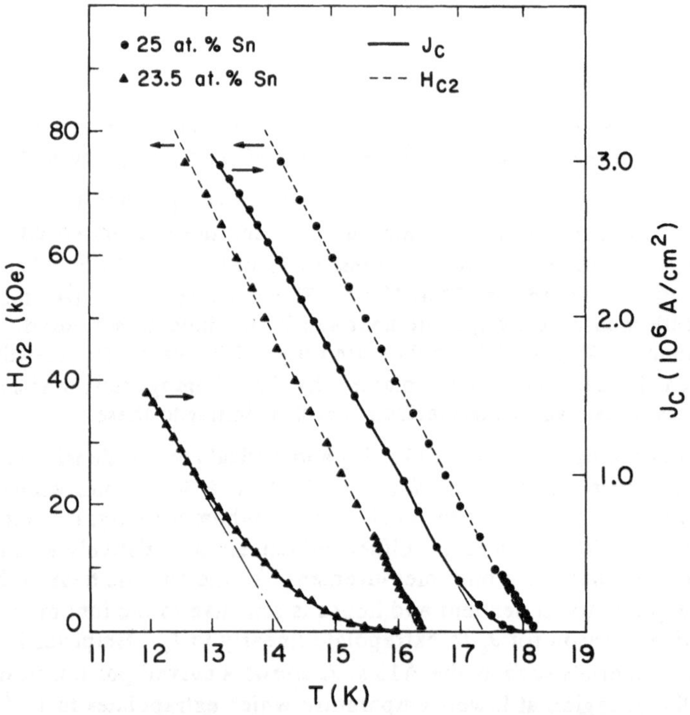

Fig. 5 Critical current and field for two samples, one on stoichiometry with a T_c of 18 K and one off with a T_c onset of 16.4 K. Critical current is shown with a dashed line, critical field with a solid. T_c^* (see text) is indicated with a dash-dot line.

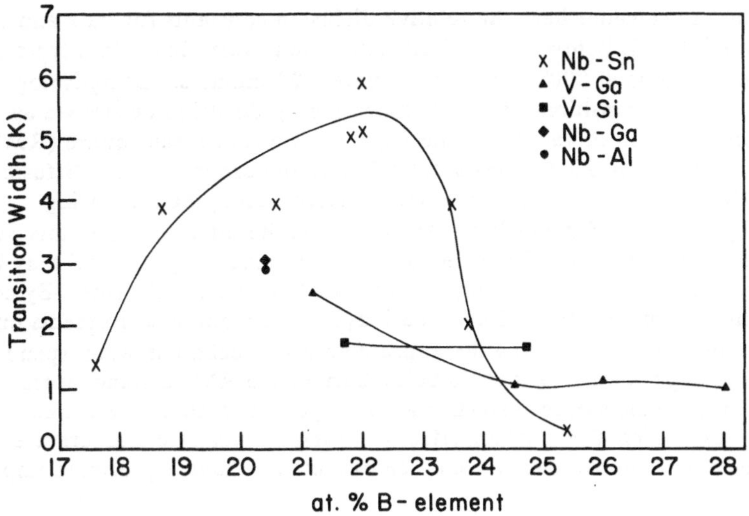

Fig. 6 Composition dependence of the width of the transition temperature measured by specific heat for the different stable A15's.

Different measurements thus reflect the inhomogeneity differently, depending both on the geometry and the length scale of both the measurement and the inhomogeneity. For example, in small devices, the inhomogeneity may show up as a non-uniformity in device characteristics, rather than as a breadth in T_c.

Figure 6 shows the composition dependence of the transition width as determined by specific heat for several of the stable A15's. All were grown on (1$\bar{1}$02) sapphire at 750°-800°C and 10-20 Å/sec. The transition width varies greatly, although Nb-Sn is the most extreme. For both V-Ga and V-Si $\frac{\partial T_c}{\partial\,(\text{at.}\%)}$ is smaller than for the Nb-based compounds, hence the transition width underestimates the relative composition spread of the V-based compounds compared to the Nb-based. As in Nb-Sn, the width of the transition in V-Ga shows a strong dependence on composition. The asymmetry in V-Ga of the composition dependence of the width with respect to the stoichiometric composition is striking. Tunneling measurements on V-Ga and also Nb-Sn document this dependence.[3,10]

MODEL

The model proposed to account for the inhomogeneity is documented elsewhere.[8] It is related to classic surface segregation and is a result of the presence of differently oriented grain surfaces. Consider first surface segregation in its simplest formulation as it might occur in Sn-deficient A15 Nb-Sn. Unlike a simple alloy, the sites are not equivalent. Compare the bond energies of Nb on a Sn site with Sn on a Sn site. For this site, only the 12 nearest-neighbor Nb bonds are significant. Presuming the Nb-Nb bond stronger than the Sn-Nb bond,[11] Sn will segregate to the surface where bonds are broken. This process will stop when the surface reaches 25 at.% Sn as extra Sn would then need to occupy Nb sites, which is energetically a very different question. The process may also be limited by the mixing entropy lost by the segregation.

Different orientation surfaces break different numbers of bonds; the segregation will thus be less on the close-packed surfaces. In the A15, the (100) surface only breaks 4 of the 12 nearest neighbor Nb bonds, while the (210) breaks 5 and the (311) breaks 6. Surface segregation has not yet been measured in under-stoichiometric Nb-Sn. Based on limited results on the stoichiometric material, it is expected to be strongly oxygen-dependent.[12,13]

It is important to consider here the relative diffusion constants for bulk and surface diffusion. Since 800°C is below 1/2 T_M, bulk diffusion is very slow. In a typical deposition, the sample is at 800°C for approximately 20 minutes. Using isotope diffusion constants for transition metals,[14] and assuming diffusion occurs via an equilibrium concentration of vacancies, the approximate root mean square (RMS) distance an atom travels in 20 minutes at 800°C is on the order of Å's. Diffusion measurements have been made on A15 V_3Ga.[15] After scaling the measuring temperatures by the ratio of the melting temperatures, the diffusion constants expected on the basis of reference 7 and 10 are similar. Hence, annealing at 800°C may be primarily used to achieve local ordering, rather than long-range diffusion. By contrast, from limited measurements made directly on Nb_3Sn[8] and from measurements on elemental transition metals,[16] during the approximately 1 second an atom spends on the surface at typical deposition rates before being buried, the RMS distance it travels is on the order of μm's. Thus, surface effects may be expected to dominate the equilibration process and factors which alter surface diffusion, such as substrate temperature, background gases and deposition rate, may be expected to have a profound impact on the film.

Consider now a growing film of a material which in equilibrium would exhibit surface segregation. In particular, consider two grains only one unit cell thick of different orientations. The different number of broken bonds will cause it to be energetically favorable to transfer Sn from the more close-packed surface to the less close-packed surface. When the next layer is deposited on top of this first layer, if the system were in equilibrium it would deplete the underlying layer. This process should continue, resulting in the bulk of the two grains containing the same concentration and the two surfaces possessing the appropriate surface excesses for each orientation. However, as discussed above, during the approximately 1 second a layer is the top layer, bulk diffusion distances are tenths of Å's, and hence no depletion of an underlying layer is possible. Bulk diffusion may of course be enhanced near the surface due to excess vacancies. It is difficult to estimate this effect. By contrast, however, surface diffusion distances are μm's in this same 1 second, and hence the Sn

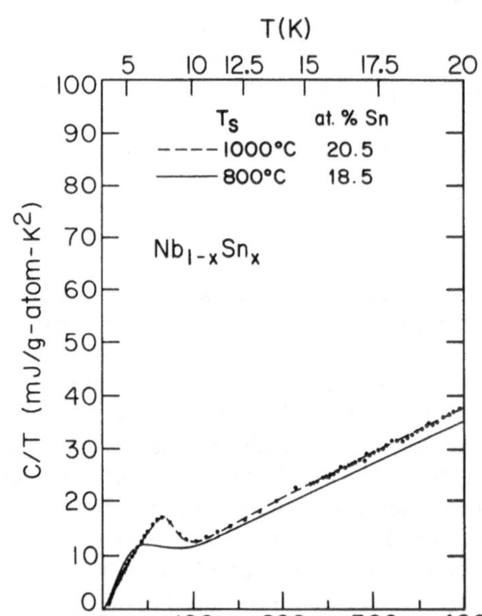

Fig. 7 Specific heat for sample grown at 1000°C. "Standard" result for T_s = 800°C shown as a solid line.

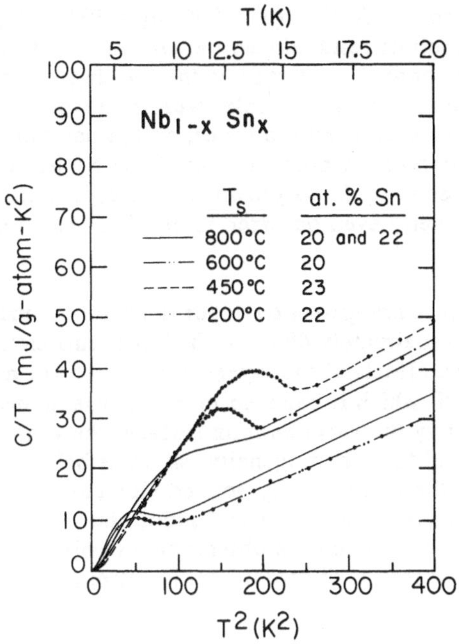

Fig. 8 Specific heat for Nb-Sn samples grown between 200°C and 800°C. "Standard" results for T_s = 800°C shown as solid curves for comparison.

transfer that occurred in the first layer can reoccur. This process will continue through the film and will result in the two grains containing different concentrations of Sn.

Returning now to the particulars of A15 film growth, in Nb-Sn the (100) grains should give Sn to the (311) grains, and in V-Ga the (100) grains should give Ga to the (210) grains. The presence of O_2 may strongly affect the segregation process and may even reverse the direction of the B-element transfer. Since O bonds strongly to Nb, Nb may be the element to segregate to the surface or to be enhanced in the surface with more broken bonds.

In practical terms, how can the inhomogeneity be reduced? Firstly, one may grow stoichiometric Nb-Sn films or one of the V-based films. Secondly, one may grow single crystal films, or even polycrystalline films but with a more preferred single orientation. Figure 7 shows the specific heat for a single crystal Nb-Sn 20.5 at.% Sn sample grown at 1000°C; the solid line shows the result from Fig. 3 for the 18.5 at.% Sn sample grown at 800°C. The transition width of the 1000°C sample is 2 K, compared with 4 K for the 800°C sample. Thirdly, one may attempt to grow polycrystalline samples of different orientation. The (100) and (311) growth of Nb-Sn on (1$\bar{1}$02) sapphire is the worst of the possible combinations. Using (0001) sapphire may result in (210) and (100) growth, which would result in less inhomogeneity.

Fourthly, if the grain size is smaller than a superconducting coherence length, which in the A15's is approximately 50 Å at T = 0, the inhomogeneity will not cause a broad transition temperature. Grain size may be reduced by lowering the deposition temperature. The resulting disordered material must be then annealed in order to reach full T_c for the composition. The minimum annealing temperature and time are not known; in our work, 800°C and 1 day were used. As discussed above, this anneal has no effect on material grown at 800°C.

The grain size of as-grown Nb-Sn crosses over from substantially less than a coherence length to substantially more between 450°C and 600°C. Figure 8 shows that the width of T_c is unchanged in a 600°C deposition, reduced markedly (to 2 K) by a

450°C deposition, and reduced still further (to 1.5 K) by a 200°C deposition. This residual width may be the result of residual disorder, in which case it could be further reduced by a longer or hotter anneal, or it may be the limiting value of the homogeneity of the deposition itself. No samples were removed from vacuum without the 800°C ordering anneal. It is believed, however, on the basis of observations of other A15's, that the 200°C deposition produced bcc phase Nb-Sn, rather than A15 phase. Without more knowledge of the as-grown structure, it is impossible to postulate whether or not the inhomogeneity even occurred in the film grown at 200°C.

CONCLUSIONS

When A15 structure samples are grown by vapor deposition, surface mobility at substrate temperatures of approximately 400°C or higher is sufficient to produce the equilibrium phase. However, effects of the layer-by-layer growth may still be seen in, for example, grain overgrowth which is believed to be a consequence of the energy of the various surfaces, specifically the vacuum/grain surface, the grain/grain surface and the grain/substrate surface, and the inhomogeneity, which is believed to be the result of grain-to-grain segregation of the more weakly bonded element. These processes are not specific to growth of A15 superconductors and should be visible elsewhere. In particular, the segregation process should be observable in polycrystalline films made from materials containing elements of different bonding strengths.

ACKNOWLEDGEMENTS

We would like to thank David Rudman, Rudi Bormann, Mac Beasley and Simon Bending for useful discussions on the topic of A15 film growth and Conyers Herring for his input on the surface segregation process. This work was supported by the AFOSR Contract No. F49620-83-C-0014 and the NSF MRL program under DMR 83- 16982.

Present address: †AT&T Bell Laboratories
Murray Hill, NJ.
* Westinghouse R and D Center
Pittsburgh, Pa.
**Also at Bell Communications Research, Inc.

REFERENCES

1. R. H. Hammond, *IEEE Trans. Magn.* MAG-11:210 (1975).
2. R. H. Hammond, *J. Vac. Sci. Technol.* 15:382 (1978).
3. D. A. Rudman, F. Hellman, R. H. Hammond, and M. R. Beasley, *J. Appl. Phys.* 55:3544 (1984).
4. D. F. Moore, Ph.D. Thesis, Stanford University (1978).
5. S. R. Early, Ph.D. Thesis, Stanford University, 1981.
6. S. R. Early, F. Hellman, J. Marshall, and T. H. Geballe, *Physica* 107B:327 (1981).
7. A. F. Marshall, F. Hellman, and B. Oh, Proceedings of the Materials Research Society, 1984 Fall Meeting.
8. F. Hellman, Ph.D. Thesis, Stanford University, 1985.
9. J. Talvacchio, Ph.D. Thesis, Stanford University (1982).
10. S. J. Bending, M. R. Beasley, and C. C. Tsuei, *Phys. Rev. B* 30:6342 (1984).
11. D. O. Welch, G. J. Dienes, O. W. Lazareth, Jr. and R. D. Hatcher, *J. Phys. Chem. Solids* 45:1225 (1984).
12. J. Talvacchio, A. I. Braginski, M. A. Janocko and S. J. Bending, *IEEE Trans. Magn.* MAG-21:521 (1985).
13. H. Ihara, Y. Kimura, H. Okumura, K. Senzaki, and S. Gonda, *Adv. in Cryo. Eng. — Mat.* 30:589 (1984).
14. P. Haasen, "Physical Metallury," Cambridge Univ. Press, Cambridge, (English translation) (1978).
15. A. Van Winkel, M. P. H. Lemmers, A. W. Weeber, H. Bakker, *Journ. Less Common Metals* 99:257 (1984).
16. G. Ehrlich, *Proc. IX IVC-V ICSS,* Madrid:1 (1983).

HIGH-T_c MoN THIN FILM SYNTHESIS

H. Ihara, K. Senzaki, Y. Kimura,
M. Hirabayashi, and N. Terada
Electrotechnical Laboratory
Niihari, Ibaraki, Japan

H. Kezuka
College of Engineering
Hosei University
Koganei, Tokyo, Japan

ABSTRACT

MoN films were prepared by several sputtering methods. Among them magnetron dc sputtering with a cup-shaped Mo target in a N_2 gas atmosphere was the most promising method of preparing well-defined B1-MoN films without hex-MoN phase. The highest T_{co} (onset of T_c) of the films was 12.7 K. The lattice constant of the B1-MoN compounds was 4.22 Å. The B1-MoN films annealed under high pressure (6 GPa) changed to hex-MoN phase mixed films with T_{co} of 15.8 K, which was the highest value among MoN compounds in the literature. The T_{co} of hex-MoN films prepared by normal dc sputtering was 13.7 K at its maximum.

The electronic structures of B1-MoN, hex-MoN and fcc-Mo_2N were observed by x-ray photoelectron spectroscopic measurement. The results showed that the density of states at the Fermi level of B1-MoN was 1.3 times as large as that of hex-MoN. This fact indicates that T_c of a high-quality B1-MoN crystal will be 21-30 K. The origin of low-T_c of the present B1-MoN films was in N-vacancies and N-interstitial defects. Optical emission spectroscopic measurements of nitrogen plasma suggested that excited N_2 molecules in the triplet state were effective in reducing N-vacancies in the B1-MoN film.

INTRODUCTION

Research for high-T_c (superconducting critical temperature) materials has concentrated on finding new nonequilibrium superconductors. B1-type transition metal compounds, like A15-type compounds, are one of the promising material groups. In general, $N(E_F)$ (density of states at the Fermi level) of B1 transition metal compounds has a tendency to increase with increasing valence electrons from 8 to 14. Some of the compounds, with large valence electron numbers between 11 and 14, are expected to have a higher T_c than that of NbN, as shown in Fig. 1. These compounds include MoN, TcN, RuN, RhN, TcC, RuC and RhC.

B1-MoN has been predicted to have a T_c of 29.3 K on the basis of energy-band calculation.[1] From empirical rules, its T_c has been estimated as 20 to 25 K.[2] Some other B1 compounds with 11 to 14 valence electrons, such as

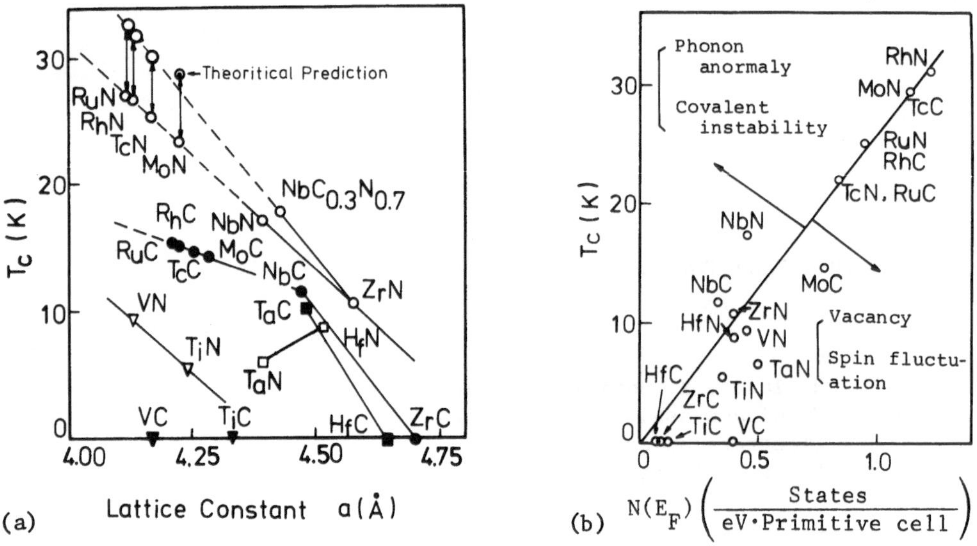

Fig. 1. Empirical rules of T_c values versus lattice constant (a) and density of states at Fermi level (b) in B1 transition metal nitrides and carbides.

TcN, RuN and RhN, have been expected to have a higher T_c than B1-MoN from empirical rules and electronic structures.[3] Even the B1-phase MoN, however, does not exist in the equilibrium phase diagram of the Mo-N system.[4] Molybdenum-mononitride exists only as a hexagonal-MoN phase. Then the synthesis of B1-MoN is difficult by the usual equilibrium preparation technique. Moreover, B1-phases of TcN, RuN, RhN, TcC, RuC and RhC compounds are in a higher grade nonequilibrium phase and would be more difficult to prepare than B1-MoN is. Then it is important to study first a preparation technique, electronic structures and superconductivities of B1-MoN in order to open up a new frontier in the nonequilibrium superconductive materials field.

Recently several attempts have been made to prepare MoN films by nonequilibrium or other techniques.[5-13] There are only four reports concerned with the preparation of the B1-phase MoN. Saur et al.[5] identified a cubic phase as the B1 phase. Its lattice constant, 4.16 Å, however, is too short, which would correspond to that of fcc Mo_2N. Linker et al.[11] have prepared B1-MoN_x films with large lattice constants of 4.20-4.26 Å and with high nitrogen concentrations ($x = 0.9 - 1.8$), but T_c of the films has been below 5 K. Wolf et al.[12] have obtained T_c values exceeding 6 K in Mo-rich $Mo_yNb_{1-y}(N_xC_{1-x})_z$ thin films. Ihara et al. have observed[13] the electronic structures of a series of B1-MoN_x films with 12.5 K at their T_{co} maximum.

In this work, the preparation method of B1 and hexagonal MoN phases was investigated by using the sputtering deposition technique and a high-pressure annealing technique. The sputtering deposition technique is the most suitable way to synthesize nonequilibrium B1-NbN films.[14] Then its technique is thought to be a promising method for the synthesis of B1-MoN films, too. The high pressure annealing technique is thought to be useful in improving the quality of the B1-MoN crystal or transforming the hexagonal-MoN phase to a B1-MoN phase with further close-packed structure.

EXPERIMENT

Several different types of sputtering deposition techniques, that is, dc or rf sputtering, magnetron or nonmagnetron sputtering, using a plane target or a cup-shaped target, N_2+Ar mixed or N_2 single sputtering gas, bias sputtering and self-epitaxial growth were attemped to prepare MoN films. Through these trials, effective sputtering methods were determined for the synthesis of B1-MoN films. In addition to this, a high pressure-annealing was applied to sputter-deposited MoN films.

The first trial for the preparation of B1-MoN films was carried out in a similar way to B1-NbN films[15] by a normal dc sputtering. A Mo plane-target was sputtered in a N_2+Ar gas mixture at a partial pressure ratio of $P_N/(P_N+P_{Ar})$=0.2~0.9. The purities of Mo metal, N_2 gas and Ar gas were 99.8 %, 5-N and 5-N, respectively. The base pressure in a preparation chamber was 2 x 10^{-5} Pa. The total pressures of the mixed gas was 50 Pa. The cathode voltage was 1000 V. The MoN films were deposited on sapphire substrates with a small inclination $(11\bar{2}0)$face at a temperature of 400~800 ℃. The temperature of the samples was measured by a infrared optical pyrometer by fixing the emissivity of MoN samples at 0.35 for 2 μm in wave length.

The second trial was carried out by a normal dc sputtering using only N_2 sputtering gas without Ar. The N_2-gas pressure was varied from 50 to 200 Pa. In this trial, the effects of bias sputtering and self-epitaxial growth were examined.

The third trial was carried out by using a cup-shaped Mo-target in magnetron sputtering mode. Outer diameter and inner diameter of the target are 42 and 38 mm, respectively and its height is 30 mm. The magnetic field was generated by SmCo magnets and changed from 100 to 900 Gauss. The N_2-gas pressure was varied from 10 to 40 Pa. The cathode voltage changed in the range of 400~1000 V. The current was fixed at 30, 60 and 120 mA.

The fourth trial was carried out to apply a high-pressure annealing technique to improve the quality of B1-MoN films.[16] This experiment was carried out using a belt-type apparatus. The MoN sample deposited on a sapphire substrate was covered by a sapphire plate and embedded in a pressure transmitting medium of sodium-chloride powder. A graphite tube surrounding the sodium-chloride powder was used as a heater. The pressure was first raised to 6 GPa, and then an ac current was passed through the heater. The annealing temperature was varied from 500 to 1400°C in the different runs. After annealing, the samples were quenched to ambient conditions.

The compositions and electronic structures of the MoN films were measured by the XPS (X-ray photoelectron spectroscopy) technique.[13] The XPS spectrometers were a Hewlett-Packard 5950A ESCA system and Vacuum Generator ESCA-LAB Mark II system equipped with a Al-Kα monochrometer source. Each of two XPS systems has a dc-sputtering chamber in order to carry out in situ XPS observation without breaking the vacuum.[17] The HP-5950A ESCA system has an auxiliary dry-box port system which enables us to transfer the samples from isolated sputtering systems to the XPS analysis chamber without exposing the samples to air. The nitrogen concentration was determined from the core level XPS spectra with reference to the high-T_c hexagonal-MoN film as a stoichiometric composition. This estimation method was consistent with the method of using the Scofield cross section[18] of N_{1s} and $Mo3d_{5/2}$ levels within an experimental error of 10%.

The crystal structure of the samples was observed by a reflected high-energy electron-diffraction (RHEED) system and an X-ray diffractometer. The lattice constants were determined by X-ray diffraction with the Cu Kα line.

The T_c of the sample was measured resistively using a conventional four-probe technique. T_{co} (onset of T_c), T_c and T_{ce} (the end of T_c) were defined

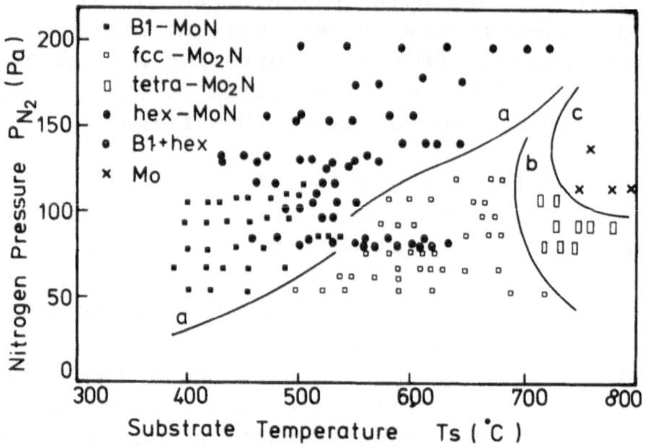

Fig. 2. Reaction diagram as a function of N_2-gas pressure and substrate temperature for Mo-N films prepared by dc sputtering.

as the 1%, 50%, and 99% points of the resistance transition. The transition width (ΔT) is the difference between T_{co} and T_{ce}. The temperature variation was measured with a calibrated germanium thermometer. The experimental error of T_c was ±0.1 K.

Optical emission spectroscopic (OES) measurement was carried out to discover what kind of nitrogen species was effective in forming the B1-MoN phase. Emission line-intensity ratios of an excited nitrogen molecule N_2^* to nitrogen ion N_2^+ were observed by optical multichannel analyser with changing N_2-gas pressure and magnetic field.

RESULTS AND DISCUSSION

(a) DC Sputtering

For the first trial, N_2+Ar mixed sputtering gas was used. At the N_2 partial pressure ratio of $P_N/(P_N+P_{Ar}) = 0.2$, which is a suitable condition for the preparation of B1-NbN films,[15] fcc-Mo_2N and metal Mo phases were observed. For a higher N_2 partial pressure ratio of 0.3 to 0.9, fcc-Mo_2N phase was observed, but B1-MoN phase was not. T_{co} and T_c of fcc-Mo_2N films ranged from 7.8 to 5.5 K and from 7.4 to below 4.2 K, respectively. The lattice constant of fcc-Mo_2N ranged from 4.11 to 4.16 Å. The composition of fcc-Mo_2N varied from $MoN_{0.5}$ to $MoN_{1.3}$. Even if the composition of the films enough reached $MoN_{1.0}$ with the increase of N_2 partial pressure, the mononitride phase (B1 and hexagonal) was not observed by using N_2+Ar mixed sputtering gas. For this reason, an N_2 sputtering gas only was used in the following preparation processes.

Figure 2 shows the Mo-N phase fraction in sputtering films as a function of N_2 sputtering gas pressure and substrate temperature. This diagram includes results from self-epitaxial growth and bias sputtering techniques. Molybdenum mononitride phases (hexagonal and B1) appeared in the region of high nitrogen pressure, P_N, and low substrate temperature, T_s, upside of line a in Fig. 2. The boundary of the hexagonal phase and the B1 phase were not distinguished clearly. Sometimes B1 phase formed in a mixed phase with hexagonal MoN. The hexagonal phase was relatively reproducible, but the B1 phase was irreproducible at temperatures above 450°C. At lower temperatures, B1 phase formed reproducibly, but it did not show superconducting property. The B1-phase films usually showed negative thermal coefficients of resistance.

Normal dc sputter produced good quality hexagonal MoN crystals with high T_c's, as shown in Fig. 3. Some of them contained partly B1-MoN phase. Superconductivity of those films, however, would be caused by hexagonal

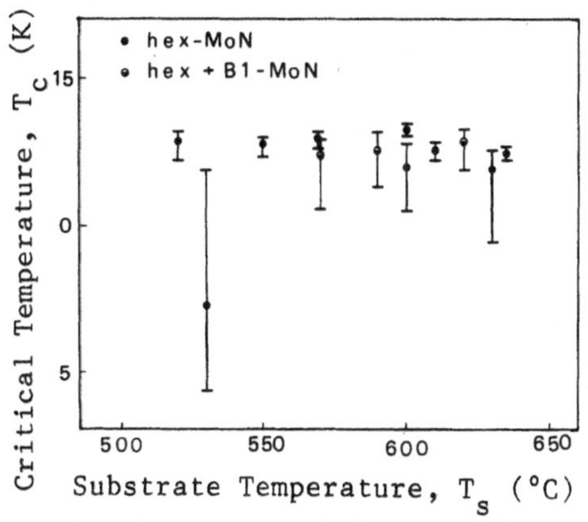

Fig. 3. T_c dependence on T_s for hexagonal MoN films prepared by normal dc sputtering.

phase and not by B1 phase. The maximum T_c of 13.2 K (T_{co} = 13.4 K) was obtained at a substrate temperature of 600°C. T_c values, however, did not strongly depend on substrate temperature. Almost all these high-T_c films were obtained as hexagonal phase appeared in the region of fcc-Mo_2N phase in Fig. 2. Although proper N_2-gas pressure was 80 Pa, high-T_c films were not always reproducible.

By the way, B1-MoN single-phase films showed no trace of superconductivity above 4.2 K. Only hexagonal-mixed B1-phase films showed high T_c between 12.9 and 12.4 K. An α-Mo film with a lattice constant of 3.185 Å showed a T_{co} of 8.3 K and a T_c of 8.1 K at their maxima. The maximum T_{co} of an fcc-Mo_2N film was 7.2 K, which was lower than the case of using N_2 and Ar mixed sputtering gas.

Molybdenum submononitride phases (fcc-Mo_2N and tetra-Mo_2N) appeared underside of the line a in Fig. 2. Fcc-Mo_2N was distinguished from B1-MoN by lattice constant, density of states of the valence band, and composition. The value of 4.19 Å was a boundary of lattice constant between fcc-Mo_2N and B1-MoN. The criterion in lattice constant corresponded well to that in density of states, but composition was not always a well-defined criterion for distinguishing between B1-MoN and tetragonal Mo_2N. Tetra-Mo_2N phase presented on a high temperature side of line b. In the higher temperature region, metal Mo films were formed.

Self-Epitaxy

Since B1-MoN phase was not formed reproducibly above 500°C, self-epitaxial growth and bias sputtering technique were attempted to stabilize the B1 phase and to fill nitrogen vacancies in B1-phase film. In order to grow B1 phase by self-epitaxial method above 500°C, B1-MoN films deposited at a low temperature around 400°C were used as substrates on which an additional MoN layer was deposited. Though the B1-MoN structure of the sublayer was kept at 600°C during deposition, the self-epitaxial growth technique did not always succeed in keeping the B1 structure of the overlayer. The overlayer film preferred to change to the fcc-Mo_2N structure. This would be caused by a similarity in lattice structure between B1-MoN and fcc-Mo_2N. That is, the fcc-Mo_2N lattice has a structure with alternate N-vacancies in the B1-MoN lattice. Actually, the self-epitaxial effect was useful in keeping the cubic structure, but it did not help to raise the T_c value of the B1 phase.

Bias Sputtering

To fill the N-vacancies in the B1-MoN or the fcc-Mo_2N lattice by impinging nitrogen ions, the bias sputtering technique was employed. Bias voltage was applied by a grid placed between target and substrate. The

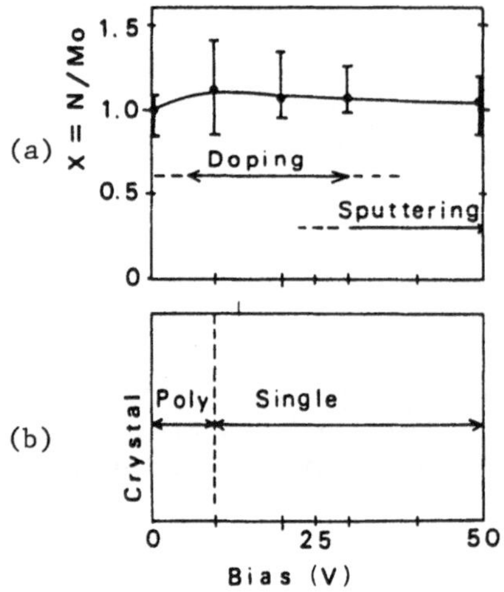

Fig. 4. The effect of bias sputtering. (a) composition versus bias; (b) crystal state versus bias.

results of bias sputtering are shown in Fig. 4. This method had a little effect in increasing nitrogen content and filling N-vacancies in MoN films. This doping effect was observed at a bias from 10 to 30 V, but resputtering of deposited film was observed at a bias above 30 V. Above 30 V in bias, glow discharge was observed between the grid and substrate, and the deposition rate reduced obviously.

A remarkable effect of bias sputtering was an improvement of crystallization. Bias above 10 V could change MoN film from polycrystal to single crystal. The RHEED pattern in Fig. 5 shows that sputtering deposition with 30-V bias formed a (111) face single crystal of the B1 phase at a substrate temperature of 500°C. This B1-MoN film itself had no superconductivity above 4.2 K. This bias sputtering was not effective in raising the T_c value of the B1 phase. The reason is this bias sputtering promotes growth of hexagonal MoN above 530°C, even if B1-MoN substrates were used for the purpose of self-epitaxial growth. On the other hand, the substrate of hexagonal MoN film was useful for stabilizing the hexagonal phase and raising its T_c value by self-epitaxial growth. Hexagonal-phase MoN films prepared by self-epitaxial growth gave reproducibly high T_c values. T_c values as a function of substrate temperatures are shown in Fig. 6 for bias-sputtered MoN films. Almost all high-T_c films were obtained by a combination of self-epitaxial growth and bias sputtering. Some of fcc-Mo_2N films had higher T_{co} than 10 K. That would be caused by small quantities of mixed B1 or hexagonal phase. The maximum T_{co} (13.8 K) and T_c (13.0 K) of hexagonal MoN film were obtained at a substrate temperature of 610°C. Maximum T_{co} increased a little, but

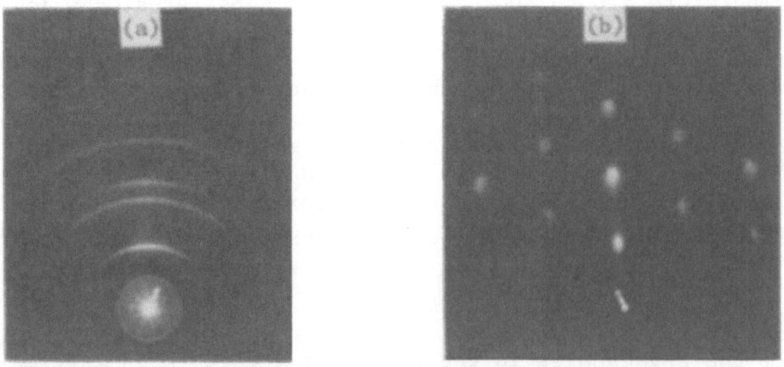

Fig. 5. RHEED patterns of B1-MoN films without (a) and with (b) 30-V bias.

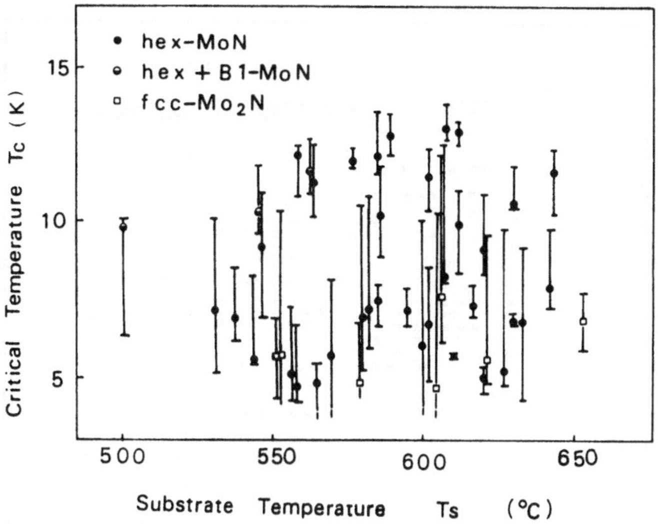

Fig. 6. Dependence of T_c on substrate temperature for MoN films prepared by bias sputtering.

maximum T_c did not in comparison with normal dc sputtering without self-epitaxial growth and bias sputtering. A merit of these self-epitaxial growth and bias sputtering techniques was an improvement in reproducibility of the hexagonal phase.

The results of the aforementioned sputtering technique suggest that ion and electron bombardment deteriorates the B1 phase and transforms it to the equilibrium phase of fcc-Mo_2N and hexagonal MoN. XPS measurement has suggested that N-vacancies are the main origin of deterioration of superconductivity in B1-MoN films.[13] Then we attempted to use a cup-shaped Mo target with a magnetron mode. The fundamental ideas were to reduce electron bombardment and to use active nitrogen species by confining nitrogen plasma magnetically and geometrically to reduce N-vacancies in the B1-MoN films.

(b) Magnetron DC Sputtering

In this sputtering deposition, a cup-shaped Mo target was used in a magnetron mode, as shown in Fig. 7. A proper magnetic field was about 300 Gauss in the center of the target. Figure 8 shows the fraction of Mo-N phase in magnetron dc-sputtering films as a function of N_2-gas pressure and substrate temperature. This method made B1-MoN form at a high substrate temperature and a low N_2-gas pressure. The formation of hexagonal MoN was suppressed considerably in contrast to bias sputtering, though hexagonal MoN appeared in a high N_2-gas pressure region. The upper limit of temperature for B1-phase formation rose from 500°C to 650°C with magnetron-mode sputtering. This high-temperature preparation led to improvement in crystal quality and increased T_c values of B1-MoN films, as shown in Figs. 8 and 9.

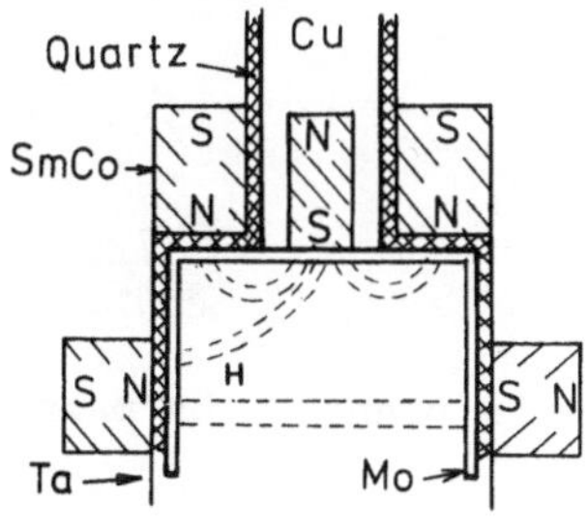

Fig. 7. Cup-shaped Mo target for magnetron sputtering.

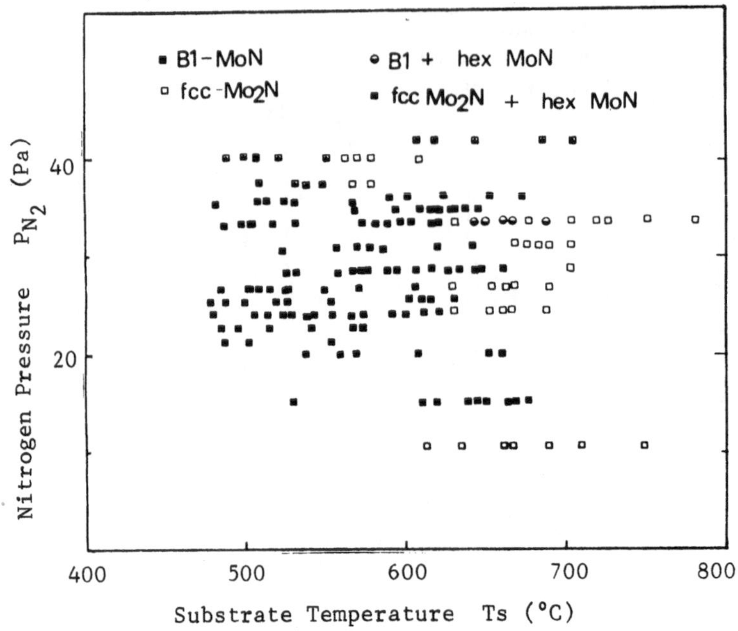

Fig. 8. Reaction diagram as a function of nitrogen pressure and substrate temperature for Mo-N films prepared by magnetron dc sputtering.

Figure 9 is an x-ray diffractometer trace of MoN film prepared at 590°C. A high-angle (400)-peak-splitting suggests a high-quality crystal structure. The film was almost single phase of B1-MoN, though a small unidentified peak was observed between (111) and (200) peaks. At least the hexagonal MoN phase was not observed at all. Preferential orientation was a (100) plane. This fact was common for all high-T_c B1-MoN films. The lattice constant of the film was 4.227 Å, which was nearly equal to the average value of B1-MoN films, as shown in Fig. 10. The lattice constant of B1-MoN film became smaller than 4.225 Å at a substrate temperature above 600°C. Then it abruptly diminished to a value of 4.16 Å above 650°C. The temperature of 650°C would be a phase-boundary temperature between B1-MoN and fcc-Mo_2N crystal. Figure 11 shows the composition of MoN film versus substrate temperature. The nitrogen content tends to decrease with temperature, though the scattering of plots was very large. This trend was not classified systematically by N_2-sputtering gas pressure. The atomic ratios (N/Mo) of almost all B1-MoN films were larger than unity, and even those of fcc-Mo_2N films were above 0.8. This means that poor bonding or decomposed

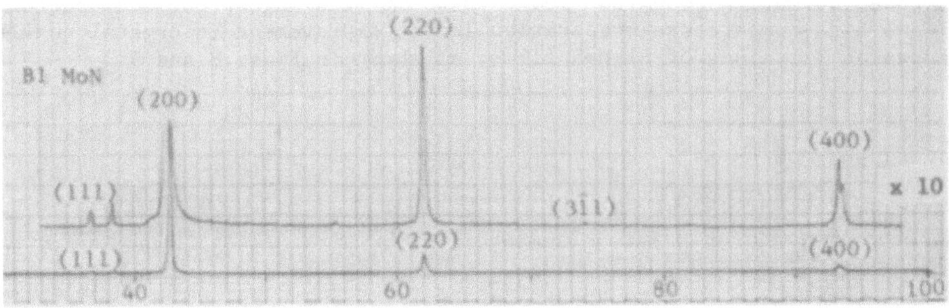

Fig. 9. X-ray diffractometer trace of B1-MoN film prepared by magnetron dc sputtering.

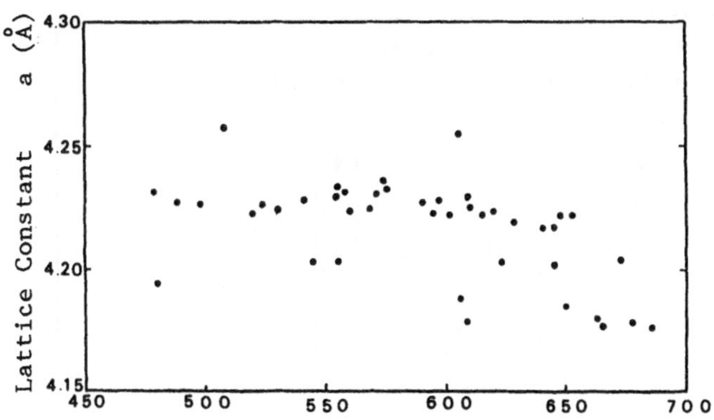

Fig. 10. Lattice constant versus substrate temperature for magnetron sputtering films.

nitrogen atoms are included in the films in large quantities. These excess nitrogen atoms would deteriorate the superconductivity of B1-MoN films.

Figure 12 shows T_c values versus substrate temperatures. T_c values increase with substrate temperature and have a peak near 650°C. The maximum T_{co} and T_c of B1-MoN film were 12.7 and 11.2 K, respectively. T_c of B1-MoN film abruptly decreased above 620°C with the formation of fcc-Mo_2N phase. Hexagonal-phase mixed MoN films had relatively high T_c values, even above 650 ℃.

The magnetron dc-sputtering with a cup-shaped Mo-target produced B1-MoN films with high T_c at temperatures from 620°C to 650°C. The normal dc-sputtering, however, could not produce B1-MoN films at a temperature above 500°C. The B1-MoN films prepared at low temperature below 570°C did not show superconductivity above 4.2 K. An important difference between the present magnetron sputtering and normal sputtering were thought to be in chemical activity of nitrogen plasma. Then optical emission spectroscopic (OES) measurement was carried out on nitrogen plasma during sputtering.

Fig.13 shows the optical emission spectra of nitrogen plasma for the present magnetron sputtering and normal sputtering. The peak intensity ratio of $N_2^*(0,0)$ peak at 337 nm to $N_2^+(0,0)$ peak at 391 nm was a convenient

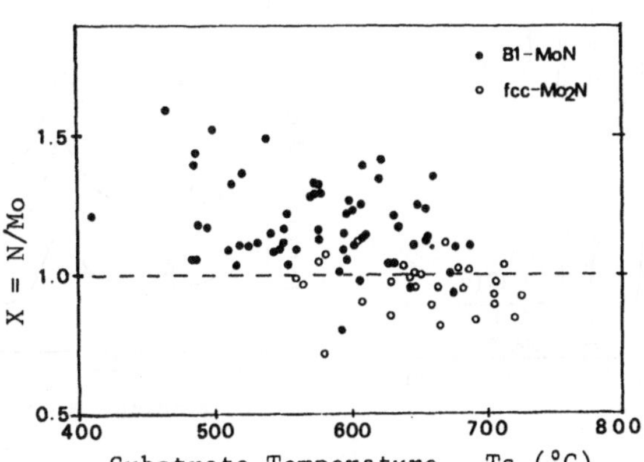

Fig. 11. Atomic ratio versus substrate temperature for magnetron sputtering MoN films

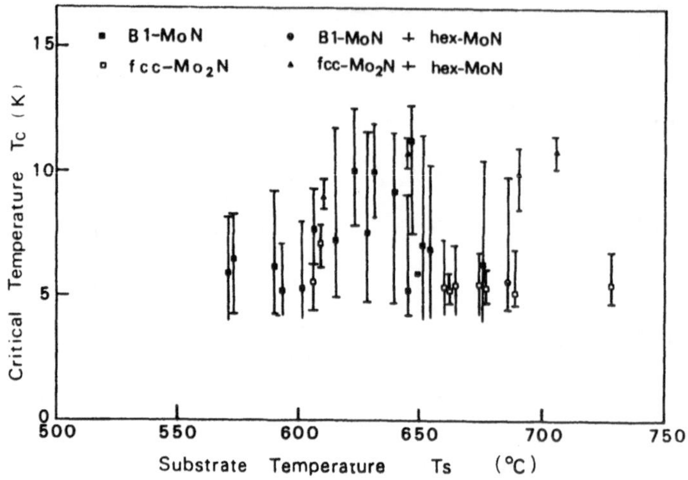

Fig. 12. Dependence of T_c on substrate temperature for MoN films prepared by magnetron dc sputtering.

index for measure of chemical activity of nitrogen plasma. The peak-intensity ratio of $I_{N_2^*}/I_{N_2^+}$ had a low value around 0.3 for normal dc sputtering and a higher value from 0.4 to 1.5 for magnetron dc sputtering. The peak-intensity ratio increased with increase of magnetic field and nitrogen gas pressure, and the reddish purple color of nitrogen plasma became opaque. The ratio correlated with the deposition rate and stability of the plasma. Then the proper peak-intensity ratio was around 0.5.

Increment of nitrogen-plasma activeness improved the electronic structure of MoN films, as shown in XPS spectra of Fig. 14. At stoichiometric composition (x = 1.0), the density of states of B1-MoN_x film manifested a characteristic structure similar to the B1 type, though it still had a non-negligible difference from the real B1 type. The spectra of $MoN_{1.3}$ film was very close to the real B1 type. Previously, the density of states of B1-$MoN_{1.0}$ films manifested an intermediate type between B1-MoN and fcc-

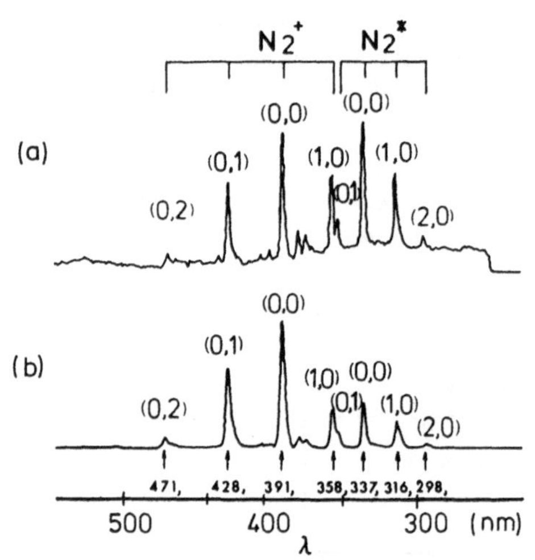

Fig. 13. Optical emission spectra of nitrogen plasma for magnetron sputtering (a) and normal sputtering (b).

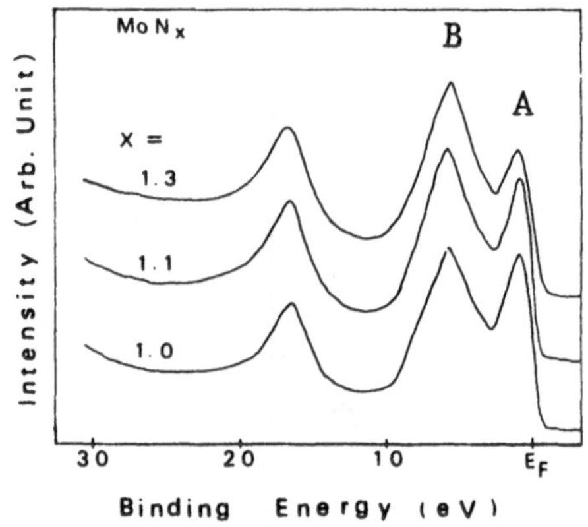

Fig. 14. Dependence of XPS spectra of valence band on nitrogen concentration for B1-MoN films.

Mo_2N.[13] Although the present electronic structure of stoichiometric B1-MoN film has been improved in comparison with the previous one,[13] it shows that N-vacancies still exist in high concentration (about 20%). Lower T_c than the predicted value and large transition width of the present B1-MoN films would be caused by the N-vacancies. It will be necessary to improve the electronic structure in order to attain a higher T_c.

(c) Rf Sputtering

Rf sputtering was carried out by using NH_3 and Ar mixed gas. This technique produced B1-MoN films reproducibly. The maximum T_{co} and T_c of the film were 12.2 and 11.5 K, respectively. The occurrence of superconductivity, however, was not reproducible.

(d) High Pressure Annealing

B-MoN films were annealed under 6 GPa at temperatures from 600°C to 1400°C. Figure 15 shows x-ray diffractometer traces of the MoN films annealed for 8 h under this pressure. The B1-MoN phase was transformed mainly to hexagonal MoN phase and partly to tetragonal Mo_2N and α-Mo phases. The B1-MoN phase disappeared at temperatures from 750 to 1100°C, but a trace of it again appeared at 1200°C. The lattice constant of the B1-MoN crystal was 4.240 Å. The B1-MoN phase again disappeared at 1400°C. The reappearance of B1-MoN phase at 1200°C depended on the duration of annealing. The B1 phase was not observed for an annealing time shorter

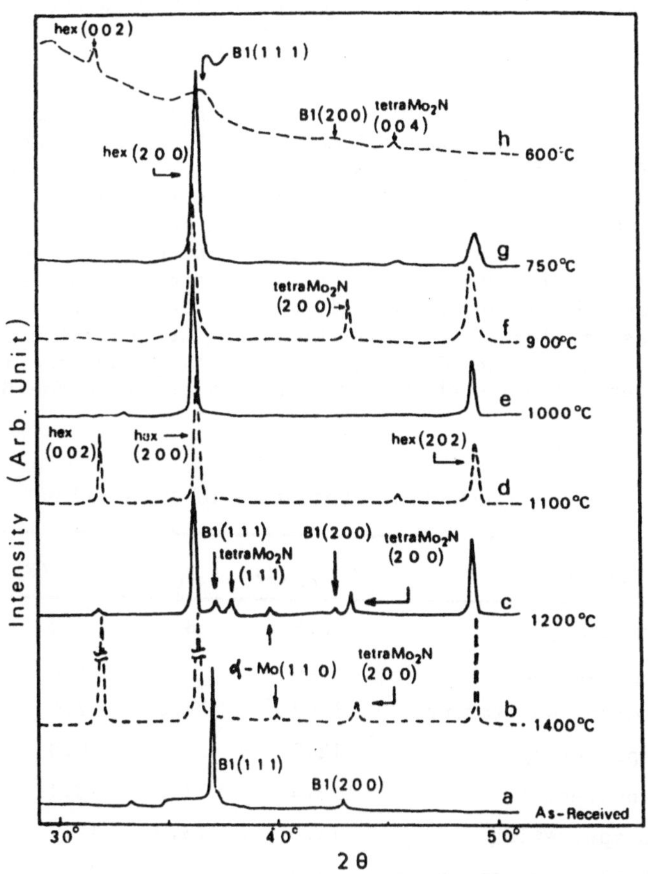

Fig. 15. X-ray diffractometer traces of MoN films annealed under a pressure of 6 GPa for 8 h. B1-MoN phase is observed at 1200°C annealing.

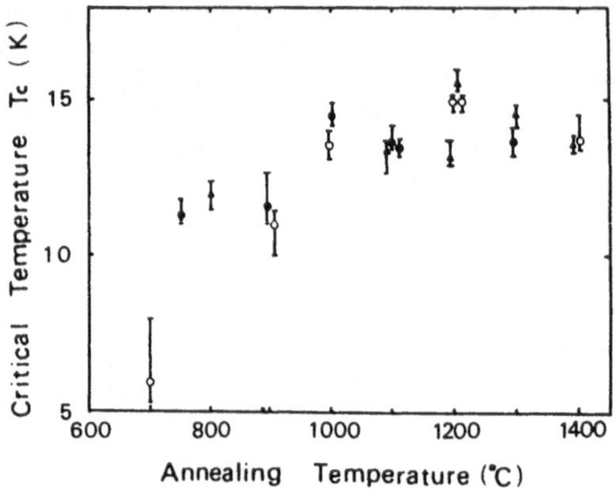

Fig. 16. Dependence of T_c on annealing temperature of B1-MoN films. The marks represent the difference of starting materials.

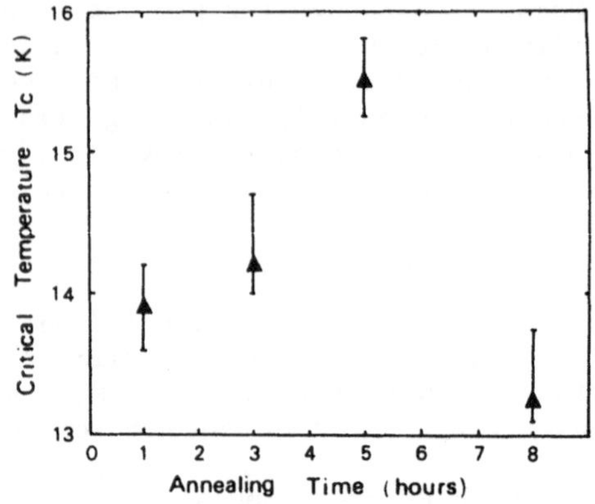

Fig. 17. Dependence of T_c on annealing time. Single-phase hexagonal MoN was obtained after 5 h. B1-MoN phase was observed after 8 h.

than 5 h. Figure 16 shows the T_c of annealed MoN films versus annealing temperature. The maximum T_{co} (15.8 K) and T_c (15.5 K) were obtained at a temperature of 1200°C. The T_c value depended on annealing time, as shown in Fig. 17. The 5-h annealing produced single-phase hexagonal-MoN with the maximum T_c. The maximum T_c value was 1.0 K higher than the value previously reported,[16] which was the highest value in the Mo-N system.

The density of states at the Fermi level [$N(E_F)$] of B1-MoN was 1.3 to 1.5 times as large as that of hexagonal MoN from XPS measurement.[13] From this result, the T_c of a high-quality B1-MoN crystal is estimated to be 21 to 30 K by using McMillan's formula[19] and the present T_c value of hexagonal MoN. In this estimation we assumed that the Debye temperature, effective coulomb repulsion, phonon density of states, and the electron-phonon matrix element, in fact all parameters except the $N(E_F)$ term in McMillan's formula, are the same in both the B1 and hexagonal phase of MoN. This assumption would be an underestimation of T_c for B1-MoN, because the effect of soft phonon usually existing in the B1 phase with high $N(E_F)$ and the increase of the Debye temperature with the change from the hexagonal to the B1 phase were neglected.

Table 1. Superconducting Properties of Mo-N Films

Phase	Process	T_{co}(K)	T_c(K)	T_{ce}(K)	ΔT(K)
α-Mo	dc-sputtering	8.3	8.1	8.0	0.3
fcc-Mo_2N	dc-sputtering	7.8	7.4	7.0	0.8
B1-MoN	magnetron sputtering	12.7	11.2	7.6	5.1
	rf-sputtering	12.2	11.5	10.8	1.4
hex-MoN	dc-sputtering	13.4	13.2	13.1	0.3
	bias sputtering	13.8	13.0	12.7	1.1
	high-pressure annealing	15.8	15.5	15.3	0.5

SUMMARY

Superconducting properties of Mo-N films were summarized in Table 1. The B1-MoN films were prepared by magnetron dc sputtering and rf sputtering. The magnetron dc-sputtering technique was the most promising method of preparing well-defined B1-MoN films without hexagonal MoN reproducibly. This would be caused by active nitrogen molecules of the excited triplet state. The highest T_{co} of the B1-MoN films was 12.7 K. The lattice constant of the B1-MoN films with high T_c values was 4.22 Å (average).

The hexagonal MoN films were prepared by normal dc sputtering and bias dc sputtering. The maximum T_{co} (13.8 K) was obtained by bias sputtering. Reproducibility of high-T_c hex-MoN films was high for the bias sputtering. This would be caused by impinging of nitrogen ions with high energy.

The B1-MoN films annealed under a pressure of 6 GPa changed to hex-MoN mixed films. The maximum T_{co} of a film was 15.8 K, which was the highest value among Mo-N systems. The B1 phase disappeared by high-pressure annealing at temperatures from 750 to 1100°C, but reappeared at 1200°C after a longer annealing time of 8 h. This fact suggests that B1-MoN is a stable phase at pressures higher than 6 GPa.

Comparison of experimental results of the B1-MoN phase and the hexagonal MoN phase enabled us to estimate the T_c of B1-MoN as 21 to 30 K. The origin of the low T_c of the present B1-MoN films was in N-vacancies and N-interstitial defects. Active nitrogen species, such as triplet-state excited molecules are thought to be effective in reducing N-vacancies and in raising the T_c of the B1-MoN crystal.

REFERENCES

1. W.E. Pickett, B.M. Klein and D.A. Papaconstantopoulos, Physica B 107 667 (1981), D. A. Papaconstantopoulos, W. E. Pikett, B. M. Klein and L. L. Bayer, Phys. Rev. B 31;752 (1985), and nature 308 494 (1984).
2. Z. You-Xiung and H. Shou-an, Solid State Commun. 45:281 (1983).
3. From the rigid band models based on the band structure of MoN[1] and NbN; D. J. Chadi, M. L. Cohen, Phys. Rev. B 10:496 (1974).
4. H. Jehn and P. Ettmayer, J. Less-Common Metals 58:85 (1978).
5. E. J. Saur, H. D. Schlechinger and L. Rinderer, IEEE. Trans. Magn. MAG-17, 1029 (1981).
6. W. W. Fuller, S. A. Walf, D. V. Gubser, E. F. Skelton and T. L. Francavilla, J. Vac. Sci. Technol. A 1, 517 (1983).
7. M. Ikebe, N. S. Kazama, Y. Muto and H. Fujimori, IEEE. Trans. Magn. MAG-19, 204 (1983).
8. D. S. Easton, E. H. Henninger, O. B. Cavin and C. C. Koch, J. Mater. Sci. 18:2126 (1983).
9. M. J. Kim, D. M. Brown and W. Katz, J. Electrochem. Soc. 130:1196 (1983).
10. S. T. Sekula, J. R. Thompson, G. M. Beardley and D. H. Loundes, J. Appl. Phys. 54:6517 (1983).
11. G. Linker, R. Smithey and O. Meyer, J. Phys. F 14 L 115 (1985).
12. S. A. Wolf, S. B. Qadui, K. E. Kihlstron, R. M. Simon, W. W. Fuller, D. van Vechter, E. F. Skelton and D. V. Gubser, IEEE. Trans. Magn. MAG-
13. H. Ihara, Y. Kimura, K. Senzaki, H. Kezuka and M. Hirabayashi, Phys. Rev. 31:3177 (1985).
14. For example; K. S. Keskar, T. Yamashita and Y. Onodera, Jpn. J. Appl. Phys. 10:370 (1971), G. Oya and Y. Onodera, J. Appl. Phys. 45:1389 (1974).
15. H. Ihara, Y. Kimura and S. Gonda, Proc. Int. Cryo. Mater. Conf. (1982, Kobe) p. 258.

16. H. Ihara, M. Hirabayashi, K. Senzaki, Y. Kimura and H. Kezuka, Phys. Rev. B 32 No. 3 (to be published on 1 August 1985).
17. H. Ihara, Y. Kimura, H. Okumura, K. Senzaki and S. Gonda, in: "Advances in Cryogenic Engineering Materials", vol 30, Plenum Press, New York, (1983) pp.589.
18. J. H. Scofield, J. Electron Spectrosc. Relat. Phenom. 8:129 (1976).
19. W. L. McMillan, Phys. Rev. 167:331 (1968).

EPITAXIAL GROWTH OF NbN ON MgO FILM

T.Yamashita, K.Hamasaki and T.Komata

Department of Electronics, The Technological University of Nagaoka, Nagaoka 949-54 Japan

ABSTRACT

A novel process for preparing epitaxial MgO/NbN bilayers has been developed using an RF magnetron sputtering in Ar/N_2 gas mixtures. MgO/NbN bilayers formed on various substrates were essentially $(200)_{MgO}//(200)_{NbN}$ oriented, and the orientation effect was independent on the crystallinity of substrate materials. With this technique, thin and ultrathin NbN films with thicknesses ranging from 10 nm to 250 nm were prepared at substrate temperature about 210 °C. These films had high critical temperatures and relatively low residual resistivity. T_c's of these films were found to depend systematically on the thickness of the predeposited MgO films. The T_c of NbN(20 nm) deposited on MgO(80 nm) films was about 15.7 K and the residual resistivity was less than 200 μΩ·cm. In addition, the calculated magnetic penetration depth, λ(0), was substantially small (λ(0)≤300 nm) for NbN films thicker than 10 nm. All-NbN Josephson tunnel junctions were prepared successfully with epitaxial MgO/NbN/MgO/NbN tetralayers. Epitaxial junctions show excellent I-V characteristics with high sum-gap voltage of more than 5 mV.

INTRODUCTION

Recently, there has been much interest in the preparation of high quality ultrathin films of niobium nitride as an electrode material for Josephson devices[1-17]. This is because NbN offeres several advantages in addition to its high critical temperature T_c~16 K[18]. The advantages offered by this material include high energy gap Δ(0)=2.6-3.1meV[3,5,16], excellent thermal cyclability, ease of preparation and thermodynamic stability of surface in regard to oxidation[15]. Josephson devices made on NbN films are also promising to achieve large value of the critical current I_o and tunneling resistance R_N product(I_oR_N) which is important for high frequency and logic applications and stable operation when small temperature fluctuation occurs.

So far, high quality NbN films thicker than 100 nm have been successfully prepared by many researchers using the technique of DC or RF magnetron sputtering at 500 °C or higher substrate temperatures[19-25]. For device processing it is most desirable that the substrate temperature is less than 300 °C or lower. In the high temperature range impurities may diffuse from the substrate or tunnel barrier materials into the superconducting films. For ultrathin NbN films less than 100 nm thick deposited at low substrate temperatures less than 300 °C, there were some unattractive features such as degradation of T_c, large magnetic penetration depth $\lambda(0) \geq 400$ nm and high normal state resistivity $\rho_o \geq 500$ μΩ·cm[26,27]. Such features are undisirable for materials used as electrodes of Josephson tunnel junctions or microbridges, because the quality of the ultrathin films strongly influences the superconducting characteristics of those devices. The large penetration depth often observed in high resistivity NbN films results in large kinetic inductances and is hard to prevent from flux trapping.

An estimate of the penetration depth is made with the transition temperature and resistivity data by using the GLAG theory, $\lambda(0) \propto (\rho_o/T_c)^{1/2}$ in the dirty limit[11,26,28,29]. According to this theory, a short $\lambda(0)$ occurs with high T_c's and low ρ_o values. In order to reduce the large penetration depth of NbN films, a composit Nb/NbN electrode has been used for the preparation of tunnel junctions[1,13]. However, high quality ultrathin films of NbN are clearly the key to prepare devices with high quality.

Recently, the authors have developed a novel technique for the deposition of ultrathin films of NbN with high critical temperatures. This technique consists of using an RF magnetron sputtered MgO films as a base on which to epitaxially grow NbN films. Although high T_c NbN films were formed using this technique, preparation conditions have not been optimized. In this paper, recent studies on epitaxial growth of MgO/NbN bilayer and MgO/NbN/MgO, or Al_2O_3/NbN tetralayers are described, and preliminary results for all-NbN tunnel junctions are discussed.

A NOVEL PROCESS FOR PREPARING HIGH QUALITY ULTRATHIN NbN FILMS

One problem in choosing a suitable substrate is that impurities may diffuse from the substrate and contaminate the films. MgO exhibits extremely high thermodynamic stability at lower than 300 °C, that is to say, its decomposition pressure is relatively low. Therefore, MgO deposited by sputtering an MgO target in pure Ar atomosphere will be in congruent evaporation, and may be stoichiometric. The hope is that the crystallite

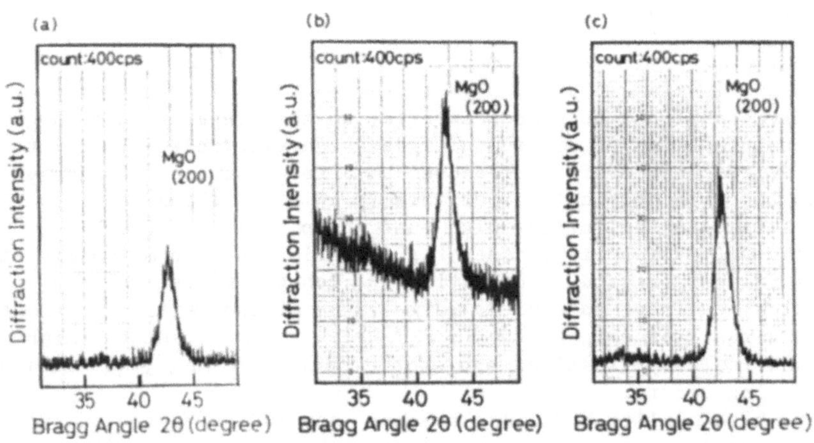

Fig.1 X-ray diffractometer traces of MgO films deposited onto various substrates. The film thickness was about 80 nm, and the substrate temperature was about 210 °C. The trace(a),(b) and (c) for a sapphire, a fused quartz, and a (100) silicon substrate, respectively.

orientation is independent on deposition conditions over a wide range, as reported in NaCl films[31].

MgO films were deposited by an RF magnetron sputtering in an multi-target sputtering system. To investigate the effect of different substrate materials on the crystallite orientation of MgO films, series of runs were conducted using different substrates. The substrates used were a (100) silicon, sapphire, and fused quartz. The background pressure before deposition was typically 2×10^{-4} Pa. MgO films were first deposited by an RF magnetron sputtering in about 4.2 Pa using a 99.99 % MgO target, and the sputtering rate was about 9.5 nm/min.

After the MgO deposition, a thin NbN layer was sequentially deposited onto the predeposited MgO layer on the substrates without breaking vacuum. NbN films were produced by reactive RF magnetron suputtering in 2.4 Pa of $Ar+(\sim10\%)N_2$ mixture. The thickness of films thicker than 40 nm were measured by a surface profiler. Using this results the sputtering rate for 5.1 W/cm^2 of RF power density was 40 ± 0.2 nm/min in $(\sim10\%)N_2$ partial pressure. The substrate temperature was about 210 °C. The thickness of films thinner than 40 nm were inferred from the sputtering rate.

CRYSTALLITE ORIENTATION OF MgO FILMS AND MgO/NbN BILAYERS

Figure 1 shows an X-ray diffractometer traces of MgO films of about 80 nm thick deposited onto various substrates. All traces clearly indicate that the crystallites of MgO films are strongly oriented (200) planes visible to $2\Theta = 75°$. In this technique, no effect of film thickness on crystallite orientation was observed in the wide range 30-160 nm. It is well known that MgO has the NaCl crystal structure, the same as δ-phase NbN. Therefore, it is possible to study the preparation conditions for epitaxial growth of NbN sequentially deposited onto predeposited MgO layers with (200) orientation.

The crystal structure of NbN films of 25 nm thick was studied as a function of the thickness of the predeposited MgO films using an X-ray diffractometer. The thickness of MgO films were in the range 0-160 nm. The substrates used were silicon wafer slices. Figure 2 illustrates the diffractometer traces for MgO/NbN bilayers deposited. For NbN deposited

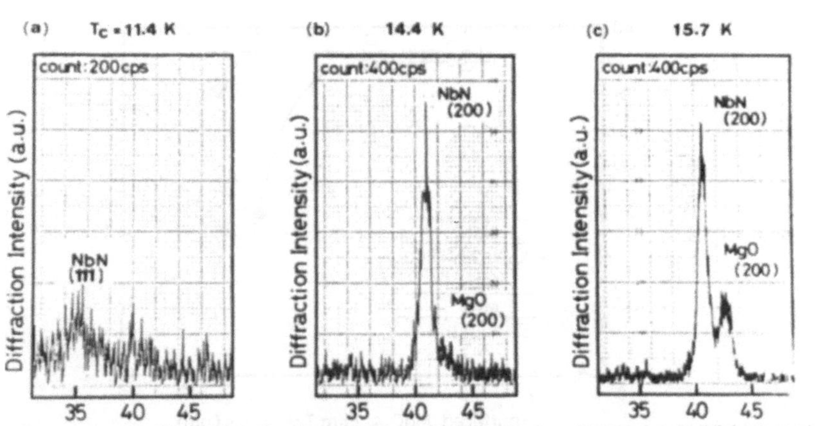

Fig.2 X-ray diffractometer traces of MgO/NbN bilayers sequentially deposited onto a silicon substrate. Trace (a) is without a predeposited MgO layer. Trace (b) and (c) are with MgO layers of 30nm and 80 nm thick, respectively.

on a silicon substrate, its crystallite orientation of δ-phase (111) plane are parallel to the substrate surface, as shown in Fig.2(a). As the thickness of MgO film was increased, we observed rapid elimination of (111) orientation of the δ-phase and growth of (200) orientation, as shown in Fig.2(b) and (c). Under these preparation conditions, single phase films with $(200)_{MgO}//(200)_{NbN}$ orientation were formed on silicon substrates. In our results, the orientation effect of MgO/NbN bilayer was independent on the crystallite of substrate materials. The grain sizes of NbN(25 nm) deposited on MgO(80 nm) were in the range 10-20 nm, with a lattice parameter close to the bulk value(0.438 nm).

Other researchers have found single δ-phase NbN with several diffraction lines when sputtered in an Ar/N_2 mixture[10,12,32-34]. In the present work only one δ-phase diffraction line for (200) reflection was observed on the diffraction traces. From this result, we can conclude the formation of epitaxial NbN films sequentially deposited onto a predeposited MgO layer. Although, for the thicker films, the epitaxial effect may not result in a high T_c compared with a polycrystalline film, the hope is that an epitaxial effect may stabilize the δ-phase (high T_c) of an ultrathin NbN film. Also, the density of defects in the ultrathin NbN film is probably reduced by the epitaxial effect similar to the annealing effect[35,36] resulting in the higher T_c and low resistivity.

CRYSTALLITE ORIENTATION OF MgO/NbN/MgO, OR Al_2O_3/NbN TETRALAYERS FOR JOSEPHSON TUNNEL JUNCTIONS

MgO and Al_2O_3 are attractive barrier materials due to their good thermodynamic stability[30]. From this point of view, MgO/NbN/MgO, or Al_2O_3/NbN tetralayers have been prepared to fabricate Josephson tunnel junctions. The normalized peak intensity of (200)/((111)+(200)) is a relative index of the crystallite preferred orientation of these tetralayers. Figure 3 is a plot of this value as a function of the thickness of thin MgO or Al overlayers on NbN base-electrodes. These thin films act as a tunnel barrier. The Al_2O_3 tunnel barrier was formed by thermal oxidization of a thin Al layer. The thickness of both NbN film electrodes was about 150 nm. The preferential orientation of (200) planes in MgO/NbN/MgO/NbN tetralayers is independent on the thickness of thin MgO overlayers on the NbN base-electrode.

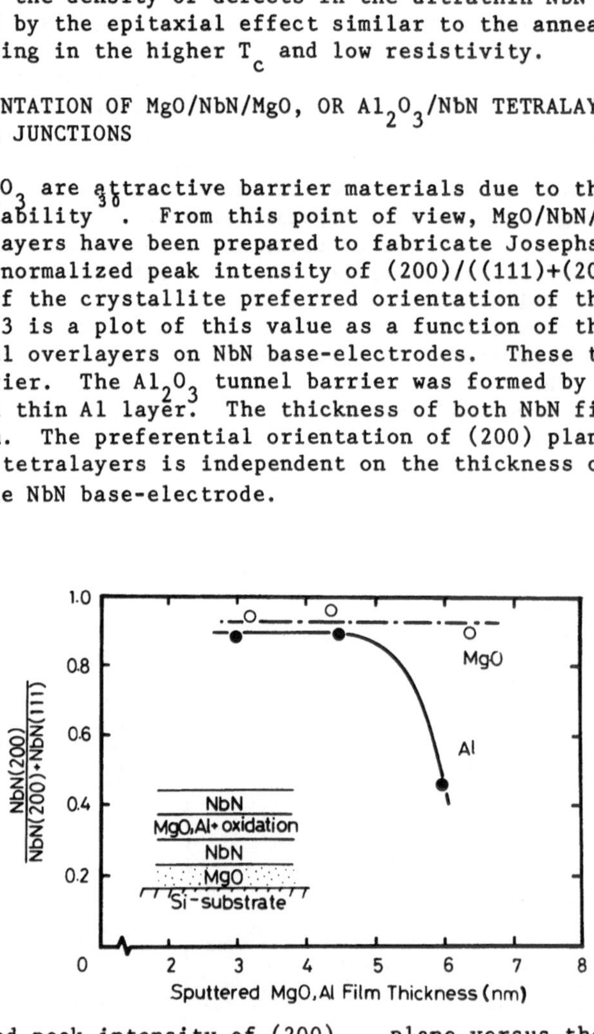

Fig.3 Normalized peak intensity of $(200)_{NbN}$ plane versus the thickness of predeposited MgO, or Al overlayers on NbN base-electrodes. The Al_2O_3 tunnel barrier was formed by thermal oxidation of a predeposited thin Al. We observe the heteroepitaxial growth for MgO, and thin Al_2O_3.

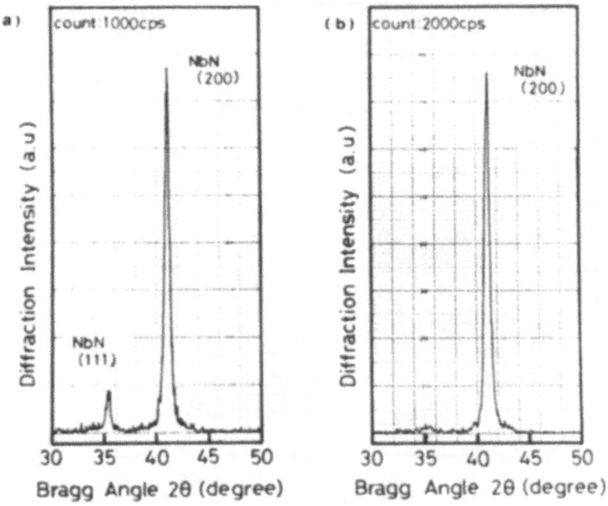

Fig.4 X-ray diffractometer traces of MgO/NbN/MgO/NbN tetralayers for different sputtering power densities. The trace (a) at 5.1 W/cm^2, and (b) at 2.6 W/cm^2 for deposition of a NbN counter-electrode.

However, the preferential orientation of MgO/NbN/Al_2O_3/NbN varies remarkable by increasing the thickness of the predeposited Al layer. For a thicker Al_2O_3, the epitaxial growth is broken because of the large lattice mismatch between Al_2O_3 and NbN.

It is concluded that the predeposited Al layer less than 4 nm in thickness is useful to form tunnel barriers without breaking the epitaxial growth. Therefore various kinds of material may be useful as a tunnel barrier if the thickness is thin enough.

In Fig.3, NbN films of counter-electrode were deposited at a high power density of about 5.1 W/cm^2. Figure 4 shows the X-ray diffractometer traces of MgO/NbN/MgO/NbN tetralayers. Unlike in Fig.4(a), the peak of (111) reflection can not be observed in Fig.4(b). Therefore, the small peak of (111) reflection in Fig.4(a) may be caused by ion damage at deposition of the NbN counter-electrode. In conclusion, the epitaxially growth technique of MgO/NbN/MgO, or Al_2O_3/NbN tetralayers on various kinds of substrates may be useful for the preparation of Josephson devices.

ELECTRICAL PROPERTIES OF EPITAXIAL NbN FILMS

The transition temperature and normal resistivity of NbN films were measured as a function of the thickness of predeposited MgO films. A superconducting transition of a typical NbN film about 20 nm thick is shown in Fig.5. Note a high transition temperature and relatively low residual resistivity indicative of high quality of the NbN film. We obtained T_c~15.7 K, RRR(=ρ_0(300 K)/ρ_0(20 K))~0.93 and the residual resistivity ρ_0(20 K)~110 μΩ·cm.

The T_c's of NbN films of about 25 nm thick is illustrated in Fig.6 as a function of the thickness of the predeposited MgO films. For NbN deposited on a silicon wafer, the T_c is about 11.5 K, and somewhat low compared with ones prepared by DC bias or DC magnetron sputtering[26,27]. When the thickness of predeposited MgO film was 80 nm, we obtained relatively high T_c~15.7 K. This result apparently concerns with the orientation effect shown in Fig.2. In the present technique, T_c's were not dependent on substrate materials when the thickness of MgO film was larger than 80 nm.

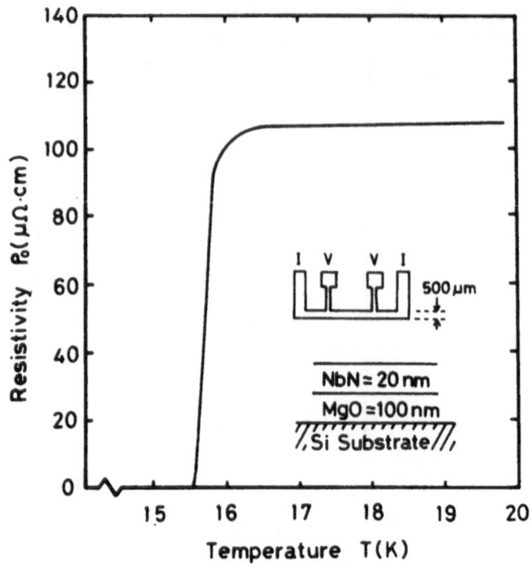

Fig.5 A typical superconducting transition in a NbN film of about 20 nm thick. The predeposited MgO layer is about 100 nm in thichness. Note a high T_c and low ρ_0(20 K).

The thickness dependences of T_c and ρ_0(20 K) for epitaxial NbN films are shown in Fig.7. The thickness of predeposited MgO films was 80-160 nm. A series of NbN films were prepared at about 210 °C and 10% partial pressure of nitrogen at an RF power density of 5.1 W/cm^2. As shown in Fig.7, T_c remains constant until the thickness of NbN is less than about 30 nm. In fact, films thicker than 30 nm could be prepared with T_c~16 K.

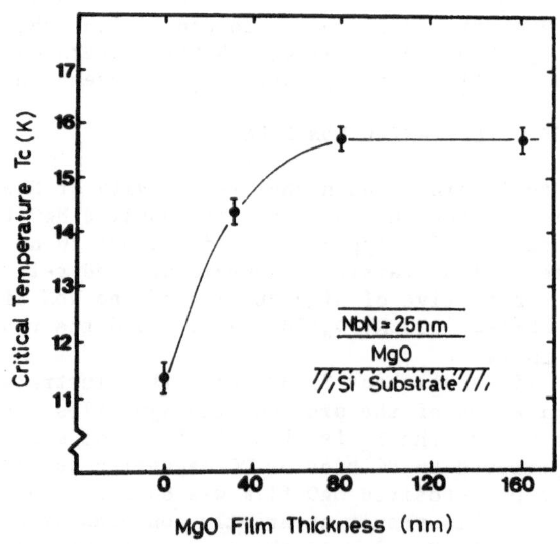

Fig.6 The transition temperature of NbN(25 nm) as a function of the thickness of the predeposited MgO films.

T_c's of NbN films of 10 nm thick were in a range of 13.5-14.7 K. The obtained T_c was relatively high compared with ones reported by other researchers[26,27]. The ρ_0 value of NbN of 10 nm thick was low$\leq$300 μΩ·cm and the best resistivity achieved was approximately 80 μΩ·cm. The extremely low resistivity for ultrathin NbN films prepared by the present technique results probably from the (200) orientation effect. The main difference from an annealing effect[35,36] is that the epitaxial NbN films with high critical temperature are prepared at very low substrate temperature about 210 °C.

MAGNETIC PENETRATION DEPTH λ(0)

The penetration depth, λ(0), sets a scale for the thickness of NbN films. Because $\lambda(0)\geq 400$ nm, films thicker than 1 μm would be necessary to obtain the minimum inductance. This is undesirable from the viewpoint of circuit patterning and step coverage. We have investigated and found that λ(0) of MgO/NbN bilayers is smaller than the values of NbN reported already.

An estimate of λ(0) was made using the transition temperature and resistivity data. The penetration depth is given by[29]

$$\lambda(0) = 2^{1/2}/\lambda_{GL}^{BCS}(0)1.33^{1/2}\varepsilon_{\lambda(0)}$$

where λ_{GL}^{BCS} is the Ginzburg-Landau penetration depth for the BCS weak coupling case in the dirty limit and $\varepsilon_{\lambda(0)}$ is the strong coupling correction by $\varepsilon_{\lambda(0)}=\varepsilon_{\Delta(0)}^{-1/2}$. $\lambda_{GL}^{BCS}(0)$ is determined by the normal state resistivity and the transition temperature[29].

$$\lambda_{GL}^{BCS}(0) = 64.2[\rho_0(\mu\Omega\cdot cm)/T_c(K)]^{1/2} \qquad (nm)$$

Using these equation and the measured values of T_c and normal state resistivity the theoretical values of the penetration depth of our epitaxial NbN films were calculated. Assuming a value of the strong coupling constant $3.53\varepsilon_{\Delta(0)}=2\Delta(0)/kT_c$ of 4.2 for our materials based on the measured value of Bacon et al and Hikita et al[5,12].

Figure 8 illustrates the calculated penetration depth as a function

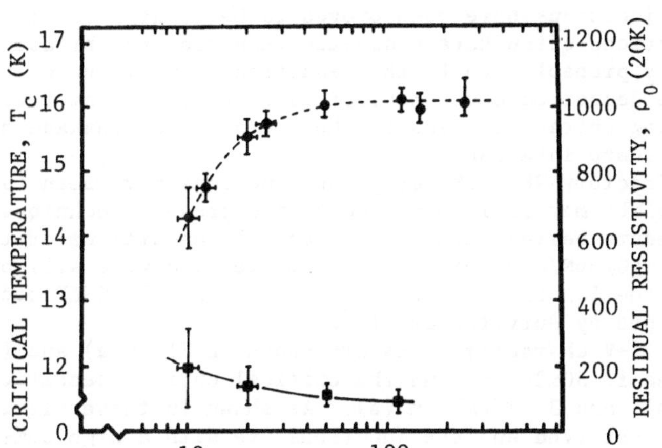

Fig.7 T_c and ρ_0 (20 K) as a function of the thickness of NbN films. The predeposited MgO layers are 80 or 160 nm in thickness.

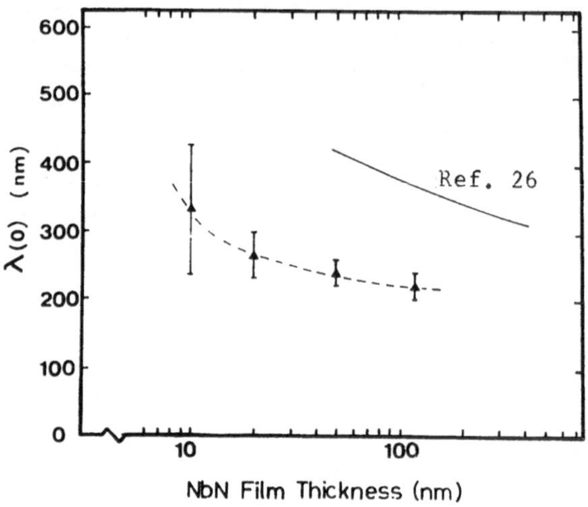

Fig.8 Film thickness dependence of λ(0) for epitaxial NbN films deposited at about 210 °C. The predeposited MgO is 80 nm thick. NbN data reported by other authors are also shown.

of the thickness of NbN. The λ(0)'s of epitaxial films are relatively small in comparison with that reported by several researchers[1,15,26]. The best quality films with ρ_0=80 μΩ•cm and T_c~14.3 K with a thickness of about 10 nm have a theoretical penetration depth of approximately 230 nm. The small λ(0) results in small kinetic inductance and will be able to avoid the use of a composit Nb/NbN electrode.

JOSEPHSON TUNNEL JUNCTIONS

Epitaxially growth MgO/NbN bilayer sandwiches described above have been successfully utilized in several applications requiring reproducible materials with high T_c, large gap voltage, and high I_oR_N product. In this section we present our preliminary results, and discuss the application for tunnel junctions.

NbN/Pb junctions have been proved to be of high quality, however, all-NbN junctions often have a drastic reduction and broadening on energy gap. This is probably due to the reduction of T_c value on the initial layer of the deposited counter-electrode. An important requirement for good tunneling characteristics is the use of a low leakage barrier material with a sharp interface.

All-refractory NbN/NbN Josephson junctions have been fabricated using high quality NbN materials prepared by the present technique. The fabrication procedure begins with the sequential depositions of the tetralayer sandwich of MgO/NbN/MgO/NbN. The substrate used were silicon wafer slices, and the junctions were fabricated using the SNEP processing technique developed by Gurvitch et al[38].

Typical I-V characteristics are shown in Fig.9(a) and (b). The junction area is 30x20 μm^2 and the critical current densities are 0.03 KA/cm^2 in (a), and 3 KA/cm^2 in (b). As shown in these figures, no "knee" structure is observed and the junctions are with a high sum-gap voltage of more than 5.0 mV. This value is larger than that for NbN/NbO_x/NbN junctions[1]. The tunnel junction results presented represent only preliminary ones. Both NbN films of base- and counter-electrode were deposited at high power density of about 5.1 W/cm^2. Therefore, the large leakage current may be caused by the imhomogeneous mixture of super or normal

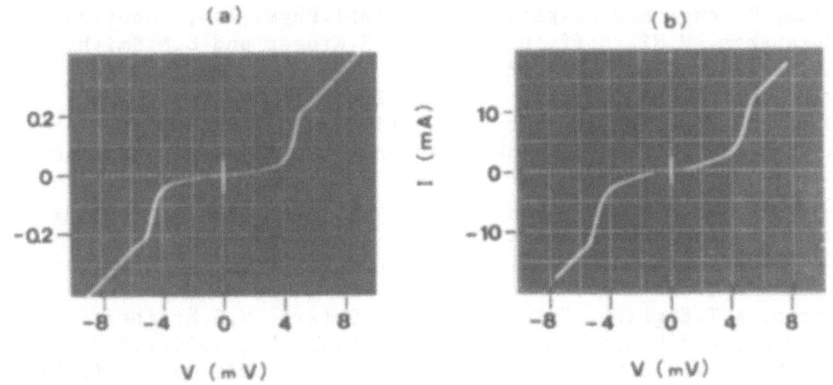

Fig.9 I-V characteristics of MgO/NbN/MgO/NbN junctions: (a) J_c=0.03 KA/cm^2 and (b) J_c=3 KA/cm^2.

conductors in the bottom layer of the NbN counter-electrode. Although sum-gap voltage has improved, further work on improving sub-gap leakage current should be done without degrading the T_c's of NbN electrodes.

CONCLUSION

High quality NbN films have been prepared successfully using a novel processing of epitaxial growth technique. The prepared MgO/NbN bilayer and MgO/NbN/MgO, or Al_2O_3/NbN tetralayer were with strongly (200) orientated. From the present study the authors conclude that, as a tunnel barrier, the various kinds of materials can be used without breaking the epitaxial growth. By optimizing the deposition conditions, we are able to routinely make NbN films of 10 nm thick with T_c~13-15 K. These films indicated relatively low resistivity ρ_0 (20 K)$\lesssim$300 μΩ·cm, and small penetration depth $\lambda(0)\leq$300 nm. The small $\lambda(0)$ results in small kinetic inductance and will be able to avoid the use of a composit electrode. The tunneling characteristics for all-NbN junctions were improved by using the epitaxial NbN films, and sum-gap voltages for MgO/NbN/MgO/NbN junctions were typically 5.0 mV. The epitaxial growth technique presented in this work will be useful for the preparation of Josephson devices, such as tunnel junctions.

ACKNOWLEDGEMENTS

The authors wish to thank S.Nagaoka and Y.Hirukawa for measurements of film thickness and Auger analysis. We would also like to thank K.Ueda for his film preparation. This work was supported by the Grant-in-Aid for "New Digital Circuit with Bridge Josephson Junctions" from the Ministry of Education of Japan.

REFERENCES

1) "Special Issue; Josephson Computer Technology", Bulletin of the Electrotechnical Laboratory, Vol.48, No.4(1984)
2) D.W.Jillie, H.Kroger, L.N.Smith, E.J.Cukauskas and M.Nisenoff, Appl.Phys.Lett., 40, 747(1982)
3) R.B.Van Dover, D.D.Bacon and W.R.Sinclair; Appl.Phys.Lett., 41, 764(1982)
4) M.Hikita, K.Takei, T.Iwata and M.Igarashi; Jpn.J.Appl.Phys., 21, L724(1982)

5) M.Hikita, K.Takei and M.Igarashi; Jpn.Appl.Phys., 54, 7066(1983)
6) E.J.Cukauskas, M.Nisenoff, D.W.Jillie, H.Kroger and L.N.Smith; IEEE Trans. Magn., MAG-19, 831(1983)
7) R.B.van Dover and D.D.Bacon; ibid., 951(1983)
8) M.T.Deen and E.D.Thompson; ibid., 954(1983)
9) M.Igarashi, M.Hikita and K.Takei; Advances in Cryogenic Engineering, 30, 535(1984)
10) E.J.Cukauskas, M.Nisenoff, H.Kroger, D.W.Jillie and L.R.Smith; Advances in Cryogenic Engineering, 30, 547(1984)
11) E.J.Cukauskas, W.L.Carter, S.B.Qadri and E.F.Skelton; IEEE Trans. Magn., MAG-21, 505(1985)
12) D.D.Bacon, A.T.English, S.Nakahara, F.G.Peters, H.Schreiber, W.R.Sinclair and R.B.van Dover; J.Appl.Phys., 54, 6509(1983)
13) R.B.van Dover and D.D.Bacon; IEEE Trans. Magn., MAG-19, 951(1983)
14) A.M.Cuculo, L.Maritato, A.Saggese and R.Vaglio; Cryogenics, 45(1984)
15) J.C.Villegier, L.Vieux-Rochaz, M.Goniche, P.Renard and M.Vabre; IEEE Trans. Magn., MAG-21, 496(1985)
16) M.Gurvitch, J.P.Remeika, J.H.Rowell, J.Geek and V.P.Lowe; IEEE Trans. Magn., MAG-21, 509(1985)
17) A.Shoji, M.Aoyagi, S.Kosaka, F.Shinoki and H.Hayakawa; Appl.Phys.Lett., 48, 1098(1985)
18) C.Kittel; "Introduction of Solid State Physics", 5th ed. (Wiley, New York, 1976)
19) J.R.Gavaler, J.K.Hulm, M.A.Janocho and C.K.Jones; J.Vac.Sci.Technol., 6, 177(1969)
20) K.S.Kesker, T.Yamashita and Y.Onodera; Jpn, J.Appl.Phys., 10, 370(1971)
21) S.A.Wolf, I.L.Siger, E.J.Cukauskas, T.L.Francavilla and E.F.Skelton; J.Vac.Sci.Technol., 17, 411(1980)
22) R.T.Kampwirth and K.E.Gray; IEEE Trans.Magn., MAG-17, 565(1981)
23) R.B.van Dover, D.D.Bacon and W.Robert Sinclair; J.Vac.Sci.Technol., A2, 1257(1984)
24) J.R.Gavaler, M.A.Janocko and C.K.Jones; J.Vac.Sci.Technol., 10, 17(1973)
25) S.A.Wolf, J.J.Kennedy and M.Nisenoff; J.Vac.Sci.Technol., 13, 145(1976)
26) S.Kubo, M.Asahi, M.Hikita and M.Igarashi; Appl.Phys.Lett., 44, 258(1984)
27) T.Goto and P.Anprung; Jpn.J.Appl.Phys., 22, 955(1983)
28) E.J.Cukauskas, W.L.Carter and S.B.Qadri; J.Appl.Phys., 57, 2538(1985)
29) T.P.Orlando, E.J.Ncniff, Jr., S.Foner and M.R.Beasley; Phys.Rev.B 19, 4545(1979)
30) T.Yamashita, K.Hamasaki, Y.Kodaira and T.Komata; IEEE Trans.Magn., MAG-21, 932(1985)
31) L.G.Schulz; J.Chem.Phys., 17, 1154(1949)
32) E.J.Cukauskas; J.Appl.Phys., 54, 1013(1983)
33) J.C.Villegier and J.C.Veler; IEEE Trans.Magn., MAG-19, 946(1983)
34) T.Akune, N.Sakamoto and Y.Shibuya; Jpn. J.Appl.Phys., 21, 772(1982)
35) Y.M.Shy, L.E.Toth and R.Somasundaram; J.Appl.Phys., 44, 5539(1973)
36) V.M.Pan and A.G.Popov; Advances in Cryogenic Engineering, 30, 571(1984)
37) J.Kwo, G.K.Wertheim, M.Gurvitch and D.N.E.Buchanan; IEEE Trans.Magn., MAG-19, 759(1983)
38) M.Gurvitch, M.A.Washington, H.A.Huggins and J.M.Rowell; IEEE Trans.Magn., 19, 791(1983)

EPITAXIAL GROWTH OF NbN FILMS *

J. R. Gavaler, J. Talvacchio, and A. I. Braginski

Westinghouse Research and Development Center

Pittsburgh, Pennsylvania

ABSTRACT

We have investigated the influence of epitaxy on the growth of the superconducting B1 phase in NbN films which were magnetron sputtered at <700°C. As is typical, even without the benefit of epitaxy or the deliberate addition of impurities, the background impurities in the sputtering environment were sufficient to stabilize the growth of this phase, which normally is unstable at these temperatures. The T_c's of these films however were less than optimum. Polycrystalline epitaxy on MgO or the addition of more impurities were both effective in raising T_c. Single crystal epitaxy on MgO (17.2K at a 700°C deposition temperature) and NbN substrates (16.4K at 90°C) produced the highest T_c's. We conclude that the stabilization of the B1 phase, by inhibiting the nucleation of the non-superconducting phase(s) via epitaxy or by the addition of impurities, is necessary to obtain optimum T_c's in NbN deposited at low temperatures.

INTRODUCTION

At present much of the research directed toward the development of a high-critical-temperature (T_c > 10K) counterelectrode for use in superconducting tunnel junctions is focussed on the B1 structure compound, NbN. To minimize damage to the junction barrier, it is desirable to deposit the counterelectrode at as low a temperature as possible. From earlier work, it was known that NbN could be reactively sputtered at temperature as low as 400 - 500°C with T_c's of ~15K [1,2] thus making it a good candidate as a counterelectrode material. Later, NbN became an even more attractive material for this application when it was found that by using magnetron sputtering, high-T_c's could be obtained at even lower temperatures.[3,4,5] Very recently Shoji, et al reported on an all NbN junction which had a finite critical current up to 14.5K. The counterelectrode in this case was rf sputtered on substrates heated to 150°C.[6] Despite these successes in growing NbN, a clear understanding of the mechanisms which permit the formation of the high T_c material at low temperatures is not available. We have been attempting to obtain this understanding by studying the effect of epitaxy and impurities on the formation of NbN. In this paper we report the results of this study.

*Supported in part by AFOSR Contract No. F49620-85-C0043

EXPERIMENT

The NbN films discussed here were grown in a UHV deposition and analytical facility which is described in detail elsewhere in this volume.[7] The films were reactively sputtered from a dc magnetron gun which employs a 5cm diameter niobium target. The target-substrate separation is approximately 11.5 cm. The background impurity level, prior to sputtering, was 1×10^{-6} Pa. The sputtering gases were various mixtures of argon and nitrogen and, in some cases, methane. Substrates included sapphire, single and polycrystalline MgO, and silicon. The polycrystalline MgO substrates were made by rf sputtering a 30 nm thick MgO layer onto sapphire in a separate deposition system. The single crystals for substrates were purchased from outside suppliers. The dimensions in each case were 0.64 x 0.64 x 0.05 cm. The structures of the films were analyzed in-situ immediately after deposition by Reflection High Energy Electron Diffraction (RHEED). After removal from the system, x-ray diffraction, T_c, and tunneling data were obtained. Critical temperature measurements were by a standard four-point resistive technique using germanium thermometry. The tunneling data were from junctions made by using NbN as the base electrode, native oxide or oxidized Al barriers, and Pb-Bi counterelectrodes.

RESULTS

We will first report results on a series of deposition experiments which were done with the substrates heated to 300°C. A relatively low sputtering current of 200 milliamps was chosen for these experiments to minimize additional heating of the substrates by the target. With these two parameters constant, a large number of films were sputtered on silicon, sapphire and polycrystalline MgO substrates using various argon pressures and argon/nitrogen ratios. In these experiments highest T_c's were obtained in films which were sputtered in a partial pressure of nitrogen of 0.133 Pa and a total $A+N_2$ pressure of 1.2 Pa. Under these conditions the sputtering voltage was 160V and the deposition rate 0.1 nm/sec. The T_c's of the NbN sputtered on silicon and sapphire substrates were similar with the maximum values being between 12 and 13K. The films on the polycrystalline MgO however were found to be approximately

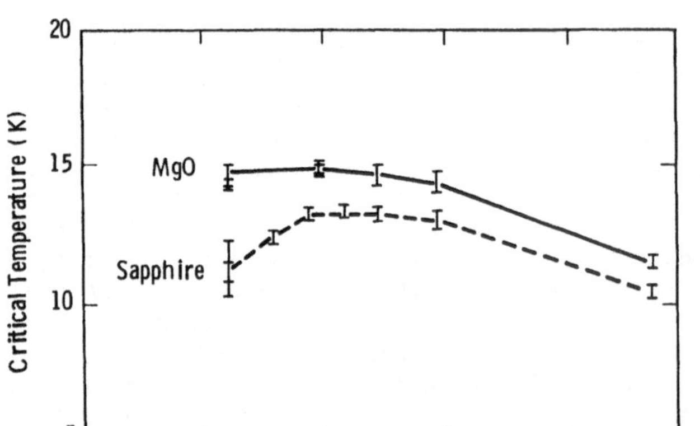

Fig. 1 Critical temperatures versus total $A + N_2$ pressure of NbN films sputtered onto sapphire and polycrystalline MgO. Substrate temperature(T_s) - 300°C, N_2 partial pressure, P_{N_2}, - 0.133 Pa, thickness - 50 nm, and rate - 0.1nm/sec.

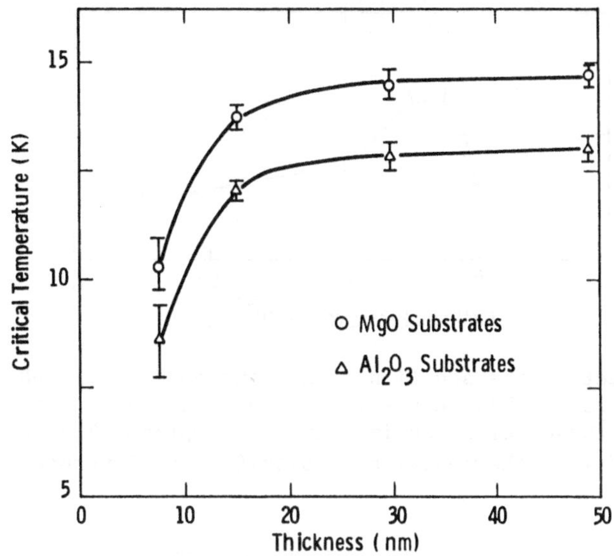

Fig. 2 T_c as a function of thickness for NbN films sputtered on polycrystalline MgO and sapphire. Argon + nitrogen pressure - 1.2 Pa, T_s = 300°C, P_{N_2} - 0.133 Pa.

2 to 3K higher. The significant data from this series of experiments are shown in Figure 1, in which T_c is plotted versus total sputtering gas pressure. The thicknesses of all of the films shown in this figure are ~50nm. Similar differences in T_c's were also observed in other NbN films. Figure 2 shows an example of some of the T_c vs. thickness results. These data are similar to those previously reported by Yamashita et al.[8] The I-V characteristics of tunnel junctions made from two of the thinnest films on the MgO substrates are shown in Figure 3. The low leakage currents, the narrow energy gaps, and gap values which give $2\Delta/k_B T_c = 4.25$, all indicate that even the very thin NbN films are homogeneous. The I-V characteristics at tunnel junctions of NbN deposited on silicon and sapphire substrates were found to be similar to those on NbN/MgO except for the expected smaller gap values. All of these tunneling data imply that any inhomogeneities which may exist in these films would have to be on a scale much smaller than the coherence length of NbN, i.e. ~5nm.

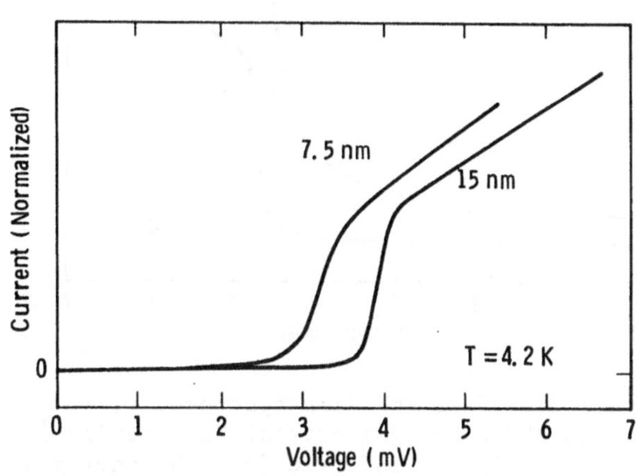

Fig. 3 I-V curves for 7.5 nm and 15 nm NbN films on polycrystalline MgO substrates.

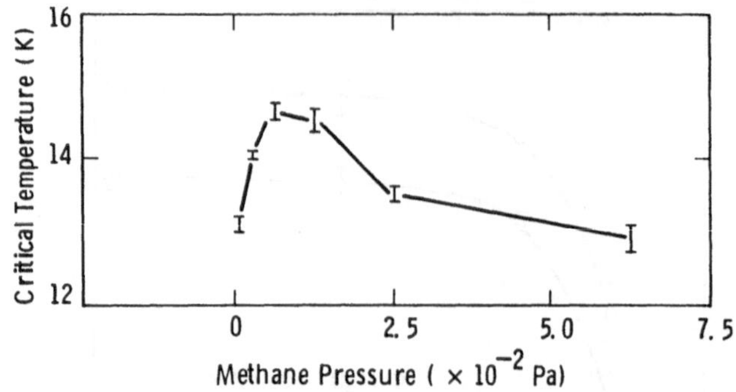

Fig. 4 The effect of the addition of methane on T_c's of NbN films on si substrates thickness - 70 nm. Sputtering condition the same as those for films in Figure 2. Optimum T_c's were obtained with the addition of 6.5×10^{-3} Pa of methane.

It has been reported that the addition of carbon (methane) to the sputtering gas is also effective in raising T_c's of NbN deposited at low temperatures.[9] We have studied the influence of methane additions to NbN at 300°C and some of the results are shown in Figure 4 where T_c is plotted as a function of methane pressure. The data on this figure are from films sputtered on silicon. As can be noted, the maximum critical temperature is ∿15K, approximately 2K higher than that obtained with no methane added. Similar data were obtained with sapphire substrates. However in the case of polycrystalline MgO on which, as was discussed, NbN with a T_c of ∿15K can be grown at 300°C, the addition of methane produced no further increase in critical temperature.

To study single crystal epitaxy, NbN films were also sputtered onto single crystal MgO and on sapphire. To achieve good epitaxial growth, the main requirement is to have a clean and damage-free surface. This requirement is easily met with both sapphire and MgO and single crystal NbN films were grown on both of these substrates. Single crystal NbN,

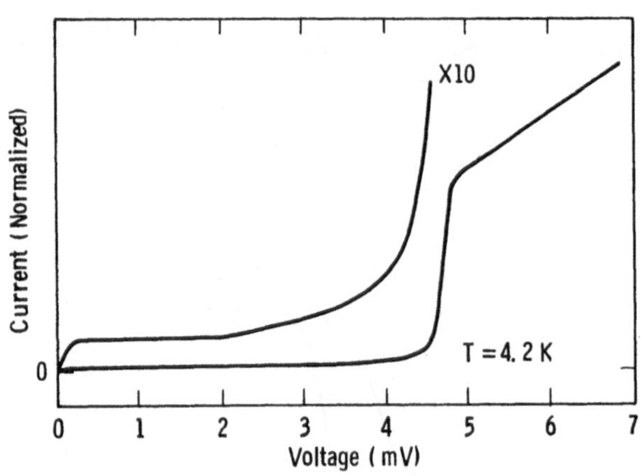

Fig. 5 I-V curve for a 50 nm NbN film sputtered at T_s = 90°C. The substrate is 100 nm single crystal NbN previously sputtered on sapphire at 700°C. T_c inferred from the energy gap = 16.4K (assuming $2\Delta/k_B T_c = 4.25$). This is the same as the resistively measured T_c of the substrate.

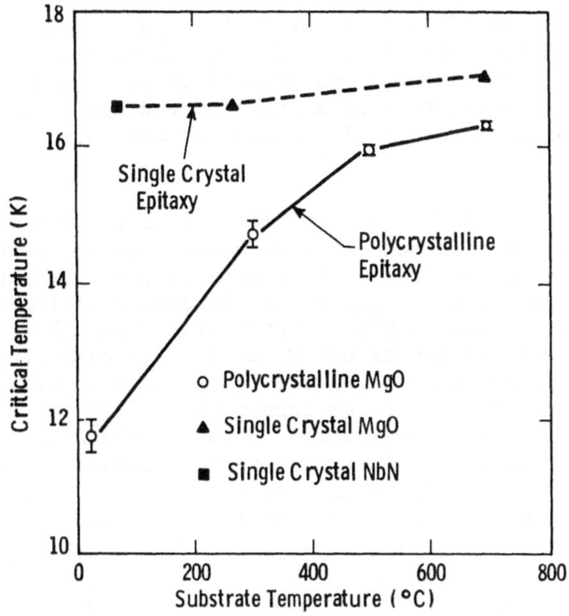

Fig. 6 A comparison of the optimum T_c values from polycrystalline and single crystal epitaxy.

as indicated by in-situ RHEED, was grown on MgO at temperatures as low as 300°C. On sapphire, temperatures of 700°C were required. More details on these deposition experiments, as well as photographs of pertinent RHEED patterns are included in a paper by Talvacchio et al.[7] Critical temperatures of the NbN deposited on the single crystal MgO substrates at 300°C reached as high as 16.3K, which is over a degree greater than the maximum value measured for NbN/polycrystalline MgO films. A T_c of 17.2K was found for a single crystal film grown on MgO at 700°C which is again our highest value obtained for that temperature. Finally, NbN was also sputtered at 90°C on an NbN single crystal film film which had previously been deposited on sapphire at 700°C. The T_c of the NbN substrate was 16.4K. The critical temperature of the 50 nm film sputtered at 90°C could not, of course, be measured resistively. The value was determined from tunneling data shown in Figure 5. From these data a T_c of $\sim$16.4 ($2\Delta/k_B T_c$ assumed to be 4.25) is inferred for the NbN overlayer which is the same as the NbN substrate and strikingly higher than that obtained without the aid of epitaxy. A comparison of the efficacy of single crystal and polycrystalline epitaxy with respect to their effect on T_c is shown in Figure 6.

DISCUSSION

As we have discussed in a previous publication,[10] extensive phase diagram studies [11-15] have shown that the NbN δ-phase, which crystallizes in the superconducting cubic B1 structure, forms only at high temperatures ($\sim$1400°C). At lower temperatures NbN crystallizes into the hexagonal δ' or ε-phases which are not superconducting. It is also well established that oxygen and carbon can stabilize the B1 phase at lower temperatures.[11-13] In this discussion, therefore, we start from the documented premise that the NbN B1 phase does not form at the temperatures used in this work without the aid of some stabilizing mechanism. We believe that our initial NbN films, grown without the aid of epitaxy or the deliberate addition of impurities, formed into the B1 structure as a result of being stabilized by the background carbon and oxygen impurities in our sputtering system. The T_c values that we obtained in these experiments concerning optimum deposition rate, sputter current, N_2 partial pressure, etc., have no significance other than that they provide some hint regarding the cleanliness of our particular sputtering environment.

There are several possible explanations why a maximum in T_c of only ~12.5K was achieved in this set of experiments. It is known that deviations from ideal stoichiometry in NbN result in lower T_c's with the decrease becoming larger with increasing nitrogen deficiencies.[16] Low T_c's could thus be due simply to not having enough N_2 in the sputtering gas. This explanation is not valid in this case since increasing the N_2 partial pressure did not improve T_c. Another possibility is that a very high level of impurities was incorporated into the films, sufficient to depress critical temperature. However, the fact that the addition of impurities in the form of methane increased T_c would seem to obviate this possibility. We believe therefore that the most probable explanation is that while the background level impurities in our system was high enough to stabilize the formation of the B1 phase, it was not sufficient to completely inhibit the nucleation of the non-superconducting δ' or ε NbN phases. We hypothesize that as the result, the B1 NbN phase, because it was forced to compete for nitrogen with these other phases at the substrate surface could not achieve ideal stoichiometry. The non-superconducting NbN phases, being more stable than the ideal stoichiometric B1 phase, tended to grow preferentially. We have tried to verify the presence of second phase growth in some of our low-T_c films sputtered at $\lesssim$ 300°C, using x-ray diffraction. The results however are inconclusive. Because of the extremely fine-grained microstructure of these films, diffraction peaks due to the major (B1) phase were found to be very broad. The fact that peaks from a possible second phase were not found may only mean that they were so weak and broad that they were not identifiable. Despite our inability to provide this piece of direct experimental evidence, we believe this "second-phase" hypothesis is valid particularly because, as will be discussed, it provides a rational framework for understanding all of our experimental results.

As mentioned, polycrystalline epitaxy on MgO and carbon additions both raised T_c's in the 300°C-sputtered NbN to similar maximum values. From these results we conclude that both of these techniques successfully stabilized a more ideal stoichiometric B1 phase, permitting it to grow at the expense of the non-superconducting phases. One of the most interesting features of the data shown in Figure 2 is that the differences in T_c's of the thicker films are the same as that of the thinner ones. This fact is perhaps the strongest piece of circumstantial evidence supporting our belief that there is a second phase in the lower-T_c films which is not present (or present at some lower level) in the epitaxial films. If this were not true, one would have the extremely difficult task of explaining how two single phase films sitting side by side could continue to grow for 50 nanometers and more with two entirely different compositions.

As shown in Figure 6 the NbN films, epitaxially grown on single crystal substrates, had the highest critical temperatures. In the context of our ideas, this result can be explained by the fact that the single crystal substrates presented an exclusively B1 surface for nucleating the B1 phase. In polycrystalline MgO, for example, preferential nucleation of the B1 phase would occur on the grains, however undesired second phase(s) could nucleate at the grain boundaries. This possibility is eliminated when using single crystal substrates.

Finally, we have shown that high-T_c films can be grown on NbN single crystals at 90°C (Fig. 5). This result, while of little practical significance, is important since it clearly demonstrates the efficacy of homoepitaxy in stabilizing the stoichiometric B1 phase. Some workers have disputed the need of impurities to grow B1 structure NbN at low temperatures because analyses of their films showed the presence of no impurities. Such results can be explained by the concept of homoepitaxy. One can easily conceive of a situation in which there are sufficient impurities present during the first few monolayers of deposition to promote the nucleation of single phase B1 film. Once nucleated, this phase could then continue to grow solely through homoepitaxial stabilization in the complete absence of any further impurities.

CONCLUDING REMARKS

Although they have been generally discussed informally more than in publications, the problems of reproducibility and the general vagaries of growing high-T_c NbN films are well known. We believe that the results and concepts presented in this paper provide a framework for understanding the mechanisms involved in the low temperature growth of this increasingly important superconducting material.

ACKNOWLEDGMENTS

We thank H. Pohl and R. Wilmer for their valuable technical assistance.

REFERENCES

1. T. Mitsuoka, T. Yamashita, T. Nakazawa, Y. Onodera, Y. Saito, and T. Anayama, J. Appl. Phys. 39:4788 (1968)
2. J. R. Gavaler, J. K. Hulm, M. A. Janocko, and C. K. Jones, J. Vac. Sci. Tech. 6: 177 (1969)
3. R. T. Kampwirth and K. E. Gray, IEEE Trans. Magn. MAG 17: 565 (1981)
4. R. B. Van Dover, D. D. Bacon, and W. R. Sinclair, Appl. Phys. Lett. 41:764 (1982)
5. J. C. Villegier, and J. C. Veler, IEEE Trans. Magn.MAG 19: 946 (1983)
6. Akira Shoji, Masahiro Aoyagi, Shin Kosaka, Fujitoshi Shinoki, and Hisao Hayakawa, Appl. Phys. Lett. 46: 1099 (1985)
7. J. Talvacchio, M. A. Janocko, J. R. Gavaler, and A. I. Braginski, This volume.
8. T. Yamashita, K. Hamasaki, Y. Kodaira, and T. Komata, IEEE Trans. Magn. MAG-21: 932 (1985)
9. E. J. Cukauskas, J. Appl. Phys. 54: 1013 (1983)
10. J. R. Gavaler, A. I. Braginski, M. Ashkin, and A. T. Santhanam, Superconductivity in d- and f- Band Metals H. Suhl and M. Brian Maple, eds. Academic Press, New York (1980) p. 25
11. G. Brauer and J. Jander, Z. Anorg. Allg. Chem. 270:160 (1952)
12. G. Brauer and R. Esselborn, Z. Anorg. Allg. Chem. 309:151 (1961)
13. G. Brauer and H. Kirner, Z. Anorg. Allg. Chem. 328:34 (1964)
14. E. V. Storms, High Temperature Science, 7:102 (1975)
15. R. W. Guard, J. W. Savage, and D. G. Swarthout, Trans. AIME 239:643 (1967)
16. T. H. Geballe, B. T. Mathias, J. P. Remeika, A. M. Clogston, V. B. Compton, J. P. Maita, and H. J. Williams, Physics 2:293 (1966)

DUAL ION-BEAM DEPOSITION OF SUPERCONDUCTING NbN FILMS

E. K. Track, L.-J. Lin, G.-J. Cui,* and D. E. Prober

Section of Applied Physics
Yale University, New Haven, CT

ABSTRACT

Superconducting NbN films have been deposited onto unheated substrates using a dual ion-beam technique. The NbN films produced have T_c up to 12 K and resistivities of $\gtrsim 150\ \mu\Omega$-cm. The substrate temperature does not exceed 100 °C. TEM analysis of these films shows a random in-plane orientation of <100 Å-size grains. Electron diffraction indicates fine-grain polycrystalline material, and a Read x-ray camera verifies this to be the high-T_c δ-phase with the B1 crystal structure. Using a native oxide barrier on these films and a $Pb_{.71}Bi_{.29}$ counterelectrode, tunnel junctions are produced with current densities of 30 A/cm^2 and V_m of 50 mV at 4.2 K. Artificial barriers have been successfully produced, using oxidized Al or Ta overlayers or AlN. The quality of the resulting I-V curves is comparable to that of junctions with native oxide barriers. The junction resistance with the artificial barriers is higher, however, and the conductance at large voltages reflects the different properties of the artificial barriers.

INTRODUCTION

After its discovery in 1941, and a period of activity in the seventies, NbN has seen a resurgence of interest in the early eighties[1-6], largely due to the potential microelectronic applications. NbN has a high T_c (16-17 K). Unlike the A-15 compounds, NbN is relatively insensitive to stress and is relatively easy to produce. It has a very high critical field, up to 50 Tesla[5]. Furthermore, there are many unanswered scientific questions, such as the origin of the high resistivity and an understanding of the electron-phonon coupling spectrum.

The technique most commonly used to fabricate NbN thin films is reactive sputtering (dc or rf, diode or magnetron)[1-3,5,6]. Such techniques employ relatively high pressures (∿1.5 Pa) and substrate heating is usually necessary to attain near-bulk T_c values. One of the few studies using a significantly different method of thin-film synthesis had a growing Nb film, deposited by electron-beam evaporation, bombarded with N_2^+ ions[7]. Broad transitions, extending from 9 K to 14.5 K, were reported.

The dual ion-beam technique was pioneered by Weissmantel[8] for fabrication of Si_3N_4 films. Later, it was applied to AlN film formation by Harper et al.[9] The dual-beam method, as shown in Fig. 1, allows independent control of the ion flux and energy during deposition, at low pressures (∿0.02 Pa) such that the mean free path is larger than typical chamber dimensions. As a result, the deposition has a directional character.

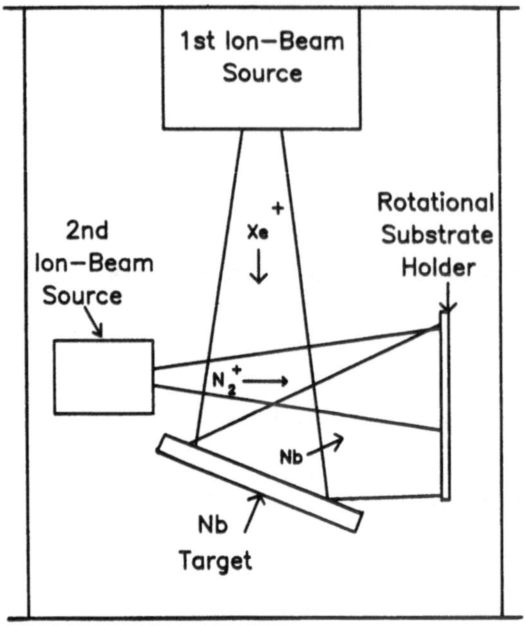

Fig 1. Sputtering configuration. The target holder holds up to four targets and can be rotated in-situ.

For producing NbN films with the dual ion-beam technique, the energy released at the surface of the growing film by the second ion beam may provide a substitute for the thermal energy used previously to attain high T_c values. A deposition method which does not require heated substrates may allow the formation of a NbN counterelectrode (i.e., for an all-NbN junction) while preserving the very thin tunnel barrier.

In this paper we report on initial success in forming δ-phase NbN films by the dual ion-beam technique. We present results on characterization of these films, and on the properties of tunnel junctions formed with these films using native oxide barriers, and also artificial barriers produced with oxidized metallic overlayers (Al,Ta) and AlN. Some of this work has been reported in a separate publication[10].

FILM PREPARATION AND PROPERTIES

The films are prepared in a diffusion-pumped vacuum chamber with a pyrex glass cylinder. The base pressure is 1-2 x 10^{-5} Pa. As shown in Fig. 1, two ion sources are fitted in the chamber. The first one is a 2.5 cm diameter source[11]. It is fixed to the top plate and faces a multiple target holder ~15 cm away. This target assembly is mounted on the bottom plate and can be rotated in-situ. The first ion source uses Xe as a sputtering gas at a flow rate of 0.5 sccm and a partial pressure of 0.013 Pa (uncompensated ion-gauge reading). This first ion source has been successfully used by itself to make high quality films of Nb and Ta[12]. In those studies, Xe was found to be the inert gas which produced Nb films with the highest T_c values. The second ion source is a 3 cm source[13] mounted with brackets to the bottom plate and oriented to face the rotary substrate holder, ~10 cm away. It uses one of the following gases, or a mixture: N_2, Kr and CH_4 at a total pressure of 0.013 Pa. We employ 0.6 cm x 0.6 cm Si(100) and Si(111) substrates. The approximate substrate temperature is measured with a thermocouple placed on the substrate holder.

With just the first ion source operating and N_2 flowing into the chamber, we obtain films which adhere poorly to the Si substrates.

Resistance measurements on these samples show multiple superconduting transitions with a maximum T_c of 11-12 K obtained at higher N_2 partial pressure (0.02 Pa). Adhesion to the substrate is improved by pre-sputter-cleaning the substrate with 500 eV Krypton ions for a few minutes.

When both ion sources are operating, we find that until the N_2^+ ion energy exceeds about 500 eV, the films display multiple transitions, and in some cases the films are resistive at 4.2 K. At present, the best films are obtained by operating the second ion source at 1500 V using either pure N_2 or N_2 mixed with Kr or CH_4 (see Table I). By using only N_2 with the second source, we found an upper limit for the beam current of 2 mA beyond which the films start displaying multiple transitions, presumably from a nitrogen-rich phase of NbN. By mixing Kr or CH_4 with N_2, the total current can be increased, allowing for more stable operation of the source, while not exceeding the critical N_2^+ arrival rate at the substrate. We find that addition of a small amount of CH_4 has the potential of improving T_c, possibly by helping to stabilize the high-T_c δ-phase of NbN[1]. In all cases the NbN deposition rate is $\sim$100 Å/min. The substrates are not intentionally heated. The temperature of the substrate holder is typically 60-80 °C.

T_c for the films deposited under optimal conditions is 11-12 K, with a transition width of 60-150 mK (see Table I). The resistivity is typically $\gtrsim$150 $\mu\Omega$-cm with higher values observed for the inhomogeneous, multiple-transition films. The residual resistance ratio (ρ(298K)/ρ(20K)) is $\sim$0.9. Fig. 2 shows an electron diffraction and TEM photograph of a 500-Å film of NbN on a 200-Å carbon film suspended on a standard 200 mesh copper TEM grid. This NbN film was made without the addition of CH_4, with the parameters of sample E. It appears to have a grain size <100 Å, with a random in-plane orientation. For a 3000-Å film made on a Si substrate using a mixture of N_2 and Kr in the second ion source (sample F), a Read X-Ray camera identifies the structure to be single-phase δ-NbN (B1 structure). Auger (AES) analyses on similar samples indicate uniform composition with a N to Nb ratio of 0.96 and traces of a few atomic percent of C and O.

Table I. Typical results for selected film depositions. First ion source is operated at 1500 V, 34 mA with P_{Xe} $\sim$15 mPa. RRR $=\rho_{298K}/\rho_{20K}$. ΔT_c is the width of the transition (10%-90%).

Sample	Parameters of Second Ion Source				Film Properties			
	V (V)	I (mA)	P_{N_2} (mPa)	P_{other} (mPa)	ρ_{20K} ($\mu\Omega$-cm)	RRR	T_c (K)	ΔT_c (K)
A	0	0	20	-	604	0.97	10.8-11.9	Multiple trans.
B	200	3.6	14	-	475	0.90	9-12	Multiple trans. R>0 at 4.2 K
C	500	0.8	13	-	870	0.88	10.3-11.7	Multiple trans.
D	1500	3	9	-	234	0.90	11.1-12.3	Two trans.
E	1500	2	13	-	206	0.80	11.6	0.08
F	1500	3.5	9	Kr: 4	214	0.82	11.5	0.09
G	1500	3	9	CH_4: 1.5	138	0.86	11.6	0.06

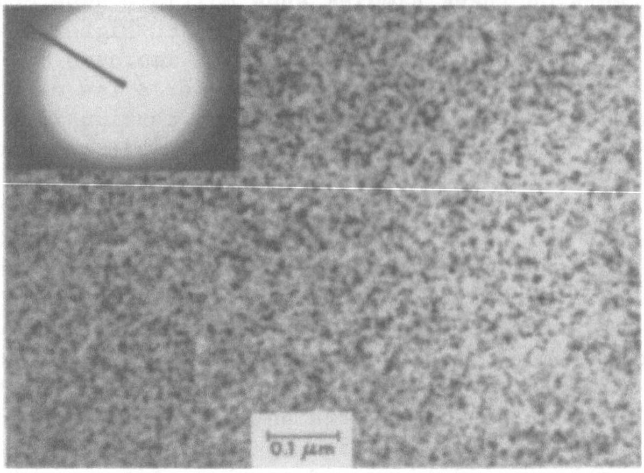

Fig 2. TEM photograph and electron diffraction pattern of a sample made with the parameters of sample E (see Table I).

TUNNEL JUNCTIONS: NATIVE-OXIDE BARRIERS

We have used films obtained with the present optimum deposition conditions (of sample G) to fabricate tunnel junctions with native oxide and with artifcial barriers, and Ag or $Pb_{.71}Bi_{.29}$ (29% Bi by weight) counterelectrodes. The junction area is defined in one direction by a window formed by evaporating 2000 Å of Ge through an 80-μm wire mask. A second mechanical mask 400-μm wide is used to define the width in the other direction during evaporation of the counterelectrode.

To form the native oxide barriers, the NbN base electrode is oxidized for up to 4 hours in room air. The counterelectrode is then evaporated in a separate system. With a $Pb_{.71}Bi_{.29}$ counterelectrode, we obtain junctions with reproducible properties and fair-quality I-V curves as shown in Fig. 3(a). At T=1.4 K, the subgap conductance at 2 mV is 1.5% of the conductance above the gap (at 5 mV). Typical junction resistances R_n are 0.2-5Ω. For one low resistance junction the maximum Josephson current (I_c) at 4.2 K was 10 mA, which gives a critical current density of 30 A/cm^2 and I_cR_n=2 mV. Here $V_m=I_cR_{sg}$=50 mV, with R_{sg} defined at 2 mV and 4.2 K. The sum gap at 1.4 K is 3.67 mV measured as the midpoint of the current rise which extends over about 0.4 mV. For $Pb_{.71}Bi_{.29}$, $\Delta(0)$=1.74 meV[14], so that for our junction, Δ_{NbN}=1.93 meV and $2\Delta/kT_c$=3.9. This is in agreement with previous findings[1-4] that NbN is a strong coupled superconductor.

OXIDIZED METALLIC OVERLAYERS

Oxidized metallic overlayers have been successfully employed to provide a high quality tunnel barrier on Nb base electrodes[15,16] and other materials for which a good native oxide barrier is not easily obtained. In-situ deposition of the metallic overlayer is essential to minimize contamination of the interface. One previous attempt to use Al overlayers on NbN films after they had been exposed to air did not yield satisfactory results[17]. It has been speculated that the Al combines with nitrogen to form a damaged layer at the NbN/Al interface[18]. It is also possible that the exposure to air may have degraded that interface.

We have used overlayers of Al, Ta, or sequentially depsosited Nb+Al or Nb+Ta on the NbN base electrode to form artificial oxide barriers. To form the thin surface layer, the N_2 gas flow is stopped, 10-15 minutes are allowed for the N_2 to be pumped out, the appropriate target is rotated into place and the overlayer is ion-beam sputter-deposited. The thickness of the

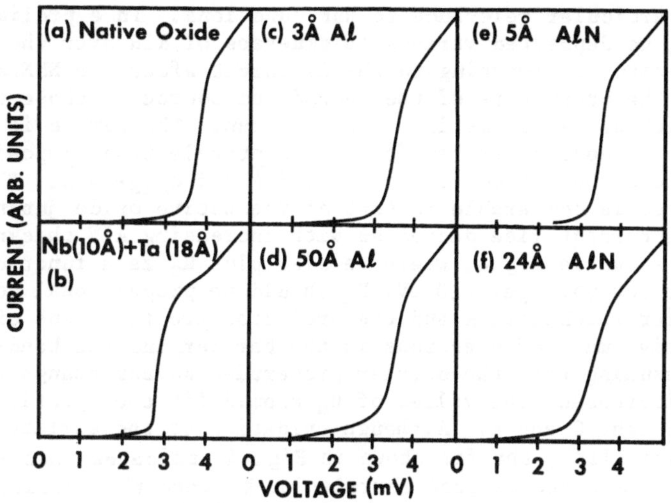

Fig 3. I-V characteristics for various junctions. a) R_n=0.3Ω b) R_n=12Ω c) R_n=92Ω d) R_n=146Ω e) R_n=4Ω f) R_n=444Ω. a) and b) are at T=1.4 K, c)-f) at T=4.2 K. Counterelectrode is $Pb_{.71}Bi_{.29}$ in all cases.

overlayer is inferred from the deposition time and the known rates of deposition. The quality of the I-V curves of these overlayer junctions, as determined by subgap leakage, is comparable to, but not better than, the native oxide junctions (see Fig. 3), but the overlayer junctions have a higher normal tunneling resistance R_n. The fact that there is no improvement of the I-V quality with overlayers may indicate that the leakage current and the width of the current rise are caused by intrinsic properties of the NbN film itself (tailing of the density of states in the gap[15], gap inhomogeneity, etc.) and are not due to defects in the barrier. In addition, the 10-15 minute exposure to the background pressure in the vacuum system may have produced imperfections at the interface, causing part of the subgap leakage observed.

Fig. 3(b) shows the I-V curve of a NbN/Nb(10 Å)+Ta(18 Å)/oxide/ $Pb_{.71}Bi_{.29}$ junction. Here the current rise occurs at a reduced bias of 2.78 mV, due to the proximity effect[19]. The proximity effect also causes the structure above the current rise. This structure displays a decrease in the conductance dI/dV, followed by an increase, with the minimum dI/dV occuring at a voltage of 3.68 mV. This voltage corresponds to the sum-gap of our native-oxide junctions, consistent with the proximity effect model[19]. Use of thinner Nb+Ta overlayers or Al or Nb+Al overlayers produces a smaller depression of the current rise and no resolved structure above the current rise.

All the Al-overlayer junctions had a similar voltage for the current rise (∿3.4 mV at 4.2 K). There was a slight increase in subgap leakage with increasing Al thickness. This may be due to Al diffusion along the grain boundaries and subsequent formation of regions of depressed T_c[18].

AlN ARTIFICIAL BARRIERS

AlN is a single-phase, wide band-gap (6.3 eV), chemically stable insulator[9]. It thus has good potential for use as an artficial tunneling barrier. Furthermore, it can be deposited by the dual ion-beam technique[9], making it of particular relevance to NbN junctions. In a preliminary investigation, we deposited various thicknesses of AlN over the NbN base electrode, in-situ, by rotating to the Al target after the NbN deposition and adjusting the parameters of the second ion source to those which produced insulating AlN films in a separate run. The sample is then transferred to a separate system for the counterelectrode deposition.

Fig 3. shows the I-V curves for d_{AlN}=5 Å and d_{AlN}=24 Å. The quality of these I-V curves is comparable to that of the native oxide junctions. The voltage of the current rise decreases with increasing AlN thickness. This dependence is shown in Fig.4, where we also plot R_n as a function of the AlN thickness. At low voltage, ∿10 mV, R_n should be proportional to de^{kd} where d is the barrier thickness, k and the prefactor are functions of the barrier height, effective mass of electrons in the barrier, and the band-gap of the barrier[20]. Assuming that the barrier properties do not change when the AlN thickness is increased, the values of R_n should fit the functional form Cde^{kd} with constant C and k. Although oxidation of the surface of the AlN layer is a possibility, the fit shown in Fig. 4 indicates that for d_{AlN} greater than ∿10 Å there is good correlation between the barrier width and d_{AlN}. For d_{AlN} = 5 Å, the lower R_n obtained could be due to incomplete coverage of the NbN base electrode yielding a mixed barrier of AlN and NbN oxide. Positions 1 and 2 correspond to the position of the junction on the substrate. The higher R_n values for position 2 are likely due to a gradient in the deposition rate of AlN on the substrate.

BARRIER PROPERTIES

The high voltage conductance of a tunnel junction reflects the properties of the tunnel barrier. The simplest means of describing the barrier is to assume that tunneling occurs through a trapezoidal WKB barrier of a given effective width (d_{eff}) and average barrier height ($\bar{\Phi}$)[21]. Such a model does not resolve the different tunnel channels which may exist in a non-uniform oxide[22], but yields an average effective barrier which would produce the same conductance as the real barrier. Using this WKB

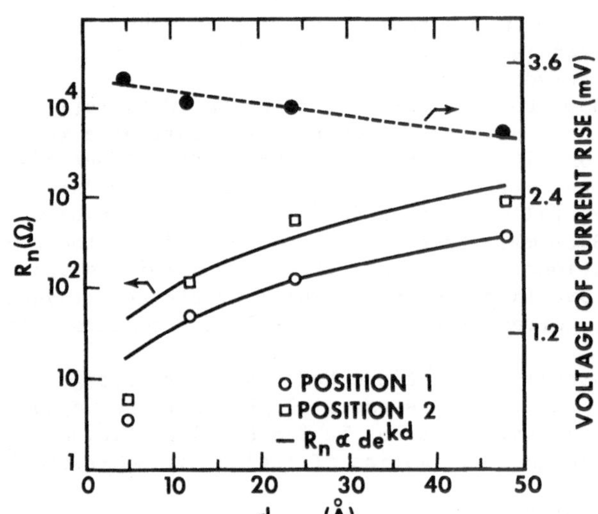

Fig 4. R_n and voltage of midpoint of current rise for AlN barrier junctions. The dashed line is a guide to the eye.

analysis, we find that NbN native oxide gives low and wide barriers ($\bar{\phi}$ = 0.2-0.3 eV, d_{eff}= 23-30 Å), whereas the Al overlayer yields the highest and narrowest barrier ($\bar{\phi}$=0.8-1.2 eV, d_{eff}=17-22 Å). These results are similar to those from previous studies of Al overlayers on Nb[23].

When this WKB analysis is applied to the AlN barriers, the inferred effective barrier width does correlate roughly with the thickness of AlN deposited: from d_{eff}= 25 Å for d_{AlN}= 5 Å, to d_{eff}= 41 Å for d_{AlN}= 48 Å. The average barrier heights obtained for AlN are $\bar{\phi}$ = 0.2-0.5 eV.

The authors thank D. W. Face for many helpful discussions, E. J. Cukauskas of the Naval Research Laboratory for the Read x-ray camera study, P. Male for the TEM study, H. Erskine for the AES study, and A. Pooley for energy dispersive x-ray analysis. This work was supported in part by ONR N00014-80-C-0855 and NSF ECS-8305000, and by the APS Chinese-American Basic Research Program (for G.-J.Cui). AES studies were conducted at the University of Minnesota NSF Regional Instrumentation Facility.

REFERENCES

* On leave from Peking University, China.

1. E. J. Cukauskas, W. L. Carter, S. B. Qadri, and E. F. Skelton,IEEE Trans. Magn., MAG-21, 505 (1985).
2. D. D. Bacon, A. T. English, S. Nakahara, F. G. Peters, H. Schreiber, W. R. Sinclair, and R. B. van Dover, J. Appl. Phys. 54, 6509 (1983).
3. J. C. Villegier, L. Vieux-Rochaz, M. Goniche, P. Renard, M. Vabre, IEEE Trans. Magn., MAG-21, 498, (1985).
4. M. Gurvitch, J. P. Remeika, J. M. Rowell, J. Geerk, and W. P. Lowe, IEEE Trans. Magn., MAG-21, 509, (1985).
5. M. Ashkin, J. R. Gavaler, J. Greggi, and M. Decroux, J. Appl. Phys. 55, 1044 (1984).
6. M. Igarashi, M. Hikita, and K. Takei, in "Advances in Cryogenic Engineering" 30, 535, Ed. by A. F. Clark and R. P. Reed, Plenum Press (1984).
7. J. J. Cuomo, J. M. E. Harper, C. R. Guamieri, D. S. Lee, L. J. Attanasio J. Angilelo, C. T. Wu, and R. H. Hammond, J. Vac. Sci. Technol. 20, 349 (1982).
8. C. Weissmantel, Thin Solid Films, 32, 11 (1976); H.-J. Herler, G. Reisse, and C. Weissmantel, ibid., 65, 233 (1980).
9. J. M. E. Harper, J. J. Cuomo, and H. T. G. Hentzell, J. Appl. Phys. 58, 550-563, (1985).
10. L.-J. Lin, E. K. Track, G.-J. Cui, and D. E. Prober. To appear in Physica B, North-Holland Publishing Co., Eds. K. A. Gschneidner and E. L. Wolf.
11. Ion Tech, Inc., Box 1388, Ft. Collins, Col.
12. D. W. Face, S. T. Ruggiero, and D. E. Prober, J. Vac. Sci. Technol. A2, 326 (1983).
13. Commonwealth Scientific Corp., Alexandria, Va.
14. R. C. Dynes and J. M. Rowell, Phys. Rev. B 11, 1884 (1975).
15. M. Gurvitch and J. Kwo, in "Advances in Cryogenic Engineering" 30, 509, Ed. by A. F. Clark and R. P. Reed, Plenum Press (1984)
16. S. T. Ruggiero, G. B. Arnold, E. K. Track and D. E. Prober, IEEE Trans. Magn., MAG-21, 850 (1985).
17. M. Gurvitch, private communication.
18. D. Van Vechten and J. F. Liebman, J. Vac. Sci. Technol. A3 1881, 1985.
19. E. L. Wolf and G. B. Arnold, Phys. Reports 91, 31 (1982).
20. K. H. Gundlach, J. Appl. Phys. 44, 5005 (1973).
21. W. F. Brinkman, R. C. Dynes, and J. M. Rowell, J. Appl. Phys. 41, 1915 (1970).
22. J. Halbritter, IEEE Trans. Magn., MAG-21, 858 (1985).
23. J. M. Rowell, M. Gurvitch, and J. Geerk, Phys. Rev. B 24, 2278 (1981).

SUPERCONDUCTING AND STRUCTURAL PROPERTIES OF RF MAGNETRON

SPUTTERED NIOBIUM NITRIDE FOR JOSEPHSON JUNCTIONS

E. J. Cukauskas and W. L. Carter

Electronics Technology Division
Naval Research Laboratory
Washington, D.C.

ABSTRACT

The material properties of rf magnetron sputtered niobium nitride prepared at a substrate temperature of 650°C have been investigated as a function of substrate rf bias. These materials exhibit a preferential crystallite orientation related to the amount of rf bias, (200) orientation for low bias and (111) for high bias. We have observed greater than expected lattice parameters of 4.46 Å for these films which we associate with a distorted fcc structure. The best films have a low temperature resistivity less than 70 μΩ-cm and transition temperatures exceeding 16 K. NbN/Si/Nb tunnel junctions have been fabricated from these films with unoxidized hydrogenated silicon barriers using the SNAP process. These junctions have V_m values exceeding 40 mV and sum gap values ~3.9 mV. These barriers are being used in all NbN tunnel junction development.

INTRODUCTION

Substantial progress has been realized over the past several years in the development of a high quality, high T_c all refractory Josephson junction technology suitable for electronic applications.[1-5] Researchers have been studying the materials, barriers, interfaces and circuit applications of NbN based Josephson junctions. At NRL we have been studying the conditions under which high quality NbN can be deposited at high and low substrate temperatures suitable for base and counter electrodes in NbN tunnel junctions.[2,6,7] We define high quality NbN as material with both high T_c (greater than 16 K) and low resistivity (less than 100 μΩ-cm) such that the material will have a reasonable penetration depth of 200-300 nm. Other researchers have been successful in attaining high T_c on substrates held at room temperature, however these materials usually have resistivities several hundred μΩ-cm and probably have very large penetration depths.[8] These materials can be successfully used in Josephson junctions using the technique of Shoji et. al. where an all NbN junction structure is sandwiched between two Nb films.[9] The Nb has a much smaller penetration depth and thus results in an acceptable magnetic field modulation characteristic of the Josephson junction.

Our goal at NRL is to develop a high quality NbN thin film technology for use in all NbN Josephson junctions. Concurrent with this research, we are studying Josephson junctions with artificial barriers suitable for high temperature deposition. Presently we are using unoxidized hydrogenated silicon barriers with Nb and NbN counter electrodes deposited at substrate temperatures up to 400°C. The base electrode material for these devices is NbN deposited by rf magnetron sputtering at 600°–850°C substrate temperatures and small amounts of rf bias.

NbN FILM PREPARATION

The NbN films used in this research were deposited on quartz substrates in an ultra high vacuum sputtering system by reactive rf magnetron sputtering at a substrate temperature of 650°C using various amounts of rf substrate bias. The complete details of the deposition system and the general film preparation techniques have been described in detail elsewhere.[6] Briefly, prior to film deposition the vacuum chamber received an 8 hour bake at 100°C. The quartz substrates were then outgassed at 750°C for 1/2 hour and sputter etched at 250 W for 20 min. in 1.3 Pa of argon. In the bias sputtering mode, the films were deposited at a total power of 750 W capacitively coupled to the 6 inch Nb target cathode and to the substrate table. For the case of no bias (grounded table), the Nb target charged to a potential of -2300V and decreased to -1500V when the bias was increased to -90V on the table. The potential to which the table charged was used as a measure of the substrate bias rather than the power density because of the ease of its measurement. All the bias sputtered films used in this study were deposited at a total gas pressure of 1.3 Pa consisting of 0.19 Pa nitrogen, 0.08 Pa methane and 1.03 Pa argon.

FILM CHARACTERIZATION

The NbN films were characterized by their transition temperature, T_c, transition width, room temperature resistivity, residual resistivity ratio, crystal structure and composition. Figure 1 illustrates the transition temperature and room temperature resistivity as a function of substrate bias potential. The transition temperature was found to be relatively independent of bias out to approximately -60 V after which T_c rapidly falls. The transition width (not illustrated) was sharp (<0.1 K) for

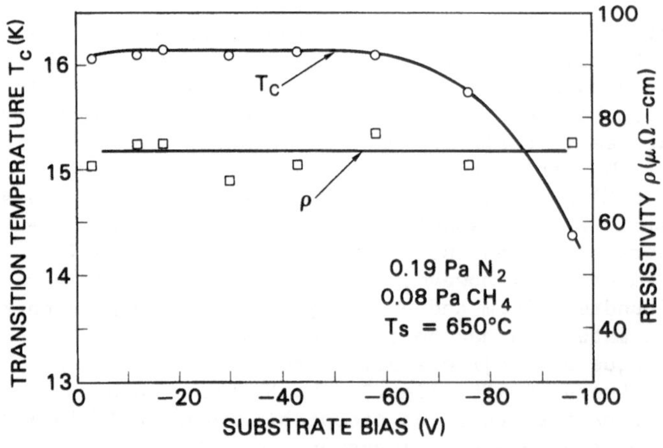

Fig. 1. The transition temperature and room temperature resistivity of rf magnetron bias sputtered NbN as a function of substrate bias. The total sputtering gas pressure was 1.3 Pa. The resistance ratio was 1 for all films.

most films except those deposited at high bias where widths as great as 0.7 K were observed. The room temperature resistivity showed no dependence on substrate bias and the residual resistivity ratio was approximately 1 for all films. This is indicative of grain boundary scattering as the predominant contribution to the resistivity and grain damage due to ion bombardment at the higher bias voltages resulting in the reduction of T_c.

The crystal structure was investigated using x-ray diffractometer scans and Read camera x-ray photographs. There was no lattice parameter dependence observed on substrate bias, however there was a strong dependence of crystallite orientation on bias. Figure 2 illustrates two diffractometer scans for films sputtered with -3 V and -78 V substrate bias. The (200) is the preferred crystallite orientation for the low bias case and the (111) for the films sputtered at the higher bias. We calculated a larger than expected lattice parameter from the Read camera data and found disagreement in the lattice parameter determined from the (111) and (200) diffraction peaks. Other researchers have also observed a larger lattice constant for NbN.[8] We associate this discrepancy with a possible

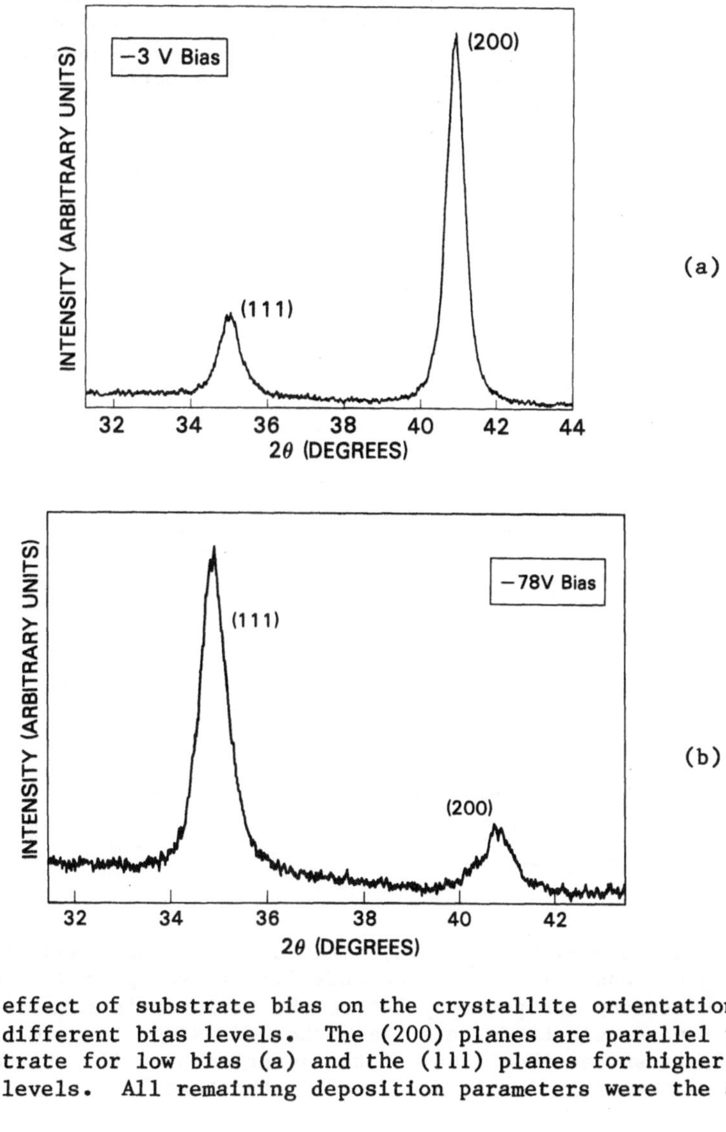

Fig. 2. The effect of substrate bias on the crystallite orientation for two different bias levels. The (200) planes are parallel to the substrate for low bias (a) and the (111) planes for higher bias (b) levels. All remaining deposition parameters were the same.

distortion of the fcc structure due to interstitial carbon or film stress. The composition of these films were determined to be $NbC_{0.4}N_{0.6}$ and independent of substrate bias.

A more extensive study of the properties of rf bias sputtered NbN will be reported in a future publication.

NIOBIUM NITRIDE-SILICON-NIOBIUM TUNNEL JUNCTIONS

Niobium nitride based tunnel junctions with hydrogenated silicon barriers were fabricated from these films using the SNAP process.[10] In the SNAP process, the junctions are isolated by anodizing the Nb counter electrode around the junction area down to the base electrode. This technique converts the Nb to a durable Nb_2O_5 insulator. The SNAP process was performed after defining the base electrode geometry which eliminates the possibility of our pogo-pin contract arrangement from penetrating through the trilayer and shorting out the structure. Figure 3(a) illustrates the top view of a 6 junction chip layout, and Figure 3(b) illustrates the device cross section. The junctions range in size from 25 to 400 μm^2 on the chip. The base electrode is rf magnetron bias sputtered at 850°C and patterned into the bar geometry by plasma etching in CF_4. The hydrogenated silicon barrier similar to that of Kroger et. al.[11] is then deposited after sputter etching at 30 watts in 1.3 Pa of 99.999% pure argon followed by the Nb counter electrode deposition without a barrier oxidation step.

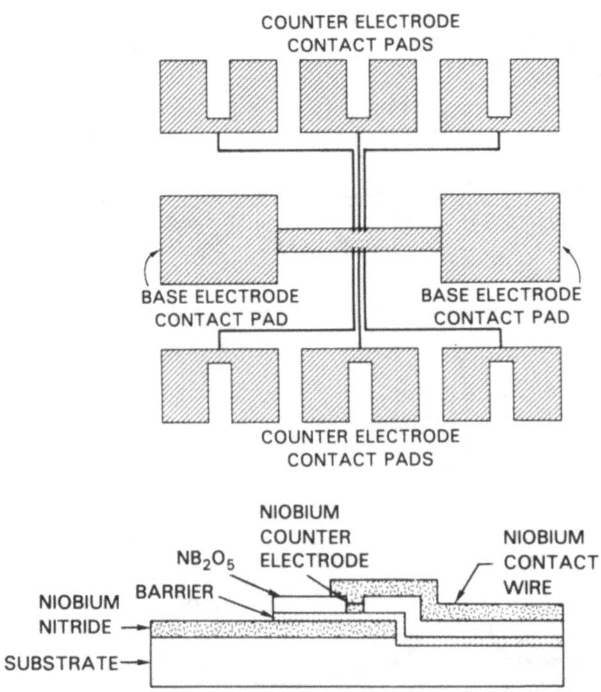

Fig. 3. The top view geometry (a) used for fabricating NbN/Si/Nb tunnel junctions. The base electrode film is sculptured into the base electrode bar geometry before the trilayer is formed. The counter electrode pads are patterned using a wet etch. In (b) a cross sectional view of the junction area illustrates the various film levels discussed in the text.

This sandwich structure is then processed using the standard SNAP process, cleaned and returned to the sputtering system for the niobium contact wire. This wire is patterned and sculptured using a wet niobium chemical etch.

We have been using these fabrication techniques in our investigation of barrier optimization preparation conditions for unoxidized amphorous silicon barriers. The barrier consists of 5-10 nm of reactively sputtered hydrogenated silicon sandwiched between two thin silicon layers approximately 1 nm thick. Figure 4 illustrates the current-voltage characteristics of a typical NbN/Si/Nb tunnel junction of area 400 μm^2 having a sum gap of 3.85 mV and R_n ~2.5Ω. We define V_m for our junctions as the product of the subgap resistance at 2 mV times 0.7 of the quasi particle current rise. This allows us to compare junction quality of high resistance junctions where the Josephson current is small or suppressed by trapped flux. The junction in Figure 4 has a V_m value of 40 mV. Using the trace of the differential resistance vs. voltage (not illustrated) we calculate the NbN and Nb gaps to be 2.72 mV and 1.13 mV respectively. The Nb gap is suppressed partially due to the Nb target being nitrided. We find that we are unable to deposit niobium with 9.2 T_c from the magnetron target without extensive presputtering.

NIOBIUM NITRIDE-SILICON-NIOBIUM NITRIDE TUNNEL JUNCTIONS

The anodization of niobium forms a tough oxide which allows it to serve as an insulating layer in the fabrication of niobium based tunnel junctions. Niobium nitride does not anodize as easily as niobium and forms a soft porous oxide layer which easily breaks away as the anodization voltage is increased. This property of niobium nitride makes the SNAP process unacceptable for junctions with a NbN counter electrode. Other researchers have used the SNEP process of Gurvitch and Kwo[12] for the fabrication of all NbN tunnel junctions. In this process, the counter electrode is selectively removed by plasma etching or reactive ion etching and replaced with an insulating layer to isolate the junctions.

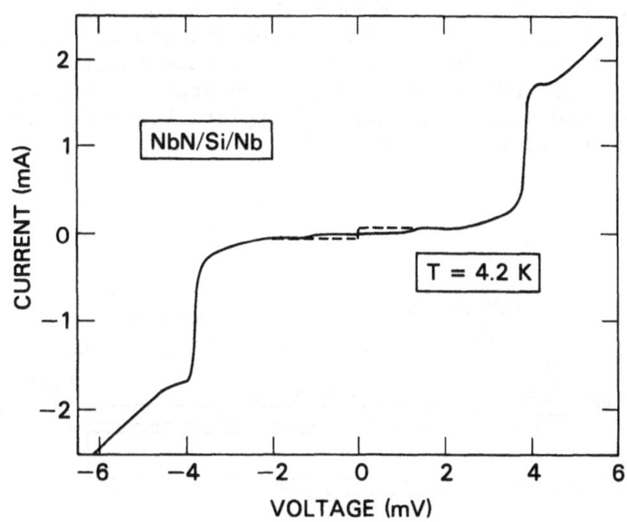

Fig. 4. The current-voltage characteristics of a NbN/Si/Nb tunnel junction with an unoxidized amorphous hydrogenated silicon barrier. The junction area is 400 μm^2 with a sum gap of 3.85 mV and R_n ~ 2.5Ω. V_m defined in the text exceeds 40 mV at 4.2K and 2 mV.

In the development of all niobium nitride tunnel junctions we chose an edge junction geometry which uses an anodized Nb film over a NbN base electrode as an insulating layer. Figure 5 illustrates the cross sectional view of the edge junction. The process starts with the deposition of a NbN/Si/Nb trilayer which is then completely anodized down through the silicon layer. The silicon layer functions as an indicator to stop the anodization process. The chip is then patterned in the base electrode geometry and plasma etched in a CF_4 and O_2 mixture down into the substrate such that a well defined edge is attained. The chip is then cleaned and returned to the sputtering system for barrier and counter electrode deposition. An unoxidized silicon barrier consisting of 5-10 nm of hydrogenated silicon sandwiched between two thin silicon layers is deposited at 200°C, 1.3 Pa pressure and 150 watts after a low power sputter etch. The counter electrode is then deposited at 200°C and in some cases at 400°C. The counter electrode is patterned using a wet chemical etch. Junction areas ranged from 2.5 μm^2 to 20 μm^2 for a 500 nm thick film.

The current-voltage characteristic of a NbN/Si/NbN tunnel junction is illustrated in Figure 6. The counter electrode for this device was deposited at a substrate temperature of 200°C using rf magnetron sputtering at a power of 750 watts and a total sputter gas pressure of 1.3 Pa. The properties of NbN deposited using the technique of rf magnetron sputtering have been reported elsewhere for several substrate temperature and reactive gas mixtures.[7] The figure of merit V_m defined above for this junction is 20 mV measured at 2 mV and 4.2 K. The junction has an area of 10 μm^2 and a resistance of 185 ohms. No Josephson current was visible and the sum gap was 4.3 mV. From the differential resistance vs. voltage curve we calculated the base electrode gap to be 2.6 mV and the counter electrode gap to be 1.7 mV indicating a substantial interface problem between the barrier and counter electrode. The counter electrode gap should be greater than 2.2 mV for a strong coupled superconductor with a T_c greater than 13 K. We also observe minima in the dV/dI curve at the base and counter electrode gaps indicating a normal conducting layer at each interface forming a parallel SIN junction.

We are attempting to improve the leakage and gap characteristics of these junctions by changing the composition of the hydrogenated silicon and the deposition conditions of both barrier and counter electrode. Some additional junctions were fabricated using the same barrier and a counter electrode deposited at 400°C. These junctions showed a higher sum gap (4.9 mV) and increased subgap leakage. A typical V_m value for such a junction is 15 mV at 4.2 K and 2 mV.

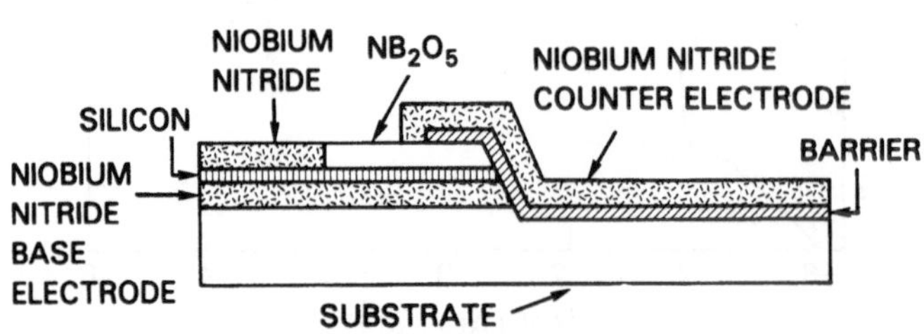

Fig. 5. The edge junction geometry used in the fabrication of NbN/Si/NbN tunnel junctions. The device uses a plasma etched anodized NbN/Si/Nb trilayer to form the junction edge. The barrier and counter electrode are deposited at elevated temperatures.

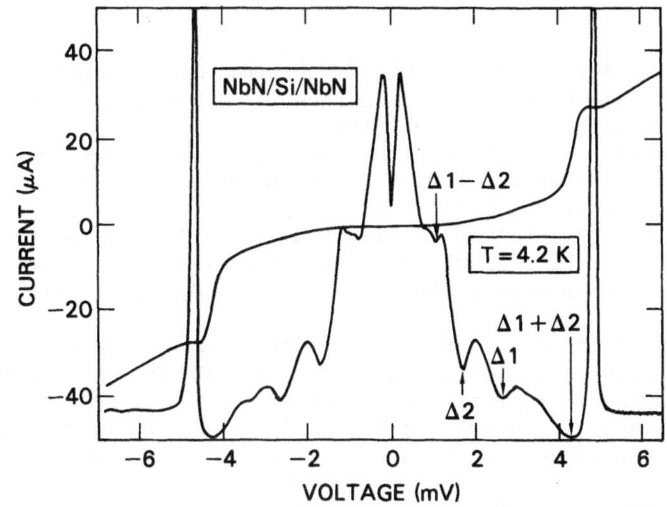

Fig. 6. The current-voltage characteristics of a NbN/Si/NbN edge tunnel junction incorporating an unoxidized amorphous hydrogenated silicon barrier. Using the differential resistance trace, we calculate the base electrode gap to be 2.6 mV and the counter electrode gap to be 1.7 mV. Each energy gap is also indicated on the trace. The junction has a resistance of 185 ohms and an area of 10 μm^2. V_m is 20 mV.

CONCLUSION

Rf magnetron reactively sputtered NbN thin films have been investigated as a function of rf substrate bias for a substrate temperature of 650°C and a fixed reactive gas partial pressure. The transition temperature for these films was approximately 16 K for bias values up to -60 V after which a rapid decrease was observed. The low temperature resistivity remained relatively constant at approximately 70 μΩ-cm for all bias levels. The major structural variation observed was the preferred crystallite orientation being (200) for low bias changing smoothly to (111) for higher bias levels. The film composition was independent of bias and found to be $NbC_{0.4}N_{0.6}$ for the gas mixture used during deposition.

The tunneling characteristics of Josephson junctions fabricated on similar films were investigated using unoxidized hydrogenated silicon barriers and niobium or niobium nitride counter electrodes. Those junctions fabricated with niobium counter electrodes were processed using the SNAP technique. The best V_m value attained was greater than 40 mV and sum gaps approached 4 mV. Junction sizes ranged from 25 to 400 μm^2. The NbN counter electrodes were deposited at 200°C and 400°C by rf magnetron sputtering. V_m values of 20 mV were attained for the lower temperature counter electrode deposition. A sum gap of 4.3 mV for 200°C and 4.9 mV for the 400°C deposition temperatures were achieved. We are presently concentrating our efforts at improving the subgap leakage by investigating the barrier composition and preparation conditions.

ACKNOWLEDGEMENTS

The authors wish to thank S. B. Qadri and Allen Lewis for their help in attaining the structure and composition data respectively and Martin Nisenoff for his many discussions. This work was partially supported by the Office of Naval Research.

REFERENCES

1. A. Shoji, M. Aoyagi, S. Kosaka, F. Shinoki and H. Hayakawa, Niobium Nitride Josephson Tunnel Junctions with Magnesium Oxide Barriers, Appl. Phys. Lett. 46, 1098 (1985).
2. E. J. Cukauskas, M. Nisenoff, H. Kroger, D. W. Jillie and L. R. Smith, All Refractory, High T_c Josephson Device Technology, Adv. Cryogenic Eng. 30, 547 (1984).
3. D. W. Jillie , H. Kroger, L. N. Smith, E. J. Cukauskas and M. Nisenoff, Niobium Nitride-Niobium Josephson Tunnel Junctions with Sputtered Amorphous Silicon Barriers, Appl. Phys. Lett. 40, 747 (1982).
4. F. Shinoki, A. Shoji, S. Kosaka, S. Takada and H. Hayakawa, Niobium Nitride Josephson Tunnel Junctions with Oxidized Amorphous Silicon Barriers, Appl. Phys. Lett. 38, 285 (1981).
5. J. C. Villegier, L. Vieux-Rochaz, M. Goniche, P. Renard and M. Vabre, NbN Tunnel Junctions, IEEE Trans. Magn. Mag-21, 498 (1985).
6. E. J. Cukauskas, The Effects of Methane in the Deposition of Superconducting Niobium Nitride Thin Films at Ambient Substrate Temperature, J. Appl. Phys. 54, 1013 (1983).
7. E. J. Cukauskas, W. L. Carter and S. B. Qadri, Superconducting and Structure Properties of Niobium Nitride Prepared by rf Magnetron Sputtering, J. Appl. Phys. 57, 2538 (1985).
8. D. D. Bacon, A. T. English, S. Nakahara, F. G. Peters, H. Schreiber, W. R. Sinclair and R. B. van Dover, Properties of NbN Thin Films Deposited on Ambient Temperature Substrates, J. Appl. Phys. 54, 6509 (1983).
9. A. Shoji, F. Shinoki, S. Kosaka, M. Aoyagi and H. Hayakawa, New Fabrication Process for Josephson Tunnel Junctions with (Niobium Nitride, Niobium) Double Layered Electrodes, Appl. Phys. Lett. 41, 1097 (1982).
10. H. Kroger, L. N. Smith and D. W. Jillie, Selective Niobium Anodization Process for Fabricating Josephson Tunnel Junctions, Appl. Phys. Lett. 39, 280 (1981).
11. H. Kroger, L. N. Smith, D. W. Jillie and J. B. Thaxter, Improved Nb-Si-Nb SNAP Devices, IEEE Trans. Magn. Mag-19, 783 (1983).
12. M. Gurvitch and J. Kwo, Tunneling and Surface Properties of Oxidized Metal Overlayers of Nb, Adv. Cryogenic Eng. 30, 509 (1984).

DEPENDENCE OF THE UPPER CRITICAL FIELD AND CRITICAL CURRENT ON RESISTIVITY IN NbN THIN FILMS

J. Y. Juang and D. A. Rudman
Department of Materials Science and Engineering
Massachusetts Institute of Technology
Cambridge, Massachusetts

R. B. van Dover, W. R. Sinclair, and D. D. Bacon
AT&T Bell Laboratories
Murray Hill, New Jersey

ABSTRACT

In this study we have measured $H_{c2}(T)$ and $J_c(4.2K)$ on a series of NbN thin films which have a fine (5 nm) equiaxed grain structure with no evidence for columnar voids. Samples with a range of critical temperatures and resistivities have been made by varying the percentage of nitrogen in argon during the deposition. The slope of the upper critical field at T_c saturates for $\rho > 250$ μΩ-cm (the calculated maximum resistivity for NbN). Despite the lack of columnar grain structure, H_{c2} is anisotropic, with $H_{c2}(\perp) > H_{c2}(//)$. At low temperatures $H_{c2}(T)$ is limited by Pauli paramagnetism ($\lambda_{so} \approx 5$) which suggests that further increases in $H_{c2}(0)$ may be possible if additional spin-orbit scattering can be induced in the material. Finally, the critical currents for these very fine-grained samples are similar to the values found by others in larger columnar-grained samples.

INTRODUCTION

The current prospects for using NbN as both a conductor in high field magnets[1,2] and as an element in superconducting electronics[3,4] has renewed the interest in this material in thin film form. These films can be made by a variety of deposition techniques, including dc and rf reactive sputtering[5,6] and ion beam deposition.[7] Independent of the deposition technique used, most NbN films display a characteristic film growth microstructure consisting of columnar grains (grain size from ten to several hundred nanometers) growing with a <111> preferred orientation. Quite often these columnar grains are separated by several nanometers, with the intergrain regions being either columnar voids or some alternate (and possibly insulating) phase.[1,8]

In general NbN films display a variety of unusual transport and superconducting properties often attributed to the specific microstructure described above. For example, high-T_c NbN has been made with film resistivities ranging from 100 to 2000 μΩ-cm. However, the maximum possible resistivity[9] calculated for bulk NbN is $\rho_{max} = 250$ μΩ-cm. In addition, these high resistivity films have residual resistivity ratios

(RRR = $\rho(300K/\rho(T_c))$ less than unity, suggesting a non-metallic or activated conduction mechanism. These transport properties are consistent with a microstructure composed of grains with resistivity ρ_{max} separated by highly resistive or insulating grain boundaries.[10]

Assuming that the grain size in NbN films is greater than the Ginzburg-Landau coherence length ($\xi_{GL}(0) \approx 3$ nm) ρ_{max} can be used to calculate a maximum initial slope of the upper critical field, $(dH_{c2}/dT)_{max}$. Using the GLAG relation for a dirty superconductor

$$dH_{c2}/dT = 4.48\gamma\rho \quad (T/K) \qquad \text{Eq. 1}$$

and assuming that the electronic coefficient of the specific heat γ in a thin film is the same as in the bulk[11,12] ($\gamma = 3.2$ erg cm^{-3}K^{-2}) we find $(dH_{c2}/dT)_{max} = 3.5$ T/K. Reported values of the slope generally fall below this limit provided the field is applied parallel to the sample, although there are exceptions.[13]

The upper critical field in NbN films is almost always anisotropic, with the perpendicular critical field $H_{c2}(\perp) > H_{c2}(//)$. The anisotropy can be as large as a factor of 1.8 near T_c. Again, this unusual behavior has been correlated with the columnar-void microstructure. A model proposed by Ashkin et al.[8] based on electrically isolated columnar grains with one grain dimension comparable to $\xi_{GL}(\approx 5$ nm) has been used to explain this behavior. This model has also been used to fit the temperature dependence of $H_{c2}(T)$ for these columnar-void samples. In general when the Werthamer, et al. (WHH) theory[14] for $H_{c2}(T)$ is applied to NbN films with a large anisotropy in H_{c2}, the theory overestimates $H_{c2}(T)$ at low temperatures (assuming no paramagnetic limiting). The Ashkin model accounts for this effect with the temperature dependence of the geometric pair-breaker used to fit the anisotropy. The temperature dependence of H_{c2} can also be fit by including the effects of paramagnetic limiting in the WHH theory, but this does not account for the anisotropy.

For this study we have used reactive magnetron sputtering to prepare a series of NbN thin films having a wide range of resistivities and critical temperatures. These samples have a fine equiaxed grain structure with no evidence for columnar voids, and yet systematically exhibit the same unusual properties described above. Therefore an explanation of these effects based on a columnar microstructure may not be applicable, and alternate origins for these features must be explored.

Table 1. Normal State and Superconducting Parameters

Sample	% N_2 in Ar/N_2	$\rho(T_c)$ ($\mu\Omega$-cm)	RRR	T_c (K)
NbN-16	10	295	0.86	11.9 ± 0.1
NbN-41A	15	195	0.889	14.1 ± 0.1
NbN-17	20	394	0.698	12.6 ± 0.15
NbN-17A	25	662	0.630	10.9 ± 0.2
NbN-18	30	558	0.625	9.9 ± 0.1
NbN-18A	35	675	0.633	8.4 ± 0.1
NbN-19	40	804	0.544	6.8 ± 0.15
NbN-19A	45	2450	0.354	5.3 ± 0.1

SAMPLE PREPARATION

The samples used in this study were made at Bell Laboratories. Details of the techniques used to make the samples have been previously reported.[5] Briefly, the NbN was made by reactive dc magnetron sputtering of a pre-conditioned Nb target in a N_2-Ar atmosphere. The single crystal sapphire substrates used were not intentionally heated and remained below 90°C during the deposition. Different samples were made by varying the percentage of N_2 in Ar while keeping all other deposition parameters fixed at their "optimum" values: total pressure during deposition = 8×10^{-3} torr, sputtering voltage = 340 V, deposition rate = 250 Å/min, and deposition time = 10 minutes. With these conditions the T_c and resistivity of the samples varied systematically with N_2 fraction, obtaining a maximum T_c of 14.1 K (and minimum resistivity of 195 μΩ-cm) at 15% N_2. The nitrogen percentage, T_c, resistivity, and residual resistivity ratio (RRR) are listed in Table 1. Note that all these samples have sharp resistive transitions. The sample with the highest T_c (Sample 41A) had the approximate composition $NbN_{.91}$ as determined by thermogravimetric analysis. Samples prepared with other N_2 fractions in the sputter gas had proportionate levels of N incorporated in the films, as determined by Auger electron spectroscopy.

The microstructure of these films has also been determined and reported.[5] TEM studies of several high-T_c samples show uniform grains approximately 5 nm in diameter. There is no evidence for either columnar grain growth or for columnar voids between grains (with resolution of 0.7 nm). From both X-ray and electron diffraction it was determined that the grains are oriented isotropically (equiaxed). This microstructure is distinctly different from the typical oriented columnar grains obtained by deposition at higher pressures (i.e. diode sputtering).

The critical fields of the samples were measured at the Francis Bitter National Magnet Laboratory. The resistive transition was measured as a function of field using a small sampling current (< 1 A/cm^2). The critical field was defined as the field at which the sample regained 50% of its normal state resistance. Typically the transition broadened with increasing field from approximately one Tesla (10% to 90%) at low field to two Tesla at the highest field. Therefore, while the selection of an alternate definition of the critical field (e.g. the 10% or 90% transition points) does shift the resulting data slightly, it does not

Table 2. Parameters for Fits to $H_{c2}(T)$ Data Using WHH Theory

Sample	Parallel dH_{c2}/dT (T/K)	Parallel $H_{c2}(0)$ (T)	Perpendicular dH_{c2}/dT (T/K)	Perpendicular $H_{c2}(0)$ (T)	λ_{so}
NbN-16	3.2	24.0	4.0	28.7	5
NbN-41A	2.5	24.4	2.8	27.3	∞
NbN-17	3.5	27.4	4.5	33.1	5
NbN-17A	4.0	25.3	5.4	29.9	3
NbN-18	3.4	23.3	3.9	26.7	∞
NbN-18A	3.8	19.1	4.5	21.6	4
NbN-19	3.5	15.2	---	---	8
NbN-19A	4.0	13.0	---	---	6

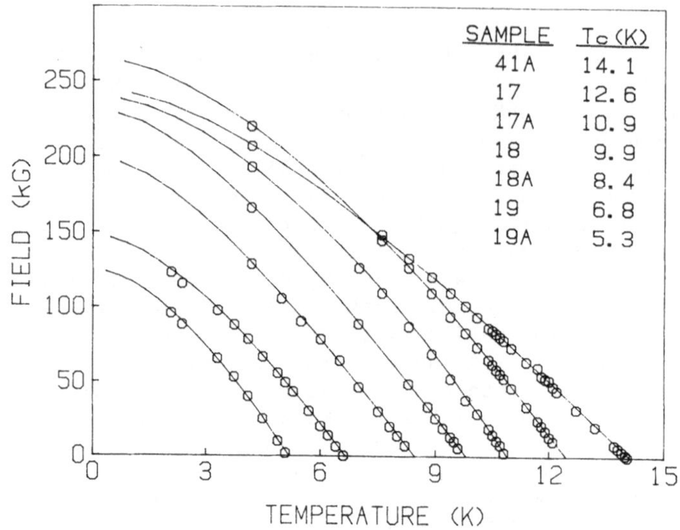

Fig. 1. Parallel upper critical field for all samples (except Sample 16). Lines are fits using the WHH theory (including paramagnetic limiting) with the parameters listed in Table 2.

significantly change the field dependence. Most samples were measured with the field applied both parallel to the film surface $H(T,//)$ and normal to the film surface, $H(T,\perp)$.

RESULTS

Figure 1 shows the temperature dependence of the parallel upper critical field $H_{c2}(T,//)$ for all the samples (except Sample 16). The data are linear near T_c for all samples, with no evidence for the positive curvature near T_c often found in inhomogeneous samples. The T_c found by extrapolating the H_{c2} data to zero field matches the resistively measured T_c to within the transition width indicated in Table 1. The lines through the data points are fits generated using the WHH theory including paramagnetic effects, and using both dH_{c2}/dT and the spin-orbit scattering constant (λ_{so}) as adjustable parameters. The values used to produce these fits are given in Table 2 along with their extrapolations to zero temperature, $H_{c2}(0)$. The absence of positive curvature in the data near T_c allows an accurate and unambiguous determination of dH_{c2}/dT and T_c. These parameters together with the low temperature (high field) data are sufficient in most samples to determine λ_{so}. The need to include paramagnetic limiting in the fitting procedure is discussed below.

As is easily seen in Figure 1, the slopes of the upper critical fields are nearly equal for all but the highest T_c sample. These slopes are shown as the circles in Figure 2, plotted as a function of measured sample resistivity. For comparison the result of Eq. 1 is shown as the dashed line in Figure 2 using a constant value for the electronic specific heat ($\gamma = 3.2 \times 10^3$ erg $cm^{-3}K^{-2}$). For low resistivity samples the data is in reasonable agreement with the theory, but for $\rho > \rho_{max}$(= 250 μΩ-cm) dH_{c2}/dT (parallel) saturates at approximately the value calculated for the corresponding maximum dH_{c2}/dT, 35 kG/K. While it is possible that γ could decrease with increasing resistivity such that dH_{c2}/dT remains constant, it seems more likely that the resistivity above ρ_{max} is due to increased grain boundary resistance and thus does not contribute to dH_{c2}/dT. It should be noted that the grain size of

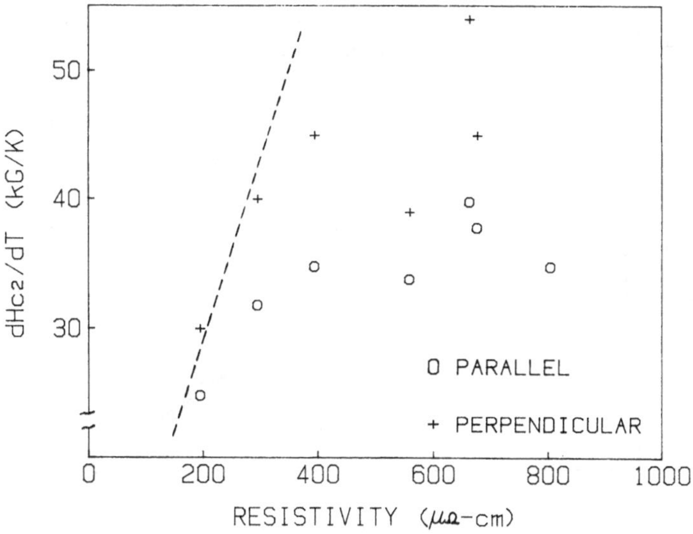

Fig. 2. Parallel (0) and perpendicular (+) upper critical field slopes as a function of sample resistivity. Dashed line is from Eq. 1 assuming a constant $\gamma = 3.2 \times 10^3$ erg cm^{-3} K^{-2}.

these films is approaching the zero temperature Ginzburg-Landau coherence length ($\xi_{GL}(0) \gtrsim 3$ nm). Thus very near T_c the grain boundary resistance may become important.

The anisotropy of $H_{c2}(T)$ has been measured and is shown for two representative samples in Figures 3 and 4. For the lowest resistivity sample (Sample 41A, Fig. 3) there is only a slight anisotropy, although $H_{c2}(\perp)$ is still greater than $H_{c2}(//)$. In contrast, Figure 4 shows the larger anisotropy typical of the high resistivity samples. Values for all samples are given in Table 2. In general the anisotropy appears to increase with increasing resistivity. This is seen in Figure 2 where dH_{c2}/dT (perpendicular) is plotted as the crosses. Despite the apparent lack of a distinct columnar-void microstructure in these samples, the anisotropic nature of H_{c2} in NbN has remained essentially unchanged. It is therefore possible that microstructure alone is insufficient to explain the anisotropy in these NbN films. More work is being done to try and clarify this issue, including measurements of the angular dependence of $H_{c2}(T)$.

In Figure 3 it is clear that the available data in both the parallel and perpendicular directions can be adequately fit by the "standard" WHH theory assuming no paramagnetic limiting ($\lambda_{so} = \infty$). This is not the case for most of the samples, represented by Figure 4, where the $\lambda_{so} = \infty$ curves (dashed lines) seriously overestimate $H_{c2}(T)$ below T_c. It proved necessary to include some degree of paramagnetic limiting to fit the data for these samples, as shown in Figure 4 by the solid lines and indicated in Table 2 by the values of λ_{so}.

These relatively large values of λ_{so} ($\lambda_{so} \sim 5$) still have a pronounced effect on $H_{c2}(T)$ due to the very large slopes of the critical field near T_c. (A large initial slope makes the orbital critical field H^*_{c2} larger than the Pauli field H_p ($\lambda_{so} = 0$).) For example, Sample 17A (Fig. 4) can only be fit by values of λ_{so} between 2 and 4. Conversely, the low initial slope found for Sample 41A makes the fitting procedure much less sensitive to the value of λ_{so} (for $\lambda_{so} > 5$), and thus $H_{c2}(T)$ does not provide a measure of λ_{so} for this sample.

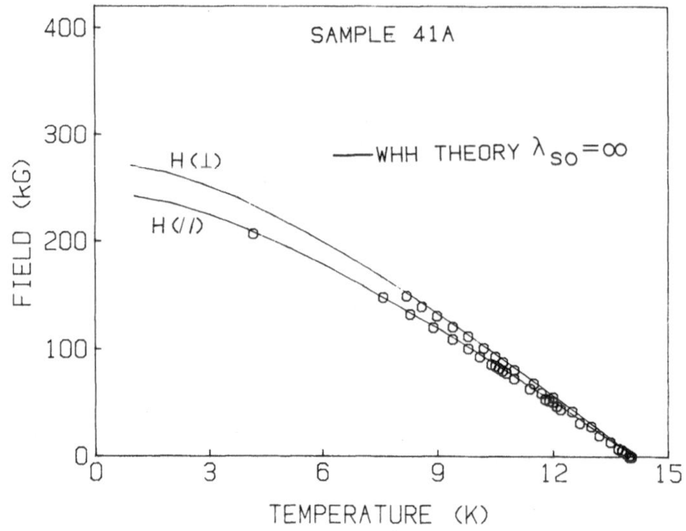

Fig. 3. Parallel and perpendicular $H_{c2}(T)$ for the lowest resistivity sample. Solid lines are theoretical fits using the parameters in Table 2.

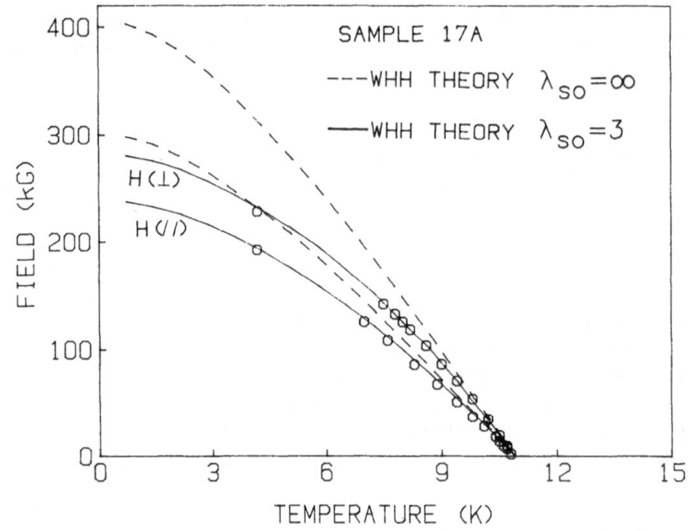

Fig. 4. Parallel and perpendicular $H_{c2}(T)$ for a representative high resistivity sample. Dashed lines are from WHH theory with no paramagnetic limiting. Solid lines are fits using parameters from Table 2.

Due to the lack of a columnar-void microstructure in these films, no attempt was made to use Ashkin's model to fit the data. However, the fact that both the parallel and perpendicular critical field data could be fit with the same value of λ_{so} suggests that these samples are paramagnetically limited. The values of λ_{so} used correspond to a spin-orbit scattering length approximately 20 times the mean free path, a value which is not physically unreasonable. If in fact NbN is paramagnetically limited, then it may be possible to increase H_{c2} for the highest critical field material, as indicated by the dashed lines in Figure 4.

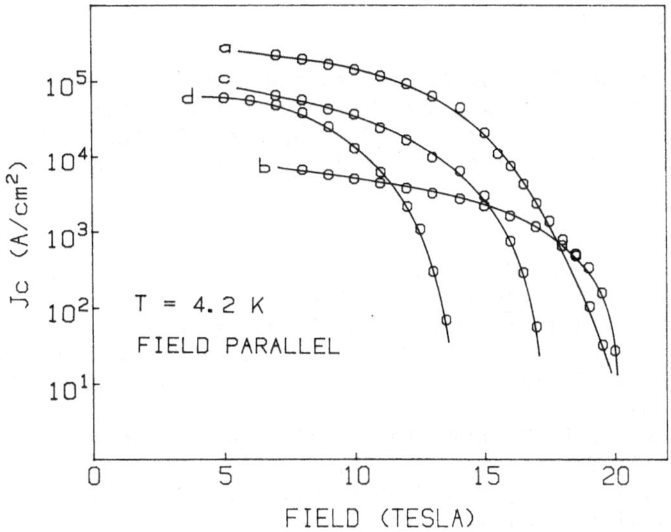

Fig. 5. Critical current measured at 4.2 K with the field parallel to the sample. Lines are only to guide the eye. The samples are: a = # 41A b = # 17, c = # 17A, and d = #. 18.

CRITICAL CURRENTS

The critical current density J_c as a function of (parallel) applied field has been measured on four selected films at liquid helium temperatures. A criterion of 0.5 μV/cm was used to determine the critical current value. The results are shown in Figure 5. The three high-J_c samples (a,c, and d) all carried $J_c > 10^4$ A/cm^2 for fields less than $0.8*H_{c2}(4.2)$ and all saturated to current densities between 10^5 and 10^6 A/cm^2 at low field. These values are similar to those found by other workers[1,16] on larger grained columnar samples. Since reducing the grain size would normally be expected to increase J_c (for grains larger than ξ_{GL}), a change to a columnar morphology for these type of samples may yield more effective pinning.

It should be noted that one of the samples in Figure 5 (curve b, Sample 17) has an order of magnitude smaller J_c than the other samples, and a distinctly different field dependence. The origins of this difference are not known at this time, but may be important since this sample has the largest $H_{c2}(4.2)$.

SUMMARY

We have measured the upper critical fields of a series of fine equiaxed-grain thin films of NbN. Several unusual properties common to other NbN thin films remain in these samples. The saturation of dH_{c2}/dT for resistivities above ρ_{max} supports the isolated-grain film model for highly resistive NbN films. These films exhibit a large anisotropy in H_{c2} despite the lack of a columnar-void microstructure. Finally it is necessary to include paramagnetic limiting in the WHH theory to fit $H_{c2}(T)$ at low temperatures. This suggests that it may be possible to increase H_{c2} in this material provided additional spin-orbit scattering can be introduced.

ACKNOWLEDGEMENTS

We would like to thank J. A. X. Alexander and T. Orlando for invaluable discussions about H_{c2}, and R. Jacowitz for assistance in data reduction. One of us (DAR) is partially supported by the National Science Foundation through the Francis Bitter National Magnet Lab.

REFERENCES

1. R. T. Kampwirth, D. W. Capone II, K. E. Gray, and A Vincens, IEEE Trans. Magn. MAG-21, 459 (1985).
2. M. Dietrich, IEEE Trans. Magn. MAG-21, 455 (1985).
3. S. Kasaka, A. Shoji, M. Aoyagi, F. Shinoki, S. Tahara, H. Ohigashi, H. Nakagawa, S. Takada, H. Hayakawa, IEEE Trans. Magn. MAG-21, 102 (1985).
4. J. C. Villegier, L. Vieux-Rochaz, M. Goniche, P. Renard and M. Vabre, IEEE Trans. Magn. MAG-21, 498 (1985).
5. D. D. Bacon, A. T. English, S. Nakahara, F. G. Peters, H. Schreiber, W. R. Sinclair, and R. B. van Dover, J. Appl. Phys. 54, 6509 (1983).
6. E. J. Cukauskas, M. Nisenoff, H. Kroger, D. W. Jillie, and L. N. Smith, in: "Advances in Cryogenic Engineering", vol. 30, 547 (1984).
7. L. J. Lin, E. K. Track, G. J. Cui, and D. E. Prober, to be published in the Proceedings of the International Conf. on the Materials and Mechanisms of Superconductivity, (1985). Physica B, North-Holland Publishing Co.
8. M. Ashkin, J. R. Gavaler, J. Greggi, and M. Decroux, J. Appl. Phys. 55, 1044 (1984).
9. R. R. Hake, Appl Phys. Lett. 10, 189 (1967).
10. M. Ashkin and J. R. Gavaler, J. Appl. Phys. 49, 2449 (1978).
11. M. P. Mathur, D. W. Deis, and J. R. Gavaler, J. Appl. Phys. 43 3158 (1972).
12. A smaller value of γ for NbN has recently been reported: $\gamma = 1.7 \times 10^3$ erg cm^{-3} K^{-2}. This will produce a correspondingly smaller value of $(dH_{c2}/dT)_{max} = 1.9$ T/K.
    G. Geibel, H. Reitschel, A. Junod, M. Pelizzone and J. Muller, J. Phys. F.: Met. Phys. 15, 405 (1985).
13. K. E. Gray, R. T. Kampwirth, D. W. Capone II, and R. Vaglio, to be published in the Proceedings of the International Conf. on Materials and Mechanisms of Superconductivity (1985), Physica B, North-Holland Publishing Co.
14. N. R. Werthamer, E. Helfand, and P. C. Hohenberg, Phys. Rev. 147 295 (1966).
15. H. Jones, O Fisher and G. Bongi, Solid State Comm. 14, 1061 (1974).
16. J. R. Gavaler, A. T. Santhanam, A. I. Braginski, M. Ashkin, and M. A. Janocko, IEEE Trans. Magn. MAG-17, 573 (1981).

HIGH FIELD PROPERTIES OF NbN RIBBON CONDUCTORS

D. W. Capone II, R. T. Kampwirth, K. E. Gray

Materials Science and Technology Division

Argonne National Laboratory,* Argonne, Illinois

ABSTRACT

In this paper we report the first high field measurements on ribbon conductors composed of NbN and Cu deposited onto Hastelloy ribbons. These ribbons are 5 to 8 cm long, 1/8" wide, and have several microns of NbN, covered with several microns of Cu, deposited onto one side. Such samples have a $T_c \sim 14.5$ K, H_{c2} (2.0 K) ~ 26 T in the parallel direction and J_c (20 T, 2.0 K) up to 3×10^4 A/cm^2.

A recent mirror fusion reactor study outlined the need for a superconducting solenoid capable of operating in the 20-24 T range. Presently, there are no commercially available superconductors with properties suitable for magnets in this range. In a program aimed at developing such a conductor, we have produced NbN films with high field properties which could satisfy this requirement. In this paper, we report the first high-field measurements on short lengths of practical ribbon conductors based on NbN. These reproducible high-field properties coupled with the demonstrated strain and radiation tolerance of NbN make it an ideal candidate for reducing the size and cost of the next generation of fusion reactor magnets.

Recently, the high field properties of thin film NbN deposited onto sapphire and Hastelloy were reported.[1] These films, deposited by d.c. reactive magnetron sputtering, were used to optimize the high field properties of NbN films. The details of this process will be reported elsewhere. In order to study the more practical aspects of our films, we have begun fabricating 1/8" wide conductors consisting of several microns of NbN deposited onto 0.002" thick Hastelloy ribbons overcoated with several microns of copper using a thermal evaporator. These ribbons are 5-7 cm long and, at present, have NbN and Cu deposited onto one side only.

Using the high-field facilities available at the Francis Bitter National Magnet Laboratory** the critical current density (J_c) has been measured as a function of applied field, for both the parallel and perpendicular directions, at two temperatures (4.2 and 2.0 K). In Figure 1, we show J_c vs. H for one of these ribbons for both the parallel and perpendicular directions. The 1 μV/cm criterion, is used to define J_c.

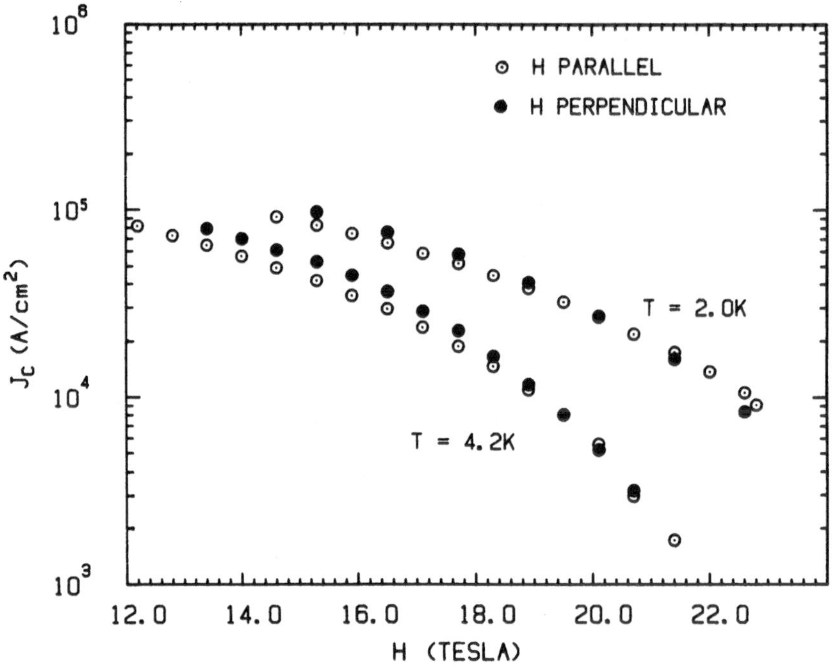

Fig. 1. Critical current density (J_c) vs. applied magnetic field (H) for one NbN ribbon conductor. The data are at T = 4.2 and 2.0 K, with the field orientation both parallel and perpendicular to the film surface.

This corresponds to an effective resistivity of about 10^{-10} Ω-cm.

Several features, common to all of our NbN samples, with important consequences for the feasibility of producing a practical conductor, are: 1) The ribbons have J_c well in excess of 1×10^4 A/cm^2 at 20 T and 2.0 K (the highest value is 3.0×10^4 A/cm^2); 2) we are able to deposit substantial thicknesses of NbN onto the Hastelloy ribbons without experiencing bonding problems (~ 17 to 20 μm); 3) we are able to deposit up to 5 μm of Cu onto these ribbons; 4) the J_c of the ribbons in the perpendicular direction is always higher than in the parallel direction; and 5) the J_c of the ribbons is the same as that measured for patterned NbN on sapphire with a Cu coating, implying that the NbN is not degraded at the edges of the ribbon.

The measured properties of our ribbons can be used to estimate the properties of a more practical conductor design. For example, a 1 cm wide ribbon with 15 μm of NbN and 5 μm of Cu deposited onto both sides, having a J_c for the NbN alone of 3.0×10^4 A/cm^2 at 20 T and 2.0 K (our best ribbon) has an overall J_c, including Cu and Hastelloy, in excess of 1×10^4 A/cm^2 at 20 T and 2.0 K. This value is well above the value (3.5×10^3 A/cm^2) set by Hoard et al.[2] in designing a 20 T magnet operating at 1.8 K. A conductor of this configuration would carry 100 A at 20 T. A cable of 10 to 20 ribbons, possibly with internal cooling, has the potential to satisfy the current requirements of large fusion magnet designs. Also, the higher J_c in the perpendicular direction should reduce the stability problems associated with the perpendicular field component at the ends of tape-wound solenoids.

The substantial increase in J_c obtained by operating at 2.0 K appears to be almost entirely due to an increase in H_{c2}, thus indicating the potential additional gains which could be achieved by increasing H_{c2} by means other than temperature. For example, the addition of appropriate third elements, such as C or Ti. It has been shown that such additions increase T_c of NbN, but, very little work has been done on their effect on H_{c2}. One of our immediate goals is to investigate the high field properties of Nb(Ti)N and Nb(C)N compounds formed by reactive sputtering from Nb(Ti) and Nb(C) targets. Another goal is to fabricate continuous lengths of NbN ribbon conductor (~ 20-70 cm) for more extensive testing of their high-field properties and suitability as practical superconducting materials. A new system designed to simultaneously deposit NbN onto both sides of moving ribbon substrates is being constructed and the results on materials produced in this system will be reported in the future.

To summarize, we have succeeded in producing short lengths of ribbon conductor based on NbN, coated with Cu, which have excellent high field superconducting properties (J_c = 1.5-3.0 × 10^4 A/cm^2 at 20 T and 2.0 K) and have potential as conductors for the next generation of superconducting magnets for fusion reactors. Future work will involve attempts to increase H_{c2} of these materials together with the fabrication of longer lengths for more relevant testing.

*The work at Argonne National Laboratory is supported by the U.S. Department of Energy Division of Basic Energy Sciences. **The Francis Bitter National Magnet Laboratory is supported by the National Science Foundation.

REFERENCES

1. R. T. Kampwirth, D. W. Capone II, K. E. Gray, A. Vicens, IEEE Trans. Magn. MAG-21:459 (1985).

2. R. W. Hoard, D. N. Cornish, R. M. Scanlan, J. P. Zbasnik, R. L. Leber, R. M. Hickman, J. D. Lee, Adv. Cryog. Eng. 29:57 (1984).

SYNTHESIS OF B1 MoN FILMS BY ION BEAM DEPOSITION

N. Terada*, M. Naoe** and Y. Hoshi***

* Electrotechnical Laboratory, Sakura-mura, Niihari-gun, Ibaraki 305, Japan
** Tokyo Institute of Technology, Oh-okayama, Meguro-ku, Tokyo 152, Japan
***Tokyo Institute of Polytechnics, Iiyama, Atsugi-shi, Kanagawa 243-02, Japan

ABSTRACT

Molybdenum nitride thin films have been synthesized by using a ion beam deposition apparatus which is able to control kinetic energy and ionization coefficient of depositing particles. And dependence of their crystal structure and superconducting properties on preparation conditions has been investigated.

Crystal structure of the films depends mainly on substrate temperature Ts and Nitrogen partial pressure P_{N2}. On the condition of high Ts and low P_{N2}, films of single B1 structure are obtained. On the other hand, lattice constant a(200) of B1 phase shows a significant change with the ionization ratio of depositing particles R_i and a little one with their kinetic energy E_i. By adjusting R_i above 90 % and E_i about 15 eV, films of single B1 phase with large a(200) about 4.25 Å, which is the predicted value for ideal B1 type MoN, are obtained even on the condition of very low P_{N2}. Superconducting critical temperature Tc shows a good correlation with the lattice constant. The films composed of single B1 crystallites with lattice constant in the range from 4.24 to 4.25 Å show high Tc up to 12.8 K, which is one of the highest values for the films of single B1 structure.

INTRODUCTION

It is theoretically predicted that molybdenum nitride of B1 type of crystal structure should have high superconducting critical temperature Tc of about 29 K[1].In order to realize such high Tc, it is necessary to synthesize the crystal without defect. However, B1-MoN has large structural instability and does not exist in thermal equibrium diagram. High dissociation pressure of nitrogen and high existance ratio of anti-bonding orbitals in this structure prevent from synthesizing well ordered crystal; theyincrease the number of nitrogen vacancies and that of nitrogen atoms which exist at interstitial site. Therefore, some kinds of non-equibrium deposition techniques with the effects which activate the reaction between molybdenum and nitrogen should be used for suppressing such unstability and obtaining high Tc specimen. Plasma deposition techniques such as reactive sputtering and plasma CVD are suitable for the purpose, since there are non-equibrium reaction processes in them. But, in the conventional methods, it is not easy to clarify growth mechanism and relationship between film properties and preparation conditions, since there are numbers of unknown and uncontrollable reaction processes. On the other hand, ion beam deposition is one of the most suitable techniques, because it has controllability not only for conventional deposition parameters but also for kinetic energy of depositing particles just before the incidence onto the substrate and the degree of their excitation. Besides, according to the study of synthesis of B1 NbN films by means of this technique, both excitation of depositing particles by ionization and acceleration of them are very effective for activating the reaction between nitrogen and transition metal, and they are useful to obtain high Tc films with good reproducibility[2].

In this study, the authors have attempted to synthesize MoN thin films of B1 type of structure by using an ion beam deposition apparatus and to investigate the dependence of their crystal structure and superconducting properties on preparation conditions, especially on the degree of excitation of depositing particles.

EXPERIMANTAL PROCEDURE

Figure 1 shows a schematic diagram of the ion beam deposition apparatus used in this study. The sputtering type of ion source is composed of two targets, magnetic coil, anode and screen grid which are arranged in the same way as that in Targets Facing type of high rate sputtering[3] system. Molybdenum disks 99.95 % in purity were used as the target. 19 holes were made through the lower target and screen grid for extracting nitrogen and molybdenum ions to deposition chamber. Total opened area of them was 1 cm^2. Kinetic energy of depositing ions just after the extraction E_i was controlled by adjusting applied anode voltage Vp; E_i corresponds to (Vp+5) eV. Since a part of ions extracted from the ion source are neutralized by the collision between atmosphere gas species and them on the way to the substrate, ionization ratio of depositing particles just before their incidence to the substrate can be controlled by varying grid-substrate spacing and/or atmosphere gas pressure P_{total}; in this study, the spacing was fixed at 2 cm, then, the ratio was about 90 % and 50 % on the condition of P_{total} below 2 mTorr and 2.5 mTorr, respectively. Specimen films about 2000 Å thick were deposited on Si(100) substrate by using argon-nitrogen mixture as sputtering gas. The ranges of film preparation conditions are listed in Table I, gas pressures were measured in deposition chamber and substrate temperature Ts was measured by using thermocouple at substrate surface Crystal structure of the films were characterized by means of electron and x-ray diffractometry. Superconducting critical temperature was determined by means of four probe method.

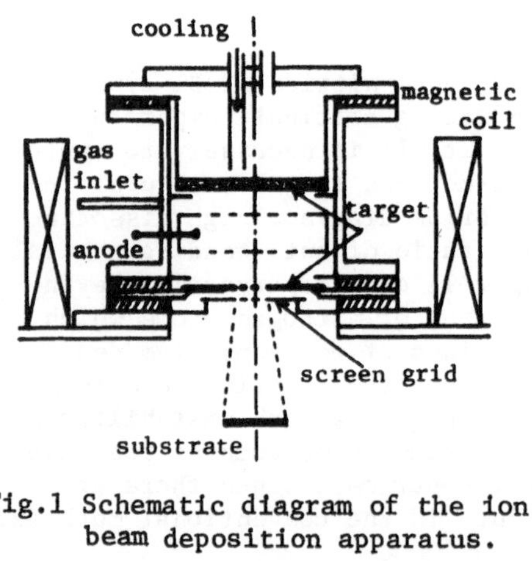

Fig.1 Schematic diagram of the ion beam deposition apparatus.

Table I Preparation conditions

Depo. rate	20	Å/min
P_{N2}	$1x10^{-4}$~$1x10^{-1}$	Torr
P_{Ar}/P_{N2}	0 ~2	
P_{total}	$1x10^{-4}$~$2x10^{-1}$	Torr
Vp	0 ~20	V
Ts	300 ~600	°C

RESULTS AND DISCUSSION

All films show good surface smoothness and structural uniformity in macroscopic scale. On the other hand, their crystal structure such as kinds of crystal phase and lattice constant changes with preparation conditions. Figure 2(a) shows the relationship between crystal structure and P_{N2}, Ts. All films in this figure were deposited on the fixed condition of Vp and P_{Ar}/P_{N2} of 10 V and 0.5, respectively. On the condition of high Ts and low P_{N2}, films composed of single B1 structure are obtained. On the other hand, on the opposite condition, films of multi-phase of B1 and hexagonal MoN are obtained. The phase boundary between the single phase region and the multi phase one shifts towards higher P_{N2} region as Ts increases. Figure 2(b) shows typical x-ray diffraction diagrams. The diffraction line from (200) plane of B1 phase in the single phase films deposited at Ts above 450°C is rather sharp. These results indicate that reaction between molybdenum and nitrogen on substrate surface is dominant in growth process of the film; dissociation of nitrogen is promoted by an increase of Ts and, contrary to this, growth of B1 phase is accelerated by it. As mentioned after, Tc of the films shows a good correlation to lattice constant of B1 type crystallites as well as their orientation. Therefore, details of preparation condition dependence of them are represented in the following.

Figure 3(a) shows the dependence of lattice constant a(200) of B1 phase on P_{N2} and kinetic energy of depositing ions E_i. All films in this figure were deposited at Ts of 500°C and showed single B1 phase In the low range of P_{N2}, a(200) increases steeply with a rise of P_{N2}. On the other

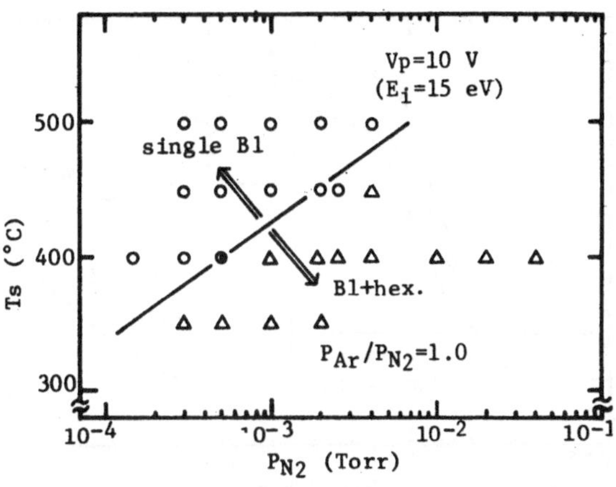

Fig.2(a) Dependence of crystal structure of the films on P_{N2} and Ts.

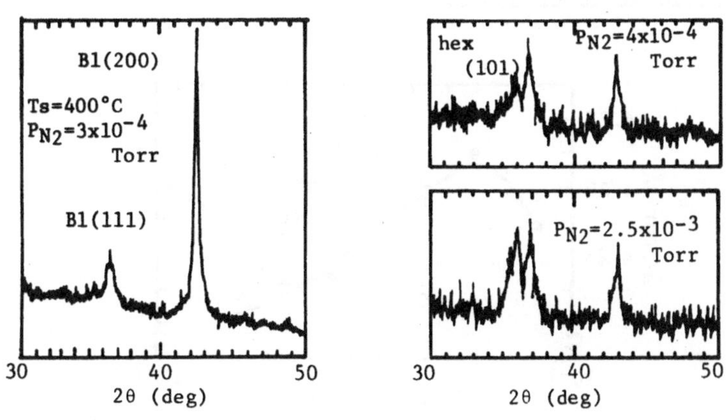

Fig.2(b) Typical x-ray diffraction diagram of the films.

hand, a(200) decreases temporarily in the middle range of P_{N2} and it increases gradually with a further in crease of P_{N2}. Since lattice constant and nitrogen content of B1 structure correlate, this result indicates the P_{N2} dependence of nitrogen content in B1 crystallites. The increase of a(200) in low P_{N2} region results from that of incidence rate of nitrogen ions to substrate. On the other hand, the decrease of a(200) in the middle range cannot be explained by it. In this range, changable deposition parameter other than incidence rate of nitrogen is ony the ratio of ionization of depositing particles, which decreases steeply as P_{N2} increases in the range from $2x10^{-3}$ to $1x10^{-2}$ Torr. Taking it into consideration, the temporary decrease of a(200) results from the decrease of the ratio. In higher range of P_{N2}, the ratio decreases more and the P_{N2} dependence of a(200) becomes similar to that of the films deposited by means of conventional reactive sputtering[4~7]. In the range of P_{N2} from $1.5x10^{-3}$ to $4x10^{-3}$ Torr, the films with large lattice constant above 4.23 Å which is close to the expected value of stoichiometric B1 MoN are obtained. It shoud be noted that this P_{N2} range is more than two order of magnitude smaller than that for such films deposited by means of conventional reactive sputtering. And curves of a(200) vs. P_{N2} shifts with the change of kinetic energy of depositing ions E_i. On the condition of E_i of 15 eV, the films with a(200) of 4.25 Å are obtained at the lowest value of P_{N2}. Then, these results indicate that the ionization of depositing particles and proper acceleration of them are very effective for activating surface reaction between molybdenum and nitrogen, and they suit for synthesizing B1 type of structure. Figure 3(b) shows P_{N2} dependence of a(200) of the films deposited at Ts of 400°C

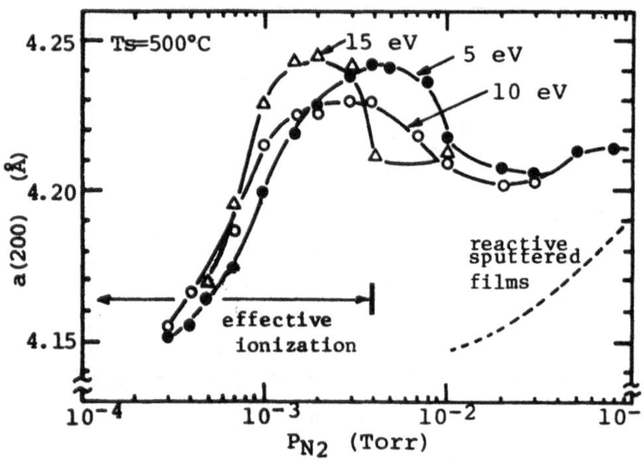

Fig.3(a) P_{N2} dependence of lattice constant of B1 phase a(200) of the films deposited at Ts of 500°C.

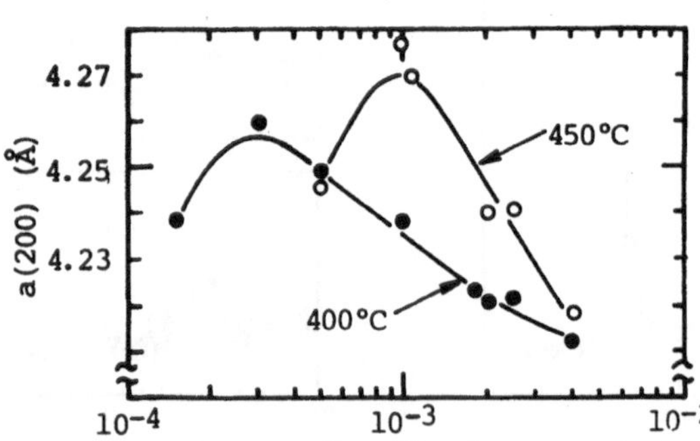

Fig.3(b) P_{N2} dependence of a(200) of the films deposited at 400 and 450°C.

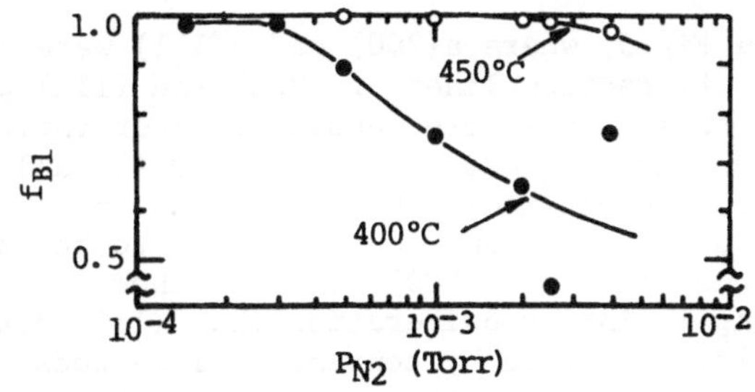

Fig.3(c) P_{N2} dependence of volume fraction of B1 phase f_{B1}.

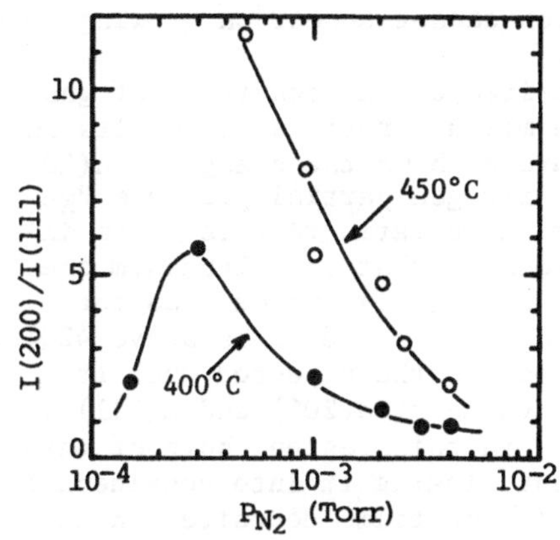

Fig.4 P_{N2} dependence of I(200)/I(111).

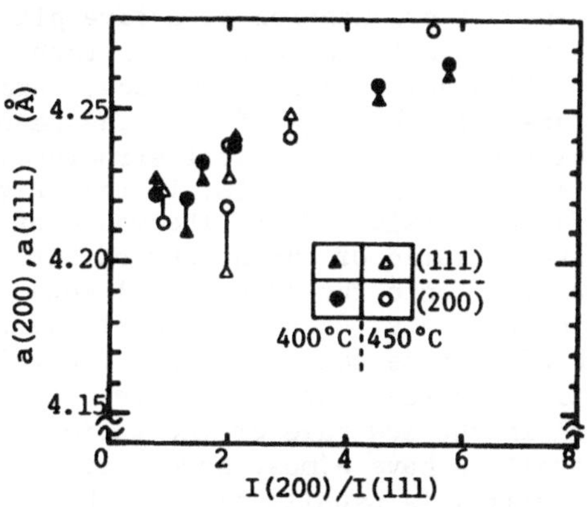

Fig.5 Relationship between a(200) and I(200)/I(111).

and 450°C. a(200) shows maxima on the P_{N2} condition of $3x10^{-4}$ and $1x10^{-3}$ Torr for the films deposited at Ts of 400°C and 450°C, respectively. And it decreases with a further increase of P_{N2}. This result from suppression of reduction of nitrogen by a decrease of Ts. The maximum of a(200) is larger than the expected value for B1 MoN. In the P_{N2} region where a(200) decreases with an increase of P_{N2}, films show multi-phase of B1 and hexagonal MoN phase and the volume fraction of the latter increases with a rise of P_{N2}. Figure 3(c) shows the P_{N2} dependence of volume fraction of B1 phase of the films plotted in Fig.3(b). The fraction f_{B1} was obtained as follows:

$$f_{B1}=I(B1)/(I(B1)+I(hex))$$

where I(B1) and I(hex) are integrated intensities of x-ray diffraction line fron B1 and hexagonal crystallites, respectively. On both conditions of Ts, f_{B1} begins to decrease at the value of P_{N2} where a(200) takes a maximum and it decreases monotonically with an increase of P_{N2}. Taking it to account that deposition parameters, such as ionization ratio of depositing particles, deposition rate and E_i don't change such low pressure region, these results indicate that it is possible to introduce nitrogen into B1 phase up to some values over stoichiometry and that the further introduction of nitrogen induces precipitation of hexagonal phase and decrease of nitrogen content in B1 phase crystallites. All films in Fig.3 were deposited on the condition of E_i around 15 eV, which is typical value of kinetic energy of depositing particles in conventional high energy deposition method such as sputtering. In sputtered films, (111) plane of B1 phase shows proper orientation[4]. Contrary to this, B1 sigle phase films have the proper orientation of (200) plane. Figure 4 shows P_{N2} dependence of I(200)/I(111). The ratio increases steeply with an increase of P_{N2} in the range where single B1 phase films are obtained. And the ratio of multi-phase films decreases steeply as P_{N2} rises. This result indicates that I(200)/I(111) and nitrogen content of B1 crystallites in the films deposited in this study correlate. For making it more clear, the relationship between I(200)/I(111) and a(200) and a(111) is shown in Fig.5, where a(200) and a(111) were calculated from the position of x-ray diffraction lines of (200) and (111) plane of B1 phase. Thoughfilms in this figure were deposited under various condition of P_{N2} and Ts, (ionization ratio R_i was fixed about 90 %), all plots look like fit to one line and lattice constant expands with an increase of I(200)/ I(111). From these results, nitrogen atoms are easy to be introduced into B1 structure on the condition that (200) plane of them is exposed as the growing surface. Taking it into consideration that (111) plane shows proper orientation in the films deposited by conventional methods, surface free energy of (200) plane may be higher than that of (111) plane under normal condition. Therefore, the authors infer that ionization of depositing particles would cause not only activation of reaction between elements but also proper orientation of active plane which cannot appear on growing surface in conventional deposition techniques.

All specimen described above were deposited on the condition of gas mixture ratio P_{Ar}/P_{N2} of 0.5. Change of the mixture ratio also results in modulation of their crystal structure. Figure 6 shows the change of a(200) and a(111) with the ratio. In this figure, nitrogen partial pressure P_{N2} as fixed at $5x10^{-4}$ Torr. (Decrease of the mixture ratio results in an increase of ratio of the number of nitrogen ions to that of molybdenum ones depositing onto the substrate.) Both a(200) and a(111) increase as the ratio decreases. On the range of the ratio below 0.5 and of Ts above 420°C, both of lattice constants take the value close to the expected one. It should be noted that difference between the value of a(200) and a(111) is small in the low rage of the ratio. That is, in this region, both of these crystallits have almost same nitrogen content. Taking it into consideration that a(111) is usually smaller than a(200) in the films deposited on the ratio of 0.5, setting of the gas mixture ratio at low region results in an increase of nitrogen content in B1 crystallites with (111) proper orientation and is useful for obtaining the single B1 phase films with uniformity in composition.

From the results in Fig.6, the (111) oriented crystallites with large lattice constant may be less stable than the (200) oriented one. Figure 7 shows the relationship between a(111) and mean crystallite size of (111) oriented crystallites $\langle D \rangle_{(111)}$. In the region of $\langle D \rangle_{(111)}$ down to 150 Å, a(111) takes the value around 4.18 Å, which is rather close to the lattice constant of fcc Mo_2N equibrium phase. And it increases steeply with a further decrease of $\langle D \rangle_{(111)}$. a(111) larger than 4.25 Å are obtained only in the range of $\langle D \rangle_{(111)}$ below 100 Å. On the other hand a(200) does not show such relation to the size of (200) oriented crystallites; by adjusting Ts and the gas miture ratio, large a(200) around 4.25 Å is obtained in the wide range of $\langle D \rangle_{(200)}$ up to 300 Å. These results also means the less stability of (111) oriented crystallites with large lattice constant;they cannot grow to large size, if nitrogen content in them were near or over stoichiometry of B1 type pf structure. Therefore, it seems to be effective

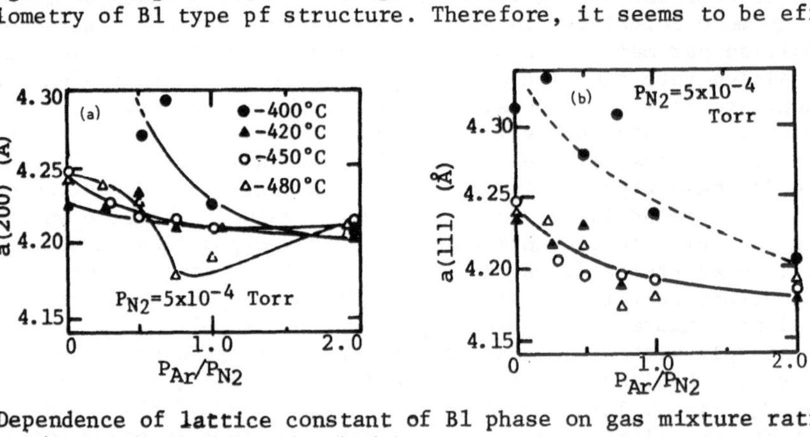

Fig.6 Dependence of lattice constant of B1 phase on gas mixture ratio P_{Ar}/P_{N2}; a(200)-(a), a(111)-(b).

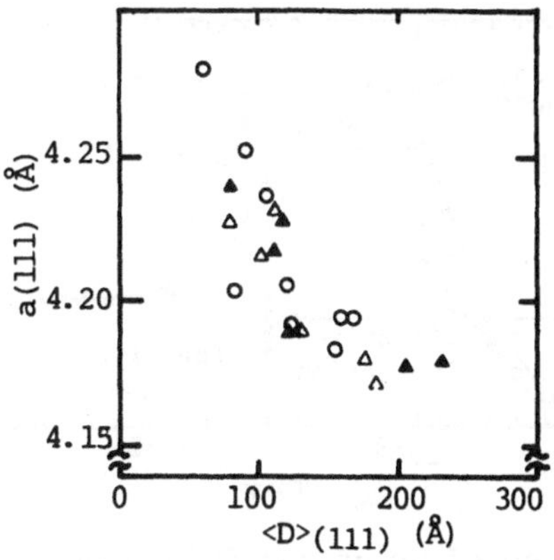

Fig.7 Relationship between a(111) and size of (111) crystallite <D>(111).

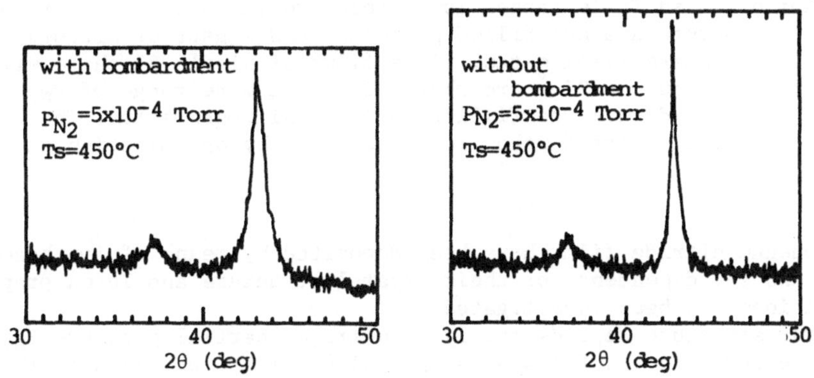

Fig.8 Dependence of diffraction diagram on high energy particle bombardment.

for synthesis of B1-MoN with better crystalinity to deposit the films with (200) proper orientation. The fact that large lattice constant can be obtained even in the large crystallites with (200) orientation means that this orientation should be effective for synthesizing high Tc films.

All specimen described before were deposited without bombardment of high energy particles. In order to investigate the effect of bombardment of such particles (E>100 eV), films were deposited with incidence of high energy nitrogen ions. Figure 8 shows the typical x-ray diffraction diagrams of the films deposited with and without the bombardment. By addition of the bombardment, peak position of (200) and (111) diffraction lines shifts toward low angle region and width of them expands. The bombardment induces a decrease of lattice constant snd suppression of growth of B1 type crystallite. This effect becomes serious in the films which should have large lattice constant if they were deposited without bombardment. These results indicate that sufficient suppression of the bombardment is necessary for obtaining well ordered B1 type MoN.

In the films composed of single B1 phase crystallites. superconducting critical temperature Tc shows good correlation to the lattice constant a(200). Figure 9 shows the P_{N2} dependence of Tc of the films deposited on the condition of Ts of 500°C and P_{Ar}/P_{N2} of 0.5. The dependence of Tc is is similar to P_{N2} dependence of a(200) as shown in Fig.3(a). Tc takes maximam in the range of P_{N2} from 1.5 to 3.0x10^{-3} Torr. The highest value of Tc is 12.8 K which is one of the highest for the films of single B1 phase. In the range of a(200) up to 4.25 Å, Tc increases with an increase of a(200). On the other hand, Tc decreases steeply and electrical resistivity in normal conducting state begins to take an apparent maximum in low temperature region with a further increase of a(200). Taking it into

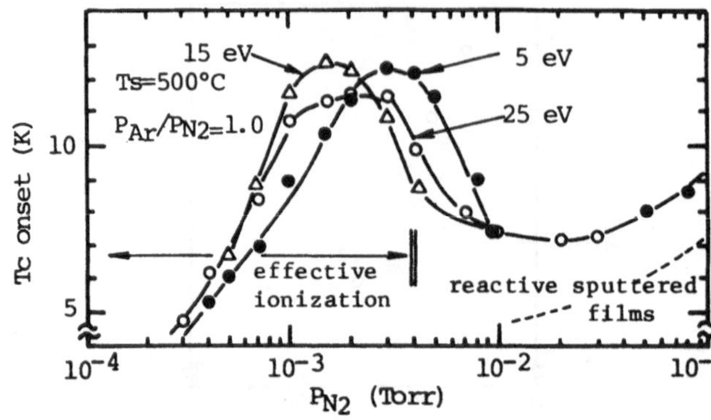

Fig.9 P_{N2} dependence of Tc of the films deposited at 500°C.

account that obtained Tc is rather lower than predicted value, nitrogen sites of B1 structure are not filled perfectly and a part of nitrogen is introduced into interstitial sites. Since films of single B1 phase with lattice constant around 4.25 Å are obtained in the wide range of P_{N2} and Ts, change of them should vary the filling ratio of nitrogen sites. Therefore, there is a possibility for further increase of Tc by optimization of them.

CONCLUSION

Molybdenum nitride films have been deposited by means of ion beam deposition. And the dependence of their crystal structure and Tc on preparation conditions has been investigated.

Crystal structure depends mainly on nitrogen partial pressure P_{N2} and substrate temperature Ts. Films of single B1 phase are obtained on the condition of low P_{N2} and high Ts. Lattice constant a(200) of B1 phase changes drastically with ionization ratio of depositing particles; in the range of the ratio of 90 %, films with large a(200) around 4.25 Å which is close to the expected value for B1-MoN are obtained even in the low range of P_{N2} below $1x10^{-3}$ Torr. This range of P_{N2} is two orders of magnitude lower than that for such films deposited by conventional methods. Thus, ionization of depositing particles is effective for activating surface reaction between molybdenum and nitrogen and for obtaining B1 phase with high nitrogen content. Tc of the films shows a good correlation to a(200) and the films with a(200) around 4.25 Å show the highest Tc of 12.8 K, which is one of the highests for films of B1
technique to obtain high Tc films of non-equibrium structure and Tc of B1 MoN shoud rise more if strict optimization of deposition conditions such as P_{N2},Ts and ionization ratio of depositing particles is performed.

REFERENCES

1. W.E.Pikett,B.M.Klein and D.A.Papaconstantopoulos;Physica,107B,667 (1981).
2. Y.Honshi,N.Terada,M.Naoe and S.Yamanaka;Advances in Cryogenic Engng., 30, 607 (1984).
3. M.Naoe,S.Yamanaka and Y.Hoshi;IEEE Trans. Magn.,MAG-16,646 (1980).
4. H.Ihara,M.Hirabayashi,K.Senzaki,Y.Kimura and H.Kezuka;Phys.Rev.,B32, (3), (1985).
5. G.Linker,R.Smithey and O.Meyer;J.Phys.,F14,L115 (1985).
6. H.Ihara,Y.Kimura,K.Senzaki,H.Kezuka and M.Hirabayashi;Phys.Rev., B31, 3177 (1985).
7. A.W.Webb,E.F.Skelton,S.B.Qadri,T.L.Francavilla,A.Onodera and K.Suito; Bull.Am.Phys.Soc.,29,484 (1984).

PREPARATION OF SUPERCONDUCTIVE MoN_x FILMS BY REACTIVE SPUTTERING

Hiroshi Yamamoto, Toshihiro Miki, and Masaichi Tanaka

College of Science and Technology
Nihon University
Funabashi-shi, Chiba, Japan

ABSTRACT

A series of MoN_x films ($0.5 \lesssim x \lesssim 2.3$) are prepared by reactive dc sputtering in the atmosphere of nitrogen. Crystalline structure and conductivity of the film obtained are discussed in relation to sputtering parameters. The crystal phase of the polycrystalline films changes from fcc Mo_2N phase ($x \lesssim 0.7$) to the cubic and hexagonal MoN mixed phase ($0.7 \lesssim x \lesssim 1.9$) as decreasing substrate temperature, T_s. A single cubic phase with a suitable lattice constant, a_o ($x \gtrsim 1.0$) shows low T_c less than 7K. Films with a single cubic structure can be grown epitaxially on (100) planes of MgO or (200) planes of Mn-Zn Ferrite in the range of T_s between 450 °C and 500 °C. A comparatively high T_c of about 11K is observed in the epitaxial film with a stoichiometrical B1 MoN phase and a_o of 0.4213nm.

INTRODUCTION

A B1 MoN crystal with the structure of NaCl is a new and unknown superconductor. It is expected theoretically[1] and/or empirically[2] to have a high superconductive critical temperature, T_c above 20K, though the B1 phase is a nonequilibrium phase. Several superconductive phases are found already in a Mo-N system; highly disordered $MoN_{0.3}$[3] with a T_c of 9K, fcc Mo_2N[4] with a T_c less than 7K, hexagonal Mo_2C phase[5] with a T_c of about 7K, and hexagonal MoN[6] with a high T_c less than 15K.

Recently Linker et al.[7] have prepared a stoichiometrical B1 MoN film with a suitable lattice constant, a_o of 0.4212nm by reactive rf sputtering. However, the T_c of the film is about 3K in contrast to a high value predicted theoretically. From X-ray photoelectron spectroscopic measurement of various MoN_x films, Ihara et al.[8] suggest that only nitrogen excess B1 MoN films ($x \gtrsim 1.3$) have a B1-type electronic structure. Then an onset temperature of T_c of a MoN_x film with a single cubic structure decreases from 12.5K to below 4.2K with increasing x from 1.1 to 1.8. They also think that vacancies or interstitial defects deteriorate a primary superior superconductivity of a B1 MoN crystal.

In this work, a series of MoN_x ($0.5 \lesssim x \lesssim 2.3$) films have been prepared by reactive dc sputtering. Films obtained are discussed with respects to crystalline structure and electrical conductivity. Especially attentions are paid to epitaxial films in addition to polycrystalline films. Polycrystalline films ($x \gtrsim 1.0$) with both a single cubic structure and a suitable a_o reveal low T_c's less than 7K, similar to the results by Linker et al.[7] On the contrary, a comparatively high T_c of about 11K is obtained in a stoichiometrical film grown on MgO crystal with a single cubic structure and a_o of 0.4213nm. From results obtained, it can be expected that a B1 MoN crystal is primarily a high T_c superconductor.

FILM PREPARATION AND MEASUREMENT

Details of the sputtering apparatus used in this work are appeared in elsewhere[9]. The feature of the sputtering beyonds to a hollow type of target which gives a high rate deposition and activation of reactive gas to proceed effectively nitriding reactions. The target is made by 0.5mm thick of Mo plates (3N) and it is a cylindrical-hollow with the inner diameter of 30mm and the height of 20mm. A top plate of the target has a hole of 5mmϕ and faces to a substrate. Sputtered particles and secondary electrons are emitted through the hole. A distance between the top plate of the target and the substrate is typically about 25mm. Substrate temperature is controled through a Cu plate holder by tungsten filaments and is monitored by a thermocouple.

Nitrogen gas of high purity is introduced into the chamber which is evacuated to the vacuum of about 1.3×10^{-4}Pa in advance. Before film deposition, presputtering is done for about 10 minutes. Typical sputtering conditions are cathode bias of about 450V; discharge power of 50W; deposition rate of about 5nm/min. Following substrates are used; Si or fused quartz for polycrystalline films and MgO, Mn-Zn Ferrite or AlN film for epitaxial films, respectively.

Film thickness is measured by an interference microscope. Crystalline structure is investigated by reflected X-ray diffraction technique. Concentration of nitrogen of the film, C_N is determined by the X-ray intensity ratio of Mo-M_α and N-$K_{\alpha 1}$ observed in electron probe microanalysis. Then the standard samples are Mo_2N films and the nitrogen excess MoN_x film of which the value of x is determined in advance from X-ray photoelectron spectroscopic spectra. Superconductive critical temperature , T_c is measured resistively by a four-probe technique. Transition width is defined as the temperature width between the 1% point (onset of T_c, T_{con}) and 99% point of the full resistive decrease. Temperature is monitored by a calibrated germanium thermometer in cryogenic region and by Au-0.07%Fe/chromel in the range of temperatures above 30K.

CRYSTALLINE STRUCTURE

Polycrystalline Films (Poly-Films)

A crystal phase changes in poly-films by nitrogen gas pressure, P_{N2}

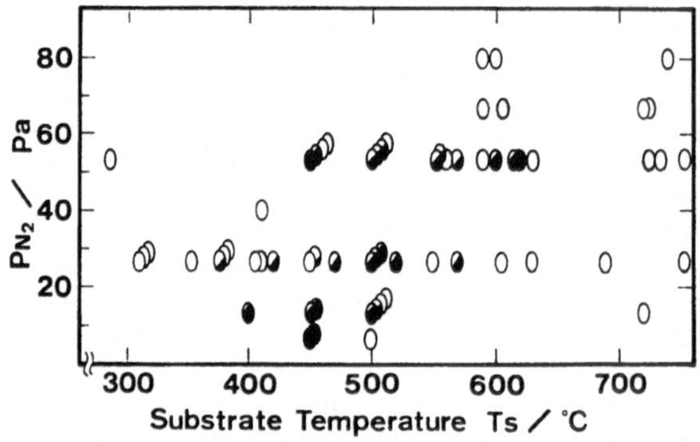

Fig. 1. Crystal phase diagram observed in the films prepared on Si, where open circles and closed circles represent a single cubic and a hexagonal MoN phase, respectively.

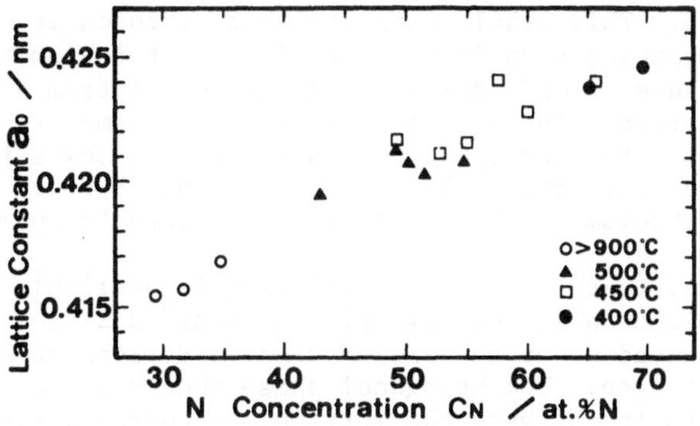

Fig. 2. Lattice constant, a_o of a cubic phase as a function of concentration of nitrogen, C_N in the films prepared under the indicated substrate temperatures.

and substrate temperature, T_s, as shown in Fig. 1. A single cubic structure appears in the range of T_s above 650 °C and below 400 °C. In the exculding range of T_s, a cubic and hexagonal MoN mixed phase is observed. The phase primarily depends on T_s rather than nitrogen gas pressure, P_{N2}.

A lattice constant of a cubic phase, a_o increases linearly with increasing concentration of nitrogen, C_N as shown in Fig. 2. The linear correlation is observed between a_o and x in the MoN_x films by Linker et al.[7] In their films, a_o attains to 0.422nm at x of 1.0, similar to the stoichiometrical film with a_o of 0.421±0.001nm in this work. A value of a_o increases with decreasing T_s. This result indicates that a thermal reaction is an important mechanism for nitriding reactions on a film growing surface. According to the linearity between a_o and C_N, a_o is an effective index to the performance of nitriding reactions. Then it is thought that a_o depends on T_s in a function of $\exp(E_a/k_BT_s)$, where E_a is an activation energy taking part in the nitriding reactions. From the point of view, $\ln a_o$ is plotted as a function of T_s^{-1} as shown in Fig. 3. It is found that a value of E_a in the range of T_s above 450 °C is about 2.5 times large as that in T_s below 450 °C. Therefore the mechanism of crystal growth seems to change at a_o of 0.422±0.001nm as a critical value. The growing process is thought to be dominated by a constant activation energy

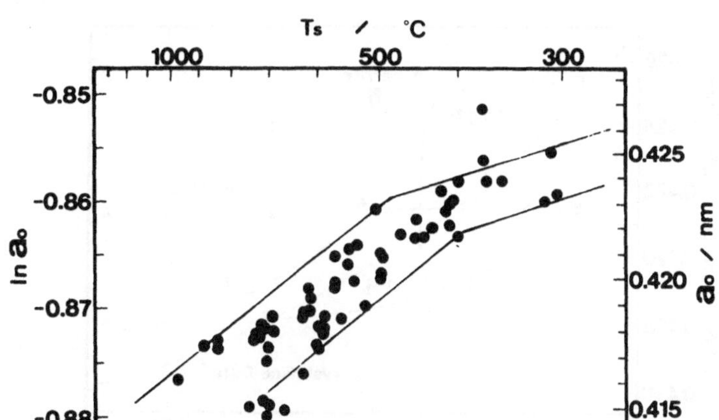

Fig. 3. Lattice constant, a_o as a function of inverse of substrate temperature, T_s^{-1} observed in polycrystalline MoN_x films.

from x=0.5 to x=1.3. This result suggests that nitrogen atoms of a few % tend to produce vacancies even in the vicinity of stoichiometry, corresponding to the change of occupation of lattice sites from a fcc Mo_2N to a B1 MoN. On the contrary, the more excess nitrogen atoms are thought to occupy interstitial sites easily with a comparatively low activation energy in the range of T_s below 450 °C. It remains, however, to be an open question what a mechanism takes part in such the defects formed by excess nitrogen atoms.

It is confirmed that a reactive sputtering is available to some degree for synthesis of a B1 MoN crystal. However, a B1 and hexagonal MoN mixed phase is obtained under the preparation conditions to give an adequate amount of nitrogen. The hexagonal phase should be suppressed as possible in order to investigate the intrinsic superconductivity of an ideal B1 MoN crystal. It is also noticed that T_s is comparatively low in the case of the syntheses of a B1 MoN crystal with a suitable a_o. The low T_s arises a serious problem in relation to a columner structure as pointed out by Linker et al.[7] As a result, a new preparation parameter must be introduced in order to proceed and/or control the nitriding reactions on a film growing surface. In this work, an epitaxial growth is examined as stated in the following section.

Epitaxial Films (Epi-Films)

An epitaxial technique is applied to suppress the growth of the hexagonal MoN phase and to proceed the growth of a B1 MoN phase. Considering for the degree of mismatch of a lattice constant, substrates used are (100) planes of MgO or (200) planes of Mn-Zn Ferrite. A c-axis oriented AlN film is also used supplementally.

Figure 4 shows a_o's as a function of T_s in the films grown on MgO and Ferrite. Typical X-ray diffraction diagrams of epi-films are shown in Fig. 5. A good epi-film has a strong diffraction peak from (200) plane of a B1 MoN phase. The value of a_o changes complicatedly by T_s in the epi-films, while poly-films show a monotonous dependence of a_o on T_s. In the case of the epi-film, a_o is quenched to the value of the substrate planes. The region of a constant a_o, that is a plateau, is a peculiar result observed in epi-films. Another feature is that the film grown on a comparatively hot substrate shows a phase separation into the two cubic phases, each of which reveals a_o corresponding to an epitaxial and a polycrystalline growth. The polycrystalline cubic phase with a_o less than

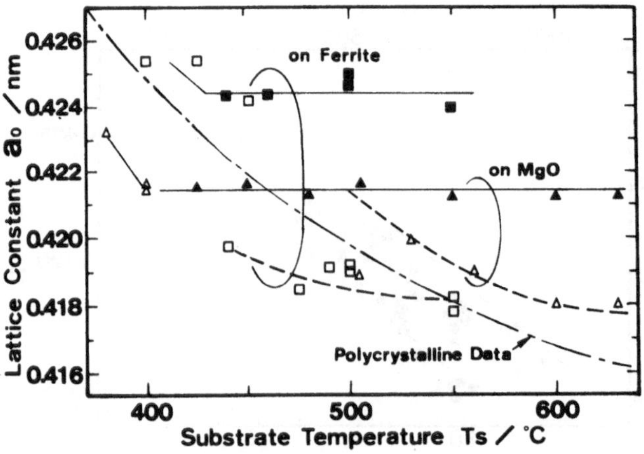

Fig. 4. Lattice constant, a_o as a function of substrate temperature, T_s in epi-films, where closed marks represent data of epi-films and open marks data of coexisting polycrystalline phase.

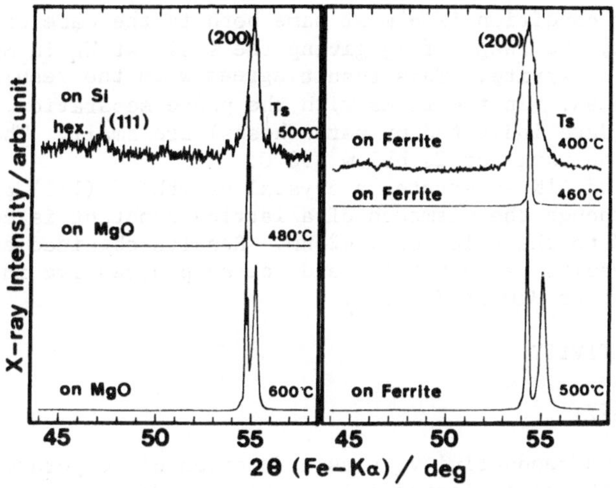

Fig. 5. Typical X-ray diffraction diagrams in sputtered MoN_x films.

0.420nm accompanies with a very small (111) diffraction peak. The magnitude of the peak intensity of the polycrystalline phase has a tendancy to increase with increasing film thickness. On the other hand, a hexagonal MoN phase in the epi-film is estimated to be a very small amount less than 10^2ppm from the intensity of X-ray diffraction peak.

The plateau value of a_o is 0.4213±0.0003nm in the film on MgO and 0.4243±0.0003nm on Ferrite, respectively. The plateau region of MgO is much wider than that of Ferrite. This result is consistent with the fact that the mechanism of nitriding reactions essentially changes through the critical value of a_o, about 0.422nm.

A concentration of nitrogen, C_N is dominated by T_s rather than a_o also in the epi-films. The stoichiometry is obtained at T_s of about 500 °C similar to the case of polycrystalline growth. Even in the plateau region of a_o, C_N or the amount of nitrogen defects changes corresponding to T_s. As a result, the optimum value of C_N on Ferrite is about 54at.%N, while stoichiometry can be obtained on MgO.

As a criterion of crystal growth, a half width of (200) peak, W_H is shown as a function of T_s in Fig. 6. The typical value of W_H is about 0.2 degree, much less than that in poly-films, about 0.5 degree. The value of

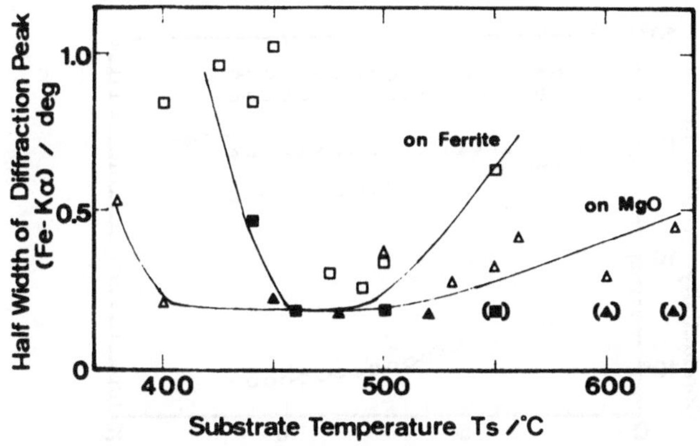

Fig. 6. Half width of (200) diffraction peak as a function of substrate temperature, T_s in the epi-films, where parentheses represent phase-separating samples and each marks mean followed in Fig. 4.

W_H in the optimum condition is almost same both in the case of MgO and Ferrite. However, the range of T_s giving the smallest W_H in MgO is much wider than that in Ferrite. This result agrees with the result mentioned above relevant to a_o. In the films with the phase separation, W_H's of the epitaxial phase (data indicated by parentheses) are small. The value of W_H increases in the range of T_s below 450 °C.

In the case of AlN substrate, a crystal growth of (111) planes is characteristic, though the mismatch of a lattice constant is comparatively large, about 3.3% to the value of 0.425nm. Results obtained are generally similar to the results of poly-films and not so progressive in the comparison with the case of MgO or Ferrite.

ELECTRICAL CONDUCTIVITY

Normal Conductivity

Figure 7 shows conductivity, σ as a function of temperature in typical MoN_x films with the indicated a_o's. Then σ(300K) at room temperature of a poly-film is distributed in the range of $4x10^5 \sim 2x10^3 \Omega^{-1}m^{-1}$ and decreases with increasing a_o. In the films with a comparatively large a_o, σ decreases with decreasing temperature, T and $\ln\sigma$ seems to decrease in proportion to $T^{-1/4}$ in low temperatures. The nonmetallic behavior of σ is interpreted by the mechanism of three-dimensional variable range hopping rather than a single activation type of conduction mechanism. Linker et al.[7] proposed the question whether columner structural defects result in a small σ and a small residual resistance ratio in MoN_x ($x \gtrsim 1.0$). From X-ray diffraction diagrams, the evident difference of structure is not recognized on comparing the films with a_o of 0.4189nm and 0.4218nm. The decrease of σ is, therefore, thought to have a close relation to the amount of effective excess nitrogen atoms. Then it should be mentioned that effective excess nitrogen atoms exist even in the apparently stoichiometrical film with a_o of about 0.421nm and that the more excess atoms increase interstitial defects to enlarge a_o as discussed in the previous section.

The stoichiometrical film on MgO reveals σ of about $10^5 \Omega^{-1}m^{-1}$ and residual resistance ratio of 1.0. It is noticed that the epi-film shows a metallic conduction. These results indicate that an epitaxial growth is available to decrease effective excess nitrogen atoms and it enables the film to approach an ideal B1 MoN crystal. On the other hand, σ of the epi-film on Ferrite seems to be influenced by a comparatively large amount of excess nitrogen atoms, though a semiconductive Ferrite prevents an

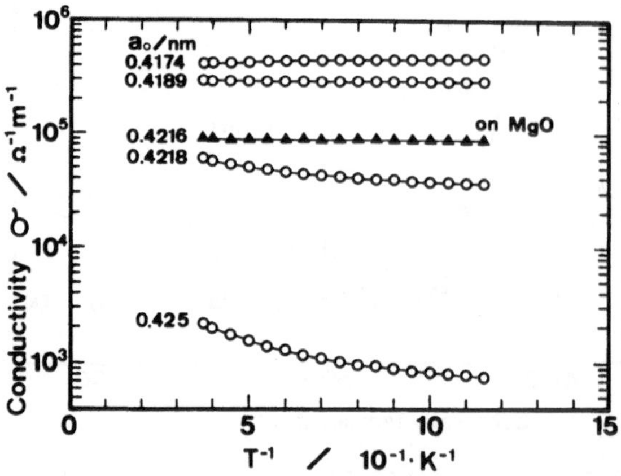

Fig. 7. Conductivity, σ as a function of inverse of temperature, T^{-1} in typical poly- and epi-films with the indicated a_o's.

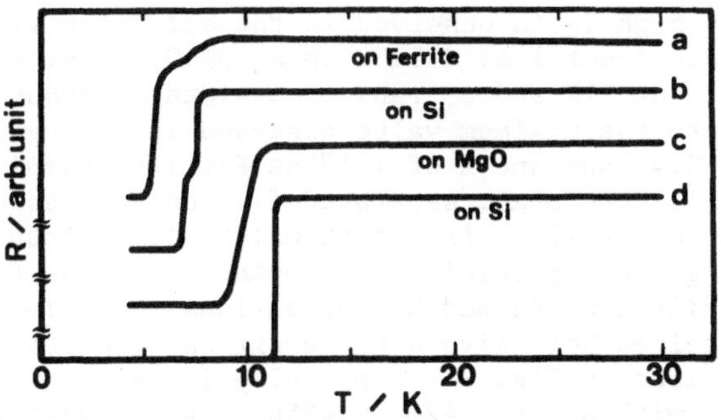

Fig. 8. Resistance, R as a function of temperature, T in typical films deposited on the indicated substrates; (a) is a datum of a single cubic phase with a_o=0.4244nm, (b) a mixed phase with a_o=0.4203nm, (c) a single cubic phase with a_o=0.4213nm, and (d) a single cubic phase with a_o=0.4196nm.

exact measurement of σ of the film. No improvement is obtained with respect to σ in the film grown on the AlN film.

Superconductivity

Temperature dependences of resistance, R of typical films are shown in Fig. 8. Also T_c's observed in this work are summarized in Fig. 9 as a function of a lattice constant, a(200) determined by only (200) diffraction peak. Since the hexagonal MoN phase has primarily a high T_c, about 15K, the data of the films with a cubic and hexagonal MoN mixed phase are neglected for simplicity.

A sharp transition with a width of about 0.4K and T_{con} of about 11K are observed in the single cubic film with a_o of 0.4196nm (sample d), of which C_N is about 37at.%N, much less than stoichiometry of a B1 phase. The mixed-phase film (sample b) reveals typically a T_{con} below 10K. On the contrary, T_{con} is at most 7K in a stoichiometrical poly-film with a B1 phase. The films with a_o above 0.421nm show very low T_c's below 4.2K. These results agree well with the results of Linker et al.[7] Most of films with a_o less than 0.419nm reveal T_{con}'s below 9K and they show characteristics expected in a fcc Mo_2N phase.

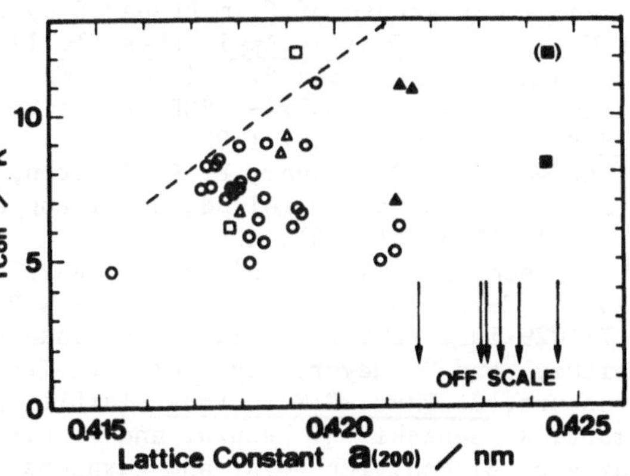

Fig. 9. Onset of T_c, T_{con} as a function of a lattice constant, $a_{(200)}$, where open circles and arrows represent poly-films with a single cubic phase, triangles the data on MgO, squares the data on Ferrite, and closed marks the case of a superior epitaxy.

A comparatively high T_c is observed in the epi-films with good reproducibility. The stoichiometrical film with a_o of 0.4213nm on MgO (sample c) shows a T_{con} of about 11K and a transition width of about 2.7K. The value of T_c belongs to the maximum value observed in a single B1 MoN film. The film with a_o of 0.4244nm and x of 1.17 on Ferrite (sample a) gives a T_{con} of about 8K, while the phase separated film on Ferrite exceptionally shows a high T_{con} of about 12K. As a conclusion, it is found that the epitaxial technique gives a possibility to obtain an ideal B1 MoN crystal which can not be easily synthesized by conventional reactive sputtering.

The results obtained here give a new empirical prediction for T_c's as shown by the broken line in Fig. 9. Especially it may be thought that the ideal B1 MoN crystal with a_o of 0.423 ~ 0.425nm has comparatively high T_c, at least 16 ~ 18K. Certainly the value of T_c predicted here is much lower than that expected theoretically, but it should be continued to optimize the preparation conditions and to clarify a primary superconductivity of a B1 MoN crystal.

SUMMARY

A series of MoN_x films have been prepared by reactive dc sputtering and discussed with respects to the crystalline structure and conduction. Polycrystalline MoN_x films are synthesized in the range of x from 0.5 to 2.3 by dominantly changing T_s. The poly-film of x of about 1.0 has typically a cubic and hexagonal MoN mixed phase. Excess nitrogen atoms result in a single cubic phase with a_o above 0.421nm, but they give a low T_c, at most 7K. Epitaxial films with a single B1 MoN phase on MgO or Ferrite show a comparatively high T_c, 11 ~ 12K. It remains, however, to be vague in regard to vacancies and/or interstitial defects caused by the excess nitrogen atoms. If nitriding reactions are controled satisfactorily, a T_c of a B1 MoN crystal will attain to the highest value in MoN system.

REFERENCES

1. W. E. Pickett, B. M. Klein, and D. A. Papaconstantopoulos, THEORETICAL PREDICTION OF MoN AS A HIGH T_c SUPERCONDUCTOR, Physica 107B:667-668 (1981).
2. Z. You-xiang and H. Shou-an, B_1-TYPE MoN, A POSSIBLE HIGH T_c SUPERCONDUCTOR, Solid State Comm. 45:281-283 (1983).
3. G. Linker and O. Meyer, SUPERCONDUCTING PROPERTIES AND STRUCTURAL TRANSFORMATIONS OF NITROGEN IMPLANTED MOLYBDENUM FILMS, Solid State Comm. 20:695-698 (1976).
4. S. Nagata and F. Shoji, Structure of Film Prepared by Low Energy Sputtering of Molybdenum, Jpn. J. Appl. Phys. 10:11-17 (1971); M. Ikebe, N. S. Kazama, Y. Muto, and H. Fujimori, Mo BASE SUPERCONDUCTING MATERIALS PREPARED BY MULTI-TARGET REACTIVE SPUTTERING, IEEE TRANS. ON MAG. MAG-19:204-207 (1983).
5. W. W. Fuller, S. A. Wolf, D. U. Gubser, E. F. Skelton, and T. L. Francavilla, Properties of a new molybdenum nitrogen phase, J. Vac. Sci. Technol. A, 1:517-519 (1983).
6. E. J. Saur, H. D. Schechinger, and L. Rinderer, PREPARATION AND SUPERCONDUCTING PROPERTIES OF MoN AND MoC IN FORM OF WIRES, IEEE TRANS. ON MAG. MAG-17:1029-1031 (1981); H. Ihara, private communication.
7. G. Linker, R. Smithey, and O. Meyer, Superconductivity in MoN films with NaCl structure, J. Phys. F:Met. Phys. 14:L115-L119 (1984).
8. H. Ihara, Y. Kimura, K. Senzaki, H. Kezuka, and M. Hirabayashi, Electronic structures of B1 MoN, fcc Mo_2N, and hexagonal MoN, Phys. Rev. B, 31 (1985), to be published.
9. H. Yamamoto, K. Ishii, and M. Tanaka, PREPARATION OF NbC_xN_{1-x} THIN FILMS BY HIGH RATE REACTIVE SPUTTERING, in: "Advances in Cryogenic Engineering-Materials," vol. 30, Plenum Press, New York (1984), pp. 623-630.

CRITICAL FIELD STUDIES OF REACTIVELY SPUTTERED AND NITRIDED NbN, VN AND V(Mo)N FILMS

J. S. Moodera, P. M. Tedrow, and R. Meservey

Francis Bitter National Magnet Laboratory
Massachusetts Institute of Technology
Cambridge, Massachusetts

ABSTRACT

We have made critical field (H_{c2}) measurements on high T_c NbN and VN films prepared either by reactive sputtering or by nitriding evaporated metal films at high temperature in an N_2 atmosphere. Good quality films were obtained having high T_c and H_{c2} by both methods. Thick sputtered NbN films (> 1000 Å) showed $H^{\perp}_{c2} > H^{\parallel}_{c2}$ whereas thinner films had $H^{\perp}_{c2} < H^{\parallel}_{c2}$. On the other hand evaporated and nitrided films displayed $H^{\parallel}_{c2} > H^{\perp}_{c2}$ for thick as well as thin films. In the case of VN thick reactively sputtered films showed small anisotropy $H^{\perp}_{c2} \gtrsim H^{\parallel}_{c2}$, while $H^{\parallel}_{c2} > H^{\perp}_{c2}$ for thin sputtered films as well as for thick nitrided films. These observations show strong columnar growth in reactively sputtered NbN films and its absence in nitrided films. Transmission electron microscope pictures support this view. We have also made measurements on VN films with a small percentage of Mo included with the aim of increasing H_{c2} by spin-orbit scattering. Werthamer-Helfand-Hohenberg theory has been applied to the H_{c2} versus T data.

INTRODUCTION

The search for superconducting materials for electronics applications has renewed the interest in the superconducting nitrides like NbN and VN because of their high transition temperature T_c, high critical magnetic field H_{c2} and ruggedness. Upper critical field studies on NbN films have established[1] that the perpendicular H_{c2} ($H^{\perp}_{c2}$) is higher than the parallel H_{c2} ($H^{\parallel}_{c2}$) in thick sputtered films. This anomalous anisotropy of H_{c2} in NbN has been attributed to the columnar growth of the films. A less-pronounced tendency to this type of anisotropy has been observed in VN. In addition, the upper critical field of VN is paramagnetically limited, similar to that of Al films.[2] It has been shown[2] that adding a thin layer of Pt over the Al films increases the spin-orbit scattering and hence increases H_{c2} values. We expect H_{c2} to increase in VN films also by the addition of high atomic number Z elements, since in VN the spin-orbit scattering rate is low.

With the above problems in mind we have investigated the critical field behavior of NbN and VN films made by reactive sputtering and by nitriding pure metal films. Also, we added a small percentage of Mo in

VN to form (V-Mo)N and a monolayer of Pt over VN with the idea of increasing the spin-orbit scattering rate.

Our studies indicate that sputtered NbN and some sputtered VN thick films have $H_{c2}^{\perp} > H_{c2}^{\parallel}$, whereas films thinner than ~ 500 Å show normal thin film behavior, i.e., $H_{c2}^{\parallel} > H_{c2}^{\perp}$. On the other hand, evaporated and nitrided films always show $H_{c2}^{\parallel} > H_{c2}^{\perp}$ irrespective of the film thickness (200 Å to ~4500 Å). The latter films are found to grow in a normal way with flat big grains, and <u>do not</u> form columns as is the case in sputtered films. Secondly, additions of Mo and Pt in VN films cause an increase in H_{c2} over that of pure VN films, tentatively indicating an increase in spin-orbit scattering by the addition of these high-Z-elements.

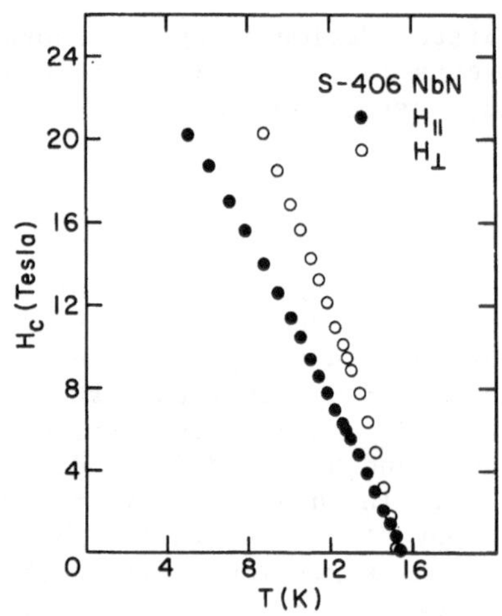

Fig. 1 Parallel and perpendicular critical magnetic field versus temperature for a thick sputtered NbN film.

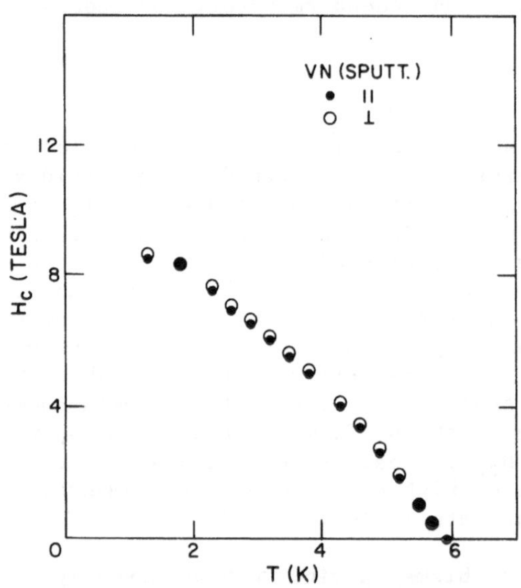

Fig. 2 Parallel and perpendicular critical magnetic field versus temperature for a sputtered VN film.

EXPERIMENTAL TECHNIQUE

The films of s-NbN and s-VN were made by rf sputtering in an $Ar:N_2$ (ratio of 8:1 for NbN and 4:1 for VN) atmosphere. (V-Mo)N films were sputtered by covering about 25% of the V target with Mo sheet. The glass or sapphire substrates were heated to $\gtrsim$ 500°C and films were deposited at a rate of about 150 Å/min.

Another set of films (e-NbN and e-VN) was made as follows:[3] metal films of Nb or V were evaporated from an e-gun source onto unheated sapphire substrates in high vacuum (10^{-6} - 10^{-7} torr). For V-Mo films, V and Mo ingots were melted together in the e-gun, and then the charge was evaporated. These films were immediately heated *in situ* in a pure N_2 atmosphere at ~ 1000°C for about 5 minutes and allowed to cool in the N_2 atmosphere. The formation of the B-1 phase of NbN and VN was established by X-ray diffractograms. Energy dispersive X-ray analysis gave the percentage of Mo in V-Mo films. In addition, films for TEM studies were made on MgO substrates.

RESULTS AND DISCUSSION

The transition temperature T_c of the films were measured resistively. Sputtered films in general showed negative temperature coefficient of resistivity TCR whereas evaporated and nitrided films had positive TCR. The T_c's of sputtered films were somewhat lower than those of the nitrided films. Critical field H_{c2} measurements were made in Bitter magnets by a 4-terminal dc resistance technique from 0.5 K to T_c.

Plots of H_{c2} versus T for thick sputtered films of NbN, VN and (V-Mo)N are shown in Figs. 1-3, respectively. For film thickness above ~ 500 Å, sputtered NbN always showed $H_{c2}^{\perp} > H_{c2}^{\parallel}$, whereas thinner film ~ 200 Å had $H_{c2}^{\parallel} > H_{c2}^{\perp}$. In the case of VN, $H_{c2}^{\perp}$ was not always higher than $H_{c2}^{\parallel}$ even for thicker tilms. The anisotropy in sputtered (V-Mo)N is as striking as in sputtered NbN.

The critical field behavior for nitrided films is different from that of the sputtered films, in that they always show $H_{c2}^{\parallel} > H_{c2}^{\perp}$. In Figs. 4 and 5, the H_{c2} variation with T is shown for evaporated NbN

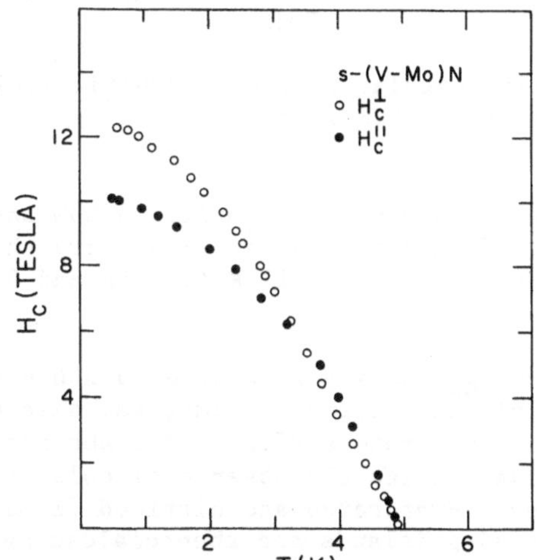

Fig. 3 Parallel and perpendicular critical magnetic field versus temperature for a sputtered V(12%Mo)N film.

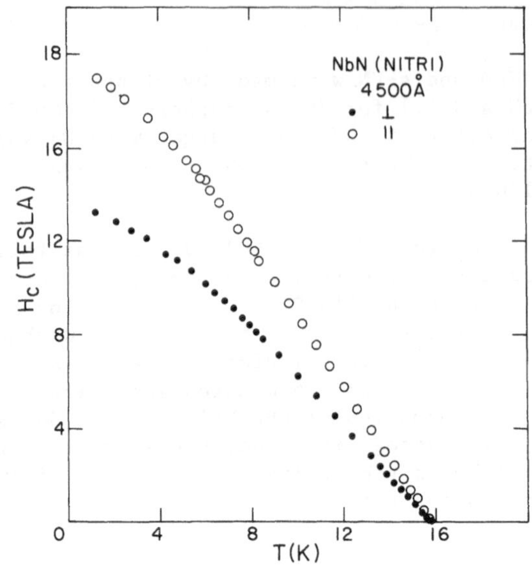

Fig. 4 Parallel and perpendicular critical magnetic field versus temperature for a thick evaporated NbN film.

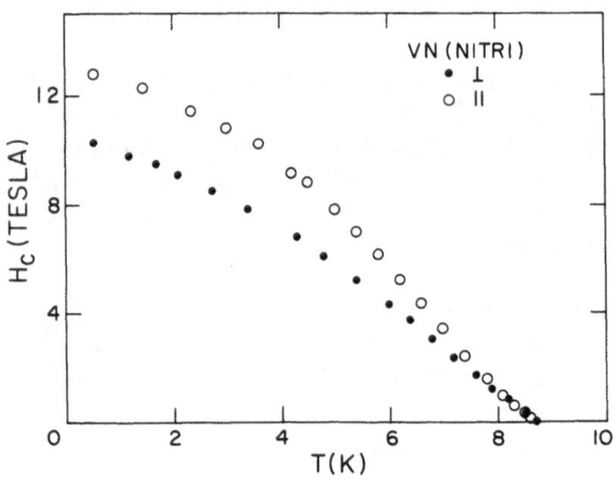

Fig. 5 Parallel and perpendicular critical magnetic field versus temperature for an evaporated VN film.

and VN. For (V-Mo)N, the behavior was similar to NbN and VN. This was found to be true for all film thicknesses. Even very thin films ($\gtrsim$ 400 Å) of these materials showed high T_c [(~ 17 K for NbN and 8.8 K for (V-Mo)N] and high H_{c2}.

The presence of the H_{c2} anomaly in sputtered NbN has been attributed to the columnar growth of the sputtered films, the axis of the column being perpendicular to the substrate surface. The absence of the H_{c2} anomaly in nitrided films implies the absence of columns in these films for all film thicknesses. Evaporated and nitrided films grow in the normal way, i.e., by forming islands and then coalescing to form flat, big grains. This is confirmed by the TEM picture shown in Fig. 6, taken for a thick film (~ 1500 Å) of e-NbN. No columnar growth of the grains could be observed. Instead, large flat grains are seen in the figure.

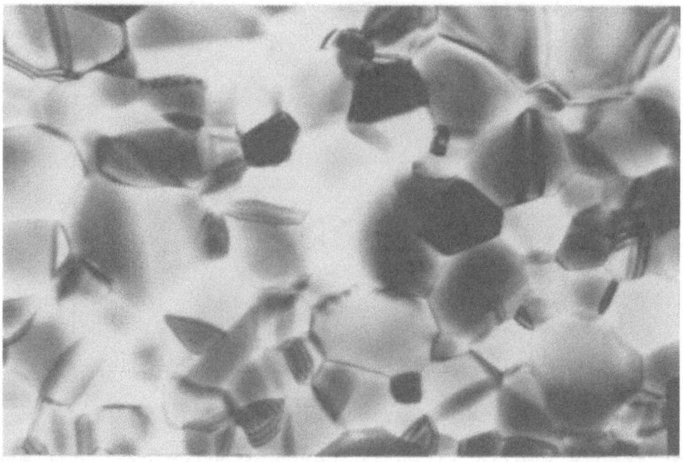

Fig. 6 TEM photograph of an evaporated and nitrided NbN film grown on a MgO substrate at 60,000X.

Tilting the specimen by 40° also did not reveal any sign of columns. Similar observations were done in e-VN, yielding the same results. A more pronounced H_{c2} anomaly, we found, occurred in general when the substrate temperature T_s during the deposition of s-VN was lower ($\tilde{>}$ 400°C), which perhaps means that at least in VN, higher mobility of the adatoms (because of higher T_s) inhibits columnar growth.

We also studied H_{c2} versus T for some films of e-VN with a 3 Å overlayer of Pt. This was done to increase the spin-orbit scattering in VN. The plot in Fig. 7 is for evaporated and nitrided VN, (V-Mo)N and VN-PT,

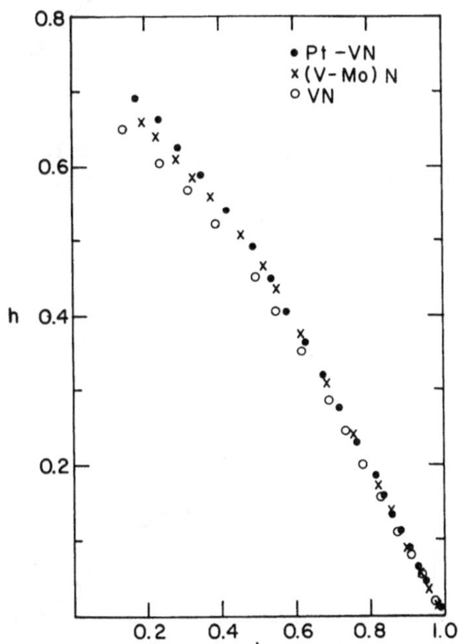

Fig. 7. Comparison of the perpendicular critical magnetic field versus temperature using reduced coordinates for evaporated and nitrided VN, V(5%Mo)N and VN-Pt films.

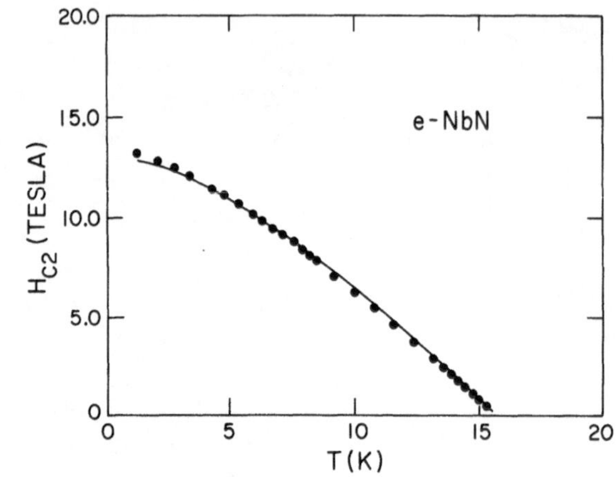

Fig. 8 Fit of theory (solid line) to perpendicular critical magnetic field data for an evaporated NbN film.

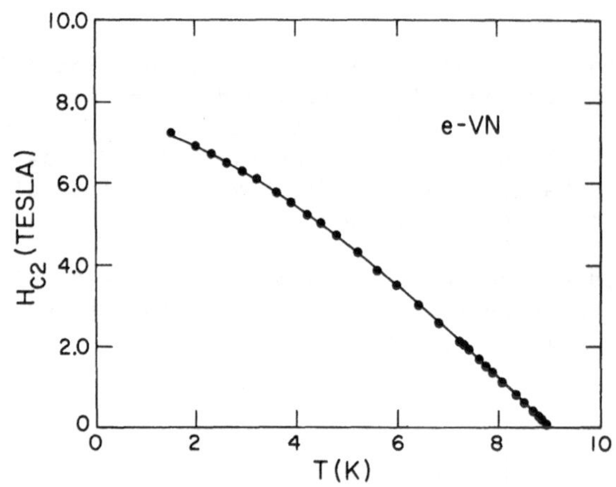

Fig. 9 Fit of theory (solid line) to perpendicular critical magnetic field data for an evaporated VN film on an MgO substrate.

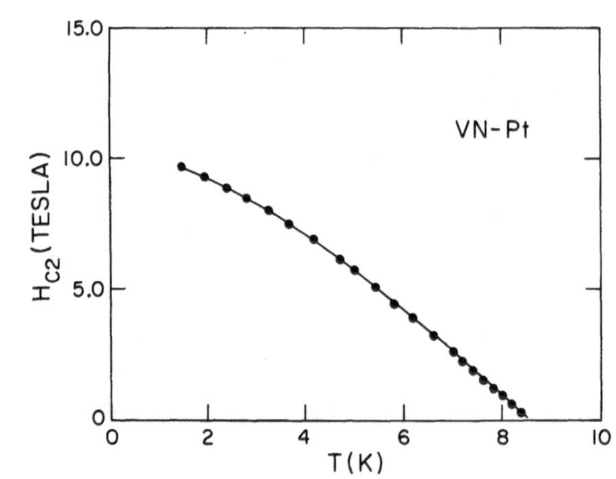

Fig. 10 Fit of theory (solid line) to perpendicular critical magnetic field data for an evaporated VN film in contact with a 3 Å film of Pt.

in reduced coordinates. The critical field has been normalized by dividing by the H_{c2} slope at T_c and by T_c. Small differences are seen for three films. The one with Pt has higher values of h than the other two, and VN has the lowest h values. The effect of Mo and Pt on VN is not as striking as Pt on Al films,[2] but the data here seem to indicate that spin-orbit scattering has increased by the addition of Mo and Pt. The amount of Mo in this (V-Mo)N film was 9 at. percent. Mo has an atomic number of 43 compared to 79 for Pt. These data, although not entirely conclusive, are consistent with the lessening of the paramagnetic limitation of the upper critical field in VN by the addition of high Z elements without unduly lowering the T_c of the superconductor.

We have attempted to fit our H_{c2} versus T data using the modified WHH theory.[4] This theory[5] takes into account both the paramagnetic limiting and the spin-orbit scattering as well as many body interactions of the quasi-particles. Unfortunately these various effects cannot be separated out just by H_{c2} versus T data. Therefore, the values of spin-orbit scattering parameter λ_{so} ($\equiv \hbar/5.6\Delta_o\tau_{so}$, where Δ_o is the superconducting order parameter and τ_{so} is the spin-orbit scattering relaxation time) and the renormalization parameter $(1 + G^o)$ (where G^o is the ℓ=0 antisymmetric Fermi liquid parameter) obtained in our fits may be unphysical, especially for the strong-coupling material NbN. Tunneling conductance measurements in the magnetic field in addition to the H_{c2} data would be required to uniquely establish the values of these two parameters. Some of the fitted curves are shown in Figs. 8-10 for e-NbN, e-VN and VN-Pt. In general theoretical fits to NbN daa were not as good as the fits to VN data. Also (V-Mo)N and VN-Pt showed higher λ_{so} compared to VN films for the best fits, which adds credibility to Fig. 7 and the discussion in the previous paragraph. Some of the typical λ_{so} values are displayed in Table I.

From the perpendicular critical field we evaluated the coherence length (ξ) for VN and NbN, which are shown in the table. In the case of NbN, the nitrided films have about twice the value of ξ compared to

TABLE I

	T_c (K)	ρ ($\mu\Omega$-cm)	R_{300}/R_{20}	$\left(\frac{dH_{c2}}{dT}\right)_{T_c}$ (Tesla/deg.)	ξ (Å)	λ_{so} [a]	$(1+G^o)$ [a]
s-NbN	<15.5	>120	0.9	2-4	45	5	1.1
s-VN	6.8		0.85	2.8	92		
e-NbN	16-17	65	1.1	1.2	83	3.0	1.6
e-VN	8-9		1.5	1.25	99	0.4	1.8
V(Mo)N	8.7		1.2	1.75			
VN-Pt	8.6		1.4	1.7		4.2	2.0

[a]These values for λ_{so} and $(1+G^o)$ give the best fit of the theory to the data. Tunneling measurements are required to determine values for these parameters independent of the theory of the critical field.

sputtered films, whereas in VN compounds the values are somewhat closer to each other. For e-NbN a value for $\xi \sim 85$ Å shows that all our films were much thicker than the coherence length. Hence the difference in $H_c^{\perp}$ and $H_c^{\parallel}$ data for 4500 Å thick film (Fig. 4) is somewhat puzzling. We presume $H_c^{\parallel}$ in this case is actually H_{c3}. This is supported by the lack of temperature dependence of the ratio $H_c^{\parallel}/H_c^{\perp} \sim 1.45$. The same argument is true for e-VN data in Fig. 5 also.

SUMMARY

We have compared critical magnetic field measurements on NbN and VN films made by reactive sputtering and by heating metal films in a N_2 atmoshpere. The latter films appear to be free of columnar growth and have slightly higher transition temperatures and lower resistivities. The spin-orbit scattering rate in VN appears to be affected by the addition of high atomic number elements.

ACKNOWLEDGMENTS

We thank M. W. Dumais for the EDAX and TEM measurements. The Francis Bitter National Magnet Laboratory is supported at MIT by the NSF. This work was preformed under AFOSR Contract No. F499620-82-K-0028.

REFERENCES

1. W. Wagner, D. Ast and J. R. Gavaler, Electronmicroscopic evidence for a columnar-void-type structure in sputtered NbN films, J. Appl. Phys. 45:465 (1974).
2. P. M. Tedrow and R. Meservey, Critical magnetic field of very thin superconducting aluminum films, Phys. Rev. B 25:171 (1982).
3. P. M. Tedrow and R. Meservey, Critical magnetic field of superconducting VN thin films, Physica 107B:527 (1981).
4. N. R. Werthamer, E. Helfand, and P. C. Hohenberg, Temperature and purity dependence of the superconducting critical field H_{c2}. III. Electron spin and spin-orbit effects, Phys. Rev. 147:295 (1966).
5. J. A. X. Alexander, T. P. Orlando, D. Rainer, and P. M. Tedrow, Theory of Fermi-liquid effects in high-field tunneling, Phys. Rev. B 31:5811 (1985).

NbTi BASED SUPERCONDUCTORS: TECHNICAL ASPECTS AND TRENDS

H. Hillmann and H. Krauth

Vacuumschmelze GmbH

Hanau, F. R. Germany

ABSTRACT

In this paper the status of industrial manufacturing technology for NbTi multifilamentary superconductors is described[1]. It focuses on the main research and development activities to enhance the performance of NbTi conductors for new or more sophisticated applications. One area concerns the optimization of the manufacturing techniques for multifilamentary strands with a large number of very thin filaments to reduce hysteresis losses and field distortions. Some applications require Cu/CuNi mixed matrix strands with a high amount of stabilization material to guarantee simultaneously low coupling losses and a high degree of stability. Large scale application of NbTi up to fields of 12 T seems to be possible by operating at 1.8 K and by using an optimized binary alloy. To have a sound basis for comparison, it is important to have a common definition of critical current and to know the influence of manufacturing steps like twisting, flattening and cabling on the critical current density. The quality factor, n, gives a global measure for the homogeneity and integrity of a strand.

INTRODUCTION

Most superconducting (s.c.) magnets built up to now (much more than 90 %) are using NbTi exclusively as the s.c. material, although many materials with much better s.c. parameters (T_c, B_{c2}) are known. This is mainly due to the excellent workability of NbTi together with copper as the stabilizing material. It is anticipated that this predominance will persist for a long time.

A true industrial application of s.c. magnets has evolved over the last decade in the field of NMR-spectroscopy and even more recently in the field of NMR-imaging (MRI). The conductors for these applications are relatively simple in terms of current and filament size ($\gtrsim$ 50 μm) and number (1 to 1000). On the other hand, long lengths and wires with a high homogeneity of conductor properties along the length are required because these magnets are operating in persistent mode and very low decay rates are needed. The manufacturing technology of these conductors is very well developed and a high quality standard has been reached to fullfill these requirements. Nevertheless, a considerable amount of research and development work is still being performed to enhance the capabilities of NbTi filamentary conductors and to qualify them for new applications. In the following sections we present some of the most important activities:

Conductors with a large number of fine filaments (< 10 μm): Fine filaments are required for applications with pulsed or oscillating field contributions, either to reduce hysteresis losses or to reduce field distortions due to residual currents inside the filaments at low excitation levels of the coils. Examples for the first case are poloidal field coils for Tokamaks and windings for superconducting turbo-generators. The second case applies for dipole and quadrupole coils in particle accelerators like the Superconducting Super Collider (SSC) in the U.S. and the Large Hadron Collider (LHC) in Europe.

Mixed matrix conductors: To reduce also the coupling losses between the filaments in a strand subjected to pulsed fields, mixed matrix (Cu/CuNi) conductors must be used in conjunction with the fine filaments. Examples are again superconducting generators and poloidal field coils.

Applications in the range of 11 T to 12 T: Due to the relatively poor values of B_{c2} and T_c, application of NbTi at 4.2 K is limited to fields of about 9 T. Quite an improvement can be reached by operating at 1.8 K. Fields up to 12.5 T were produced in laboratory magnets. This can be of interest for projects like LHC. Mostly ternary alloys like NbTiTa and NbTiHf were proposed for this type of application. But it seems that the use of optimized binary alloys is more recommendable from a technological and economical point of view.

Definition of critical current and factors influencing the critical current density: It is important to have a common definition of critical current either in terms of electrical field, E (0.1 to 1 μV/cm), or of resistivity, ρ (10^{-14} to 10^{-12} Ωm) of the superconductor in order to allow a sound comparison of measured critical current densities. Optimized laboratory scale conductors typically exhibit larger values than reproduceably achievable production conductors. Factors influencing or degrading the current density are strand diameter, twist pitch, flattening and cabling. Empirically the internal quality of a strand can be characterized by the quality factor, n, given by the slope of the log ρ versus log j curve.

CONDUCTORS CONTAINING LARGE NUMBERS OF FINE FILAMENTS

In the case of multipole coils for accelerators and of turbogenerator filament diameters in the range of 5 μm to 1 μm or below, up to 20,000 or more filaments in one strand are necessary. This requires modifications of the manufacturing process.

Billet Assembly and Treatment before Extrusion

The generally applied billet assembly process for reaching large filling factors uses hexagon-shaped tubes containing NbTi rods[2]. This technique can be applied down to rod diameters of about 2 mm. This results in a maximum number of $\sim$ 5000 filaments for a copper to superconductor area ratio of 1.8 and a billet diameter of 250 mm. For higher filament numbers round shaped single cores may be assembled and isostatically compacted before extrusion, to avoid buckling of single filaments when applying an uniaxial pressure.

In the case of hydrostatic extrusion isostatic compaction is integrated into the extrusion process. But in this case the billet diameter is reduced to about 170 mm. This means that the filament number is limited to about 20,000 when using 1 mm overall diameter single cores, which seems to be the lower limit for assembling straightened rods. For reaching much higher filament numbers and applying a single extrusion step, an alternative assembly technique using a woven structure can be used[21]. For very large numbers of filaments the multiple extrusion technique must be used.

Compatibility between Mechanical Properties and Optimization Heat Treatment

To reach high current densities intermediate annealing is required for the precipitation of pinning-active α-Ti[3,4,5]. The final current density depends on the integrated time, the number and the temperature of these heat treatments. Due to the precipitation of α - Ti the mechanical properties of NbTi are degraded. In addition, each annealing process creates a small layer of Ti_2Cu on the filaments which will be disconnected and crumpled during the following drawing operations. This means new unreacted areas will be ready for forming new layers during the next annealing process. Thus the volume of intermetallic components will increase with the number of intermediate annealing processes for comparable integrated annealing time. This means generally that there is strong competition between the mechanical properties and conductor current density. Additional care must be applied to the grain size of the NbTi used. Large grain sizes lead to better workability but to low final current density. Therefore the grain sizes should be individually chosen for a special conductor design.

Reduction of Surface Interaction between NbTi and Matrix

For bronze/niobium or copper/niobium composite conductors 1-µm-diameter filaments are well established because no chemical interaction between copper and niobium occurs by heat treatments during working down. In the case of NbTi-based conductors, either Ti- and Cu-diffusion takes place and mechanical alloying caused by imperfect NbTi-surfaces can occur[1]. To reduce or avoid the formation of CuTi intermetallics, the following measures can be taken:

- Avoid surface defects.
- Keep the extrusion temperature as low as possible.
- Limit the optimization heat treatments.
- Apply diffusion barriers.

Figure 1 shows a conductor with 12 µm filaments being optimized to 2,500 A/mm^2 at 5 T. The filaments show only very small intermetallic compound particles in the range of 1/10 of the filament diameter. No sausaging was observed and the current density decrease when increasing the sensitivity by one decade was only about 5 %.

Figure 2 shows 10 µm filaments of hydrostatically extruded Cu/CuNi mixed matrix conductors with Cu next to the filaments which were processed differently. Conductor 'a' got heat treatments at 650° C for 1 h to anneal the CuNi which obviously must be avoided, whereas conductor 'b' got a long optimization heat treatment. Conductor 'c', on the other hand, got only short heat treatments. This conductor was also drawn down to 1 µm filament diameter without degrading the filament quality (Fig. 2 d). Current densities for 10 µm, as for 1 µm, filaments of above 1300 A/mm^2 were obtained by applying a final optimization heat treatment. With intermediate optimization heat treatments, current densities were increased, e.g., to 1950 A/mm^2 for a 3-µm-filament diameter.

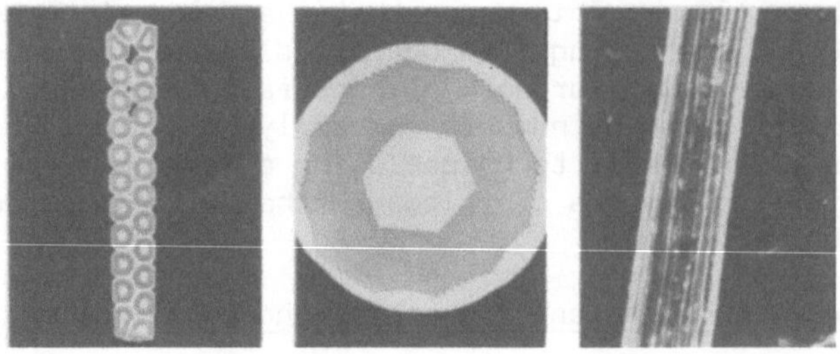

a) b) c)

Fig. 1. Prototype cable for the HERA dipoles: a) keystoned cable with 24 strands; b) strand with 1700 filaments à 12 μm, j_c = 2500 A/mm^2; c) filament quality with low amount of intermetallic particles.

Insertion of diffusion barriers is well known for Nb_3Sn conductors. In the case of NbTi they can also be applied. A suitable material is niobium. But the high additional costs make it economically unattractive. Another material which also forms intermetallic phases with Ti but with much lower diffusion rates is CuNi.

In the case of submicron filamentary conductors[6,7] the Cu resistivity in the filament region is degraded by Ti or Ni diffusion into the Cu and is due to the size effect. On the other hand at very low dimensions, the Cu may become superconductive due to the proximity effect[7]. Additionally the comparable mechanical properties of CuNi and NbTi provide an enhanced workability to submicron filaments, so that CuNi is recommended for these applications.

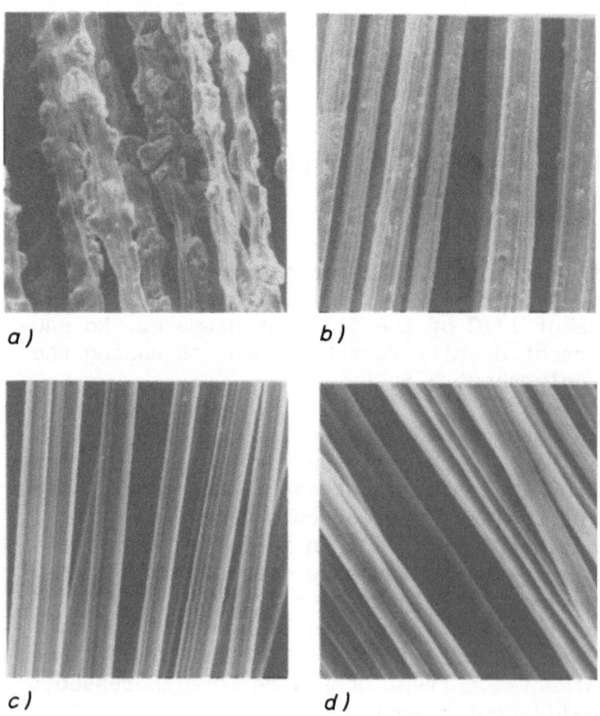

Fig. 2.
Typical filament quality at different diameters and heat treatments

a) 10 μm, long time optimization with additional annealing treatment at 650° C for 1 h;

b) 10 μm, long time optimization heat treatment;

c) 10 μm, final optimization heat treatment;

d) 1 μm, final optimization heat treatment.

MIXED MATRIX CONDUCTORS WITH LARGE COPPER AREA

For applications requiring both a large amount of Cu to get a high degree of stabilization and very low ac losses, the transverse resistivity of the matrix must be increased. This is done by introducing CuNi barriers to subdivide the Cu inside the strand and by using a CuNi outer shell. Special care must be applied in the design and the processing to maintain the workability of these three-component composites. Figures 3 and 4 show two versions of this type of conductors containing about 1000 filaments. In Figure 3 can be seen the cross sections after extrusion and at wire diameters of 1 mm (design value, 10 µm filament diameter) and 0.1 mm, respectively. The change in the internal geometry is due to nonaxial (i.e., radial and circumferential) flow during drawing operations. The filaments at these stages were already shown in Figs. 2 c and d. The example shown in Fig. 3 has a Cu outer shell but also strands with a CuNi shell could be fabricated. The measured time constants at 1 mm diameter were 140 µs for a CuNi shell and 380 µs for a Cu shell. For the conductor shown in Fig. 4 with a CuNi shell, a time constant of 105 µs was measured.

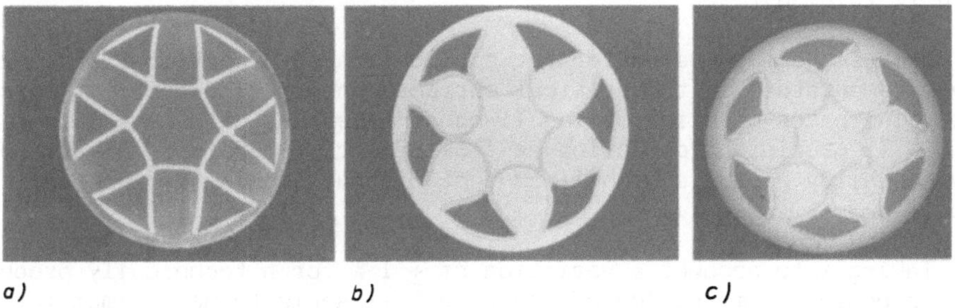

Fig. 3. Mixed matrix conductor for the POLO project at KfK - Karlsruhe (F.R.G.)
Cu/CuNi/NbTi ≃ 6/3/1, 1000 filaments, Cu outer shell

a) after hydrostatic extrusion (50 mm diameter);
b) at final dimensions (1 mm diameter, 10 µm filaments);
c) drawn down to 0.1 mm diameter with CuNi outer shell.

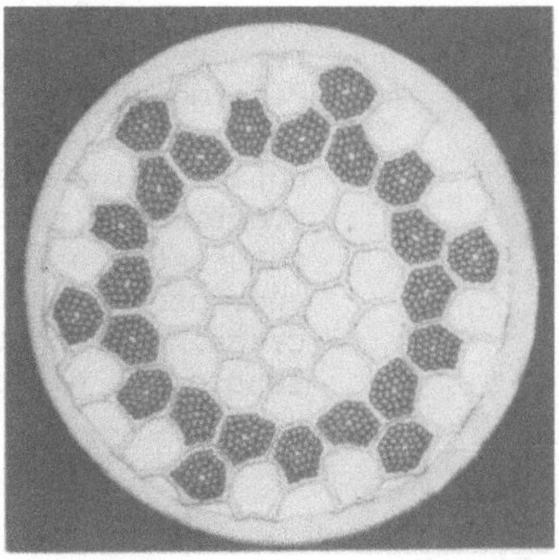

Fig. 4.

Mixed matrix conductor for POLO with double extrusion technique, 24 x 36 filaments, Cu/CuNi/NbTi ≈ 4.5/2.5/1, with CuNi outer shell.

CONDUCTORS FOR HIGH FIELD APPLICATIONS UP TO 12 T AT 1.8 K

For high field applications at 1.8 K ternary Nb-Ti-Ta and Nb-Ti-Hf alloys have been investigated. It was shown that the increase in B_{c2} for ternary alloys is only 0.2 T to 0.5 T at 4.2 K and 1 T at 2 K as compared with the binary alloy[9,10]. This is shown in Fig. 5 for the Nb-Ti-Ta system at 4.2 K. The increase of the gain in B_{c2} for the ternaries at low temperature is due to their lower critical temperature because, in thermodynamic approximation, $B_{c2}(T) = B_{c2}(0)\cdot[1-(T/T_c)^2]$.

On the other hand the production of ternary alloy rods is much more expensive and mechanical properties (i.e., workability) are deteriorated. The question arises whether the critical current density requirement for 1.8 K application can be reached with binary alloys and which composition should be chosen.

A view on the Nb-Ti binary side of Fig. 5 shows that for technical alloys both T_c and B_{c2} (4.2 K) are decreasing with increasing Ti-content. Because T_c decreases much faster, there may be a relatively pronounced peak of B_{c2} at lower temperature at a composition with about 48 to 49 w/o Ti (Fig. 6).

However the main parameter for application at a certain field and a certain temperature is the critical current density. The critical current density increases with increasing Ti-content because of higher potential for α - Ti precipitation acting as flux pinners. Therefore we have to choose an alloy having a high Ti-content without loosing in upper critical field.

Taking into account a variation of $\pm$ 1 % for a technically produced alloy, a nominal composition of about 48 % Ti seems to be optimum for the envisaged applications. Not many laboratories have the facilities to perform j_c measurements at 1.8 K and these measurements are relatively expensive. It seems therefore adequate to determine the gain, ΔB, for j_c = const. when going from 4.2 K to 1.8 K. Values taken from the available data are $\Delta B \simeq 3$ to 3.3 T for binary alloys, 3.6 to 3.8 for NbTiHf

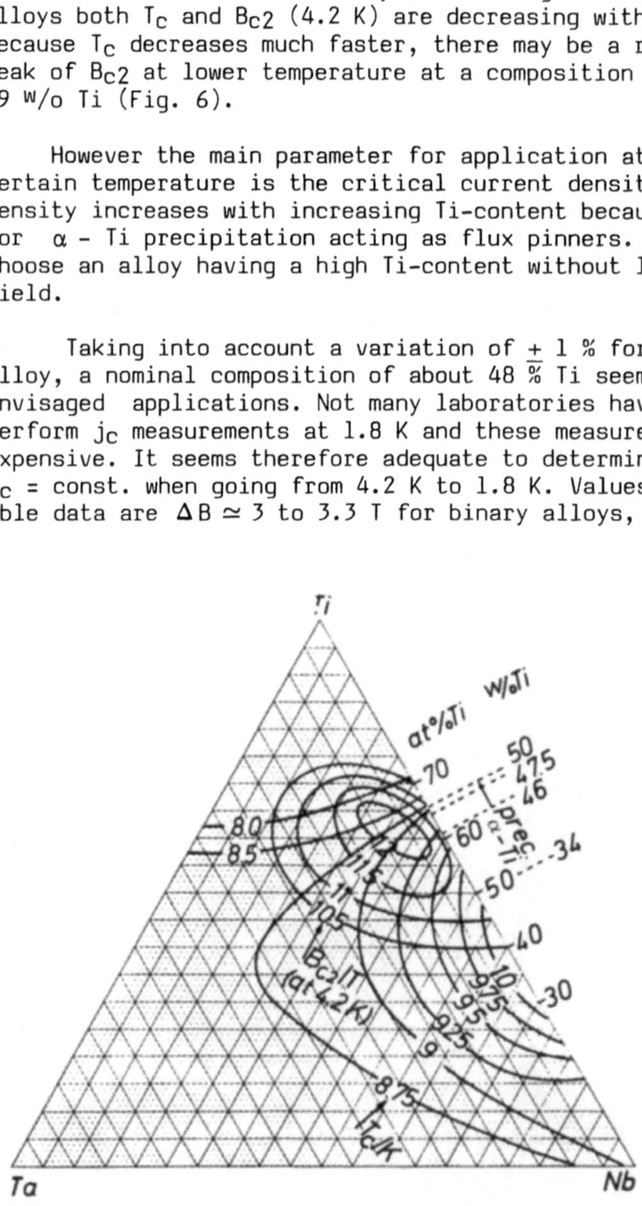

Fig. 5.
Lines of constant T_c and B_{c2} (4.2 K) for the ternary NbTiTa system. Pinning-active α - Ti precipitation during optimization heat treatment starts at about 46 w/o Ti and increases with increasing Ti-content leading to higher j_c values.

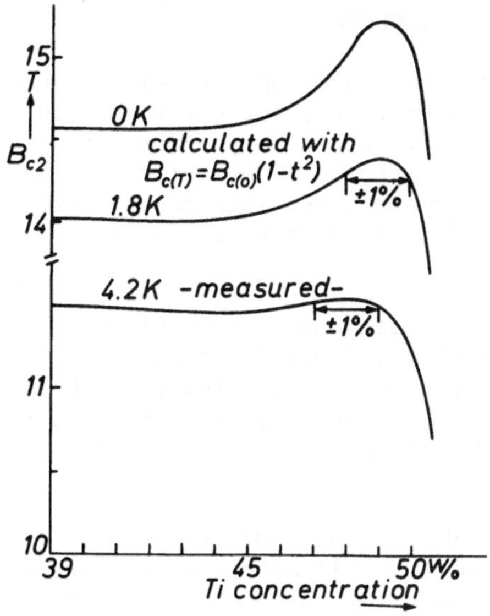

Fig. 6.

Measured and estimated B_{c2} values at 4.2 K and at reduced temperature, respectively. The composition of a technical alloy typically is specified with a tolerance of about $\pm$ 1 %.

and 3.6 to 4.0 T for NbTiTa. Fig. 7 shows critical current densities for different alloys either measured directly at 1.8 K or extrapolated by the above rule.

More exact measurements for binary alloys with well defined Ti-content have to be performed on j_c (1.8 K) and ΔB-values to end up with an optimally treated binary alloy. It seems obvious that such a binary would be preferable to all ternary alloys because of the above mentioned reasons.

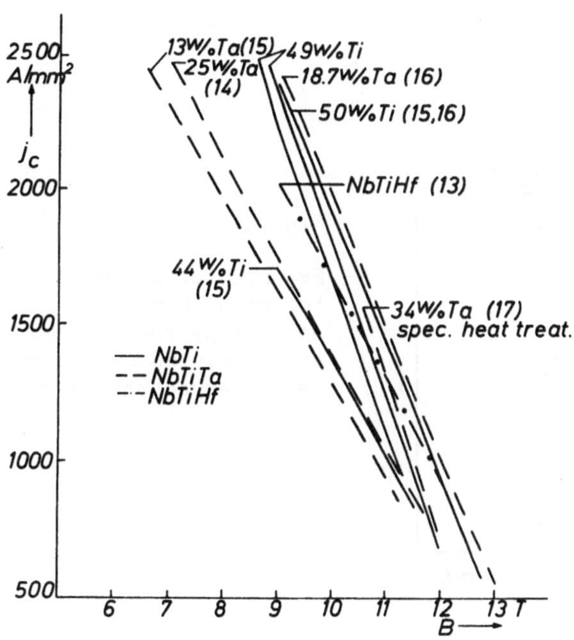

Fig. 7.

Compilation of published critical current densities of binary NbTi and ternary (NbTiTa and NbTiHf) alloys. Compositions shown are nominal values. The two curves labeled 49 w/o Ti concern typical standard production conductors at VAC measured at 4.2 K. Extrapolation to 1.8 K was made by the ΔB - rule given in the text.

DEFINITION OF CRITICAL CURRENT AND DEGRADING INFLUENCES

The costs for superconducting magnets depend very sensitively on the needed quantity of superconducting material. Therefore current densities as high as possible are required. Published current density values on optimized laboratory scale conductors concern generally untwisted conductors and may not be applicable to technical conductors. When specifying conductors for technical applications, the following degrading influences must be regarded:

Twist: Twisting causes straining of the filaments, decreasing their cross section and influencing the microstructure. Therefore a current density degradation must be expected. Values depend on the twist pitch relative to the strand diameter and whether a conductor is twisted in the hard or annealed condition and in one or more steps. Observed values for tight pitches ($lp \lesssim 12$ D) are 3 - 5 %.

Strand diameter: The optimum current density depends very sensitively on the integrated cold work degree of the NbTi-material that means the microstructure of the filaments. Main parameters are subband density, number, size and shape of normal conducting precipitates. For comparable grain sizes of the starting material this means that the critical current density depends on the final conductor diameter and decreases with increasing diameter.

Influence of flattening: Conductor flattening by profiling can cause an anisotropy and a degradation[18]. The anisotropy factor is defined by the ratio $j_{c\perp}/j_{c\parallel}$, the degradation factor by $j_{c\perp}/j_{co}$. In general, the anisotropy factor depends on the aspect ratio and is independent on the induction, B. The critical current density, $j_{c\perp}$ with the induction, B, perpendicular to the flat area of the conductor generally is lower than the critical current density of the round conductor. In the case of anomalous anisotropy, both $j_{c\perp}$ and $j_{c\parallel}$ can be increased by flattening. Figure 8 shows an example of a conductor with anomalous anisotropy factor including the starting round shape condition.

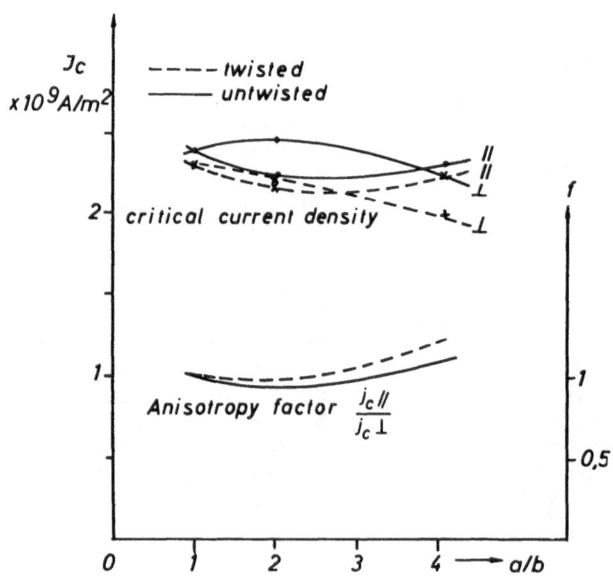

Fig. 8.

Example for a conductor with anomalous anisotropy for twisted and untwisted condition (a/b = aspect ratio).

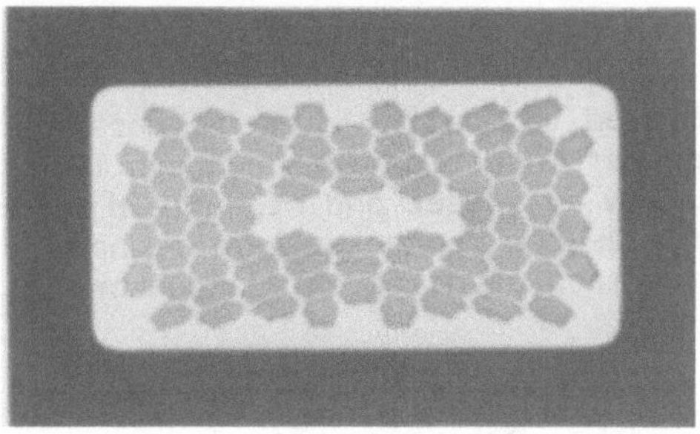

Fig. 9.

Monolithic conductor ($2.8 \cdot 5.6\ mm^2$) for TORE SUPRA with 10,000 fil. in a mixed matrix showing the deformation pattern due to profiling. It is supposed that the flat pinning centers are reoriented in a similar way, producing the observed type of anisotropy.

Conductor flattening changes the microstructure within the filaments, mainly the distribution and shape of pinning-active α-Ti precipitates. It was shown that the current density anisotropy depends on the degree of optimization. Anomalous anisotropy can be observed if the conductor is flattened in the optimized or overoptimized conductor. In multifilamentary conductors the filaments and therefore their microstructure are flattened differently, according to their size, their position inside the cross section and the applied flattening treatment. Twisted conductors behave differently from untwisted ones.

The existence of anomalous anisotropy shows that the microstructure can be improved even after current optimization by round shape drawing. The explanation for this may be connected to the fact that α - Ti precipitates can be deformed by cold working to flat noodles, as was reported recently[5]. In the round shaped condition the flat area directions of these noodles are randomly distributed over the cross section. Flattening causes a preferred orientation of the flat area of α - Ti precipitates having a higher degree of longitudinal area components along the flux lines which are expected to be more pinning active. This is analogous to the deformation of a filament conductor as shown in Fig. 9.

Influence of cabling: Flat cables which are mainly used for high current capacity generally show a current degradation of the individual strands of 3 - 18 %. This degradation is caused by the anisotropy of the flattened strands and the cross sectional change of the strands. In addition, a cable degradation ($j_{c\perp}$ compared with the mean j_c of the used strand material) and a cable anisotropy ($j_{c\perp}$ compared with $j_{c\parallel}$) exists[19]. The values for the cable shown in Fig. 1 were 10 % and 9 % ($j_{c\parallel} > j_{c\perp}$), respectively. The observed effects seem to be a combination of strand anisotropy and degradation, of the cable transposition angle (increasing $j_{c\parallel}$ by about 3 %) and of different local field enhancements due to the conductor self field for different field orientations.

Measuring sensitivity and quality factor: The measured critical current density depends on the applied criterion which may be an electrical field level or resistivity level. It was shown years ago[20], that the transition can be fitted by the equation $\rho \sim j^n$. The quality factor n describes the steepness of the transition curve. It can be estimated by changing the measuring sensitivity by a factor of 10. To date there is no systematic experimental work concerning the physical and microstructural meaning of n. But obviously it reflects the integrity and homogeneity of the filaments over the conductor length and over the cross section. It depends on the filament diameter and number, the Cu area and the applied external field. Typical values for high quality conductors at 4.2 K are 40 to 50 at 5 T and 25 to 35 at 8 T. Cabling leads to a degradation of n to about 20 at 5 T.

CONCLUSIONS

NbTi conductors still can be improved to qualify them for more sophisticated or new applications. Areas of investigations concern fine filaments, large numbers of filaments, low ac losses and high field applications. A common procedure to characterize the conductor performance is required to give a sound basis for comparison of different conductors.

ACKNOWLEDGEMENT

The authors gratefully acknowledge the effort and help of many collegues at Vacuumschmelze who contributed to the work and to the results reported here. They also would especially like to express their thanks for helpful discussions and for providing test results to U. Jeske, K. P. Jüngst, C. Schmidt, P. Turowski (KfK), S. Wolff (DESY), B. Turck (CEA), D. Leroy, R. Perin (CERN), C. L. Goodzeit, A. F. Greene, W. B. Sampson (BNL) and D. C. Larbalastier (UW).

REFERENCES

1. H. Hillmann, in:"Superconductor Materials Science-Metallurgy, Fabrication and Applications", S. Foner and B. B. Schwartz, eds., Plenum Press, New York (1981), p. 275.
2. H. R. Weber, in:"Extrusion", G. Lang et al., eds., Deutsche Gesellschaft für Metallkunde e. V., Oberursel/F.R.G. (1981), p. 271.
3. I. Pfeiffer and H. Hillmann, Acta Met. 16: 1429 (1968).
4. D. C. Larbalestier, loc. cit. 1., p. 133.
5. D. C. Larbalestier and A. W. West, Acta Met. 33: 1781 (1984).
6. P. Dubots, A. Février, J. C. Renard, J. C. Goyer, H. G. Ky, Journ. Physique 45: C1-467 (1968).
7. I. Hlásnik, S. Takács, V. P. Burjak, M. Majoroś, J. Krajcik, L. Krempasky, M. Polak, M. Jergel, T. A. Korneeva, O. N. Mironova, I. Ivan, private communication (to be published in Cryogenics, 1985).
8. H. Dittrich, S. Förster, A. Hofmann, U. Jeske, H. Katheder, A. Khalil, H. Krauth, W. Lehmann, W. Nick, A. Nyilas, A. Ulbricht, in: "Proc. 10th Symp. Fus. Eng., Philadelphia 1983, Dec. 5-9", IEEE, New York (1983), Vol. 2, p. 1350.
9. M. Suenaga, K. M. Ralls, Journ. Appl. Phys. 40: 4457 (1969).
10. G. D. Hawksworth, D. C. Larbalestier, Adv. Cryo. Eng. 26: 479 (1980).
11. E. W. Collings, "A Sourcebook of Titanium Alloy Superconductivity", Plenum Press, New York (1983).
12. H. Hillmann, K. J. Best, IEEE Trans. Mag. 13: 1568 (1977).
13. Y. Yamada, S. Murase, H. Wada, K. Tachikawa, Cryogenics 23: 670 (1983).
14. H. R. Segal, T. M. Hrycaj, J. J. Stekly, T. A. de Winter, IEEE Trans. Mag. 17: 53 (1981).
15. M. Wake, T. Shintomi, M. Kobayashi, H. Hirabayashi, in: "High Field Magnetism" (M. Date ed.), North-Holland 1983, p. 339.
16. M. Wake, T. Shintomi, M. Kobayashi, K. Tsuchiya, H. Hirabayashi, H. Wada, K. Inoue, K. Itoh, K. Tachikawa, IEEE Trans. Mag. 19: 552 (1983).
17. E. Olzi, R. Rondelli, P. Bruzzone, Cryogenics 24: 115 (1984).
18. K. J. Best, P. Genevey, H. Hillmann, L. Krempasky, M. Polak, E. Turck, IEEE Trans. Mag., 15: 765 (1979).
19. M. Garber and W. B. Sampson, IEEE Trans. Mag. 21: 336 (1985).
20. F. Voelker, Particle Accelerators, 1: 205 (1970).
21. H. Hillmann, Patent pending.

SUPERCONDUCTING MATERIALS FOR THE SSC*

R. Scanlan, J. Royet, and C. E. Taylor

Lawrence Berkeley Laboratory
University of California
Berkeley, California

ABSTRACT

The proposed Superconducting Supercollider presents several new challenges with regard to materials for dipole magnets. One design requires a NbTi superconductor with J_c (5T) greater than 2400 A/mm², whereas the Tevatron recently completed at Fermilab required a J_c (5T) $\geq$ 1800 A/mm². In addition, the high field design requires a conductor with a filament diameter of about 2.5 μm, if correction coils are to be eliminated. Finally, the high field design utilizes a 30-strand cable which again is a significant increase from the 23-strand cable used in the Tevatron. This paper describes the results of recent R and D programs aimed at meeting the stringent material requirements for the SSC.

INTRODUCTION

A review of the SSC design and program status is presented in another paper in these proceedings.[1] Presently, two dipole magnet designs are under active development. The high-field design option, referred to as Design D, requires a 30-strand cable for the outer layer and a 23 strand cable for the inner layer. The low field design, referred to as Design C, requires a 24-strand main cable and a 12-strand correction coil cable. Since the conductor requirements for the low field design are the same as those for the high field inner cable, the discussion here will focus on the superconductor requirements for the Design D dipoles. The conductor parameters for the Design D cables are listed in Table I.

A Reference Designs Study for the Superconducting Super Collider (SSC) was prepared for DOE during Feb.-May 1984. During this study the choice of a target critical current density for NbTi was discussed at length. A conservative choice would have been the value used for the Doubler/Saver magnets and for CBA, i.e., 1800 A/mm² at 5T. However, recent results[2] suggested that considerable improvements should be possible and these improvements would result in a substantial cost savings for the magnet system.[3] Consequently, a target value of 2400 A/mm² at 5T was chosen.

---

*This work was supported by the Director, Office of Energy Research, Office of High Energy and Nuclear Physics, High Energy Physics Division, U.S. Dept. of Energy, under Contract No. DE-AC03-76SF00098.

TABLE I

STRAND PARAMETERS	INNER	OUTER
Alloy	Nb 46.5 ± 1.5 wt % Ti High Homogeneity Grade or Equivalent	
Size	0.81 ± .01 mm	.65 ± .01 mm
Critical Current (5T, 4.2K)	535 A (minimum)	282 A (minimum)
Critical Current (8T, 4.2K)	212 A (minimum)	112 A (minimum)
Copper to Superconductor Ratio	1.3 ± 0.1:1	1.8 ± 0.1:1
Twist Pitch	.8 twists/cm	.8 twists/cm
Filament Size*	< 23 μm	< 23 μm
Minimum Length	760 m	760 m
CABLE PARAMETERS	**INNER**	**OUTER**
Number of Strands	23 strands	30 strands
Width	6.76 ± .03 mm	9.73 ± .03 mm
Mid-Thickness	1.458 ± .015 mm	1.166 ± .015 mm
Keystone Angle	1.67°	1.24°
Twist Pitch	74 mm	69 mm
Critical Current (5T, 4.2K)	10453 A (minimum)	7202 A (minimum)
Critical Current (8T, 4.2K)	4142 A (minimum)	2860 A (minimum)
Minimum Length	1082 m	1347 m

*Most Desirable Size is Approximately 2.5 μm.

During the 14 months since that decision, a number of NbTi billets have been produced; the results show that the J_c value chosen can be met with confidence by a number of U.S. and foreign manufacturers.

Another materials challenge posed by one of the reference design dipoles was the use of a wide cable with 30 strands which represents a step up in difficulty from the 23 strand Doubler/Saver and CBA cables. Once again, results of the SSC R&D program have shown that this cable can be manufactured with a high degree of confidence.

The present thrust of superconducting materials R&D in support of the SSC program is the development of fine filament NbTi as a replacement for the 20 μm filament material presently being used. This work is still in progress; however, initial results are quite encouraging.

As these examples show, the SSC design requirements are providing new impetus for the development of improved superconducting materials. The technical details of these and other SSC R&D programs will be presented.

IMPROVEMENTS IN NbTi CURRENT DENSITY

The "New Era" for high critical current density NbTi was initiated in 1982 with the report by the Baoji group of a J_c = 3900 A/mm² at 5T[(2)] (using a more sensitive criterion for J_c, Larbalestier confirmed a value of about 3450 A/mm²). This announcement stimulated a new interest in binary NbTi alloys in the U.S., in particular by Larbalestier and coworkers at U. Wisc. This group made an extensive analysis of conductors being produced in the U.S. and found that the composition of the NbTi alloy was quite inhomogeneous. They concluded that this lack of homogeneity prevented these alloys from responding effectively to the multiple heat treatments used by the Baoji group.[4,5] After a series of discussions with the NbTi alloy manufacturer (Teledyne Wah-Chang, Albany) a collaborative experiment aimed at testing these ideas was begun in August, 1983.

In this experiment a 10-inch billet (Billet 5183) was ordered by LBL. A special lot of high homogeneity alloy was purchased from TWCA and provided to IGC for processing. After extrusion, the material was divided into two lots - one for processing by IGC using their standard commercial process and the other to be held until Larbalestier could complete a J_c-optimization study and suggest an alternate treatment. As seen in Table II, Billet 5183 processed by conventional techniques produced an improved J_c (about 2300 A/mm² compared with about 2000 A/mm² for the best Doubler/ CBA material). This result (see Billets 5198-1 and 5198-2 in Table II) was verified on two additional billets procured by LBL and processed by IGC while Larbalestier was completing his optimization studies. Larbalestier[6] recommended a new processing schedule and IGC processed the remainder of Billet 5183 (designated 5183-2) with this schedule. As seen in Table II, the J_c values improved significantly (from 2365 A/mm² to 2645 A/mm² for the 0.6 mm diam. strand). Based on these results, LBL ordered two additional billets (5210-1 and -2) and FNAL ordered five billets (5209-1 through 5) from IGC in July, 1984. This material was delivered in January, 1985; the J_c values in all cases exceeded our target specification value of 2400 A/mm². The only problem encountered was a switch of extruded rod labels by the extruder; as a result IGC processed the billets with Larbalestier's heat treat schedule, but not at the specified wire sizes, for billets 5210-1 and -2.

The final material order for Design D dipoles was placed in November, 1984, after competitive bidding won by IGC. IGC delivered 373 kgm (inner layer material) and 373 kgm (outer layer material) in April, 1985, (see Table II); the J_c (5T) values are 2509 A/mm² for inner and 2719 A/mm² for outer layer material. With the exception of Billet 5210-2, all outer

TABLE II

Designation	Strand Diameter (mm)	J_c Value** $(10^{-12}\Omega\text{-cm})$ 5T	8T	Intermediate Heat Treatment	Final Heat Treatment
5183 - 1 (Cu/SC = 1.36/1)	.8 .65	2280 2365	930 1015	IGC	IGC
5198 - 1 (Cu/SC + 1.35/1)	.8	2238	885	IGC	IGC
5198 - 2 (Cu/SC = 2.05/1)	.65	2273	880	IGC	IGC
5183 - 2 (Cu/SC = 1.36/1)	.8 .65	2545 2645	1030 1070	U. Wisc. (3 X 40 hr at 375°C)	260°C
5210 - 1 (Cu/SC = 1.25/1)	.8	2505	1034	*	260°C
5210 - 2 (Cu/SC = 1.80/1	.65	2435	1070	*	260°C
5212 - 3 (Cu/SC - 1.77/1	.65	2717	1146	U. Wisc.	260°C
5212 - 1 (Cu/SC = 1.28/1	.8	2509	1083	U. Wisc.	260°C

*U. Wisc. heat treatment with slight variation.
**Data obtained by Larbalestier and co-workers, U. Wisc.

layer material processed with the new heat treatment (3 x 40 hr at 375°C) has a significantly higher J_c than inner layer material. The FNAL material (Billets 5209-1 through 5209-5) is equivalent to outer layer material and also is consistent with this observation. A possible explanation for this behavior is that the additional cold working after extrusion in the case of the outer layer material is beneficial in improving the J_c; if true, this result suggests that it may be possible to get somewhat higher J_c values in the fine filament conductors which also contain more cold working.

Another favorable result from these procurements has been the piece lengths compared with Doubler/CBA experience. The longer piece length greatly facilitates cabling and simplifies testing and quality control.

FINE-FILAMENT NbTi R&D

In addition to the SSC requirement of high current density it is desirable to use conductors with a fine filament size[3] in order to reduce field distortion at low fields due to magnetization effects. If a conductor can be fabricated with filament diameters of 2.5 µm or less, correction coils presently required for each dipole magnet could be eliminated. Another correction approach currently being evaluated would require a filament size in the 5-6 µm range.[7] During the past few months, significant progress has been made in establishing the technical feasibility of fine-filament NbTi (see Fig. 1). These results show that fine filaments are feasible; however, in each case the quantity produced was small and the process was not production scale (with the exception of the Supercon billet discussed below). We will now discuss several problems

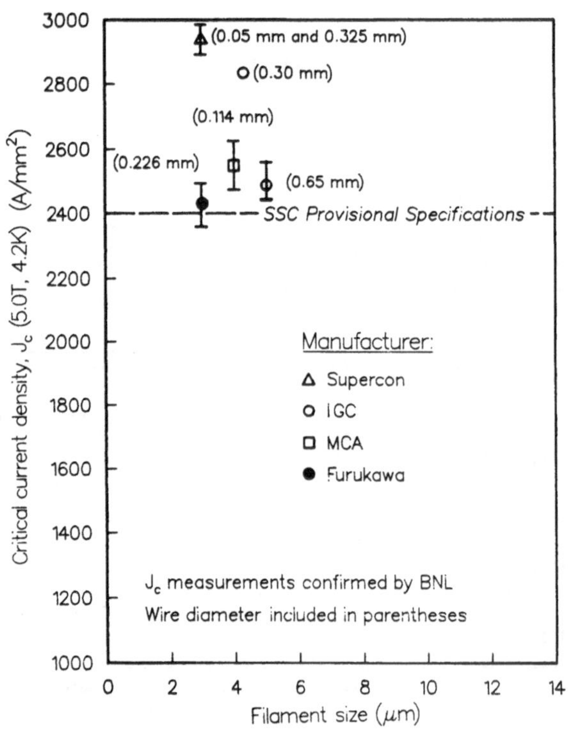

Fig. 1. Recent results show that high J_c values can be achieved in fine filament NbTi. The Supercon result for the 0.325 mm wire with 3 µm filaments was produced from a 305 mm diameter billet. The other results are from billets smaller than conventional production size billets.

which must be solved in the large scale manufacture of fine-filament NbTi and several R and D programs that are in progress.

Conventional production of NbTi superconductor consists of a hot extrusion (500-600°C) of NbTi rods in a copper matrix. During this extrusion and the prior heating of the billet, a layer of titanium-copper intermetallic compound, perhaps 1-2 μm thick, can form around the filaments.[8,9,10] This brittle intermetallic layer does not co-reduce and thus can become nearly equal to the filament diameter at final wire size; this results in extensive filament breakage and sometimes strand breakage. This problem can be eliminated by enclosing the NbTi rods at extrusion size in a barrier material, such as Nb or Ta, which prevents the titanium-copper intermetallic formation.[8,11,12] This barrier need only be 0.1 to 0.2 mm thick, and will be reduced to an insignificant fraction of the filament cross section at final filament size.

Another problem can arise from the introduction of foreign particles during the billet preparation operations. Any "dirt" consisting of micron size particles or any inclusions of this size in the NbTi rods or the copper components can result in filament breakage at the final wire size. This type of problem is insidious since processing may proceed successfully until the final wire size in approached. Also,the size of inclusion which is tolerable depends upon the desired filament size, e.g., a one micron diameter inclusion is acceptable for a 20 micron filament, but not for a 2 micron filament. This problem can be minimized by careful selection of raw materials and by clean room practice in billet assembly.

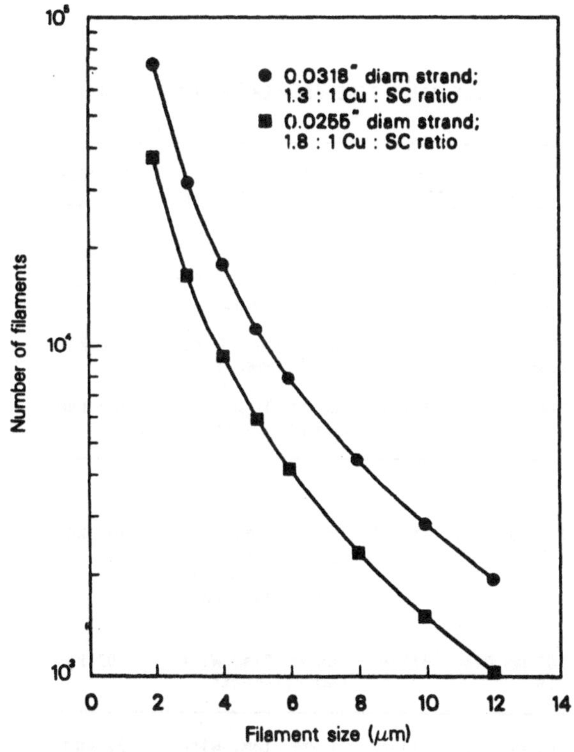

Fig. 2. A plot showing the number of filaments vs. desired filament size for the two Design D dipole conductors. To achieve a 2 μm filament size, approximately 40,000 filaments are necessary for the outer layer conductor, and over 70,000 filaments are necessary for the inner layer conductor.

When a large number of rods are stacked in a billet, as is necessary to achieve fine filaments, (Fig. 2), a large void fraction is present, and this can lead to non-uniform reduction in the extrusion step. The filaments are necked down locally and this also leads to filament breakage. This problem can be eliminated by compacting the billet before extrusion.

When these potential problems are eliminated by proper processing and quality control, there is no metallurgical reason why a J_c value of greater than 2400 A/mm² cannot be achieved in filaments approximately 2 μm in diameter. In fact, the increased total reduction in area of the NbTi filaments may mean that it is possible to introduce more heat treat/cold work cycles and hence raise the value of J_c.

These potential problems and the proposed solutions were discussed with several superconducting material manufacturers between December, 1983 and August, 1984. In September, 1984, both IGC and Supercon responded with proposals to investigate the production of high J_c, fine-filament NbTi. The deliverable items include material for J_c optimization studies and also material for construction of model magnets. The final reports will include an economic analysis of the fabrication method. The details of these projects are listed in Table III, and the current status of these efforts are reported in other papers at this conference.[13,14]

A practical problem to be solved in producing fine filament NbTi for the SSC Design D configuration is to devise a method of stacking a large number of elements to produce the end product (see Fig. 2). There are at least three promising approaches, which will be discussed below.

TABLE III - FINE FILAMENT R AND D EFFORTS

ORGANIZATION	BILLET SIZE AND Cu:SC RATIO	FINAL WIRE SIZE AND FILAMENT SIZE	QUANTITY	DELIVERY DATE
IGC				
PHASE I	152 mm Diam. Billet 1.8:1	.65 mm Diam. Wire 4μm Filaments	45 kgm	July 1985
PHASE II	254 mm Diam. Billet 1.3:1	.8 mm Diam. Wire 2μm Filaments	182 kgm	Dec. 1985
	254 mm Diam. Billet 1.8:1	.65 mm Diam. Wire 2μm Filaments	182 kgm	Dec. 1985
SUPERCON				
PHASE I	305 mm Diam. Billet 1.3:1	.8 mm Diam. Wire 8μm Filaments .27 mm Diam. Wire For 2-Level Cable 2μm Filaments	182 kgm	Aug. 1985
PHASE II	305 mm Diam. Billet 1.8:1	.65 mm Diam Wire 6μm Filaments	182 kgm	Oct. 1985
LBL HYDROSTATIC BILLETS	165 mm Diam. Billet 1.3:1	.8 mm Diam. Wire 2.5μm Filaments	91 kgm	Oct. 1985
	165 mm Diam. Billet 1.8:1	.65 mm Diam. Wire 2.5μm Filaments	91 kgm	Oct. 1985
	165 mm Diam. Billet 1.8:1	.65 mm Diam. Wire 2.5μm Filaments	91 kgm	Nov. 1985

First, one can stack a large number of rods in a single billet, and this is the approach being investigated by Supercon (see Table III). To date, they have stacked a 305 mm billet with 4164 rods and completed the extrusion successfully. Small amounts of this material have been processed to .8 mm diam. wire (8 μm filament size) and .3 mm diam. wire (3.0 μm filament size). The results are excellent. The filaments show no signs of degradation and the J_c (5T) = 2950 A/mm² for material with 3.0 μm filaments. A decision on final configuration will be made after results are complete on this optimization study and on two-level cable experiments in progress (see discussion below). The second phase of the Supercon study will consist of another 305 mm diam. billet with a 1.8:1 Cu:SC ratio and a filament number yet to be specified. If Supercon can demonstrate an acceptable stacking scheme, a Phase III billet will contain approximately 40,000 filaments and yield a final filament size of 2 μm.

IGC proposed to investigate a double stack-hot extrusion approach in which a number of rods are extruded in the first billet and then this material is stacked in a second billet. The Phase I extrusions consisted of an initial 152 mm diam. billet with 7 NbTi rods and a second 152 mm billet with 7 x 858 NbTi filaments. Approximately 45 kgm of this material has been drawn to final wire size and will be cabled in order to produce samples for I_c and magnetization measurements. Initial results indicate that this material will produce a J_c = 2450 A/mm² in 0.6 mm wire with 5 μm diam. filaments. It appears that some filament sausaging is occurring and this limits the J_c; this is discussed in more detail in Ref. 14. IGC is proceeding with Phase II (see Table III) and have procured the raw material for two 254 mm first stage billets. Material from these billets will then be restacked to produce a 254 mm billet of inner layer and a 254 mm billet of outer layer material. This wire should be ready for cabling in December, 1985. If this phase is successful, there will be enough cable for one 16-m model dipole and several 1-m models.

A promising alternative to the use of conventional hot extrusion with diffusion barriers is cold hydrostatic extrusion. Production size hydrostatic presses providing toll extrusion services are available in Europe (but not in the U.S.), and the costs are competitive with conventional extrusion. The maximum billet diameter is 165 mm, but billets to 1600 mm long can be extruded. Hence, one can process approximately the same weight of material using hydrostatic extrusion as can be produced from a 250 mm diam. conventional extrusion. However, the yield of useful material can be much higher in the hydrostatic extrusion case because of reduced end losses. This factor is especially important in a double extrusion process.

In order to evaluate, both technically and economically, the potential of hydrostatic extrusion for producing fine filament NbTi, three billets have been assembled for hydrostatic extrusion (see Table III). The elements for stacking the first two billets are being prepared using a bundle and draw approach. NbTi rods are clad with Cu, then 19 of these rods are loaded into another Cu tube, drawn and bundled to form the billet stacking elements. The third billet will be assembled with NbTi rods containing a diffusion barrier for comparison with the other two billets without barriers.

At the conclusion of this R&D program in December, 1985, we will have a data base on both cost and technical feasibility of various fine filament options. This will allow the SSC management to evaluate the fine filament option and to begin incorporating fine filament NbTi into the SSC long range plan.

CABLE FABRICATION

The effort to develop the cables required for SSC Design D dipoles has been proceeding along two paths - cabling experiments at LBL and process improvements at New England Electric Wire (NEEW). An experimental cabling machine has been constructed at LBL that can produce long continuous lengths of cable (up to about 1500 m. with the present spool system) at production speeds e.g., 4 m/min. In addition, the machine has several features not found on conventional machines, but essential for developing the optimum cabling parameters. These include variable planetary motion for the supply spools, precise tension control for the individual strands, capacity for 36 spools, and easy adjustment of cable twist pitch or cabling direction.

Several trial runs were made at NEEW between February and November, 1984. These trials were disappointing, especially for the 30-strand outer cable. Many crossovers occurred and only about 140 m. could be produced before crossovers recurred. In order to determine whether we were approaching some practical limit on strand number with the 30-strand cable, we attempted a 36-strand cable at LBL and made a successful cable. Additional trials on the LBL experimental cabling machine showed that two conditions contributed to crossovers - uneven tension from strand to strand and a small mandrel diameter (6.4 mm). When these two conditions were corrected at NEEW, the crossover problem disappeared. After these changes were made, a total of approximately 3600 m of cable have been made at LBL at a speed of 3-4 m/min. and a yield of over 95%. The increased yield is due (1) to improved wire lengths and quality, and (2) to improved cabling parameters. Cross sections of the 23-strand and 30-strand cables are shown in Fig. 3 and Fig. 4. At this time, we feel that the Design D cable can be made for the same cost, or perhaps somewhat less, than the Tevatron cable.

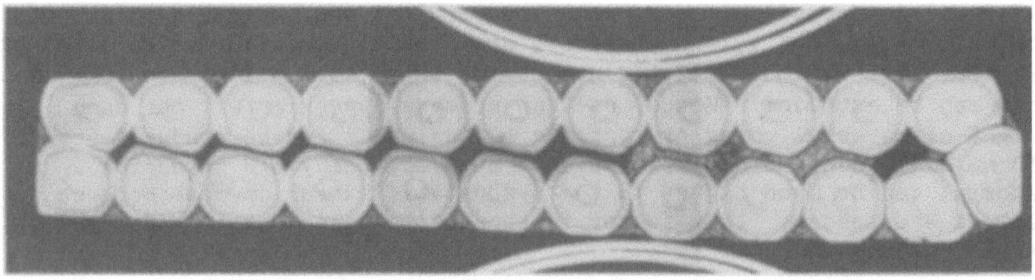

Fig. 3. Cross section of Design D dipole inner layer cable with 23 strands of 0.8 mm diameter. Note that the keystoning results in increased strand deformation at the narrow edge of the cable.

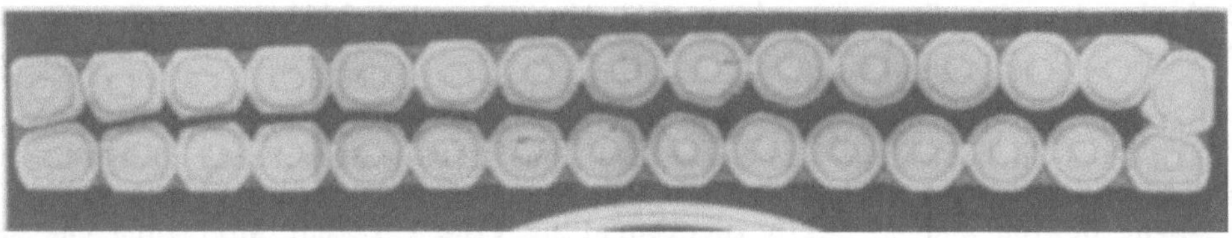

Fig. 4. Cross section of Design D dipole outer layer cable with 30 strands of 0.65 mm diameter.

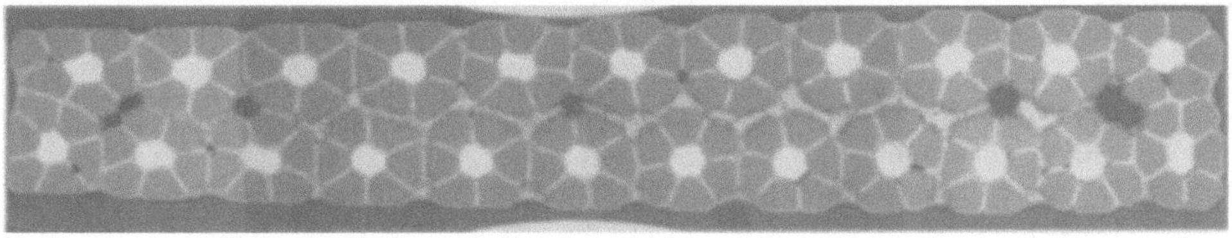

Fig. 5. Cross section of an experimental two-level cable made from Isabelle-type strands. This experiment demonstrates that a two-level cable substitute for the conventional cable shown in Fig. 3 can be made.

Currently, we are investigating several new cables which could have advantages for SSC dipoles. These include two-level cables, internal wedge cables, and internal flat cables. The two-level cable is of interest from the standpoint of fine filaments and increased flexibility. For example, we can use the 4,164 filament material being produced by Supercon as a .8 mm diam. strand with 8 μm diam. filaments, or we can reduce the wire diam. to .28 mm (filament diam. = 2.8 μm), fabricate a 7 element cable, and then fabricate a 23 strand cable from these elements. At LBL, we recently produced a mechanical model of this cable using surplus Isabelle strand material (Fig. 5). The cabling was completed without problems and at a typical production line speed of 3 m/min. This cable will be wound into 1-m coils in order to evaluate its applicability for flush-end magnets. As soon as the new material is completed by Supercon, we will repeat this experiment with high-J_c strand material and make electrical measurements.

CONCLUSIONS

We now have a substantial data base from these production-size billets (15 billets for a total weight of approximately 2270 kgm, including FNAL billets), and several conclusions can be drawn:

(1) The interim SSC specification value for J_c(5T) of 2400 A/mm^2 can be met in industrial scale production. This performance has been demonstrated by a number of U.S. conductor manufacturers.

(2) The specification of high homogeneity NbTi appears to reduce the spread in J_c values (although a more stringent test of this hypothesis, based on a larger data base, is necessary).

(3) The use of high homogeneity NbTi has resulted in extremely long piece lengths.

As a result of cabling experiments at LBL and at NEEW, we feel that the 30 and 23-strand cables planned for the SSC Design D dipoles can be produced in a reliable and cost-effective manner.

ACKNOWLEDGMENTS

The improved conductors described in this paper are the result of an ongoing collaboration involving D. C. Larbalesteir and coworkers at the University of Wisconsin, M. Suenaga and W. Sampson's group at Brookhaven National Laboratory, and several groups in industry including Teledyne Wah Chang Albany, Intermagnetics General Corp., and Supercon, Inc. We wish to acknowledge their contributions and support.

REFERENCES

1. M. Tigner, Where is the Superconducting Supercollider (SSC) today? in: "Advances in Cryogenic Engineering," vol. 31, Plenum Press, New York (1986).
2. Li Chengren, Wu Xiao-zu, and Zhou Nong, IEEE Trans. Magn. MAG 19:284 (1983).
3. C. E. Taylor, in: "Proceedings, 1985 Particle Accelerator Conference," Vancouver, B.C. (May 1985).
4. D. C. Larbalestier and A. W. West, The metallurgical and superconducting properties of niobium titanium alloys, Annales de Chimie - Francaises Science des Materiaux 9:813 (1984).
5. D. C. Larbalestier, Towards a microstructural description of the superconducting properties, IEEE Trans. Magn. MAG-21:257 (1985).
6. D. C. Larbalestier, A. W. West, W. Starch, W. Warnes, P. Lee, W. K. McDonald, P. O'Larey, K. Hemachalam, B. Zeitlin, R. Scanlan, and C. Taylor, High critical current densities in industrial scale composites made from high homogeneity Nb 46.5 Ti, IEEE Trans. Magn. MAG-21:265 (1985).
7. B. C. Brown and H. E. Fisk, A technique to minimize persistent current multipoles in superconducting accelerator magnets, in: "Proceedings of the 1984 Summer Study on the Design and Utilization of the Superconducting Super Collider," Snowmass, Colorado, June 23-July 13 (1984), p. 336.
8. P. Dubots et al., in: "Proceedings, ICEC 8," (1980), p. 505.
9. M. Garber, M. Suenaga, W. B. Sampson, and R. L. Sabatini, Effect of CuTi compound formation on the characterisitics of NbTi accelerator magnet wire, in: "Proceedings, 1985 Particle Accelerator Conference," Vancouver, B.C. (May 1985).
10. D. C. Larbalestier, L. Chengren, W. Starch, P. J. Lee, Limitation of critical current density by intermetallic formation in fine filament Nb-Ti superconductors, IEEE Trans. Magn. MAG-21:39 (1985).
11. M. T. Taylor, C. Graeme-Barber, A. C. Barber, and R. P. Reed, Cryogenics 11, 224-226 (1971).
12. E. Gregory, Manufacture of superconducting materials, R. W. Meyerhoff, ed., American Society for Metals, Metals Park, Ohio (1977), pp. 1-16.
13. T. S. Kreilick, E. Gregory, and J. Wong, Fine filamentary NbTi superconducting wires, in: "Advances in Cryogenic Engineering--Materials," vol. 32, Plenum Press, New York (1986).
14. K. Hemachalam, C. G. King, B. A. Zeitlin, and R. M. Scanlan, Fabrication and characterization of fine filaments of NbTi in a copper matrix, in: "Advances in Cryogenic Engineering--Materials," vol. 32, Plenum Press, New York (1986).

CRITICAL CURRENT STUDIES ON FINE FILAMENTARY NbTi ACCELERATOR WIRES*

M. Garber, M. Suenaga, W.B. Sampson, and R.L. Sabatini

Brookhaven National Laboratory, Upton, New York

ABSTRACT

The magnets for the Superconducting Super Collider, a high energy proton colliding beam accelerator, require a superconductor with very high current density ($>$ 2400 A/mm^2 at 5 T) and very small filaments ($<$ 2 μm in diameter). Previous work has shown that by controlling the formation of Cu_4Ti compound particles on the filament surfaces it is possible to make fine filamentary NbTi wire with high critical current density. The performance of multi-filamentary wire is characterized by the current density and the quantity "n" which describes the superconducting-normal transition. Micrographs of wires having high J_c and high n show smooth, uniform filaments. Recently wires of very high critical current and high n have been produced in experimental quantities by commercial manufacturers.

INTRODUCTION

Magnet design requirements of the Superconducting Super Collider Project (SSC)[1] have recently stimulated two impressive improvements in NbTi multifilamentary wire. Larbalestier and coworkers have shown that critical current densities in the vicinity of 3000 A/mm^2 (at 5 Tesla and 4.2 K) can be attained by means of a series of anneal-strain cycles during the reduction of starting billet to final wire.[2] This is more than 50% greater than the mean current density of the wire used in the Fermilab Tevatron.[3] It is also well in excess of the original SSC specification of 2400 A/mm^2. A more appropriate specification might now be 1800 A/mm^2 at 6.5T.

A second requirement of the SSC design is that the diameter of the NbTi filaments be in the range 2-3 μm, considerably smaller than was common previously; the filaments in Tevatron wire are 9 μm. It has been pointed out that uniform small diameter filaments can be produced only if care is taken to control the formation of intermetallic compound (Cu_4Ti) particles on the filament surface during processing.[4] Magnetization measurements have shown that the transport current in wires having non-uniform filaments may be drastically reduced without altering the intrinsic current density.[5]

*Work supported by the U.S. Department of Energy

In this paper, further electrical and metallurgical characterizations of this problem are presented.

CRITICAL CURRENT AND FILAMENT UNIFORMITY

In order to compare results for different wires it is usual to quote a critical current density, $J_c \equiv I_c/A_s$, where I_c is the measured critical current and A_s is the total superconductor cross section. In wires having large filament diameters, $d \gtrsim 20$ μm, say, the filament cross section is usually constant enough as a function of length that J_c is equal to the microscopic or intrinsic critical current density. However, variations in filament cross section, such as sausaging, pinching, breaks, etc., cause I_c to be reduced.

The reduction in I_c due to filament non-uniformity is accompanied by a broadening of the superconducting-normal state transition. This transition is frequently represented by the equation

$$\rho = VA/I = \text{const} \cdot I^n$$

where V is the voltage per unit length, A is the total wire cross section area, and I is the measuring current in the sample.[6] The critical current is defined as the current at which $\rho = 10^{-12}$ ohm cm. This definition is based on the practical fact that for this value of the resistivity the dissipation in large scale devices is acceptably small or negligible.[7] The n-value is inversely related to the width of the V-I transition.

Figure 1 illustrates these comments by showing results for a series of wires which are drawn to successively smaller diameters. The starting wire was a BNL CBA Project (CBA) wire in which the n-value was relatively large and in which the 9 μm filaments were seen to be smooth and uniform as observed using scanning electron microscopy, Fig. 2A. As the wire is drawn down, surface imperfections, which are not important at large filament size, gradually produce significant variations in filament cross section, as illustrated in Fig. 2B. Correlated with this are decreases in I_c and n.

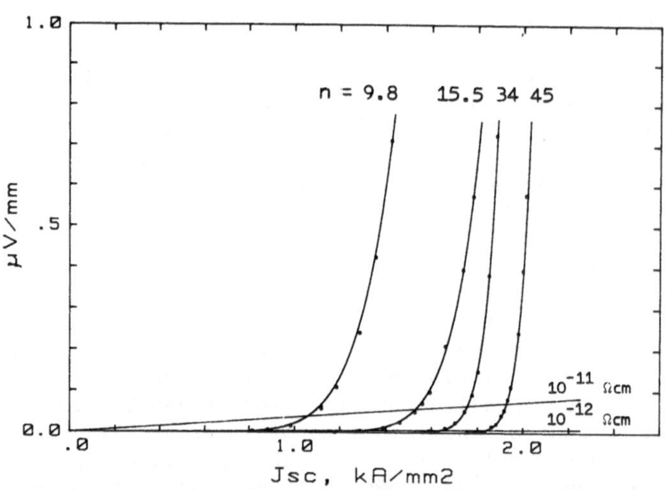

Fig. 1 Superconducting-normal transitions for a series of drawn wires. Mean filament diameters are, right to left, 9.0, 4.6, 2.3, and 1.3 μm, respectively. CBA starting wire, Cu/SC ratio = 1.7, T = 4.23K, B = 5T.

Fig. 2A SEM micrograph of 9 µm filaments in CBA wire used in the draw down experiment. J_c = 2000 A/mm^2, n = 40 at T = 4.23K, B = 5T.

Fig. 2B SEM micrograph of filaments after reduction to mean diameter 1.3 µm (wire diameter = .004").

The same effect is observed within a group of wires which are nominally the same in composition and metallurgical treatment but in which the prevention of Cu_4Ti particle formation is inconsistent. A statistical distribution of I_c's and n-values results, and these are correlated as shown in Fig. 3. This figure gives data for CBA wires obtained from one particular manufacturer. In this graph J_c is plotted as a function of n; the latter is an index of geometric filament quality.

Photomicrographs of filaments of high n, high I_c and of low n, low I_c wires are shown in Figs. 4A and 4B, respectively. Examination of the necked regions at higher magnification indicates that these result from the presence of compound particles.[4] Similar results have been observed in the newer, very high J_c wire produced by multiple anneal-strain cycles, and in monolithic conductors produced for NMR medical apparatus. The filaments in the latter conductors are generally large, typically 30-50 µm, and n-values between 50 and 100, at 5T, are commonly observed.

VARIATION OF n-VALUE WITH APPLIED FIELD

The shape of the superconducting-normal state transition varies with the external field. In high quality multifilamentary wire n varies over the entire range of applied field, going to zero at the critical field. In low n-value material on the other hand n is relatively independent of field. This is illustrated in Fig. 5, which shows n-values for three CBA wires as a function of field. As before, SEM photomicrographs show a correlation between smooth, uniform filaments and high n-values, and moderate to strong filament non-uniformity in the low n wire.

In studying multifilamentary Nb_3Sn wire, Suenaga et al have observed that in wires in which n is low the n vs B curve is flat over most of the field range and joins the curve for high n wire near the

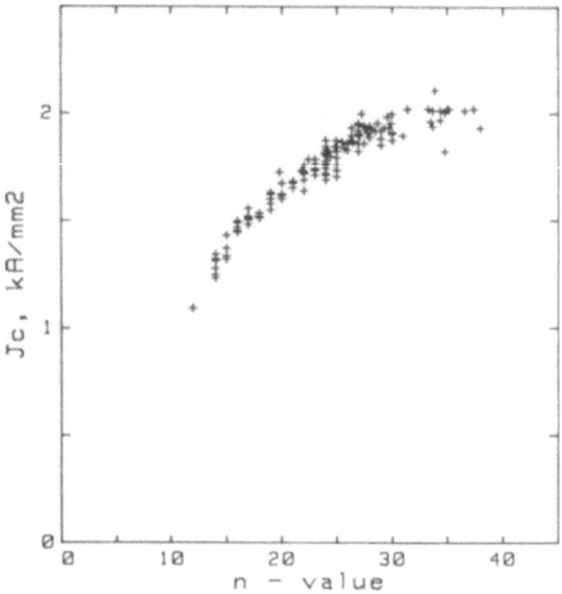

Fig. 3 J_c vs n for CBA wires of one manufacturer. Cu/SC ratio = 1.7. T = 4.23K, B = 5T.

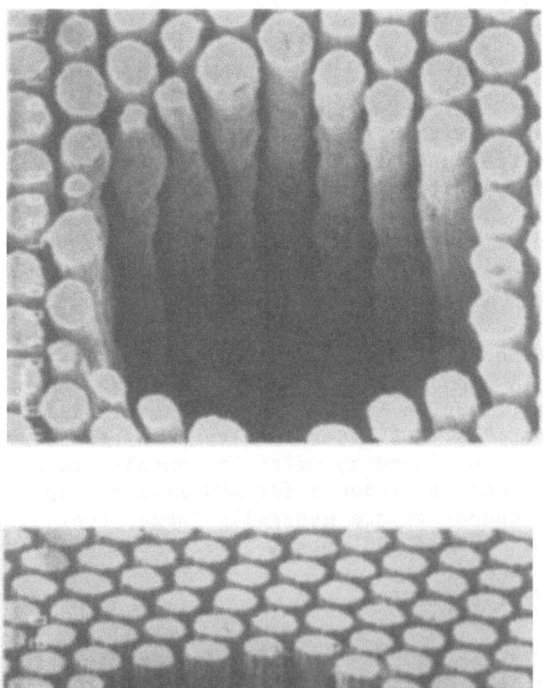

Fig. 4A SEM micrograph of filaments of low quality wire. n = 14, J_c = 1400 A/mm^2.

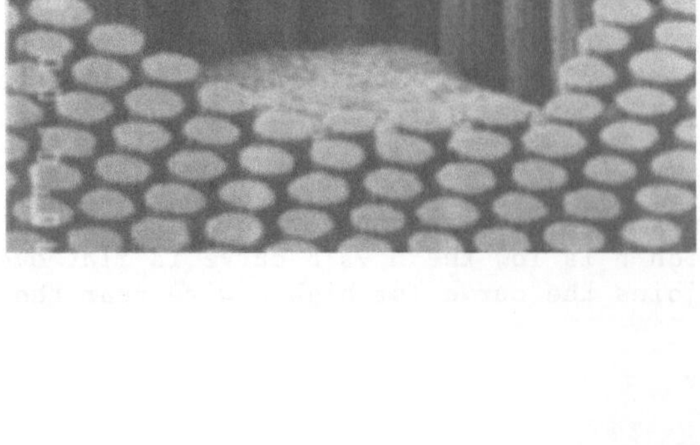

Fig. 4B SEM micrograph of high quality CBA wire. n = 40, J_c = 2000 A/mm^2.

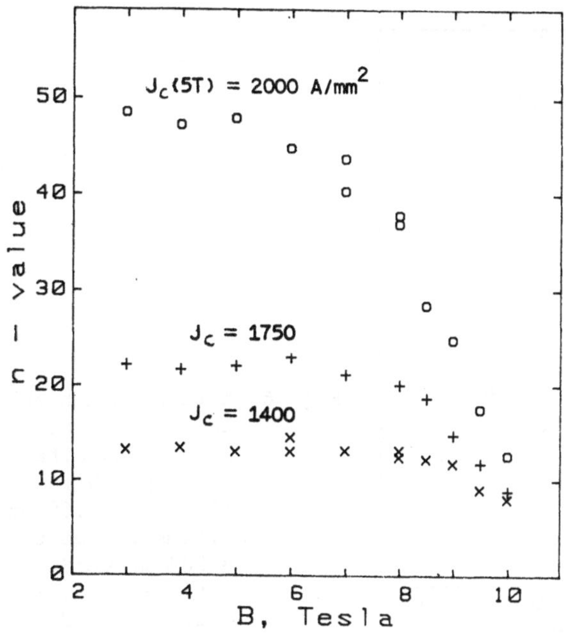

Fig. 5 n vs B for CBA wires of varying filament uniformity.

critical field.[8] Furthermore, when the temperature was lowered from 4.2K to 1.8K, the critical currents of both types increased by the same percentage. The value of n stayed constant for the low n wire but increased for the high n wire (∿50% at B = 16T, e.g.). These observations in Nb_3Sn and the present data suggest that J_c is limited in low n wire by geometrical factors, such as filament non-uniformity, while in high n-wire it is determined by superconducting properties, such as flux pinning strength.

W. Warnes and D. Larbalestier have independently arrived at similar conclusions regarding the correlation between filament uniformity and n-value behavior.[9] For a uniform monofilament NbTi sample they observe the strongly varying type of n(B) curve with n-values in excess of 100 at low fields. In multifilamentary samples, more slowly varying or flat n(B) curves are observed depending upon the degree of sausaging or non-uniformity of the filaments.

EFFECT OF MECHANICAL DAMAGE

Flattened cables of multifilamentary wire usually have lower I_c's and n-values than those of the pre-cabled wire.[10] In keystoned cables, the packing factor of the wire in the thin edge region is frequently around 100%, so the possibility of mechanical damage is not small. Wires tested before cabling and after removal from a damaged cable show a pattern of behavior similar to that described above. This is illustrated in Fig. 6. The NbTi in this example is of the newer high J_c, high homogeneity type; the pattern of behavior is the same as for the older material. In this case, the filament degradation is localized to the regions of the wire which pass around the cable edge. The appear ance of the filament distortion is also different, as shown in Fig. 7. Scanlan and co workers observed the formation of slip lines on the 200 μm filaments of a large, square cross section, monolithic conductor due to the final sizing steps.[11]

Although the example cited above is a drastic case of mechanical damage on the filaments influencing the n-value, less drastic deformation such as light rolling of a wire can also reduce n. In

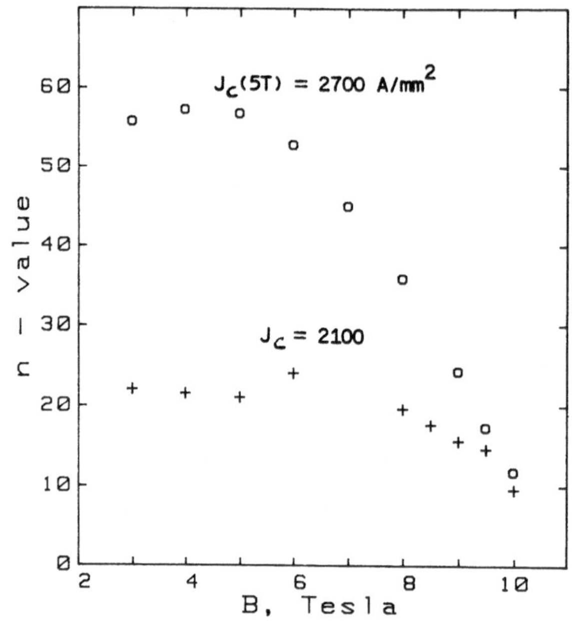

Fig. 6 n vs B in a high homogeneity wire before and after cabling. Wire diameter = .65 mm, filament diameter = 20 μm, Cu/SC ratio = 40, T = 4.23K, B = 5T.

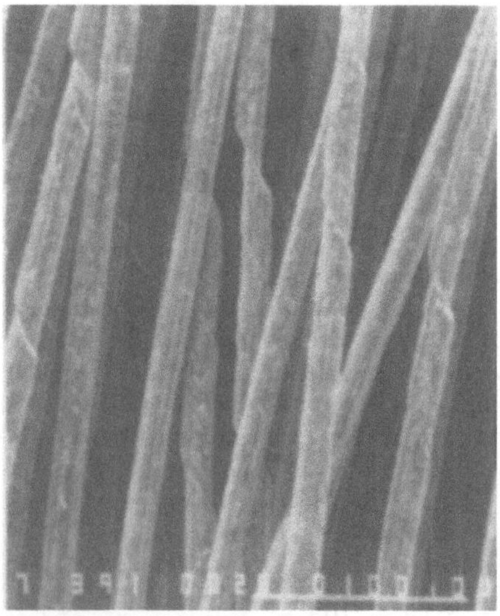

Fig. 7 SEM micrograph showing filament damage in thin edge region.

general, when a wire is deformed to make an aspected conductor, the value of its J_c becomes anisotropic; i.e., J_c for B parallel is greater than that for B perpendicular to the flat face. Correspondingly, if the original wire n-value is large, in the flattened wire the parallel n-value is larger than the perpendicular.

ORIGIN OF THE n-VALUE

The inverse correlation between n-value and filament non-uniformity indicates that some form of current sharing is responsible for the observed behavior. Suenaga[12] has suggested that a model similar to that used to describe resisitivity of in situ superconductors[13] may be appropriate. However, a detailed theory of the n-value is lacking, nor is it clear that the correct functional dependence is $\rho \sim I^n$.

In an example of very fine filament wire in which the matrix is cupronickel, the n-B curve is as shown in Fig. 8. The data in this figure are for wire having 0.6 μm filaments.[14] Photomicrographs show them to be smooth and clean. J_c is low, ~ 900 A/mm^2 at 5T, and the n-values are low. In this case the curve is not flat even though n is low. It is suspected that filament breakage arising from sources other than compound particles is responsible for this behavior.

It is possible to have a wire in which n is high and Ic is low: we have measured a conductor with 3.5 μm filaments which had an n-value in excess of 100 at 5T but current density of 500 A/mm^2. However, it is not possible to have a high Ic, low n wire because of the large cross section area variations associated with low n.

RECENT DEVELOPMENTS

Using the newly devised methods of improving J_c, and proprietary techniques for suppressing compound formation, a number of manufacturers have produced high J_c wires with filament diameters below 5 μm. Some of these have been made available to us in connection with the SSC R and D program and are listed in the following table.

Recent Developmental Fine Filament NbTi Wire

Wire Size mm	No. of Filaments	Cu/SC Ratio	Filament Size μm	J_c* A/mm^2	n	Manufacturer
.23	3000	1.1	2.9	2400	25	Furukawa
.29	1950	1.4	4.3	2800	32	I.G.C.
.11	367	1.1	4.0	2500	31	M.C.A.
.04	127	0.75	2.7	2950	45	Supercon
.33	4164	1.2	3.5	2950	45	Supercon

*Determined from transport current

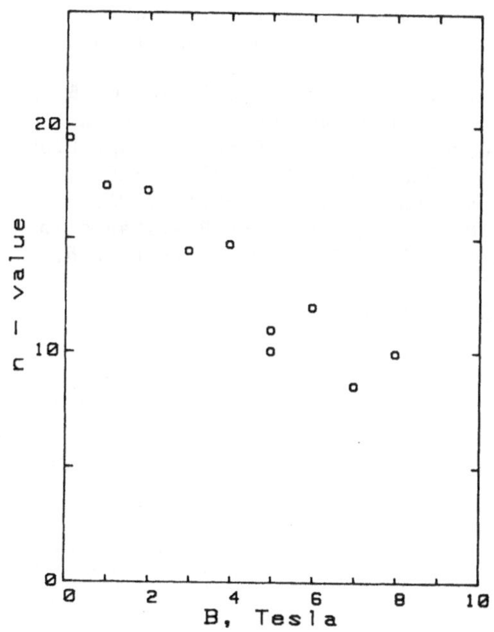

Fig. 8 n vs B for a 0.12 mm wire containing 15000 0.6 μm filaments. The filaments are in a cupronickel matrix which has a stabilizing copper core.[15] T = 4.23K B = 5T.

ACKNOWLEDGEMENTS

Wire samples have been provided by many manufacturers in this country, Europe, and Japan and we thank them collectively. R. Scanlan of LBL has provided samples of the new, high homogeneity wire and cable. We thank E. Sperry, J. D'Ambra, and F. Perez for technical help.

REFERENCES

1. Report of the Reference Designs Study Group on the Superconducting Super Collider, May 8, 1984.
2. D. C. Larbalestier, A. W. West, W. Starch, W. Warnes, P. Lee, W. K. McDonald, P. O'Larey, K. Hemachalam, B. Zeitlin, R. Scanlan and C. Taylor, "High Critical Current Densities in Industrial Scale Composites Made from High Homogeneity Nb46.5Ti", IEEE Trans., MAG-21, 269, 1985.
3. M. J. Tannenbaum, M. Garber, and W. B. Sampson, "Correlation of Superconductor Strand, Cable and Diople Critical Currents in CBA Magnets," IEEE Trans., MAG-19, 1357, 1983.
4. M. Garber, M. Suengaga, W. B. Sampson, and R. L. Sabatini, "Effect of Cu_4Ti Compound Formation on the Characteristics of NbTi Accelerator Magnet Wire," IEEE Trans. NS-32, 3681, 1985.
5. A. K. Ghosh, private communication. See also A. K. Ghosh, K. E. Robins and W. B. Sampson, "Magnetization Measurements on Multifilamentary Nb_3Sn and NbTi Conductors," IEEE Trans., MAG-21, 328, 1985 and these Proceedings.
6. M. Garber, W. B. Sampson, and M. J. Tannenbaum, "Critical Current Measurements on Superconductor Cables," IEEE Trans., MAG-19, 720, 1983.
7. W. B. Sampson, R. B. Britton, P. F. Dahl, A. D. McInturff, G. H. Morgan, and K. E. Robins, "Superconducting Synchrotron Magnets," Particle Accelerators, 1, 173, 1970.
8. M. Suenaga, K. Tsuchiya, N. Higuchi, and K. Tachikawa, "Superconducting Critical Current Densities of Commercial Multifilamentary Nb_3Sn(Ti) Wires by the Bronze Process," Cryogenics, 25, 123, 1985.
9. W. H. Warnes and D. C. Larbalestier, unpublished, presented at NbTi Workshops, Berkeley, January 1985 and Madison, July 1985.
10. M. Garber and W. B. Sampson, "Critical Current Anistropy in NbTi Cables," IEEE Trans., MAG-21, 336, 1985.
11. R. M. Scanlan, J. E. Johnston, P. A. Waide, B. A. Zeitlin, G. B. Smith, and C. T. Nelson, "Manufacturing and Quality Assurance for the MFTF Superconductor Core," Proc. 8th Symp. on Eng. Prob. of Fusion Res., IEEE Publ. No. 79CH1441-5, V.1, 260, 1979.
12. M. Suenaga, Talk to NbTi Workshop January 1985, Berkeley, CA.
13. A. Davidson, M. R. Beasley, and M Tinkham. "Remnant Resistance in Tsuei's Composite Superconductors," IEEE Trans., MAG-11, 276, 1975.
14. This wire was made by Alsthom, Belfort, France.

THE GROWTH OF INTERMETALLIC COMPOUNDS AT A COPPER-NIOBIUM-TITANIUM INTERFACE

D. C. Larbalestier*, P. J. Lee and R. W. Samuel*

Applied Superconductivity Center
University of Wisconsin-Madison
Madison, Wisconsin

ABSTRACT

The diffusion process between Cu and Nb-Ti has been studied in a well-bonded composite. Interdiffusion studies were performed between 500° and 650°C and three distinct layers were seen. Two of these were intermetallic compounds of approximate composition Ti_2Cu_7 and $(Ti_xNb_{1-x})Cu$, where x is ~ 0.5. A significant diffused region of Cu with grain boundary precipitation was seen in the NbTi solid solution. The growth rates of the layers were quite well-behaved and diffusion parameters for the process are presented.

INTRODUCTION

It has been known for some time that intermetallic compounds can be formed by reaction between Cu and NbTi.[1,2,3] It has also been known that particles of the intermetallic limit the size to which NbTi filaments can be drawn. The important influence of extrusion conditions on the intermetallic formation has been described by Hillmann[2]: increasing the extrusion temperature from 600° to 650°C greatly increased the particle size, producing markedly sausaged filaments.

The consequences of this intermetallic reaction have generally been quite manageable. It seems probable that the whole Fermilab production program (approximately 1000 0.25 m billets) had moderately sausaged filaments, the prime cause of the sausaging being intermetallic particles. The nominal filament size of the Fermilab filaments (8 to 10 μm) is right in the range where filament sausaging becomes strongly catalyzed by 1 to 2 μm intermetallic particles. However, the critical current requirement of the conductor was relatively moderate (~ 1850 A/mm^2 at 5 T, 10^{-14} Ωm) so that few conductor lengths had to be rejected for this reason.

Recent researches of our group have been aimed at understanding and greatly improving the current density of niobium titanium composites.[4,5,6] An important part of our studies has been the separation of intrinsic limits on the J_c (i.e. fluxoid-microstructure interactions) from extrinsic limits, such as chemical inhomogeneities[4,5] and intermetallic

*Also Department of Metallurgical and Mineral Engineering.

formation.[6] The requirements of the planned Superconducting Supercollider (SSC) have added considerable stimulus to such studies: in particular the need for high J_c (greater than 2400 A/mm^2 at 5 T) and fine (~ 3 μm) filaments. Recent experiments have shown that the Cu-NbTi reaction is a crucial obstacle standing in the way of fabricating such composites.[6,7]

We have had a number of experiments to study this problem underway in the last year. One set of experiments investigated the influence of extrusion conditions by working with composites of Nb 46.5 wt% Ti which had no extrusion whatsoever.[6] Only precipitation heat treatments at 375°C of 10 and 40 hours duration were given. The results of these experiments were quite striking. For example, six heat treatments of 40 hours at 375°C produced a critical current density of 3190 A/mm^2 (5 T) in a 19 μm filament composite but only 1320 A/mm^2 in a 3 μm filament composite. Reducing the heat treatment to 6 x 10 hours at 375°C produced a critical current density of 2705 A/mm^2 (20 μm filaments) and 2640 A/mm^2 in the 4 μm diameter composite. The differences between the two composites was due to the degree of sausaging produced by the heat treatments. The longer heat treatments produced 1 - 3 μm particles of intermetallic. This produced only minor sausaging in the 20 μm filaments but major sausaging in the 4 μm filaments. The particles were much smaller for the 10 hour heat treatments and the filaments were consequently not greatly sausaged even at 4 μm diameter. These experiments clearly showed that the extrusion at 550° to 650°C is not the only source of intermetallic. Our conclusion was that there is a potential for sausaging introduced by the low temperature precipitation heat treatments designed to raise the critical current density at any filament size below 10 μm.

The present paper amplifies these earlier experiments by providing basic information about the intermetallic compound and its growth in the temperature range from 500° to 650°C.

EXPERIMENTAL DESIGN

Basic Diffusion Kinetics

The composite used for our experiments was a 361 filament composite of nominal Nb 46.5 wt% Ti fabricated in our own laboratories solely by cold drawing techniques. The alloy rod was of high homogeneity grade and had an analyzed composition of Nb 47 wt% Ti. The oxygen content was 500 ± 50 ppm and it met the Fermilab specification in all respects. The copper to Nb-Ti ratio of the composite was 1.58:1. A cross-section of the composite is shown in Fig. 1.

Interdiffusion experiments were carried out at 500°, 550°, 600° and 650°C on sections of the composite at a wire size of 1.7 mm. The filament diameter was then about 55 microns and the interfilament distance 10 - 15 μm. Our aim was to have a well bonded composite of well shaped filaments in which relatively thick layers of intermetallic could be grown without exhausting the copper or NbTi sources. This was designed to facilitate layer thickness measurements by light microscopy and phase identification by the electron probe.

Metallographic Examination

Layer thicknesses were measured from several light micrographs of the reacted layer, the measurements being calibrated against a stage micrometer. Each quoted thickness is an average of about 20 measurements. The measurements of the inner layer showed greater scatter than the outer layer. Measurements were made on plane regions of the reaction

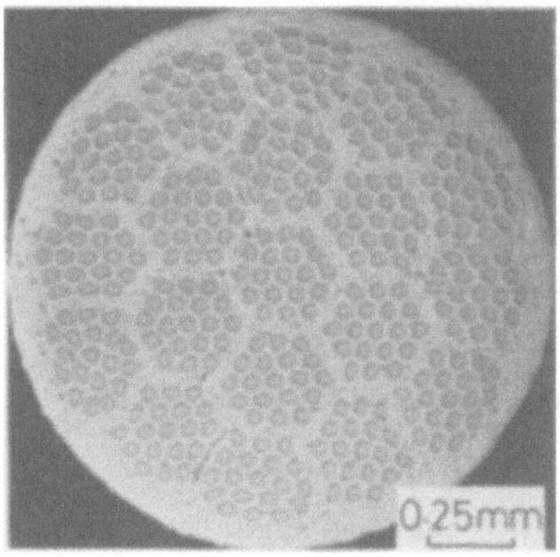

Fig. 1. Overall view of composite.

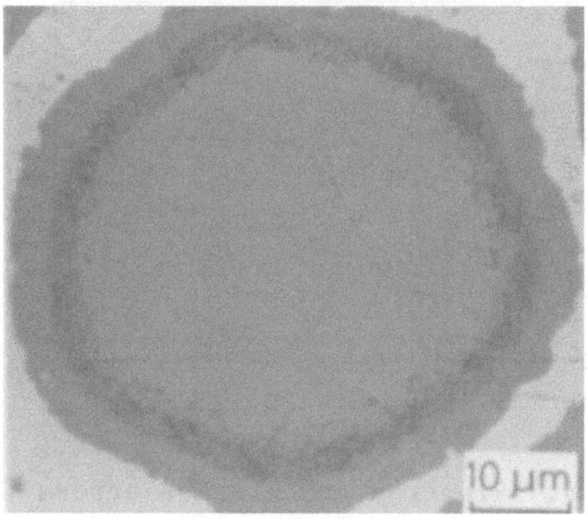

Fig. 2. Light micrograph of filament reacted 20hrs/650°C.

front (see Fig. 2) without significant indentation or curvature. The different layers were revealed by etching. An etch of 5 HF, 5 HNO_3, 90 H_2O (parts by volume) was found to distinguish the layers.

Chemical analysis of reacted layers was performed with an ARL microprobe, a JEOL 35C scanning electron microscope and a PEI 595 scanning Auger microprobe (SAM). A limited amount of TEM and STEM analysis was also performed.

RESULTS

Phase Identification

Cursory examination of the thicker layers showed there to be more than one phase present. The samples reacted for 20 and 100 hours at 650°C were accordingly selected for further study. Figures 2 through 5 show the samples examined by light microscopy, SEM, the electron probe and the SAM. Three layers were clearly distinguished in the SEM, the probe and the SAM; the outer two layers were not however well distinguished by light microscopy.

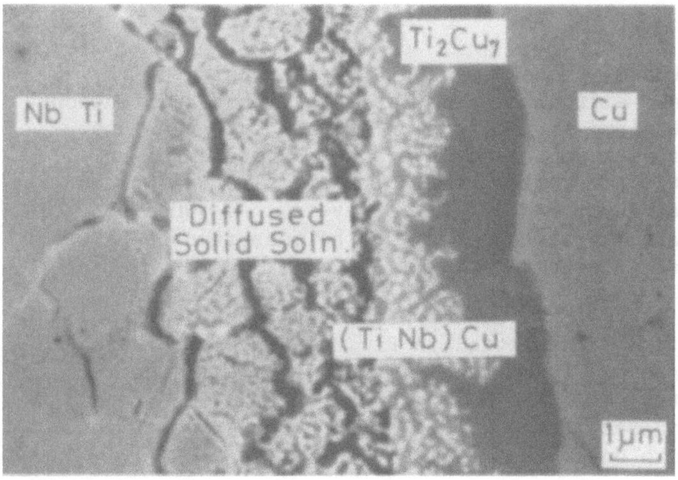

Fig. 3. Back scatter scanning electron micrograph of filament reacted 20hrs/650°C.

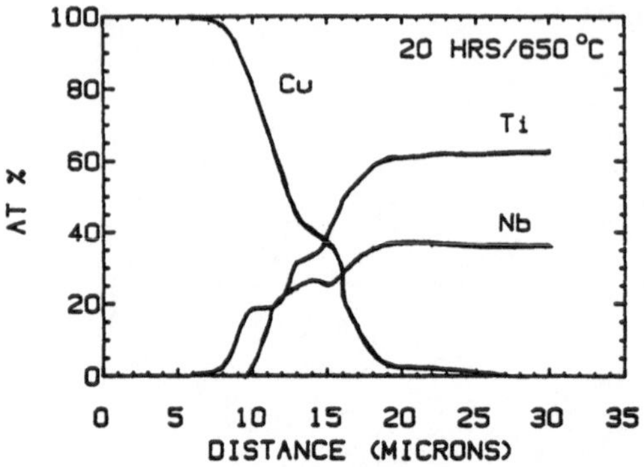

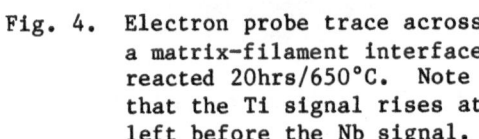

Fig. 4. Electron probe trace across a matrix-filament interface reacted 20hrs/650°C. Note that the Ti signal rises at left before the Nb signal.

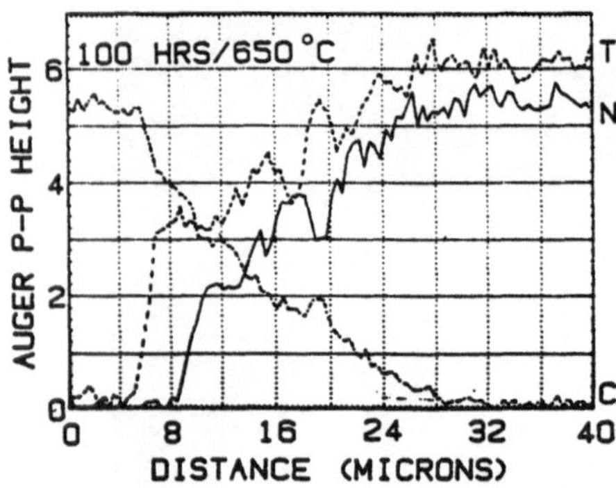

Fig. 5. Scanning Auger microprobe trace across a matrix-filament interface reacted 100hrs/650°C.

Quantitative phase identification was most accurately made with the electron probe. However, the spatial resolution of the electron probe for quantitative work is uncertain for layers of less than 5 μm thickness and it was found useful to supplement the probe traces with SAM line scans. The information provided by the SAM is only semi-quantitative but the spatial resolution is of order 0.25 μm. The SAM line scan in Fig. 5 clearly shows three distinct layers, the outer being Nb free, and the inner two layers containing all three elements. The outer two layers have regions of approximately constant composition, indicative of fixed composition compounds, while the innermost layer has a continuously variable composition. The SEM and the light micrographs (Figs. 2 and 3) suggest that this inner layer contains grain boundary phases but their size was too small for quantitative identification. A significant copper signal was found 15 μm into the filament given a reaction of 100 hours at 650°C (Fig. 4).

Diffusion Kinetics

Layer thickness results are plotted against the square root of time in Fig. 6. All measurements were found to be well described by the relationship:

$$x = (Dt)^{1/2}$$

where x is the layer thickness, t the time and D the diffusion parameter. We report D as a diffusion parameter rather than a diffusion coefficient since it is obtained from diffusion through more than one layer. Since the two outer intermetallic layers could not be reliably distinguished in the light microscope, the results are plotted for the outer two layers, the inner layer and all three layers. At 500°C none of the layers could be separately distinguished and the measurements only refer to the whole layer at this temperature.

Table 1 collects the diffusion equation parameters obtained from a least squares analysis of the data plotted in Fig. 6. The correlation coefficients were always above 97% and generally above 99%, indicating the good nature of the fits. Measured diffusion parameters are of the order 10 to 1000 x 10^{-18} m^2/sec. Figure 7 is an Arrhenius plot of the data, assuming a temperature dependence of the form

$$D(T) = D_0 \exp -(Q/RT)$$

where D_0 is a constant, R the gas constant and Q the activation energy for layer growth. The data of Fig. 6 produced reasonable straight line fits (Fig. 7) from which the activation energy and D_0 parameters shown in Table 2 were derived. The inner layer is seen to have an activation energy about half that of the outer layers (117 versus 205 kJ/mol). The relative growth rates of the layers changed between 650° and 600°C. At 650°C the outer layers grew faster but at 600°C and below the inner layer had the faster growth rate.

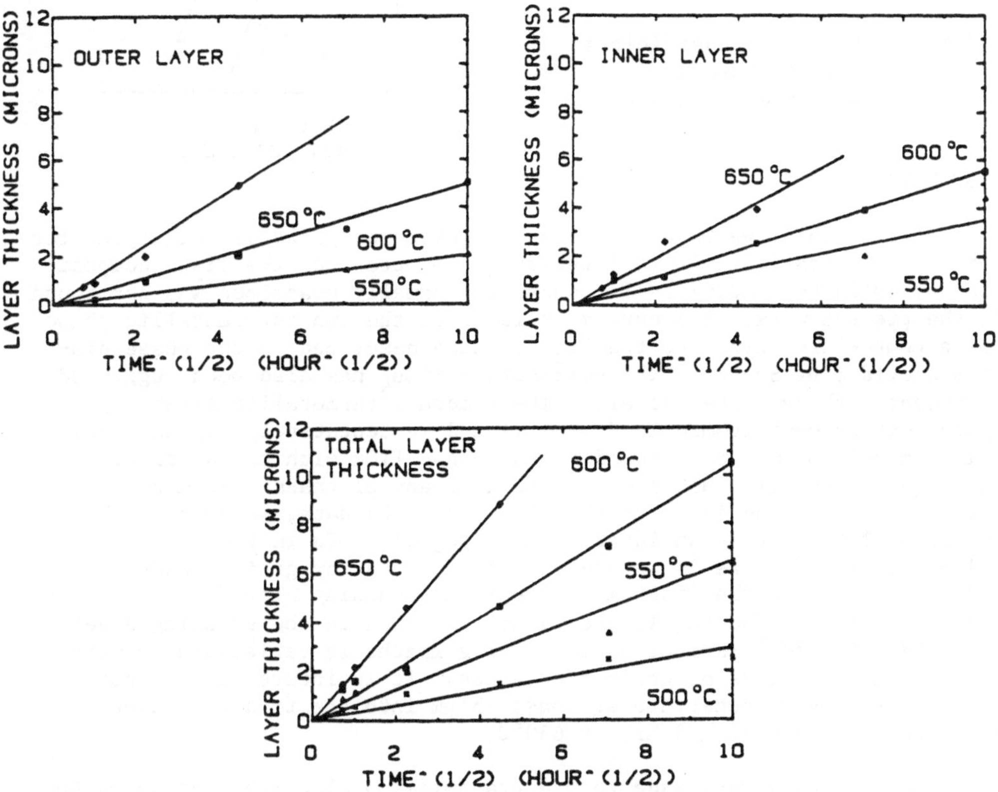

Fig. 6. Layer thickness vs. $(\text{time})^{1/2}$ plots for the outer intermetallic layer, the inner diffused layer, and all three layers.

Table 1. Diffusion Parameter Data

Temperature	Diffusion Parameters (m^2/sec)		
(°C)	Outer two layers	Inner layer	All three layers
650	344×10^{-18}	204×10^{-18}	1080×10^{-18}
600	72.1×10^{-18}	85.3×10^{-18}	299×10^{-18}
550	13.4×10^{-18}	80.3×10^{-18}	107×10^{-18}
500	--	--	17.7×10^{-18}

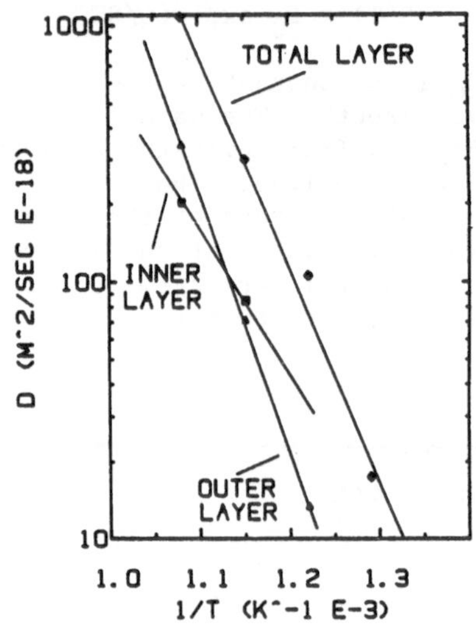

Fig. 7. Diffusion coefficients vs. the inverse absolute temperature.

DISCUSSION

At this point we have relied on the electron probe microanalyzer for phase identification. It is interesting to note that the Ti_2Cu composition by which the intermetallic phase is normally described was not found in the present work. The outermost layer of the two intermetallic phases has a composition close to the Ti_2Cu_7 phase described in the phase diagram provided by Shunk.[8] The composition $TiCu_4$ has also been suggested in recent work by Garber et al.[7] The second intermetallic layer is approximately equi-atomic in Nb and Ti and we tentatively suggest that this composition may correspond to the phase TiCu with the Nb freely substituting for Ti. Based on the lack of any plateaus of composition seen in the scanning Auger or the electron probe data, we propose that the innermost layer is an interdiffused region of Cu in the bcc NbTi solid solution. There is strong visual evidence for grain boundary precipitation from the light microscopy and scanning electron microscopy (Figs. 2 and 3). The Nb, Ti and Cu traces shown in the scanning Auger data (Fig. 5) show distinct evidence of sympathetic variations, possibly being caused by crossing grain boundaries. It is interesting to note that Cu appears to penetrate at least 15 μm into the filament given a reaction treatment of 100 hrs at 650°C.

Some attempts were made to use transmission electron microscopy to investigate the layer formed at lower reaction temperatures. However it was found difficult to thin the intermetallic layer. In one case we were able to examine the layer found on a composite reacted for 10 hrs at 405°C. Nodules of intermetallic approximately 100 nm thick were seen after this treatment. EDX analysis showed that only Cu and Ti were found in this layer. There was little sign either of any grain boundary precipitation typical of that seen in Fig. 2 or Fig. 3. There is some possibility therefore that the stable phases existing at 400°C are not the stable phases observed in our investigations at 500° to 650°C.

Table 2. Diffusion Equation Parameters

Layer	D_0 (m^2/sec)	Q (kJ/mol)
Outer two layers	137×10^{-6}	205
Inner layer	819×10^{-12}	117
All three layers	1.09×10^{-6}	159

Table 3. Predicted Layer Thicknesses for Important Conditions of Superconductor Fabrication. (Data generated from information in Table 2).

Temperature (°C)	Time (hr)	Intermetallic Layers	Solid Solution Layer	All Layers
300	3	0.6 nm	14 nm*	6 nm
375	40	24 nm	210 nm*	155 nm
	160	49 nm	420 nm*	310 nm
550	1	0.22 μm	0.33 μm	0.56 μm
	3	0.38 μm	0.58 μm	0.98 μm
600	1	0.52 μm	0.54 μm	1.1 μm
	3	0.90 μm	0.94 μm	1.9 μm
650	1	1.1 μm	0.84 μm	2.0 μm
	3	1.9 μm	1.45 μm	3.4 μm

*These figures are believed to be too high. The calculated total layer thickness is believed to be an upper limit, based on limited TEM data.

The growth rate data shown in Fig. 6 is rather well behaved and appear to yield good Arrhenius plots for the various layers (Fig. 7). The activation energies (Table 2) cannot be taken too rigorously, since they are composites for more than one phase. The trend is reasonable however. The outer intermetallic layers have the reasonably high value of 205 kJ/mol which is consistent with diffusion through an intermetallic compound, while the inner layer has only about half this value (117 kJ/mol). This latter value is consistent with diffusion through a solid solution, particularly one where there is marked evidence of preferential grain boundary diffusion.

An attempt to use the data for assessing the composite fabrication process is shown in Table 3. In this table, we have calculated the layer thicknesses for various times and temperatures of interest. Extrusion is commonly carried out between 550° and 650°C and extrusion preheats of 1 to 3 hours are frequently used. Reducing the extrusion conditions from 3 hrs at 650°C to 1 hr at 550°C is certainly likely to be beneficial, the intermetallic thickness declining from 1.9 to 0.22 μm (3.4 to 0.56 μm if the whole layer is considered).

Optimization heat treatments for Nb 46.5 wt% Ti are frequently carried out at 375°C, three such heat treatments not being uncommon.[5] For 40 hour treatments, about 24 nm of intermetallic (approximately 155 nm of total layer) should form with each heat treatment. The layer falls to negligibly low levels at 300°C.

Other points need to be considered, however. These experiments were carried out on a well bonded composite and they are likely therefore to overestimate the layer growth in a normal extrusion, where there is no metallurgical bond and there may be some void space in the billet. It should be accurate for a second stage extrusion however, where a metallurgical bond was previously formed. A second point concerns the behavior of the layer during drawing. This breaks up and is redistributed non-uniformly (a typical coverage is 10% or less at final size) on the filament surfaces. It seems likely that some process of agglomeration occurs, the fine particles of intermetallic being captured at asperities, other obstacles and larger particles. Figure 8 shows an admittedly extreme case of a composite like that of Fig. 1, given nine

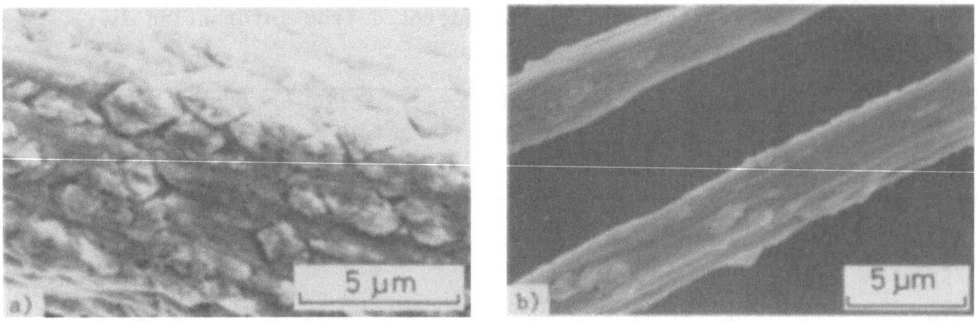

Fig. 8. Scanning electron micrographs of a filament from composite CB1233 (ref. 6); a) immediately after the last HT, b) after further drawing by a strain of ~ 3.

heat treatments of 40 hrs at 375°C.[6] It also had no extrusion and the heat treatments were spaced by a strain of about 1.15. Figure 8a shows the filament surface immediately after the ninth heat treatment. Large particles of 1 to 2 µm diameter are imbedded in the filament. On drawing by a further strain of about 3 (Fig. 8b), the filaments have obviously sausaged but the nodules remain about the same size. Table 3 suggests that only about 25 nm of intermetallic should have grown at each heat treatment (or perhaps 150 nm if it is accepted that the bcc interdiffused layer may also break off). Some type of agglomeration appears necessary to explain the large particles seen in Fig. 8.

ACKNOWLEDGMENTS

We are grateful to H. DeVries, J. Goddard, D. Moffat, R. Noll and H. So for experimental assistance. The SAM data was taken on the NSF Regional Instrumentation Center Facility at the University of Minnesota. Helpful discussions with Ron Scanlan, M. Suenaga and M. Wells are gratefully acknowledged. The work was supported by DOE - Division of High Energy Physics.

REFERENCES

1. E. Gregory in Manufacture of Superconducting Materials, p.1, Ed. R. W. Meyerhoff, American Soc. for Metals, Metals Park, Ohio, 1977.
2. H. Hillmann in Superconductor Materials Science, p. 275, Ed. S. Foner, B. B. Schwartz, Plenum Press, NY, 1982.
3. P. Dubots, J. Maldy, J. C. Renard, J. Goyer, H. Nithart, J. L. Sabrie, Proc. of ICEC-8, p. 505, IPC Science and Technology Process, London, 1980.
4. D. C. Larbalestier, A. W. West, Acta Met. 32, 1871 (1984).
5. D. C. Larbalestier, A. W. West, W. Starch, W. Warnes, P. Lee, W. K. McDonald, P. M. O'Larey, K. Hemachalam, B. Zeitlin, R. Scanlan, C. Taylor, IEEE Trans. on Magnetics 21, 269 (1985).
6. D. C. Larbalestier, Li Chengren, W. Starch and P. J. Lee, Limitation of critical current density by intermetallic formation in fine filament Nb-Ti superconductors. To appear IEEE Trans. on Nuclear Science, 1985.
7. M. Garber, M. Suenaga, W. B. Sampson, R. L. Sabatini, Effect of CuTi compound formation on the characteristics of NbTi accelerator magnet wire. To appear IEEE Trans. on Nuclear Science, 1985.
8. F. A. Shunk, Constitution of Binary Alloys, p. 296, McGraw Hill, NY, 1969.

DEVELOPMENT OF MULTIFILAMENTARY NbTi AND Nb_3Sn COMPOSITE

CONDUCTORS WITH VERY FINE FILAMENTS

T. Ogasawara, Y. Kubota, K. Yasohama, Y. Oda*,
T. Makiura** and H. Okon
College of Science and Technology
Nihon University
Kanda-Surugadai, Chiyoda-ku, Tokyo, Japan

ABSTRACT

A NbTi multifilamentary composite conductor with about 10,000 filaments has been manufactured in long lengths. A filament diameter of 0.52 μm, a twist pitch of 1.13 mm, a strand diameter of 0.1 mm and a Cu/CuNi mixed matrix result in strongly reduced a.c. losses. The hysteresis loss and the coupling loss are 73 kW/m^3 and 56 kW/m^3 for a 50 Hz magnetic field with an amplitude of 1.5 T. From three strands a conductor was formed with a twist pitch of 2.4 mm. Several small coils were wound and operated at 50 Hz. One of the coils generated a maximum field of 1.52 T(center) at an operating current of the same size as the static critical current. Similarily the construction of a Nb_3Sn multifilamentary composite conductor with about 280,000 sub-micron filaments for a.c. use was tried.

INTRODUCTION

A large section of recent research in high field superconductors has concentrated on multifilamentary composite (MFC) conductors to be used for a.c. or pulsed fields. In 1983, the highest rate for a superconducting pulsed coil[1] was an average ramp rate of 200 T/s with a maximum magnetic field of 4 T. However, research on a.c. applications is yet in its beginnings. In 1982, we produced a low loss NbTi MFC conductor[2] with a fine filament of 1 μm, however, it was found out soon that this conductor had no economical benefits when operated in fields of 50 Hz with amplitudes in excess of 1 T. In 1983, Hlåsnik[3] and Dubots et al.[4] fabricated a NbTi multifilamentary composite conductor with a fine filament of 0.5 μm diameter and succeeded in the 50 Hz operation of a small coil wound from this conductor, but their operational current did not reach I_c.

We have tried to achieved the following objectives: 1) to design and fabricate a highly stable, low loss NbTi MFC conductor, 2) to verify the feasibility of a 50 Hz operation of a small coil wound from the new NbTi MFC conductor and 3) to investigate the feasibility of fabricating a Nb_3Sn MFC conductor for a.c. use.

Present address; * Showa Electric Wire & Cable Co., Kawasaki-shi, Kanagawa, Japan
** Toshiba Engineering, Shinagawa-ku, Tokyo, Japan

DESIGN AND FABRICATION OF CONDUCTORS

To reduce the a.c. losses in MFC conductors, the filament diameter should be smaller than 1 μm, the twist pitch length must be smaller than 10 times the conductor diameter, and the copper matrix must be replaced with high electrical resistivity materials such as CuNi. If the dynamic stability has to be increased yet further, in addition to the above requirements, the following two points must be taken into consideration:

- the heat flux due to the a.c. losses should be much smaller than the cold-end recovery heat flux Q_r;
- the thermal characteristic time must be shorter than a quarter period of the a.c. field.

We have been able to fabricate and test a NbTi MFC conductor for a.c. use which fulfills these requirements. The conductors used in this study were fabricated by Showa Electric Wire & Cable Co.

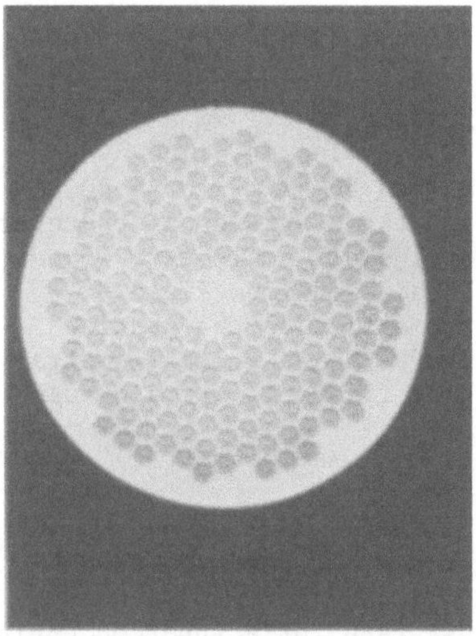

(a)

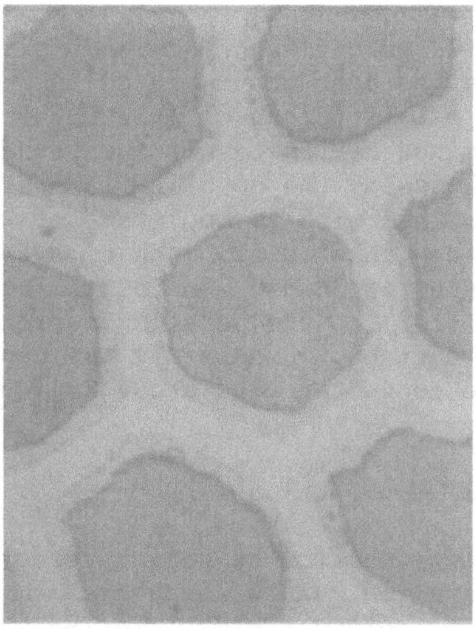

(b)

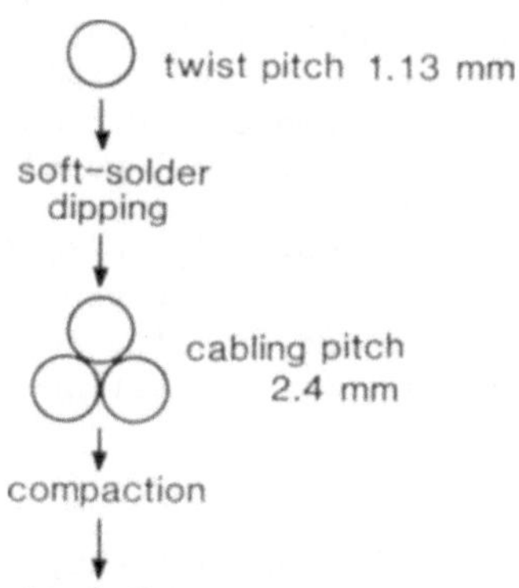

Fig. 1. Cross sectional view of the NbTi conductor
(a) basic strand.
(b) enlarged picture, dark filaments embedded in Cu/CuNi mixed matrix.
(c) cable construction of the triplex.

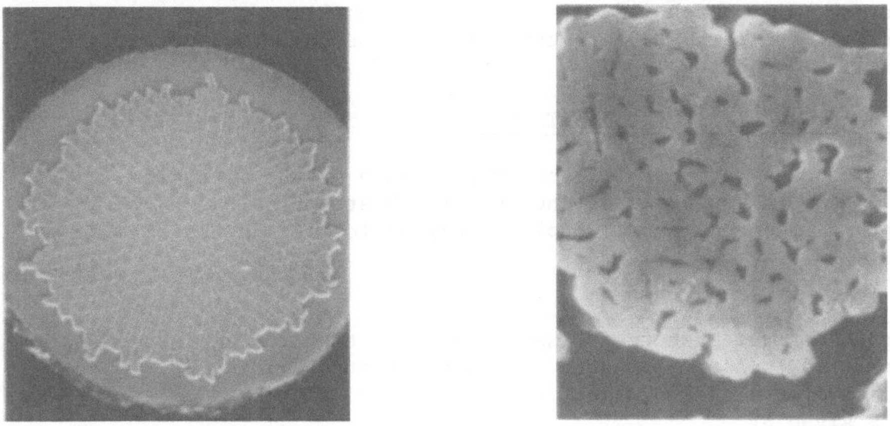

(a) (b)

Fig. 2. Cross sectional views of the Nb_3Sn conductor
(a) basic strand.
(b) enlarged puture, light regions are Nb_3Sn filaments embedded in CuSn matrix.

For the Nb_3Sn conductor, the external tin diffusion method was employed because it is more economical for the fabrication of Nb_3Sn conductors with very fine filaments than the bronze method. Tin-plating and insulation were done at our laboratory.

Structure of the Conductors

A cross sectional view of the NbTi basic strand is given in Figs. 1-(a) and (b). A Cu/CuNi mixed matrix is used each of the NbTi filaments (0.52 μm) is surrounded by a copper stabilizer and embedded in the CuNi matrix. There are 9720 filaments. The diameter of the strand, D is 0.103 mm and the twist pitch, ℓ_p, 1.13 mm. Three of the strands are cabled into a final conductor with a pitch of 2.4 mm as shown in Fig. 1-(c).

Fig. 2-(a) shows a cross sectional view of the Nb_3Sn conductor which has been reacted at 600°C for 5 days. Originally it contained 278,425 filaments of a diameter of 0.11 μm within a bronze matrix. In the heat-treated conductor, it was found that the filaments became coarsened and within the first stacks of bundles, the filaments were all tangled together as shown in Fig. 2-(b). Thus the effective diameter of the Nb_3Sn filaments in the finished conductor is much larger than 0.11 μm. It was measured later and found to be 4 μm. The conductor diameter is

Table 1. The parameters of the two strands.

NbTi composite conductor		Nb_3Sn composite conductor	
composite diameter(mm)	0.106	composite diameter(mm)	0.11
insulator	formvar	insulator	varnish
number of filaments	54 x 84 = 9720	number of filaments	925x301= 278,425
filament diameter(μm)	0.52	filament diameter(μm)	0.11
twist pitch(mm)	1.13	twist pitch(mm)	1
matrix between filaments	Cu/CuNi	ratio of Nb_3Sn/CuSn	1/2
ratio of NbTi/Cu/CuNi	1/0.35/2.88	stabilizer	no
outer sheath	CuNi	heat treatment	600°C x 5 days

0.11 mm and the twist pitch 1.0 mm. No copper stabilizer was included in this conductor. The parameters of the NbTi and the Nb_3Sn strands are given in Table 1.

A.C. Losses and Stability of the NbTi Conductor

The a.c. losses per unit time and unit volume of the composite conductor, P, can be expressed as the sum of a hysteresis loss P_h in the superconducting filaments and a coupling current loss P_c in the filamentary zone,

$$P = P_h + P_c. \quad (1)$$

First, we consider the hysteresis loss. Assuming that the amplitude of the applied field is large compared to the penetration field, H_p, the hysteresis loss for a cyclic magnetic field orthogonal to the circular conductor is given by

$$P_h = \frac{8}{3\pi} f \, d \, \lambda \, J_c \, B_m \quad (2)$$

where d is the superconducting filament diameter, λ the filling factor of the filaments, f the frequency of the a.c. magnetic field and B_m the amplitude of the magnetic field. For our NbTi conductor, eq. (2) gives $P_h = 33$ kW/m^3 for the following set of parameters: f = 50 Hz, d = 0.52×10^{-6}m, $\lambda J_c = 10^9$ A/m^2 and $B_m = 1.5$ T. The estimated value of P_h compares faborably with the lowest hysteresis losses reported in the literature[5].

The eddy current loss[6] P_c can be expressed as

$$P_c = \frac{B_m{}^2}{M_o} \frac{\omega^2 \tau}{1 + \omega^2 \tau^2} \quad (3)$$

where ω is the radian frequency μ_0 the permeability in a vacuum, τ is the time constant[6] for the coupling loss and defined as

$$\tau = \frac{1}{2} \frac{\mu_0}{\rho_e} \left(\frac{\ell_p}{2\pi}\right)^2 \quad (4)$$

where ρ_e is the transverse resistivity and ℓ_p is the twist pitch of the filaments. Here, we use the resistivity of CuNi as an approximation for ρ_e of the mixed matrix in our NbTi conductor. $\rho_e = 1.25 \times 10^{-7}$ Ωm at 4.2 K for Cu-10 wt.%Ni. For this ρ_e, eq. (4) gives $\tau = 0.16$ sec and from eq. (3), the eddy current loss P_c becomes 28 kW/m^3.

From eq. (1) the a.c. losses of the NbTi conductor are evaluated as 61 kW/m^3.

In the following, we estimate the stability of the conductor in a.c. fields. The cold-end recovery heat flux Q_r represents a stability condition for the normal zones in the d.c. mode. It may also be used as an indicator of the stability in time-varying fields as long as the heat flux due to the a.c. losses is much smaller than Q_r. For a.c. losses of 1.6 W/m^2 while the cold-end recovery heat flux Q_r[7] is, 10^4 W/m^2, as measured with variotis NbTi MFC conductor. In addition to the above requirement, the heat generated within the conductor due to the a.c. losses must be quickely removed in order to prevent a temperature rise of the conductor. An estimate is given by the characteristic time $\Delta t, = 4D^2(\pi^2 D_{th})^{-1}$ where D is a diameter of the conductor and D_{th} the average thermal diffusivity of the mixed matrix. The requirement is fulfilled as long as Δt is shorter than a quarter period of the a.c. field. If we assume that D_{th} of the mixed matrix is dominated by that of the CuNi matrix, which is 10^{-3} m^2 sec^{-1} for Cu-10 wt.%Ni, Δt becomes 5 sec. This is much smaller than a quarter of the AC period. (5 ms)

The hysteresis and a.c. losses of the NbTi MFC conductor were measured in another experiment details of which have been published elsewhere. The hysteresis loss of the basic strand at a 1.5 T peak field of 50 Hz was found to be 73 kW/m^3, which is twice as large as the estimate

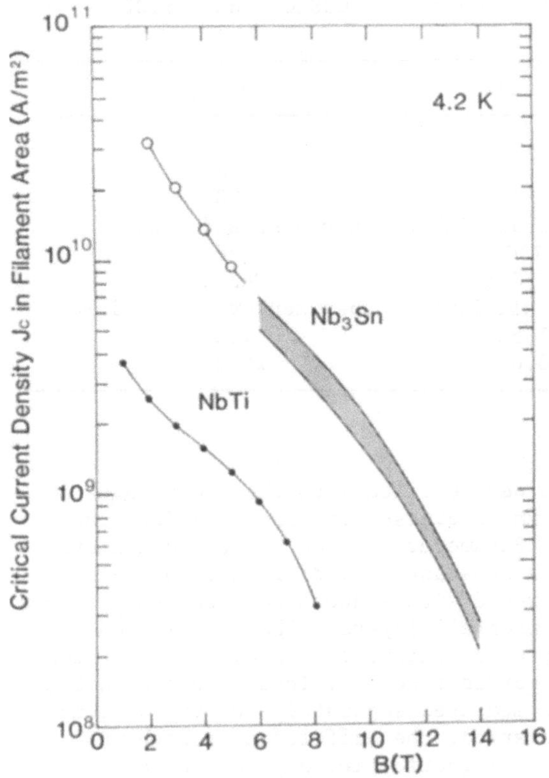

Fig. 3. The critical current density, J_c, of strands vs. magnetic field.

$P_h = 33$ kW/m^3. We surmise that the larger loss results from a lack of uniformity in the filament diameters, and/or a poor estimate of the current density. The measured eddy current loss is 56 kW/m^3, contrary to 28 kW/m^3 estimated from eq. (3). The discrepancy seems to arise from our adoption of the electrical resistivity of CuNi, instead of the true resistivity 8 T of the mixed matrix, in eq. (4). However, in view of the difficulty of calculating ρ_e of the mixed matrix, the discrepancy can be considered as acceptable.

In a cabled conductor, the strands become also electromagnetically coupled by way of the solder into which the strands have been dipped during the fabrication process. The coupling loss at 1.5 T was measured to be 55 kW/m^3.

CRITICAL CURRENT DENSITY IN STATIC FIELDS

The critical current densities J_c of short samples of the NbTi and Nb_3Sn conductors are shown in Fig. 3. Here J_c is defined as the critical current (1 μV/cm criterion) divided by the total cross sectional area of the superconducting filaments. For our NbTi conductor, J_c is almost the same as that of other conductors with filaments of a diameter larger than 0.5 μm. For the Nb_3Sn conductor, the shaded areas represent sample-to-sample variations in J_c. In the measurements below 6 T, two copper wires of 0.1 mm in diameter were used as a shunt to prevent a burn-out of the samples.

50 Hz QUENCHING CURRENTS AND LOSSES FOR COILS

Several small coils were wound from the three-strand cabled conductors onto FRP bobbins with a diameter of 10 mm and a length of 20 mm. For

Table 2 Parameters of the two coils, C3S201 and C3S102

coil	C3S201	C3S102
inner diameter(mm)	10	10
outer diameter(mm)	37	18
axial length (mm)	20	20
type of conductor	three strand cable	three strand cable
conductor length(m)	120	38
spacer/thickness(mm)	FRP/0.5	FRP/0.5
arrangement of spacers	between all layers	between every 2nd layer
maximum field (T)	1.52	1.13
maximum current (A)	20.8	25.3

cooling purposes, a number of holes were drilled into the end-flanges of the bobbins and FRP spacers of 0.5 mm thickness were inserted into the windings to form cooling channels. Parameters of two typical coils are given in Table 2. The two coils are substantially different in their cooling conditions; coil C3S102 is wound with spacers between every second layer, coil C3S201 with spacers between all layers. The a.c. losses of the coils at 50 Hz were measured by a calorimetric method. In order to eliminate background losses and to obtain true a.c. losses in the coils, evaporation rates of liquid He were measured separately for the a.c. mode operation and for the d.c. mode operation; the difference of both was considered to be due to a.c. losses. Magnetic fields generated by the coils were measured by pick-up coils inserted in the bore.

The 50 Hz peak quenching currents of the coils are given in Fig. 4. For C3S201, the peak quenching current is 20.8 A, which is equal to the d.c. mode quenching current and nearly the same as the short sample critical current. At a peak current of 20.8 A, the maximum magnetic field is to 1.52 T, which was confirmed by the pick-up coil. Coil C3S102 is inferior in cooling condition to coil C3S201, only half of the total

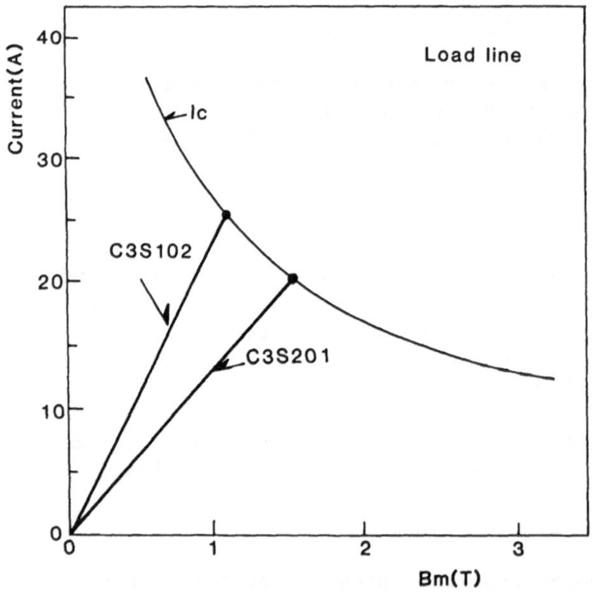

Fig. 4. Test results of the coils, C3S201 and C3S102. Solid circles represent the maximum operation currents in the a.c. mode

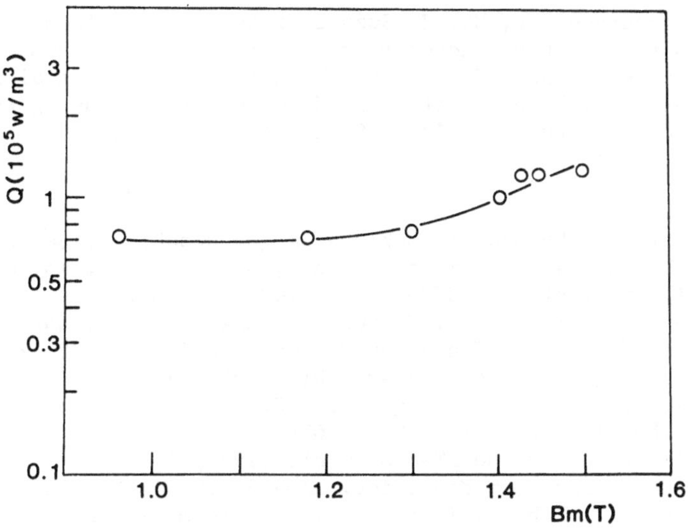

Fig. 5. The a.c. losses of the coil C3S201 vs. the peak field.

surface of the conductor of coil C3S102 is in free contact to liquid He. However, the 50 Hz peak quenching current of coil C3S102 reaches 25.3 A equal to the d.c. mode quenching current as well the short sample critical current. This means that the conductor temperature rise due to the a.c. losses is negligible even in the coil with the poor cooling. In other words, the conductor has a high a.c. stability. The maximum field at 25.3 A is 1.13 T, and for this field, the average rate of the field variation is 226 T/s.

The 50 Hz a.c. losses of the coil, C3S201, are presented in Fig. 5 as a function of the a.c. peak fields. The loss increases gradually with the peak field, and at 1.52 T the loss becomes 130 kW/m^3. This value is even better than the 140 kW/m^3 loss measured in low loss conductors at 1.5 T by Dubots et al.

SUMMARY

MFC conductor of NbTi with a low a.c. loss and a high a.c. stability at power frequencies has been successfully fabricated. Several small coils with different cooling conditions are tested by feeding 50 Hz a.c. current. All the coils have been operated at 50 Hz up to a peak current very close to the quenching current of the d.c. mode operation and also to the short sample critical current in static fields. A maximum peak field of 1.52 T was attained in one of the coils, and at this field the a.c. loss was measured to be 130 kW/m^3. For the Nb_3Sn MFC conductor, the filaments seem to be connected and the effective diameter becomes about 4 ~ 5 μm. Further improvements are required before a Nb_3Sn MFC conductors with a sub-micron filament can be produced.

ACKNOWLEDGEMENTS

This work is in part supported by Grant-in-Aid for Fusion Research, Science and Culture, Ministry of Education in Japan.

The authers would like to thank the late professor K. Yasukōchi for continuing encouragement, Mr. E. Suzuki of Showa Electric Wire & Cable Co. for production of the conductor, Mr. T. Noguchi of Vacuum Metallurgical Co. Ltd. for tin-plating and professor L. Böesten for critical reading of the manuscript. The high magnetic fields were obtained using the magnet of the Research Institute for Iron, Steel and other Metals.

REFERENCES

1. T. Ogasawara, Y. Kubota, T. Makiura, K. Yasukōchi, T. Satow, M. Iwamoto, K. Toyoda, H. Momota, K. Sato, S. Yamada, K. Koyama and T. Onishi, Proc. of the ICEC-9 Kobe (1982) 339-342.
2. T. Ogasawara, Y. Kubota, T. Makiura, T. Akachi, T. Hisanari, Y. Oda and K. Yasukōchi, IEEE Trans. on Magnetic, MAG-19 (1982) 248-251.
3. I. Hlásnik, Proc. of the MT-8 Grenoble (1983) 459
4. P. Dubots, A. Février, J.C. Renard, J.C. Goyer and Hoang Gia Ky, Proc. of the MT-8 Grenoble (1983) 467-470.
5. I. Hlásnik, Proc. of the MT-8 Grenoble (1983) 459-466.
6. G. Ries, IEEE Trans. on Magnetics, MAG-13 (1977) 524-526.
7. T. Ogasawara, Y. Kubota, T. Makiura, R. Akihama and K. Yasukōchi, Proc. of the US-Japan Workshop on Superconductive Energy Storage (1981) 372.

FABRICATION AND CHARACTERIZATION

OF FINE FILAMENTS OF NbTi IN A COPPER MATRIX

K. Hemachalam, C. G.King, B. A. Zeitlin
Intermagnetics General Corporation
Waterbury, Connecticut

R. M. Scanlan
Lawrence Berkeley Laboratory
Berkeley, California

ABSTRACT

Practical multifilamentary Nb-46.5 Wt. % Ti/Cu composites have been made by a double extrusion process. The composites contain up to 6000 filaments with a diameter of 2 to 6 μm at the final wire size. Through careful attention to the billet design and thermo-mechanical processing, the formation of intermetallics (Cu-Ti-Nb) is virtually halted. The intermetallic precipitates, when allowed to form at the filament surface, interfere with the uniform reduction and give rise to poor filament quality; including filament breaks and reduction in critical current density, Jc. The integrity of the present fine filaments is studied with SEM and compared with that of conventionally processed material. The Jc, as a function of the filament size, is investigated over a transverse magnetic field of up to 8 Tesla. The value of 'n' in $\rho=kI^n$ is measured and the results are compared to those obtained for similar M.F. wires currently under study at other institutions. It is hoped that the fine filamentary wire produced by the double extrusion process will greatly reduce the magnetization which is responsible for field distortions in the High Energy Physics program applications.

INTRODUCTION

The desire for niobium titanium filaments in the micron to sub-micron range has been manifest since the realization that magnetization, and hence A.C. losses were dependent linearly on filament size. Progress on producing conductors with the required filament size and matrix resistivity has been slow, due to both market and technical factors. This has severely restricted the application of superconductivity in devices such as motors, generators, transformers, and fast ramped or pulsed devices. Recent progress[1] has shown that 1.3 to .12 μm filament conductors can be made in a laboratory environment.

The technical challenges to produce micron size filaments in production quantities focus on two major areas; alloy uniformity and reproduceability, and fabrication technology. The conductor manufacturer has to be confident that the alloy will reproducibly deform, without significant reduction in current density, to micron sizes. Fabrication technology requires understanding and optimization of multiple extrusion and elimination of intermetallic formation.

The high energy physics community is challenging the past technology by requiring a new accelerator twenty times the energy of the Fermi Tevatron, the Superconducting Super Collider (SSC). In addition, basic understanding and improvements in the alloy and processing[2] have contributed significantly to the present technology. This is due in part to the foresight of the Department of Energy in funding the University of Wisconsin Applied Superconductivity Center and the excellent work and cooperative spirit the Center has developed in working with industry and the national laboratories.

The high field option for the SSC's dipole magnets requires a current density of a minimum of 2400 A/mm^2 in a filament of 2.5 μm diameter[3]. The Tevatron, the first truly large scale application of superconductivity, required conductor of 9 μm filaments and a current density of 1800 A/mm^2. To minimize the magnet cost the aperature of the SSC is 40 mm versus 75 mm in the Tevatron. The combination of smaller aperature, higher current density, and a lower injection field requires the filament size to be 2.5 μm if the sextapole uniformity is to be met without expensive distributed correction windings.

This paper describes progress in an industrial scale towards conductor designed for the high field option of the SSC and, just as importantly, opens new possibilities for economic application of superconductors to A.C. devices.

COMPOSITE FABRICATION

Design Philosophy

For the high-field options of the SSC design, two superconductors, 0.808 mm dia. and 0.648 mm dia. with a Cu:Sc ratio of 1.3:1 and 1.8:1, respectively, are being considered[4]. The filament size in the conductors will have to be far below 10 μm to reduce magnetization. If a 2 μm size is chosen for the two wires, the number of filaments required will be 71,100 and 37,600 respectively. In conventional billet assembly, for conductors with a few thousand filaments, billets are assembled by stacking arrays of hexagonal copper tubes which are filled with NbTi rods. For fine filament conductors, this approach can not be employed economically.

In order to produce billets containing the large numbers of filaments described above, a double extrusion method is used, whereby a restack element is manufactured in the first extrusion, drawn to a specified restack element size, straightened and cut to length. This restack element is assembled into a second stage billet which is extruded and then drawn to the final product wire. In this study, two methods of first stage billet design were tried to make fine filament conductor. The first method, the single filament restack method, is applicable for production conductors with 5 μm filament diameters or greater. For production of fine filament conductors, with filament diameters in the 2 to 4 μm range, a multiple filament restack element is used.

Single Filament Restack Billet (0509)

The first stage billet used to manufacture this product wire was a copper clad monofilament composite. The alloy used was a 76 mm diameter ingot with a composition of Nb 46.5 Wt. % Ti. After extrusion, this

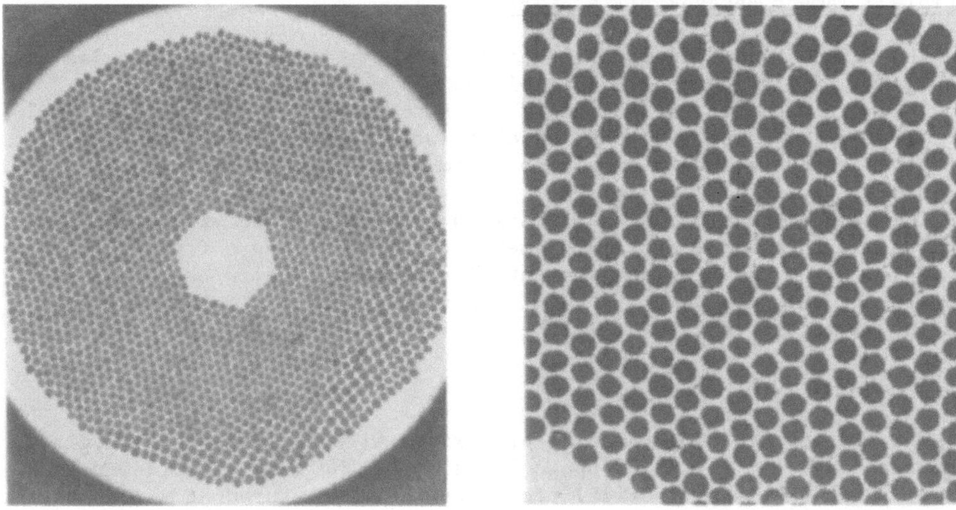

Fig. 1. Cross section of conductor made from single filament restack billet 0509 and a closeup view of the NbTi filaments.

single filament restack element was drawn to final size, straightened, cut to length and then stacked into a 89 mm diameter second stage billet. This billet was assembled to yield a conductor with 1953 filaments and a Cu:Sc ratio of 1.4:1. This billet was to serve as a preproduction model to investigate optimal process parameters necessary for fine filaments in the range of 2 to 6 µm. In order to reach the filament size range of interest, the conductor had to be drawn to 0.152 and 0.471 mm diameter, respectively. Cross sectional views of the conductor from billet 0509 are shown in Fig. 1. For the proposed SSC conductors to have 2 to 5 µm filaments, significantly larger numbers of filaments have to be assembled into the second stage billet. This can be accomplished by using a larger second stage billet or by incorporating a multiple filament restack element in the first stage billet design.

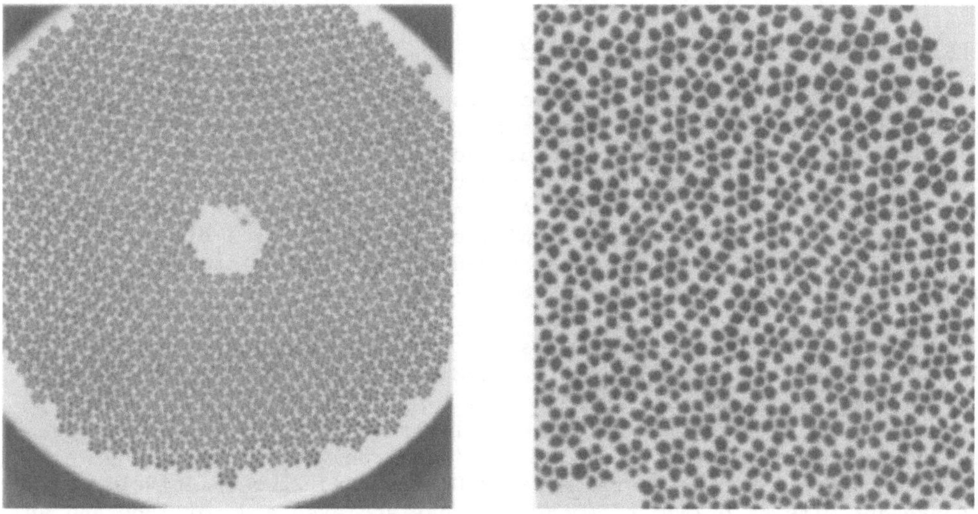

Fig. 2. Cross sectional views of 6006 filament conductor from billet 5914.

Multiple Filament Restack Billet (5914)

The first stage billet consisted of a copper billet containing seven Nb 46.5 Wt. % Ti alloy rods. This billet was extruded, drawn to a restack element diameter, straightened and cut to length. 858 multiple filament restack elements were then stacked into a 150 mm second stage billet to yield a conductor containing 6006 filaments, shown in Fig. 2.

The second stage billet was designed specifically to make a 1.8:1 conductor, with a final overall diameter of 0.648 mm and a final filament diameter of 5 μm. During development of an optimized draw/heat treatment schedule for this conductor, several sub size conductors were made, having filament diameters of 2 to 5 μm, the results of which are reported below.

RESULTS AND DISCUSSION

The critical current capacity of various wire samples was tested at transverse magnetic fields of up to 8 Tesla. A one meter sample was inductively wound on a cylindrical probe and Ic was measured at a resistivity of 1×10^{-14} Ωm. Particular attention was given to record the superconductive-to-normal transition of the V-I characteristic. This was done to be able to calculate the value of 'n' in the relationship[5],

$$\rho = kI^n$$

where ρ = resistivity
I = current
k = constant.

The value of 'n' is derived from critical currents at two convenient resistivities such as 1×10^{-13} Ωm and 1×10^{-14} Ωm. The value of 'n' is a measure of filament quality both in terms of geometric integrity and chemical/microstructural homogeneity. A large 'n' is obtained from an abrupt superconductive-to-normal transition and it is an indicator of uniform filaments. On the other hand, broad transitions yield low values of 'n' which relate to poor filament quality often characterized by the presence of extensive intermetallic compounds on the surface of filaments, filament sausaging, and even fracturing. The results of the short sample testing and further discussion of 'n' are presented separately for the two types of superconductive wire below.

Single Filament Restack Billet (0509)

The critical current density as a function of magnetic field is shown in Fig. 3 for three different wire sizes. The measured values of 'n' at 5 Tesla for the three wires are tabulated at the bottom of the figure. It can be seen that both Jc and 'n' are the highest for 0.471 mm diameter wire containing 6.2 μm filaments. As the filament diameter is decreasing, the Jc and 'n' are diminishing in spite of the fact that a different optimal heat treat processing has been given to each of the samples. The filament quality was examined in a Scanning Electron Microscope (SEM). Typical filaments of 0.152 mm diameter wire are shown in Fig. 4. The filaments are free of intermetallics and they are relatively uniform along the length. For comparison, an SEM picture of the filaments obtained in a production wire is shown in Fig. 5. This wire contains 2070 filaments of 9 μm size and it represents a large quantity of wire manufactured for the Fermi National Accelerator (Tevatron). Typical Jc and 'n' values at 5 Tesla for this wire were 1850 A/mm^2 and 22 respectively. The present wire at 6.2 μm filament size has achieved a Jc of 2884 and an 'n' value of 40 at 5 Tesla which are substantially better than the Tevatron conductor. The enhanced properties are considered to be the result of elimination of Cu-Ti-Nb intermetallic formation and recent improvements in the microstructure[6] of the NbTi alloy.

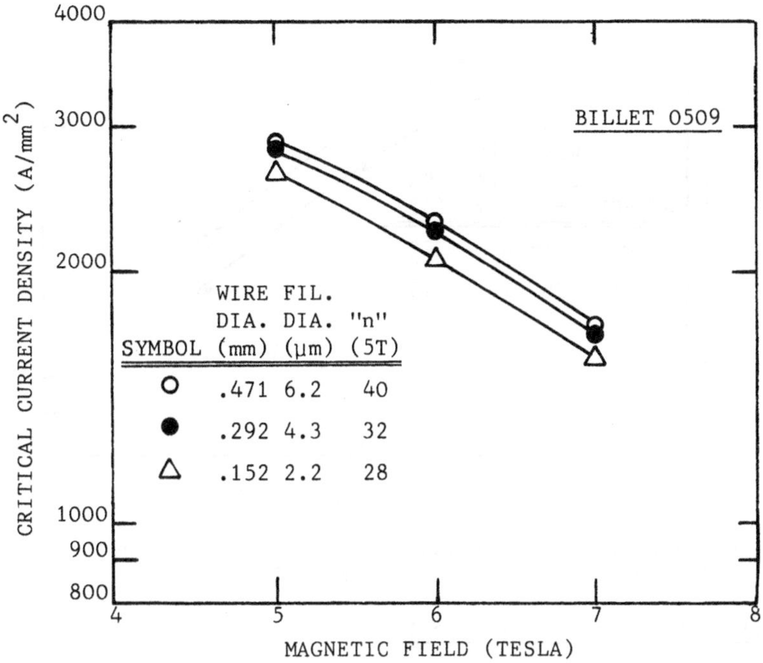

Fig. 3. Critical current density as a function of magnetic field and measured values of 'n' for three wires.

Fig. 4. SEM picture of typical 2.3 μm filaments from billet 0509.

Fig. 5. SEM picture of 9 μm filaments in a production conductor made for the Tevatron. Note intermetallics on filament surface.

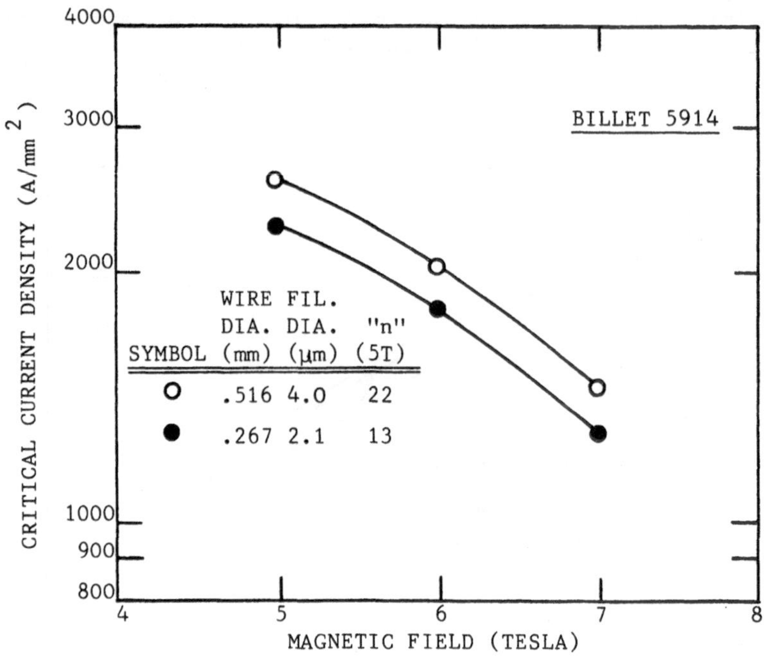

Fig. 6. Critical current density as a function of magnetic field and values of 'n' for two wires of billet 5914.

Multiple Filament Restack Billet (5914)

The superconducting properties for two optimized wires made from the 6006 filament billet are shown in Fig. 6. The best 5 Tesla Jc and 'n' values are 2595 A/mm^2 and 22, respectively, obtained for 4 μm filaments. This Jc is approximately 8% lower than for billet 0509 at corresponding filament size. The value of 'n' for the 5914 samples is also lower in comparison to 0509 samples. The filaments can be qualitatively examined from SEM pictures shown in Fig. 7. Although the surface is free of

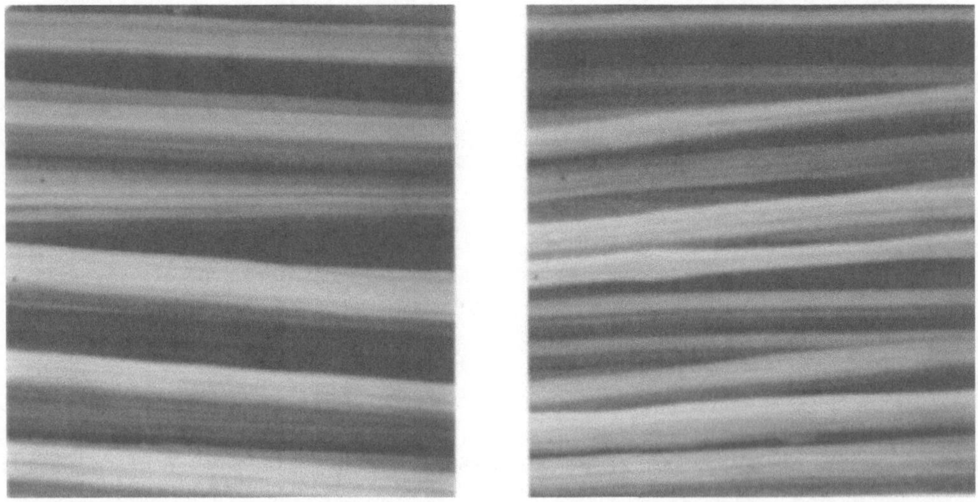

Fig. 7. SEM pictures of 5.3 μm and 3.3 μm filaments from billet 5914. Sausaging can be clearly seen in the smaller filaments.

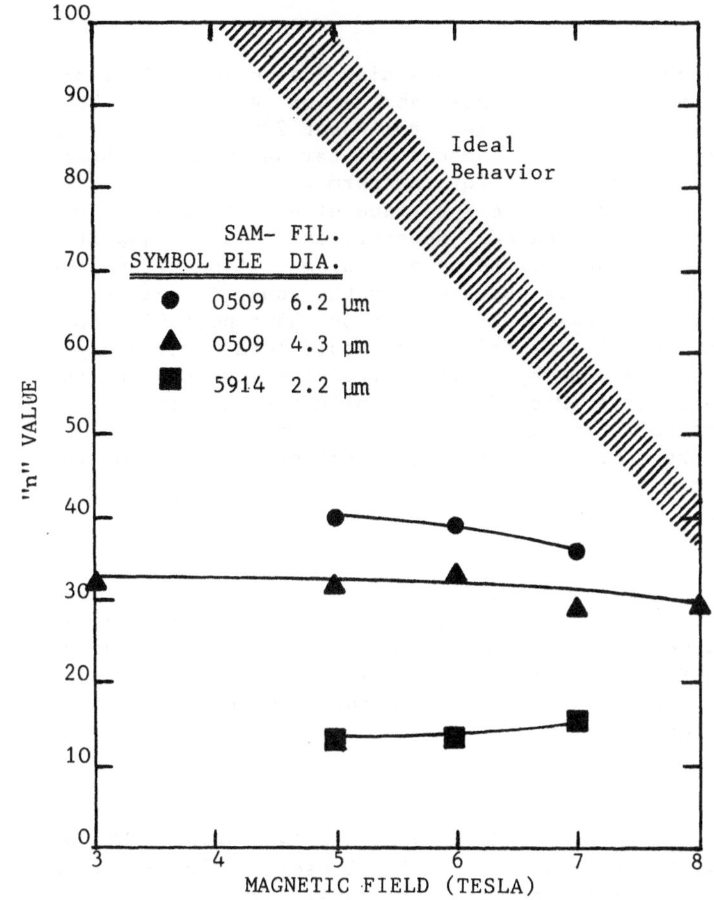

Fig. 8. Plot of value of 'n' as a function of magnetic field for some of the wires studied.

intermetallics, the filaments from 5914 samples show a definite sign of sausaging especially at the smaller size. Since the necked down sections of the filaments have less cross sectional area for current transport, they will enter current sharing mode and eventually drive the overall conductor normal before the undamaged filament regions reach their critical limit. Without sausaged regions in the filaments, implying a high level of filament integrity, a much higher Jc can be expected.

It is interesting to note that the 5 Tesla Jc of the wires at a resistivity of 1×10^{-13} Ωm is relatively high (eg. 2880 A/mm^2 for the 0.516 mm diameter wire). This indicates that there is a potential in these wires for improvement in Jc by controlling the filament sausaging. The value of 'n' measured for the present test samples is plotted as a function of magnetic field in Fig. 8. The shaded area represents an idealized behavior expected of NbTi. This is in fact observed for monofilament wires where the filament is uniform[7]. The non-ideal behavior of the fine filament conductors is believed to be largely due to the existence of geometric irregularities in the filaments. The microstructural/chemical inhomogenieties present inside the body of the filaments are also expected to play a roll in determining the value of 'n' especially at the fine filament sizes studied. Thus, for future superconductors involving fine filaments, the conductor design and the processing techniques have to be specifically aimed at further improving the filament integrity.

CONCLUSIONS

Practical superconductors containing 2 to 6 μm NbTi filaments have been processed using conventional fabrication techniques. The critical current density at 5 Tesla varies from about 2300 to 2900 A/mm^2 depending on the filament size. These numbers are far superior to those obtained for production conductor made for the Fermilab Tevatron. The Cu-Ti-Nb intermetallic formation at the interface of NbTi filaments and Cu matrix has been completely eliminated. However, the filament quality with respect to sausaging continues to be a problem at fine filament diameters below about 4 μm. This is evidenced by the lower 'n' values in the low 20's and below. The present investigation shows potential for further improvement in the value of 'n' which should lead to even higher Jc values. This can be done by paying particular attention to the design and process parameters which directly influence the filament quality, and the microstructural/chemical aspects of the NbTi alloy rod. More work is definitely in order to establish the effect of alloy microstructure on the value of 'n'.

ACKNOWLEDGEMENTS

The authors wish to thank M. Kovic and D. Pollock for manuscript preparation; R. Hlava, T. Keene and T. O'Connor for their help in conductor processing; T. Baran and C. Meeks for Ic testing, and D. Hazelton and K. Pickard for SEM work. The research on billet 0509 was internally funded by IGC. Billet 5914 was manufactured for LBL under a DOE/SSC conductor development contract.

REFERENCES

1. P. Dubots, A. Fevrier, JC. Renard, JP. Tavergnier, NbTi wires with ultra-fine filaments for 50-60 H2 use. Influence of the filament diameter upon losses, in IEEE Trans. Magn. Vol. MAG 21, (1985) pp. 177-180.
2. D. C. Larbalestier, A. W. West, W. Starch, W. Warnes, P. Lee, W. K. McDonald, P. O'Larey, K. Hemachalam, B. Zeitlin, R. Scanlan, C. Taylor, IEEE Trans. Magn. Vol. MAG 21, (1985) pp. 269-272
3. C. L. Taylor from 4th NbTi Workshop, Madison, WI, July 1985.
4. R. M. Scanlan, Superconducting Materials for the supercollider, 1985 ICMC proceedings.
5. M. Garber, W. B. Sampson and M. J. Tannenbaum, critical current measurements on Superconducting Cables, in IEEE Trans. Magn. Vol. MAG 19, No. 3, (1983).
6. D. C. Larbalestier, Towards a microstructural description of the Superconducting properties, IEEE Trans. Magn. Vol. MAG 21, No. 2 (1985) pp. 257-261.
7. W. H. Warnes, D. C. Larbelestier, from 4th NbTi Workshop, Madison, WI, July 1985.

FINE FILAMENTARY NbTi SUPERCONDUCTING WIRES

T. S. Kreilick, E. Gregory, and J. Wong

Supercon, Inc.

Shrewsbury, Massachusetts

ABSTRACT

The present interest in NbTi composite conductors with very fine filaments has developed because of the requirements of the Superconducting Super Collider. The Fermilab Tevatron utilizes wire with 8 μm diameter filaments; the SSC may require wire with less than 3 μm diameter filaments. The reduction in filament diameter means that, while Fermilab wire required 2,000 filaments, a similar size wire for the SSC will require 40,000 or more filaments. Conventional techniques of billet assembly will no longer suffice.

Recent improvements in billet design, material preparation, and billet assembly techniques have resulted in a production size NbTi billet containing 4,164 filaments. Extensive development of thermomechanical processing schedules has enabled us to optimize the critical current density in this material to a distinctly higher level than has been previously reported. Experiments have been started to explore the feasibility of scaling this material to a 40,000 filament conductor.

INTRODUCTION

Superconducting wire manufacturers were offered the Gordian knot at the Second NbTi Workshop held in Madison, Wisconsin during August of 1984. The problem was to produce fine filaments of NbTi less than 3 μm in diameter and capable of exhibiting a J_c greater than 2,500 A/mm^2 at 5 T. This meant that the filaments had to be "unsausaged" and devoid of copper titanium intermetallic on their surface.

Until recently the magnet designers for high energy physics and other applications, with the exception of some working on rotating machinery, have been satisfied with 10 μm diameter filaments. However, filaments of this relatively large size would cause undesirable magnetization effects at injection fields in the SSC. Historically, NbTi in a copper matrix, when processed below 10 μm in diameter, has exhibited greatly reduced quality,[1] and this has resulted in a lower current carrying capability. Values of 1,800 to 2,000 A/mm^2 were standard in Fermilab conductor produced by the wire industry for a number of years. Uneven filament cross sections were prevalent in some of the high current density fine wires, but data on its extent has only recently become generally known.[2] The situation has now changed, and a diagnostic tool for filament quality is being widely used. This is the shape of the resistive transition in the I-V curve as expressed by the "n" value (where n is defined by R = constant x I^n). High n values (greater than 30) are generally indicative of good filament quality, but the value is highly dependent in field and other factors and should be used with discretion.

Inhibition of the growth of the intermetallic compound of copper and titanium on the filaments, while not the only factor in the improvement in the J_c, is a necessary one. In 10 μm diameter filaments, the size of the nonductile intermetallic is frequently the same order of magnitude as that of the filaments themselves, and their breakage, and that of the wire will soon occur if further reduction is carried out.

It has been demonstrated recently that formation of the intermetallic compound takes place not only at conventional extrusion temperatures (525-650 K), but also during subsequent heat treatments designed to optimize J_c values.[3]

These factors point to the desirability of the addition of a diffusion barrier to prevent the formation of this intermetallic compound. Such a barrier allows much more flexibility in the temperatures and times of the heat treatments used during processing. The barrier technology has been alluded to previously[4] and was published as early as 1971 for NbZr.[5] Diffusion barrier techniques have been used in the past to produce high quality filaments in double extrusions, and references 6 and 7 describe material with 10,285 filaments, each of which was 1.2 μm in diameter. While this material was not optimized for current density, it is believed that if present high homogeneity NbTi had been used it is probable that high current density wire could have been produced.

EXPERIMENTAL PROCEDURES

Three methods suggest themselves for the production of fine filamentary NbTi that will meet the requirements of the SSC design:

1) cabled cable[8]
2) single stack
3) double extrusion

Cabled-cable Approach

While initial problems have been encountered in the development of the cabled-cable approach, considerable effort has gone into improving it, and it remains a very viable option.[9] This is particularly true, as our present discussion shows, since high quality fine 4,000 filament NbTi is available in the correct wire sizes.

Single Stack Approach

Supercon, Inc., has been funded by the Department of Energy (DOE), through the Lawrence Berkeley Laboratory (LBL), to make a production quantity (360 kg) of fine filament high-current density material. This would then be processed to two diameters, 0.81 mm (0.0318 in) and 0.65 mm (0.0255 in) suitable for the inner and outer cables used in the prototype SSC magnets of the "D" design. Since cost is a very important factor in the magnets, our approach was to put as many filaments into a single-stage extrusion billet as was feasible using relatively conventional methods. This avoids a costly double extrusion step which leads to very poor yields of the highly expensive NbTi raw material. The single-stage process also avoids the alternative cold stacking step, which can lead to poor bonding, wire breakage and, at best, increased costs.

For these reasons, we proposed to stack two 305 mm (12 in.) diameter billets with approximately 4,100 filaments each. One billet would have a 1.3:1 Cu:SC ratio and would yield 8 μm filaments in a wire of 0.81 mm (0.0318 in.) and the other, a Cu:SC ratio of 1.8:1 would yield 6 μm filaments in a 0.65 mm (0.0255 in.) wire.

In this paper we report on the 1.3:1 ratio material only, i.e., Phase I of the contract. We have recently been authorized to proceed with Phase II, which is a similar billet with a Cu:SC ratio of 1.8:1.

To prevent the growth of the intermetallic compound and to maintain filament quality, a diffusion barrier was added to each individual filament. Figure 1 shows several such filaments in a cross section of the material. The Cu matrix and the NbTi core have been chemically etched to display the diffusion barrier.

Trial Billet to Optimize Processing Parameters

In order to reveal any flaws in the processing parameters and to optimize cold working and heat treating schedules for the 305 mm (12 in.) diameter billet, we carried out some preliminary experiments.

A small amount of barrier clad, NbTi monofilament was taken from the same starting materials as the large billet. The material was drawn to an appropriate size, cut to length, and restacked in a 51.0 mm (2 in.) diameter extrusion billet. This billet contained 127 filaments. To achieve the desired filament diameter of 2.0 μm, the trial billet had to be drawn to a wire diameter of 0.0381 mm (0.0015 in.). Critical current densities in this material are listed in Table 1. The geometrically uniform filaments with clean, smooth surfaces are shown in Figure 2 (the Cu was removed chemically).

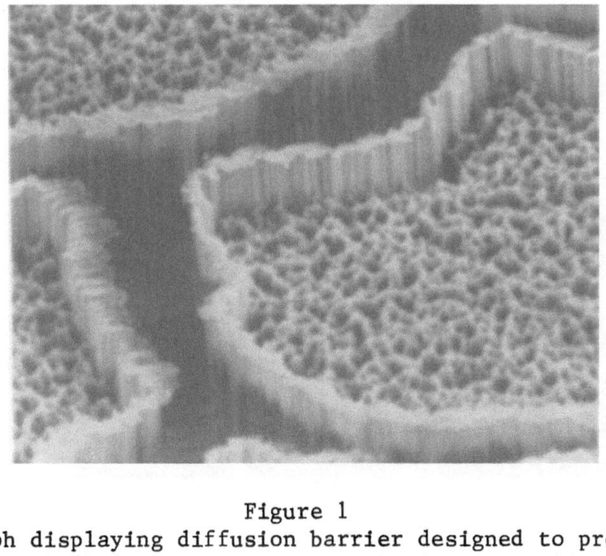

Figure 1
SEM Micrograph displaying diffusion barrier designed to prevent Cu-Ti intermetallic formation.

Table 1.

Wire Ø (mm)	Fila. Ø (μm)	B (T)	I_c(A)	n	I_q(A)	J_cA(mm^2)
0.0508	3.6	3	--	--	--	4287 (e)
		5	43.5	45	45	3113
		6	35.3	49	36	2526
		8	17.6	33	19	1259
0.0381	2.0	3	--	--	--	4483 (e)
		5	47.4	48.5	51	3157
0.0236	1.2	3	0.826	27.7	--	3299
		4	0.688	32.8	--	2749
		5	0.589	32.3	--	2352
		6	0.492	25.7	--	1963

Note: All values reported are confirmed by Brookhaven National Laboratory.
(e) = extrapolated. J_c measured at 4.2 K, I_c = p Ω cm

As these trial experiments showed encouraging results, we then proceeded to process the large billet. Material was drawn to two different sizes: one where it could be used in a single cable with 8 μm filaments and another, where, by employing a cabled-cable approach, it could be used to make a trial magnet with a conductor containing 2.5 μm diameter filaments. The properties of the material made under these two conditions are given in Table 2.

Standard high homogeneity Nb 46.5 w/o Ti alloy was used as opposed to 50 w/o Ti alloy reported upon recently.[10] Whether or not properties can be still further improved or costs reduced by changing the alloy remains to be proven.

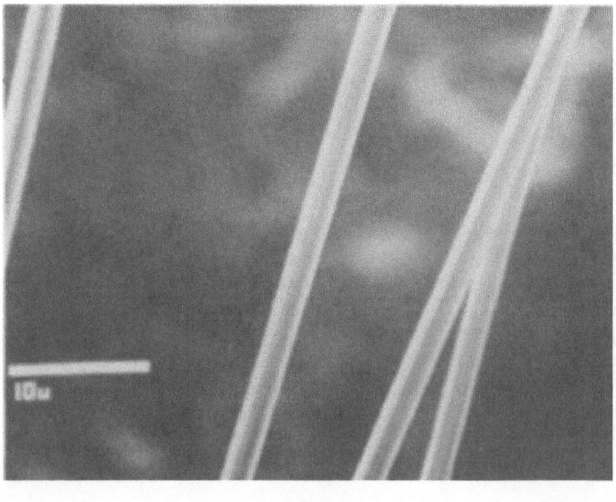

Figure 2
SEM micrograph of filaments chemically removed from 0.0236 mm wire.

Table 2.

	Alloy composition	HI HO Nb 46.5 w/o Ti
	Cu:SC Ratio	1.20:1
	# of Filaments	4164

Wire Ø (mm)	Fila. Ø (μm)	B (T)	I_c (A)	n	I_q (A)	J_c (A/mm^2)
0.813	8.8	5	654	57.3	664	2773
		6	534	38.8	554	2264
		7	417	54.8	434	1768
		8	295.5	43.8	314	1250
0.737	7.9	5	--	--	--	2933 (e)
		6	463.2	53.7	469	2391
		7	358.2	37.6	372	1849
		8	--	--	--	1307 (e)
0.358	3.9	3	196.7	42	207	4295
		5	136.7	37	146	2985
0.325	3.5	3	159	50	168	4207
		5	112	46	120	2963
0.267	2.9	3	91.4	29	99	3588
		5	65.4	37	71	2574

Note: Material has been optimized for 8 and 4 μm filaments respectively. All values reported are confirmed by Brookhaven National Laboratory.
(e) = extrapolated J_c measured at 4.2 K, I_c= p Ω cm

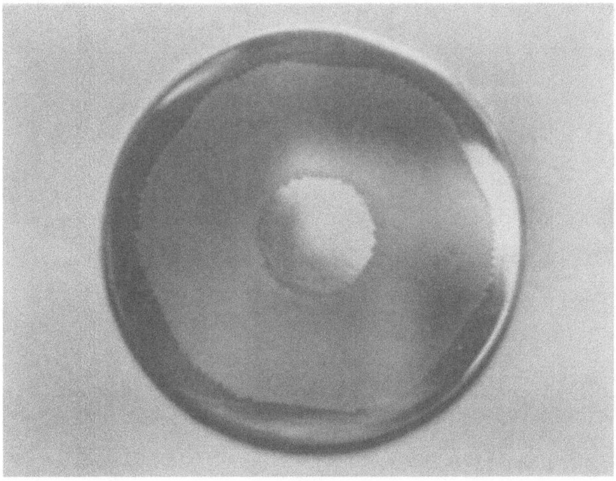

Figure 3
Cross section of single stack extrusion billet (4164 filaments)

Double Extrusion

More, barrier-clad NbTi monofilament of the type put into the 4,164 filament billet was used to make a small two-stage extrusion billet. While, as stated earlier, the high cost of extrusion and the loss of starting materials usually make this approach economically prohibitive, it was decided to explore the technical feasibility of making a high current density 40,000 filament material, in order to understand some of the problems involved.

The first stage consisted of 139 filaments assembled in a 51.0 mm (2 in.) diameter extrusion billet. This billet was then extruded at a ratio of 16:1, drawn to a suitable size, and restacked. The second extrusion billet was also 51.0 mm (2 in.) in diameter and contained 289 rods of the earlier extruded material and, therefore, a total of 40,171 filaments with a Cu:SC ratio of 0.90:1. A partial cross section exhibiting uniform filament distribution is shown in Figure 4. This material has been drawn to a diameter of 0.254 mm, without heat treatment, to establish a baseline of drawability. The filaments at this wire diameter are approximately 0.75 μm. To date, this material has not been fully processed and the J_c has not been optimized.

CONCLUSIONS

1. Experimental results demonstrate that critical current values of greater than 3000 A/mm^2 (4.2 K, I_c=ρ Ω cm) are attainable in high homogeneity Nb 46.5 w/o Ti fine filaments on the order of 2 to 3μm diameter, so long as filaments of high quality are maintained.

2. Geometrically uniform "unsausaged", fine filaments, devoid of copper-titanium intermetallic compounds are achieved by the addition of a diffusion barrier around each filament. This is a controllable process which allows greater freedom of heat treatment times and temperatures. The cost effectiveness of this technology appears to depend on the relative costs of diffusion barrier raw materials and assembly versus the design and

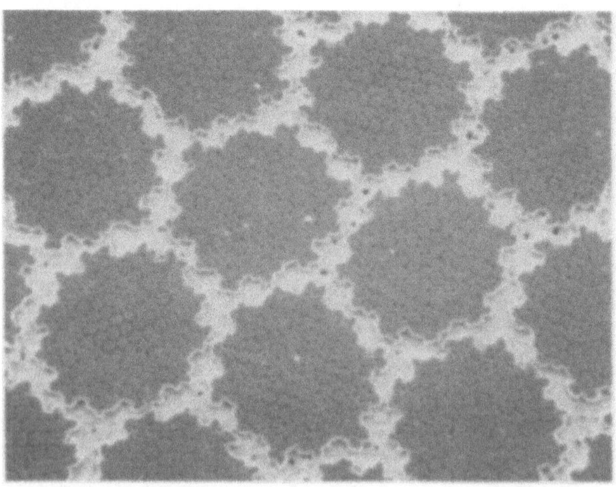

Figure 4
SEM micrograph showing a cross section of double extrusion billet (40,171 filaments).

fabrication of internal correction coils for large accelerator magnets.

3. Double extrusion of barrier clad NbTi has yielded a composite containing 40,171 submicron diameter filaments of high quality. The current-carrying capacity of this material remains to be optimized.

4. High J_c values are no longer just a laboratory phenomenon. Production quantities of high current density multifilamentary Nb 46.5 w/o Ti superconductor, with J_c approaching 3000 A/mm^2 at 5 T, have been produced by Supercon, Inc.

ACKNOWLEDGEMENTS

The financial support of Lawrence Berkeley Laboratory and of the Small Business Innovative Research Administration, both under the auspices of the U.S. Department of Energy, is gratefully acknowledged. The expertise and assistance of W.B. Sampson and his staff at BNL for critical current measurements is greatly appreciated. Credit for much of the success of these experiments goes to J. Bishop and the rest of the staff and technical support group of Supercon, Inc.

REFERENCES

1. H. Hillman, "Superconductor Materials Science" S. Foner, B.B. Schwartz, eds. Plenum Press, NY (1981), p. 275.
2. M. Garber, M. Suenaga, W.B. Sampson, and R.L. Sabatini, Effect of Cu4Ti Compound Formation on the Characteristics of NbTi Accelerator Magnet Wire, IEEE Trans. Nucl. Sci. (1985).
3. D.C. Larbalestier, Li Chengren, W. Starch, and P.J. Lee, Limitation of Critical Current Density by Intermetallic Formation in Fine Filament NbTi Superconductors, IEEE Trans. Nucl. Sci. (1985).
4. E. Gregory, "Manufacture of Superconducting Materials", American Society for Metals Materials and Metal Working Technical Series, R.W. Meyerhoff, ed., American Society for Metals, Metals Park, Ohio (1977), p.1.
5. M.T. Taylor, C. Graeme-Barber, A.C. Barber, and R.B. Reed, Co-processed Nb-25% Zr/Cu Composite, Research and Technical Notes, Cryogenics, p. 224 (June 1971).
6. T. Ogasawara, Y. Kubota, T. Makiura, T. Akachi, T. Hisanari, Y. Oda, and K. Yasukochi, IEEE Trans. Magn., MAG-19 (3) (1983).
7. W.J. Carr, Jr., and G.R. Wagner, "Advances in Cryogenic Engineering Materials", Vol. 30, Plenum Press, NY (1984) p. 923.
8. E. Gregory, E. Adam, S. Hong, W. Marancik, P. Sanger, and C. Spencer, Recent Developments in Superconductors for Large Magnets, IEEE Trans. Magn., MAG-17 (5) (1981).
9. W. Sampson, Brookhaven National Laboratory, private communication.
10. Y. Tanaka, M. Ikeda and H. Tanaka, Superconducting Wire with Small Filaments for SSC Magnets, IEEE Trans. Nucl. Sci., (1985).

NbTi ULTRAFINE FILAMENT WIRES FOR 50/60 HERTZ USE

A. Fevrier, P. Dubots, and J. C. Renard
Laboratoires de Marcoussis - CR - C.G.E.
Marcoussis, France

Y. Laumond, Hoang Gia Ky, and J. L. Sabrie
Alsthom
Belfort, France

ABSTRACT

Thanks to technological progress carried out in our Company for the last ten years, we have succeeded in manufacturing long lengths of NbTi ultra-fine filament wires in which 50 Hertz losses are strongly reduced. These results have been obtained by manufacturing the wires with a Cu-30 At % Ni matrix between the filaments, by reducing the filament diameter well below .5 micron and by reducing the twist pitch length to 4 times the wire diameter.

These new wires, which open very new prospects for industrial applications of superconductivity in fast pulsed magnets and in large 50/60 Hertz equipments, herald promises of a technological revolution in electrotechnology and very likely in high power electronics.

In this paper we present the state-of-the-art of NbTi ultra-fine filament wires and we present the reflections of Alsthom and Laboratoires de Marcoussis which lead to reconsider electric machine design on account of technical and econominal potentials of these new superconducting wires.

INTRODUCTION

In 1983, the development, by Alsthom and by les Laboratoires de Marcoussis (LdM), of NbTi ultra-fine filament wires with a high transverse resistivity matrix, has allowed to open very new prospects for industrial applications of superconductivity in fast pulsed magnets and in large 50/60 Hertz equipments.[1, 2].

For about fifteen years, several designs of superconducting conductors for fast pulsed magnets and for A.C. devices have been proposed[3,4,5]. But it takes many years to begin controlling industrial manufacturing of such conductors [6,7].

In the fields of electrotechnology and of high power electronics, these new conductors constitutes a first step forward a deep technological evolution. The goal of this paper is to present the state-of-the-art of

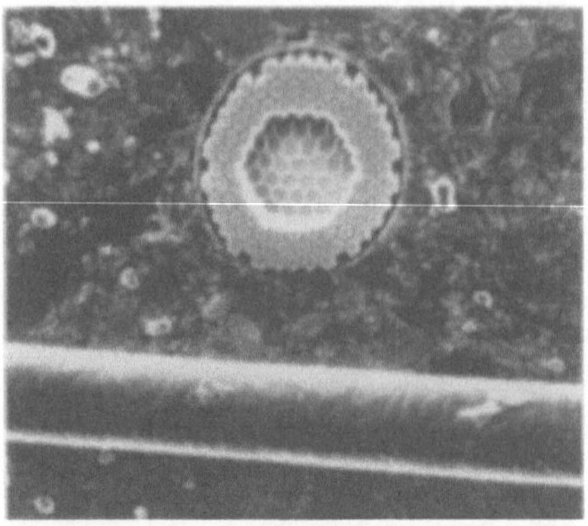

a) Wire section compared to a human hair. Wire diameter = .12 mm

b) Detail of the filaments (defects are undissolved matrix). Filament diameter = .12 micron.

Fig. 1. SEM views of the 254 100 filament wire.

NbTi ultra-fine filament wires and to discuss Alsthom and LdM reflections at the dawn of the technological evolution these new conductors start up.

CONDUCTOR MANUFACTURING

Studies relating to the design and to the manufacturing of a conductor for the rotor of a superconducting generator have allowed Alsthom and LdM to gain the know-how for manufacturing wires which comprise from several thousand to several ten thousand 5 to 10 micron-diameter NbTi filaments in a Cu-Cu 30 At % Ni mixed matrix[6]. These wires have already given consistently fine performances under pulsed conditions, in good agreement with our calculations, which also attest that we have mastered loss computations[8]. Loss computations show it is possible to design conductors for A.C. use if filament and wire diameters, twist pitch length and transverse electric conductivity of the matrix can be sufficiently reduced.

With the aim of developing such conductors we have started a multiphase program in order to determine technological and industrial limits :

a) In a first phase, ended in 1983[1], our purpose was to manufacture a wire in which the filaments are embedded in a Cu 30 At % Ni matrix. Such a wire of .12 mm in diameter and which comprises 14 496 .5 micron filaments

Table 1. Critical current density and critical current of .5 micron 14 496 filament wires.

B (T)	0,5	2	5	8
J (10^9 A/m^2)	10.25	4.5	2	.745
I (A)	39.6	17.35	7.7	2.88

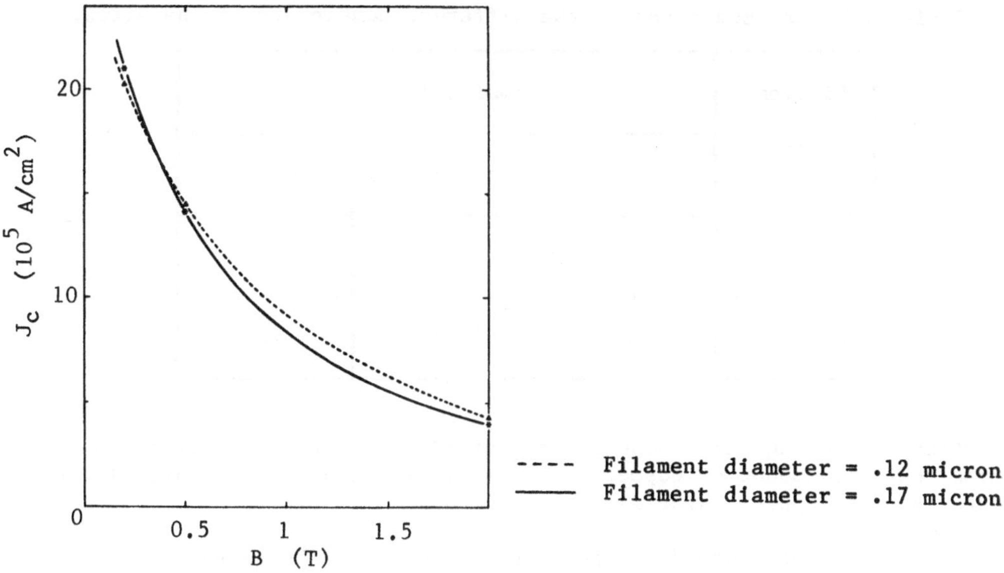

Fig. 2 Typical values of critical current densities obtained on ultra-fine filament wires in the low magnetic field range.

and a Cu-CuNi central core is manufactured by Alsthom in an industrial scale.

b) In a second phase, we tried to reduce, as much as possible, the filament diameter and the twist pitch length :

- .08 micron filaments have been obtained in a .08 mm wire which comprise 254 100 filaments[7]. Fig. 1 shows a section of this type of wire and a SEM view of the filaments on which neither surface defect nor necking are observable.

- twist pitch length to wire diameter ratio has been reduced to a value very close to 4 which is well below the common value of 7.

c) In a third phase, the work is directed towards improvement of critical current density values. For ultra-fine filament wires, specific problems are encountered. Among them, the strain undergone by the differents metals is very high, in particular for NbTi, and for filament diameter lower than .5 micron surface effects may be important.

- As far as .5 micron 14 496 filament wires are concerned, critical current densities are now very close to those of standard wires (See table I).

- As far as 254 100 filament wires are concerned, very high critical current densities have been obtained in the low magnetic field range (See. Fig. 2) and work is in progress in the high magnetic field range.

d) In a fourth phase, we study the magnetic stability of these wires. With ultra-fine filaments, we were able to get a good magnetic stability although the filaments are embedded in a Cu 30 At % Ni matrix and the pure copper is restricted to the central core divided-up by Cu 30 At % Ni barriers. Table 2 gives the normalized areas of the different components of these wires and fig. 3 shows their ranges of magnetic stability as a

Table 2. Normalized areas of the different components of the wires.

Filament Number	Normalized areas		
	Cu	Cu 30 At % Ni	Nb Ti
14 496	.253	.413	.334
254 100	.222	.535	.243

function of the magnetic induction and of filament and wire diameters. We think that the amount of copper can be again reduced while maintaining a wide range of magnetic stability.

By the end of these first four phases, we should have technical knowledge to help conductor design for different applications. We have grouped these conductors in three main families which satify the various kinds of applications envisioned to-date (See. next paragraph).

Two new phases could then be undertaken :

e) In phase five we would set the manufacturing conditions of model conductors for each kind of them (See. next paragraph). Depending on the considered applications, these conductors could be used at different filament and wire diameters.

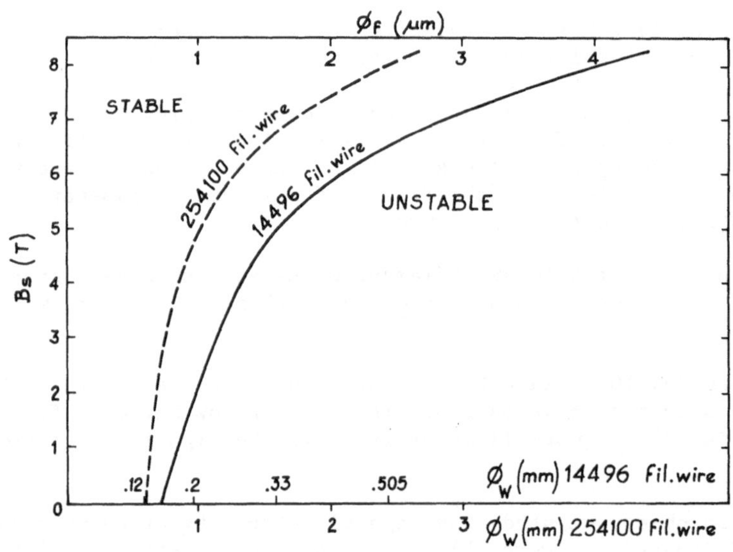

Fig. 3. Stability domains of 14 496 filament wire and of 254 100 filament wire as a function of the magnetic induction, B_s, and of filament and wire diameters, $\emptyset_f$ and $\emptyset_w$.

f) The sixth phase would establish the manufacturing conditions of cabled conductors able to carry thousands of amps which are necessitated by the main considered applications (See. next paragraph).

APPLICATIONS: CONDUCTOR DESIGN

The design of a conductor for any superconducting equipment necessitates to carry out a fine analysis of time-variations of losses and temperature in the conductor, which arise from variations of the current, of the magnetic field components and of cooling conditions. From 1975 to 1982, we set up a computer program for the calculation of losses in monolithic and cabled conductors which undergo any kinds of variation of the magnetic induction, of the current and of the temperature[8,9]. This program, which has been thoroughly validated for conductors with filament diameter larger than 1 micron, allows us to compute losses within an accuracy of 15 % and to calculate conductor temperature within an accuracy of a few .01 K for "Roebel" type cabled conductors. In 1984, we experimentally showed that, for a first kind of ultra-fine filament wires, hysteretic losses do not show a linear decrease with filament diameter for diameters smaller than about 1 micron[10]. To begin with, the loss computer program has been modified by using an empirical formula to take this new effect into account but work is in progress in order to analyse hysteretic losses in superconducting filaments in the sub-micron range.

According to the considered application, the type of equipment, the considered part of the equipment and electromagnetic and thermal working conditions, conductors may be very different from each other. We think superconducting equipments may be classified according to three main types of working conditions to which three main families of conductors are respectively associated and which are very different in their design :

- A first family of conductors for D.C. applications but which are able to undergo transient conditions without quenching. For this family, conductor enthalpy plays an important part in the design.

- A second family of conductors for 50/60 Hertz applications. For this family, conductor design is mainly ruled by A.C. losses.

- A third family of conductors which are alternately used in the superconducting and in the normal state, like in superconducting rectifiers. For this family, conductor design is mainly ruled by the normal state electrical resistance of the conductor.

<u>Conductors for D.C. Applications</u>

This family of conductors, is defined by the following working conditions :

- Main working under D.C. or quasi-D.C. conditions and with the highest overall current density in the conductor.
- Transient conditions without conductor quenching.

Alsthom already produces this kind of conductors in an industrial scale.

Application fields for these conductors are rotor windings of generators and motors, Superconducting Magnetic Energy Storage (SMES) for electric power network stabilization or for peak shaving in utility, magnets for fusion power reactors and high energy physics...

The conductors of this family are cabled from several wires the diameter of which is around 1 mm. The wires comprise several thousand NbTi filaments the diameter of which ranges from 1 micron to a few microns and which are embedded in a copper matrix or in a copper/copper-nickel mixed matrix. Enthalpy of the conductor and of the surrounding medium plays a very important part in the design of these conductors.

The design and description of a rotor winding conductor for a 250 MW generator[11] are presented to illustrate this conductor family. This conductor has been designed so that it does not quench when it undergoes the most severe fault condition. In the present case, the most severe fault condition is encountered on the straight part of the rotor winding during a three-phase short-circuit followed by a fault clearance and a resynchronization. This fault is the most severe for the conductor since high current and magnetic field peaks occur but also because resynchronization low frequency (1-2 Hertz) magnetic field oscillations are imperfectly shielded by the damper. Fig. 4 (a) presents current variations which show a peak at the short-circuit clearance time followed by resynchronization oscillations. The maximum value of the magnetic field is reached on the outer diameter of the rotor winding where the armatùre feed-back effect adds to the magnetic field increase due to the rotor winding current increase. Fig. 4 (b) presents magnetic induction variations.

The optimal conductor has been designed by using our loss and winding temperature computer program. This conductor, of "Roebel" type, is cooled along the edge and the overall current density is equal to $1.5\ 10^8\ A/m^2$ for normal working conditions. It comprises 19 insulated wires of 1.16 mm in diameter. Each wire, with a CuNi outer sheath, comprises 13 068 5.25 micron-diameter NbTi filaments in a Cu/CuNi mixed matrix surrounding a Cu/GuNi central core. This optimal structure has been determined so that the current never exceeds 70 % of the critical current during the cleared short-circuit and so that the winding temperature rise be enough low in case of an unexpected quenching. Fig. 4 (c) and 4 (d) present variations of the current to critical current ratio and of the conductor temperature during the first 1.5 s of the cleared short-circuit. The maximum value of I/I_c is .68 and the maximum reached temperature is 4.85 K. At last, fig. 4 (e) which presents variation of losses in the conductor, shows that a considerable part of losses is removed by enthalpy.

Ultra-fine filament conductors lead to a simplificaton of the generator which can be built without damper[2]. In this case, the conductor will comprise a larger number of a few .1 mm-diameter wires, and each of them will comprise several ten thousand filaments which diameter is smaller than 1 micron. Such a conductor will be able to undergo time-variations of the magnetic induction of several ten Tesla per second with low A.C. losses.

Conductors for 50/60 Hertz applications

This family of conductors is defined by the fact that AC losses due to 50/60 Hertz magnetic field and current variations must be as low as possible.

Application fields for these conductors are armature windings of generators and motors, power transformer windings, magnetic switch for Superconducting Electric Valves (SEV)...

To lower A.C. losses to a level which makes ultra-fine filament conductors attractive for designers from an economic point of view, we worked in the two following directions[7] :

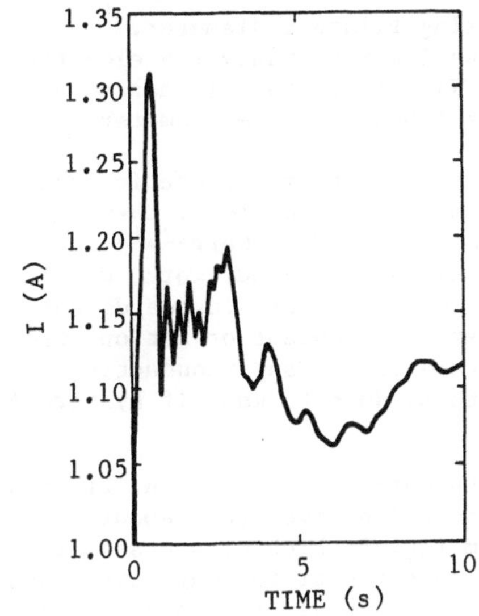

a) Current variations

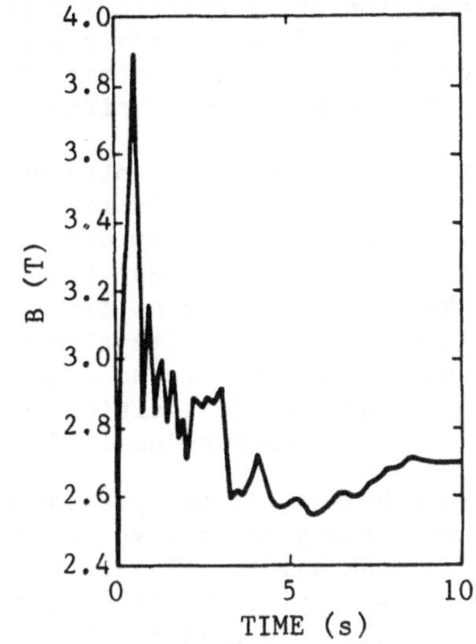

b) Magnetic induction variations at the critical point of the winding.

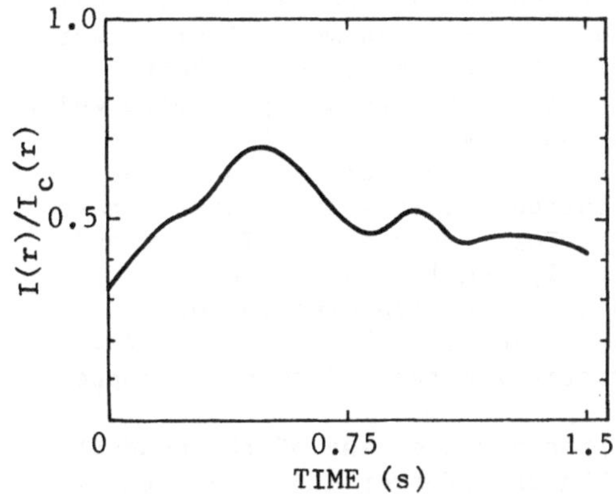

c) Variations of the fault current to critical current ratio

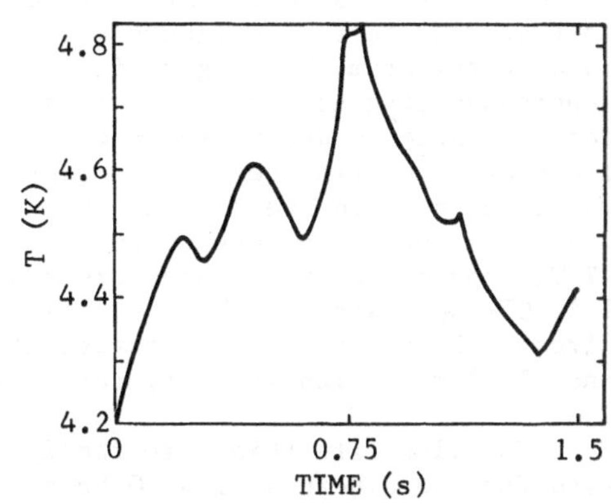

d) Conductor temperature variations

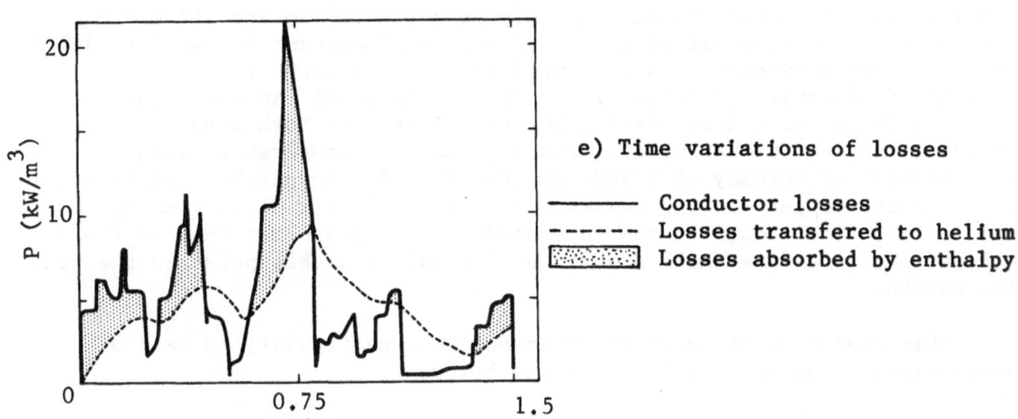

e) Time variations of losses

Fig. 4. Behaviour of a superconducting rotor winding during a three-phase short-circuit followed by a fault clearance and a resynchronization

- reduction of hysteretic losses by lowering filament diameter.
- reduction of eddy current losses by lowering the transverse electric conductivity of the matrix (Cu 30 At % Ni between the filaments) and by lowering the twist pitch length to 4 times the wire diameter.

Up to now we have not obtained a sharp reduction of hysteretic losses as expected from the Bean's model[10] for filament diameter lower than about .5 micron. This behaviour has been analyzed by W.J. Carr[12,13] in term of surface effects. But from this analysis and from our work, we cannot know if the filament diameter from which hysteretic losses do not have a linear decrease with filament diameter is dependent or not on wire manufacturing conditions, or even on the nature of the superconducting material. Considerable additional work has to be done to know if hysteretic losses could be still reduced.

Nevertheless, in the present state-of-the-art, our ultra-fine filament wires are already attractive for magnetic induction lower than about 1.5 Tesla peak value. This can be shown from fig. 5 which presents 4.21 K 50 Hertz losses of our best conductors as a function of the magnetic induction peak value. On this figure we also drew loss curves for copper wires carrying different current densities (peak values) ; in order to make a comparison with superconducting wires, ambient temperature copper losses have been normalized to 4.21 K by using a refrigerator efficiency of 1/500. Losses are expressed in watt per Amp-meter generated which is a practical unit of measure for designers who want to make an assessment of different conductors. From this figure it is shown that, at 1.5 Tesla, 50 Hertz superconducting conductor losses are equal to losses in a copper wire which carry 15 A/mm^2 - peak value - which is a common value of ambient temperature current density in copper wires. For smaller values of the magnetic induction we have to lower the current density in copper to have losses as low as in superconducting wires. For instance at 1 T, .5 T and .2 T, superconducting losses are respectively equal to 5 10^{-5} W/Am, 1.4 10^{-5} W/Am and 3 10^{-6} W/Am, values which can be achieved in copper wires with current densities respectively as low as 2.9 A/mm^2, .82 A/mm^2 and .18 A/mm^2 which are very far from practical current density in copper.

To illustrate this wire family, we presente the case of the conductors used for the windings of a 50 Hertz-220 kVA transformer under construction in our laboratories. Both conductors are made from 254 100 filament wires which have a Cu-CuNi central core and in which filaments are embedded in a Cu 30 At % Ni matrix. The primary conductor is a cable which comprises 6 insulated superconducting wires stranded around a CuNi insulated wire. Wire and filament diameters are respectively equal to .17 mm and .17 micron. Conductor length is equal to 160 m and the rated current is equal to 349 A. The secondary conductor is a .155 mm-diameter wire with .155 micron-diameter filaments. Its length is equal to 1013 m and the rated current is equal to 55 A. At nominal rating 50 Hertz losses in both windings should be equal to .4 W at 4.21 K, i.e. 200 W at ambient temperature with a refrigerator efficiency of 1/500. In the case of a comparable conventional transformer, copper winding losses are equal to 1733 W. Thus winding losses should be 8 times lower in the superconducting transformer than in the conventional one we shall experimentally validate this point in the next few months.

This kind of wires have other advantages, particulary in case of short-circuit, which will be presented below.

Conductors for rectifiers

This family of conductors is defined by the fact that the overall normal state electric resistivity must be as high as possible.

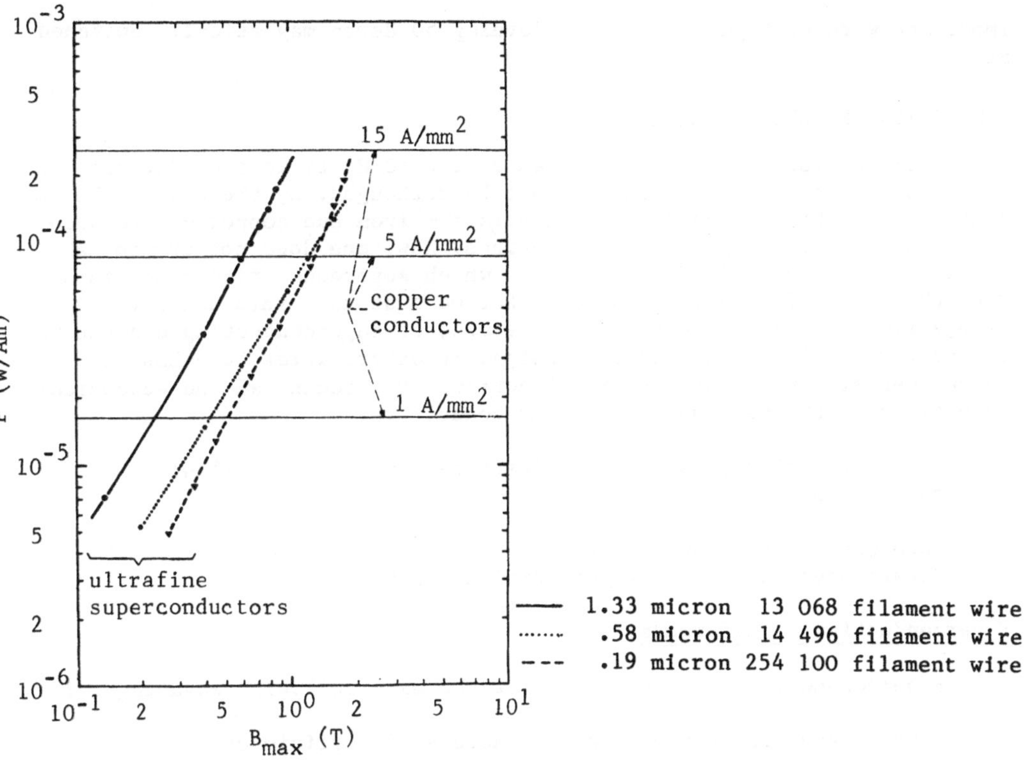

Fig. 5. 50 Hz losses in our best conductors compared to losses in copper conductors.

Application fields for these conductors are thermally switched Superconducting Electric Valves - SEV - (SEV using NbTi composites), magnetically switched SEV (which need to use wires made of a low critical magnetic field superconductor : $B_c < 1$ Tesla), fault current limiters...

For this family of conductors, the main problem to solve is to obtain a good magnetic stability with filaments in a low conducting matrix like Cu-30 At % Ni. In this case a good magnetic stability can be obtained by reducing filament and wire diameters. This can be shown from fig. 5 which presents magnetic stability ranges of 14 496 and 254 100 filament wires. For a given magnetic induction value, magnetic stability of the 254 100 filament wire is obtained for smaller filament diameter than for the 14 496 filament wire since the diameter of the first wire is higher than the second one for the same filament diameter. Work is in progress to reduce the amount of copper in the central core, nay to suppress it.

To illustrate this conductor family we shall say a few words about Superconducting Electric Valves (SEV). These SEV, which presents symetrical current-voltage characteristics, are of controled turn-on and turn-off type and in some ways they combine GTO thyristor and Triac properties. Fig. 6 presents a SEV current-voltage characteristic in comparison with a thyristor one : $J_c = J_c(B,T)$ is the critical current density of the SEV conductor. In the off-state, $J_c = 0$ and the characteristic is given by $E = \rho J$, in which ρ is the overall normal state electric resistivity of the SEV conductor. In the on-state, the characteristic is given by $E = 0$ for $|J| < J_s$.

In the present state-of-the-art, we think it is possible to built 98 % efficiency SEV with a specific power of 400 W per cm^3 of conductor. But important work must yet be done to develop 50 Hertz magnetically switched SEV.

NEW DESIGN OF ELECTRIC MACHINES

From the very new prospects open by low 50 Hertz loss conductors, the design of electrotechnic machines must be rethought. By the end of the last century, magnetic and conducting materials - iron and copper - have deeply stood out in the design of these machines. Now, one does not try to make conventional machine "cryo-copies", in which superconducting conductors take the place of copper conductors, but we must take into account cryogenics and the very particular features of superconducting conductors at the very beginning of machine design. In others words, one has to reconsider electrotechnic machine functions from technical and economical potentials of the new superconducting conductors.

This point of view shall be illustrated by the two following particular exemples :

- Superconducting Transformers
- Superconducting AC Generator Complex (GRASC)

Superconducting Transformers

Superconducting transformers offer the main following advantages :

- Loss reduction which leads to increase in efficiency
- Significant weight reduction which results from low 50 Hertz losses of the new conductors which allows to increase amper-turns and to reduce iron core area.

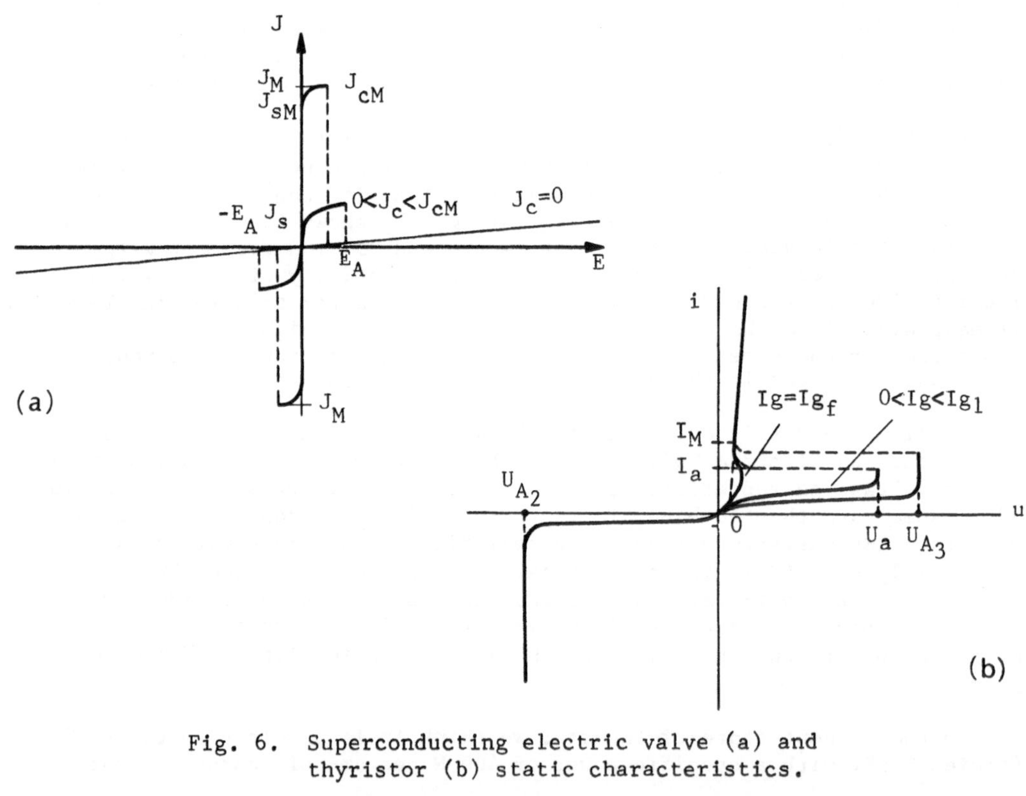

Fig. 6. Superconducting electric valve (a) and thyristor (b) static characteristics.

- Considerable simplification of mechanical design problems which results from a small increase of magnetic forces in case of short-circuit fault since the high normal state resistivity of the the conductor limits overcurents to a few 10 % of the rated current.

To illustrate these avantages, we briefly describe the general design and main features of the 50 Hertz single-phase transformer now under construction :

- Nominal rating : 220 kVA
- Nominal primary voltage : 660 V
- Nominal secondary voltage : 4000 V, with a cos φ value equal to .8
- Windings are cooled by natural convection in a 4.21 K helium bath

Fig. 7 presents total losses of 220 kVA conventional and superconducting transformers as a function of the iron core area. Using state-of-the-art ultra-fine filament wires allows to increase Amper-turns and to reduce the iron core area by a factor of six. In comparison with a comparable conventional transformer, the iron core area reduction leads to the following gains : the iron core weight is equal to 52 kg instead of 610 kg, iron core losses are equal to 100 W instead of 700 W and the iron core exciting current in the low voltage winding is equal to .36 Amp instead of 1.2 Amps. A further reduction of A.C. losses of superconducting conductors will lead to additional reduction of the iron core area (dashed curve on fig. 7 which corresponds to a reduction of A.C. losses by a factor of 3).

The general design of the transformer is given on fig. 8. The near ambient temperature iron core, of conventional shape, is situated in a fiber glass reinforced composite inner vessel and it is vacuum insulated from the helium bath. The iron core is cooled by means of an inner channel.

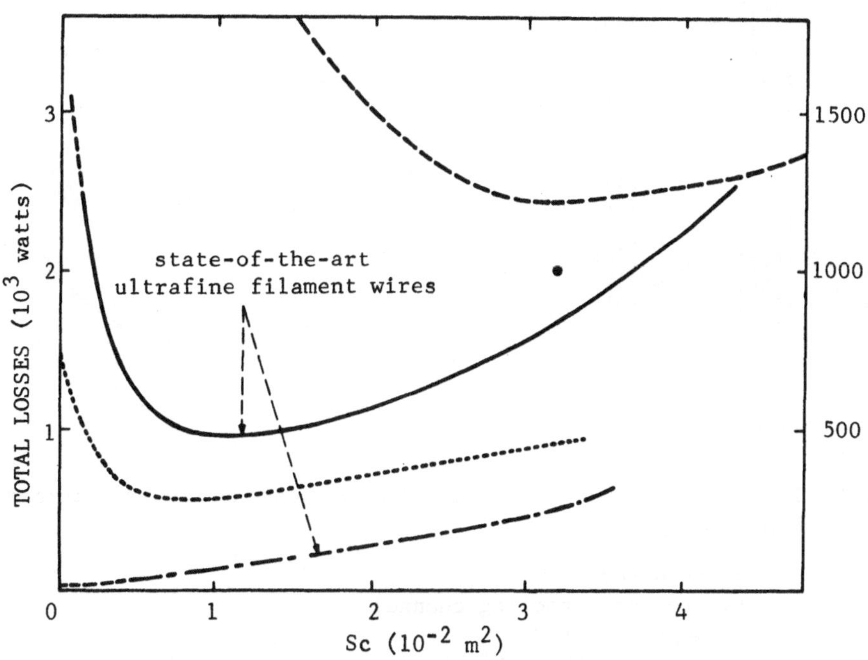

- Conventional transformer : losses (—–), weight (●)
- Superconducting transformer : losses (——, - - - - -), weight (-——-)

Fig. 7. Total losses and weight of 220 kVA conventional and superconducting transformers as a function of the iron core area S_f.

Inter-leaved low and high voltage windings, which are made from the conductors described in the preceding paragraph, are wound around the inner vessel by means of half-cut formers. Windings weight is about 10 kg instead of 150 kg for copper windings of a comparable conventional transformer.

The theoretical X_T reactance of this transformer is equal to 7.5 %. For a comparable conventional transformer the short-circuit overcurrent would be equal to 15 or 20 times the rated current which leads to transient magnetic forces 200 or 400 times higher than the nominal ones. In the present superconducting transformer, beyond the conductor critical current, the theoretical total electric resistance seen from the primary winding is equal to 6.6 Ω which limits the primary short-circuit current to about 500 A, i.e to 1.45 times the nominal one. Fig. 9 illustrates theoretical variations of primary and secondary currents in case of a secondary short-circuit fault.

Table 3 summarizes the main features of the superconducting transformer and of a comparable conventional one. As it can be seen, the most important part of losses comes from the inner vessel and current leads.

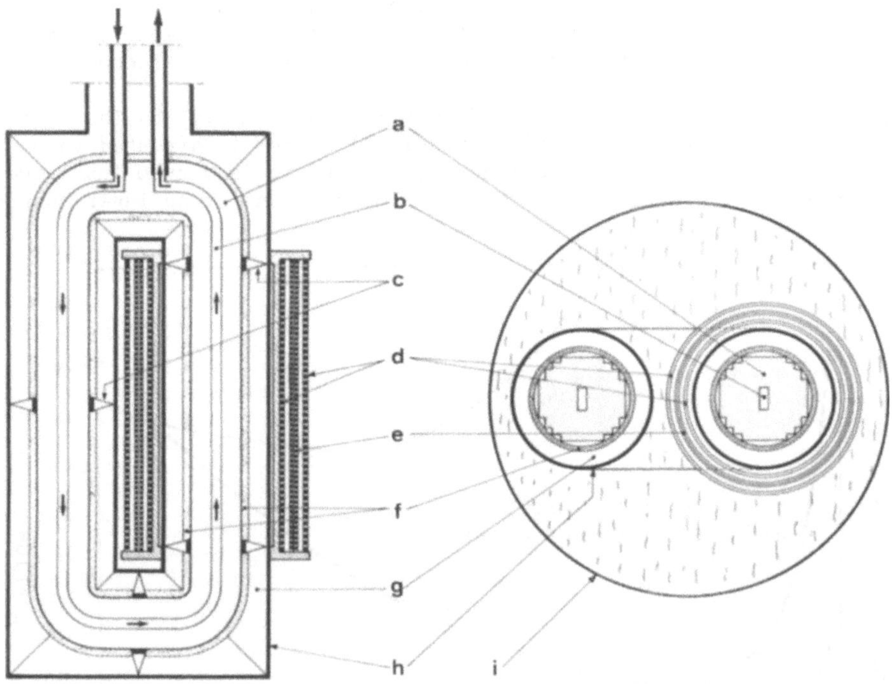

Fig. 8. Schematic diagram of the 220 kVA superconducting transformer under construction.

a. iron core
b. iron core cooling channel
c. support struts
d. low voltage windings
e. high voltage winding
f. radiation shield
g. vacuum gap
h. fiber glass reinforced composite inner vessel
i. cryostat cold vessel.

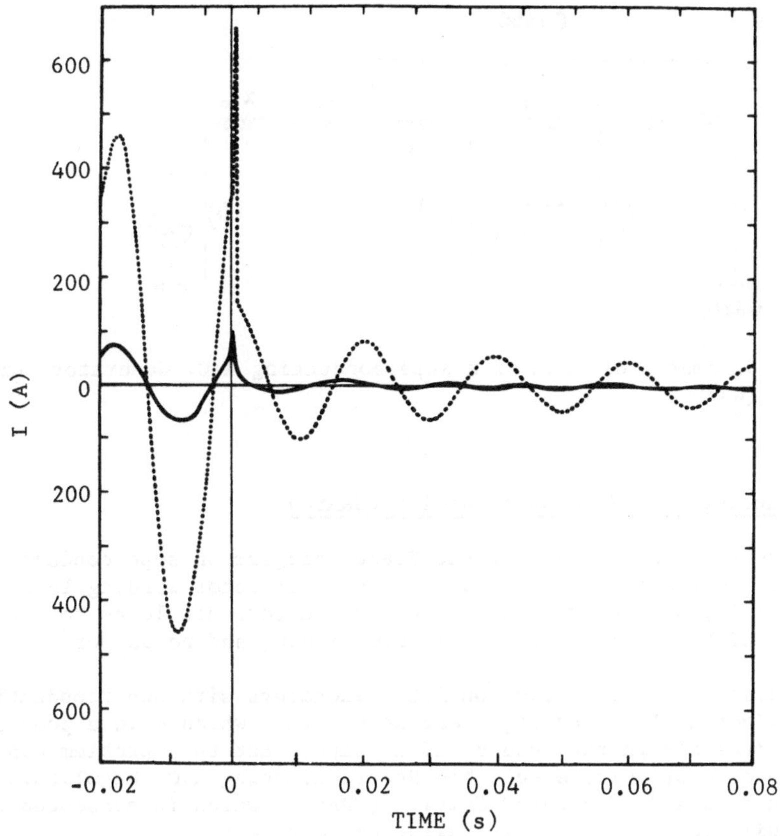

Fig. 9. Theoretical variations of primary (........) and secondary (——) currents in case of a secondary short-circuit fault.

Table 3 - Main features of the 220 kVA superconducting transformer and of a conventional one

Main features	Superconducting transformer	Conventional transformer
. Nominal iron induction	1.8 Tesla	1.5 Tesla
. Iron core losses	100 W	694 W
. Ambient temperature winding losses (full load, Cos φ = 1)	.4 W x 500*	1733 W
. Ambient temperature inner vessel and current lead losses	2 W x 500*	-
. Total weight	100 kg	1000 kg
. Total losses	1300 W	2400 W
. Efficiency	99,4 %	98,9 %
. Short-circuit current normalized to rated one	1,5	15

(*) Refrigerator efficiency : 1/500

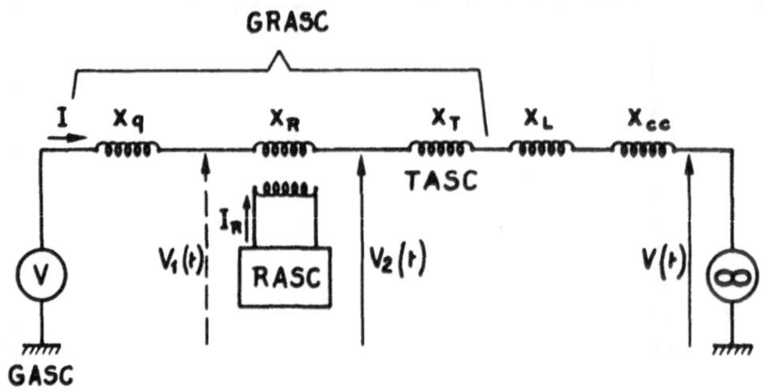

Fig. 10. Schematic diagram of a superconducting A.C. Generator Complex (GRASC)

Superconducting A.C. Generator Complex (GRASC)

During the last few years, the French program on superconducting A.C. generators[14] was exlusively planned to increase rotor winding load. But feasibility of low 50/60 Hertz loss conductors now leads to A.C. generators with superconducting armature winding and no damper[2].

Preliminary investigation on A.C. generators with superconducting and armature windings leads to high reactance values which give a poor stability, particularly in the absence of a damper. But this problem can be easily overcome by using a suitable Superconducting A.C. Regulator. Thus, a Superconducting A.C. Generator Complex (GRASC), which is presented on fig. 10, will comprise the main following components :

- A Superconducting Rotor and Armature A.C. Generator (GASC)
- A Superconducting A.C. Regulator (RASC) which uses a small SMES unit.
- A Superconducting Output Power Transformer (TASC).

In the case of a high-voltage short-circuit fault, the whole complex is maintained in the superconducting state thanks to the high value of the generator transient reactance (≃ 300 %). In case of a low voltage short-circuit fault, the current is limited by the high normal state electrical resistance of the conductor.

The Superconducting A.C. Regulator (RASC) shall in particular fulfill the following purposes :

- voltage adjustment
- static stability control
- Torque control, to avoid shaft oscillations.

A precise network analysis will lead to specify the functions and the design of the RASC from the following unrelated parameters : shaft speed of rotation, wattfull and wattless output powers and network voltage. Otherwise this RASC could also lead to operating cost savings on the turbine, on the generator and on the transformer.

From our preliminary economical analysis, the break-even point of Superconducting A.C. Generator Complex (GRASC) should be well below 1000 MW which is the accepted one for Superconducting Rotor A.C. Generator. At the present time, we are designing a 250 MW GRASC and a small scale experimental program (20 kW) is conducted in collaboration with the CRTBT-CNRS-Grenoble to examine GRASC features which are crucial to its viability.

Other Applications

As a matter of interest we mention some other applications of ultra-fine filament superconducting conductors :

a) Generators and Motors : The use of superconductors for slowly time-varying current windings allows to design light and compact synchronous machines[15]. It is possible to reach 25 MW per metric ton instead of 2.5 in case of electric nuclear plant equipment.

b) Fault Current Limiters, Power Filters : The high normal state electric resistivity of these conductors leads to allow their use in fault current limiters set up-stream of conventional breakers. Such fault current limiters should reduce the size of overhead transmission line surge-protecting equipements. The fault current limiting function can be provided by a superconducting transformer the secondary windings of which are shorted for the rated current while primary windings are serie-connected in network phases.

If the secondary short-circuit of the superconducting transformer is substituted by a frequency-varying impendance, one can make a power filtrer to suppress low and high frequencies. Such a power filter should allow machine or network part descoupling to avoid harmfull interferences.

CONCLUSIONS

In 1983-1984, superconductivity took a turn for the better on account for the two following occurences :

- The establishment by Alsthom and Laboratoires de Marcoussis of design and manufacturing parameter for superconducting conductors which can be economically used in fast pulsed magnet and in large 50/60 Hertz equipments.
- The production in an industrial scale of NMR imaging superconducting magnets.

These two advances herald promises of a technological revolution in electrotechnology and very likely in high power electronics.

Electric machines designed from the unique aspect of ultra-fine filament conductors are one of the important economic stake for the next 10 years. Our group is embarked on a new technological route with the two following principal aims :

- Improvement of the properties of the three conductor families itemized.
- Design and building of model machines to demonstrate our analyses.

ACKNOWLEDGMENTS

The authors thank J. Gerard, G. Bottini, D. Legat, F. Saint-Saens, J.P. Tavergnier and W. Weber for their assistance in obtaining the reported experimental results.

REFERENCES

1. P. Dubots, A. Février and al., "8th International Conference on magnet Technology", Journal de Physique, Sup.1, (1984), pp.C1-467-470.
2. J.L. Sabrié, "8th International Conference on magnet Technology", Journal de Physique, Sup.1, (1984), pp. C1-717-720.
3. A. Février, Thèse, Université de Paris-Sud (1973).
4. I. Hlasnik, IEEE Trans. Magn. MAG 17 : 2261-2269 (1981).
5. T. Ogasawara and al., IEEE Trans. Magn. MAG 19 : 248-251 (1983).

6. P. Dubots and al, "Proceedings of the Eighth International Cryogenic Engineering Conference" England (1980), pp 505-508.
7. P. Dubots and al., "Proceedings of the tenth international cryogenic Engineering Conference" England (1984), pp 610-615.
8. A. Février, Cryogenics April 1983 : 185-200.
9. A. Février, J.C. Renard, IEEE Trans. Magn. MAG 17 : 224-227(1981).
10. P. Dubots, A. Février, and al., IEEE Trans. Magn. MAG 21 : 177-180 (1985).
11. J.L. Duchateau, A. Février, Y. Laumond, "Stability of superconductors in helium I and helium II", Int. Inst. Refr. Paris (1981) pp. 269-283.
12. W.J. Carr Jr., G.R. Wagner, "Advances in Cryogenic Engineering" Vol.30, (1984) pp. 923-930.
13. W.J. Carr Jr., IEEE Trans. Magn. MAG 21 : 355-357 (1985).
14. J.L. Sabrié, J. Goyer, IEEE Trans. Magn. MAG 19 : 529-531 (1983).
15. B.B. Gamble, T.A. Kein, "Advances in Cryogenic Engineering", Vol. 25, (1980) pp. 127-136.

COUPLING LOSSES IN SUPERCONDUCTING CABLES

T. Okada, K. Takahata, and S. Nishijima

ISIR Osaka University
Ibaraki, Osaka, Japan

ABSTRACT

Measurements of the coupling losses have been made on the several types of cables by changing the cable twist pitch and direction systematically in order to optimize the cable structures. A cable twist opposite in direction to the strand twist led to higher losses than those with identical twist directions. The twist pitch dependence of the losses could be interpreted on the basis of Campbell's theory. The results indicate that the losses may be reduced by optimizing the cable twist pitch. The influences of the volume fraction of superconductors in the cable and the arrangement have also been examined.

INTRODUCTION

Multistage cables have been fabricated with multifilamentary superconducting wires aiming at the large scale pulsed magnets. To improve the stability, the copper or the aluminum stabilizers are often assembled together with the wires. The cables are soldered in order to ensure stable current sharing as well as the mechanical rigidity of the cables. Previous researchers[1-3] have shown that good electrical contacts between strands cause an increase in ac losses because of increased interstrand coupling. It is not, therefore, desirable to solder the cables completely. For this reason, ordinary cables have been soldered only at the first level of cable construction.

Campbell estimated the coupling losses of the first level cables by assuming that the cable consists of a large number of strands.[4] Quantitative estimation of the actual losses in the cables might not be possible using the theory, since the practical cables consist of a limited number of strands (three to seven in this study).

The purpose of this study is to measure the coupling losses of practical cables and to compare the data with the calculated losses. The influences of the volume fraction of strands and the arrangement on the losses were investigated. The effects of the cable twist direction, twist pitch, and electrical conductivity of the solder were also examined.

SAMPLE AND EXPERIMENTAL METHOD

The basic superconducting strands were two types of multifilamentary Nb-Ti-Zr-Ta composites with copper matrices. The specifications of the strands are shown in Table 1. Strand B has an additional outer copper sheath, which is 30 μm thick.

In Table 2, the parameters of the cabled samples investigated in this work are listed. The two types of cables, which are three- and seven-strand cables, were twisted with various pitch lengths and filled with three kinds of solders. The cables were made with a twist direction identical to or opposite that of strand. In the seven-strand cable, copper wire with the same diameter as the strand was located at the center of the cable. The

Table 1. Superconducting Strand Specifications

	Without Sheath (A)	With Sheath (B)
Conductor	Nb-Ti-Zr-Ta	
Diameter, mm	0.32	0.35
Cu/SC ratio	0.88	1.3
Filament dia., μm	30	
Number of filament	61	
Twist pitch, mm	10 (R.H.)	

Table 2. Cable Sample Parameters

Cable Sample	Number of Strands	Strand	Solder	Cable Twist Pitch (mm)
#1	3 (3 SC wires)	A	40Pb-60Sn	Varied
#2	7 (6 SC wires and 1 Cu wire)	A	40Pb-60Sn	Varied
#3	3	B	40Pb-60Sn	Varied
#4	7	B	40Pb-60Sn	Varied
#5	7	B	38Pb-60Sn-2Ag	23(L.H.)
#6	7	B	48Pb-50Sn-2Cu	23(L.H.)

Table 3. Effect of Solder Resistivity

Sample Cable	Solder	Resistivity (Ω·m) at LNT	$\tau_c/(\tau_s \cdot f)$
#4	40Pb-60Sn	2.32×10^{-8}	3.7
#5	38Pb-60Sn-2Ag	1.66×10^{-8}	5.2
#6	48Pb-50Sn-2Cu	1.98×10^{-8}	4.5

electrical resistivity of solders was measured at liquid nitrogen temperature (LNT) and is given in Table 3.

The losses of the cabled samples were determined as the area of M-H curves, which were obtained by an electromagnetic measurement. A block diagram of the measuring system is illustrated in Fig. 1. The measurements of losses were made under transverse magnetic fields. The 15-mm-long samples were set perpendicular to the applied fields in the saddle-shaped pickup coil. A trapezoidal-shaped field changing with time was applied to the specimen, under a background maximum field of 3.2 T and with a sweep rate from 0.65 to 6.3 T/s, and the ac losses were measured.

Coupling losses, P_c, can be expressed as follows:

$$P_c = \frac{\tau}{\mu_o} \cdot \dot{B}^2 \qquad (1)$$

where μ_o is the magnetic permeability in vacuum and τ, the time constant characterizing the coupling loss. To compare the interstrand coupling loss with the interfilamentary loss, the ratio of the cable to the strand loss was evaluated by $\tau_c/(\tau_s \cdot f)$, where τ_c and τ_s are the time constants for cable and strand and f is the volume fraction of the cable occupied by the strands.

The $\tau_c/(\tau_s \cdot f)$ corresponds to the ratio of the cable loss to the sum of the strands loss. Hence, if $\tau_c/(\tau_s \cdot f) = 1$, the cable losses are equivalent to those in the case where strands are insulated from each other and no interstrand coupling occurs.

RESULTS

Figure 2 shows P_c versus $\dot{B}^2$ curves for Sample #1 and the basic strand A. Two cabled samples were twisted with the same pitch of 14 mm in different directions. It should be noted that the interstrand coupling losses depend on the cable twist direction. In these measurements, the losses in the cable that is twisted in the same direction as the basic strand (right hand direction) are significantly lower than the one with the opposite twist.

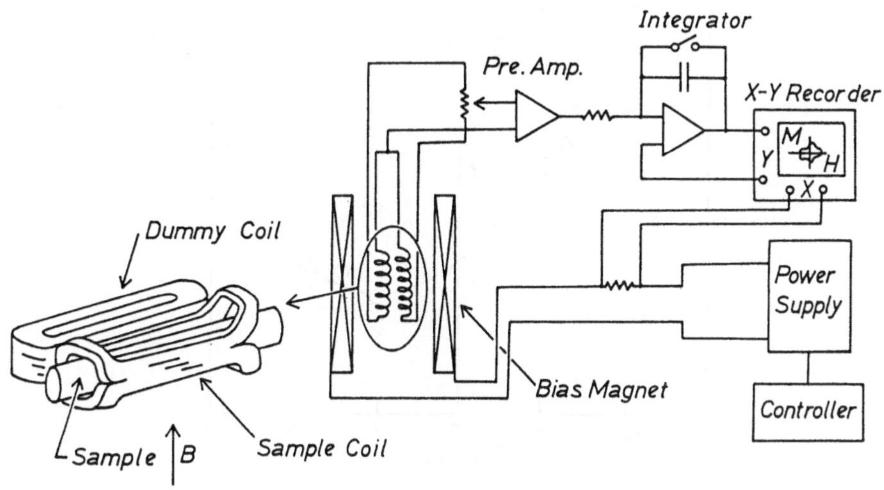

Fig. 1. Block diagram of the measuring system.

To clarify the effect of the cable twist direction and cable twist pitch, the coupling losses in Samples #1, #2, #3, and #4 are shown in Fig. 3 and Fig. 4 as a function of the cable twist pitch. The $\tau_c/(\tau_s \cdot f)$ is plotted against the ratio of the cable twist pitch to strand pitch, ℓ_c/ℓ_s. To indicate the cable twist direction, ℓ_c is given a negative sign where cable and strand twist directions are opposite. The twist pitch dependence

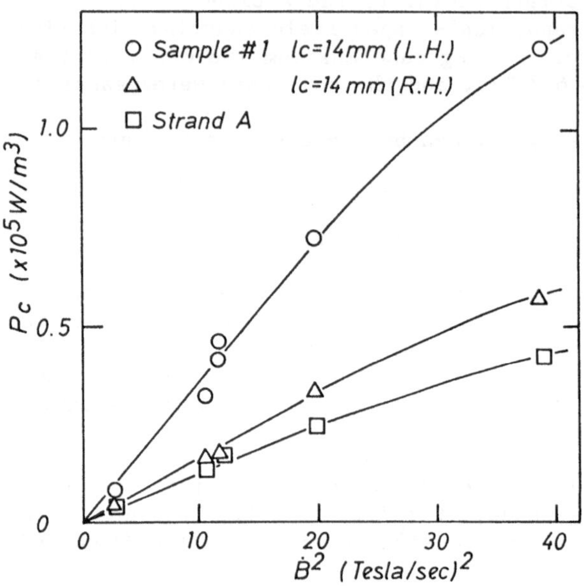

Fig. 2. P_c versus $\dot{B}^2$ curves for Sample #1 and the basic strand.

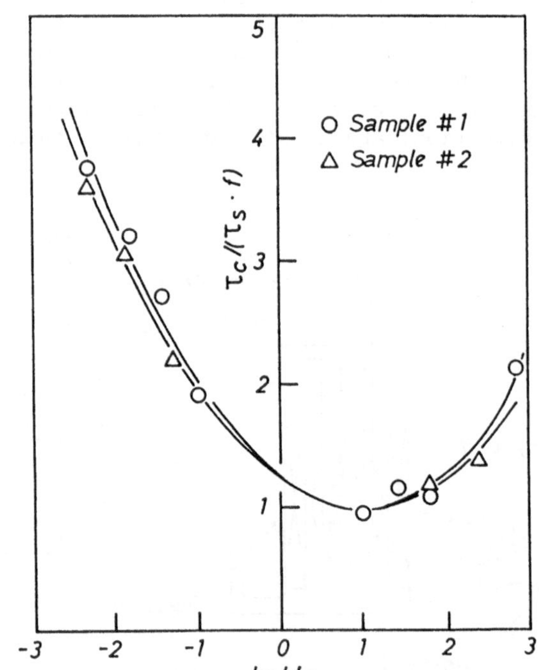

Fig. 3. Twist pitch dependency of the cable losses for Samples #1 and #2.

on the coupling losses is shown as a curve of the second degree for all cabled samples. It is noticeable that the losses in the cable are not minimized at zero twist pitch of cable ($\ell_c = 0$). For Samples #1 and #2 produced with the strand without copper sheath (strand A), the optimum ℓ_c, which minimizes the losses, is approximately identical to ℓ_s, as shown in Fig. 3. For Samples #3 and #4, which are made of the strand with a copper sheath (strand B), the optimum ℓ_c is approximately half of ℓ_s (see Fig. 4). Note that twisting the cable in the direction opposite to that of the strand twist always results in higher losses.

By comparing the data in Fig. 3 with those in Fig. 4, it is found that the outer copper sheath affects the losses. By attaching the copper sheath to the strands, the cable twist pitch dependence of loss becomes relatively strong. The losses in the three-strand cable #1 and the seven-strand cable #2 are nearly identical to those of strand A. However, the losses in the three-strand cable #3 with strand B are much larger than those in the seven-strand cable #4 with strand B.

In Table 3, the measurement data for Samples #4, #5, and #6 are shown. All of them have an identical cable pitch, ℓ_c, of -23 mm. The three-strand samples are filled with the different solders, as shown in Table 3. It is clear that the solder with high resistivity reduces the losses.

DISCUSSION

The coupling losses, P_c, in the multistrand cable have been calculated by Campbell as follows:

$$P_c = \frac{\sigma_s \ell_s^2 \dot{B}^2 f}{4\pi^2} \left\{ 1 + \left[\frac{p^2}{f} - 1 + \frac{2(p-1)^2}{1-f} \right] \sigma \right\} \qquad (2)$$

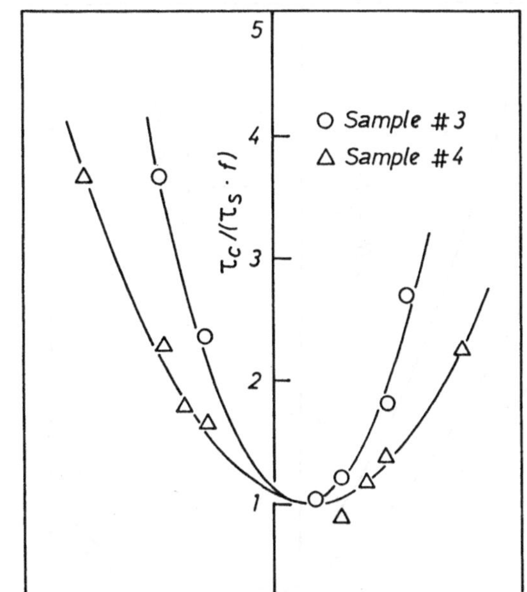

Fig. 4. Twist pitch dependency of the cable losses for Samples #3 and #4.

where p is the pitch ratio, ℓ_c/ℓ_s, and σ is the ratio of the electrical conductivity in the cable matrix to the mean transverse conductivity in the strand, σ_c/σ_s. Thus, the $\tau_c/(\tau_s \cdot f)$ is given by

$$\frac{\tau_c}{\tau_s \cdot f} = 1 + \left[\frac{p^2}{f} - 1 + \frac{2(p-1)^2}{1-f}\right]\sigma. \tag{3}$$

The σ_s, which is determined from the time constant for the strand, was 0.88 nΩ·m. The σ was calculated to be approximately 0.1. The $\tau_c/(\tau_s \cdot f)$ versus p curves, which were calculated from eq. (3), are shown in Fig. 5. It can be seen that an optimum pitch ratio exists where the twist directions of the cable and the strand are identical. These results are consistent with the experiment.

The optimum pitch ratio, p_o, determined from eq. (3), is written as

$$p_o = 2f/(1+f). \tag{4}$$

The p_o varies from 0 to +1, which is also consistent with the experiment.

It should be noted, however, that the losses are overestimated by the equations. The difference may be attributed to the assumption that the cable consists of a large number of strands.

The shifts of p_o when the outer copper sheath is attached to the strand can be explained by eq. (4). The volume fraction, f, is reduced owing to the copper sheath. It could be understood that the shift of p_o depends on f. The σ_c increases owing to the copper sheath. From eq. (3), increasing σ_c may increase the effects of parameters p and f on losses. It can be explained in terms of eq. (3) that, owing to the copper sheath, the

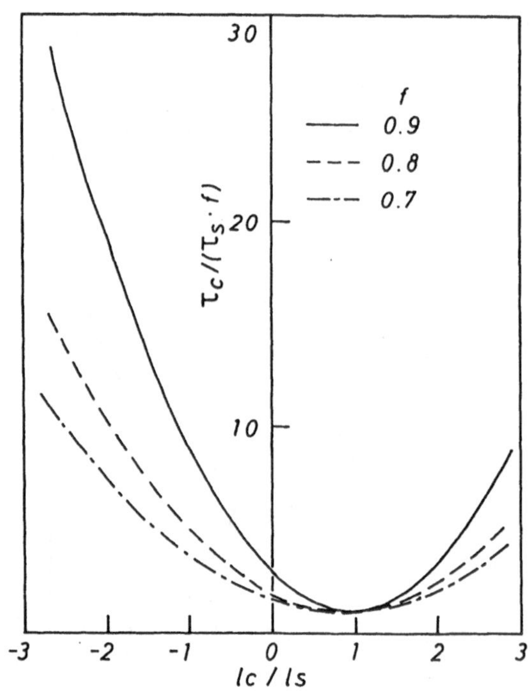

Fig. 5. Calculated twist pitch dependency of the cable losses.

losses depend more strongly on p, and the three-strand cables create higher losses than the seven-strand cables.

CONCLUSIONS

From coupling loss measurements on the several types of cables with changing the cable twist pitch, twist direction, and electrical resistivity in the filled solder, the following conclusions were drawn:

(1) Cable losses increase in proportion to $(\ell_c - \ell_{co})^2$, where ℓ_c is the cable twist pitch and ℓ_{co} is the ℓ_c that minimizes the losses. The loss minimum occurs when cable and strand twisting direction are identical. The ℓ_{co} is not zero and smaller than the strand twist pitch. As a result, the cable twist direction opposite to that of the strand twist leads to higher losses than when twist directions are identical. The interstrand coupling losses might be negligibly small by optimizing the cable twist pitch.

(2) The configuration of cable also affects the losses. The losses are increased by inserting a low resistivity material in a gap between strands, such as low resistivity solder or an outer copper sheath. In this case, the losses depend strongly on the twist pitch and volume fraction of the filament bundle region.

(3) These feature can be explained to some extent on the basis of Campbell's theory. The losses, however, are overestimated quantitatively by the theory.

ACKNOWLEDGMENTS

The authors are grateful to all members of the Low Temperature Center, Osaka University for their help in the experiment using liquid helium. This work is partly supported by Grant in Aid for Scientific Research No. 60050044, Ministry of Education in Japan.

REFERENCES

1. K. Kwasnitza and I. Horvath, IEEE Trans. Magn. MAG-17:2278 (1981).
2. K. Kwasnitza and I. Horvath, Cryogenics 23:9 (1983).
3. S. H. Kim, J. W. Dawson, D. G. McGhee, and S. S. Shen, in: "Advances in Cryogenic Engineering -- Materials," vol. 30, Plenum Press, New York (1984), p. 939.
4. A. M. Campbell, Cryogenics 20:651 (1980).

A.C. LOSSES IN MULTIFILAMENTARY COMPOSITE SUPERCONDUCTING STRANDS AND CABLES*

T.M. Mower and Y. Iwasa

Plasma Fusion Center
Massachusetts Institute of Technology
Cambridge, Massachusetts

ABSTRACT

A.C. losses in multifilamentary composite superconducting strands and cables have been measured in adiabatic conditions for transverse field sweep rates up to 70 $T\,s^{-1}$. Measurements were performed on NbTi and Nb_3Sn conductors of several configurations and surface preparations: single strands, soldered strands and cables of varying degrees of compaction composed of bare strands, strands with CuNi barriers and strands with chrome plating. To achieve the adiabatic measurements, the test conductor was placed in a chamber which was evacuated and then subjected to an exponentially decaying field. By measuring the induced temperature rise of the test conductor the A.C. losses are computed from cryogenic enthalpy plots. Loss data are characterized in terms of effective coupling current time constants.

INTRODUCTION

A.C. losses within single multifilamentary strands may be reasonably predicted with analytical loss models by summing the hysteresis loss and the filament coupling loss. Determination of losses in cabled superconducting strands, however, presents a challenging problem. Among the parameters which are unique to cabled superconductors (ICCS) are conductor geometry and configuration, cable compaction and strand surface preparation. Strand-to-strand coupling occurs in cables when the pressure between strands reduces the contact resistance enough to allow induced currents to pass across strand interfaces, thus dissipating additional power. At large field sweep rates the strand coupling loss may, depending upon cable compaction, dominate the total A.C. loss. It is likely that the development of ICCS conductors will prove their merit for application to tokamaks and other large superconducting coils which will be subjected to rapid field changes. Consequently, magnet system designers will require more quantitative information about the magnitudes of A.C. losses and their important parametric dependence within ICCS conductors. The ultimate goal of this research is to provide a better understanding of the nature and magnitude of A.C. losses in cabled superconductors.

* Research sponsored by the Office of Magnetic Fusion Energy, U.S. Department of Energy, under Contract No. DE-AC02-78ET-51013. Based on part of a thesis submitted by T. Mower in partial fulfillment of requirements for the degree of Master of Science in Mechanical Engineering at the Massachusetts Institute of Technology, September, 1985.

EXPERIMENTAL PROGRAM

Motivation for the experimental work of this research was initiated by the need to determine losses in the ICCS conductor of the poloidal field coils for a proposed superconducting tokamak fusion device at M.I.T. The initial step taken was to establish an experimental technique which could measure losses in conductors subjected to field sweep rates (~30 $T s^{-1}$) relevant to the proposed Alcator DCT machine.

Pulse Field

To achieve an applied field sweep in the form of an exponential decay, a superconducting pulse magnet and a dump resistor in parallel were placed in series with a current supply and an SCR/capacitor controlled switch. The magnet would be charged with the switch in the closed position, drawing all of the current through the superconducting coil. When the SCR was commutated open, the current flowing within the coil would be discharged exponentially through the dump resistor with time constant $\tau = L/R$. The values of resistance were selected to produce time constants of 9, 32, 117, 231, 540 and 1250 ms. Using a pulse coil which produced an average flux density of 1.6 T at an operating current of 500 A, a maximum B/τ of 70 $T s^{-1}$ could be achieved without exceeding a safe coil terminal voltage of 600 V.

Adiabatic Chamber

The sensitivity of common techniques (magnetization and calorimetry) used to measure A.C. losses of conductors in isothermal conditions is proportional to the mass of specimen. To achieve high sensitivity in a compact apparatus, losses in this research were meausered in adiabatic conditions. An annular stainless steel vacuum chamber was placed within the pulse magnet to provide an adiabatic environment for specimen conductors. Gold/chromel thermocouples, with a calibrated sensitivity of 14.4 $\mu V/K$ in the range of 4.2 K to 9 K, were used to measure the induced temperature increase of the sample. The reference junction was achieved by passing the thermocouple leads from the vacuum to liquid helium via a cryogenic ceramic feed-thru. The sample was isolated from the vacuum chamber by thin nylon spacers, resulting in a heat transfer from the sample of ~0.2 mW. The ratio of sample mass to non-dissipative mass in contact with the sample (thermocouples and spacers) was ~80:1. Figure 1 shows a typical field pulse and the resulting temperature increase of a specimen. The extended temperature response for the same event as in Figure 1 is shown in Figure 2, where it may be observed that the sample cooled from a peak of ~8 K to ~6 K in 6 minutes. Considering the extremely low specific heats (less than 1 mJ/g) in this temperature range, the temperature response indicates a very adiabatic environment indeed.

Loss Determination

The dissipation per unit volume, Q, could be easily determined from computed composite enthalpy plots by subtracting the enthalpy at 4.2 K from the enthalpy at the measured peak temperature:

$$Q = H(T_{final}) - H(4.2\text{K}). \tag{1}$$

$H(T)$ is the enthalpy of the specimen conductor determined by volume weighting the specific heats of constituents (filaments, matrix, sheath and solder). The experimental error, a function of the heat transfer and heat capacity of the spacers and thermocouples, is calculated to be less than 10%. Any error introduced by the physical system will be negative in sense, thus it is judged that the method errs on the low side of correct loss magnitudes. Excellent agreement of calculated single multifilamentary strand losses with measured losses provides confidence in the technique.

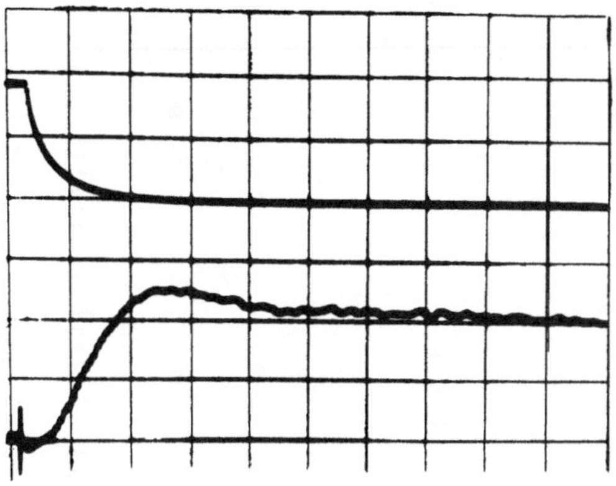

Fig. 1. Oscillogram Showing Temperature Response: Upper trace, coil current decay, 300 A/div.; Lower trace, thermocouple output, 25 μV/div. ($\simeq$1.7 K/div.); Time scale, $\simeq$80 ms/div.

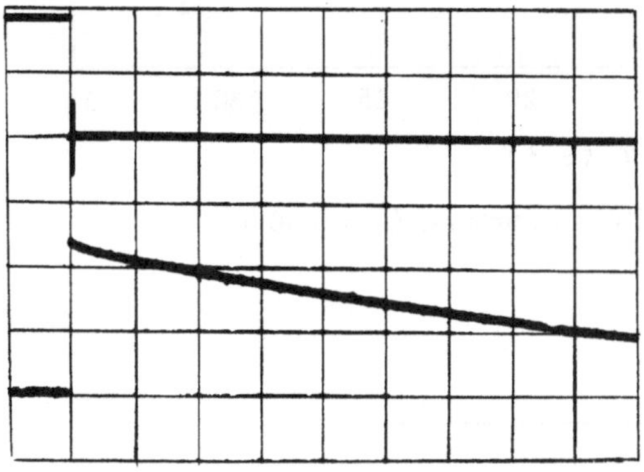

Fig. 2. Oscillogram Showing Adiabatic Behavior: Upper trace, coil current; Lower trace, thermocouple output, 25 μV/div. ($\simeq$1.7 K/div.); Time scale, $\simeq$40 s/div. (Same event as in Figure 1.)

Test Conductors

Three independent groups of specimen conductors were chosen for loss measurements for different purposes.

(1) BNL NbTi: During the early stages of this research, configurations of the BNL "Isabelle" NbTi composite wire with a surface layer of CuNi were being considered for the Alcator DCT poloidal coils. Consequently, loss measurements were performed on the basic Isabelle strand, the BNL "Cluster" (nine Isabelle strands soldered around a copper core), and a sub-size sheathed cable of 81 strands compacted to ~3% void.

(2) MCA NbTi: To determine the effect of compaction on A.C. losses in ICCS cables an MCA, 0.7-mm diameter NbTi wire was chosen for the basic strand. Cables composed of 27 twisted MCA strands were sheathed in stainless steel tubes and drawn to the appropriate diameters to yield void percentages of 55%, 35%, 27% and 24%. Loss measurements were performed on cables with bare strands to determine the loss dependence upon degree of compaction. To determine the effect of chrome plating on coupling losses, strands were plated with ~ 0.8μm of hard chrome prior to compaction in sheaths.

(3) Oxford/AIRCO Nb_3Sn : Since the primary purpose of chrome plating is to reduce sintering in Nb_3Sn cables, loss measurements were also performed on Nb_3Sn single strands and 27-strand cables, both with bare strands and with chrome plated strands.

Results

Figures 3-6 display a portion of the loss measurements obtained in this research. The total dissipation Q, which is the sum of the hysteresis loss, filament coupling loss and strand coupling loss, is given per unit volume of strand. Figures 7-10 illustrate normalized losses and effective coupling time constants.

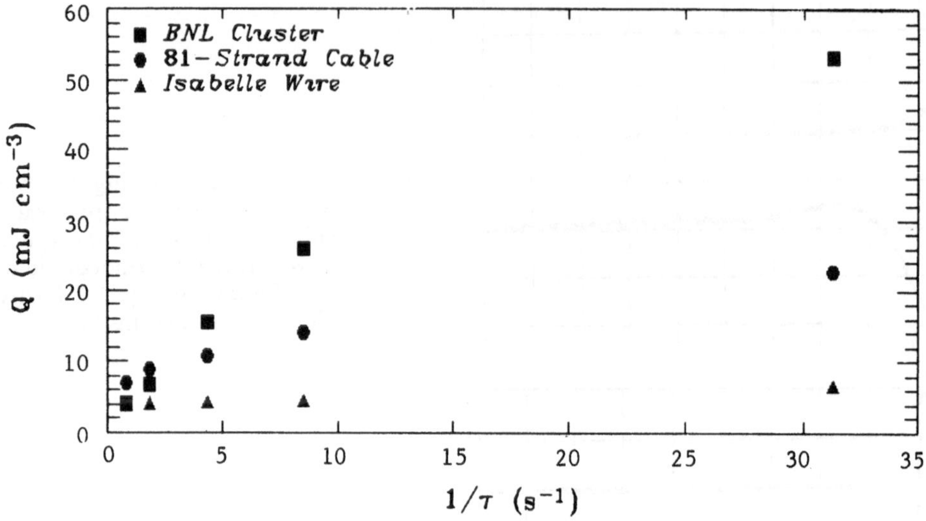

Fig. 3. BNL NbTi composite conductors, B_m=0.96 T.

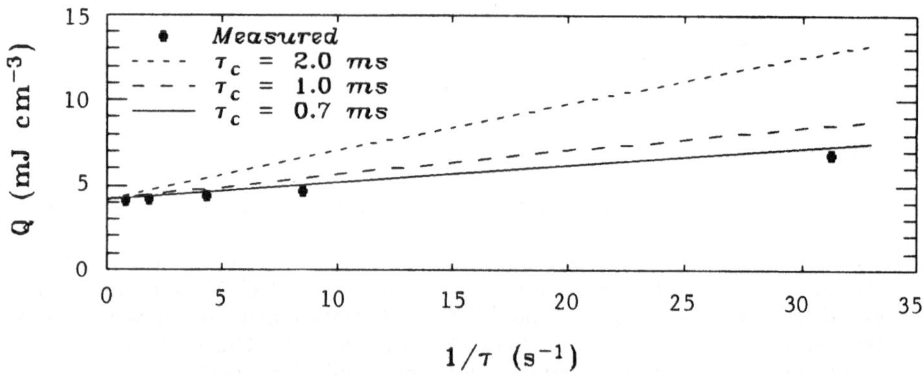

Fig. 4. BNL Isabelle, measured and computed.

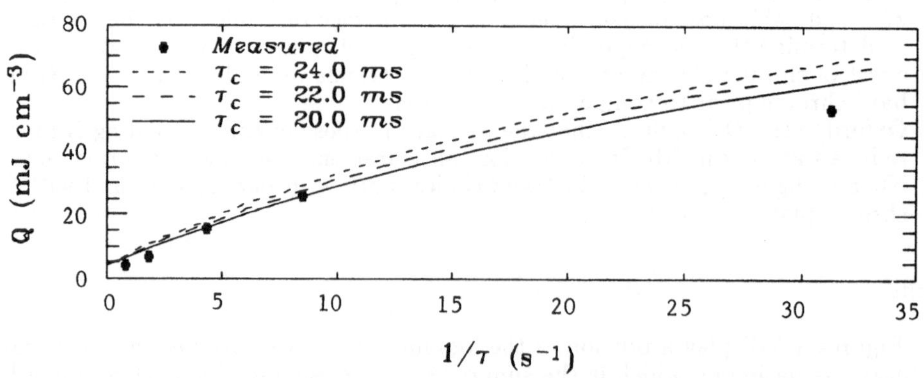

Fig. 5. BNL Cluster, measured and computed.

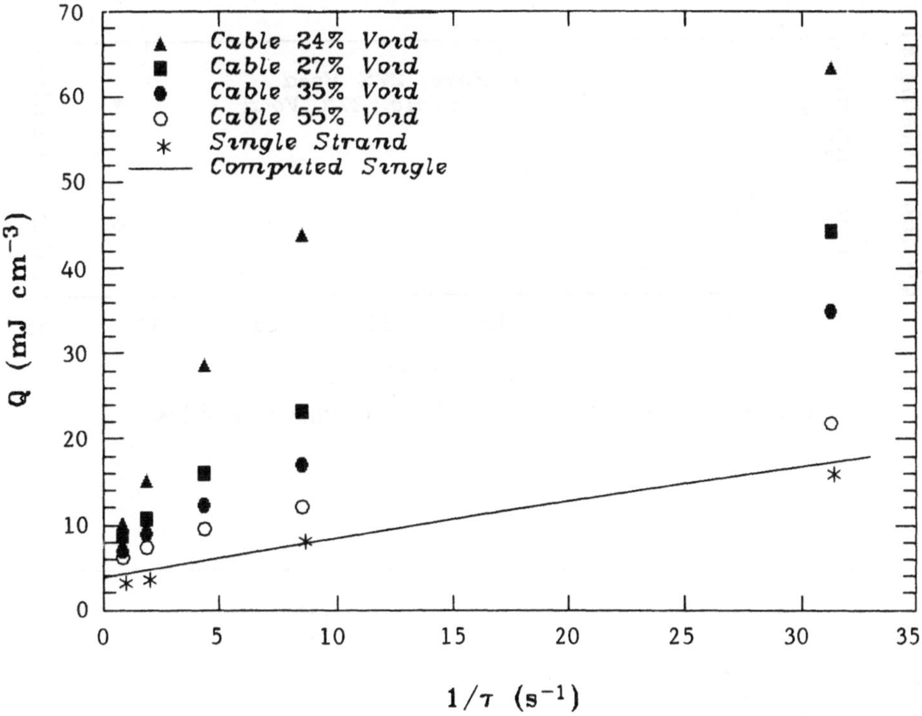

Fig. 6. MCA NbTi composite conductors, B_m=0.96 T.

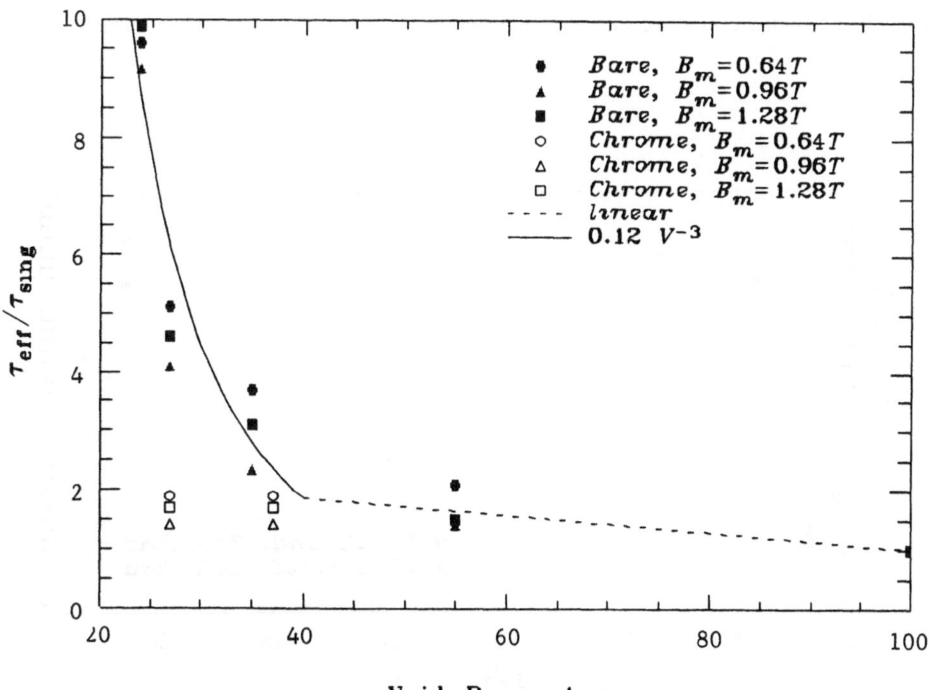

Fig. 7. Effective coupling time constant of 27-strand MCA NbTi cables.

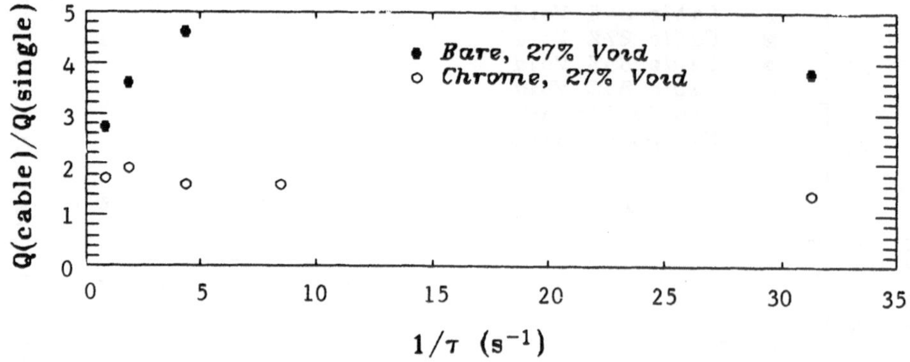

Fig. 8. Total cable loss normalized to single strand loss.

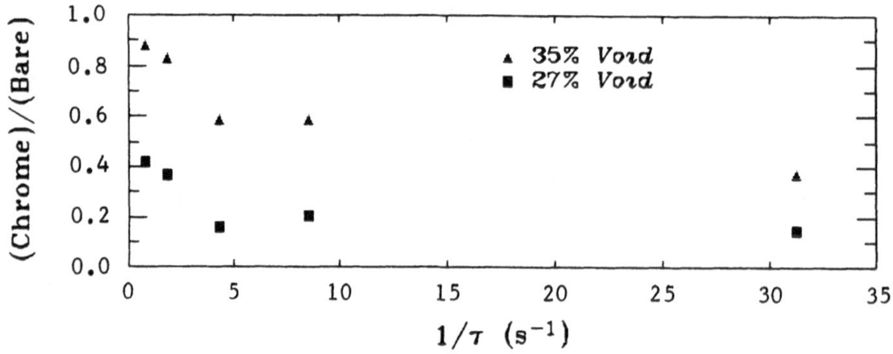

Fig. 9. Strand coupling loss in chrome cables normalized to bare cables.

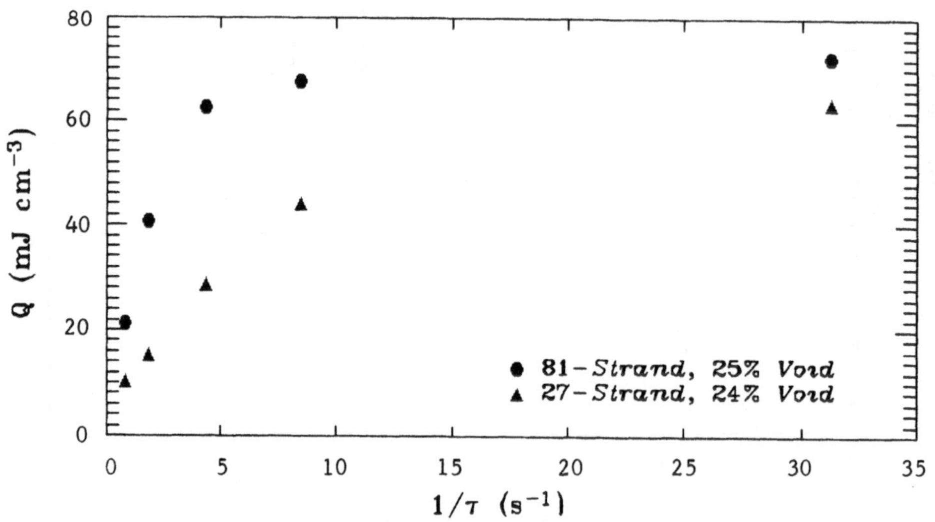

Fig. 10. MCA 27 and 81-strand cables, B_m=0.96 T.

DISCUSSION OF RESULTS

A plot of adiabatic dissipation (Q) per unit volume of superconducting strand *vs.* nominal frequency ($1/\tau$) is shown in Figure 3 for the BNL conductors subjected to a field decay from a peak flux density of 0.96 T. On such a plot, the y-intercept is equal to the magnitude of the intrinsic hysteresis loss and the gradient is equal to the filamentary coupling loss as a function of $1/\tau$ where τ is equal to the time constant of the applied field sweep. Inspection of Figure 3 shows, as expected, that on a unit volume basis the hysteresis loss in the single Isabelle wire is identical to that in the cluster of nine soldered to a copper core. The coupling loss, however, is far greater in the cluster as a result of the induced currents which are allowed to flow between adjacent strands and across the copper core due to the increased area of contact provided by the solder. The 81-strand cable, though highly compacted (~ 3% void), exhibits an intermediary coupling loss reflecting the successful inhibition of strand coupling by the outer shell of CuNi on the Isabelle wire.

For an applied magnetic flux density which decays exponentially with time constant τ from a peak value of B_m, the coupling loss per unit volume of superconductor may be expressed by[1]

$$Q_{cpl} = \frac{B_m^2}{\mu_o} \frac{\tau_c}{\tau + \tau_c} . \tag{2}$$

The natural decay time constant, τ_c, of the induced filament surface currents may be given by

$$\tau_c = \frac{\mu_o}{2\rho_e} \left(\frac{L_p}{2\pi} \right)^2 , \tag{3}$$

where L_p is the filament twist pitch length and ρ_e is the effective transverse matrix resistivity.[2] An approximation for the effective matrix resistivity is given by $\rho_e = \rho_m (1 + \lambda)/(1 - \lambda)$.[3] The above loss expressions may be multiplied by λ, the volume fraction of filaments within a strand to yield the loss per unit volume of strand.

Applying the above equations to the Isabelle strand, with $\rho_m \sim 2.4 \times 10^{-10} \Omega\,\text{m}$, L_p=10 mm and λ=0.4, τ_c is calculated to be about 5 ms. Based upon this value, the computed coupling loss far exceeds the measured magnitudes. Figure 4 depicts measured losses of the Isabelle strand and calculated losses for time constants of 2, 1 and 0.7 ms. From this plot it is apparent that the coupling time constant of the specimen was actually quite near 0.7 ms. It is hypothesized that the conductor tested may have had a contaminated copper matrix or that some of the CuNi barrier may have penetrated into the bulk between filaments, thus increasing the effective matrix resistivity. A similar procedure is used in Figure 5 to determine that the effective coupling time constant of the BNL Cluster is about 20 ms.

Loss measurements were performed on 27-strand, MCA NbTi cables with varying degrees of compaction subjected to field decays with peak values of from 0.6 T to 1.6 T. The results from sweeps with B_m=0.96 T are shown in Figure 6. It may be observed that excellent agreement exists between the measured single strand values and the computed curve (τ_c= 3.6 ms). As expected, the losses increase with increasing compaction reflecting the enhanced strand-to-strand coupling resulting from increasing pressure and contact area between strands. This increasing loss becomes nonlinear at the higher sweep rates (and, more noticeably, at higher peak fields) due to two factors. First, Eq. 3 indicates that the coupling loss becomes less linear with $1/\tau$ as the coupling time constant τ_c becomes larger. Notice, for example, the curve in Figure 5 for τ_c=20 ms compared to the line in Figure 4 for τ_c=0.7 ms. The second nonlinear influence results from the adiabatic nature of this experiment. As losses increase, the specimen temperature is driven closer to T_c, and in fact, portions of the interior may be normal. Consequently the induced filament surface currents may quickly collapse and adiabatic "clamping" of the loss results.

Data for field sweeps from peak flux densities of 0.64, 0.96 and 1.28 T were plotted on graphs similar to Figure 6. From these plots the effective coupling time constants were determined from the loss gradients of measurements on cables with 55%, 35%, 27% and 24% void fractions. The effective time constants (τ_{eff}), normalized to the single strand coupling constant (τ_c) of 3.6 ms *vs.* the cable void percentage are plotted in Figure 7. A void percentage of 100% corresponds to a single strand which is isolated, while a void of zero percent would be a solid strand with the diameter of the cable. From the case of a very loose cable down to about 40% or 50% void it appears that only moderate strand-to-strand coupling exists. Below 40% void the strand coupling quickly increases and becomes the dominant loss mechanism. A simple curve fit suggests that the total cable loss grows as $\sim 1/(\text{void})^3$ below void fractions of 0.4.

The effective cable coupling constant for the chrome plated cables is also plotted on the graph of Figure 7. The loss measured in chrome cables compacted to 37% void was identical to that measured in 27% void chrome cables. Apparently the surface resistance of chrome is sufficient to effectively minimize strand coupling regardless of degree of compaction. Figure 8 indicates that the total loss in chrome cables is lower by a factor of two to three at all sweep rates. The strand coupling loss, equal to the total loss minus the single strand loss (all losses are per unit volume of strand), is depicted in Figure 9. This plot shows that the strand coupling in chrome cables can be as low as 20% of the strand coupling in bare cables. Loss measurements on both bare and chrome Nb_3Sn cables, not included here due to space limitation, also indicate similar reductions in strand coupling due to chrome plating.

To address the issue of applicability of sub-sized cable measurements to full-sized conductors, measurements were performed on an 81-strand, MCA NbTi cable compacted to 25% void. The results of sweeps from B_m=0.96 T are shown in Figure 10 for both the 27-strand cables and the 81-strand cable. This plot indicates that the effective coupling constant in the larger cable is indeed larger by about a factor of two. Further experimentation and analysis needs to be performed to adequately determine size effect.

CONCLUSIONS

An experimental technique has been established to accurately measure the adiabatic A.C. loss in sub-sized superconducting cables. Though superconducting applications are typically isothermal, the research reported here provides useful information. Effective coupling time constants were determined as a function of cable compaction. Below void fractions of 40%, the coupling time constant (and hence total A.C. loss) of bare cables was observed to grow as the inverse cube of the void fraction. It has been shown that chrome plating on individual strands dramatically reduces the strand coupling which is the dominant loss in highly compacted bare cables. Further research should be done to determine the effect of cable size on the total loss. A more detailed account of the research summarized in this paper will be published at a later date.

REFERENCES

1. Wilson, Martin N., Superconducting Magnets, Oxford University Press, New York, 1983, p.171
2. Wilson, Martin N., Superconducting Magnets, Oxford University Press, New York, 1983, p.180
3. Carr, Jr., W.J., A.C. Loss and Macroscopic Theory of Superconductors, Gordon and Breach, N.Y., 1983, pp.148-151

AC LOSSES IN Nb-Ti MEASURED BY MAGNETIZATION AND COMPLEX SUSCEPTIBILITY

R. B. Goldfarb and A. F. Clark
Electromagnetic Technology Division
National Bureau of Standards
Boulder, Colorado

ABSTRACT

DC magnetization and complex ac susceptibility were measured at 4 K as functions of longitudinal dc field for a multifilamentary Nb-Ti superconductor with no transport current. Minor hysteresis loops were obtained in the dc measurements. The full-penetration field, H_p, a function of applied field, H, was deduced directly for each minor loop. The values for H_p were fit to the Kim-type equation, $H_p(H) = H_p(0)/(1+H/H_k)$, where $H_p(0)$ and H_k are constants. The minor hysteresis-loop areas gave losses that were in excellent agreement with Carr's theoretical critical-state equation, $W = (4\mu_0 H_0 H_p/3)(1-H_p/2H_0)$, where H_0 is the maximum applied field for each loop.

An expression was obtained for the ideal reversible differential susceptibility: $\chi_{rev} = \phi_0/8\pi\mu_0(H-H_{c1})\lambda^2$, where ϕ_0 is the flux quantum, H_{c1} is the lower critical field, and λ is the penetration depth. H_{c1} and λ for the sample were deduced from the shape of the major hysteresis loop. Clem's theoretical expressions for the real (χ') and imaginary (χ'') components of ac susceptibility are functions of χ_{rev}, H_p, and ac field amplitude, h. The predicted susceptibilities based on these expressions were in good agreement with measured curves of χ' and χ'' as functions of h and H. The measured χ' and χ'' were independent of frequency up to 1 kHz, as expected when bulk hysteresis is the primary loss mechanism.

INTRODUCTION

In an earlier work[1] we discussed the relationship between dc magnetization and ac susceptibility in a type-II superconductor. That work described how magnetic measurements provide information on hysteresis losses. In this paper, as in Ref. 1, we examine magnetization and susceptibility for longitudinal fields and no transport current. Here, however, we obtain the full-penetration fields, H_p, directly from the hysteresis loops rather than estimate H_p from measurements of critical current density, J_c. Also, we derive an expression for the reversible susceptibility, χ_{rev}, rather than using the experimental susceptibility, χ_{dc}, as an approximation. A superconducting wire whose low-field magnetization approached a reversible curve was selected for study.[2] Several minor hysteresis loops were obtained in addition to the major loop. The susceptibility curves were more nearly reversible and virtually independent of frequency, as expected from theory.

Table 1. Characteristics of Nb-Ti Wire

Cross section bare wire:	0.63 × 0.88 mm
Twist length:	1.7 cm
Cu/Nb-Ti volume ratio:	0.91
Number of filaments:	240
Filament radius:	19.3 μm
Density of Nb-Ti alloy:	6.20 g/cm^3

EXPERIMENT

The magnetization measurements were made with a vibrating-sample magnetometer (VSM) at 4 K. Magnetization was computed as magnetic moment per unit volume of Nb-Ti. The volume of the Cu matrix was not included.

It is useful to compare the advantages of the VSM method vis-à-vis the popular integration method of Fietz.[3] The VSM method (1) is useful for small samples, (2) is a dc measurement, not sensitive to coupling losses, (3) does not require precise pick-up coil balance when the applied field is stepped, and (4) is not subject to integrator drift. The integration method (1) measures frequency dependences, (2) detects flux jumps, and (3) is easily adaptable to measurements with transport current.

The experimental methods are more fully described in Ref. 1. The sample in this study is different and its length is 3.0 rather than 1.5 cm. The purpose of a longer sample was to avoid significant end effects. The characteristics of the wire are given in Table 1. This is the same wire as sample 8 in Ref. 2.

DC HYSTERESIS LOOP

Full-Penetration Field

The major and minor hysteresis loops are shown in Fig. 1. The full-penetration field, H_p, a function of applied field, H, may be estimated directly for each minor loop.[2] As the loops are traversed, the filaments go from full penetration in one direction to full penetration in the other direction. Therefore, at a high-field end of a loop, H_p is approximately one-half the field required to reverse the magnetization. The reversal field, $2H_p$, for the major loop is shown in Fig. 1.

The well-known Kim model for critical current density is[4]

$$J_c(H) = J_c(0)/(1+H/H_k) , \tag{1}$$

where $J_c(0)$ and H_k are constants. In the critical-state model, for a field applied axially,

$$H_p = J_c r , \tag{2}$$

where r is the filament radius. We did a linear least-squares fit of each loop's H_p to the expression

$$H_p(H) = H_p(0)/(1+H/H_k) , \tag{3}$$

where $H_p(0)$ is a constant equal to $J_c(0)r$. The fit was excellent. We obtained $H_p(0)$ = 140 kA/m (1.76 kOe) and H_k = 1.33 MA/m (16.7 kOe).

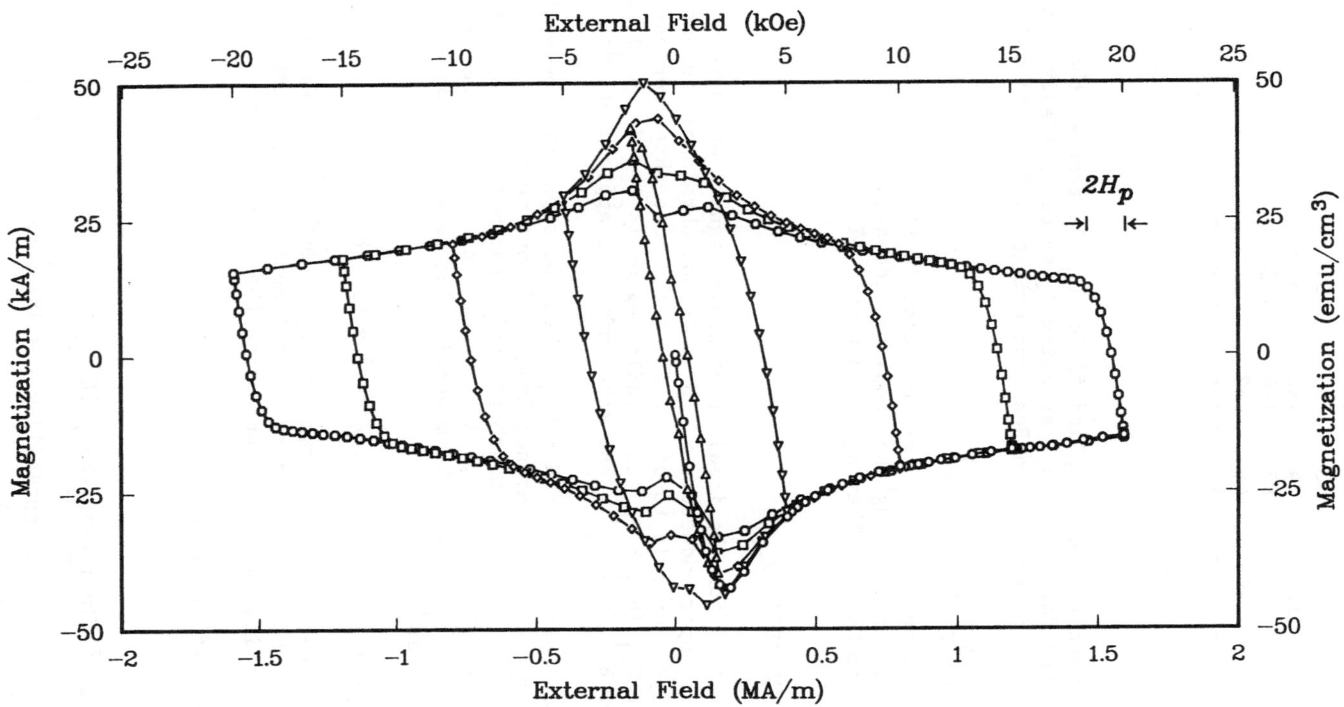

Fig. 1. Major and minor hysteresis loops for a multifilamentary Nb–Ti superconductor. DC magnetization is plotted vs. longitudinal applied field.

Table 2. Comparison of Losses Obtained from Eq. 4 and Loop Area.

H_0 (kA/m)	H_p	Eq. 4 (kJ/m^3)	Area
1591	63	165	159
1194	74	143	136
796	87	110	105
398	108	62	58

Hysteresis Loss

For the case when the peak applied field, H_0, is greater than H_p, Carr derived an expression for the loss, assuming J_c to be spatially constant and field independent in the London-Bean approximation.[5] In SI units,

$$W = (4\mu_0 H_0 H_p/3)(1-H_p/2H_0) \quad . \tag{4}$$

Clearly H_0 is much larger than H_p for the loops in Fig. 1. Losses calculated by numerically integrating the loops are compared in Table 2 to losses obtained from Eq. 4. They agree to within 6%.

REVERSIBLE SUSCEPTIBILITY χ_{rev}

An exact equation for the ideal, reversible magnetization-versus-field curve of a type-II superconductor, $H > H_{c1}$, is not obtainable in analytic form. The usual procedure is to divide the field into three regions: immediately above H_{c1} ($H_{c1} \lesssim H$), the middle region ($H_{c1} < H < H_{c2}$), and immediately below H_{c2} ($H \lesssim H_{c2}$), and make judicious approximations.

One tractable case is $H_{c1} < H < H_{c2}$, when the separation between flux vortices is larger than the coherence length, ξ, but much smaller than the penetration depth, λ. Thus the Ginzburg-Landau parameter $\kappa \equiv \lambda/\xi$ is much greater than 1. For Nb-Ti at 4 K,[6] $\lambda \cong 394$ nm and $\xi \cong 5$ nm, and this case is applicable. To first order, the magnetization as a function of magnetic field strength M(H) is[7]

$$M = (\phi_0/8\pi\mu_0\lambda^2)\ \{\ln[4\pi\mu_0(H-H_{c1})\lambda^2/\phi_0] + \alpha\} - H_{c1} \ , \tag{5}$$

where ϕ_0 is the magnetic flux quantum = h/2e = 2.068 × 10^{-15} Wb ($T\cdot m^2$) and α is a constant on the order of unity. The functional dependence of M(H) can be seen by dropping the α term and using the relationship[8]

$$H_{c1} = (\phi_0/4\pi\mu_0\lambda^2)\ \ln(\lambda/\xi) \ . \tag{6}$$

Thus,

$$M = -(\phi_0/8\pi\mu_0\lambda^2)\ \ln[\phi_0/4\pi\mu_0(H-H_{c1})\xi^2] \ . \tag{7}$$

The volume magnetic susceptibility χ = dM/dH can be obtained from either Eq. 5 or 7 and is simply

$$\chi = \phi_0/8\pi\mu_0(H-H_{c1})\lambda^2 \ . \tag{8}$$

We take this χ to be equal to the ideal reversible susceptibility χ_{rev} in the range $H_{c1} < H < H_{c2}$. For $H < H_{c1}$, $\chi_{rev} = -1$. χ_{rev} represents the susceptibility under conditions of thermodynamic equilibrium.

Our goal is to obtain numerical values for χ_{rev} for our wire. The differential slope of the third-quadrant branch of the major magnetization curve (Fig. 1), from 0.2 to 1.6 MA/m (2.5 to 20 kOe), was fit to Eq. 8 using linear least squares. While this portion of the magnetization curve itself is not reversible, we expect that its slope will simulate χ_{rev}. For the adjustable parameters H_{c1} and λ, we obtain 92 kA/m (1 kOe) and 78 nm, respectively. These values are not too unreasonable, though this fit risks overestimating H_{c1} with values closer to $H_p(0)$. Substituting into Eq. 8 we get χ_{rev} for this wire for $H_{c1} < H < H_{c2}$ (SI units):

$$\chi_{rev} = 10772/(H-91751) \; . \tag{9}$$

This equation will be used in the next section to calculate χ' and χ''. An alternate approach would be to use textbook values of H_{c1} and λ in Eq. 8.[9] This would give slightly different values of χ' and χ'' for $H \gtrsim H_{c1}$. For larger fields, the differences are negligible.

MEASURED AND IDEAL χ' AND χ''

The real and imaginary components of susceptibility, χ' and χ'', were measured as functions of frequency (10, 100 and 1000 Hz), sinusoidal ac field amplitude, h [11 and 56 kA/m (140 and 700 Oe)], and dc bias field, H [0 - 1.6 MA/m (0 - 20 kOe)]. The susceptibilities were independent of frequency to within 3%. Since χ'' is a measure of the losses and hysteresis is known to be frequency independent, bulk hysteresis is probably the primary loss mechanism. Frequency-dependent eddy-current and coupling losses are likely insignificant for this sample.

Figures 2 and 3 show χ' and χ'' measured at 10 Hz. The initial and decreasing-field branches are plotted. Some irreversibility may be seen. In Fig. 3, h is on the order of H_p. As shown by Clem,[10] χ' and χ'' as functions of H may be predicted from h, H_p, and χ_{rev}. For our cases, $h < H_p$, the equations are:[1,10]

$$\chi' = (1+\chi_{rev})(h/H_p - 5h^2/16H_p^{\,2}) - 1 \; , \tag{10}$$

$$\chi'' = (1+\chi_{rev})(4h/H_p - 2h^2/H_p^{\,2})/3\pi \; . \tag{11}$$

Using the expressions for H_p and χ_{rev} (Eqs. 3 and 9), we computed the theoretical curves of χ' and χ'' for this wire for ac field amplitudes of 11 and 56 kA/m (140 and 700 Oe). They are shown in Figs. 4 and 5. They compare quite favorably with the actual curves in Figs. 2 and 3. For $H < H_{c1}$, $\chi' = -1$ and $\chi'' = 0$. There are discontinuities at H_{c1}, as expected. In the actual curves these are rounded owing to flux pinning.

CONCLUSION

DC magnetization curves and measurements of complex susceptibility provide information on hysteretic losses in multifilamentary superconductors such as Nb-Ti. These measurements may be performed on small samples. Interpretation of the data yields information on full-penetration field, critical current density, lower critical field, and penetration depth. The frequency independence of susceptibility suggests bulk hysteresis as the primary loss mechanism.

ACKNOWLEDGMENT

This work was sponsored by the Air Force Office of Scientific Research.

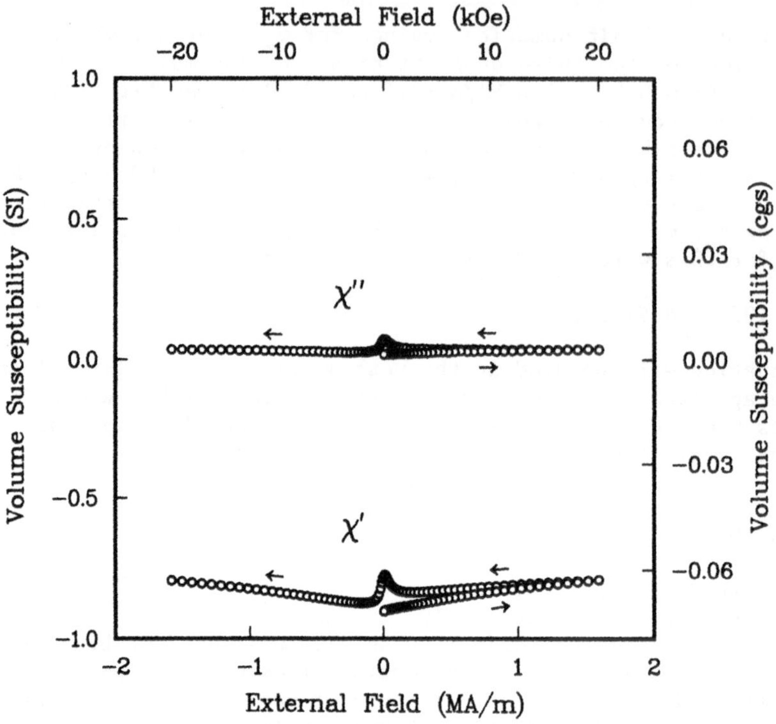

Fig. 2. Complex ac susceptibility as a function of dc bias field. The ac field amplitude is 11 kA/m (140 Oe) at 10 Hz.

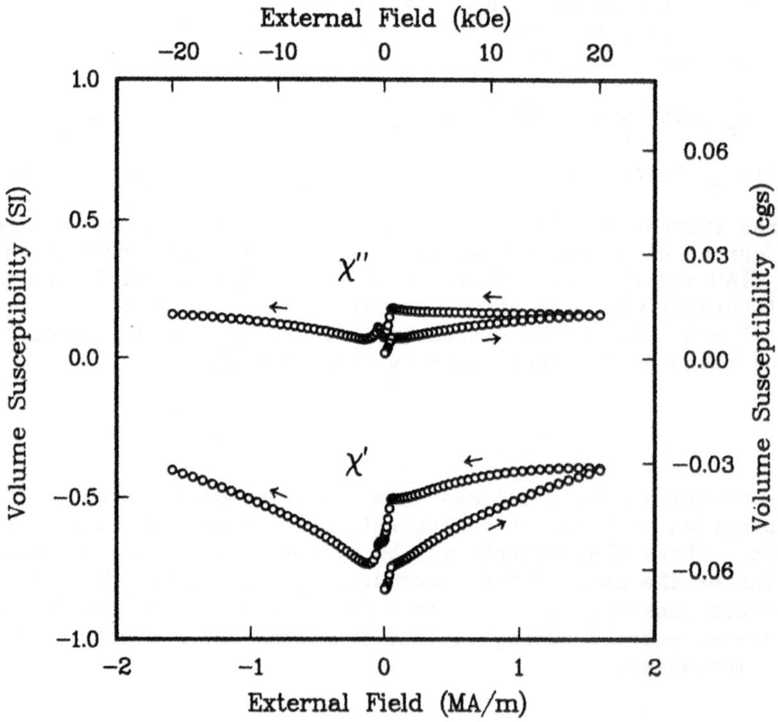

Fig. 3. Complex ac susceptibility as a function of dc bias field. The ac field amplitude is 56 kA/m (700 Oe) at 10 Hz.

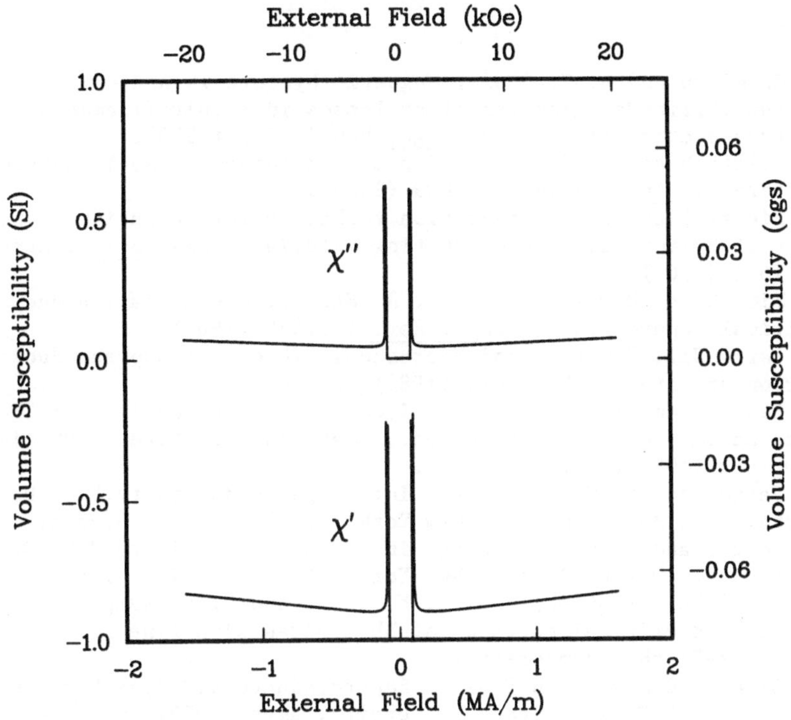

Fig. 4. Theoretical complex susceptibility vs. dc bias field for an ac field amplitude of 11 kA/m (140 Oe).

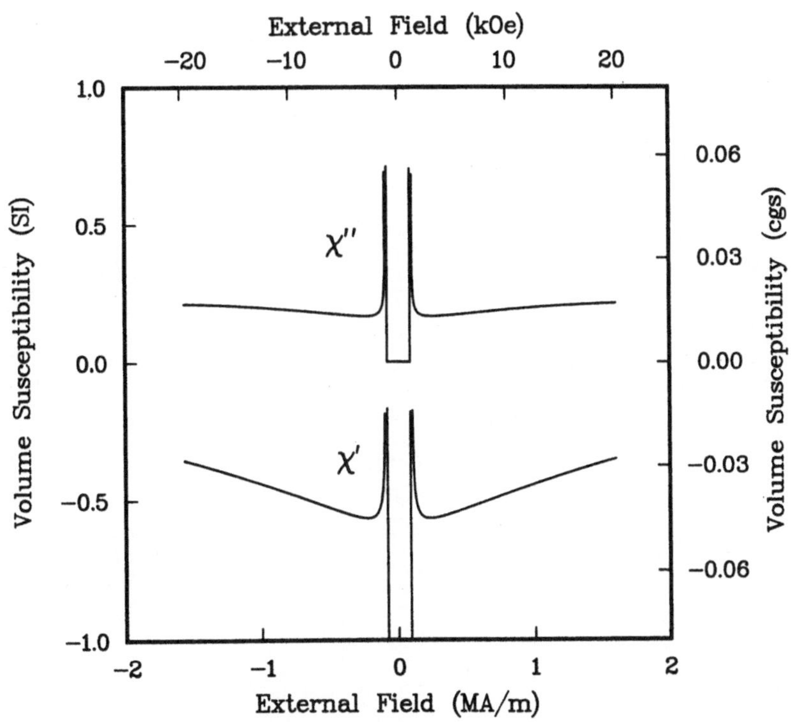

Fig. 5. Theoretical complex susceptibility vs. dc bias field for an ac field amplitude of 56 kA/m (700 Oe).

REFERENCES

1. R. B. Goldfarb and A. F. Clark, Magnetic hysteresis and complex susceptibility as measures of ac losses in a multifilamentary NbTi superconductor, IEEE Trans. Magn. MAG-21:332 (1985).
2. R. B. Goldfarb and A. F. Clark, Hysteretic losses in Nb-Ti superconductors, J. Appl. Phys. 57:3809 (1985).
3. W. A. Fietz, Electronic integration technique for measuring magnetization of hysteretic superconducting materials, Rev. Sci. Instrum. 36:1621 (1965).
4. Y. B. Kim, C. F. Hempstead, and A. R. Strnad, Magnetization and critical supercurrents, Phys. Rev. 129:528 (1963).
5. W. J. Carr, Jr., "AC Loss and Macroscopic Theory of Superconductors," Gordon and Breach, New York (1983), p. 70.
6. M. R. Beasley and T. P. Orlando, cited by R. J. Donnelly, in: "Physics Vade Mecum," H. L. Anderson, ed., American Institute of Physics, New York (1981), ch. 7, p. 121.
7. A. L. Fetter and P. C. Hohenberg, in: "Superconductivity," R. D. Parks, ed., Marcel Dekker, New York (1969), ch. 14, p. 838.
8. A. L. Fetter and P. C. Hohenberg, in: "Superconductivity," R. D. Parks, ed., Marcel Dekker, New York (1969), ch. 14, p. 843.
9. Cf. $H_{c1} \cong 8$ kA/m (100 Oe), in: A. K. Ghosh and W. B. Sampson, Magnetization and critical currents of NbTi wires with fine filaments, paper DZ-7, this conference.
10. J. R. Clem, "AC Losses in Type-II Superconductors," Ames Lab. Tech. Rept. IS-M 280, Iowa State University, Ames (1979), pp. 23-24.

THE EXACT SOLUTION OF THE ELECTROMAGNETIC FIELD CONFIGURATION IN MULTIFILAMENTARY WIRE IN A TIME-DEPENDENT FIELD

P. C. Rem, D. Dijkstra, F. P. H. van Beckum,
and L. J. M. van de Klundert

Twente University of Technology
Enschede, The Netherlands

ABSTRACT

In some macroscopic applications of superconductivity, multifilamentary wires are subjected to large time-dependent magnetic fields. To estimate the losses and the performance under such conditions, it is essential to have a thorough understanding of the electromagnetic field inside such a wire.

In the past few years several analytical models[1,2] have yielded approximate solutions which are valid for a part of the parameter range. Recently[5,6] the exact quasi-static solution was extended to the entire frequency range. In this paper we present the general numerical solution of the electromagnetic field in a multifilamentary wire in a (periodic) time-dependent magnetic field.

INTRODUCTION

The problem of finding the solution of Maxwell's equations for the case of a multifilamentary wire in a time-dependent applied field has been discussed by several authors. Purely analytical approaches by G. Ries[1] and A. M. Campbell[2] yielded solutions which are fundamentally correct for infinite twist pitch, L_p, in the low frequency limit. T. Ogasawara et al.[3] incorporated the phenomenon of saturation into their model using a slab model to describe the shielding. Furthermore, they neglected the transverse current in the saturated regions. K. Kanbara et al.[4] derived a set of two differential equations for the induced magnetic field, assuming elliptical shielding. Basing their model on that of T. Ogasawara et al., they treated the reaction field as a homogeneous field.

In a previous paper[5] we solved Maxwell's equations for the stationary case. It was found that the profiles of the unsaturated regions are not quite elliptical. This result was confirmed by D. Cyazinski and B. Turck, using a network model.[6] They extended the solution to the entire frequency range and found a peculiar effect at intermediate saturation (β_s = 0.36). Furthermore, it was realized that the reaction field is homogeneous in the unsaturated region only and that the transverse current is not negligible. This means that a description of the reaction field by an ordinary differential equation can only be very approximate. None of the aforementioned authors took into account the resistivity of the filaments due to the external field effect.

In this paper we will present some simple solutions for the stationary case and discuss the results obtained from a numerical model which solves the general problem of a multifilamentary wire in a timedependent field.

DEFINITION OF THE PROBLEM

We will treat the problem within the Carr-model[7], i.e., we will consider the wire as an anisotropic medium in which the current density $\vec{j}$ consists of two parts, the ohmic conduction $\vec{\vec{\sigma}}.\vec{E}$ and the superconducting current density j_s in the direction of the filaments $\vec{p}$.

$$\vec{j} = \vec{\vec{\sigma}}.\vec{E} + j_s\vec{p} \qquad 1$$

Four dimensionless parameters characterize the problem, apart from the particular shape of the applied field B_x^a as a function of time. We choose the x-axis in the direction of the applied field.

$$\hat{\alpha} = \frac{\omega \hat{B}_x^a L_p \bar{\sigma}}{2\pi \bar{j}_c} , \quad \beta = \left(\frac{L_p}{2\pi R}\right)^2 , \quad \gamma = \frac{\mu_0 \bar{j}_c R}{\hat{B}_x^a} , \quad \delta = \frac{\pi R_f}{4R} . \qquad 2$$

$\hat{B}_x^a$ is the amplitude of the applied field, $\bar{j}_c$ and $\bar{\sigma}$ are mean values for the effective critical current density and the ohmic conductivity, R_f is the radius of the filaments and R is the radius of the composite. The parameter $\hat{\alpha}$ relates the transverse current density to the longitudinal current density, β is a geometry factor, γ relates the penetration field of the wire to the external field and δ is a measure of the resistance of the filaments due to the external field effect. Instead of using the physical electromagnetic field $\vec{\tilde{E}}$, $\vec{\tilde{B}}$ and $\vec{\tilde{j}}$ we define dimensionless quantities by:

$$\vec{E} = \frac{2\pi \vec{\tilde{E}}}{\omega \hat{B}_x^a L_p} , \quad \vec{B} = \frac{\vec{\tilde{B}}^i}{\mu_0 \bar{j}_c R} , \quad \vec{j} = \vec{\tilde{j}}/\bar{j}_c . \qquad 3$$

$\vec{\tilde{B}}^i$ is the magnetic field due to the currents inside the wire.

Furthermore we define dimensionless space and time variables:

$$\rho = r/R \ , \ \tau = \omega t, \qquad 4$$

and local dimensionless material properties:

$$j_c = j_c/\bar{j}_c \ , \ \vec{\vec{\sigma}} = \vec{\vec{\tilde{\sigma}}}/\bar{\sigma}. \qquad 5$$

Usually, $\hat{B}_x^a$ will be much larger than the penetration field of the filaments. In this case j_s, the effective superconducting current density is given by:

$$j_s = \begin{cases} j_c \ \mathrm{sign}\ (\vec{E}.\vec{p}) & |\vec{E}.\vec{p}| \ge \dfrac{\delta|f(\tau)|}{\sqrt{\beta}} \\ j_c \dfrac{\sqrt{\beta}(\vec{E}.\vec{p})}{\delta|f(\tau)|} & \text{otherwise} \end{cases} \qquad 6$$

where $f(\tau) = \partial_\tau B_x^a/\hat{B}_x^a$ and $\vec{p} = (\rho\vec{e}_\phi + \sqrt{\beta}\vec{e}_z)/\sqrt{\beta + \rho^2}$. Fig. 1 shows the

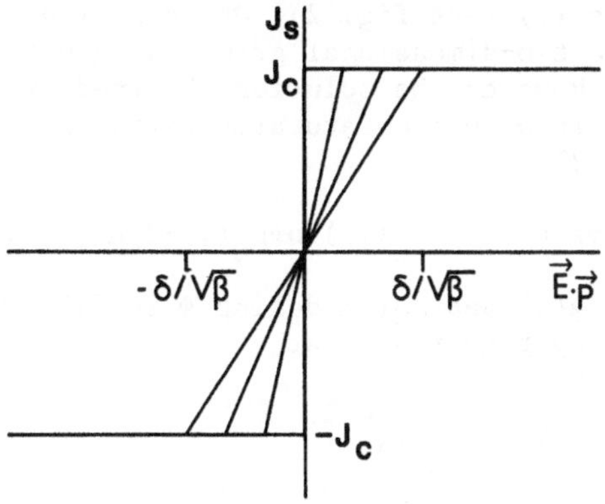

Fig. 1. The dependence of j_s on $\vec{E}.\vec{p}$ for $0 \le |f(\tau)| \le 1$.

dependence of j_s on $\vec{E}.\vec{p}$. Recently[8,9] this constituitive equation was confirmed by experiments performed in our group. Using these definitions, the Maxwell equations read:

$$\nabla\times\vec{E} = -(f(\tau)\vec{e}_x + \gamma\partial_\tau\vec{B})/\sqrt{\beta} \qquad 7$$

$$\nabla\times\vec{B} = \alpha\vec{\vec{\sigma}}\cdot\vec{E} + j_s\vec{p} \qquad 8$$

THE STATIONARY CASE

If either the currents inside the conductor are stationary or the penetration field of the conductor is negligible with respect to the external field, the second term on the right hand side of 7 vanishes and equations 7, 8 can be solved separately. The general solution of 7 is:

$$\vec{E} = f(\tau)[-\vec{e}_x - \frac{y}{\sqrt{\beta}}\vec{e}_z + \nabla\Phi] \qquad 9$$

We will first show a simple solution of 8 for $\delta = 0$ and then discuss the more general case $\delta \neq 0$.

The case $\delta = 0$.

If the filaments are very thin, the resistance due to the external field effect may be neglected. In this case the first equation of 6 is valid everywhere in the conductor. One can see from 6 and 9 that $\Phi = 0$ in the central unsaturated part (0) of the conductor and that Φ depends only

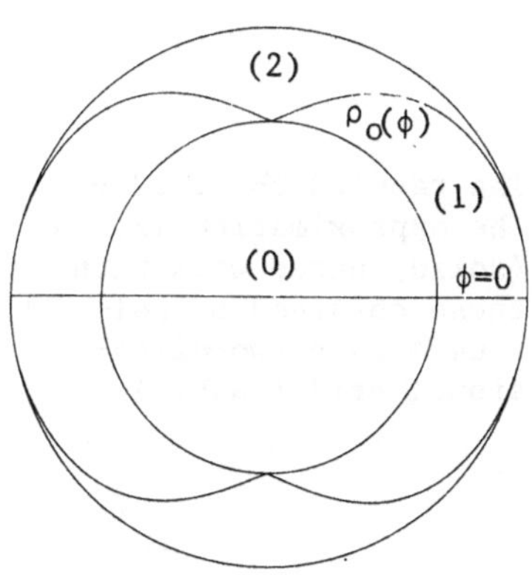

Fig. 2. A schematic view of the saturation in multifilamentary wires, due to a perpendicular field change. In (2) the filaments are saturated.

on ρ in the remaining unsaturated parts (1) (see fig. 2). One can find Φ in the saturated parts (2) by setting up a two-dimensional grid, using either a continuous model or a network model. However the solution obtained in this way shows that Φ is almost linear in ρ in the saturated parts (2). Therefor a good approximation of Φ in (2) is:

$$\Phi(\rho,\phi) = g(\phi) + (\rho-\rho_0(\phi))\cos\phi + (\rho-\rho_0(\phi))(\rho+\rho_0(\phi)-2)h(\phi) \quad 10$$

Since Φ depends only on ρ in (1), $g(\phi)$ and $\rho_0(\phi)$ define Φ in (1). We can now solve for $g(\phi)$, $\rho_0(\phi)$ and $h(\phi)$ by requiring that

- E be continuous at $\rho_0(\phi)$: $\partial_\phi\Phi|_{\rho=\rho_0(\phi)} = 0$
- $|j_s| = j_c$ at $\rho_0(\phi)$
- $\int_0^1 j_\phi d\rho = 0$ (conservation of current, equation 8).

These conditions lead to three ordinary differential equations for g,h and ρ_0. For $j_c = 1$ and $\vec{\sigma} = \vec{U}$ these equations are:

$$g' - \rho_0'(\cos\phi - 2h(1-\rho_0)) = 0 \quad 11$$

$$\hat{\alpha}|f(\tau)|\phi\partial_\phi(\rho_0(\cos\phi - 2h(1-\rho_0))) - \partial_\phi\sqrt{\rho_0^2+\beta} = 0 \quad 12$$

$$[(1-\rho_0)(\frac{-3}{2} + \frac{\rho_0}{2}) + \rho_0(\rho_0-2)\ln\rho_0]h' = \phi\rho_0(\cos\phi-2h(1-\rho_0)) + $$
$$+ \rho_0\sin\phi(\ln\rho_0-1) + \frac{1}{\alpha|f(\tau)|}(\sqrt{1+\beta} - \sqrt{\rho_0^2+\beta}) \quad 13$$

Notice that 12 and 13 may be solved first and g can be integrated afterwards. One can integrate these ordinary differential equations by finite-difference techniques. In fig. 3 the results are compared to those obtained by using a two-dimensional grid. If terms of the order $(1-\rho_0^2)$ are neglected with respect to terms of the order 1, we can immediately solve 12 and 13 for ρ_0 and h:

$$\sqrt{1+\beta} - \sqrt{\rho_0^2+\beta} = \hat{\alpha}|f(\tau)|(\sin\phi-\phi\cos\phi) \quad 14$$

$$h = \frac{-\cos\phi}{2\rho_0} \quad 15$$

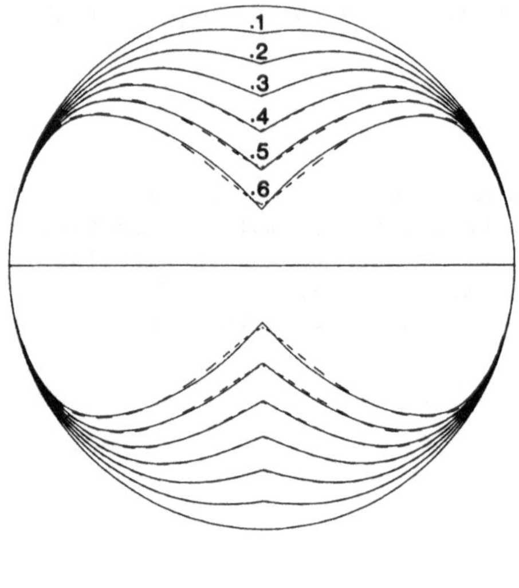

Fig. 3. The results obtained by the approximation 12, 13 (solid) agree well with those obtained by solving Φ on a two-dimensional grid (dashed).

and the corresponding coupling loss is:

$$P(\tau) = \frac{\tilde{P}(\tau)}{(\omega \hat{B}_x^a R^2)^2 \sigma\beta} = f^2(\tau)[\pi(1+\frac{1}{4\beta})-4(\frac{8}{9}-\frac{\pi}{6})\hat{\alpha}|f(\tau)|\sqrt{1+\beta}] \qquad 16$$

The case $\delta \neq 0$.

If the filaments have a finite diameter ($\delta \neq 0$), equation 8 can be solved for small values of $\hat{\alpha}\sqrt{\beta}$ (as long as no saturation occurs). In this case the current density is linear in $\vec{E}$ everywhere in the conductor according to 6. If $j_c = 1$ and $\overset{\Rightarrow}{\sigma} = \overset{\Rightarrow}{U}$ it follows that:

$$\begin{aligned}
E_\rho &= (p_\nu(\tfrac{\rho}{\sqrt{\beta}})-1)f(\tau)\cos\phi, \\
E_\phi &= (q_\nu(\tfrac{\rho}{\sqrt{\beta}})+1)f(\tau)\sin\phi, \\
E_z &= \frac{-\rho}{\sqrt{\beta}} f(\tau)\sin\phi \quad , \\
p_\nu(x) &= \frac{P_{\nu-1}(1+2x^2)+((4\nu+2)x^2-1)P_\nu(1+2x^2)}{\beta x^2[P_{\nu-1}(1+\frac{2}{\beta})+(\frac{4\nu+2}{\beta}-1)P_\nu(1+\frac{2}{\beta})]} \\
q_\nu(x) &= \frac{P_{\nu-1}(1+2x^2)-(1+2x^2)P_\nu(1+2x^2)}{\beta x^2[P_{\nu-1}(1+\frac{2}{\beta})+(\frac{4\nu+2}{\beta}-1)P_\nu(1+\frac{2}{\beta})]}
\end{aligned} \qquad 17$$

and $\qquad (2\nu+1)^2 = 1 + \sqrt{\beta}/(\alpha\delta|f(\tau)|)$

(P_ν is the Legendre-function).

The loss as a function of time is:

$$P(\tau) = \int_{\rho=0}^{1}\int_{\phi=0}^{2\pi} ((2\nu+1)^2 E_{/\!/}^2 + E_\rho^2 + E_\perp^2)\rho\, d\rho\, d\phi. \qquad 18$$

In fig. 5 $P(\tau)$ averaged over one cycle is shown as a function of $\hat{\alpha}$ and δ.

THE GENERAL (TIMEDEPENDENT) CASE

To solve equations 6, 7 and 8 for general $\hat{\alpha},\beta,\gamma,\delta$ and $f(\tau)$ we use a finite difference scheme on a two-dimensional grid, representing the cross-section of the wire. Fig. 4 shows the relative positions of the variables in a small (3*3 cells) version of the grid. Notice that not all components of the electromagnetic field share the same positions in the grid. The discretization we employed is based on the integral representation of the Maxwell equations. Thanks to symmetry only half of the cross-section is needed for the calculation. As an example the contours representing the z-components of $\nabla.\vec{B} = 0$, $\nabla\times\vec{E} = -\partial_\tau\vec{B}$ (solid) and $\nabla.\vec{j} = 0$, $\nabla\times\vec{B} = \vec{j}$ (dashed) are indicated in fig. 4.

On the circular part of the boundary the conditions

$$E_\rho = 0, \quad B_z = 0 \qquad 19$$

are imposed. The remaining conditions, which link the magnetic field inside the wire to the vacuum solution are expressed in an integral representation using $\nabla.\vec{B} = 0$ at the boundary. A slight disadvantage of the grid is that not all components of the electromagnetic field needed for the calculation of Poynting s vector are available at the boundary. The local dimensionless dissipation is calculated in two parts:

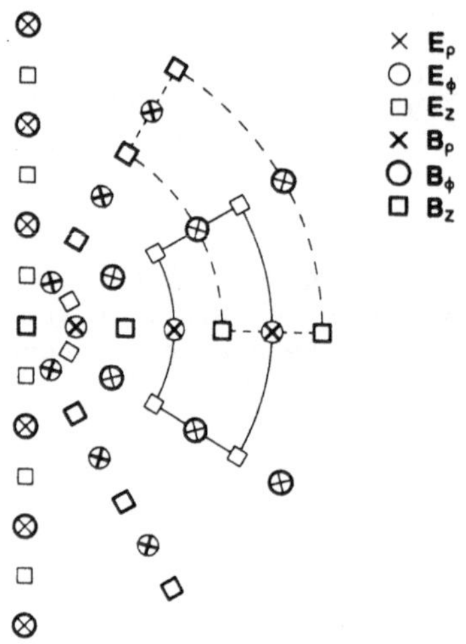

Fig. 4. The positions of the components of the electromagnetic field on the grid are chosen in a natural way with respect to the Maxwell equations and boundary conditions.

$$P_c = \frac{(2\pi)^2 \tilde{P}_c}{\bar{\sigma}(\omega \hat{B}^a_x L_p)^2} = \frac{\vec{j}\cdot\vec{E}}{\alpha} \qquad 20$$

$$P_m = \frac{4\tilde{P}_m}{\pi R_f \bar{j}_c \omega \hat{B}^a_x} = \begin{cases} 0 & |\vec{E}\cdot\vec{p}| \geq \dfrac{\delta |f(\tau)|}{\sqrt{\beta}} \\ \dfrac{16\gamma}{3\pi^2} j_c \left|\dfrac{\partial B_\perp}{\partial \tau}\right| (1-(j_s/j_c)^2) & \text{otherwise} \end{cases} \qquad 21$$

P_m is the local loss due to magnetization of the filaments. Furthermore we calculated the fraction of the critical current which can be carried DC without loss (except for the loss due to the filament resistivity):

$$i^{max}_{DC} = \int_{\rho=0}^{1} \int_{\phi=0}^{2\pi} (j_c - j_s)\sqrt{\frac{\beta}{\beta+\rho^2}}\, \rho d\rho d\phi \Big/ \int_{\rho=0}^{1} \int_{\phi=0}^{2\pi} j_c \sqrt{\frac{\beta}{\beta+\rho^2}}\, \rho d\rho d\phi \qquad 22$$

THE RESULTS

The actual numerical calculation was performed on a grid with $n_\rho * n_\phi$ = 12*24 cells and takes about 10 min. of C.P.U.-time on a Cyber 205 for one set of parameters. The calculations were carried out for $\beta = 25$, $\delta = .025$, $.02<\alpha<.16$, $\gamma = 0.5, 1.0, 2.0$ and a sinusoidal field $f(\tau) = \sin(\tau)$. In fig. 5 the results are compared with those of the stationary models (γ=0).

For small values of γ the saturation varies in time like a succession of stationary solutions except for a shift in phase which is linear in γ. In fact this is what one might expect from a rough comparison with the solution by G. Ries[1]:

$$B^i \propto \partial_\tau B^a_x / (1 + j\alpha\sqrt{\beta}\gamma/2)$$

For larger values of γ, however, there are two circulations of current in opposite directions (see fig. 6). Clearly, it is not possible to expand such a solution from the low frequency limit. Fig. 7 shows the superconducting current density j_s for the same time ($\tau = 19\pi/128$) and the same parameters ($\alpha = .16$, $\gamma = 2.0$).

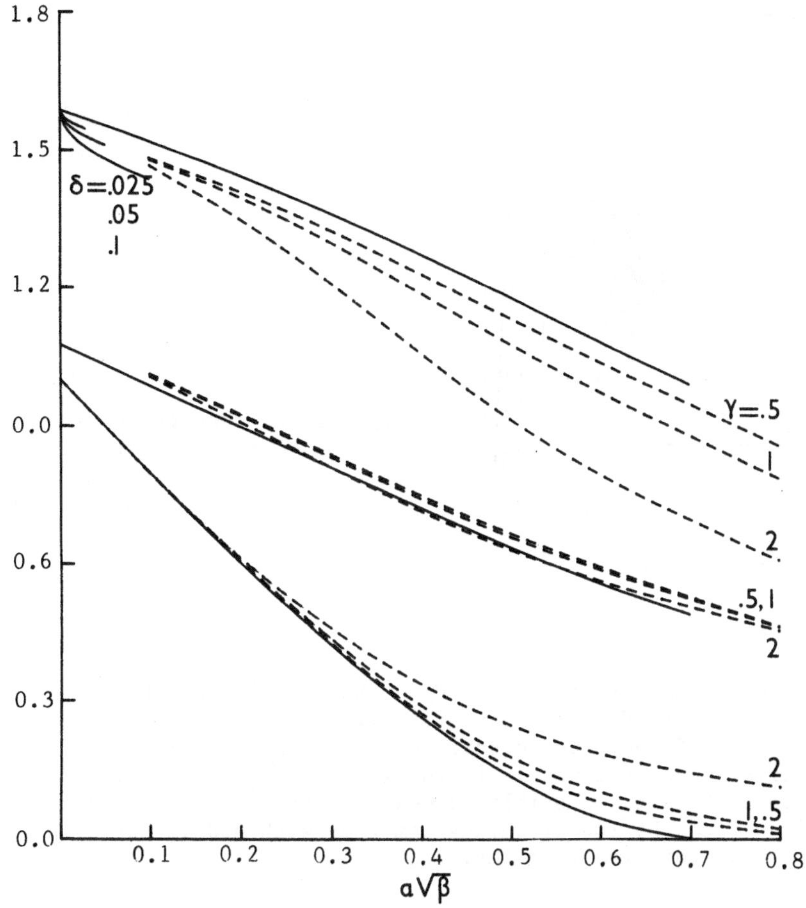

Fig. 5. Results for the coupling loss $(2\pi)^2 P/(\sigma(\omega B_x^a L_p R)^2)$, (upper curves), the loss due to filament magnetization $4P/(\pi R_f j_c \omega B_x^a R^2)$, (middle curves), where P denotes the corresponding loss per length of wire and averaged over a cycle of a sinusoidal applied field, and the maximum loss-free DC current (see equation 22, lower curves), obtained by numerical integration of equations 6, 7 and 8 (dashed, $\beta = 25$, $\delta = 0.025$) are compared with results from the quasistatic models (solid, $\beta = 25$, $\gamma = 0$). The three curves in the upper left part of the graph are calculated according to equations 17, 18. The curves stop at the onset of saturation.

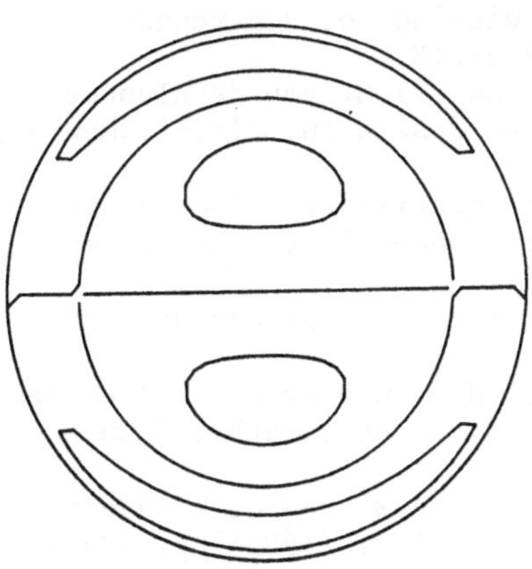

Fig. 6. The streamlines of the current density in the cross-section of a wire in a sinusoidal applied field show two modes of circulation for $\alpha = .16$, $\beta = 25$, $\gamma = 2$ and $\delta = 0.025$. In fact for higher values of γ more modes of circulation show up, in contrast to the quasistatic model.

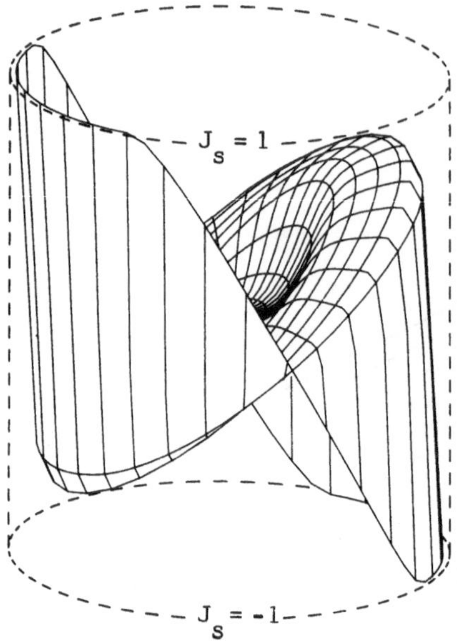

Fig. 7. The superconducting current density, for the same time and parameters as fig. 6, is shown as a function of ρ and ϕ.

CONCLUSIONS

Using modern computers (like the Cyber 205) it is possible to calculate the detailed behaviour of multifilamentary wires under realistic AC-conditions. The results of the model described in this paper agree with those obtained by timeindependent models in the limits $\hat{\alpha}\sqrt{\beta}\gamma \rightarrow 0$. It is clear from the results that it is not possible to describe the whole parameter-range with expansions from the low frequency limit or with models based on the assumption of a uniform reaction field.

REFERENCES

1. G. Ries, AC-losses in multifilamentary superconductors at technical frequencies, in: "IEEE Transactions on Magnetics", vol. MAG-13 (1977), pp. 524-526.
2. A. M. Campbell, A general treatment of losses in multifilamentary superconductors, Cryogenics 22:3-16 (1982).
3. T. Ogasawara, M. Itoh, Y. Kubota, K. Kanbara, Y. Takahashi, K. Yasohama and K. Yasukochi, Transient field losses in multifilamentary composite conductors carrying transport currents, in "IEEE Transactions on Magnetics", vol. MAG-17 (1981), pp. 967-970.
4. K. Kanbara, Y. Kubota and T. Ogasawara, Transverse field loss of a twisted multifilamentary round wire in windings of superconducting magnets, in "PROC. ICEC", (1981), pp. 715-718.
5. P. C. Rem, F.P.H. van Beckum, D. Dijkstra and L.J.M. van de Klundert, A contribution to the understanding of AC losses in multifilamentary wires, J. de Physique C1-471 (1984).
6. D. Cyazinski and B. Turck, Theoretical and experimental study of the saturation of a superconducting composite under fast changing magnetic field, Cryogenics 24:507-514 (1984).
7. W. J. Carr, Electromagnetic theory for filamentary superconductors, Physical Rev. B 11:1547-1554 (1975).
8. P. C. Rem, J.L. de Reuver, D. Dijkstra, F.P.H. van Beckum, "The external field effect", Memorandum No. 478, Dep. of App. Math., Twente University of Technology.
9. J. L. de Reuver, G.B.J. Mulder, P.C. Rem and L.J.M. van de Klundert in "IEEE Transactions on Magnetism", vol. MAG-21 (1985) pp. 173-176.

AC LOSSES FOR IN SITU SUPERCONDUCTING MICROCOMPOSITES

V. M. Pan, S. I. Mukhin, V. S. Flis, M. G. Vasilenko,
V. I. Latysheva, and E. N. Litvinenko
Institute of Metal Physics
Kiev, USSR

L. M. Fisher and V. M. Dzugutov
V. I. Lenin's All-Union Electrical Engineering Institute
Moscow, USSR

ABSTRACT

Metallurgical linking of the filaments and their proximity coupling through the normal metal matrix in in situ prepared superconducting composites cause greater ac losses than those observed for bronze process materials. Twisting of in situ wire, provided twist length is sufficiently small, results in a limitation of screening supercurrent density by the critical supercurrent density across the filament direction before the twisting. Supercurrent density across the filament direction is determined by both cross-linking and proximity coupling of filaments. Proximity supercurrents between the filaments can be greatly reduced by a high enough magnetic field, so that the separate contribution of metallurgical links can be determined. In fact, this is done by comparison of the magnetization hysteretic curves for the wires before and after twisting.

Experimental results are presented for in situ composites with a superconductor concentration above the percolation threshold. Magnetization of the sample due to proximity coupling currents was nearly twice that due to the currents through metallurgical links. Reduction of the proximity currents in in situ composites can reduce ac losses by several times.

INTRODUCTION

In this work, the structural and superconducting properties were investigated for in situ Cu-Nb microcomposites with a high concentration of Nb (35 wt.%) by measuring and subsequent analysis of the magnetic field dependencies of magnetization for samples with different twist lengths.

For our investigation, two considerations were crucial: (a) screening supercurrent density in heavily twisted wires is determined by the supercurrent density across the filaments direction (J_c) except near the wire axis and (b) the portion of J_c due to the proximity effect is highly sensitive to the external magnetic field. This results in a strong dependence of the magnetization of the twisted wire on magnetic field under some conditions discussed below. This enables one to examine the metallurgical connectivity of in situ superconducting microcomposites by using their hysteretic magnetization curves for different values of twist length.

UNTWISTED SAMPLES: EXPERIMENT AND INTERPRETATION

To produce Cu-Nb ingots, vacuum-melted copper and electron-beam-melted niobium were used. The Cu-35 wt.% Nb ingots were remelted in an induction furnace and cast in a copper crystallizer with subsequent hydroextrusion and drawing into wire with different amounts of deformation, $R = 10^3 - 10^4$. Magnetization was measured at liquid helium temperatures by vibrating a sample magnetometer in the transverse (to the coil plane) ac magnetic field with various periods (16 s, 64 s, and 128 s) and amplitudes (from 0.03 T up to 0.45 T).

Hysteretic Magnetization Curves

Magnetization curves are presented in Fig. 1. The inset shows (for comparison) analogous magnetization curves for continuous multifilamentary superconducting Nb-Ti wire with a similar external diameter. In the last case, magnetization curves for different field amplitudes obviously have coincident parts, which correspond to recurrent flux profiles when the field magnitude exceeds its full penetration value for a given sample. The maximum magnetization value then corresponds to the full penetration of the field. Similar magnetization patterns were obtained only for the thinnest of our in situ samples. In contrast, magnetization curves for the thickest in situ sample (Fig. 1) do possess maxima, but have no coincident parts. None of the magnetization curves have coincident parts or regions of steep variation with the field.

The features described above can be simply explained by allowing for the suppression of supercurrent density in the external magnetic field. To figure this, we assume the following conditions:

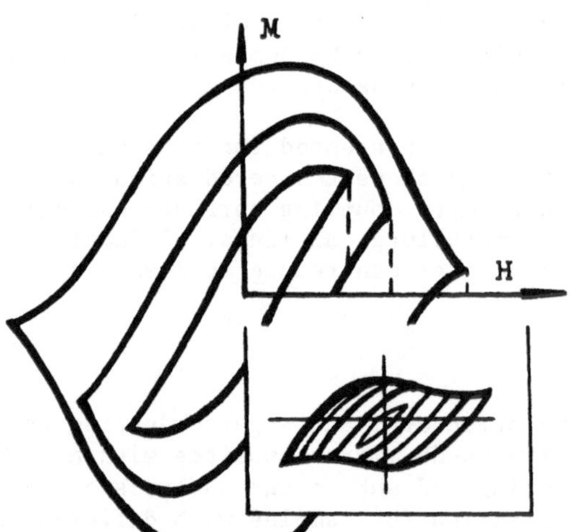

Fig. 1. Magnetization vs. magnetic field curves for in situ Cu-Nb superconducting sample with a diameter of 0.292 mm. Inset: M vs. H curves for continuous multifilamentary superconducting Nb-Ti wire of similar diameter.

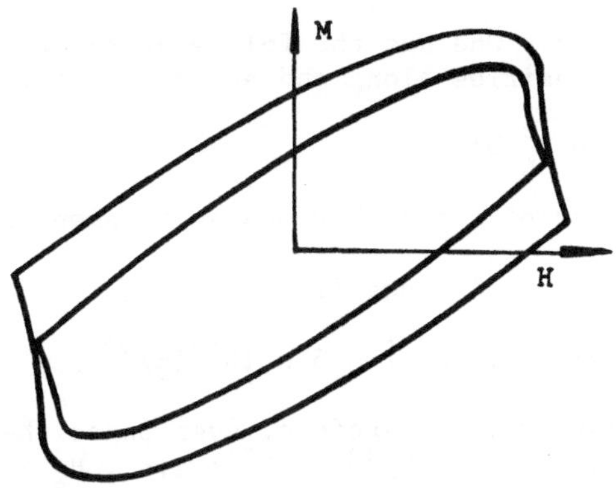

Fig. 2. Variation of M vs. H for the model system described in the text; $H_s < J_{co}D$ and $J_{co}/J_{cl} = 10$; $H_s/2J_{co}D = 1/3$.

$$J = \begin{cases} J_{co}; & H < H_s \\ J_{cl}; & H > H_s \end{cases} \qquad (1)$$

where $J_{cl} < J_{co}$ and H_s is estimated below. Figure 2 shows magnetization versus field plots calculated in frame of Eq. (1) for a superconducting layer with a 2D thickness in the parallel magnetic field. Here the ratio J_{co}/J_{cl} equals 10, and magnetic field is scaled in the units of H_s. The plot presented in Fig. 2 is for the case $H_s < J_{co}D$, i.e., when the magnetic field approaches H_s before it fully penetrates the layer. Here the maxima of the magnetization curves appear to be due to suppression of the screening current flow as the field increases but not due to full penetration of the field. On the contrary, when $H_s > J_{co}D$, magnetization first saturates when the magnetic field fully penetrates the sample. After that, magnetization decreases as the field exceeds H_s until the field exceeds H_s throughout the volume of the layer. The result is that magnetization saturates a second time. The ratio of the two saturation values of the magnetization equals [Eq. (1)] precisely the current densities ratio J_{co}/J_{cl}. The calculated magnetization versus field curves in Fig. 2 are qualitatively similar to those presented in Fig. 1, except for the presence of the regions of great variation of magnetization in Fig. 2. This discrepancy is discussed in the next section.

Suppression of Proximity Current by Magnetic Field

The screening supercurrent in the magnetic field is reduced mainly due to suppression of the proximity currents between the filaments as well as to the reduction of the supercurrent density inside each filament as the field approaches the upper critical value, H_{c2}, for the superconductor material. Proximity current density, J_p, in the normal matrix depends exponentially on magnetic field,[1] H:

$$J_p = J_1 \exp(-D_n K_n). \qquad (2)$$

where, for high enough fields,

$$K_n = 2(eH/hc)^{1/2}, \qquad (3)$$

J_1 is a constant, D_n is the thickness of the normal layer between two bulk superconductors, c is the velocity of light, e is the electron charge, and h is Planck's constant. At a given amount of the sample deformation ($R \sim 10^4$), one takes $D_n \gtrsim L_n \simeq 10^3$ Å, where L_n is diffusion length of the Cooper pair in the copper matrix. One has the following relation between the critical supercurrent densities along and across the filament axis:[2]

$$J_p = J_c D'/w, \tag{4}$$

where D' is the thickness of the ribbon-shaped filament and w is its width.

Using eqs. (2) through (4), one finds:

$$(J_1 w/J_c D') \exp\left[-4.5 \times 10^{-2} (H)^{1/2}\right] \lesssim 1 \tag{5}$$

The factor w/D' is about 10^2.[3] Therefore, even under the condition $J_1 \sim J_c$, eq. (5) gives: $\exp\left[-4.5 \times 10^{-2} (H)^{1/2}\right] < 10^{-2}$; i.e., $H_s \gtrsim 0.1 - 1$ T. In such fields, the suppression of proximity currents becomes substantial, but since this field interval is close to the H_{c2} value for niobium filaments, their internal supercurrent density decreases simultaneously. Thus, the large ratio w/D' masks the suppression of proximity currents in our untwisted samples until the field approaches a H_{c2} value at which intrafilamentary supercurrent also decreases sharply. This results in the absence of regions of great variation in the magnetization curves in Fig. 1 within the experimental interval of field variation, but one cannot extract the separate contribution of the screening supercurrent density due to the metallurgical linking for untwisted in situ samples in the transverse magnetic field.

TWISTED IN SITU WIRES: THEORY AND EXPERIMENT

Twisting a cylindrical superconducting sample (of diameter D) with anisotropic critical supercurrent density (twist length equals L) results in the following radial distribution of the longitudinal supercurrent density:[2]

$$J_z = J_c/2pr; \quad r \geqslant r_o \tag{6}$$

$$J_z = fJ_c/2; \quad r < r_o \tag{7}$$

where r is the radial coordinate, $r_o = J_{c\perp}/J_c fp$, $p = \pi/L$, and f is the fraction of superconductor in the unit volume of the sample. For a tight twist, the ratio of the magnetization values of twisted and untwisted samples in the external magnetic field, H, equals:

$$M(H,L)/M\,(H,L = \infty) \sim J_{c\perp}(H)L/J_c(H)fD\pi < 1 \tag{8}$$

for the case of full penetration.

To determine the field dependence of the transverse critical current density, $J_{c\perp}(H)$, we consider the model system of parallel superconducting filaments (filament diameter equals D_o) embedded in a normal metal matrix and distributed at random over the wire cross section. The filaments are interconnected randomly by a number of metallurgical links. Then the average critical supercurrent density, $J_{c\perp}$ for such a model may be represented as follows:

$$J_{c\perp}(H) = (f/2)P_m J_c(H) + J_w(H) \tag{9}$$

where P_m is the volume ratio of the filaments connected into an infinite cluster to the number of filaments in the wire; $J_w(H)$ is the critical density of the proximity supercurrent between the filaments. As is shown in Ref. 1, at high enough magnetic field H [an estimate for copper gives:

$H > H_o \sim 0.01$ T, since the diffusion coefficient for copper D_N, is about 0.15 m^2s^{-1} and the temperature ($< T_c$) is about 10 K] one has:

$$I_{ij} = I_o L_o \exp[-2D_{ij}(eH/hc)^{1/2}] \tag{10}$$

where I_{ij} is the critical proximity current between the superconducting filaments i and j (of the length L_o) and D_{ij} is the thickness of the normal metal layer between them. Since D_{ij} has a random distribution ($D_{ij} \gtrsim$ 1000 Å for the samples considered), using the results of percolation theory[4,5] for the case of the cross section of the wire normal to filament direction, one finds:

$$2\pi B_c^{1/3}(\pi D_o^2/4f)^{1/2} J_w = I_o \exp[-2(eH/hc)^{1/2}\, 0.95(\pi D_o^2/4f)^{1/2}] \equiv$$

$$I_o \exp(-B_c) \tag{11}$$

Substituting eq. (11) into eq. (8), one finds:

$$I_1 = \left.\frac{M(H,L)}{M(H,L=\infty)}\right|_{H<<H_1} \sim \frac{L}{fD\pi}\left[\frac{f}{2}P_m + \frac{J_w(H)}{J_c(H)}\right] \tag{12}$$

$$I_2 = \left.\frac{M(H,L)}{M(H,L=\infty)}\right|_{H>>H_1} \sim \frac{L}{fD\pi}\left(\frac{f}{2}P_m\right) \tag{13}$$

where H_1 is determined by: $B_c > 1$ (for $f \sim 0.4$), i.e., $H_1 \gtrsim 0.01 - 0.1$ T.

Hysteretic magnetization curves obtained from Ref. 6 are presented in Fig. 3. The outside loop was obtained for an untwisted Cu-40 wt.% Nb in situ superconducting microcomposite with a 0.3-mm diameter at T = 4.2 K. The inside loop in Fig. 3 was obtained for a sample with a twist length of 1.8 mm (the other parameters were unchanged;[6] the deformation amount (R) of both samples equals 1.8×10^4). Extracting I_1 and I_2 from Fig. 3, one finds: $P_m \sim 0.2$ and $J_w/J_c \sim 0.1$ at $H << H_1$. Thus, it follows from eq. (12) and the above estimates that the proximity current contribution to the

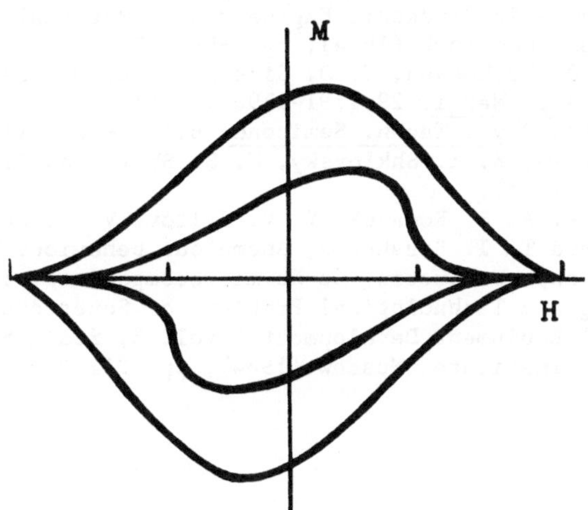

Fig. 3. M vs. H for two in situ Cu-Nb microcomposite samples of 0.3-mm diameter. The outside loop corresponds to twist length (L) of ∞; the inside one corresponds to L = 1.8 mm.

magnetization of the in situ superconducting sample with a superconductor concentration above the percolation threshold can exceed by several times the contribution due to metallurgical links.

CONCLUSIONS

Interfilamentary coupling in the in situ prepared superconducting composites can be investigated by measuring the magnetization of twisted wires in a magnetic field. Twisting is used here to distinguish the contributions to the screening supercurrent density that are due to two coupling mechanisms: metallurgical linking and proximity coupling through the copper (the last mechanism is more sensitive to external magnetic fields far from the H_{c2} value). For this method to be effective, the following conditions should be met: (a) twist should be tight enough so that the screening supercurrent density is substantially dependent on the supercurrent density across the filaments [(see eqs. (6) and (7)]; (b) the full penetration field should be less than field H_1, which substantially suppresses the proximity currents in the sample [see text following eq. (13)]; (c) H_1 should be much less than the upper critical field, H_{c2}, of the superconductor material in order that the dependence $J_c(H)$ is much smoother than $J_{c\perp}(H)$ at $H \sim H_1$; (d) the metallurgical connectivity, P_m, should be small enough (<<1) to distinguish the field-dependent contribution to the magnetization due to the proximity currents from the background magnetization due to the supercurrent through the metallurgical links.

So, the estimates obtained for P_m and J_w/J_c show that substantial suppression of proximity currents through the normal metal matrix (by external magnetic field or magnetic impurities in the matrix) can reduce the ac losses for in situ microcomposites by several times. One should allow for this possibility in efficiency estimates of the utilization of in situ microcomposites in superconducting technology.

REFERENCES

1. T. Y. Hsiang and D. K. Finnemore, Phys. Rev. B 22:154-163 (1980).
2. W. J. Carr, Jr., Electromagnetic theory for in situ superconductors, in: "Advances in Cryogenic Engineering - Materials," vol. 30, Plenum Press, New York (1984), pp. 899-907.
3. J. J. Sue, J. D. Verhoeven, E. D. Gibson, J. E. Ostenson, and D. K. Finnemore, Acta Metal. 29:1791-1796 (1981).
4. B. I. Shklovsky, Phys. Techn. Semicond. 6:1197-1226 (1972).
5. M. E. Levinshtein, B. I. Shklovsky, M. S. Shur, and A. L. Efros, JETP 69:386-392 (1975).
6. A. L. Rakhmanov, A. S. Romanuk, V. V. Jeltov, V. D. Jemerihkin, I. A. Kijansky, and T. I. Brushkova, Anomalous behaviour of magnetization of in situ superconducting Cu-Nb microcomposites, in: "Scientific Engineering and Technological Problems of Superconducting Electro-energetical Equipment Development," vol. 3, State Scientific-Research Energetical Institute, Moscow (1984), pp. 212-213.

MAGNETIC BEHAVIOR OF A VERY FINE FILAMENT

CONTINUOUS SUPERCONDUCTOR*

W. J. Carr, Jr. and G. R. Wagner

Westinghouse R&D Center
Pittsburgh, Pennsylvania

ABSTRACT

The magnetic moment of an untwisted multifilamentary superconductor consisting of 0.12 µm diameter Nb filaments in a copper matrix was studied as a function of applied magnetic field. The wire contains 72,102 filaments arranged in 61 bundles of 1182 each. In a transverse magnetic field the wire behaves as a solid superconductor yielding a moment characteristic of the wire diameter as expected for a long untwisted wire. But in a parallel field the moment is much smaller and is interpreted as arising from strong proximity coupled filaments within a bundle, with weak coupling between bundles. A Meissner effect was observed for the individual filaments but not for the bundles or wires.

INTRODUCTION AND SUMMARY

A study has been made of the magnetic moment in both weak and strong magnetic fields of a multifilament superconductor having 0.12 µm diameter filaments, presumed to be continuous through the wire. The conductor consists of Nb filaments embedded in a Cu matrix with the configuration shown in Fig. 1, except that the outer Cu and the Nb barrier have been etched away. The filaments are arranged in bundles with 61 bundles of 1182 filaments in the wire, giving a total of 72,102 filaments. Neither the filaments nor the bundles are twisted. Each bundle has a core of Sn but this fact is considered to be unimportant in the analysis. All measurements in the superconducting state were made at 4.2°K. The wire was obtained from Intermagnetics General Corporation.[1] The magnetic moment was measured with a Foner-type vibrating sample magnetometer.

Since the separation between filaments in a bundle is the order of 0.05 µm it is to be expected that the filaments are strongly coupled together via the proximity effect in the copper,[2] and for some purposes the bundle may be expected to act as a single large filament. It is found that for perpendicular magnetic fields the bundles can also be considered as coupled together, and the magnetic moment is determined by the wire diameter. For parallel fields each bundle acts as a single filament, but

*Supported by the Air Force Aero Propulsion Laboratory, Contract No. F33615-81-C-2040.

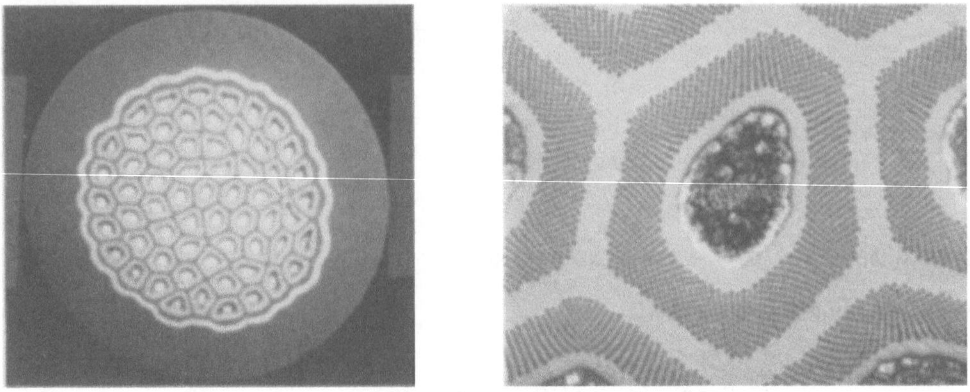

(a) (b)

Fig. 1. (a) Conductor cross-section showing 61 bundles of filaments with Sn central core, diffusion barrier, and outer stabilizing copper. (b) A bundle at higher magnification showing individual filaments.

the bundles are only weakly coupled and the magnetic moment is much smaller. In all magnetic fields a weak partial Meissner effect is observed for the parallel field case, but it occurs only in the filaments.

BEHAVIOR IN A TRANSVERSE FIELD

The initial slope of the magnetization curve as a function of applied magnetic field[3] is shown in Fig. 2, where the ordinate corresponds to -4π times the magnetic moment divided by the total volume V of the sample. For a solid superconductor, simple theory would predict an initial slope of two for this plot. Since, to good approximation, the measured curve fits this theoretical curve, it follows that the filamentary nature of the sample has no effect on the measured curve: i.e., the magnetic moment is the same as that for a solid superconductor with an equivalent current density. Thus the bundles are coupled together in addition to the filaments. This result is not entirely surprising, since for a long untwisted wire in a transverse field, the currents tend to flow parallel to the filaments and bundles, and these currents have the whole length of the wire to cross over from one bundle to another. Even a very small transverse supercurrent in the Cu matrix between bundles will lead to the observed result. In terms of the effective longitudinal current density j_c for the wire, the critical state model gives for the magnetic moment[4]

$$\frac{m}{V} = \frac{0.2}{3\pi} d\, j_c \qquad (1)$$

for full penetration of the current into the wire, where d is now the diameter of the wire (71×10^{-4} cm). The peak in the curve of Fig. 2 is taken to correspond to the point of full penetration, and using the measured value of m/V one obtains $j_c = 1.32 \times 10^6$ amps/cm^2. With the approximation $j_c = 0.23\, j_{cf}$ where 0.23 is the fraction of superconductor in the wire, the value obtained for the critical current density j_{cf} of the filaments at H = 1500 Oe is

$$j_{cf} = 5.7 \times 10^6 \qquad (2)$$

in good agreement with the value measured for this wire as shown in Fig. 3.

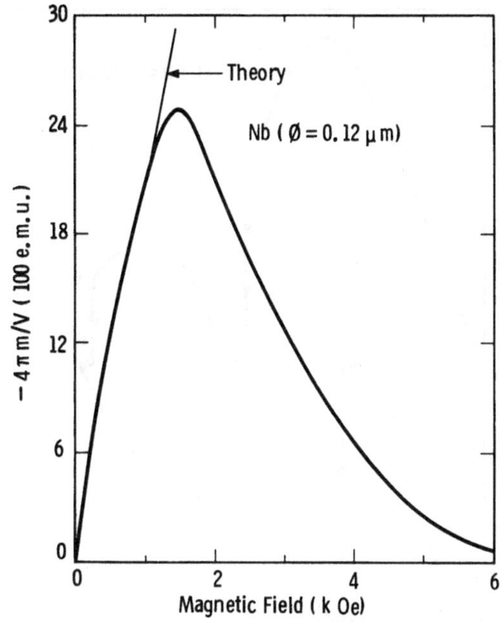

Fig. 2. Initial magnetization as a function of the applied transverse magnetic field, where m is the magnetic moment and V the sample volume.

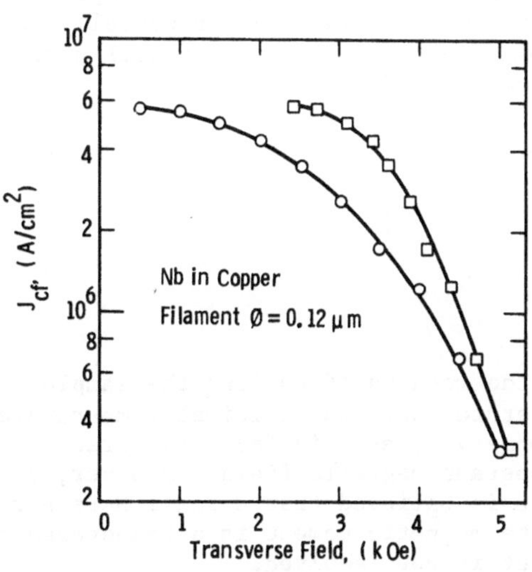

Fig. 3. Measured critical current density in the filaments vs. the applied transverse field. The circles are as measured and the squares are corrected for self field.

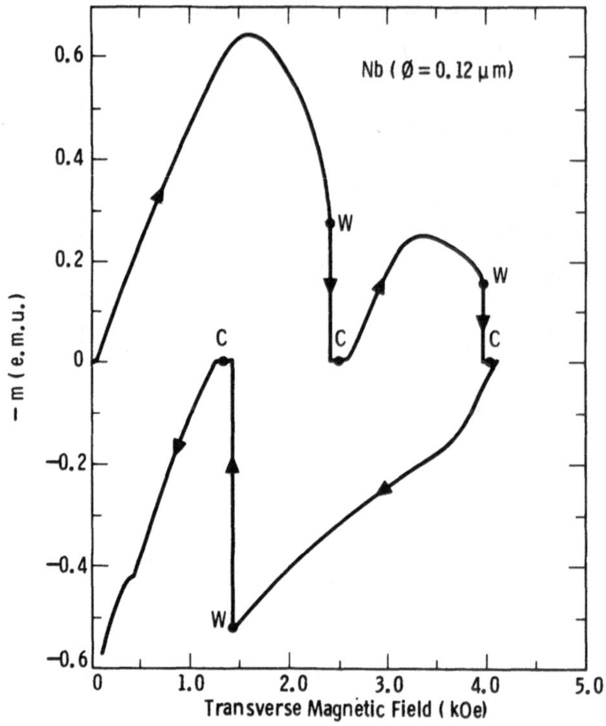

Fig. 4. Measured magnetic moment in a transverse magnetic field. At points indicated by W the sample was warmed above T_c at constant field. At points indicated by C the sample was recooled to 4.2°K at constant field. While the sample temperature was above T_c, the recorder pen was displaced horizontally to properly observe the effect of cooling, so the horizontal portions have no significance.

Figure 4 shows the results of cooling the sample in various magnetic fields from a temperature above the critical temperature T_c. No bulk Meissner effect is observed, and, in fact, no effect at all is observed from cooling in a constant magnetic field. However, as discussed in the following section, it is believed that a small Meissner effect exists in the filaments, but the magnetic moment in a transverse field is so large that this small effect is not resolved.

BEHAVIOR IN A PARALLEL FIELD

A sample was prepared for parallel field measurements by cutting the wire into one centimeter lengths and using 77 segments to form a sample of nearly the same volume as that used for the transverse field measurements. The initial magnetization curve is shown in Fig. 5, where two peaks in the curve are observed, one at m = −0.057 e.m.u., H = 400 Oe; with the other at m = −0.084, H = 3000. It is observed that the magnetic moments tend to be an order of magnitude smaller than those for the transverse field. The interpretation assumed for this curve is the following: (1) each bundle acts like a single large superconducting filament which becomes fully penetrated at the lower peak, H = 400 Oe; and (2) bulk supercurrents also circulate around larger loops (Fig. 6) due to migration along favorable paths between bundles. The latter are presumably determined by some randomness in the spacing, and these current loops produce the additional moment leading to the second peak. One may check the first assumption as follows. The field for full penetration of a hollow filament of outer diameter d_{so} and inner diameter d_{si} is[5]

$$H_p = 0.2\pi\, j_c(d_{so} - d_{si}) \tag{3}$$

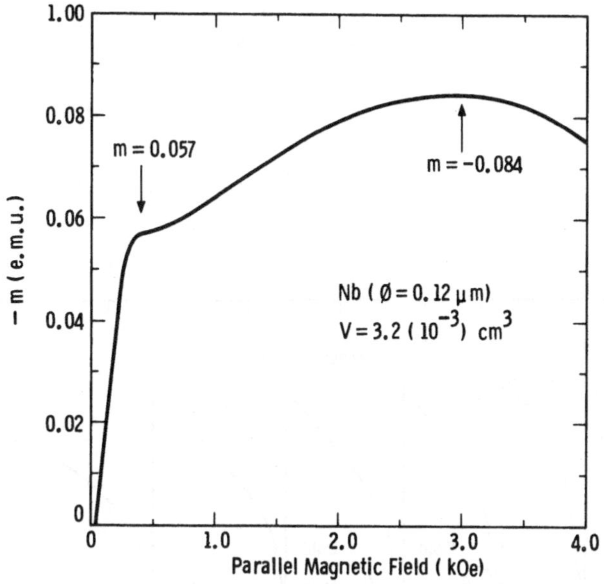

Fig. 5. Initial magnetization measured as a function of the applied field parallel to the filaments.

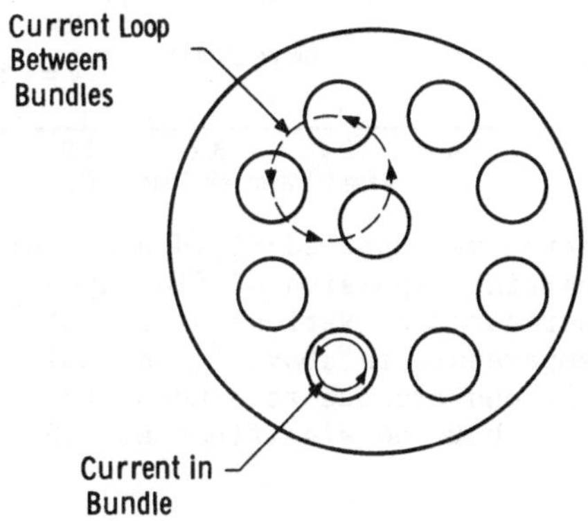

Fig. 6. Schematic picture indicating a current loop within a bundle and a larger loop between bundles.

with j_c the current density in the filament. For the hollow filament formed by the bundle of Fig. 1, $d_{so} - d_{si} \approx 3.2 \times 10^{-4}$ cm, and for $H_p = 400$

$$j_c = 2 \times 10^6 \ A/cm^2 \tag{4}$$

for the critical current density circulating in the part of the bundle occupied by filaments. This value is smaller than the longitudinal critical current density given by (2), as expected, since the circumferential current density must flow, in part, through the matrix. However the two have the same order of magnitude. It is easily shown that the magnetic moment due to the critical current j_c of a fully penetrated cylinder in a longitudinal field is

$$m = \frac{0.1 \ j_c \ d}{6} V \ . \tag{5}$$

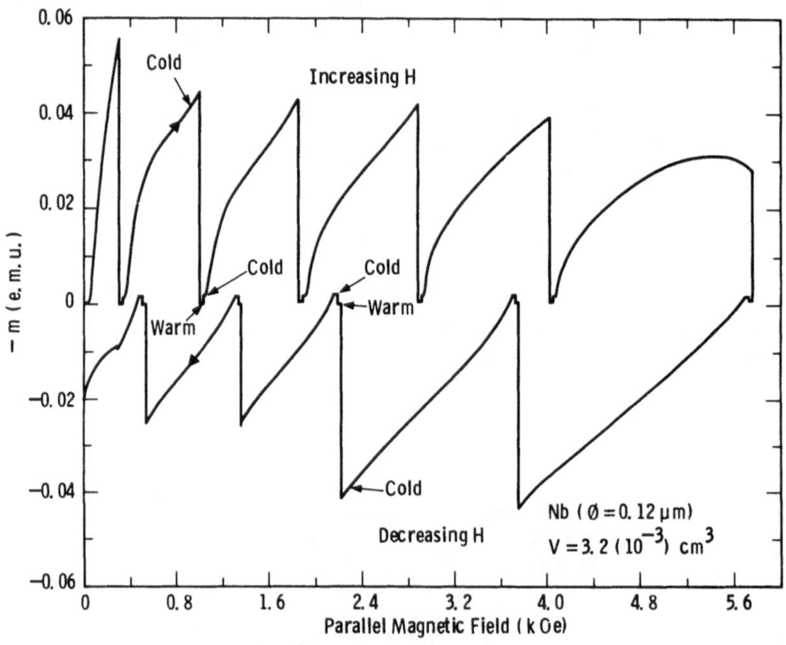

Fig. 7. The magnetization vs. applied field parallel to the filaments showing the partial expulsion of flux upon cooling through the critical temperatures at various values of field. Warm indicates the sample temperature is above T_c and cold is 4.2°K. The horizontal portions are due to a mechanical displacement of the recorder pen and have no significance. (See Fig. 4 caption)

Thus for a hollow strand

$$m = \frac{0.1\ j_c}{6} \left[d_{so} \frac{\pi}{4} d_{so}^2\ \ell - d_{si} \frac{\pi}{4} d_{si}^2\ \ell\right]$$

$$= \frac{0.1\ j_c}{6} d_{so}\ V_s\left(1 - \frac{d_{si}^3}{d_{so}^3}\right) \tag{6}$$

where ℓ is the total length of the strand and V_s is the volume. One obtains using (4)

$$m = 2.75 \times 10^{-8}\ j_c \tag{7}$$

$$= 0.055$$

in excellent agreement with the measured value 0.057. The second assumption is more difficult to verify since the larger current loops involve a distribution of current paths.

Figure 7 shows an investigation of the Meissner effect in various parallel applied fields. It is clear that no bulk Meissner effect occurs in the wire or bundles, but a small effect is observed at all fields, which is attributed to a partial Meissner effect in the filaments. The change in magnetic moment observed on cooling through the critical temperature is $\Delta m \sim 2 \times 10^{-3}$ and the total volume of filaments is 0.67×10^{-3}. One finds that

$$\frac{4\pi\ \Delta m}{V_f} \sim 38 \tag{8}$$

in e.m.u., which is less than the lower critical field H_{c1}, as expected.

REFERENCES AND FOOTNOTES

1. This conductor was manufactured by IGC to be subsequently heat treated to form Nb_3Sn by the internal tin technique. The results reported here are for the unreacted state with the outer copper stabilizer and diffusion barrier etched away leaving Nb filaments in a copper matrix with the tin core in the bundles. We thank Dr. G. M. Ozeryansky of IGC for supplying the wire.
2. It is interesting to compare these results with the hysteresis losses in similar fine filament conductors reported by P. Dubots, A. Fevrier, J. C. Renard and J. P. Tavergnier at the Applied Superconductivity Conference, San Diego (1984). In the latter the matrix consists of a Cu-Ni alloy which has a much weaker proximity effect, and filamentary behavior exists down to a smaller filament diameter.
3. The sample consisted of a wire coiled on a mandrel. A small correction was made to the applied magnetic field to account for the field of neighboring turns.
4. W. J. Carr Jr. and G. R. Wagner, Ad. in Cryogenic Eng., Ed. A. F. Clark and R. P. Reed, Plenum, New York, Vol. 30, 923 (1984).
5. W. J. Carr Jr., AC Loss and Macroscopic Theory of Superconductors, Gordon and Breach, New York (1983).

MAGNETIZATION AND CRITICAL CURRENTS OF NbTi WIRES WITH FINE FILAMENTS*

A. K. Ghosh and W. B. Sampson

Brookhaven National Laboratory
Upton, New York

ABSTRACT

In high energy accelerators such as the SSC, the magnetization of the superconductor is an important component in determining the harmonic fields at injection (~0.3T). In an effort to reduce these residual fields, interest has focused on NbTi conductors with fine filaments which are expected to have a reduced magnetization as dictated by the critical state model. With this in view, the magnetization and critical currents were measured at 4.3K for a set of NbTi wires with filament diameters, d, ranging from 1.0 to 5.0 microns. The data show that, although the magnetization scales linearly with d, it does not do so with the product $J_c d$ for d less than 3 μm. However, at these d values, the critical transport current density, J_c, of NbTi was observed to decrease rapidly as a function of d. The origin of this J_c degradation and its effect on the scaling of magnetization within the framework of the critical state model is explored. We also examine the question of the observed asymmetry of the hysteretic magnetization.

INTRODUCTION

The effects of magnetization currents in the superconductor of accelerator dipoles appear at low fields as a non linear transfer function and systematic sextupole and higher multipole fields which are hysteretic in nature. These magnetization fields, which were first measured extensively by Sampson et al.[1] in early ISABELLE dipoles, dominate the systematic field errors at injection which must be corrected by trim coils. For the Superconducting Super Collider (SSC) which is envisioned to have an acceleration range of 1-20 TeV, the injection field is nominally 0.3T for the high field SSC Reference Design D[2]. At present prototype Design D dipoles which are being wound with NbTi cables with a large filament size (~ 20 μm), exhibit significant magnetization fields at 0.3T. This has motivated conductor R&D in an effort to produce NbTi wires, with high critical current density and small filament size (~ 2 to 5 micron), which are expected to show a significantly lower magnetization.

* Work supported by the U. S. Department of Energy

We report here magnetization and critical current measurements on NbTi conductors with filament diameters ranging from 5 to 1 micron. These data provided the basis for testing the scaling of magnetization with current density and filament diameter.

EXPERIMENTAL DETAILS

The average magnetization, M, of several multifilamentary NbTi samples were determined in these experiments as a function of applied transverse magnetic field, H, and magnetic history at 4.35K. Measurement of changes of flux through appropriate pickup coils during changes in H were integrated to yield a complete set of values of M as a function of H. The field dependence of the critical transport current I_c was determined in a separate experiment[3].

The details of the magnetization apparatus and the measurement techniques are given elsewhere[4]. The test samples were in the form of a stack of 30 cm long pieces of wire of volume ~ 2-3 cm^3, held together by epoxy and fiberglass and mounted in a non-magnetic holder. Since the measurement involves taking the difference of the measuring coil signal with and without the sample, the net susceptibility of the holder and the material other than the superconductor influences the shape of the hysteretic loop at high fields. Data were taken with an external ramp rate of ~ 20 mT/s. No eddy currents were detected for these rates.

Samples

A description of the NbTi samples are given in Table I. The IGC, MCA and Furukawa wire which have been manufactured recently represent possible future conductor for the SSC magnet program and some of the metallurgical details and critical current measurements are to be found in a report by Garber et al.[5] The 0.22 mm Furukawa wire was drawn without any heat treatment to the smaller sizes. The OST conductors manufactured in 1981 as a qualification test for the Tore II Supra project have a Cu+CuNi matrix and twist lengths ranging from 12 mm to 1.1 mm. The first four OST wires were drawn down from the same billet whereas the 0.20 mm wire was from a different billet. Eddy current loss measurements[6] have been reported for conductor similar to the 0.20 mm wire, while Carr and Wagner[7] have studied the magnetic hysteresis loop for the 0.367 mm wire.

Analysis of Measurements

The magnetization measurements gave reproducible hysteresis curves after the samples were cycled once to high fields. To compare conductors, the "width" of the hysteresis loop defined as $2\,\mu_o M$ was determined at the field of interest.

$$2\mu_o M = \mu_o\left(M^- - M^+\right)$$

where M^- (M^+) denotes the increasing (decreasing) applied field branch of the magnetization loop. In the critical state model this is related to the critical current density for fields greater than the full penetration field Hp, as

$$2\mu_o\, M(H) = 2\mu_o\, \frac{2}{3\pi}\, \nu\, J_c\,(H)\, d \tag{1}$$

where ν is the volume fraction of NbTi, d is the nominal filament diameter and J_c is the current density in the NbTi. To facilitate comparison of conductors with different copper-superconductor ratio we also define

$$2\mu_o\, M_s = 2\mu_o\, M/\nu \tag{2}$$

Table 1. Sample Description

Manufacturer	Wire Dia. (mm)	Vol.fraction of NbTi	# of filaments	Filament Dia.(μm)
1. IGC	0.29	0.42	1953	4.3
2. MCA	0.114	0.48	367	4.0
3. Furukawa	0.226	0.48	3000	2.9
	0.142	0.48		1.8
	0.114	0.48		1.5
4. OST (Airco)	0.91	0.31	10285	5.0
	0.57	0.31		3.1
	0.455	0.31		2.5
	0.36	0.31		2.0
	0.20	0.31		1.1

The asymmetry of the magnetization loop for $H > H_p$ can be characterized by

$$2\mu_o M_a = \mu_o (M^- + M^+) \tag{3}$$

I_c, the critical current defined for a conductor resistivity, ρ, of $10^{-14}\Omega$-m, was measured in transverse external fields from 0 to 8T. The resistive transition is characterized by I_c and a "quality factor" n defined empirically by the relation[3,5]

$$\rho = 10^{-14} (I/I_c)^n \tag{4}$$

A transport J_c is calculated using the known Cu/SC ratio and quoted at 5T in units of A/mm^2.

RESULTS AND DISCUSSION

In Fig. 1 the measured J_c at 5T and the width of the magnetization loop for unit volume of superconductor, $2\mu_o M_s$, at 1 Tesla are plotted as a function of filament diameter. The data show that although the J_c drops rapidly for $d < 3$ microns, the magnetization scales linearly with d. In fact this linear trend can be shown to extend up to filament sizes as large as ~ 23 μm[8]. This indicates that for those conductors of interest to the SSC magnet program (i.e. J_c (5T) $\geq$ 2400 A/mm^2), the systematic error fields produced in the dipoles would be proportionally reduced as the filament size went from 23 to ~ 3.0 μm.

Validity of the Critical State Model

Table 2 lists some of the measured quantities. J_c^M is the critical current density calculated from the magnetization data using Eq. 1. J_c^* is the transport critical current density which has been corrected for self field using the relation $(\mu_o H)_{sf} = 10^{-7}$ (2I/R) where R is the wire radius. In Fig. 2 the ratio J_c^*/J_c^M at 1 Tesla is plotted against filament diameter. At first glance the data would seem to indicate that for $d < 3$ μm the magnetization is larger than what would one expect from the critical state model. However, we argue that for $d < 3$ μm the transport critical current is not intrinsic to the superconductor. The reason that unusually high M values are observed in fine filament wires of low current capability is due to the fact that magnetization is unaffected by filament breakage or "necking", whereas transport current is directly affected by such geometric defects. Thus a sample with severe filament necking will have a larger magnetization than would be expected from the measured current. In

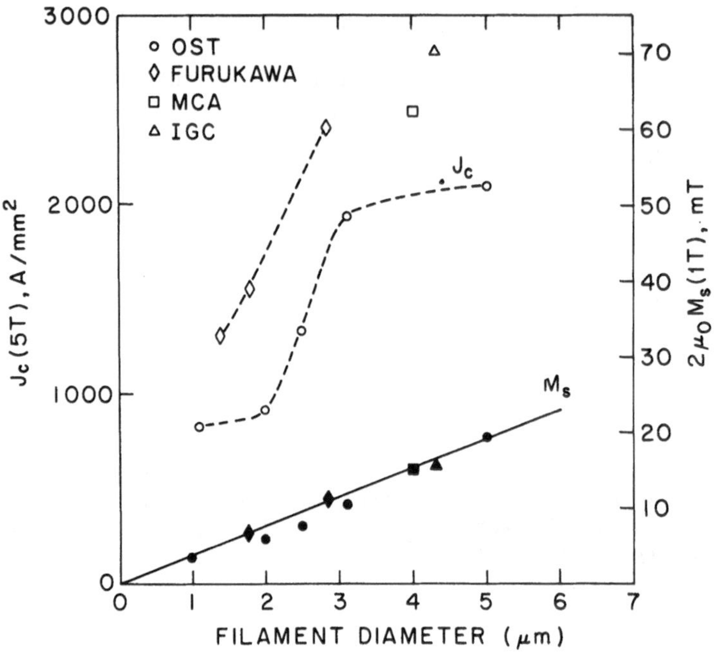

Fig. 1. Plot of $J_c(5T)$ and $2\mu_o\ M_s(1T)$ versus filament diameter The open symbols are the J_c data. The closed symbols give the magnetization "width" at 1 Tesla.

Table 2. Critical Current and Magnetization Data

Sample	$J_c(5T)$ A/mm²	n at 5T	$2\ \mu_oM(1T)$ mT	$J_c^M(IT)$ A/mm²	J_c^*/J_c^M
IGC-0.29	2835	31.5	6.5	6710	-
MCA-0.114	2497	31.1	7.1	6980	0.98
FUR-0.226	2381	24.5	5.1	6770	1.00
-0.142	1580	18.1	3.1	6740	.64
-0.114	1320	15.5	-	-	-
OST-0.91	2100	17.6	6.0	7190	1.00
-0.57	1950	10.8	3.1	6000	1.05
-0.455	1330	5.5	2.3	5513	0.75
-0.36	920	4.5	1.8	5393	0.55
-0.20	830	9.5	1.0	5450	0.48

fact, it is not the magnetization that is high but the current that is low. We have observed that the magnetization of cable samples is not affected by compaction even when significant degradation has been induced by excessive rolling. In one example the cable transport current was reduced by 30% while the magnetization was identical to an equivalent volume of uncabled wire. This is because the bulk of the superconductor is unaffected but the filaments are serrated at edges reducing the transport current[5]. In the following section, the cause of the J_c degradation in fine filament NbTi is discussed.

J_c Degradation for $d < 3$ μm

Garber et al.[5,9] have shown that high J_c wire have filaments with smooth surfaces and uniform cross sections. Filaments in low J_c wires show Cu_4Ti compound particles 1 ~ 2 μm size on their surfaces and large variations in cross section (i.e. necking). Scanning electron microscope (SEM) photographs of the Furukawa wires shows the presence of these nodules of Cu-Ti on the filament surface which would account for the J_c degradation as this wire is drawn down.

In the case of the OST wires, evidence of filament damage or necking is reflected in the reduced n values characterizing the resistive transition (see Table 2). In Fig. 2 the dashed curve is the J_c(5T) normalized to the J_c(5T) of the 0.91mm wire, which shows the degradation in critical current as the wire is drawn down. Although these wires had Nb barriers around each filament to prevent Cu-Ti compound formation extensive necking was observed in SEM micrographs, with the 0.20 mm wire showing a lot of filament breakage. At this wire size, 1 micron ZrO_2 inclusions were found at the necking sites. Contamination of the alloy with ZrO_2 from grinding wheels used in a centerless grinding operations likely led to this inclusion and the degradation of the wires for $d < 3$ μm.

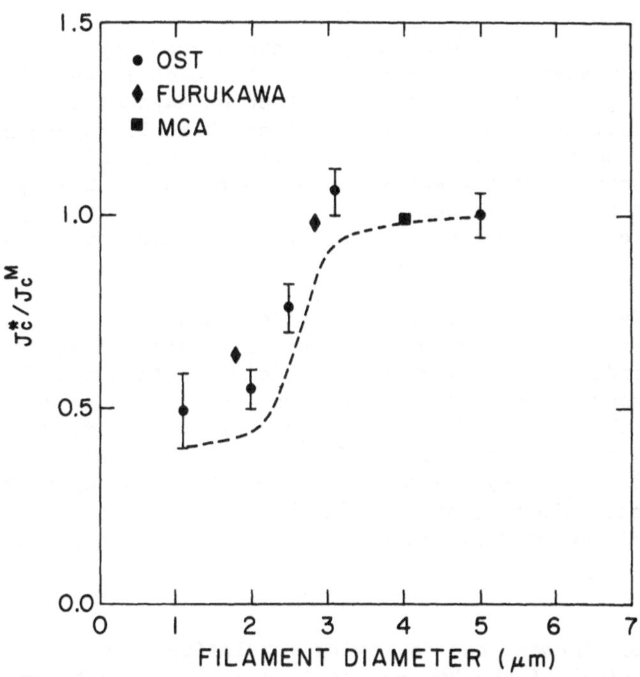

Fig. 2 Plot of J_c^*/J_c^M at 1 Tesla versus filament diameter

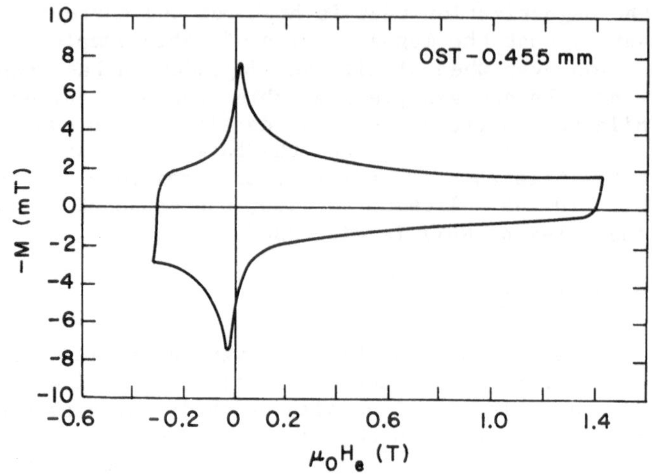

Fig. 3 Magnetization data for the OST-0.455 mm wire.

Asymmetry of the Magnetization Loop

Carr and Wagner[7] showed that for fine filaments the magnetization hysteresis has an up-down asymmetry. This they attribute to a surface current which is non zero (in the positive field region) for positive $\frac{dH}{dt}$, i.e. the M^+ branch, and is zero for $-\frac{dH}{dt}$, i.e. the M branch. This asymmetry was observed in the fine filament wires reported here an example of which is shown in Fig. 3. In this case most of the normal material susceptibility was eliminated by taking the difference of the signal with the conductor at 4.35K and when heated to ~ 20K without removing the sample from the measuring coils. The asymmetric term $2\mu_o M_a$ is ~ 1 mT at an external field of 1 Tesla and is similar in magnitude to the 1.1 μm filament wire.

Although the critical state model predicts up-down symmetry, data in the literature show asymmetry at high fields for even large filament diameter superconductors[10]. It is argued that an asymmetric contribution describing the equilibrium magnetization of any Type II superconductor should be present at all fields. This equilibrium magnetization M_e is described by the models of Abrikosov and Ginzburg-Landau theory[11]. It is assumed that this term is independent of the critical state magnetization and depends only on the magnitude of the applied field and is therefore reversible.

For extreme Type II (κ >>1) superconductors and for $H \gg H_{c1}$, the Abrikosov solution of the "London model" gives an analytical expression for M_e for a cylinder in parallel applied field[11]. Although the applied field in these experiments is transverse, nevertheless one can estimate the magnitude of this term. Using an upper critical field, H_{c2}=11.2T,[12] and heat capacity data of Nb-46.5 wt.%Ti,[13], the following values are estimated: $\kappa = 38$, H_{c1}(4.3K) ~ 10mT, λ, the penetration depth, = 2.5 x 10^{-7}m. Using these values M_e at 1 Tesla is calculated to be ~ 1.7mT. The observed magnitude of M_a is consistent with this interpretation.

CONCLUSIONS

From these experiments, the following conclusions are drawn: (1) at low fields the "width" of the hysteresis loop scales with filament diameter down to ~ 1 micron. This indicates that the magnetization effects seen in the current SSC Design D prototype magnets would be reduced by an order of magnitude by using the fine filament conductors with high J_c at 5T. These wires represent the next generation of SSC conductors. (2) The J_c degradation seen in wires with filament diameter < 3 μm are believed to be the result of local "necking" of the filaments caused by the drawing down of filaments with Cu-Ti compound on their surface or with micron size inclusions. These geometric defects of the filaments do not degrade the magnetization properties which reflect the intrinsic critical current of the filaments. (3) The asymmetry of the M-H curve can be explained by the presence of the reversible equilibrium magnetization of Type II superconductors and should be present even in wires with large filaments. This contribution to the total magnetization is assumed to be independent of the hysteretic contribution of the critical state model, and is apparent when the critical state contribution is small due to fine filaments.

ACKNOWLEDGMENTS

Samples of NbTi wire supplied by the manufacturers are greatfully acknowledged. We thank M. Garber and K. E. Robins for useful discussions and A. Wirszyla, E. Sperry and E. Ostheimer for technical assistance. We also thank R. L. Sabitini for the SEM analysis.

REFERENCES

1. W. B. Sampson, P. F. Dahl, A. D. McInturff, K. E. Robins, and E. J. Bleser, IEEE Trans. on Magnetics 15:114 (1979).
2. P. Dahl et al., "Performance of Two 4.5m Dipoles for Reference Design D", submitted to the 9th Int. Conf. on Magnet Technolgy, Zurich, 1985.
3. M. Garber, W. B. Sampson and M. J. Tannenbaum, IEEE Trans. on Magnetics 19:720 (1983).
4. A. K. Ghosh, K. E. Robins and W. B. Sampson, IEEE Trans. on Magnetics 21:328 (1985).
5. M. Garber, M. Suenaga, W. B. Sampson and R. L. Sabitini, this conference.
6. T. Ogadsawara, Y. Kubota, T. Makiru, T. Akachi, T. Hisanari, Y. Oda and K. Yasukōchi, IEEE Trans. on Magnetics 19:248 (1983).
7. W. J. Carr and G. R. Wagner, in "Advances in Cryogenic Engineering - Materials", Vol. 30, Plenum Press, N.Y. (1984), pp. 923-930.
8. A. K. Ghosh, W. B. Sampson and P. Wanderer, submitted to the 1985 Particle Accelerator Conference, Vancouver, B.C.
9. M. Garber, M. Suenaga, W. B. Sampson and R. L. Sabatini, "Effect of Cu_4Ti Compound Formation on the Characteristics of NbTi Accelerator Magnet Wire", submitted to the 1985 Particle Accelerator Conf., Vancouver, B. C.
10. W. A. Fietz, M. R. Beasley, J. Silcox and W. W. Webb, Phys. Rev. 136:A335 (1964).
11. A. L. Fetter and P. C. Hoenberg in "Superconductivity", R. D. Parks ed., Mercel Dekker, N.Y. (1969), pp. 817-924.
12. D. G. Hawksworth and D. C. Larbalestier, IEEE Trans. on Magnetics 17:49-52 (1981).
13. S. A. Elrod, J. R. Miller and L. Dresner in "Advances in Cryogenics Engineering-Materials" Vol. 28, Plenum Press, N.Y., pp. 601 (1981).

STRAIN EFFECTS IN "IN SITU" PROCESSED Nb_3Sn CONDUCTORS

M. Fukumoto, K. Katagiri, T. Okada and K. Koyanagi
ISIR, Osaka University, Ibaraki, Osaka, Japan

K. Yasohama
College of Science and Technology, Nihon University
Tokyo, Japan

ABSTRACT

The strain effects and mechanical properties in "in situ" processed Nb_3Sn conductors have been investigated in relation to the parameters specifying the fabrication conditions. The effects of these parameters are discussed in order to provide a guide for the optimization of fabrication. The parameters are [1] Nb content in Cu-Nb alloy (25, 30 and 35 w%Nb), [2] total reduction ratio R.R. of cross-sectional area (R.R.=4700, 40000, and 250000), [3] presence/absence of Cu stabilizer with Ta barrier, and [4] heat treatment time (at 550 C for 1, 3.5, 7 and 14.2 day).

INTRODUCTION

The "in situ" formed Nb_3Sn conductors provide higher mechanical strength and higher strain tolerance in critical current than conventional Nb_3Sn conductors[1-3], and seem desirable for large scale dc magnets in 8-14 Tesla range[3-4]. The characteristics of "in situ" conductors depend on fabrication conditions. For the purpose of the fabrication of the conductors optimized to their applications, the relation between the characteristics and fabrication conditions should be understood. We investigated strain effects and mechanical properties for "in situ" Nb_3Sn conductors,varying some parameters related to the fabrication conditions. The parameters are described in the next section.

EXPERIMENT

Specimens examined are "in situ" formed Nb_3Sn conductors fabricated by means of consumable arc melting and external Sn diffusion method[5]. The final size of all the specimens are 0.31 or 0.33 mm in diameter. Specimens fabricated in various process are prepared. The varied fabrication parameters are Nb content in initial Cu-Nb alloy, total reduction ratio R.R. of cross-sectional area in swaging and drawing process, presence/absence of pure copper stabilizer with tantalum barrier, and heat treatment time. Specimen characteristics and structure are presented in Table 1 and Fig. 1, respectively. The electroplated Sn content to Cu(excluding stabilizer) is 14w% for all the specimens. The

Table 1. Specimen Characteristics

#	Nbw%	R.R.	Cu Stabilizer	Ta Barrier	Heat Treatment	
1	25	4,700	26 vol%	11 vol%	550°C x	1 day
1	25	4,700	26	11	550	3.5
1	25	4,700	26	11	550	7
1	25	4,700	26	11	550	14.2
2	30	4,700	26	11	550	7
3	35	4,700	26	11	550	7
4	25	250,000	0	0	550	7
5	30	40,000	0	0	550	7

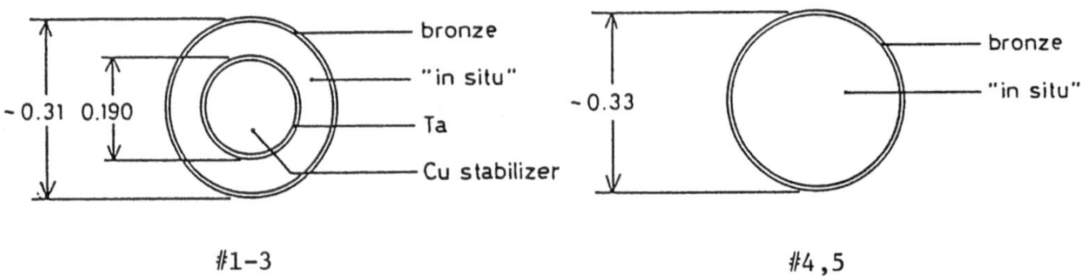

Fig. 1 Specimen structure.

filaments in the heat treated specimens have been observed by SEM, and are found to be approximately 0.1 μm x 1 μm in size. The filament size shows little dependence on R.R. and Nb content. The conductors with higher reduction ratio require the Cu-Nb ingot of larger diameter. The size of dendritic precipitates of Nb in Cu is the larger in the larger ingot because cooling rate decreases as the diameter of ingot is increased. This may be the reason of the little dependence of the final filament size on R.R. The comparison between specimen #1 (7 day) and 4, or #2 and 5 may, therefore, provide the information of the influence of Cu stabilizer and Ta barrier rather than that of RR.

The apparatus for the measurements of strain effects and mechanical properties has been reported[6]. Critical current I_c is measured under magnetic flux density of 6 T supplied by a split pair magnet. The gauge length which is the distance between two current terminals is 16.5 cm. The accuracy of strain measurement is within 0.01 % strain. Voltage taps are soldered to a specimen with a distance of 6 cm. The I_c is defined as the current where the potential gradient of 0.5 μV/cm is detected. Upper critical magnetic flux density B_{c2} at zero strain has been measured by use of 23 T hybrid magnet HM-2 of Tohoku university. Applied field is varied with supplying constant current of 0.2 A to the specimen. The B_{c2} is defined as the flux density where the potential gradient of 5μV/cm is detected. The total length of the specimen is 15 mm and the potential taps are soldered with a distance of 5 mm.

RESULTS AND DISCUSSION

Critical Current

In Fig. 2 a) the dependence of critical current I_c on reaction time is shown for specimen #1. The I_c increases with heat treatment time and reaches the maximum value in the range from 7 to 14.2 days. This result

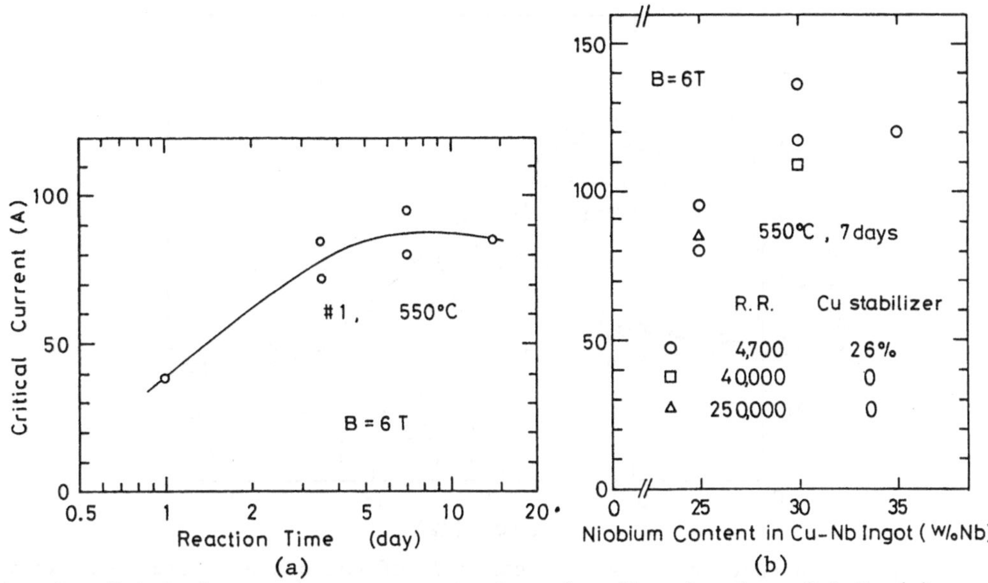

Fig. 2 Critical current at zero strain under flux density of 6 T. (a) Heat treatment time dependence for specimen #1. (b) Effect of Nb content and Cu stabilizer with Ta barrier for specimens heat treated for 7 days.

is supported by EDX (energy dispersion X-ray spectroscopy) analysis. The line scan profile of Sn concentration steeply slopes down from outside to inside across the cross-section of the specimen heat treated for 1 day, whereas it is nearly flat for the specimen of 7 day heat treatment. Figure 2 b) shows critical currents for all the specimens heat treated for 7 days. The I_c increases with Nb content among the specimens which have Cu stabilizer. The I_c is comparable between the specimens with and without Cu stabilizer, although the specimen without stabilizer has larger amount of Nb filaments than the specimen with the stabilizer. This means that the sufficient amount of Sn diffusion into Cu matrix is restricted only within the depth which is comparable with the thickness of the cylindrical "in situ" alloy in the specimens with the stabilizer. The SEM observation of the fracture surface of specimens without stabilizer (#4 and #5) reveals the presence of ductile core of about 1/3 thickness of wire diameter.

Strain Effects

Figure 3 presents the strain dependence of critical current with the parameter of heat treatment time for #1 specimen. The strain induced change in critical current was found to be reversible up to the fracture strain in all these examinations. The strain ε_m where critical current reaches its maximum value is little dependent on heat treatment time, and is 0.44 %. In bronze processed Nb_3Sn conductors, ε_m decreases with an increase in heat treatment time[7]. The filament thickness in these "in situ" specimens is as thin as 0.1 μm. The heat treatment time longer than 1 day presumably gives little change to the thickness of Nb-Sn compound layer, but varies only the composition of Nb-Sn system because of their thin filaments. Strain dependence of critical current is shown in Fig. 4 for specimen #1-5 heat treated for 7 days. The change in I_c seems reversible up to fracture for all these specimens as well. The fracture strains are in the order: #1 > #2,3; #1 > #4; and #2 > #5. These facts suggest that the presence of Cu stabilizer or Ta barrier, and higher bronze ratio (=bronze/Nb) improve the ductility of wire. In specimen #1 the

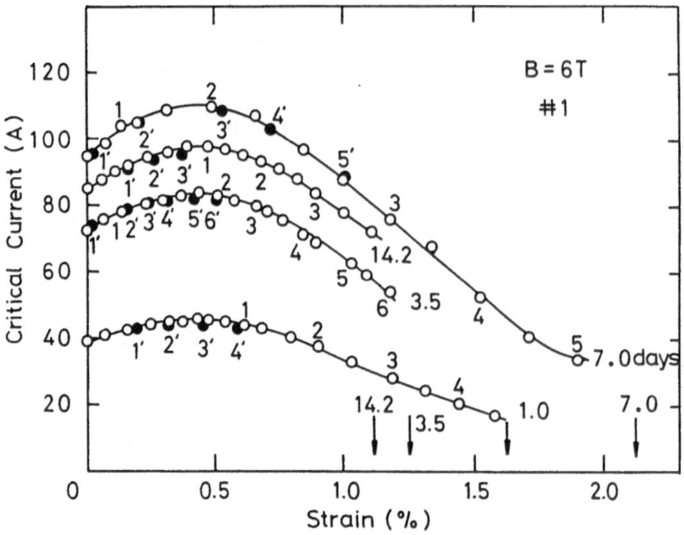

Fig. 3 Strain effects on I_c for specimen #1 of various heat treatment time. Open and closed circles show loaded and unloaded data, respectively. Arrows present fracture strains. These symbols are also used in Fig. 4.

fracture strain is 2.1%, and the reversibility of the change in critical current is preserved up to 1.4 % in intrinsic strain. The SEM observation for the filaments of the fractured specimen removing bronze matrix by HNO_3 shows few damages in the filaments[6]. This result means that the filaments in "in situ" conductors have high strain tolerance. This may be due to extremely fine filaments. It is noticeable that Nb_3Sn filaments are capable of being deformed up to strain as high as 1.4 % without plastic deformation. This result may be related to the reduction of Young's modulus caused by cooling down from room temperature to 4.2 K where martensitic transformation takes place[8].

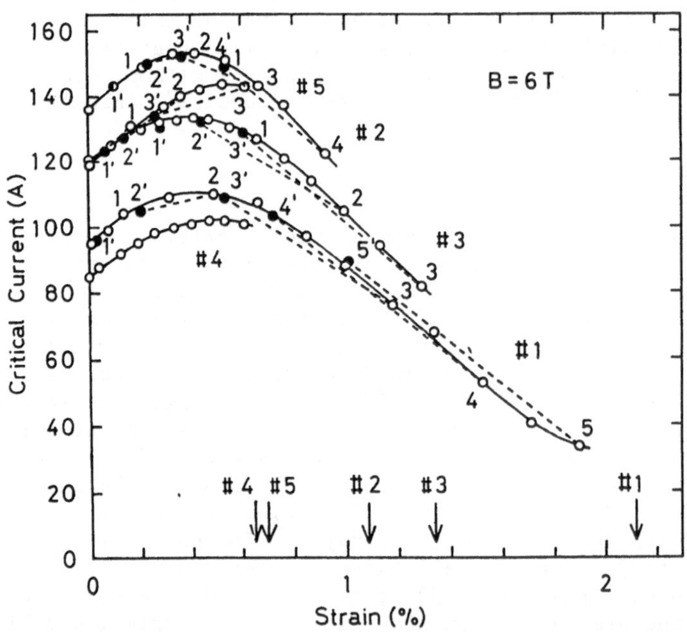

Fig. 4 Strain effect on I_c for specimens heat treated for 7 days.

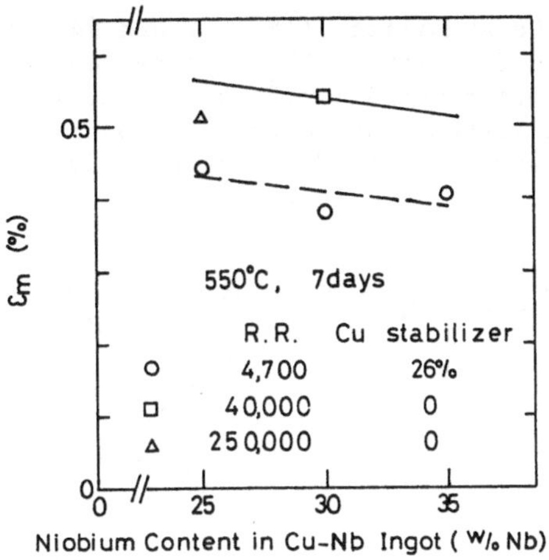

Fig. 5 Effect of Nb content and Cu stabilizer with Ta barrier on the strain ε_m where I_c reaches its maximum value.

Figure 5 presents ε_m of these specimens. Solid and dashed curves are obtained from calculation based on simple rule of mixtures for bronze-Nb_3Sn and bronze-Nb_3Sn-Ta systems, respectively. In the latter case the presence of Cu stabilizer is neglected. Pure Cu contributes little to the thermal compression of Nb_3Sn filaments because of its low yield strength. Experimental data fit these predictive curves with an exception of specimen #4. These results of experiments and calculation indicate that thermal compression of Nb_3Sn filaments is reduced with an increase in Nb content and by the presence of Ta barrier which has smaller thermal expansion coefficient and higher Young's modulus than Nb_3Sn and bronze.

In order to compare the strain sensitivity of critical current I_c among all the specimens examined, plots of I_c normalized to their maximum values I_{cm} versus intrinsic strain ε_0 ($=\varepsilon-\varepsilon_m$) are presented in Fig. 6. The curves for specimens heat treated for 7 and 14.2 days coincide with each other. As heat treatment time is reduced from 7 to 1 day, the curve becomes steeper, that is, strain sensitivity of I_c increases. This means that, the optimum heat treatment time is 7 days or more from the standpoint of both magnitude of I_c and strain effects on I_c. Strain sensitivity of specimens #1-5 of 7 day heat treatment is close to each other as shown in Fig. 6 (b). More precise comparison, however, leads to the following relative magnitude in the sensitivity: #2 > #1 > #3, #1 > #4, and #2 > #5. This result is related to the maximum upper critical flux density B_{c2m} to be mentioned later.

Mechanical Properties

Stress-strain relations at liquid helium temperature are shown in Fig. 7. Yield strength of 0.2 % offset decreases with an increase in heat treatment time. This result is interpreted in terms of the reduction in residual Sn concentration in bronze matrix and the consequent decrease in the yield strength of the matrix with reaction time. The stress-strain curves of the specimens of heat treatment time longer than 3.5 days are close to each other. This suggests that Nb filaments are sufficiently reacted with Sn, and Sn concentration in bronze matrix is almost constant among the specimens heat treated for longer than 3.5 days. The comparison among the specimens heat treated for 7 days are shown in Fig. 7 (b). Yield and fracture strength are excellently high for all the specimens examined in comparison with a bronze processed commercial Nb_3Sn[9]. The yield strength is improved with an increase in Nb content (#2 > #1 and #5 >

#4), and by absence of Cu stabilizer (#5 > #2 and #4 > #1). As Nb content is increased, the volume fraction of Nb_3Sn increases and inter-filament spacing decreases The former means increase in fraction of high yield stress material and the latter enhancement of constraints against plastic deformation of bronze matrix. Yield strength of pure Cu is extremely

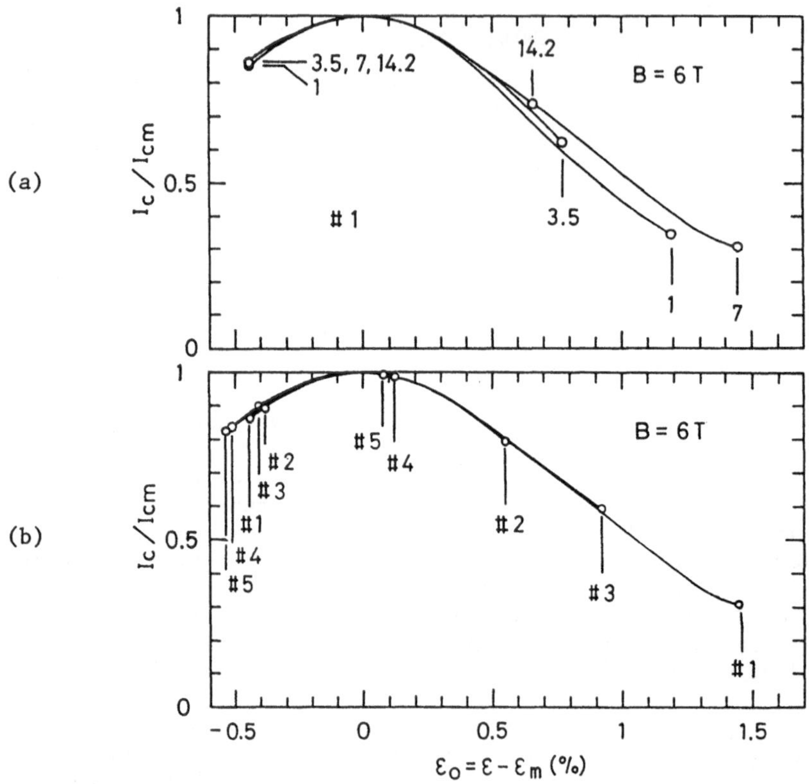

Fig. 6 Normalized I_c versus intrinsic strain for specimen #1 (a), for specimen #1-5 heat treated for 7 days. (b)

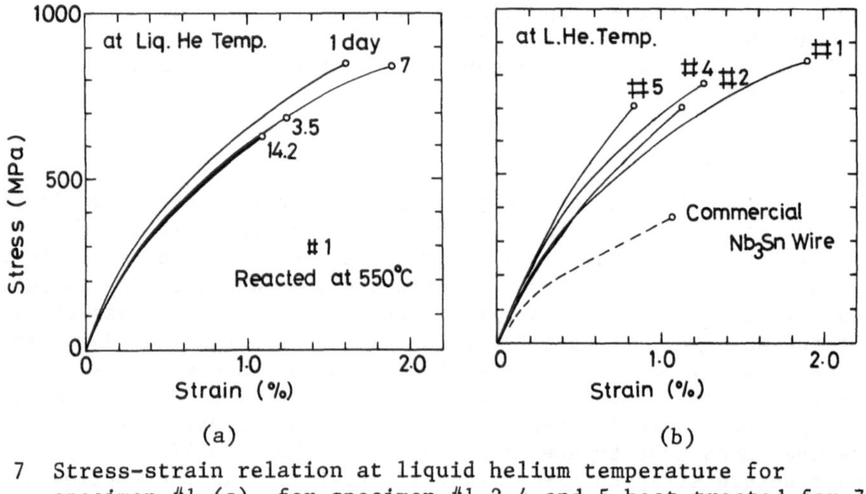

Fig. 7 Stress-strain relation at liquid helium temperature for specimen #1 (a), for specimen #1,2,4 and 5 heat treated for 7 days (b).

lower than that of bronze. These facts explain the results described above.

Upper critical flux density B_{c2} at zero strain and the maximum value B_{c2m} in B_{c2} vs. strain curve are shown in Fig. 8 for specimen #1-3, and 5. The value of B_{c2m} is obtained based on the B_{c2} at zero strain and ε_m by means of the empirical formula presented by Ekin[10]:

$$B_{c2}(\varepsilon)=B_{c2m}[1-900(\varepsilon_m-\varepsilon)^{1.7}] \quad (\text{for } \varepsilon > \varepsilon_m).$$

This formula is proposed for bronze processed and fully reacted Nb_3Sn conductors. We assume this relation applicable to the "in situ" specimens, then B_{c2m} is obtained as following formula:

$$B_{c2m}=B_{c2}(0)/(1-900\varepsilon_m^{1.7}).$$

The magnitude of B_{c2m} is in the order: #5 > #1 = #3 > #2. Neither Nb content dependence nor the effect of the presence of Cu stabilizer/Ta barrier in Fig.8 has been interpreted yet. The order is similar to that of strain sensitivity of I_c. This suggests that strain sensitivity is principally dependent on the magnitude of B_{c2m}. It is predicted that B_{c2m} increases with heat treatment time because of the decrease in the strain sensitivity with heat treatment time.

SUMMARY

The investigation has been carried out on the influence of Nb content in initial Nb-Cu alloy, heat treatment time, and the presence/absence of Cu stabilizer with Ta barrier on critical characteristics, strain effects and mechanical properties in "in situ" Nb_3Sn conductors. The results are summarized in Table 2. This table provides a guide for the fabrication condition which is to be selected depending on the objects of practical application of the conductor.

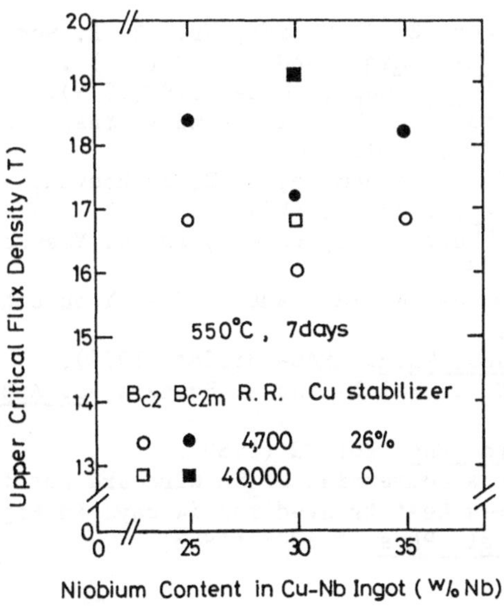

Fig. 8 Upper critical flux density at zero strain (B_{c2}) and at ε_m(B_{c2m}).

Table 2. Summary of Examinations

Properties	Heat Treatment Time ↑	Nb w% ↑	R.R. ↑	Cu/Ta ↑
I_c	↑	↑	Supposed Little Effect	Little Effect
B_{c2m}	Supposed ↑	Min. at 30 w%	Supposed Little Effect	↓
ε_m	Little Effect	↓	Supposed Little Effect	↓
Strain Sensitivity	↓	Max. at 30 w%	Supposed Little Effect	↑
Yield Strength	↓	↑	Supposed Little Effect	↓
Ductility		↓	Supposed Little Effect	↑

ACKNOWLEDGEMENT

The authors are grateful to Dr. J. Yamamoto, Mr. Y. Tsuji, Mr. Y. Wakisaka and H. Makiyama of the Low Temperature Center of Osaka University for facilitating the experiments using liquid helium. The measurement of upper critical flux density was performed at High Field Laboratory for Superconducting Materials RIISOM, of Tohoku University by use of 23 T hybrid magnet HM-2. The authors are, therefore, grateful to Professors Y. Muto, K. Noto and the staffs of RIISOM for giving us the opportunity to obtain the high field data. The authors are indebted to Mr. A. Kreeda and Mr. T. Ishibashi of EM Lab. in ISIR for the EDX analysis.

This work is in part supported by Grant in Aid for Scientific Research No. 59050044 and 60050044, Ministry of Education in Japan.

REFERENCES

1. R. Roberge, S. Foner, E. J. McNiff, Jr., B. B. Schwartz, and J. L. Fihey, IEEE Trans. Magn. MAG-13:683 (1977).
2. J. W. Ekin, IEEE Trans. Magn. MAG-14:197(1979).
3. T. Okada, M. Fukumoto, K. Yasohama, and K. Yasukochi, IEEE Trans. Magn. MAG-21:775 (1985).
4. J. E. Ostenson, D. K. Finnemore, J. D. Verhoeven, and E. D. Gibson, Appl. Phys. Lett. 37:662 (1980).
5. K. Yasohama, H. Ohkubo, T. Ogasawara, and K. Yasukochi, Adv. Cryo. Eng. 28:555 (1982).
6. M. Fukumoto, T. Okada, K. Yasohama, and K. Yasukochi, Adv. Cryo. Eng. 30:867 (1984).
7. G. Rupp, IEEE Trans. Magn. MAG-13:1565 (1977).
8. J. F. Bussiere, D. O. Welch, and M. Suenaga, J. Appl. Phys. 51:1024 (1980).
9. G. Rupp, Adv. Cryo. Eng. 26:522 (1980).
   The curves for a commercial Nb_3Sn wire are obtained from the data for the specimen heat treated for 64 days in Fig. 6 of this reference.
10. J. W. Ekin, J. Appl. Phys. 54:303 (1983).

FATIGUE DAMAGE IN Nb_3Sn CONDUCTORS

K. Katagiri, M. Fukumoto, and T. Okada
ISIR Osaka University, Ibaraki, Osaka, Japan

R. Onodera and R. Aoki
Kyushu University, Fukuoka, Japan

ABSTRACT

Changes in critical current, I_c, caused by cyclic straining have been investigated in a multifilamentary Nb_3Sn composite and two types of "in situ" Nb_3Sn wires, i.e., one fabricated in Jelly Roll method and the other cast specimen reacted using the external tin diffusion technique. No degradation in I_c was found in the wires cyclically strained at a strain level up to ε_{irrev}, the monotonic strain above which irreversible drop in I_c occurs. The I_c decreased cumulatively with the cycle of straining when the strain level is higher than ε_{irrev}. In the external diffusion wire, ε_{irrev} was very close to fracture strain.

In the Jelly Roll wire, plastic strain range (width of hysteresis loop) decreased with the number of strain cycles until 10^2 cycles, then it saturated to certain value depending on the stress amplitude. Electrical resistivity, however, increased continuously beyond 10^5 cycles.

The characteristics change caused by fatigue is discussed in connection with damages in Nb_3Sn filament and the cyclic deformation behavior of constituents in the composite.

INTRODUCTION

The strain tolerance of A-15 type superconducting materials, such as Nb_3Sn and V_3Ga have been a major concern for developers of large scale high field magnets. Consequently, many studies of the strain effects on critical current have been made for a wide variety of composite Nb_3Sn conductors.[1-6] In practical operation of the magnets, however, conductors are subjected not only to monotonic (static) strain but also cyclic strain. Cumulative damage can be essential in some situations. Ekin[2] has presented measurements on the effects of cyclic stress in NbTi and bronze processed Nb_3Sn multifilamentary composite. He pointed out that the degradation in I_c can occur in the latter if the peak value of cyclic strain exceeds certain level which is determined by a test for static strain effect.

The "in situ" processed conductors have been proved to provide higher mechanical properties and excellent strain characteristics of I_c as compared with conventional multifilamentary composites.[4,6] In this paper, fatigue damage which result in degradation of I_c is examined in two types of "in situ" processed wires and a multifilamentary composite. Changes in mechanical behavior and electrical resistivity are also investigated.

EXPERIMENT

Specimen

Specimens employed in this study were two types of "in situ" wires; A, fabricated in Jelly Roll method and B, cast specimen reacted using the external tin diffusion technique. For comparison, bronze processed multifilamentary composite, C, was also tested. Principal specifications are shown in Table 1. More detailed description on wire A and B has been given in Refs.7 and 6, respectively.

Apparatus and Fatigue Test

The I_c was measured at 4.2K as a function of tensile strain level at various stages of strain cycles under a transverse magnetic flux density of 6 T. Total length of a specimen is 24.5 cm. Each end of the specimen is soldered to a copper block with 4 cm length which serves as a current terminal as well as a grip for transferring tensile load to the specimen. Voltage tap spacing is 6 cm and I_c is determined within 1% error using criterions of 0.5 μV/cm for wire A, B and 1.0 μV/cm for C. The strain is measured using a 2-strain gage extensometer with accuracy of 0.01%. Details of the apparatus has been given in Ref. 6. Specimens are subjected to repeated strain in constant peak strain mode at a strain rate of approximately 5×10^{-2}/sec.

Long term fatigue test in constant load mode is also conducted on wire A using a Baldwin type fatigue testing machine at a frequency of 16 Hz in liquid He.[8] The gage length of the specimen is 44 mm and load ratio being 0.167. During the fatigue test plastic strain range is measured using hysteresis loop, and electrical resistivity just above the critical temperature of Nb_3Sn through four probe method interrupting fatigue run and raising temperature above the critical temperature.

Table 1. Nb_3Sn Wire Specifications

Wire	Size (mm dia.)	Number of fils.	Fil. size (μm)	Heat treat.	Composition (% volume fraction)
A: "In situ" Jelly Roll	1.0		0.1x1	793K 2 day + 823K 8 day	31 Cu-25 w%Nb 23 Bronze (Cu-10w%Sn) 16 Cu stabilizer; center 30 Cu-30 w%Ni sleeve
B: "In situ" External diffusion	0.31		0.1x1	823K 7 day	62 Bronze + Nb_3Sn (Cu-25w%Nb-14w%Sn) 27 Cu stabilizer; center 11 Ta barrier
C: Bronze process multi-filament	0.31	745	5 dia.	973K 5 day	62 Bronze + Nb + Nb_3Sn (bronze/Nb = 2, bronze: Cu-13w%Sn) 30 Cu stabilizer; periphery 8 Nb barrier

Microscopic damage caused by straining in the Nb_3Sn layer was observed using SEM. The CuNi and bronze layers in the wire were dissolved using 30% HNO_3. The fracture surface was also examined to analyse fatigue fracture mechanisms.

RESULTS AND DISCUSSION

Monotonic Strain Effects on Critical Current

The monotonic strain dependence of I_c at 6 T is presented in Fig. 1 (a), 2 (a) and 3 (a), for wire A, B, and C, respectively. The I_c in the "in situ" wire A increases with strain up to the maximum value (ε_m^c) at the strain of 0.50%, then decreases with further strain. In Fig. 1 (a), I_c values obtained after unloading are presented as well (8', for example). It is noticeable that the alteration in I_c induced by strain is reversible up to a strain as high as 1.6%, or more. The strain beyond which the change in I_c is irreversible (ε_{irrev}) is at least larger than 1.6%. In this wire, fracture strain, ε_f, is >11 % and is very high as compared with the other wires. This seems to be a consequence of higher bronze ratio and the presence of highly ductile CuNi sleeve. The stress-strain relation shows that strain increases with fine serration and no eventual hardening in the high strain region. Extensive multiple filament fractures in the wire are presumed in that region. The I_c vs. strain characteristics in wire B is much the less similar to wire A; high ε_m and ε_{irrev}, the strain beyond which I_c in unloaded condition does not fall on the curve for that in loaded condition. The ε_{irrev} in this wire is almost equal to ε_f. This means fracture in this wire is controlled by fracture of Nb_3Sn filaments.

In the case of wire C, however, ε_m is small and ε_{irrev} exists far below ε_f. Further, ε_f itself is smaller than those in wire A and B. All characteristic strain values obtained are tabulated in Table 2.

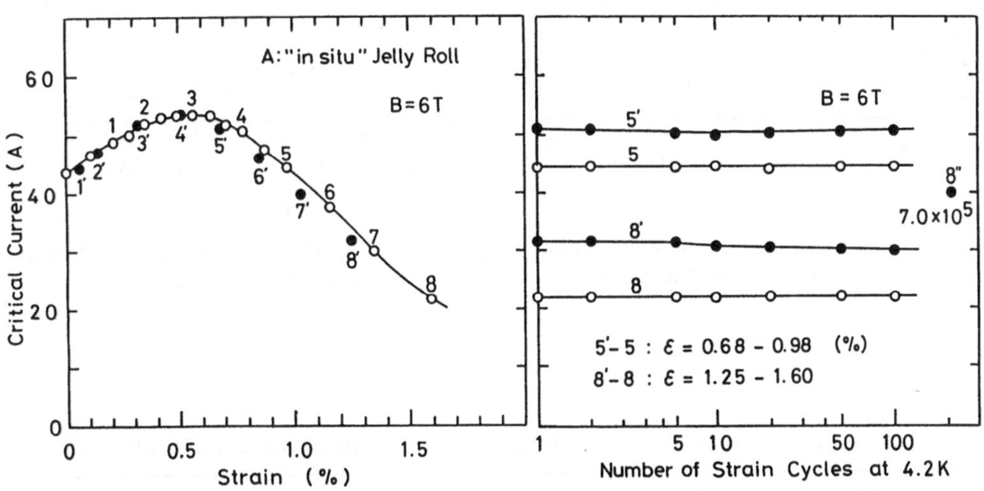

(a) (b)

Fig. 1 Critical current as a function of strain (a) and number of strain cycles (b) in the Jelly Roll "in situ" Nb_3Sn wire. [Open circles indicate I_c at loaded condition and solid one at unloaded condition corresponding to the same number; for 8" see the text. The same symbol is used hereafter.]

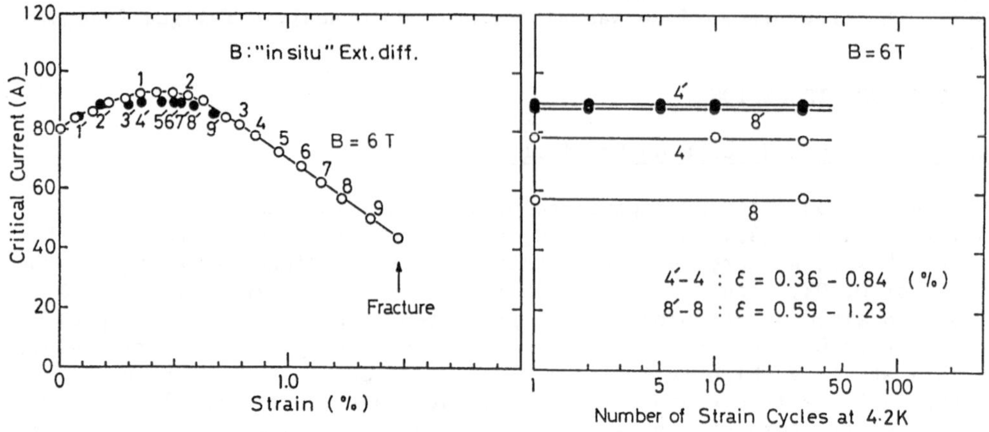

(a) (b)

Fig. 2 Critical current as a function of strain (a) and number of strain cycles (b) in the externally tin diffused "in situ" Nb_3Sn wire.

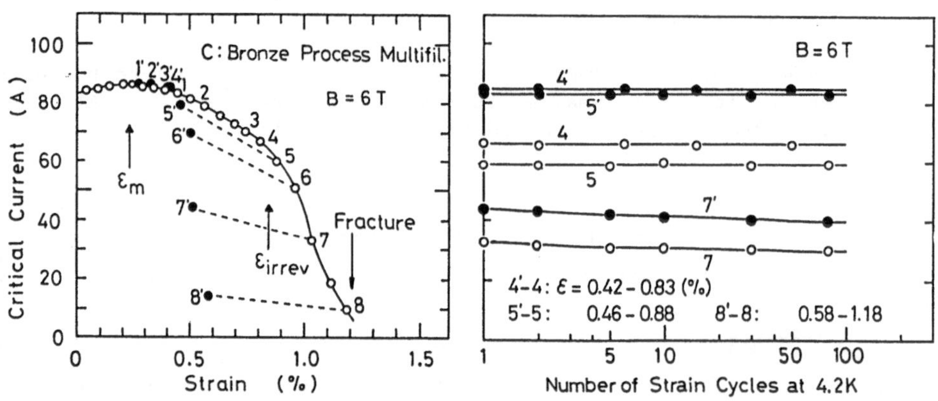

(a) (b)

Fig. 3 Critical current as a function of strain (a) and number of strain cycles (b) in the multifilamentary Nb_3Sn composite.

Table 2. Characteristic strain (%) in monotonic strain effect

Wire	ε_m	ε_{irrev}	$\varepsilon_{0irrev} = \varepsilon_{irrev} - \varepsilon_m$	ε_f
A:"In situ" Jelly Roll	0.50	>1.60	>1.10	>11
B:"In situ" External diff.	0.45	1.50	1.05	1.50
C: Bronze process multifil.	0.20	0.85	0.65	1.20

Cyclic Strain Effects on Critical Current

The I_c values during strain cycling at constant peak strain in wire A are shown in Fig. 1 (b), for loaded and unloaded condition, respectively. Strain levels at loading and unloading condition correspond to the numbers indicated in Fig 1 (a). No change in the I_c was found within the scatter of data. Slight decrease in I_c for unloaded condition can be attributed to increase in the strain caused by cyclic strain hardening under constant peak strain. In Fig. 1 (b), the I_c value at unloaded condition in the specimen subsequently fatigued at a constant stress by 7.0×10^5 cycles is included (8"). The peak stress (380 MPa) in this cycling is set so as to the peak strain is equal to that in the previous 100 strain cycling using saturated cyclic stress-strain relation. Marked recovery in I_c has occurred. It can be a consequence of stress redistrbution in the wire as is reported by Cogan et al.[5] A control test without the long term stress cycling, however, revealed that major part (about 50 %) of the recovery, should be ascribed to temperature rise in the specimen (up to approximately 500K) experienced during sample re-setting for I_c measurement. No degradation in I_c with the number of cycles at a strain close to ε_f is seen in the case of wire B (Fig. 2 (b)). Consequently, it can be safely concluded that there is no cyclic strain effect on I_c in "in situ" processed Nb_3Sn wire, if the peak strain is less than ε_{irrev} in monotonic straining.

In the case of wire C (Fig. 3 (b)), strain cycling effect can not be seen within the range of strain $\varepsilon<\varepsilon_{irrev}$. Degradation of I_c occurs if the cyclic strain exceeds the limit. This is consistent with similar results previously reported by Ekin,[2] although one aspect of degradation in the high strain region is somewhat different: the decrease in I_c saturates within 10 cycles in his wire. This may be ascribed to the difference in conductor composition.

In order to examine microstructural aspect of ε_{irrev} in "in situ" wire, Nb_3Sn layer in wire C was examined using SEM. Fig. 4 (a) shows surface appearance in Nb_3Sn layer observed in the specimen shown in Fig. 1 (b), i.e., fatigued to 100 cycles at 1.6% peak strain, which is still within the range of reversible strain. No crack is observed with the exception of few microcracks of the order of less than 2μm in length. This is in

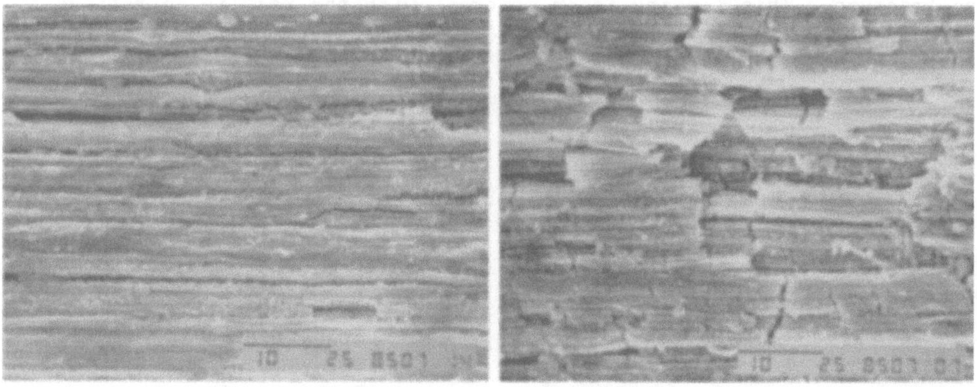

(a) (b)

Fig. 4 Surface appearance in Nb_3Sn layer in Jelly Roll wire, (a) fatigued to 100 cycles within reversible strain range [the same sample shown in Fig. 3], and (b) fatigued to failure [σ=470 MPa, $N_f=5.8\times10^5$ cycles].

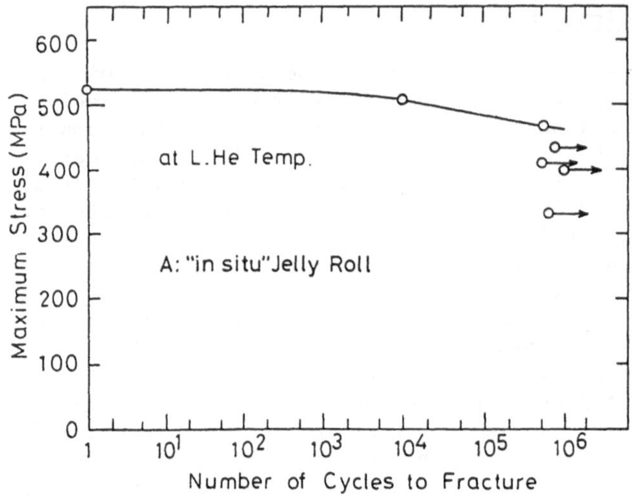

Fig.5 Peak stress vs. number of cycles to fracture in Jelly Roll wire. [Arrows indicate specimens interrupted without fracture.]

contrast with the appearance of the specimen fatigue fractured at a peak stress of σ=470 MPa, after N_f=5.8x10^5 cycles to be mentioned in the next section. Cracks of various sizes cover the entire surface of the sample (see Fig. 4(b)). We previously presented that there was no cracking in another "in situ" wire B even at fracture in tensile test.[6] These results are consistent with each other in the sense that eventual crack formation is not found so long as strain is less than ε_{irrev} in both monotonic and cyclic situations.

S-N Curve of "in situ" Jelly Roll Nb_3Sn Wire

Wire A was chosen as a vehicle to get information on long-term fatigue characteristics of "in situ" wires to be described in the following sections. Results of constant peak load fatigue tests at 4.2K are shown in Fig. 5. It seems noteworthy that the fatigue endurance limit criterion of 10^6 cycles is close to the ultimate tensile strength.

Fractographic examination revealed the absence of Stage 1 (early stage) facets in the periphery of the wire, indicating a crack initiation site within the interior of the sample (Fig. 6). This is consistent with the result mentioned above; the Nb_3Sn layer appears to have been seriously damaged before fracture. This finding is in contrast with the results on fatigue at cryogenic temperatures in the multifilamentary NbTi composite, in which cracks initiated at the CuNi sleeve surface were associated with inclusions.[8]

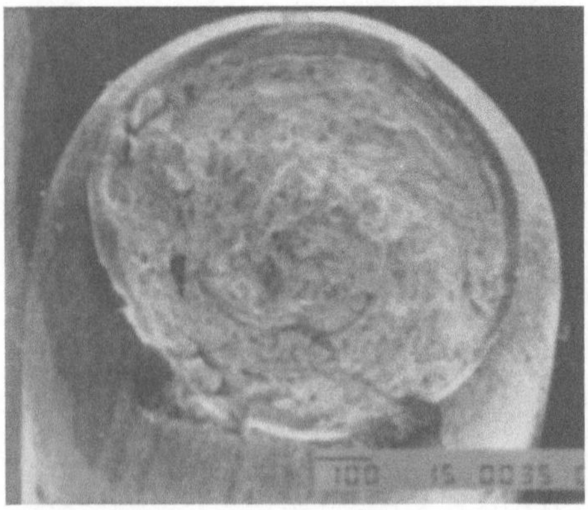

Fig. 6 Fractograph showing initiation of fatigue crack within the interior of the Jelly Roll wire [Two depressed regions in the periphery appear to be artefacts induced at unloading after fracture].

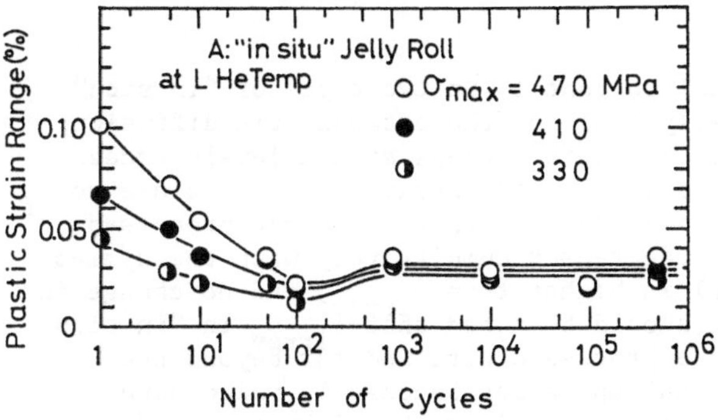

Fig. 7 Cyclic stress hardening behavior of Jelly Roll wire.

Cyclic Stress Hardening

The plastic strain range defined as the width of the hysteresis loop at the mean stress level decreased monotonically during the early stage of fatigue test until approximately 10^2 cycles and remained constant after that (Fig. 7). We can expect cyclic hardening in ductile constituents of the wire, and consequently, stress redistribution in the wire during the early stage. The I_c in the loaded condition during constant peak load cycling is inferred to increase on account of the decrease in peak strain caused by hardening coupled with reverse yielding. This is in contrast with the behavior of I_c in the loaded condition during constant peak strain cycling, as shown in Figs. 1-3; there was no change in I_c with the number of cycles.

In the saturated strain stage, an almost reversible motion of mobile dislocations containing jogs continuously creates point defects in the copper stabilizer, resulting in an increase in the resistivity, to be mentioned in the next section.

Changes in Electrical Resistivity

The residual resistivity in the copper stabilizer just above the critical temperature in wire A changed during the fatigue run, as shown in Fig. 8. Because the resistivity in CuNi, bronze, and Nb_3Sn is higher than that in copper by a factor of more than 100, their contribution to the wire resistivity change is neglected in the first approximation. The resistivity increased monotonically with the number of cycles. It still increased beyond 10^5 cycles, although the rate of increase tended to decrease. The resistivity was almost doubled by fatigue of 10^6 cycles at 330 MPa, which caused a decrease in I_c by about 19% in the loaded condition. Similar resistivity changes in long-term fatigue tests of NbTi wire have been reported.[2,9]

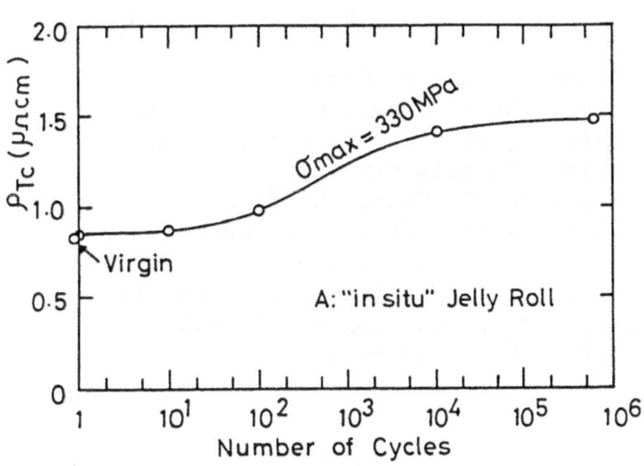

Fig. 8 Degradation of the residual resistance ρ at transition temperature as a function of fatigue cycles in Jelly Roll wire.

CONCLUSIONS

It has been shown that critical current I_c in two types of "in situ" Nb_3Sn conductors fabricated by Jelly Roll and the external tin diffusion method does not decrease by cyclic straining unless strain levels exceed certain value. In the case of the latter, the strain level is close to ε_f and also ε_{irrev}. These results coincide with that in bronze processed multifilamentary wire, in which I_c decreases cumulatively with the cycles of straining when the strain level is higher than ε_{irrev}, and no change in I_c within low strain region. It should be noted that ε_{irrev} in "in situ" wires is higher than that in bronze processed wire and far beyond the strain to be encountered in practical operation because I_c value have already degraded there. Microscopic damage in Nb_3Sn filaments in "in situ" wire appears closely associated with ε_{irrev}.

Fatigue endurance limit in Jelly Roll "in situ" wire is close to ultimate tensile strength. Plastic strain range decreases with the number of strain cycles whereas electrical resistivity increases even beyond 10^5 cycles. Perhaps the most important consequence of fatigue damage in long term operation of "in situ" wire is to be the degradation of the resistivity in copper and the accompanying loss of stability as is the case in NbTi composites.

ACKNOWLEDGEMENTS

The authors are grateful to Dr. J. Yamamoto and the staffs at Low Temp. Center in Osaka University for facilitating measurement at liquid He temperature. They also would like to thank Dr. K. Yasohama for providing "in situ" sample B and Prof. T. Anayama Japanese reference multifilamentary Nb_3Sn sample C. Experimental assistance by Mr. K. Saito and Mr. K. Koyanagi is acknowledged. This work is in part supported by Grant in Aid for Scientific Research No. 60050044 and 60050004, Ministry of Education, Japan.

REFERENCES

1. G. Rupp, IEEE Trans. Mag. MAG-13, 1565 (1977).
2. J. W. Ekin, in: "Acvances in Cryogenic Engineering," vol. 24, Plenum Press, New York (1978), pp. 306-316.
3. T. Luhman, M. Suenaga, D. O. Welch and K. Kaiho, IEEE Trans. Magn. MAG-15: 699-702 (1979).
4. J. W. Ekin, in: "Advances in Cryogenic Engineering - Materials," vol. 30, Plenum Press, New York (1984), pp.823-836.
5. S. F. Cogan, J. D. Klein, S. Kwon, H. Landis and R. M. Rose, IEEE Trans. Magn. MAG-19: 917-919 (1983).
6. M. Fukumoto, T. Okada, K. Yasohama and K. Yasukochi, in: "Advances in Cryogenic Engineering - Materials," vol. 30, Plenum Press, New York (1984), pp.867-874.
7. R. Aoki, N. Mori, H. Iiboshi, Y. Matsubara, R. Onodera, and F. Sumiyoshi, in: "Proc. Int. Cryog. Mats. Conf., Kobe," Butterworths, Surry (1982), pp.191-194. [The present sample is not exactly the same as one described in this paper.]
8. K. Katagiri, S. Nishijima, T. Nishiura, T. Okada, H. Takei, M. Yokota, Y. Murakami and Y. Inuishi, in: "Proc. Int. Cryog. Mats. Conf., Kobe," Butterworths, Surry (1982), pp.191-194.
9. K. Katagiri, K. Koyanagi, M. Fukumoto, T. Nishiura, T. Okada and M. Nagata, in: "Advances in Cryogenic Engineering - Materials," vol. 30, Plenum Press, New York (1984), 461-468.

THE EFFECT OF ASPECT RATIO ON CRITICAL CURRENT IN MULTIFILAMENTARY SUPERCONDUCTORS*

L. F. Goodrich, W. P. Dube, E. S. Pittman, and A. F. Clark

National Bureau of Standards
Boulder, Colorado

ABSTRACT

Experimental data and discussion are presented on the critical current of straight superconductors as a function of the orientation of a perpendicular applied magnetic field. Commercial, multifilamentary NbTi and Nb_3Sn samples were measured in a radial access magnet that allowed an arbitrary angle setting. The change in critical current was measured at different magnetic fields to scale the effect for use in a standard test method. For a NbTi sample, the critical current with the magnetic field parallel to the wider face of the conductor is higher than that with the perpendicular orientation. The effect can be as high as 40% for a NbTi sample with an aspect ratio of six. The effect in Nb_3Sn is opposite that in NbTi. A discussion of the most likely cause of the effect, which accounts for the difference between NbTi and Nb_3Sn, is given.

INTRODUCTION

The aspect ratio effect is defined as the variation of critical current, I_c, with sample-field orientation in a constant applied magnetic field perpendicular to the wire axis. The aspect ratio of a conductor or filament is defined as the ratio of the cross sectional dimensions (wide/narrow). The emphasis of this study was measurement of how the critical current varied with the angle of constant applied magnetic field. The most likely cause of the aspect ratio effect is the asymmetric deformation of the filaments and, thus, the orientation of the flux-pinning sites.

The aspect ratio effect on critical current is important for most applications where a monolithic rectangular conductor is used. It may also be of importance for understanding the critical current of cable conductors that have been deformed during compaction. The difference in I_c can be 20% for a NbTi conductor with an aspect ratio of two (40% for a ratio of six). The present American Society for Testing and Materials (ASTM) critical current standard test method[1] suggests a measurement orientation with the magnetic field parallel to the wider face of a rectangular sample unless otherwise specified, but this orientation is difficult to obtain in the

* Work supported in part by the Department of Energy through the Office of Fusion Energy and the Division of High Energy Physics.
Contribution of NBS, not subject to copyright.

hairpin geometry. Also, in some applications, such as the end turns of a dipole magnet, might require the measurement in the other (or both) orientations.

EXPERIMENT

The critical current measurements reported here were made using a straight sample geometry and a radial access magnet. Samples are identified in Table 1. The critical current was defined as the current at which the electric field strength was 0.2 μV/cm. The sample was centered in a holder, and four conductors (symmetrically located on a 2-cm diameter circle) were used for the return current path. This arrangement reduced the effect of the magnetic field from the return current path and reduced the net torque on the sample cryostat. The sample was held in, and electrically isolated from, a machined brass rod holder with epoxy. The rest of the structure was fiberglass epoxy.

Two experimental problems were identified and their effects on precision and accuracy reduced. The first was voltage noise, as large as several μV, introduced by slight motion of the sample during acquisition of the voltage-current (V-I) curve. The net force on the sample cryostat is not zero if the current return path does not traverse exactly the same magnetic field profile as the sample path. Additionally, stray Lorentz forces and mechanical vibrations can cause relative motion of the magnet and sample. The noise problem was reduced to the 0.1 μV level with the addition of a band brake device that was mounted on the top of the magnet. This brake could be tightened around the sample cryostat with a room temperature screw actuator, thus restricting the relative motion of the magnet and sample.

The second problem was persistent currents in the four parallel return wires, because the wires were soldered together at each end and the magnetic flux was changed each time the angle was changed.[2] These persistent currents would cause a time dependent, partial shielding of the sample from the applied magnetic field. The decay time constant was reduced from 25 s to 5 s by adding a small resistance (brass shims, approximately 50 nΩ) on the end of each lead. These resistors also helped to ensure that the current was shared more evenly among the return wires. This reduced the net torque and effects of the magnetic field of the return path to a negligible level.

The orientation of the sample was changed by rotating the sample cryostat relative to the background magnet. A pin in a series of circumferential holes at the top of the cryostat and the band brake mentioned above were used to hold the orientation. The zero angle was defined as the orientation with the magnetic field parallel to the wider side of the conductor. The zero position in the raw data was only approximate because of variation in how the sample was mounted in the holder.

Two experiments were conducted to test the apparatus and the effect. The first experiment was to measure I_c of a large (I_c 1200 A at 8 T) round wire as a function of angle from 0° to 360°. The measured critical currents were independent of angle within the experimental precision of ±1%. This result indicated that the apparatus had no artificial angular variation due to the applied magnetic field adding or subtracting from the magnetic field of the return leads.

The second experiment was a comparison of measurements made on a monolithic conductor and on a conductor composed of two of these monoliths soldered together along their length. This comparison was not ideal, but it allowed a direct comparison of aspect ratio effect and critical current. There were no indications of current-sharing problems in the I_c measurements.

Table 1. Physical Parameters of the Samples Measured. Critical Current at 8 T for NbTi and 10 T for Nb_3Sn.

Sample number	Type	Conductor aspect	Filament aspect	I_c, A	ΔI_c, %	Cu/non Cu ratio
1	NbTi	round	∿ 1.0	1215	< 1	2
2	NbTi	1.28	∿ 1.11	114	5	1.8
3	NbTi	1.93	1.28 to 1.41	350	19 to 29	1.25
4	NbTi	2.0	1.15 to 1.17	700	19.5	2
5	NbTi	2.0	∿ 1.24	600	17.5	5
6	NbTi	6.0	1.61 to 1.93	2350	42	5
7	Nb_3Sn	2.0	∿ 1.92	42	15.5	1.63
8	Nb_3Sn	2.0	∿ 3.15	1020	19	?

Within the experimental precision, I_c of the two-monolith conductor was twice that of the single monolith and the aspect ratio effect was the same percentage of I_c. These results suggested that the demagnetization factor of the overall conductor did not have a significant effect.

DATA

The general dependence of I_c on angle was similar for all samples measured. It was more broad at 0° than at 90° and fell between $\sin^2\theta$ and $\sin^4\theta$. Typical data are shown on Fig. 1 with $\sin^2\theta$ and $\sin^4\theta$ for reference. The filaments were not perfectly aligned with the surface of the conductor; some were tilted slightly on either side of alignment. This caused the curve to be flat around 0° and is the most likely cause of the small peaks observed in some of the samples (see Fig. 2) on either side of 0°.

The shape of I_c with angle scales approximately as a percentage of I_c for the magnetic field range of 2 to 10 T (see Figs. 2, 3, and 4). Obviously, the effect at zero magnetic field must be zero and should approach zero continuously. In general, the effect at 2 T was a little lower percentage than at higher magnetic fields. The lack of a strong field dependence suggests that the fluxoid-core size and fluxoid spacing does not have a large effect on the resulting pinning force.

The aspect ratio effect in Nb_3Sn (see Fig. 4) is opposite that of NbTi, but it is comparable in magnitude for similar conductor aspect ratios. Two types of multifilamentary, bronze process Nb_3Sn conductors were measured, one made using the conventional process and the other with the jelly-roll process. Results were consistent between these two conductors. A Nb_3Sn tape measured previously[3] had an aspect ratio effect in the same direction as these multifilamentary Nb_3Sn conductors.

The amount of asymmetric filament deformation, filament aspect ratio, was determined from photographs of the wire cross sections. A large number of filaments (in some cases all of the filaments) were averaged by summing the ratio of width to thickness and then dividing by the number of filaments to be averaged. These ratios are listed in Table 1.

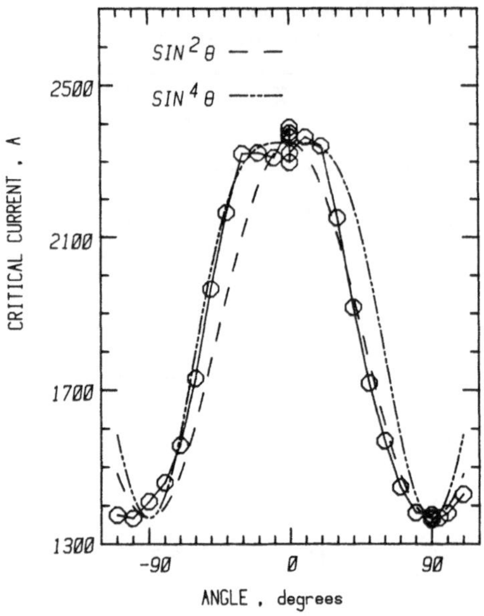

Fig. 1. I_c versus angle for a NbTi conductor with an aspect ratio of 6, sample #6.

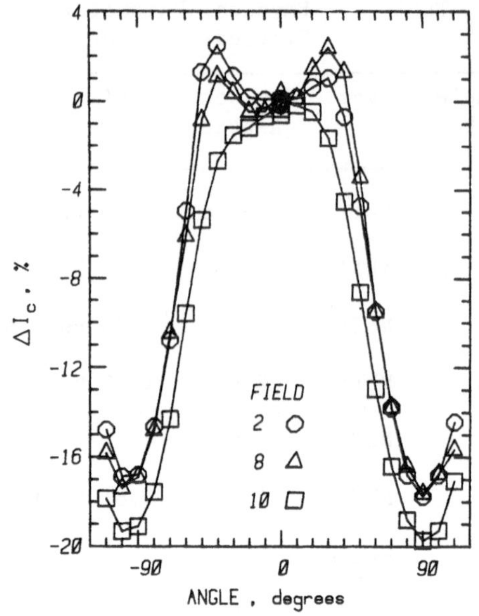

Fig. 2. Percentage change in I_c versus angle for NbTi sample #5.

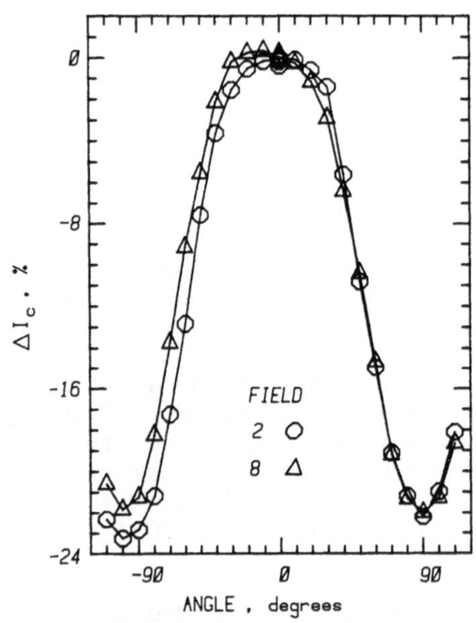

Fig. 3. Percentage change in I_c versus angle for NbTi sample #3.

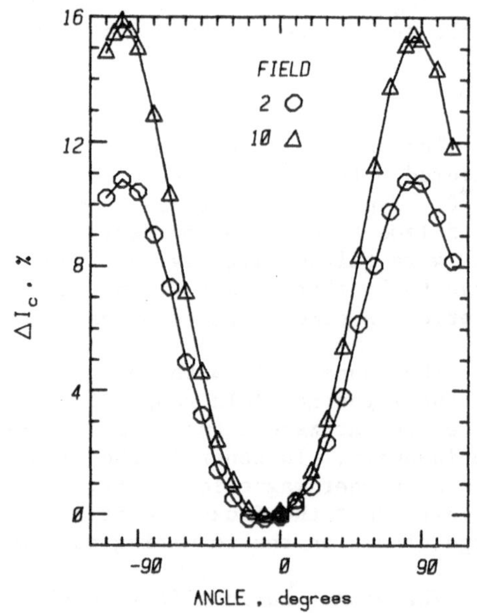

Fig. 4. Percentage change in I_c versus angle for Nb_3Sn sample #7.

Some filament aspect ratios of some wires varied by more than a factor of 2 for cross sections separated by less than a few centimeters along the length of the wire. Two such cross sections of the same wire are shown in Fig. 5. Further observation indicated that this variation might be occurring periodically along the conductor. As a result of this, another sample was prepared with a number of closely spaced voltage taps to look at short range variation in I_c and aspect ratio effect. Figure 6 is a plot of I_c versus position along the wire using adjacent voltage taps separated by about 0.23 cm. Because of the close tap spacing (low voltage with E_c = 0.2 μV/cm) and I_c variation (quench current determined by the lowest I_c) a number of repeat determinations were made.

The variation in I_c was about 12% for the 0° data and 10% for the 90° data. The variation also looked periodic for both 0° and 90°. There was a high point for both the 0° and 90° data, starting from the left, approximately every third point. There was also a variation in the aspect ratio effect along the wire. Consider the third position: the 0° point was relatively low and the 90° point was relatively high, this resulted in a small aspect ratio effect. The situation was reversed for the fifth position, which results in a large aspect ratio effect and alternates along the wire. The aspect ratio effect varied from 19 to 29%.

The cause of this periodic variation in filament shape, I_c, and aspect ratio was most likely the combination of filament twist and hexagonal filament array.[4] Filament deformation was extreme when the flats of the hexagonal filament array were parallel to the minor dimension, where the filaments were on top of each other. A transverse view of the filaments etched from the matrix indicated that the wavelength of the extreme filament deformation was about 0.28 cm, which was about 1/6 of the twist length. The length of the flattened filament was about 0.1 cm. The fact that the tap separation was about the same as the deformation wavelength, apparently resulted in a beating of the two, forming a third artificial wavelength in the variation of I_c. The measured I_c values were consistent with a 0.28 cm wavelength but the actual amplitude of the variation may be larger. Measurements with a smaller tap separation, on the order of 0.05 cm, would have been very difficult because of the extremely low voltages.

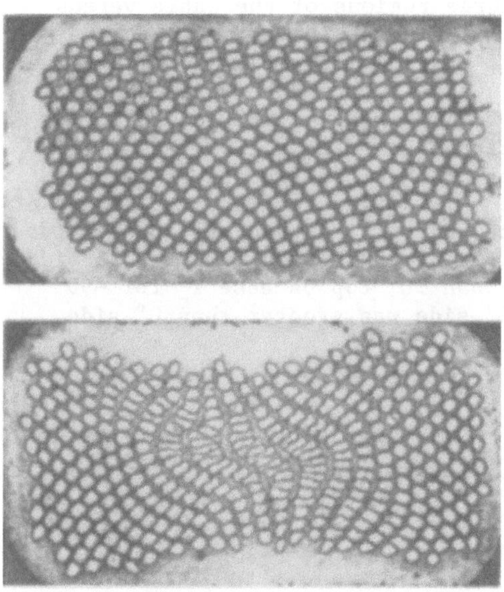

Fig. 5. Photographs of two different cross sections of the same NbTi wire, sample #3.

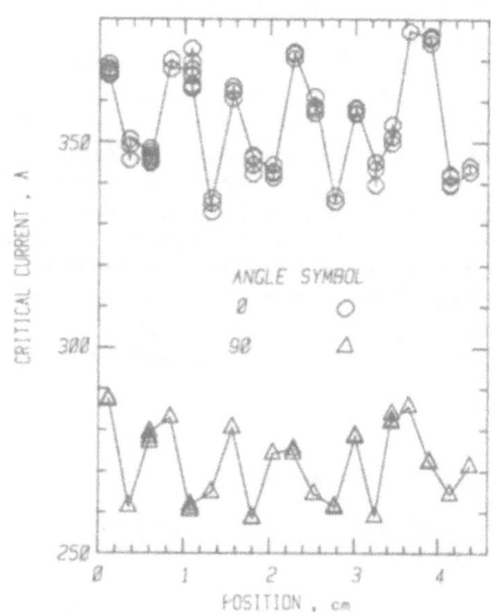

Fig. 6. Variation of I_c versus position for 0° and 90° at 8 T and Ec =0.2 μV/cm for sample #3.

The cross sections of other rectangular conductors were examined more closely considering this periodic filament deformation. All of the others measured showed some change in deformation as indicated by a change in the filament array. Two cross sections of the same wire are shown in Fig. 7. Notice the change in the shape of the center copper island and difference in filament deformation. Voltage tap and microscope studies did not reveal another case as extreme as Fig. 5, but there was always some variation. The physical parameter that may have been unique to the sample shown in Fig. 5 was the low copper to superconductor ratio 1.25/1.

Some measurements were made on a seven-strand NbTi cable, which had been compacted into a rectangular cross section. A photograph of the cable cross section is given in Fig. 8. A strand changes from almost round in the middle top position to deformed at the corner and side positions of the cable. Despite current sharing among the strands, V-I curves on short sections of individual strands gave some indication of the local I_c.

A number of voltage taps were placed along individual strands to measure I_c for different cable positions. Three regions of the cable were measured: an "overall" region encompassing two cable twist lengths of a strand, a "top" region encompassing only the portion of a strand on top of the cable, and a "side" region encompassing the corners and side portions. The shorthand notation for these three regions from here on will be: overall, top, and side. Voltage taps for the top and side measurements were placed on the top and bottom (wider faces) near the corner of the cable on a given strand. One of these voltage taps was used for both the top and side measurements. This placement caused most of the corner region to be included in the side measurement. Thus, all strand regions were included in either the top or the side measurements.

The cable I_c data had a combination of aspect ratio and self field effects. The self field of the conductor vectorially added to the applied magnetic field, which resulted in different magnetic fields for different angles and cable positions. The I_c of three regions of the cable versus angle are given in Fig. 9. The overall I_c, circles, had some structure approximately every 45° but the variation was less than 2%. Combining the aspect and self field effects, the top I_c as a function of angle should have been ordered in magnitude as follows; 180° > 0° > -90° and 90°. The side I_c should have been ordered in magnitude as follows; -90° > 90° > 180° and 0°. In fact, these experimental data and data on two other strands were consistent with this ordering. The only difference was a splitting of the lower critical currents. This was caused by a mixture of top and side characteristics due to the finite width of the transition from top to side. Notice that I_c for -90° top was greater than 90° top because I_c for -90° side was at a maximum. Also, I_c for 180° side was greater than 0° side because I_c for 180° top was at a maximum.

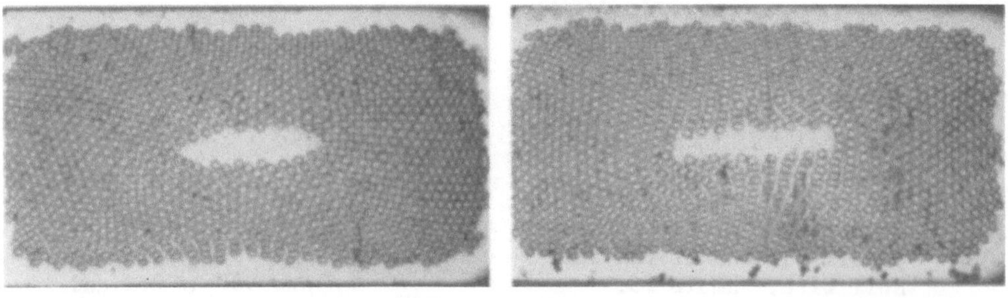

Fig. 7. Photographs of two different cross sections of sample #5.

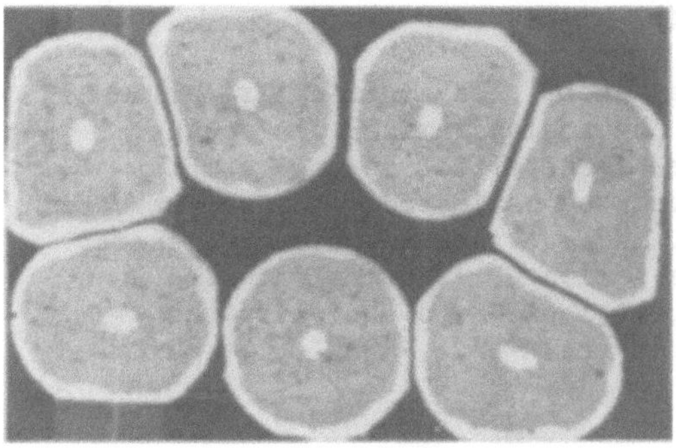

Fig. 8. Photograph of the cross section of a NbTi cable.

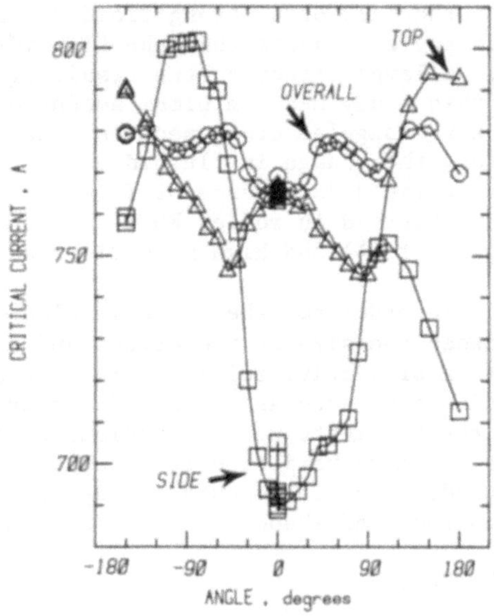

Fig. 9. I_c versus angle for different regions of a NbTi cable, see Fig. 8.

The size of the aspect ratio and self field effects can be estimated with the assumption that the self field effect can be removed by averaging the critical currents taken 180° apart. Using this assumption at 8 T, the top has an aspect ratio effect of 8% and a self field effect of 3%. The side at 8 T has an aspect ratio effect of 14% and a self field effect of 6%. At 4 T, the aspect ratio and self field effects respectively are: top 6% and 5%; side 19% and 10%.

DISCUSSION

The most likely cause of the aspect ratio effect is the asymmetric deformation of the filaments and thus the orientation of the flux-pinning sites. If the magnetic fluxoid is more closely aligned with the pinning site, the pinning force will be larger than in the nonaligned orientation.[5] Extending to the whole conductor cross section, the total pinning force will be anisotropic. In the case of NbTi, the pinning sites (precipitates, dislocations, and impurities) have a preferred orientation parallel to the wider side of the filament (the wider side of the overall conductor) thus the higher critical current with the magnetic field in that orientation. For Nb_3Sn, the pinning sites (grain boundaries) have a preferred orientation perpendicular to the wider side of the filament, due to columnar grain growth; therefore, Nb_3Sn has the opposite orientation effect. Even though there is a mixture of equiaxial and columnar grain growth, the presence of some columnar grains apparently causes the total pinning force to be asymmetric.

The effect in Nb_3Sn is opposite that of NbTi, but it is comparable in magnitude for similar conductor aspect ratios. Due to the relative hardness, the Nb filaments in the bronze matrix (Nb_3Sn) is deformed more than the NbTi filaments in the copper matrix. However, as mentioned above, the mechanism is different for the two materials and the resulting effect on I_c is similar.

The lack of a strong field dependence of the percentage change in I_c with angle suggests that the fluxoid-core size and fluxoid spacing do not have a large effect on the resulting pinning force[6]. The samples measured in this study had a minimum amount of area reduction in going from a round to a rectangular cross section. However, for thin, highly deformed pinning sites, the change in fluxoid-core size with magnetic field could cause a strong field dependence of I_c with angle for NbTi. A sign reversal has been observed in rolled NbTi conductors[7] between low magnetic fields (I_c higher at 0°) and high magnetic fields (I_c higher at 90°).

In order to make these results more general, an empirical relationship between the size of the effect and some physical parameter was sought. With a predictive relationship, the size of the effect in more complex conductors, such as compacted cables, could be estimated. The aspect ratio of the filaments was the obvious choice. However, as shown above, there was too much variation for even a simple estimate. There might also be a dependence on the manufacturing technique. These data exemplify the magnitude and shape of the aspect ratio effect.

CONCLUSIONS

The aspect ratio has a significant effect on the critical current of rectangular monolithic and compacted cable NbTi and Nb_3Sn conductors. A 20% change in critical current with angle in a constant applied magnetic field is typical. The most likely cause of the aspect ratio effect is the asymmetric deformation of the filaments and thus the orientation of the flux-pinning sites. The effect in Nb_3Sn is opposite that of NbTi due to the different kinds of pinning sites in the two materials. Aspect ratio and self field effects can be observed in I_c measurements on a cable, which would account for some of the degradation of I_c from strand to cable. Large variations in filament deformation, critical current, and aspect ratio can be observed in rectangular conductors along the length of a wire.

ACKNOWLEDGMENTS

The authors extend their thanks to M. S. Allen for drafting and cryostat construction; F. R. Fickett for being an always available sounding board; J. W. Ekin for discussions of flux pinning; R. G. Benson for making sample holders; R. A. Shenko for preparing this paper; and to the rest of the Superconductors and Magnetic Materials Group.

REFERENCES

1. Standard Test Method for D-C Critical Current of Composite Superconductors, Annual Book of ASTM Standards, ASTM B714-82, Part 02.03, pp. 595-98, American Society for Testing and Materials, Philadelphia, Pennsylvania (1983).
2. S. L. Wipf, Los Alamos National Laboratory, Los Alamos, New Mexico, private communication.
3. L. F. Goodrich and F. R. Fickett, "Critical current measurements: a compendium of experimental results," Cryogenics 22: 222-241, 1982.
4. C. King, Intermagnetics General Corporation, Guilderland, New York, private communication.
5. A. M. Campbell and J. E. Evetts, "Flux vortices and transport currents in type II superconductors," Adv. Phys. 21: 199-429 (1972).
6. J. W. Ekin, "Critical currents in granular superconductors," Phys. Rev. B 12: 2676-2681 (1975).
7. M. Suenaga, Brookhaven National Lab., Upton, New York, private communication.

PERFORMANCE OF Nb_3Sn-Cu IN SITU CONDUCTOR IN A SUPERCONDUCTING MAGNET

J. E. Ostenson, D. K. Finnemore, E. D. Gibson,
J. D. Verhoeven, and D. L. Bruhwiler

Ames Laboratory, USDOE
Iowa State University
Ames, Iowa

ABSTRACT

A small superconducting magnet was constructed from in situ prepared Nb_3Sn-Cu wire in order to test the suitability of this conductor material for use either as an insert magnet at 4.2K in the 10T range or as a stand alone magnet operating on a closed cycle refrigerator in the 10 to 14K range. The conductor was a single strand untwisted wire made by the Sn core process and stabilized with a jacket of pure Cu. Initial tests in a pool of liquid helium showed that the magnet critical current reached 90% of short sample performance with no evidence of training or changes in the field-current relation on repeated cycling even with repeated quenches. Subsequent tests on a close cycle refrigerator showed that the magnet was remarkably stable and gave a very reproducible magnetic field-current relation.

INTRODUCTION

In situ composites based on Nb_3Sn-Cu are of considerable interest as a conductor material for certain classes of superconducting magnets because the materials have a low reaction temperature of 550°C and can operate effectively in the temperature range from 10 to 12K where closed cycle refrigerators can conveniently provide the cooling. Ohkubo and coworkers[1] have constructed moderate scale test magnets using both in situ tape and in situ wire. They find very encouraging results at 4.2K. The purpose of the work reported here is to explore the performance of in situ magnets in the 4 to 14K temperature range.

EXPERIMENT

Conductor for this magnet was fabricated as a single strand wire by the tin core process reported previously.[2] A 10 kilogram billet of dendritic Cu 20 wt% Nb was cast by consumable arc melting. A 5 cm diameter cylinder was machined from the casting, jacketed in Cu and extruded as a tube having an outer diameter of 1.3 cm and an inner diameter of 0.5 cm. A rod of Sn 5 wt% Cu was inserted in the core and the assembly was drawn to 0.66 mm diameter wire by Supercon, Inc. Approximately 41% of the cross section area was pure Cu stabilizer, 15% was the bronze core and the remaining 45% was in situ material. A layer of S-glass insulation was applied by New England Electric Wire Co. to give a final wire diameter of 0.79 mm. The wire was wound onto a stainless steel coil form to give a coil having 3.8 cm i.d., 8.9 cm o.d. and 4.4 cm height. There were approximately 1910 total turns on the coil giving an inductance of about 0.14H and a field to current ratio of 30 mT/A. The stored energy at a typical operating current of 100A would be 700J.

The magnet was reacted at 400°C for 14 days to diffuse Sn into the in situ composite and at 550°C for 9 1/2 days to react to form Nb_3Sn. A series of short samples reacted for various times at 550°C showed that the reaction beyond 5 days gave no further increases in overall critical current density, J_c.

After reaction an inspection of the wire with an optical microscope revealed that only about 30% of the in situ material had reacted to form Nb_3Sn. There was a region of the in situ material near the Sn core where the Cu was almost totally depleted and the Nb_3Sn filament structure cross linked very substantially. This then formed a nearly monolithic Nb_3Sn barrier which inhibited further Sn diffusion. This serious cross linking was not observed in previous studies with 0.25 mm diameter wire. Even with this very serious problem of reaction, the wire at 4.2K carried 300A at 5T, 100A at 8T and 12A at 12T. The total area of the wire including Cu jacket and bronze core was 3.4×10^{-3} cm^2 so 100A implies a current density of 2.9×10^4 A/cm^2 at 8T. The data between 5 and 12T were linear on a Kramer[3] plot of $J_c^{1/2}H^{1/4}$ vs. H and an extrapolation gives H_{c2} of 14.5T.

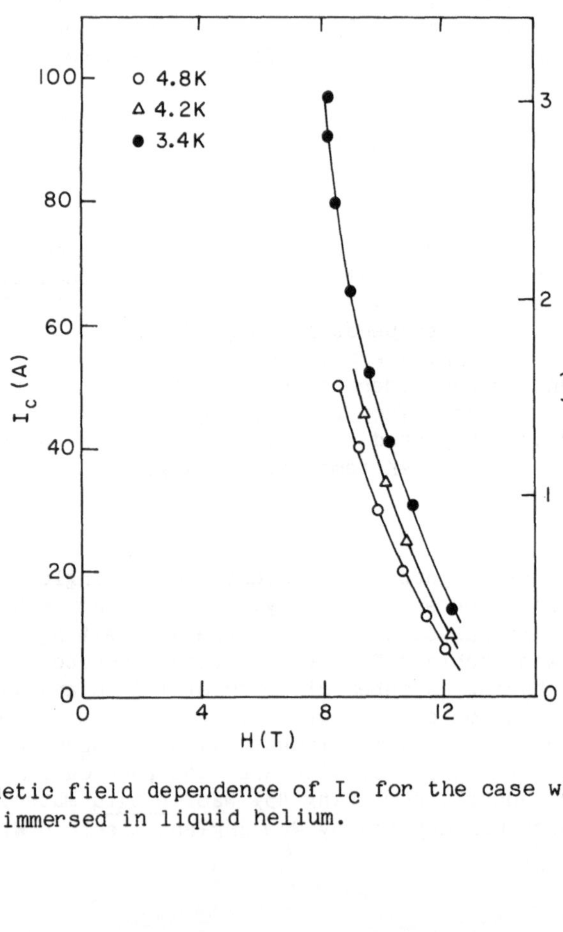

Fig. 1. Magnetic field dependence of I_c for the case where the magnet was immersed in liquid helium.

Tests of the magnet were made in 3 different cryostats. The high field performance as an insert magnet was tested at the National Magnet Laboratory; the high temperature performance for currents up to 20A was tested on a closed cycle refrigerator; the performance over the full range of temperature and currents up to 100A was tested in a specially designed dewar in which the current leads fed through a helium bath but the magnet was isolated in a vacuum space. Temperatures were measured with carbon glass thermometers and magnetic fields were measured with a Cu magnetoresistance probe.

RESULTS AND DISCUSSION

The first test of the magnet was a study of the magnetic field dependence of the critical current, I_c, with the magnet immersed in liquid helium. These were carried out at the National Magnet Laboratory in fields up to 12T. At 4.2K with no field from the Bitter magnet the superconducting magnet was cycled to 100A (3.1T self-field) many times with the current-field relation reproducible to 1%, the accuracy of the Cu magnetoresistance field probe. There were no signs of flux jumping. The Bitter magnet was then set at fields ranging from 8T to 12T and the data shown on Fig. 1 were taken. Critical currents for the magnet are within about 10% of short sample predictions at 4.2K. For magnet quenches at 100A, the stored energy is 700J. Hence a quench would blow off about a quarter of a liter of helium and the magnet would return to equilibrium in about 3 minutes. Also shown are the I_c vs H curves for 3.3 and 4.8K, temperatures which could be obtained by pumping or pressurizing the liquid helium.

After completion of the 4.2K measurements, the magnet was mounted on a closed cycle refrigerator to test performance of the magnet when there was no liquid helium to protect the magnet during a quench. No special attempt was made to build special high current leads so the measurements were limited to currents less than 20A. Critical currents measured in this apparatus are shown by the open triangles in Fig. 2. Because Nb_3Sn has a T_c of 18K it was a bit of a surprise that the critical currents extrapolate to an effective transition temperature of about 15.5K. The resistive transition for the magnet was then measured.

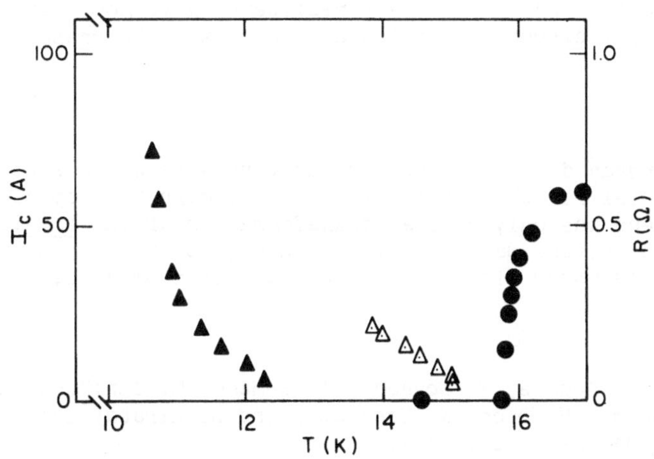

Fig. 2. Temperature dependence of the magnet critical current in the closed cycle refrigerator (open triangles) and variable temperature cryostat (solid triangles). The resistive transition at a measuring current of 0.1A is shown by the solid circles.

Results, shown by the solid circles of Fig. 2, indicate a T_c of about 15.5K in good agreement with the I_c vs T extrapolation. The refrigerator would easily cool the magnet below 10K but there was no convenient way to get 100A into the magnet without completely rebuilding the closed cycle refrigerator which did not belong to our group.

To test the magnet in a vacuum space with currents up to 100A, a metal cryostat was built so that the current leads fed through the liquid helium and into the vacuum through glass to metal seals having a 6.3 mm diameter Cu rod as feed throughs for each lead. Pressure contacts (Cu to Cu) and Pb-Sn solder joints were used to connect the current leads to the magnet.

The temperature dependence of the critical current for values up to 100A are shown by the solid triangles in Fig. 2. The magnet clearly changed characteristics between these two measurements in that there is a clear shift to lower temperatures from the open triangle data of Fig. 2. The same carbon glass thermometer was used in both experiments. Probably the most interesting feature of these data is the very steep rise in I_c vs T below 11K. The magnet behaves very well as a 3T magnet for all temperatures below 10K. Quenches of the magnet at currents in the 50 to 100A range cause the temperature of a thermometer mounted on the outside edge of the magnet to rise to about 18K. The temperature distribution across the magnet was not measured.

The shift in I_c vs T illustrated by the difference between the open triangles and solid triangles in Fig. 2 could arise from a number of sources. The magnet could have been damaged in one of its many remountings. It could have been damaged in a quench. In any event, once we started taking the solid triangle data of Fig. 2, the data were quite reproducible and no further damage was apparent, even for quenches at 100A or 700J.

The ac losses for the entire magnet assembly were measured as a function of frequency, amplitude of sweep and temperature. It was found that the losses increase linearly with frequency indicating that hysteretic losses dominate for low fields. Typical ac losses for sweep rates of 0.07T/sec were 0.6mW at 0.3T and less than 0.1mW at 1.2T. As presently configured, the contact resistance gave more loss than the ac losses. A full discussion of the ac losses will be presented elsewhere.

CONCLUSIONS

In situ prepared Cu-Nb_3Sn superconducting wire was a satisfactory conductor material for the fabrication of a 3T magnet to operate at 10K. Even though there was only partial transformation of the _in situ_ composite to Nb_3Sn, the magnet still reached 3T operation in the temperature range where closed cycle refrigerators operate.

REFERENCES

1. H. Ohkubo, M. Kodera, T. Noguchi, T. Kumano, M. Ichihara, E. Suzuki, K. Yasohama, H. Okon, K. Yasukochi and H. Hirabayashi, IEEE Trans on Mag. MAG-21, 312 (1985).

2. J. D. Verhoeven, E. D. Gibson, C. V. Owen, J. E. Ostenson and D. K. Finnemore, Appl. Phys. Lett. 35, 270 (1979).

3. E. J. Kramer, J. Appl. Phys. 44, 1360 (1973).

HEAT CAPACITY OF SUPERCONDUCTING MATERIALS

G. R. Stewart

Materials Science and Technology Division
Los Alamos National Laboratory
Los Alamos, New Mexico

ABSTRACT

Heat capacity as a technique for exploring the fundamental physical properties, as well as the more materials-oriented properties, of superconductors is discussed. From such measurements can be obtained the electronic density of states at the Fermi energy, the Debye temperature, an approximation for the superconducting energy gap, the electron-phonon coupling strength, the thermodynamic critical field, the approximate percentage of the sample that superconducts and at what temperature, and an idea of the lattice order. Data for A-15 Nb_3Ge, V_3Si, and Nb_3Si are used as examples in the discussion.

INTRODUCTION

Low temperature specific heat (LTSH) measurements of a superconductor ($0 < T \leq 2\ T_c$), where T_c is the superconducting transition temperature, can be useful in two different ways. From a physics viewpoint, such data is extremely rich in content, giving one (with approximations discussed below) the electronic density of states at the Fermi energy, N(0), the lattice stiffness via the Debye temperature, Θ_D, an idea of the superconducting energy gap, Δ, a measure of the strength of the electron-phonon coupling via the parameter λ calculated from Θ_D and T_c via McMillan's formula[1] (or Rowell's formula[2]), another measure of the coupling strength in the normalized discontinuity in the specific heat C, i.e. $\Delta C/\gamma T_c$ (where γ comes from $C = \gamma T + \beta T^3$), the thermodynamic critical field, $H_c(0)$, and the presence in the temperature range of measurement of any other (e.g. structural) transitions. From a materials point of view, LTSH data for superconductors tell what fraction of a sample is superconducting (thus preventing minority second phase resistive and inductive indications of superconductivity from falsely being assigned as characteristic of the majority phase), how well ordered the lattice of a given sample is based on the width of the ΔC, and approximately what fraction of the sample goes superconducting at what temperature. In addition, when one adds the capability of measuring LTSH in field, one determines γ (and therefore N(0), $\Delta C/\gamma T_c$, and $H_c(0)$) more accurately, as well as learning (although there are easier methods) the upper critical field, H_{c2}.

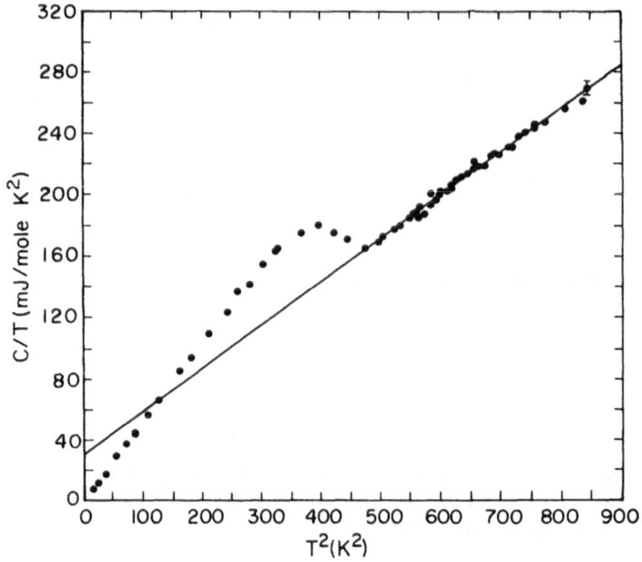

Fig. 1. Low temperature specific heat of single phase A-15 Nb_3Ge prepared by chemical vapor depostion. The 3 K broad transition into the superconducting state is due to inhomogeneities in the sample.

Using this (almost) natural separation of viewpoints (basic physical properties and materials properties) as an organizational basis, we discuss below the utility of LTSH measurements.

PHYSICAL PROPERTIES

As an example of LTSH data, the specific heat[3] of the high T_c A15 structure superconductor Nb_3Ge, $T_c = 21.8$ K, is shown in Figure 1.

Above T_c,

$$C = \gamma T + \beta T^3 \qquad (1)$$

where

$$N(0)(1 + \lambda) = 0.1061\ \gamma \qquad (2)$$

and

$$\Theta_D = \left(\frac{1944 \times r}{\beta}\right)^{1/3} \times 10 \qquad (3)$$

with C in units of mJ/mole-K^2 and r the number of atoms per formula unit, in this case four. Normally materials obey eq. 1 up to a minimum temperature of $\Theta_D/50$, and often as high as $\Theta_D/25$ before the lattice specific heat, C_L, deviates from the simple Debye T^3 law ($C_L = \beta T^3$). We see in the case of Nb_3Ge that eq. 1 is obeyed (i.e. C/T versus T^2 is a straight line with slope β and intercept γ) to almost 30 K, while from eq. 3 the slope of the line in Fig. 1 gives $\Theta_D = 302$ K. Thus, in some materials, the Debye law for C_L is followed up to $\Theta_D/10$, although for most materials a T^5 term has already entered above $\Theta_D/25$. In the case of a high T_c superconductor, non-Debye-like behavior above T_c makes a determination of γ more difficult, since then at least a 3 term computer fit to the normal state specific heat, C_n, is required to extrapolate C_n below T_c to $T = 0$ to determine γ, and the higher the T_c the further the extrapolation. Since superconductors have second-order transitions, i.e. the entropy S is continuous at T_c, one can use the relation $S_{normal}(T_c) = S_{superconductor}(T_c)$ to check the extrapolation of C_n, since one measures C_s and therefore $S_s(T_c) = \int_o^{T_c} \frac{C_s}{T} dT$ is known. As long as an applied magnetic field does not change C_n, an improved knowledge of C_n (and therefore γ, N(0), $\Delta C/\gamma T_c$, and $H_c(0)$) can be had by depressing T_c and measuring C_n below T_c ($H = 0$).

Fortunately in the case of Nb_3Ge the simple fit of the C_n data gives the correct $S_n(T_c)$ (i.e. = measured $S_s(T_c)$), since the critical field of Nb_3Ge is relatively high, i.e. quite a large field (∿ 25T) would be needed to suppress T_c to 4 K.

Knowing for Nb_3Ge that $T_c = 21.8$ K, and $\Theta_D = 302$ K, one uses McMillan's formula[1] if $\lambda < 1$ and Rowell's[2] if $\lambda > 1$, to find $\lambda \sim 1.7$ for Nb_3Ge, in agreement with tunneling measurements.[4] Knowing $\gamma = 30$ mJ/mole-K^2, eq. 2 then gives us N(0) $\simeq$ 1.7 states/eV-atom (both spin directions.) In comparison to other superconductors,[5] e.g. elemental Nb, this is not a large value for such a high T_c ($T_c \sim e^{-1/N(0)}$), indicating that the source of the high T_c in Nb_3Ge is more connected with a strong electron-phonon coupling.

From

$$\int_o^{T_c} (C_s - C_n) dT = \frac{VH_c^2(0)}{8\pi} \tag{4}$$

one determines[3] that $H_c(0) = 4500$ for Nb_3Ge. Since the superconducting transition measured in Figure 1 is somewhat broad, the integral in eq. 4 is clearly decreased by this broadening, i.e. $C_s - C_n$ near T_c is smaller than it would be for "ideal" Nb_3Ge. Extrapolating C_s upwards for such a sharp transition gives $H_c(0) \cong 5380$ Gauss.

From these values of $H_c(0)$ for Nb_3Ge we may obtain $2\Delta/kT_c$ via

$$\frac{2\Delta}{\kappa T_c} = \frac{H_c(0)}{T_c} \left(\frac{2\pi V}{3\gamma}\right)^{1/2} \tag{5}$$

where the molar volume of Nb_3Ge is[3] 40.89 cm^3/mole. One obtains 3.5 for $H_c(0) = 4500$ G and $\frac{2\Delta}{k T_c} = 4.2$ for $H_c(0) = 5380$ G. The latter number is in good agreement with tunneling results.[6] Other ways exist to obtain $\frac{2\Delta}{kT_c}$ values from specific heat, including plotting ln(C_s -

C_L) (i.e. $\ell n\ C_{electronic,\ T<Tc}$) versus T_c/T. Since $C_s - C_L$ ($\equiv C_{es}$) goes as $e^{-\Delta(T)/kT}$, one may obtain an idea from such data of the temperature dependent energy gap. One problem for some materials is that C_L is not Debye like above T_c, or is known to change effective Θ_D for $T < T_c$, making the subtraction $C_s - C_L$ for $T < T_c$ somewhat inaccurate. For a discussion, see ref. 7.

As an example of LTSH of a superconductor with a sharp transition, as well as the utility of field data, results for transforming and non-transforming V_3Si are shown in Figures 2 and 3. At one time there existed a controversy[8-10] (of interest from a physics viewpoint, see ref. 10) over which form of V_3Si had a larger dressed density of states, cubic ("nontransforming") or tetragonal (note the anomaly at 20 K in the "transforming" V_3Si, Fig. 2) V_3Si. Due to the high T_c, plus the large change in C_p above T_c due to the martensitic transformation, the answer was difficult to resolve without suppressing T_c by applied magnetic field. As seen in Figures 2 and 3, the field data imply that the non-transforming sample has the larger γ. (As a technical problem, note that even a 19T field does not totally suppress superconductivity in the temperature range of measurement (Fig. 3.) in these high critical field A-15 superconductors.)

In zero field, we see from Fig. 2 what a sharp ΔC looks like, with $\Delta C/\gamma T_c \sim 1.8$ for cubic V_3Si. Since BCS predicts $\frac{\Delta C}{\gamma T_c} = 1.43$ for weak coupled superconductors, this result is a sign of strong coupling in V_3Si.

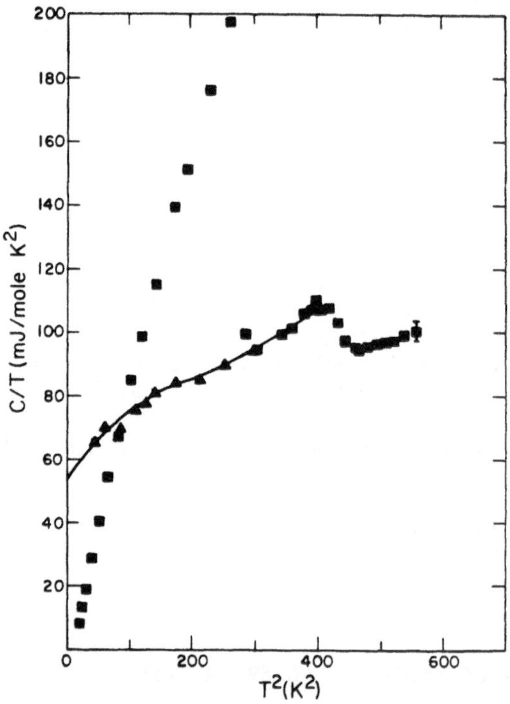

Fig. 2. LTSH of a rather good sample (4 single crystals) of transforming V_3Si, where the quality of the sample is identified with the sharpness and size of the specific heat anomanly at the martensitic transformation, $T \sim 20$ K. The 18 T data (triangles), which give a $\gamma = 53$ mJ/mole-K^2, extend only down to 6.7 K for technical reasons involving difficulties in the high field measurements.

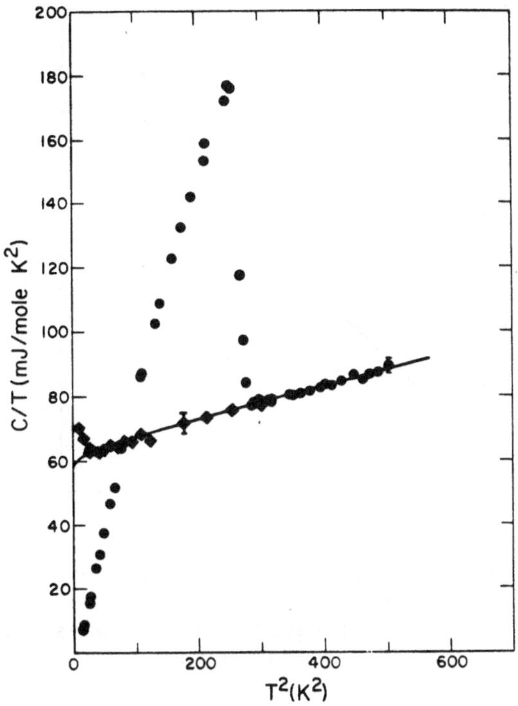

Fig. 3. LTSH of A-15 (cubic) non-transforming V_3Si, both in zero (dots) and applied field (diamonds, H = 18T except below 7 K, where H = 19T). The field data give γ = 58 mJ/mole-K^2. A-15 V_3Si is stable with temperature until almost its melting point. Thus, the lattice order, upon which the transition width depends critically, of Nb_3Ge deposited at 900°C is seen in Figure 1 to be not nearly as perfect as that seen in V_3Si formed at a 1000°C higher temperature. Specific heat, being a bulk measure of superconductivity, is almost a unique method for determining this transition width correctly.

MATERIALS PROPERTIES

The LTSH of A-15 (cubic) Nb_3Ge (Figure 1) obviously has a much broader superconducting transition than that of A-15 (cubic) V_3Si. In fact, these data show that some bulk fraction of the Nb_3Ge is still not yet superconducting 3 K below T_c^{onset} = 21.8 K. The reason for this is the lack of stability of the A-15 phase in Nb_3Ge, so that special meta-stable preparation techniques are required, in this case chemical vapor deposition ("CVD"). Attempts[11] at optimization of the materials preparation parameters (gas flow rates, substrate temperature) were unsuccessful at narrowing the transition width, although even this negative result was an important example of the utility of LTSH as a materials optimization tool.

Figure 4 shows[12] the LTSH of explosively compressed A-15 structure Nb_3Si. These data were of several uses. First and foremost, they showed conclusively that the A-15 majority phase (determined by x-rays), produced by the < 1 μsec of applied pressure ∿ 1 Mbar, was in fact a bulk superconductor at 18 K. Thus, worries over a minority phase, possibly formed on the surface between Nb and N and C from the explosive, being responsible for the inductively observed superconductivity were dispelled. Second, the LTSH data, see Fig. 4, indicated a rather sharp transition in this most unstable of known A-15 high

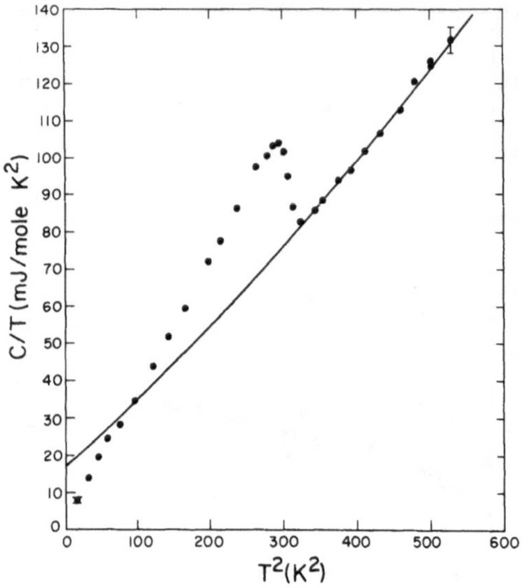

Fig. 4. LTSH of A-15 Nb_3Si produced by explosively compression, where ∿ 50-60% of the sample is the A-15 phase, as determined by x-ray measurements. The sharpness of ΔC has been a subject of some discussion, see text.

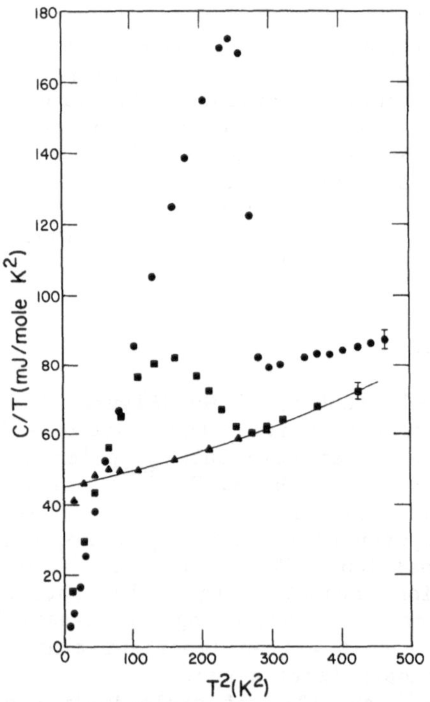

Fig. 5. LTSH of as prepared (dots) and explosively compressed (squares) V_3Si, where the explosive compression was done under conditions similar to those for the A-15 Nb_3Si sample, Figure 4. Field data (triangles) for the explosively compressed sample show a decreased γ value from the as prepared material, as expected from long range order considerations.

T_c superconductors. That such good lattice ordering (e.g. better than in Nb_3Ge, see Fig. 1) could take place during the explosive compression was surprising. Another possible explanation[12] was that the T_c for "ideal" well ordered A-15 Nb_3Si was in fact much higher, with the narrow transition observed (Fig. 4) indicative of a highly damaged superconductor with a significantly depressed T_c.

In order to investigate this second possibility, well ordered A-15 V_3Si was explosively compressed in a similar manner and the before and after material compared[13] via LTSH, see Fig. 5. One sees clearly that explosively compressed A-15 V_3Si has a broader transition than similarly prepared A-15 Nb_3Si, lending at least some support to the hope that A-15 Nb_3Si, as so far prepared, is in fact highly disordered and has a T_c significantly below that of ideal A-15 Nb_3Si.

CONCLUSIONS

We have seen that LTSH can be a useful measurement technique for materials characterization. Knowledge of fundamental properties, such as N(0) and λ, is a help in understanding superconductivity in materials, and an aid in choosing future directions for materials development. As a process control instrument, LTSH is certainly a useful tool, albeit to this point one involving several days effort per sample. Recent advances[14] in measurement technique have produced LTSH data from sub-milligram thin films, and work in progress[15] will allow even small masses to be measured. Thus, the modern trend toward thin film preparation of superconductors has not left one of the more useful characterization tools behind.

ACKNOWLEDGMENTS

Work at Los Alamos performed under the auspices of the US DOE. This manuscript was written while the author was a guest at the Institut für Technische Physik, Kernforschungszentrum, West Germany. Their hospitality is gratefully acknowledged.

REFERENCES

1. W. L. McMillan, Phys. Rev. 167, 331 (1968).
2. J. M. Rowell, Solid State Comm. 19, 1131 (1976).
3. G. R. Stewart, L. R. Newkirk, and F. A. Valencia, Solid State Comm. 26, 417 (1978).
4. K. E. Kihlstrom and T. H. Geballe, Phys. Rev. B24, 4101 (1981).
5. G. R. Stewart, in Superconductivity in d- and f- Band metals 1982 (ed. by W. Buckel and W. Weber, Kernforschungszentrum Karlsruhe, 1982), p. 81.
6. J. M. Rowell and P. H. Schmidt, Applied Phys. Lett. 29, 622 (1976).
7. G. R. Stewart and A. L. Giorgi, Phys. Rev. B17, 3534 (1978).
8. C. C. Huang, A. M. Goldman, and L. E. Toth, Solid State Comm. 33, 581 (1980).
9. A. Junod and J. Muller, Solid State Comm. 36, 721 (1980).
10. G. R. Stewart and B. L. Brandt, Phys. Rev. B 29, 3908 (1984).
11. G. R. Stewart, L. R. Newkirk, and F. A. Valencia, unpublished.
12. G. R. Stewart, B. Olinger, and L. R. Newkirk, Solid State Comm. 39, 5 (1981).
13. G. R. Stewart, B. Olinger, and L. R. Newkirk, Phys. Rev. B 31, 2704 (1985).
14. S. R. Early, F. Hellman, J. Marshall and T. H. Geballe, Physica 107B, 327 (1981).
15. A. M. Goldman, private communication.

IRRADIATION DAMAGE IN SUPERCONDUCTORS*

Harald W. Weber

Atominstitut der Österreichischen Universitäten
Vienna, Austria

ABSTRACT

An overview of recent neutron irradiation experiments on superconductors is presented. Special attention is paid to those materials which are considered as candidates for fusion reactor magnets, and to an irradiation environment which is closely related to the magnet operating conditions over the lifetime of a fusion reactor. The results on a variety of NbTi superconductors may be summarized as follows: 1) the influence of irradiation temperature is very small, if thermal cycles following low temperature irradiations are included; 2) the influence of the metallurgical starting conditions becomes very small at high fields ($\geq$ 7 T) indicating saturation effects for flux pinning and and an increasing influence of H_{c2} on critical current densities j_c; 3) the overall degradation of j_c does not exceed 20% over the lifetime of a fusion magnet. The experimental results on A15 superconductors, in particular Nb_3Sn with small additions of Ti, show the normal increase of critical current densities with neutron fluence followed by a sharp decrease of j_c. However, the peak of j_c is shifted to much lower fluences in the alloyed samples as compared to pure Nb_3Sn. Finally, a comment on the resistivity change of "magnet" copper is made indicating that a considerable increase of resistivity ($\sim$60% at 8 T) has to be taken into account, if "reasonable" operating conditions of the magnet (i.e. only occasional annealing cycles to room temperature) are taken as a basis.

INTRODUCTION

Neutron irradiation experiments on superconductors are made mainly for two purposes: firstly, radiation damage in carefully selected materials, which is introduced under special radiation environments, represents an ideal tool for studying flux pinning mechanisms[1,2] or the influence of disorder on some primary superconductive properties[3]. Secondly, the progress of concepts for the confinement of a fusion plasma with superconducting magnets has spurred experimental investigations of radiation damage in those materials, which are considered as candidates for the construction of these coils. Pursuing the literature of the past few years[4-7] a clear trend to irradiation conditions, which are closely related to the operating conditions of a fusion magnet over its lifetime, will be observed. It is the purpose of the present paper to emphasize this particular

*Invited paper

aspect of neutron irradiation work and to review the progress achieved recently.

In the following sections I wish to present results on a careful characterization of various neutron irradiation facilities and compare them with a typical neutron spectrum expected for the coil location in a fusion reactor. Then I will discuss results on the critical current densities as a function of neutron fluence obtained on a variety of alloy and compound superconductors. In the final section some conclusions and requirements for further experiments will be presented and a remark on the radiation induced resistivity change of the copper stabilizer made.

NEUTRON SPECTRA AND DAMAGE ENERGY CROSS SECTIONS

Radiation damage in high-j_c superconducting materials is caused primarily by fast neutrons through the production of defect cascades, minor contributions (which are usually neglected) stem from transmutations and Frenkel pairs induced mainly by thermal neutrons and the corresponding increase of normal state resistivity. Therefore, neutron flux densities or their time integral, the neutron fluence, to which a superconductor is exposed, are usually quoted in terms of the fast neutron component only, i.e. for neutron energies greater than 0.1 or 1 MeV, respectively. In order to compare irradiation experiments carried out at different laboratories and to judge the relevance of these studies with respect to calculated fusion neutron spectra at the magnet location, a detailed knowledge of the neutron energy distribution in each irradiation facility is required. In order to meet these requirements, neutron dosimetry experiments were made at several facilities[8-10] and evaluated in terms of neutron energy distributions and various damage parameters in metals using advanced computer codes[11,12].

As an example, results on three different neutron sources, which will be referred to later on, and a fusion spectrum[13] are shown in Fig. 1. Except for the monoenergetic 14 MeV neutron source (RTNS-II, Livermore) the fission (TRIGA, Vienna), spallation (IPNS, Argonne) and fusion (STARFIRE) spectra show distinct similarities at high neutron energies ($E > 0.1$ MeV), but a slight enhancement of the fusion neutron flux density towards 14 MeV. From this comparison it seems to be obvious that simulation experiments carried out in fission or spallation sources have to be supplemented by 14 MeV irradiations. On the other hand, each spectrum can be characterized by a damage energy E_D, i.e. by an average energy transferred to each atom at a certain neutron fluence. Now, if we prove experimentally, that the superconductive parameter of interest (e.g. the critical current density j_c, the transition temperature T_c, etc.) scales with damage energy, any spectrum can be used to obtain the desired information. First results of this type will be presented later and in more detail in a separate paper at this conference[14]. A summary of damage energy cross sections, which have to be multiplied by the neutron fluence in order to obtain E_D, is presented for selected elements in Table 1, the corresponding calculations were made using the computer code SPECTER[12]. The results demonstrate the similarities between the fission and spallation spectra on the one hand and the fusion spectrum on the other hand, in a very clear way.

As a concluding remark the operating conditions and the radiation environment of the magnets should be specified. It is generally agreed upon that the maximum lifetime fluence at the magnet windings will be around 4×10^{22} neutrons per m^2 ($E > 0.1$ MeV). Radiation damage will be produced at the operating temperature of the magnet, i.e. near 5 K, but several thermal cycles to room temperature will occur on the occasion of plant shut-downs. Thus, simulation experiments on superconducting magnet components should be extended well into the 10^{22} m^{-2}-fluence range and meet the

Table 1. Damage energy cross sections (keV-barns) for selected elements and four different neutron spectra

ELEMENT	TRIGA	IPNS	RTNS	FUSION
Niobium	70.2	62.0	280.4	67.9
Titanium	81.7	64.9	248.5	66.5
Tin*	76.2	54.0	242.1	62.6
Tantalum	46.6	38.6	219.8	45.1
Copper	70.1	59.6	300.8	70.9

*not included in SPECTER, best estimate.

temperature requirements specified above. All fluences, quoted in the present paper will refer to energies greater than 0.1 MeV.

RESULTS

Alloys: NbTi and NbTaTi

Concerning recent irradiation work on the alloy superconductor NbTi, previous research by our group [7,15-17] was taken up, extended into the high-field region (8 - 10.5 T) and adopted to meet the irradiation environment specified for appropriate simulation experiments in the last section. The low temperature irradiations were made at the Radiation Effects Facility (REF) of the spallation source IPNS, additional ambient temperature irradiations were made at IPNS as well as at the 14 MeV source RTNS-II in Livermore. A survey of the materials used is given in Table 2. Because of the large number of samples investigated and because of the lack of an in-situ high-field magnet for the critical current measurements, the following procedure was adopted for the low temperature irradiations: irradiate the samples at 5 K to a certain neutron fluence, store them at 77 K for

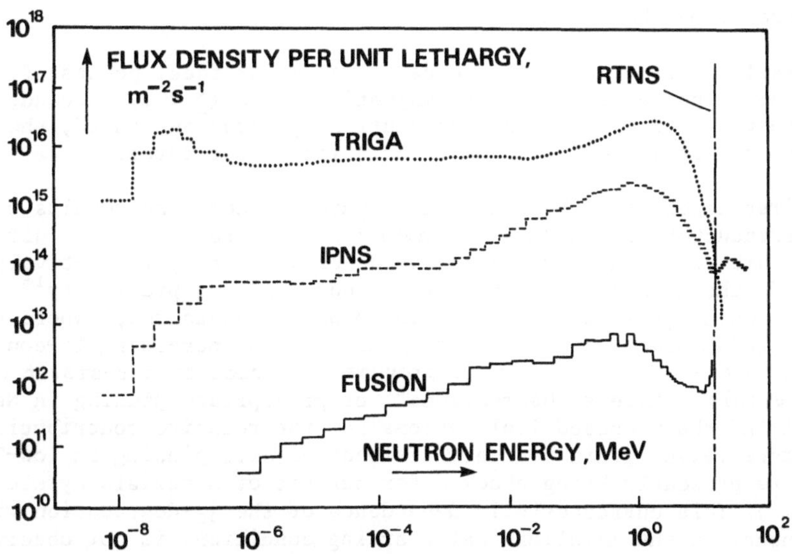

Fig. 1. Neutron flux density distribution for four different neutron sources: fission reactor (TRIGA, Vienna), spallation source (IPNS, Argonne), DT-source (RTNS-II, Livermore), fusion spectrum at the magnet location (STARFIRE).

Table 2. Summary of material parameters for NbTi-superconductors

A) Single-core conductors (Cu: superconductor = 1:1) supplied by BBC Brown Boveri & Cie, Zürich, Switzerland

Alloy Composition	Annealing Temperature, ^{0}C (Annealing Time: 14 hours)	Degree of Final Cold Work (fcw), %	Total Number of Samples
42 wt%Ti	320,350,370,400,430	0,46,71,84,91	11*
49 wt%Ti	320,350,370,400,430	0,46,71,84,91	11*
54 wt%Ti	320,350,370,400,430	0,46,71,84,91	11*

*Each sample is characterized by one particular thermomechanical treatment; e.g.: 370 ^{0}C, 91 %fcw.

B) Multifilamentary conductors

Alloy Composition	Cu: Superconductors	Number of Filaments, Filament Diameter	Number of Samples	Supplied by
49 wt%Ti	2.4 : 1	48, 43 μm	2	BBC,S-48-3
46.5 wt%Ti	1.2 : 1	330, 17 μm	1	BBC,S-300-1
17 wt%Ta, 41 wt%Ti	1.44 : 1	361, 7.5 μm	1	Univ. of Wisconsin

10 days, anneal them to room temperature for 30 hours, measure the critical current densities and the stabilizer resistivity at 4.2 K, store them again at 77 K until the next run in the radiation effects facility and so on. Clearly, under these conditions we do not obtain information on the radiation damage prevailing at the operating temperature of the magnet prior to a shut-down, which may add between 5 and 10%[18,19] to the j_c-degradation, but obtain results on the change of critical current densities following the (simulated) shut-down.

The results will be presented under three major focal points: First, the radiation response of j_c at high magnetic fields ($\geq$ 7 T); second, the radiation response of j_c to different neutron spectra; and third, the radiation response of j_c to different irradiation temperatures.

The first aspect is illustrated in Figure 2, where the results on three superconductors of identical thermomechanical treatment, but different Ti-content, are compared at magnetic fields of 5 and 8 T, respectively. At low fields the data display again the trend observed previously[15,7] and discussed in much detail in terms of flux pinning mechanisms, where the increasing predominance of precipitate pinning with increasing Ti-concentration was invoked to explain the improvement of radiation resistance in high-Ti materials. (Recent observations[20] of precipitate pinning in Nb-46.5 wt% Ti and the proposed implications for the relative contribution of precipitate versus grain boundary/dislocation cell pinning in low-Ti materials are presently being checked for our set of materials by electron microscopy[21]). This characteristic dependence of the j_c-degradation with neutron fluence on the metallurgical starting conditions is not observed at high fields. This statement holds in a very general form as long as the upper critical fields H_{c2} of the samples are comparable. This fact is emphasized by the results shown in Figure 3: H_{c2} of samples 4 and 5 is almost identical (10.8 and 10.7 T, respectively), whereas H_{c2} of sample 6 is only 9.8 T. If we correct for this difference by referring to the same

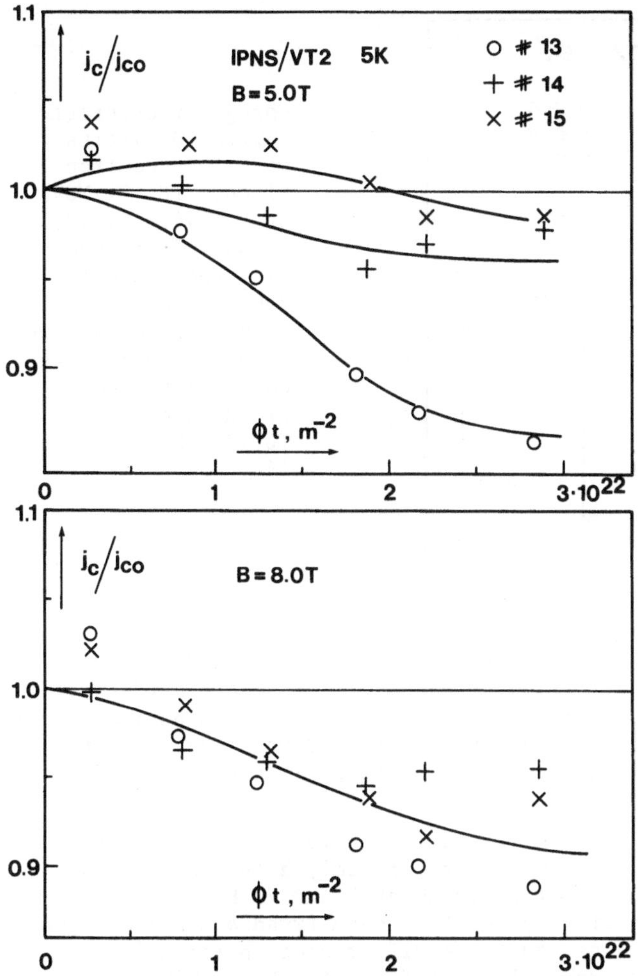

Fig. 2. Change of critical current densities with fast neutron fluence (E >0.1MeV) at 5 and 8 T. #13,14,15: 42,49,54 wt%Ti; annealing temperature: 350°C; final cold work: 71%.

reduced field $b = B/B_{c2}$ ($\sim$ 0.74) instead of the applied field B(= 8T), the data show again a universal behavior.

These results and preliminary data[22] on the scaling of pinning forces in unirradiated materials near H_{c2} strongly suggest that in these high-j_c materials saturation effects for flux pinning occur and the magnitude of j_c is determined essentially by the magnitude of H_{c2} (and/or the elastic response of the flux line lattice). Unfortunately, at the present stage no H_{c2}-data of the irradiated samples are available[21], which inhibits a final interpretation of this interesting effect.

It should be pointed out furthermore, that the degradation of j_c is always found to be larger at 8 T than in low fields (2 - 10%); shows a tendency to level off in the fluence range of interest for fusion applications and does, in general, not exceed 20%. All of these features are displayed clearly by a conductor of technical relevance (S-300-1), which was used for the Swiss LCT-coil[17]

The second aspect I wish to present refers to a test of the damage energy concept, i.e. the feasibility of scaling the damage in the superconductor produced by different neutron spectra. For this purpose three identical sets of 17 samples were irradiated at ambient temperature in the TRIGA reactor (Vienna), the spallation source IPNS (Argonne) and the 14 MeV-Source RTNS-II (Livermore) and the critical current densities measured as a function of neutron fluence. Based on the damage energy cross sections for Nb and Ti presented in Table 1, appropriate values

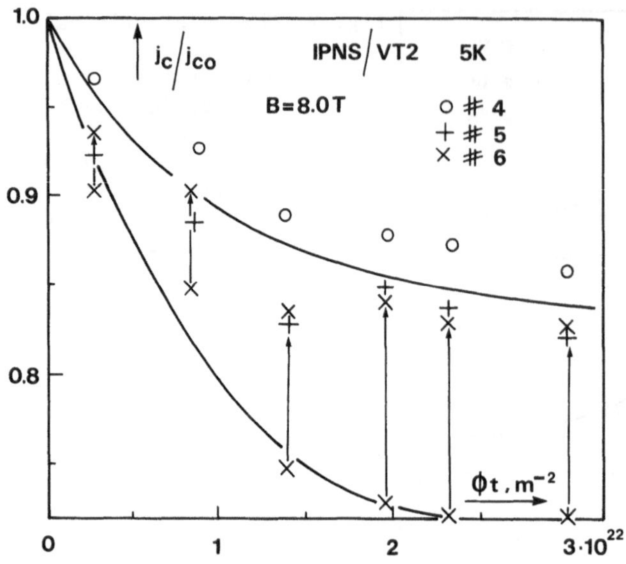

Fig. 3. Change of critical current densities with fast neutron fluence (E > 0.1MeV) at 8 T. #4,5,6: 42, 49,54 wt%Ti; annealing temperature: 350°C; final cold work: 0%. The arrows refer to results at constant reduced field, b = 0.74 (B = 7.3 T for sample #6).

for each alloy and each neutron spectrum were calculated according to the atomic compositions of the materials. We find[14], that all the data agree to within experimental accuracy, although the neutron fluences differ by up to a factor of more than four. This result is considered to be most valuable for reactor design purposes and proves in a direct way, that the damage mechanisms in the superconductor remain unchanged even upon changing the neutron spectrum dramatically.

The third aspect of this work refers to the influence of temperature maintained during irradiation. Although there are only a few data on identical samples available, which were irradiated at 5 and 77 K, respectively, and again subjected to an annealing cycle after each irradiation step, the general trend of results indicates, that the differences in radiation damage after the thermal cycle are negligible (Fig. 4). Even the ambient temperature irradiations lead to comparable results in most cases. This behavior can be understood qualitatively in terms of the defect production processes and the few in-situ irradiation/annealing experiments published for NbTi (e.g.[12]). The original defect produced by a fast neutron is a displacement cascade, which remains stable at low temperatures (4.2 to ≈ 50 K), starts to disintegrate with increasing temperature and finally forms small clouds of point defects at room temperature. The total amount of j_c-recovery observed at 77 K after 5 K-irradiation amounts only to about 2%[18]. Hence, the similarity between results obtained after 5 K and 77 K irradiation and after including a thermal cycle to room temperature, is not surprising. At ambient temperature irradiation the disintegration of cascades occurs in-situ. The close agreement of these results with the low temperature irradiation/annealing data seems to indicate, that the final damage of the superconductor is similar in all cases.

As a concluding remark of this section, irradiation experiments on the ternary alloy NbTaTi should be mentioned. It is favored for some advanced magnet designs because of the increase of H_{c2} at low temperatures (∿ 2 K) compared to the binary alloys and the corresponding accessibility of the 12 T range at good critical current densities[23]. The wire included in the present study (cf. Table 2B) was supplied by D.C.Larbalestier and had a nominal composition of 17 wt%Ta and 41 wt%Ti (j_c at 5 and 8 T: 1.91 and 0.76 x 10^9 Am^{-2}, respectively). Low temperature neutron irradiation up to a fluence of 1.4 x 10^{22} m^{-2} (E > 0.1 MeV) showed significant <u>increases</u> of j_c with fluence (+7% at 5 T, +19% at 7.5 T). In view of the large number of alloy samples investigated so far we conclude that the defect structure

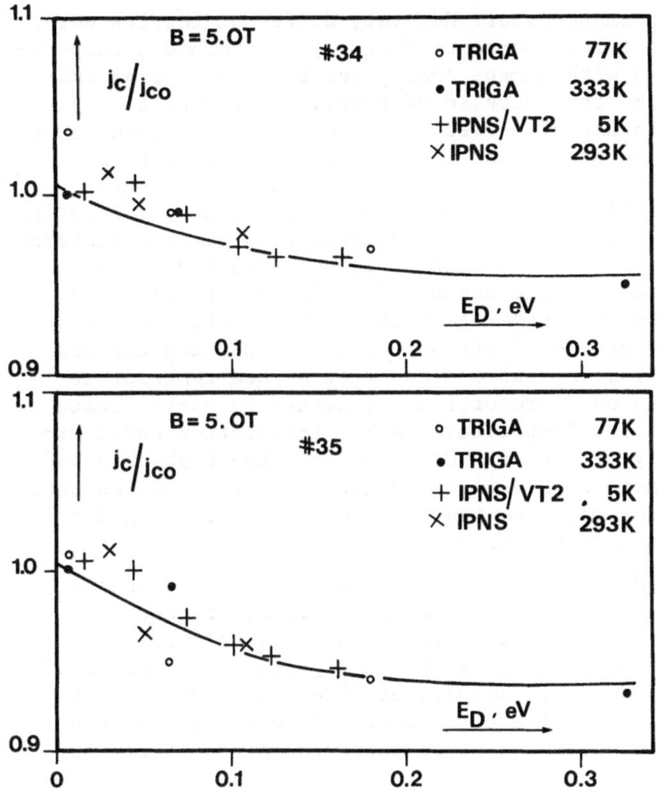

Fig. 4. Change of critical current densities at 5 T with damage energy per atom. Multifilamentary conductors S-48-3 (#34,35;49 wt%Ti). Irradiation conditions: TRIGA reactor (77 K, 333 K), IPNS (5 K, 293 K).

of this particular wire was not optimized. It is, however, safe to assume that the radiation response of this alloy will, in general, be comparable to the one characteristic of binary alloys.

Compounds: Nb_3Sn and $(NbTi)_3Sn$

Concerning recent irradiation work on A15 superconductors, which would fit into the scope of the present paper, not a lot of activity can be reported on. However, both of the investigations to be discussed in more detail have provided interesting results, which should stimulate further work for their clarification.

The first study refers to a monofilamentary Nb_3Sn-bronze conductor, which was irradiated at ambient and low temperatures with 14 MeV-neutrons[24] and at ambient temperature to very high fluences in the Brookhaven High Flux Beam Reactor[25,26]. The decrease of transition temperature is found to scale with damage energy independently of the irradiation temperature, which conforms to the understanding that radiation damage in the A15's is

Table 3. Summary of material parameters for Nb_3Sn superconductors

Compound	Code	Description	Thermal treatment	Bronze: Nb	Comment
Nb_3Sn	NBSN1	Monofil.cond. filament diameter: 95 μm	700°C, 96 h	12.5:1	Refs.24-26
(Nb-0.8 wt%Ti)$_3$Sn	#38	Multifil.cond.	625°C,139.5 h	3:1	Supplied by
(Nb-0.8 wt%Ti)$_3$Sn	#39	∿10.000 filam.	700°C,28.5 h	3:1	University
(Nb-1.5 wt%Ti)$_3$Sn	#40	fil.size:	625°C,150 h	3:1	of
(Nb-1.5 wt%Ti)$_3$Sn	#41	∿1.3 x 5 μm^2	700°C,14.5 h	3:1	Wisconsin

dominated by changes in long-range order. However, different results were obtained for the change of critical currents (Fig. 5): the ambient temperature irradiations scale again with damage energy and show the normal initial increase of j_c (caused by the increase of normal state resistivity ρ_n and, hence, H_{c2}, with disorder) followed by a steep decrease (caused by the predominance of the T_c-depression and the change in the density of states). On the contrary, the low temperature irradiations show a much steeper increase of j_c persisting up to considerably higher damage energies and amounting to about twice the enhancement observed upon ambient temperature irradiation. Furthermore, this enhancement is not annealed out completely after a final room temperature anneal. In order to explain this discrepancy the authors of Ref.[24] stipulate that low temperature irradiation leads to the production of a siginificant amount of pinning centers (cf. also[27]), most of which remain stable even after a room temperature anneal. On the other hand, an over-proportional increase in normal state resistivity in the course of low temperature 14 MeV irradiation could explain the initial discrepancy between the data, but not the high dose and annealing result. Clearly, more work will be needed to clarify all aspects of this data, but the need for more low temperature irradiations on A15 superconductors has been demonstrated unambiguously.

The second study was made by our group on several alloyed conductors of the type $(Nb_{1-x}Ti_x)_3Sn$, which were provided by D.B.Smathers and D.C. Larbalestier from the University of Wisconsin. Much interest[28] has focused on these materials recently because of the significant increase of H_{c2} achieved by the addition of small amounts of Ti of the order of 0.5 - 1.5 wt%. In fact, H_{c2} at 4.2 K was found to be 22.5 T in a compound containing 0.8 wt%Ti and reacted at 700 °C as compared to 18.3 - 18.5 T for pure Nb_3Sn.

The irradiations were made again at low temperatures (REF, IPNS) and included a thermal cycle to room temperature after each irradiation step. Because of the shut-down of the radiation effects facility at Argonne only the low-fluence range up to 1×10^{22} m^{-2} (E >0.1 MeV) was covered. Results on a conductor containing 1.5 wt%Ti are shown in Fig. 6. We observe again an increase of critical current densities followed by a rather sharp decrease, but the maximum enhancement of j_c occurs at a neutron fluence,

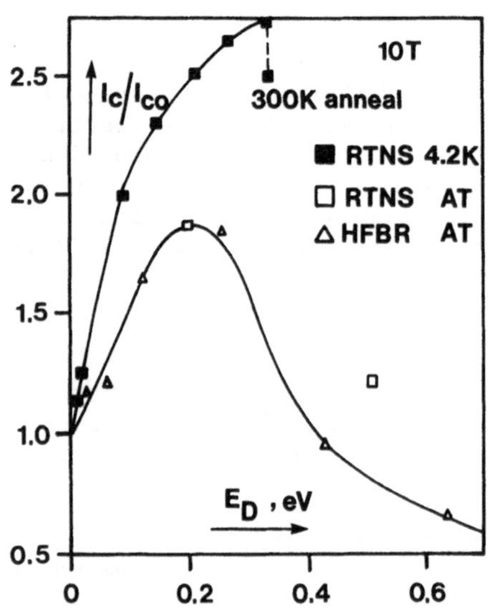

Fig. 5. Change of critical currents with damage energy per atom. Monofilamentary conductor NBSN1[24].

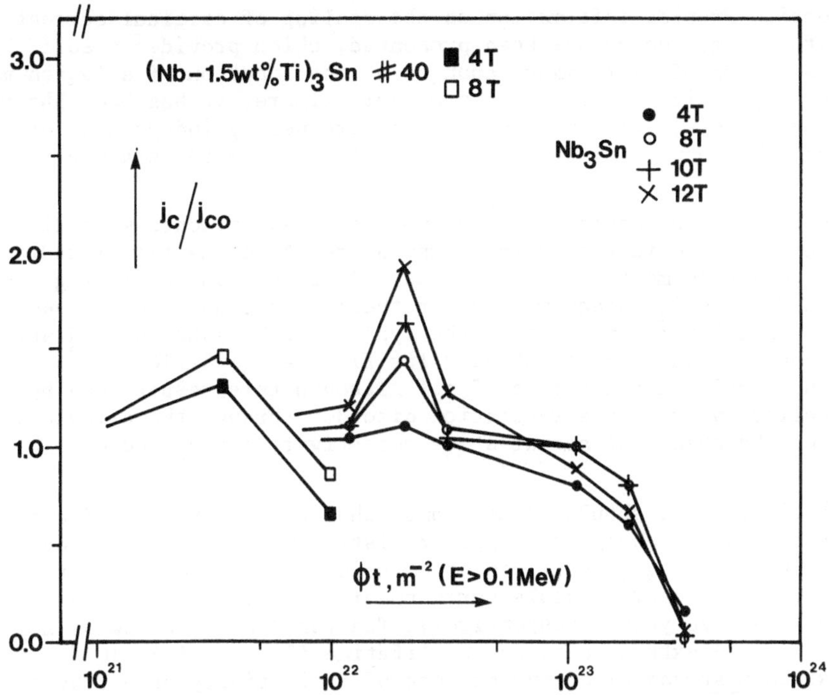

Fig. 6. Change of critical current densities with neutron fluence. The scaling is not completely accurate because of slight differences in damage energy cross sections. Low temperature irradiations on an alloyed conductor are compared to ambient temperature irradiations of pure Nb_3Sn[29].

which is lower by a factor of almost 6 compared to pure Nb_3Sn[29]. An explanation of these results in terms of the disorder induced increase of H_{c2} seems to be quite straightforward: whereas a neutron fluence of about 2×10^{22} m^{-2} ($E > 0.1$ MeV) is needed, in order to achieve the maximum enhancement of H_{c2} by about 12% in pure Nb_3Sn (data on NBSN1,[30]), the preirradiation enhancement of H_{c2} in the alloyed conductor already amounts to ∿13% (#40: $\mu_0 H_{c2}(4.2) \approx 21$ T,[31]). Hence, an almost immediate decrease of performance with neutron fluence is to be expected. Of course, more detailed experiments, especially on the change of H_{c2} with neutron fluence and on a broader spectrum of alloyed samples, will be needed to clarify all aspects of the present data.

CONCLUSIONS

The main results of the investigations presented in this paper may be summarized as follows.

For NbTi superconductors, the extension of experiments into the high field range revealed the existence of a rather uniform radiation response for metallurgically different materials, in marked contrast to the low field results. Further work on the radiation induced change of H_{c2} and the metallurgical microstructure will be needed to discuss these observations in more detail. Precise information on the scaling of critical current changes with damage energy has been presented, which provides a solid basis for the evaluation of performance changes at any location of a fusion magnet and for any fusion neutron spectrum. Furthermore, it has been shown that the changes of j_c with neutron fluence are nearly independent of irradiation temperature, if an annealing cycle to room temperature is included.

Concerning Nb_3Sn-superconductors two recent experiments were discussed. The first one refers to low temperature 14 MeV neutron irradiation of pure Nb_3Sn and seems to indicate, that radiation induced pinning centers play a role for the j_c-enhancement under these conditions. The second showed that in alloyed compounds of the type $(Nb_{1-x}Ti_x)_3Sn$ the radiation induced j_c-enhancement occured at very low fluences ($\sim 3 \times 10^{21}$ m^{-2}) followed by the usual sharp decrease of j_c. Although this result can be explained qualitatively by preirradiation disorder through the alloying, more data on the change of normal state resistivity and H_{c2} seem to be desirable.

In conclusion, it should be mentioned that a complete set of data on the radiation induced change of copper resistivity has been obtained, which simulates the operating conditions of a fusion magnet over 20 years and includes room temperature anneals corresponding to plant shut-downs after 2,6,9,12.5,15 and 20 years, respectively. The results on the Swiss LCT-conductor (Fig. 7) extrapolated to the lifetime fluence of 4×10^{22} m^{-2} show a maximum post-annealing enhancement of resisitivity at 8 T by a factor of 1.22. However, if we include the in-situ (low fluence) experiments by Klabunde and Coltman[32] on similar "magnet" copper into the extrapolation, the final pre-annealing enhancement of resistivity will amount to 1.6 at 8 T. Although extrapolation errors should be allowed for, it is fairly obvious that present design limits for the permissible resistivity increase of the copper stabilizer at 8 T (25%) can not be reconciled with a reasonable operating schedule of a fusion plant.

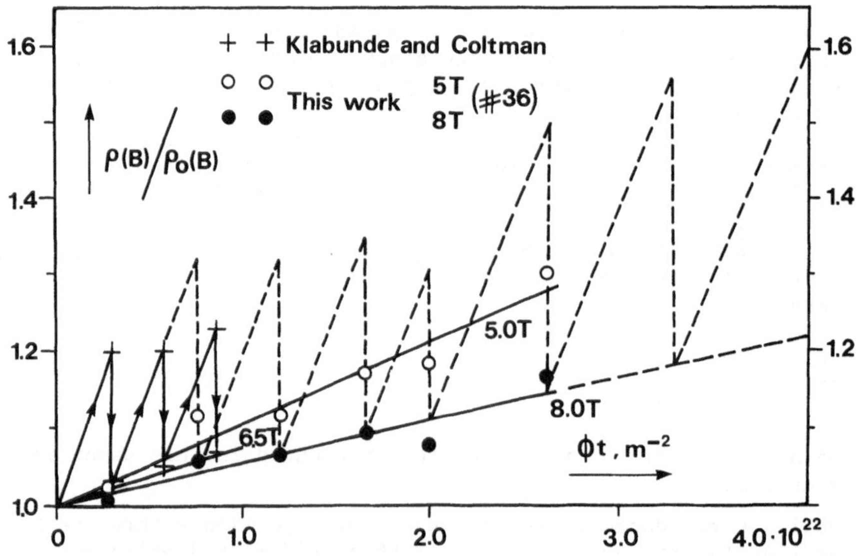

Fig. 7. Complete simulation cycle (dashed lines) for the change of stabilizer resistivity with neutron fluence over the plant lifetime. The in-situ low fluence data of Ref. 32 (+++) are used to extrapolate the present post-annealing data at 8 T (oo) to a fluence of 4×10^{22} m^{-2} (E > 0.1 MeV).

ACKNOWLEDGEMENTS

It is a pleasure to acknowledge the close and most fruitful cooperation with Dr.Peter Hahn. I wish to thank R.C.Birtcher, B.S.Brown and L.R. Greenwood, Argonne, M.W.Guinan, Livermore, D.C.Larbalestier, Madison, C.L. Snead, jr., Brookhaven, and R.R.Coltman, jr., Oak Ridge for valuable discussions and for providing data prior to publication. This work was supported in part by "Bundesministerium für Wissenschaft und Forschung", Wien.

REFERENCES

1. H. R. Kerchner, D. K. Christen, C. E. Klabunde, S. T. Sekula, and R. R. Coltman, jr., Low temperature irradiation study of flux line pinning in type-II superconductors, Phys.Rev. B27: 5467 - 5478 (1983).
2. R. Meier-Hirmer, H. Küpfer, and H. Scheurer, Experimental study on the collective theory of flux pinning in a high-field superconductor, Phys.Rev. B31: 183 - 192 (1985).
3. P. Müller, G. Wallner, W. Weck, and F. Rullier-Albenque, Ion and neutron damage studies of A-15 superconductors, J.Nucl.Mat. 108 & 109: 585 - 592 (1982).
4. A. R. Sweedler, D. E. Cox, and S. Moehlecke, Neutron irradiation of superconducting compounds, J.Nucl.Mat. 72: 50 - 69 (1978).
5. S. T. Sekula, Effect of irradiation on the critical currents of alloy and compound superconductors, J.Nucl.Mat. 72: 91 - 113 (1978).
6. A. R. Sweedler, C. L. Snead, and D. E. Cox, Irradiation effects in superconducting materials, in: "Treatise on Materials Science and Technology", Vol. 14, Metallurgy of Superconducting Materials, T. Luhmann and D. Dew-Hughes, eds., Academic Press, New York (1979), pp. 349 - 426.
7. H. W. Weber, Neutron irradiation effects on alloy superconductors, J.Nucl.Mat. 108 & 109: 572 - 584 (1982).
8. L. R. Greenwood, Review of source characterization for fusion materials irradiations, BNL-NCS-51245, Brookhaven National Laboratory, Upton, New York (1980), p. 75.
9. L. R. Greenwood, Neutron source characterization and radiation damage calculations for material studies, J.Nucl.Mat. 108 & 109: 21 - 27 (1982).
10. H. W. Weber, H. Böck, E. Unfried, and L. R. Greenwood, Neutron dosimetry and damage calculations for the TRIGA MARK-II reactor in Vienna, J.Nucl.Mat., submitted for publication.
11. F. G. Perey, Least squares dosimetry unfolding: the program STAY'SL, ORNL/TM-6062, Oak Ridge National Laboratory, Oak Ridge, Tennessee (1977); modified by L. R. Greenwood (1979).
12. L. R. Greenwood and R. K. Smither, SPECTER: neutron damage calculations for materials irradiations, ANL/FPP/TM-197, Argonne National Laboratory, Argonne, Illinois (1985).
13. C. C. Baker and M. A. Abdou, eds., STARFIRE - A commercial tokamak fusion power plant study, ANL/FPP-80-1, Argonne National Laboratory, Argonne, Illinois (1980)
14. P. Hahn, H. W. Weber, R. C. Birtcher, B. S. Brown, L. R. Greenwood, and M. W. Guinan, Neutron irradiation of superconductors and damage energy scaling of different neutron spectra, This volume, paper GP-7.
15. F. Nardai, H. W. Weber, and R. K. Maix, Neutron irradiation of a broad spectrum of NbTi superconductors, Cryogenics 21: 223 - 233 (1981)
16. H. W. Weber, F. Nardai, C. Schwinghammer, and R. K. Maix, Neutron irradiation of NbTi with different flux pinning structures, in: "Advances in Cryogenic Engineering-Materials", vol. 28, Plenum Publishing

Corporation, New York (1982), pp. 329 - 335.
17. P. Hahn, H. Hoch, H. W. Weber, R. C. Birtcher, and B. S. Brown, 5 K neutron irradiation and thermal cycling of NbTi superconductors, in Proc. LT-17, U. Eckern, A. Schmid, W. Weber, and H. Wühl, eds., Elsevier Science Publishers B.V., Amsterdam (1984), pp. 1297 - 1298.
18. M. Söll, S. L. Wipf, and G. Vogl, Change in critical current of superconducting NbTi by neutron irradiation, in: "Proc. of the 1972 Applied Superconductivity Conference", IEEE Publ. 72 CHO682-5-TABSC (1972), pp. 434 - 439.
19. C. L. Snead, jr., Low-temperature 30 GeV-proton effects on critical properties of type-II superconducting filamentary conductors, J.Nucl.Mat. 72: 190 - 197 (1978).
20. D. C. Larbalestier and A. W. West, New perspectives on flux pinning in Nb-Ti composite superconductors, Acta metall. 32: 1871 - 1881 (1984).
21. H. Hoch, H. W. Weber, and E. Hörl, work in progress.
22. H. Hoch, Kritische Ströme in NbTi-Supraleitern bis H_{c2}, Diploma Thesis, Technische Universität, Wien (1983), unpublished.
23. D. C. Larbalestier, Niobium-titanium superconducting materials, in: "Superconductor Materials Science", S. Foner, B. B. Schwartz, eds., Plenum Publishing Corporation, New York (1981), pp. 133 - 199.
24. M. W. Guinan, R. A. Van Konynenburg, and J. B. Mitchell, Effects of low-temperature fusion neutron irradiation on critical properties of a monofilament niobium-tin superconductor, UCID-20048, Lawrence Livermore National Laboratory, Livermore, California (1984).
25. C. L. Snead, jr., and M. Suenaga, Synergism between strain and neutron irradiation in filamentary Nb_3Sn conductors, Appl.Phys.Lett 37: 659 - 661 (1980).
26. C. L. Snead, jr., D. M. Parkin, and M. W. Guinan, High energy neutron damage in Nb_3Sn: changes in critical properties and damage energy analysis, J.Nucl.Mat. 103 & 104: 749 - 754 (1981).
27. H. Küpfer, R. Meier-Hirmer, and T. Reichert, Field dependent change of the critical current density in neutron irradiated A15 superconductors with grain boudary pinning, J.Appl.Phys. 51: 1121 - 1126 (1980).
28. D. B. Smathers, K. H. Marken, P. J. Lee, D. C. Larbalestier, N. K. McDonald, and P. M. O'Larey, Properties of idealized designs of Nb_3Sn composites, IEEE Trans.Magn. MAG-21: 1133 - 1136 (1985).
29. C. L. Snead, jr., and D. M. Parkin, Effect of neutron irradiation on the critical current of Nb_3Sn at high fields, Nucl.Technol. 29: 264 (1976).
30. C. L. Snead, jr., private communication.
31. D. C. Larbalestier, private communication.
32. C. E. Klabunde and R. R. Coltman, jr., The magnetoresistivity of copper irradiated at 4.4 K by spallation neutrons, DOE/ER-0113/3, U.S. Department of Energy, Washington, D.C., (1984).

NEUTRON IRRADIATION OF SUPERCONDUCTORS AND DAMAGE ENERGY SCALING OF DIFFERENT NEUTRON SPECTRA

Peter A. Hahn* and Harald W. Weber
Atominstitut der Österreichischen Universitäten
Vienna, Austria

Michael W. Guinan
Lawrence Livermore National Laboratory, Livermore
California

Robert C. Birtcher, Bruce S. Brown and
Lawrence R. Greenwood
Argonne National Laboratory, Argonne, Illinois

ABSTRACT

Three different neutron sources were used to irradiate identical sets of NbTi superconductors up to about half the lifetime dose of a superconducting magnet in a fusion reactor. Based on a careful source characterization of the TRIGA Mark-II reactor in Vienna, the spallation neutron source IPNS at Argonne and the 14 MeV neutron source RTNS-II at Livermore, the damage energy cross sections were calculated for four different types of NbTi alloys (42, 46.5, 49 and 54 wt% Ti). The experimental results on the variations of critical current densities J_c with neutron dose are found to scale within the experimental uncertainties with the appropriate damage energy cross sections. This first explicit proof of damage energy scaling for J_c-variations in superconductors is considered to be most valuable for the evaluation of radiation damage in superconductors under fusion reactor conditions.

INTRODUCTION

The primary objective of this work was to determine the effects of neutron irradiation obtained from different sources on the critical current density J_c in commercially manufactured NbTi superconductors suited for magnet construction in future fusion reactors. The crucial question raised by magnet designers, whether the results from various irradiation experiments obtained so far can be directly applied to the superconductor at the magnet location, suggested the experimental verification of the damage energy concept developed in recent years. This method is increasingly used to characterize and compare the results of different neutron irradiation studies.

*Present address: Nuclear Chemistry Division, Lawrence Livermore National Laboratory, Livermore, California

The changes of the critical parameters of a superconductor, such as T_c, H_{c2} and J_c during irradiation are primarily due to the total damage deposited in the material. The calculation of this quantity, measured in displacements per atom (dpa) or more practically eV/atom, requires a complete knowledge of the neutron spectrum, the fluence obtained during irradiation, the differential neutron cross sections and the primary recoil distribution of the irradiated material. In metals, a significant contribution to the total damage is caused by high energy neutrons; hence, the flux of several irradiation facilities is frequently quoted for energies greater than 0.1 or greater than 1.0 MeV, respectively. A large fraction of the incident neutron energy goes into electronic excitation, activation and thermal vibrations of the lattice atoms rather than into production of displacement cascades. Moreover, the majority of produced defects, may they be point defects or entire cascades, are subject to recombination during the process of formation, thus the temperature maintained during irradiation affects the number of "surviving" defects. An adequate model of the interaction between neutrons and the atoms of the irradiated metal has been discussed previously.[1,2]

Based on the knowledge of differential elastic and inelastic neutron scattering cross sections,[3] $\sigma(E)$, the primary recoil energy distribution $T(E)$ for each individual energy-flux group of a given neutron spectrum has been calculated using the SPECTER-computer code.[4] The total displacement energy cross section $\langle\sigma \cdot T\rangle$ averaged over all neutron energies can be expressed as

$$\frac{\int \sigma(E) \cdot T(E) \cdot \frac{d\phi}{dE}\, dE}{\int \frac{d\phi}{dE}\, dE} \qquad (1)$$

Multiplication of this quantity with the actual fluence obtained during irradiation yields the average energy per atom deposited for defect production. As indicated earlier, the fluence (flux ϕ x irrradiation time t) may be specified for $E > 0.1$ or $E > 1.0$ MeV, respectively. The resulting product, however, remains constant for the damage energy cross sections are scaled accordingly.

If there is more than one element in the irradiated sample, which is the case in the investigated NbTi alloys, one has to calculate the contribution of each element to the total damage energy cross section. Linear scaling with the atomic percentage c_i is employed during summation:

$$\langle\sigma \cdot T\rangle_{\text{Alloy}} = \sum_{\text{Elements}} \langle\sigma \cdot T\rangle_i \cdot c_i \qquad (2)$$

More advanced theoretical models for binary systems[5] indicate that in the present case, this simplification does not represent a significant source of error. All irradiations of the NbTi superconductors described in this work have been carried out at room temperature. Irradiation experiments at 5.0 K with a subsequent annealing cycle to 300 K have been completed and will be published.[6]

EXPERIMENT

Materials

All NbTi conductors were manufactured by Brown Boveri & CIE/Switzerland. Two identical sets containing 14 single core conductors and three multifilamentary conductors have been selected for the irradiation programs at

IPNS/Argonne and RTNS-II/Livermore. A detailed characterization of all irradiated materials is given in Ref. 7. One of the multifilamentary conductors has been developed for the LCT (Large Coil Test) program currently carried out at Oak Ridge National Laboratory. An additional set has been irradiated earlier in the TRIGA Mark-II reactor in Vienna.[7] All single core conductors have undergone an annealing step at 400°C prior to final cold work which was varied (0 - 91%) as well as the Ti concentration (42, 49 and 54 wt% corresponding to 58, 65 and 70 at% respectively). Whereas the Ti-rich samples (54 wt% Ti) exhibit significantly higher current densities in manetic fields up to 6 T than materials with 42 wt% Ti, their performance at the projected field of 8 T is declining rapidly because of their lower H_{c2}. For the present standard conductors, including the LCT-conductor, a composition of 46.5 wt% (63 at%) Ti has been chosen. The single core conductors were all 0.4 mm in diameter, their copper to superconductor ratio (with respect to volume) was about 1:1. All samples were prepared to meet the requirements of the experimental setup measuring J_c and the spatial constraints in the irradiation facilities. Straight pieces, 47 mm in length, were cut and polished to remove the oxide layer from the copper stabilizer. Thin potential leads were soldered onto each sample with a relatively small gauge length of 6 mm to avoid excessive fluence gradients along the sample arising from the strong inhomogenious flux distribution inherent to all accelerator based neutron sources.

Neutron Irradiation

One set of samples has been irradiated with 14.6 MeV neutrons at RTNS-II in 3 cycles to fluences of 2.22×10^{21}, 4.26×10^{21} and 5.82×10^{21} n/m^2 averaged over all specimens. Small Nb foils attached to the samples were used to monitor the fluences. After each cycle the whole batch was shipped to Argonne for J_c-measurements. Before the second and third cycle the positions of the samples have been rotated with respect to the beam spot of the target in order to compensate the differences in neutron fluence between individual samples.

Another set has been irradiated at ambient temperature at IPNS in 3 cycles up to fluences ($E > 0.1$ MeV) of 0.53×10^{22}, 0.86×10^{22} and 1.96×10^{22} n/m^2 with spallation neutrons. These numbers are based on an activation analysis of Ni dosimetry wires attached to the samples during irradiation.

The results on an identical set of materials irradiated in the TRIGA-reactor were already reported by Nardai et al.[7] and are used for comparison. Although there have been six irradiation cycles carried out up to a total fluence of about 1.75×10^{23} n/m^2, only data for the first three cycles, 8.8×10^{20}, 8.8×10^{21} and 4.4×10^{22} n/m^2 ($E > 0.1$ MeV), are quoted in this work because the results at higher fluences exceed the range of data obtained from the experiments at RTNS-II and IPNS. It should be noted that the absolute errors in determining the neutron fluence are typically estimated at 10%.

J_c-Measurements

After each irradiation cycle the critical current densities have been measured at 4.2 K as a function of the magnetic field (perpendicular to the wire axis) in the range between 1.0 and 8.0 T in steps of 0.5 T. Each sample was mounted free from mechanical stress onto a sample holder in a manner to prevent any movement caused by the Lorentz force when magnetic fields were present during J_c measurement. A standard four probe technique was employed to determine J_c applying a voltage criterion of 50 μV/cm. The implications of "defining" J_c by means of a certain voltage criterion have been discussed earlier.[8] In the case of single core conductors, this does not represent a major problem for their transition from the superconducting to the normal state occurs quite rapidly in contrast to multifilamentary conductors with a rather smooth I-V characteristic near J_c.

ANALYSIS

Neutron spectra

Based on elastic and inelastic scattering cross sections, i.e. nuclear data for (n,xn), (n,α) and (n, γ), reactions listed in ENDF-B/V,[3] the differential fluxes $d\phi/dE$ in all 100 energy groups have been calculated using the STAYS'L computer code.[9] An activation analysis[10] of the central irradiation thimble of the TRIGA Mark-II reactor in Vienna with subsequent STAYS'L evaluation revealed no significant deviation from the spectrum originally specified by the contractor General Atomics/San Diego. The neutron spectra of the irradiation facilities used are illustrated in Fig. 1. where the flux per unit lethargy, $d\phi/d\ln E$, is plotted on a linear scale. In this representation the area under the spectrum or sections thereof is directly proportional to the total number of neutrons or its fraction within a specified energy range.

In characterizing the neutron spectra of the three facilities, one has to deal with different source strengths which are summarized in Table 1. The spectra have been normalized in a way that the total of all group fluxes is equal to 1.0 which is useful if one is comparing the percentage of neutrons within a particular energy range between different neutron sources regardless of their total flux. In addition, the theoretically modelled spectra at the magnet location for the STARFIRE[11] and MARS (Mirror Advanced Reactor Study)[12] designs are presented in Fig. 2., viewing the regions with the least amount of shielding between the plasma and the superconductor, which can be regarded as a worst case assumption.

Damage energy cross sections averaged over the entire neutron spectrum or scaled to E > 0.1 or E > 1.0 MeV respectively, have been calculated for pure Nb, Ti and several alloys. The numbers listed in Table 2. are applicable to fluences quoted for E > 0.1 MeV. The damage energy cross sections

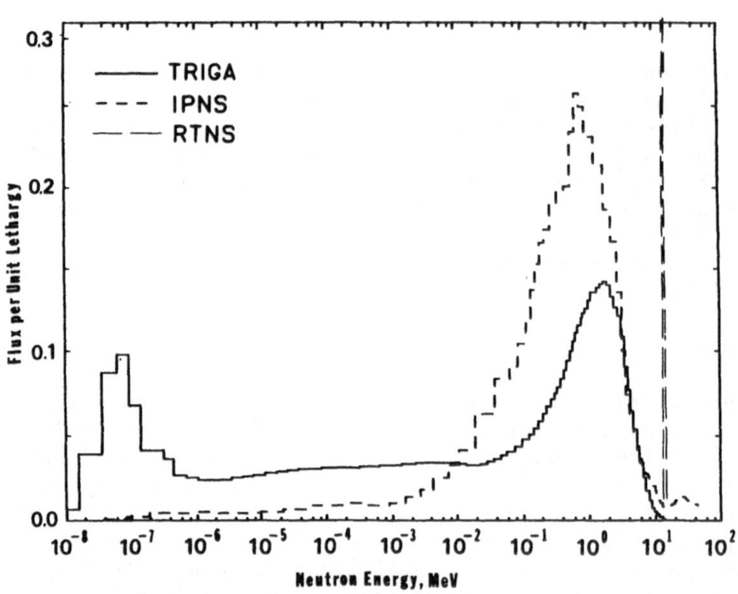

Fig. 1. Spectra of the neutron sources used during the irradiations.

Table 1. Flux densities of different neutron sources and calculated fusion reactor designs, Units in $n/m^2 \cdot s$.

Source/ design	Total	E > 0.1 MeV	E > 1.0 MeV
TRIGA-reactor, Vienna[a]	$1.9x10^{17}$	$7.5x10^{16}$ (38.8%)	$3.9x10^{16}$ (20.3%)
IPNS/REF[b]	$1.0x10^{16}$	$6.9x10^{15}$ (71.3%)	$2.7x10^{15}$ (27.5%)
RTNS-II[c]	----------$1.15x10^{16}$ of 14.6 MeV neutrons----------		
MARS[d]	$1.2x10^{14}$	$6.0x10^{13}$ (50.0%)	$1.1x10^{13}$ (9.0%)
STARFIRE[e]	$3.1x10^{13}$	$1.7x10^{13}$ (55.0%)	$5.0x10^{12}$ (16.0%)

[a]At 250 kW reactor power.
[b]At a beam current of 8 μA of 450 MeV protons.
[c]At beam spot under operation with 150 mA d^+-beam, 380 keV.
[d]In Yin-Yang region assuming 4.3 MW/m^2 neutron wall loading at blanket of center cell.
[e]Assuming 3.6 MW/m^2 neutron wall loading.

for both fusion reactor designs are found to be slightly smaller than those of the TRIGA-spectrum and are very close to the numbers obtained for IPNS.

Results

The results of all J_c-measurements prior and after the irradiations have been recorded and are available in files under VAX/VMS. The selection of data presented here primarily emphasizes the validity of damage energy scaling and also indicates the effect of neutron irradiation on materials with different Ti-concentration and degree of final cold work (fcw). A more detailed analysis is given in Ref. 6.

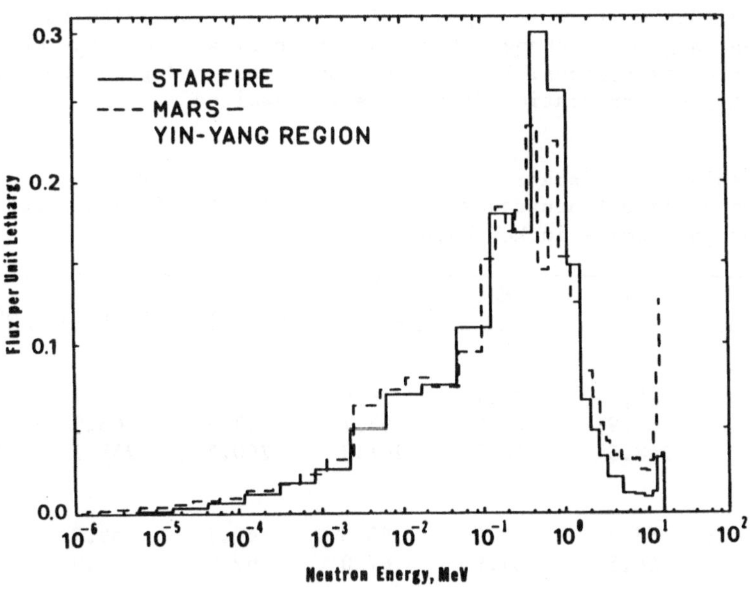

Fig. 2. Calculated neutron spectra at the magnet location for two fusion reactor designs.

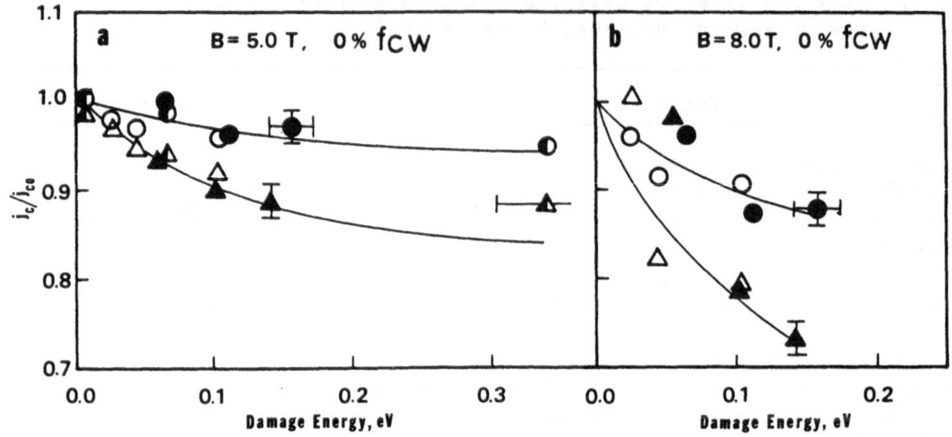

Fig. 3a. Fractional change of the critical current density J_c at 4.2 K and 5.0 T as a function of the deposited damage energy for Nb-42 wt% Ti(○◐●) and Nb-54 wt% Ti (△◭▲) with no fcw. The light symbols represent the results of the irradiation at IPNS/REF, the shaded ones those of the TRIGA-reactor, and the bold ones those of the 14 MeV neutron irradiation at RTNS-II

3b. Equivalent to Fig. 3a. but for B = 8.0 T.

The fractional changes of J_c as a function of the damage energy at 4.2 K and 5.0 T for all 3 irradiations are illustrated in Fig. 3a. The results for both Nb-42 wt% Ti (○◐●) and Nb-54 wt% Ti (△◭▲) fit a smooth line within the experimental uncertainties of the dosimetry results (+-10%) and of the quantity J_c/J_{co} which depends slightly on the absolute value of the critical current I_c (typically 2%). The reasons for the observed differences in J_c-degradation between Nb-42 wt% Ti and Nb-54 wt% Ti are not completely understood at this time, particularly in consideration of the lacking final cold work in these materials which is required for a well defined microstructure and efficient flux pinning. As displayed in Fig. 3b., this trend was found to be enhanced by a factor of about 3.0 at higher fields (8.0 T) where the performance of Nb-54 wt% Ti dropped to 70% of its value prior to irradiation. We assume that a gradual change of the upper critical field H_{c2} during neutron irradiation is responsible. H_{c2} has been measured prior to irradiation, corresponding post-irradiation data are still awaited.

Table 2. Displacement damage energy cross sections $\langle \sigma \cdot T \rangle_{E} > 0.1$ MeV for different neutron sources and calculated fusion reactor designs, units in keV barn

Source Design	Ti	Nb	Nb-42wt%Ti	Nb-46.5wt%Ti	Nb-49wt%Ti	Nb-54wt%Ti
TRIGA-reactor,						
Vienna	81.5	70.7	76.9	77.4	77.7	78.2
IPNS/REF	64.9	62.0	63.7	63.8	63.9	64.1
RTNS-II	248.5	280.4	261.9	260.3	259.7	258.1
MARS,						
Yin-Yang region	58.3	62.9	60.3	60.1	59.9	59.7
STARFIRE	66.5	67.9	67.0	67.0	67.0	66.8

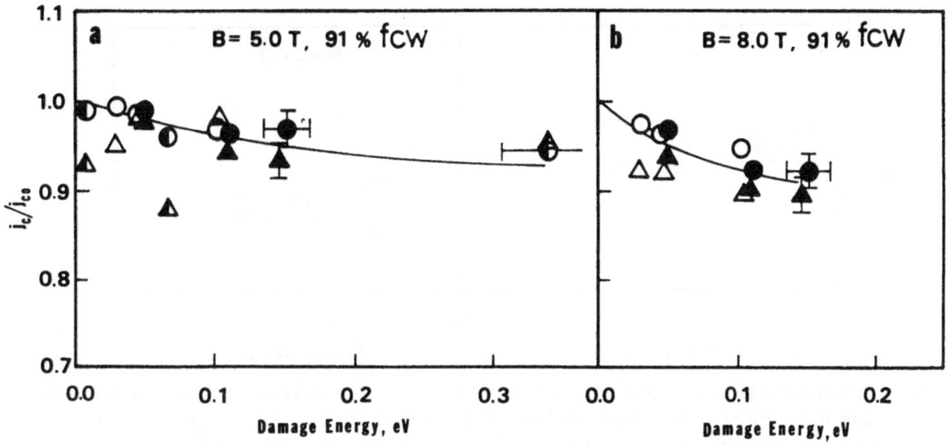

Fig. 4a. Fractional change of the critical current density J_c at 4.2 K and 5.0 T as a function of the deposited damage energy for heavily cold worked Nb-42 wt% Ti(○◐●) and Nb-54 wt% Ti (△◭▲) the light symbols exhibit the results of the irradiation at IPNS/REF, the shaded ones those of the TRIGA-reactor and the bold ones those of the 14 MeV neutron irradiation at RTNS-II.
4b. Equivalent to Fig. 4a. but for B = 8.0 T.

In contrast, the response to neutron irradiation of heavily cold worked superconductors depicted in Figs. 4a. and 4b. is significantly smaller. Apparently their structure with a high density of dislocation cells and cell walls which is primarily responsible for an improvement in the flux pinnning mechanism and consequently in J_c, is equally affected by irradiation in both materials. Considering that the absolute increase in the normal state resistivity ρ_o is roughly equal in both cold worked (Figs. 4a. and 4b.) and non-cold worked NbTi (Figs. 3a. and 3b.), one expects higher J_c-degradation in materials with 0% fcw. The change of J_c can be written as

$$\frac{\Delta J_c}{J_{co}} = -c_1 \frac{\Delta\rho}{\rho_o} + c_2 \frac{\Delta H_{c2}}{H_{c2}} + c_3 \frac{\Delta T_c}{T_c} \qquad (3)$$

with J_{co}, ρ_o, H_{c2} and T_c representing the values prior to irradiation. Our experimental observations demonstrate that NbTi with 0% fcw and initially lower ρ_o are more affected by irradiation. A more detailed analysis of the effects of neutron irradiation with respect to the metallurgical parameters of NbTi superconductors will be presented later. For practical applications, the absolute value of J_c prior to irradiation is as important as their performance during neutron irradiation. In cold worked (91%) Nb-54 wt% Ti, J_c was found to be 1.7×10^9 A/m^2 at 5.0 T and 0.61×10^9 A/m^2 at 8.0 T, whereas the same material with 0% fcw yielded 8.3×10^8 and 2.2×10^8 A/m^2 respectively.

As further evidence for the validity of damage energy scaling, we present the results on the Swiss LCT-conductor in Fig. 5. At 8.0 T, the highest magnetic field in which NbTi superconductors are expected to be employed in projected magnet designs for fusion reactors, its performance decreases roughly 10% after half the lifetime fluence.

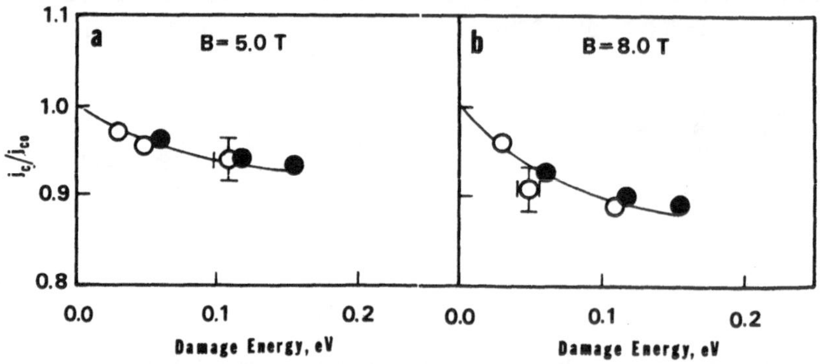

Fig. 5. Fractional change of J_c at 4.2 K as a function of the deposited damage energy for the Swiss LCT-conductor. The light symbols represent the result of the irradiation at IPNS/REF, the bold ones those of the 14 MeV irradiation at RTNS.

CONCLUSIONS

In general, the results of all three irradiation experiments on NbTi superconductors were found to agree when the obtained fluences are scaled with the appropriate damage energy cross sections. This experimental observation enables magnet designers to precisely determine the effects of neutron irradiation in magnet components since the effects of a fusion spectrum can be inferred from irradiations in other spectra. Considering technological applications of NbTi on a large scale, materials with multiple thermomechanical treatment are suited best for they have both higher initial critical current density and are least effected by irradiation.

ACKNOWLEDGEMENTS

The authors are gratefully indebted to the technical staff of IPNS and RTNS-II for their support during the experiments and Ruth M. Nuckolls for handling the dosimetry. This work was performed under the auspices of the U.S. Department of Energy by Lawrence Livermore National Laboratory under contract No. W-7405-Eng-48 and by the Federal Ministry of Science and Research, Wien, Austria.

REFERENCES

1. L. R. Greenwood, J. Nucl. Mater. 108 & 109: 21-27 (1982).
2. C. Lehmann in: "Interaction of Radiation with Solids and Elementary Defect Production," North Holland Publishing Company, Amsterdam (1977), pp 88-92, pp 105-214.
3. "Evaluated Nuclear Data File, Version V," National Neutron Cross Section Center, Brookhaven National Laboratory (1979).
4. L. R. Greenwood and R. K. Smither, ANL/FPP/TM-197, Argonne National Laboratory, Argonne, IL 60439 (1985).
5. D. M. Parkin and C. A. Coulter, J. Nucl. Mater. 85 & 86: 611 (1979).
6. P. A. Hahn, Thesis, Technical University Vienna (1984), unpublished.
7. F. Nardai, H. W. Weber and R. K. Maix, Cryogenics 21: 223-233 (1981).
8. A. F. Clark and J. W. Ekin, IEEE Trans. Mag. 13 (1977).
9. F. G. Perey, "Least Squares Dosimetry Unfolding: The Program STAYSL," ORNL-TM-6062 (1977); modified by L. R. Greenwood (1979).
10. H. W. Weber, H. Böck, E. Unfried and L. R. Greenwood, to be published in J. Nucl. Mater.
11. "STARFIRE - A Commercial Tokamak Fusion Power Plant Study," C. C. Baker and M. A. Abdou, eds., ANL/FFP-80-1, vol. 1 & 2 (1980).
12. "Mirror Advanced Reactor Study," UCRL-53480 (1984).

DISORDERING MECHANISM IN IRRADIATED AND

QUENCHED A15 TYPE COMPOUNDS

R. Flükiger

Kernforschungszentrum Karlsruhe
Institut für Technische Physik
Karlsruhe, Federal Republic of Germany

ABSTRACT

A new diffusion mechanism describing the changes of the long range order parameter in A15 type compounds after both quenching from high temperatures or low temperature irradiation with high energy particles is presented. It is based on the occupation of a small concentration of non-equilibrium or "virtual" sites centered halfway between two neighbouring A atoms on 6c sites, arising from the instability of a single 6c vacancy recently found by Welch and coworkers by pair potential calculations. After low temperature irradiation, the occupation of this interstitial site creates the necessary conditions for A↔B site exchanges over several interatomic distances by focused replacement collision sequences. At occupied virtual sites, atomic "overlapping" occurs not only between A atoms on the chains or between A and B atoms (due to deviations from perfect ordering), but also between B atoms on BBB sequences. The latter are retained after low temperature irradiation only and are responsible for the observed lattice expansion and static displacements.

INTRODUCTION

The variation of the superconducting transition temperature, T_c, after irradiation with high energy particles or after fast quenching has been studied for various A15 type compounds in the last decade[1]. From the wealth of published data, it follows that in high T_c A15 type compounds, e.g. V_3Si, Nb_3Ge, Nb_3Sn, ... a strong decrease of T_c is always observed after irradiation doses exceeding certain values (10^{18} neutrons/cm^2), 10^{14} sulfur ions/cm or 10^{15} He ions/cm) for irradiation temperatures $T_{irr} \leq 150^{o}C$. A decrease of T_c after quenching was observed in V_3Au[2,3], Nb_3Pt[4,5], Nb_3Al[6,7], Nb_3Ga[8,9]. In analogy to the quenching case, the initial decrease of T_c at low doses is generally attributed to a decrease of the long range atomic order parameter, S. The first experimental evidence for the decrease of S in A15 type compounds after irradiation was furnished by Sweedler et al.[1] on the basis of neutron diffraction measurements on neutron irradiated Nb_3Al. More recently, a decrease of S in the same compound has also been reported by Schneider et al.[10] after irradiation with H^+ and Ne^+ ions of several hundreds keV. A competing mechanism to this picture of homogeneous disordering has been advanced by Pande[11], who postulated that the formation of disordered microregions (or depleted zones) of sizes below the coherence length (i.e. $\leq$ 5.0 nm) could lead to a decrease of T_c by proximity effects.

Due to the complexity of the problem, involving simultaneously additional effects as lattice expansion[1] and static atomic displacements[12], it is, however, difficult to decide whether homogeneous (disordering) or inhomogeneous (depleted zones) effects have a dominant influence on the initial decrease of T_c in irradiated A15 type compounds. The arguments given in Refs. 1 and 5 strongly support the hypothesis of a homogeneous decrease of the long range order parameter over the whole sample volume. They are confirmed in a more detailed analysis carried out recently by the author[13]. The question arises: how is it possible that a homogeneous decrease of S implying site exchanges over several lattice spacings can occur during irradiations at temperatures $T \leq 150^{\circ}C$, where no noticeable thermal diffusion takes place. It is the aim of the present paper to show that disordering in irradiated A15 type compounds occurs by focused replacement collision sequences, which are only possible in the particular configuration of the irradiated A15 structure. A model is proposed on the basis of occupied interstitial (or "virtual") sites, placed midway between two 6c sites. As recently shown by Welch et al.[15], this configuration is energetically more favourable than a 6c vacancy. It will be shown that this picture explains as well the occurrence of the lattice expansion as of static atomic displacements in irradiated A15 type compounds.

THE "VIRTUAL" LATTICE SITE IN THE A15 STRUCTURE

In the tightly packed A15 structure, the site exchanges required for a change in atomic ordering, at high as well as at low temperature irradiation, are ordinarily assumed to occur by a vacancy diffusion mechanism. In a bcc lattice, such a mechanism would be described by an atom jumping into a neighbouring vacancy, thus creating a new vacant site. Atoms and vacancies are expected to undergo a large number of position exchanges, in order that a very small number of vacant lattice sites ($\sim 10^{-3}$) would be sufficient to induce a substantial diffusion by vacancies. In the A15 structure, however, the diffusion mechanism is expected to be considerably more complex than for the simple bcc structure, essentially due to the covalent bonding[16] between two A atoms lying on neighbouring 6c sites (in the perfectly ordered case, the A atoms of the A15 type compound A_3B are at the 6c sites, the B atoms at the cubic 2a sites). This leads to AA interatomic distances being considerably shorter than the sum of two A atomic radii, and thus to highly nonspherical shapes for the atoms lying on the 6c sites.

Welch et al.[15] have recently shown by means of pair potential calculations that such an individual vacancy of an A atom on a 6 c site is unstable. They found that the state of lower energy corresponds to a configuration where one of the two A atoms adjacent to the 6c vacancy is shifted towards a new site which is equidistant from the next two A neighbours (see Fig. 1a). This kind of vacancy can either be called "split vacancy" , as proposed by Welch et al.[15] or "negative crowdion", following the terminology of Seeger[17]. It has to be noted that the corresponding new site is not an equilibrium site of the A15 structure, but coincides with the region of overlapping between two neighbouring A atoms on 6c sites and will thus be called "virtual" site in the following. Two possibilities of site exchange in the A15 structure are illustrated in Fig. 2. In Fig. 2a, the jump of a B atom into a 6c site is shown, which simultaneously requires the motion of an A atom from a virtual site to an equilibrium 6c site. In Fig. 2b, an A atom jumps into a B vacancy, followed by the motion of an adjacent A atom into the vacant virtual site in both cases, it appears that the diffusion mechanism in

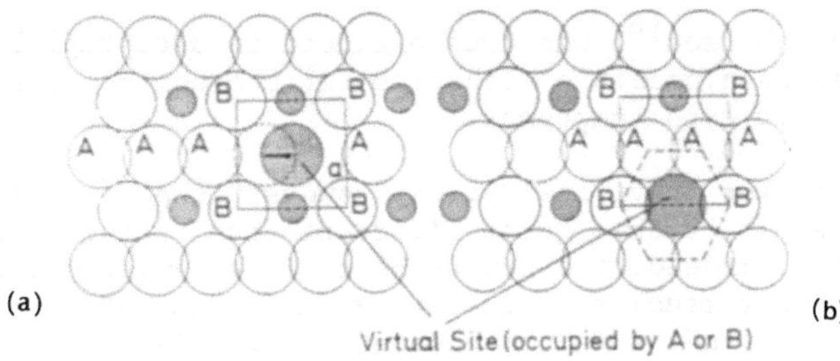

Fig. 1. Occupation of the virtual site in the A15 structure as a consequence of the instability of single 6c vacancies: a) in a chain parallel to the image plane, b) in a chain perpendicular to the image plane, showing the hexagonal arrangement of the complex around this site. The small circles correspond to the overlapping region between two A atoms belonging to chains perpendicular to the { 100 } plane. a is the lattice parameter.

the A15 structure comprises at least two steps, in contrast to the one-step mechanism for bcc structures. As suggested by Welch et al..[15] various multi-step processes can also be imagined, leading to even more possibilities for site exchanges by diffusion mechanism in the A15 structure. In the present case of the A15 structure, the situation is complicated by the fact that a jump into the virtual site requires simulaneously a rearrangement of more than one neighbour atom.

FOCUSED REPLACEMENT COLLISION SEQUENCES IN DAMAGED A15 CRYSTALS

As first recognized by Seeger[17], the transport energy in an irradiated crystal is focused along certain crystallographical directions of the lattice. The so-called "focused replacement collision sequences" can carry both energy and matter, thus transporting the interstitial atoms of Frenkel pairs formed during the primary collision events several interatomic distances away from their associated vacancy. In the A15 structure, replacement collision sequences are possible in the < 100 > and the < 111 > direction, but do not produce A↔B site exchanges, these directions containing only one type of atom, A or B, respectively. The only focusing direction where collision sequences could in principle produce A↔B site exchanges is the <102> direction. In the unirradiated state, however, the atomic sequence in the < 102 > direction is • ABA•ABA• , the space between two ABA sequences being occupied by the region of overlap between two A atoms belonging to perpendicular chains (small circles in Fig. 3). This renders the occurrence of A↔B site exchanges quite unlikely, as pointed

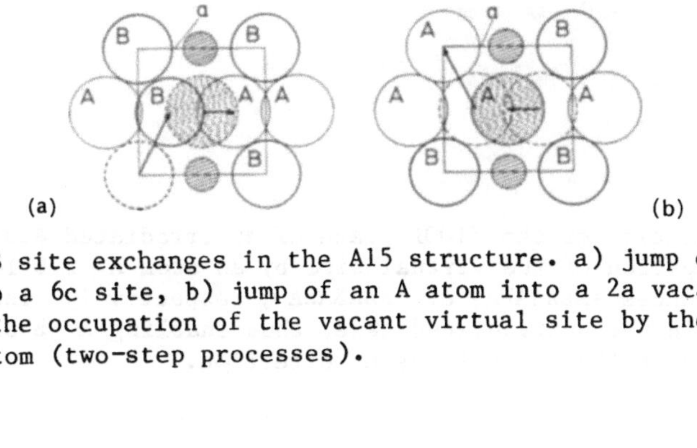

Fig. 2. A↔B site exchanges in the A15 structure. a) jump of a B atom into a 6c site, b) jump of an A atom into a 2a vacancy, followed by the occupation of the vacant virtual site by the neighbouring A atom (two-step processes).

out by Pande[18] and Schulson[19], who thus excluded focused replacement collision sequences as a possible mechanism for the homogeneous decrease of S in irradiated A15 type compounds.

With the occupation of virtual sites, however, the situation in irradiated A15 crystals becomes quite different from that encountered prior to irradiation. Indeed, the new sequence in the <102> direction including an occupied virtual site is now •ABA•ABAAABA•ABA• or •ABA•ABABABA•ABA• , depending on the occupation of this site by an A or a B atom, as shown in Fig. 3. Due to the transport of matter caused by the large number of focusing replacement collision (each entering particle removes up to 100 atoms), the site exchanges between normal and virtual sites will occur at a rapid sequence, thus leading to a high mobility of the atoms at low temperatures, even for very small concentrations of occupied virtual sites. The virtual sites will alternatively be occupied by either A or B atoms, thus constituting a "bridge" between ABA sequences and allowing A↔B exchanges over several interatomic distances. This is the necessary condition for a homogeneous decrease of the degree of ordering over the whole crystal volume after irradiation of an A15 type compound at low temperatures.

THE VARIATION OF THE LATTICE PARAMETER AFTER IRRADIATION

The lattice expansion after irradiation observed in A15 type compounds[1] seems to be connected with particularities of this crystal structure. Indeed, a comparison shows that the transition elements V, Nb and Mo do not exhibit a lattice expansion after irradiation, except when they contain impurities[21].

A very important conclusion can be drawn from Fig. 1b showing that the occupation of the virtual site not only leads to very short interatomic distances and thus to overlapping between A and B atoms, but also between B atoms. As will be shown in the following, the occurrence of BBB sequences (retained by low temperature irradiation only) could furnish the key for understanding the causes of the lattice expansion and the static displacement observed in this class of materials. It is particularly interesting to study the consequence of BBB sequences if B is a nontransition element.

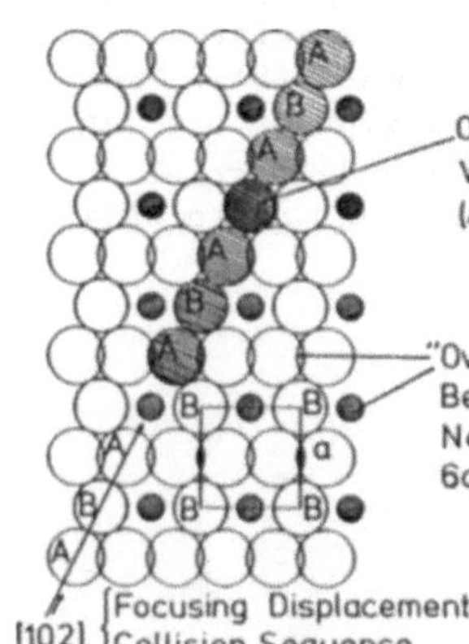

Fig. 3. Representation of the {100} plane of an irradiated A15 lattice. The occupation of the virtual site by an atom A or B leads to the sequences •ABAAABA• or •ABABABA•, respectively, instead of •ABA as in the unirradiated case, thus enabling A↔B site exchanges in the <102> focusing direction.

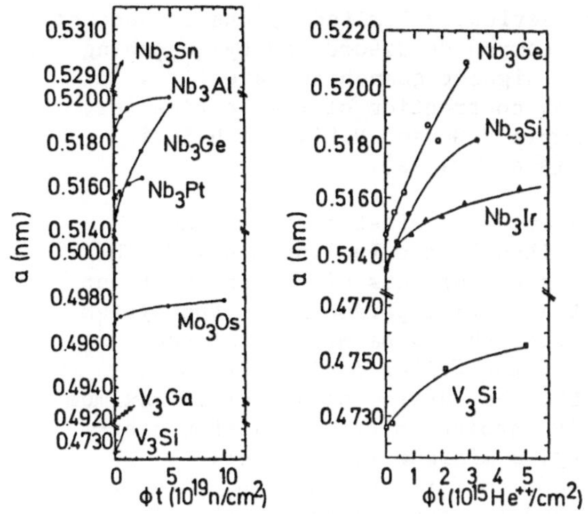

Fig. 4. Lattice parameters of various A15 compounds after irradiation. a) neutrons (E>1 MeV): ● Sweedler et al.[1], ▲ Cox and Tarvin[22], b) He ions (E = 300 keV): ○ Pflüger and Meyer[23], ● Haase and Rudzicka[24], □ Meyer and Linker[25] and △ Schneider and Linker[26].

The observed increase of the lattice parameter in several A15 type compounds is represented in Fig. 4a for neutron and in Fig. 4b for He ion irradiation. In spite of the small number of analyzed systems, it can be concluded that the increase a is considerably smaller if the B atom is a transition element, e.g. Os, Ir or Pt. The largest lattice expansion is observed on systems containing nontransition B elements, as Ge, Si or Sn, independently on the A element. It is interesting that Nb_3Ge[23] and Nb_3Si[24] in Fig. 4b show the same increase Δa with He dose, in spite of the very different atomic radii of Ge and Si. The same conclusion can be drawn from a comparison between Mo_3Ge[27] and Mo_3Si[28] after irradiation with ^{32}S ions, both showing an expansion $\Delta a/a$ exceeding 1.1%. This shows convincingly that there is no simple correlation betwen the lattice expansion rate in irradiated A15 type compounds and the ratio of the radii of the constituents: The value of Δa is rather due to the electronic charge distribution of the B element, which influences the repulsion between the atoms in the complex around the virtual site, the radius ratio being of secondary importance only.

Spherical Atomic Shapes and their Limitations

In spite of the fact that the shapes of the A atoms in the A15 structure are highly nonspherical, there have been several attempts to calculate the A15 lattice parameters using spherical (geometrical) approximations. Two geometrical approaches, the Pauling[20] and the Geller[29] model, based on spherical shapes of the atoms A and B, predict the A15 lattice parameters with an accuracy better than 1% without any consideration of the respective electronic configurations. The present considerations about the lattice parameter changes as a function of increasing damage and disorder, as well as of compositional changes will show the limitations of these geometrical models.

The observed lattice expansion in A15 type compounds after irradiation has been attributed to the effect of increased disorder[1,4], the slope $d(\Delta a)/d(\phi t)$ being assumed to depend on the difference between the atomic radii, characterized by the ratio r_A^G/r_B^G, where r_A^G and r_B^G are the Geller radii of the atoms A and B. This hypothesis was mainly based on the Geller model[21], which postulates atomic "contact" between spherical A and B atoms, the lattice parameter being obtained by the formula $a = \frac{4}{\sqrt{5}}(r_A^G + r_B^G)$. Taking into account site exchanges due to disordering, this formula can be modified to $\Delta a = a(S\neq 1) - a(S=1) = \frac{8}{\sqrt{5}}(1-r_a)(r_A^G - r_B^G)$, where r_a is the probability of occupying 6c sites by A atoms. The validity of this formula is, however, seriously limited by the absence of lattice parameter changes in A15 type compounds disordered by quenching from high temperatures, even after the highest quenching rates[2,30]. In addition, this formula would predict a contraction of the lattice for A15 systems where $r_A^G < r_B^G$, as for example V_3Ga and V_3Si, which is in contrast to the experimental results (see Fig. 4a).

The invariance of the lattice parameter against order parameter changes, $\Delta a(S) = 0$, is correctly described by the Pauling model[20]. The Geller model[29] also fails in describing the effects of stoichiometry on the lattice parameter. For example, the lattice parameter in the system $Nb_{1-\beta}Sn_{\beta}$ as a function of the Sn content should be minimum at the stoichiometric composition, $\beta = 0.25$, if the Geller radius for Sn, 0.144 nm, would be correct. In reality, the lattice parameter is maximum at $\beta = 0.25$ (a = 0.5289 mm). Similarly, stoichiometric V_3Si should exhibit a maximum of a, while in reality, a minimum is observed: a = 0.4724 nm for $\beta = 0.25$ and a = 0.4736 nm for $\beta = 0.20$. In spite of the spherical approximation, the Pauling model thus describes the situation in A15 type compounds far better than the Geller model which should not longer be used.

Deviation from Sphericity and the Electronic Configuration

Based on the proposed virtual site exchange mechanism leading to AB or BB "overlapping", the variation of the lattice parameter in irradiated A15 type compounds can be interpreted as reflecting the electrostatic repulsion would be smallest between two transition elements, intermediate between a transition and a nontransition element and maximum between nontransition elements. It is possible to estimate the effect of AB repulsion on the lattice parameter by studying the variation of the latter as a function of composition in A15 systems where the phase field extends to both sides of stoichiometry. This is the case in the four systems Nb - Ir, Nb - Pt, V - Pt and V - Ga, where the A15 phase is stable within the composition ranges $0.22 \leq \beta \leq 0.28$, $0.20 \leq \beta \leq 0.30$, $0.19 \leq \beta \leq 0.325$ and $0.18 \leq \beta \leq 0.32$, respectively. At compositions $\beta > 0.25$, the number of B atoms exceeds that of available 2a sites, and the quantity (β - 0.25) of B atoms will thus be located on 6c sites, giving rise to AB "overlapping." The repulsive effect of the presence of B atoms on the chain sites can be visualized by the different slopes, $da/d\beta$, at both sides of the stoichiometric composition. It can be seen from Fig. 5 that there is a lattice expansion contribution at $\beta > 0.25$ in the three systems Nb - Pt, V - Pt, and V - Ga, regardless of the respective sizes of A and B atoms.

At $\beta = 0.30$, which corresponds to 6.7% of 6c sites occupied by B atoms or to Bragg-Williams long range order parameters $S_a = 0.78$ and $S_b = 1$, the measured value of a is larger than that obtained by extrapolation from the data below $\beta = 0.25$. The positive difference, Δa, corresponding to a lattice expansion, is 0.0003 nm for Nb_3Pt, 0.0003 nm for V_3Pt and 0.0007 nm for V_3Ga, while no deviation could be observed for Nb_3Ir (see Fig. 5). The positive change of the slope $da/d\beta$ for $\beta > 0.25$ in Nb-Pt and in V-Pt can also be found by applying the Geller model[29], as pointed out by Moehlecke et al.[31], while no change is predicted by the Pauling model[20]. However, the agreement with the Geller model is accidental, since for V_3Ga, this model would predict a contraction instead of the observed expansion. In reality, no geometrical model is able to explain correctly the additional lattice expansion at $\beta > 0.25$ shown in Fig. 5, which is attributed here to the occupation of virtual sites.

The lattice expansion $\Delta a = \Delta a(\beta)$ for V_3Ga in Fig. 5 exceeds the corresponding ones for the Pt based systems by more than a factor two. Indeed, there are now interactions between a nontransition and a tran-

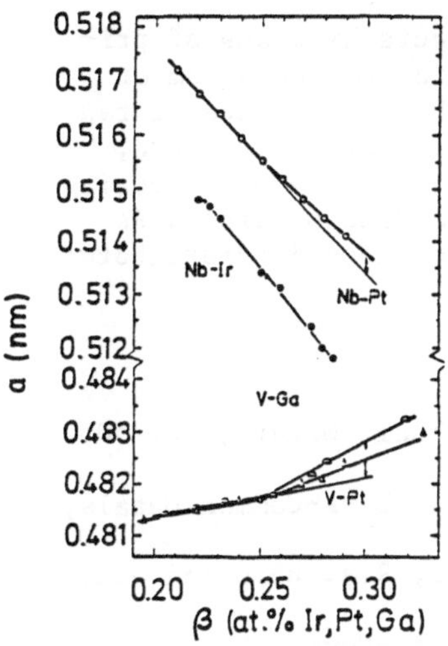

Fig. 5. Lattice parameter vs. composition, β, in the systems Nb-Ir(●, Ref.4), V-Pt(Δ, Ref. 13) and V-Ga(○, Ref.33).

sition element (represented by Ga_{6c} and V_{6c}) and between two nontransition elements (Ga_{6c} and Ga_{2a}), the distances $V_{6c}-Ga_{6c} \approx 0.2409$ nm and $Ga_{6c}-Ga_{2a} = 0.2693$ nm being both shorter than the corresponding sums $r_V + r_{Ga} = 0.2713$ nm and $2 \times r_{Ga} = 0.2762$ nm, respectively. From the small deviations of $\Delta a = \Delta a(\beta)$ from linearity in Fig. 5, it can be understood why no lattice parameter changes have so far been observed in quench disordered A15 type compounds.

The occurrence of static displacements of the atoms from their equilibrium positions observed in several A15 type compounds after irradiation[12] is also related to the proposed virtual site exchange mechanism. As illustrated in Fig. 1b, an occupied virtual site causes a perturbation of the potential in the surrounding hexagonal complex, which not only leads to an increase of the lattice parameter, but also to displacements from the equilibrium atomic positions for this group of atoms. A comparison between the known static displacement data shows that the lattice parameter expansion, a, as well as the static displacements, $\langle u^2 \rangle$, are enhanced when heavier incident particles are chosen. For example, irradiation of Nb_3Al up to saturation of T_c (Ref. 5) yielded for 300 keV H^+ ions a value $\langle u^2 \rangle = 5\times10^{-4}$ nm^2, while 700 keV N^+ ions yielded $\langle u^2 \rangle = 1\times10^{-4}$ nm^2.

COMPARISON BETWEEN IRRADIATED AND QUENCHED A15 TYPE COMPOUNDS

It appears that both processes are influenced by the number of occupied virtual sites, the time of occupation being very different in both cases. A high temperature, the occupation of the virtual site has the character of an intermediate state in the course of two-step or multistep processes connected with the particular diffusion mechanism in A15 type compounds. Since the time of occupation of a virtual site is expected to be very short (depending on temperature), in order to retain a significant amount of occupied virtual sites the total cooling time during the quenching process should be shorter than the occupation time. From quenching experiments on V_3Si, where a lowering of T_c is only observed at cooling rates well above 10^5 °C/s[32], the occupation time at high temperature can be estimated to be less than 10^{-3} s, the probability of retaining occupied virtual sites after quenching being thus very low. The main difference between quenching and irradiation resides in the formation process of the vacancies. In quenched crystals, the occurrence of vacancies is a consequence of the anharmonic thermal vibrations of the atoms around their equilibrium lattice site at a given temperature. In irradiated crystals, the formation of vacancies occurs by means of primary collision events. There is also a difference in the mechanism of disordering: In the thermal case, disordering is the consequence of random site exchange, the driving force being a function of temperature, while site exchange due to high energy irradiation occurs along focusing directions, the driving energy being due to the incident particle. As discussed above, however, both cases lead to a homogeneously distributed degree of ordering throughout the whole crystal.

REFERENCES

1. A. R. Sweedler, D.E. Cox and S. Moehlecke, J. Nucl. Mater., 72, 50 (1978).
2. R. Flükiger, J.L. Staudenmann and P. Fischer, J. Less-Common Metals, 50, 253 (1976).
3. E. C. Van Reuth, R.M. Waterstrat, R.D. Blaugher, R.A. Hein and J.E. Cox, Proc. LT 10, 137 (1967).
4. S. Moehlecke, D.E. Cox and A.R. Sweedler, J. Less-Common Metals, 62, 111 (1978).
5. R. Flükiger, unpublished. The results were made available to Moehlecke et al.[4] for including in their paper.
6. A. R. Sweedler and D.E. Cox, Phys. Rev., B12, 147 (1975).
7. R. Flükiger, J.L. Jorda, A. Junod and P. Fischer, Appl. Phys. Comm., 1, 9 (1981).
8. G. W. Webb and J.J. Engelhardt, IEEE Trans.Magn., MAG-11, 208 (1975).
9. R. Flükiger and J.L. Jorda, Sol. State Comm., 22, 109 (1977).
10 U. Schneider, G. Linker and O. Meyer, J. Low Temp. Phys., 47, 439 (1982).
11. C. S. Pande, Sol. State Comm., 24, 241 (1977).
12. L. R. Testardi, J.M. Poate and H.J. Levinstein, Phys. Rev., B15, 2570 (1977), O. Meyer and B. Seeber, Sol. State Comm., 22, 603 (1977).
13. R. Flükiger, to be published.
14. R. Flükiger, Proc. LT 17, p. 609 (1984).
15. D. O. Welch, G.J. Dienes, O.W. Lazareth and R.D. Hatcher, IEEE Trans. Magn., MAG-19, 889 (1983).
16. J. L. Staudenmann, Solid State Comm., 23, 121 (1977).
17. A. Seeger, in "Radiation Damage in Solids", edited by Int. Atomic Energy Agency, Vienna, 1962, Vol. 1, p. 101.
18. C. S. Pande, Phys. Stat. Sol., (a) 52, 687 (1979)
19. E. M. Schulson, J. Nucl. Mater., 83, 239 (1979).
20. L. Pauling, Acta Cryst., 10, 374 (1957).
21. G. Linker, J. Nucl. Mater., 72, 275 (1978), G. Linker, Radiation Effects, 47, 225 (1980).
22. D. E. Cox and J.A. Tarvin, Phys. Rev., B18, 22 (1978).
23. J. Pflüger and O. Meyer, Sol. State Comm., 32, 1143 (1979).
24. E. L. Haase and J. Ruzicka, J. Low Temp. Physics, 47, 461 (1982).
25. O. Meyer and G. Linker, J. Low Temp. Physics, 38, 747 (1980).
26. R. Schneider and G. Linker, to be published.
27. M. Lehmann, H. Adrian, J. Bieter, P. Müller, C. Nöltscher, G. Saemann-Ischenko, Phys. Lett., 87A, 369 (1982).
28. M. Lehmann and G. Saemann-Ischenko, Phys. Lett., 87A, 369 (1982).
29. S. Geller, Acta Cryst., 9, 885 (1956).
30. R. Flükiger, S. Foner and E.J. McNiff, Jr., in "Superconductivity in d and f Band Metals", Ed. H. Suhl and M.B. Maple, Academic Pres, New York 1980, p. 26.
31. S. Moehlecke, D.E. Cox and A.R. Sweedler, Sol. State Commun., 23, 703 (1970)
32. B. Pannetier, T.H. Geballe, R.H. Hammond and J.F. Gibbons, Physica, 107B, 471 (1981).
33. T. LL. Francavilla, B.N. Das, D.V. Gubser, R.A. Meussner and S.T. Sekula, J. Nucl. Mater., 72, 203 (1978).

THE RELATIONSHIP BETWEEN CRITICAL CURRENT AND MICROSTRUCTURE OF INTERNAL TIN WIRE

D.R. Dietderich, W.V. Hassenzahl, and J.W. Morris, Jr.

Lawrence Berkeley Laboratory and
Department of Materials Science and Mineral Engineering,
University of California, Berkeley
Berkeley, California

ABSTRACT

Prior work on internal tin wire has shown that an increase in critical current results when the Nb_3Sn reaction temperatures (650-730°C) are preceded by low temperature diffusion heat treatments that distribute the tin. These heat treatments produce a more uniform tin distribution through the niobium filament array before substantial Nb_3Sn formation has occurred. Heat treatments as long as 19 days have been proposed as the optimal heat treatment for the conductor. However, it is possible to substantially reduce the low temperature heat treatment time while retaining the same high critical current. The success of shortened heat treatments may be interpreted on the basis of the Cu-Sn reaction, diffusion kinetics and the Nb_3Sn growth kinetics.

INTRODUCTION AND SUMMARY

The internal tin process[1,2] is a fabrication technique that makes possible the increase of the fraction of Nb_3Sn in a conductor and thus increases the overall critical current. The process starts with a Cu-Nb-Sn composite that can be deformed to final size without intermediate anneals since no bronze is present and the elemental units co-deform well. In addition, the J_c of the Nb_3Sn produced by this process is high.[3,4,5] Thus the critical current of the internal tin wire is enhanced in that both the quantity and the quality of Nb_3Sn are improved in comparison to bronze-processed wires. A representative conductor of this type, which has a Ta barrier between the Cu-Nb-Sn composite and the Cu stabilizer, is shown in Figure 1.

Many different elaborate heat treatment schedules have been proposed for internal tin wire.[4,5,6,7] The highest critical currents are achieved with multiple heat treatments at different temperatures with total heat treatment times of up to 20 days. Our earlier work[8] indicates that the total heat treatment time may be reduced to about 12 days with no reduction in performance by altering the low temperature tin diffusion steps. The shortened heat treatment still includes the 218 hour heat treatment at 580°C, which has been used in several reaction schemes; however, this step has not been shown to be essential.

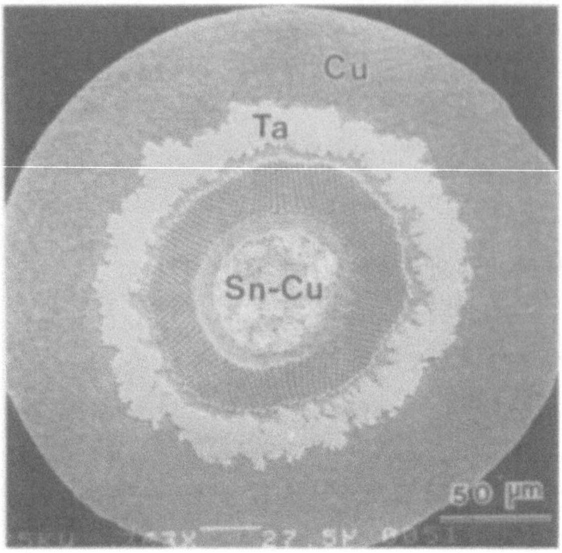

Fig. 1. SEM micrograph of an internal tin wire.

The selection of alternate, shortened heat treatments to form the Nb_3Sn phase here is motivated by the improved performance of a bronze-processed conductor that received a low-temperature heat treatment followed by a shorter high-temperature heat treatment.[9] The best heat treatment for the bronze wire was 700°C/4 days followed by 730°C/2 days. These temperature and time combinations are inappropriate for the internal tin wire since overaging will occur after 4 days at 700°C.[8] Nevertheless, the concept can be applied to the internal tin wire. The low-high scheme is incorporated into most internal tin heat treatment schedules in the form of a treatment at either 580°C or 625°C followed by a treatment at or above 700°C.

The present work was undertaken to simplify and shorten the heat treatment of the internal tin wire without degrading performance and to clarify and quantify the microstructural sources of the observed critical current density. The principal results of the work are that a high overall J_c can be obtained with a short (2-3days), intermediate-temperature (580°C) heat treatment prior to 700°C, or with a 0.1°C/minute temperature ramp from room temperature to 700°C. The high J_c seems to result from the fine equiaxed grained Nb_3Sn region which forms at intermediate temperatures.

EXPERIMENTAL PROCEDURE

Two internal tin wires fabricated by Intermagnetics General Corporation were utilized in this study. One, a 1.73mm (0.068in.) wire, which was used for metallography and x-ray analysis, contained 61 sub-elements with nominal diameter of 0.12mm (0.005in.). The other wire, a 0.27mm (0.0105in.) wire (Fig. 1), was produced with a Ta barrier and Cu stabilizer specifically for critical current measurements. The internal dimensions and filament size of this wire approximated those in the large conductor. The Ta barrier and Cu stabilizer of the sub-element accounted for approximately 15% and 60% of the cross sectional area of the wire respectively.

Heat treatments were done in sealed quartz tubes back-filled with argon such that the pressure would approximate ambient pressure at the highest temperatures (650-700°C). To ensure that no tin was lost, the conductors' ends were sealed by melting until a small alloy bead was created. At least 20mm at each end of the 0.27mm diameter conductor was removed prior to critical current measurements. The heat treatment notation and the corresponding temperature/time combinations are given in Table 1. The effects of two temperature ramps, 1.0 and 0.1°C/minute from room temperature to 700°C, were also investigated.

All the critical current measurements were done in a transverse magnetic field with a potential criterion of 0.1μV/mm or less. At least five samples were measured for heat treatments with a high I_c (700°C/4d, 0.1°C/min.+700°C/1d, IVB+700°C/1d, and I+700°C/1d) while a minimum of three tests were performed on low I_c samples (IVA, IVB, and I). A four point probe arrangement was used with 30mm long samples. The voltage taps were placed 5mm apart in the middle of the sample, equidistant from the current contacts. The overall J_c is calculated from the active core area (i.e. area within the Ta barrier) which is $1.3 \times 10^{-2} mm^2$. The areas were measured from SEM micrographs using a digitizing tablet. To correct for magnification error, the active core areas were scaled by the ratio of the wire area obtained from the micrographs to the wire area determined by a physical measurement of the wire diameter.

Table 1. Heat Treatment Notation

IVA	IVB	I
380°C/2d + 580°C/3d	380°C/2d + 580°C/2d + 625°C/1d	380°C/2d + 580°C/9d

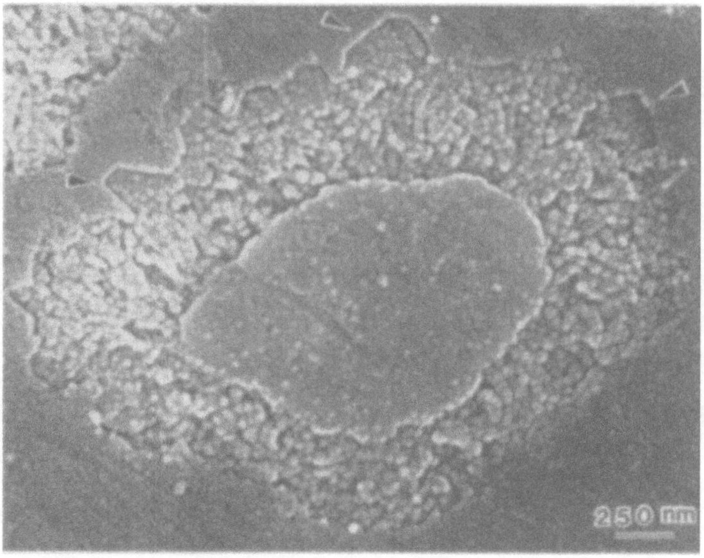

Fig. 2. SEM micrograph of a reacted filament that has received heat treatment I. The sample has been etched to reveal the Nb_3Sn. Note the large grains at the filament periphery.

TEM observations were performed on longitudinal sections of the 1.73mm conductor. The samples were thinned mechanically to about 0.025mm and then ion milled.

RESULTS AND DISCUSSION

Many different heat treatments have been proposed as the optimum for internal tin wires. However, there is a limited amount of work relating conductor performance to the reaction temperatures and times. Microstructural characterization of the conductor led to the development of heat treatments that included only a short low temperature step.[8] In fact, long heat treatments at 580°C may be detrimental to the magnetization properties of the conductor by promoting filament bonding (i.e. bridging). Figures 2 and 3 illustrate the positive and negative aspects of a long 580°C treatment. Figure 2 is a high magnification micrograph of an individual filament that has been etched to reveal the Nb_3Sn. At the filament periphery large grains of 200 to 300nm in size are visible. Some filaments have been observed to grow together when adjacent filaments have large grain extrusions in close proximity. These large grains can also be seen in the upper portion of the TEM micrograph in Figure 3. The same micrograph also illustrates a beneficial aspect of the 580°C treatment, which is the fine equiaxed grains (40-45nm) produced at this intermediate temperature.

Prior work has shown that a multi-phase bronze microstructure is produced after 2 days at 580°C.[8] The microstructure consisted of the α

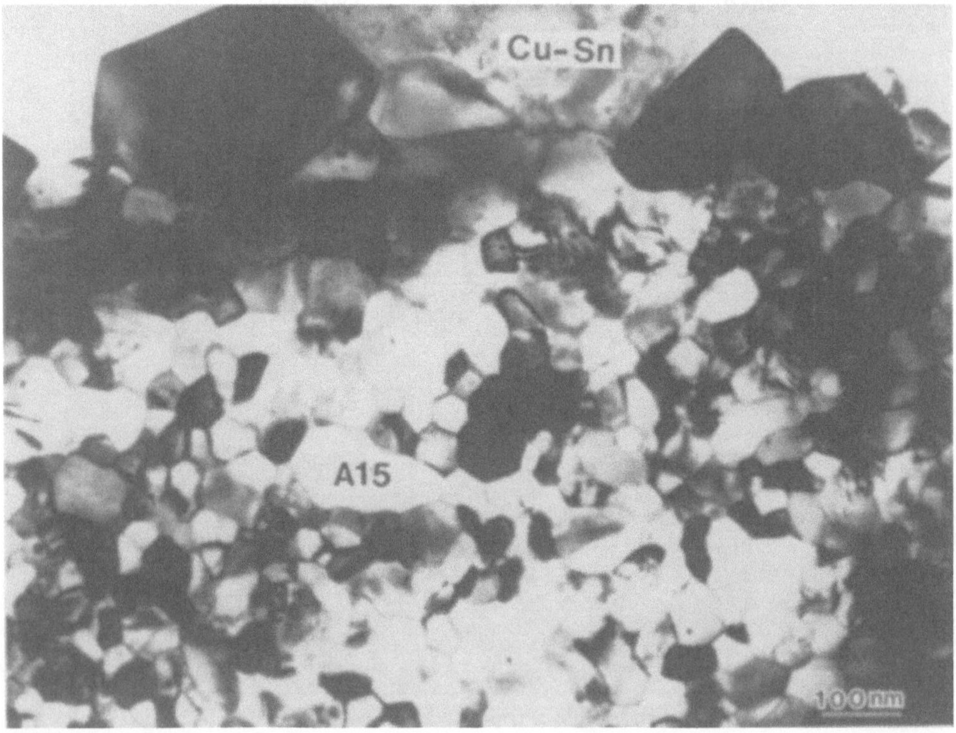

Fig. 3. TEM micrograph from a sample with heat treatment I. Note the large grains (200-300nm) at the bronze interface and the small grained (40-45nm) region.

phase (Cu-Sn solid solution) and decomposed γ phase ($\alpha + \delta$). The bronze adjacent to the Nb filaments is predominantly α but contains isolated regions of decomposed γ. Energy dispersive x-ray spectroscopy indicates that the Sn distribution throughout the α phase is relatively uniform. The analyses show that the tin concentration in the bronze on either side of the Nb filament array is ~7.5at.%Sn. Recent observations show that at this point in the heat treatment the Nb filaments are already about 40-50% converted to Nb_3Sn. The extent of the filament reaction and the good tin distribution suggested that this was the appropriate time in the heat treatment schedule to increase the temperature. This led to heat treatment schedules IVA and IVB.

Alternative heat treatments which employ a slow temperature ramp to 700°C were designed to take advantage of the variation with temperature of the Nb_3Sn formation rate and the diffusivity of tin in the Cu-Sn system. Nb_3Sn does not readily form until a temperature of ~500°C yet available data on Sn diffusion in Cu[10] suggest that diffusion of Sn is relatively fast at this temperature. Therefore, it was thought that a slow ramp would produce a redistribution of tin at intermediate temperatures and provide a "low-high" temperature heat treatment for the Nb_3Sn.

The overall critical current density at 10T and 4.2K at various stages in the heat treatments described above is shown in Figure 4. The data in the lower portion of the graph are the critical current densities of wires that are partially reacted using schedules IVA, IVB, and I. The data in the upper portion of the graph are the critical currents of the wires after an additional heat treatment at 700°C. Data from reference 8 for raising the temperature directly to 700°C is included for comparison. All the samples that received an intermediate temperature (580-625°C) heat treatment for more than 2 to 3 days by either a temperature ramp or

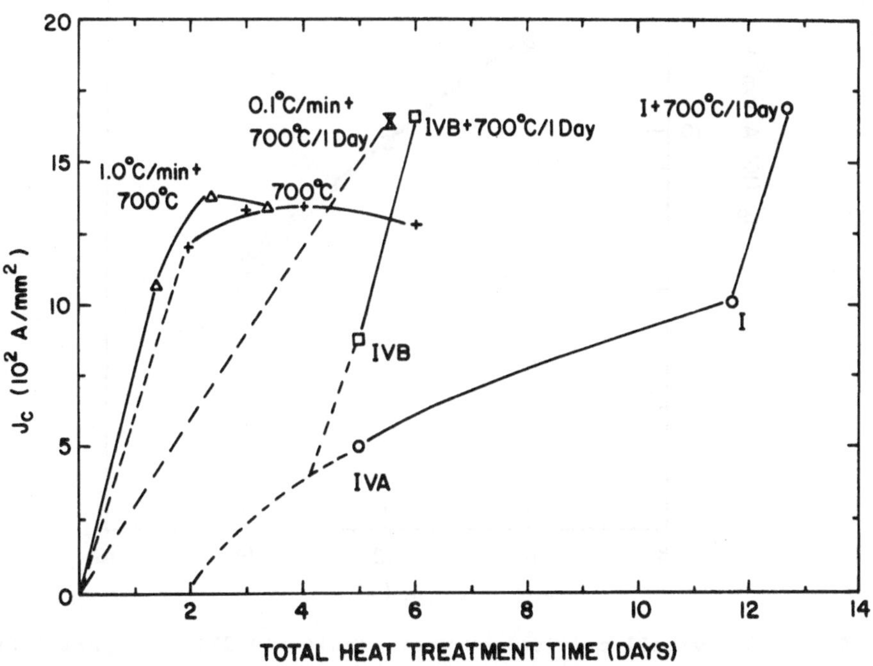

Fig. 4. The overall critical current density (10T, 4.2K) vs. the total heat treatment time the wire received.

constant temperature had a higher critical current density than those taken directly to the high reaction temperature. The heat treatment times for samples IVA, IVB, and I include the 2 day treatment at 380°C. Figure 5 shows the J_c variation with field for the three best heat treatments. There is no significant difference between these heat treatments at fields in the range 10 to 14T.

The high critical current densities measured here agree favorably with those measured by Suenaga et al. on long barrel samples of a similar wire.[11] Although different testing techniques were used, the overall critical currents shown in Fig. 5 for heat treatment I+700°C/1 day agree to within 15% of sample A in Fig. 3 of reference 11. There are at least two possible reasons for the higher J_c values given in this study (see also comments at end of text). One is the the less sensitive voltage criterion, 0.1 μV/mm as opposed to 0.01 μV/mm. However, the large n value of the conductor at 10T (35-40) has the consequence that this difference in technique can cause at most a 5% change in the critical current. The second reason is the different wire dimensions of the two conductors. Each wire must receive a different heat treatment. Minor differences in the resulting final microstructure could cause differences in J_c.

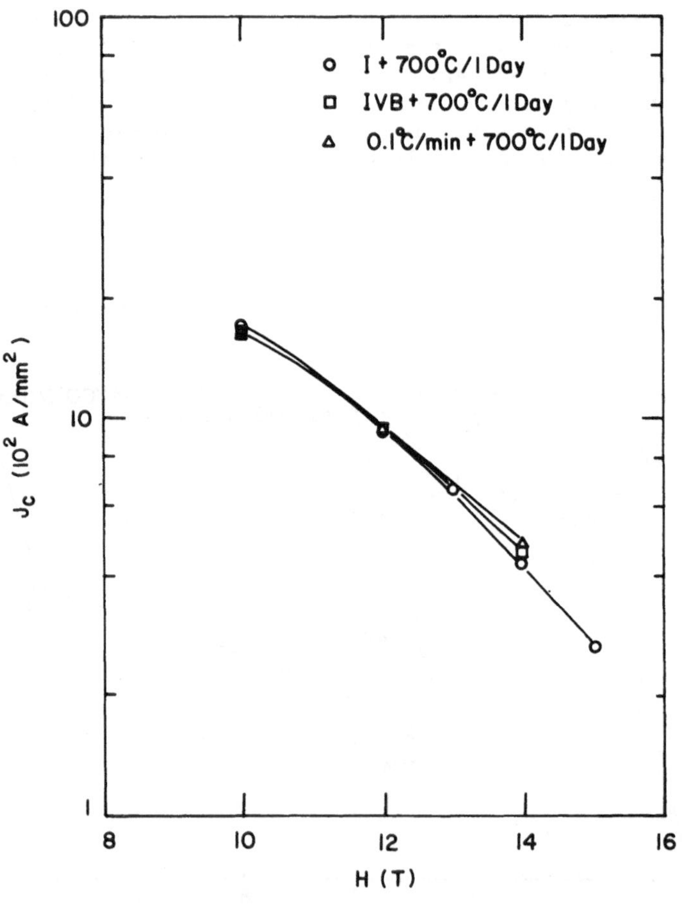

Fig. 5. The overall critical current density (4.2K) vs. applied field.

Our testing procedure, which uses very short samples, provides the accuracy and speed needed to map out a sizeable heat treatment matrix for a conductor. Although the barrel technique may be more precise and reflect the current densities expected in operating magnets more accurately, it is not as well suited to a general survey of heat treatments.

The high critical currents obtained here after slow ramps to temperature make the internal tin conductor a promising choice for large wind-and-react magnets. It may not be feasible to heat treat a large magnet uniformly in discrete temperature steps because of the time required for the temperature to stabilize throughout the magnet. By contrast, a slow ramp is a convenient treatment for a large magnet. The ramping treatment is effective since it combines time at low temperature, which distributes the tin and promotes formation of fine grained Nb_3Sn, and time at high temperature, which improves the stoichiometry of the superconductor.

CONCLUSIONS

Several shortened heat treatments produce a high overall J_c at 10T. The shortest of these reduces the total heat treatment time from approximately 19 days to about 6 days. To achieve the highest possible critical current a low-high temperature sequence is required. Some time at an intermediate temperature is necessary to obtain the fine equiaxed grain structure required for high critical currents. A temperature ramp is a practical and effective low-high heat treatment scheme.

ACKNOWLEDGMENTS

The authors would like to especially thank J. Holthuis for assistance during all stages of the research and D. Tribula and J. Glazer for in-depth discussion of fundamental aspects of this work. We would also like to thank Dr. K. Hemachalam and Dr. G.M. Ozeransky of Intermagnetics General Corporation for the special fabrication of the samples used for critical current measurements, and to L. Rubin and his associates at the National Magnet Laboratory for advice and assistance with the critical current measurements. The critical currents measurements were performed at the Francis Bitter National Magnet Laboratory, supported by the National Science Foundation. This work was jointly sponsored by the Director, Office of Energy Research, Office of Basic Energy Sciences, Materials Sciences Division, and by the Director, Division of High Energy Physics, U.S. Department of Energy under Contract No. DE-AC03-76SF00098.

COMMENTS

A potential source of critical current density variation is strain in the Nb_3Sn. A compressive strain in the Nb_3Sn will reduce the critical current density. The origin of the compressive strain is the thermal contraction differences of the wire components. Altering the amount of any component (Ta, Nb_3Sn, and Cu-Sn) will change the strain in the Nb_3Sn. Since the conductor used in this study has a larger Ta diffusion barrier than that of a typical commercial wire (~15% vs. ~5%) less compressive strain is expected, and thus a higher J_c. This difference can account for about a 10-15% increase in J_c at 10T in the conductor used in this study over a commercial wire. However, the conductor referred to in reference 11 also has a diffusion barrier that accounts for about 15% of the wire cross section. Therefore, this implies that it and the wire in this study have similar compressive strains, thus, making the amount of Ta unable to account for the difference in the critical current density.

REFERENCES

1. M. Suenaga, Metallurgy of Continuous Filamentary A15 Superconductors, in: "Superconductor Materials Science," ed. S. Foner and B. Schwartz, Plenum Press, New York, (1981) p. 201.

2. Y. Hashimoto, K. Yoshizaki, and M. Tanaka, Processing and Properties of Superconducting Nb_3Sn Filamentary Wires, in: "London ICEC Proceedings," IPC Science and Technology Press, London (1974) p. 332.

3. K. Yoshizaki, O. Taguchi, F. Fujiwara, M. Imaizumi, Y. Hashimoto, K. Wakamoto, T. Yamada and T. Satow, Properties of Multifilamentary Nb_3Sn Wires Processed by Internal Tin Diffusion, in: "Advances in Cryogenic Engineering," vol. 28, (1982) p. 380.

4. R.E. Schwall, G.M. Ozeryansky, D.W. Hazelton, Properties and Performance of High Current Density Sn-Core Process MF Nb_3Sn, IEEE Trans. Magn., MAG-19:1135 (1983).

5. S.F. Cogan, J.D. Klein, S. Kwon, H. Landis and R.M. Rose, On The Mechanical Properties of Sn-Core Processed Nb_3Sn Filamentary Composites, IEEE Trans. Magn., MAG-19:917 (1983).

6. N. Higuchi, K. Tsuchiya, C.J. Klamut, and M. Suenaga, Superconducting Properties of Nb_3Sn Multifilamentary Wires Fabricated by Internal Tin Process, in: "Advances in Cryogenic Engineering," vol. 30, (1984) p. 739.

7. B.A. Zeitlin, G. Ozeryansky, K. Hemachalam, An Overview of the IGC Internal Tin Nb_3Sn Conductor, IEEE Trans. Magn., MAG-21:293 (1985).

8. D.R. Dietderich, J. Glazer, C. Lea, W.V. Hassenzahl and J.W. Morris, Jr., The Critical Current Density and Microstructural State of an Internal Tin Multifilamentary Superconducting Wire, IEEE Trans. Magn., MAG-21:297 (1985).

9. I.W. Wu, D.R. Dietderich, J.T. Holthuis, M. Hong, W.V. Hassenzahl, and J.W. Morris, Jr., The Microstructure and Critical Current Characteristic of a Bronze-Processed Multifilamentary Nb_3Sn Superconducting Wire, J. Appl. Phys., 54(12), (1983), p.7139.

10. H. Oikawa, A. Hisoi, Interdiffusion in Cu-Sn Solid Solutions, Scr. Metall., 9(8), (1975), p.823.

11. M. Suenaga, C.J. Klamut, N. Higuchi and T. Kuroda, Properties of Ti Alloyed Multifilamentary Nb_3Sn Wires by Internal Tin Process, IEEE Trans. Magn., MAG-21:305 (1985).

CRITICAL CURRENTS AND FLUX-PINNING PROPERTIES IN $PbMo_6S_8$ SUPERCONDUCTORS

K. Noto, *K. Hamasaki, *T.Komata, *T.Saito, *T.Yamashita, and K. Watanabe

The Research Institute for Iron, Steel and Other Metals, Tohoku University, Sendai 980, Japan

*Department of Electronics, Technological University of Nagaoka, Nagaoka 949-54, Japan

ABSTRACT

We report on recent investigations of the fabrication of $PbMo_6S_8$ tapes prepared using an electro-plating and diffusion technique. Measurement of critical currents was carried out in either 16.5 T superconducting magnet, or 20.5 T (HM-3) and 23 T (HM-2) hybrid magnets in High Field Laboratory for Superconducting Materials RIISOM, Tohoku University in Japan. After optimal reaction the critical temperature measured by a resistive method reaches about 14.5 K, and a transition width was less than 0.5 K. Critical current densities are relatively high in a low magnetic field, and reached 1×10^4 A/cm^2 in a 15 T magnetic field at 4.2 K. However, J_c's in a strong magnetic field above 15 T are still insufficient for practical applications. In some of these samples pinning force densities were found to obey a simple scaling law, and in most cases the scaling is not obey. Poor J_c-B characteristics in a strong magnetic field may be due to inhomogeneities of this compound.

INTRODUCTION

Among the Chevrel-phase compounds, $PbMo_6S_8$ is very interesting as a superconducting material due to its extremely high critical magnetic field and substantially high critical temperature.[1,2] Therefore, this compound is one of the most promising materials for the production of steady-state magnetic field above 20 T. Since the discovery of the high critical field of $PbMo_6S_8$ compounds, many investigations reporting critical current densities in this type of thin films, wires, tapes and bulk samples have been published.[3-11] However, a large enough J_c for practical use has not yet been obtained for this type compound.

Recently pioneering work on the fabrication of $PbMo_6S_8$ wires has been carried out by Seeber et al.[12,13] They successfully produced Mo-$PbMo_6S_8$ wires using industrial methods for hot drawing of molybdenum. Although by this methods it is possible to produce very long wires, the critical current density of such a wire was $J_c=3\times10^3$ A/cm^2 at 6.5 T and 4.2 K.[12] In this paper we report recent investigations on the fabrication of $PbMo_6S_8$ tapes using an electro-plating and diffusion method. Systematic studies of T_c and J_c are carried out. We also discuss recent J_c-result measured in a 20.5 T and a 25 T hybrid magnets, and the pinning properties in this type compound.

SAMPLE CHARACTERIZATION

$PbMo_6S_8$ tapes were prepared by an electro-plating and diffusion technique.[15] The molybdenum tape plated with lead was sealed in an evacuated quartz tube, and heat treated in sulfur vapour from MoS_2 for 4-70 hours at about 1050 °C. The heating-up time from room temperature to 1000 °C was either 30, or 180 min. After this process, samples were then cooled down slowly to room temperature.

Figure 1 illustrates the typical superconducting transition of $PbMo_6S_8$ samples reacted for different reaction period. The resistively measured critical temperature were relatively high, and typically 14.5 K (midpoint transition). The resistive transition width was less than 0.5 K for most samples reacted within the time range 14-70 hrs, as shown in Fig.1. We can see an optimum reaction time is about 14 hr. Note the degradation of T_c for shorter reaction time. The graph showing the reaction period dependence (Fig.1) indicate that 14 hr does not have any particular disadvantage as compared with 42, or 70 hrs. In addition T_c's was strongly affected by the reaction temperature in the present fabrication technique. The high T_c's resulted from a reaction at temperature higher than 1000 °C. Reaction temperature below 1000 °C leads to a reduction of T_c value. This is because the reaction is very slow and must be performed above 1000 °C to get complete reaction within reasonable time.

Figure 2 is a scanning electron microscope view of the typical fractured surface of our tapes reacted for 14, or 70 hrs. (a) is for 4 hr, and (b) for 70 hr. The final product was a tape with a $PbMo_6S_8$ layer about 200 µm thick. A finely divided structure of average grain size 0.5 µm was observed in Fig.2(a). On the other hand, for longer reaction time the grain size is larger than about 5 µm, as can be seen in (b). We can see the crystal grows as increasing in the reaction time. It has been shown that in A-15 compounds, such as Nb_3Sn, the flux pinning caused by the small grain size is required to obtain high J_c values.[16] This may also be true of $PbMo_6S_8$ compound, therefore a small grain size with the ideal composition may be required to obtain higher J_c values.

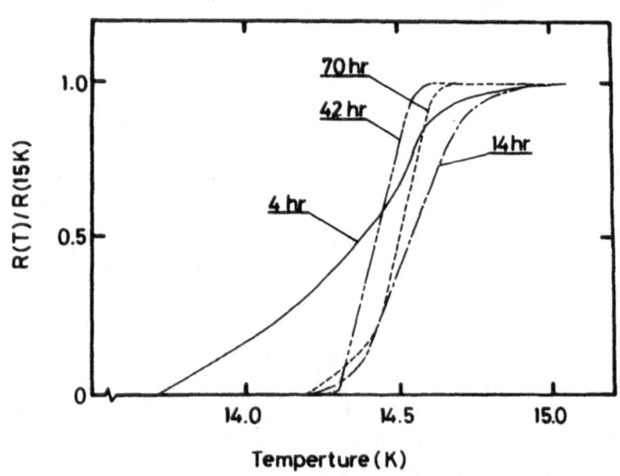

Fig.1 Curves of superconducting transition in four $PbMo_6S_8$ specimens. Reaction temperature is about 1050 °C.

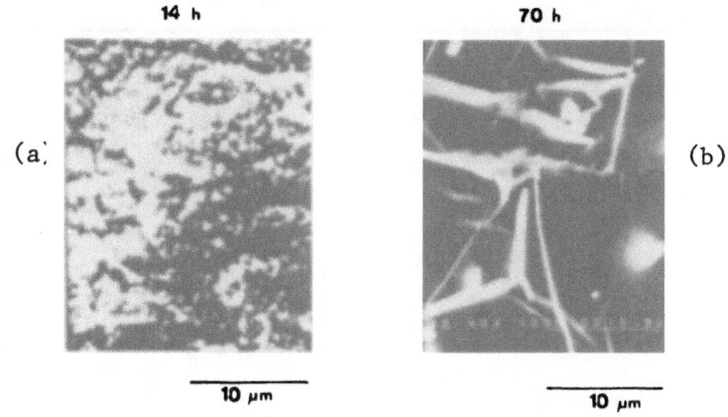

Fig.2 Scanning electron micrographs of typical fractured surface of $PbMo_6S_8$ tapes with different average grain sizes: (a) D $\leq$ 1 μm, (b) D $\cong$ 10 μm. Reaction time is 14 h for (a) and 70 h for (b), respectively.

CRITICAL CURRENTS

The current-voltage characteristics taken at 20 T are shown in Fig.3 for three specimens. As can be seen from the figure tapes indicate a sharp transition from the superconducting to normal state. The typical examples of critical current densities are shown in Fig.4 as a function of applied magnetic field in field up to 23 T. Samples were prepared for different reaction period. All measurements were in liquid helium to avoid heating effects. Critical currents were measured with a 16.5 T superconducting magnet, and a 20.5 T (HM-3) and a 23 T (HM-2) Hybrid magnets in High field Laboratory for Superconducting Materials RIISOM, Tohoku University in Japan.

The critical current densities of these samples is strongly dependent on the reaction time. The samples reacted for 4 hr shows large J_c's approximately a factor 2 higher than one for 42 hours over most of the magnetic field range. Although in all samples J_c's are relatively high and reaches to 10^5 A/cm^2 in a weak magnetic field, in a strong field the graph shows the almost linear decrease in J_c's against applied field. The J_c's for a sample reacted for 4 hr was about 1×10^4 A/cm^2 at 15 T, and about 3×10^3 A/cm^2 at 23 T. J_c's in a strong magnetic field above 15 T are still insufficient for practical applications.

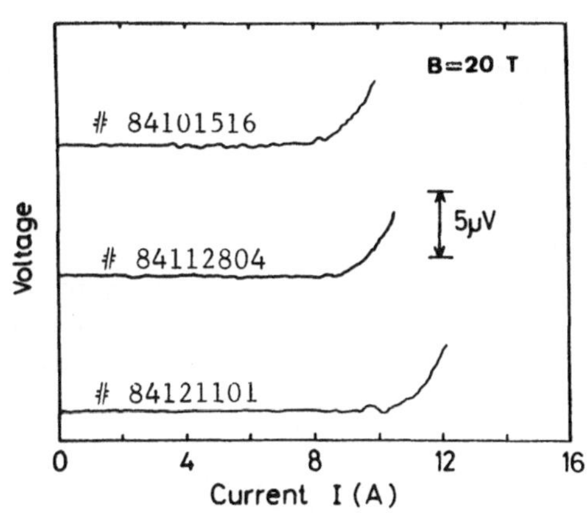

Fig.3 I-V curves for three $PbMo_6S_8$ specimens taken at 20 T. Reaction was performed at about 1050 °C for 14 hours.

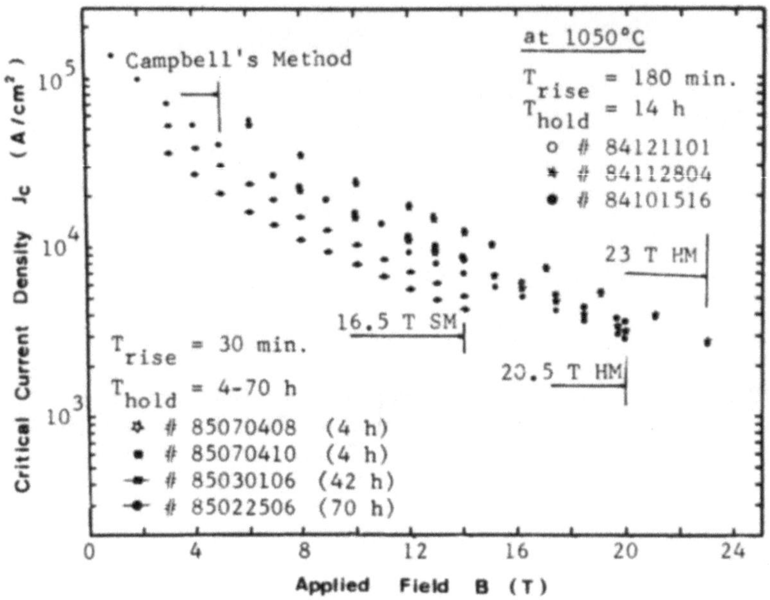

Fig.4 Critical current densities as a function of an external magnetic field B. J_c's in the range of 1- 5 T was measured by Matsushita et al.[18] using Campbell's method.

The variation of pinning force, $F_p = J_c \times B$, with applied magnetic field for different reaction period is shown in Fig.5. The pinning force increases for smaller grain sizes (see Figure 2). The measured T_c values suggest that the B_{c2} for these samples should be above 50 T.[2] Using B_{c2}=50 T, the pinning force maximum is seen to fall at $b(=B/B_{c2}) < 0.1$ for all specimens. In these specimens, pinning force results are not consistent with a major feature of Kramer equation, i.e., that the pinning force maximum is b=0.2 and moves to lower values of b as the pinning force increases. This is probably due to inhomogeneities in our materials. In some samples, we observed the large b value of about 0.2 for relatively short reaction period.[15] Unfortunately, high quality samples with large b value can not be reproducibly prepared at the present time. For a more

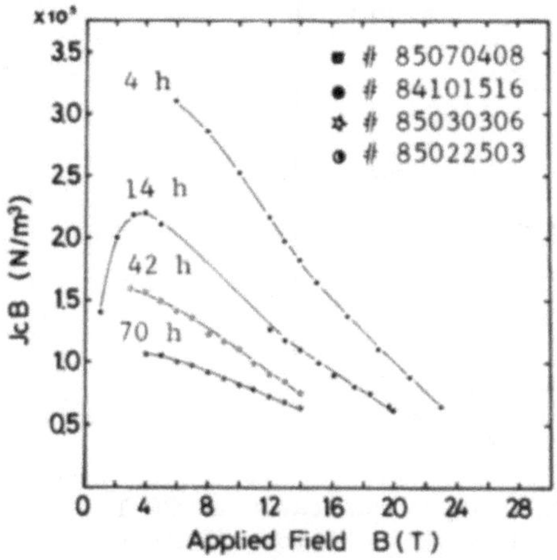

Fig.5 Plots of J_cB versus B for $PbMo_6S_8$ samples.

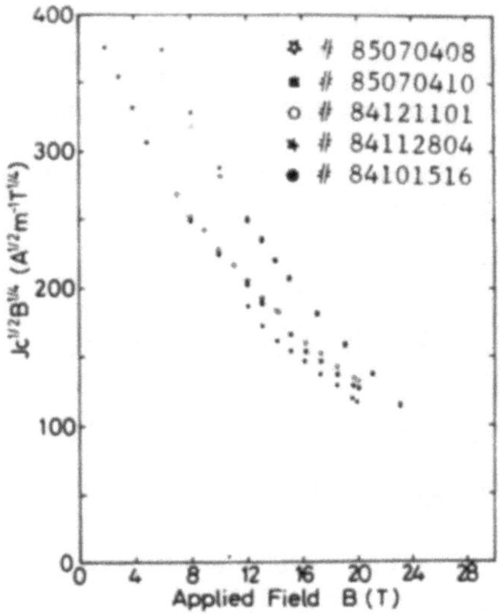

Fig.6 Plots of $J_c^{1/2}B^{1/4}$ versus B for $PbMo_6S_8$ samples.

detailed comparison with flux pinning theories, data on B_{c2} and J_c as a function of temperature and field are necessary.

According to Kramer's law, $J_c^{1/2}B^{1/4}$ product is given as, $J_c^{1/2}B^{1/4} \propto 1-(B/B_{c2})$.[17] Therefore, using only high field data the plot of this product on applied field should then be a straight line intercepting the abscissa at $B=B_{c2}^*$. Figure 6 illustrates such plots for several samples. The data in the 4-23 T range are not linear on this plot. The positive curvature at low field is seen for all samples. The extrapolated values of B_{c2}^* in the range of 30-35 T are very different from the evaluated B_{c2} with data on T_c. The curves in Fig.6 can thus be viewes as reflecting the existence of at least two separate flux-pinning mechanisms. The pinning law in its simplest form $F_p \propto b^{1/2}(1-b)^2$ is no longer valied at lower fields. The reason of the poor J_c-B characteristic, or pinning properties is not clear at the present time. This may well be due to inhomogeneities in our materials. A more detailed analysis of these results is in progress. More recently, Ta matrix wires with high J_c's were successfully fabricated by Kubo et al.[19] Using Mo,[12,13] or Ta [19] matrix wires, the superconducting magnets with field above 20 T and more can be expected in the future.

CONCLUSION

The superconducting $PbMo_6S_8$ tapes have been produced with an electro= plating and diffusion method, and critical currents have been measured up to 23 T. The resistively measured critical temperatures were relatively high, and typically 14.5 K (midpoint transition). The J_c values were about 1×10^4 A/cm^2 at 15 T, and 3×10^3 A/cm^2 at 23 T. Optimal reaction conditions improved J_c's in a weak magnetic field. However, J_c's in a strong magnetic field above 15 T are still in sufficient for practical applications. The poor J_c-B characteristics may be caused by inhomogeneities in our materials. We note that much work has still to be done in order to elucidate the flux-pinning mechanism in $PbMo_6S_8$ superconductors.

ACKNOWLEDGEMENTS

This work was financially supported by a Grant-in-Aid for Fusion Research from the Ministry of Education of Japan. The authours wish to thank K.Yoshizaki for his assistance in T_c measurements, and T.Matsushita for high useful information.

REFERENCES

(1) Ø.Fisher; Appl. Phys., 16, 1 (1978)
(2) S.Foner, E.J.Mcniff,Jr., and E.J.Alexander; IEEE Trans. on Magn., MAG-11, No.2, 155 (1975)
(3) M.Decroux, Ø.Fisher and R.Chevrel; Cryogenics, Vol.17, 291 (1977)
(4) N.E.Alekseevskii, N.M.Dobrovolskii, D.Eckert and V.T.Tsevro; J. Low Temp. Phys., Vol.29, 565 (1977)
(5) N.E.Alekseevekii, M.Glinski, N.M.Dobrovolskii and V.I.Tsebro; JETP Lett. Vol.23, 413 (1976)
(6) T.Luhman and Dew-Hughes; J. Appl. Phys., Vol.49, 336 (1978)
(7) S.A.Alterovitz, J.A.Woolam, L.Kammerdiner; Philos. Mag., Vol.38, 619 (1978)
(8) S.A.Alterovitz and J.A.Woolam; "Ternary Superconductor", New York, North Holland 113 (1981)
(9) B.Seeber, C.Rossel and Ø.Fisher; "Ternary Superconductor", New York, North Holland 119 (1981)
(10) C.Rossel; Ph.D Thesis No.2019, University of Geneva (1981)
(11) K.Hamasaki, T.Inoue, T.Yamashita, T.Komata and T.Sasaki; Appl. Phys. Lett., Vol.41, 667 (1982)
(12) B.Seeber, C.Rossel, Ø.Fisher and W.Glaetzle; IEEE Trans. on Magn., MAG-19, 402 (1983)
(13) C.Rossel and Ø.Fisher; J. Phy. F: Met. Phy., Vol.14, 455 (1984)
(14) K.Hamasaki, T.Yamashita, T.Komata, K.Noto and K.Watanabe; Adv. Cryog. Emg., Vol.30, 715 (1983)
(15) K.Hamasaki, K.Hirata, T.Yamashita, T.Komata, K.Noto and K.Watanabe; IEEE Trans. on Magn., MAG-21, 471 (1985)
(16) R.M.Scanlan, W.A.Fietz and E.F.Koch; J. Appl. Phys. Vol.46, 2244 (1975)
(17) E.J.Kramer; J. Appl. Phys. Vol.44, 1360 (1973)
(18) T.Matsushita et al.; Proc. Japan-US workshop. Superconducting Mat. for Fusion., 232 (1984)
(19) Y.Kubo, K.Yoshizaki, F.Fujiwara and Y.Hashimoto; Adv. Cryog. Eng., Vol.32, (1985) (to be published)

FLUX FLOW DYNAMICS IN SUPERCONDUCTING FILMS

V. M. Pan, G. G. Kaminsky, V. G. Prokhorov, and
K. G. Tretiatchenko

Institute of Metal Physics
Ukrainian Academy of Sciences
Kiev, USSR

ABSTRACT

Current-voltage characteristics (CVC) and their first-order derivatives have been measured for niobium and Pb-22% Bi alloy films in magnetic fields in the temperature range 0.2 to 0.5 T_c. CVC derivatives were found to have a peak that the authors attributed to the incoherent movement of flux lines in the weakly supercritical range due to their interaction with pins. Peak amplitude and position strongly depend on the magnetic field. This effect occurred for all magnetic fields higher than 0.2 H_{c2}, temperatures, and specimen orientations. A model of the distribution function of pins qualitatively accounts for experimental data obtained. The theoretical description must be modified by taking into account the interaction between flux lines.

INTRODUCTION

Most experimental investigations on the resistive state of superconducting films in magnetic fields are aimed at determining the nature of nonlinearity of the current-voltage characteristic (CVC) in the range of weakly supercritical range, i.e., when $I - I_c \ll I_c$, followed by the viscous flux flow state. However, there is no common viewpoint on this effect. The mechanism of complete current-induced destruction of superconductivity is not fully understood yet. On the other hand, a number of theoretical studies concerning flux line lattice (FLL) dynamics in type II superconductors have been published recently.[4,5] They need to be proven experimentally.

Investigation of peculiarities in the CVC of niobium and Pb-22% Bi alloy films in the mixed state was the main object of the present work. An attempt was made to describe our new experimental data theoretically.

SAMPLES AND EXPERIMENTAL PROCEDURE

Niobium films under investigation were prepared by electron beam evaporation in a vacuum ($p \leq 10^{-4}$ Pa). Lead-bismuth films were prepared by the usual evaporation method with direct heating. Sapphire was used as the substrate material. The substrate temperature during niobium deposition was about 850°C. The Pb-Bi alloy was deposited at room temperature. The deposition rate was approximately 10 Å/s. All films were 90-μm wide and

1.86-mm long. The film thickness ranged from 0.1 to 1.0 μm. The point $R(T) = R_N/2$ was taken as the critical temperature. The CVC curves and their first-order derivatives were recorded in liquid helium for two orientations relative to the magnetic field direction, $\vec{H}$: $\vec{H}$ normal to the film surface and parallel to the film surface. Transport current direction always remained perpendicular to the H vector. The current level corresponding to a voltage of about 1 μV/cm was taken as the critical value, I_c. First-order derivatives of CVC were recorded at a modulation frequency of about 300 Hz; the modulating current was approximately 100 μA. Other experimental details can be found elsewhere.[6]

EXPERIMENTAL RESULTS

Figure 1 shows[7] that the shape of the CVC curves of niobium films are rather complicated; the shape strongly depends on the value of applied magnetic field. In the high magnetic field range, a strongly marked nonlinear section is observed. With a decrease in magnetic field, the CVC curves become steeper, and the transition of a film to the normal state is abrupt, as when there is no magnetic field.

CVC peculiarities are seen more distinctly in plots of their first-order derivatives (Fig. 2), which reveal three specific sections characterized by different V(I) dependences. In the weakly supercritical range under high magnetic fields, the increase in voltage with an increase in the transport current can be described by the law $V \sim (I - I_c)^n$, where the value of the exponent n changes with the applied magnetic field. At a certain current value, dV/dI reaches a maximum followed by a region that can be interpreted as a viscous coherent flux flow regime. A further increase in current results in a sudden increase of resistance corresponding to the transition of the film to the normal state. This process is accompanied by intensive boiling of the liquid helium, resulting in random jumps of the voltage. If the experiments had been performed in gaseous helium, smooth curves would have been obtained, in accordance with Ref. 8. But under these conditions, there was no confidence in the constancy of the temperature and the correctness of its measurement. At the temperature just above the λ-point, overheating was strong enough to prevent observation of the first peak.

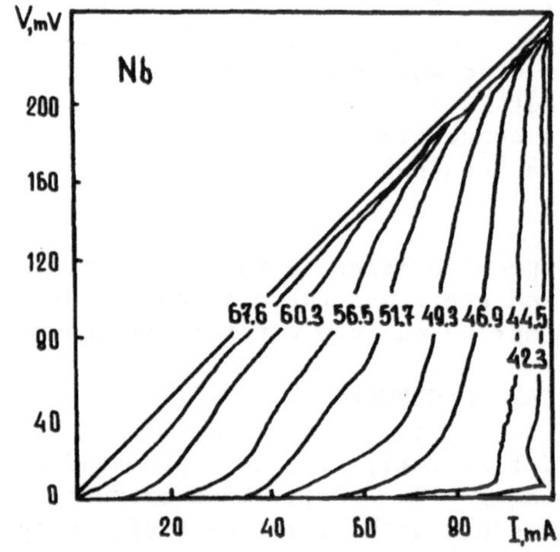

Fig. 1. CVC curves of niobium films at T = 4.2 K in different magnetic fields (x 10^4 A/m).

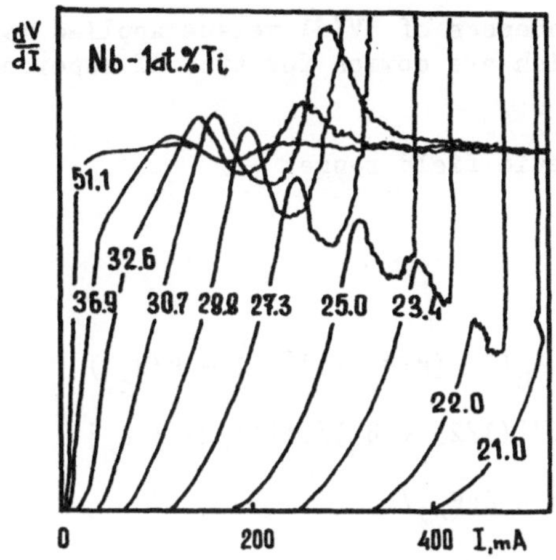

Fig. 2. CVC first-order derivatives, dV/dI, for different magnetic fields.

The shape of the experimental CVC first-order derivative curves did not change dramatically for measurements taken in parallel magnetic fields or at different temperatures.

It was interesting to find some correlation between the CVC curve shape and a certain intrinsic property of the test samples. From these experimental dV/dI curves it can be seen that for all films there is characteristic magnetic field $H_{n=2}$, at which CVC is very close to the curve $V \sim (I - I_c)^2$. Figure 3 shows that the value of $H_{n=2}$ correlates with the film resistivity. On the other hand, the volume pinning force in this magnetic field, $F'_p = F_p(H_{n=2})$, is approximately constant for all films.

Such behaviour indicates a direct correlation between pinning forces and the V(I) law at the initial part of the curve. Thus, we can conclude that the CVC curve shape characterizing moving magnetic flux dynamics is determined mainly by the interaction forces between the flux line ensemble and pin system.

Dependence of maximum volume pinning force, F_p^{max}, versus film resistivity is also shown in Fig. 3 for comparison. As expected, a tendency of F_p^{max} to increase with ρ_o was observed in this case.

Figure 2 shows that at a certain value of the current, I*, dV/dI curves have a peak that could be interpreted as a transition from one magnetic flux movement regime (referred to later as incoherent) to another (viscous flow) at which the whole flux moves with the same velocity. Generally, in ideal type II superconductors with low pin density, the viscous flux flow region is characterized by a constant resistance value the CVC first-order derivative curves. For niobium films, the viscous flux flow region is very narrow and just a minimum in the dV/dI curves, or in some cases, it is absent. This is the case because of high current density during the measurements; hence, Joulean heating of the film occurs at the very beginning of the coherent flux line movement resulting in complete destruction of superconductivity. However, assuming the resistance at the minimum of dV/dI curves to be equal to the viscous flux flow resistance,

R_f, experimental dependences of dV/dI versus applied magnetic field can be obtained (Fig. 4), which are common for type II superconductors with strong pins.

In the weak magnetic field range:[4]

$$R_f/R_N \sim H/H_{c2}(0) \quad (1)$$

and for high fields:

$$R_f/R_N = 1 - 2.49\, L_D(T)\, [\kappa/\kappa_2(T)]^2\, (1 - H/H_{c2})$$

$$L_D(T) = 2 + \{b_o\, \Psi''[(1/2) + b_o]/\Psi'[(1/2) + b_o]\} \quad (2)$$

$$b_o = \varepsilon_o/4\pi T;\ \varepsilon_o = 2\, DeH_{c2}/c$$

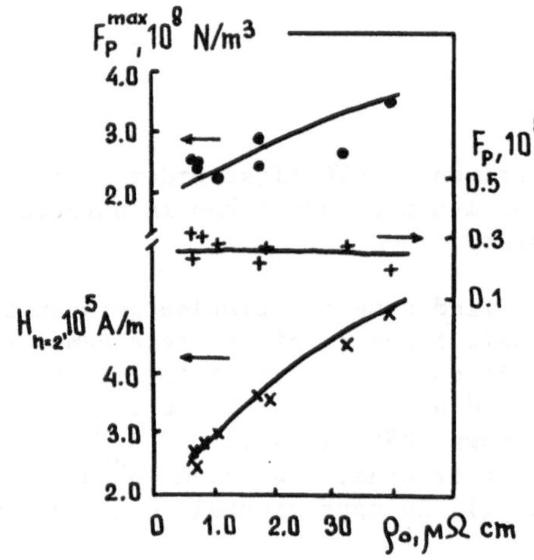

Fig. 3. Maximum pinning force, F_p^{max}, characteristic magnetic field, $H_{n=2}$, and corresponding pinning force, F_p', versus film resistivity.

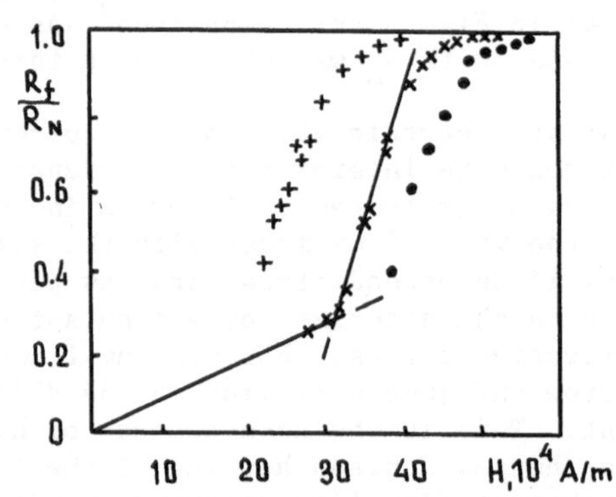

Fig. 4. Viscous flux flow resistivity versus magnetic field for niobium films with different ρ_o.

where Ψ', Ψ" are Ψ-function derivatives and D is the diffusion coefficient. R_f/R_N dependences corresponding to equations (1) and (2) for measured values of H_{c2} and T are shown in Fig. 4 with solid lines. Therefore, the incremental resistance at the minimum of the dV/dI curves is believed to correspond to the viscous flux flow resistance value.

Figure 5 shows that the dV/dI peak position characterizing a change in the moving flux dynamics depends on the applied magnetic field value. This dependence can be described empirically with the law $I^* \sim I_o^* (H_{c2}/H)^{5/2}$, i.e., the dynamic transition current value decreases with an increase of applied magnetic field and approaches the value I_o^*; the more impure the film, the less its I_o^* value.

Note that this effect occurs only for superconducting films with strong pins. Measurements on Pb-22% Bi alloy where the F_p^{max} value was two orders smaller than that of niobium films have shown that the dV/dI peak vanished in the high current range at $H/H_{c2} \lesssim 0.2$, and the transitions to coherent flux flow were montonic.

DISCUSSION

The problem of theoretically describing superconducting film CVC in magnetic fields is not solved yet. The nature of the moving flux line ensemble is not completely clear: Is it a fluid or an ordered structure? How are the net dynamic and static pinning forces related? How are they affected by the real pinning potential?

Baixeras and Fournet[9] were the first to assume that the CVC curve shape is determined by the elementary force distribution of pins in material. However, they assumed a relatively artificial form for this distribution. Their distribution function depends on certain general characteristics of flux line ensemble and material. Its physical meaning appeared to be determined only by those pins for which the effective pinning force was nonzero; i.e., this function could be considered as a product of the distribution function of pins and the effective number of flux lines that are pinned with a given force. Moreover, the flux line ensemble was assumed to be fluid.

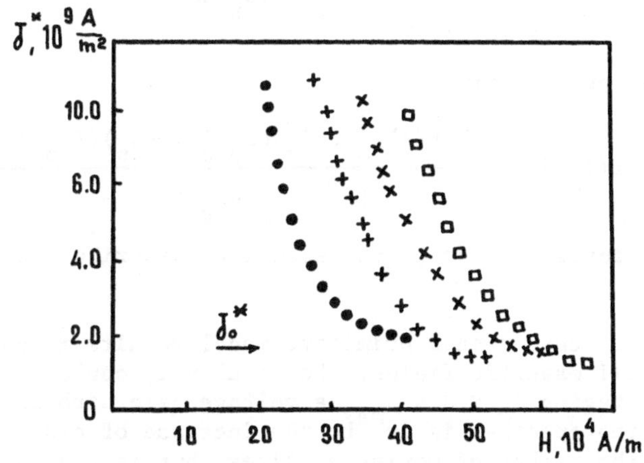

Fig. 5. Peak position of dV/dI versus magnetic field for niobium films with different resistivities.

Magradze et al.[10] have developed the concept of elementary force distribution of pins. They have shown that a fluid flux line ensemble assumption is not obligatory; i.e., all considerations are more or less valid for the case of the flux line lattice. They proved that when one flux line interacts simultaneously with a great number of pins, the distribution function is a Gaussian normal distribution function and therefore can be described with three parameters: f_o, the mean value; σ, dispersion; and A, the normalization factor.

The effectively pinned flux line number $g(f_i)$ is assumed to be nonzero for $f_i > JH$. This assumption is equivalent to the statement that for $J < J_c$ (J_c = critical current density) only rearrangement of flux lines occurs within the film to reach the maximum pinning force. Response to this rearrangement was shown to be few orders of magnitude smaller that the flux flow state and could hardly be observed experimentally.

Magradze et al.[10] made an attempt to take the elasticity of the flux line lattice into account and to evaluate its influence. Nevertheless, some facts from this paper are apparently proven, and some statements appear to be invalid. An important factor was not taken into account in all previous investigations. When a flux line moves, it interacts with all pins and not only with the strongest pins, as is the case just before reaching the critical current; i.e., the dynamic pinning force always differs from the static value even for a very low vortex velocity. It means that the vortex velocity effect on the elementary pinning force was not taken into account. We have made an attempt to establish a more correct foundation for the CVC curve and to explain the data obtained in Ref. 7. The present report is a continuation of that work.

In accordance with Ref. 10 and assuming a flux line to be affected by all pins with $J_p < J$ in terms of current,

for the mean flux line retardation force we have:

$$\langle J_p \rangle_J = \frac{\int_o^J f(J_p) J_p dJ_p}{\int_o^\infty f(J_p) dJ_p} \tag{3}$$

and for the viscous flux flow model for CVC:

$$V \sim \frac{\int_o^J f(J_p) dJ_p}{\int_o^\infty f(J_p) dJ_p} \int_o^i (J-J_p)\, f(J_p) dJ \tag{4}$$

and for CVC first-order derivative:

$$dV/dJ \sim \int_{J_c}^J f(J_p) dJ_p + \frac{f(J) \int_o^J f(J_p) dJ_p \int_o^J (J-J_p) f(J_p) dJ_p}{[\int_o^\infty f(J_p) dJ_p]^2} \tag{5}$$

Dependences corresponding to the step-like distribution function are shown in Fig. 6.

It can be seen that even a primitive model results in peaks of CVC derivatives for all magnetic fields. Physically it could be understood as follows: at the region $J_c < J < J_{max}$ a voltage gain with an increase of current through the specimen is due to the increase of drag force and to the increase in the number of moving vortices, but in the current range $J > J_{max}$ it is due to the drag force increase only.

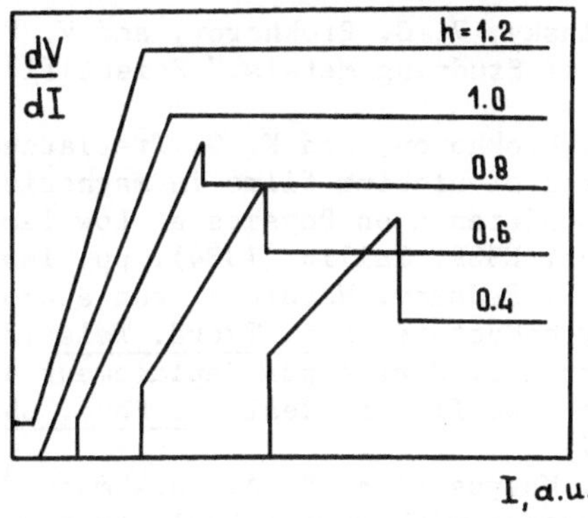

Fig. 6. Model CVC first-order derivative curves with the assumption of step-like distribution function of pins.

Note that the CVC derivative at point J_c has a jump. The explanation for this is: When the current approaches J_c from above, the number of moving flux lines, n_m, approaches zero, but their mean velocity does not reach zero. Hence $d\bar{V}/dJ(J_c) \sim f(J_c)\bar{v}(J_c)$ and a jump is absent only when $\bar{v}(J_c) \equiv 0$. This might be due to interaction between flux lines. This point remains for future studies.

CONCLUSION

Current-voltage characteristics (CVC) and their first-order derivatives have been measured for niobium and Pb-22% Bi alloy films at magnetic fields at the temperature range 0.2 to 0.5 T. CVC derivatives were found to have a peak that the authors attributed to the incoherent movement of flux lines in the weakly supercritical range due to their interaction with pins. Peak amplitude and position strongly depend on the magnetic field and temperature. The effect occurred for all magnetic fields higher than 0.2 H_{c2}, specimen orientations, and temperatures. Just above the λ-point, peaks were obscured by the overheating of the film.

A model of the distribution function of pins qualitively accounts for experimental data obtained. The theoretical description must be modified by taking into account the interaction between flux lines.

REFERENCES

1. T. Ōgushi and V. Shibuya, Flux flow in type I and II superconducting films, J. Phys. Soc. Japan 32:400-415 (1972).
2. R. G. Jones, E. H. Rhoderick, and A. C. Rose-Innes, Nonlinearity in the voltage-current characteristic of type-2 superconductor, Phys. Lett. 24A:318-319 (1967).
3. S. M. Wasim and N. H. Zeboni, Nonlinear resistivity and dissipation in the mixed state of superconducting Nb and V, Solid State Commun. 6:69-71 (1968).
4. A. E. Larkin and Yu. N. Obchinikov, Nonlinear conductivity of superconductors in the compound state, Zh. Eksp. Teor. Fiz. 68:1915-1927 (1975).
5. B. Yu. Block and S. V. Lempitsky, Influence of magnetic fields on the resistive condition of wide superconducting films, Fiz. Tverd. Tela 26:457-461 (1984).

6. V. M. Pan, G. G. Kaminsky, V. G. Prokhorov, and V. F. Taborov, in: "Physical Methods of Studying Metals," Scientific Council, Kiev (1981), pp. 36-41.
7. G. G. Kaminsky, V. G. Prokhorov, and K. G. Tretiatchenko, Peculiar resistivity features of niobium films in magnetic fields, in: "23rd All-Soviet Conference on Physics at Low Temperatures," Academy of Sciences, ESSR, Tallin (1984), pp. 166-167.
8. A. P. Smirnov and N. F. Fedorov, Nature of the anomolous resistive condition in superconductors, Fiz. Tverd. Tela 13:796-798 (1971).
9. J. Baxieras and G. Fournet, Pertes par deplacement de vortex dans un superconducteur de type II non ideal, J. Phys. Chem. Solids 28:1541-1547 (1967).
10. O. V. Magradze, L. V. Matyushkina, V. A. Shukhman, Vortex motion in type II superconductors with controllable type defects, J. Low. Temp. Phys. 55:475-494 (1984).

THE PINNING FORCE AND CRITICAL CURRENT IN NbTi SUPERCONDUCTING WIRE WITH PLATE-LIKE α-Ti PRECIPITATES

Wang Keguang, Zhang Tingjie, Wu Xiaozu, Li Chengren,
Zhou Nong

Baoji Institute for Non-Ferrous Metal Research
Baoji, Shaanxi, The People's Republic of China

ABSTRACT

Analytical expressions for the pinning force density F_p and the critical current density J_c of NbTi wire having plate-like α-Ti precipitates are derived from Matsushita's statistical theory and the local free-energy density function. These expressions show that both F_p and J_c are an increasing function of the volume fraction of α-Ti plates, and proportional to the square root of the average thickness of the plates. When the average separation of α-Ti plates is about equal to the spacing of the fluxoids F_p and J_c have their maxima. The calculated results from our expressions are compared with measurements. The calculated and measured results are in rather good agreement.

INTRODUCTION

Improving the technology, the critical current density of NbTi superconductors has been strikingly raised, recent J_c values up to 2000 - 3400 A/mm^2 (5 T, 4.2 K) have been achieved.[1] West and Larbalestier[2,3] have measured the critical parameters and microstructures of several kinds of NbTi conductors and discussed the effects of microstructures on J_c values. These investigations show that the inverse sub-band size dependence of J_c does not hold in two phase composites, but that the morphology of α-Ti precipitates play a crucial role in raising the J_c value in NbTi alloys. The large equiaxed α-Ti precipitates lead to low critical current density, but the plate-like precipitate having suitable thickness and separation can lead to high J_c. The equiaxed α-Ti precipitates in low J_c NbTi conductors are generally located at the boundaries between sub-bands. The micromorphology of high J_c NbTi multifilamentary wire is completely different from that of low J_c. The diameters of sub-bands are about 30 - 50 nm. After multiple drawing, thermal treatment and final cold deformation all of the α-Ti precipitates are drawn out into ribbons along the wire axis. In the photograph of the transverse section of the wire, these ribbons of α-Ti particle have a rippled-plate morphology, their thickness is about 3 - 7 nm having the same order of magnitude as the coherence length in NbTi alloy (5 nm). The width and separation of the plates are about 20 - 50 nm, depending on the chosen process technology.[2,4]

As it is well known that the most reliable theoretical approach for this purpose is statistical theory and these were used for relatively

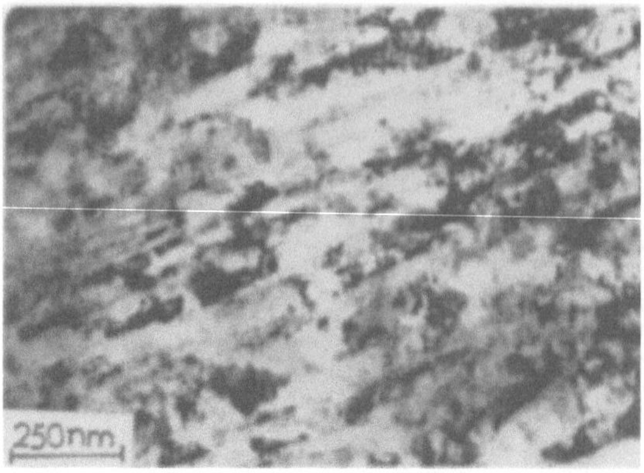

Fig. 1. Electron micrograph of a longitudinal section in the wire CH661.

weak pins. For superconducting materials with strong pins, the pinning force density is experimentally known to obey a linear summation model. But the theoretical foundation of the model has not been clarified. By the aid of the statistical theory for strong pins, Matsushita derived an expression of the pinning force density analogous to that of the linear summation model. This gives the dependence of the pinning force density on microstructures.[5]

In this paper, plate-like α-Ti precipitates are looked upon as the strong pinning centers. Analytical expressions for the pinning force density and critical current density are derived with the aid of Matsushita's statistical theory and the Ginzburg-Landau local free energy density function.

RESULTS

The Expressions of the Pinning Force and the Critical Current Density

If the volume that a flux line intercepts α-Ti plates is denoted by U. The pinning force of the action on this flux line is given by

$$P = - U\left(\frac{dE}{dx}\right)_{max} = - Up_f \qquad (1)$$

Fig. 2. Electron micrograph of a transverse section in the wire CH661.

where E is the local free-energy density and x is the site coordinate. The E given by Campbell and Evetts[6] is as

$$E = \frac{1}{2} \mu_o H_c^2 [- |\psi|^4 + \xi^2 \nabla^2 |\psi|^2] \quad (2)$$

where $|\psi|^2$ is the local order parameter and H_c the thermodynamic critical field. In the case of one-dimension, $|\psi|$ can be written as

$$|\psi|^2 = \frac{2}{3} (1 - b) (1 - \cos \frac{2}{a_f} x) \quad (3)$$

where $a_f = (\frac{2}{\sqrt{3}} \frac{\phi_o}{B})^{1/2}$ is the fluxoid spacing. From eqs. (1) - (3) we have an equation for P_f

$$P_f = \frac{8\pi}{9} \frac{\mu_o H_c^2}{a_f} [(1+D)(1-b)^2 + \frac{3\sqrt{3}}{4} \pi b (1-b)] \times (1-D^2)^{1/2}$$

$$= \frac{1}{a_f} f(b)$$

where $b = \frac{B}{B_{c2}}$ is the reduced field, and

$$D = \frac{1}{4} \{[(1 + \frac{3\sqrt{3}}{4} \pi \frac{b}{1-b})^2 + 8]^{1/2} + (1 + \frac{3\sqrt{3}}{4} \pi \frac{b}{1-b})\} \quad (5)$$

the value of U depends on the morphology of the α-Ti precipitates and the relative orientation between the precipitate and flux line. We assume that the length of plate-like α-Ti precipitate along the wire axis is much greater than the fluxoid spacing, the average thickness and width of the plates in the transverse section of the wire are denoted by t and d, respectively, and the average separation of α-Ti plates is L. When the magnetic field is perpendicular to the axis of the wire, from the point of view of the statistics, the flux lines can only go through α-Ti plates at an angle to t or d. The probabilities for these two cases are $\frac{1}{\pi}$ and $(1 - \frac{1}{\pi})$, respectively. If we let Ut and Ud be the volumes of α-Ti plate occupied by a flux line in the above two cases, and n_f and n_d be the number of flux lines entering into the unit length of a α-Ti plate, respectively, we roughly have $U_t = t_d a_f$,

$$U_d = \frac{\sqrt{3}}{2} t a_f^2 , \quad n_t = \frac{1}{a_f} \text{ and } n_d = \frac{2\sqrt{3}}{3} \frac{d}{a_f} .$$

On Matsushita's strong pinning theory,[5] when a large pinning center interacts with a lot of flux lines, the contribution of the center to the force is $P = \sqrt{n} P_f$, where n is the number of the flux lines interacting with the center. Thus, the pinning force of a α-Ti plate with random orientation in unit length at the wire axis can be written as

$$P_u = [\frac{1}{2} n_f^{1/2} U_f + \frac{1}{2} (1 - \frac{1}{\pi}) n_d^{1/2} U_d] P_f (b) \quad (6)$$

where the factor 1/2 in second term of eq. (6) is introduced because a larger pinning center has a lower effective condensation energy density than a smaller center with size less than ξ.[5]

When the average separation of α-Ti plates is less than that of the flux lines, that is high pin density, every flux line can simultaneously interact with a lot of plates. In this case, we choose $V = L_c a_f$ as the statistical average unit of the volume to obtain pinning force density. Here $L_c = 4\pi a_f$ is the cutoff length of the flux lines. The number of α-Ti plates in this volume unit is $N = a_f L_c n_o$, where $n_o = 1/L^2$ is the concentration of pinning center. From Matsushita's theory, pinning force density can be expressed as $F_p = \frac{\nu g}{V} \sqrt{N}\, P_v$ where g is the statistical constant.

On the other hand, when the average separation of α-Ti plates is larger than that of flux lines a_f, that is low pin density, we choose $V = L^2$ as statistical average unit of the volume to derive the force density. In this case, pinning force density can be written as

$$F_p = \frac{1}{V} P_v$$

Thus, the expression of the pinning force density in NbTi alloy wire with plate-like α-Ti precipitates is

$$F_p = \begin{cases} \frac{Lg}{V} \sqrt{V}\, P_v \; ; & L < a_f \quad (7a) \\ \frac{1}{V} P_v \; ; & L > a_f \quad (7b) \end{cases}$$

Substituting eq. (6) into the eq. (7) and reducing eq. (7) we obtain

$$F_p = \begin{cases} \left[0.1592 \frac{SL}{a_f^{3/2}} + 0.3172 \frac{(St)^{1/2}}{a_f}\right] f(b) & L < a_f \quad (8a) \\ \left[0.3183 \frac{S}{a_f^{1/2}} + 0.3172 \frac{(St)^{1/2}}{L}\right] f(b) & L > a_f \quad (8b) \end{cases}$$

where $s = dt/L^2$ is the volume fraction of the precipitate particle in the conductors, f(b) is given by eq. (4),

$$f(b) = \frac{82}{9} \mu_o H_c^2 \left[(1+D)(1-b)^2 + \frac{3\sqrt{3}}{4} \pi b(1-b)\right](1-D^2)^{1/2}$$

D is only a function of b (see eq. (5)).

From $J_c = F_p/B$, the critical current density J_c is given by

$$J_c = \begin{cases} \left[0.1592 \frac{L}{a_f^{3/2}} + 0.3172 \frac{(St)^{1/2}}{a_f}\right] \frac{f(b)}{B} \; ; & L < a_f \quad (9a) \\ \left[0.3183 \frac{S}{a_f^{1/2}} + 0.3172 \frac{(St)^{1/2}}{L}\right] \frac{f(b)}{B} \; ; & L > a_f \quad (9b) \end{cases}$$

COMPARISON WITH EXPERIMENT AND DISCUSSION

Equations (8) and (9) show that F_p and J_c are very complicated functions of magnetic field and the microstructure parameters S, T and L. It is not possible that the contribution of the field to F_p and J_c are completely distinguished from that of the microstructure because the part of the square brackets in eqs. (8) and (9) depends not only on the structure parameters, and the spacing of flux line. Therefore, discussing the dependence of F_p and J_c on the field, we have to do with the structural parameters, and vice versa.

The Dependence of F_p and J_c on the Magnetic Field

The NbTi 50 conducting wire CH661 worked in Baoji, China has quite high values of F_p and J_c in midfield, for example, it has the F_p value of 18.5 GN/m^3 at about 4 T and 4.2 K and the J_c value has reached 3460 A/mm^2. Its microstructure parameters are L = 7 nm, S = 0.24, L = 31 nm, the upper critical field is about 9.5 T at 4.2 K and 12.2 T at 2.2 K.[1] The J_c values vs. the field are carefully measured in the full field. The L value of 31 nm corresponds to the spacing of flux line at 2.5 T. According to the eqs. (8) and (9), when the applied field is lower than 2.5 T, the values of F_p and J_c should be calculated from eqs. (8a) and (9a), and when the field is grater than 2.5 T. The values of F_p and J_c should be calculated from eqs. (8b) and (9b) respectively. Figs. 3 and 4 give the results of the calculations and measurements of F_p vs. B and J_c vs. B.

The Dependence of F_p and J_c on the Microstructure

Both Eqs. (8) and (9) show that F_p and J_c are increasing functions of the volume fraction S, and are proportional to the square root of the thickness to of the α-Ti particles, when the separation L is greater than a_f, F_p and J_c are an inverse function of the separation L, when the separation L is Less than a_f, they are proportional to L, when L is equal to a_f, F_p and J_c have the maxima, giving $g\nu = 2\sqrt{\pi}$.

The J_c values of some NbTi conductors have been calculated from eq. (9) at 5 T, 4.2 K. The comparison of the calculated results with the

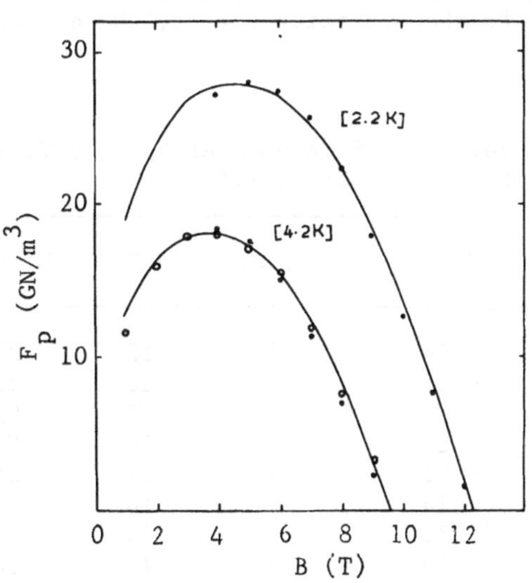

Fig. 3. The pinning force density vs. B. curves.

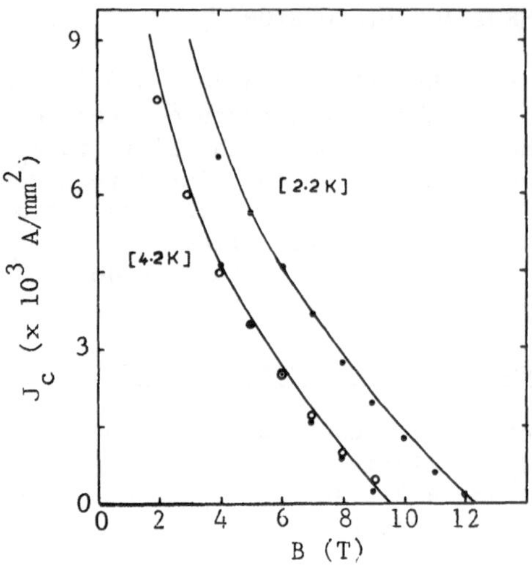

Fig. 4. The critical current density vs. field curve.

with the measurements are listed in Table 1. The microstructure details are given at the same time.

The J_c values computed in eq. (9) have a good agreement with the measurements. The investigations by Larbalestier et al.[3] show an approximately inverse relationship between the plate separation and J_c at 5 T, 4.2 K, when the plate separation becomes less than about 20 nm, however the J_c begins to decline. This result implies that J_c has a maximum when L is about 20 nm. Accounting for the fact that the separation of 20 nm is approximately equal to the space of flux line at 5 T. It is just as prospective result of eq. (9) that the maximum of J_c appears at about 20 nm at 5 T. Fig. 5 shows the J_c vs. L curve at 5 T, 4.2 K. The curve is calculated in eq. (9). The parameters used are as follows: H_{c2} = 10.2 T, t = 4 nm, s = 0.18 which are typical values for high J_c NbTi conductors. The measurement data in Fig. 5 come from ref. 1. The departure of the calculating curve from the measurements is because the conductors have different microstructure parameters from above typical parameters. Kramer and Freyhardt[8] have theoretically demonstrated that due to the proximity effect, the elementary interaction

Table 1. The J_c Values of Some NbTi Alloy at 5 T, 4.2 K and Microstructure Parameters.*

Sample	H_{c2}(T)	L(nm) +	t(nm)	S(%)	J_c(A/mm^2)	J_c(A/mm^2)cal.
F187	10.4	22	3	17	2525	2854
F233	10.4	25	4	12	2214	2214
F245	10.4	31	6	20	2340	3240
F220	10.4	27	3	14	2315	2233
F275	10.4	31	4	14	2170	2244
B879	10.4	24	5	14	2460	2687
CH661	9.5	31	7	24	3428	3476
BA98345	10.2	19	4	18	2610	2789

*The measurement data in the table are from ref. 7.

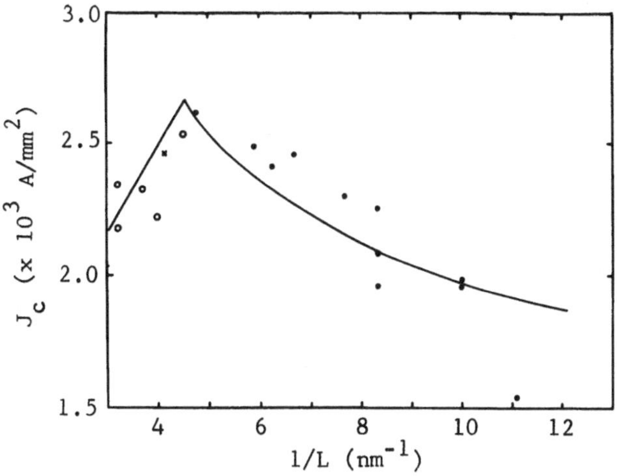

Fig. 5. The critical current density vs. 1/L at 5 T and 4.2 K.

force for small normal conducting precipitate particles is related to the smallest dimension of the particle, it is the more so when the dimension is much less than the BCS coherence length, the effect is not discussed in this paper. In addition, the microstructural parameters can affect the J_c in other ways.[4] A further study is necessary, even so, the relationship between J_c and microstructure yet has a better agreement with measurement. On the basis of eq. (9), if the volume fraction and thickness of a-Ti plate are properly increased, and the separation of α-Ti plate is is adjusted to about the spacing of flux lines in the applied field, the best critical current density should be obtained.

ACKNOWLEDGMENTS

We wish to thank Dr. D. Larbalestier and Zhou Lian for their valuable help in carrying out critical current and microstructure parameter measurements of our superconducting wire.

REFERENCES

1. Zhou Lian et al., Properties of NbTi50 superconducting composite wire, in "Journal de Physique", colloque C1, supplement au n°1, Tome 45, janvier 1984, pp. c1-437 - c1-440.
2. A. W. West and D. C. Larbalestier, IEEE Trans. Mag., vol. Mag-19, No. 3 (1983), pp. 754-757.
3. D. C. Larbalestier, Invited paper VI at 1983 Particle Accelerator Conference, IEEE Trans. on Nuclear Science, 30, 3299 (1983).
4. Zhang Tingjie et al., ICEC10 (1984), p. 700.
5. T. Matsushita, Japan J. Appl. Phys., vol. 20 (1981), pp. 1153-1159 and pp. 1955-1966.
6. A. M. Campbell and J. E. Evetts, Adv. in Physics, vol. 21, no. 90 (1972).
7. See also A. W. West, D. C. Larbalestier, IEEE Trans. Mag., vol. Mag-19, (1983), p. 548.
8. E. J. Kramer and H. C. Freyhardt, J. Appl. Phys., vol. 51, No. 9 (1980), pp. 4930-4938.

THE DEVELOPMENT OF MICROSTRUCTURE IN MULTIFILAMENTARY BRONZE ROUTE A15 COMPOSITES

E.R. Wallach and J.E. Evetts

University of Cambridge
Department of Metallurgy and Materials Science
Cambridge, England

ABSTRACT

Although prototype multicomponent A15 composites continue to show steady improvement in properties, the optimisation of the design, fabrication and heat treatment of these wires still proceeds on a semi-empirical basis. Two major problems confront anyone setting out to design a new composite or modify an existing design. Firstly, although the main elements of microstructure, composition, degree of order and stress state are now appreciated, there is still a lack of quantitative understanding of the relation of superconducting properties to the details of the microstructure. Secondly, even if an optimum microstructure could be defined, understanding of how it should be developed during fabrication and heat treatment is at best rudimentary. This review has three parts. Initially, aspects of A15 formation and microstructure on which there is general agreement are summarised and then discussed. This is followed by a critical assessment of different models of microstructural development. Finally, the most important unresolved problems are highlighted. Thus the review includes a discussion of pre-reaction layers, grain morphology, growth mechanisms and kinetics, stress distribution and the main effects of alloying additions; where appropriate, these are related to superconducting properties.

INTRODUCTION

The engineer developing a high-field magnet as part of a complex cryogenic system needs to work with a conductor specification expressed in terms of the bulk superconducting and mechanical properties of available high field superconducting composites. Such bulk engineering properties are inevitably averages over both the filaments of the composite and the superconducting layer within each individual filament, and it has become increasingly clear that local variations of properties within commercial superconducting composites are very large indeed. Thus properties such as the critical temperature, T_c, and upper critical field, B_{c2}, may vary by up to 25%, while other properties such as the critical current density, J_c, and Young's modulus may vary by 100–500%. One consequence of these local inhomogeneities is that a composite becomes difficult to characterise in terms of bulk parameters[1]. For example, a resistive measurement of T_c will sample the regions with highest T_c within the composite, a critical current measurement at high fields samples regions with the highest B_{c2}, while the low field critical current depends less on the value of B_{c2} and more on the local grain size. Since these regions are unlikely to coincide in a technological composite, the problem of optimising a conductor by reference only to its bulk engineering properties becomes intractible. This is the main practical reason why a detailed understanding of the development of microstructure and its relation to superconducting properties is essential.

A complete microstructural description includes the local composition variation, the defect state and state of order, the grain size and morphology and the local strain state of the A15 lattice. Whilst transmission electron microscopy (TEM) imaging and analysis can elucidate many features of microstructure, there are other features that may only be determined indirectly. Most important is the state of stress which can only be deduced through an understanding of the process of evolution of the microstructure. Only if there is a quantitative understanding of how the microstructure develops can one predict the effect of changes in composite design, fabrication and heat treatment on microstructure and superconducting properties; this then can be used to optimise the engineering or in-service performance of the superconductor.

A review of this length cannot be comprehensive; reference should be made to other recent reviews[1-4]. In this paper, the accepted aspects of A15 microstructural development are summarised so that unresolved problems can be put in perspective.

A15 FORMATION AND MICROSTRUCTURE: FACTS

Several approaches to composite design are currently of interest. However, the fundamental considerations that determine the development of A15 microstructure are generally common to the different designs. Thus the primary concern of this review is the formation of Nb_3Sn by a solid-state diffusion reaction of Nb with α-phase CuSn bronzes. The major variants on the bronze route employ different schemes for entraining Sn during fabrication and for dispensing Sn during compound growth. Their aim is to avoid two particular disadvantages of the conventional bronze route. The first is the inherently low value of the ratio of Nb cross-section to the overall conductor cross-section, because the maximum Sn content in the bronze is 13.5wt%. The second is the requirement for intermediate bronze anneals after each 60% reduction in area cross-section.

The main stages of microstructure development in multifilamentary bronze route composites are pre-reaction anneals and then the reaction heat-treatment. Although the subject of close study for more than a decade, it is only recently that TEM studies have provided a consistent picture of the filament microstructure at various stages of the layer reaction[4-8]. This is because the submicron scale of the inhomogeneous grain structure of the A15 layer dictates that only transverse TEM data will reveal unambiguously the microstructure details across the diffusion layer and at the boundaries with bronze and Nb core. Caution has to be exercised in interpreting transverse SEM micrographs of fractured filaments as there is evidence that a fracture surface accentuates the columnar appearance of the grain structure[9].

Pre-reaction

Two microstructural changes of fundamental importance occur during composite fabrication prior to the main heat-treatment reaction. Firstly, the initially equiaxed structure of the Nb filaments develops a marked <011>-fiber texture with two of the four <111> directions oriented favourably for slip parallel to the fiber axis[10,11]. In addition, filaments are observed to develop pronounced slip steps on the (011) plane parallel to the filament axis

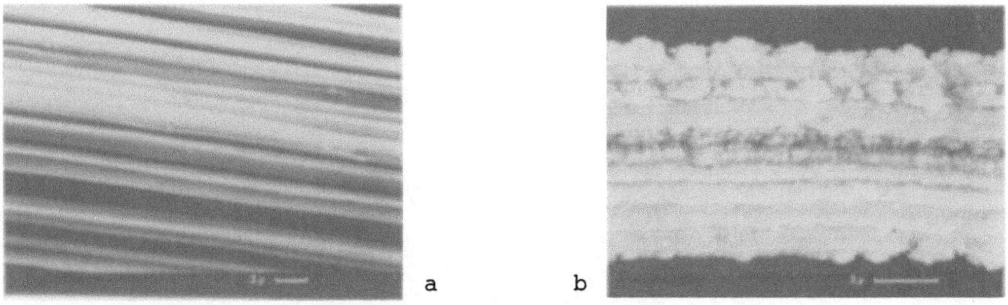

Fig.1 Nb filaments from Wah Chang MJR composites - matrix etched away.
(a) Nb-Cu composite PBI, thus no pre-reaction A15.
(b) Nb-bronze composite M88, thus Nb filaments with pre-reaction A15.

that contains both active slip directions; Bevk et al.[11] attribute these steps to the off-axis stress components during drawing. As a consequence, the filament shape distorts and displays a characteristic angular aspect. Secondly, each low-temperature bronze anneal, typically at 450°C, results in the formation of a small amount of A15 phase. Since twenty or more such anneals may be necessary during the composite fabrication, considerable amounts of A15 phase are formed (see Fig.1). This layer is referred to as the "pre-reaction" layer; it is irregular and fragmented since the brittle A15 phase formed at each anneal is fractured and redistributed during subsequent drawing stages. The pre-reacted layer can grow to more than 0.5μm and there is strong evidence that it seriously degrades the critical current of the composite[4,12].

The Main Diffusion Reaction

The Nb filaments may typically have diameters of 2-5μm and, depending on composite design (availability of Sn), all can be converted to Nb_3Sn during reaction heat-treatments of about 100 hrs at 700°C. The formation of Nb_3Sn is accompanied by a 37% volume expansion which, in practice, occurs as a 37% increase in area[13]. (Because of contraction by the bronze, the overall composite changes by only 4%.) Thus although the layer growth rate is quite small, quite large strains can be introduced. At the growth front of the A15 layer, there is always a region or band of columnar grains. This region is already well established after as little as 12 minutes at 700°C (Fig.2). The columnar grains, with a mean diameter in the region of 30-50nm, grow into the Nb layer with the occasional nucleation of a new grain, and initially extend from the pre-reaction layer to the growth front at the Nb core. At a later stage, a new region consisting of equiaxed grains forms and separates the pre-reaction layer from the columnar region. The columnar grains have length:diameter ratios between 4:1 and 10:1 while the diameters of the new equiaxed grains are comparable to those of the columnar grains. As layer growth continues, the equiaxed region grows in extent, while the columnar region remains of roughly constant thickness, perhaps decreasing gradually. There is some uncertainty as to the factors that determine the extent of the columnar region. There is clearly a strong dependence on the filament geometry (the columnar region is far more extensive in tape-like filaments) and some workers have also suggested a correlation with the tin concentration in the bronze[14]. Models for the development of this unusual duplex microstructure are discussed below.

Additional processes, continuing throughout the layer growth phase and also after the diffusion reaction is complete when all the Nb is consumed, are

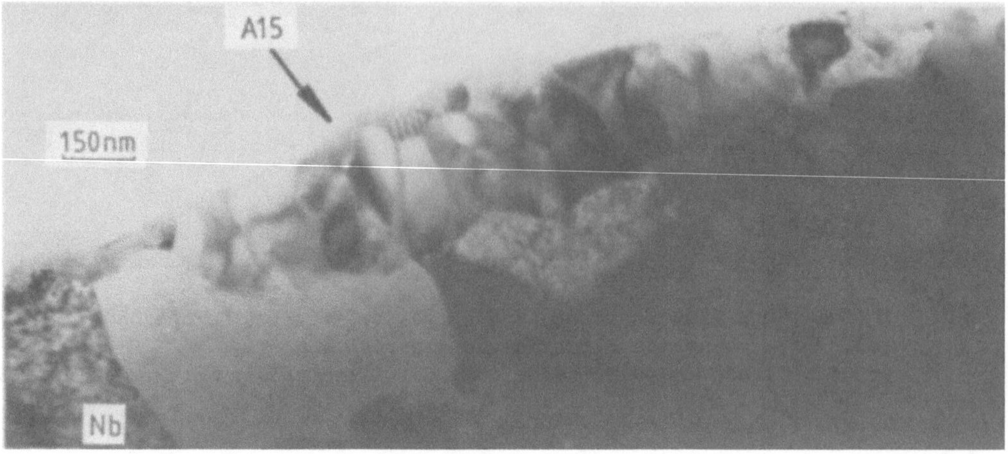

Fig.2 TEM micrograph of multifilamentary Nb-Ta composite after a reaction heat-treatment at 700°C of 12 minutes.

grain growth, compound ordering, adjustment of composition profiles and stress relief. All of these processes are of importance for the superconducting properties. The composition and state of stress determine the local variation of primary superconducting parameters T_c, B_{c2} and the Ginzburg-Landau parameter, while the critical current for a given set of primary parameters depends on the size and morphology of the grain structure[1].

A15 FORMATION AND MICROSTRUCTURE: DISCUSSION

Strain

The above facts are generally accepted. The following considers briefly aspects of microstructural development that are less well understood and commences with the effects of strain. The A15 structure displays anomalous elastic behaviour that make its superconducting properties highly sensitive to its strain state. For instance a strain of 0.3% can reduce the critical current by 40% at 0.75 B_{c2}[15]. When the compound maintains cubic symmetry, the superconducting properties are optimum; thus deviatoric stress causes much more severe degradation than pure hydrostatic pressure[16]. This behaviour has its origin in the electronic structure of the A15 phase with a very high density of states at the Fermi surface. There are three main sources of strain in a reacted layer in a multifilamentary composite. Firstly, there is axial compression imposed on the filaments by the bronze matrix through differential thermal contraction. Secondly, high levels of strain are introduced into the layer during layer growth by the 37% volume expansion[13]. Finally, certain Nb_3Sn compositions are prone to a low temperature second-order martensitic transformation[16-17]. In single crystals, the untransformed cubic and transformed tetragonal phases have a difference in T_c of 0.4K, and B_{c2} values at 4.2K of 26T and 21.5T respectively[17]. The practical effect of all three of these distortions is best viewed by considering the anomalous elastic modulus of a polycrystalline A15 phase of the appropriate composition. Pure stoichiometric Nb_3Sn displays a Young's modulus that drops by a factor of 2.5 from 300K to 4.2K; this material clearly is highly susceptible to degradation by deviatoric stress. Off-stoichiometric Nb_3Sn and compounds with ternary additions do not soften to the same extent and so are far less susceptible to strain degradation[18].

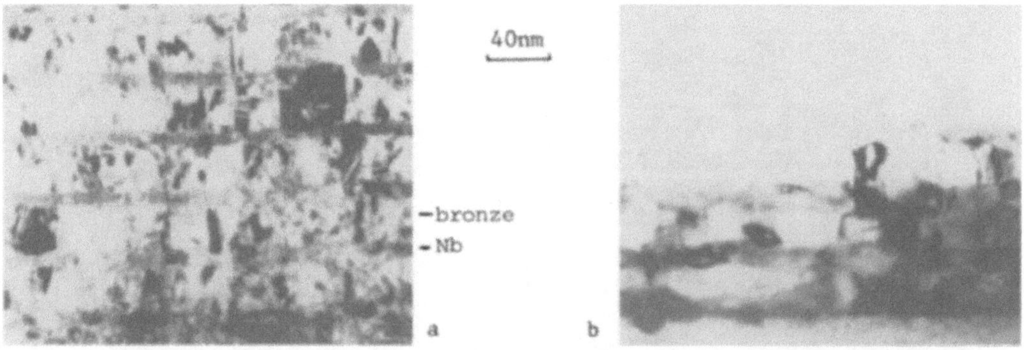

Fig.3 TEM micrographs of multi-layered thin-film composites (ref.19).
(a) Unreacted. (b) Almost fully reacted after 2 hours at 450°C.

Nucleation

The formation of Nb_3Sn from Nb and bronze is a nucleation and growth phenomenum yet the nucleation stage appears to have received only scant attention in the literature compared with growth studies. Given the importance of small grain size, e.g. in optimising J_c in medium applied fields (10–14T), and since grain size is dependent on nucleation, this aspect of Nb_3Sn formation is now considered. Nucleation generally occurs heterogeneously, for example on dislocation arrays and grain or sub-grain boundaries which intersect the original Nb-bronze interface. Thus the grain sizes and dislocation densities within both Nb and bronze may influence the grain size of the initial Nb_3Sn which is formed during pre-reaction anneals. This Nb_3Sn may influence the grain size of that subsequently grown during reaction heat-treatments. The bronze grain size after reaction heat-treatments of more than 24 hours at temperatures above 650°C is of the order of 1–2µm; in the heavily worked state, its grain size will be considerably less and its dislocation density high. In contrast, the Nb has a grain size of 0.1µm after deformation and is not expected to recrystallise given its high melting point. Thus the grain size of the pre-reaction Nb_3Sn grains may well be consistent with the grain size of, or dislocation sub-structure within, the deformed bronze. Some evidence for this hypothesis can be seen from preliminary results of Somekh and Pugh[19] who examined the initial growth of Nb_3Sn in sputtered multi-layered thin-film composites using TEM and X-ray diffraction (XRD). The grain size of the Nb_3Sn nuclei in the heat-treated multilayer appears to match that of grains in the bronze layers (Fig.3). Another factor that will influence the density of nuclei is surface (or interface) energy which can be related to the degree of epitaxy that occurs between the Nb_3Sn and the Nb or bronze substrate. The extent of this epitaxy will depend on the strain energy arising from the misfit between the substrate and Nb_3Sn lattice planes. The lack of clear and consistent orientation relationships between the Nb and columnar Nb_3Sn grains in TEM studies of filaments suggests any initial epitaxy must occur only with the bronze.

Growth

For heat-treatments of time t at a constant temperature, the diffusion distance, x, is expressed in terms of a growth law of the form $x = kt^{n}$. Values of the exponent n can be predicted from theoretical analyses and depend, for thin-film diffusion couples, on the precise path taken by the diffusing atom[20]; values of n for diffusion control can be between 0.5 and 0.25. Higher values, tending to 1, reflect conditions under which a reaction

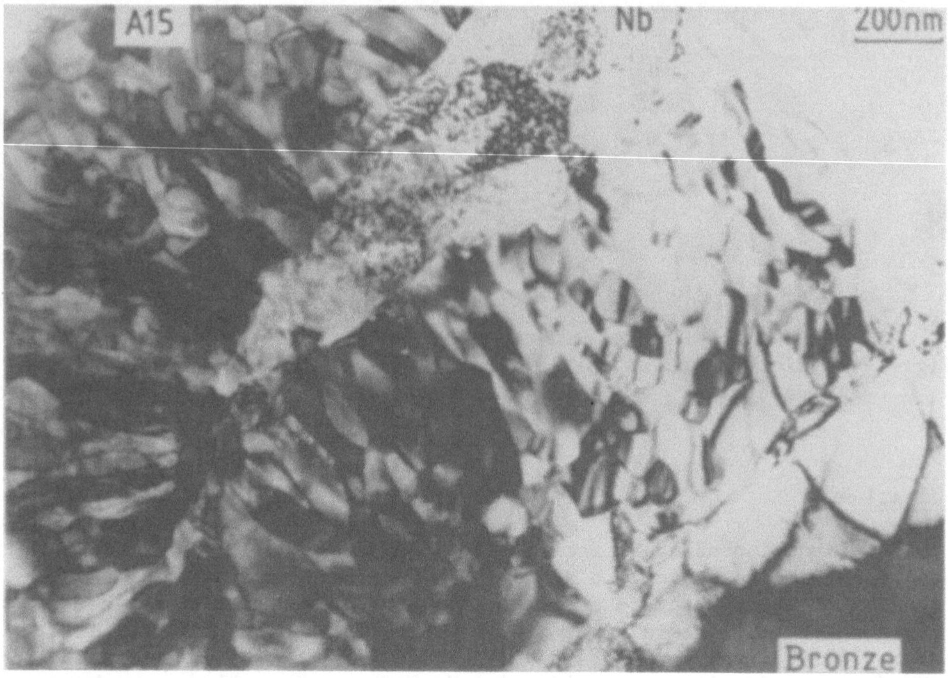

Fig.4 TEM micrograph of Wah Chang MJR composite M88 after a reaction heat-treatment at 650°C for 49 hours.

at a growing interface is rate controlling rather than flux to the interface; in practice, this occurs when a growing layer is very thin or if the layer becomes extensively cracked[21]. The analyses by Farrell et al.[21-22] predict a value of n of 0.5 for growth of the Nb_3Sn layer, assuming the diffusing Sn atoms are transported down grain boundaries in the Nb layer with negligible diffusion sideways into non-stoichiometric grains, as is generally accepted. This value was shown to fall to 0.4 when there was a competing grain growth mechanism within the Nb_3Sn layer which resulted in a time-dependent decrease in the flux of Sn to the growing interface. Experimentally, n has been determined for multifilamentary internal-route composites by a number of investigators and values generally between 0.1 and 0.4 have been reported for heat-treatments between 650 and 800°C[7,13,22-24].

The first point to consider in the analysis and experimental determinations of n is the implicit assumption that the Nb_3Sn layer can be regarded as a uniform structure comprising grains which do not change morphology. In fact, TEM studies[5-8] have shown this is not so (see Fig.4). Thus values of n change as the Nb_3Sn layer grows and the grain morphology alters. A second and more serious point that has not been accounted for is the variation in rates of growth of filaments which are different distances from the main Sn supply (bronze reservoir in either wire core or circumference) in multifilamentary composite wires. While the overall Sn concentration has been recognised as affecting layer growth[14], the consequences of considerable variations in the availability of Sn to filaments at different positions within a composite has not been discussed until recently[7,25]. It has been shown that, for Nb filaments away from the Sn reservoir, three regimes of Nb_3Sn growth can be found according to the availability of Sn[7]. Initially, rapid growth occurs using Sn around individual filaments but, as Sn depletion occurs, slow growth (n falling to values of 0.1) commences. This

Fig.5 TEM micrograph of Wah Chang MJR composite M88 after a reaction heat-treatment at 650°C for 49 hours.

continues until Sn becomes available from the reservoir when n increases again to around 0.4. In contrast, filaments near to the tin reservoir grow with a single time exponent (near to 0.4 as predicted by Farrell et al.). Thus any analyses on multifilamentary wires must consider and allow for the availability of tin, i.e. filament location. The route taken by diffusing Sn atoms is believed to be down the few grain boundaries in the large grains of bronze and along the many A15-bronze interfaces. Hence depletion of the diffusing Sn will occur by the growth of A15 layers in filaments nearer the bronze reservoir. This continues until these A15 layers are reasonably thick and only then will the Sn become available to filaments further from the reservoir. The different thicknesses of Nb_3Sn formed in various locations in a wire after the same heat-treatment will cause the superconducting properties to be inhomogeneous.

Kidson, as early as 1961, commented on the dangers of trying to ascribe a physical significance to an activation energy in a multi-component system in which a number of simultaneous kinetically-controlled events were occurring; it is not possible to do so unless one event is dominant and remains so[26]. Activation energies for the formation of Nb_3Sn layers have been determined experimentally and have been reported to be approximately 220 $kJmol^{-1}$ (2.3 eV $atom^{-1}$)[7,27]. Given the variation in grain morphology and the dependence of layer growth on filament location, the reported values for activation energy of layer growth are of little fundamental significance.

Finally, the importance of two other experimental results has not previously been emphasised. Firstly, the Nb_3Sn-Nb interface is essentially planar (see Fig.5) despite the variety of orientations of both small Nb_3Sn and large Nb grains along the interface. Secondly, TEM studies have shown that the columnar grains do not have a common texture. Both results suggest that the orientation of the Nb grains is not important with regard to nucleation and/or growth because many columnar grains are nucleated and grow at similar rates from any Nb grain, the growth faces of which have all orientations within the <011> zones. The planar interface will also be favoured due to interface diffusion and the hoop strains in the region of the growth front.

Additives

Thus far our attention has been directed to the binary Nb_3Sn composite. However, the pressure to optimise conductors with ever higher critical currents at high magnetic fields has produced an irresistible trend to more complex alloys with ternary and quaternary additions. A dozen or more elemental additions have been investigated to varying extents with most attention being given to additions of Ta, Ti, Ga and Mg. Such additives can act beneficially in two different ways. Firstly, they can become incorporated into the A15 phase, thereby improving superconducting or elastic properties. Secondly, they can change the kinetic process of microstructure development, e.g. affecting the nucleation density, layer growth rate, grain morphology or grain size. Most additives act in both ways; Mg is perhaps an exception in that it appears to reduce grain coarsening in the growing layer without having a significant effect on primary superconducting properties[9].

A major role of ternary additions is to raise B_{c2}; there is now general agreement that B_{c2} enhancement of an unstrained A15 phase depends predominantly on the normal state resistivity which controls the Ginzburg-Landau coherence length[3,28]. A maximum in B_{c2} occurs for a resistivity of about 35 μΩ cm; additives as disparate as Ni, Ta, Ti and Hf produce similar improvements in B_{c2} when expressed in terms of the resistivity change produced. Additives generally lead to a gradual depression of T_c at low concentrations and this depression accelerates if a resistivity of about 40 μΩ cm is exceeded[3]. Only Ga significantly raises T_c, by more than 0.6K, although there is evidence that 2at%Ta increases T_c by 0.2K[3,29]. Of more importance in practical composites is the effect of additives on T_c in off-stoichiometric compounds. Suenaga has shown that the decrease in T_c with decreasing Sn content is more gradual for most of the ternaries than for the binary composite[3].

Ternary additions have a major effect on the elastic moduli of the A15 phase[18] and the importance of this for superconducting properties has been underestimated in the literature. If Young's modulus at 4.2K can be increased by a factor of two (e.g. by adding 6at%Ta), then the strain degradation in a composite is roughly half that of an equivalent binary composite. If the effect of an additive on the elastic moduli can be fully described, this probably would also take into account any effects that depend on the second order martensitic transformation[17]. It is clear, however, that across a diffusion layer the composition, stoichiometry and state of stress are likely to vary significantly. Furthermore there is accumulating evidence that the A15 stoichiometry and the concentration of additives can vary strongly near the grain boundaries; this is likely to be of most importance for flux pinning at low fields[3,30].

The role of additives in the development of microstructure is more complex and less well understood. Certain facts are reported but it is not yet possible to predict theoretically what might be the effect of a given additive and its optimum concentration. Additions may be made to the bronze or to the Nb; usually solubility limits and effects on drawability determine which is preferred. Thus Ta additions are made to the Nb because drawability can actually improve, while Ti is added to core or matrix even though when added to the latter the amount of Ti in the A15 layer is less. The proportion of an additive incorporated in the A15 layer and the corresponding effects on the amount of Sn in the layer, general stoichiometry within the ternary compound and kinetics of layer growth depend in a complex way on thermodynamic factors. The Gibbs free energy determines equilibrium phase boundaries and therefore the stoichiometry expected, whereas gradients of activity of the different components in Cu and Nb determine the driving forces for diffusion. Ga is a good example of an additive whose activity in Cu profoundly influences Sn diffusion and the A15 layer growth rate[31]. Additives may also change grain boundary energy and structure which can affect grain boundary diffusivity and/or the driving force for grain coarsening rate. Both can alter layer growth kinetics and final morphology of grains in the A15 layer. For instance, although Ta and Ti both enhance the layer growth rate, Ta causes grain refinement whereas Ti results in a coarser grain morphology; perhaps Ti increases grain boundary diffusivity to a greater extent than it decreases grain boundary energy. In contrast, the reduction in grain coarsening rate by Mg could result from a marked decrease in grain boundary energy.

In the above, some correlations between metallurgical structure and superconducting properties have been made. However, the precise mechanisms by which the additives affect the thermodynamics and kinetics of A15 layer growth are still not fully understood and further work is needed.

MODELS FOR THE DEVELOPMENT OF MICROSTRUCTURE

Although measurements of layer growth kinetics, activation energies and Nb_3Sn grain sizes have been reported, observation rather than prediction has tended to dominate. Thus, despite the need for a quantitative model for microstructural development, as outlined in the Introduction, such a model does not exist. Several qualitative models have been recently proposed to explain the grain morphology observed within the Nb_3Sn layer and from these the basis of a quantitative model now has appeared.[3] This section summarises the various models and suggests a basis for modelling pre-reaction.

Okuda et al.[14] correlated the formation of equiaxed A15 grains with the availability of Sn at the Nb-A15 growth front. Small equiaxed grains were associated with a high Sn content, which it was suggested would favour the nucleation of new grains, while the columnar structure resulted from less Sn at the growth front. This was used to explain the variation in columnar to equiaxed ratio at different regions within the same multifilamentary composite; filaments near the bronze reservoir have more equiaxed grains than those further away. In fact, the columnar grains are observed by TEM to reach a critical length (which is not accounted for in the model of Okuda et al.) and so the proportion of equiaxed grains increases as the layer thickens. A further problem for this model is the TEM observation that after just 12 minutes at 700°C columnar grains are observed (see Fig.2), i.e. they form when the Sn concentration is highest. Thus the model of Okuda et al. seems less convincing than those that follow.

The role of stress in controlling the grain morphology has been discussed qualitatively both by Cave and Weir[13], and by Wu et al.[6]. Cave and Weir demonstrated that the A15 layers form without cracking, contrary to an earlier suggestion by Sanger[32], despite the 37% increase in volume from the original Nb volume. The strain induced by this volume expansion far exceeds the tensile yield strain of the Nb_3Sn, which is less than 2%. Thus it was proposed that these large strains are relieved during layer growth by a relaxation process. The tensile hoop strain in the columnar grains, due to their being forced outwards by the 37% expansion in area, increases on moving away from the Nb growth front until its magnitude is such that dislocations nucleate and separate the tips of the columnar grains to form equiaxed grains. These grains will then rotate and/or slide to relieve the stress and, given the low strain rates associated with the slow layer growth rates, will continue to do so under very low loads, a characteristic of microcrystalline materials[33]. Wu et al.[6] suggested an identical mechanism with support from TEM micrographs which showed both low-angle boundaries at the tips of columnar grains and dislocations. The observed time and temperature dependence of microstructure, especially of the fine-grained equiaxed region, was discussed with particular emphasis on dislocation mobility; the greater their mobility, the easier the breakdown of columnar grains.

The role of stress has been analysed further in a quantitative analysis by Cave et al.[34]. The elastic stress distributions, hence strains, across

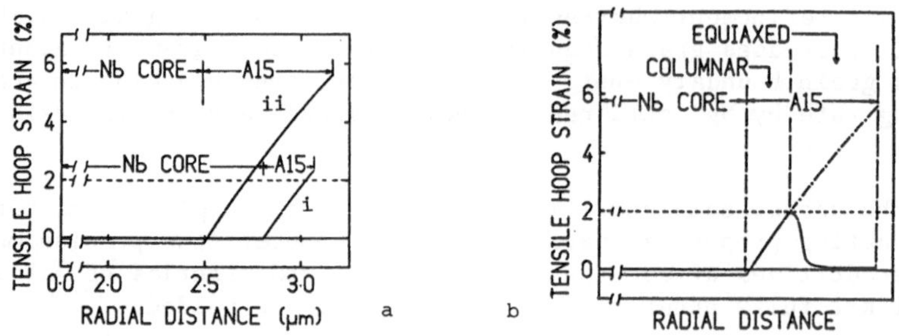

Fig.6 Variation of hoop strain with radial distance across a reacted filament.
(a) Predicted hoop strain from linear elasticity model for two different A15 thicknesses of (i) 0.3μm and (ii) 0.7μm.
(b) Schematic hoop strain expected resulting allowing for columnar to equiaxed transition.

partially-transformed Nb filaments have been described using a simplified plane-strain linear elasticity model. Typical results are shown in Fig.6 for a Nb filament initially of 3μm radius. The calculated hoop strain can be seen to be slightly compressive in the unreacted Nb core whereas in the A15 layer it becomes tensile and rises steeply. In theory, strains of approximately 3.3 and 5.2% for layer thicknesses of 0.3 and 0.7μm respectively at the bronze interface would be reached. In practice, these strains exceed the 2% tensile yield strain of the A15 and so cracking would occur unless the stress relief mechanism proposed above operates. Thus in Fig.6 the intercept at 2% elastic strain denotes a critical layer thickness above which the original columnar grains break down to form equiaxed grains; there is good agreement with experimental observations of the length of the columnar region (around 0.4μm at 700°C). The schematic diagram in Fig.6 shows the hoop strain expected across a filament allowing for the decrease in stress in the equiaxed region due to grain rotation and/or sliding. The analysis also predicts a decrease in columnar length both as growth proceeds and as the effective Nb filament radius decreases. This can be extended to non-circular filaments for which the columnar length will be greatest where the curvature is least; TEM studies on transverse sections on MJR composites support this. The observed severe depression in B_{c2} values for tape-like filaments[35], also is consistent with a high level of strain in the layer as predicted by the analysis.

The mechanisms and kinetics of formation of grains comprising the pre-reaction layer have not been modelled to-date. Some initial ideas are now proposed. It is evident (see Fig.1) that the Nb filaments do deform on definite slip planes, as discussed by Bevk et al.[11]. During the subsequent pre-reaction anneal, any such freshly-exposed Nb surfaces will be in contact with bronze of high Sn concentration and rapid A15 formation will occur. The XRD results of Somekh and Pugh[19] on multi-layered thin-film composites have shown on an Arrhenius plot that the effective diffusion coefficients at low pre-reaction temperatures fall on the same straight line extrapolated from higher main reaction temperatures - see Fig.7. (Incidentally, this result confirms that "short-circuit" diffusion rather than volume or bulk diffusion is the mechanism for material transport during the main reaction heat-treatment.) Using their data, estimates of the thickness (about 0.5μm) expected after the twenty or so pre-reaction anneals can be made and there is good agreement with SEM and TEM observations. Also, the A15 layers that form

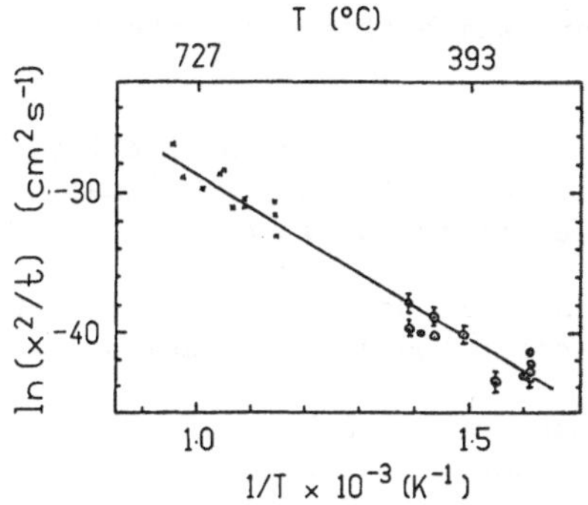

Fig.7 Variation of effective diffusion coefficient with temperature. Data presented for multi-layered thin-film composites at low temperatures[19] and for multi-filamentary composites at high temperatures[7,14,22].

well-defined slip bands seen in Fig.1 have very fine grain sizes; sizes of around 10-20nm are predicted from the data in Fig.7 and these values are consistent with TEM observations. Further reduction of a pre-reacted Nb filament leads to a break-up of these fine grains and some redistribution. Assuming a reduction in area of 60%, there would be an approximate increase in surface area of 58%, although this assumes cylindrical symmetry which clearly is not the case in practice. However, the deformation of the Nb filaments is inhomogeneous, occuring at the slip bands, and most of the increase in area would be expected on the slip bands. Therefore, any existing A15 pre-reaction grains will tend to be rotated/bunched up and not spread extensively along the new Nb interface. This may account for the very irregular appearance of pre-reaction grains at locations other than the slip bands. Clearly, this is a topic for further development, especially since pre-reaction grains may affect the diameters of columnar grains in the subsequent reaction heat-treatment.

Existing models have accounted for the microstructure observed in single filaments. In the future, it is necessary to model microstructural development in the entire composite by considering pre-reaction, the flux of Sn to individual filaments at different locations and the use of both diffusion coefficients and concentration or activity gradients.

CONCLUSIONS

The development of microstructure in multifilamentary A15 composite superconductors has many interlocking aspects. Although a unified quantitative description is still lacking, very substantial progress has been made since 1980. On the basis of the above discussion, it is possible to draw some important conclusions and also point to major unresolved problems. Dealing first with the former, one may state:

1. For the formation of Nb_3Sn, there is no evidence for any change in diffusion mechanism at temperatures as low as 350°C and so at the commonly-used reaction heat-treatment temperatures (around 750°C) "short-circuit" diffusion predominates.

2. The almost planar A15 layer growth front across many grains of Nb and Nb_3Sn indicates that growth is dominated by interface diffusion and is effectively independent of crystal orientation.

3. The linear elastic model for stress development in the growing A15 layer predicts a large tensile hoop stress superimposed on a somewhat smaller overall dilational strain. The columnar region of the A15 layer is expected, therefore, to have a strongly depressed B_{c2}.

4. The dilational strain is likely to enhance diffusion in the columnar

region and assist the nucleation of dislocations which then cause the formation of equiaxed grains.

5. Since the bronze grain size during reaction is determined by the interfilament spacing, the interface between the bronze and A15 layer becomes an important diffusion path for Sn from any bronze reservoirs in the composite.

6. The effect of additives on the A15 grain size is determined by a combination of changes in grain boundary energy and grain boundary diffusion rates.

7. The pre-reaction layer is extensive because, after each drawing stage, the freshly exposed Nb slip steps are able to react with a high tin bronze during the following pre-reaction anneal.

Some of the major unresolved problems are considered to be:

1. What mechanism determines the nucleation of the A15 phase at the bronze/Nb interface? There is circumstantial evidence that the nucleation sites are determined by the bronze.

2. Does the nucleation stage or the pre-reacted layer determine the columnar grain diameters?

3. How do specific additives affect the thermodynamics and kinetics of layer growth, e.g. the activity gradients for diffusion, the A15 grain boundary energy and the diffusion coefficient for Sn in the A15 phase.

4. A better understanding is required of composition profiles across the A15 layer and of unusual composition variations in the vicinity of grain boundaries within the A15 layer.

5. A self-consistent diffusion model is required to describe Sn supply, the Sn concentration and layer growth from filament to filament throughout a multifilamentary composite.

6. The quantitative stress model predicting the nucleation of equiaxed grains from the ends of the columnar grains furthest from the Nb interface requires further experimental verification.

7. A full treatment of the properties of an inhomogenous A15 filament must include the strong local variation in Young's modulus and stress in order to account for superconducting properties.

ACKNOWLEDGEMENTS

We would like to thank Dr J.R. Cave, Dr. B. Glowacki, Dr N.J. Pugh, Dr R.E. Somekh and J.L.M. Robertson for helpful discussion and collaborations. Support from the Science and Engineering Research Council and the Dept. of Metallurgy and Materials Science, University of Cambridge, is acknowledged.

REFERENCES

1. J.E. Evetts, IEEE Trans. Mag. MAG-19:1109-1119 (1983).
2. M. Suenaga, in "Superconductor Materials Science", S. Foner and B.B. Schwartz, eds., Plenum Press, New York (1981), pp.201-274.
3. M. Suenaga, IEEE Trans. Mag. MAG-21:1122-1128 (1985).
4. D.C. Larbalestier, IEEE Trans. Mag. MAG-21:257-264 (1985).

5. W. Schelb, J. Mat. Sci. 16: 2575-2582 (1981).
6. I.W. Wu, D.R. Diederich, J.T.Holthius, M. Hong, W.V. Hassenzahl & J.W. Morris Jr., J. Appl. Phys. 54: 7139-7152 (1983).
7. N.J. Pugh, J.L.M. Robertson, E.R. Wallach, J.R. Cave, R.E. Somekh & J.E. Evetts, IEEE Trans. Mag. MAG-21: 1129-1132 (1985).
8. N.J. Pugh, J.E. Evetts and E.R. Wallach, J. Mat. Sci., 20 (Dec,1985).
9. P.E. Johnson-Walls, D.R. Dieterich, W.V. Hassenzahl and J.W. Morris Jr., IEEE Trans. Mag. MAG-21: 1137-1139 (1985).
10. W.F. Hosford Jr., Trans. Met. Soc. AIME 230: 12 (1964).
11. J. Bevk, J.P. Harbison and J.L. Bell, J. Appl. Phys. 49: 6031-6038 (1979)
12. D.B. Smathers, K.R. Marken, D.C. Larbalestier and R.M. Scanlan, IEEE Trans. Mag. MAG-19: 1417 (1983).
13. J.R. Cave and C.A.F. Weir, IEEE Trans. Mag. MAG-19: 1120-1123 (1983).
14. S. Okuda, M. Suenaga and R.L. Sabatini, J. Appl. Phys. 54: 289-302 (1983).
15. J.W. Ekin, in "Filamentary A15 Superconductors", M. Suenaga & A.F. Clark, eds., Plenum Press, New York (1980), pp187-203.
16. D.O. Welch, Adv. in Cryogenic Eng. 30: 671-682 (1984).
17. R. Roberge, H. Le Huy and S. Foner, IEEE Trans. Mag. MAG-21: 811-814 (1984).
18. J.F. Bussiere, B. Faucher, C.L. Sneed and M. Suenaga, Adv. in Cryogenic Eng. Mat. 28: 453-460 (1984).
19. R.E. Somekh and N.J. Pugh (to be published).
20. J.E.E. Baglin and J.M. Poate, in "Thin Films - Interdiffusion and Reactions", J.M. Poate, K.N. Tu and J.W. Mayer, eds., J. Wiley, New York (1978), pp.305-357.
21. H.H. Farrell and G.H. Gilmer and M. Suenaga, Thin Solid Films 25: 253-264 (1975).
22. H.H. Farrell, G.H. Gilmer and M. Suenaga, J. Appl. Phys. 45: 4025-4035, (1974).
23. D. Salathe and K. Kwasnitza, J. de Physique 45: C1/489-493 (1984).
24. S.K. Agarwal, S.B. Samanta, K.V. Batra and A.V. Narlikar, J. Mat. Sci. 19: 2057-2063 (1984).
25. D.B. Smathers, K.R. Marken, P.J. Lee, D.C. Larbalestier, W.K. McDonald & P.M. O'Larey, IEEE Trans. Mag. MAG-21: 1133-1136 (1985).
26. G.V. Kidson, J. Nucl. Mat. 3: 21-29 (1961).
27. D.C. Larbalestier et al., IEEE Trans. Mag. MAG-11: 247 (1975).
28. R. Bormann, D.-Y. Yu, R.H. Hammond, T.H. Geballe, S. Foner and E.J. McNiff, IEEE Trans. Mag. MAG-21: 1140-1143.
29. J. Wecker, R. Bormann and H.C. Freyhardt, J. Appl. Phys. 54: 3318-3324 (1983).
30. M. Suenaga and W. Jansen, Appl. Phys. Lett. 43: 791 (1983).
31. S. Kwon, S.F. Cogan, J.D. Klein and R.M. Rose, J. Appl. Phys. 4: 1008-1012 (1983).
32. P.A. Sanger, E. Adam, E. Ioriatti and S. Richards, IEEE Trans. Mag. MAG-17: 666 (1981).
33. N.J. Grant in "Rapidly Quenched Metals", S. Steeb and H. Warlimont, eds., North-Holland, Amsterdam (1985), Vol.1, pp.14-22.
34. J.R. Cave, E.R. Wallach and J.E. Evetts (to be published).
35. M. Suenaga and D.O. Welch, J. Appl. Phys. 53: 5111-5115 (1982).

CHARACTERIZATION OF BULK AND MULTIFILAMENTARY Nb_3Sn AND Nb_3Al BY DIFFRACTOMETRIC AND RESISTIVE MEASUREMENTS

R. Flükiger, W. Goldacker, and R. Isernhagen

Kernforschungszentrum Karlsruhe
Institut für Technische Physik
Karlsruhe, Federal Republic of Germany

ABSTRACT

A series of bronze-processed Nb_3Sn wires (binary and alloyed) and of powder metallurgically processed Nb_3Al filaments were characterized by resistivity measurements. The composition profile across the layer was determined by Auger spectroscopy. Particular attention was given to the determination of the global composition profiles in both Nb_3Sn and Nb_3Al by means of diffractometric analysis in the stress-free state. The correlation between the lattice parameter and the chemical composition was the basis for a careful analysis of the shape of several strong diffraction lines. This relatively simple method gives a reliable picture of the overall composition profile. We have found that the indication of the average lattice parameter alone does not give a complete description of the real state of the layer because B_{c2} is strongly correlated to variations in the composition. Therefore, the indication of the composition profile is necessary for understanding superconducting properties of the layers.

INTRODUCTION

The A15-type materials Nb_3Sn and Nb_3Al both fulfill the technical and economical conditions for use in high field magnets. For several years, Nb_3Sn multifilamentary wires have been commercially available, and Nb_3Al is actually passing from the laboratory to the industrial level. The critical current density of Nb_3Sn wires prepared by several techniques, including those with ternary element additions, has been gradually improved and seems to be not too far away from its optimum value, while the J_c of Nb_3Al is still subject to further improvements.

In order to characterize Nb_3Sn and Nb_3Al in view of the improvement of their current-carrying capacity, the physical metallurgical properties of the bulk materials have to be known. X-ray and resistive measurements have been found to be very useful in characterizing Nb_3Sn wires. An attempt is made in this paper to extend their use to Nb_3Al wires. For this purpose, the variation of T_c, B_{c2}, ρ_o (the electrical resistivity just above T_c) and a (lattice parameter) as a function of composition and atomic ordering is first reviewed for both Nb_3Sn and Nb_3Al. The variations in the lattice parameter and in the thermal vibrational amplitude up to 1000 K are compared between these systems. Finally, the critical current density, J_c, of bronze Nb_3Sn is compared with that of powder-metallurgical (PM) Nb_3Al wires.

It is found that most physical metallurgical properties of bulk Nb_3Sn and Nb_3Al are the same as those in the multifilamentary state. The only exception, the enhanced lattice parameter of PM-processed Nb_3Al, may be attributed to the interstitial oxygen content arising from the oxides originally adsorbed on the very large Nb and Al powder surfaces.

THE EQUILIBRIUM PHASE DIAGRAMS

The stoichiometric composition in the A15 phase Nb_3Sn is stable below 1200°C.[1,2] A more recent work of Shiffman and Bailey[3] indicates a highly temperature-dependent behavior of the low Sn limit, compositions below 23 at.% being stable at high temperatures only. However, new measurements[4] on the same samples previously used by Devantay et al.,[2] originally homogenized at 1800°C and annealed 14 days at 1000°C after a supplementary heat treatment of 14 days at 1000°C, confirmed the stability of 20 at.% Sn at this temperature in agreement with the earlier phase diagrams.[1,2] Low temperature X-ray diffraction measurements on the samples with 20 and 23 at.% Sn did not reveal any particular features; the diffraction lines maintained their shape down to 10 K. The lattice parameters (a) at 10 K for the 20, 23, and 24.5 at.% Sn compositions were determined to be 0.52751, 0.52769, and 0.5281 nm, respectively. This variation is very similar to that observed at 300 K, represented in Fig. 1.

The phase field of Nb_3Al, established by Jorda et al.,[5] is characterized by a strongly temperature-dependent Al-rich phase boundary, varying from 21 at.% Al at 1000°C to 22 at.% Al at 1500°C and finally, 25 at.% Al at 1940°C, the eutectic temperature. Thus, values of $T_c \cong 14$ K and $B_{c2} \cong 15$ to 20 T would be expected for Nb_3Al wires reacted below 1000°C (see Fig. 2). The reported values, $T_c = 16$ K and $B_{c2} = 23$ T, however, do not correspond to the equilibrium Al content.

PHYSICAL PROPERTIES OF Nb_3Sn AND Nb_3Al AS A FUNCTION OF COMPOSITION

The variation of T_c and $B_{c2}(0)$ in Nb_3Sn and Nb_3Al, plotted in Fig. 2, shows a linear behavior of both properties up to 23 at.% Sn or Al. With respect to all other known A15-type superconductors, however, an anomalous behavior is observed when approaching the stoichiometric composition. The "saturation" in T_c is explained as follows: The perfectly ordered state in Nb_3Sn[13] is thought to create the necessary conditions for the cubic-tetragonal martensitic transformation for Sn contents exceeding 24.5 at.% Sn.[2]

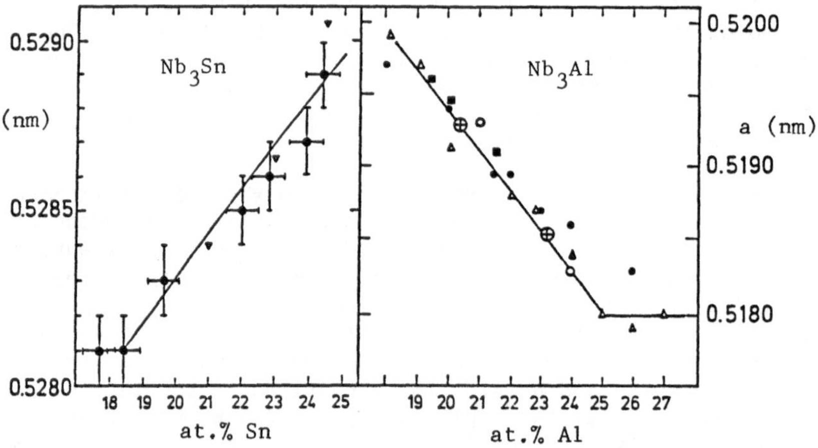

Fig. 1. Lattice parameter vs. composition in Nb_3Sn[2] and Nb_3Al.[5,6]

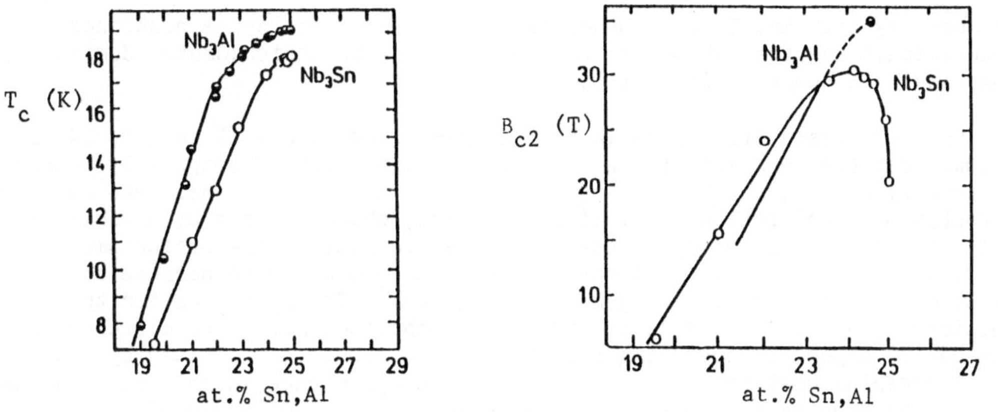

Fig. 2. Variation of T_c and $B_{c2}(0)$ as a function of composition in the A15 systems Nb_3Sn and Nb_3Al. a) T_c: Refs. 2,6; b) B_{c2}: Refs. 7-12.

The extrapolation of T_c to 25 at.% Sn for a hypothetical, cubic and stoichiometric sample would yield 19 K, i.e. nearly 1 K higher than the reported value for tetragonal stoichiometric Nb_3Sn. This is indirectly confirmed by the slight increase of T_c in alloyed Nb_3Sn occurring simultaneously with the suppression of the martensitic transformation, as shown for the additives Al,[14] Ti,[15,16] Ta[17,18] and others. The case of Nb_3Al is quite different from Nb_3Sn, since this compound exhibits a marked deviation from perfect ordering. As reported by Flükiger et al.,[6] the order parameter decreases from $S_a = 0.97$ to 0.95 for a compositional change from $\beta = 0.231$ to $\beta = 0.245$, the error being $\Delta S_a = \pm 0.02$. Thus, the saturation of T_c could indeed be correlated to a decreasing order parameter approaching the metastable stoichiometric composition Nb_3Al, but this is still uncertain.[6]

The variation $B_{c2}(0)$ vs. β in Fig. 2 follows the corresponding variation of T_c vs. β in the linear part only, i.e., up to 23 at.%. The particular behavior of $B_{c2}(0)$ above 23 at.% in Nb_3Sn can be explained by the variation of the electrical resistivity, which is itself a function of both composition and atomic ordering. The variation ρ_o vs. β for the systems Nb_3Sn,[2] Nb_3Ge,[19] V_3Si[20] and Nb_3Al[20] is illustrated in Fig. 3. Thus, the maximum of $B_{c2}(0)$ for Nb_3Sn in Fig. 2 can be explained on the basis of the equation $B_{c2}(0) \sim \rho_o \gamma T_c$ as the result of two opposite effects: i) below 23 at.% Sn, the increase of T_c and γ is dominant over the decrease of ρ_o, and $B_{c2}(0)$ increases with β, and

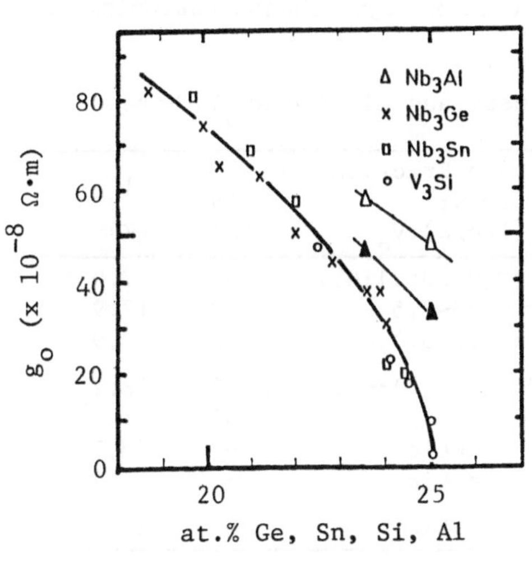

Fig. 3. ρ_o vs. composition in various A15 compounds. Nb_3Al: ▲ quenched, Δ reordered at 750°C.[20]

ii) above 23 at.% Sn, T_c (and probably also γ) shows the above mentioned "saturation," and the strong decrease of ρ_0 close to stoichiometry dominates, thus causing a decrease of $B_{c2}(0)$.

From the resistivity data in Fig. 3, described in detail in Ref. 20, it follows that the behavior of the electrical resistivity of completely ordered A15-type systems like V_3Si and Nb_3Sn (Nb_3Ge has also to be considered as being perfectly ordered) follows a similar variation, characterized by a particularly strong decrease of ρ_0 in the immediate vicinity of the stoichiometric composition. The system Nb_3Al does not fit the common curve because of its deviation from perfect ordering mentioned above.[6] The ρ_0 values for the compositions 23.5 and 25 at.% Al in Fig. 3 were obtained after two different heat treatments, argon jet quenching from 1940°C and quenching followed by prolonged ordering heat treatments at 750°C. The difference in the order parameter, S_a, between these two states lies between ΔS_a = 0.01 and 0.02.

ANALYSIS OF MULTIFILAMENTARY WIRES

Nb_3Sn

Occasionally, Nb_3Sn wire samples have been characterized by the value of the lattice parameter. However, it is obvious that the indication of a lattice parameter in an inherently inhomogeneous A15 layer will only be meaningful in cases where the Nb filaments are completely or nearly completely reacted to Nb_3Sn. For example, a 19-core bronze-processed Nb_3Sn wire with core diameters of 60 μm and A15 layer thicknesses of 12 μm exhibited an average lattice parameter, $\bar{a}$, of 0.52835 nm after being treated 64 h at 700°C.[22] The same wire of 0.6-mm diameter was subsequently rolled to a tape of 0.1-mm thickness. After 290 h at 700°C, the filaments were completely reacted, which led to considerably narrower X-ray lines and to a lattice parameter of 0.52888 nm,[16] i.e., considerably closer to the value at stoichiometry, 0.5292 nm. In particular, the contributions arising from the low Sn contents were drastically reduced.

Goldacker and Flükiger[16] have recently shown that X-ray diffraction can be used for determining a very precise picture of the composition profile in a Nb_3Sn layer. Semiquantitative line shape considerations can give detailed information about the global composition profile. This also holds for low temperature diffractometry: for bronze-processes binary Nb_3Sn wires, a successive decrease of the reaction temperature from 800°C to 700°C led to a decreasing tetragonal volume fraction, from 75% to 50%[16] (see Table 1). Results of a study on position and shape of the X-ray diffraction lines on these samples enabled us to correlate a shift of the mean composition toward lower Sn contents for the reaction at 700°C to a significant broadening of

Table 1. Diffractometric Results on Binary and Alloyed Nb_3Sn Wires

Wire Sample (at.%)	Reaction Conditions (°C/h)	a (300 K) (nm)	Volume Fraction at 10 K: $V_t(1-c/a):V_c$	T_M (K)	T_c (K)
Nb_3Sn	800/70	0.52892	0.75(0.0052):0.25	43	17.9
Nb_3Sn	750/138	0.52888	~0.5/~0.5	~43	17.9
Nb_3Sn	700/290	0.52888	0.5(0.0057):0.5	~43	17.9
+1.7 Ta	800/70	0.52888	~0.5(broad):~0.5	~20 -50	18.2
+2.8 Ta	800/70	0.52884	cubic	–	18.1
+4.3 Ta	800/70	0.52880	cubic	–	17.0
+1.3 Ti	800/70	0.52873	cubic	–	18.0

the composition distribution; i.e., the cubic volume fraction was higher for the sample reacted at 700°C than for the sample reacted at 800°C.

X-ray diffractometry and resistive measurements can also be used for characterizing alloyed Nb_3Sn wires with Ti and Ta additions. It is seen from Table 1 that the cubic-tetragonal phase transition is suppressed for 1.3 at.% Ti and for ≥ 2.8 at.% Ta. The wire with the lowest Ta content, 1.7 at.% Ta, exhibited about 50% of tetragonal phase and is, therefore, located just at the cubic/tetragonal phase boundary. This sample exhibited a broad distribution of (1 - c/a) values. These data can be compared to the $(Nb_{1-x}Ta_x)_3Sn$ phase diagram established earlier[23] on the basis of sintered samples. The deviation from stoichiometry with increasing Ta content seems to be more pronounced for the bronze-processed wires, which is attributed to the very different temperatures of phase formation in both cases.[16,23] Indeed, the comparison of the lattice parameters of Ta-alloyed Nb_3Sn wires (see Fig. 4) with those measured on arc melted samples[24] suggests that the suppression of the phase transformation is not only due to the increasing Ta content, but also to the increasing deviation from stoichiometry. This is also in agreement with the observations of Tafto et al.[25] The same argument holds for Ti-alloyed Nb_3Sn wires, where the deviation from stoichiometry is even stronger than for Ta additions.[15]

The effect of alloying ternary elements to Nb_3Sn is to enhance the upper critical magnetic field, as shown by Ekin,[27] Suenaga,[28] Sekine et al.,[29] and Drost et al.[26] It was expected that the observed increase of B^*_{c2} (the value obtained by the Kramer extrapolation at 4.2 K) would be caused by an increase of the electrical resisitivity. This hypothesis was first confirmed by Drost et al.,[26] who measured ρ_o on fully reacted A15 filaments of Ta, Ti, and Ni+Zn alloyed 19-core bronze-processed wires (the same wires as listed in Table 1). The results, also represented in Fig. 4, show convincingly that the value of ρ_o, which is 0.16 μΩ·m for the binary Nb_3Sn, increases up to values close to 0.40 μΩ·m for alloyed Nb_3Sn filaments. It follows from a comparison with the corresponding values of J_c at high fields that the optimum values of J_c are obtained for ρ_o values around 0.35 μΩ·m. Such ρ_o values are reached for 3.6 at.% Ta and 1.3 at.% Ti in the A15 layer, as shown by Auger spectroscopy.[26] The very narrow X-ray lines of the completely reacted layers[16] indicate that the concentration profiles are quite narrow, a necessary condition for comparing the ρ_o values between bulk and filamentary Nb_3Sn. Indeed, it immediately follows from the highly nonlinear behavior of ρ_o vs. β in Fig. 2 that in a nonhomogenous sample, the contributions to ρ_o arising from the low Sn contents lead to too high values of the electrical resistivity.

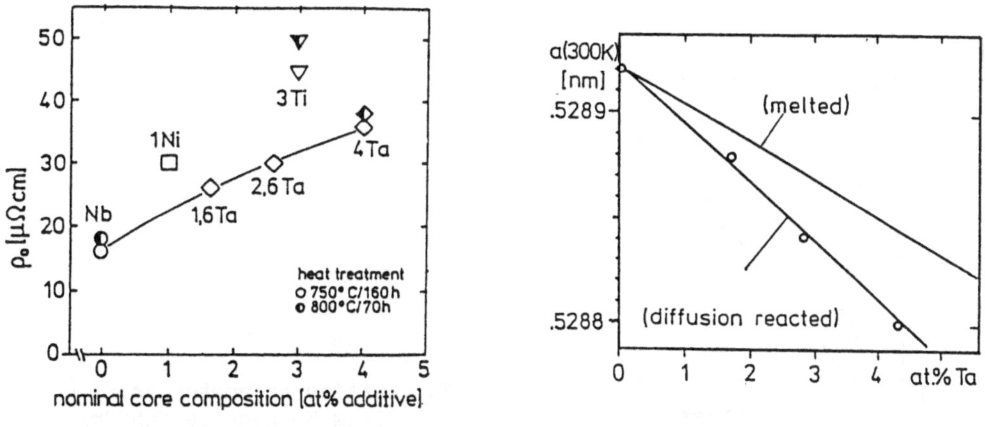

Fig. 4. Variation of ρ_o[26] and a[16] in Ta alloyed Nb_3Sn multifilamentary wires. The values for Ti and Ni+Zn additions are also added.[26]

$\underline{Nb_3Al}$

In contrast to Nb_3Sn, Nb_3Al wires cannot be prepared by means of the bronze diffusion reaction,[30] owing to the formation of two intermetallic phases, the μ phase Nb_2CuAl and the C14 Laves type phase $Nb(Cu,Al)_2$.[31] For this reason, Nb_3Al wires are formed by a direct reaction between Nb and Al_3 at temperatures below 1000°C without the presence of Cu. This reaction mechanism has some similarities to the infiltration technique[32] in Nb_3Sn, where Nb and Sn are reacted together at 900°C. This is a relatively high temperature when compared with 650 to 700°C, the current reaction temperature range for Nb_3Sn wires using the diffusion reaction process, i.e., with Cu as a "catalyzer." Thus, it is not surprising that the reaction temperature for Nb_3Al lies above 850°C. The analogy between infiltrated Nb_3Sn and Nb_3Al is, however, not complete, since the latter only forms at these temperatures if the Nb and Al dimensions are well below 10^{-6}m. This was discovered by Ceresara et al.,[33] who rolled Nb and Al foils together, which were drawn to fine wires and reacted. Larson et al.[34] used a particular kind of PM approach, the "mechanical alloying" for producing Nb_3Al wires with J_c values comparable to those of Ceresara et al.,[33] which were, however, much lower than those of Nb_3Sn wires.

The decisive improvement of J_c in Nb_3Al wires came with the application of the "cold PM" technique by Akihama et al.,[35] Foner et al.,[36] and Thieme et al.[37] The cold PM technique had first been applied on Nb_3Sn by Flükiger et al.[38] Mixtures of fine Nb and Cu powders are extruded and drawn to fine wires which are Sn plated and reacted. Niobium and copper is an ideal material combination, the plastic deformation behavior being very similar. The substitution of Cu by Al, however, changes drastically the deformation behavior of the composite, which becomes much more difficult to control. From the experiences with Nb_3Sn,[38] it was clear that only Nb with low oxygen content, i.e., highly ductile Nb, would be convenient. The necessity of a cold deformation process can be illustrated by the lattice parameter variation of hydride-dehydride Nb powders of ≤40 μm size as a function of temperature, shown in Fig. 5. This experiment[4] was carried out under 100 mm of argon pressure on Nb with a large surface and thus a large amount of adsorbed oxygen. Hot extraction experiments revealed a total (dissolved + adsorbed) oxygen content of 1200 wt. ppm (or 1200 μg oxygen per g Nb). Figure 5 demonstrates how the oxygen at the Nb surface is gradually dissolved in Nb with increasing temperature. The interstitially dissolved oxygen leads to a lattice expansion; the lattice parameter at 300 K increases from 0.3302 nm before to 0.3308 nm

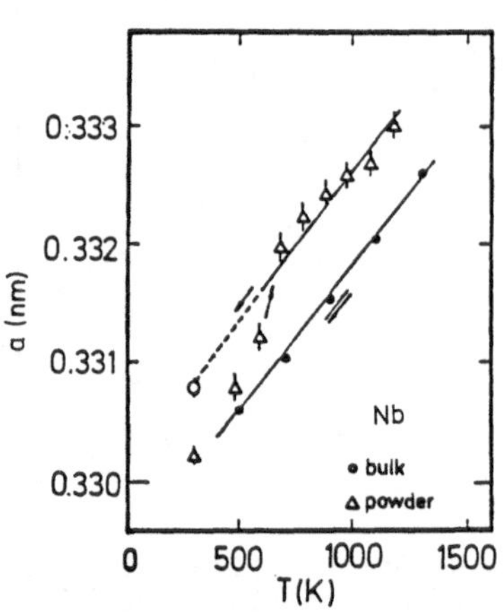

Fig. 5. Lattice parameter vs. T on 40 μm Nb hydride-dehydride powders (p = 10 kPa)

after the experiment. The oxygen starts to enter into the Nb at temperatures between 500 to 600 K. Above 700 K, the linear thermal expansion is the same as that for pure, bulk Nb,[39] indicating that a certain equilibrium has been reached. This experiment shows that at any time during the deformation process the temperature should exceed certain limits.

Preparation of Nb_3Al Wires

A series of Nb_3Al wires were prepared at the Kernforschungszentrum by using the cold PM technique, but choosing a different approach than Thieme et al.,[37] the extrusion constituting the first step, rather than compacting the powders by a swaging procedure. As starting materials, different Nb powders were used: REP (rotating electrode) powders 500 μm in size and various hydride-dehydride powders with sizes between 125 and 200 μm and different oxygen contents. The aluminum powders consisted of inert gas-atomized powders of 20 μm. In agreement with Thieme et al.,[37] it was found that the REP Nb with <200 ppm oxygen had by far the best deformation behavior, wire drawing being possible without breakage down to an areal reduction factor, R, of 1.5×10^5. However, because the starting size was too large, even this large R factor was not sufficient to reach reasonably small Nb fibers, thus leading to relatively small J_c values. Thus, in the following, only the results and analyses on hydride-dehydride powders are meaningful.

The deformability of the hydride-dehydride Nb powders was found to depend on the content of interstitial impurities, which determined the limits of the wire drawing process. Depending on the impurity content, the drawing limit varied between $R = 3 \times 10^3$ and $>10^5$ (the lower value was found for the Nb powders described in Fig. 5). The fabrication process consists of a series of mixing, evacuating, compacting, and extruding steps, which can be repeated once or more, followed by swaging, rolling and drawing. The wires had no diffusion barrier between the Nb-Al composite and the external copper.

The cross section of a twice-extruded 18-core Nb-8wt.% Al wire with a diameter of 1 mm corresponding to a reduction factor, R, of 1×10^5 is represented in Fig. 6. The initial powder sizes were 125 μm for the Nb and 20 μm for the Al powder particles. This picture shows the effect of swaging on the external Nb-Al composite cores. The same sample was further deformed to $R = 3.4 \times 10^5$, the Cu/Nb-Al ratio remaining constant at 1:1. All the reaction heat treatments for the present work were carried out at 900°C for 1 h, i.e., the samples had not yet been submitted to the ordering heat treatment at 750°C, the optimization being not yet completed.

J_c Measurements

The J_c measurements of several Nb_3Al wires after reaction at 900°C for 1 h are represented in Fig. 7, together with corresponding values of Thieme et al.[40] after the same heat treatment and after an additional ordering heat treatment. When comparing different J_c data on Nb_3Al wires prepared by the cold PM process, it is necessary to know the R factor as well as the initial Nb powder size. Under the assumption that the really important property is the thickness of the Nb fibers, it should be possible to define normalized R factors with respect to the initial Nb powder size. Since Thieme et al.[37] used Nb powders of 80 μm size, our R factors should be normalized, i.e., the value $R_{125} = 3.4 \times 10^5$ would become $R_{80} = 1.4 \times 10^5$. A detailed study of the data of Thieme et al.[37,40] and Akihama et al.[35] showed that this scaling of R is roughly correct. The scaling also holds for the very large REP Nb powders, with R up to 2×10^7, but yielding low values of J_c, which nevertheless agree with the values in Ref. 37 after normalization, $R_{80} = 5 \times 10^4$. The present reactions were restricted to 900°C, higher temperatures not seeming ideal.

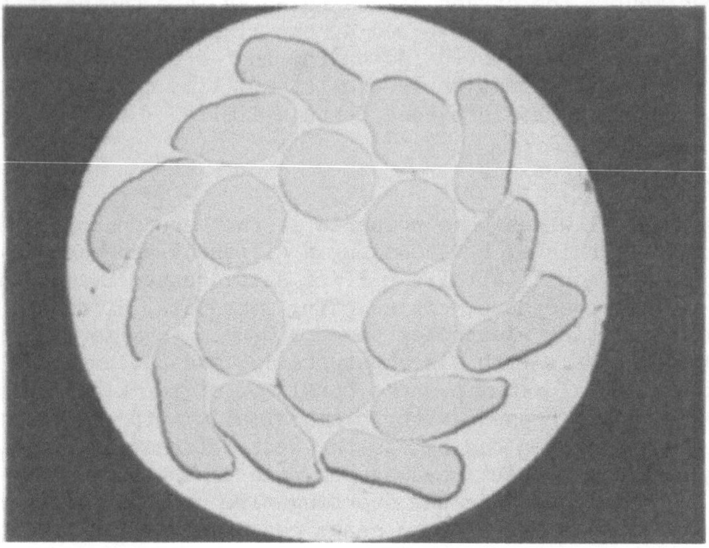

Fig. 6. Cross section of a Nb_3Al wire. Diameter: 1 mm; $R_{125} = 1 \times 10^5$.

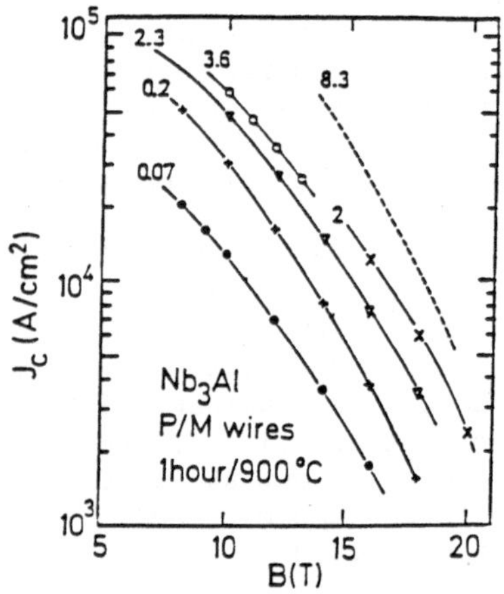

Fig. 7. J_c of Nb_3Al wires with different R_{125} factors. Dashed line: Data of Thieme et al.[37,40] after additional ordering heat treatment (3 days at 750°C).

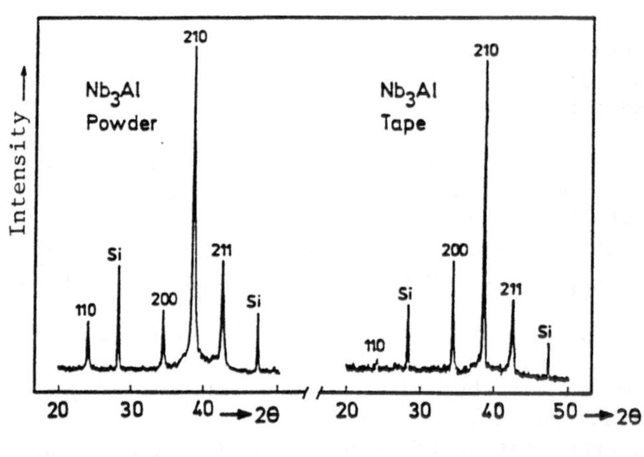

Fig. 8. X-ray powder diffraction patterns of a Nb_3Al tape ($R = 3 \times 10^5$), a) powdered and b) measured on the radial tape surface ($CuK_{\alpha\perp}$, 300 K).

Analysis of Nb_3Al Wires: The Lattice Parameter Paradox

The X-ray diffraction patterns reveal the presence of additional phases Nb_2Al, the σ phase, and unreacted Nb. The amount of these additional phases decreases considerably when either the R factor or the reaction temperature increases, in agreement with Thieme et al.[37] As shown in Fig. 8, there is a marked texture in the A15 filaments. This was determined by X-ray diffraction measurements on Nb_3Al tapes obtained by rolling the wires mentioned above to tapes of 0.100-mm thickness, corresponding roughly to a reduction factor of 3 x 10^5.

There is an aspect common to Nb_3Sn and Nb_3Al wires: The cold working of Nb. As shown by Togano et al.[41] by X-ray pole figure analysis, this results in a texture of the unreacted Nb (well-known for bcc materials) that can be described by the orientation (100) [011], where [011] indicates the orientation along the longitudinal wire axis and (100) the radial crystallographical planes in the commonly used nomenclature. Small deviations from this texturing behavior were found by neutron diffractometry on various commercial Nb_3Sn wires; they are attributed to individual differences in the cold-working procedure. After reaction, the Nb_3Sn layer was also found to be textured in the same way, (100) [011], which was proposed to be initiated by the nearly parallel relationship $(001)_{bcc} \parallel (001)_{A15}$.[41] This behavior of the Nb_3Sn layer is expressed by the enhanced intensity of the (200) reflection and a partly suppressed (211) reflection in the X-ray diffraction pattern on the radial surface of an etched Nb_3Sn tape when comparing with the powdered texture-free material. This behavior is strongly similar to the diffraction patterns of a Nb_3Sn tape and powdered Nb_3Al filaments, as shown in Fig. 8. This is an indication for an oriented growth mechanism of the Nb_3Al phase during the reaction process. Of course, the analogy between Nb_3Al and Nb_3Sn (and V_3Ga) is not complete, the microstructure in Nb_3Al being different, the grains being much smaller (<100 nm), as reported by King et al.[42]

It is remarkable that the variation of R_{125} from 2 x 10^4 to 1.4 x 10^5 as well as the enhancement of the reaction temperature from 900 to 1080°C had little effect on the lattice parameter, which was mostly situated at 0.5198 nm (the largest value after 16 h/900°C was 0.5204 nm). This lattice parameter is the same as those previously reported by Ceresara et al.[33] and Thieme et al.[37] However, a comparison with Figs. 1 and 2 shows that this lattice parameter corresponds to an Al content of 17.5 at.%, where T_c would be of the order of 6 K, which is in contrast to the measured value of 16 K for the onset of T_c. To check if this is a volume effect, a specific heat investigation on Nb_3Al filaments after different reductions and reaction temperatures was undertaken. The main result of this work, which will be published soon, is that there are ordinarily two calorimetric transitions, one starting at 16 K, attributed to the A15 phase, and a second one starting at 9 K, arising from the unreacted Nb. The A15 transition is quite broad, as expected from the width of the corresponding X-ray lines. There is no doubt that the enhancement of J_c values in Nb_3Al wires is correlated to a further narrowing of the calorimetric superconducting transition and to the almost complete disappearance of neighboring phases.

The above remarks have shown that there is really a paradox: Nb_3Al wires have a T_c value corresponding to 21 to 22 at.% Al, but the lattice parameter corresponds to ~18 at.% Al. Knowing that in Nb_3Sn all measured physical metallurgical properties are the same in the bulk as well as in the multifilamentary state, the situation in Nb_3Al is indeed intriguing, the most questionable value being that of the lattice parameter. Thus, the question arises: what factors contribute to the lattice expansion? Assuming that an Al content of 18 at.% cannot be responsible for T_c values around 16 K, there is practically only one explanation for the lattice expansion in Nb_3Al wires: interstitially dissolved oxygen. This possibility is realistic; considerable oxygen is adsorbed at the large Nb and Al surfaces. The difficulty resides in proving that the oxygen in Nb_3Al is really interstitially dissolved and not just present at the grain boundaries. For this purpose, high temperature X-ray studies were undertaken at the Kernforschungszentrum Karlsruhe.[43]

Two different experiments were undertaken, the first one starting with bulk samples, powdered to 40-μm size, and the second one with powdered Nb_3Al filaments with $R_{125} = 1.5 \times 10^5$, reacted 1 h at 900°C. The results for the bulk samples with $\beta = 0.21$ and 0.246, plotted in Fig. 9, show an excellent reproducibility below 1023 K. The measured linear thermal expansion coefficients between 300 and 1000 K are $\alpha = 8.61 \times 10^{-6} K^{-1}$ for Nb_3Al and $\alpha = 7.69 \times 10^{-6} K^{-1}$ for Nb_3Sn. The upper temperature limit for the experiments in Fig. 9 is 1073 K (compared with 1373 K for Nb_3Sn). Above this temperature, all line intensities decrease. At the same time, the Nb line appears, indicating Al losses (Fig. 10). After cooling down from cycle I, the $Nb_{.754}Al_{.246}$ sample had a lattice parameter of 0.5195 nm, i.e., considerably higher than the initial value, 0.5182 nm, which would correspond to 20 at.% Al. A second cycle on the same sample (cycle II in Fig. 9) shows reversibility below 1000 K. The question arises whether this lattice expansion is uniquely due to Al losses or if some additional oxygen or other gases (present in the apparatus) have been dissolved in the A15 structure at high temperature.

The second experiment is complementary to the first one, since i) the Nb_3Al filaments already contain oxygen (wherever it may be located) and ii) the measurements were now performed under vacuum (p = 1 mPa). At 1123 K, the Nb line appears again (Fig. 10), as for the bulk sample. However, after cooling down from a temperature cycle up to 1123 K, the lattice parameter was now considerably reduced, from the starting value of 0.5198 nm, to 0.5189 nm after one cycle, and to 0.5187 nm after a second cycle. This considerable lattice contraction of 0.0011 nm was observed in spite of possible Al losses, which would tend to enhance the lattice parameter value. Since Si powder had been added to the powdered Nb_3Al filaments for calibration purposes, it was first thought that this would have caused the decrease of a. After excluding this possibility, the following explanation can be given: In contrast to the situation in the wire, where the Nb_3Al filaments are surrounded by Monel[37] or Cu (as in Ref. 33 or in the present case), the exposition of the same filaments after crushing to powders of ≤30 μm size to high vacuum favors the elimination of the interstitially dissolved gases,

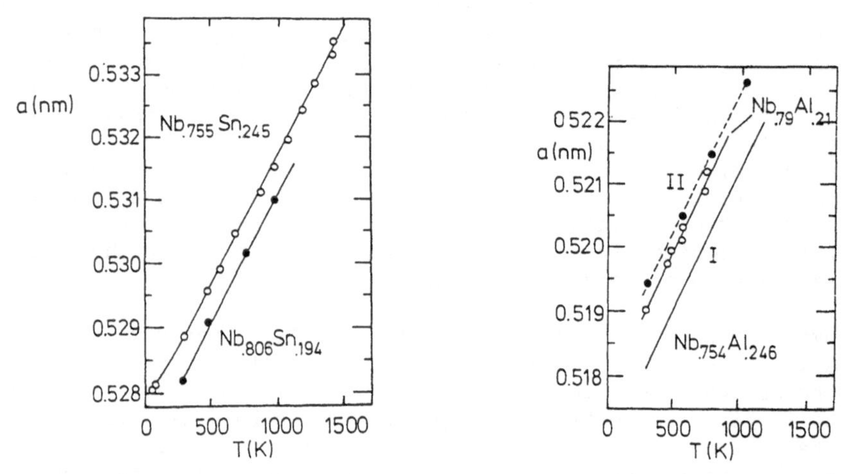

Fig. 9. Lattice parameter vs. T on various Nb_3Sn and Nb_3Al powdered bulk samples. I and II are the data for the first and second cycle of the same $Nb_{.754}Al_{.246}$ sample (p = 10 kPa argon).

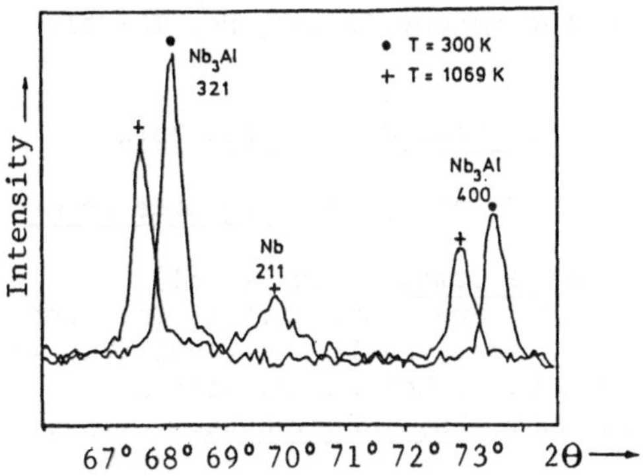

Fig. 10. Occurrence of Nb peaks in powdered $Nb_{.754}Al_{.246}$ at 1069 K.

e.g., oxygen, nitrogen, or hydrogen. Thus, the lattice parameter after cooling down has to be considered as lying close to the true value of the A15 lattice parameter. It corresponds to ~22 at.% Al, as follows from Fig. 1. Although this agrees well with the T_c value of 16 K (see Fig. 2), it is premature to establish a definitive correlation. It could be, indeed, that further vacuum annealing would lead to even smaller values of a, corresponding to higher Al contents. The possible effect of interstitial gases on T_c also needs to be investigated. A more detailed work about these high temperature diffraction experiments will be published soon.[43]

CONCLUSION

In the present work, it is shown that the physical metallurgical properties (T_c, B_{c2}, ρ_o, a) of bulk Nb_3Sn are essentially the same as in the multifilamentary state. Although the information for Nb_3Al is not as complete as that for Nb_3Sn, it can now be stated that T_c and the lattice parameter in bulk and multifilamentary Nb_3Al are also the same, provided that perturbations like interstitially dissolved gases (such as oxygen) can be eliminated. Estimations using the Kramer law indicate that B_{c2} of bulk Nb_3Al could be correlated to that in the multifilamentary state. Preliminary measurements of ρ_o in multifilamentary Nb_3Al indicate the correct tendency when compared to Fig. 3 (RRR ~ 2), but it must first be ascertained that the unreacted Nb is completely eliminated before reliable ρ_o value can be obtained. We have found that the A15 phase in Nb_3Al wires (formed by a nonequilibrium process, according to Vandenberg et al.[44] and Bormann et al.[45]) has a composition close to 23 at.% Al, thus exceeding the equilibrium concentration by 1.5 at.% Al. The possible role of interstitially dissolved oxygen on this effect is being investigated. The J_c values of wires prepared by a modified cold-PM process are comparable with those of Thieme et al.,[37] if one considers that the present wires were not optimized. Different improvements, such as diffusion barriers, the use of smaller powders, the achievement of higher R values and additional ordering heat treatments at 750°C, are expected to enhance the actual J_c values. The present investigation confirms Nb_3Al as a promising material for high-field applications.

ACKNOWLEDGMENTS

The authors would like to thank Prof. Wühl for the specific heat measurements and Drs. W. Schauer and W. Specking for the J_c measurements.

REFERENCES

1. J. P. Charlesworth, I. Macphail, and P. E. Madsen, J. Mater. Sci. 5:580 (1970).
2. H. Devantay, J. L. Jorda, M. Decroux, J. Muller, and R. Flükiger, J. Mater. Sci. 16: 2145 (1981).

3. R. A. Schiffman and D. M. Bailey, Ames Laboratory Report, No. IS-J615, (1984).
4. Present work
5. J. L. Jorda, R. Flükiger, and J. Muller, J. Less-Common Met. 75:227 (1980).
6. R. Flükiger, J. L. Jorda, A. Junod and P. Fischer, Appl. Physics Comm. 1:9 (1981).
7. S. Foner and E. J. McNiff, Jr., Solid State Comm. 39:959 (1981).
8. R. Flükiger, W. Schauer, W. Goldacker, in: "Superconductivity in d and f Band Metals," W. Buckel and W. Weber, eds., Karlsruhe (1982), p. 41.
9. A. J. Arko, D. H. Lowndes, F. A. Müller, L. W. Roland, J. Wolfrat, H. W. Myron, R. M. Müller, and G. W. Webb, Phys. Rev. Lett. 40:1590 (1978).
10. J. Kwo, R. H. Hammond and T. H. Geballe, J. Appl. Phys. 51:1726 (1980).
11. T. P. Orlando, Phys. Rev. B19:4545 (1979).
12. S. Foner, in: High Field Magnetism," North-Holland (1982), p. 133.
13. R. Flükiger, R. Isernhagen, W. Goldacker, and W. Specking, Adv. Cryo. Eng. 30:851 (1984).
14. L. J. Vieland and A. W. Wicklund, Phys. Lett. 34A:43 (1971).
15. K. Tachikawa, H. Sekine and Y. Iijima, J. Appl. Phys. 53:5354 (1982).
16. W. Goldacker and R. Flükiger, Materials and mechanisms of superconductivity, to be published in Physica.
17. W. Goldacker and R. Flükiger, IEEE Trans. Magn., MAG-21:807 (1985).
18. R. Roberge, H. Lehuy, and S. Foner, Phys. Lett. 82A:259 (1981).
19. K. E. Kihlström, R. H. Hammond, J. Talvacchio, T. H. Geballe, A. K. Green and V. Rehn, J. Appl. Phys. 53:8907 (1982).
20. R. Flükiger, J. L. Jorda, and J. Muller, to be published.
21. R. Flükiger, Ann. Chim. (Paris), Sci. Materiaux 9:841 (1984).
22. E. Drost, R. Flükiger and W. Specking, Cryogenics 24:622 (1984).
23. R. Flükiger, Adv. Cryo. Eng. 28:399 (1982).
24. W. Kunz and E. Saur, Z. Phys. 189:401 (1966).
25. J. Tafto, M. Suenaga, and D. O. Welch, J. Appl. Phys. 55:4330 (1984).
26. E. Drost, W. Specking and R. Flükiger, IEEE Trans. Magn. MAG-21:281 (1985).
27. J. W. Ekin, IEEE Trans. Magn. MAG-17:658 (1981).
28. M. Suenaga, in "Superconducting Material Science - Metallurgy, Fabrication and Application," S. Foner and B. B. Schwartz, eds. Plenum (1981).
29. H. Sekine, K. Tachikawa, and Y. Iwasa, Appl. Phys. Lett. 35:472 (1979).
30. J. D. Livingston, Krist. Tech. 13:1379 (1978).
31. C. R. Hunt and A. Rahman, Z. Metallk. 59:701 (1968).
32. K. Hemachalam and M. R. Pickus, IEEE Trans. Magn. MAG-13:466 (1977).
33. S. Ceresara, M. V. Ricci, N. Sacchetti, and G. Sacerdoti, IEEE Trans. Magn. MAG-11:263 (1975).
34. J. M. Larson, T. S. Luhman, and H. F. Merrick, Proc. Int. Conf., Port Chester, New York 8-10 Nov. 1976).
35. R. Akihama, R. J. Murphy, and S. Foner, Appl. Phys. Lett. 37:1107 (1980).
36. S. Foner, Prog. Powder Metall. 38:107 (1982).
37. C. L. H. Thieme, S. Pourrahimi, B. B. Schwartz, and S. Foner, Appl. Phys. Lett. 44:260 (1984).
38. R. Flükiger, S. Foner, E. J. McNiff, Jr., and B. B. Schwartz, Appl. Phys. Lett. 34:763 (1979).
39. J. B. Conway, A. C. Loskamp, Trans. AIME 236:702 (1966).
40. C. L. H. Thieme, H. Zhang, J. Otubo, S. Pourrahimi, B. B. Schwartz, and S. Foner, IEEE Trans. Magn. MAG-19:567 (1983).
41. K. Togano and K. Tachikawa, J. Appl. Phys. 50:3495 (1979).
42. W. King, C. L. H. Thieme, and S. Foner, IEEE Trans. Magn. MAG-21 (1985).
43. W. Goldacker and R. Flükiger, to be published.
44. J. M. Vandenberg, M. Hong, R. A. Hamm, and M. Gurvitch, J. Appl. Phys, 58:618 (1985).
45. R. Bormann, H. U. Krebs, and A. O. Kent, Adv. Cryo. Eng. 32 (1986).

DEVELOPMENT OF A15 CONDUCTORS IN THE USSR

V. M. Pan and V. S. Flis
Institute of Metal Physics
Kiev, USSR

V. A. Bashilov, V. A. Blisnyuk
Experimental Design Bureau "Horizon"
Moscow, USSR

V. P. Buryak
Physics Engineering Institute
Donetsk, USSR

ABSTRACT

Results of the development of multifilamentary and microcomposite superconducting materials for a large-scale manufacturing of A15 conductors at the "Horizon" plant of the Ministry of Energy of the USSR are presented. These materials will be used in the construction of GJ superconducting magnetic energy storage systems and high-field test facility. The three main areas of research are: (1) development of large-scale fabrication of the multifilamentary and microcomposite Nb_3Sn conductors; (2) development of the hydrostatic extrusion techniques and equipment suitable for ingots and billets as well as rods and wires; (3) development of methods to produce multifilamentary conductors with alloyed Nb_3Sn using an internal tin source for high-field magnetic system applications. Advantages of hydrostatic extrusion are discussed as well as the properties of the microcomposite Nb_3Sn conductors produced.

INTRODUCTION

The higher complexity of A15 multifilamentary conductor fabrication and its relatively high cost[1] are compensated by their ability to function in high magnetic fields (~16T). Since the stored energy in a coil increases in proportion to the volume and square of the magnetic field, it is much more beneficial to operate at higher fields than to increase the inductive storage volume of magnetic systems. The Experimental Design Bureau (EDB) "Horizon" of the Ministry of Energy of the USSR is currently involved in constructing superconducting magnetic energy storage systems with stored energy in the GJ range. The EDB "Horizon" plant was built for the large-scale commercial fabrication of superconducting wires and conductors required by these projects; it has facilities to develop and test materials and to optimize their properties.

Currently, the most widely used method of producing commercial multifilamentary A15 conductors is the selective solid state reactive diffusion method, which is usually called the "bronze process."[2] The deformation process used in this method is hot pressing and drawing. However, a single reduction of cross-sectional area cannot exceed 15 to 20% by drawing, and this process requires many die changes. As a result, many drawing machines and expensive dies are needed, and a significant amount of material is wasted during wire fabrication. Also, in drawing the wire, the tensile stress applied during deformation tends to break the filaments in multifilamentary conductors.

In many cases, hot pressing in the initial stage is an undesirable and unacceptable method of deformation. Technological processes are needed to treat the low plasticity structural constituents of multifilamentary and microcomposite A15 conductors and to avoid preliminary heating before deformation. An application of high hydrostatic pressure during deformation, that is, hydropressing,[3,4] appeared to be very successful. Improved conditions of contact friction at the point of deformation, an application of dies with small cone angles, and an absence of high tensile stresses during hydropressing ensure a high uniformity of deformation and a large number of correctly distributed, very thin filaments with no filament breakage.

Use of the Nb_3Sn conductors in high-field magnetic systems will require an increase in their high-field critical current densities as well as a reduction in their cost. If the optimized conventional bronze process (with 13 wt.% tin bronze and approximately 1-μm filament diameter) is used, it may be possible to achieve a J_c (12 T) of 5×10^4 A/cm^2 considering the area occupied by filaments and bronze,[5] although the usual J_c (12 T) is somewhat lower than (3 to 4) $\times 10^4$ A/cm^2.[6] However, achieving such high J_c (12 T) values in large-scale production is too expensive and difficult, and the probability of filament breakage is very high during drawing.[7] Therefore, alternative methods to produce the microcomposite A15 superconducting materials have been developed. Now in situ and powder metallurgy processed A15 conductors with discontinuous filaments are commonly used (see, for example, Ref. 8). In another new approach, bronze is enriched with tin without a deterioration of its technological plasticity (processes using an external[2] or internal[9] tin source).

However Nb_3Sn conductors could not be considered for use in magnet systems with operational fields of more than 15 to 16 T until it had been proved that their upper critical field, $H_{c2}(0)$, could be greater than 30 T.[10] By then, the multifilamentary conductors with alloyed Nb_3Sn had been developed with high enough J_c at fields greater than 16 T (see, for example, Ref. 8). The use of alloyed Nb_3Sn combined with an internal tin reservoir seems to be especially promising.

In this paper, results of research in three areas are presented:

1. Development of large-scale production of A15 conductors at EDB "Horizon".

2. Development of the hydroextrusion technique and associated equipment at the Donetsk Physics Engineering Institute.

3. Development of multifilamentary conductors with alloyed Nb_3Sn having an internal tin source at the Institute of Metal Physics.

MULTIFILAMENTARY AND MICROCOMPOSITE A15 CONDUCTORS AT EDB "HORIZON"

A15 multifilamentary conductors have been developed by the conventional bronze process; A15 microcomposite conductors with discontinuous fiber structure have been produced by the in situ and powder metallurgy processes; and superconducting Nb_3Sn tape has been made by the diffusion process. The fabrication technique of multifilamentary Nb_3Sn wires usually includes subsequent deformation of the composite assemblage. Ingots with 13 wt.% tin bronze are used for the first assemblage. The initial deformation of the billets is accomplished by a horizontal hydroextrusion press with a force of 24.5 MN. The operational pressure in the container is as high as 1.4 GPa. The press, made by the Swedish company ASEA, is fully automated, computer controlled, and supplied by industrial robots; it is capable of treating ingots and billets of up to 100 kg mass. Before the use of hydroextrusion, the billets could be heated to 600°C.

Subsequent deformation of rods is accomplished by either rod or drawing mills: first, by a driving chain mill and then by reel mills for multiple drawing. The process used for rod deformation depends on the type and structure of the conductor and on its technical requirements. To produce a conductor with a rectangular cross section, it is necessary to use roller drawing mills or flattening mills.

Circular cross-section Nb_3Sn conductors with from 37 to more than 10,000 filaments (diameters up to 1 μm) and flattened multifilamentary Nb_3Sn tape with a cross-sectional area of 0.1 to 0.2 cm^2 have been fabricated at the EDB "Horizon" plant. The conductors are stabilized with 10 to 50% copper. Their overall critical current densities are 4 to 7 x 10^4 A/cm^2 in a transverse magnetic field with an induction of 10 T at 4.2 K. Several different magnetic systems with operational fields up to 10 T have been manufactured from these materials at EDB.

However, the principal superconducting material activity there is the development of the microcomposite in situ and powder metallurgy processed Nb_3Sn conductors, which are expected to be used in superconducting magnetic energy storage. Wires of 0.3- to 1.0-mm diameter and tape with a 0.1-mm x 5-mm cross-sectional area have been manufactured and tested. The microstructure of the powder metallurgy processed Nb_3Sn wire with easily distinguishable discontinuous Nb_3Sn fibers linked to each other is shown in Fig. 1.

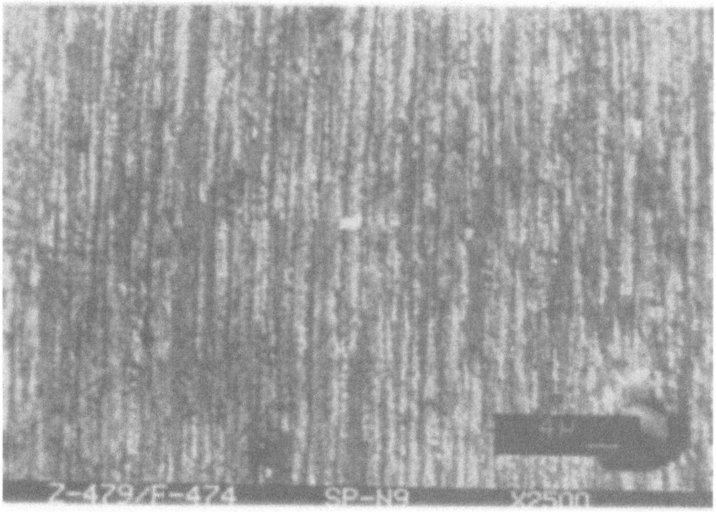

Fig. 1. Powder metallurgy processed Nb_3Sn wire microstructure (SEM), longitudinal section (matrix is slightly etched away); f_s = 0.4; R = 1.5 x 10^4.

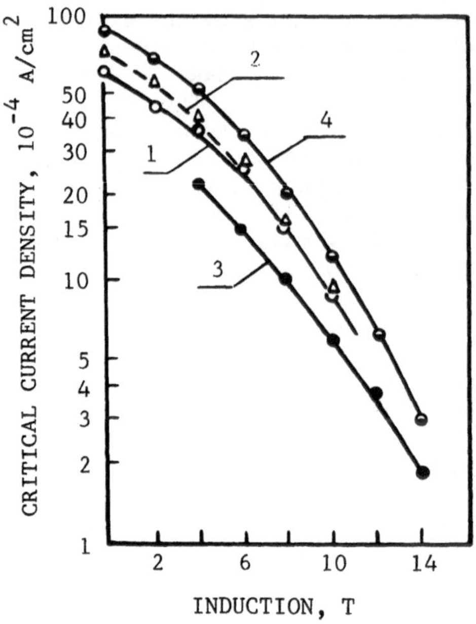

Fig. 2. Critical current densities versus induction of magnetic field at 4.2 K (referred to the cross-sectional area occupied by $Nb_3Sn/Nb+CuSn$). 1 - powder metallurgy processed Nb_3Sn wires; 2 - in situ processed Nb_3Sn wires; 3 - the best commercial conventional bronze processed Nb_3Sn wires;[5,6] 4 - the best in situ processed Nb_3Sn wires (see, for example, Ref. 8).

The overall critical current densities of these microcomposite Nb_3Sn conductors are 10 and 3×10^4 A/cm^2 in transverse fields of 10 and 14 T, respectively (when superconductor volume fraction, f_s, equals 0.4). The average values of the overall critical current densities versus magnetic field induction for the microcomposite Nb_3Sn materials produced at EDB "Horizon" are shown in Fig. 2. For comparison, $J_c(B)$ curves for the best Nb_3Sn conductors discussed in Refs. 5, 6, and 8 are also shown. Completed studies of the critical parameters and ac losses versus cross-sectional area reduction, R; superconductor volume fraction, f_s; influence of additives to the copper matrix; and heat treatment regimes indicate the feasibility of semicommercial fabrication of microcomposite Nb_3Sn materials having high current-carrying ability and sufficiently low level of ac losses.

By calorimetric measurements of the hysteretic losses in transverse magnetic fields with an induction, B_o, of 4 T and a peak value of the alternating component, B_m, of 0.2 ± 0.6 T (frequency = 0.1 Hz), it was found that the behavior of microcomposite Nb_3Sn conductors with $f_s \geq 0.3$ ($R = 5 \times 10^3$) is significantly different from those with $f_s \leq 0.2$ ($R = 5 \times 10^3$). The former possess high ac losses like bulk superconductors; the latter, in contrast, have low ac losses, and from this viewpoint, are competitive with conventional multifilamentary conductors with filament diameters of 1 to 2 μm. The measured ac losses are given in Table 1.

The level of properties achieved for these materials combined with their comparatively low cost is the basis for recommending their application in current-carrying systems of superconducting magnetic energy storage.

With regard to the emerging A15 materials, conductors based on Nb_3Ge, Nb_3Al, and $Nb_3(Al,Ge)$ have been developed. Long superconducting tape with a Nb_3Ge layer on a molybdenum substrate has been produced by the chemical vapor

Table 1. AC Losses of the Powder Metallurgy Processed Nb_3Sn Conductors (mJ/cm^3 cycle)

f_s	R	B_m, T				
		0.2	0.3	0.4	0.5	0.6
0.15	5000	0.2	-	0.29	0.46	4.3
0.30	5000	16.3	30	45.2	66.7	72.1
0.30	5000	17.3	-	45.9	64.4	80.2

deposition process. The critical temperature of these tapes is as high as 20 to 21.5 K. Techniques for manufacturing Nb_3Al tapes and multifilamentary wires have been developed as well. The main feature of the process is the application of two-stage annealing--a very short, high-temperature anneal followed by a long, low-temperature anneal.

DEVELOPMENT OF TECHNIQUE AND APPARATUS FOR HYDROEXTRUSION OF SUPERCONDUCTING WIRES

The peculiarities of the hydrostatic pressing process are the reason that high-quality wire of any diameter (with cross-sectional area reduction many times more than that achieved by the usual drawing method) can be fabricated by this method. The equipment for hydrostatic pressing is more complicated than that for conventional drawing. However, the significantly higher quality wire produced by this method makes it competitive. At the present, commercial equipment for hydropressing extremely long wire is not available, and so hydrostatic apparatus for this purpose had to be designed and built at the Donetsk Physics Engineering Institute of the Academy of Sciences of the Ukrainian SSR. The technology and equipment for complete wire fabrication, from initial billet to a final diameter of 0.3 mm, are expected to be developed.

The experimental investigation of hydroextrusion processes of billets, rods, and wires evolved into the following basic stages: (1) hydrostatic

Fig. 3. The 2-GPa hydropressing container for 100-mm diameter billets.

pressing of the billets (assemblages) from 70 to 120 mm in diameter to rods 30 mm in diameter by a vertical-type apparatus; (2) hydropressing of rods to wire blanks 7 mm in diameter by a horizontal-type apparatus; (3) hydrostatic pressing of wire blanks to a diameter of 0.3 mm.

The commercial vertical-type presses, with forces of 6.2 to 9.8 MN, appear to be the most convenient to use in the first stage. Special containers have been designed and manufactured at the Donetsk Institute for hydrostatic pressing with operational pressures up to 1.4 GPa (Fig. 3).

For hydroextrusion of rods, a machine with semicontinuous action has been designed and built. The principle of its action is: The rod blank has to be periodically fed through a clamping seal into the container when the liquid pressure inside equals atmospheric pressure. When the pressure inside the

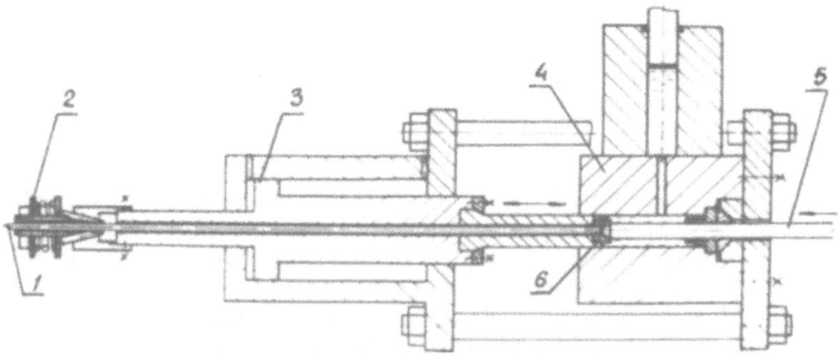

(a)

(b)

Fig. 4. Machine for semicontinuous hydrostatic pressing of 30-mm diameter rods. (a) Machine components: 1 - pressed rod; 2 - clamp; 3 - 2-MN press; 4 - high-pressure block; 5 - rod blank; 6 - die. (b) General view of the machine.

container reaches the operational level, the clamping seal sets to work holding the rod blank, and a working stoke of a plunger with die occurs, resulting in hydropressing. At the end of the working stroke, the pressure is released, the plunger with die moves back to its primary position, and at the same time feeds the next part of the rod blank into the container (Fig. 4). In this way, semicontinuous processing is accomplished by alternate hydropressing of the die and feeding of the rod blank. The apparatus is capable of forging rods of unlimited length at pressures up to 1.0 GPa and diameters of 30 to 7 mm. The length of the rod treated during a working stoke is 300 mm. The pressing rate is up to 0.024 m/s. The output of the machine is up to 27 m/h. Because pressing metals, alloys, and composites (σ_B = 295 to 785 MPa) with this machine can accomplish up to a 50% degree of deformation at a time, a single hydropressing appears to take the place of five or six drawing operations, and it simplifies pressing procedures, improves physical-mechanical properties, and increases production of suitable wire.

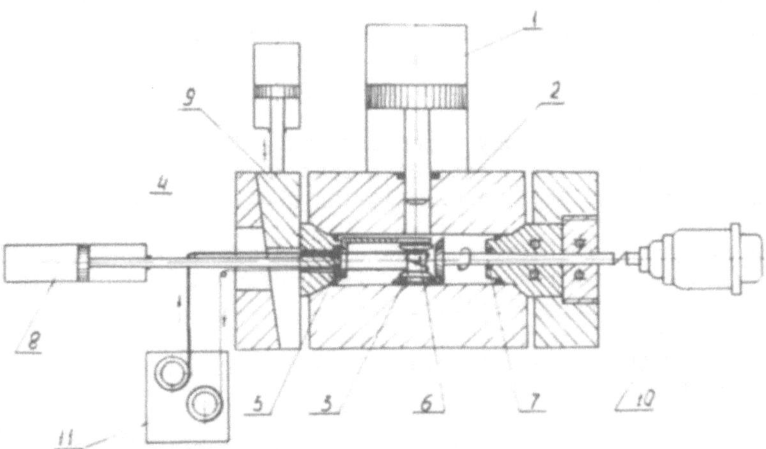

(a)

(b)

Fig. 5. Machine for continuous hydrostatic pressing of 2-mm-diameter wires. (a) Machine components: 1 - 2-MN press; 2 - high-pressure block; 3 - contact block; 4 - sealing die; 5 - deforming die; 6 - retracting reel; 7 - reel rotating mechanism; 8 - contact block shifting mechanism; 9 - clamping device; 10 - electric driver; 11 - receiving device. (b) General view of the machine.

The last stage of wire fabrication by hydropressing involves deformation of wire blank many times longer than the container. So it is necessary to process by passing the wire continuously through the high pressure chamber and die. The principle of continuous hydropressing accompanied by drawing is: the wire blank is continuously drawn into the chamber by a rotating reel and then is forced out by hydrostatic pressure through the working die (Fig. 5). This method of deformation of wire results in a cross-sectional area reduction several times that which would be obtained by drawing. To achieve continuous hydrostatic pressing accompanied by drawing, a commercial test machine was constructed at the Donetsk Institute. Wires can be fabricated with up to a 50% degree of deformation and an operational liquid specific pressure less than the yield strength of the material. The pressing rate is determined by the rate of the receiving device, which simultaneously exerts a slight tensile force at the front point of the wire. The tensile force is less than half the material yield strength and practically vanishes during the static regime of forging. This machine is intended for multifilamentary wire fabrication with diameters from 2 to 0.5 mm at operational liquid pressures up to 0.8 GPa. The receiving device can apply a tensile force up to 735 N and receive the finished wire at a rate up to 5 m/s.

HIGH-FIELD MULTIFILAMENTARY ALLOYED Nb_3Sn CONDUCTORS

To develop the multifilamentary superconducting Nb_3Sn wire, which will be used as a winding for the inner coil of a test magnetic system with magnetic field inductions up to 18 T, the following plan was chosen:

- Manufacture of the basic strand, which consists of 48 rods in the first stage and seven tin cores and is surrounded by a niobium diffusion barrier and a copper stabilizer.

- Manufacture of the multifilamentary (Nb_3Sn) conductor assemblage, consisting of 55 basic strands.

At first, three kinds of bimetallic billets were made: In the first, a pure copper matrix and pure electron-beam-melted niobium cores were used. In the second, the copper matrix was alloyed with 0.35 wt.% titanium. In the third, copper alloy matrix (0.35 wt.% titanium) and niobium cores alloyed with 0.6 wt.% titanium were used. The copper-to-niobium ratio in the first bimetallic billet was to 3:1. In the 70-mm-diameter copper matrix, seven

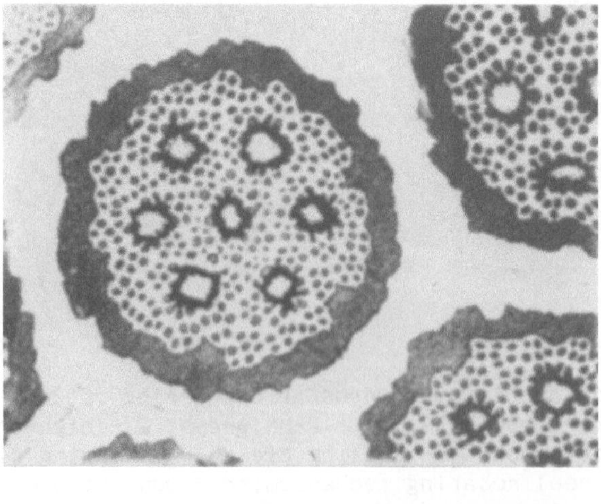

Fig. 6. Structural constituents of the multifilamentary Nb_3Sn alloy wire: basic 336-filament strand with internal tin sources.

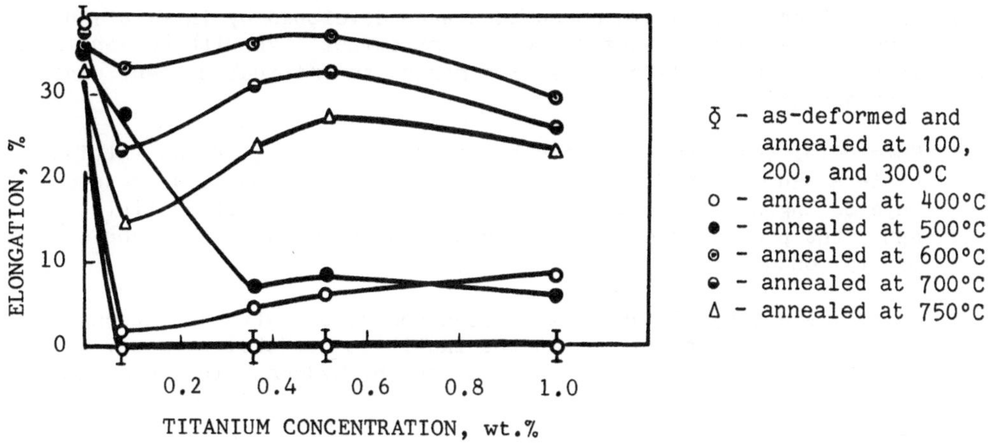

Fig. 7. Relative elongation versus titanium content of the copper matrix; deformation degree - 55%; annealing time = 1 h

holes were drilled into which niobium rods were inserted. The billets assembled in this way were extruded by hydrostatic pressing to a diameter of 14 mm with a single deformation degree of approximately 80% and then to a diameter of 6.2 mm with a single deformation degree of about 30%. In the next stage, the rods of the first bimetallic seven-filament assemblage were used to prepare the second billet having eventually 336 niobium filaments. The second (basic) strand of multifilamentary wire was made in the following way: Forty-eight rods of the first assemblage and seven 6.4-mm diameter tin rods (the tin was premelted together with 7 wt.% copper) were put together and inserted into the niobium tube. The latter serves as a diffusion barrier during the subsequent thermal diffusion treatment to prevent the tin from penetrating the surrounding stabilizing copper. The second billet was extruded by hydrostatic pressing from a diameter of 70 mm to 7 mm. The single deformation degree amounted to 55% and 30% at diameter reductions of 70 to 14 mm and 14 to 7 mm, respectively. The rods of the basic strand with 336 niobium filaments were used to fabricate the third assemblage of multifilamentary Nb_3Sn wire containing 18,480 filaments (Fig. 6). The tin content in the basic strand, considering the bronze to be formed during annealing, should amount to 14.4 wt.%.

It is necessary to call attention to the difficulties in fabricating the titanium alloyed copper matrix. Titanium additives strengthen copper and sharply reduce its plasticity. Some results of investigations on the mechanical properties of alloyed copper after deformation and annealing are shown in Fig. 7. it is clear enough from Fig. 7 that only alloys annealed at 600°C recover their plasticity. However, such high-temperature annealing could not be used because of the danger of melting and leakage of tin from internal sources. However, use of the hydrostatic pressing process substantially increases the deformability of this copper-titanium alloy, and in spite of the difficulties, the multifilamentary wire was produced.

The $J_c(B)$ measurements for the multifilamentary alloyed Nb_3Sn wires should be completed soon and published elsewhere.

CONCLUSIONS

- Large-scale fabrication of A15 conductors with the application of hydrostatic pressing at least for the initial stage of deformation is being developed at the experimental plant of EDB "Horizon."

- At EDB "Horizon," Nb_3Sn conductors were developed by in situ and powder metallurgy processes, which are competitive with conventional multifilamentary Nb_3Sn wires with regard to critical current densities and ac losses.

- Commercial test equipment for hydrostatic pressing of the initial billets and long rods and wires was designed and built at Donetsk Physics Engineering Institute; these machines made it possible to apply the promising hydrostatic extrusion process to the fabrication of A15 conductors.

- Multifilamentary alloyed Nb_3Sn-based conductors with an internal tin source are being developed at the Institute of Metal Physics; the deformation of these conductors is done by hydrostatic pressing.

REFERENCES

1. E. Gregory, E. Adam, W. Marancik, S. Richards, and P. Sanger, in: "Proceedings of the International Cryogenic Materials Conference, Kobe, Japan," Butterworths, London (1983), pp. 305-308.
2. M. Suenaga, in "Superconductor Materials Science," S. Foner and B. B. Schwartz, eds., Plenum, New York (1981), p. 201.
3. V. P. Buryak, in: "Physika i tekhnika vysokikh davleniy," 3rd edition, Naukova Dumka, Kiev (1981), p. 73.
4. E. G. Smith, R. J. Fiorentino, E. W. Collings, and F. J. Jelinek, IEEE Trans. Magn. Mag-15:91 (1979).
5. C. H. Rosner, B. A. Zeitlin, R. E. Schwall, M. S. Walker, and G. M. Ozeryansky, in: "Filamentary A15 Superconductors," M. Suenaga and A. F. Clark, eds, Plenum, New York (1980), p. 69.
6. P. A. Sanger, E. Adams, E. Ioriatti, and S. Richards, IEEE Trans. Magn. MAG-17:666 (1981)
7. D. K. Smathers, K. R. Marken, D. C. Larbalestier, and R. M. Scanlan, IEEE Trans. Magn. Mag-19:1417 (1983).
8. V. M. Pan, V. G. Prokhorov, A. S. Shpigel, "Metallophysika sverkhprovodnikov," Naukova Dumka, Kiev (1984), 192 pp.
9. K. Yoshizaki, O. Taguchi, F. Fujiwara, M. Imaizumi, Y. Hashimoto, K. Wakamoto, T. Yamada, and T. Satow, in: "Proceedings of the International Cryogenic Materials Conference, Kobe, Japan," Butterworths, London (1983), p. 380.
10. V. M. Pan, and A. G. Popov, in: "18 vsesoyuznoye soveshchaniye po physike nizkikh temperatur (NT-18), Kiev, 16-20 sentyabrya," Kiev (1974), p. 493.

MULTIFILAMENTARY $(Nb,Ti)_3Sn$ CONDUCTORS FOR 15-T-CLASS MAGNET APPLICATION

K. Tachikawa
National Research Institute for Metals
1-2-1 Sengen, Sakura-mura, Niihari-gun, Ibaraki 305, Japan

K. Kamata* and H. Moriai**
*Metal Research Laboratory, Hitachi Cable, Ltd.
**Tsuchiura Works, Hitachi Cable, Ltd.
3550 Kidamari-cho, Tsuchiura-shi, Ibaraki 300, Japan

N. Tada
Hitachi Research Laboratory, Hitachi Ltd.
4026 Kuji-machi, Hitachi-shi, Ibaraki 319-12, Japan

T. Fujinaga and R. Saito
Hitachi Works, Hitachi Ltd.
3-1-1 Saiwai-cho, Hitachi-shi, Ibaraki 317, Japan

ABSTRACT

Ti-bronze multifilamentary Nb_3Sn conductors have been successfully fabricated in full production scale through triple extrusion and drawing process, for a 15 T-class magnet. The heat treatment condition has been optimized using reduced size small capacity conductors. The 2.84 mm wide and 0.55 mm thick 5 µm-diam 31 x 361-core Nb/Cu-7.5at%Sn-0.4at%Ti rectangular shaped conductor with the aspect ratio of 5.2 shows an overall critical current density J_c(overall) of 300 A/mm^2 at 15 T in magnetic field parallel to the flat surface of the specimen after optimum heat treatment. A 190 mm winding inner diameter magnet has been constructed from double pancakes wound from 9.5 mm wide and 1.8 mm thick 5 µm-diam 349 x 361-core Ti-bronze Nb_3Sn conductors heat treated at 660°C for 200 hr. The $(Nb,Ti)_3Sn$ intermediate coil is connected in series to the NbTi outer coil. The magnet has been successfully excited to a central magnetic field of 14.2 T in 28 min., and held at that field for 30 min. The magnet has been designed to have an enough margin for generating 15 T.

INTRODUCTION

A high-field magnet capable of generating a central magnetic field over 18 T at 4.2 K in a clear bore of 30 mm is planned to install at NRIM.[1] The back up coils of the magnet consist of an outer solenoid wound from multifilamentary NbTi conductor and an intermediate double pancake coil wound from multifilamentary Nb_3Sn conductors. These two coils are connected in series. The back up magnet is designed to operate at a central magnetic field of 14 T in a 180 mm-diam clear bore. Thus, it was required to improve the high-field overall critical current density J_c(overall) for multifilamentary bronze-processed Nb_3Sn conductors, because the currently available pure Nb_3Sn conductors show rapid decrease in J_c(overall)'s with increasing magnetic field over 12 T.

We have previously shown that a small amount of titanium addition to the bronze matrix of bronze-processed multifilamentary Nb_3Sn conductors is most effective for the purpose, among the various alloying elements.[2,3,4] It has been also revealed that critical current I_c in parallel field ($H_{\parallel}$) drops rather drastically with increasing aspect ratio up to 4 and it decreases slightly with increasing aspect ratio over 4, though I_c in perpendicular field ($H_{\perp}$) increases gradually with increasing aspect ratio, for rectangular shaped multifilamentary Ti-bronze Nb_3Sn conductors.[3,4] It has been decided that the double pancakes for the intermediate coil are to be wound from 9.5 mm wide and 1.8 mm thick rectangular shaped multifilamentary Ti-bronze Nb_3Sn conductors. In this paper, the details of manufacturing and evaluating the multifilamentary Ti-bronze Nb_3Sn conductors for the intermediate coil, and also the excitation test result for the assembled back up magnet are reported.

OPTIMIZATION OF PROCESS PARAMETERS IN DOUBLE EXTRUDED Ti-BRONZE Nb_3Sn CONDUCTORS

Structure and Mechanical Properties of Ti-Bronze

Since it has been revealed that a small amount of titanium addition to the bronze matrix is most effective for enhancing the critical current density J_c(overall) of multifilamentary Nb_3Sn conductors[2,3,4], we examined the structure and mechanical properties of Ti doped Cu-7.5at%Sn ingot, in order to establish the fabrication procedure for Ti-bronze multifilamentary Nb_3Sn conductors in full production scale. A Cu-7.5at%Sn-0.4at%Ti ingot of about 150 kg charge was cast into 160 mm-diam mold after melting it in graphite crucible in vacuum, and was heat treated at 750°C for 50 hr for homogenization. Small δ phase particles were observed in the Cu-7.5Sn-0.4Ti ingot, even after the homogenization heat treatment, while they disappear completely in the Cu-7.5Sn alloy with no titanium addition after the same heat treatment.

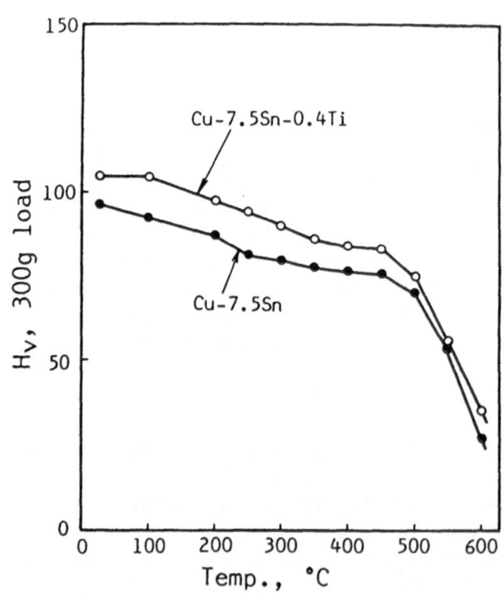

Fig. 1. High temperature hardness for Cu-7.5Sn-0.4Ti and Cu-7.5Sn alloys.

Table 1 Conductor Specifications

Specifications	Round Wire		Rectangular Wire	
Composition (at%)	Nb/Cu-7.5Sn-0.4Ti		Nb/Cu-7.5Sn-0.4Ti	
Barrier	Nb		Nb	
Size (mm)	ϕ1.39	ϕ0.69	1.84x0.80	2.84x0.55
Core	ϕ5μmx361x31	ϕ5μmx361x7	ϕ5μmx361x31	
Cu/non Cu	1.2	0.77	1.2	
Bz/Core	2.5		2.5	

Fig. 1 shows the high temperature hardness for the alloys. Usually, we practice triple warm hydrostatic extrusions at temperature from 300°C to 400°C for fabricating practical multifilamentary Nb_3Sn conductors. Extrusions of Nb/Cu-7.5Sn-0.4Ti single-core billets have been successfully made, though Cu-7.5Sn-0.4Ti alloy keeps 10-15 higher Vickers hardness values than those for Cu-7.5Sn alloy, up to 450°C. It has been shown from the measurement of annealing characteritics for cold worked Cu-7.5Sn-0.4Ti and Cu-7.5Sn alloys that both alloys fully recrystallize after 1 hr annealing at 500°C. The grains for the fully annealed Cu-7.5Sn alloy have been observed to be significantly refined by the 0.4at%Ti addition.

Conductor Preparation

Table 1 shows the conductor specifications prepared in this study. 1.39 mm-diam Nb/Cu-7.5Sn-0.4Ti wire with 5 μm-diam 31 x 361-cores was used for optimizing the heat treatment condition. Nb/Cu-7.5Sn-0.4Ti rectangular wires with 5 μm-diam 31 x 361-cores were for evaluating the I_c anisotropy. Uniaxial strain dependence of I_c was also measured using 0.69 mm-diam Nb/Cu-7.5Sn-0.4Ti wire with 5 μm-diam 7 x 361-cores after heat treatment.[5] All the conductors listed in Table 1 were prepared from double extruded 361-core composites surrounded by a niobium barrier inside a stabilizing Cu. Extrusions were done in full production scale at a temperature from 300°C to 400°C, utilizing 4,000 tonnage capacity hydrostatic extrusion press. Billet size is 160 mm in diameter and 1,200 mm in length, typically. A Nb/Cu-7.5Sn-0.4Ti single-core billet with bronze/core volume ratio of 2.5 was extruded to a 40 mm-diam rod. A 'submulti' billet was assembled from hexagonal shaped 361 Nb/Cu-7.5Sn-0.4Ti single-core rods surrounded by a niobium barrier in a stabilizing OFC(Oxygen Free Copper) tube. This 'submulti' billet was extruded and drawn with intermediate annealings at 550°C for 1 hr. 31 or 7 'submulti' wires were inserted into OFC tubes. The resulting composites were then drawn to the specifications listed in Table 1.

Superconducting Current-Carrying Capacities

Fig. 2 (a) shows J_c(overall)-H curves for double extruded 1.39 mm-diam multifilamentary Nb/Cu-7.5Sn-0.4Ti wires with 5 μm-diam 31 x 361-cores after the quoted heat treatments. All the J_c(overall)'s up to 16 T increase slightly with decreasing heat treatment temperature from 690°C to 650°C. A J_c(overall) for the wire heat treated at 650°C for 200 hr exceeds 3.5 x 10^4 A/cm^2 at 15 T. It has been also measured that the J_c(overall)'s for the wire with 5 μm-diam cores decrease if the heat treatment temperature is lowered below 650°C. Fig. 2 (b) shows the J_c(overall)-$H_{\parallel}$ curves for the multifilamentary 1.39 mm-diam wire, and the 1.84 x 0.80 mm and 2.84 x 0.55 mm rectangular wires with the aspect ratio of 2.3 and 5.2, respectively, prepared through double extrusions and heat treated at 690°C for 200 hr.

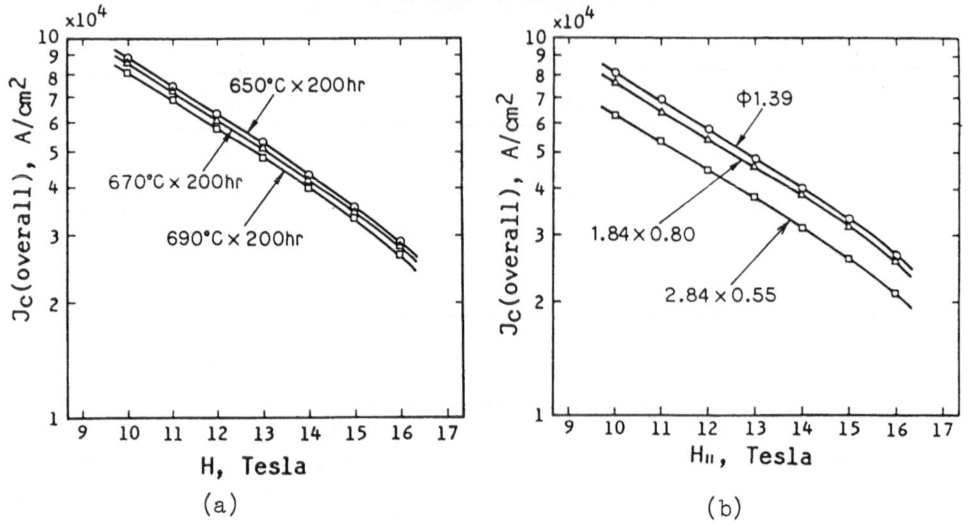

Fig. 2. J_c(overall)-H curves at 4.2 K for the multifilamentary 5 μm-diam 31 x 361-core Nb/Cu-7.5Sn-0.4Ti conductors fabricated through double extrusion and drawing process.
(a) 1.39 mm-diam round wires heat treated at the quoted temperatures for 200 hr.
(b) 1.39 mm-diam round wire, and 1.84 x 0.80 mm and 2.84 x 0.55 mm rectangular wires heat treated at 690°C for 200 hr.

The degree of the aspect ratio dependence of the I_c anisotropy for the double extruded rectangular shaped Nb/Cu-7.5Sn-0.4Ti wires with the aspect ratio of 2.3 and 5.2 is in good agreement with that for the wires prepared through drawing process only, after the same heat treatment, as is shown in the former report.[3,4]

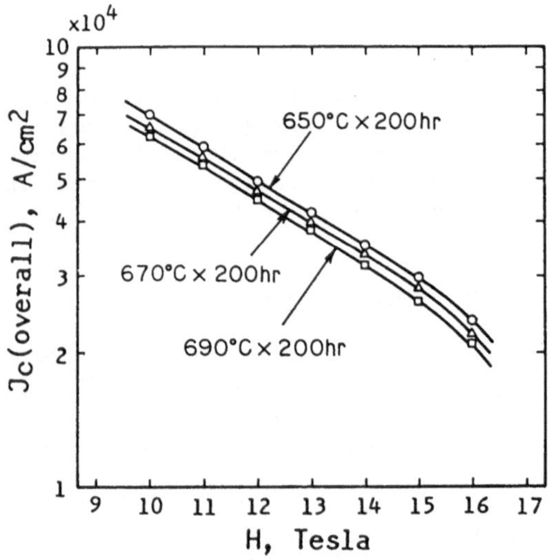

Fig. 3. J_c(overall)-$H_{\parallel}$ curves at 4.2 K for double extruded 2.84 x 0.55 mm rectangular shaped multifilamentary Nb/Cu-7.5Sn-0.4Ti wires with 5 μm-diam 31 x 361-cores heat treated at the quoted temperatures for 200 hr.

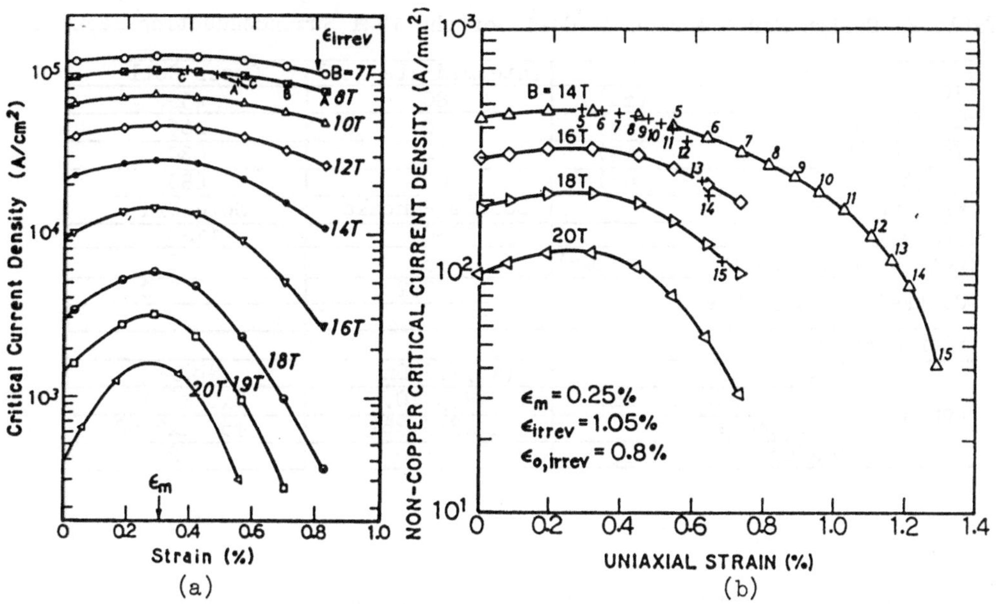

Fig. 4. Uniaxial strain dependece of J_c(overall) at 4.2 K for multifilamentary Nb_3Sn wires.[5]
(a) Pure Nb_3Sn wire
(b) Nb/Cu-7.5Sn-0.4Ti wire

Fig. 3 shows J_c(overall)-$H_\parallel$ curves for double extruded 2.84 x 0.55 mm rectangular shaped multifilamentary Nb/Cu-7.5Sn-0.4Ti wires with 5 μm-diam 31 x 361-cores and the aspect ratio of 5.2 after the quoted heat treatments. All the J_c(overall)'s up to 16 T for the rectangular wires also increase slightly with decreasing heat treatment temperature from 690°C to 650°C, in the same manner as those for round wires shown in Fig. 2 (a). It is revealed from Fig. 2 and Fig. 3 that the degree of the I_c anisotropy for the rectangular wires decreases with decreasing heat treatment temperature. The ratio of the J_c(overall) ($H_\parallel$) for the rectangular wire with the aspect ratio of 5.2 to the J_c(overall) for the round wire increases from 0.78 to 0.84 with decreasing heat treatment temperature from 690°C to 650°C.

Strain Dependence of Critical Current

Fig. 4 (b) shows a J_c(overall) vs. uniaxial strain characteritics at magnetic fields from 14 T to 20 T for a 0.69 mm-outer diam and 5 μm-diam 7 x 361-core multifilamentary Nb/Cu-7.5Sn-0.4Ti conductor with a bronze/core volume ratio of 2.5 and a Cu/non Cu volume ratio of 0.77, heat treated at 650°C for 200 hr.[5] A J_c(overall) vs. uniaxial strain characteritics for a pure multifilamentary Nb_3Sn conductor is also shown in Fig. 4 (a), as a reference.[5] The compressive prestrain ε_m for the Ti-bronze Nb_3Sn wire is 0.25%. It is seen that J_c(overall) at ε_m exceeds 3 x 10^4 A/cm^2 at 16 T and 1 x 10^4 A/cm^2 at 20 T for the Ti-bronze wire. It is observed that the uniaxial strain sensitivity J_c(overall) for a pure multifilamentary Nb_3Sn wire is greatly reduced by the small amount of titanium addition to the bronze matrix. The intrinsic tensile irreversible strain limit for the Ti-bronze Nb_3Sn wire showed a much higher value of 0.8% as compared to 0.5% for the pure multifilamentary Nb_3Sn wire. These improvements in the strain sensitivity of J_c(overall) and the intrinsic irreversible strain limit for the Ti-bronze multifilamentary Nb_3Sn conductor may be due to the enhancement in H_{c2} as a result of titanium incorporation into the Nb_3Sn layer and to the suppression of Kirkendall void formation during the heat treatment resulting from the titanium addition to the bronze matrix.

Table 2 Design for the Back up Coils of the 18 T Superconducting Magnet.

Item			Intermediate Coil	Outer Coil
Clear Bore		(mm)	180	455
Winding Inner Diam		(mm)	190	465
Winding Outer Diam		(mm)	422	630
Length		(mm)	665	680
Winding Method			Double Pancake	Solenoid
Current		(A)	1180	1180
Coil Current Density		(A/mm^2)	55	85
Central Magnetic Field		(T)	7.2	7.0
			14.2	
Stored Energy		(MJ)	6	
Conductor	Material		$(Nb,Ti)_3Sn$	NbTi
	Cross-Section	(mm)	1.8x9.5 R0.5	2x6 R0.5
	Filaments		φ5μmx349x361	φ50μmx2050
	Cu/non Cu		1	2

MANUFACTURE OF PRACTICAL Ti-BRONZE Nb_3Sn CONDUCTORS

Manufacturing and Evaluating the Practical Ti-Bronze Nb_3Sn Conductors

Table 2 shows the design for the back up coils of a 18 T superconducting magnet . The back up coils consist of an outer solenoid wound from multifilamentary NbTi conductors and an intermediate double pancake coil wound from multifilamentary Ti-bronze Nb_3Sn conductors. These two coils are connected in series. The back up magnet is designed to operate at a central magnetic field of 14 T in a clear bore of 180 mm in diameter. The multifilamentary Nb/Cu-7.5Sn-0.4Ti conductor has a cross-section of 9.5 mm in width and 1.8 mm in thickness and 5 μm-diam 349 x 361-cores. The Cu/non Cu volume ratio is 1.0 and the bronze/core volume ratio is 2.5. 5 μm-diam 361 niobium cores are embedded in Ti-bronze matrix surrounded by a niobium barrier inside a stabilizing OFC. This 9.5 mm wide and 1.8 mm thick Ti-bronze Nb_3Sn conductor with the aspect ratio of 5.2 has been successfully fabricated in full production scale through triple extrusion and drawing process.

Fig. 5 shows a cross-section for the rectangular shaped multifilamentary Ti-bronze Nb_3Sn conductor. The heat treatment condition for the 9.5 mm wide and 1.8 mm thick practical Ti-bronze Nb_3Sn conductors has been decided to practice at 660°C for 200 hr, referring the J_c(overall)-$H_{\parallel}$ curves in Fig. 3 for the reduced size 2.84 mm wide and 0.55 mm thick Ti-bronze Nb_3Sn conductors with the same aspect ratio of 5.2 as that for the practical ones. The 9.5 mm wide and 1.8 mm thick practical Ti-bronze Nb_3Sn conductors have I_c's over 1,600 Amperes at 16.3 T ($H_{\parallel}$) from the measurement for the half wide specimens cut from the heat treated conductors.

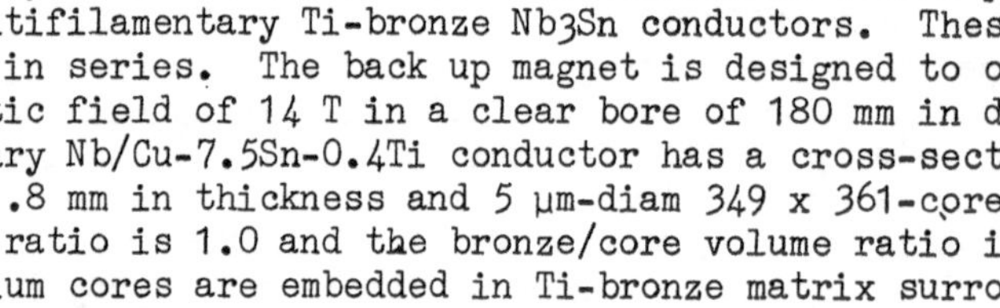

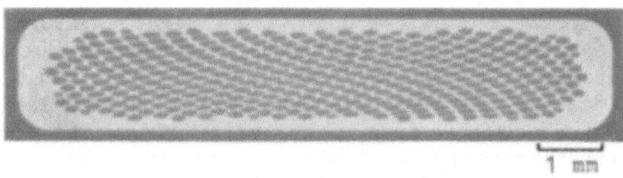

Fig. 5. A cross-section of a 9.5 mm wide and 1.8 mm thick 5 μm-diam 349 x 361-core Nb/Cu-7.5Sn-0.4Ti conductor fabricated in full production scale through triple extrusion and drawing process.

Fig. 6. An outer view of the assembled back up magnet with a clear bore of 180 mm in diameter.

Construction and Test of a 15 T-Class Magnet

The $(Nb,Ti)_3Sn$ coils were wound in a double pancake manner. Turn to turn insulation was made by helically wrapping double layer 45 μm-thick Kapton sheets overlapping half width around the heat treated conductors. Layer to layer one by inserting 1 mm thick glass fiber epoxy sheet, between

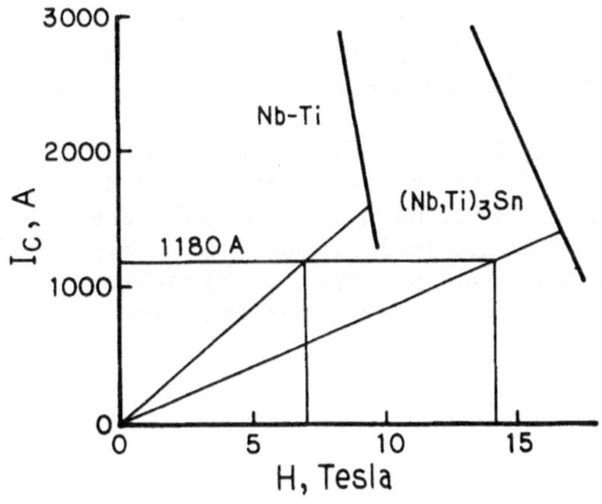

Fig. 7. Calculated excitation load lines for the back up coils.

pancakes. 32 double pancakes were stacked and connected on the outer radius. The assembled $(Nb,Ti)_3Sn$ intermediate coil was connected in series to the NbTi outer coil. Fig. 6 shows the outer view of the assembled back up magnet. Fig. 7 shows the calculated load lines for the magnet. It is designed that the back up coil generates a central magnetic field of 14.2 T at the current supply of 1,180 Amperes. The NbTi outer coil generates a central magnetic field of 7.0 T at this current supply. The calculated total expanding electromagnetic force loaded to the outermost layer conductors in radial direction is 3,340 ton and the total compressive force loaded to the central pancakes in axial direction is 236 ton, at the current supply of 1,180 Amperes. The back up magnet has been successfully excited to 14.2 T at the sweeping rate of 0.5 T/min., and held for 30 min., at that field. It is shown in Fig. 7 that the magnet still has an enough margin for generating 15 T.

CONCLUSION

The 9.5 mm wide and 1.8 mm thick rectangular shaped multifilamentary Ti-bronze Nb_3Sn conductors with 5 µm-diam 349 x 361-cores have been successfully fabricated in full production scale through tripple extrusion and drawing process, and heat treated at 660°C for 200 hr. These $(Nb,Ti)_3Sn$ conductors showed I_c's over 1,600 Amperes at 16.3 T ($H_{\parallel}$). The heat treatment condition was optimized and the I_c anisotropy was made clear using reduced size 5 µm-diam 31 x 361-core $(Nb,Ti)_3Sn$ conductors prepared through double extrusions. It was also revealed that both the strain sensitivity of J_c(overall) and the intrinsic irreversible strain limit for multifilamentary Nb_3Sn conductor are greatly improved by the small amount of titanium addition to the bronze matrix.

The back up magnet for the 18 T magnet consists of an outer solenoid wound from multifilamentary NbTi conductors and an intermediate double pancake coil wound from multifilamentary $(Nb,Ti)_3Sn$ conductors, with a clear bore of 180 mm in diameter. The outer solenoid coil and and the intermediate double pancake coil are connected in series. We have successfully excited the back up magnet to a central magnetic field of 14.2 T in 28 min., at the current supply of 1,180 Amperes and held it at that field for 30 min.

REFERENCES

1. K. Tachikawa, H. Sekine, K. Itoh, K. Kamata, K. Noguchi, N. Tada T. Fujinaga and R. Saito, Titanium Bronze Multifilamentary Nb_3Sn Conductors, Proceedings of Japan-US Workshop on High-Field Superconducting Materials for Fusion, p. 17 (Tsukuba, 1984)
2. K. Tachikawa, H. Sekine and Y. Iijima, Composite-Processed Nb_3Sn with Titanium Addition to the Matrix, J. Appl. Phys., 53, 5354 (1982)
3. K. Kamata, N. Tada, K. Itoh and K. Tachikawa, High-Field Current-Carrying Capacities of 'Titanium Bronze' Processed Multifilamentary Nb_3Sn Conductors with Pure and Alloy Cores, Adv. in Cryogen. Engr., 30, 771 (1984)
4. K. Kamata, H. Moriai, N. Tada, T. Fujinaga, K. Itoh and K. Tachikawa, Manufacturing of Titanium Bronze Processed Multifilamentary Nb_3Sn Conductors, IEEE Trans. Magn., to be published.
5. J. W. Ekin, Electro-Mechanical Properties of High Field Superconductors, Proceedings of Japan-US Workshop on High-Field Superconducting Materials for Fusion, p. 153 (Tsukuba, 1984)

OBSERVATION OF THE MICROSTRUCTURE IN BRONZE-PROCESSED MULTIFILAMENTARY Nb_3Sn TAPES

Liu Chunfang, Wang Keguang, Zhang Tingjie,
Yang Zhaoling, and Zhang Shudong

Baoji Institute for Nonferrous Metals Research
Baoji, Shaanxi
People's Republic of China

ABSTRACT

The microstructure of the longitudinal and transverse sections of multifilamentary Nb_3Sn tapes in which 0.35 at.% Ti is added to the Cu-Sn matrix was observed by TEM and SEM. A kind of grain boundary-like network exists at the Nb_3Sn filamentary interface near the Cu-Sn matrix consisting of coarse Nb_3Sn grains. It is not evident that the size of the network changes with aging time. This paper presents the relationship between the average size of the Nb_3Sn grains and the heat-treatment program and also the Nb_3Sn layer thickness as a function of aging time. A small quantity of Nb_3Sn is already present in specimens cold-worked and annealed at 450°C.

INTRODUCTION

It has been reported that the addition of Ti to the Nb core or to the Cu-Sn matrix of multifilamentary Nb_3Sn conductors fabricated by the bronze process greatly increases the critical current density, J_c, in high fields.[1-4] We have fabricated multifilamentary Nb_3Sn tapes in which 0.35 at.% Ti was added to the Cu-Sn matrix in order to use a prereacted conductor for a high-field test magnet. Overall J_c of about 8.6×10^3 A/cm² at 20 T (⊥) and 4.2 K; 2.64×10^4 A/cm² at 16 T (⊥) and 4.2 K were obtained for tapes heat-treated at 700°C for 200 h. The critical current density of A-15 Nb_3Sn material is strongly dependent on the microstructure, such as grain size, grain morphology, and stoichiometry. The purpose of the present work is to investigate the microstructure of multifilamentary Nb_3Sn tapes in order to understand the growth characteristics of the Nb_3Sn diffusion layer and to look for a suitable heat-treatment program in tape specimens with Ti additions.

EXPERIMENTAL PROCEDURE

Nb_3Sn composites consisting of a Cu-7.4 at.% Sn-0.35 at.% Ti alloy matrix, 912 pure Nb cores, a tantalum diffusion barrier, and a Cu sheath were fabricated into 5 mm x 0.25 mm tape. Specifications of the conductor are shown in Table 1.

The tapes were heat-treated at 700°C and 750°C for 48 to 200 h. The critical currents, I_c, were measured at 4.2 K in magnetic fields up to 20 T at the High Magnetic Field Laboratory in Grenoble, with a criterion of 0.1

Table 1. Specifications of the Conductor

Dimensions	5 mm x 0.25 mm
Number of filaments	912
Cu/non-Cu ratio	0.25
Bronze/Nb	3.4:1

μV/cm. The thicknesses of the Nb_3Sn layers were measured by optical microscope and SEM on the cross sections of the tapes. The Nb_3Sn grain morphology was observed by SEM on the fracture surface of the tape. The average grain size was determined by applying the line-intercept method on at least 50 grains using TEM and SEM. SEM-EDAX was also performed to obtain the content of the coarse Nb_3Sn grains of which the grain boundary-like network consists.

RESULTS AND DISCUSSION

It was found that a kind of grain boundary-like network exists at the Nb_3Sn filamentary interface near the Cu-Sn matrix. A typical micrograph of a sample heat-treated at 700°C for 175 h is shown in Fig. 1, where one can see coarse Nb_3Sn grains coalescing into the grain boundary-like network, which surrounds a lot of small Nb_3Sn grains. The size of the network did not change obviously with heat-treating time. Examination by SEM-EDAX indicated that the average Sn content of the coarse Nb_3Sn at the network was 6 percent higher than the fine Nb_3Sn grains surrounded by the network. The microstructure of the longitudinal section of the bare Nb filaments removed from the Cu-Sn matrix was investigated by SEM. The fibroid area with a high Sn content was observed for a bare Nb filament without heat treatment. As the bare Nb filament was heat-treated, for example at 700°C for 150 h, the grain boundary-like network was not observed. The result measured by SEM-EDAX shows that a small quantity of Nb_3Sn was already present at the Nb filament interface near the Cu-Sn matrix before heat treatment. It is possible that it formed during the cold working and annealings at 450°C for 1 h. In general, the Nb_3Sn grains coalesce into

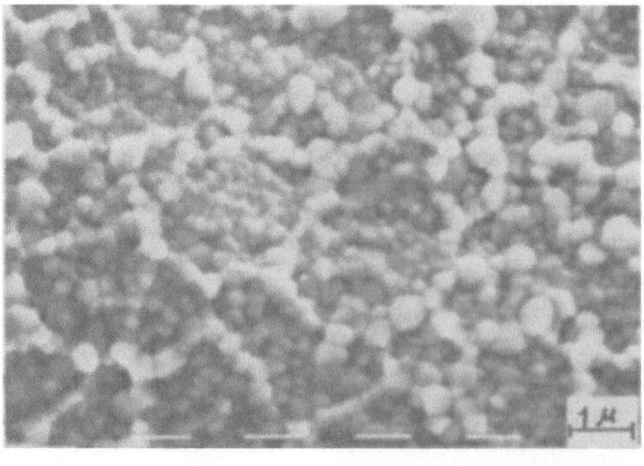

Fig. 1. Scanning electron micrograph of the longitudinal interface of a Nb_3Sn filament reacted at 700°C for 175 h.

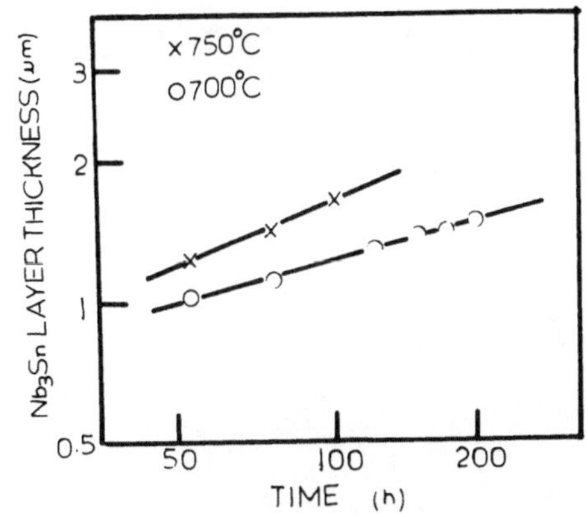

Fig. 2. Nb_3Sn layer thickness versus heat treatment time at 700°C and 750°C.

inconsecutive fibres. For the bare Nb filament, x-ray diffraction analysis indicated that there is no Nb_3Sn phase within this measurement capability.

In general, the Nb_3Sn phase forms first in the grain boundaries of the Nb filaments because the grain boundaries are in the area where defects (such as dislocations, vacancies, and impurities) concentrate, and the energy to form new phases is lower in the grain boundaries. The bronze matrix is a Sn source; during the diffusion reaction forming Nb_3Sn, Sn atoms are constantly transferred to the reacted interface. For diffusion of Sn atoms in the bronze matrix, the diffusivity coefficient along the grain boundary of the bronze is much larger than that of bulk diffusion. Therefore, it is possible that the Sn-rich grain boundary-like network existing at the interface of the Nb_3Sn filament near the Cu-Sn matrix is dependent on the structure (e.g., grain boundary) of the unreacted bronze and Nb filament and on the changes in structure during heat treatment.

Figure 2 shows the time dependence of the Nb_3Sn layer thickness at 750°C and 700°C. The time dependence of the thickness can be expressed approximately by the relation $Y - t^n$, where Y is the thickness, t is the time, and n is a constant. For the specimen heat-treated at 700°C, the value of n is about 0.26; for the specimen heat-treated at 750°C, n is about 0.41. Observation indicates that the Nb_3Sn layer thickness is dependent on the position of the filaments in the tape. It has been previously reported that the growth law of the Nb_3Sn layer is dependent on

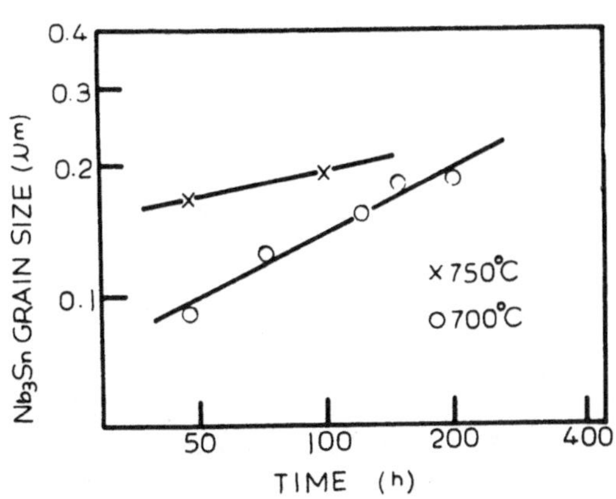

Fig. 3. Nb_3Sn grain size versus heat treatment time at 700°C and 750°C.

Fig. 4. Typical transmission electron micrograph of the Nb_3Sn grains reacted at 700°C for 150 h.

several factors, such as bronze/Nb, Sn composition in the bronze matrix, number of filaments, and shape of filaments.[5-7] The measured results of the present work were obtained for Nb_3Sn filaments located in the middle of the tapes.

Figure 3 shows the average Nb_3Sn grain size versus heat treatment time at 700°C and 750°C. It can be expressed approximately by the relation $L - t^m$, where L is the average Nb_3Sn grain size, t is time, and m is a constant. The value of m is about 0.48 for specimens heat-treated at 700°C and about 0.18 for those heat-treated at 750°C. Introducing the observed m = 0.48 (at 700°C) and m = 0.18 (at 750°C) into the equation deduced by Farrell et al.,[8,9] n = 0.26 (at 750°C) and n = 0.41 (at 750°C), respectively, were obtained. The results are identical with the observed values in the present work.

The average grain size measured by TEM was less than that measured by SEM. The grain size is dependent on the position in the tape. The data in Fig. 3 were obtained from the results measured by SEM. Figure 4 shows a typical micrograph of the Nb_3Sn grains obtained by TEM.

The observation indicates that for the multifilamentary Nb_3Sn tapes, a better conductor with a high critical current in high fields and usable mechanical properties can be obtained through a longer heat treatment at 700°C (e.g., 200 h). The measured results of critical current density are shown in Fig. 5.

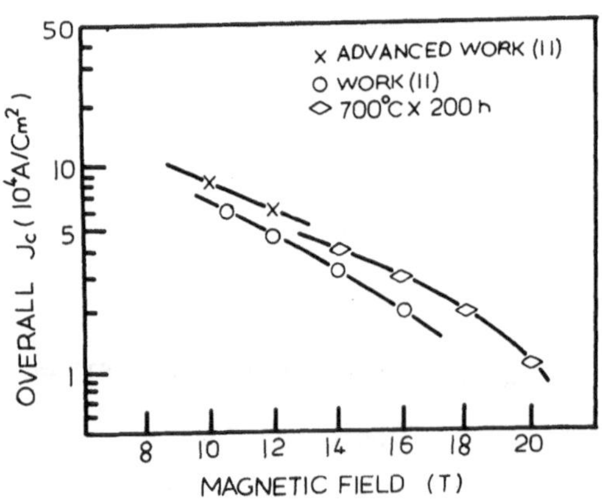

Fig. 5. Overall J_c versus magnetic field curves for the Nb_3Sn tapes.

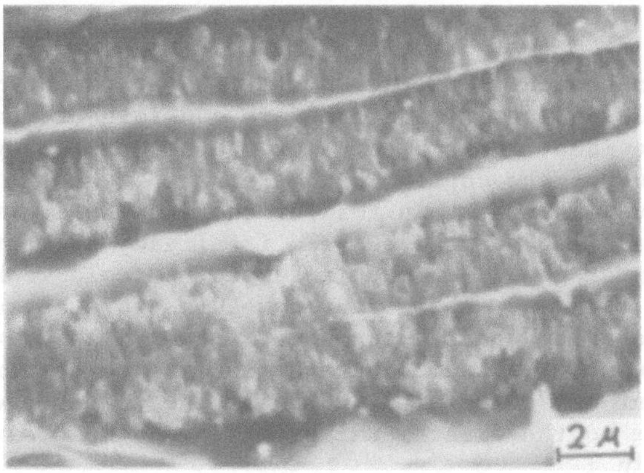

Fig. 6. Scanning electron micrograph of the Nb_3Sn grain structure reacted at 700°C for 150 h.

The grain morphology of the Nb_3Sn layer in the tapes was observed by SEM. In this study, the observed results were essentially the same as the research.[10] As shown in Fig. 6, the inner shell is made up of columnar grains that radiate from the Nb interface. The columnar grains grow on the fine equiaxial grains. In the present work, the coarse grains, of which the grain boundary-like network consists, were formed primarily during the diffusion processes.

Figure 7 shows the measured results of the critical current density, J_c, for the U-shape specimens and the solenoid-shape specimens. During the measurement process, magnetic fields are perpendicular to the tape surface for the U-shape specimens and are parallel to the tape surface for the solenoid-shape specimens. The result indicates that the I_c measured in the perpendicular field is considerably larger than that measured in the parallel field. As shown in Fig. 6, the Nb_3Sn grains grow perpendicular to the tape surface. The difference in the grain boundary density due to the directional growth of Nb_3Sn grains may account for the anisotropy in I_c.

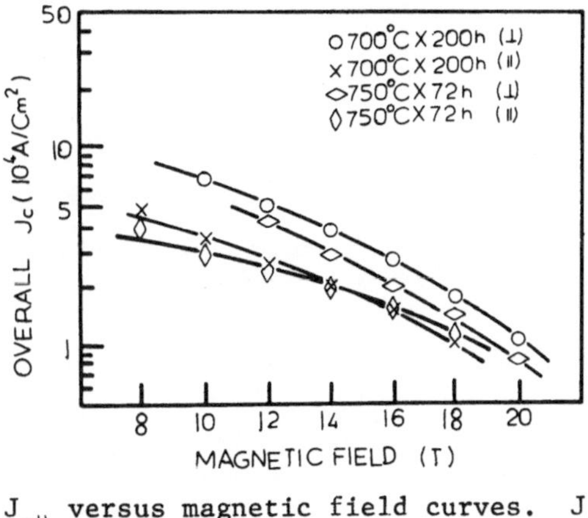

Fig. 7. $J_{c\perp}$ and $J_{c\parallel}$ versus magnetic field curves. $J_{c\perp}$ and $J_{c\parallel}$ are the critical current measured in magnetic fields perpendicular and parallel to the tape surface, respectively.

SUMMARY

A kind of grain boundary-like network exists at the Nb_3Sn filamentary interface near the Cu-Sn matrix. The network consists of coarse Nb_3Sn grains. It is not evident that the size of the network changes with aging time. The Sn content of the coarse Nb_3Sn grains, of which the grain boundary-like network consists, is more than that of the fine Nb_3Sn grains surrounded by the network. A small quantity of Nb_3Sn is already present at the Nb filament interface near the Cu-Sn matrix during the cold working and annealings at 450°C for 1 h. The measured results for the multifilamentary Nb_3Sn tapes indicate that a better conductor with a high critical current in high fields and usable mechanical properties can be obtained through a longer heat treatment at 700°C (e.g., 200 h). The difference in the grain boundary density due to the directional growth of Nb_3Sn grain may account for the anisotropy in I_c.

ACKNOWLEDGMENT

The critical currents were measured by Mr. Wang Jingrung and Mr. Teng Xinkang at the High Magnetic Field Laboratory in Grenoble. We would like to express our thanks to them.

REFERENCES

1. K. Tachikawa, T. Asano, and T. Takeuchi, Appl. Phys. Lett. 39:766 (1981).
2. M. Suenaga, S. Okuda, R. Sabatini, K. Itoh, and T. S. Luhman, Adv. Cryog. Eng. 28:319 (1982).
3. K. Kamata, N. Tada, K. Itoh, and K. Tachikawa, IEEE Trans. Magn. MAG-19(3):1433 (1983).
4. H. Sekine, Y. Iijima, K. Itoh, K. Tachikawa, Y. Tanaka, and Y. Furuto, IEEE Trans. Magn. MAG-19(3):1429 (1983).
5. M. Suenaga, O. Horigami, and T. S. Luhman, Appl. Phys. Lett. 25:624 (1974).
6. M. Suenaga, O. Horigami, and T. S. Luhman, J. Appl. Phys. 45:4049 (1974).
7. K. Togano and K. Tachikawa, J. Less-Common Met. 68:15 (1979).
8. H. H. Farrell, G. H. Gilmar, and M. Suenaga, J. Appl. Phys. 45:4025 (1974).
9. H. H. Farrell, G. H. Gilmar, and M. Suenaga, Thin Solid Films 25:253 (1975).
10. I. W. Wu, D. R. Dietderich, et. al., J. Appl. Phys. 54:7193 (1983).
11. K. Noto, K. Watanabe, et. al., J. Phys. (Paris) 45 (1984).

EFFECT OF STARTING MATERIALS AND PROCESSING VARIABLES FOR THE PRODUCTION OF DISCONTINUOUS-FILAMENT Nb_3Sn WIRE

P.L. Upadhyay and D. Dew-Hughes

Department of Engineering Science
Oxford University
Oxford, U.K.

ABSTRACT

Discontinuous multifilamentary wires of Nb_3Sn have been prepared from compacted mixtures of 30 wt. % Nb in Cu, extruded, drawn, annealed, tin plated and reacted. Processing variables include starting materials, extrusion ratio and extrusion temperature. Continuous lengths of wire could be satisfactorily produced from compacts of either ultra-pure Nb (VPN ~ 95 kg mm^{-2}) and Cu powder or from centrifugal arc-cast Nb spheroids (VPN ~ 120 kg mm^{-2}) and tough pitch Cu powder. After a total area reduction of 10^4:1, the latter materials resulted in long, unbroken, highly regular filaments of Nb ~ 6μm in diameter. The high degree of perfection of these filaments is due in part to the uniformity of the initial spheroids, compared to the highly irregular hydride-dehydride Nb powder. However their greater hardness requires that the spheroids be co-processed in a less-pure Cu matrix. Critical currents were measured on helical specimens involving more than 1m length of wire, in fields up to 15T at 4.2 K, after reaction for various times at different temperatures. Overall current densities of 3×10^8 Am^{-2} were obtained at 12T in the best samples. Further reductions are expected to produce material with improved current densities.

INTRODUCTION

In recent years, alternate technologies have been developed for the production of multifilamehtary wires of Nb_3Sn. The most important of these are the in-situ method[1] and the powder process[2-5], both of which produce discontinuous filamentary conductors. The cold[2,3] and hot[4,5] powder processes produce conductors with superconductoring critical current densities comparable to those of commercial MF conductors of Nb_3Sn. More recently, the cold process is being used for the production of large quantities of this material using large sized Nb powders and additional Sn cores in rebundled powder-processed wires[6,7].

This paper reports the powder-metallurgical fabrication of Nb_3Sn wires using hydride-dehydride Nb powder and spherical Nb particles[3]. A wide range of processing conditions have been used to show the improved performance of the latter.

EXPERIMENTAL DETAILS

Processing of Powders

For the fabrication of Cu-Nb composites, 3 different powder mixtures of 30 wt% Nb in Cu were prepared using a high purity hydride-dehydride Nb powder (Wah-Chang Teledyne, Albany, Oregon) and Nb spheriods prepared by the Centrifugal Shot Casting technique developed and patended by UKAEA Harwell. This technique produces spherical particles of uniform shape and size (~ 300 µm diameter for this work). Since the Nb rod used to prepare the spheroids was not the purest available zone refined Nb, the microhardness of the spheriods (VPN ~ 120Kg mm^{-2}) was higher that that of ultra purity hydride-dehydride powder (VPN ~ 95Kg mm^{-2}) The powder particles and the spheroids are shown in figures 1 and 2 respectively.

The hydride Nb was mixed with high-purity precipitated copper powder and the Nb spheroids were mixed with both the high-purity Cu and a less pure but harder tough pitch Cu powder. The mixtures were compacted in OFHC copper containers of outer diameters 35 mm and 25.4 mm. The evacuated, sealed billets were extruded at $300^{0}C$, $500^{0}C$ and $900^{0}C$ to a 6.3 mm diameter rod. The extruded compacts were annealed under vacuum at $975^{0}C$ ~ for 3 hours and subsequently cold-drawn to the final wire size.

Successfully processed wires were electro-plated with Sn and heat-treated in a temperature range from $550^{0}C$ to $800^{0}C$ for different times ranging from 1 day up to 3 weeks. Heat-treatment conditions were optimised by examining the diffusion reacted layers of Nb_3Sn under the optical microscope.

J_c Measurements

The system used for measuring critical currents at the Clarendon Laboratory has been described previously[8]. The sample is helically wound (3 mm pitch) on a 19 mm diameter stainless steel tube with Cu electrical contacts pressed into either end. The wire is reacted after mounting in order to avoid damage to the Nb_3Sn filaments and is then firmly fixed onto the holder by dipping in a soft-solder bath. The current is passed through about 1.2 metres of the wire and the voltage generated is measured across the central 5 turns, i.e. about 30 cm of the conductor length. Critical currents were measured at 4.2 K in magnetic fields up to 15 tesla in the Clarendon Laoratory 16 Tesla Superconducting Magnet. The criterion for J_c was a voltage generation of 0.1 µv across 1 cm of the wire.

Fig. 1 Particles of hydride dehydride Nb

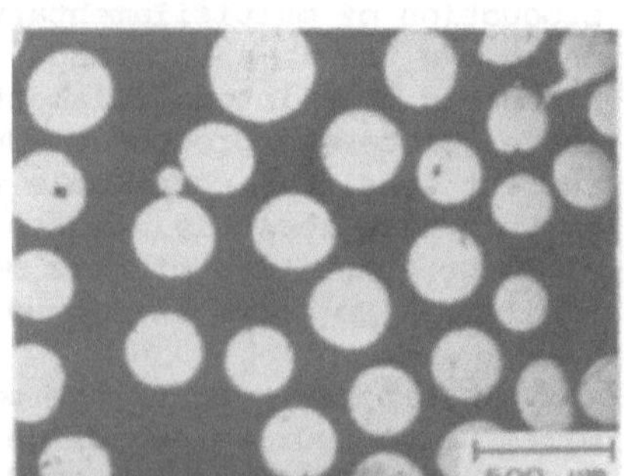

Fig. 2 Nb Spheroids

RESULTS AND DISCUSSION

Powder Characteristics

Two of the three powder-mixtures used were successfully processed to obtain several metres of continuous lengths of the conductor. Nb spheroids in the high purity Cu matrix remained undeformed in all the extruded compacts of this mixture, whereas the spheroids were deformed to give a fine filamentary structure when processed with the tough pitch Cu. The initial higher value of microhardness of Nb spheroids (VPN = 120 kg mm^{-2}) makes them compatible for processing with the harder TP Cu but not with the much softer, high purity, Cu.

The filametary structure for the composites of spheroids and tough pitch Cu mixture was very uniform for the compacts extruded at 300°C (cold extrusion, Fig (3)) and 500°C(warm extrusion). The Nb spheroids underwent different degrees of deformation at different stages of size reduction in the composites extruded at 900°C (hot extrusion) resulting in a large variation in size of the filaments (Fig 4).

The processing of the hydride-dehydride Nb was possible with the high-purity Cu powder but only when the compacted mixture was extruded at 300°C. At higher extrusion temperatures (500°C and 900°C) the Nb particles could not be deformed because of solution-hardening due to absorbtion of oxygen. The microhardness of Nb increases from 95 to 155 and 280 kg mm^{-2} in compacts extruded at 500°C and 900°C respectively (Table 1). This increase makes the Nb particles so hard that they will not deform during cold drawing of the extruded compacts as is also the case for the hot-extruded compact of Nb spheroids and TPCu powder. The compact extruded at 300°C was successfully drawn into long continuous wires although the filaments (Fig 5) are not as uniform and well defined as those in Nb shperoids + TP Cu mixture (Fig 3)

The striking difference in the filamentary structure of the two successfully processed powder mixtures, Nb spheroids + TP Cu and hydride-dehydride Nb and high purity Cu, is believed to be in part due to the initial shape of the Nb particles. Because the spheroids are perfectly spherical they yield fine, straight and regular filaments distributed uniformly in the Cu matrix, whereas the irregularly shaped flake-like particles of the hydride-dehydride powder produce irregular filaments of varying thickness and non-uniform distribution. However, Zhang et al[9] appear to find no difference between powders and spheroids of niobium. They do not quote purity or microhardness. From the description of their copper constituent it is probably less pure and harder, than the ultra high purity copper used here.

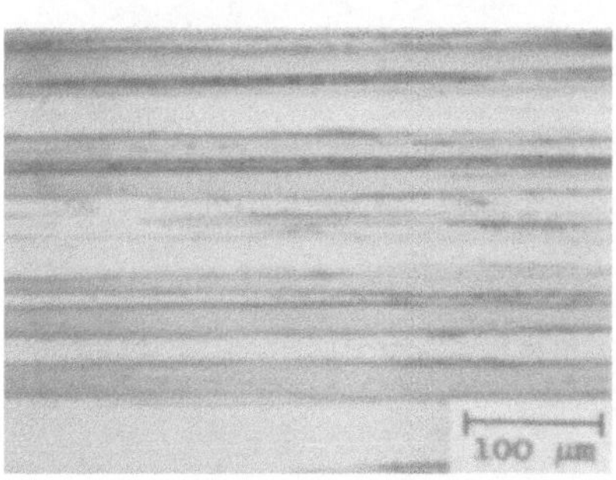

Fig. 3 Nb spheriods and TPCu, cold extruded and drawn to 0.7mm diameter wire

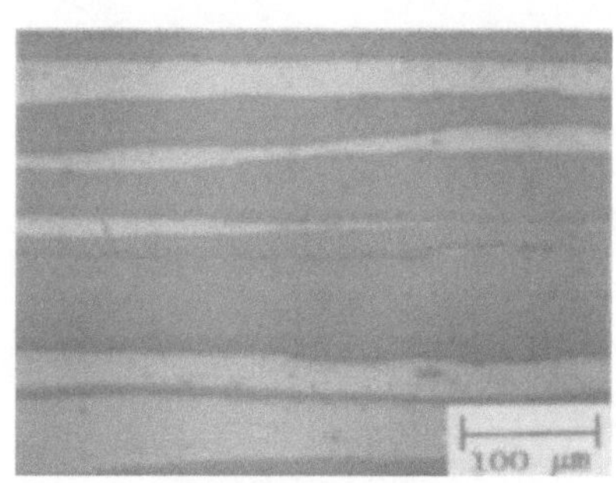

Fig. 4 Nb spheroids and TPCu hot extruded and drawn to 0.7mm diameter wire.

Table 1. Processing conditions and current densities

Powder Mixture	Extrusion Temp	Micro Hardness (VPN kg mm^{-2})	Heat Treatment	J_c at 4.2 K (Am^{-2}) 5T	12T
Nb spheroids	300°C	140	725°C-7days	$1.5x10^9$	$7x10^7$
+ tough pitch	500°C	120	700°C-7days	$2x10^9$	$3x10^8$
Cu	900°C	170	750°C-7days	$4x10^8$	$7x10^7$
Hydride	300°C	140	725°C-7days	$9x10^8$	$4x10^7$
dehydride Nb	500°C	155	*	*	*
+ High Purity Cu	900°C	280	*	*	*

The spherical shape of the spheroids also helps in reducing the effect of oxygen absorption during extrusion at high temperatures. There is a very small increase in microhardness of Nb spheroids after extrusion at 900° C (from 120 to 170 kg mm^{-2}) and none at 500°C (Table 1), whereas for the hydride-dehydride powder the effect is serious enough to make processing impossible. The flat flake-like particles of this powder, which are essentially of 2 dimensional nature, have a much higher surface area to volume ratio as compared to the spheroids and are much more susceptible to contamination due to impurites at high temperature[3]. The processing of these powders requries low temperature of extrusion[4] or additions of a third element to absorb preferentially the oxygen and leave the Nb deformable at high extrusion temperatures[5].

Heat-treatment Optimisation and Critical Current Densities

The optimum temperature and duration of reaction for the formation of Nb_3Sn was found to vary for different wires as shown in Table 1. The temperature needed for complete reaction of the filaments are higher than those generally used for the discontinuous type of material(range 550-650°C) THe reduction ratio of ~ 10^4:1 in these composites produces a final filament diameter of 6 μm. This is greater than the sub-micron sizes reported by other workers. Higher temperatures are probably needed for complete reaction of these larger filaments. The optimum temperature for the composites prepared from a 35 mm billet, thus having the highest

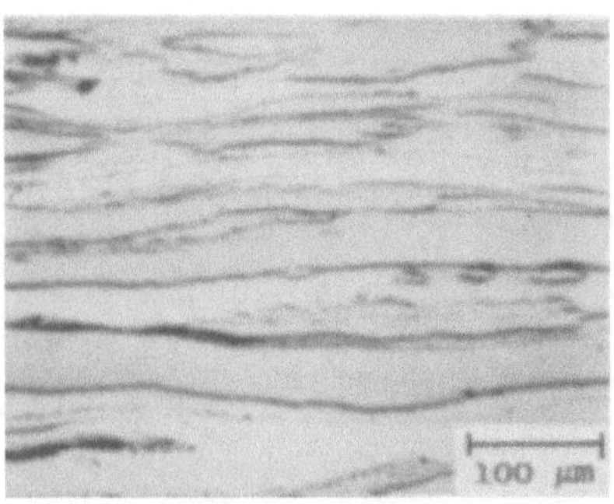

Fig. 5 Hydride-dehydride Nb and high purity Cu, cold-extruded and drawn to 0.7mm diameter wire.

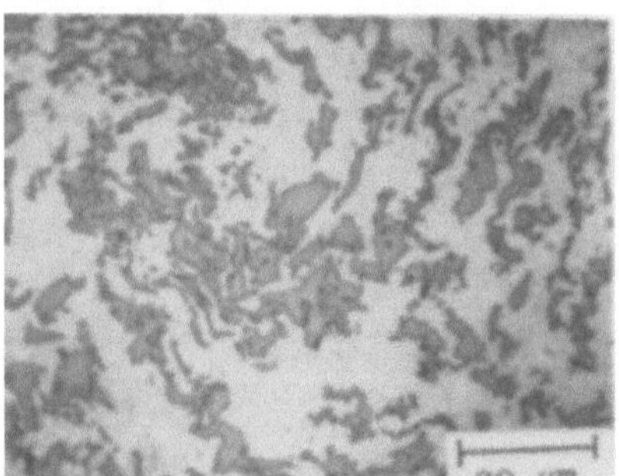

Fig. 6 Completely reacted filaments of Nb_3Sn

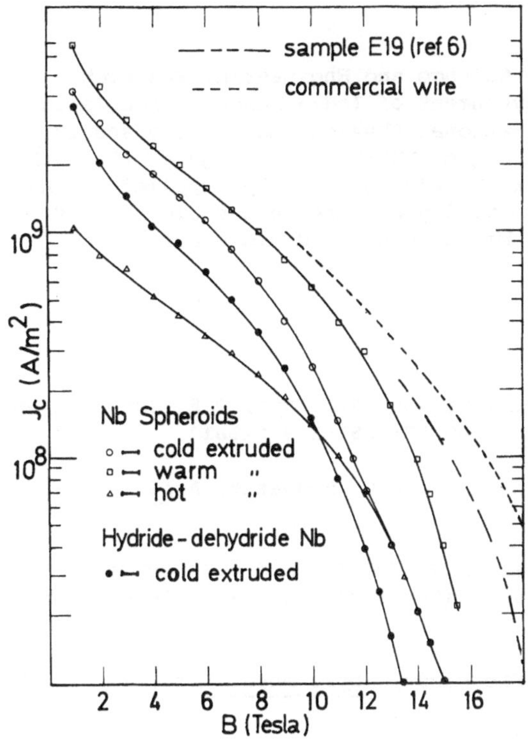

Fig. 7. Current density versus applied magnetic induction. Dotted curves are for a commercial Nb_3Sn wire and one of the best reported powder processed wires.

reduction ratio of 2 x 10^4:1 was lower (700°C - 7 days) than that for the composites of 25 mm billet (725°C-7days) with a reduction ratio of 10^4:1.

Higher reduction ratios can also lead to higher J_c values. Fine filaments require lower temperatures and shorter times for complete reaction and also large reduction in size gives more cold work in the filaments, all these lead to finer grain size in Nb_3Sn resulting in high J_c values. However, close spacing of filaments which accompanies a reduction in filament size, increases the strength of the matrix, raises the prestress in the filaments, and causes a reduction in J_c at high fields[10]

J_c values of the completely reacted (Fig 6) wires produced from different powder mixtures are shown in Fig 7. The values, although, lower than the ones reported for discontinuous material and commercial MF Nb_3Sn, can be improved by increasing the Nb content and reducing the temperature of reaction. The best wire, Nb spheroids plus TP copper, extruded at 500°C, has current densities which approach those of Pourrahimi et al[11] extruded at the same temperature and containing 36% Nb.

CONCLUSIONS

The powder materials, Nb spheroids and tough pitch Cu are easier to process and can carry current densities close to the commercial MF Nb_3Sn.

Prospects for improvement include;

(i) Use of greater reduction ratio to achieve finer filaments and higher J_c.
(ii) increase volume fraction of Nb to 40 wt %
(iii) use of zone-refined Nb as starting material for spheroids.
(iv) replace tough pitch Cu by Cu alloyed with elements such as Mg[12] and Zn[13] which are known to enhance growth rate and thus refine grain-size of Nb_3Sn.

ACKNOWLEDGMENTS

This work was supported by the Science and Engineering Research Council. PLU was supported by a Government of India Research Scholarship. Extrusions were carried out at the National Physical Laboratory and IMI Ltd. Dr. J.A. Lee and Mr. D. Armstrong of UKAEA, Harwell gave help with the powder and the wire drawing. Dr. R. Flukiger kindly supplied the high purity niobium powder. The superconducting measurements could not have been made without the assistance of Mr. H. Jones and his staff at the Clarendon Laboratory.

REFERENCES

1) J.D. Verhoeven, F.A. Schmidt, E.D. Gibson, J.J. Sue, J.E. Ostenson and D.K. Finnemore, IEEE Trans. Magn. MAG 17: 251-254 (1981).

2) R. Flukiger, S. Foner, E.J. McNiff, and B.B. Schwartz, Appl. Phys. Lett. 34: 763-766 (1979).

3) R. Flukiger, R. Akihama, S. Foner, E.J. McNiff, and B.B. Schwartz, in "Advances in Cryogenic Engineering Materials", Vol. 26, Plenum Press, New York (1980), pp 337-342.

4) R. Borman, L. Schultz and H.C. Freyhardt, Appl. Phys. Lett. 32: 79-81 (1978).

5) H.C. Freyhardt, R. Borman and K. Mrovice, in "Filamentary A-15 Superconductors", M. Suenaga and A.F. Clark, eds., Plenum Press, New York, (1980), pp 289-298.

6) J. Otubo, S. Pourrahimi, C.L.H. Thieme, H. Zhang, B.B. Schwartz and S. Foner, IEEE Trans. Magn. MAG 19: 764-768 (1983).

7) J. Otubo, S. Pourrahimi, H. Zhang, C.L.H. Thieme and S. Foner, Appl. Phys. Lett. 42: 469-471 (1983).

8) P.A. Hudson, F.C. Yin, and H. Jones, IEEE Trans. Magn. MAG 19: 903-906 (1983).

9) H. Zhang, S. Pourrahimi, J. Otubo, C.L.H. Thieme, B.B. Schwartz and S. Foner, IEEE Trans. Magn. MAG 19: 769-772 (1983).

10) R. Flukiger, W. Specking and P. Turowski, presented at the "MT-9 Conference", Zurich, August 1985 (to be published).

11) S. Pourrahimi, C.L.H. Thieme, B.B. Schwartz and S. Foner, IEEE Trans. Magn. MAG 21: 764-767 (1985).

12) K. Togano, Y. Asano, and K. Tachikawa, J. Less. Common Metals 68: 15-22 (1979).

13) E. Drost, W. Specking and R. Flukiger, IEEE Trans. Magn. MAG 21: 281-284, (1985).

CHARACTERIZATION STUDIES OF A FULLY REACTED HIGH BRONZE-TO-NIOBIUM RATIO FILAMENTARY Nb_3Sn COMPOSITE

K. R. Marken, S.-J. Kwon, P. J. Lee and D. C. Larbalestier*

Applied Superconductivity Center
University of Wisconsin-Madison
Madison, Wisconsin

ABSTRACT

An attempt has been made to increase the chemical homogeneity of Nb_3Sn layers in a conventional bronze-process composite with excess Sn through extended time reaction heat treatments at high temperature (> 750°C). This study has included measurements of J_c, B_{c2}, local chemical composition, and grain size and morphology. The local bronze/Nb ratio has been found to influence the local chemical composition even after extensive reaction times. The dependence of B_{c2} on local composition variation is uncertain. Critical current densities at high magnetic fields have been improved up to 50% over those obtained for heat treatments at 700°C. These increases are attributed to increased values of B_{c2}, but they are accompanied by a reduction in J_c below 12 T. Drastically reduced J_c is found with overaging, despite a quite high value of B_{c2}. Grain size is found to increase strongly with heat treatment temperature and a large variation is found above 750°C. Equiaxed morphologies have been found at all temperatures between 650°C and 800°C.

INTRODUCTION

The critical current density (J_c) of bronze-process Nb_3Sn composites in low magnetic fields depends directly on grain size and morphology, the peak pinning force being proportional to grain boundary area or the inverse grain size.[1] In high fields the pinning force (and hence J_c) are less sensitive to the microstructure, and the upper critical field (B_{c2}) is important in determining J_c.[2] B_{c2} has been shown to vary with crystallographic order and chemical composition in thin Nb_3Sn films[3] and in bulk samples.[4] However, there is little direct evidence to substantiate a variation of B_{c2} with chemical composition in bronze-process composites.

Because Nb_3Sn layers in these wires are grown by solid-state diffusion, a Sn concentration gradient across the layer is typically present.[5] This varying composition, together with the small thickness of the layers (< 3 μm), makes quantification of the chemical state difficult. Some success in measuring the relative composition across these

*Also Department of Metallurgical and Mineral Engineering.

layers has been achieved, both with the scanning Auger microprobe (SAM)[5-7] and with x-ray analysis on the scanning transmission electron microscope (STEM).[8] However, there are problems in obtaining absolute concentrations from both of these techniques.

An additional problem involved with a quantitative characterization of these composites is that the local bronze/Nb ratio can vary significantly with position in a wire.[9] This variation can affect the local grain size and morphology, as well as the local composition of the layer. This presents problems for the correlation of high resolution analysis of the microstructural and chemical state with measurements of the superconducting properties, which are generally interpreted as bulk averages for the wire.[10] Convincing correlations must depend on an appropriate average of the locally varying structure and composition. A first step toward such a process may be taken by identifying and characterizing the local environment in which the high resolution analysis is made.

In fact, the situation appears to be even more complicated. Evetts has pointed out that common techniques for measurement of T_c, B_{c2}, and J_c do not generally give bulk averages.[11] He suggests that the appropriate program to pursue is the development of higher resolution techniques for the determination of local superconducting property variation within a layer.[11]

This work is a subset of a broader program to characterize the local composition and morphology, and to explore the extent to which such microstructural measurements can be correlated with conventional measurements of J_c and B_{c2}. A series of extended time heat treatments at a range of temperatures has been given to an excess Sn composite in an attempt to react it as fully as possible and to achieve as uniform a composition as possible. We report here some of the results to date.

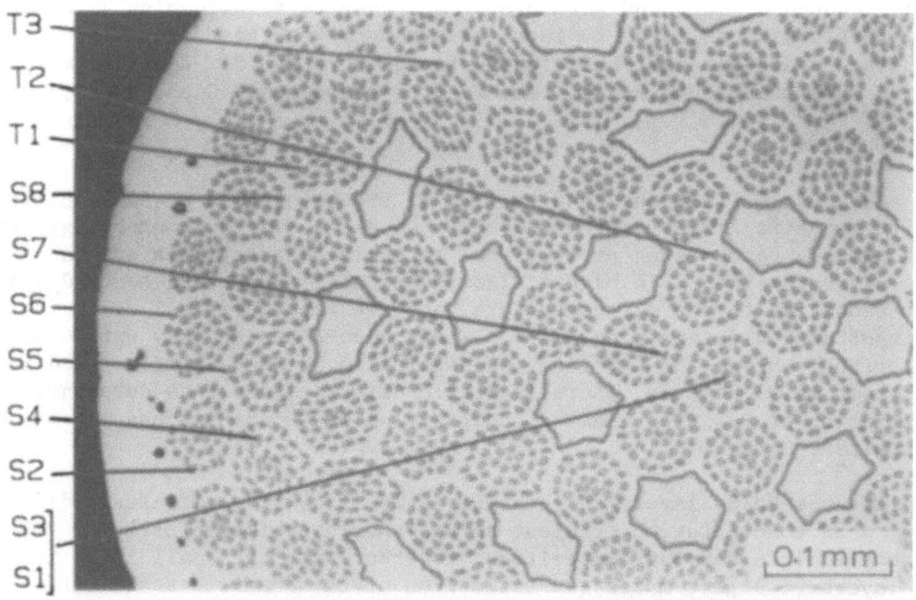

Fig. 1. Segment of the cross section of a wire reacted for 191hr/800C. Filament labels show the location of filaments studied with SAM and TEM (compare Table 2, Fig. 5).

EXPERIMENTAL DETAILS

The Composite and Its Reaction Heat Treatment

The composite discussed here was fabricated at AERE, Harwell, England, and is a conventional bronze composite with a bronze/Nb ratio design of 4.1:1 and a bronze composition of 14 w/o Sn. Full reaction of the Nb to stoichiometric Nb_3Sn from a 13.5 w/o Sn bronze matrix requires a ratio of 3:1. Fig. 1 shows a segment of a cross section of the wire at 1 mm diameter. The stabilizing copper is internally located in 24 larger filaments protected by Nb diffusion barriers. The 5143 filaments have an average diameter of 7 μm and are grouped in bundles of 37. A series of reaction heat treatments have been done in 1 atmosphere of Ar for various times and temperatures from 24 to 1000 hours and 650 to 800°C.

Superconducting Properties

I_c was measured resistively at 4.2 K on 0.6 m samples wound helically on 35 mm diameter stainless steel barrels. The tap length varied from 0.11 to 0.33 m. The criterion for I_c was 10 μV/m.

B_{c2} was measured resistively on 10 mm samples at 4.2 K. Measuring current densities varied between 0.01 and 1.2 A/mm^2 over the wire cross section. This corresponds to estimated Nb_3Sn current densities between 0.06 and 6 A/mm^2. We also calculated values of the critical field, B^*_{c2}, from the $J_c(B)$ data using the Kramer $J^{1/2}B^{1/4}$ extrapolation.

Concentration Profiles

Reacted filaments were chemically analyzed for Nb, Sn, Cu, O, and C using line scans and multiplexed point analyses on a Physical Electronics 595 scanning Auger microprobe. The samples were polished transverse sections of whole wires. Experimental details were similar to those previously reported.[5] The major difference from our past experiments was that these results were obtained while continuously ion-etching at a very slow rate (~ 5 Å/min on a Ta_2O_5 standard). This reduced the effect of residual oxygen and carbon in the system, which can strongly affect the sensitivity factors of Nb and/or Sn. This problem was particular to the machine used, since the sample chamber pressure was higher than normal, ranging between 6 and 10 x 10^{-10} Torr. The results obtained on a thick layer tape standard showed less statistical error and greater reproducibility from run to run using this technique. The disadvantage of data acquisition during sputtering was decreased signal-to-noise ratio. Since higher beam currents were required to obtain good signal-to-noise ratios in a reasonable time, some spatial resolution was lost. The probe size used for these measurements was ~ 0.3 μm.

Microstructure

The grain size and morphology were studied using a JEOL 100B or 200CX transmission electron microscope. Transverse section foils were thinned electrolytically. Details of the sample preparation are similar to those used previously for transverse sections of filamentary NbTi composites.[12] Average grain sizes were determined by the linear intercept method.

RESULTS

Attempts to fully react the filaments in this composite have been only partially successful. Fig. 1 shows that the outermost filaments in the composite are reacted to a greater extent than the inner filaments.

Nb_3Sn area measurements are being made but are unavailable at this time. We can comment that complete reaction in the outermost filaments of the wire has been achieved at all reaction temperatures between 650 and 800°C. Nearly complete reaction in the filaments at the wire center required 636 hours at 800°C.

Results for the critical current density (bronze plus Nb area) are shown in Fig. 2. These results are for those reaction times that produced essentially complete reaction at the outside of the wire at 650, 700, 750 and 800°C. Additionally we show results for a 636 hr/800°C sample. Above 16 T J_c increases with higher reaction temperature. At 16 T the J_c for 191 hr/800°C is 118 A/mm^2, 7% more than with 672 hr/700°C. At 19 T the J_c is 22 A/mm^2, for an increase of 83%. At fields below 10 - 12 T the trend is reversed, with lower reaction temperature resulting in higher J_c. An obvious exception is the 636 hr/800°C sample, which had quite low J_c at all fields.

In Fig. 3 the Kramer extrapolations to B^*_{c2} are plotted. Some of the lower field data is not linear; our extrapolations emphasize the linear high field region, as the model predicts linearity only at fields well above the field at which the pinning force peaks (0.2 B_{c2}).[2]

Fig. 4 shows the resistive B_{c2} as a function of measuring current density in the whole wire. The inset shows the definition of B_{c2} used for these results. B_{c2} increased with decreased measuring current. In most cases the slope of the B_{c2}-J curve decreased with decreasing current as well, the 650°C sample being an exception. Table 1 lists the values obtained at the lowest measuring currents and compares these with the extrapolated B^*_{c2} values. The two sets of values are consistent in rank, with the resistive values varying from 2.1 to 2.8 Tesla higher than the extrapolations. In both cases B_{c2} increases with increasing reaction temperature. The highest value is for the 636 hr/800°C sample, which reaches 24.9 T. The transition widths in Table 1 are from B_{c2} to the

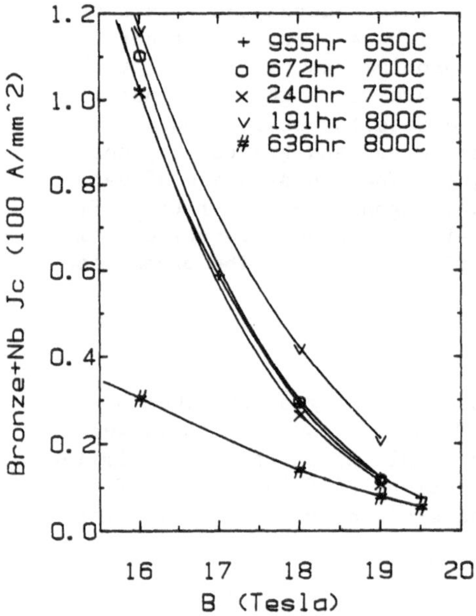

Fig. 2. Critical current density in the bronze + Nb area. Criterion for I_c is 10 μV/m.

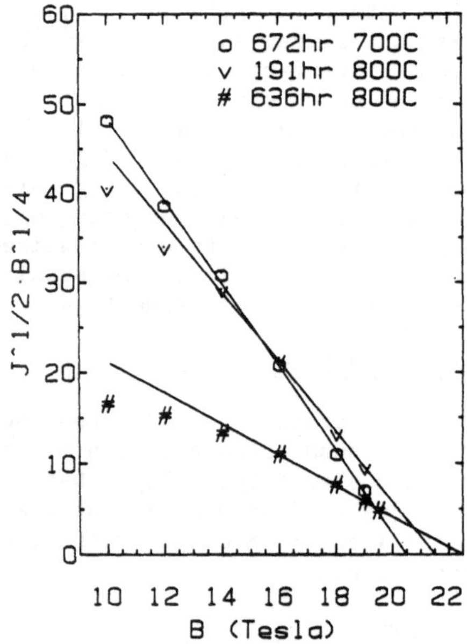

Fig. 3. Kramer plot showing extrapolated B^*_{c2} for three of the samples studied.

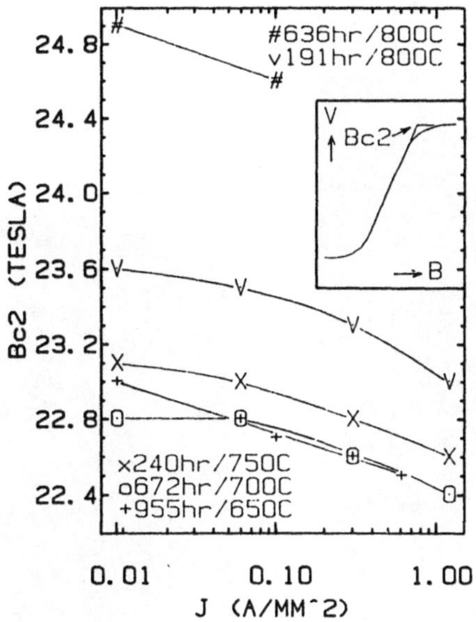

Fig. 4. Upper critical field measured resistively as a function of measuring current density. The inset shows the definition of B_{c2} used.

Table 1. Comparison of Critical Field Values Measured Resistively with Extrapolated Values

Sample	B_{c2} (T)	$\Delta B_{c2}/2$ (T)	B^*_{c2} (T)
955hr/650C	23.0	0.7	20.2
672hr/700C	22.8	0.4	20.5
240hr/750C	23.1	0.5	20.5
191hr/800C	23.6	0.4	21.5
636hr/800C	24.9	0.5	22.3

Table 2. Maximum Sn Contents and Estimated Gradients Found in Fully (Right) and Partially Reacted (Left) Filaments.*

Sample	955hr 650°C	672hr 700°C	240hr 750°C	191hr 800°C	955hr 650°C	672hr 700°C	240hr 750°C	191hr 800°C
Fil.#	S1	S3	S5	S7	S2	S4	S6	S8
Max Sn at%	23.0	23.4	26.0	21.0	23.0	25.3	26.0	25.9
Gradient at%/µm	5	1	3	2	4	0	0	0.3

*Locations of these filaments in the wire are shown in Fig. 1, referenced by filament number.

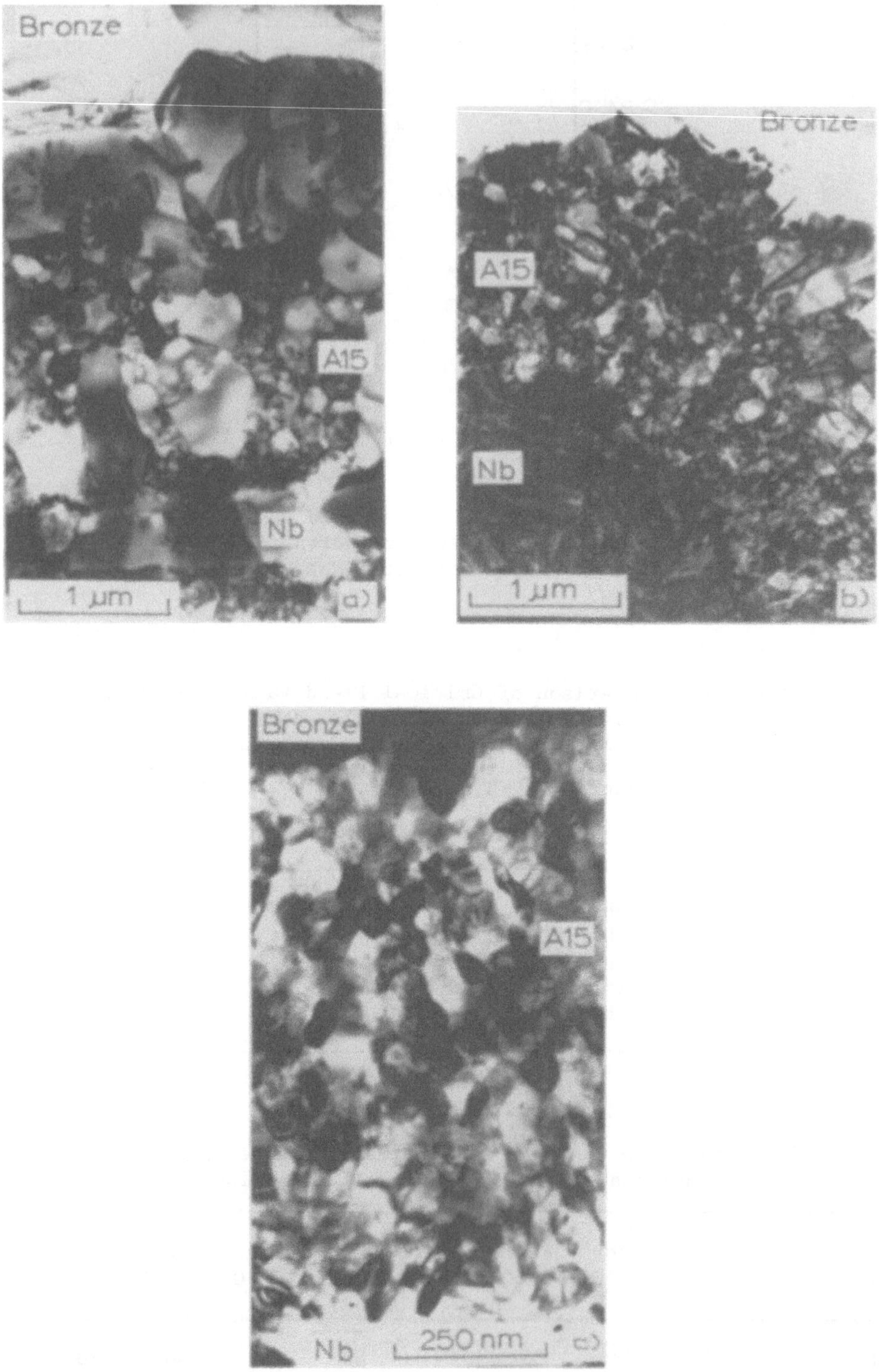

Fig. 5. Transverse section TEM micrographs showing the complete reaction layer at three temperatures. Locations of these filaments in the wire are shown in Fig. 1, referenced by Fil. # a) Fil. T1, 120hr/800°C; b) Fil. T2, 240hr/750°C; c) Fil. T3, 955hr/650°C.

midpoint, as defined by the half voltage point. These widths are all close to 0.5 T.

Preliminary composition measurements were made on one filament at the wire edge and one at the wire center for each sample. Locations of measured filaments are shown in Fig. 1. Results are summarized as a function of temperature in Table 2. The maximum Sn concentrations measured are shown; in all cases these were measured nearest the bronze interface. The tabulated gradients are estimates only, since in most cases point analyses were not done at the minimum Sn concentration points. The filaments from the wire center had unreacted Nb cores, strong composition gradients, and low maximum Sn concentrations, regardless of reaction temperature. In contrast, the filaments at the wire edge were completely reacted (except for the 650°C sample) and the line scans showed relatively constant Sn concentration across the layers, with maximum Sn contents between 25 and 26 a/o.

Typical grain morphologies found in this composite are shown in Fig. 5 in TEM micrographs from samples reacted for 955 hr/650°C, 240 hr/750°C, and 120 hr/800°C. These are transverse sections which show the entire reaction layer. The location of these filaments in the composites is given in Fig. 1. An equiaxed morphology was found in this composite at all reaction temperatures from 650°C to 800°C. A few aspected grains were observed, occurring at random locations within the layers. Average grain size measurements are presented in Table 3. The size increased strongly with increasing temperature; the spread in size increased with temperature also.

DISCUSSION

The radial dependence of the extent of reaction is explained by the radial bronze/Nb ratio variation of the wire. There is a significant Sn reservoir in the extra bronze at the outside of the wire, and the fully reacted filaments are all found within a few rows of the outside. In all cases the central filaments are the least fully reacted. There is also a variation in bronze/Nb ratio from outside to inside each bundle. However, the extent of reaction does not appear to vary significantly within a bundle at these long reaction times.

The capability of the SAM to probe whole wire cross sections is a real advantage for the study of these radial variations. The exact location in the composite of the filament under study is apparent, and the layer can be analyzed along the direction of growth. The initial composition results, while not extensive enough to be statistically representative of the whole sample, do suggest the range of variation within a

Table 3. Average Grain Size Results

Sample	Size (nm)
955hr/650°C	60
672hr/700°C	115
240hr/750°C	102
120hr/800°C	187
636hr/800°C	542

composite for a given heat treatment. The results suggest that this range is about the same for temperatures between 700°C and 800°C at these extended times. All of these samples had high, uniform Sn concentrations across outer filaments, and lower concentrations with a strong gradient in the inner filaments. This means that the bronze/Nb ratio variation across the wire strongly affects the local composition, even after extended time and high temperature heat treatments. It is not clear why the measured Sn concentration reaches 26% Sn. This may be a standardization problem; however, Smathers has also measured high Sn concentrations at the bronze-Nb_3Sn interface.[13]

There is a consistent difference between the measured and extrapolated B_{c2} values, the measured values being about 3 T higher. A difference of 2 T was found by Suenaga and Welch in a bronze process monofilamentary wire.[16] They suggested this difference was due in part to composition gradients across the Nb_3Sn layer or to variations in strain. In our samples there is only an approximate correlation between the B_{c2} values and the composition measurements. B_{c2} is almost constant from 650 to 700°C (22.8 - 23.1 T), rising to 23.6 and 24.9 T at 800°C. The Sn compositon is low at 650°C (23 at%), the maximum value being fairly constant between 25.3 - 26.0 at% Sn above 700°C. It seems unlikely then that radial composition variation alone can account for the 3 T difference. It is possible that this difference is due in part to the distribution of critical current values in the wires, caused by structural or chemical variations along the length of the filaments. A sensitive measurement of I_c probes the low end of this distribution. On the other hand, a low current density resistive B_{c2} measurement samples the best material present.

The grain morphology is notable for its equiaxed nature over a wide temperature range (650 - 800°C). Transverse sections allowed complete layers to be examined along the growth direction. The results of this study are quite different from recent reports of two and three shell layer morphologies.[8,14] The difference may be due to the high bronze/Nb ratio of this composite. However, we have also found uniform columnar layers in a composite with a low bronze/Nb ratio.[15] We have found no dependence of morphology on position in this composite. The grain size measurements are complicated by the wide variation in size throughout the layer, especially at temperatures above 750°C. However, the average sizes measured are comparable to averages measured by West and Rawlings on a similar composite.[1]

The J_c results suggest that B_{c2} is a major determinant of high field current density for most typical reaction conditions. The increased high field J_c of the 191 hr/800°C sample is, however, at the expense of J_c below 15 T, indicating that heat treatment for optimum J_c is different for different field ranges. The low J_c in the 636 hr/800°C sample indicates overaging is possible in this composite. The sample shows no signs of damage, having sharp transitions and little I_c variation along its length. Since B_{c2} is high, the implication is that the large grain size has depressed J_c even at very high fields.

CONCLUSIONS

1. Transverse wire sections are important for the study of composites with radial structure and property variations.
2. Even after extended time, high temperature heat treatments the bronze/Nb ratio variation across the wire has influenced the local composition of this excess Sn composite.
3. The grain morphology of this composite is equiaxed across the reaction layer, independent of location in the wire and reaction temperature.

4. Overaging is possible in this high bronze/Nb ratio composite. This appears to occur for grain sizes between 0.19 and 0.54 μm.

ACKNOWLEDGMENTS

We are grateful to E. Labourdeth for experimental help, J. A. Lee (AERE Harwell) for the composite and D. B. Smathers for many discussions on Auger and other techniques. The SAM work was performed at the NSF Regional Instrumentation Facility, University of Minnesota and the high field work at the National Magnet Laboratory, MIT. The work has been supported by DOE - Office of Fusion Energy.

REFERENCES

1. A. W. West and R. D. Rawlings, J. Mat. Sci. 12:1862-1868 (1977).
2. E. J. Kramer, J. Appl. Phys. 44:1360-1370 (1973).
3. T. P. Orlando, J. A. Alexander, S. J. Bending, E. J. McNiff, Jr., and S. Foner, IEEE Trans. Mag. Mag-17:368-369 (1981).
4. H. Devantay, J. L. Jorda, M. Decroux, J. Muller, R. Flukiger, J. Mat. Sci. 16:2145-2153 (1981).
5. D. B. Smathers, K. R. Marken, D. C. Larbalestier, and J. Evans, IEEE Trans. Mag. Mag-19:1421 (1983).
6. M. Suenaga and W. Jansen, Appl. Phys. Lett. 43:791-793 (1983).
7. J. E. Drost, R. Flukiger, and W. Specking, Cryogenics 24:622-628 (1984).
8. I. W. Wu, D. R. Dietrich, J. T. Holthuis, M. Hong, W. V. Hassenzahl, and J. W. Morris, Jr., J. Appl. Phys. 54:7139-7152 (1983).
9. D. B. Smathers, K. R. Marken, P. J. Lee, D. C. Larbalestier, W. K. McDonald, and P. M. O'Leary, IEEE Trans. Mag. Mag-21:1133-1136 (1985).
10. D. C. Larbalestier, IEEE Trans. Mag. Mag-21:274-284 (1985).
11. J. E. Evetts, IEEE Trans. Mag. Mag-19:1109 (1983).
12. A. W. West and D. C. Larbalestier, Met. Trans. A 15A:843-852 (1984).
13. D. B. Smathers, Ph.D. thesis, U. Wisconsin-Madison (1982).
14. W. Schelb, J. Mat. Sci. 16:2575-2582 (1981).
15. K. R. Marken, A. W. West, P. J. Lee, D. B. Smathers, D. C. Larbalestier, to be published.
16. M. Suenaga and D. O. Welch, in "Filamentary A15 Superconductors", ed. M. Suenaga and A. F. Clark, p. 131 (1980).

EFFECTS OF HOT ISOSTATIC PRESSING ON THE SUPERCONDUCTING PROPERTIES OF (Nb, Ti)$_3$Sn MULTIFILAMENTARY WIRES

Y. Monju,* T. Fukutsuka* and T. Horiuchi**

* New Materials Division
** Asada Research Laboratory
Kobe Steel Ltd., Kobe, Japan

ABSTRACT

The effects of Hot Isostatic Pressing (HIP) on the superconducting properties of (Nb, Ti)$_3$Sn multifilamentary wires were studied. The (Nb, Ti)$_3$Sn wires, having a bronze (Cu-13wt%Sn-0.3wt%Ti) to niobium ratio of 2.8, were manufactured by the conventional bronze method using the hot hydrostatic extrusion. After the diffusion heat treatment, the (Nb, Ti)$_3$Sn wires were HIP treated in a high pressure argon gas atmosphere of 130 MPa in a temperature range from 600 to 750°C. The critical current measured in the transverse magnetic field of up to 23 tesla indicates that the optimum temperature of HIP treatment on (Nb, Ti)$_3$Sn was less than 600°C.

INTRODUCTION

Effects of Hot Isostatic Pressing (HIP) on the stress sensitivity of Nb_3Sn multifilamentary wires were reported[1-3] in ICMC. It was found that Nb_3Sn wires HIP treated at 825°C for 2 hours at 130 MPa were able to be strained up to 1.5% without any degradation in critical current. Also, it had already been shown that the HIP treatment could be useful in extending the handling limits and bending strain limits of conventionally processed Nb_3Sn superconductors. This is because a very large enhancement of the irreversible strain limit was measured on the HIP treated Nb_3Sn conductors.

In recent years, many studies have been reported that the titanium addition to Cu-13wt%Sn matrix improves the critical current density at high field. It is, thus, interesting to apply the HIP treatment to (Nb, Ti)$_3$Sn wires. This study was carried out to optimize the HIP treatment conditions and to improve the strain sensitivity on multifilamentary (Nb, Ti)$_3$Sn wires.

EXPERIMENTAL PROCEDURES

Two kinds of (Nb, Ti)$_3$Sn multifilamentary wires were manufactured by the bronze method. One was internal copper stabilized wire (specimen A) and the other was external copper stabilized wire (specimen B). The details and cross-sections of wires are shown in Table 1 and Fig. 1. In Fig. 1(a), composite billets composed of 6174 (=126x7x7) filaments were bundled in Cu jackets with a stabilizing copper core and a Ta barrier. However, in Fig. 1(b), composite billets composed of 10285 (=187x55) filaments were bundled with an external OFC Cu jacket with a Ta barrier.

These billets have a bronze (Cu-13wt%Sn-0.3wt%Ti) to niobium ratio of 2.9 and 2.8, and a unit weight of 5 kg and 50 kg respectively. The billets were sealed by an electron beam and, at a temperature of 670°C extruded by hot hydrostatic 400 and 1,650 ton presses to rods of outside diameters of 19 mm and 40 mm respectively.

These two kind of rods were cold drawn to the outside diameter of 0.86 mm and 0.70 mm with intermediate annealings after every several passes. The cold drawn wires were diffusion heat treated in a vacuum at 600 ~ 690°C for 100 ~ 250 hours and some of these wires were HIP treated at 600 ~ 750°C for two hours under 130 MPa argon gas atmosphere. When these wires were HIP treated, hydrostatic pressure was applied after the wires reached predetermined temperatures.

Table 1. Specifications of Composite Wires

Specimen	Number of Filaments	Diameter of Wire	Filament Diameter	Bronze/Nb
A	6174 (=126x7x7)	0.70 mmϕ	3.4 μmϕ	2.9
B	10285 (=187x55)	0.86 mmϕ	3.5 μmϕ	2.8

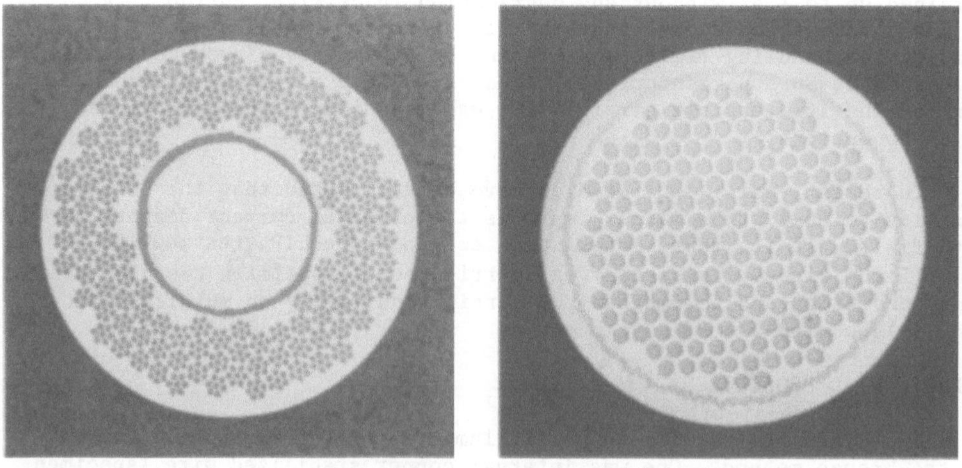

(a) (b)

Fig. 1 Entire cross-sections of specimen A:(a) and specimen B:(b)

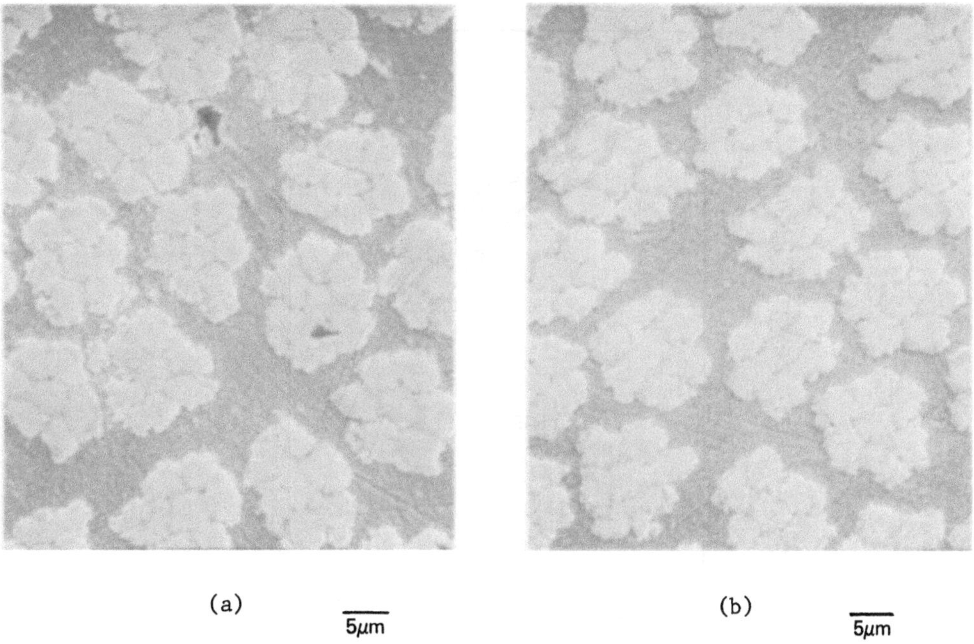

Fig. 2 Cross-sections of $(Nb, Ti)_3Sn$ filaments for specimen A without HIP:(a) and with HIP:(b) treatment at 600°C

Critical currents in the transverse magnetic field of up to 23T were measured by using the hybrid magnet at Tohoku University. The critical current was determined by the criterion : $V=0.5x10^{-6}V/cm$. The grain structures were observed by a scanning electron microscope.

RESULTS

Fig. 2 shows the cross-sections of $(Nb, Ti)_3Sn$ filaments for specimen A with and without HIP treatment at 600°C. Many voids are observed in the $(Nb, Ti)_3Sn$ layers in the specimen without HIP treatment. These voids were eliminated in the specimen with HIP treatment. These results suggest that the HIP treatment at 600°C is suitable on the $(Nb, Ti)_3Sn$ wires.

Critical current densities of specimen A vs. the transverse magnetic field of up to 23 tesla are shown in Fig. 3. Diffusion heat treatment time at 690°C is 200 hours. The critical current densities of specimen A treated at 600°C by HIP did not change, but a large decrease was measured at 660°C. Fig. 4 shows the critical current ratio of specimen A with HIP treatment to the same without HIP treatment. The critical current ratio of the HIP treated specimen decreased gradually as the magnetic field increased. This result with $(Nb, Ti)_3Sn$ wires is completely contrary to the results with Nb_3Sn wires in that, in the latter the critical current ratio of the sample with HIP to that without HIP increased gradually as the magnetic field increased.[4]

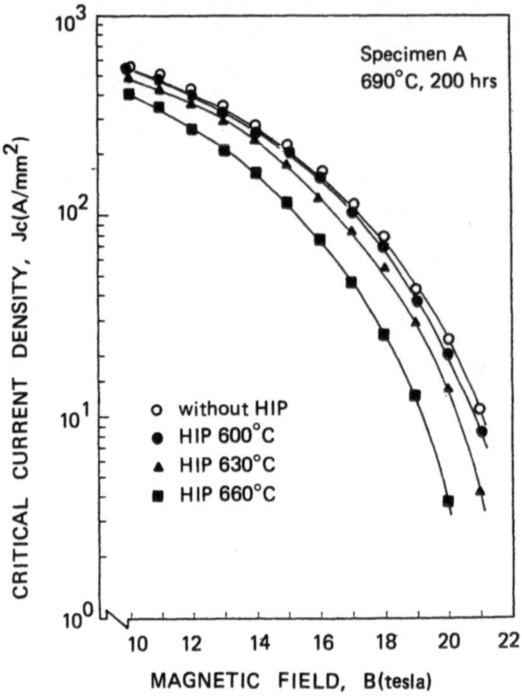

Fig. 3 Critical current density vs. magnetic field for specimen A diffusion heat treated at 690°C for 200 hours.

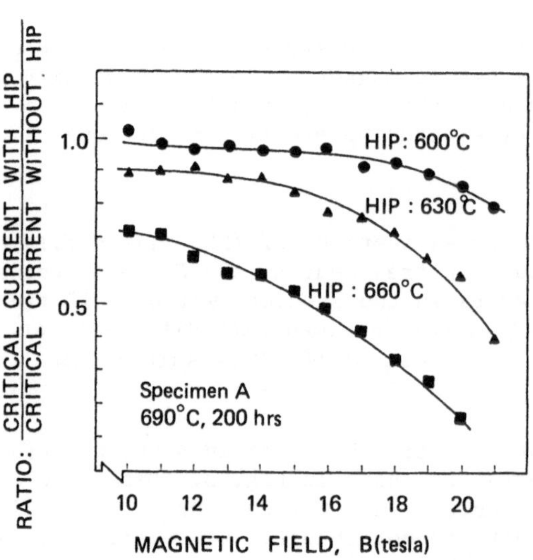

Fig. 4 The critical current ratio of the specimens with HIP treatment to the same without HIP treatment vs. magnetic field. Diffusion heat treatment for specimen A is 690°C for 200 hours.

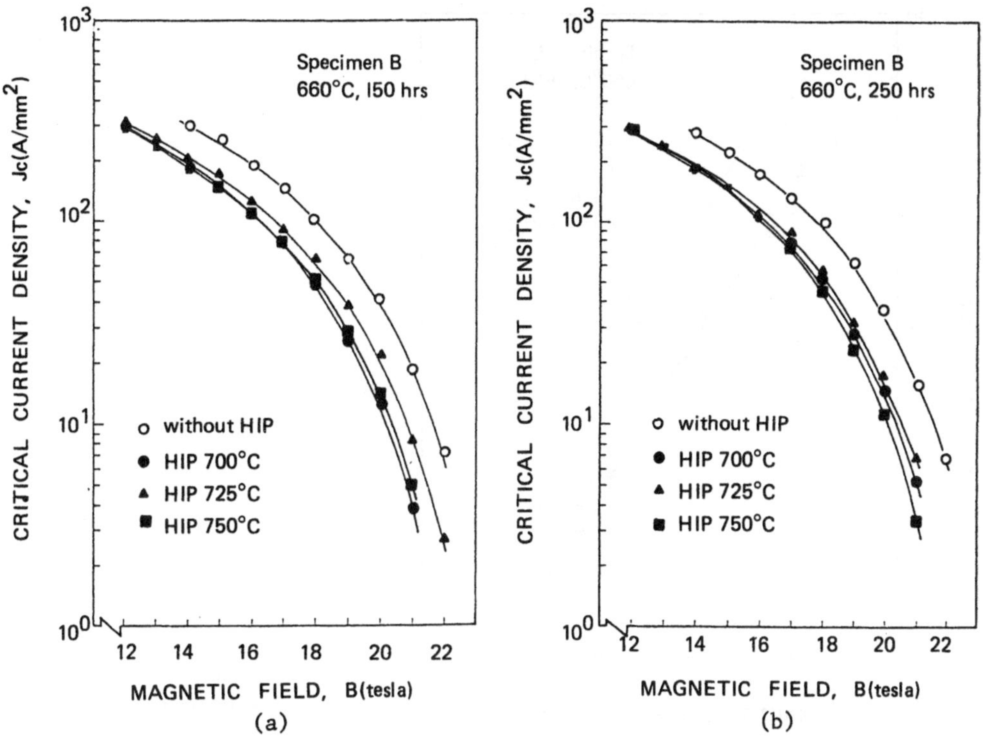

Fig. 5 Critical current density vs. magnetic field for specimen B diffusion heat treated at 660°C for 150 hours:(a), and for 250 hours:(b).

Fig. 5 shows the critical current density vs. the transverse magnetic field up to 23 tesla for specimen B. Diffusion heat treatment time at 660°C is 150 hours for Fig. 5(a), and 250 hours for Fig. 5(b). The critical current density of specimen B heat treated at 660°C increases as the heat treatment time increases. When these specimens were HIP treated at 700 ~ 750°C for two hours, the critical current densities of both samples decreased 30 ~ 50% over the entire magnetic field. Between the HIP treatment temperatures of 700 and 750°C after diffusion heat treatment times of 150 hours and 250 hours at 660°C, differences of critical current density were not observed. This tendency is the same for specimen B heat treated at 690°C.

Fig. 6 shows Kramer's plot of critical current for specimen B. It can be seen that these samples obey Kramer's law very well. Hc_2 values of specimen B with HIP estimated by extrapolation were 0.5 ~ 1.2 tesla lower than those of specimen B without HIP. These results show that the Hc_2 value of $(Nb, Ti)_3Sn$ wires would be decreased by HIP treatment at 700 ~ 750°C. These results concerning void elimination, critical current and critical magnetic field properties show that the optimum temperature of HIP treatment on $(Nb, Ti)_3Sn$ is less than 600°C.

Fig. 7 shows SEM photographs of specimen B. It was found that specimen B without HIP (a) showing the highest critical current density has a $(Nb, Ti)_3Sn$ grain size of about 0.2 µm and a clear grain boundary. Specimen B with HIP (b) has a same grain size of about 0.2 µm, but the grain boundary of specimen B is not clear. This suggests that the cause of degradation in critical current density is not the difference of the grain size but the difference of the grain boundary.

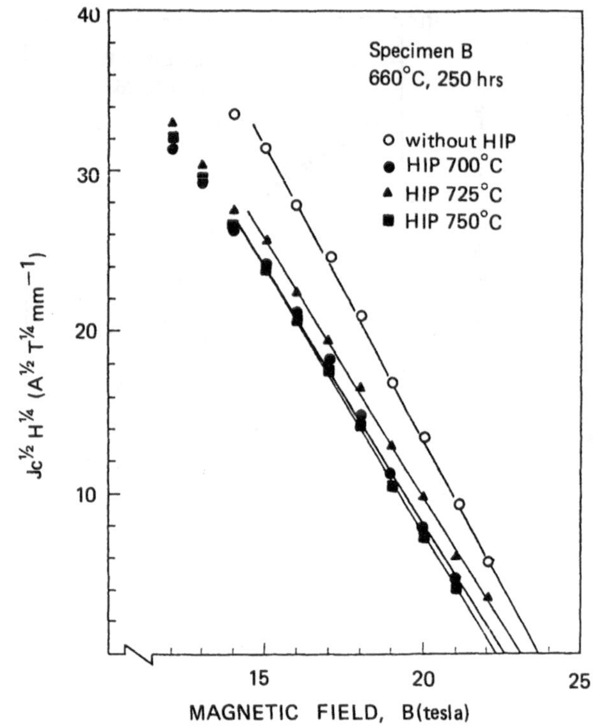

Fig. 6 Kramer's plot for specimen B

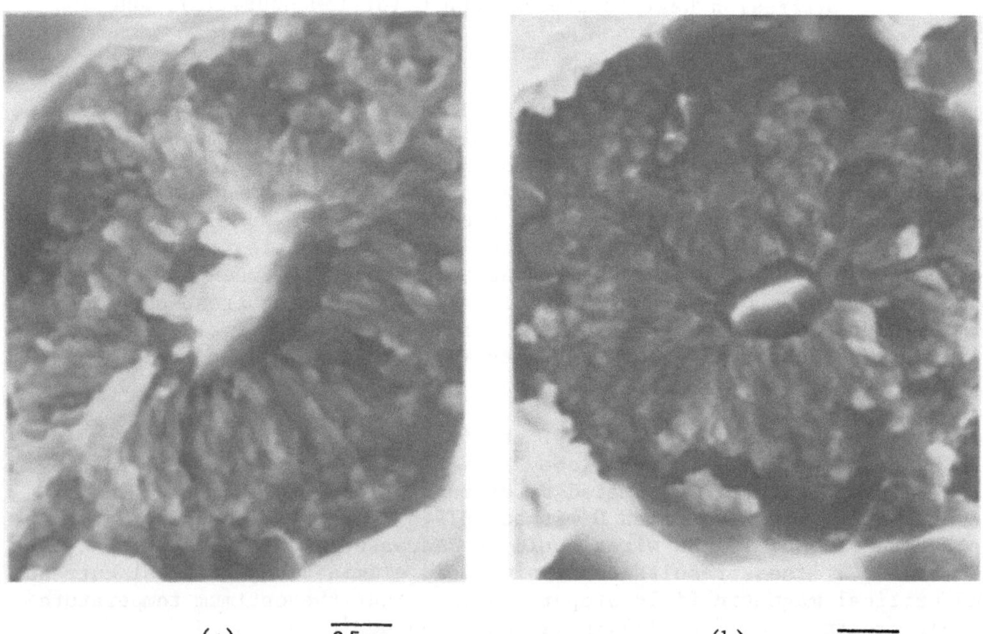

Fig. 7 SEM photograph of specimen B without HIP:(a), with HIP:(b).

Fig. 8 shows the mechanical properties of specimen A with and without HIP treatment measured at room temperature. The application of HIP treatment causes the increase in tensile strength and a reduction in elongation. This behavior is the same as that of Nb_3Sn and can be explained as a result of the improvement in the adherence of the $(Nb, Ti)_3Sn$ filaments to the bronze matrix due to HIP treatment.

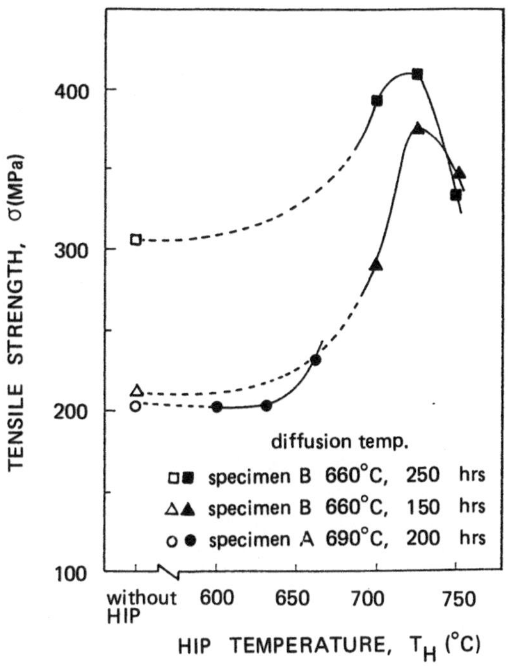

Fig. 8

Tensile strength at room temperature of the specimen A and B vs. HIP treatment temperature.

CONCLUSIONS

Effects of HIP on the (Nb, Ti)$_3$Sn wires were studied. When the HIP treatment temperatures increased from 600 to 750°C, the critical current density and critical magnetic field gradually decreased. The grain size was almost unchanged. However, Kirkendahl voids were varnished out and the mechanical properties improved. It is clear that the optimum temperature of HIP treatment for (Nb, Ti)$_3$Sn is lower than 600°C, and for (Nb, Ti)$_3$Sn wires which are HIP treated under optimum conditions, the critical current under strain should be measured.

ACKNOWLEDGEMENT

It is a pleasure to acknowledge the hospitality of the members of the High Field Laboratory for Superconducting Materials, Tohoku University.

REFERENCES

1. T. Fukutsuka, Y. Monju, I. Tatara, Y. Maeda, M. Moritoki and T. Fujikawa, "Effects of hot isostatic pressing on the superconducting properties of Nb_3Sn multifilamentary wires", in: "Proceedings of the International Cryogenic Materials Conference, KOBE, JAPAN, 11-14 May 1982" Butterworth, Great Britain (1982), pp21-24.

2. T. Fukutsuka, T. Horiuchi, Y. Monju, I. Tatara, Y. Maeda and M. Moritoki, "Effects of hot isostatic pressing on the superconducting properties of Nb_3Sn multifilamentary wires", in: "Advances in Cryogenic Engineering — Materials", vol. 30, Plenum Press, New York (1984), pp891-898

3. J.W. Ekin, "Electro-Mechanical Properties of High Field Superconductors", in: "Proceedings of Japan-US workshop on high-field superconducting Materials for fashion, December 10-12, 1984" JAPAN (1985), pp153-155.

4. Ibid.

THE DIFFUSION PROCESS IN Nb_3Sn-Cu SUPERCONDUCTING WIRE MADE BY THE EXTERNAL TIN METHOD

J. D. Verhoeven, C. C. Cheng and E. D. Gibson

Ames Laboratory and M.S.E. Dept.
Iowa State University
Ames, Iowa

ABSTRACT

The external Sn process for producing Nb_3Sn-Cu superconducting wire generally employs a solid state diffusion process for the Sn diffusion step. The latter process is reviewed and shown to have limitations at larger wire sizes due to both long time requirements and a blister formation problem. New experimental data are presented on wire geometries and compared to previous results on sheet geometries. It is concluded that the blister formation problem results from nucleation of rows of voids at the ε/α interface during stage I of the process. The void and blister formation problem is found to be more severe with a wire versus a sheet geometry, which is consistent with the idea that phase transformation induced tensile stresses play a key role in nucleation of the void rows. Methods for minimizing the blister problem, and implications for the Sn core process are discussed.

INTRODUCTION

In the past two decades or so several schemes have evolved for preparing Nb_3Sn-Cu superconducting wire which involves production of a composite Nb-Cu wire followed by diffusion of Sn from a pure Sn source, located at either the surface or core of the wire, to produce Nb_3Sn. In this paper we are concerned with the external Sn process, but the results have some implications for the internal process. A problem with the utilization of a pure Sn source, as opposed to a bronze Sn source, is that the Sn melts at 232°C and temperatures above this value are required for the diffusing step. After melting, the Sn metal may flow under the action of either capillary forces and/or gravitational forces [1,2] and this has deleterious effects on wire quality. The solid state diffusion process proposed by Benz et al. [3] overcomes this difficulty and is now generally utilized with external Sn processing. There are some serious limitations with this process which will be demonstrated here.

The process may be analyzed with the aid of the Cu-Sn phase diagram and Fig. 1. In stage I the sample is held at 200 to 220°C for a time sufficient to convert a Sn layer of thickness W_{Sn} into a bilayer of $\eta + \varepsilon$ having thickness W_I. The sample may now be heated to 415°C without melting the η layer, so stage II is carried out at 300 to 400°C. At 300 to 350°C one expects the $\eta + \varepsilon$ bilayer to convert to an ε layer of

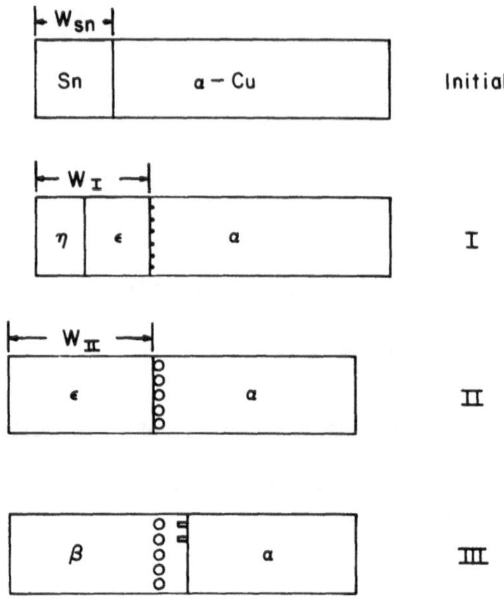

Fig. 1. The 3 stages of the solid state diffusion process.

thickness W_{II}, while at 350 to 400°C it should convert to a bilayer of $\delta + \varepsilon$. Since the $\delta + \varepsilon$ compositions melt at temperatures above 700°C the stage III step is carried out at 650 to 700°C to achieve rapid diffusion rates. During stage III the $\delta + \varepsilon$ layer is quickly (generally in a matter of minutes) converted to a β phase layer which achieves a maximum thickness, shown in III on Fig. 1, and then decreases in thickness as Sn continues to diffuse into the Cu phase. Hence, a series of intermetallic layers are formed on the Cu phase which achieve a maximum thickness in stage III and then decrease to zero thickness as time proceeds.

A complication has been found to occur in the process which is demonstrated in Fig. 2. Under certain processing conditions, such as those given in Fig. 2, blisters form at the Cu interface, causing the Sn-rich intermetallic layers to completely break away from the Cu substrate. Hence, the practical utilization of the solid state process requires answers to two important questions: For a given Sn layer thickness,

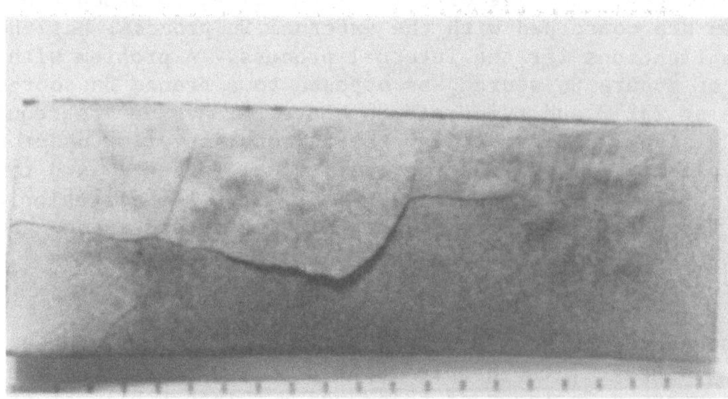

Fig. 2. Severe blister formation and nearly complete delamination. A 9 μm Sn layer treated by stage I, 200°C/16d and stage II, 400°C/28d. Scale: 1 mm/division.

(1) how long must the sample be held at stages I and II to assure absence of melting in the following stage, and (2) how can one avoid the blistering problem.

We have recently reported[4,5] a series of experiments on sheet samples in which both of these problems were studied. These results will first be reviewed and then some additional results on wire samples will be presented which indicate that the blistering is more severe for the wire geometry.

SHEET GEOMETRY

Analysis[4] of the kinetics of the layer growth on sheet geometries during the 3 stages has shown that the time to complete stage I is by far the longest in the process. The analysis indicates that a Sn layer of thickness W_{Sn} should be gone after a stage I time at 200°C given as,

$$t_I = 0.14\ W_{Sn}^{\ 2}$$

where W_{Sn} is in μm and t_I is in days. Experiments give rough agreement with this equation which predicts t_I values of 14 to 126d for Sn layer thicknesses ranging from 10 to 30 μm. At near maximum stage I temperatures, 220°C, the equation becomes $t_I = 0.075\ W_{Sn}^{\ 2}$ and stage I times of 7.5 to 67 days are required for W_{Sn} ranging from 10 to 30 μm. It is clear that stage I times become prohibitively long for Sn layer thicknesses much above 10 μm or so. These results demonstrate that there are limits on the wire size for which the solid state process is applicable. For example, if one desires a final bronze/Nb_3Sn ratio of 3.0 a maximum value of W_{Sn} = 10 μm limits the Cu-Nb wire size to 0.21 mm while W_{Sn} = 20 μm limits it to 0.42 mm. These numbers assume complete conversion of Nb to Nb_3Sn and a final bronze composition of 3 wt.% Sn.

The blister problem in the solid state process has been traced[5] to the formation of voids during the diffusion induced mass transport. As illustrated in Fig. 3 for a 6h 340°C treatment, in which stage II was not yet complete because some η phase remained, rows of voids appeared upon the α-Cu side of ε/α interface. During stage II the voids grew in size and were present both when δ did not form, 340 and 370°C, and when it did form, 400°C. Throughout stage II and whether or not δ phase formed the

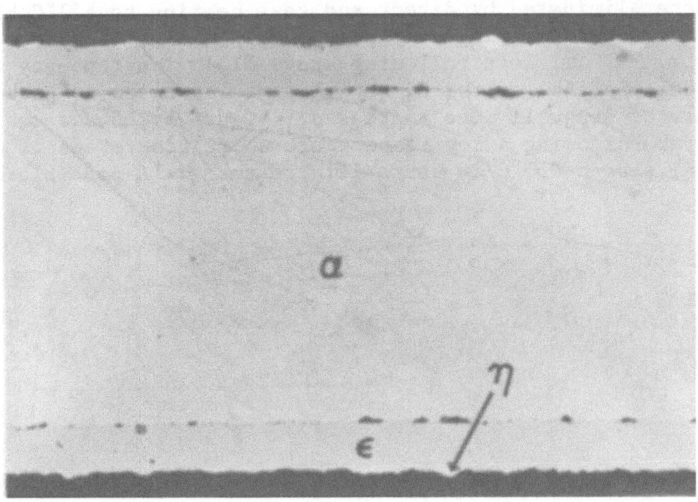

Fig. 3. The ε + η layer remaining after stage I, 200°C/20d, plus stage II, 340°C/6h, on a sheet sample of 10 μm original Sn thickness. Unetched, orig. mag. = 400x.

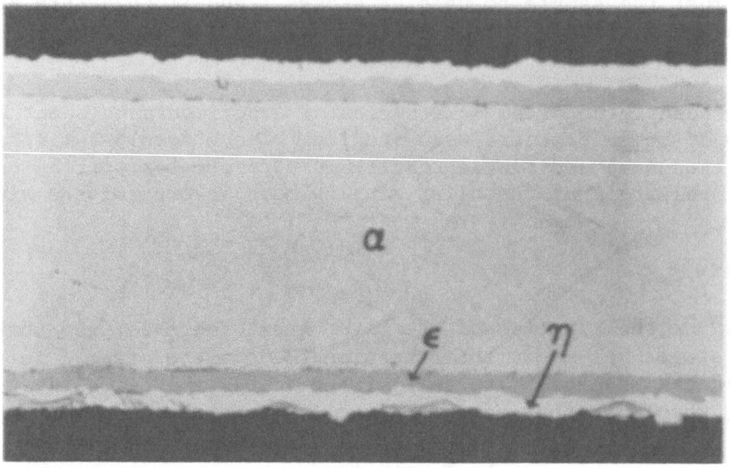

Fig. 4. The $\varepsilon + \eta$ layer formed in stage I, 200°C/20d from a 10 μm thick Sn layer on a sheet sample. Unetched, orig. mag. = 400x.

voids remained on the α-Cu side of the ε(or δ)/α interface. There was some qualitative evidence that the void growth was enhanced in regions where δ formed but void formation clearly did not require δ formation. Following stage I treatment quite small voids were observed on the α-Cu side of the ε/α interface as illustrated in Fig. 4. It was difficult to be sure that the void features observed in Fig. 4 (which were also examined in an SEM) were not produced in the polishing process, but careful metallographic work gave strong evidence that these features were voids formed in stage I. Hence, as illustrated schematically on Fig. 1 it was found that void rows nucleated in the α-Cu upon the ε/α interface during stage I, and then during stage II they grew in size and migrated with the interface. During stage III the β/α interface was observed to migrate past the void rows into the α-Cu as illustrated in Fig. 1. In this stage a new set of voids sometimes appeared. These voids were elongated in the direction of the growing β/α interface and appeared to be forming as depressions in the moving interface. Whether or not blisters formed was found to be dependent upon the time of the stage II treatment. If stage II were eliminated by direct and fast heating to 650°C blisters did not occur. (Note: melting of η phase was not a problem here with the thin layer of η, ~7 μm, left following stage II.) The tendency to form blisters and delamination of the Su rich layers increased as both the time and temperature of stage II were increased. Figure 5 shows a small discrete blister following a low temperature short time stage II treatment, 340°C/6h, plus a 650°C/2m stage III treatment. A possible blister

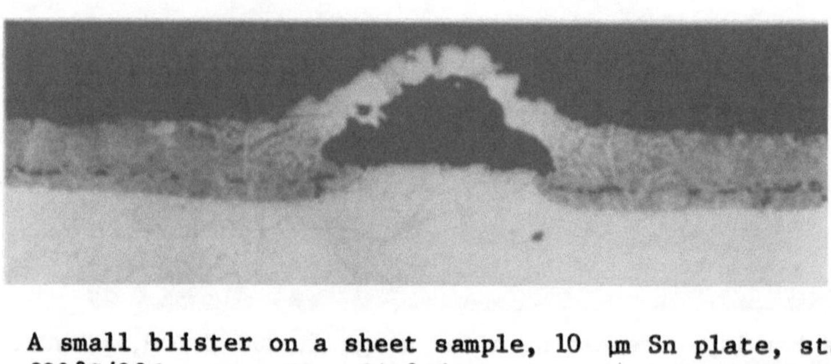

Fig. 5. A small blister on a sheet sample, 10 μm Sn plate, stage I = 200°C/23d, stage II = 340°C/6h and stage III = 650°C/2m. Grard's etch, org. mag. = 200x.

formation mechanism may be understood from this picture. The void rows extending away from each side of the base of the blister were formed during stage II and then during stage III the β/α interface migrated past the void row into the Cu everywhere except at the base of the blister. A vapor gap must have formed at the base, either by void coalescence or formation of connecting cracks. A given volume of Cu will expand when it is converted to β phase. Hence, as the β phase grew into the Cu around the periphery of the vapor gap it expanded laterally thereby compressing the isolated ε layer above the vapor gap and caused it to bulge upward under the compressive force. A short remnant of the blue ε phase can be seen at the center top of the blister cap in Fig. 5. It is surrounded by δ phases and the β phase layer extending outward from the blister has decomposed on cooling to a 2-phase microstructure of δ + α. Discrete blisters as shown in Fig. 5 were generally observed at stage II temperatures of 340°C. More often at the higher temperatures of both 370 and 400°C complete delamination of the ε layer was found, as illustrated in Fig. 2. The delamination occurred both when δ phase formed, 400°C, and when it did not, 370°C and 340°C.

The blister problem is clearly related to the void row formation and therefore it is important to determine the mechanism of void formation during stage I. This question has been discussed at length in the previous study[5] and it is believed to be due in large measure to the development of tensile stresses in the α-Cu at the ε/α interface. Such stresses could arise from two mechanisms: (1) Stresses would result from the steep Sn concentration gradient in the Cu required by the very low diffusion coefficient at the stage I temperature of 200°C ($D \sim 10^{-20}$ cm^2/s). (2) A given volume of Cu will expand by a factor of roughly 1.6 when converted to ε phase and this could lead to a lateral expansion, as demonstrated in Fig. 5 for β phase formation, with resultant tensile stresses in the Cu. If these phase transformation induced tensile stresses do play a key role

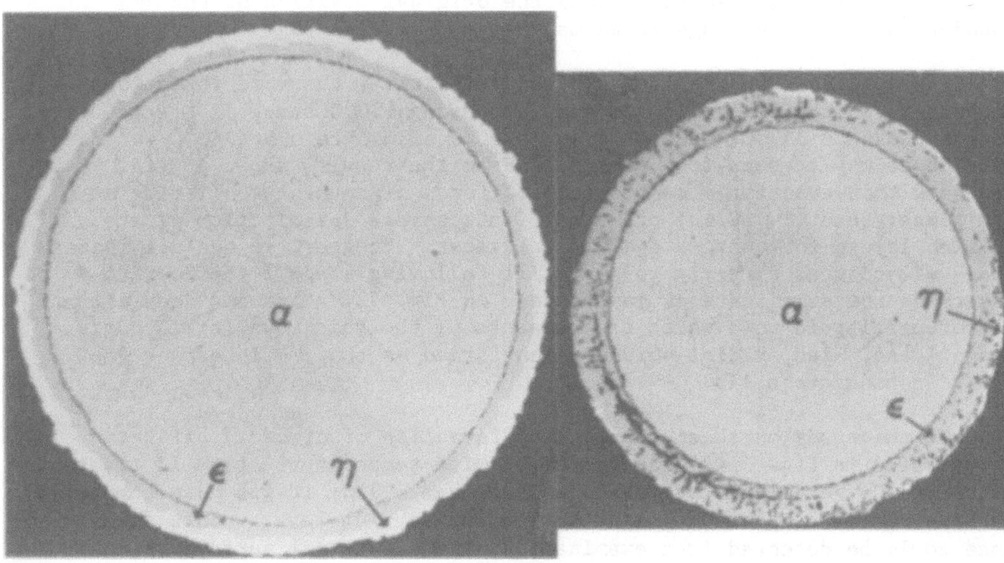

(A) (B)

Fig. 6. (A) Unetched transverse section after stage I, 200°C/20d. 0.25 mm wire, orig. mag. = 300x. (B) Unetched transverse section after stage II, 340°C/6h. 0.25 mm wire, orig. mag. = 250x.

in void nucleation in the sheet geometries, then one would expect void formation to be even more of a problem in the wire geometry. As the ε layer grows inward on a wire surface its lateral expansion would tend to produce a radial tensile strain at the ε/α interface. Also, as the ε layer thickens the corresponding volume increase at the growing ε/α interface region should produce circumferential tensile stains in the outer regions of the ε layer. Both of these effects might lead to enhanced void formation in a wire geometry.

WIRE GEOMETRY

A series of experiments has been carried out on wire samples under conditions similar to the previous study on sheet samples. Three different Cu wire diameters, 0.25, 1 and 3 mm were studied. The wires were drawn from C101 Cu and cleaned in a 1:1HNO_3:H_2O solution for 30s. They were plated with Sn from a sodium stannate bath and to insure uniformity of plate thickness the wires were rotated during the plating operation. Based on weight gain measurements the plate thicknesses varied from 10.1 to 10.4 μm. The Sn plated wires were cut to 2.5 cm lengths, sealed in 5 mm quartz tubes under helium and given a stage I treatment of 20d at 200°C. Batches of wire were then examined after stage I, after stage II treatments of 340°C/6h or 400°C/6h, and after stage III treatments of 650°C/2m both with and without preceding stage II treatments.

Figure 6A presents a transverse section of a typical result following the stage I treatment for a 0.25 mm diameter wire. The significant void density of Fig. 6A was consistently found with the wire samples, whereas for sheet samples the void density was generally smaller, as illustrated in Fig. 4. A more definite difference for wires versus sheet samples may be seen by comparing Figs. 6B and 3. The ε phase in all wire samples was consistently found to contain a large area fraction of voids throughout as shown in Fig. 6B, whereas it was rare to find voids within ε in sheet samples. This effect was most pronounced for wires of the smallest diameter. In addition, for the 0.25 mm wire, circumferential cavities were frequently observed at roughly the original position of the η/ε boundary produced in stage I, as may be seen in Fig. 6B.

As expected from results on sheet geometries, the void formation effects were more severe in the samples having 400°C stage II treatments versus those at 340°C. The minimum time to eliminate η at 340°C is just over 6h as may be seen from Fig. 6B. From the results shown in Fig. 7A one sees that even for a sample treated for a minimum stage II time at a low temperature (340°C/6h) one still finds severe delamination of the Sn-rich layers following stage III treatment. However, if one eliminates stage II and goes directly to stage III following stage I the Sn-rich layers do not delaminate as may be seen in Fig. 7B. Here one sees circumferential void rows which are remnants of the void rows formed during stage I and, also, radial voids, which formed as the β/α interface grew inward during stage III.

In the study on sheet samples the formation of discrete blisters illustrated in Fig. 5 was favored at the low temperature stage II treatment, 340°C. At higher temperatures, 370 and 400°C, it was more common to observe nearly complete delamination of the Sn rich layer. This difference could be detected from examination of the external surface after stage II or short time stage III experiments. Over both delaminated regions and discrete blisters a color difference was apparent due to the lack of diffusive mass transport. For example, after stage II the color difference was blue over delaminated regions versus grey over intact regions, while after short time stage III it was grey versus bronze. Some differences were observed with the wire samples as a function of diameter.

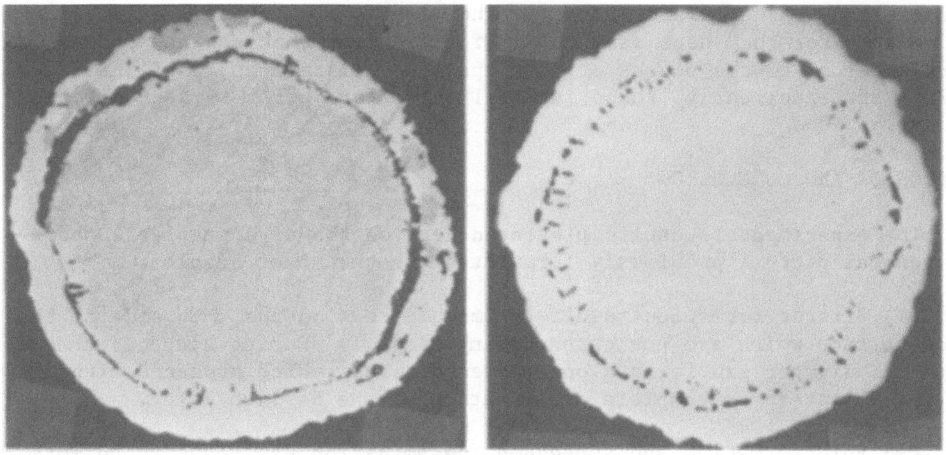

(A) (B)

Fig. 7. (A) Unetched transverse section after stage II; 340 °C/6h plus stage III 650 °C/2m. (B) Unetched transverse section after stage III; 650 °C/2m, no stage II 0.25 mm wire, orig. mag. = 250x.

Discrete blister formation was rarely observed on the 0.25 mm wire samples by external examination or on transverse sections and a nearly complete delamination, as illustrated in Fig. 7A, was generally found. In the 1 and 3 mm wire samples discrete blister formation was easily detected from external surface examination and Fig. 8 presents sectioned views of typical blisters for these two wire sizes. Apparently the blister sizes are sufficient that when they form under the conditions of these experiments in the smaller 0.25 mm wire, they extended nearly around the entire wire

(A)

(B)

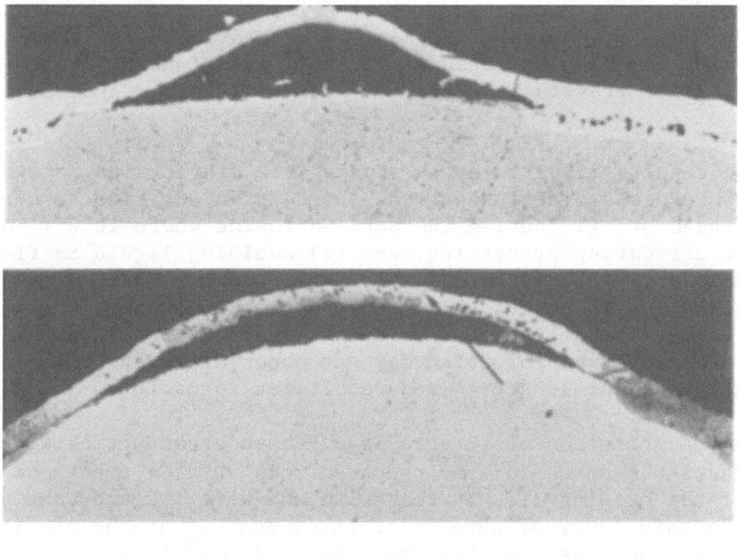

Fig. 8. (A) 3 mm wire, unetched, stage II = 340 °C/6h, stage III = 650 °C/2m. (B) 1 mm wire, Grard's etch, stage II = 400 °C/6h, stage III = 6550 °C/2m. Orig. mag. = 200x. Arrow at ε/δ boundary.

circumference. The blister form for the wire samples appears quite similar to that for the sheet samples of Fig. 5 except that one now often sees voids internal to the phases of the separated blister cap, as illustrated in Fig. 8B. Apparently, these internal voids form prior to separation of the cap.

DISCUSSION AND CONCLUSIONS

The experimental results presented here on the wire samples support the general picture previously formulated[5] on the sheet samples.

(1) Blister formation results from void rows on the α-Cu side of the ε/α interface which are nucleated during stage I. During stage II and/or stage III a vapor gap forms along these void rows which produces either discrete blisters or complete delamination of the Sn rich layer.

(2) Void formation and subsequent delamination was found to be more of a problem on wire samples versus sheet samples. More voids are present at the ε/α interface following stage I and, in addition, voids are now formed internally within the ε phase during stage II. Tensile transformation stresses are expected to be more severe in the wire geometry and can account for this enhanced void formation.

(3) The blister/delamination problem may be avoided even in the wire samples by eliminating stage II, although one is left with fairly coarse voids in the outer bronze regions following stage III treatments. Such voids will not be detrimental to mechanical properties as they are contained in a ductile bronze phase after stage III.

(4) It seems unlikely that the external Sn process will be effective on wire samples of diameter much larger than 0.2 mm because of the long times required for stage I of the solid state diffusion process. In addition, the possibility of avoiding blistering by eliminating stage II is decreased as the Sn layer increases because the η layer thickness produced in stage I will increase, and flow of liquid η will become a problem when stage II is omitted. It has been shown[1] that Sn flow problems induced by capillary forces result when Sn layers are melted because molten Sn does not wet the η phase. However, if η phase is melted following stage II one would not expect this problem because molten Sn solutions have been shown[6] to wet the ε-phase. Gravity induced flow would still be a problem however.

(5) These results illustrate that there is a distinct advantage of the internal Sn process over the external process in addition to the two obvious advantages of, (1) having the soft Sn inside where it will not become scraped off during processing, and (2) avoiding liquid Sn flow because the molten tin can be held in place by the surrounding Cu-Nb material. Blistering is not expected to be a problem because the η, ε and β phases all form internally where the stresses are compressive. In the core process, the void row formation may not occur and even if it does the compressive stresses should act to reduce blister formation.

(6) In the fabrication of in situ Nb_3Sn-Cu superconducting wire one generally encases the Nb-Cu in situ prepared rods in a Cu sheath to inhibit galling of the drawing dies by the Nb in the wire. Hence, the final wire contains a thin layer of pure Cu of thickness W_{Cu} around the in situ Cu-Nb core. The above results show that the void rows leading to blistering occur at a depth W_{II} which one may calculate from a mass balance, W_{Sn}, and the ratio of the η to ε thickness which is given in reference 4. If $W_{II} > W_{Cu}$ the void rows should appear within the in situ matrix, while if $W_{II} < W_{Cu}$ the void rows will appear within the Cu.

Copper clad in situ sheet was prepared by rolling, and samples were plated with a 10 µm thickness of Sn. By rolling to various thicknesses the Cu cladding thickness was adjusted to provide W_{II}/W_{Cu} = 0.84, 1.7 and 2.8. As illustrated in Fig. 9, following a solid state diffusion process severe blister formation was observed in the sample with W_{II}/W_{Cu} = 0.84 while no blisters occurred in the W_{II}/W_{Cu} = 1.7 and 2.8 samples, but void rows appeared within the in situ material. Apparently the mechanical support of the Nb fibers prevented blister formation in the latter cases. These results indicate that the proper design of cladding thickness the blister problem can be controlled somewhat on in situ Cu-Nb materials. Similar control might also be possible on conventional Cu-Nb rod materials.

ACKNOWLEDGMENTS

The authors would like to acknowledge helpful discussions with R. K. Trivedi and D. T. Peterson, and the experimental assistance of H. Baker in the metallographic work. This work was done at the Ames Laboratory, which is operated for the U.S. Dept. of Energy at Iowa State University, Contract No. W-7405-Eng-82, supported by the Director of Energy Research, Office of Basic Energy Sciences.

(A)

(B)

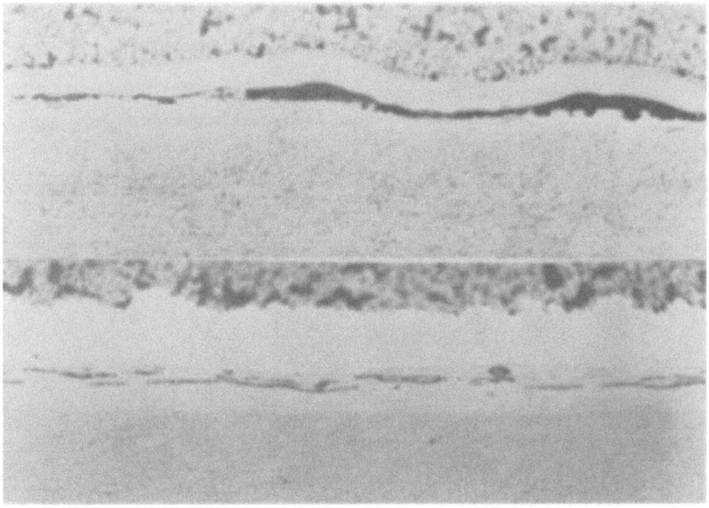

Fig. 9. As polished section of Sn plated Cu clad in situ sheet after a stage I of 200°C/29d, stage II of 400°C/10d and stage III of 650°C/1h. Orig. mag. = 200x. (A) W_{II}/W_{Cu} = 0.84. (B) W_{I}/W_{Cu} = 2.8.

REFERENCES

1. J. D. Verhoeven, E. D. Gibson, and C. C. Cheng, Appl. Phys. Lett. 40:87 (1982).
2. S. F. Cogan, S. Kwon, J. D. Klein and R. M. Rose, IEEE Trans. Mag. MAG-19:1139 (1983).
3. H. Benz, I. Horvath, K. Kwasnitza, R. X. Maix and G. Mayer, Cryogenics 19:435 (1979).
4. J. D. Verhoeven, K. Heimes and A. Efron, submitted JAP (1985).
5. J. D. Verhoeven, A. Efron, E. D. Gibson and C. C. Cheng, submitted JAP (1985).
6. H. Fidos and H. Schreiner, Z. Metallk. 61:225 (1970).

THE STRUCTURE AND SUPERCONDUCTING PROPERTIES OF MULTIFILAMENTARY Nb_3Sn WIRES PREPARED BY INTERNAL Sn DIFFUSION PROCESS USING Sn-Ti CORES

S. Miyashita*, K. Yoshizaki*, Y. Hashimoto*, K. Itoh** and K. Tachikawa**

*Materials and Engineering Lab. Mitsubishi Electric Corp.
Sagamihara, Japan
**National Research Institute for Metals
Ibaraki, Japan

ABSTRACT

The multifilamentary Nb_3Sn superconducting wires with the titanium addition have been fabricated by internal tin diffusion process to improve critical current density (J_c) at high fields (>12T). The wires fabricated have seven modules in a copper matrix. The module is composed of the Sn-2wt%Ti alloy core instead of the tin core at the center and 90 niobium filaments located by three layers around the Sn-Ti core.

The wire heat-treated at 675°C for 168hr shows a high overall J_c of $7.5x10^4 A/cm^2$ at 12T and $3x10^4 A/cm^2$ at 16T. The J_c is further improved by increasing titanium content in the core, and the wire with Sn-2.5wt%Ti core has a J_c of $8.2x10^4 A/cm^2$ at 12T. The observation by electron probe microanalizer (EPMA) indicates that the amount of titanium dissolved in the Nb_3Sn layer depends on the location of filaments. The filaments close to the center of a module have maximum titanium amount of ~0.5wt% titanium in the Nb_3Sn layers.

The results indicate that Nb_3Sn wire using Sn-Ti cores is one of the most favorable superconductors for high field applications.

INTRODUCTION

The Nb_3Sn superconducting wire has good superconducting properties at high fields. Today the Nb_3Sn multifilamentary wire is applied to high field magnets for laboratory use, and to the technical applications for magnetic fusion, accelerator, superconducting magnetic energy storage (SMES), high resolution nuclear magnetic resonance and so on. In these applications, the conductor that has high current densities at fields from 10T to 15T is required for the magnetic operation.

Recently, various researches to improve superconducting properties by third element additions to Nb_3Sn wires have been carried out by many workers.[1,2]

It is noted that internal tin diffusion process is very economical in manufacturing Nb_3Sn wires. In our previous paper,[3] it was indicated that the indium addition to a tin core and the titanium

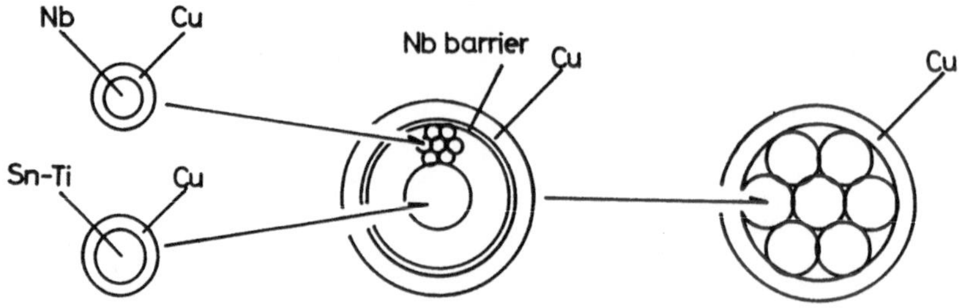

Fig. 1. Fabrication process for the internal tin diffusion processed titanium addition Nb_3Sn wire.

addition to a copper matrix were effective to improve J_c in the Nb_3Sn wires made by internal tin diffusion process.

In this study, the titanium addition to the tin core was attempted in this process and measurements of superconducting properties and microstructure observations were performed.

EXPERIMENTAL PROCEDURE

Figure 1 shows the fabrication process of the Nb_3Sn wire with Sn-Ti cores. The wire has seven modules in a copper matrix. The module is composed of the Sn-2wt%Ti alloy core instead of the tin core at the center and 90 niobium filaments located in three layers around the Sn-Ti core. The composition estimated as bronze matrix inside the barriers was Cu-19.4wt%Sn-0.4wt%Ti.

To investigate the effect of the filament size on the J_c, the wires with 8.5μm, 5.6μm and 3.7μm filaments were prepared by reducing the wire diameter, respectively. The wire specifications are shown in Table 1. These wires were very easily drawn to final sizes without any intermediate annealings and heat-treated to form Nb_3Sn compound at 650°C-775°C for 25-300hr.

Critical current (I_c) and critical temperature (T_c) were measured by a four probe resistive method. I_c measurements were performed by using the 17.5T superconducting magnet at NRIM.[4] The J_c was calculated by dividing I_c by the cross-section area of the wire excluding that of the stabilizing copper. T_c was defined as the midpoint of the transition.

Metallurgical studies of Nb_3Sn compound layers were performed by optical microscopy, scanning electron microscopy (SEM) and EPMA.

Table 1. The specifications of Nb_3Sn wires fabricated by internal tin diffusion process using Sn-Ti cores.

Wire Diameter (mm)	0.76	0.50	0.33
Filament Diameter (μm)	8.5	5.6	3.7
Barrier Diameter (mm)	0.146	0.096	0.063
No. of Filaments	630		
Barrier Material	Nb		
Cu Ratio	2.5		
Matrix Composition(wt%)	Cu-19.4Sn-0.4Ti		

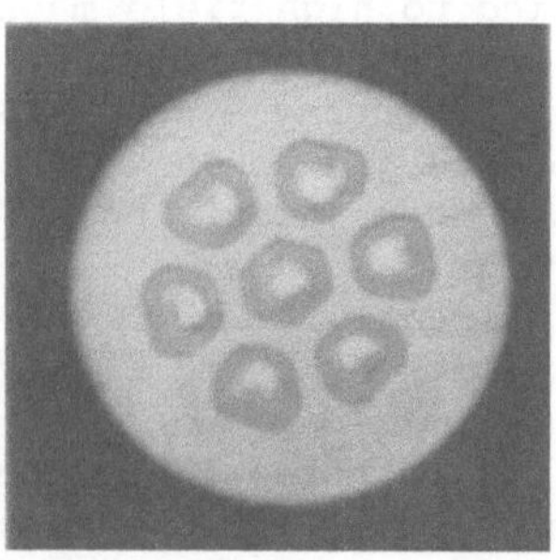

RESULTS AND DISCUSSION

CRITICAL CURRENT DENSITY

Figure 2 shows the heat-treatment dependence of J_c for the titanium added wires (0.5mm in wire diameter). The reaction times that give maximum J_c's at each temperature are chosen. In this figure, the J_c increases with decreasing heat-treatment temperature down to 675°C. The wire reacted at 675°C for 168hr has a high J_c of $3x10^4 A/cm^2$ at 16T, 4.2K.

Figure 3 shows the filament size dependence of J_c at each field for the titanium addition wires. The wire with 3.7μm (0.33mm in wire diameter), 5.6μm (0.5mm) and 8.5μm filament diameters (0.76mm) were reacted at 675°C for 168hr. The wire with finer filaments had a high J_c at low and mid fields. The J_c of the 3.7μm filament wire is lower than those of 5.6μm and 8.5μm filament wires at high fields over 12T.

UPPER CRITICAL FIELD AND CRITICAL TEMPERATURE

The B_{c2}* calculated by Kramer's scaling law[5] is increased with increasing filament diameter. The 3.7μm and the 8.5μm filament wires reacted at 675°C for 168hr have B_{c2}*'s of 21T and 23T, respectively. The B_{c2}* is increased from 23T to 26T with increasing heat-treatment temperature from 675°C to 775°C in the wire with 8.5μm filaments.

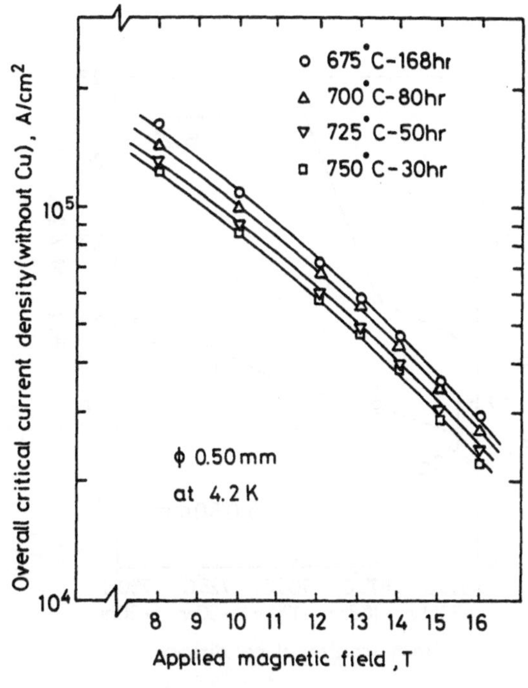

Fig. 2. Overall critical current densities for wires with titanium addition reacted at different temperatures as a function of magnetic fields at 4.2K.

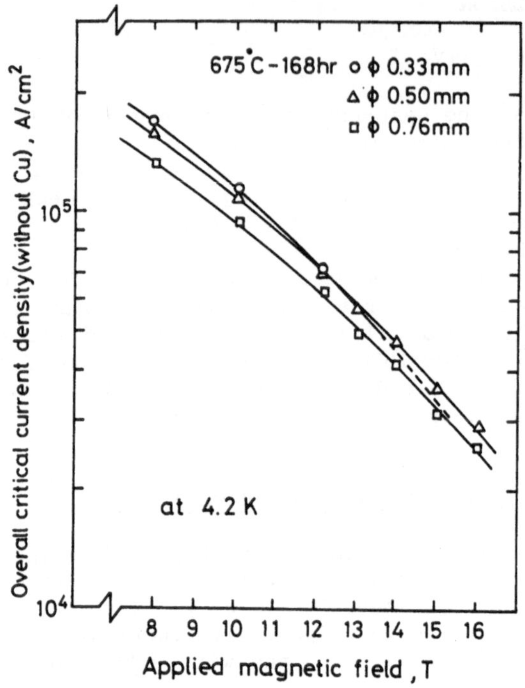

Fig. 3. Overall critical current densities for wires with titanium addition reacted at 675°C for 168hr with 3.7μm filament (0.33mm diameter), 5.6μm (0.50mm) and 8.5μm (0.76mm) as a function of magnetic field at 4.2K.

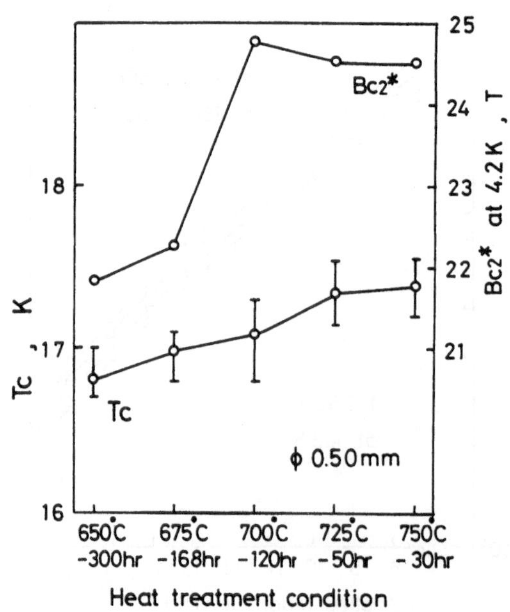

Fig. 4. The critical magnetic field and the critical temperature of the 0.5mm wires as a function of heat-treatment temperature.

The T_c is increased with increasing heat-treatment temperature. The T_c's of the wire reacted at 675°C for 168hr and 775°C for 25hr were 16.9K and 17.4K, respectively.

Figure 4 shows the heat-treatment dependence of B_{c2}^* and T_c. There appears to be fairly good correlation between T_c and B_{c2}^*.

METALLURGICAL ASPECTS

The copper matrix of the wire turns to the bronze without the Nb_3Sn formation by the heat-treatment below 600°C. It is indicated by EPMA that the tin content in the matrix is 16~18wt% after the heat-treatment. It is very difficult to draw the wire with such a high tin bronze matrix in bronze process because of the existence of β phase. Internal tin diffusion process realizes the fabrication of high tin bronze wires. In this process the constituent metals of a wire before reaction are quite ductile ones, i.e., niobium, copper and tin, so this process does not require intermediate annealing.[6,7] The high tin bronze matrix is effective to increase the Nb_3Sn formation rate and shorten the reaction time compared with the low tin bronze matrix. The tin content in the bronze matrix influences to the Nb_3Sn grain size. Since the nucleation rate is higher than grain growth rate in such a high tin content, layers of finer equiaxed Nb_3Sn grain are formed.[8] This is why the wire by internal tin diffusion process has high J_c at ~12T compared with the wire by the bronze process.

On the other hand, titanium segragates at the interface between niobium filaments and the bronze after the heat-treatment at under 600°C, and almost no titanium is found in the bronze matrix. The titanium content is decreased from the filament-bronze interface close to the center of the wire toward the outer filament-bronze interface (close to niobium barrier). It is considered that the cause of such a titanium gradient lies in the difference of the diffusion length and the difficulty of decomposition of the Sn-Ti intermetallic compound. The Sn-Ti compound already forms in the tin matrix when the Sn-Ti alloy ingot is fabricated.

Figure 5 shows SEM photograph and the X-ray emission Ti(Kα) image for the cross-section of the wire after heat-treatment (675°C for 168hr). The titanium content of 0.5wt% was analized in the Nb_3Sn layer at the filament close to the center of the wire, but the titanium content was decreased to 0.2wt% at the Nb_3Sn of the niobium barrier.

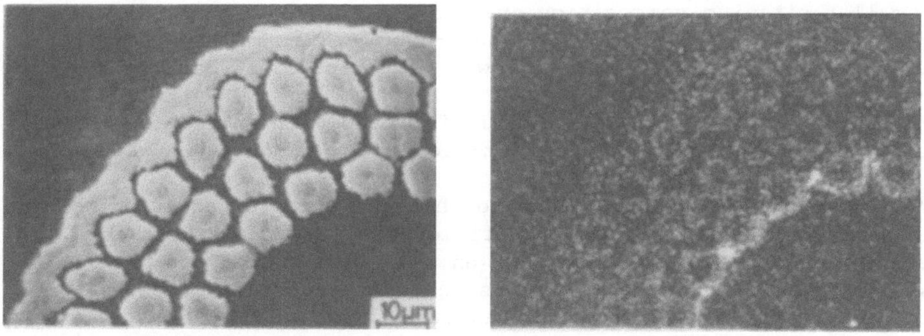

Fig. 5. The SEM photograph and the X-ray emission Ti(Kα) image of the cross-section of the wire reacted at 675°C for 168hr.

Figure 6 shows the titanium content profile in the cross-section heat-treated at 650°C, 700°C and 750°C. The heat-treatment at low temperature for long time (650°C-168hr) gives smaller titanium composition gradient than that of heat-treatment at high temperature for short time (750°C-30hr). This may be the reason why the wire heat-treated at lower temperature has better superconducting properties.

Figure 7 shows the titanium content profile of the cross-section heat-treated at 650°C for 168hr-300hr. Although the heat-treatment at 650°C for 168hr gives a steeper composition gradient of titanium than that of the heat-treatment at 650°C for 300hr, longer heat-treatment time yields inferior J_c because of the grain coarsening.

HIGH TITANIUM CONTENT WIRES

It is reported the J_c at high fields depends on the titanium content in the Nb_3Sn layer formed in the bronze processed wire. The Nb_3Sn wires with Sn-2.5wt%Ti and Sn-3.7wt%Ti cores were then fabricated to investigate the effect of titanium content.

The titanium content in the 0.76mm wires reacted at 700°C for 80hr influences on Jc properties. The J_c and B_{c2}* of the wires are improved with increasing titanium content in tin cores from 2wt% to 2.5wt% and 3.7wt%. The wire with Sn-2.5wt%Ti cores has a high J_c of $8.2 \times 10^4 A/cm^2$ at 12T. In the wire with Sn-3.7wt%Ti cores, the gradient of the J_c-B curve becomes smaller and therefore this wire has better superconducting properties at high fields.

CONCLUSIONS

The internal tin diffusion processed Nb_3Sn wire with Sn-2wt%Ti core has a good workability without intermediate annealing.

The J_c is increased with decreasing heat-treatment temperature and filament diameter. The wire with 5.6μm filaments reacted at 675°C for 168hr has a high overall J_c of $3 \times 10^4 A/cm^2$ at 16T at 4.2K.

The B_{c2}* of the wire is increased with increasing heat-treatment temperature and increasing filament diameter. The wire with 3.7μm filaments and with 8.5μm filaments reacted at 675°C for 168hr has B_{c2}* of 21T and 23T, respectively.

The T_c is increased by increasing heat-treatment temperature and there appears to be fairly good correlation between T_c and B_{c2}*.

The titanium content of 0.5wt% was analized in the Nb_3Sn layer at the filament close to the center of the wire, but the titanium content was decreased to 0.2wt% at the Nb_3Sn of the niobium barrier.

The condition of low temperature for long time (650°C-168hr) gives smaller titanium composition gradient than that of the condition of high temperature for short time (750°C-30hr). This may be the reason why the wire heat-treated at lower temperature has better superconducting properties.

Furthermore, the J_c of the wires is improved with increasing titanium content in tin cores from 2wt% to 2.5wt% and 3.7wt%.

The results of this study indicate that the titanium addition to the tin core increases the J_c at high fields similar to the titanium addition to the copper matrix. The Nb_3Sn wire by internal tin diffusion process using a Sn-Ti core is one of the most favorable superconductors for generating high magnetic fields.

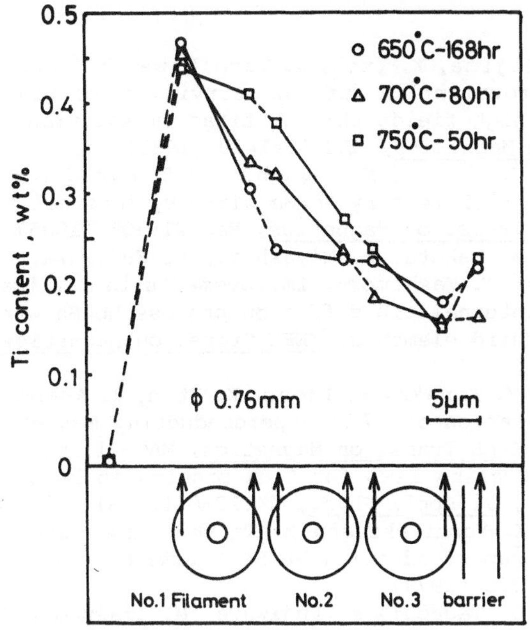

Fig. 6. The titanium content in the Nb_3Sn layers as a function of filament position in the wire reacted at 650°C, 700°C and 750°C.

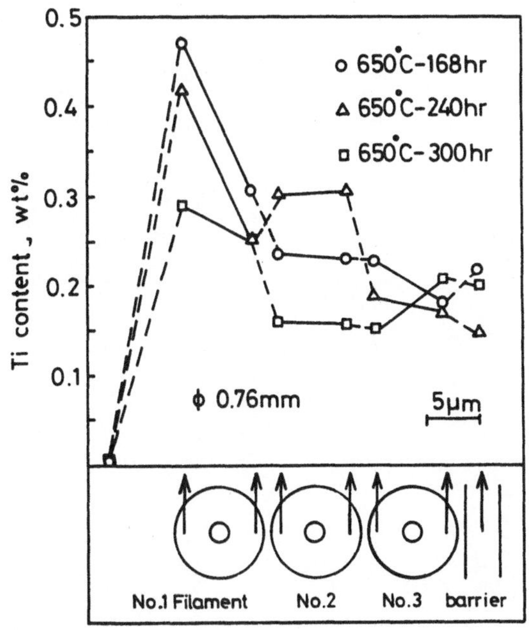

Fig. 7. The titanium content in the Nb_3Sn layers as a function of filament positions in the wire reacted 650°C for 168hr-300hr.

REFERENCES

1. H. Sekine, Y. Iijima, K. Itoh, K. Tachikawa, Y. Tanaka and Y. Furuto, Improvement in current-carrying capacities of Nb_3Sn composites in high fields through titanium addition to the matrix, IEEE Trans. on Magnetics, MAG-19:1429 (1983)
2. M. Suenaga, C. J. Klamut, N. Higuchi and T. Kuroda, Properties of Ti alloyed multifilamentary Nb_3Sn wires by internal tin process, IEEE Trans. on Magnetics, Mag-21:305 (1985)
3. K. Yoshizaki, M. Wakata, S. Miyashita, F. Fujiwara, O. Taguchi, M. Imaizumi and Y. Hashimoto, Improvements in critical current densities of internal tin diffusion process Nb_3Sn wires by additions of third elements, IEEE Trans. on Magnetics, MAG-21: 301 (1985)
4. K. Tachikawa, Y. Tanaka, K. Inoue, K. Itoh, T. Asano and Y. Iijima, Operation of 17.5T superconducting magnet system in the last 8 years, IEEE Trans. on Magnetics, MAG-21:1048 (1985)
5. E. J. Kramer, Scaling laws for flux pinning in hard superconductors, J. Appl. Phys., 44:1360 (1973)
6. Y. Hashimoto, K. Yoshizaki and M. Tanaka, Processing and properties of superconducting Nb_3Sn filamentary wires, Proc. ICMC-5, 332 (1974)
7. K. Yoshizaki, O. Taguchi, F. Fujiwara, M. Imaizumi, Y. Hashimoto, K. Wakamoto, T. Yamada and T. Satow, Properties of multifilamentary Nb_3Sn wires processed by internal tin diffusion, Proc. ICMC-9, 380 (1982)
8. M. Suenaga, Properties of pure and alloyed Nb_3Sn by solid-state diffusion process, Proc. ICMC-9, 61 (1982)

INTERNAL TIN PROCESS Nb_3Sn SUPERCONDUCTORS FOR 18 TESLA*

D.W. Hazelton, G.M. Ozeryansky, M.S. Walker
B.A. Zeitlin, K. Hemachalam

Intermagnetics General Corporation
Guilderland, New York

and

E.N.C. Dalder, L. Summers

Lawrence Livermore National Laboratory
Livermore, California

ABSTRACT

The internal tin process has been established as a cost effective and reliable method for the manufacture of practical high current density Nb_3Sn superconductor.[1-3] This paper describes fabrication and test results to date for internal tin process conductor elements that are being developed for the Lawrence Livermore National Laboratory for fusion applications at 18 Tesla and beyond. The filaments of the conductors are niobium or alloyed niobium reacted in a bronze matrix formed from copper and tin or doped tin elements. Ti, Mg and Ta are among the alloying elements being explored. Critical current measurements, and scanning electron microscope results are presented for variously reacted conductor elements.

INTRODUCTION

This research was undertaken to develop a data base for the development of a Nb_3Sn high field superconductor utilizing the internal tin process. This superconductor is targeted for use by Lawrence Livermore National Laboratory for fusion energy applications at 18T and beyond,[4] such as the MFTF-B+ Upgrade "plug" magnet now under consideration.

The internal tin process, detailed elsewhere,[5] involves co-reduction of copper, niobium, and tin based components to final size. This configuration is then heat treated in two major stages, a lower temperature stage to form bronze from the copper and tin components and a higher temperature stage to form Nb_3Sn. The resultant Nb_3Sn filaments in a bronze matrix are contained within a diffusion barrier surrounded by copper stabilizer material.

In recent years, research work by several sources[6-15] has shown that the addition of certain dopants to the Nb filaments and/or the bronze matrix dramatically improves the critical current density (J_c) and upper

Table 1. Parameters Investigated

Nb Filament Dopants:	Nb, Nb-1.2 w/o Ti, Nb-7.5 w/o Ta
Sn Core Dopants:	Sn, Sn-2.1 w/o Mg
Residual Sn Percentages:	6%, 8% by weight
Local Cu:Nb Ratio:	1.1:1, 1.2:1, 1.5:1

critical field (H_{c2}) of the conductor over undoped materials, particularly at higher fields. The addition of dopants has also been shown to reduce the strain sensitivity of the resultant Nb_3Sn material.[16]

Conductor elements fabricated were prepared using the basic internal tin process with the following modifications added to enhance H_{c2} and J_c of the material. A list of the parameters investigated, is given in Table 1.

- o Alloying of the niobium filaments
- o Alloying of the tin core reservoir
- o Variation of the residual tin content of the bronze after reaction Residual tin content is defined as the residual weight percentage of Sn in the bronze after complete reaction of the Nb_3Sn filaments
- o Variation of the local Cu:Nb ratio to investigate tin diffusion paths, rates, and gradients

Preparation of Doped Nb_3Sn

Five Cu-$Nb_{1-w}X_w$ tubular extrusion billets were prepared as detailed in reference 5. The doping of the Nb filaments was done during melting of the material. The local copper to niobium filament ratio was determined by varying the amount of copper surrounding each filament. These billets were then tubularly extruded from approximately 140 mm to approximately 63.5 mm outer diameter. Tin core rods of the compositions listed in Table 1 were prepared by casting the tin or tin alloy into copper tubes.

Sections of the extruded billets were filled with the variously alloyed Sn cartridges. These sections were then placed in Ta and Cu tubing, serving as a diffusion barrier and stabilizer respectively. The residual tin percentages of the bronze were fixed by varying the copper to tin ratio of the material inside the diffusion barrier. The material was co-drawn to a final wire diameter of 0.254 mm resulting in a filament diameter of approximately 6.2 μm.

The combination of the parameters of filament dopant, Sn core dopant, local Cu:Nb ratio, and residual tin content resulted in a possible total of 36 different parameter combinations. Of these 36, 15 of the more promising candidates were prepared and selected for heat treatment, scanning electron microscopy studies and J_c measurements. The composition of these 15 candidate materials are given in Table 2. The component volume percentages of the two types of elements differentiated by residual Sn content are given in Table 3.

Table 2. Element Compositions

Element	Filament	Core	% Residual Sn	Local Cu:Nb
1	Nb(Ti)	Sn	6	1.2:1
2	Nb(Ti)	Sn(Mg)	6	1.2:1
3	Nb(Ti)	Sn	8	1.2:1
4	Nb(Ti)	Sn(Mg)	8	1.2:1
5	Nb(Ti)	Sn	6	1.5:1
6	Nb(Ti)	Sn	8	1.5:1
7	Nb(Ti)	Sn	6	1.1:1
8	Nb(Ta)	Sn	6	1.2:1
9	Nb(Ta)	Sn(Mg)	6	1.2:1
10	Nb(Ta)	Sn	8	1.2:1
11	Nb(Ta)	Sn(Mg)	8	1.2:1
12	Nb	Sn	6	1.2:1
13	Nb	Sn(Mg)	6	1.2:1
14	Nb	Sn	8	1.2:1
15	Nb	Sn(Mg)	8	1.2:1

Heat Treatment

Prior experience with internal tin process Nb_3Sn materials has indicated that splitting the heat treatment cycle into a low temperature "homogenization" stage and a higher temperature "reaction" stage results in a morphological structure producing higher J_c values than a single high temperature heat treatment. In order to facilitate comparisons of the effects of the various conductor parameters, a single homogenization heat treatment was utilized with a range of reaction heat treatments. Tests were conducted on the material to determine a homogenization heat treatment by which a uniform bronze composition was developed throughout the matrix as well as generating the formation of some Nb_3Sn on the filaments. The resultant homogenization heat treatment was 100 h at 200°C, 24 h at 325°C, and 200 h at 580°C. Up to eight final reaction heat treatments, listed in Table 4, were then used.

Four approximately 200 mm long samples of each material listed in Table 2 were placed in four holed ceramic holders to prevent bonding between samples during heat treatment. The sample ends were then sealed

Table 3. Component Volume Percentages

	Low Residual Sn		High Residual Sn	
	As Drawn	As Reacted	As Drawn	As Reacted
Cu Stabilizer	54.49	55.35	54.50	55.41
Ta Barrier	8.40	8.53	8.40	8.54
Cu Matrix	23.67	-	23.11	-
Sn Cores	5.84	-	6.38	-
Nb Filaments	7.60	-	7.61	-
BZ Matrix	-	25.62	-	25.53
Nb_3Sn Filaments*	-	10.50	-	10.52

* Assumes 100% reacted

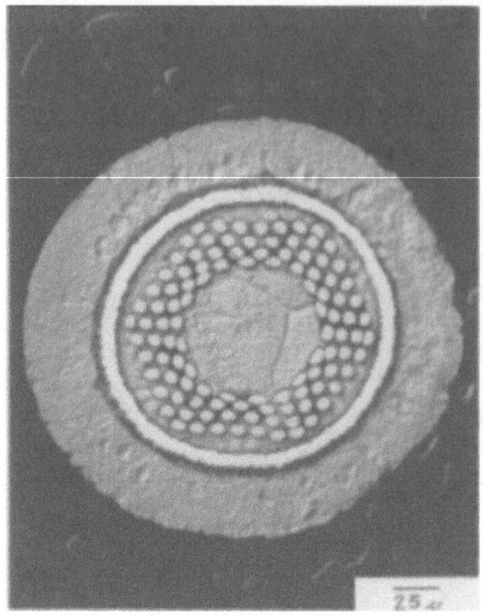

Fig. 1. Reacted $(NbTi)_3(SnMg)$ conductor Element 5, heat treatment B. Etched to expose filament pattern.

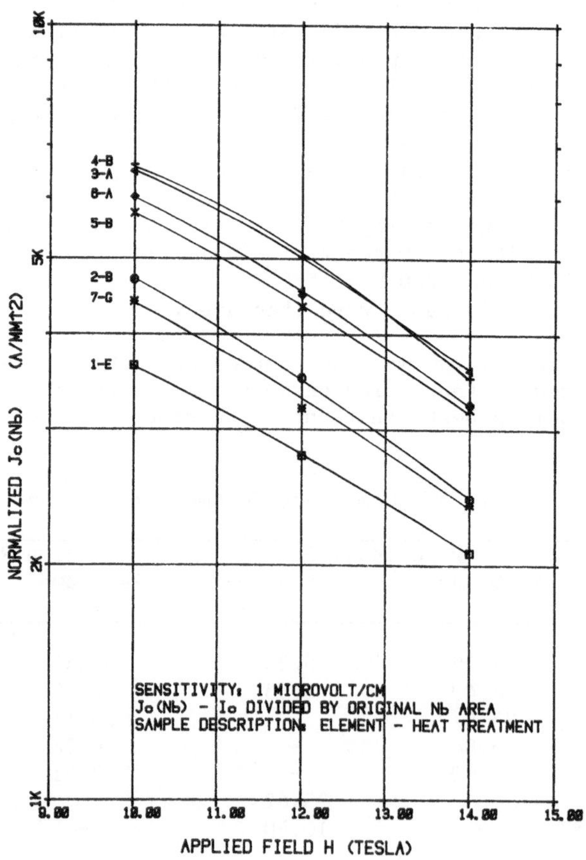

Fig. 2. Jc vs H measurements on Ti doped Nb_3Sn elements. Normalized to original Nb area.

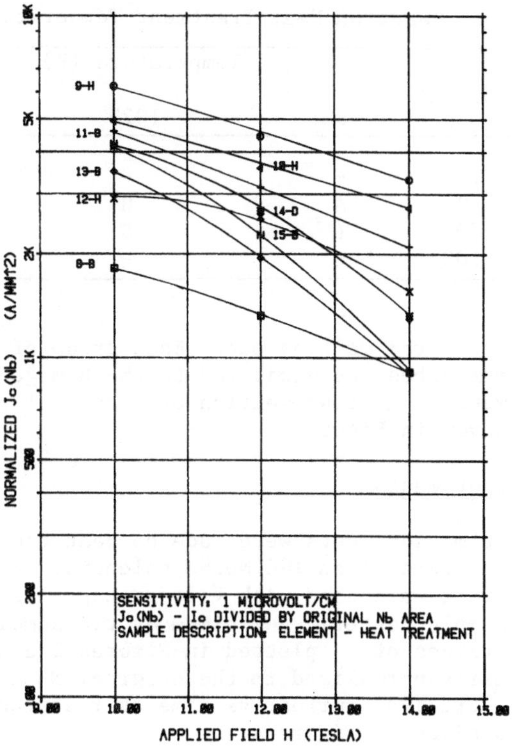

Fig. 3. Jc vs H measurements on Ta and undoped Nb_3Sn elements. Normalized to original Nb area.

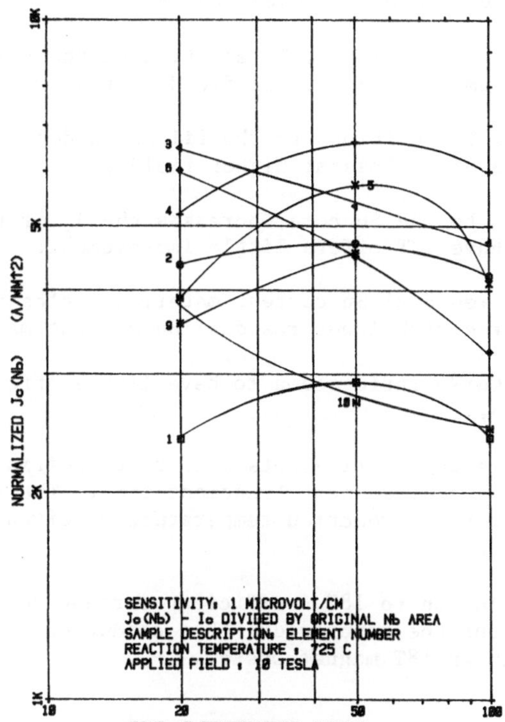

Fig. 4. Jc vs T, t measurements on selected Nb_3Sn elements at 10T. Normalized over original Nb area.

Table 4. Reaction Heat Treatment Temperatures

		Temperature (K)		
		998	1023	1048
Time (Hours)	10	-	-	D
	20	A	H	E
	50	B	G	F
	100	C	-	-

to prevent the Sn cores from melting out. Each group of four was sealed under vacuum in quartz tubes and subjected to the homogenization and reaction heat treatments. A cross-section of a typical conductor element after reaction are given in Figure 1.

Critical Current Measurements

Critical current measurements were made on heat treated samples in fields of 8-14T in the bore of an IGC Nb_3Sn solenoid. Straight samples approximately 50 mm long were used with voltage taps located approximately 12 mm apart at the sample center. A minimum of two samples per specimen were measured. The values of J_c plotted in Figures 2 and 3 represent the critical current density normalized to the original Nb area for selected conductor elements. Critical current vs time at reaction temperatures for selected elements is given in Figure 4.

CONCLUSIONS

Based on the data obtained, the following conclusions can be made of the influences of the parameters investigated.

1. For all samples measured, Ti addition to the filaments offers the best improvement in J_c at all fields investigated.

2. Both Ti and Ta additions to the filaments decrease the severity of the rate of J_c loss at higher fields.

3. Mg addition to the Sn core increases the J_c of the material up to 12T, but above 12T offers little improvement.

4. The higher residual Sn content material yields higher J_c values when compared with lower residual Sn content material.

5. The local Cu:Nb ratio seems to have little effect on J_c performance.

6. Ti additions to the filaments and/or Mg additions to the Sn core shorten the required heat treatment time. The additions also seem to lower the reaction temperature required for peak J_c performance.

7. By extrapolation to 18T, sufficient current density appears to be available for the manufacture of a conductor for that could be utilized in an 18T magnet system.

ACKNOWLEDGEMENTS

* Work supported by the U.S. Department of Energy, Lawrence Livermore National Laboratory under Contract #4279305.

REFERENCES

1. R.E. Schwall, G.M. Ozeryansky, D.W. Hazelton, S.F. Cogan, R.M. Rose, Properties and Performance of High Current Density Sn-Core Process MF Nb_3Sn, in: "IEEE Trans. on Magnetics" vol. MAG-19, No. 3 (1983).
2. K. Yoshizaki, O. Taguchi, F. Fujiwara, M. Imaizumi, M. Wakata, Nb_3Sn Superconducting Cables Processed by Internal Tin Diffusion, in: "IEEE Trans. on Magnetics", vol. MAG-19, No. 3 (1983)
3. N. Higuchi, K. Tsuchiya, C.J. Klamut, M. Suenaga, Superconducting Properties of Nb_3Sn Multifilamentary Wires Fabricated by Internal Tin Process, in: "Advances in Cryogenic Engineering - Materials", vol. 30, Plenum Press, New York (1984).
4. R.W. Hoard, D.W. Cornish, R.M. Scanlan, J.P. Zbasnik, R.L. Leber, R.B. Hickman, J.D. Lee, "Ultra High Field Superconducting Magnets", UCID-19853, Lawrence Livermore Nationary Laboratory, Livermore, California (1983).
5. B.A. Zeitlin, G.M. Ozeryansky, K. Hemachalam, An Overview of the IGC Internal Tin Nb_3Sn Conductor, in: "IEEE Trans on Magnetics", vol. MAG-12, No. 2 (1985).
6. K. Tachikawa, Recent Developments in Filamentary Compound Superconductors", in: "Advances in Cryogenic Engineering - Materials", vol. 28 Plenum Press, New York (1982).
7. J.D. Livingston, Effect of Ta Additions to Bronze Processed Superconductors, in: "IEEE Trans. on Magnetics", vol. MAG-14, No. 5 (1979).
8. M. Suenaga, S. Okuda, R. Sabatini, K. Itoh, T.S. Luhman, Superconducting Properties of $(NbTi)_3Sn$ Wires Fabricated by the Bronze Process, in: "Advances in Cryogenic Engineering - Materials", vol. 28, Plenum Press, New York (1982).
9. Tachikawa, T. Takeuchi, T. Asano, Y. Iijima, H. Sekine, J.T. Holthuis, Effects of the IVa Element Additions on Composite - Processed Nb_3Sn, in: "Advances in Cryogenic Engineering - Materials" vol. 28, Plenum Press, New York (1982).
10. I.W. Wu, D.R. Dietderich, P.E. Johnson-Walls, J. Glazen, J.T. Holthuis, W.V. Hassenzahl, J.W. Morris, The Critical Current Density of Bronze Processed Multifilamentary Nb_3Sn Wires with Magnesium Addition to the Matrix, preprint from Applied Physical Letters paper - Fall 1984.
11. R. Bormann, Phase Relationships and Superconducting Properties of Ternary Systems Used in the Bronze Process, in: "Advances in Cryogenic Engineering - Materials", vol. 30 Plenum Press, New York (1984).
12. N. Tada, K. Aihara, R. Saito, Y. Kazawa, T. Suzuki, K. Kamata, K.Noguchi Development of Hitachi's Multifilamentary Nb_3Sn and $(NbTi)_3Sn$ Superconductors, preprint from 2nd US-Japan Workshop on High Field Superconducting Materials for Fusion, San Diego, California, May 1983.
13. K. Kamata, N. Tada, K. Itoh, K. Tachikawa, High Field Current Carrying Capacities of Titanium Bronze Processed Multifilamentary Nb_3Sn Conductors with Pure and Alloy Cores, in: "Advances in Cryogenic Engineering - Materials", vol. 30 Plenum Press, New York (1984).
14. R. Roberge, H. LeHuy, S. Foner, Effects of Added Elements and Strain on the Martensitic Phase Transformation of Nb_3Sn, Physics Letters, 82A-5:259-262 (1981).
15. O. Horigami, T. Luhman, C.S. Pande, M. Suenaga, Superconducting Properties of $Nb_3(Sn_{1-x}Ga_x)$ by a Solid State Diffusion Process, Applied Physics Letters, 28-12:738-740 (1976).
16. J.W. Ekin, Strain Effects in Superconducting Compounds, in: "Advances in Cryogenic Engineering - Materials", vol. 30 Plenum Press, New York (1984).

INTERNAL-TIN-PROCESSED Nb_3Sn MULTIFILAMENTARY WIRES ALLOYED WITH Mg, Zn+Ni, AND Ti THROUGH THE Sn CORE

T. Kuroda, M. Suenaga, C. J. Klamut, and R. Sabatini

Brookhaven National Laboratory
Upton, New York

ABSTRACT

The influence on superconducting properties by alloying Nb_3Sn with an alloyed Sn core in internal-tin-processed multifilamentary Nb_3Sn wires was investigated for alloying elements, Mg, Ni+Zn, and Ti. The results indicate that the critical current densities, J_c of the Mg, Ni, and Ni+Zn-alloyed wires were essentially identical to or less than those for the unalloyed control wires although small amounts of Mg, Ni and Zn were found in the Nb_3Sn filaments. However, additions of Ti increased the J_c of the wire considerably at intermediate as well as very high-magnetic fields. Results of detailed characterization of these wires for superconducting and metallurgical properties will be reported.

INTRODUCTION

The bronze process for fabrication of multifilamentary Nb_3Sn wires has been developed to the point such that a number of small as well as large high-field magnets were successfully constructed.[1] However, the annealing process required for the bronze matrix during the drawing operation is still very time consuming and costly. In order to avoid this step of wire fabrication, the internal-tin process was developed[2] and appears to be a viable alternative process for production of Nb_3Sn multifilamentary wires.[3] Recently, it was shown that additions of third elements to Nb_3Sn by the bronze process were very effective in increasing values of the critical magnetic field H_{c2}[4,5] and thus improving the values of critical current density J_c at very high-magnetic fields.[6-10] Accordingly, similar elemental additions (In[11] and Ti[11,12]) to Nb_3Sn multifilamentary Nb_3Sn wires by the internal-tin process were attempted to improve the values of J_c at high-magnetic fields. These results indicated that Ti can be introduced in the Cu matrix or the Sn core and in both cases significant increases in J_c at high fields were observed. In the case of the In addition via the Sn core, J_c was shown to improve over the entire range of magnetic fields in comparison with that for pure Nb_3Sn wire.

From the studies of alloying in the bronze process, there are other elements (Mg[10] and Ni+Zn[9]) which were reported to have beneficial effects on superconducting properties. However, both of these elemental additions are considered not suitable for fabrication of practical wires by the bronze process due to excessive hardening of the bronze matrix and reduction of Sn content in the bronze, respectively. These problems may be circumvented if these elements were introduced in Sn of the internal-tin process. Thus, we examined the above possibilities and the results are reported here.

We also studied the influence of Ti additions in the Sn core at a higher level (Sn-3 wt% Ti) than in the previous study[12] (Sn-1.5 wt% Ti) on the critical-current densities of an internal-tin-processed wire. This result is also discussed here.

SPECIMEN PREPARATION

In order to study the influence of alloying the Sn core with Mg, Ni, Ni+Zn, and Ti on the critical-current density, a Cu-Nb-Sn composite (called a subelement) produced by Intermagnetics General Corporation was used for this study. This subelement consists of an annular Cu matrix region filled with 1182 Nb filaments and of a central core region which is filled with Sn and occupies ~20% in area. For the purpose of this study, the Sn cores from several ~30 cm long sections of the subelements were replaced with alloyed Sn cores containing the various elements, Mg, Ni, Ni+Zn, and Ti. The amounts of alloying elements in each core are indicated in Table 1.

In order to fabricate wires having sufficiently small filaments, the individual subelements were inserted into a Ta tube used for the Sn diffusion barrier and these in turn were placed inside a pure Cu tube used for the stabilizer. Each composite was then drawn down to 0.63 mm except for a length of wire with the Sn-Ti core which was drawn to 0.16 mm. (See Fig.1 for an example of the cross section of the wire.) At this stage the filaments were ~5 μm and ~1 μm, respectively.

For measurements of critical current, sections of these wires were heat treated on spirally grooved stainless heat treating forms. The heat treatment consisted of 1 day at 340°C + 8 days at 580°C and additional heat treating at higher temperatures, 700, 725, 750 and 800°C for various

Table 1. Compositions of Sn Cores and Nb_3Sn Filaments

Wire	Compositions of Sn Cores (wt%)		Compositions of Filaments[a]	
	Nominal	Measured[b]	(at.%)	Heat Treatment[c]
A	Cu(?)	-	Nb-23.6 Sn	750°C/48 h
B	1 Mg	1.2 Mg	-	-
C	2 Mg	2.0 Mg	Nb-21.1 Sn-0.1 Mg	750°C/48 h
D	1 Ni	1.1 Ni	-	-
E	1 Ni	0.97 Ni	Nb-21.7 Sn-0.2 Ni	800°C/60 h
	-23 Zn	-22.3 Zn	-0.3 Zn	-
F	3 Ti	3.7 Ti	Nb-23.7 Sn-1.5 Ti	750°C/48 h

[a]In this table, the amount of Cu (1-2 at.%) was excluded from the above chemical compositions and renormalized since it is known that essentially all of the Cu in Nb_3Sn was at grain boundaries rather than incorporated into the bulk of Nb_3Sn.
[b]These were measured by atomic absorption.
[c]This heat treatment is after all of the wires received an identical preheat treatment of 340°C/24 h + 580°C/192 h.

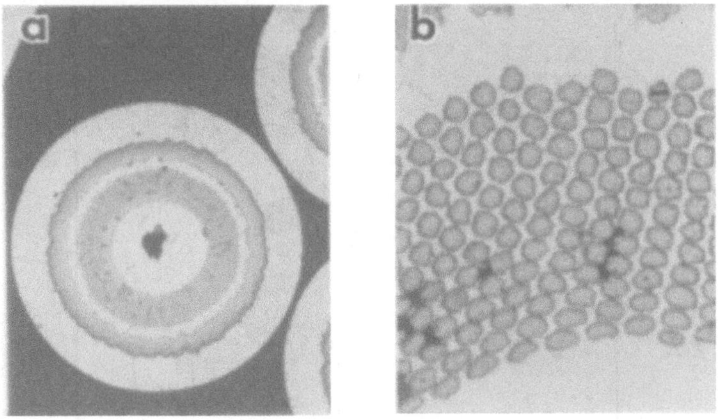

Fig. 1. Optical micrographs of a reacted wire showing (a) the entire cross section and (b) a magnified view of the Nb_3Sn filaments.

durations depending on the wire. The high-temperature heat-treatment condition for each wire is listed in Table 2.

METALLURGICAL CHARACTERIZATION

In order to study the effects of alloying on superconducting properties, the chemical composition and the grain size of alloyed Nb_3Sn filaments in these wires were measured. The chemical compositions were measured using a JEOL JXA-35 scanning electron microscope with a wave length dispersive X-ray detector. The results are listed in Table 1 for representative Nb_3Sn wires. Two wires containing a core of Sn-1 wt% Ni and Sn-1 wt% Mg were not measured for chemical composition since the amount of Mg was too small in Nb_3Sn and the Nb_3Sn layer was too thin for the Ni-alloyed wire. The atomic percent composition was given, excluding a small amount (~1 at.%) of Cu in the filaments. This was done since, as shown previously, most of the Cu in the filaments is located at the grain boundaries rather than incorporated in the lattice of Nb_3Sn.[13]

In contrast to the earlier study which reported relatively large amounts of Mg (~3.5 at.%) was incorporated into the bronze processed Nb_3Sn layer,[10] we found very little Mg in Nb_3Sn. However, a small amount of Mg in Nb_3Sn appears to reduce Sn significantly from the stoichiometric value (25 at.%). A similar reduction of Sn in Nb_3Sn was also observed when a small amount of Zr (~0.1 at.%) was incorporated into a Nb_3Sn layer.[5] Although both Ni and Zn are incorporated into Nb_3Sn filaments, the amounts are still very small, and these elemental additions also depress the Sn content in Nb_3Sn. The amount of Ti in the filaments varied considerably (0.7 to 2.5 at.%) depending on the location of the filaments. Those filaments on the outside rim contained the smallest amount of Ti and those near the Sn core contained the largest. In all cases, due to a higher concentration of Sn near the core, the growth of the compound is faster at the inner rings than on the outside rings.

In all of the wires the grain size of the alloyed Nb_3Sn filaments was observed from the fracture surfaces of the wires using a scanning electron microscope (JEOL JXA-35). The grains of alloyed Nb_3Sn were approximately equal (0.2-0.5 µm) to those for the pure Nb_3Sn except those for the Sn-1 Ni core when all of the wires were heat treated at 750°C for 48 h. The Nb_3Sn grains which were alloyed with a Sn-1 Ni core were approximately 5 times larger than those for the other alloys.

Table 2. Critical-Current Densities (Exclude Cu and the Barrier)

	Specimen Identification	Heat Treatment[a]	Critical-Current Densities (A/mm^2)[b] 10 T	14 T	18 T	Sharpness Factor "n"[c] 10 T	14 T	18 T
A	Pure	700°C/96 h	1211	370	31	-	29	10
B	Sn-1 Mg	700°C/96 h	812	241	24	23	17.5	8
		750°C/48 h	868	312	40	32	17	9
C	Sn-2 Mg	700°C/96 h	973	299	31	50	25	10
		750°C/48 h	759	270	40	26	12	7
D	Sn-1 Ni	750°C/48 h	512	193	15.6	51	23	8
		800°C/24 h	450	219	39	72	31	13
E	Sn-1 Ni-23 Zn	750°C/48 h	639	256	41.6	11	8	11
		800°C/24 h	440	217	65.5	10	9	12
F	Sn-3 Ti	750°C/48 h	928	496	218	70	58	38
		800°C/24 h	408	242	148	29	29	27
		625°C/144 h[d]	1650	-	-	56	-	-

[a]All specimens received an identical preheat treatment: 340°C/24 h + 580°C/192 h.
[b]The critical current criterion was 0.01 μV/mm.
[c]n is defined as $V \sim I^n$
[d]This wire had the Nb filaments of ~1 μm rather than ~5 μm for other wires and received a preheat treatment: 340°C/24 h + 580°C/24 h.

However, when Ni was added to Sn with Zn, the grains were reduced to the size equal to or somewhat smaller than those in other alloys. These grains were more equiaxed than the others.

SUPERCONDUCTING PROPERTIES

Superconducting critical currents I_c as well as critical temperature T_c of these wires were measured to determine the influence of the alloy cores of internal-tin processed Nb_3Sn wires. T_c was measured by placing two to three pieces of approximately 5 mm wires into a balanced pair of secondary coils. The onset (90% of the change in the inductive signal) critical temperature for pure Nb_3Sn was 17.65 and 17.8 K for wires heat treated for 48 h at 750°C and 24 h at 800°C, respectively. All of the alloy additions reduced T_c (onset) by 0.1 ~ 0.2 K. It is interesting to note that reduction of T_c due to alloying, particularly with Ni+Zn and Mg, is quite small in spite of a very large reduction in Sn content due to the additions of these elements (see Table 1). A similar observation[5] was also made when Nb_3Sn by the bronze process was alloyed with Zr, and it is not understood why T_c does not drop more with an observed large decrease in the Sn content.[5]

Superconducting critical currents I_c were measured on helically wound wires using the technique which was previously described.[14] The specimen was ~600 mm long and the voltage taps were placed in the middle 200 mm. The critical current produced 2 μV across the 200 mm and the measurements were performed using a 19 or 20 T Bitter solenoid at the Francis Bitter Magnet National Laboratory as well as in a 11 T superconducting magnet.

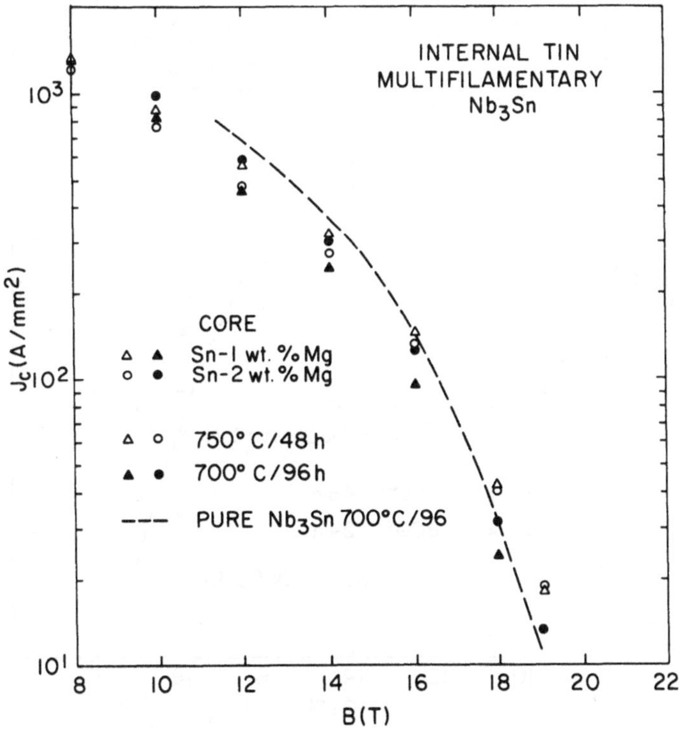

Fig. 2. Critical-current density vs magnetic field for Mg alloyed Nb_3Sn wires.

The results of the measurements on three types of alloyed wires as well as for pure Nb_3Sn wire are shown in Figs. 2-4 and in Table 2. (J_c is calculated for the area inside the barrier.) These results point out that additions of Mg and Ni or Ni+Zn (Figs. 3 and 4) into the Sn core are not very effective in improving critical-current density J_c in contrast to earlier reports for the bronze processed Nb_3Sn wires.[9,10] In both cases, the amounts of alloying elements in Nb_3Sn filaments by the internal-tin process are substantially lower than those reported for the bronze process. However, the combined addition of Zn and Ni to the core appears to put small amounts of Zn and Ni in the Nb_3Sn lattice and to raise the critical field H_{c2}, thus the values of J_c at very high fields was increased by a small amount in comparison with the addition of Ti as described below.

In contrast to the above results, very large improvements in the values of J_c were measured on wires with a 3 wt% Ti addition to the Sn core, and the results were compared with the best values of J_c which were measured in the bronze-process $Nb_3Sn(Ti)$ wires.[14,15] As shown in Fig. 4, J_c in the wire containing ~5 μm Nb filaments by the internal-tin process was as high as the best bronze-processed wire with a similar filament size of Nb_3Sn. Furthermore, when the Nb filaments were reduced to ~1.2 μm before reaction, it was possible to achieve very high-current density at the medium field range (~10 T) as shown in Table 2. Thus, it was demonstrated that by an addition of ~3 wt% Ti to the Sn core, multifilamentary Nb_3Sn by the internal-tin process can produce wires with one of the highest values of J_c at intermediate as well as very high magnetic fields by choosing a proper filament size and a corresponding heat treatment.

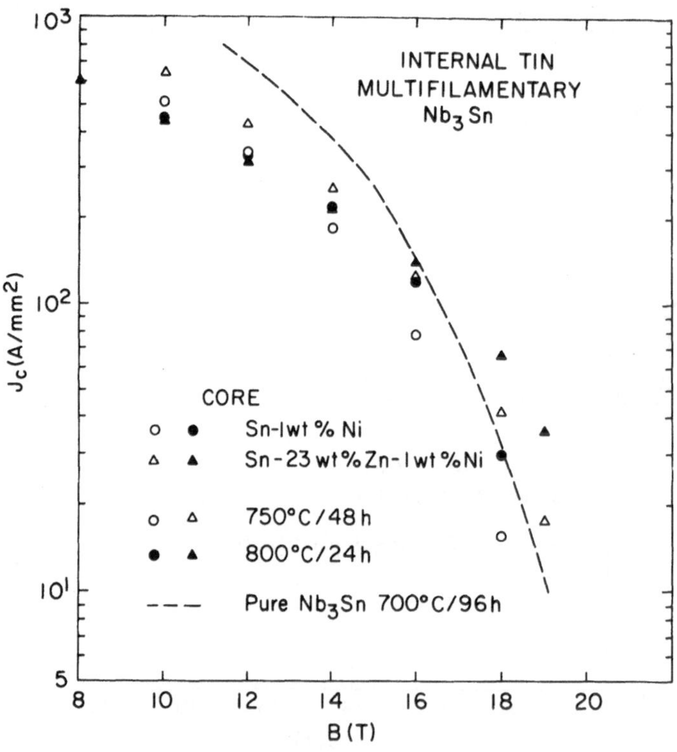

Fig. 3. Critical-current density vs magnetic field for Ni and Ni+Zn alloyed Nb_3Sn wires.

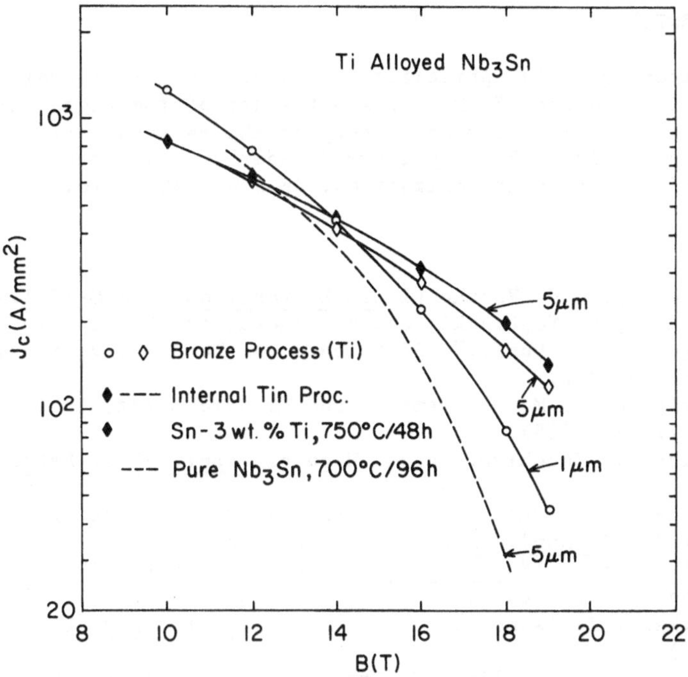

Fig. 4. Critical-current density vs magnetic field for the Ti alloyed Nb_3Sn wires by the internal-tin process in comparison with the similar Ti Nb_3Sn wires by the bronze process.

Also listed in Table 2 are the values of n(B) which is defined as $V \sim I^n$ where V and I are the voltage and the transport current in the wire at the superconducting to the normal transition state in a given magnetic field. As pointed out previously,[14,16,17] the exponential factor n is a convenient means for quantifying the sharpness of the transition. Also, it was shown that the low values of n and J_c are associated with nonuniformity in superconducting filaments.[14] Furthermore, the origin of the low n[18] was suggested to be due to the premature current sharing among nonuniform filaments and this could be described by an extension of a model to estimate remanent resistivity in <u>in situ</u> wires.[19,20] This particular approach to associate the value of n to physically observable factors was quite successful in describing the sharpness of the transition in many cases of the bronze-processed Nb_3Sn[14] and NbTi wires.[21,22] However, in the present wires, uniformity of the Nb filaments in each of the wires are presumably identical. As shown in Table 2, great variations in the values of n are measured. Thus, the nonuniformity in Nb_3Sn filaments is not a main controlling factor in determining the value of n in these wires.

Another possible factor is the limit in superconductivity in the filaments, i.e., at high-magnetic field or temperature, pinning strength of the filaments become weak and flux flow which may result in lowering of n. However, there is not a consistent trend between the values of J_c and n among the present wires with regard to the values of H_{c2}. It is shown that the value of n is not always an indication in uniformity of the filaments and there appears to be other unknown factors strongly influencing the value, and it requires further study to clarify this puzzle.

ACKNOWLEDGMENTS

The authors greatly appreciate the expert technical assistance from A. Cendrowski, D. Horne, F. Perez, and the use of the high-field magnets at the Francis Bitter Magnet Laboratory at the Massachusetts Institute of Technology. This work was performed under the auspices of the U.S. Department of Energy under Contract No. DE-AC02-76CH00016.

REFERENCES

1. See for example, K. Yasukochi, IEEE Trans. Magn. MAG-19:194 (1983).
2. Y. Hashimoto, K. Yoshizaki, and M. Tanaka, in: "Proc. 5th Intern. Cryogenic Eng. Conf.," IPC Science and Technology Press, London, (1974), p. 332.
3. B. A. Zeitlin, G. M. Ozeryansky, and K. Hemachalam, IEEE Trans. Magn. MAG-21:293 (1985).
4. M. Suenaga, K. Tsuchiya, and N. Higuchi, Appl. Phys. Lett. 49:919 (1984).
5. M. Suenaga, D. O. Welch, R. L. Sabatini, O. F. Kammerer, and S. Okuda, J. Appl. Phys., to be published.
6. J. D. Livingston, IEEE Trans. on Magn. MAG-14:611 (1978).
7. H. Sekine and K. Tachikawa, Appl. Phys. Lett. 35:472 (1979).
8. K. Tachikawa, T. Asano, and T. Takenchi, Appl. Phys. Lett. 39:766 (1981).
9. R. Flükiger, W. Speckling, E. Drost, and L. Oddi, in: "Proc. of ICMC,", K. Tachikawa and A. F. Clark, eds., Butterworth, London, (1982), p. 75; R. Flükiger, E. Drost, W. Goldacker, and W. Speckling, IEEE Trans. on Magn. MAG-19:1441 (1983); E. Drost, R. Flükiger and W. Speckling, Cryogenics 24:622 (1984).
10. I. W. Wu, D. R. Dietderich, J. T. Holthuis, W. V. Hassenzahl, and J. W. Morris, Jr., IEEE Trans. on Magn. MAG-19:1437 (1983); I. W. Wu, D. R. Dietderich, P. E. Johnson-Walls, J. Glaser, J. T. Holthuis, W. V. Hassenzahl, and J. W. Morris, Jr., Appl. Phys. Lett. 45:792 (1984).
11. K. Yoshizaki, M. Wakata, S. Hiyashita, F. Fujiwara, O. Taguchi, M. Imaizumi, and Y. Hashimoto, IEEE Trans. on Magn. MAG-21:301 (1985).
12. M. Suenaga, C. J. Klamut, N. Higuchi, and T. Kuroda, IEEE Trans. on Magn. MAG-21:305 (1985).
13. M. Suenaga and W. Jansen, Appl. Phys. Lett. 43:791 (1983).
14. M. Suenaga, K. Tsuchiya, N. Higuchi, and K. Tachikawa, Cryogenics 25:123 (1985).
15. Y. Tanaka, M. Idada, and H. Tanaka, to be published in: "Proc. of 1985 Particle Accelerator Conf."
16. M. Garber, W. B. Sampson, and M. J. Tannenbaum, IEEE Trans. on Magn. MAG-19:720 (1982).
17. L. F. Goodrich and F. R. Frickett, Cryogenics 22:225 (1982).
18. M. Suenaga, unpublished.
19. A. Davidson, M. R. Beasley, and M. Tinkham, IEEE Trans. on Magn. MAG-11:276 (1975).
20. W. J. Carr, Jr., Adv. Cryogenic Eng. 30:899 (1984).
21. M. Garber, M. Suenaga, W. B. Sampson, and R. L. Sabatini, to be published in: "Proc. 1985 Particle Accelerator Conf."
22. M. Garber, M. Suenaga, W. B. Sampson, and R. L. Sabatini, in: this Proceedings.

NEW METHODS FOR OBTAINING METASTABLE METALLIC ALLOYS

Frans Spaepen

Division of Applied Sciences
Harvard University
Cambridge, Massachusetts

ABSTRACT

The thermodynamic and kinetic conditions for metastable phase formation in three new areas are discussed: melt quenching by picosecond pulsed laser irradiation, amorphous phase formation by interdiffusion of elemental crystalline metals, and the non-equilibrium extension of the latter process by ion bombardment or mechanical deformation.

INTRODUCTION

Metastable phases are generally created by raising the free energy of the (stable) starting materials, followed by "quenching", or de-energization, which allows the system to transform into or through any of the lower free energy states. At the end of the quench, the system can remain in a metastable, or even unstable state if the atomic mobility is low enough. It is then *kinetically* constrained to a configurationally frozen state. The initial energization methods include heating to form a vapor or melt, dissolution, ion bombardment, as in sputtering or ion implantation, or plastic deformation. An extensive review of these processes, their attendent quench rates, and the configurational freezing phenomenon has been given by Turnbull in his 1980 Campbell Memorial lecture.[1] In this paper, some new methods, developed over the last five years, are discussed. They are: metallic alloy formation by picosecond pulsed laser quenching, formation of amorphous metallic phases by interdiffusion and reaction of the elemental metals, and the enhancement of the latter process by ion bombardment or plastic deformation.

PICOSECOND PULSED LASER QUENCHING

Pulsed laser irradiation of a crystal surface is a quenching method that allows confinement of the energy deposition to a very thin surface layer. This layer is melted and, as a result of the very large thermal gradient, is subsequently quenched at rates of 10^{12} K s^{-1} or higher. The mechanism of pulsed laser quenching has been analyzed in detail in a number of publications.[2,3] The process can be divided into three stages, as illustrated by Figure 1 for a specific example.

In the first stage, while the pulse is on, the energy is deposited in a layer of thickness α^{-1} (α: absorption coefficient). Since the transfer

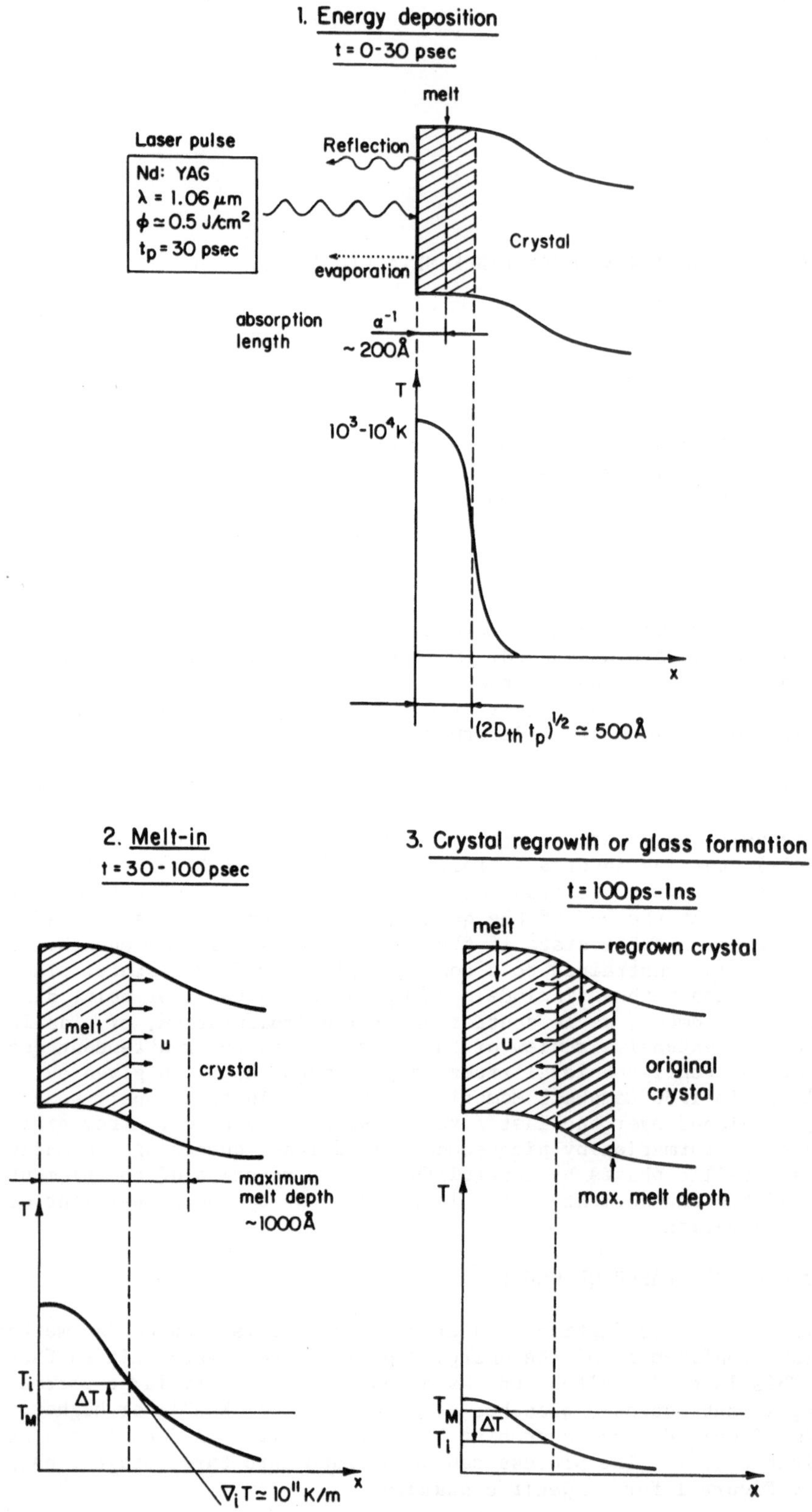

Figure 1: Schematic representation of the successive stages of the picosecond pulsed laser quenching process.

of the energy from the electrons, which interact with the laser, to the lattice occurs on a timescale of about 1 ps,[4] the energy deposition process may be treated as a thermal one. The thermal diffusion length over the duration of the pulse is $\ell_T = (D_{th}t_p)^{1/2}$ (D_{th}: thermal diffusivity, t_p: pulse duration). As discussed by Bloembergen,[2] for irradiation of metals with a pulse of a picosecond or longer ℓ_T is greater than α^{-1}. In such an experiment, therefore, the minimum thickness of the initial molten layer and the maximum thermal gradients are determined by the length scale ℓ_T. It is worth noting that for a pulse length less than 1 ps, ℓ_T is less than α^{-1}, which means that the quench rate in this regime becomes independent of the pulse length. Femtosecond laser pulses, although very useful for time-resolved probing of the irradiation process, therefore do not produce higher quench rates than picosecond pulses.

In the second stage of the process illustrated in Figure 1, which starts after the end of the pulse, more of the underlying crystal is melted, until the overheat of the melt is spent. The crystal-melt interface moves in until its temperature drops to the equilibrium melting temperature.

In the third stage of Figure 1, the interface temperature falls below the equilibrium melting temperature and the crystal-melt interface reverses direction. If the crystal growth is fast enough, the entire melt is consumed by the regrowth process; otherwise, the melt transforms into a glass. The lifetime of the melt, τ, under the conditions of Figure 1, is known from both heat flow calculations[2,3] and time-resolved reflectivity measurements[5-7] to be about 1 ns. This immediately shows that the cooling rate in this process must be very high: $\dot{T} \approx 10^3\ K/\tau = 10^{12}\ K\ s^{-1}$. Since the cooling rates in "conventional" melt-quenching techniques, such as melt spinning, are only about $10^6\ K\ s^{-1}$, it is not surprising that the picosecond laser quenching technique can produce many more metastable phases.

The short lifetime of the melt creates a special experimental problem in alloy formation. Since the mixing length in the liquid, $(D\tau)^{1/2}$, is only about 3nm, the alloy components must be mixed on this scale in the starting material. This can be accomplished, for example, by ion implantation or coevaporation. A particularly convenient method is to deposit alternate layers of the elemental metals, each on the order of 1 nm thick, to form a multilayered (compositionally modulated) film about 100 nm thick (the maximum melt depth). This type of deposition can conveniently be performed by any multiple source apparatus, such as DC[8,9] or ion beam[10] sputtering. An additional advantage of preparing the starting material as a thin film is that it can be supported on a dissolvable backing, which facilitates TEM observation after irradiation.

The short time scale of the picosecond laser quenching process also puts restrictions on the type of phase transformations that can occur. If the crystal regrowth (stage 3 of Figure 1) is to consume the entire melted layer thickness ($\sim$ 100 nm), the crystal growth velocity must be on the order of $100\ nm/\tau = 100\ m\ s^{-1}$. Since the diffusive speed in the liquid phase, D/λ (D: liquid diffusivity $\sim 10^{-8} - 10^{-9}\ m^2s^{-1}$; λ: interatomic distance), is on the order of $10\ m\ s^{-1}$, no solute redistribution can occur. The only type of transformation that can occur, therefore, is *partitionless* solidification, either by growth of a crystal of the same composition as the melt, or, if this is thermodynamically prohibited, by the formation of a glass. Figure 2 illustrates the thermodynamic conditions for partitionless crystallization on a schematic phase diagram, containing two primary solid solutions (α,δ) and two compounds (β,γ). The free energy diagram shows that at temperature T_1 a partitionless transition from the liquid (ℓ) to the δ-phase is only possible for compositions $x > x_{o,\delta\ell}$. The locus of the points $x_{o,\delta\ell}$ for each temperature forms the $T_{o,\delta\ell}$ line. Partitionless crystallization to

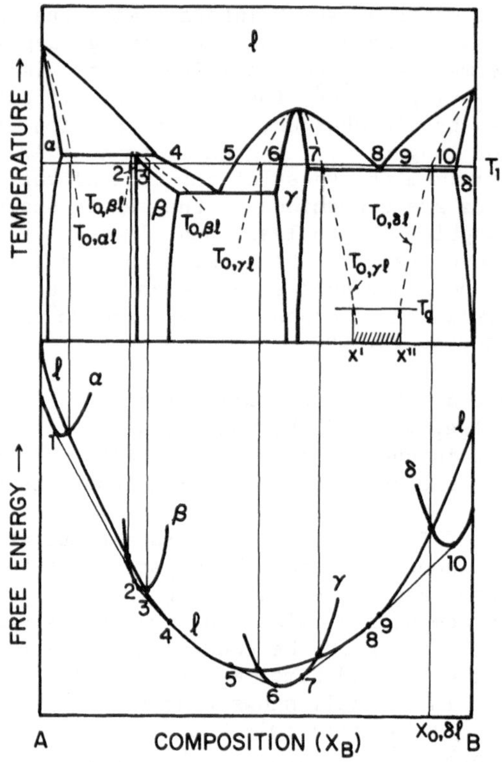

Figure 2: Schematic phase diagram (top), and corresponding free energy diagram at temperature T_1 (bottom), illustrating the construction of the T_o-lines for the primary solutions and the intermetallic compounds. In the dashed composition range between x' and x" only glass formation is possible under conditions of partitionless solidification. The numbers correspond to the same compositions in the top and bottom diagrams.

the δ-phase can only occur below this T_o- line. Similar T_o-lines are shown for all the other crystallization processes. At a low enough temperature, T_g, the rate of atomic transport in the liquid becomes small enough to preclude any transformation. The liquid is then configurationally frozen and becomes a glass. For a composition $x' < x < x''$ on Figure 2, no partitionless crystallization is possible, and a glass is formed in picosecond laser quenching. The T_o-lines, however, only delineate a *sufficient* condition for glass formation: most picosecond laser quenching experiments show glass formation even below the T_o-lines. It is therefore necessary to consider also the *kinetic* aspects of the crystal growth process.

As discussed by Spaepen and Turnbull,[11] the crystal growth velocity, u, can be written as:

$$u = fk\lambda \ [1 - \exp(\frac{\Delta G}{RT_i})] \qquad (1)$$

where k: atom jump frequency across the crystal-melt interface; f: fraction of crystal surface sites that can incorporate a new atom; $f \approx 1$ for metals; ΔG: difference in free energy between crystal and melt (negative); T_i: interface temperature.

If T_i is only moderately below the melting temperature T_M, this equation can be linearized:

$$u = k\lambda\beta \frac{T_M - T_i}{T_i} \tag{2}$$

where: $\beta = \Delta S_f/R \approx 1$ for metals (ΔS_f: molar entropy of fusion).

The crystal growth rate is also determined by the rate of removal of the heat of transformation

$$u_{th} = \frac{\kappa \bar{V} \nabla_i T}{\Delta H_f} \tag{3}$$

where: κ: thermal conductivity; $\bar{V}$: molar volume; $\nabla_i T$: temperature gradient at the interface; ΔH_f: molar enthalpy of fusion.

In picosecond laser quenching $\nabla_i T$ is very large, and u_{th} is on the order of $100\ m\ s^{-1}$. The interface temperature is then found by combining equations (2) and (3):

$$\frac{u_{th}}{k\lambda} = \frac{T_M - T_i}{T_i} \tag{4}$$

In pure metals and some dilute solutions, the jump frequency k is the thermal vibration frequency, and the maximum growth velocity, $u_{max} = k\lambda$, is close to the speed of sound.[12] In this regime, crystal growth is *collision-limited*, and even in picosecond laser quenching u_{th} is much smaller than $k\lambda$. Equation (4) then shows that the undercooling $\Delta T = T_M - T_i$ is small. The presence of large crystals (1 μm lateral diameter) following picosecond laser quenching experiments on pure metals or dilute alloys[13,14] allows an experimental estimate of a minimum velocity of $(.5\ \mu m)/\tau = 500\ m\ s^{-1}$ for collision limited growth.

In concentrated alloys, the jump frequency is the diffusional one, and the maximum growth velocity is then $u_{max} = k\lambda = D/\lambda$, which is on the order of $10\ m\ s^{-1}$. In this regime, crystal growth is *diffusion-limited*, and equation (4) shows that large undercoolings are possible, so that glasses can be formed. The formation of glasses below the T_0-line in many systems[15-17] shows that partitionless crystallization can be a diffusion-limited process. Even though no *long-range* compositional changes are allowed, it may still be necessary to change the short range chemical ordering, which requires changes of nearest neighbors. For example, most intermetallic compounds are not formed in picosecond laser quenching,[16,17] presumably due to either their large unit cell or ordered structure. Similarly the growth of primary solid solutions can also be diffusion-controlled if the short range order in the liquid is different from that in the crystal. For example in the Fe-B system, the nearest neighbor environment of a B-atom in the amorphous phase is known to consist of 9 Fe atoms in a trigonal prismatic arrangement,[18] which is very different from the environment around a B interstitial in a Fe(B) bcc solid solution.

That Fe-B glasses can be produced by picosecond laser quenching with as little as 5 at.% B,[15,19] far below the T_0-line, can be explained if one-and-a-half coordination shell around each B requires diffusional rearrangement upon crystallization. The minimum solute content to form metal-metal glasses, such as Ni-Nb, is higher (10-20 at.%),[17] possibly due to a lesser degree of interaction between solute and solvent atoms.

The concentrated alloys that do crystallize in picosecond laser quenching are therefore those that have simple crystal structures. For example, disordered fcc Cu-Co has been obtained at equiatomic composition.[14] This is to be expected, since the crystal growth kinetics of a *disordered* fcc alloy are likely to be similar to those of a pure fcc crystal.

A final illustration of the power of picosecond laser quenching is the production of fcc solid solutions in the Nb-Si system over the range 10-27 at.% Si.[20] So far, this phase had only been prepared at 22 at.% Si by shock compression.[21]

FORMATION OF AMORPHOUS ALLOYS BY SOLID STATE REACTION OF THE ELEMENTAL METALS

Figure 3 shows a schematic phase diagram with two primary solid solutions (α,γ) and an intermetallic compound β. At the temperature T_1, the β phase is the stable one between compositions x_4 and x_5; in the ranges $x_1 - x_4$ and $x_5 - x_8$, two-phase equilibrium between β and, respectively, α and γ is established. If the presence of the β phase is kinetically suppressed, however, a different (metastable) equilibrium is established: between compositions x_3 and x_6, the liquid phase occurs,

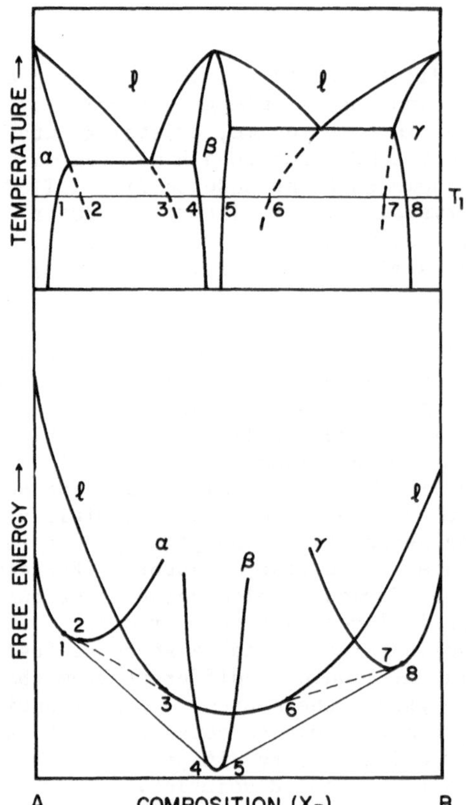

Figure 3: Schematic phase diagram (top), and corresponding free energy diagram at temperature T_1 (bottom). The solid lines represent stable equilibrium (with the β phase present). The dashed lines represent the metastable extension in the absence of the β phase. The numbers correspond to the same compositions in the top and bottom diagrams.

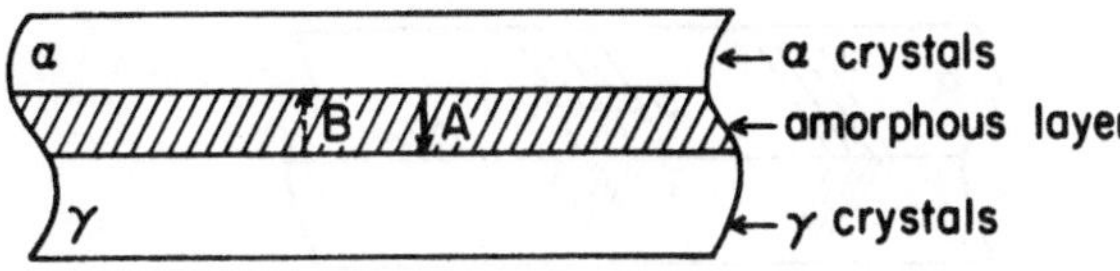

Figure 4: Schematic diagram of the amorphous phase formation in a bilayer of the elemental crystalline metals A and B from Figure 3. A is the dominant diffusing species through the amorphous layer.

and in the ranges $x_2 - x_3$ and $x_6 - x_7$, the liquid is in (metastable) equilibrium with, respectively, α and γ. These metastable phase fields can be estimated conveniently by extending the liquidus and solidus lines on the phase diagram (dashed lines on Figure 3). If the temperature T_1 is below the glass transition temperature of the liquid, the (solid) amorphous phase can be in metastable equilibrium with the primary solid solution.

Since the rate of atomic transport in the amorphous phase can be reasonably high with negligible crystallization occurring,[22] it is possible to establish this metastable equilibrium by interdiffusion of the elemental components, as Johnson and co-workers demonstrated for a number of systems.[23-30] A typical bilayer experiment is schematically shown in Figure 4. At the start of the experiment, only pure A and B are present. The bilayer is annealed at a temperature that is high enough to allow atomic transport, but low enough to prevent rapid crystallization. Initially some B atoms dissolve into α, and A atoms into γ, and after compositions x_2 or, respectively, x_1 are reached, an amorphous layer begins to form. In metastable equilibrium the degree of completion of the crystalline-to-amorphous ($c \rightarrow a$) reaction depends on the average composition. In the Au-La system, Schwartz and Johnson[23] found fairly good agreement between the relative amounts of crystal and amorphous phase observed experimentally, and those estimated from the free energy curves of the phases. An extrapolation of the liquidus lines of the primary Au and La phases to the reaction temperature gives qualitatively similar phase fields.

In order for metastable equilibrium to be established, diffusional equilibrium must occur throughout the sample. In the mechanism shown in Figure 4, the rate-limiting process is the diffusion in the amorphous phase. In order for rapid crystallization not to occur, the diffusivity in the amorphous phase must be less than $10^{-20}\ m^2s^{-1}$. At this limit, no crystallization is observed for annealing times up to 10^4 s.[22] This corresponds to a diffusion distance, $(Dt)^{1/2}$, on the order of 10 nm. Due to the rough scaling of the diffusivity and the crystallization rate, this distance is fairly independent of temperature. In order to increase the amount of material transformed, the crystal-to-amorphous reaction experiments are usually performed on multilayered films with a repeat length on the order of 10 nm. It should be kept in mind that even after completion of the reaction the multilayers may still not be entirely equilibrated, and still contain concentration gradients.

Factors that favor formation of an amorphous phase by the $c \rightarrow a$ reaction include:

(i) A large negative heat of mixing in the liquid/amorphous (ℓ/a) phase. It is clear that this lowers the free energy of the ℓ/a phase relative to the pure elements and their primary solutions, which makes it more likely for the ℓ/a phase to be in metastable equilibrium at a temperature that is low enough for negligible crystallization to occur. A

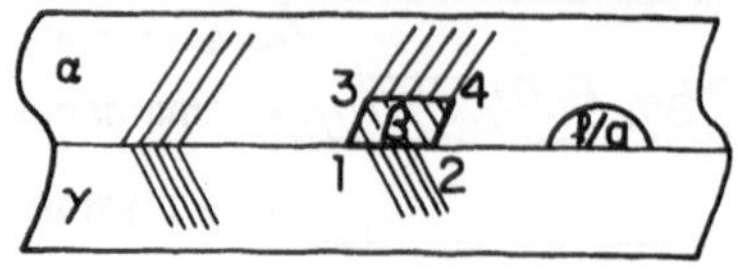

Figure 5: Schematic diagram illustrating that the nucleation of the liquid/amorphous (ℓ/a) phase is kinetically favored over that of the intermetallic compounds (see text).

large negative heat of mixing also favors formation of intermetallic compounds, which must be suppressed kinetically, as discussed below. Experimentally, the $c \to a$ reaction has indeed only been observed in systems with a large negative heat of mixing. It is interesting that Ni/Nb, Ni/Zr and Co/Nb multilayers (2 nm repeat length) produced by DC sputtering are amorphous, whereas Ni/Mo and Co/Mo multilayers produced by the same method are crystalline.[16,17] The heat of formation of the latter systems is indeed much less negative.[31]

(ii) A difference in the diffusion rates of the two species in the amorphous phase. Many of the diffusion couples that yield the amorphous phase are known to exhibit an asymmetry in their crystalline diffusion behavior: A diffuses considerably more rapidly in crystalline B than vice-versa. Cheng et al.[32] demonstrated that Ni was the dominant moving species in the Ni-Zr $c \to a$ reaction by Rutherford backscattering measurements of the relative motion of a W marker and the $c - a$ interface in a Ni/Zr bilayer. Schröder et al.,[33] using cross-sectional TEM, showed that Co was the dominant moving species in the $c \to a$ reaction in a Co/Zr multilayer from the asymmetrical thickening of the amorphous layer. One possible explanation of the importance of the diffusional asymmetry for the $c \to a$ reaction may be that the formation of the amorphous phase from the primary solid solution requires only very short range atomic rearrangements that can be accomplished by the motion of one species only, whereas the formation of the intermetallic compound, which often has a more specific stoechiometry, a large unit cell and strict chemical order, may require the cooperative motion of both species.

(iii) The nucleation of the amorphous phase at a grain boundary is kinetically favored over that of a crystalline intermetallic compound. As illustrated schematically in Figure 5, the β nucleus at the α/γ grain boundary can form a well-matching (low energy) interface (1 - 2) with one phase, but then only an incoherent one with the other phase (3 - 4). Since the surface tension of a crystal-melt interface is much lower than that of an incoherent crystalline interface, the nucleation of the ℓ/a phase is generally favored. The TEM observations of Schröder et al.[33] showed that the amorphous-crystalline interface remained parallel throughout the interdiffusion reaction. This means that nucleation of the amorphous phase was indeed copious along the original boundaries. Their observations also support the notion that diffusion through the amorphous phase (as in Figure 4) is the rate-limiting process in the $c \to a$ reaction.

Finally, it should be noted that the $c \to a$ transition was first observed as a result of a reaction of gaseous hydrogen with the crystalline intermetallic compound Zr_3Rh, to form amorphous $Zr_3RhH_{5.5}$.[34]

NON-EQUILIBRIUM EFFECTS

In the previous section, the formation of the amorphous phase under metastable equilibrium conditions was considered. In a simple eutectic system, such as that of Figure 6, the $c \to a$ reaction can therefore not occur under

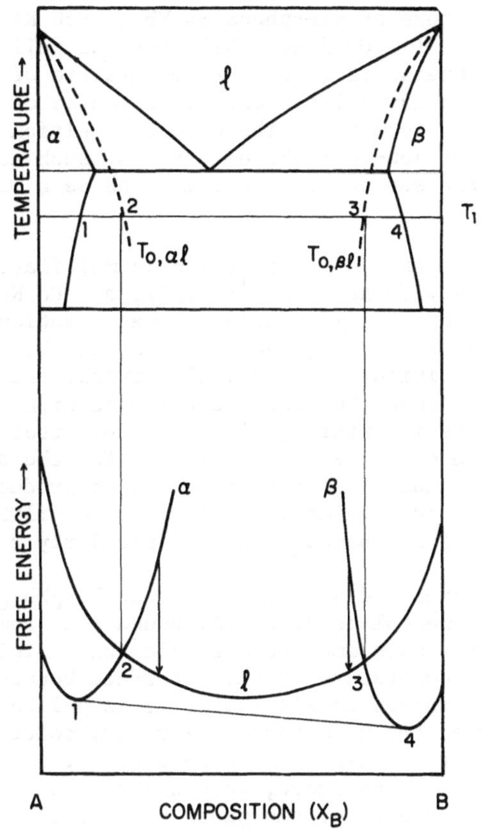

Figure 6: Schematic simple eutectic phase diagram (top), and corresponding free energy diagram at temperature T_1 (bottom), illustrating the non-equilibrium conditions that allow the crystalline-to-amorphous reaction to occur below the eutectic temperature. The numbers correspond to the same compositions in the top and bottom diagrams.

these conditions: at a temperature, T_1, below the eutectic temperature, the free energy of the ℓ/a phase is always higher than that of the primary solid solution, α and β, in equilibrium (compositions x_1 and x_2). However, if the free energy of the α and β solutions can somehow be raised above that of the ℓ/a phase, the $c \rightarrow a$ transition becomes thermodynamically possible. For example, if the solid solutions are made supersaturated past the T_o points (compositions x_2 and x_3), partitionless $c \rightarrow a$ transitions are possible (see arrows on Figure 6). This supersaturation can be accomplished, for example by ion bombardment, either by direct implantation[35] or as a result of substrate bias in DC sputtering,[8,9,17] or by large amounts of plastic deformation, as in mechanical alloying.

That non-equilibrium effects must occur in the $c \rightarrow a$ reaction is clear from a number of observations. In the Ni-Nb system, multilayered films (period ~2 nm), produced by DC bias sputtering of the elemental metals from alternating targets, were *entirely* amorphous in their as-prepared form from average film compositions between 40 and 79 at.% Ni.[17,37] That diffusional equilibrium was incomplete in these films was clear from the remaining compositional inhomogeneity, observed by the X-ray diffraction satellite resulting from the composition modulation.[8,9,38–40] Furthermore, extrapolation of the solidus and liquidus of the primary solutions to the glass transition temperature of amorphous Ni-Nb (~ 950 K) gives a very narrow composition range (~ 60 ± 5 at.% Ni) for the fully amorphous phase field. The reaction temperature (i.e. the temperature of the water-cooled substrate) was also far below all reasonable estimates of the (ℓ/a)-primary solutions metastable eutectic. The wide composition range can therefore only be explained by non-equilibrium effects, such as mixing induced by the bombardment of the substrate by plasma ions as a result of the bias voltage.

As mentioned earlier, the $c \rightarrow a$ reaction in multilayer sputtering is only observed in systems such as Ni/Nb,[17] Ni/Zr, and Co/Nb [16] that satisfy the favorable criteria for the diffusional $c \rightarrow a$ reaction discussed above. Other elemental multilayer systems, such as Co/Mo and Ni/Mo, produced under identical sputtering conditions, are entirely crystalline.[16] This may mean that the $c \rightarrow a$ reaction under DC bias sputter conditions is formed by interdiffusion as well as ion mixing, although the latter process would also be enhanced by the large negative heat of mixing for the amorphous phase. Entirely amorphous Cu/Zr multilayers[41,42] have been produced by ion beam sputter deposition[10] of the elemental metals. Non-equilibrium effects can arise here from the substrate bombardment by specularly reflected ions.

Koch et al.[36] have prepared a fully amorphous $Ni_{60}Nb_{40}$ alloy by mechanical alloying of the elemental powders. Although this composition corresponds to the narrow metastable amorphous single phase field discussed above, and although the local temperature at the interface between the deforming powder particles can be raised considerably, it is unlikely that the narrow temperature range between the metastable ℓ-primary solutions eutectic temperature and the rapid crystallization temperature of the amorphous alloy is consistently reached. Furthermore, Koch and collaborators have recently extended the composition range for amorphous phase formation in this system considerably.[43] It is therefore clear that non-equilibrium effects must play an important role in the $c \rightarrow a$ reaction during mechanical alloying. That the mechanical work extended in the process is sufficient to produce free energy changes in the primary solid solutions, such as those shown in Figure 6, can be seen from the following simple estimate. The required free energy increase in the primary solutions is on the order of $\Delta S_f(T_e - T_{mix})$ (ΔS_f: entropy of fusion, $\sim R$ for metals; T_e: eutectic temperature, T_{mix}: mixing temperature), which is on the order of a few kJ per mole. The mechanical work per mole expended in plastic deformation is $\sigma \varepsilon \overline{V}$ (σ: flow stress, $\geqslant 10^{-3}\,\mu$ in work-hardened metals; μ: shear modulus; ε: strain; $\overline{V}$: molar volume). A plastic strain on the order of 10, which is much less than the total strain in mechanical alloying, is sufficient to provide these few kJ per mole.

ACKNOWLEDGEMENTS

It is a pleasure to acknowledge many useful discussions with David Turnbull on all aspects of this paper. I am grateful to C.C. Koch and W.L. Johnson for making their latest results available prior to publication. The reported work on picosecond laser quenching has been carried out in collaboration with C.J. Lin and W.K. Wang, and has been supported by the Office of Naval Research under contract N00014-K-83-0030.

REFERENCES

1. D. Turnbull, Met. Trans. A 12:695 (1981).
2. N. Bloembergen, in "Laser-Solid Interactions and Laser Processing", S.D. Ferris, H.J. Leamy and J.M. Poate, eds., AIP, New York (1979), p. 1.
3. F. Spaepen and C.J. Lin, in "Amorphous Metals and Non-Equilibrium Processing", M. Von Allmen, ed., Les Editions de Physique, Les Ulis, France (1984), p. 65.

4. For a review see W.L. Brown, Mat. Res. Soc. Symp. Proc., vol. 23, 9 (1984).
5. J.M. Liu, R. Yen, H. Kurz, and N. Bloembergen, Appl. Phys. Lett. 39: 755 (1981).
6. P.H. Bucksbaum and J. Bokor, Mat. Res. Soc. Symp. Proc., vol. 13, 51 (1983).
7. C.A. MacDonald, A.M. Malvezzi, and F. Spaepen, Mat. Res. Soc. Symp. Proc., to be published.
8. M.P. Rosenblum, F. Spaepen, and D. Turnbull, Appl. Phys. Lett. 37:184 (1982).
9. A.L. Greer, C.J. Lin, and F. Spaepen, Proc. 4th Int. Conf. on Rapidly Quenched Metals, Jap. Inst. Metals, Sendai (1982), p. 567.
10. F. Spaepen, A.L. Greer, K.F. Kelton, and J.L. Bell, Rev. Sci. Inst. 56:1343 (1985).
11. F. Spaepen and D. Turnbull, in "Laser Processing of Semiconductors", J.M. Poate and J.W. Mayer, eds., Academic, New York (1982), p. 15.
12. S.R. Coriell and D. Turnbull, Acta Metall. 30:2135 (1982).
13. C.J. Lin and F. Spaepen, in "Chemistry and Physics of Rapidly Solidified Materials", B.J. Berkowitz and R.O. Scattergood, eds., TMS-AIME, New York (1983), p. 273.
14. C.J. Lin and F. Spaepen, Mat. Res. Soc. Symp. Proc., vol. 28, 75 (1984).
15. C.J. Lin and F. Spaepen, Appl. Phys. Lett. 41:721 (1982).
16. C.J. Lin, F. Spaepen, and D. Turnbull, J. Non-Cryst. Solids 61/62: 767 (1984).
17. C.J. Lin and F. Spaepen, "Nickel-Niobium Alloys Obtained by Picosecond Pulsed Laser Quenching", submitted to Acta Metall.
18. For a review see J.M. Dubois and G. LeCaer, Acta Metall. 32:2101 (1984).
19. C.J. Lin and F. Spaepen, Scripta Metall. 17:1259 (1983).
20. W.K. Wang and F. Spaepen, submitted to J. Appl. Phys.
21. W.K. Wang, Y. Syono, G. Goto, H. Iwasaki, A. Inoue, and T. Masumoto, Scripta Metall. 15:1313 (1981).
22. For a review, see B. Cantor and R.W. Cahn, in "Amorphous Metallic Alloys", F.E. Luborsky, ed., Butterworths, London (1983), p. 487.
23. R.B. Schwarz and W.L. Johnson, Phys. Rev. Lett. 51:415 (1983).
24. R.B. Schwarz, K.L. Wong, W.L. Johnson, and B.M. Clemens, J. Non-Cryst. Solids 61/62:129 (1984).
25. M. Atzmon, J.D. Verhoeven, E.D. Gibson, and W.L. Johnson, Appl. Phys. Lett. 45:1052 (1984).
26. W.L. Johnson, M. Atzmon, M. Van Rossum, B.P. Dolgin, and X.L. Yeh, Proc. 5th Int. Conf. on Rapidly Quenched Metals, S. Steeb and H. Warlimont, eds., North-Holland, Amsterdam (1985), p. 1515.
27. B.P. Dolgin and W.L. Johnson, "Kinetics of the Formation of an Amorphous Layer During a Solid State Reaction", DOE Report ER10870-146, (1984).
28. B.M. Clemens, W.L. Johnson, and R.B. Schwarz, J. Non-Cryst. Solids 61/62:817 (1984).
29. M. Van Rossum, M.-A. Nicolet, and W.L. Johnson, Phys. Rev. B 29:5498 (1984).
30. M. Atzmon, J.D. Verhoeven, E.D. Gibson, and W.L. Johnson, in Proc. 5th Int. Conf. on Rapidly Quenched Metals, S. Steeb and H. Warlimont, North-Holland, Amsterdam (1985), p. 1561.
31. A compilation of heats of formation is given by A.R. Miedema, Philips Tech. Rev. 36:218 (1976).
32. Y.T. Cheng, W.L. Johnson, and M.-A. Nicolet, Appl. Phys. Lett. Sept. (1985) in press.
33. H. Schröder, K. Samwer, and U. Köster, Phys. Rev. Lett. 54:197 (1985).
34. X.L. Yeh, K. Samwer, and W.L. Johnson, Appl. Phys. Lett. 42:242 (1983).
35. W.L. Johnson, Y.T. Cheng, M. Van Rossum, and M.-A. Nicolet, Nucl. Instr. Meth. B 7/8:657 (1985).

36. C.C. Koch, O.B. Cavin, C.G. McKamey, and J.O. Scarbrough, Appl. Phys. Lett. 43:1017 (1983).
37. C.J. Lin, Ph.D. Thesis, Harvard University (1983).
38. A. Guinier, "X-Ray Diffraction", Freeman, San Francisco (1963), pp. 209-212, 279-281.
39. D. de Fontaine, in "Local Atomic Arrangements Studied by X-Ray Diffraction", J.B. Cohen and J.E. Hilliard, eds., Gordon and Breach, New York (1966), Chapter 2.
40. H.E. Cook and J.E. Hilliard, J. Appl. Phys. 40:2191 (1969).
41. T. Mizoguchi and F. Spaepen, Mat. Res. Soc. Symp. Proc., to be published.
42. R.C. Cammarata, Ph.D. Thesis, Harvard University (1985).
43. C.C. Koch, private communication.

PRACTICAL PROCESSING OF A-15 MULTIFILAMENTARY SUPERCONDUCTING WIRE FROM POWDERS: Nb_3Al AND Nb_3Sn^+

S. Foner,[++] C.L.H. Thieme, S. Pourrahimi
and B.B. Schwartz

Francis Bitter National Magnet Laboratory*
and Plasma Fusion Center
Massachusetts Institute of Technology
Cambridge, Massachusetts

ABSTRACT

Several approaches to fabricating Nb_3Al superconducting wires have been published in the last decade. These processes include: the Swiss roll, rapid quenching, mechanical alloying and powder metallurgy processing. This paper will address powder metallurgy (P/M) processing of Nb_3Al and Nb_3Sn. The recent measurements of critical current densities at high field for P/M processed Nb-Al gave $J_c > 10^4 A/cm^2$ at 19T and 4.2K. Good strain tolerance, low reaction temperatures, and low ac losses at high field all indicate that the P/M process for multifilamentary Nb-Al should lead to practical, high performance, superconducting wires for high fields. Significant improvements in powder metallurgy processing of Cu-Nb-Sn conductors have also been attained with low areal reductions, extrusion processing and with internal Sn cores. Increased Nb content increased J_c more than proportionally, and Ti additives yield very high J_c's at high field. Thus, powder metallurgically processed Nb_3Al and Nb_3Sn are promising for high field applications.

INTRODUCTION

The Nb-based A-15 superconductors, such as Nb_3Sn, Nb_3Al, Nb_3Ga, Nb_3Ge, $Nb_3(GeAl)$, and V based A-15 superconductors, such as V_3Ga and V_3Si, all have promise as practical high field superconducting materials, because of their high upper critical fields, H_{c2}, and high critical temperatures, T_c (see Figure 1). Of the Nb_3X compounds, only Nb_3Sn has been manufactured commercially as a multifilamentary wire for high field applications, yet Nb_3Sn has the lowest H_{c2} and T_c of any of the stoichiometric Nb_3X compounds above. The reasons that Nb_3Sn is the most advanced material are: 1) It forms the stoichiometric composition rather easily as can be ascertained from its phase diagram; 2) Relatively low temperatures (between 550 and 800°C) are required for reactions between the Nb and Sn to form Nb_3Sn; 3) Nb_3Sn has a relatively high H_{c2} and T_c compared to that

+ Supported by the Department of Energy
++ Also the Department of Physics
* Supported by the National Science Foundation

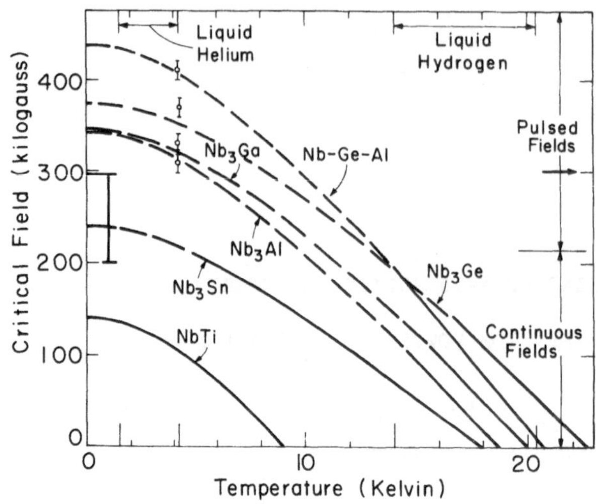

Fig. 1. Upper critical field, H_{c2} versus temperature,T, for several Nb_3X materials (after Ref. 1, see also Ref. 2). The spread in H_{c2} for Nb_3Sn is based on the more recent data. The arrow on the right ordinate indicates the advance that superconductors have made in a Hybrid magnet which achieves over 30T.

of Nb-Ti; and 4) Practical processes have been developed over the last 20 years. Both the bronze process and more recently the Sn-core process have been demonstrated for practical large scale superconductors.

Recently interest has developed in higher field superconductors for applications in mirror fusion devices. However, lower field applications would also profit from improved alternative practical processes. The major materials development problems for high field superconductors can be divided into two main areas: 1) Is there a better method of fabricating Nb_3Sn which is less expensive, more reliable, and more adaptable to varied configurations, and has equal or better high field properties? and 2) Are there other materials which can be fabricated into practical wires using commercial processing which exceed the properties of currently available Nb_3Sn multifilamentary wire? In this paper we will concentrate on cold powder metallurgical processing for fabrication of multifilamentary Nb_3Al. We will first review other methods being investigated for fabricating wires of Nb_3Al. In addition we will briefly review some recent developments in P/M processing of Nb_3Sn multifilamentary wires. In order to compensate for space limitations, an extensive reference list is included where more details of earlier work can be found.

Review of Fabrication Processes for Nb-Al

In 1958, Wood et al.[3] discovered that the A-15 compound Nb_3Al was superconducting with a transition temperature of 17.5K. Soon thereafter Willens et al.[4] prepared samples of Nb_3Al from compacted mixed powders. After heat treatment, a T_c of 18.8K was measured. Foner[2] et al. measured a critical field of 30 T at 4.2K, and recently Jorda et al.[5] gave evidence for T_c of Nb_3Al as high as 19.1K.

Although the H_{c2} and T_c for Nb_3Al seemed very attractive, the measurement of critical current density on rapidly quenched Nb_3Al by Bevk and Lo[6] suggested the potential vast superiority of Nb_3Al. Their measurements showed exceptionally high critical current densities of the A-15 layer ($J_c > 10^6 A/cm^2$ at 20T). This high J_c justified renewed efforts to fabricate multifilamentary Nb-Al wire.

The formation of Nb_3Al is difficult. The phase diagram of Nb-Al indicates that the stable A-15 Nb-Al will be nonstoichiometric over almost the entire temperature range, and that stable compounds Nb_2Al and $NbAl_3$ can form.[7] These are stable up to high temperatures and their formation is much faster than the A-15 phase at temperatures[8] below, say, 1100°C. Hunt et al.[9] showed that in the Nb-Al-Cu system, stable (Nb,Cu,Al) phases exclude the formation of A-15. This makes the use of an Al bronze technique for Nb_3Al virtually impossible. Various attempts to fabricate Nb-Al superconductors include the Swiss (jelly) roll technique, infiltration process, rapid quenching, mechanical alloying and powder metallurgy processing among others. For the Swiss roll technique, thin layers of rolled Nb foils and Al foils were prepared by Ceresara et al.[10,11] The composite was inserted into a hollow copper cylinder, rod-rolled, wire drawn then heat treated. The formation of Nb-Al was critically dependent on heat treatment and the thickness of the final Al layers. A critical current above $10^5 A/cm^2$ was observed at 6.4T.

The infiltration process has been reviewed by Pickus et al.[12] Here, a sintered porous Nb-powder compact is immersed in liquid Al(or Sn) after which areal reduction by rolling, wire drawing, etc. is carried out. This article also describes quenching of BCC Nb-Al solutions done more recently by Hong et al.[13] In this process superconducting materials are made by deforming the quenched supersaturated solutions of Nb-Al, and then precipitating the A-15 phase by aging at intermediate temperatures. Initial measurements yielded moderate critical current densities at high field. There is a delicate balance between the A-15 grain size, superconductor continuity and volume fraction of the Nb_3Al formed with heat treatment.

Recently Togano et al.[14] have reported fabrication of composite tapes of Nb(AlGe) using a liquid quenching technique, and this is reviewed in this proceedings. A powder metallurgy process for Nb(Al,Ge) was discussed in 1973 by Lohberg et al.[15]

The development of our powder metallurgy process for Nb-Al followed earlier work on other A-15 materials. Early work using brittle powders was initiated by the group at the Netherlands Energy Research Foundation, ECN, for Nb_3Sn, V_3Ga, and V_3Si.[16] Hot powder metallurgy has been developed by the group at the University of Goettingen.[17] Billets from 2 to 8 kg have been prepared and third-element additives can be incorporated.

Earlier, there had been an interest in the use of mechanical alloying[18] for producing Nb-Al powders through high energy impact ball milling. Some recent results in our laboratory are presented in the latter part of this paper. The hope is to achieve more stoichiometry in the Nb_3Al compound, which should lead to higher H_{c2} and J_c.

The success of the in situ processing of Cu-Nb-Sn and Cu-V-Ga[19] led to the investigation of cold P/M processing using powders of Cu and Nb. Initial tests by Flukiger et al.[20] demonstrated that cold processing was feasible starting with 40μm hydride-dehydride Nb and Cu powders. Compaction of the powder in a Be-Cu tube, swaging, and wire drawing followed by external Sn plating, diffusion, and reaction at 650° C, yielded submicron filaments of Nb_3Sn with good strain tolerance and with high J_c values at high field, as observed in our in situ processed materials. Soon after the success of the powder metallurgy process for Nb_3Sn, Akihama et al.[21] made Nb-Al superconducting wire starting from powders. Further progress of the powder metallurgy processed multifilamentary Nb-Al wire was given by Thieme et al.[22] and for Nb_3Sn by Pourrahimi et al.[23] In what follows we report on the most recent results for powder metallurgy processing of Nb-Al, some preliminary results on mechanical alloying for Nb-Al powders, and powder processing of Cu-Nb-Sn with Ti additions employing extrusion and Sn core processing.

Nb-Al Powder Processed Superconducting Materials

We found that Nb-6to8wt%Al is an optimal composition for high J_c P/M processed[24] multifilamentary wire. Furthermore, improved J_c is obtained for increased effective areal reduction. Fig. 2 shows results for J_c as a function of field for various samples and heat treatment. An overall $J_c > 10^4$ A/cm^2 at 19 T is achieved for an areal reduction of 3.4×10^5 obtained by swaging and wire drawing.[25]

Despite the large J_c achieved by the P/M process for Nb-Al, it is clear that we may expect a large increase in J_c at high fields if processing can be improved further. The highest values of T_c and H_{c2}(4.2K) of the P/M processed Nb-Al in Fig. 2 were 16K and 24.5T, respectively, whereas bulk values are much higher.[2] Our results show that independent of powder size, the one common feature is that J_c continues to increase with increased areal reduction, R, and there is no evidence of saturation of J_c versus R. This is illustrated in Fig. 3. The symbols are connected by lines to guide the eye. (In contrast, saturation of J_c versus R has been reported for both in situ[26] and P/M processed Nb_3Sn.) A consistent explanation for this behavior could be that the reacted Nb-Al layer is quite thin and that larger R leads to a larger surface area of Nb-Al. To examine this feature in the P/M processed wires, analytical electron microscopy investigations of the microscopic structure was initiated.[25,27]

Low field (2mT) magnetic moment measurements versus T made on short sections (<5mm) of a wide variety of P/M processed Nb-Al wires show no simple correlation with J_c, e.g., some wires with high J_c had lower T_c's

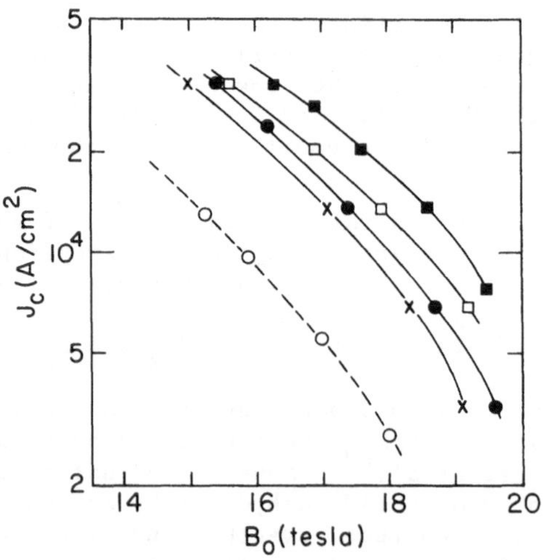

Fig. 2. Improved overall critical current densities, J_c, versus applied field, B_o, for Nb-8wt%Al wires using fine powders and $R=3.4 \times 10^5$ with varying heat treaments (upper solid curves). Lower dashed curve used larger powders and $R=8 \times 10^4$ (after Ref. 25).

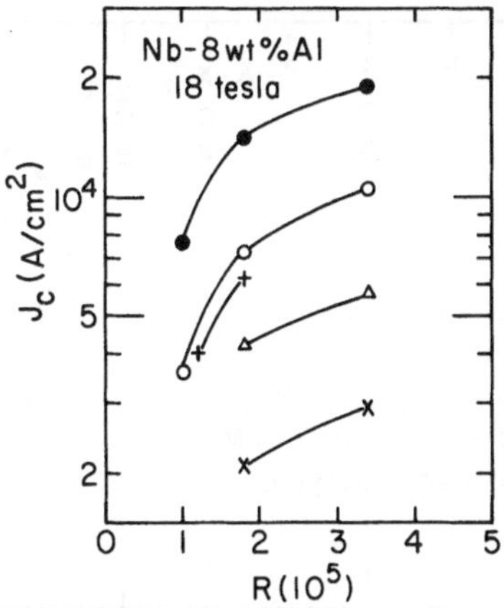

Fig. 3. Overall critical current density, J_c, versus areal reduction ratio, R, for a variety of Nb-8wt%Al wires at 4.2K and 18 T. The curves are to guide the eye for each wire. The powder sizes in μm for Nb and Al respectively are: x 80,50; Δ150,5; + 60,5; o 80,5;   80,5.

or broad T_c transitions and some resistive wires had a high T_c. It should be noted that J_c is a severe measure of a superconductor, requiring continuity, pinning, homogeneity, etc. as well as high H_{c2} and T_c. That an increase of T_c is not produced by an increase in R is illustrated in Fig. 4 where the relative moment versus temperature is shown for sections of P/M processed wires with R = 0.97 and 3.4×10^5. Because the data is essentially identical, each of the sets of data are displaced vertically as indicated for clarity. All the materials had T_c's≤16K.

Recent efforts in the development of the P/M process for Nb-Al[22,23] have been directed toward extrusion technology, examination of the limits of J_c with further increased R, effects of powder size, increased area of contact using Al flakes, and the effects[22] of additives. The major improvement is achieved with increased areal reduction R. Mechanical alloying with high impact ball milling was also initiated to determine whether improved Nb-Al alloys can be produced. Some of the best ball milled powders have magnetic properties similar to the P/M processed wires. This is shown in Fig. 4, where the mechanical alloyed powder has a somewhat broader transition and an onset of $T_c \simeq$ 16K. However, fabricated wires using these powders have not yielded high J_c's at high field so far. These results are consistent with the earlier work.[18] As an alternative approach, Prof. D. Rudman at MIT has started a program with rapidly quenched powders.

Areas for improvement of the P/M process for Nb-Al involve: 1) An understanding of the formation of the A-15 phase on a microscopic basis; 2) Means to achieve more nearly stoichiometric Nb_3Al with highest T_c and H_{c2} in order to achieve even higher J_c; and 3) Ways to achieve high J_c with low areal reductions. Of course, much further improvement in practical wires can be achieved if the P/M process can be modified for the higher H_{c2}, Nb_3X compounds.

Practical fabrication of exceptionally high J_c Nb-Al superconducting conductors made by the P/M process remains quite promising. We believe

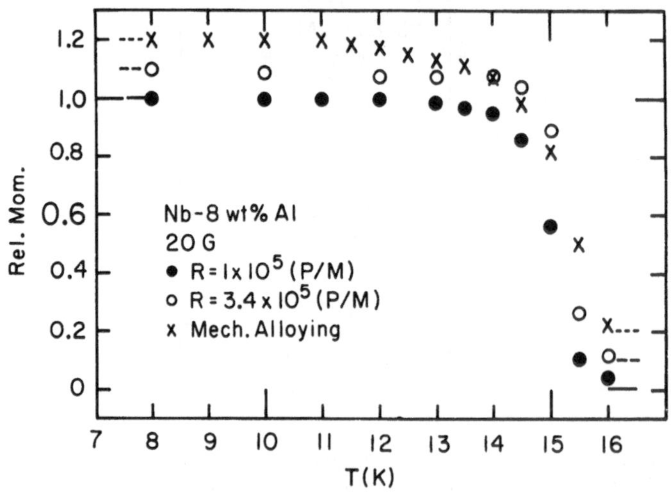

Fig. 4. Relative moment versus temperature for Nb-8 wt% Al at 2mT for sections of P/M processed wires (open and solid circles) and for high energy ball milled powder (x). Sets of data are displaced by 0.1 for clarity as indicated.

that further improvements can be achieved with a detailed microscopic understanding of the process and have initiated a basic research program using model bilayer and multilayer Nb/Al thin films to explore this area.

Nb_3Sn Powder Metallurgy Processing

Powder metallurgy processing of Cu-Nb-Sn has been successful in achieving high values of J_c at high field and in serving as a vehicle for development of the Nb-Al P/M process. It allows fabrication of multifilamentary fine fiber wires with complete Nb conversion, good strain tolerance and high J_c (as in the in situ process). The size of the fibers can be controlled by choosing appropriate powder sizes and cold processing can be used. Progress on scaleup, multistrand processing, tin core processing, etc. using swaging and wire drawing is presented in Ref. 28 and references cited therein.

Recent work has been directed toward increasing J_c for highest field applications and development of practical processing technology, which is transferrable to industry. Model small scale cold hydrostatic extrusion and industrial hot extrusion have been successful.[29,23] In addition low areal reductions (R ~ 2000), were used to produce relatively thick fibers corresponding to those of conventional bronze processed wires with low prestress. This resulted in improved J_c at high field for P/M processed wires.[23] More than linear increases in J_c were achieved by increasing the Nb content. Finally, we find a continuing improvement in J_c by additions of Ti. A particularly successful approach done in collaboration with M. Suenaga of Brookhaven National Laboratory uses Sn-Ti alloyed cores in the P/M process. An example of wires with Sn core versus Sn core with added Ti is shown in Fig. 5 for P/M processed wire.[30,31] Values of $J_c > 10^4$ A/cm^2 at 20 T and 4.2K and 3×10^4 A/cm^2 at 18 T have been measured, and J_c continues to increase with added Ti up to 3wt%Ti in the Sn core. Thus the P/M processed Cu-Nb-Sn is also promising for very high field applications. Addition of Sn cores after extrusion prior to wire drawing was also successful.

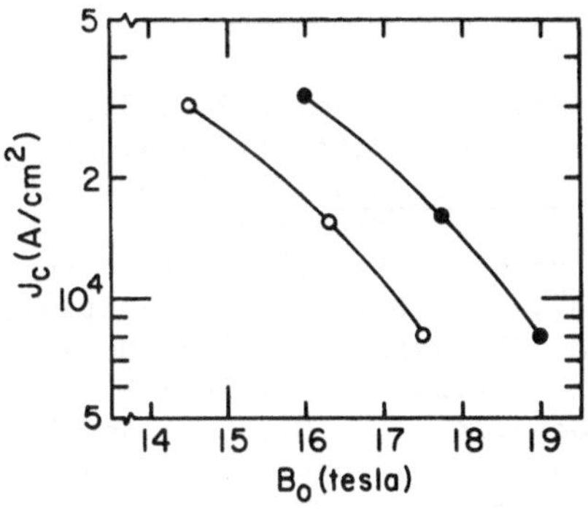

Fig. 5. Effect of Ti addition to Sn core on J_c at 4.2K versus B_o for P/M processed Cu-45wt%Nb wires. The wires were hydrostatically extruded and wire drawn with R=2000 and with: • Sn-1.5wt%Ti core; o Sn-5wt%Cu core.

The P/M process for Nb_3Sn has many advantages over conventional processing including: 1) Processing can be done on small or large scale; 2) The fiber size can be controlled by choice of powder size and areal reduction; 3) The Nb content can be varied easily so that high J_c's are obtained at very highest fields, and 4) Use of Sn cores, barriers, multifilament arrays, etc. has been demonstrated. However, our P/M processing requires ductile powders with no foreign matter. Almost all the approaches to fabrication have been demonstrated in our Laboratory and many are directly transferrable to industry. Thus the P/M process for Cu-Nb-Sn appears to be quite near commercial processing and we foresee no obstacles to large scale fabrication.

Areas being examined are: 1) Optimization of Ti additions to the Sn core processing so that higher J_c's at high fields are attained; and 2) Means to introduce Ga additions to the Cu-Nb-Sn P/M process to improve on H_{c2} (and consequently J_c) as suggested by Ref. 32.

General Features of P/M Processed Wires

As with the in situ processed wires, one expects that the interconnections between fibers would lead to increased losses. thus the effective fiber size, d_{eff} is measured to be between the wire diameter and the individual fiber size.[33,34] Low frequency losses at high fields[34] show that these losses can be reduced by twisting and by using multistrand configurations. The measured low frequency losses were purely hysteretic and d_{eff} was independent of field so that extension of the loss measurements to low fields allowed determination of the very high J_c's at low field of P/M processed wires. For Nb-Al P/M processed wires a d_{eff}<3μm was observed.[34] Long lengths of multifilamentary Sn core P/M processed Cu-Nb-Sn and P/M processed Nb-Al incorporated in small magnets tested to 15 T gave short sample characteristics[35] and upper limits of resistivity indicating that uniform long lengths can be fabricated with good high field performance. Another measure of homogeneity and connectivity in conventional superconductors is the current-voltage characteristic[31] where $E=kI^n$. Preliminary measurements on P/M processed wires at high field show exponents which are consistent with those observed for continuous filament wires.[31] Extension of this work will be published elsewhere.

REFERENCES

1. S. Foner, "Colloq. Int. CNRS" No. 242 (1975), pp.423-429.
2. Source references for these materials are: Nb_3Sn: S. Foner and E.J. McNiff, Jr., Solid State Commun. 39:959-964 (1981); Nb-Ge-Al: S. Foner, E.J.McNiff, Jr., B.J.Matthias, T.H. Geballe, R.H. Willens, and E. Corenzwit, "Proceedings 11th International Conference on Low Temperature Physics", University of St. Andrews, Scotland (1968) Eds. J.F.Allen, D.M. Finlayson, and D.M. McCall, Vol.II, pp.1025-1032 and Phys. Lett. 31A:349-350 (1970); Nb_3Al: S. Foner, E.J. McNiff, Jr., T.H. Geballe, R.H. Willens, and E. Buehler, Physica 55:534-539 (1971); Nb_3Ga: S. Foner, E.J. McNiff, Jr., L.J. Vieland, A. Wicklund, R.E.Miller, and G.W. Webb, IEEE, Pub. No. 72CH-0682-5-TABSC, pp.404-406 (1972); Nb_3Ge: S. Foner, E.J. McNiff, Jr., J.R. Gavaler, and M.A. Janocko, Phys. Lett. A47:485-486 (1974); V_3Ga: R. Flükiger, S. Foner, and E.J. McNiff, Jr., "Superconductivity in d- and f-Band Metals", eds. H. Suhl and M.B. Maple, Academic Press, NY, (1980), pp.265-271; S. Foner, E.J. McNiff, Jr., S. Moehlecke, and A.R. Sweedler, Solid State Commun. 39:773-776 (1981); V_3Si: S. Foner and E.J. McNiff, Jr., Phys. Lett. A58:318-320 (1976).
3. E.A. Wood, V.B. Compton, B.T. Matthias and E. Corenzwit, Acta. Cryst. 11:604-606 (1958).
4. R. H. Willens, T.H. Geballe, A.C. Gossard, J.P. Maita, A. Menth, G.W. Hull,Jr., and R.R. Soden, Solid State Commun. 7:837-841 (1969).
5. J.L. Jorda, R. Flükiger, A. Junod, and J. Muller, IEEE Trans. on Magn., MAG-17:557-560 (1981).
6. K. Lo, J. Bevk and D. Turnbull, J. Appl. Phys. 48: 2597-2600 (1977); J. Bevk, "Rapidly Quenched Metals", Third International Conference (Ed. B. Cantor) Vol. 2:17-20 (1978).
7. J.L. Jorda, R. Flükiger and J. Muller, J. Less Common Metals 75:227-239 (1980).
8. G. Slama, A. Vignes, J. Less Common. Met. 29:189-202 (1972).
9. C.R. Hunt and A. Raman, Zeit. für Met. 59:701-707 (1968).
10. S. Ceresara, M.V. Ricci, N. Sacchetti and G. Sacerdoti, IEEE Trans. on Magn., Vol. MAG-11:263-265 (1975).
11. B. Annaratone, R. Bruzzese, S. Ceresara, G. Giordano. G. Psotti, V. Pericoli-Ridolfini, G. Pitto, M.V. Ricci, N. Sacchetti, and M. Spadoni, "Proceedings 8th International Cryogenic Engrineering Conference (ICEC-8)",pp. 424-428 (1980), Ed. C. Rizzuto, Genoa; IEEE Trans. on Magn. MAG-17:1000-1001 (1981).
12. M.R. Pickus, J.T. Holthuis and M. Rosen,"Filamentary A-15 Superconductors", Eds. M. Suenaga and A.F. Clark, Plenum Publ. Corp. New York (1980), pp.331-353.
13. M. Hong, D.R. Dietderich, I.W. Wu and J.W. Morris, Jr., IEEE Trans. on Magn. MAG-17:278-281 (1981).
14. K. Togano, H. Kumakura, Y. Yoshida, and K. Tachikawa, IEEE Trans. Magn. MAG-21:463-466 (1985).
15. R. Lohberg, T.W. Eagar, I.M. Puffer, and R.M. Rose, Appl. Phys. Lett. 22:69-71 (1973).
16. J.D. Elen, J.W. Schinkel, A.C.A. Van Wees, C.A.M. van Beijnen, E.M. Hornsveld, T. Stahhe, H.J. Veringa and A. Verkaik, IEEE Trans. on Magn. MAG-17:1002-1005 (1981) and references cited therein.
17. R. Bormann, L. Schultz and H.C. Freyhardt, Appl. Phys. Lett. 32:79-81 (1978); R. Bormann and H.C. Freyhardt Appl. Phys. Lett. 35:944-946 (1979) and IEEE Trans. on Magn. MAG-17:270-273 (1981).
18. J.M. Larsen, T.S. Luhman, and H.F. Merrick, "Manufacture of Superconducting Materials", Ed. R.W. Meyerhoff, American Society of Metals, Metals Park (1977), pp.154-163.
19. S. Foner, E.J. McNiff, Jr., B.B. Schwartz, and R. Roberge, Appl. Phys. Lett. 31:853-854 (1977); R. Roberge and S. Foner, "Filamentary

A-15 Superconductors" Eds. M. Suenaga and A.F. Clark, Plenum Publishing Corporation New York, NY (1980), pp. 241-257; R. Roberge, J. LeHuy, J-L Fihey and S. Foner, "Proceedings of International Cryogenic and Engineering Materials Conference"., Kobe, Japan, Eds. K. Tachikawa and A.F. Clark, Butterworth Scientific Ltd. (1983), pp.331-334 and references cited therein.

20. R. Flükiger, S. Foner, E.J. McNiff, Jr., and B.B. Schwartz, Appl. Phys. Lett. 34:763-766 (1979); R. Flükiger, R. Akihama, S. Foner, E.J. McNiff, Jr., and B.B. Schwartz, Appl. Phys. Lett. 35:810-812 (1979).
21. R. Akihama, R.J. Murphy and S. Foner, Appl. Phys. Lett. 37:1107-1109 (1980); R. Akihama, R.J. Murphy and S. Foner, IEEE Trans. on Magn. MAG-17:274-277 (1981).
22. C.L.H. Thieme, S. Pourrahimi, B.B. Schwartz and S. Foner, IEEE Trans. on Magn. MAG-21:756-759 (1985).
23. S. Pourrahimi, C.L.H. Thieme, B.B. Schwartz and S. Foner, IEEE Trans. on Magn. MAG-21:764-767 (1985).
24. C.L.H. Thieme, H. Zhang, J. Otubo, S. Pourrahimi, B.B. Schwartz and S. Foner, IEEE Trans. on Magn. MAG-19:567-569 (1983).
25. C.L.H. Thieme, S. Pourrahimi, B.B. Schwartz and S. Foner, Appl. Phys. Lett. 44:260-262 (1984).
26. R. Roberge, Chapter 6, "Superconducting Materials Science: Metallurgy, Fabrication and Applications", Eds. S. Foner and B.B. Schwartz, Plenum Publishing Corporation, New York, NY, (1981) pp.389-453; H. Zhang, S. Pourrahimi, J. Otubo, C.L.H. Thieme, B.B. Schwartz and S. Foner, IEEE Trans. Magn. MAG-19:769-772 (1983) and references cited in these papers.
27. W.E. King,C.L.H. Thieme and S. Foner, IEEE Trans. Magn. MAG-21:815-818 (1985).
28. J. Otubo, S. Pourrahimi, C.L.H. Thieme, H. Zhang, B.B. Schwartz, and S. Foner, IEEE Trans. Magn. MAG-19:764-768 (1983); H. Zhang, S. Pourrahimi, J. Otubo, C.L.H. Thieme, B.B. Schwartz, and S. Foner, IEEE Trans. Magn. MAG-19:769-772 (1983).
29. S. Pourrahimi, C.L.H. Thieme, S. Foner, and R.J. Murphy, Appl. Phys. Lett. 43:1070-1072 (1983).
30. See also Ref. 19 and work to be published.
31. For recent work on Ti additions to Nb or the bronze for continuous fiber bronze processed conductors see M. Suenaga, K. Tsuchiya, N. Higuchi, and K. Tachikawa, Cryogenics 25:123-128 (1985).
32. R. Bormann, D-Y Yu, R.H. Hammond, T.H. Geballe, S. Foner, and E.J. McNiff, Jr., IEEE Trans. Magn. MAG-21:1140-1143 (1985).
33. T.P.Orlando, C.B. Braun, B.B. Schwartz, S. Foner and W.K. McDonald, IEEE Trans. Magn. MAG-19:435-438 (1983).
34. A.J. Zaleski, T.P. Orlando, A. Zieba, B.B. Schwartz and S. Foner, J. Appl. Phys. 56:3278-3283 (1984).
35. A.J. Zaleski and S. Foner, J. Appl. Phys. Lett. 44:1098-1100 (1984).

THE FORMATION OF THE METASTABLE PHASE Nb_3Al BY A SOLID STATE REACTION

R. Bormann,[+*] H. U. Krebs,[+] and A. D. Kent[*]

[+]Institut für Metallphysik
Universität Göttingen , FRG

[*]Department of Applied Physics
Stanford University, USA

ABSTRACT

The formation of the stoichiometric Nb_3Al phase by a solid state reaction of thin Nb/Al multilayers is basically determined by the intermediate formation of a metastable $\underline{Nb}Al$ solid solution. If the nucleation of Nb_2Al is avoided, the homogeneity range of the A15 phase can be extended metastably up to about 28 at.% Al. The temperature dependence of the metastable Al-rich A15 phase boundary can be determined by the respective (free) enthalpy curves of the A15 and the bcc phase calculated from the phase diagram.

INTRODUCTION

The formation of metastable alloys is of interest particularly in the case of superconductive compounds since the high-T_c A15 phases Nb_3Al, Nb_3Ga, and Nb_3Ge in the stoichiometric composition are metastable under usual preparation conditions. A variety of processes have been used to synthesize these materials, for example, condensation techniques, chemical vapor deposition, and rapid quenching from the melt and solid state reactions. Nb_3Al tapes and wires with favorable superconducting properties can be prepared either from the metastable bcc structure[1-4] or from the pure elements.[5] However, the thermodynamic origin of the formation of the metastable composition is still not completely understood.

The transformation from the bcc $\underline{Nb}Al$ solid solution into the A15 phase is determined by the particular shape of the (free) enthalpy curves of the phases, as described in Ref. 3. In this article we report the thermodynamic process involved in the solid state reaction between Nb and Al.

For technical applications wires are prepared either by a powder-metallurgical[5] or a jelly-roll[6] technique. Favorable superconducting properties are achieved only if the composites are highly deformed, leading to a layer thickness of less than 100 nm. The A15 phase is formed by a heat treatment at temperatures of about 1000°C. In order to study the thermodynamic processes involved in the formation of the metastable compositions, we prepared a model system of Nb and Al multilayers. It is possible to monitor the reactions between Nb and Al layers with an X-ray diffractometer equipped with a heated sample holder. In this way we have been able to monitor the formation of different metastable phases as a function of temperature and time. The results are discussed on the basis of the (free) enthalpy curves for the various phases involved in the formation of A15 Nb_3Al.

EXPERIMENTAL PROCEDURE

The Nb/Al multilayered composites were prepared by sequential sputtering of Nb (purity 99.9+%) and Al (99.999%) onto single crystal sapphire substrates at ambient temperature. The background pressure before the admission of Ar was better than $4 \cdot 10^{-5}$ Pa. Samples with an overall composition of 25 at.% were prepared with a Nb/Al layer thickness of 100/30.7 nm and 30/9.2 nm. In addition, off-stoichiometry samples with an overall concentration of 23 at.% Al were prepared with a Nb/Al layer thickness of 100/27.6 nm. The total thickness of the composites was about 1 µm. For the X-ray diffraction measurements, the samples were mounted on a heated Cu block inside a vacuum chamber (background pressure less than $1 \cdot 10^{-2}$ Pa). Typically, the temperature was held constant for about 20 min during the recording of the diffraction pattern and then increased by steps of 50° or 100° to 800°C. The diffraction patterns measured at high temperature were used for the determination of the phases, while their lattice constants were detected on samples cooled to room temperature. After the X-ray analysis, the critical temperature, T_c, of the sample was determined by an inductive probe. The T_c was defined as the midpoint of the normal to superconducting transition.

RESULTS

From the Nb/Al phase diagram,[7] the intermetallic compounds Nb_3Al (A15 phase), Nb_2Al (σ phase) and $NbAl_3$ can be expected during the reaction of Nb and Al. However, owing to the different time scales of their formation, not all of the phases occur simultaneously. The small diffusion lengths involved in their synthetically prepared, layered structures enable us to seperate the different intermediate steps kinetically and to prepare the metastable phases in a bulk form.

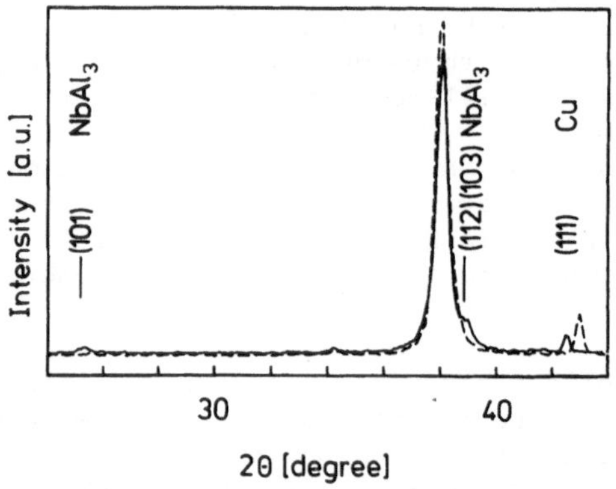

Fig. 1. Diffraction pattern of the Nb/Al multilayers taken at room temperature (dashed line) and at T= 650°C (solid line).

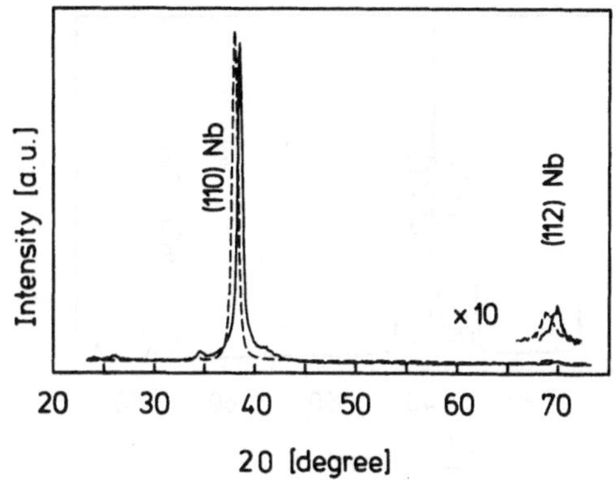

Fig. 2 Diffraction pattern of Nb/Al multilayers (both taken at room temperature) before (dashed line) and after a heat treatment of 2 h at 700°C (solid line), indicating the formation of a single phase solid solution. (The small peak at about $2\theta = 34°$ corresponds to the K_β-(110) Nb line.)

The formation of the Al-rich $NbAl_3$ compound has been observed at temperatures as low as 300°C. While increasing the temperature in steps of 50°C every 20 min all of the Al is transformed to $NbAl_3$ at the melting point of Al (660°C). In addition, the solution of Al in Nb is observed. This intermediate step is shown in Fig. 1. The small peak at $2\theta \approx 25.4°$ and the shoulder at the right side of the (110) Nb peak correspond to the (101) and (112)/(103) lines of the $NbAl_3$ phase, respectively. The intensity of the (110) Nb peak is decreased slightly and is shifted to higher 2θ values, despite the influence of thermal expansion visible qualitatively at the (111) Cu peak in the case of the Cu substrate holder. While increasing the temperature to 700°C, the $NbAl_3$ phase reacts with the NbAl, increasing its solubility range metastably. After 2 h at 700°C all the Al is dissolved in the NbAl solid solution in the case of the 30-nm-thick Nb layers. Some of the samples were cooled down in order to determine the decrease of the lattice parameter of Nb due to the solution of Al (Fig. 2). Previous measurements of bcc NbAl samples co-evaporated onto cold sapphire substrates[3] indicate that a shift in lattice constant of 0.004 nm corresponds to a solution of Al in the range 20 to 25 at.%. Also, this supports the observation that prior to the nucleation of the A15 phase, a single-phase structure is formed that consists of a metastable NbAl solid solution (see also Fig. 2). However, our measurements indicate that this can only be achieved if the Nb layer is smaller than about 30 nm. For coarser microstructures, the thickness of $NbAl_3$ formed is larger and its reaction with Nb cannot be completed before the transformation of bcc into A15 phase occurs. The existing concentration gradient in these samples leads to an inhomogeneous A15 phase and also to the formation of the Nb_2Al-σ phase. For thin layered structures a homogeneous bcc-NbAl solid solution can be formed, which transforms into the A15 phase, increasing its compositional range metastably towards higher Al concentrations. The formation of the σ phase can be avoided, at least up to 25 at.% Al (Fig. 3).[3] The critical temperature of the A15 phase formed from the multilayered structures reaches only 16 K, even for the samples with thin Nb layers, which may be due to the relatively low reaction temperature of 800°C. In addition, the rather broad transition and the relatively large lattice constant a_o = 0.5184 nm indicate that a slight evaporation of Al from our samples has occurred.

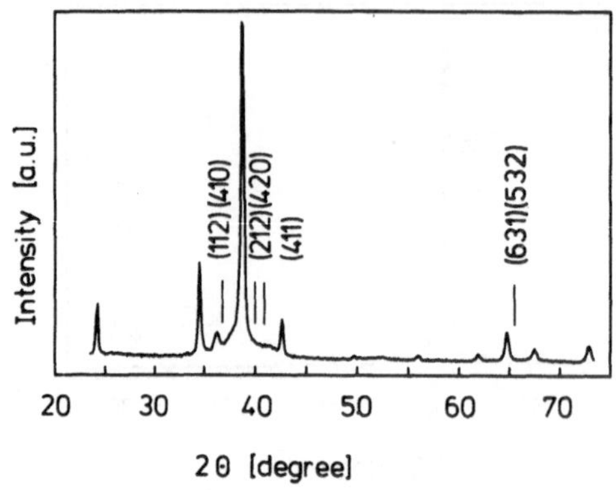

Fig. 3 Diffraction pattern of the A15 phase transformed from the bcc solid solution by a heat treatment of 80 min at 800°C. The positions of the most intense lines of the Nb_2Al-σ phase are indicated.

DISCUSSION

The formation of the A15 phase with metastable Al-rich concentrations by a solid state reaction of thin Nb/Al multilayers can be divided into three steps:

1. At low temperatures, the Al reacts with Nb forming the $NbAl_3$ phase and a solid solution:

$$Nb + Al \longrightarrow \underline{Nb}Al + NbAl_3$$

The amount of the $NbAl_3$ phase formed can be decreased by higher heating rates and by microstructures with thinner layers.

2. Increasing the temperature, the solution of Al in Nb is enhanced at the expense of the $NbAl_3$ phase. The homogeneity range of the bcc $\underline{Nb}Al$ is extended metastably:

$$\underline{Nb}Al + NbAl_3 \longrightarrow \underline{Nb}Al$$

3. For Nb layer thicknesses smaller than 30 nm, this reaction can be separated from the final transformation, where the Nb_3Al nucleates from the $\underline{Nb}Al$ solid solution:

$$\underline{Nb}Al \longrightarrow Nb_3Al$$

An A15 phase with a composition close to stoichiometry can be achieved only if the formation of the σ phase can be avoided.

The different steps of the reaction can be explained by the particular (free) enthalpy curves of the occurring phases $\underline{Nb}Al$, Nb_3Al and $NbAl_3$. These were calculated from the phase diagram and from (free) enthalpy values, measured with a galvanic cell.[8] However, in spite of Ref. 9, the Nb-Al phase diagram determined by Jordan et al.[7] was used. In addition, the concentration dependence of the (free) enthalpy of the intermetallic compounds Nb_3Al and Nb_2Al were calculated by assuming a subregular behavior, with enthalpy and entropy being temperature dependent. The fitted (free) enthalpy curves agree with those calculated by Lukas[10] for concentrations

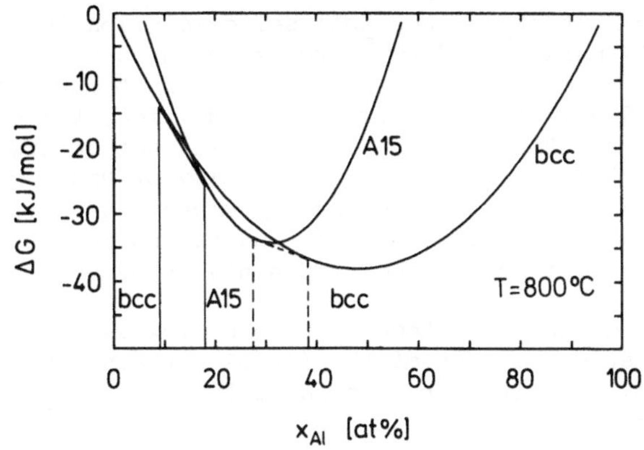

Fig. 4 (free) enthalpy curves of the A15 and the bcc NbAl solid solution at T = 800°C. The metastable equilibrium is indicated by the dashed line.

close to the equilibrium homogeneity range of the respective phase. In Fig. 4, the (free) enthalpy curves at T = 800°C of the bcc and the A15 phase are shown. The thermodynamically stable bcc and the liquid structure were chosen as a reference point at x_{Al} = 0 and 100, respectively. The maximum driving force for the nucleation of the A15 phase occurs at about 24 to 25 at.% Al, in agreement with the picture that the A15 phase is most stable close to the stoichiometric composition. If the chemical potentials of the components in each phase are defined in the metastable composition range similar to the equilibrium range, a metastable equilibrium between the two phases can be constructed (Fig. 4). This is synonymous with the assumption that a local minimum of the (free) enthalpy is reached if the chemical potentials of each component in the two phases are equal. In addition, metastable phase diagrams can be calculated for different temperature ranges, when taking into account only the occurring phases (Fig. 5). This approach for temperatures between 300 and about 700°C

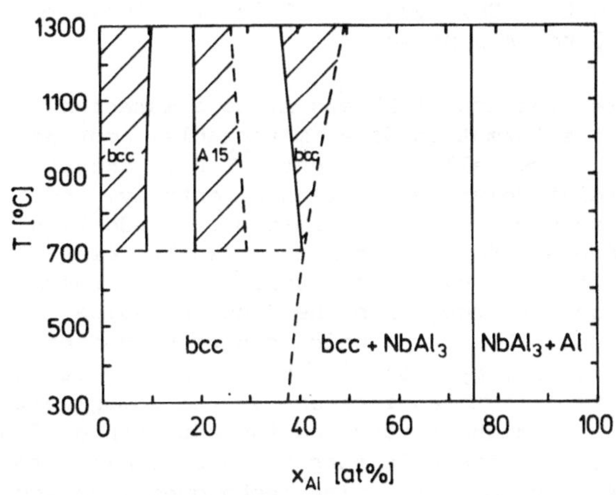

Fig. 5 Nb-Al metastable phase diagram calculated form the (free) enthalpy curves for temperatures between 300 and 1300°C. Below 700°C only the bcc solid solution and the $NbAl_3$ phase are taken into account. For T > 700°C, the formation of the A15 phase is also considered.

leads to an enhancement of the bcc NbAl solid solution to about 40 at.% Al. This metastable phase boundary is rather hard to determine accurately, owing to the error inherent in extrapolating the bcc (free) enthalpy curve far to the 50 at.% Al composition. In addition, the equilibria between the liquid and the $NbAl_3$ phase are still not determined completely, leading to an uncertainty of the (free) enthalpy of $NbAl_3$ at low temperatures. Nevertheless, the metastable phase boundary agrees rather well with the observation that the NbAl solid solution can be increased up to 42 at.% Al by co-evaporation of the components onto cold substrates.[3]

The formation of $NbAl_3$ at temperatures as low as 300°C might be explained by the drastic decrease of the (free) enthalpy of the liquid Al upon dissolving Nb, thus leading to a partially molten AlNb alloy during the formation of the $NbAl_3$ phase.

At temperatures higher than 750°C, the formation of the A15 phase must be taken into account for the calculation of the metastable phase diagram (Fig. 5). If the nucleation of the σ phase can be avoided, the homogeneity range of the A15 phase can be extended up to about 28 at.% Al (T = 1000°C) in agreement with results obtained from homogeneous NbAl samples. However, for Al concentration higher than about 27 at.%, the σ phase has been observed at temperatures around 1000°C, limiting the homogeneity range of the A15 phase. Nevertheless, it is clear from Fig. 4 and 5 that the formation of the metastable A15 phase in thin Nb/Al layered structures is due to the difference in the kinetics of phase formation bypassing the nucleation of the σ phase.

The precipitation of the Nb_3Al compound from the NbAl solid solution is determined by the shape of the respective (free) enthalpy curves, especiallly in the off-stoichiometric region,[3] where a non-polymorphous transformation occurs owing to the higher driving forces. An advantage of Nb_3Al with respect to other metastable A15 compounds might also be that the A15 curve exhibits a slope equal to that of the bcc curve at about 24 to 25 at.% Al, favoring a homogeneous, polymorphous transformation.

In addition, the σ phase can be energetically suppressed with respect to the A15 phase by the addition of third elements, such as Ge and Si, as pointed out by Dew Hughes.[11] This also leads to an extended homogeneity range of the A15 phase.

The solid state reaction of Nb and Al to a metastable NbAl solution is very similar to the formation of a (metastable) amorphous alloy by a heat treatment of the crystalline elements.[12] Both phases, the bcc and the amorphous, have a relatively high configural entropy, and therefore, an enhanced number of possibilities of formation. In addition, the energy of the respective interphase boundaries might also be relatively small, decreasing the activation energy of the nucleation. Furthermore, the A15 phase has also a higher entropy than the σ phase indicated by the shift of the Al-rich A15 phase boundary, which suggests that phases with low entropy can be kinetically bypassed in solid state reactions. At low temperatures, this results in an increase in the homogeneity range of the equilibrium phases with high entropy and the occurrence of amorphous alloys. The sequence of metastable phases may be similar to the formation of different phases by vapor deposition techniques when increasing the substrate temperatures. However, the homogeneity range of the phases might not be determined by the metastable phase diagram calculated from the (free) enthalpy curves, because the equilibrium condition may not to be valid in these processes.

CONCLUSION

The formation of the Nb_3Al A15 phase by a solid state reaction of Nb and Al is very similar to the previously investigated[3] transformation of metastable bcc NbAl solid solution into the A15 structure. If the thickness of the Nb layer is less than about 30 nm, the NbAl solid solution can be formed prior to the nucleation of Nb_3Al. To avoid an unfavorable concentration gradient, a homogeneous bcc phase should be formed first by a low temperature (≈ 700°C) heat treatment. An increase in temperatures is then used to nucleate the A15 phase.

The reaction temperature of the A15 phase should be as high as possible owing to its increased stability with respect to the σ phase. However, the nucleation of the σ phase must be avoided in order to extend the homogeneity range of the A15 phase metastably. In addition, the transformation of the σ phase can be suppressed by third elements,[11] which lower the driving force of nucleation. Other metastable A15 phases, like Nb_3Ge, should also be formed by a solid state reaction of Nb and Ge multilayered structure. In this system the amorphous phase is more stable than the bcc solid solution as indicated by investigations on co-evaporated NbGe samples.[3] It is, therefore, expected that an amorphous phase will be formed by a heat treatment of Nb/Ge multilayers before nucleating the A15 compound. However, owing to the increased stability of the σ phase with respect to Nb_3Ge, the metastable homogeneity range of the A15 phase transformed from an amorphous NbGe alloy could only be extended up to 22 to 23 at.% Ge. Similar results are expected for Nb/Ge multilayers and are under investigation.

ACKNOWLEDGEMENTS

We gratefully appreciate the valuable discussion with P. Haasen. This work was supported by the Deutsche Forschungsgemeinschaft. One of the authors (ADK) wishes to acknowledge support from an IBM fellowship during the course of investigation.

REFERENCES

1. L. Kammerdiener and H. L. Luo, J. Appl. Phys. 43:4728 (1972).
2. G. W. Webb, Appl. Phys. Lett. 32:73 (1978).
3. R. Bormann, D.-Y. Yu, R. H. Hammond, A.Marshall, and T. H. Geballe, in: "Rapidly Quenched Metals V", S. Steeb and H. Warlimont, eds., North Holland (1985), p. 879.
4. K. Togano, T. Takenchi and K. Tachikawa, Appl. Phys. Lett. 41:199 (1982).
5. C. L. H. Thieme, H. Zang, J. Otubo, S. Pourrahimi, B. B. Schwartz and S. Foner, IEEE Trans. Magn. MAG - 19:567 (1981).
6. B. Annaratone, K. Bruzzese, S. Ceresava, V. Pericoli-Ridolfini, G. Pitto, and N. Sachetti, IEEE Trans. Magn. MAG - 17:1000 (1981).
7. J. L. Jorda, R. Flükiger, and J. Muller, J. Less Common Met. 5:227 (1980).
8. G. A. Malets, Vest. Akad. Nauk SSR Ser. Khim Navuk 6:127 (1974).
9. L. Kaufmann and H. Nesor, CALPHAD 2:325 (1978).
10. H. L. Lukas, DGM Hauptversammlung, Stuttgart (1985), to be published.
11. D. Dew Hughes, Cryogenics 15:435 (1975).
12. R. B. Schwarz and W. L. Johnson, Phys. Rev. Lett. 51:415 (1983).

PROCESSING OF Nb_3Al AND OTHER EMERGING SUPERCONDUCTORS

Kyoji Tachikawa

Tsukuba Laboratories
National Research Instutute for Metals
Ibaraki, Japan

ABSTRACT

The development of new fabrication processes is essential for practical use of advanced superconductors. The powder metallurgy process has improved the fabrication of Nb_3Al and $PbMo_6S_8$ wires. The continuous liquid quenching and laser beam irradiation techniques have been successfully applied to the synthesis of Nb_3Al, $Nb_3(Al,Ge)$, and Nb_3Ga. The chemical vapor deposition process has increased reliability in the fabrication of Nb_3Ge tapes. The sputtering process is useful for the synthesis of NbN with high superconducting performance. The composite process as well as the liquid quenching process is adaptable for the fabrication of $V_2(Hf,Zr)$ wires and tapes. The present status of the development of these fabrication processes is described.

INTRODUCTION

To date, multifilamentary Nb-Ti conductors have been widely used for applications up to 9 T. As the second generation of high-field superconductors, multifilamentary Nb_3Sn, $(Nb,Ti)_3Sn$ and V_3Ga have entered the commercial stage, and 15-T-class magnets have been successfully constructed using these superconductors.

Now, Nb_3Al and other superconducting materials having performances superior to that of Nb_3Sn or V_3Ga are being developed as the third generation superconductors. However, these advanced superconductors can not be synthesized by the bronze process, which has been successfully adapted to Nb_3Sn and V_3Ga. Thus, the development of new fabrication processes is urgently needed. In this article, recent developments in the processing of Nb_3Al and other advanced superconducting materials are reviewed. The discussion, however, is limited to materials for conductor applications and does not include those for film applications.

Nb_3Al PREPARED BY THE POWDER METALLURGY PROCESS

The phase diagram of Nb-Al shows that the stable A15 phase is rather off-stoichiometric at low temperatures.[1] Ceresara et al. fabricated a Nb_3Al wire from a composite prepared by superimposing and winding Nb and Al foils around a Cu cylinder.[2] They heat-treated the composite wire at relatively

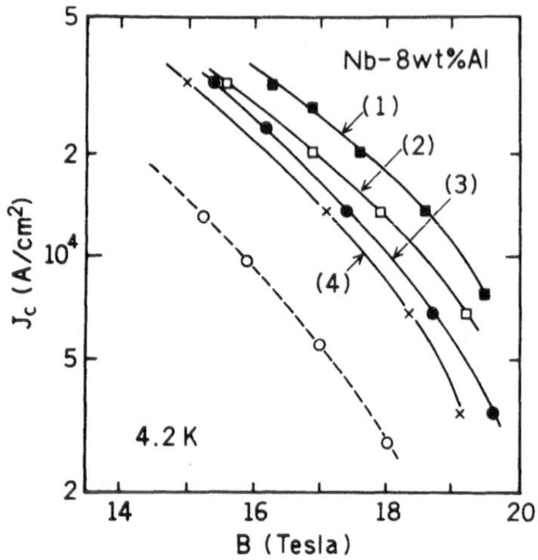

Fig. 1
Overall J_c versus applied field for the PM processed Nb-8wt%Al wire with fine powders and nominal areal reductions R of 3.4×10^5. (1) heat treated at 1100°C for 15 min plus at 750°C for 4 days. (2) 1000°C for 5 min plus 750°C for 4 days. (3) 900°C for 40 min plus 750°C for 3 days. (4) 800°C for 8 hr.[4] The dashed curve is earlier data from Ref. 3 with larger powders and R of 8×10^4.

low temperatures below 1000°C, and obtained a T_c of 15.56 K and $J_c(Nb_3Al)$ of 1.5×10^5 A/cm^2 at 6T.

Then, Foner and his group developed a powder metallurgy process for the Nb_3Al wire fabrication,[3] and succeeded in increasing the overall J_c by a factor of ~10 during these few years.[4] In their process, hydride-de-hydride Nb powder and Al powder are mixed with a composition range of 2-10wt%Al, compacted in a Cu-alloy jacket, swaged and drawn into a wire. The wire is heat treated at temperatures between 700°C and 1100°C. The overall J_c for the PM processed Nb_3Al wires is increased by using Nb powder of smaller grain size, increasing the total areal reduction of the wire, optimizing the Al percentage, and modifying the heat treatment condition.[4] Fig. 1 shows overall J_c versus applied field curves for some of improved Nb-8wt%Al wires and compares them with earlier data for the PM processed Nb-Al wire.[4]

TEM studies on the longitudinal cross sections of the PM processed Nb-Al wire were made by King et al,[5] and three different phases of unreacted Nb, A15 Nb_3Al and Nb_2Al were identified. According to their studies, filamentary A15 Nb_3Al layers of ~1μm thick and ~80μm long are formed adjacent to the unreacted Nb layers, and these A15 layers contain very fine grains of 100-300nm in size.

Small magnets have been wound by Foner et al using the PM processed Nb-Al wire and tested in background fields up to 15T. The overall J_c measured for the magnet is comparable with that for short samples.[6] The results on the PM processed Nb-Al wires suggest that these Nb-Al wires are one of the most promising conductors for generating fields over 19T.

Nb_3Al AND $Nb_3(Al,Ge)$ PREPARED BY LIQUID QUENCHING

The rapid quenching technique from the liquid state is potentially useful for the preparation of Nb_3Al or other advanced superconducting materials. Kammerdiner et al[7] succeeded in quenching small samples with Nb-Al bcc phase containing Al up to 35at%. The quenched samples were then annealed to produce A15 phase with the highest T_c of 18.6K. Bevk et al[8] prepared tape samples of Nb_3Al containing 1.5at%Si by the liquid quenching technique, which showed A15 structure at the quenched state. The grain size of the A15 phase is extremely small and J_c at 4.2K exceeds 1.8×10^6 A/cm^2 in fields up to 15T.

Recently, Togano et al[9] demonstrated that the liquid quenching on a hot substrate could achieve better wettability between the melt of high-melting-point materials and the substrate, resulting in a larger cooling rate than that for the conventional quenching techniques. Using this technique, a supersaturated bcc phase of a $Nb_3(Al,Ge)$ ternary alloy was successfully formed and transformed into a fine-grained A15 phase with high J_c.[10,11] Based on these results, they developed a new continuous liquid quenching apparatus equipped with a high-speed moving hot substrate, and obtained composite tapes composed of the quenched alloy and the substrate.[12] Fig. 2 shows a photo of the continuous liquid quenching apparatus. The alloy button of $\sim$10 gr is melted by RF levitation melting in the upper chamber which has been evacuated and backfilled with argon gas. The molten alloy is then dropped into a nozzle by cutting off RF current and ejected as a jet onto a high-speed moving copper tape in the lower vacuum chamber. The copper tape has been pre-heated to a proper temperature and driven at a velocity of 10-20m/sec on a Hastelloy tape which serves as a reinforcement. The ejected molten alloy spreads as a uniform thin layer, adhering tightly to the copper tape, forming a composite tape. Fig. 3 shows an example of the resulting composite tape.

The X-ray diffraction pattern of $Nb_3(Al_{0.9}Ge_{0.1})$ quenched on a hot copper substrate indicates the formation of a supersaturated bcc solid solution of ternary $Nb_3(Al_{0.9}Ge_{0.1})$ alloy. After an annealing above 800°C, the X-ray pattern of the quenched layer shows relatively broad lines of the A15 majar phase. The A15 $Nb_3(Al,Ge)$ phase converted from the bcc phase has an extremely small grain size of a few hundred angstroms, which leads to a high J_c in high magnetic fields. The $Nb_3(Al_{0.9}Ge_{0.1})$ tape sample annealed at 850°C for 7h shows an excellent high field performance, namely its J_c at 4.2K exceeds $10^5 A/cm^2$ in fields up to 20T.

Advantages of the liquid quenching on a moving substrate of heated copper are, (1) improved cooling rate over the conventional liquid quenching, (2) good bonding between the quenched alloy layer and the copper substrate, which enables the direct use of substrate as a stabilizing material, and (3) simplicity of the process to produce composite superconductors very rapidly.

Fig. 2
Continuous liquid quenching apparatus.[12]

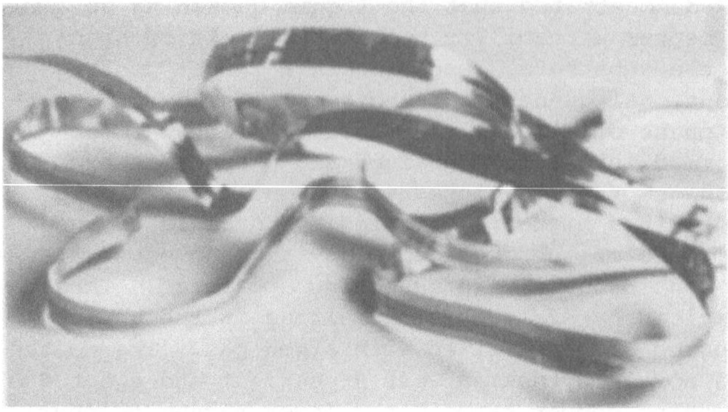

Fig. 3
SC/Cu composite tape prepared by the liquid quenching on a hot Cu substrate.[12]

SYNTHESIS BY LASER BEAM IRRADIATION

Although laser alloying has become of much interest as a new material processing method, only a few attempts have been made to apply this technique to superconductor fabrication. Irradiation experiments of short pulse laser have been performed for V-Si[13] and Mo-N(C)[14] in order to study the effect of resultant rapid quenching on the superconducting properties of these materials.

Recently, a research program to develop advanced superconducting materials utilizing the irradiation of high power density beams including both laser and electron beam has been started at NRIM. A series of preliminary investigations on Nb_3Ga and Nb_3Al using a CW-CO_2 laser have been carried out in collaboration with Toshiba.[15]

As a starting material for Nb_3Ga synthesis, the Nb tape coated with $NbGa_3$ compound layer was prepared by passing the Nb tape through a molten Ga bath and subsequently through a pipe-shaped furnace. Laser irradiation was performed in an argon gas atmosphere using a CW-CO_2 laser with a spot size of $\sim 3mm^2$ at power levels of 0.3-1.2kW. During the irradiation, the tape was moved at a velocity of 20cm/sec. High power density of the CO_2 laser beam and high speed moving of the material bring about a very quick reaction between the components and a subsequent rapid cooling. The result is the formation of a very fine structure composed of A15 Nb_3Ga and α-Nb.

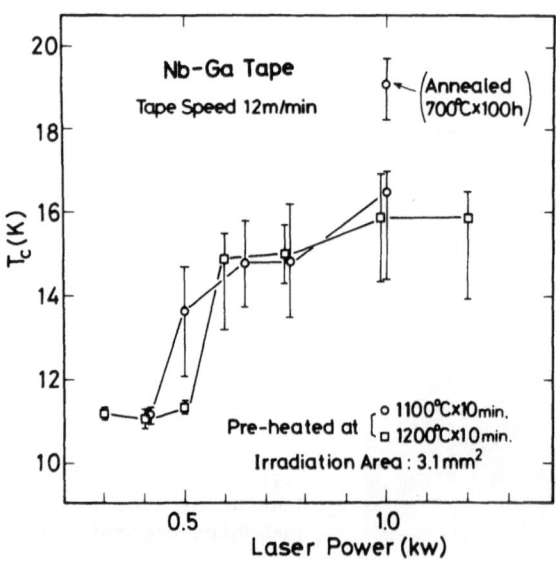

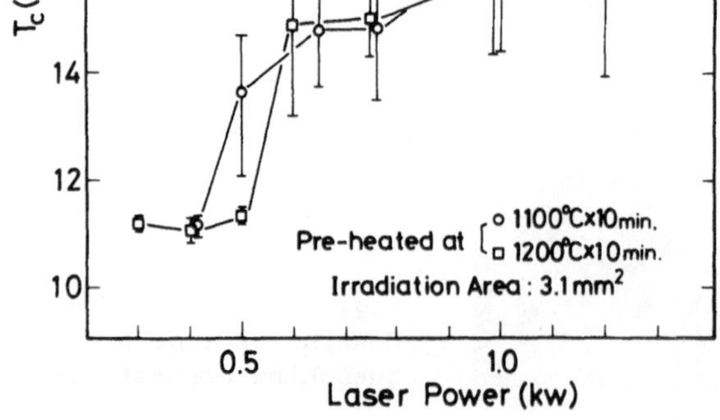

Fig. 4
T_c of Nb_3Ga versus irradiated laser power.[15]

Fig. 4 shows the T_c as a function of the laser power. High T_c values are obtained at power levels above 0.6kW, indicating the formation of stoichiometric Nb_3Ga stable only at elevated temperatures. Annealing at 700°C was carried out for the sample irradiated at a power of 1.0kW, and resulted in a significant increase in T_c from 17.0K to 19.7K.

Laser irradiation has been also performed for the synthesis of Nb_3Al superconductor. Starting material was prepared by conventional powder metallurgy method, i.e., the mixture of high purity Nb and Al powders was cold worked into a composite tape. The maximum T_c obtained after the irradiation is 16.5K, which can be increased to 18.2K by subsequent annealing at 700°C. This value is higher than that of PM processed Nb-Al wire conventionally heat treated,[3] probably due to the retension of more stoichiometric A15 phase by quenching from an elevated temperature. For some specimens, critical current measurement was carried out, and a J_c(for reacted area) of $3x10^4 A/cm^2$ was obtained at 4.2K and 15T.

These results suggest that the continuous laser irradiation onto a high speed moving material is a useful method for fabricating advanced superconductors such as Nb_3Ga, Nb_3Al and $Nb_3(Al,Ge)$.

Nb_3Ge PREPARED BY CHEMICAL VAPOR DEPOSITION

Since the discovery of its high T_c exceeding the liquid hydrogen temperature,[16] the Nb_3Ge compound has been a prominent candidate material for applications, such as superconducting transmission lines and high-field magnets. Studies for producing practical Nb_3Ge conductors are mainly based on the chemical vapor deposition (CVD) process which has a larger film deposition rate than sputtering or vacuum evaporation techniques.[17,18,19] The formation of the Nb-Ge phases occurs on the heated substrate tape by hydrogen reduction of a vapor mixture of $NbCl_5$ and $GeCl_4$ at an atmospheric pressure. Both chlorides are usually prepared by direct metal chlorination. As listed in Table 1, this process may include many parameters which affect both film growth features and superconducting properties.

Nb_3Ge with a stoichiometric composition and a high T_c is normally unstable at low temperatures. Several mechanisms for the stabilization of the high-T_c Nb_3Ge phase have been proposed. Recently, it has been pointed out that the high-T_c Nb_3Ge mostly coexists with the tetragonal Nb_5Ge_3,[20] and can be formed from the Nb_5Ge_3 phase via a heteroepitaxial growth process.[21] The A15 Nb_3Ge with a high-T_c above 20K usually exhibits smaller lattice parameters between 0.515-0.514nm. The J_c of the Nb_3Ge at high magnetic fields has been reported to be sensitive to the amount of incorporated second phase particles[18] and the grain size of Nb_3Ge.[21]

Many previous reports suggest that the optimization of the process parameters listed in Table 1 and the precise control of them is most essential for the preparation of practical high-T_c and high-J_c Nb_3Ge conductors.

Table 1. CVD Process Parameters for the Nb_3Ge Tape

Gas flow rate and mixing ratio	Temperatures
Cl_2, H_2, He(or Ar)	Chlorination furnace
$H_2/(NbCl_5+GeCl_4)$	Reaction furnace
Gas dilution ratio	Substrate tape
He(or Ar)/Cl_2	Gas flow pattern around
for Nb and Ge chlorination	substrate tape
Tape moving speed	Substrate material

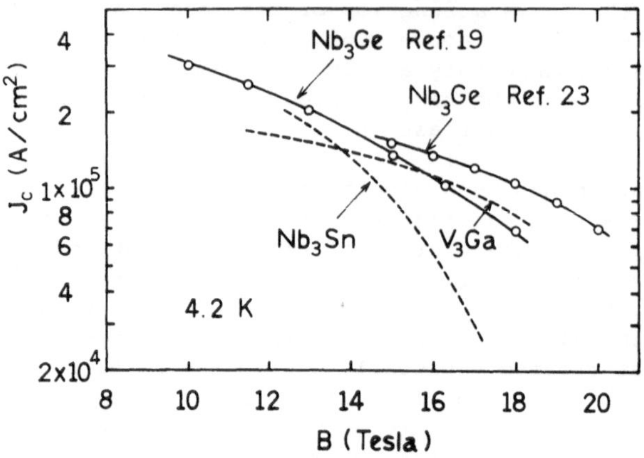

Fig. 5. Typical J_c-B curves of CVD Nb_3Ge in high fields taken from quoted references.

Recently, the formation of the Nb_3Ge layer in a reproducible fashion has been reported,[21,22] and a J_c as high as $7x10^4 A/cm^2$ at 20T and 4.2K has been obtained for the Nb_3Ge layers, as shown in Fig. 5.

NbN PREPARED BY SPUTTERING

It has been reported that the reactive sputtered NbN film can exhibit a T_c of 17.3K,[24] a B_{c2} of 50T,[25] and a J_c of $10^6 A/cm^2$ at 20T.[25] Moreover, NbN films are relatively insensitive to mechanical strain[26] and neutron irradiation.[27] Thus, NbN may be one of the most promising alternative superconductors to Nb_3Sn for high field magnet use.

Kampwirth et al[28] have been working on NbN tape conductors. Their 2.5µm thick films by DC magnetron sputtering show a J_c of about $10^4 A/cm^2$ at 20T. Another unique work has been done by Dietrich et al,[29,30] who deposited NbN on high tensile carbon fiber bundles by DC magnetron sputtering. The carbon fiber substrate is suitable for winding magnets of high stress tolerance and easy handling. Moreover, the carbon fiber deposition may lead to the multifilamentary NbN conductor. Their 0.5µm thick films sputtered on the carbon fiber show a J_c of about $10^5 A/cm^2$ at 13T.[29]

In Japan, Suzuki et al[31] have produced NbN films ~3000Å thick by RF sputtering, and obtained a J_c of $10^5 A/cm^2$ in a perpendicular field of 20T with a considerably small J_c of $4x10^2 A/cm^2$ in the same but parallel field. This J_c anisotropy has been also found by Yamada et al[32] in NbN films RF magnetron sputtered on the AlN substrate at a sufficiently low temperature.

In the above cases,[31,32] the NbN films have columner structures and this may be the reason why the anisotropy appears, as was first discussed by Gavaler et al.[25] However, the microstructure of the NbN film varies and granular structures have been also reported.[33] What causes this microstructural difference is still unclear.

The J_c of NbN degrades with increasing film thickness. Fig. 6 summarizes the dependence of J_c on the film thickness for NbN conductors so far obtained.[25,28,30,32] The J_c of the NbN film is rapidly decreased with increasing film thickness, and this is the most serious problem in developing a practical NbN conductor. Stacking NbN layers may be one of the approaches to sustain large critical currents. The anisotropy in J_c with respect to the field direction is also illustrated in Fig. 6.

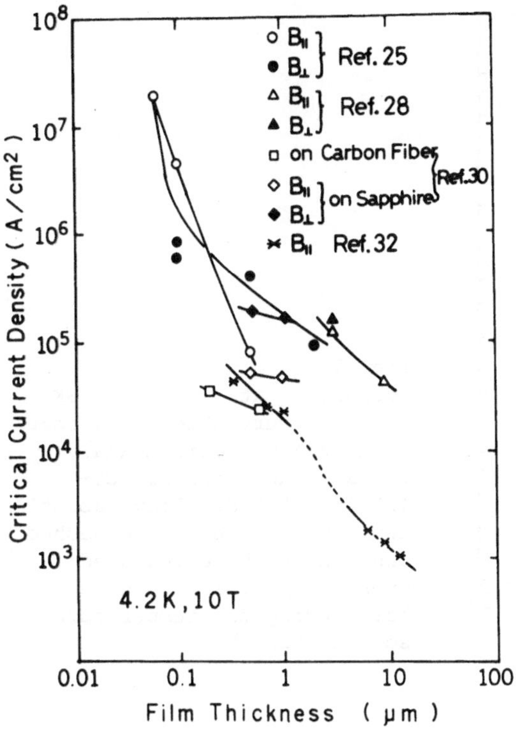

Fig. 6
The dependence of J_c on film thickness for NbN sputtered films. The data are summarized from quoted references.

Akune et al[34] quenched the NbN films with a liquid N_2 spray immediately after the deposition. Through this method the growth of the equilibrium phase is depressed and that of the phase with a high T_c is encouraged instead. Although several problems like J_c degradation with increasing film thickness remain unsolved, recent results suggest the feasibility of the NbN as the high field superconductor.

V_2(Hf,Zr) PREPARED BY COMPOSITE PROCESS AND LIQUID QUENCHING

V_2Hf-based Laves phase compounds are attractive superconductors especially for fusion magnet use due to their promising combination of high B_{c2} exceeding 20T at 4.2K[35] and insensitiveness to both neutron irradiation[36] and mechanical strain.[37] Additions of third elements such as Nb, Ta, Zr, Cr, and Mo are known to improve both T_c and B_{c2} of V_2Hf.[38,39] Among these V_2Hf-based Laves phase compounds, a pseudobinary compound V_2(Hf,Zr), is practically most interesting. Studies have been carried out on the superconducting properties of V_2(Hf,Zr) multifilamentary wires[40] and those of liquid-quenched V-Hf-Zr alloy tapes containing Laves phase compounds.[41,12]

Multifilamentary V_2(Hf,Zr) wires can be fabricated by the composite process where a composite of a V-1at%Hf alloy matrix and Zr-Hf alloy cores are cold drawn into wires, and then reacted to form Laves phase compound layers between the matrix and the cores by the diffusion reaction. 1634-core multifilamentary wires have been successfully fabricated for the composites with Zr-Hf cores containing Hf less than 60at%.[40]

A T_c of 9.9K, a B_{c2}(4.2K) of 22.1T, and an overall J_c(4.2K) of 1×10^4 A/cm^2 at 17T are obtained for multifilamentary wires with Zr-(40-50)at%Hf alloy cores reacted at 950°C-1025°C for 20-50h. These values of B_{c2} and overall J_c are comparable to those of commercial Nb_3Sn wires at 4.2K. The

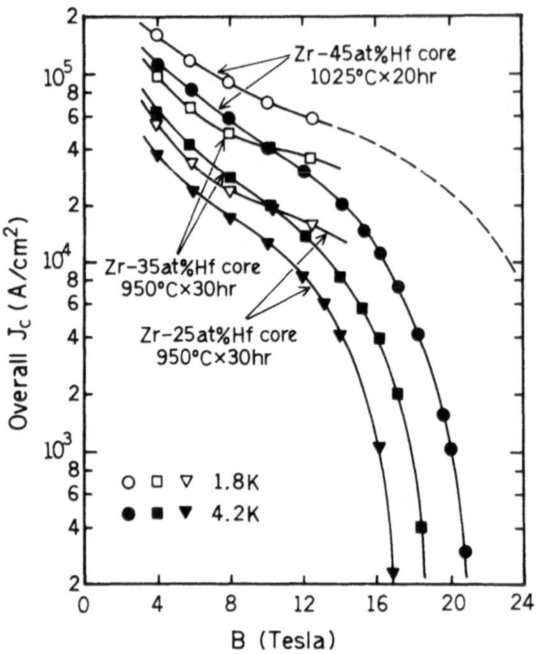

Fig. 7
Overall J_c at 4.2K and 1.8K versus B curves for 1634-core V_2(Hf,Zr) multifilamentary wires. The cores are Zr-(25,35,45)at%Hf alloys about 10μm in diameter. The dashed line shows the calculated overall J_c(1.8K) versus B curve using the temperature scaling law.[40]

J_c of the V_2(Hf,Zr) wires are rapidly increased with reducing temperature and become twice as large as those of commercial Nb_3Sn wires at 1.8K and 15T as shown in Fig. 7. The enhanced J_c at reduced temperatures may be caused by the rapid increase in B_{c2}. The B_{c2} of the Zr-50at%Hf core wires measured by a pulsed magnet is about 28T at 2.0K. Using the temperature scaling law of the pinning density, the estimated value of overall J_c at 1.8K is $2x10^4$A/cm^2 at 20T. These V_2(Hf,Zr) multifilamentary wires may be used for manufacturing magnets generating magnetic fields above 18T in a pressurized liquid HeII environment.

Meanwhile, Tenhover studied on V-Hf-Zr amorphous alloys produced by a liquid quenching technique and found that J_c values larger than 10^5A/cm^2 are obtained at 4.2K and 15T by heat treating the alloys to precipitate crystalline Laves phase.[41] Recently, a long composite tape of a V_{40}(Hf,Zr)$_{60}$ amorphous layer and a copper substrate was produced by the continuous liquid quenching technique.[12] The cross section of the composite tape is shown in Fig. 8. The amorphous phase is converted to the Laves phase and the β phase by an annealing at 600°C-900°C. The composite tape shows a T_c of 8.9K and an overall J_c(10T, 4.2K) of $5x10^4$A/cm^2. The optimization of alloy composition and process parameters will lead better critical values in the liquid quenched Laves phase composite tape.

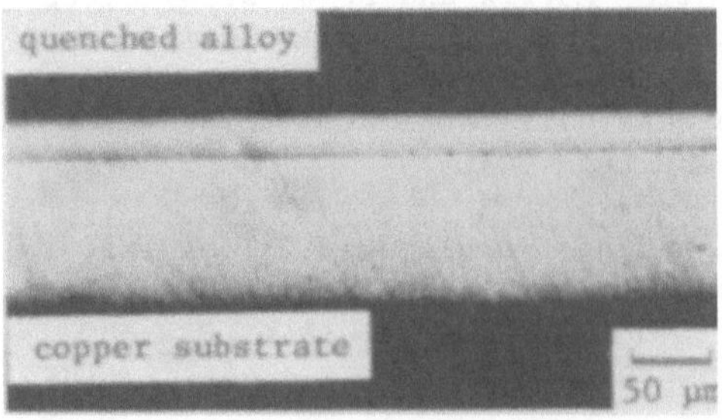

Fig. 8
Cross section of the V_2(Hf,Zr)/Cu composite tape prepared by the continuous liquid quenching.[12]

As is well known, some of Mo containing Chevrel phase compounds have the highest B_{c2} among superconductors. Especially, $PbMo_6S_8$ has become of much interest since its B_{c2} exceeds 50T at 4.2K and its T_c is around 15K.[42] Thus, various fabrication techniques have been studied for this Chevrel phase superconductor to attain a practically sufficient J_c. Among them the gas phase diffusion process and the powder metallurgy process are attractive from the industrial point of view.

In the gas phase diffusion process proposed by Decroux et al[43] and Aleksevskii et al,[44] a Mo wire is reacted with S to form a layer of Mo sulfides, and then exposed to Pb vapor at temperatures between 950°C and 1100°C. Hamasaki et al[45] has recently developed a modified process where they claim a J_c of $1.8 \times 10^4 A/cm^2$ at 14T. In this process, a Mo wire is electroplated with Pb and reacted in S vapour from MoS_2 at 1050°C for 14h. The process has an advantage that a $PbMo_6S_8$ single phase can be obtained, which is the key to high T_c, B_{c2}, and J_c, although the control of the reaction condition along a long wire is not easy.

In the powder metallurgy process, a mixture of Mo, MoS_2 and PbS powders are packed in a Ta or Mo tube which seals up the S. The composite is drawn to a final size and heat-treated to form $PbMo_6S_8$ compound in its core. With the Ta sheath it is easy to draw the composite at room temperature, but undesirable Ta sulfides are formed at the heat treatment. The Mo sheath is favourable in view of the reaction with S, but the composite with the Mo sheath can be drawn only above 600°C. Seeber et al[46] have compared J_c for the composite wire with the Mo sheath with that for the hot-pressed bulk sample. J_c at 6.5T for the wire is nearly one order smaller than that for the bulk sample. The degradation of J_c in the composite wire may be attributed to cracks in the compound layer induced by the thermal contraction. The refinement of microstructures may be effective in attaining higher J_c.

Yamasaki et al[47] have reported a J_c of $10^4 A/cm^2$ at 8T for their powder metallurgy processed wire using higher S contents and a Mo sheath, and reacted at 950°C for 10h. Kubo et al[48] have developed wires with a Ta sheath and a Cu stabilizer, heated them at 1000°C for 2h under an isostatic

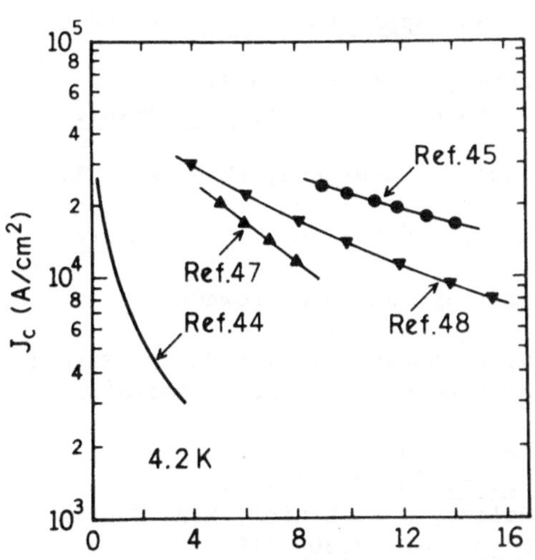

Fig. 9
Typical J_c-B curves of $PbMo_6S_8$ superconductors taken from quoted references.

pressure of 2000kg/cm^2, and obtained a J_c of $8x10^3A/cm^2$ at 15.5T and a T_c of 14.2K. Since the formed compound layer contains other phases besides $PbMo_6S_8$, J_c may be further improved if the fraction of $PbMo_6S_8$ is increased. Typical J_c-B characteristics of $PbMo_6S_8$ conductors are summarized in Fig. 9. Up to now, several attempts of forth element addition have been made to improve J_c properties of PM processed $PbMo_6S_8$ conductors.

Ekin et al[49] have examined strain effects on J_c and B_{c2} of $PbMo_6S_8$ conductors fabricated by the gas phase diffusion process and found mechanical behaviours similar to those of A15 compounds. In conclusion, the J_c so far obtained for the $PbMo_6S_8$ conductor is still insufficient for high field magnet use, although recent results are very encouraging.

SUMMARY

A lot of challenges are being proceeded on the processing of advanced superconductors throughout the world as mentioned briefly here. The author hopes that a breakthrough will be realized from these challenges in the near future.

ACKNOWLEDGMENTS

The author wishes to express his sincere thanks for preparing this review to Drs. K. Togano, Y. Tanaka, K. Inoue, H. Wada, and K. Itoh of National Research Institute for Metals, Mr. Y. Yamada of Toshiba, and Mr. K. Yoshizaki of Mitsubishi Electric Co.

REFERENCES

1. J.L. Jorda, R. Flukiger, and J. Muller, J. Less Common Metals 75:227 (1980).
2. S. Ceresara, M.V. Ricci, N. Saccheti, and G. Sacerdoti, IEEE Trans. Magn. MAG-11:263 (1974).
3. K. Akihama, R.J. Murphy, and S. Foner, ibid. MAG-17:274 (1981).
4. C.L.H. Thieme, S. Pourrahimi, B.B. Schwartz, and S. Foner, Appl. Phys. Lett. 44:260 (1984).
5. W.E. King, C.L.H. Thieme, and S. Foner, IEEE Trans. Magn. MAG-21:815 (1985).
6. C.L.H. Thieme, S. Pourrahimi, B.B. Schwartz, and S. Foner, ibid. MAG-21:756 (1985).
7. L. Kammerdiner and H.L. Luo, J. Appl. Phys. 43:4128 (1972).
8. K. Lo, J. Bevk, and D. Turnbull, ibid. 48:2597 (1977).
9. K. Togano, H. Kumakura, and K. Tachikawa, Appl. Phys. lett. 40:84 (1982).
10. K. Togano, T. Takeuchi, and K. Tachikawa, ibid. 41:199 (1982).
11. K. Togano, H. Kumakura, T. Takeuchi, and K. Tachikawa, IEEE Trans. Magn. MAG-19:414 (1983).
12. K. Togano, H. Kumakura, Y. Yoshida, and K. Tachikawa, ibid. MAG-21: 463 (1985).
13. B. Stritzker, B.R. Appleton, C.W. White, and S.S. Lau, Solid State Communications, 41:321 (1982).
14. S.T. Sekura, J.R. Thompson, G.M. Beardsley, and D.H. Lowndes, J. Appl. Phys. 54:6517 (1983).
15. K. Kumakura, K. Togano, K. Tachikawa, S. Murase, Y. Yamada, M. Sasaki, and H. Nakamura, "Proc. of 33rd Meeting of Cryogenic Association of Japan", p6 (1985).
16. J.R. Gavaler, Appl. Phys. Lett. 23:480 (1973).
17. H. Kawamura and K. Tachikawa, Phys. Lett. 50A:29 (1974).
18. A.I. Braginski, J.R. Gavaler, G.W. Roland, M.R. Daniel, M.A. Janocko, and A.T. Santhanam, IEEE Trans. Magn. MAG-13:300 (1977).

19. M.P. Maley, L.R. Newkirk, J.D. Thompson, and F.A. Valencis, ibid. MAG-17:533 (1981)
20. L. Li, B. Zhao, P. Zhou, S. Guo, and Y. Zhao, J. Low Tem. Phys. 45: 287 (1981).
21. F. Weiss, O. Demolliens, R. Mader, J.P. Senateur, and R. Fruchart, J. Physique 45:1137 (1984).
22. T. Asano, Y. Tanaka, and K. Tachikawa, to be published in Cryogenics.
23. O. Demolliens, private communication.
24. K.S. Kesker, T. Yamashita, and Y. Onodera, Jpn. J. Appl. Phys. 10:370 (1971).
25. J.R. Gavaler, A.T. Santhanam, A.I. Braginski, M. Ashkin, and M.A. Janocko, IEEE Trans. Magn. MAG-17:573 (1981).
26. J.W. Ekin, J.R. Gavaler, and J. Gregg, Appl. Phys. Lett. 41:996(1982).
27. D. Dew-Hughes, and R. Jones, ibid. 36:856 (1980).
28. R.T. Kampwirth, D.W. Capone II, K.E. Gray, and A. Vicens, IEEE Trans. Magn. MAG-21:459 (1985).
29. M. Dietrich, ibid. :p455
30. S. Ohshima, M. Dietrich, and G. Linker, J. Appl. Phys. 57:890 (1985).
31. M. Suzuki, M. Baba, M. Takahata, and T. Anayama, "Proc. of 33rd Meeting of Cryogenic Association of Japan",p14 (1985).
32. Y. Yamada, S. Murase, S. Takatsu, K. Shibuki, and T. Sadahiro, ibid.: p15.
33. D.D. Bacon, A.T. English, S. Nakahara, F.G. Peters, H. Schreiber, W.R. Sinclair, and R.B. van Dover, J. Appl. Phys. 54:6509 (1983).
34. T. Akune, N. Sakamoto, and Y. Shibuya, Jpn. J. Appl. Phys. 23:184 (1984).
35. K. Inoue, K. Tachikawa, and Y. Iwasa, Appl. Phys. Lett. 18:235(1971).
36. B.S. Brown, J.W. Hafstron, and T.E. Klippert, J. Appl. Phys. 48:1759 (1977).
37. H. Wada, K. Inoue, K. Tachikawa, and J.W. Ekin, Appl. Phys. Lett. 40: 844 (1983).
38. K. Inoue and K. Tachikawa, "Proc. of the Applied Superconductivity Conference", IEEE Cat. No. 72CHO682TABSC, p415 (1972).
39. Z. Li, Z. Zhi-tao, X. Yun-hui, C. Lin-fu, and Y. Dao-le, "Proc. of ICMC", Butterworth & Co., p219 (1982).
40. K. Inoue, T. Kuroda, and K. Tachikawa, IEEE Trans. Magn. MAG-21:467 (1985).
41. M. Tenhover, ibid. MAG-17:1021 (1981).
42. Ø. Fischer, R. Odermatt, G. Bongi, H. Jones, R. Chevrel, and M. Sergent, Phys. Lett. 45A:87 (1973).
43. M. Decroux, Ø. Fischer, and R. Chevrel, Cryogenics 17:291 (1977).
44. N.E. Alekseevskii, M. Glinski, N.M. Dobrovol'skii, and V.I. Tsebro, JETP Lett. 23:412 (1976).
45. K. Hamasaki, K. Hirata, T. Yamashita, T. Komata, K. Noto, and K. Watanabe, IEEE Trans. Magn. MAG-21:471 (1985).
46. B. Seeber, C. Rossel, Ø. Fischer, and W. Glaetzel, ibid. MAG-19:402 (1983).
47. H. Yamasaki and Y. Kimura, "Proc. of JAPAN-US Workshop on High-Field Field Superconducting Materials for Fusion", (1985) p118.
48. Y. Kubo, K. Yoshizaki, and F. Fujiwara, "Proc. of 33rd Meeting of Cryogenic Association of Japan", (1985) p13.
49. J.W. Ekin, T. Yamashita, and K. Hamasaki, IEEE Trans. Magn. MAG-21:474 (1985).

THE EFFECTS OF HIGH REACTION TEMPERATURES ON THE T_c OF MULTIFILAMENTARY Nb-Al WIRES

D. Abukay and L. Rinderer

Université de Lausanne
Institut de Physique Expérimentale
Lausanne, Switzerland

ABSTRACT

Powder metallurgy processed multifilamentary Nb-8 wt.% Al wires were heat-treated at temperatures >2000°C for very short times. Values of T_c ranging from 12 to 20 K have been resistively measured from these wires after annealing them at 700 to 750°C for 48 to 185 h. The $B_{c2}(T)$ values of the same samples were also measured, and the values of their magnetic slope $(-dB_{c2}/dT)$ at T_c varied in the range 2.38 to 2.58 T/K.

INTRODUCTION

In several recent publications[1-3] it has been demonstrated that the powder metallurgy (P/M) processing of multifilamentary (MF) fine-fibre high-field superconductors composed of Nb-Al is quite feasible and adaptable to large-scale fabrication. However, when the P/M processed MF Nb-Al wires are heated below 1000°C to form the A15 phase, they always yield a critical temperature, T_c, in the range 15 to 16 K. This is still higher than the value of 12 K measured for the bulk samples prepared in the same temperature ranges.[4] The origin of this enhancement of T_c is not yet understood. For superior performance at very high fields, a high T_c is also required, since, as described in GLAG theory, it is related to the upper critical field, B_{c2} (at T = 0 K).

In this work, we have studied the effects of high reaction temperatures on the T_c of P/M processed MF Nb-8 wt.% Al wires to explore the practical limits achievable on such composite structures. Surprisingly, we have measured extraordinarily high critical temperatures from the samples heated to temperatures >2000°C for short times followed by a low-temperature anneal.

EXPERIMENT

Niobium powder ($<2 \times 10^{-4}$ m) and Al powder ($<4 \times 10^{-5}$ m) were dry mixed at a stoichiometric ratio of 8 wt.% Al. The mixture was put into a Cu tube at a tap density. This tube, having a Ta barrier with a wall thickness of 0.5 mm already inserted tightly inside, had an outer diamter of 20 mm. The composite was evacuated, sealed, and swaged. A nominal size reduction ratio of about 250 was obtained, and a final diameter of 1 mm was measured from the strands after the Cu outer jacket was removed in a HNO_3 bath.

Table 1. The Results of T_c and B_{c2} (T) Measurements of Multifilamentary Nb-Al Wires Reacted at 2000°C

Sample Number	Reaction Time (s)	Annealing		T_c (K) before Annealing		T_c (K) after Annealing		$\left(-\frac{dB_{c2}}{dT}\right)$ (TK^{-1})	B_{c2}* (T)
		Time (h)	Temperature (°C)	Onset	Midpoint	Onset	Midpoint		
S1	6.5	185	700	—	—	19.10	19.05	2.4	31.70
S2	3.0	96	750	—	—	20.32	20.11		
		185	700			19.20	19.15	2.53	33.35
S3	1.0	185	700	17.17	16.90	19.30	19.25	2.58	34.25
S4	3.2	185	700	13.50	10.85	19.67	19.59	2.64	35.51
S5	5.0	185	700	—	—	19.05	18.93	2.38	31.15
S6	12.0	185	700	17.60	17.52	19.15	19.10	2.48	32.70
S7	40.0	48	750	16.66	16.34	20.08	19.30	2.46	33.00

*The value of B_{c2} at T = 0 K.

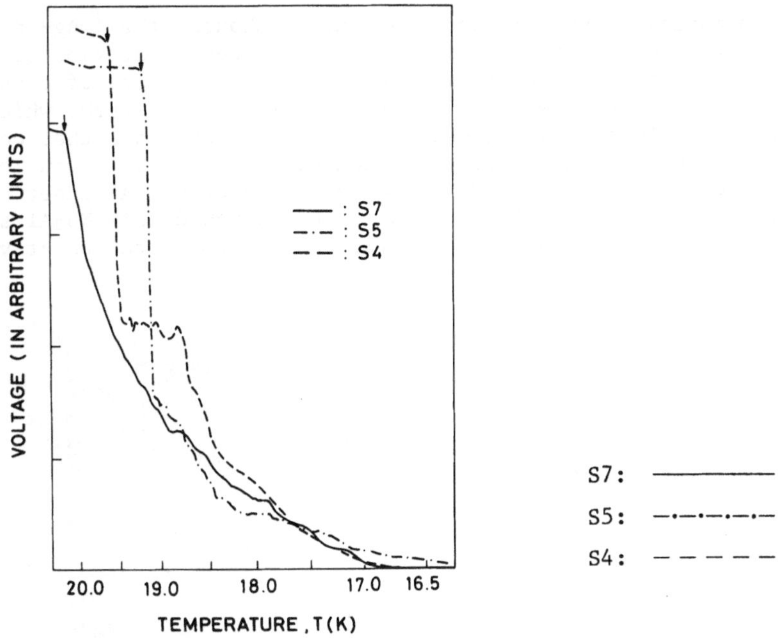

Fig. 1. Typical resistive transition curves of Nb 8 wt.% Al samples in this work after annealing for 47-185 h at 700-750°C. The arrows indicate the onset points.

The samples were cut 0.02 to 0.03 m in length and were heated to temperatures >2000°C in a vacuum of 13.33 mPa (10^{-4} torr) by passing an electric current through them from a stabilised dc current source for a short time. The temperatures of the samples were monitored with an optical pyrometer; the correction for the emissivity of Ta was considered. Prior to this heating procedure, to eliminate the possiblity of contamination by the remaining organic residues left from former processing, the samples were cleaned by treating them both in a Na-based basic solution and in an acid bath (1 part HNO_3, 1 part HF and water). The superconducting parameters, i.e., T_c and $B_{c2}(T)$, were resitively measured from the samples in a superconducting 10-T coil by taking only the midpoints of their transition curves as a criterion.

RESULTS AND DISCUSSION

The results of the present experiment are summarized in Table 1. We observed that the values of T_c of the samples increased to their maximum, ranging from 19 to 20 K, after annealing at 700 to 750°C for 48 to 185 h. Figure 1 shows typical transition curves recorded during this experiment. The general feature of these curves is that the transition width at lower temperatures is rather large, about 2 to 2.5 K, indicating the inhomogeneity of the samples. The transition width of some becomes very narrow, about 0.5 K, upon reaching the onset point. That sharp transition may arise from some small A15-phase region with the highest T_c in the sample, which is homogeneous and well-ordered after annealing. Figure 2 shows a typical back-scattered electron (BSE) picture of the cross section of a sample obtained by an x-ray spectrometer; its Al distribution, as seen by a scanning electron microprobe analyser, is shown in Fig. 3. The shapes of the Nb filaments are highly irregular and rather flattened, typical of a swaging process, and they are rather thick since the size reduction ratio was small. An onset temperature of 20.3 K represents the highest value reported so far for Nb_3Al. It appears that P/M processing of MF wires favors the direct formation of the A15 phase and also enhances the long-range atomic ordering processes during annealing. This may be attributed to the fact that during the size reduction processes a large amount of crystal defects is introduced into the Nb filaments as well as a texture, which may have also been developed along them. The metallic interface that occurs between Nb and Al under these conditions then may enhance the diffusion kinetics of the Nb-Al couple to form the Nb_3Al compound directly. An onset of 19.1 K was reported by Flükiger et al.[5] for an Ar jet-quenched bulk Nb_3Al sample after a long period of annealing. However, in our work, the onset values of

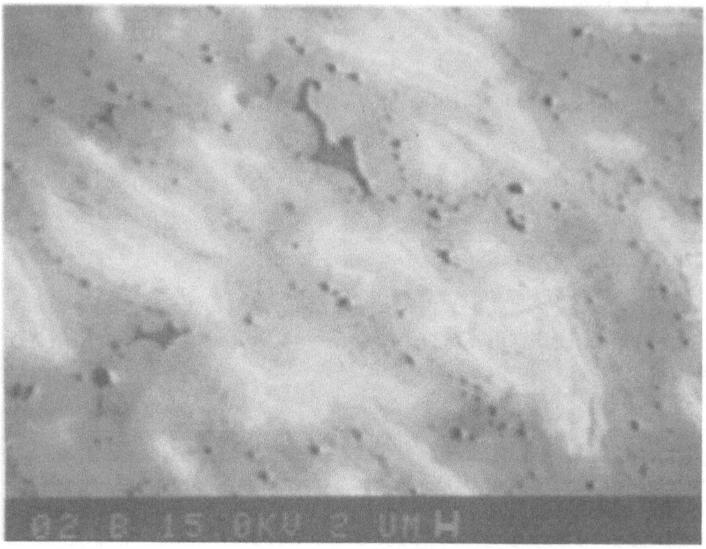

Fig. 2. Backscattered electron image showing a cross section of a Nb_3Al wire used in this work.

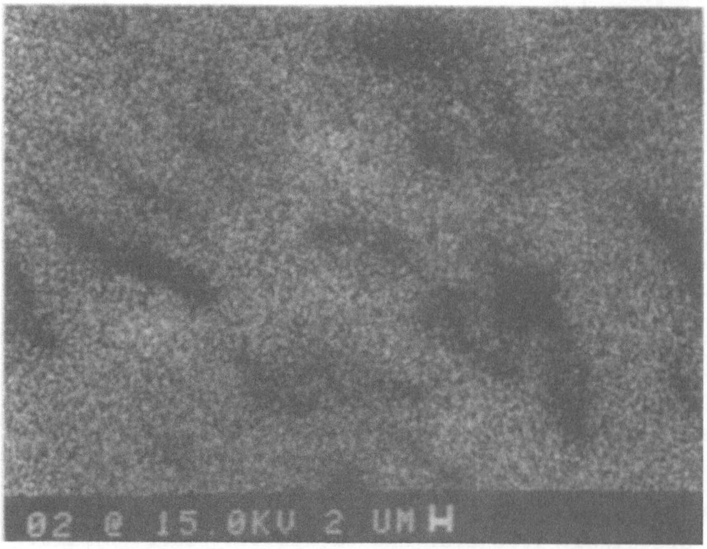

Fig. 3. Al distribution in the cross section of the sample shown in Fig. 2.

the same order or even higher were reached after 48 h of annealing. Annealing longer than 185 h did not significantly improve the T_c values already reached. An important remark here is that the cooling rate of our samples in this experiment was quite low, simply cooling by radiation. This obviously did not prevent segregation, which was quite evident in this compound. From the phase diagram of Nb-Al, we think that if the cooling rates of the samples were efficiently improved, then both their stoichiometry and homogeneity would be superior to the present ones.[6] The upper critical field measurements are shown in Fig. 4. The slope $(-dB_{c2}/dT)$ at T_c had an upward curvature below 1 T in the neighborhood of T_c at zero field for all samples. This behavior, we believe, is a simple manifestation of the large inhomogeneity of the samples. In this case, in calculating B_{c2} (T = 0 K) according to the GLAG formula, T_c was determined by extrapolating the linear part of the slope in the B_{c2} vs. T plots until it intersected the horizontal axis. The values of the slopes determined in that way varied between 2.38 T/K and 2.58 T/K, which are close to the value of 2.5 T/K reported by Foner et al. for a single-phase bulk Nb_3Al sample.[7]

CONCLUSIONS

The results of our present work have shown that a range of 19 to 20 K for the T_c of P/M processed MF Nb-Al wire is practically obtainable, which introduces a new practical limit for this compound for its technological use. It also implies that there is still much to know about the behavior of P/M processed MF Nb-Al wires for further optimization of their superconducting parameters, as well as for a better comprehension of their superconductivity. Now our remaining concern is to achieve a high J_c, at least on the order of the value reported by Lo et al.[8] To do that, one must discover how to impede the grain-growth processes in the A15 phase during

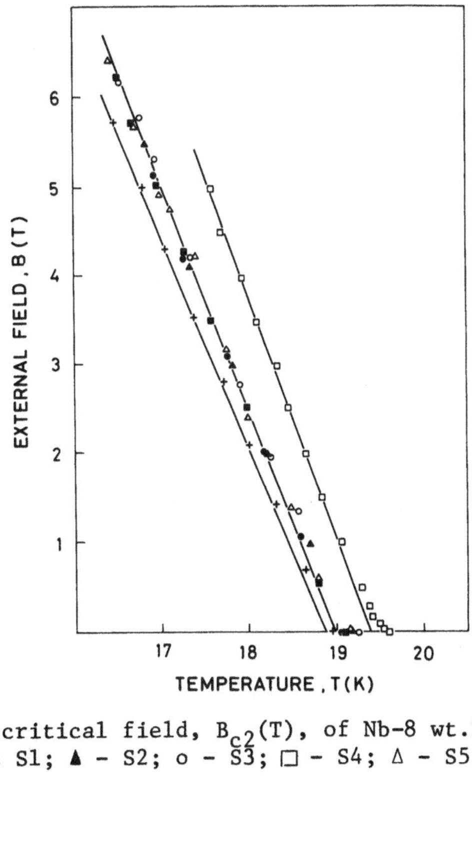

Fig. 4. The upper critical field, $B_{c2}(T)$, of Nb-8 wt.% Al samples. ● - sample S1; ▲ - S2; o - S3; □ - S4; Δ - S5; + - S6; ■ - S7.

cooling of the samples from elevated temperatures. Hence, convenient, fast cooling techniques must be employed for this type of processing. We have already begun such work in our laboratory.

ACKNOWLEDGMENTS

The authors wish to thank Mr. R. Groux for his assistance on the metallurgical processes and to Mr. G. Burri for his work on the scanning electron microprobe analysis of the samples. This work was supported by the Fonds National Suisse de la Recherche Scientifique.

REFERENCES

1. C. L. H. Thieme, H. Zhang, J. Otubo, S. Pourrahimi, B. B. Schwartz, and S. Foner, IEEE Trans. Magn. MAG-19:567-569 (1983).
2. C. L. H. Thieme, S. Pourrahimi, B. B. Schwartz, and S. Foner, Appl. Phys. Lett. 44:260-262 (1984).
3. C. L. H. Thieme, S. Pourrahimi, B. B. Schwartz, and S. Foner, IEEE Trans. Magn. MAG-21:756-759 (1985).
4. J. L. Jorda, R. Flükiger, A. Junod, and J. Muller, IEEE Trans. Magn. MAG-17:557 (1981).
5. R. Flükiger, J. L. Jorda, A. Junod, and P. Fischer, Appl. Phys. Comm. 1:9-30 (1981).
6. J. L. Jorda, R. Flükiger, and J. Muller, J. Less Common Met. 75:227-239 (1980).
7. S. Foner, E. J. McNiff, T. H. Geballe, R. H. Willens, and E. Buehler, Physica 55:534-539 (1971).
8. K. Lo, J. Bevk, and D. Turnbull, J. Appl. Phys. 48:2597-2600 (1977).

A FLEXIBLE A-15 SUPERCONDUCTING TAPE IN THE $Nb_3(Al\ Si\ B)_1$ SYSTEM

Mireille Treuil Clapp and Donglu Shi

Mechanical Engineering Department
University of Massachusetts
Amherst, Massachusetts

ABSTRACT

Melt spinning was used to rapidly solidify $Nb_3\ (Al\ Si\ B)_1$ alloys. It was found that without B, a minimum of 18 at% Si was required to form an amorphous ribbon which resulted in a low superconducting transition temperature, T_c, of the annealed A-15 product. By the addition of 0.5 at% B, the necessary Si content was reduced to 12 at%, thus greatly increasing T_c of the annealed A-15. The annealed metallic glass was far more flexible than the A-15 ribbon obtained by quenching directly into the crystalline state. Total percent bend strain to fracture was increased 30 times from 0.1% to 3%. The flexibility was increased by grain refinement. This is consistent with behavior previously observed by the authors in the $Ti_3\ Nb_6\ Mo_3\ Si_4$ system.[1]

INTRODUCTION

The A-15 structure frequently exhibits high values of the three critical parameters for superconductivity; however, it is difficult to process into wires for practical applications because it is inherently very brittle. By a novel processing technique, we discovered a flexible A-15 superconducting tape in the $Ti_3Nb_6Mo_3Si_4$ alloy system.[1] Rapid solidification via melt spinning was used to produce amorphous ribbons which were subsequently annealed into fine grained flexible A-15 tapes. The superconducting transition temperature T_c was however low (~ 4 K). We have searched for similar flexible behavior in A-15 alloys with T_c's sufficiently high to have practical importance. Results of our work on the Nb_3 (Al Si B) system are presented here

EXPERIMENTAL PROCEDURE

Samples were prepared from high purity elemental powders: Nb 99.8, Al 99.9, Si 99.999, B 99.9 wt% metallic purity. The powders were weighed out stoichiometrically on a microbalance; the total weight of the samples was between 1 and 2 g. They were compacted under high pressure (80 Mpa) with a hydraulic press, and arc melted in a titanium-gettered argon atmosphere using a noncomsumable tungsten electrode. For the purposes of annealing,t he samples were cleaned in acetone, wrapped in Nb foil, and placed in a quartz tube which was sealed off under a vacuum of ~ 10^{-6} Torr.

Melt spinning was used to rapidly solidify the alloys at rates approaching 10^6 K/s. The apparatus[2] consisted of a vacuum chamber with an argon inlet, an induction coil and a copper disk 24 cm in diameter. A stream of molten metal was ejected from the nozzle of a crucible onto the surface of the wheel rotating with tangential velocities up to 70 m/s. To avoid sample contamination from the crucible, a suspension technique was used.[1,2,3]

Crystal structure was analyzed by use of an x-ray diffractometer with Ni-filtered CuKα radiation. Grain size was determined from broadening of the x-ray peaks using the Scherrer formula

$$t = \frac{0.9\ \lambda}{B \cos \theta} \tag{1}$$

(λ = x-ray wave length, B = broadening of peak at half maximum in radians, t = diameter of grain, θ = Bragg angle) B was calculated using Warren's method.

$$B = (B^2_M - B_S^2)^{1/2} \tag{2}$$

(B_M = width of a broadened peak, B_S = width of a standard peak from a large grained sample.) B_S was determined from samples that had been annealed at 750 C for 48 hours and had grain sizes greater than 1000 Å.

Superconducting transition temperatures were monitored using a conventional four-point probe resistive method. Temperature was measured by a calibrated germanium thermistor mounted in close proximity to the sample. The temperature range examined was from 4.2 to 25 K.

The total bend strain to fracture of the ribbons was determined from a bend testing device. The samples were attached between the jaws of a vise which were slowly closed at a rate of 0.033 mm/s by a constant speed motor. When the sample fractured the distance between the vise jaws was measured. It was assumed that the ribbon bent in a semi circle and the strain was calculated from

$$\text{strain} = \frac{t}{d} \tag{3}$$

(t = thickness of sample, d = distance between vise jaws)

RESULTS

The critical first step was to obtain an amorphous ribbon by melt spinning. The critical second step was to obtain a fine grained A-15 with a reasonably high T_c by annealing. Both of these are strongly dependent on composition, so a variety of alloys was initially investigated.

It was found that it was necessary to add significant amounts of Si to Nb_3Al to form an amorphous ribbon. As can be seen in Fig. 1, a minimum of 18 at% Si was required. The addition of a small amount of B (0.5 at%) greatly enhanced the glass forming tendency, reducing the necessary amount of Si to 12 at% (Fig. 1).

The effects of Si content on T_c of $Nb_{99.5-(x+y)}Al_xSi_yB_{0.5}$ can be seen in Fig. 2. The dots are for the A-15 structure obtained by quenching from the liquid directly into the crystalline state. An ordering anneal could increase T_c by 2 K. The crosses are for the A-15 obtained by annealing the amorphous ribbons at 750°C for 48 hours. By comparing our results with those of Bevk,[4] it can be seen that they are in agreement and that

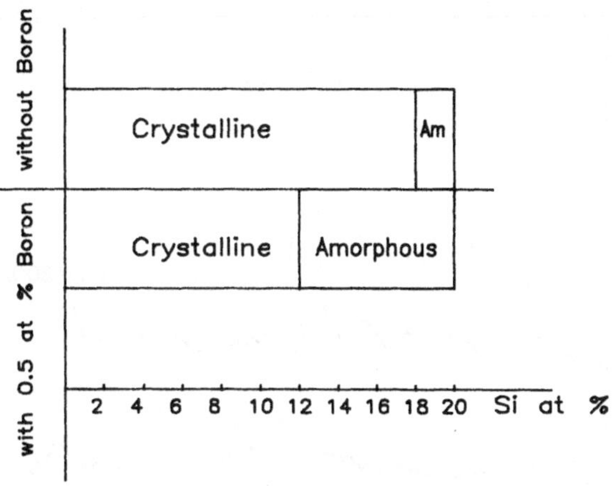

Fig. 1 Ranges of amorphous formation for the $Nb_{99.5-(x+y)}\,Al_xSi_yB_{0.5}$ and $Nb_{100-(x+y)}\,Al_xSi_y$ alloys.

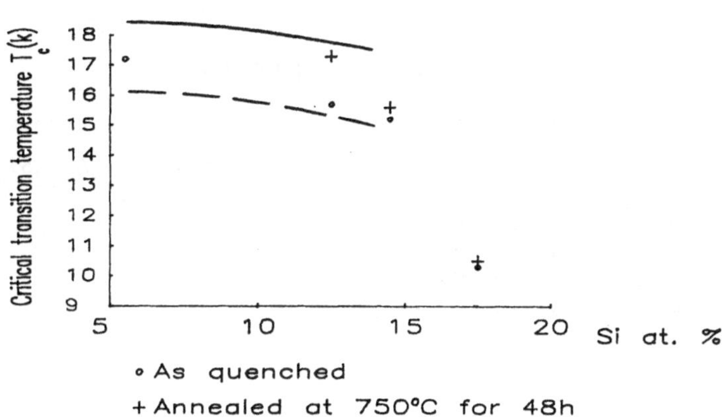

Fig. 2 Critical transition temperature vs. silicon content

$Nb_{99.5-(x+y)}Al_xSi_yB_{0.5}$: ∘ as quenched crystalline

\+ amorphous, followed by anneal at 750°c for 48 h.

$Nb_{73}Al_{27-x}Si_x$ (Bevk[4]) -------- quenched crystalline

——— quenched crystalline plus ordering anneal

the Boron has not adversely affected T_c. The composition chosen for further study was $Nb_{73}\,Al_{12}\,Si_{14.5}\,B_{0.5}$, this being a relatively easy alloy to make amorphous and having a reasonably high T_c after annealing.

If the quenching rates were moderate, the melt spun $Nb_{73}\,Al_{12}\,Si_{14.5}\,B_{0.5}$ alloys would solidify into single phase A-15 brittle fragments of ribbons up to 10 mm in length. By increasing the quenching rates, long ductile amorphous ribbons up to 250 mm in length were obtained; typical widths and thicknesses were 0.8 to 1.5 mm and 0.015 to 0.022 mm. The following melt spinning parameters were used: wheel speed 57 m/s (4,500 rpm) to 70 m/s (5,500 rpm), orifice diameter 0.6 mm to 1.0 mm, argon ejection pressure 71 KPa (10.3 psi) to 102 KPa (14.8 psi).

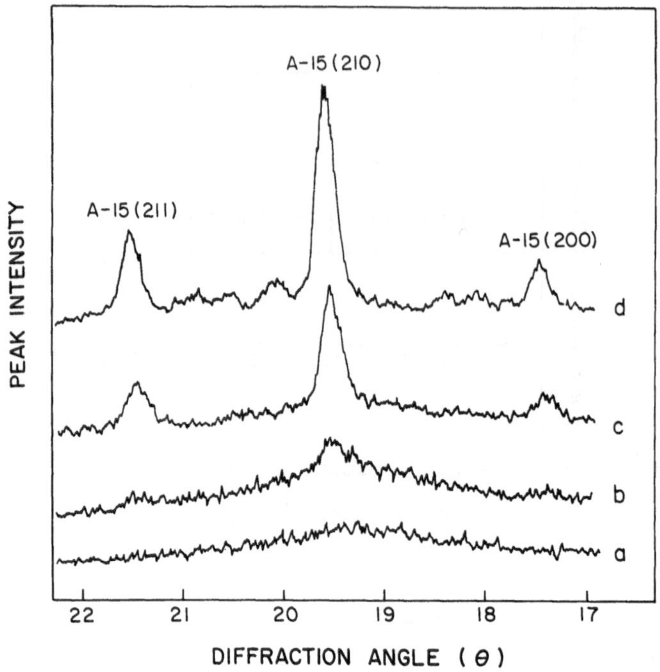

Fig. 3 Portions of x-ray diffraction plot showing the effect of annealing on the structure of amorphous $Nb_{73}Al_{12}Si_{14.5}B_{0.5}$.

a: as quenched amorphous
b: annealed at 620°c, 24 hs
c: annealed at 700°C, 24 hs
d: annealed at 750°C, 24 hs

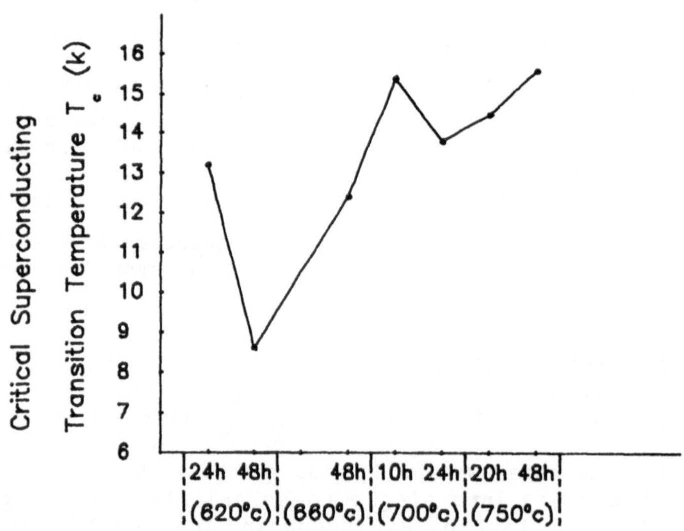

Fig. 4 Critical temperature vs. annealing temperature and time for $Nb_{73}Al_{12}Si_{14.5}B_{0.5}$.

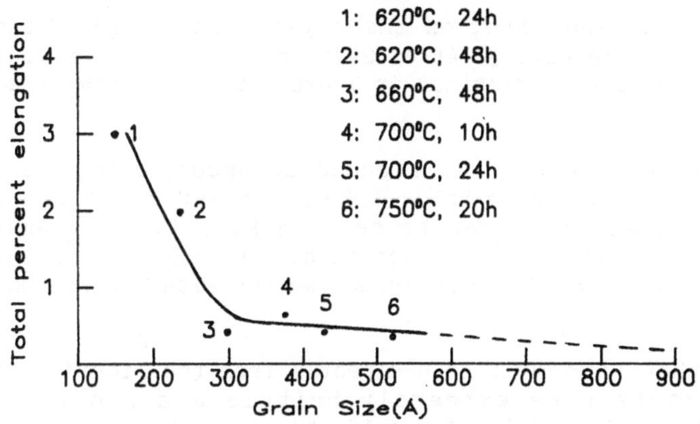

Fig. 5 Total percent elongation vs. grain size for $Nb_{73}Al_{12}Si_{14.5}B_{0.5}$.

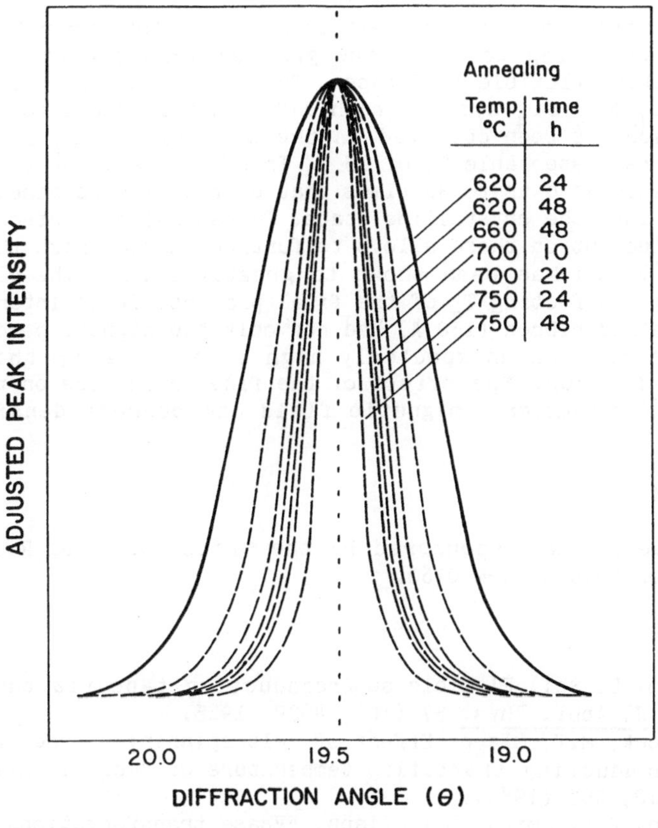

Fig. 6 Effect of annealing on the breadth of the (210) A-15 x-ray peak.

The effect of annealing on the crystal structure of the amorphous ribbons can be seen in Fig. 3. After 24 hours at 620°C, small A-15 peaks appeared and became very distinct at 700°C. At 750°C, some traces of Ti_3P peaks emerged.

The amorphous ribbons had a broad superconducting transition that began at 7 K and continued to below 4.2 K. The effect of annealing temperatures and times on T_c can be seen in Fig. 4. The transitions were sharp. A maximum T_c of 15.5 K was obtained. The high T_c of 13.2 K after annealing at 620°C for 24 hours is noteworthy although not understood at this time.

Samples that were melt spun directly from the liquid into the crystalline A-15 state were extremely brittle and had total percent elongations of less than 0.1%. The A-15 structure that was obtained by annealing the amorphous ribbons was much more flexible, the highest percent elongation being 3%, a thirty times increase in bend strain to fracture. It is worth noting that the percent elongations of 0.1 and 3.0 correspond respectively to fracture diameters of 20 and 0.7 mm. The effect of grain size on % elongation can be seen in Fig. 5. The grain sizes were determined from broadening of the three strongest A-15 x-ray and averages were taken. The effect of annealing on the breadth of the strongest A-15 peak can be seen in Fig. 6. The % elongation values are the averages of three separate bend tests.

DISCUSSION AND CONCLUSIONS

The processing technique of melt spinning to form amorphous ribbons followed by annealing to form fine grained A-15's continues to show promise for making flexible A-15 tapes. The effect was initially observed in the $Ti_3Nb_6Mo_3Si_4$ system which unfortunately had a low T_c of 4 K.[1] Similar behavior has been observed in the $Nb_{73}Al_{12}Si_{14.5}B_{0.5}$ system with a much more respectable T_c of 15 K. In both cases, the bend strain to fracture was increased up to 30 times that of the A-15 obtained by quenching directly into the crystalline state. Transmission electron microscopy is being carried out on these alloys to further characterize the increased flexibility. Boron has been shown to greatly enhance the glass forming tendency without affecting T_c of the final product. It is interesting that the finest grained ribbon (150 Å) had not only the highest percent elongation (3%) but also an unexpectedly high T_c of 13.2 K; this will be investigated further. The effect of the fine grain size on the critical superconducting parameters, magnetic field and current density will be studied.

ACKNOWLEDGMENT

This research was sponsored by the Office of Naval Research under contract number N00014-85-K-0160.

REFERENCES

1. M.T. Clapp, D. Shi, Flexible superconducting tape via the amorphous state, J. Appl. Phys. 57 (10), 4672, 1985.
2. L.R. Gresock, M.T. Clapp, Effect of melt spinning of the structure and superconducting transition temperature of "Nb_3Si" alloys, Mater. Lett. 12, 492 (1984).
3. B.S. Smith, R.A. Smith, M.T. Clapp, "Phase transformations in Solids," (North Holland Elsevier, New York, 1984), pp. 829-834.
4. J. Bevk, Superconducting properties of liquid-quenched Nb_3 (Al Si) and Nb_3 (Al, Ge), "Rapidly Quenched Metals III," Vol. 2, p. 17.

SUPERCONDUCTING MATERIALS RESEARCH AND DEVELOPMENT IN CHINA

Lin Li

Institute of Physics
Chinese Academy of Sciences
Beijing, China

ABSTRACT

Critical temperatures and critical current densities of multifilamentary Nb-Ti and Nb_3Sn superconductors are presented, along with some transverse electron microscope observations. Superconducting magnets made with such conductors have maximum fields of 8 T for multifilamentary Nb-Ti and 15 T for multifilamentary Nb_3Sn. Research on modified bronze processes by the addition of a third element in processing the multifilamentary Nb_3Sn composite superconductors is described, together with some analytical and low temperature measurement results. The practical potential for other A15 compounds, such as chemical-vapor-deposited Nb_3Ge and sputtered Nb_3Ge, is discussed.

INTRODUCTION

The Institute of Physics of the Chinese Academy of Sciences started the production of cryogens for low temperature superconductivity research and built the first Chinese helium liquefier (a piston type) in 1959.

Since the sixties, research has been devoted to high-field superconducting materials and the superconducting tunneling effect. Industrial research laboratories took part in the development of practical superconducting materials. The Institute of Physics, together with some departments of physics in the universities, expended their efforts on the superconducting tunneling effect, new materials with more favorable transition temperatures, and the flux pinning mechanisms in type II superconductors.

MULTIFILAMENTARY Nb-Ti COMPOSITES

The General Research Institute of Nonferrous Metals, together with Baoji Institute for Nonferrous Metals and the Shanghai Institute of Nonferrous Metals, turned to the development of multifilamentary Nb-Ti composite conductors. At present the laboratory uses superconducting magnets made with such conductors, whose maximum field is 8 T. Mostly Nb-50% Ti* conductors are in use, but some experimental Nb-46.5% Ti conductors are also being investigated. The general characteristics of the Nb-Ti/Cu conductors in production are: filament diameter: around 20 μm; number of filaments: from tens to

*wt.% is used throughout, unless otherwise specified.

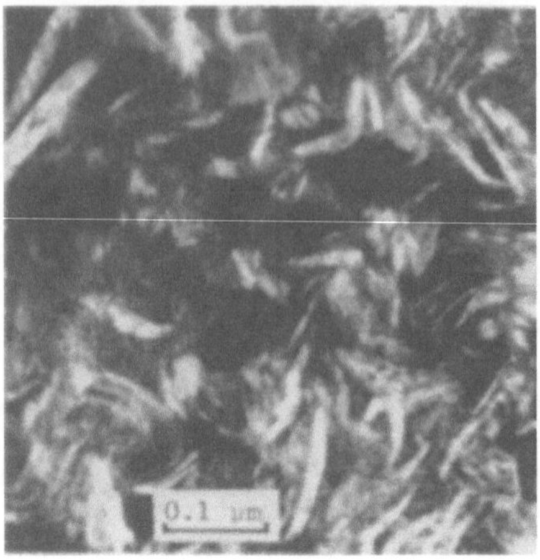

Fig. 1. Longitudinal TEM micrograph of a Baoji wire.

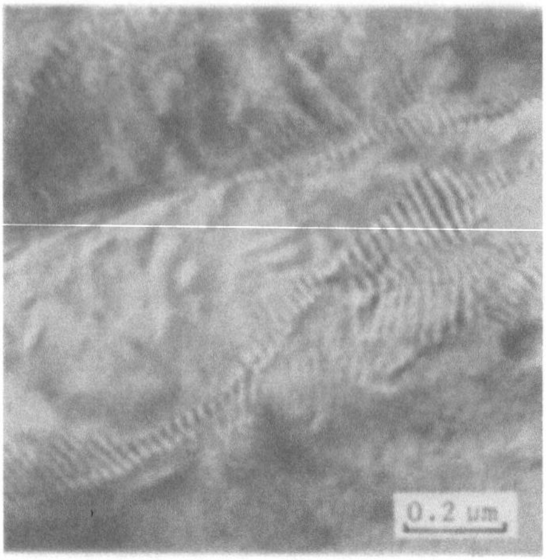

Fig. 2. Transverse TEM micrograph of a Baoji wire.

several hundreds; critical current density (J_c) at 4.2 K and 5 T: 0.15 to 0.20 MA/cm^2 for round wires of diameter < 1 mm, and 0.13 to 0.17 MA/cm^2 for ribbons of cross section > 3 mm^2.

Research on the flux pinning mechanisms and improvement in the J_c of multifilamentary Nb-Ti composites by successive deformation and aging is being carried out. Figures 1 and 2 are electron micrographs obtained by us from the Baoji composite wire. The results correspond to those of A. W. West and D. C. Larbalestier:[1] that the pinning effect is due to α-Ti precipitated at the subband boundaries.

$\underline{Nb_3Sn}$

A technique for chemical vapor deposition (CVD) of Nb_3Sn was developed by collaboration of the Research Institute of Mining and Metallurgy in Changsha, the General Research Institute of Nonferrous Metals, and the Institute of Physics. Hastelloy tapes of 40-μm thickness were used; they were passed through a reaction chamber at a high speed--up to 100 m/h. This high speed, together with the addition of C, resulted in the deposition of fine grains of A15 Nb_3Sn on the tape and, hence, a high J_c.[2] Nb_3Sn layers were around 8 to 10 μm on each side of the tape, and Cu shields 30- to 60-μm thick were deposited on the surfaces after reaction. The CVD tape provided the bulk of the material used for winding magnets for laboratory use with fields up to 10 T.

The measurements of critical current of a CVD Nb_3Sn tape and a modified diffusion-processed Nb_3Sn tape at 4.2 K (Research Institute for Nonferrous Metals, Baoji) in high magnetic fields (~22 T) show that these two Nb_3Sn tapes have good superconducting properties. Their critical current densities at 4.2 K are:[3] diffusion-processed Nb_3Sn tape: 0.30 MA/cm^2 (12 T) and 0.14 MA/cm^2 (15 T); CVD Nb_3Sn tape: 0.29 MA/cm^2 (12 T) and 0.060 MA/cm^2 (15 T).

The main research efforts now are on the development of multifilamentary Nb_3Sn composite conductors. The bronze process, in which filaments of Nb are drawn down in a Cu-Sn bronze matrix and reacted to produce the A15 compound Nb_3Sn, is now the established production method for the fully stabilized, high-current-density, multifilamentary superconducting wire with which 15-T laboratory magnets are produced by the Baoji group.

Table 1

Sample No.*	T_c (K)	ΔT_c (K)	$H_{c2}(0)$T
CNS-D-19	17.66	0.20	21.06
CNSTi-0.1%-19	17.75	0.17	23.01
CNSTi-0.2%-19	17.82	0.14	23.61
CNSTi-0.3%-19	17.96	0.17	25.90
CNSTi-0.4%-19	17.96	0.18	26.35
CNSIn-2%-19	17.66	0.28	21.42
CNSIn-6.7%-19	17.58	0.17	21.57

*CNSTi-0.1%-19 means the addition of 0.1 wt.% Ti to Nb_3Sn.
CNSIn-2%-19 means the addition of 2 wt.% In to Nb_3Sn.

A research group at the Institute of Metallurgy in Shanghai is working on three modified bronze processes in which a third element is added:

1. Method of micro-multifilamentary Nb/Cu extrusion with high-Sn central diffusion. This technique is being developed to modify the bronze process for manufacturing Nb_3Sn compound composite wires. The multifilamentary Nb/Cu tube is made by extrusion in three stages. A 91% Sn-Cu rod is inserted in the multifilamentary Nb/Cu composite tube and drawn together to a wire of 0.3- to 1-mm diameter without intermediate annealing. The Nb content in the wire is 33.6%, and the average Sn content in the Cu-Sn matrix is 21.3%. The 0.3-mm-diameter sample (9,894 filaments, 1.7-µm-thick Nb_3Sn, reacted at 700°C for 24 h) has a J_c of 0.32 MA/cm^2 (4.2 K, 6 T), a T_c of 17.53 K, and a lattice parameter of 5.28Å.

2. Addition of Ti and In to Sn-rich (Cu-35% Sn) bronze matrix. It was found that different elements affect the growth rate of Nb_3Sn and its superconducting properties to a different extent. Table 1 shows the critical temperature (T_c) and the upper critical field at 0 K [$H_{c2}(0)$] of the samples after heat treatment at 700°C for 156 h.

The growth rate of Nb_3Sn is not affected very much when the Ti or In content is low, such as in samples CNSTi-0.1%-19 and CNSIn-2.1%-19. Figure 3 shows the growth rate curves of three samples, CNS-D-19, CNSTi-0.3%-19, and CNSIn-6.7%-19, when heat-treated at 700°C. Evidently, Ti has a greater effect than In. The distribution of different elements across sample CNSTi-0.3%-19 when heat-treated at 700°C for 200 h was analyzed by electron microprobe line scan. It was found that after heat treatment, Ti diffused completely into the $(Nb,Ti)_3Sn$ layer while only a small amount of In diffused into the $Nb_3(Sn,In)$ layer. The rest remained in the matrix. The grain size of A15 Nb_3Sn with the addition of Ti is very fine. The J_c values (4.2 K) of CNSTi-0.3%-19 and CNSIn-6.7%-19 under a pulsed magnetic field of 15 T (1 pulse = 10 ms) are 60 kA/cm^2 and 25 kA/cm^2, respectively.

3. Addition of Ce and enrichment with Sn. Alloys rich in Sn were obtained by melting together pure Cu, Sn, and Ce. The nominal compositions were Sn-4.1% Cu-0.5% Ce and Sn-4.3% Cu-2% Ce. The samples were first heat-treated at 200° to 400°C for homogenization of the Cu and Sn-rich alloy within the Nb tube, and then heat-treated at 650° to 800°C for 5 to 300 h to form Nb_3Sn. However, the Ce addition did not increase the growth rate of Nb_3Sn, which is different from that of the above-mentioned Ti alloy. An electron microprobe line scan along the radius of the Nb tube found that Ce

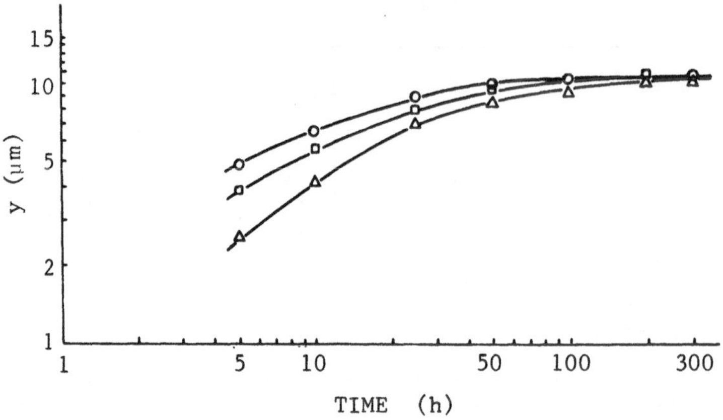

Fig. 3. The growth rate curves of reacted Nb_3Sn:
△ - CNS-D-19; □ - CNSIn-6.7%-19; ○ - CNSTi-0.3%-19.

had definitely diffused into the Nb_3Sn layer. However, the superconducting properties, such as T_c, were not affected, but in most cases $H_{c2}(0)$ decreased. The effect of Ce in Nb_3Sn needs further study.

OTHER A15 COMPOUNDS

In 1974, work on V_3Ga fabrication was carried out by the General Institute of Nonferrous Metals in Beijing. Test coils using wind-and-react techniques were constructed. Research is continuing on the modified bronze method, adding small amounts of light metals or rare earth elements. It was found that some of the elements diffused into the V_3Ga/matrix interface, which induced the V/Ga ratio to become near stoichiometric in composition. The T_c increased from 15 to 15.50 K; the J_c (5 T, 4.2 K) of single-filament V_3Ga (1.0 mm) increased to twice its original value. The overall J_c of the 15-filament (1.1-mm) V_3Ga (6 T, 4.2 K) was 42.1 kA/cm^2. The stability of the conductor was much improved by using a 40 vol.% oxygen-free Cu shield over the pure Nb core.

Preparation of Nb_3Ge is being done by the CVD, coevaporation, and sputtering methods. For the CVD Nb_3Ge tape, new deposition methods are being investigated to improve the J_c of the Nb_3Ge, with the goal of producing a potentially practical superconductor. Studies of coevaporation and sputtering of Nb_3Ge films concentrate more on the fundamental aspects, such as the formation and stabilization of the high-T_c metastable A15 Nb_3Ge phase and the origin of high T_c.

The CVD Nb_3Ge tape samples were prepared similar to the Nb_3Sn tapes. Hastelloy tapes of about 40-μm thickness were used. They were passed through a reaction chamber at a speed of 50 to 70 m/h. Multilayers of Nb-Ge were deposited on each side of the tape to a thickness of around 2 μm. Finally, a Cu shield about 40-μm thick was electroplated onto the surfaces after reaction. The uniformity of deposited Nb-Ge compositions for different heating powers and Ar/H_2+Cl_2 ratios were examined by line scans of an electron microprobe. It was found that a heating power of 350 to 400 W and a Ar/H_2+Cl_2 ratio of 4 to 6 greatly improved the uniformity of the deposited Nb-Ge. Examination by scanning electron microscope of the fractured tapes (prepared under the same conditions) showed that the grain sizes of Nb_3Ge are relatively uniform and small at the middle of the tape and coarse at the edge of the tape. Further work is being done to improve the uniformity of the grain structure and the critical current.

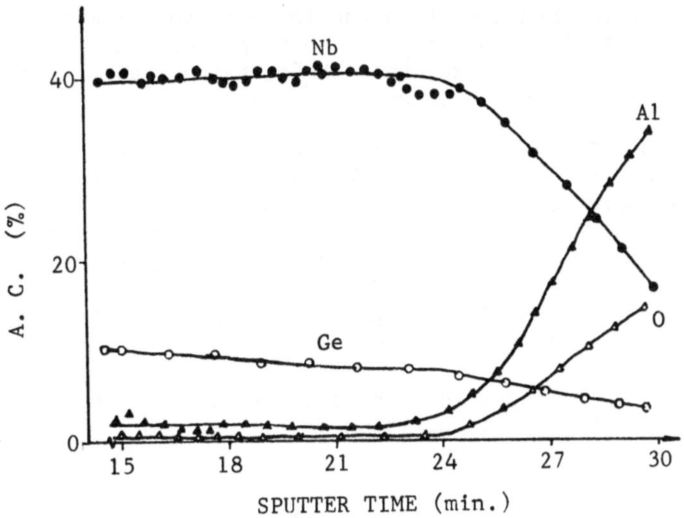

Fig. 4. Auger depth profile of a Nb-Ge sputtered sample with T_c = 23 K.

We started to investigate the high-T_c metastable Nb_3Ge in 1979. Film samples were prepared by the getter sputtering method in a conventional vacuum system. The usual high-Ar pressures and low sputtering voltages were used. In our experiment, the best deposition temperatures were 900° to 980°C. We obtained films with a T_c of 23 K in 1980.[4] However, in the high-T_c films the A15 Nb_3Ge was always accompanied by some second-phase Nb_5Ge_3. We did not find the sudden increase in oxygen content at the film-substrate interface (Fig. 4). We measured the T_c values of samples prepared with the help of Dr. Somehk of Cambridge, U.K., in which a small amount of oxygen was intentionally added during film deposition by the induction method.[5] In the sputtered samples with a single phase, A15 Nb_3Ge had broad transitions and showed several steps in the curves, whereas the samples containing A15 Nb_3Ge and Nb_5Ge_3 exhibited a smooth and sharp transition, as shown in Fig. 5. This indicates that the so-called single-phase A15 samples actually contain more than one A15 phase, and only a small fraction of them has a small lattice parameter with a high T_c. But in the samples with A15 Nb_3Ge plus Nb_5Ge_3, the

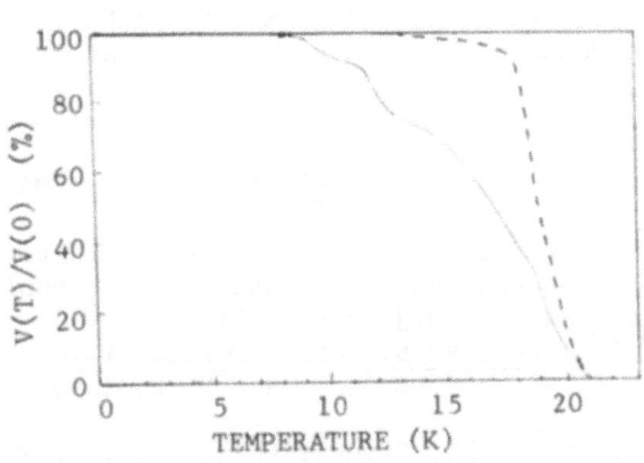

Fig. 5. Inductive transition of two-phase sample (dashed line) and a single-phase sample (solid line).

A15 phase has a more ordered structure and is near stoichiometric in composition, approaching that of the high-T_c A15 Nb_3Ge phase. This experimental result supports our previous proposal that Nb_5Ge_3 has a stabilizing effect on the metastable high-T_c A15 Nb_3Ge phase.

The preparation of high-field Nb_3Al conductors by the powder metallurgy process has been carried out for some time by a group at the Institute of Metallurgy in Shanghai. The process consists of mechanically mixing high-purity Al powder (9-40 μm) and hydride-dehydride Nb powder (150 μm), vacuum sealing in a Cu tube, swaging, and wire drawing down to an area reduction ratio of 2,500 to 20,000. Although the J_c obtained was rather low, the experimental results are still useful for improving the powder metallurgical process. The low temperature measurements indicate:

1. J_c (14 T, 4.2 K) appeared to have a maximum with the increase of the heat-treatment temperatures; the position of the maximum varied with the initial particle size of Al powder.

2. In the range 12 to 17 T, $H^{1/4}J^{1/2}$ and H have a linear relationship. It is possible to use the Kramer scaling law to describe the flux pinning mechanism.

After reaction at 800°C, the composite was composed of the solid solution (b.c.c.) of Nb-rich Nb-Al, with small amounts of A15 compound deficient in Al. With fine Al powder, the A15 compound formed more easily; however, with further heat treatment, the Al content decreased in the A15 compound. It is postulated that by using fine Al powder and a high reduction ratio, it is possible to increase the reaction between Nb and Al, and thus increase the number of ultrafine A15 compound fibers. Owing to the strong coupling action, J_c could be increased at high fields.

CONCLUSION

Some progress has been achieved in applied superconductivity over the past twenty years in the People's Republic of China. Superconducting materials and magnet technology have grown from nothing to its present stage. Current work emphasizes more basic and detailed studies on the mechanisms of pinning effects of type II superconductors and the exploration of new, high-T_c materials. Applied research on superconducting materials, such as laboratory-scale multifilamentary Nb-Ti and Nb_3Sn composite wires, is underway.

ACKNOWLEDGMENT

Most of the material was collected from the 5th National Conference on High Parameter Superconductors, held in Baoji on August 22-25, 1984.

REFERENCES

1. A. W. West and D. C. Larbalestier, IEEE Trans. Magn. MAG-19:548-551 (1983).
2. Research group of Nb_3Sn tape of Changsha Research Institute of Mining and Metallurgy and of General Research Institute for Nonferrous Metals, IEEE Trans. Magn. MAG-17:1013-1016 (1981).
3. L. Zhou and C.-G. Cui, Acta Phys. Temp. Humilis Sinica 7(3):43-46 (1985).
4. L. Li, B.-R. Zhao, P. Zhou, S.-Q. Guo, and Y.-X. Zhao, J. Low Temp. Phys. 45:287-294 (1981).
5. B.-R. Zhao, J. Gao, and L. Li, Chinese Phys. Lett. 2:53-54 (1985).

THE STABILIZATION OF HIGH T_c A15 Nb_3Ge PHASE IN SPUTTERED

Nb-Ge FILMS BY (T)Nb_5Ge_3

B. R. Zhao, J. Gao, and L. Li

Institute of Physics
Academia Sinica
Beijing, China

ABSTRACT

It has been pointed out in our previous experiments that the high-T_c (23 K) Nb-Ge films always contained some Nb_5Ge_3 with a Nb/Ge < 3. The Auger depth profiles did not show a sudden increase of oxygen at the film-substrate interface. It is found in our present work by inductive measurements of T_c that the sputtered Nb-Ge films are heterogeneous, and only a small fraction of the A15 phase is of the high-T_c nature. TEM examinations showed streaky dispersion of (T)Nb_5Ge_3 in A15 grains. These results provide further support that (T)Nb_5Ge_3 has a stabilization effect on the high-T_c A15 phase. However, formation of the high-T_c phase in chemical vapor deposited Nb-Ge samples is different from that of the sputtered films.

INTRODUCTION

It is well recognized that A15 Nb_3Ge films remain the highest T_c superconductors known. Although high-T_c Nb_3Ge can be easily obtained by vapor-quenched thin-film techniques, the mechanism of the formation and stabilization of the metastable high-T_c A15 phase is still not clear. This situation impedes the development of technologically important high-T_c superconducting material.

Reports from different laboratories indicate that the presence of oxygen is crucial for the formation and stabilization of the metastable high-T_c A15 Nb_3Ge phase.[1-4] In our previous work,[5] we proposed that a certain amount of Nb_5Ge_3 may play a role in stabilization of the metastable A15 Nb_3Ge phase, which also leads to high T_c of the deposited films. In this study, experiments were conducted on Nb-Ge films prepared in the Department of Metallurgy and Materials Sciences, University of Cambridge, U.K. with the help of Dr. R. E. Somehk and on chemical vapor deposited (CVD) films prepared in the Institute of Mining and Metallurgy, Changsha. The results of T_c transitional curves by resistive and inductive methods together with the observation of the film samples by electron microscope are compared with the sputtered samples prepared in our laboratory to elucidate the influence of oxygen and Nb_5Ge_3 on the formation of high-T_c A15 phase.

EXPERIMENT

The sputtered samples were prepared in the dc getter-sputtering system (MKII) previously described.[5] Small amounts of oxygen (∿50 ppm) were added to each batch of samples. A layer of about 1000 Å of amorphous film was sputtered first on the cold substrates before heating up the substrates to about 850°C. For batch numbers 337, 340, and 341, the sputtering conditions were the same, except that 337 was annealed for 12 h at the deposition temperature, 340 was annealed for 3 h at 50°C higher than the deposition temperature, and 341 was annealed for 3 h at 40°C lower than the deposition temperature. X-ray diffraction was used for phase determination; the results showed that annealing after sputtering has a profound effect on phase formation: for sample 337 only the strong peaks of A15(210)(211)(200) were present, sample 340 showed some fraction of the tetragonal Nb_5Ge_3, and sample 341 showed fractions of both hexagonal and tetragonal Nb_5Ge_3. The Nb and Ge composition of the deposited films were determined by electron microanalyzer.

The CVD Nb_3Ge tape samples were prepared by procedures similar to those used for CVD Nb_3Sn tapes.[6] Hastelloy tapes of 40 μm thickness were used; they were passed through a reaction chamber at a speed of 50 to 70 m/h, and multilayers of Nb-Ge were deposited on each side of the tape to a thickness of about 4 μm. Finally, Cu shields about 40 μm thick were deposited on the surfaces after reaction. The uniformity of deposited Nb-Ge compositions for different heating powers and Ar/H_2+Cl_2 ratios were examined by scanning electron microscope. The most uniformly deposited Nb-Ge layer was achieved with a heating power of 350 to 400 W and an Ar/H_2+Cl_2 ratio of 4 to 6, such as in samples G207-6 and G207-11. We chose these samples for our experiments, although their overall J_c is only about 5.8 x 10^5 A/cm^2.

The T_c transitional curves were measured with calibrated Ge resistors both by the standard four-probe resistance method and the inductive method. We designed the inductive measuring device ourselves;[7] it is a very sensitive detector of the superconductive transition in thin film samples. Some representative samples were examined with a transmission electron microscope.

Table 1. The Transitional Temperatures of Sputtered and CVD Samples Measured Resistively

Sample Number	Nb/Ge	Phase Structure	Thickness, x 10^4 Å	T_c, K (onset)	T_c, K (mid)	ΔT_c, K
16	2.16	A15+Nb_5Ge_3	0.6	23.00	22.22	1.40
19	2.20	A15+w.Nb_5Ge_3	0.5	23.00		2.50
26	2.87	A15+w.Nb_5Ge_3	0.5	22.40		1.22
337-9	2.97	A15	2.38	21.30	20.90	0.91
337-10	2.93	A15	2.62	22.35	21.84	1.30
337-11	3.04	A15	2.42	21.84	21.36	1.38
340-8	2.62	A15+Nb_5Ge_3(T)	1.04	23.47	22.95	1.29
340-9	2.73	A15+Nb_5Ge_3(T)	1.23	23.36	22.69	1.66
341-8	3.21	A15+Nb_5Ge_3(T,H)	1.01	22.44	21.84	1.00
341-9	3.24	A15+Nb_5Ge_3(T,H)	0.87	21.84	21.21	1.09
G207-6	3.16	A15 (5.156 Å)		19.52	19.20	0.50
G207-7	2.95	A15 (5.155 Å)		19.46	18.96	0.86
G207-11	3.68	A15 (5.164 Å)		19.39	18.96	0.96

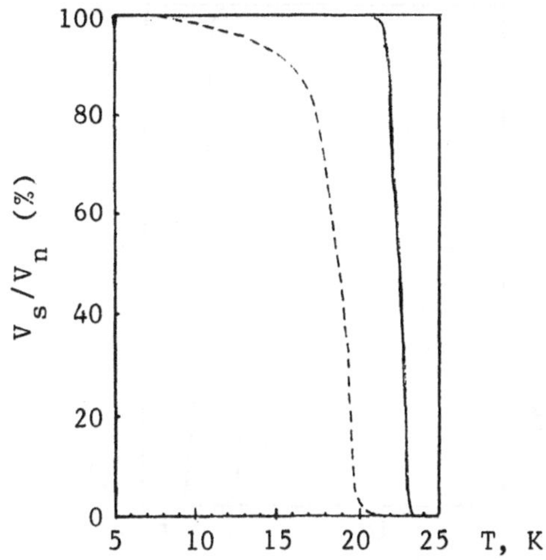

Fig. 1. T_c transitional curves by resistive and inductive methods of sample 340-8. Solid line: resistive curve; dashed line: inductive curve.

RESULTS

The T_c values, measured by the standard four-probe resistive method, of the samples made in the University of Cambridge, U.K. and the CVD tape samples prepared in Changsha are compared with those made in our laboratory in Table 1. From the sputtered samples with high T_c (onset), it is evident that all contained some Nb_5Ge_3 and Nb/Ge < 3, and the ΔT_c's are rather broad. This reflects the heterogeneous nature of the films. From the experimental results, the T_c's of the samples measured by the inductive method are all lower than those measured by the resistive method (Fig. 1). It appears that the volume fractions of the high-T_c regions are relatively small in the sputtered samples, perhaps less than 10%. Samples with single-phase A15 Nb_3Ge have broad transitions and showed several "steps" in the curves, whereas the samples containing A15 Nb_3Ge and a small amount of Nb_5Ge_3 have smoother and sharper transitions (Fig. 2). This also shows that the so-called single-phase A15 samples actually contained more than one A15 phase; only a small fraction of the A15 phase is of the high-T_c nature. However, the CVD samples behave differently; Fig. 3 shows the resistive and inductive transitional curves of a single-phase A15 sample (G207-6). Although the

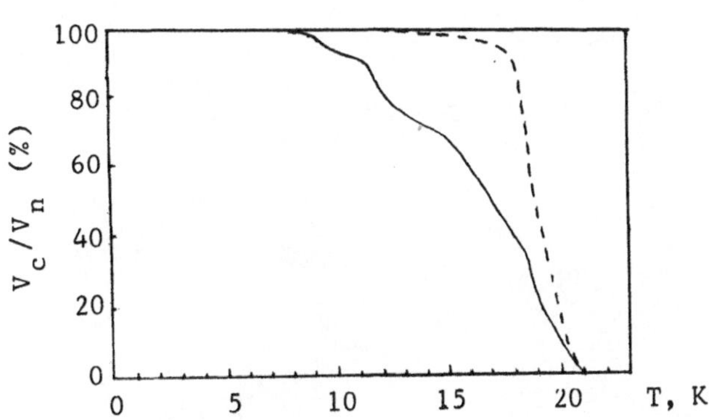

Fig. 2. Inductive transition of a single-phase sample (solid line) and of a two-phase sample (dashed line).

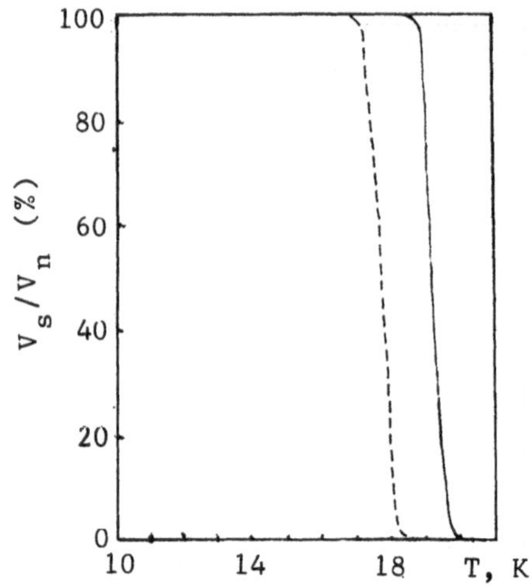

Fig. 3. T_c transitional curves by resistive and inductive methods of CVD sample G207-6. Solid line: resistive curve; dashed line: inductive curve.

T_c is about 2 K lower by the inductive method than that by the resistive method, the transition is rather sharp with no "steps" in the curve.

TEM examinations of the sputtered samples containing two phases showed that Nb_5Ge_3 was present as a fine streaky structure within the A15 grains (Figs. 4 and 5). These results are comparable to the electron micrograph of the high-T_c films obtained by Yin et al.[8]

DISCUSSION

From the experimental results, it is quite clear that the Nb-Ge films obtained by sputtering are heterogeneous in nature. It is very hard to say if the film is nonhomogenous through its depth, with low-T_c material at the bottom. Comparing the results from the samples with and without the addition of oxygen, the highest T_c values are still obtained from films containing A15 plus a small amount of tetragonal Nb_5Ge_3 and with $Nb/Ge < 3$. TEM examination showed that some of the Nb Ge phase crystallized with tetragonal Nb_5Ge_3 as eutectic structure; this structure may have formed directly from

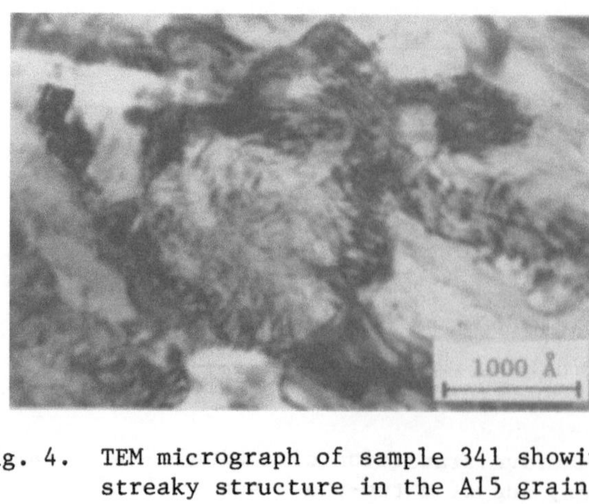

Fig. 4. TEM micrograph of sample 341 showing streaky structure in the A15 grain.

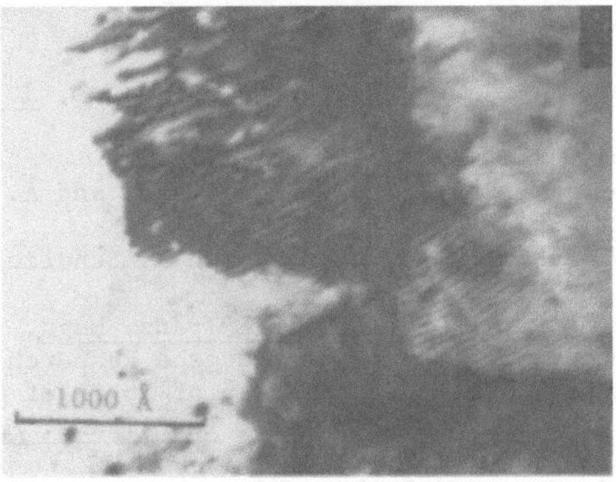

Fig. 5. TEM micrograph of sample 340 showing eutectic structure of Nb_5Ge_3 with Nb_3Ge.

examination showed that some of the Nb_3Ge phase crystallized with tetragonal Nb_5Ge_3 as eutectic structure; this structure may have formed directly from film deposition on the hot substrate. On the other hand, a few of the high-T_c two-phase samples showed striae stacks within some of the A15 grains.[9] This could be a feature of a material that had been formed by a "martensitic" transformation. We postulate that the Ge atoms may have partially substituted the Nb atoms and formed some Ge-rich A15 grains; they may have nucleated in the initially deposited films; and a "martensitic" transformation may occur at a temperature near T_c owing to their close crystallographic relationship. However, this phenomenon is drastically sample dependent and is not so likely to happen as in Nb_3Sn or V_3Si.

We agree with Weiss et al.[10] that high-T_c Nb_3Ge can be formed through a heteropitaxial process between the A15 and the competing $(T)Nb_5Ge_3$ phase. The mechanism of transformation can be summarized as follows:

$$4\ Nb_3Ge = Nb_{12}Ge_4 \xrightarrow{\text{substitution}} Nb_{10}Ge_2Ge_4 = 2\ Nb_5Ge_3$$

Repulsive forces between Ge atoms lead to a Hyde rotation of blocks, which gives the tetragonal structure of Nb_5Ge_3. In such a mechanism, the gain in interfacial free energy obtained by epitaxy contriubtes to lower the free energy of formation of the A15 phase. Thus it allows the formation of a nearly stoichiometric Nb_3Ge with a small lattice parameter of 5.14 Å.

But the formation of Nb_3Ge phase in CVD films is different from that of the sputtered films; the TEM micrographs show equiaxed A15 grains similar to those of A15 Nb_3Sn. It can be concluded, at the moment, that irrespective of the addition of oxygen, small amounts of the second-phase Nb_5Ge_3 still play an important part in stabilizing the high-T_c A15 Nb_3Ge phase in sputtered films, but this conclusion cannot be applied to CVD films. Work is being continued to clarify the essence of the low-T_c phase in sputtered films.

ACKNOWLEDGMENTS

One of us (L. Li) gratefully acknowledges the help of Dr. R. E. Somekh for preparing the samples.

REFERENCES

1. J. R. Gavaler, M. A. Janocko, and C. K. Jones, J. Appl. Phys. 45:3009 (1974).
2. R. E. Somekh, Philos. Mag. B 37:6 (1978).
3. J. R. Gavaler, M. Ashkin, A. I. Braginski, and A. T. Santhanam, Appl. Phys. Lett. 33:359 (1978).
4. B. Krevet, W. Schauer, F. Wuchner, and K. Schulze, Appl. Phys. Lett. 36:704 (1980).
5. R. E. Somekh and J. E. Evetts, IEEE Trans. Magn. MAG-15:494 (1979).
6. Research Group of Nb_3Sn Tape of Changsha Research Institute of Mining and Metallurgy and Research Group of Nb_3Sn of General Research Institute for Nonferrous Metals, IEEE Trans. Magn. MAG-17:1013 (1981).
7. B. R. Zhao, J. Gao, and L. Li, Chin. Phys. Lett. 2:53 (1985).
8. D. Yin, W. Schauer, and F. Wuchner, IEEE Trans. Magn. MAG-19:276 (1983).
9. L. Li, B. R. Zhao, D. Y. Yang, X. C. Chen, and P. Zhou, Acta Physica Temperaturae Humilis Sinica 4:291 (1982).
10. F. Weiss, O. Demolliens, R. Madar, J. P. Senateur, and R. Fruchart, J. Phys. 45:1137 (1984).

SUPERCONDUCTING PROPERTIES OF CHEVREL-PHASE $PbMo_6S_8$ WIRES BY AN IMPROVED POWDER PROCESS

Y. Kubo, K. Yoshizaki, F. Fujiwara and Y. Hashimoto

Materials & Electronic Devices Laboratory
Mitsubishi Electric Corp., Sagamihara, Kanagawa, Japan

ABSTRACT

A Chevrel-phase $PbMo_6S_8$ superconductor is one of the most promising superconducting wires for very high field magnets. In this study, $PbMo_6S_8$ superconducting wires fabricated by using a Ta barrier in a powder process have been heat-treated under high hydrostatic pressure for producing $PbMo_6S_8$ compound. And then superconductivity measurements and microstructure observations have been performed. The wires are very easily drawn to a final size at room temperature, and have a critical current density (Jc) for the compound of $1.7 \times 10^4 A/cm^2$ at 12T. The typical transition temperature (Tc) and its width, ΔTc are 14.5K and 0.6K, respectively. Furthermore, EPMA and PSPC-type X-ray Microdiffractometer observations indicate Mo, MoO_2 precipitates and unreacted MoS_2 in the compound core inside the Ta barrier. Consequently, the $PbMo_6S_8$ wires with excellent Jc at high fields can be fabricated by using a Ta barrier and heat-treating under hydrostatic pressure.

INTRODUCTION

A Chevrel-phase $PbMo_6S_8$ superconductor has a very high upper critical field (Bc2) and is one of the most promising superconducting wires for high field magnets applied to fusion, NMR Microscopy and so on, if the most serious drawback of low Jc can be overcome.

Many fabrication processes have been attempted to improve the Jc.[1-7] Among them the powder process is advantageous from the viewpoint of an industrial process for a practical conductor.

In this paper, we report on superconductivities and microstructures of $PbMo_6S_8$ wires by an improved powder process.

FABRICATION PROCESS

In the powder process,[5-7] Mo, Ta, Nb and Ag have been used as a barrier for sulphur sealing. Among them we chose a Ta barrier because of having an excellent cold-workability and not reacting with the $PbMo_6S_8$ compound.

In our process, the heat-treatment under hydrostatic pressure was performed to achieve following two objectives.

(1) To decrease the porosity in the compound core.

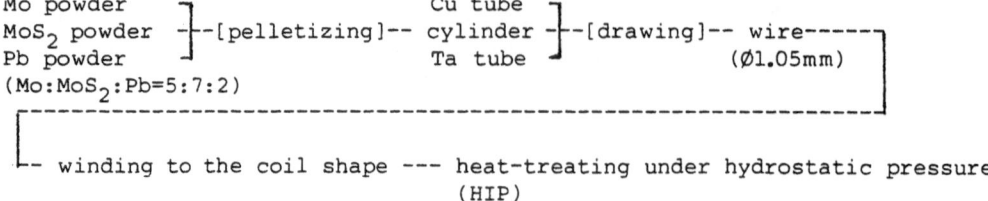

Mo powder, MoS_2 powder, Pb powder ($Mo:MoS_2:Pb=5:7:2$) --[pelletizing]-- cylinder (Cu tube, Ta tube) --[drawing]-- wire (Ø1.05mm)

-- winding to the coil shape --- heat-treating under hydrostatic pressure (HIP)

Fig. 1. Fabrication process of the $PbMo_6S_8$ wire.

(2) To prevent a Ta barrier from cracks caused by the volume expansion of the core at the compound formation during the heat treatment.[8]

The wire was fabricated as follows (shown in Fig. 1). First, starting powders (Mo, Pb and MoS_2 are mixed in the ratio 5:2:7 by atom) were pressed into a cylinder hydrostatically. The composite was assembled by putting the cylinders into a Ta barrier tube and a stabilizing Cu tube, and then was easily drawn to a final size of 1.05mm in diameter at room temperature. The wire shows excellent cold-workability and is never broken down. And then the wire was heat-treated under high hydrostatic pressure by using hot isostatic press (HIP). Heat treatments were performed at 850-1050°C, for 1-3hr under pressures of 1000-2000kg/cm^2.

A cross sectional SEM micrograph of the reacted wire is shown in Fig. 2. In this figure, the $PbMo_6S_8$ compound is formed at the center core of the wire by heat-treatment, and its surrounding white part is a Ta barrier, and the outermost layer is the stabilizing Cu.

In this process, a Pb powder is used instead of a PbS powder, because Pb has the plasticity and homogenity of the composition in the core can be get by melting of Pb at the heat treatment temperature.

RESULTS & DISCUSSION

SUPERCONDUCTING PROPERTIES

Tc measurements were carried out by a conventional four-probe resistive method, and Pt-Co resistive thermometer was used to measure sample temperatures. The typical Tc (onset) and ΔTc are 14.5K and 0.6K, respectively. This value is as high as Tc obtained in other processes.

The wires for Ic measurements were prepared as follows. A 400mm long sample of wires was wound on a 38mm diameter, and the coil shaped sample was then heat-treated to form the Chevrel-phase compound under the high hydrostatic pressure by HIP. The measurements of Ic were carried out by a four-probe resistive method under applied field up to 15.5T, and a current-transfer was not found in V-I characteristics.

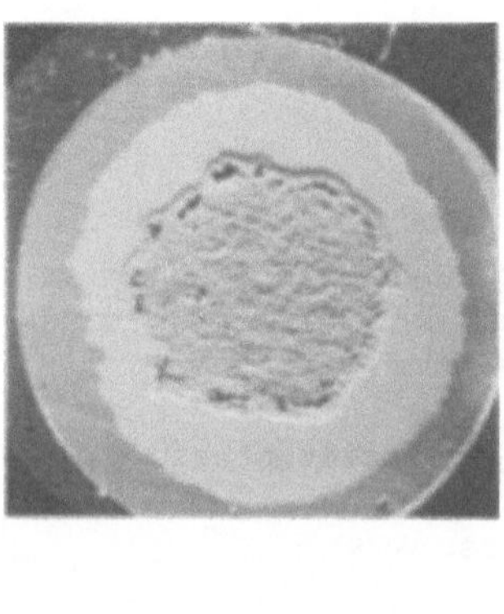

Fig. 2. A cross sectional SEM micrograph of a $PbMo_6S_8$ wire after reaction (X90)

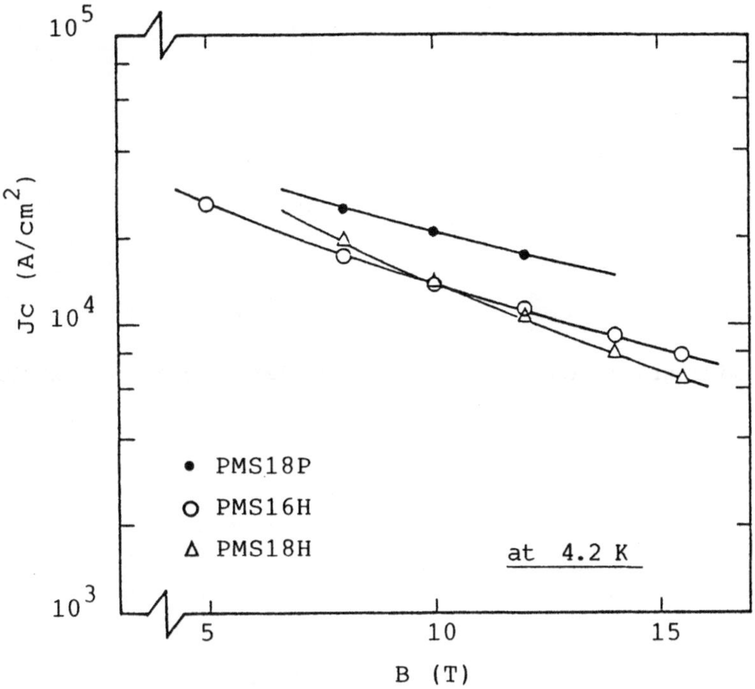

Fig. 3. Critical current density Jc (for the area of the compound core) versus applied magnetic field B at 4.2K for $PbMo_6S_8$ wires by this process.

The Jc's for the area of the compound core as a function of applied fields are shown in Fig. 3. In this figure, sample PMS18P treated at 1000°C for 2hr under the pressure of $2000kg/cm^2$ has a Jc of $1.7x10^4A/cm^2$ at 12T, and sample PMS16H has a Jc of $8x10^3A/cm^2$ at 15.5T. The Jc's obtained are higher than inductively measured one for a hot-pressed bulk sample.[7]

The Bc_2^* calculated by Kramer's scaling law[9] is a very high value of 45.2T in sample PMS16H. Furthermore, Jc's extrapolated by Kramer's scaling law in sample PMS16H are higher than those of a tipical Ti addition Nb_3Sn wire at fields above 20T.

These results show that the $PbMo_6S_8$ compound formed in this process has excellent superconductivities.

MICROSTRUCTURE OBSERVATIONS

The microstructure observations of the wires were carried out by SEM, EPMA and PSPC-type X-ray Microdiffractometer (PSPC-XMD). It was confirmed that the compound layer observed at the interface between the compound core and the Ta barrier was a Ta-sulphide by EPMA. But the existence of Ta element was not observed in the compound core by EPMA. This is one of the reason why good Tc and Jc properties are obtained in spite of using a Ta barrier. It was observed that the longitudinal cross section had almost the same microstructure as the transverse cross section.

Detailed observations by EPMA confirmed that the compound core was composed of three regions, which were the light gray part (A), the medium gray part (B) and the dark gray part (C), as shown in Fig. 4-(1), (3). It is known that the lightness of a composition image of EPMA is varied by the atomic weight of the element. Composition profiles by EPMA along the white line in Fig. 4-(1), (3) are shown in Fig. 4-(2), (4), respectively. And the results of elemental analysis in these three regions are also

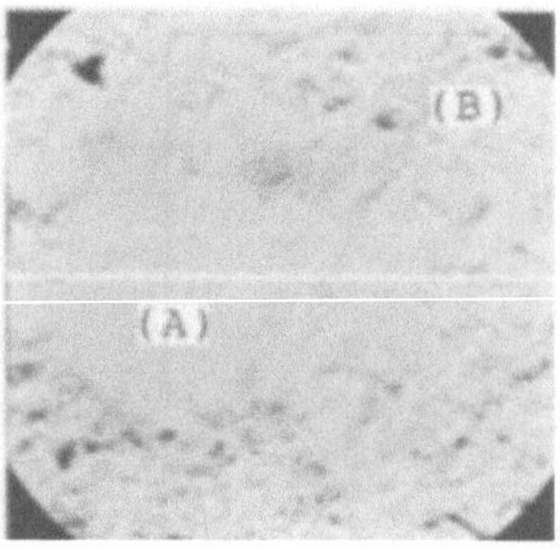

(1) EPMA micrograph (X2000)

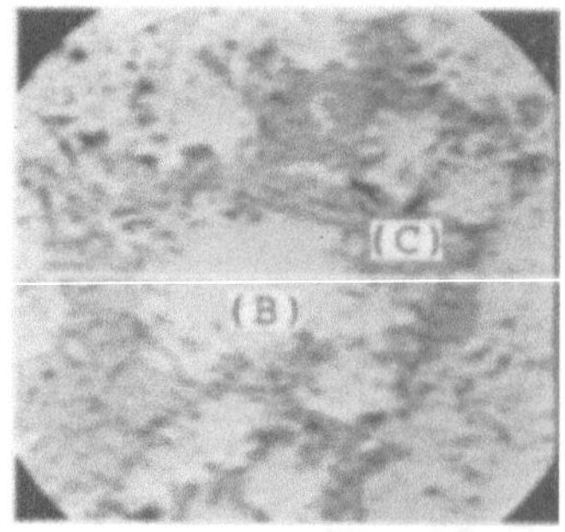

(3) EPMA micrograph (X2000)

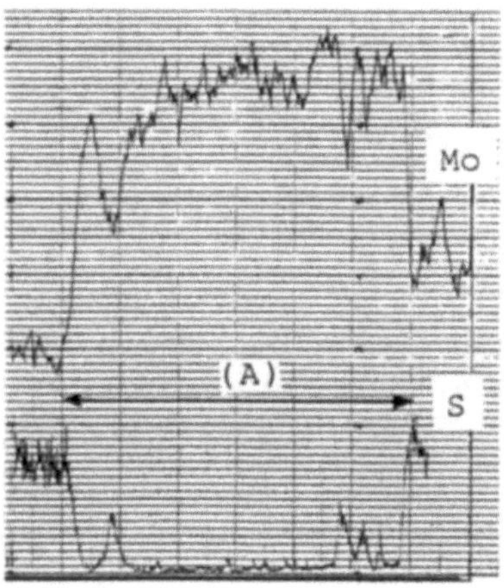

(2) Composition profiles along the white line in Fig. 4-(1)

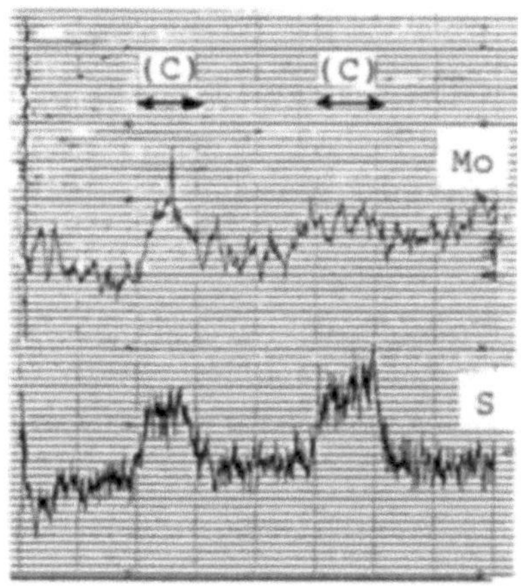

(4) Composition profiles along the white line in Fig. 4-(3)

Fig. 4 EPMA micrographs and composition profiles of the $PbMo_6S_8$ compound

shown in Fig. 5. In this figure, (A), (B) and (C) regions were a Mo precipitate, a $PbMo_6S_8$ compound and a sulphur rich Mo-sulphide which was supposed to be MoS_2, respectively.

Furthermore, phases of the core inside the Ta barrier were identified by PSPC-XMD. The region irradiated by X-ray at the compound core was 50 µm in diameter. A X-ray diffraction pattern of the sample heat-treated at the condition of 1000°C for 2hr under the pressure of 2000kg/cm^2 is shown in Fig. 6. The analysis of X-ray diffraction patterns indicated the existence of phases of $PbMo_6S_8$, Mo and MoO_2. The MoS_2 phase besides them was also found in some of samples. The Mo precipitates are not the unreacted starting Mo powder but the one formed by the decomposition of MoS_2 [$MoS_2 \rightarrow 2S + Mo$] in the reaction process since the precipitates are large in size. The $PbMo_6S_8$ compound is not more than 50% in volume at the maximum by estimation. This is why less volume fraction of the $PbMo_6S_8$ compound is due to a lack of sulphur caused by Ta sulphide formation and the composition of starting powders (Pb:Mo:S=1:6:7). The further improvement in the Jc will be able to be given by increasing the quantity of sulphur element in starting powders.

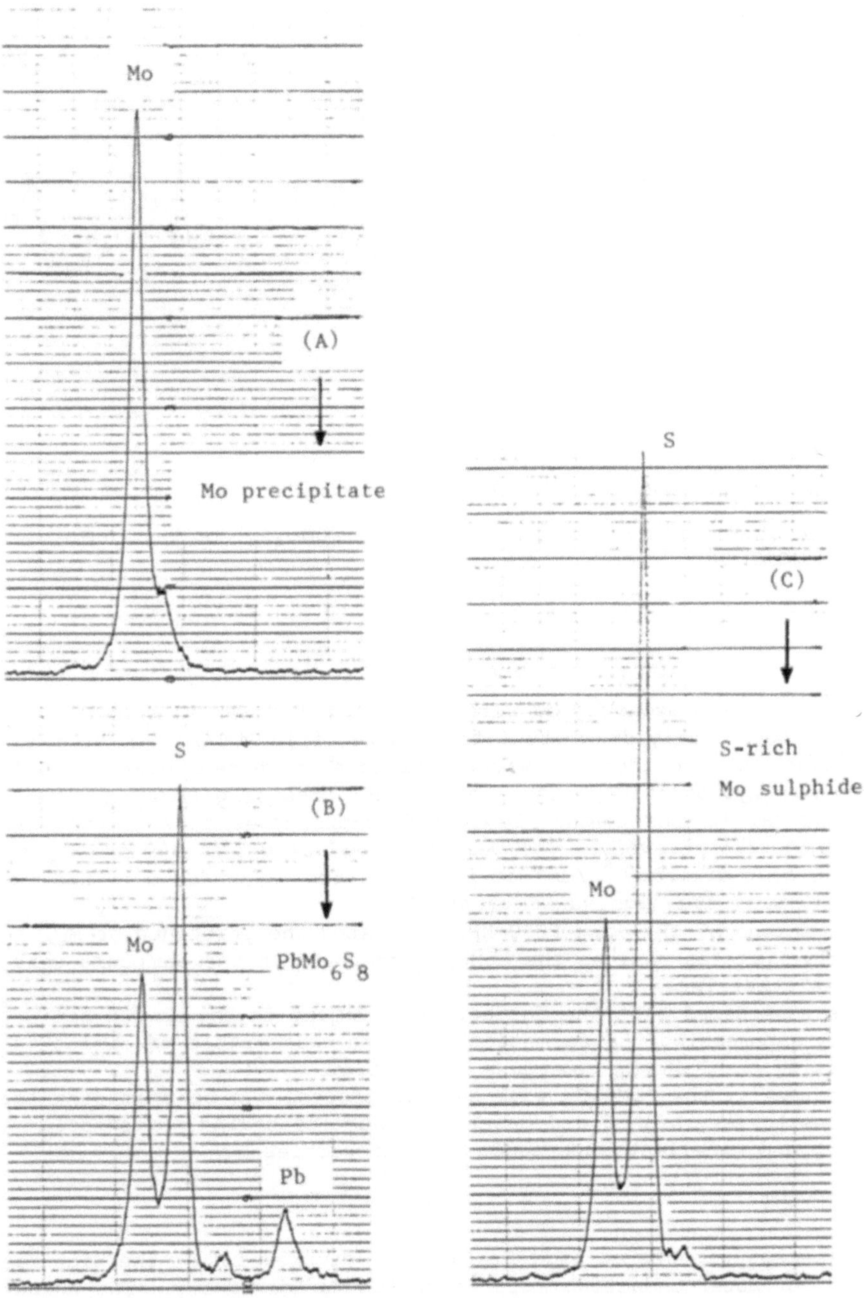

Fig. 5. Characteristic X-ray spectra of three regions (A,B,C) shown in Fig. 4 in the $PbMo_6S_8$ compound core by EPMA

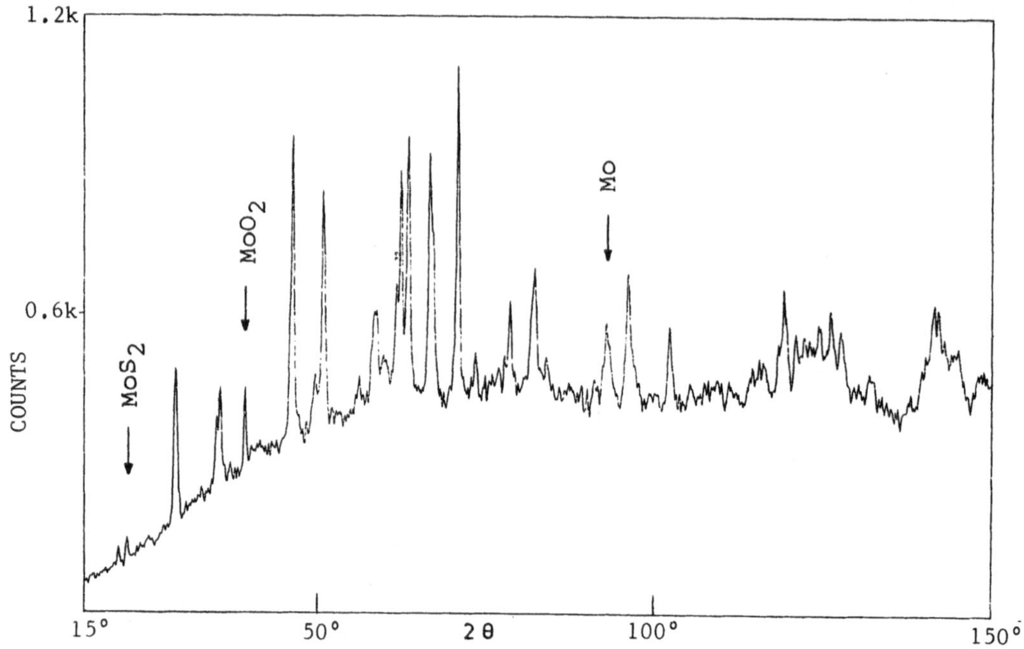

Fig. 6. A X-ray diffraction pattern of the $PbMo_6S_8$ compound core by PSPC-XMD

CONCLUSIONS

The $PbMo_6S_8$ superconducting wires with a Ta barrier in a powder process have been easily drawn to a final size of 1.05mm in diameter at room temperature. And then the wires fabricated have been heat-treated under the high hydrostatic pressure by HIP for producing $PbMo_6S_8$ compound.

The typical Tc (onset) and ΔTc by a four-probe resistive method are 14.5K and 0.6K, respectively. The wire treated at 1000°C and $2000 kg/cm^2$ for 2hr has a Jc of $1.7 \times 10^4 A/cm^2$ at 12T. The Bc_2^* calculated by Kramer's scaling law is a very high value of 45.2T.

Observations by EPMA and PSPC-XMD indicated the existence of Mo, MoO_2 and MoS_2 precipitates besides the $PbMo_6S_8$ compound in the compound core inside the Ta barrier.

It is considered that the further improvement in the Jc can be achieved by suppressing the formation of those undesirable precipitates and increasing the $PbMo_6S_8$ compound in quantity.

Consequently, the $PbMo_6S_8$ wire by this process can be promising for applications to very high field magnets because of having a good cold-workability and excellent superconducting properties.

ACKNOWLEDGEMENT

The Ic measurements at high fields were carried out at the High Field Laboratory for Superconducting Materials, Tohoku University. The authers are grateful to Drs. Y. Muto, K. Noto and K. Watanabe.

REFERENCES

1. M. Decroux, Ø. Fischer and R. Chevrel, Superconducting wires of $PbMo_6S_8$, Cryogenics 17:291-294 (1977)
2. N. E. Alekseevskii, M. Glinski, N. M. Dobrovol'skii and V. I. Tsebro,

Critical currents of certain superconducting molybdenum sulfides, JETP Lett. 23:412-414 (1976)

3. K. Hamasaki, T. Inoue, T. Yamashita and T. Komata, Critical currents in the Chevrel-phase lead molybdenum sulfide thin films, Appl. Phys. Lett. 41:667-669 (1982)
4. K. Hamasaki, K. Hirata, T. Yamashita, T. Komata, K. Noto and K. Watanabe, Critical currents of superconducting $PbMo_6S_8$ tapes, IEEE Trans. on Mag. MAG-21:471-473 (1985)
5. T. Luhman and Dew-Hughes, Superconducting wires of $PbMo_{5.1}S_6$ by a powder technique, J. Appl. Phys. 49(2):936-938 (1978)
6. B. Seeber, C. Possel and Ø. Fischer, $PbMo_6S_8$: A new genegation of superconducting wires?, in: "Ternary Superconductors," Elsevier North Holland, New York (1981), pp. 119-124.
7. B. Seeber, C. Rossel, Ø. Fischer and W. Giaetzle, Ingestigation of the properties of $PbMo_6S_8$ powder processed wires, IEEE Trans. on Mag., MAG-19:402-405 (1983)
8. N. E. Alekseevskii, N. M. Dobrol'skii, D. Eckert and V. I. Tsebro, Investigation of critical currents of ternary molybdenum sulfides, J. Low Temp. Phys. 29:565-572(1977)
9. E. J. Kramer, Scaling laws for flux pinning in hard superconductors, J. Appl. Phys. 44:1360-1370(1973)

CRITICAL FIELD MEASUREMENTS ON VERY THIN FILMS OF

V-Ga SUPERCONDUCTORS

J. E. Tkaczyk, P. M. Tedrow

Francis Bitter National Magnet Laboratory
Massachusetts Institute of Technology
Cambridge, Massachusetts

ABSTRACT

V_3Ga thin films, 500, 20, 10 and 8 nm thick have been made for the purpose of studying the role of spin-orbit scattering in the theory of high field superconductivity. Either 1 percent dopings or thin layers of high-atomic-number elements have been added to these samples in an attempt to increase the spin-orbit scattering rate in this paramagnetically limited material. Critical fields are presented and material parameters are obtained by fitting to a modified Werthamer-Helfand-Hohenberg (WHH) theory. The uniqueness of the parameters obtained in this manner is discussed in light of Fermi-liquid effects. Comparisons are drawn with bulk V_3Ga.

INTRODUCTION

V_3Ga is an A15 superconductor with a critical field at zero temperature of about 23 tesla.[1] The critical field is limited by the paramagnetism of the normal state. Theory predicts that the introduction of spin-orbit scattering can lead to a higher critical field. It is believed that spin-orbit scattering is a rapidly increasing function of atomic number.[2] This is apparently the reason for low spin-orbit scattering in V_3Ga. It has been proposed that the introduction of a small quantity (about 1%) of large Z elements uniformly dispersed into the material, or as thin layers in contact with thin films of V_3Ga, will increase the critical field. It has been shown that only eighty percent of a monolayer of Pt deposited onto a thin, aluminum film more than doubles the critical field.[3] However, such properties as critical field and superconducting transition temperature are sensitive to the degree of long-range order in the A15 crystal structure.[4] Therefore, it is expected that results similar to that in aluminum may be difficult to achieve in V_3Ga. The study of V_3Ga in particular is motivated by the current interest in its use as a material for making very high field, superconducting magnets.

The theory describing thin film superconductors in large magnetic fields has been under recent scrutiny.[5,6] Early attempts to determine the spin-orbit scattering rate, b_F[7], were made by fitting critical field data to the theory due to Werthamer, Helfand, and Hohenberg.[8] Orbital pairbreaking effects were parameterized with the slope of the critical field curves at T_c. For Nb_3Sn values for b_F were found which implied that spin-orbit scattering was stronger than transport scattering. It was suggested that Fermi-liquid effects arising from electron-electron and electron-phonon interactions implicit in the theory had been ignored. Inclusion of such effects ameliorated the disagreement between theory and experiment. Fermi-liquid effects are accounted for in the theory by the inclusion of a renormalization factor, $(1+G^{o})$, where G^{o} is the $\ell=0$ anti-symmetric Fermi-liquid parameter.[5,9] With the introduction of this new parameter into the theory of critical fields it was found that fitting only critical field data was not sufficient, to define a unique set of material parameters. The technique of spin-polarized tunneling[10] provides a direct measurement of b_F. Tunneling measurements on aluminum have produced values of b_F which differ by a factor of four from earlier values obtained by fitting critical fields.[3] It has been proposed that fitting both critical field and tunneling data provides the necessary and sufficient information required to obtain the material parameters uniquely.[9] The strong paramagnetic limiting and yet high critical field of V_3Ga makes it an interesting material in which to test the theory.

500 nm V_3Ga FILMS

Previous work[11] on V_3Ga films 500 nm thick has included assessing the effect on critical field of 1 to 2% elemental impurities of Pb, Pt, Sn, Nb and Ta. These films made at Stanford University were formed by electron beam codeposition of V,Ga and the impurity onto sapphire substrates heated to 750°C. The simultaneous production of samples with a range of composition yielded comparisons of impurity doped films both on and off stoichiometry. It was found that in off stoichiometry V-Ga (Ga rich) the 1% Nb samples had higher T_c's than the undoped off stoichiometry material. This was a significant effect; at the highest gallium concentration produced, 30%, the T_c of the Nb doped sample was 11.75 K as compared to 8.6 K for the undoped V-Ga. However, in the stoichiometrical material the doped samples had lower Tc's than the pure V_3Ga.

This zero-field effect on T_c indicates that the impurities are influencing the V_3Ga A15 structure. However, critical fields and tunneling data show no evidence of significantly increased spin-orbit scattering.

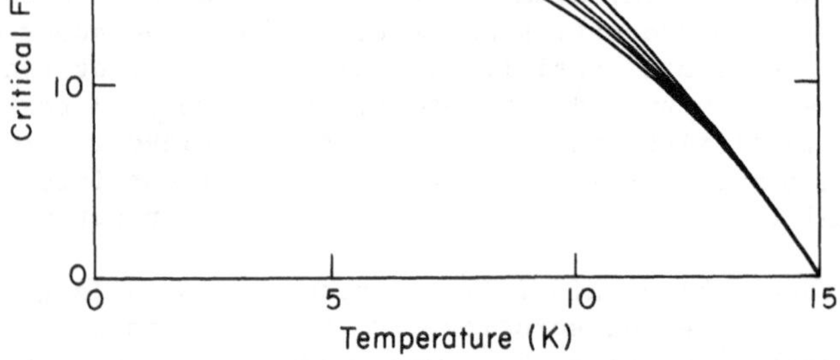

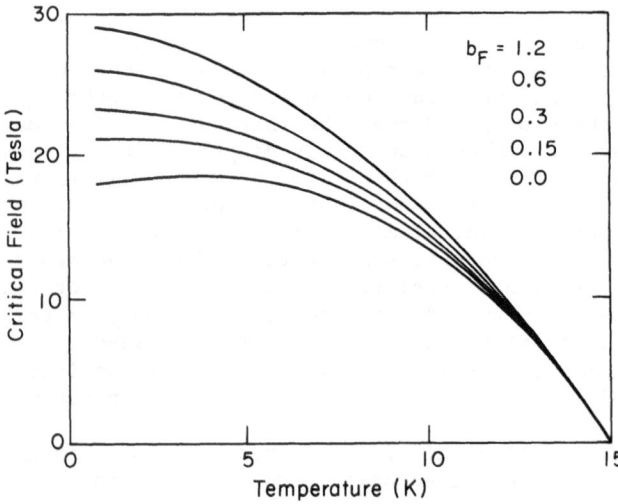

Fig. 1 Critical fields generated from the theory for parameters typical of bulk V_3Ga (T_c = 15 K, $|dH/dT|_{T_c}$=4.0 T/K, b_F=0.3) and for other values of b_F.

In fact, when plotted in reduced variables, as in Fig. 2, the critical fields are almost identical. Critical field measurements were made up to 20 Tesla with the field oriented perpendicular to the surface of the film. Parameters obtained by fitting the critical fields are given in Table I. The critical fields of some of those samples listed in Table I are shown in Fig. 3. Fermi-liquid effects were neglected for these fits by setting $G^{\circ}=0$. The fits were optimized by minimizing the statistical varience utilizing a computer gradient search. The similarity in the parameters derived from the critical field indicates no field dependent perturbation due to the impurities. This is unexpected in light of the zero-field effect on T_c. Since substrate temperature was optimized for V_3Ga crystal growth, the possibility arises that the third element is excluded from such growth and thus migrates to the grain boundaries. In the non-stoichiometric, Ga-rich, material, the Nb dopant may participate in such growth reducing distortion of the A15 lattice and thus raising the T_c. Nb forms a superconducting A15 compound with gallium with a reported T_c of 20.3 K.[4] ESCA studies of these samples were inconclusive because of a lack of standards with which to compare observed energy shifts and also because of lack of the necessary sensitivity for detecting 1 percent effects.

Studies of bulk V-Ga have correlated the reduction in T_c as composition moves away from stoichiometry with the associated disordering of the A15 lattice. A similar decrease in T_c was observed for the undoped 500 nm thick V-Ga. In order to reduce possible diffusion of the third element and the resulting disordering of the A15 lattice, which could reduce T_c, a new strategy was applied. As already mentioned the placement of sub-monolayer coverings of Pt on 4 nm Al films was successful at significantly increasing the spin-orbit scattering rate and critical field. Given a spin mean free path of 50 nm in V_3Ga one may expect to see some similar effect for V_3Ga films of thickness less than 10 nm.

THINNEST FILMS

Films of V_3Ga of thicknesses 20, 10 and 8 nm thick have been made using techniques similar to those used for making the 500 nm thick films.

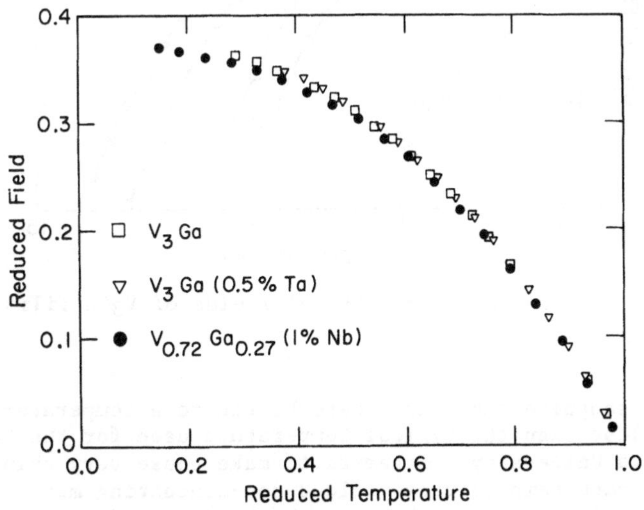

Fig. 2 Perpendicular critical fields of 500 nm thick V-Ga with and without high Z elements. The reduced field is obtained by dividing H_{c2} by its $|dH/dT|_{T_c}$ and by T_c. The reduced temperature is T/T_c.

Table 1. Characteristics of V_3Ga

Sample	T_c (K)	dT_c (K)	$(dH/dT)_{T=T_c}$ (T/K)	b_F [a]	ξ nm	Variance $T^2 \times 10^{-2}$
Bulk [b]	15.1	0.23	4.3	0.22	2.7	
500 nm, pure	13.90		3.9	0.26	3.0	0.59
500 nm, 1% Nb	10.70		4.0	0.24	3.3	0.84
500 nm, 0.5% Ta	14.51		2.9	0.25	2.9	0.94
20 nm	12.60	0.40	4.3	0.25	2.9	0.58
10 nm, 6 nm Al	10.55	0.75	5.4	0.23	2.8	0.38
8 nm, 6 nm Al	9.50	1.50	5.8	0.22	2.9	1.20
8 nm, 0.5 nm Ta, 6 nm Al	9.95	1.00	5.8	0.20	2.9	1.39

[a]See reference 7.

[b]Values obtained from reference 1.

Values of T_c, $(dH/dT)_{T=T_c}$ and b_F were taken from critical field fits. Fermi-liquid effects were neglected by taking $G^\circ=0$. The statistical variance of the fits is given in the last column. The value of dT_c is the 90-10 width of the zero-field transition.

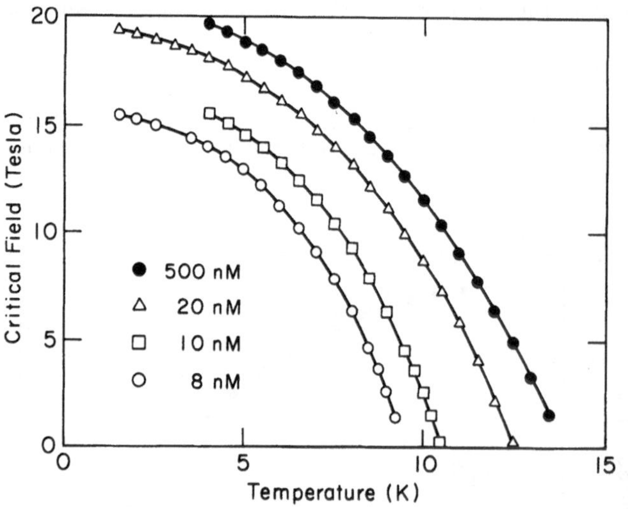

Fig. 3 Perpendicular critical fields of V_3Ga films.

The ¼ x ½ inch sapphire substrates were heated to a temperature of 500°C. This is 250°C less than the typical temperature used for the samples made at Stanford University. Attempts to make these very thin samples at higher substrate temperatures yielded non-conducting material. Three substrates were placed in line between the separated V and Ga sources for each evaporation. Comparison of the resulting films allows one to achieve stoichiometry by adjusting the respective rates to obtain the maximum T_c. The deposition rate was typically 0.6 nm/sec. and pressure during the evaporation was in the high 10^{-8} torr range. Aluminum was evaporated over some of the V_3Ga films as a protective layer. Any heavy element overlayers were evaporated before this covering. Substrate temperature during these overlayer evaporations was below 200°C. An aluminum thickness as much as 3 to 6 nm was required to achieve high transition temperatures in the thinnest V_3Ga films. The relative amounts of Al metal and oxide and the mechanism for this effect is not known at present. However, one may suspect that the Al getters oxygen from the V_3Ga film surface and from grain boundaries. A system of masks and shutters has been developed for the simultaneous deposition of two films onto the same substrate. These can be made to differ only by the presence of a third element over-layer on one. Such a pair of films would allow an unambiguous determination of the effect, or lack thereof, which the third element has on V_3Ga. A few samples with alternating layers of V_3Ga and the third element have been made. The substrate temperature during the fabrication of these structures was kept between 500 and 550°C.

DISCUSSION

As has been observed previously in other A15 films,[5] thin film samples with higher zero-field transition temperatures are found to have higher residual resistivity ratio, lower resistivity and narrower transitions. However, there has been too much scatter in these properties to quantify the correlation at present. Variations in residual pressure, annealing times, cooling rates and the presence of mask shadows or compositional variations at edges could each have their own effect on such properties. Emphasis has been placed on maximizing T_c and sample quality by varying substrate temperature and film composition.

The parameters in Table I and the critical field curves show that the zero-field transition temperature decreases with decreasing film thickness. Presumably, the factors influencing the change in T_c are diffusive contaminants, especially oxygen, and strain at the substrate-film interface. Localization may also play a role in the decrease of T_c. The field dependent properties of these films are remarkably similar. The slope at T_c increases with decreasing T_c in such a way that the coherence length remains approximately constant. The pair of 8 nm films listed in Table I were made on the same substrate. In these films the width of the constant temperature, ramped field transition narrowed when the temperature was decreased. This suggests a strong fluctuation contribution to the dT_c's listed in Table I. Of the 8 nm thick samples the one with a 0.5 nm Ta overlayer has the higher T_c and yet similarity in the values of the coherence length and b_F show that plots of their critical fields in reduced variables are nearly identical. Similarly, in pairs of 100 nm V_3Ga films differing only by 1 nm in the aluminum overlayer thickness, the sample with the greater coverage of aluminum had the higher T_c. However, a comparison of the critical field showed no field dependent effect due to additional aluminum. Samples with alternating layers of V_3Ga and the impurity showed changes in T_c, but no field dependent effect.

At present the absence of the effect which these sample preparation methods were expected to have on the spin-orbit scattering rate is not understood. The spin mean free path $\ell_{so}=\xi_o/b_F$ is of the order of 50 nm in V_3Ga. Surfaces are thought to be very effective in spin scattering.[12] From the available data, it appears that either the mechanism of spin-orbit scattering is different in a complex transition metal compound from that in a simple metal like Al, or there is some microstructural explanation. It is expected that work on very thin films and layered

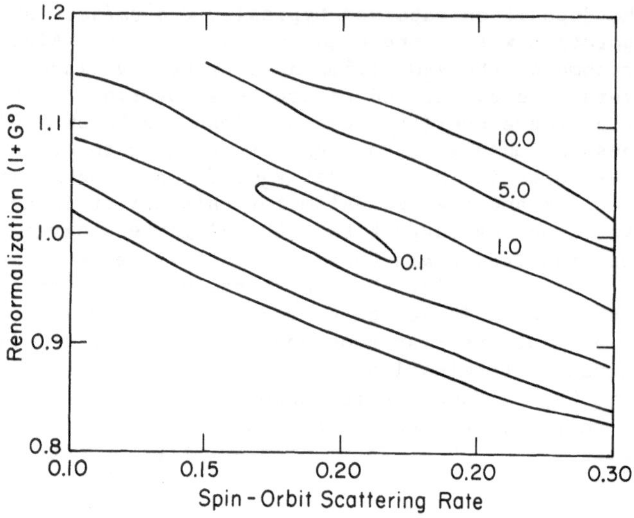

Fig. 4 Level curves of the variance minimized by adusting $|dH/dT|_{T_c}$. Details as to the construction and meaning of this figure are found in the text. The spin-orbit scattering refers to the value of b_F.

structures including spin-polarized tunneling measurements will shed light on this question.

Finally, an important point should be made concerning the uniqueness of the critical field fits. The possible presence of Fermi-liquid effects forces one to use tunneling measurements for a correct quantitative analysis of these films. The final Figure makes this clear. Twenty points of critical field "data" resembling that of V_3Ga were generated from the theory using the following parameters: T_c=10 K, slope at T_c=4.0 T/K, b_F=0.2, G^o=0. Then a new choice of b_F and G^o were made and held fixed. Critical field curves were generated for different values of slope at T_c until the statistical variance with respect to the "data" was minimized. In this way the best fit to the "data" was obtained for each point in a 100 point grid in the parameter space formed by b_F and G^o. The level curves of the variance minimized by varying the slope alone are shown in Fig. 4. Values of variance found in real data listed in Table 1 show that one may trade off b_F for G^o and still achieve a good fit to the theory. This result indicates the need for tunneling data to determine the values of G^o and b_F uniquely. In a similar analysis, minimizing the variance with respect to either b_F or G^o shows that the slope at T_c, and therefore the orbital depairing parameter, is in fact well determined by critical field data.

CONCLUSION

It has been shown that the critical field behavior of V_3Ga films down to 8 nm in thickness resembles that of the bulk material except for a depression of the T_c with decreasing film thickness. Although changes in T_c were observed when these films were doped, or covered with third elements, no field dependent effects have been seen which are attributable to the impurity. This is in spite of the care taken to form two samples on the same substrate differing only in the presence of the impurity. In addition, the inclusion of Fermi-liquid effects in the theory of the critical magnetic field necessitates the use of tunneling data to make quantitative statements concerning the parameters.

ACKNOWLEDGMENTS

This work is supported by Department of Energy Grant DE-FG02-84ER45-094. The Francis Bitter National Magnet Laboratory is supported at MIT by the National Science Foundation.

REFERENCES

1. S. Foner and E. J. McNiff, Jr., S. Moehlecke, and A. R. Sweedler, Effects of composition on the upper critical field of V-Ga, Solid State Commun. 39:773 (1981).
2. A. A. Abrikosov and L. P. Gorkov, Zh Eksp. Teor. Fiz. 42:1088 (1962); [Sov. Phys. JETP, 15:752 (1962)]; R. Meservey and P. M. Tedrow, Phys. Rev. Lett. 41:805 (1978).
3. P. M. Tedrow and R. Meservey, Critical magnetic field on very thin superconducting aluminum films, Phys. Rev. B. 25:171 (1982).
4. D. Dew-Hughes, Superconducting A-15 compounds: A review, Cryogenics, 15:435 (1975).
5. T. P. Orlando, E. J. McNiff, Jr., S. Foner, and M. R. Beasley, Critical fields, Pauli paramagnetic limiting, and material parameters of Nb_3Sn and V_3Si, Phys. Rev. B. 19:4545 (1979).
6. T. P. Orlando and M. R. Beasley, Pauli limiting and the possibility of spin fluctuations in the A 15 superconductors, Phys, Rev. Lett. 46:1598 (1981).
7. The spin-orbit scattering parameter, b_F, used here may be converted to λ_{so} by multiplying b_F by the factor, 1.123.
8. N. R. Werthamer, E. Helfand, and P. C. Hohenberg, Phys. Rev. 147:295 (1966).
9. J. A. X. Alexander, T. P. Orlando, D. Rainer, and P. M. Tedrow, Theory of Fermi-liquid effects in high-field tunneling, Phys. Rev. B. 31:5811 (1985).
10. P. M. Tedrow, J. S. Moodera, and R. Meservey, Resolution of the spin states of a superconductor by tunneling from a ferromagnet, Solid State Commun. 44:587 (1982).
11. P. M. Tedrow, J. E. Tkaczyk, R. Meservey, S. J. Bending and R. Hammond, Critical magnetic field of V_3Ga thin films with third element additions, IEEE Trans. Magn. 21:1144 (1985).
12. R. Meservey and P. M. Tedrow, Surface Relaxation Times of Conduction electron spins in superconductors and normal metals, Phys. Rev. Lett. 41:805 (1978).

ELECTRON TUNNELING INTO SUPERCONDUCTING FILAMENTS: DEPTH PROFILINC THE ENERGY GAP OF NbTi FILAMENTS FROM MAGNET WIRES.*

John Moreland, J. W. Ekin, and L. F. Goodrich

Electromagnetic Technology Division
National Bureau of Standards
Boulder, Colorado

ABSTRACT

Squeezable electron tunneling (SET) junctions consisting of superconducting NbTi filaments (extracted from magnet wires) and sputtered Nb thin-film counter electrodes were used to determine the energy gap at the surface of the filaments. The current versus voltage curves of junctions immersed in liquid helium at 4 K were measured for a series of filaments taken from the same wire. Each filament had been etched to remove a surface layer of varying thickness so that the energy gap could be determined as a function of depth into the surface of an "average" filament. It was found that some manufacturing processes yield filaments having surface layers with reduced energy gaps of 0.4 meV compared to measured interior bulk values ranging from 1.2 to 1.3 meV.

INTRODUCTION

Multifilamentary-NbTi superconductors are used extensively in the construction of large-scale magnets. This is primarily due to the relative ease of manufacture and the strain insensitivity of the superconducting properties of NbTi filaments. With the proper heat treatment, NbTi alloys precipitate Ti rich inclusions within the material. It is generally accepted that these precipitates act as vortex pinning sites that give rise to the large critical currents observed in NbTi filaments. A good understanding of the alloy metallurgy is therefore vital to developing better high-current superconductors.[1,2] Perhaps equally important are the effects of various steps in manufacturing processes on microscopic intrinsic superconducting parameters. In particular, the magnitude of the superconducting energy gap, Δ, directly affects T_c,[3] H_{c2},[4] and along with alloy metallurgy, determines J_c.[5]

An electron tunneling experiment is an excellent probe of the fundamental parameters that govern the superconducting state, providing a direct measurement of the energy gap. This is because the conductance of a superconducting tunnel junction as a function of applied bias (often

*Contribution of the National Bureau of Standards, not subject to copyright.

referred to as a tunneling spectrum) is directly related to the density of states of thermal excitations in a superconductor as a function of energy. Recently it has been shown that electron tunneling experiments can be performed using ultrastable electromechanical transducers to space two electrodes about 1 nanometer apart, stable to within a few picometers.[6] In this way the oxide barrier previously required for tunneling measurements has been eliminated, providing an alternative method for obtaining tunneling spectra. The concept of mechanically adjustable barriers is attractive because it is relatively straightforward to make a tunneling contact to any conductor including bulk materials. In particular, we have shown in an earlier paper that a superconducting squeezable-electron tunneling (SET) junction can be constructed between a filament and a thin-film counter electrode in liquid helium at 4 K.[7]

Tunneling electrons probe a superconducting surface to a depth of about one BCS coherence length. It should therefore be possible to delinitate the superconducting properties of a material as a function of depth to within submicron distances. Presented here, for example, are results obtained when the SET-junction technique mentioned above was applied to NbTi superconductors. Individual filaments were removed from multifilamentary NbTi superconductors by chemically etching away the surrounding copper stabilizer. The surfaces of many such filaments were then etched different amounts to obtain the "depth profile" of the energy gap of the "average" filament within a given conductor.

APPARATUS

The experimental apparatus used to "squeeze" filament SET junctions is described in earlier papers.[6,7] Essentially the filaments were positioned above a 0.2 µm-thick Nb counter electrode sputtered onto a flexible Si substrate with 1 µm-thick evaporated SiO spacers. An electromagnetic squeezer adjusted the bow in the substrate and thus the distance between the filament and the Nb counter electrode. In this way tunneling barriers a few nanometers thick could be obtained with stability sufficient for tunneling spectroscopy of energy gaps. The squeezer was mounted on a probe designed for rapid measurements at 4 K in a liquid-helium storage Dewar. It was possible to cool, perform a tunneling measurement, and warm within a 30-minute time span.

The I-V curve was monitored over the range from -5 to +5 mV as the junction was mechanically adjusted in liquid helium to obtain a tunneling characteristic. Four-terminal contact was made to the junction. Superconducting counter electrodes were used because they gave rise to distinct structure in the I-V curves that provided a quality check for the tunneling barrier even if the sample electrodes were normal. Also, it was easier to detect small energy gaps for the sample electrodes as shifts in the I-V characteristic using a superconducting counter electrode, compared to shifts of a broad zero-bias inflection in a relatively featureless curve using a normal counter electrode.

Curves of dynamic conductance (dI/dV) versus voltage were measured with a tunneling spectrometer that utilized a feedback circuit to keep a constant modulating voltage (250 µV, 100 Hz) across the junction. This signal was superposed onto the slowly swept dc voltage (± 5 mV, sweep rate = 100 µV/s). The conductance was proportional to the first-harmonic signal measured with a lock-in amplifier. For two superconducting electrodes the sum of the energy gaps, Δ_1 and Δ_2, can be estimated from the location of the peaks in the conductance spectrum of the junction using the formula

$$e\bar{V} = 2\Delta_1 + 2\Delta_2. \qquad (1)$$

$\bar{V}$ is the voltage separating the peaks in the conductance spectrum that occur near the current inflections in the I-V curve and e is the electron charge. This procedure is used instead of measuring $\Delta_1 + \Delta_2$ from an apparent zero bias so that small preamplifier offsets need not be considered in the energy gap measurements.

The energy gap of the Nb thin-film counter electrodes was determined by measuring the tunneling energy gap of a SET junction with two identical thin films and found to be 1.4 meV. One of these calibrated counter electrodes was then transferred to the filament-tunneling apparatus. The contribution from the counter electrode ($2\Delta_{film}$ = 2.8 meV) was subtracted from the sum of the energy gaps to obtain the energy gap of the filaments ($2\Delta_{filament}$). Typically the data lead to measurements precise to within ± 0.1 meV.

Tunneling contacts were nondestructive in that the magnitude of the tunneling energy gap was independent of the resistance for a given junction setting as long as the junction was not shorted. Since a nondestructive tunneling contact was made between the filament and the film and care was taken to avoid moisture condensation between runs, the counter-electrode films could be reused many times.

FILAMENT SAMPLES

Individual NbTi filaments from 20 to 30 μm in diameter (filaments with diameters less than 20 μm were difficult to mount in the apparatus) were removed from multifilamentary-NbTi superconductors by etching away the surrounding Cu matrix in a 50/50 solution of nitric acid and water at room temperature. Subsequent etches to remove varying-thickness surface layers from a single filament were performed in freshly prepared solutions consisting of 4 parts hydrofluoric acid, 36 parts nitric acid, and 60 parts water at room temperature. The etch rate was determined for several filaments etched different amounts using a scanning electron microscope (SEM) to compare adjacent etched and unetched portions of the filaments. The etch rate was constant over the measured etching range from 0 to 10 μm. It had a value of 50 ± 5 nm/s, which was about the same as the etch rate for pure Nb.

Typical SEM micrographs of etched NbTi filaments are shown in figure 1. The filaments in figures 1a, 1b, and 1c were taken from a wire that was made using a conventional manufacturing process that included high temperature extrusion followed by cold working and a final Cu annealing step. This wire was made in the early 1970's and may not be representative of conductors fabricated using current technology. The small nodules a few microns in size on the surface of the unetched filament (Fig. 1a) have been attributed to a Cu-NbTi surface layer that forms during the hot extrusion and subsequently breaks up into debris during cold working.[2,8] After a short etch (20 s) these nodules are removed from the filament surface (Fig. 1b). Further etching serves not only to reduce the filament diameter but also polishes the filament surface, removing striations acquired during the wire-drawing process (Fig. 1c).

Other filaments tested include those extracted from "cold-process" wire previously developed for experiments aimed at understanding mixture and alloy formation and their effect on NbTi superconductors.[8] This wire was not optimized for a high J_c. It was processed without any high-temperature extrusion, was drawn at low temperatures, and was given no heat treatments during processing. SEM micrographs of unetched filaments of this kind were devoid of Cu-NbTi nodules (fig. 1d).

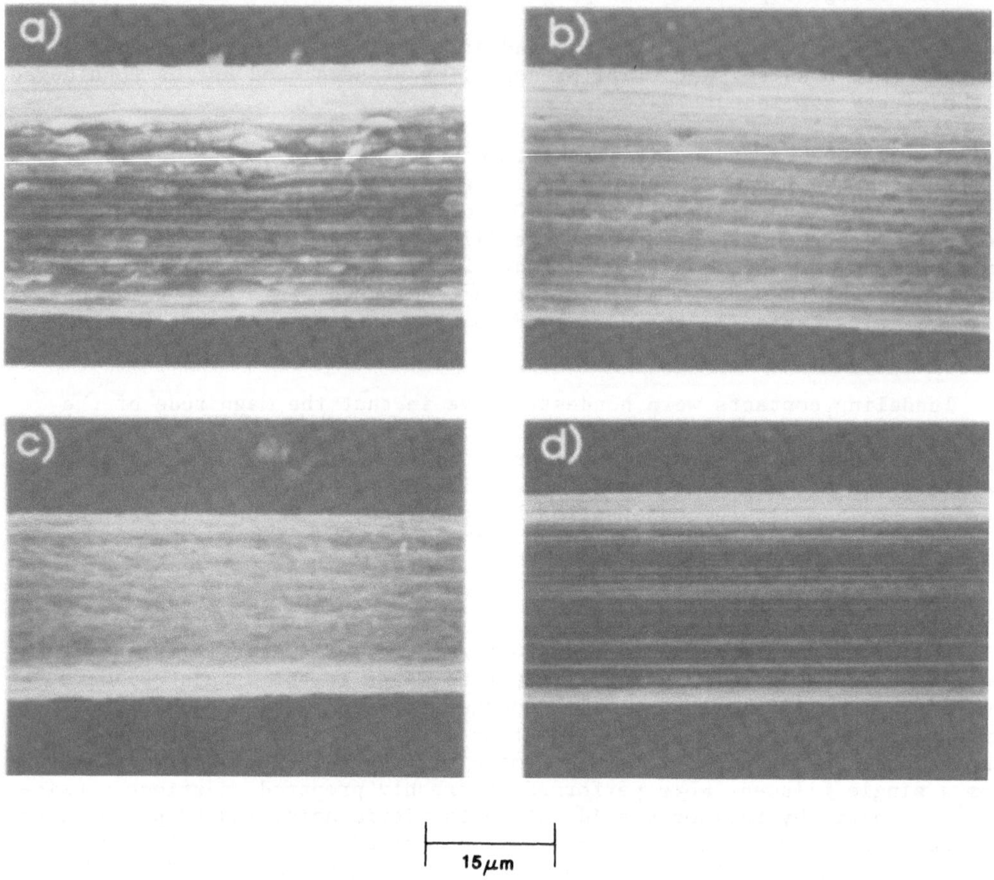

Fig. 1 Electron micrographs of various NbTi filaments.
a) Unetched conventionally processed filament with Cu-NbTi nodules.
b) Conventionally processed filament after 1 μm etch.
c) Conventionally processed filament after 6 μm etch.
d) Unetched "cold-process" developmental filament.

DEPTH-PROFILE RESULTS

Figure 2 shows the tunneling spectra (conductance versus voltage) of a series of NbTi-filament SET junctions with Nb thin-film counter electrodes immersed in liquid helium at 4 K. Each filament had been etched for a different time. The etching depths are based on the measured etch rate of 50 nm/s mentioned above. The peaks defining the gap edges in each of the curves in Fig. 2 are used to determine the energy gap of the filaments.

Data derived from curves like those shown in figure 2 are presented in profile format in figure 3. Here we plot the filament energy gaps derived from the conductance curves as a function of estimated etching depth for two different kinds of wire. The dashed lines are included to aid the viewer's eye. Each data point represents a local measurement of the energy gap. The variability of the data taken at a given etch depth may be an indication of the variability between filaments. It may also, however, be due to changes that occur along the surface of a filament.

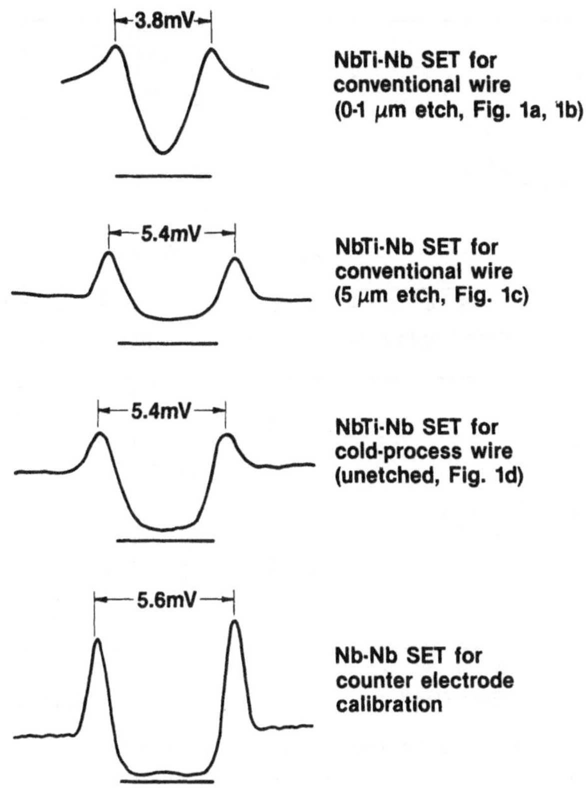

Fig. 2 Tunneling spectra (conductance versus voltage) of various NbTi-filament/Nb-film SET junctions in liquid helium at 4 K.

The results for wire manufactured using the conventional method are shown in figure 3a. In general, eV increases with increasing etching times from a value as low as 3.2 meV to about 5.4 meV, corresponding to filament energy gaps that range from 0.4 meV at the filament surface to 1.3 meV in the filament interior (see equation 1). There appears to be a 3- to 4-μm thick layer at the surface of an unetched filament with a substantially reduced energy gap compared to the highest interior bulk value.

Figure 3b, on the other hand, shows the profile of a wire that had undergone cold drawing without any heat treatments or extrusion, thus avoiding the formation of Cu-NbTi mixtures. The energy gap is relatively constant as a function of depth having a value of about 1.2 to 1.3 meV even at the unetched-filament surface.

Also included in figure 3 are the corresponding energy-dispersive X-ray microanalysis results that show the weight percent Cu as a function of filament etch depth for both types of wire. A discernable difference (dashed line) from the instrumental background signal (solid line) is apparent in Fig. 3c for conventionally processed filaments that had undergone an etch of 4 μm or less. [The dashed line is shown for reference only; the signal to noise ratio is about 2, which does not permit the detailed shape of the Cu profile to be determined.] In contrast, very little Cu above background levels could be detected in the cold-process filaments as shown in figure 3d.

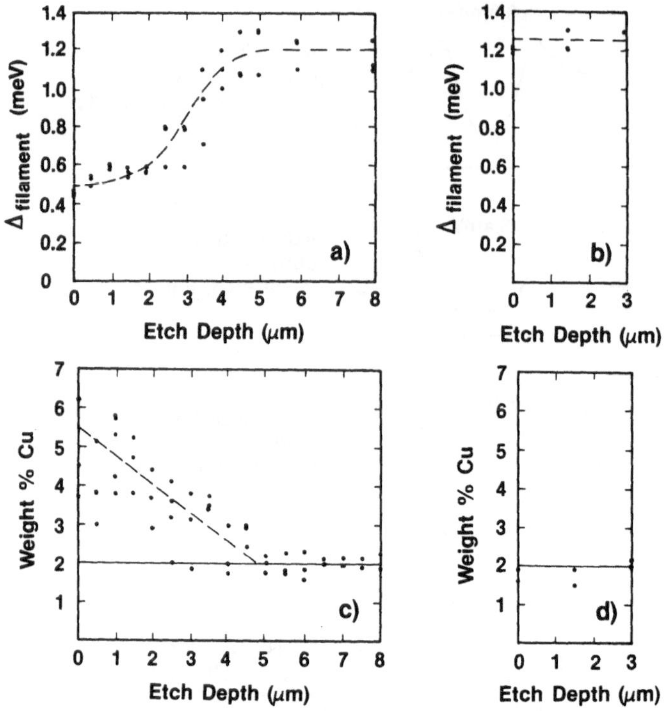

Fig. 3 Depth profiling results for NbTi filaments from magnet wires: $\Delta_{filament}$ as a function of estimated etch depth for an average filament extracted from (a) conventionally processed and (b) developmental cold-processed wires (T = 4 K). Weight percent Cu determined by X-ray microanalysis as a function of estimated etch depth for an average filament extracted from (c) conventionally processed and (d) developmental cold-processed wires. The instrumental Cu background signal was approximately 2%.

The experiment was repeated with both the conventionally-processed and the cold-processed wires etched simultaneously in the same hydrofluoric-nitric-water bath. The same X-ray microanalytic Cu concentration results were obtained as before. This indicates that the difference between the two types of samples was not simply an artifact of surface contamination introduced when the filaments were etched in different baths of the same solution.

Energy-dispersive X-ray microanalysis is a semi-quantitative technique that yields average weight percent of an element over the area illuminated by the SEM beam (in our case about 10 μm^2). Further, the penetration depth of the 25 keV beam used in these studies is about 1 μm. In comparison, tunneling electrons probe only to a depth of about one BCS coherence length (about 10 nm for NbTi) over a very small SET junction area (probably less than 1 μm^2).

CONCLUSION

We have demonstrated a new depth profiling technique for studying the energy gap in practical superconductors. The results show that a NbTi conductor made by conventional manufacturing methods of the early 1970's had filament surface layers with energy gaps as low as 0.4 meV at 4 K. The low gap region extended 3-5 μm into a 30 μm filament. NbTi wires that had not undergone high-temperature processing showed very little energy-gap reduction at the filament surface (Δ 1.3 meV at 4 K).

The tunneling data combined with the X-ray microanalysis on the conventionally-processed filaments suggests that there is a correlation between the presence of Cu and the observed low energy gaps near the filament surface. The amount of Cu detected is very small, less than about 3 wt % above the instrumentational background signal. It is unlikely that it is simply an artifact of SEM electron penetration through a thin (< 1 μm) high-Cu content intermetallic layer. This is because these small Cu concentrations are measured at etching times well beyond the point where a thin intermetallic layer should have been chemically removed. We conclude, therefore, that a low-concentration Cu diffusion layer is introduced during processing. It is unclear at this time if this layer is particular to the early 1970's conductor which we tested, or whether it occurs more generally. Further studies will be needed to accurately clarify its nature and origin.

Other phenomena, including surface mechanical damage or oxygen contamination introduced during conventional processing may also be important. In addition, the etching process used to remove filament surface layers may affect the outcome of a given profile measurement. The fact that $\Delta_{filament}$ does not decrease for the cold-process wire and increases in the conventionalprocess wire with further etching shows that at least the etch is not detrimental to the energy gap. It may, however, change the Cu concentration from the original diffusion profile. A more conclusive and more quantitative study might therefore include tunneling into bulk alloys purposely contaminated with known amounts of Cu.

Depending on the current flowing through and the magnetic field within the filaments of a magnet wire, the effective area of each filament may be substantially reduced by the presence of a surface layer with a degraded energy gap. Considering that many applications require superconducting magnets to be operated near their critical parameters, the observed energy-gap degradation could have significant effects (especially at high fields near H_{c2}). It is therefore important to isolate the manufacturing steps that lead to the formation of a degraded-superconductor layer.

ACKNOWLEDGMENTS

Partial support was provided by the Department of Energy. Support was also provided to one of us (J.M.) as a National Research Council Postdoctoral Associate. Helpful discussions with R. Scanlan, A. West, D. Larbalestier, M. Suenaga, M. Garber, and P. Purtscher are gratefully acknowledged.

REFERENCES

1. D. C. Larbalestier and A. W. West, New perspectives on flux pinning in niobium-titanium composite superconductors, Acta Metall. 32:1871-1881 (1984).
2. M. Garber, M. Suenaga, W. B. Sampson and R. L. Sabatini, Effect of Cu_4Ti compound formation on the characteristics of NbTi accelerator magnet wire, to be published in IEEE Trans. on Nuclear Science, 1985; D. C. Larbalestier, Li Chengren, W. Starch, and P. J. Lee, "Limitation of critical current density by intermetallic formation in fine filament Nb-Ti superconductors," to be published in IEEE Trans. on Nuclear Science, 1985; see also papers CZ-2 and CZ-3 of this conference proceedings and the references cited therein.

3. J. Bardeen, L. N. Cooper and J. R. Schrieffer, "Theory of Superconductivity", Phys. Rev. 108:1175-1204 (1957).
4. A. A. Abrikosov, On the magnetic properties of superconductors of the second type, Sov. Phys. JETP 5:1442-1452 (1957).
5. R. Hampshire, J. Sutton and M. T. Taylor, Effect of temperature on the critical current density of Nb-44 wt. percent Ti alloy, in: "British Cryogenics Council Conference on Low Temperatures and Electrical Power", Program preprint IIR, Royal Society, London, England (1969), pp. 69-76.
6. J. Moreland and P. K. Hansma, Electromagnetic squeezer for compressing squeezable electron tunneling junctions, Rev. Sci. Instrum. 55:399-403 (1984).
7. J. Moreland and J. W. Ekin, Electron tunneling into superconducting filaments using mechanically adjustable barriers, Appl. Phys. Lett. 47:175-177 (1985).
8. A. W. West, Rensselaer Polytechnic Institute, Troy, New York, private communication.

AUTHOR INDEX

MATERIALS INDEX

SUBJECT INDEX